1999 ANNUAL BOOK OF ASTM STANDARDS

SECTION 4

Construction

VOLUME

04.01 Cement; Lime; Gypsum

Includes standards of the following committees:

C-1 on Cement
C-7 on Lime
C-11 on Gypsum and Related Building Materials and Systems

Stock #: S040199

ASTM 100 Barr Harbor Drive, West Conshohocken, PA 19428 ■ (610) 832-9500 ■ fax: 610-832-9555
■ email: service@astm.org ■ website: http://www.astm.org

Editorial Staff

Director, Technical Publishing Services:
Roberta A. Storer

Manager, Standards Publications:
Joan L. Cornillot
Paula C. Fazio-Fluehr

Senior Indexer:
H. Joel Shupak

Editors:
Robert F. Allen
Nicole C. Baldini
Patricia E. Donofrio
Elizabeth L. Gutman
Eleanor Keefe
Joanne G. Kramer
Christine M. Leinweber
Vernice A. Mayer
Patricia A. McGee
Kathleen A. Peters
Susan Sandler
Todd J. Sandler
Richard F. Wilhelm

Editorial Assistants:
Audrey Lewandowski
Emilie A. Whealen

ISBN 0-8031-2712-X (set)
ISBN 0-8031-2646-8 (section)
ISBN 0-8031-2647-6 (volume)

Photocopy Rights

Printed in Easton, MD, U.S.A.

Foreword

Organized in 1898, ASTM has grown into one of the largest voluntary standards development systems in the world. ASTM is a not-for-profit organization which provides a forum for producers, users, ultimate consumers, and those having a general interest (representatives of government and academia) to meet on common ground and write standards for materials, products, systems, and services.

From the work of 130 standards-writing committees, ASTM publishes more than 10,700 standards each year. These standards and other related technical information are used throughout the world.

ASTM Headquarters has no technical research or testing facilities; such work is done voluntarily by 31,100 technically qualified ASTM members located throughout the world. Membership in the Society is open to all concerned within the fields in which ASTM is active. A membership application may be found at the back of this volume. Additional information may be obtained from Member and Committee Services, ASTM, 100 Barr Harbor Drive, West Conshohocken, PA 19428; tel. (610) 832-9693.

1999 Annual Book of ASTM Standards

The 1999 *Annual Book of ASTM Standards* consists of 73 volumes, divided among 16 sections, of which this volume is one. It contains approved ASTM standards, provisional standards, and related material. These terms are defined as follows in the *Regulations Governing ASTM Technical Committees:*

Categories:

standard, n—as used in ASTM, a document that has been developed and established within the consensus principles of the Society and that meets the approval requirements of ASTM procedures and regulations.

standard, adj.—as used in ASTM, a descriptive used in titles of test methods, specifications, and other documents to indicate consensus approval in accordance with ASTM procedures and regulations.

provisional standard—a document published for a limited period of time by the Society to meet a demand for more rapid issuance of specific documents, such as an emergency situation, regulatory requirements, or other special circumstances.

Discussion—Provisional standards are not full consensus documents because they require subcommittee consensus only. (See the *Regulations Governing ASTM Technical Committees*, Section 14.)

Types:

The various types of ASTM documents are to provide a flexibility of form, communication, and usage for both the technical committees and the myriad users of ASTM documents. The type of ASTM document that is developed and titled is based on the technical content and intended use, not on the degree of consensus achieved. The two categories of ASTM documents (standard and provisional standard) can be of the following forms and types:

classification—a systematic arrangement or division of materials, products, systems, or services into groups based on similar characteristics such as origin, composition, properties, or use.

guide—a compendium of information or series of options that do not recommend a specific course of action.
Discussion—A guide increases the awareness of information and approaches in a given subject area.

practice—a definitive set of instructions for performing one or more specific operations or functions that does not produce a test result.
Discussion—Examples of practices include, but are not limited to: application, assessment, cleaning, collection, decontamination, inspection, installation, preparation, sampling, screening, and training.

specification—an explicit set of requirements to be satisfied by a material, product, system, or service.
Discussion—Examples of specifications include, but are not limited to, requirements for: physical, mechanical, or chemical properties, and safety, quality, or performance criteria. A specification identifies the test methods for determining whether each of the requirements is satisfied.

terminology—a document comprising definitions of terms; descriptions of terms; explanations of symbols, abbreviations, or acronyms.

test method—a definitive procedure that produces a test result.

Discussion—Examples of test methods include, but are not limited to: identification, measurement, and evaluation of one or more qualities, characteristics, or properties. A precision and bias statement shall be reported at the end of a test method. (Refer to Section A21 on Precision and Bias.)

A new edition of the Book of Standards is published annually because of additions of new standards and significant revisions to existing standards. Approximately 30 % of each volume is new or revised. Each volume contains all actions approved by the Society at least six months before the publication date. New and revised standards approved by the Society between the annual editions of any given volume are made available as separate copies. Users are cautioned to follow the most current issue of a standard except when a specific edition of a standard is cited, for example, as in a contract.

Development and Use of ASTM Standards

ASTM believes that technically competent standards result when a full consensus of all concerned parties is achieved and rigorous due process procedures are followed. This philosophy and standards development system ensures technically competent standards having the highest credibility when critically examined and used as the basis for commercial, legal, or regulatory actions.

ASTM standards are developed voluntarily and used voluntarily. Standards become legally binding only when a government body references them in regulations, or when they are cited in a contract. Any item that is produced and marked as conforming to an ASTM standard must meet all applicable requirements of that standard.

ASTM standards are used by thousands of individuals, companies, and agencies. Purchasers and sellers incorporate standards into contracts; scientists and engineers use them in laboratories; architects and designers use them in plans; government agencies reference them in codes, regulations, and laws; and many others refer to standards for guidance.

Consideration of Comments on ASTM Standards

An ASTM standard is subject to revision at any time by the responsible technical committee and must be reviewed every five years and if not revised, either reapproved or withdrawn. Your comments are invited either for revision of any standard or for the development of new standards and should be addressed to ASTM Headquarters. Your comments will receive careful consideration at a meeting of the responsible technical committee, which you may attend. If you feel that your comments have not received a fair hearing, you should make your views known to the ASTM Committee on Standards, 100 Barr Harbor Drive, West Conshohocken, PA 19428.

Using the Annual Book of ASTM Standards

The standards are assembled in each volume in alphanumeric sequence of their ASTM designation numbers. Volumes 03.06, 04.09, 05.03, 05.04, 05.05, and 06.03 are assembled first by committee, then in alphanumeric sequence. Each volume has a table of contents, listing the standards in alphanumeric sequence by ASTM designation; and a list by subjects, categorizing the standards according to subject. A subject index of the standards in each volume appears at the back of each volume.

Availability of Individual Standards

Each ASTM standard is available as a separate copy from ASTM. Standards can be ordered from the ASTM website at http://www.astm.org, in the store section. Standards can also be ordered from Customer Services at (610) 832-9585, Monday through Friday, 8:00 AM - 5:00 PM Eastern Time.

Caveat Statements and Policies in Standards

ASTM caveat statements on Safety Hazards and Fire Hazards are required to appear in standards where appropriate. They are located in the scope section of applicable standards. The caveats on General Statement of ASTM Policy and Patents are contained in all standards and located at the end of each standard. For more information on the caveats see Section F2 of the *Form and Style for ASTM Standards.*

ANNUAL BOOK OF ASTM STANDARDS

Listed by Section and Volume

Section 1—Iron and Steel Products

Volume 01.01 Steel—Piping, Tubing, Fittings
Volume 01.02 Ferrous Castings; Ferroalloys
Volume 01.03 Steel—Plate, Sheet, Strip, Wire; Stainless Steel Bar
Volume 01.04 Steel—Structural, Reinforcing, Pressure Vessel, Railway
Volume 01.05 Steel—Bars, Forgings, Bearing, Chain, Springs
Volume 01.06 Coated Steel Products
Volume 01.07 Ships and Marine Technology

Section 2—Nonferrous Metal Products

Volume 02.01 Copper and Copper Alloys
Volume 02.02 Aluminum and Magnesium Alloys
Volume 02.03 Electrical Conductors
Volume 02.04 Nonferrous Metals—Nickel, Cobalt, Lead, Tin, Zinc, Cadmium, Precious, Reactive, Refractory Metals and Alloys Materials for Thermostats, Electrical Heating and Resistance, Contacts, and Connectors
Volume 02.05 Metallic and Inorganic Coatings; Metal Powders, Sintered P/M Structural Parts

Section 3—Metals Test Methods and Analytical Procedures

Volume 03.01 Metals—Mechanical Testing; Elevated and Low-Temperature Tests; Metallography
Volume 03.02 Wear and Erosion; Metal Corrosion
Volume 03.03 Nondestructive Testing
Volume 03.04 Magnetic Properties
Volume 03.05 Analytical Chemistry for Metals, Ores, and Related Materials (I): C 571 - E 354
Volume 03.06 Analytical Chemistry for Metals, Ores, and Related Materials (II): E 356 - latest; Molecular Spectroscopy; Surface Analysis

Section 4—Construction

Volume 04.01 Cement; Lime; Gypsum
Volume 04.02 Concrete and Aggregates
Volume 04.03 Road and Paving Materials; Vehicle-Pavement Systems
Volume 04.04 Roofing, Waterproofing, and Bituminous Materials
Volume 04.05 Chemical-Resistant Nonmetallic Materials; Vitrified Clay Pipe, Concrete, Fiber-Cement Products; Mortars and Grouts; Masonry
Volume 04.06 Thermal Insulation; Environmental Acoustics
Volume 04.07 Building Seals and Sealants; Fire Standards; Dimension Stone
Volume 04.08 Soil and Rock (I): D 420 - D 4914
Volume 04.09 Soil and Rock (II): D 4943 - latest; Geosynthetics
Volume 04.10 Wood
Volume 04.11 Building Constructions

Section 5—Petroleum Products, Lubricants, and Fossil Fuels

Volume 05.01 Petroleum Products and Lubricants (I): D 56 - D 2596
Volume 05.02 Petroleum Products and Lubricants (II): D 2597 - D 4927
Volume 05.03 Petroleum Products and Lubricants (III): D 4928 - latest
Volume 05.04 Test Methods for Rating Motor, Diesel, and Aviation Fuels; Catalysts
Volume 05.05 Gaseous Fuels; Coal and Coke

Section 6—Paints, Related Coatings, and Aromatics

Volume 06.01 Paint—Tests for Chemical, Physical, and Optical Properties; Appearance
Volume 06.02 Paint—Products and Applications; Protective Coatings; Pipeline Coatings
Volume 06.03 Paint—Pigments, Drying Oils, Polymers, Resins, Naval Stores, Cellulosic Esters, and Ink Vehicles
Volume 06.04 Paint—Solvents; Aromatic Hydrocarbons

LISTED BY SECTION AND VOLUME

Section 7—Textiles
Volume 07.01 Textiles (I): D 76 - D 3218
Volume 07.02 Textiles (II): D 3333 - latest

Section 8—Plastics
Volume 08.01 Plastics (I): D 256 - D 2343
Volume 08.02 Plastics (II): D 2383 - D 4322
Volume 08.03 Plastics (III): D 4329 - latest
Volume 08.04 Plastic Pipe and Building Products

Section 9—Rubber
Volume 09.01 Rubber, Natural and Synthetic—General Test Methods; Carbon Black
Volume 09.02 Rubber Products, Industrial—Specifications and Related Test Methods; Gaskets; Tires

Section 10—Electrical Insulation and Electronics
Volume 10.01 Electrical Insulation (I)—D 69 - D 2484
Volume 10.02 Electrical Insulation (II)—D 2518 - latest
Volume 10.03 Electrical Insulating Liquids and Gases; Electrical Protective Equipment
Volume 10.04 Electronics (I)
Volume 10.05 Electronics (II)

Section 11—Water and Environmental Technology
Volume 11.01 Water (I)
Volume 11.02 Water (II)
Volume 11.03 Atmospheric Analysis; Occupational Health and Safety; Protective Clothing
Volume 11.04 Environmental Assessment; Hazardous Substances and Oil Spill Response; Waste Management; Environmental Risk Management
Volume 11.05 Biological Effects and Environmental Fate; Biotechnology; Pesticides

Section 12—Nuclear, Solar, and Geothermal Energy
Volume 12.01 Nuclear Energy (I)
Volume 12.02 Nuclear (II), Solar, and Geothermal Energy

Section 13—Medical Devices and Services
Volume 13.01 Medical Devices; Emergency Medical Services

Section 14—General Methods and Instrumentation
Volume 14.01 Healthcare Informatics; Computerized Systems and Chemical and Material Information
Volume 14.02 General Test Methods; Chromatography; Forensic Sciences; Terminology; Conformity Assessment; Statistical Methods
Volume 14.03 Temperature Measurement
Volume 14.04 Laboratory Apparatus, Degradation of Materials, Filtration, SI, Oxygen Fire Safety

Section 15—General Products, Chemical Specialties, and End Use Products
Volume 15.01 Refractories; Carbon and Graphite Products; Activated Carbon; Advanced Ceramics
Volume 15.02 Glass; Ceramic Whitewares
Volume 15.03 Space Simulation; Aerospace and Aircraft; High Modulus Fibers and Composites
Volume 15.04 Soaps and Other Detergents; Polishes; Leather; Resilient Floor Coverings
Volume 15.05 Engine Coolants; Halogenated Organic Solvents and Fire Extinguishing Agents; Industrial Chemicals
Volume 15.06 Adhesives
Volume 15.07 End Use Products
Volume 15.08 Fasteners
Volume 15.09 Paper; Packaging; Flexible Barrier Materials; Business Imaging Products

Section 00—Index
Volume 00.01 Subject Index and Alphanumeric List

ANNUAL BOOK OF ASTM STANDARDS

Listed by Subjects

SUBJECTS	VOLUME
Acoustics, Environmental	04.06
Activated Carbon	15.01
Adhesives	15.06
Advanced Ceramics	15.01
Aerosols	15.09
Aerospace Industry Methods	15.03
Aggregates	04.02, 04.03
Aluminum and Aluminum Alloys	02.02
Amusement Rides and Devices	15.07
Analytical Atomic Spectroscopy	03.05, 03.06
Anesthetic and Respiratory Equipment	13.01
Appearance of Materials	06.01
Aromatic Hydrocarbons and Related Chemicals	06.04
Atmospheric Analysis	11.03
Biological Effects and Environmental Fate	11.05
Biotechnology	11.05
Bituminous Materials	04.03
Building Constructions	04.11
Building Seals and Sealants	04.07
Business Imaging Products	15.09
Carbon Black	09.01
Carbon Products, Manufactured	15.01
Cast Iron	01.02
Catalysts	05.04
Cellulose	06.02
Cement	04.01
Hydraulic	04.01
Rubber	09.01
Ceramic Materials	15.02
Advanced Ceramics	15.01
Ceramic Whitewares	15.02
Ceramics for Electronics	15.02
Porcelain Enamel	02.05
Chemical Analysis of Metals	03.05, 03.06
Chemical-Resistant Nonmetallic Materials	04.05
Chemicals, Industrial	15.05
Chromatography	14.02
Closures	15.09
Coal and Coke	05.05
Compatibility and Sensitivity of Materials in Oxygen-Enriched Atmospheres	14.04
Computerized Systems and Chemical and Material Information	14.01
Concrete and Concrete Aggregates	04.02
Concrete Pipe and Tile	04.05
Concrete Products, Precast	04.05
Concrete Reinforcing Steel	01.04
Conformity Assessment	14.02
Consumer Products	15.07
Containers:	
Aerosol	15.09
Glass	15.02
Shipping	15.09
Coolants, Engine	15.05
Copper and Copper Alloys	02.01
Corrosion, Metal	03.02
Detention and Correctional Facilities	15.07
Die-Cast Metals	02.01, 02.02, 02.04
Dimension Stone	04.07
Dosimetry	12.02
Ductile Iron	01.02
Durability of Nonmetallic Materials:	
General Practices	14.02
Pipeline Coatings	06.02
Electrical Conductors	02.03
Electrical Contacts and Connectors	02.04
Electrical Insulating Materials	10.01, 10.02, 10.03
Electrical Protective Equipment for Workers	10.03
Electronics	10.04, 10.05
Emergency Medical Services	13.01
Emission Spectroscopy	03.05, 03.06
Environmental Assessment	11.04
Environmental Risk Management	11.05
Erosion and Wear	03.02
Exposure Tests	14.02
Fasteners	15.08
Fatigue	03.01
Fences	01.06
Ferroalloys	01.02
Ferrous Castings	01.02
Fiber-Cement Products	04.05
Filtration	14.04
Fire Standards	04.07
Flexible Barrier Materials	15.09
Food Service Equipment	15.07
Footwear, Safety and Traction for	15.07
Forensic Sciences	14.02
Fracture Testing	03.01
Gaseous Fuels	05.05
Gaskets	09.02
Geosynthetics and Related Products	04.09
Geothermal Resources and Energy	12.02
Glass	15.02
Graphite Products, Manufactured	15.01
Graphite Products, Nuclear	12.02
Grouts for Unit Masonry	04.05
Gypsum	04.01
Halogenated Organic Solvents and Fire Extinguishing Agents	15.05
Hazard Potential of Chemicals	14.02
Hazardous Substances and Oil Spill Response	11.04
Healthcare Informatics	14.01

LISTED BY SUBJECTS

SUBJECTS	VOLUME
Healthcare Services and Equipment	13.01
High Modulus Fibers and Composites	15.03
Imaging Products, Business	15.09
Index (for all volumes)	00.01
Industrial Chemicals	15.05
Iron Castings	01.02
Knock Test Manual	05.04
Laboratory Apparatus	14.04
Leather	15.04
Lime	04.01
Magnesium and Magnesium Alloys	02.02
Magnetic Properties	03.04
Malleable Iron	01.02
Marine Technology	01.07
Masonry Units	04.05
Meat and Poultry	15.07
Medical and Surgical Materials and Devices	13.01
Metallic and Inorganic Coatings	02.05
Metallography	03.01
Metal Powders	02.05
Metals, Chemical Analysis	03.05, 03.06
Metals, Effect of Temperature on Properties	03.01
Metals, Physical and Mechanical Testing	03.01
Molecular Spectroscopy	03.06
Mortars for Unit Masonry	04.05
Naval Stores	06.03
Nickel and Nickel Alloys	02.04
Nondestructive Testing	03.03
Nonferrous Metals, General	02.04
General Test Methods	14.02
Nuclear Materials	12.01, 12.02
Occupational Health and Safety	11.03
Oil Spill Response, Hazardous Substances	11.04
Ores, Metal Bearing, Sampling and Analysis	03.05, 03.06
Orthotics, External Prosthetics, and Mobility Aids	13.01
Packaging	15.09
Paint and Related Coatings and Materials:	
Pigments, Resins, and Polymers	06.03
Products and Applications	06.02
Solvents	06.04
Tests for Chemical, Physical, and Optical Properties	06.01
Paper	15.09
Particle and Spray Organization	14.02
Pesticides	11.05
Petroleum Products and Lubricants	05.01, 05.02, 05.03, 05.04
Plastic Pipe and Building Products	08.04
Plastics	08.01, 08.02, 08.03
Polishes	15.04
Porcelain Enamel	02.05
Pressure Vessel Plate and Forgings	01.04
Products Liability Litigation, Technical Aspects of	14.02
Protective Clothing	11.03
Protective Coating and Lining Work for Power Generation Facilities	06.02
Protective Equipment, Electrical, for Workers	10.03
Quality and Statistics	14.02
Radioisotopes and Radiation Effects	12.02
Reactive and Refractory Metals	02.04
Refractories	15.01
Resilient Floor Coverings	15.04
Road and Paving Materials	04.03
Robotics	14.01
Roofing, Waterproofing, and Bituminous Materials	04.04
Rubber	09.01, 09.02
Search and Rescue	13.01
Security Systems and Equipment	15.07
Sensory Evaluation of Materials and Products	15.07
Ships	01.07
SI Practice	14.04
Sintered P/M Structural Parts	02.05
Skiing, Snow	15.07
Soaps	15.04
Soil and Rock	04.08, 04.09
Solar Energy Conversion	12.02
Solvents	06.04
Space Simulation	15.03
Spectroscopy	03.06
Sports Equipment and Facilities	15.07
Statistical Methods	14.02
Steel:	
Bars	01.05
Bearing Steel	01.05
Bolting	01.01
Castings	01.02
Chain	01.05
Concrete Reinforcing	01.04
Detention and Correctional Facilities	04.07
Fasteners	15.08
Forgings	01.04, 01.05
Galvanized	01.06
Piping, Tubing, and Fittings	01.01
Plate, Sheet, and Strip	01.03
Pressure Vessel Plate and Forgings	01.04
Rails, Wheels, and Tires	01.04
Springs	01.05
Stainless Steel	01.01, 01.02, 01.03, 01.04, 01.05
Structural Steel	01.04
Wire	01.03
Surface Analysis	03.06
Surgical Materials and Devices	13.01
Temperature Measurement	14.03
Terminology	14.02
Textiles	07.01, 07.02
Thermal Insulation	04.06
Thermal Measurements	14.02
Thermocouples	14.03
Thermostats, Electrical Heating and Resistance, Contacts, and Connectors	02.04
Tires	09.02
Traveled Surface Characteristics	04.03
Vacuum Cleaners	15.07
Vehicle-Pavement Systems	04.03
Vitrified Clay Pipe	04.05
Waste Management	11.04
Water	11.01, 11.02
Wear and Erosion	03.02
Weathering and Durability	14.04
Wood	04.10

Contents

1999 ANNUAL BOOK OF ASTM STANDARDS, VOLUME 04.01

CEMENT; LIME; GYPSUM AND RELATED BUILDING MATERIALS AND SYSTEMS (INCLUDING MANUAL OF CEMENT TESTING)

A complete Subject Index begins on p. 609

This contents includes only those standards included in Volume 04.01 and those standards that appeared previously that have been superseded or discontinued within the past three years. Since the standards in this book are arranged in alphanumeric sequence, no page numbers are given in this contents.

In the serial designations prefixed to the following titles, the number following the dash indicates the year of adoption as standard or, in the case of revision, the year of last revision. Thus, standards adopted or revised during the year 1999 have as their final number, 99. A letter following this number indicates more than one revision during that year, that is 99a indicates the second revision in 1999, 99b the third revision, etc. Standards that have been reapproved without change are indicated by the year of last reapproval in parentheses as part of the designation number, for example, (1999). A superscript epsilon indicates an editorial change since the last revision or reapproval—$\epsilon 1$ for the first change, $\epsilon 2$ for the second change, etc.

§C 5 – 79 (1997) Specifications for Quicklime for Structural Purposes
C 11 – 98 Terminology Relating to Gypsum and Related Building Materials and Systems
C 22/C 22M – 96$^{\epsilon 1}$ Specification for Gypsum
§C 25 – 98 Test Methods for Chemical Analysis of Limestone, Quicklime, and Hydrated Lime
§C 28 – 96$^{\epsilon 1}$ Specification for Gypsum Plasters
§C 35 – 95 Specification for Inorganic Aggregates for Use in Gypsum Plaster
§C 36 – 97 Specification for Gypsum Wallboard
§C 37 – 95 Specification for Gypsum Lath
§C 50 – 94 Practice for Sampling, Inspection, Packing, and Marking of Lime and Limestone Products
§C 51 – 98 Terminology Relating to Lime and Limestone (As used by the Industry)
§C 59 – 95 Specification for Gypsum Casting and Molding Plaster
§C 61 – 95 Specification for Gypsum Keene's Cement
§C 79/C 79M – 97 Specification for Gypsum Sheathing Board
§C 91 – 98 Specification for Masonry Cement
§C 109/C 109M – 99 Test Method for Compressive Strength of Hydraulic Cement Mortars (Using 2-in. or [50-mm] Cube Specimens)
§C 110 – 98 Test Methods for Physical Testing of Quicklime, Hydrated Lime, and Limestone
C 114 – 99 Test Methods for Chemical Analysis of Hydraulic Cement
C 115 – 96a Test Method for Fineness of Portland Cement by the Turbidimeter
C 141 – 97 Specification for Hydraulic Hydrated Lime for Structural Purposes
§C 150 – 98 Specification for Portland Cement
C 151 – 98a Test Method for Autoclave Expansion of Portland Cement
C 183 – 97 Practice for Sampling and the Amount of Testing of Hydraulic Cement
C 184 – 94$^{\epsilon 1}$ Test Method for Fineness of Hydraulic Cement by the 150-μm (No. 100) and 75-μm (No. 200) Sieves
§C 185 – 99 Test Method for Air Content of Hydraulic Cement Mortar
C 186 – 98 Test Method for Heat of Hydration of Hydraulic Cement
C 187 – 98 Test Method for Normal Consistency of Hydraulic Cement
C 188 – 95 Test Method for Density of Hydraulic Cement
C 191 – 99 Test Method for Time of Setting of Hydraulic Cement by Vicat Needle
C 204 – 96a Test Method for Fineness of Hydraulic Cement by Air Permeability Apparatus
§C 206 – 84 (1997) Specification for Finishing Hydrated Lime
§C 207 – 91 (1997) Specification for Hydrated Lime for Masonry Purposes
C 219 – 99 Terminology Relating to Hydraulic Cement
§C 226 – 96 Specification for Air-Entraining Additions for Use in the Manufacture of Air-Entraining Hydraulic Cement
C 230 – 98 Specification for Flow Table for Use in Tests of Hydraulic Cement
C 243 – 95 Test Method for Bleeding of Cement Pastes and Mortars
C 265 – 99 Test Method for Calcium Sulfate in Hydrated Portland Cement Mortar
C 266 – 99 Test Method for Time of Setting of Hydraulic-Cement Paste by Gillmore Needles
C 305 – 94 Practice for Mechanical Mixing of Hydraulic Cement Pastes and Mortars of Plastic Consistency

§ Approved for use by agencies of the Department of Defense. Consult the DoD Index of Specifications and Standards for the specific year of issue which has been adopted by the Department of Defense.

CONTENTS

§C 317 – 98	Specification for Gypsum Concrete
C 318/C 318M – 96	Specification for Gypsum Formboard
C 348 – 97	Test Method for Flexural Strength of Hydraulic Cement Mortars
C 349 – 97	Test Method for Compressive Strength of Hydraulic Cement Mortars (Using Portions of Prisms Broken in Flexure)
C 359 – 99	Test Method for Early Stiffening of Portland Cement (Mortar Method)
C 400 – 98	Test Methods for Quicklime and Hydrated Lime for Neutralization of Waste Acid
C 430 – 96	Test Method for Fineness of Hydraulic Cement by the 45-μm (No. 325) Sieve
§C 442 – 98	Specification for Gypsum Backing Board and Coreboard
C 451 – 99	Test Method for Early Stiffening of Hydraulic Cement (Paste Method)
C 452 – 95	Test Method for Potential Expansion of Portland Cement Mortars Exposed to Sulfate
C 465 – 99	Specification for Processing Additions for Use in the Manufacture of Hydraulic Cements
C 471M – 96	Test Methods for Chemical Analysis of Gypsum and Gypsum Products [Metric]
§C 472 – 98	Test Methods for Physical Testing of Gypsum, Gypsum Plasters and Gypsum Concrete
C 473 – 97$^{\epsilon1}$	Test Methods for Physical Testing of Gypsum Panel Products
C 474 – 97	Test Methods for Joint Treatment Materials for Gypsum Board Construction
§C 475 – 94	Specification for Joint Compound and Joint Tape for Finishing Gypsum Board
C 490 – 97	Practice for Use of Apparatus for the Determination of Length Change of Hardened Cement Paste, Mortar, and Concrete
C 511 – 98	Specification for Moist Cabinets, Moist Rooms, and Water Storage Tanks Used in the Testing of Hydraulic Cements and Concretes
§C 514 – 96	Specification for Nails for the Application of Gypsum Board
C 563 – 96	Test Method for Optimum SO_3 in Hydraulic Cement Using 24-h Compressive Strength
§C 587 – 97	Specification for Gypsum Veneer Plaster
§C 588 – 95a	Specification for Gypsum Base for Veneer Plasters
C 593 – 95	Specification for Fly Ash and Other Pozzolans for Use with Lime
§C 595 – 98	Specification for Blended Hydraulic Cements
C 596 – 96	Test Method for Drying Shrinkage of Mortar Containing Portland Cement
§C 602 – 95a	Specification for Agricultural Liming Materials
§C 630/C 630M – 96a	Specification for Water-Resistant Gypsum Backing Board
C 631 – 95a	Specification for Bonding Compounds for Interior Gypsum Plastering
C 645 – 98	Specification for Nonstructural Steel Framing Members
C 688 – 96	Specification for Functional Additions for Use in Hydraulic Cements
C 706 – 98	Specification for Limestone for Animal Feed Use
C 737 – 97	Specification for Limestone for Dusting of Coal Mines
§C 754 – 98a	Specification for Installation of Steel Framing Members to Receive Screw-Attached Gypsum Panel Products
§C 778 – 99	Specification for Standard Sand
C 786 – 96	Test Method for Fineness of Hydraulic Cement and Raw Materials by the 300-μm (No. 50), 150-μm (No. 100), and 75-μm (No. 200) Sieves by Wet Methods
C 806 – 95	Test Method for Restrained Expansion of Expansive Cement Mortar
C 807 – 99	Test Method for Time of Setting of Hydraulic Cement Mortar by Modified Vicat Needle
C 821 – 78 (1995)	Specification for Lime for Use with Pozzolans
§C 840 – 98	Specification for Application and Finishing of Gypsum Board
§C 841 – 98	Specification for Installation of Interior Lathing and Furring
C 842 – 98	Specification for Application of Interior Gypsum Plaster
§C 843 – 98	Specification for Application of Gypsum Veneer Plaster
§C 844 – 98a	Specification for Application of Gypsum Base to Receive Gypsum Veneer Plaster
C 845 – 96	Specification for Expansive Hydraulic Cement
§C 847 – 95	Specification for Metal Lath
C 897 – 96	Specification for Aggregate for Job-Mixed Portland Cement-Based Plasters
C 911 – 96	Specification for Quicklime, Hydrated Lime, and Limestone for Chemical Uses
C 917 – 98	Test Method for Evaluation of Cement Strength Uniformity from a Single Source
C 926 – 98a	Specification for Application of Portland Cement-Based Plaster
§C 931/C 931M – 98	Specification for Exterior Gypsum Soffit Board
§C 932 – 98a	Specification for Surface-Applied Bonding Compounds for Exterior Plastering
C 933 – 96a	Specification for Welded Wire Lath
C 954 – 98	Specification for Steel Drill Screws for the Application of Gypsum Panel Products or Metal Plaster Bases to Steel Studs from 0.033 in. (0.84 mm) to 0.112 in. (2.84 mm) in Thickness
C 955 – 98	Specification for Load-Bearing (Transverse and Axial) Steel Studs, Runners (Track), and Bracing or Bridging for Screw Application of Gypsum Panel Products and Metal Plaster Bases
C 956 – 97	Specification for Installation of Cast-in-Place Reinforced Gypsum Concrete
C 959 – 91	Specification for Radiant Heat Fill (Discontinued 1996†)
§C 960/C 960M – 97	Specification for Predecorated Gypsum Board
C 977 – 95	Specification for Quicklime and Hydrated Lime for Soil Stabilization

† Although this standard has been officially withdrawn from society approval, a brief description is included for information only.

CONTENTS

§C 1002 – 98 Specification for Steel Drill Screws for the Application of Gypsum Panel Products or Metal Plaster Bases
C 1005 – 97 Specification for Weights and Weighing Devices for Use in the Physical Testing of Hydraulic Cements
C 1007 – 98$^{\epsilon 1}$ Specification for the Installation of Load-Bearing (Transverse and Axial) Steel Studs and Related Accessories
C 1012 – 95a Test Method for Length Change of Hydraulic-Cement Mortars Exposed to a Sulfate Solution
C 1032 – 96 Specification for Woven Wire Plaster Base
C 1038 – 95 Test Method for Expansion of Portland Cement Mortar Bars Stored in Water
C 1047 – 98 Specification for Accessories for Gypsum Wallboard and Gypsum Veneer Base
§C 1063 – 98 Specification for Installation of Lathing and Furring to Receive Interior and Exterior Portland Cement-Based Plaster
C 1097 – 95 Specification for Hydrated Lime for Use in Asphaltic Concrete Mixtures
C 1157 – 98a Performance Specification for Hydraulic Cement
C 1164 – 92 (1997)$^{\epsilon 1}$ Practice for Evaluation of Limestone or Lime Uniformity From a Single Source
C 1177/C 1177M – 96 Specification for Glass Mat Gypsum Substrate for Use as Sheathing
C 1178/C 1178M – 96 Specification for Glass Mat Water-Resistant Gypsum Backing Board
C 1222 – 97 Practice for Evaluation of Laboratories Testing Hydraulic Cement
C 1264/C 1264M – 95 Specification for Sampling, Inspection, Rejection, Certification, Packaging, Marking, Shipping, Handling, and Storage of Gypsum Board
C 1271 – 94a Test Method for X-ray Spectrometric Analysis of Lime and Limestone
C 1278/C 1278M – 96 Specification for Fiber-Reinforced Gypsum Panels
C 1280 – 98 Specification for Application of Gypsum Sheathing
C 1301 – 95 Test Method for Major and Trace Elements in Limestone and Lime by Inductively Coupled Plasma-Atomic Emission Spectroscopy (ICP) and Atomic Absorption (AA)
C 1318 – 95 Test Method for Determination of Total Neutralizing Capability and Dissolved Calcium and Magnesium Oxide in Lime for Flue Gas Desulferization (FGD)
C 1328 – 98 Specification for Plastic (Stucco) Cement
C 1329 – 98 Specification for Mortar Cement
C 1355/C 1355M – 96 Specification for Glass Fiber Reinforced Gypsum Composites
C 1356 – 96 Test Method for the Quantitative Determination of Phases in Portland Cement Clinker by Microscopial Point-Count Procedure
C 1365 – 98 Test Method for Determination of the Proportion of Phases in Portland Cement and Portland-Cement Clinker Using X-Ray Powder Diffraction Analysis
C 1381 – 97 Specification for Molded Glass Fiber Reinforced Gypsum Parts
C 1395/C 1395M – 98 Specification for Gypsum Ceiling Board
C 1396/C 1396M – 98 Specification for Gypsum Board
C 1397 – 98 Practice for Application of Class PB Exterior Insulation and Finish Systems
PS 81 – 98$^{\epsilon 1}$ Provisional Test Method for Length Change of Mortar Bars Caused by Reaction Between Hydraulic Cement and Alkali-Reactive Aggregate

RELATED MATERIAL

PAGE

Manual of Cement Testing 585
SI Quick Reference Guide 603
Metric Practice Guide for Cement 607
Index 609
ASTM Membership Application Blank

List by Subjects

1999 ANNUAL BOOK OF ASTM STANDARDS, VOLUME 04.01

CEMENT; LIME; GYPSUM AND RELATED BUILDING MATERIALS AND SYSTEMS (INCLUDING MANUAL OF CEMENT TESTING)

A complete Subject Index begins on p. 609

Since the standards in this book are arranged in alphanumeric order, no page numbers are included in this list by subjects. The standards listed in italics are related documents included for information only and do not appear in this volume.

CEMENT

Specifications for:

§C 226 – 96	Air-Entraining Additions for Use in the Manufacture of Air-Entraining Hydraulic Cement
§C 595 – 98	Blended Hydraulic Cements
C 1157 – 98a	Hydraulic Cement, Performance
C 845 – 96	Expansive Hydraulic Cement
C 230 – 98	Flow Table for Use in Tests of Hydraulic Cement
C 688 – 96	Functional Additions for Use in Hydraulic Cements
§C 91 – 98	Masonry Cement
C 511 – 98	Moist Cabinets, Moist Rooms, and Water Storage Tanks Used in the Testing of Hydraulic Cements and Concretes
§C 150 – 98	Portland Cement
C 465 – 99	Processing Additions for Use in the Manufacture of Hydraulic Cements
§C 778 – 99	Standard Sand
C 1005 – 97	Weights and Weighing Devices for Use in the Physical Testing of Hydraulic Cements
C 1328 – 98	Plastic (Stucco) Cement
C 1329 – 98	Mortar Cement

Test Methods for:

§C 185 – 99	Air Content of Hydraulic Cement Mortar
C 151 – 98a	Autoclave Expansion of Portland Cement
C 243 – 95	Bleeding of Cement Pastes and Mortars
C 265 – 99	Calcium Sulfate in Hydrated Portland Cement Mortar
C 114 – 99	Chemical Analysis of Hydraulic Cement
§C 109/C 109M – 99	Compressive Strength of Hydraulic Cement Mortars (Using 2-in. or [50-mm] Cube Specimens)
C 349 – 97	Compressive Strength of Hydraulic Cement Mortars (Using Portions of Prisms Broken in Flexure)
C 188 – 95	Density of Hydraulic Cement
C 1365 – 98	Determination of the Proportion of Phases in Portland Cement Clinker Using X-Ray Powder Diffraction Analysis
C 596 – 96	Drying Shrinkage of Mortar Containing Portland Cement
C 359 – 99	Early Stiffening of Portland Cement (Mortar Method)
C 451 – 99	Early Stiffening of Hydraulic Cement (Paste Method)
C 917 – 98	Evaluation of Cement Strength Uniformity from a Single Source
C 1038 – 95	Expansion of Portland Cement Mortar Bars Stored in Water
C 452 – 95	Expansion, Potential, of Portland Cement Mortars Exposed to Sulfate
C 786 – 96	Fineness of Hydraulic Cement and Raw Materials by the 300-μm (No. 50), 150-μm (No. 100), and 75-μm (No. 200) Sieves by Wet Methods
C 184 – 94$^{\epsilon 1}$	Fineness of Hydraulic Cement by the 150-μm (No. 100) and 75-μm (No. 200) Sieves
C 430 – 96	Fineness of Hydraulic Cement by the 45-μm (No. 325) Sieve
C 204 – 96a	Fineness of Hydraulic Cement by Air Permeability Apparatus
C 115 – 96a	Fineness of Portland Cement by the Turbidimeter
C 348 – 97	Flexural Strength of Hydraulic Cement Mortars
C 186 – 98	Heat of Hydration of Hydraulic Cement
C 1012 – 95a	Length Change of Hydraulic-Cement Mortars Exposed to a Sulfate Solution

§ Approved for use by agencies of the Department of Defense. Consult the DoD Index of Specifications and Standards for the specific year of issue which has been adopted by the Department of Defense.

LIST BY SUBJECTS

Test Methods for:

PS 81 – 98$^{\epsilon 1}$	Length Change of Mortar Bars Caused by Reaction Between Hydraulic Cement and Alkali-Reactive Aggregate
C 187 – 98	Normal Consistency of Hydraulic Cement
C 563 – 96	Optimum SO_3 in Hydraulic Cement Using 24-h Compressive Strength
C 1356 – 96	Quantitative Determination of Phases in Portland Cement Clinker by Microscopial Point-Count Procedure
C 806 – 95	Restrained Expansion of Expansive Cement Mortar
C 266 – 99	Time of Setting of Hydraulic-Cement Paste by Gillmore Needles
C 191 – 92	Time of Setting of Hydraulic Cement by Vicat Needle
C 807 – 99	Time of Setting of Hydraulic Cement Mortar by Modified Vicat Needle

Terminology Relating to:

C 219 – 99	Hydraulic Cement

Practices for:

C 490 – 97	Apparatus for the Determination of Length Change of Hardened Cement Paste, Mortar, and Concrete, Use of
C 1222 – 97	Evaluation of Laboratories Testing Hydraulic Cement
C 305 – 94	Mechanical Mixing of Hydraulic Cement Pastes and Mortars of Plastic Consistency
C 183 – 97	Sampling and the Amount of Testing of Hydraulic Cement

Related Materials:

Manual of Cement Testing

LIME

Specifications for:

§C 602 – 95a	Agricultural Liming Materials
§C 206 – 84 (1997)	Finishing Hydrated Lime
C 593 – 95	Fly Ash and Other Pozzolans for Use with Lime
§C 207 – 91 (1997)	Hydrated Lime for Masonry Purposes
C 141 – 97	Hydraulic Hydrated Lime for Structural Purposes
§C 35 – 95	Inorganic Aggregates for Use in Gypsum Plaster
C 821 – 78 (1995)	Lime for Use with Pozzolans
C 706 – 98	Limestone for Animal Feed Use
C 737 – 97	Limestone for Dusting of Coal Mines
C 977 – 95	Quicklime and Hydrated Lime for Soil Stabilization
§C 5 – 79 (1997)	Quicklime for Structural Purposes
C 911 – 96	Quicklime, Hydrated Lime, and Limestone for Chemical Uses
C 1097 – 95	Hydrated Lime for Use in Asphaltic Concrete Mixtures

Test Methods for:

§C 25 – 98	Chemical Analysis of Limestone, Quicklime, and Hydrated Lime
C 400 – 98	Neutralization of Waste Acid, Testing Quicklime and Hydrated Lime for
§C 110 – 98	Physical Testing of Quicklime, Hydrated Lime, and Limestone
§C 50 – 94	Sampling, Inspection, Packing, and Marking of Lime and Limestone Products
C 1271 – 94a	X-ray Spectrometric Analysis of Lime and Limestone
C 1301 – 95	Major and Trace Elements in Limestone and Lime by Inductively Coupled Plasma-Atomic Emission Spectroscopy (ICP) and Atomic Absorption (AA)
C 1318 – 95	Determination of Total Neutralizing Capability and Dissolved Calcium and Magnesium Oxide in Lime for Flue Gas Desulferization (FGD)
E 119 – 98	*Fire Tests of Building Construction and Materials (see Vol 04.07)*

Terminology Relating to:

§C 51 – 98	Lime and Limestone (As used by the Industry)

GYPSUM AND RELATED BUILDING MATERIALS AND SYSTEMS

Specifications for:

C 1047 – 98	Accessories for Gypsum Wallboard and Gypsum Veneer Base
C 897 – 96	Aggregate for Job-Mixed Portland Cement-Based Plasters
§C 840 – 98	Application and Finishing of Gypsum Board
C 1397 – 98	Application of Class PB Exterior Insulation and Finish Systems
§C 844 – 98a	Application of Gypsum Base to Receive Gypsum Veneer Plaster
C 1280 – 98	Application of Gypsum Sheathing Board

LIST BY SUBJECTS

Specifications for:

Designation	Title
§C 843 – 98	Application of Gypsum Veneer Plaster
C 842 – 98	Application of Interior Gypsum Plaster
C 926 – 98a	Application of Portland Cement-Based Plaster
C 631 – 95a	Bonding Compounds for Interior Plastering
§C 931/C 931M – 98	Exterior Gypsum Soffit Board
C 1278/C 1278M – 96	Fiber-Reinforced Gypsum Panels
C 1355/C 1355M – 96	Glass Fiber Reinforced Gypsum Composites
C 1177/C 1177M – 96	Glass Mat Gypsum Substrate for Use as Sheathing
C 1178/C 1178M – 96	Glass Mat Water-Resistant Gypsum Backing Board
C 22/C 22M – 96$^{\epsilon 1}$	Gypsum
§C 442 – 98	Gypsum Backing Board and Coreboard
§C 588 – 95a	Gypsum Base for Veneer Plasters
C 1396/C 1396M – 98	Gypsum Board
§C 475 – 94	Gypsum Board Finishing, Joint Compound and Joint Tape for
§C 59 – 95	Gypsum Casting and Molding Plaster
C 1395/C 1395M – 98	Gypsum Ceiling Board
§C 317 – 98	Gypsum Concrete
§C 61 – 95	Gypsum Keene's Cement
§C 37 – 95	Gypsum Lath
§C 28 – 96$^{\epsilon 1}$	Gypsum Plasters
§C 79/C 79M – 97	Gypsum Sheathing Board
§C 587 – 97	Gypsum Veneer Plaster
§C 36 – 97	Gypsum Wallboard
§C 35 – 95	Inorganic Aggregates for Use in Gypsum Plaster
C 956 – 97	Installation of Cast-in-Place Reinforced Gypsum Concrete
§C 841 – 98	Installation of Interior Lathing and Furring
C 1063 – 98	Installation of Lathing and Furring to Receive Interior and Exterior Portland Cement-Based Plaster
C 1007 – 98$^{\epsilon 1}$	Installation of Load-Bearing (Transverse and Axial) Steel Studs and Related Accessories
§C 754 – 98a	Installation of Steel Framing Members to Receive Screw-Attached Gypsum Panel Products
C 955 – 98	Load-Bearing (Transverse and Axial) Steel Studs, Runners (Track), and Bracing or Bridging for Screw Application of Gypsum Panel Products and Metal Plaster Bases
§C 847 – 95	Metal Lath
C 1381 – 97	Molded Glass Fiber Reinforced Gypsum Parts
§C 514 – 96	Nails for the Application of Gypsum Board
C 645 – 98	Nonstructural Steel Framing Members
§C 960/C 960M – 97	Predecorated Gypsum Board
C 959 – 91	Radiant Heat Fill (Discontinued 1996†)
C 1264/C 1264M – 95	Sampling, Inspection, Rejection, Certification, Packaging, Marking, Handling and Storage of Gypsum Board
§C 1002 – 98	Steel Drill Screws for the Application of Gypsum Panel Products or Metal Plaster Bases
C 954 – 98	Steel Drill Screws for the Application of Gypsum Panel Products or Metal Plaster Bases to Steel Studs from 0.033 in. (0.84 mm) to 0.112 in. (2.84 mm) in Thickness
§C 932 – 98a	Surface-Applied Bonding Compounds for Exterior Plastering
§C 630/C 630M – 96a	Water-Resistant Gypsum Backing Board
C 933 – 96a	Welded Wire Lath
C 1032 – 96	Woven Wire Plaster Base

Test Methods for:

Designation	Title
C 471M – 96	Chemical Analysis of Gypsum and Gypsum Products
C 474 – 97	Joint Treatment Materials for Gypsum Board Construction
C 473 – 97$^{\epsilon 1}$	Physical Testing of Gypsum Panel Products
§C 472 – 98	Physical Testing of Gypsum, Gypsum Plasters and Gypsum Concrete
E 119 – 98	*Fire Tests of Building Construction and Materials (see Vol 04.07)*

Terminology Relating to:

Designation	Title
C 11 – 98	Gypsum and Related Building Materials and Systems

GENERAL METHODS OF TESTING

Specification for:

Designation	Title
E 171 – 94 (1998)	*Standard Atmospheres for Conditioning and Testing Flexible Barrier Materials (see Vol 15.09)*

† Although this standard has been officially withdrawn from Society approval, a brief description is included for information only.

LIST BY SUBJECTS

Test Methods for:

E 337 – 84 (1996)ε1	*Measuring Humidity with a Psychrometer (The Measurement of Wet- and Dry-Bulb Temperatures) (see Vol 11.03)*
E 96 – 95	*Water Vapor Transmission of Materials (see Vol 04.06)*

Practices for:

E 74 – 95	*Calibration of Force-Measuring Instruments for Verifying the Force Indication of Testing Machines (see Vol 03.01)*
E 4 – 99	*Force Verification of Testing Machines (see Vol 03.01)*
E 83 – 98	*Verification and Classification of Extensometers (see Vol 03.01)*

METRIC PRACTICE

SI Quick Reference Guide (see Related Material section)

Metric Practice Guide for:

Cement (see Related Material section)

Designation: C 5 – 79 (Reapproved 1997)

Standard Specification for Quicklime for Structural Purposes[1]

This standard is issued under the fixed designation C 5; the number immediately following the designation indicates the year of original adoption or, in the case of revision, the year of last revision. A number in parentheses indicates the year of last reapproval. A superscript epsilon (ϵ) indicates an editorial change since the last revision or reapproval.

This specification has been approved for use by agencies of the Department of Defense. Consult the DoD Index of Specifications and Standards for the specific year of issue which has been adopted by the Department of Defense.

1. Scope

1.1 This specification covers all classes of quicklime such as crushed lime, granular lime, ground lime, lump lime, pebble lime, and pulverized lime, used for structural purposes.

2. Referenced Documents

2.1 *ASTM Standards:*

C 25 Test Methods for Chemical Analysis of Limestone, Quicklime, and Hydrated Lime[2]

C 50 Practice for Sampling, Inspection, Packing, and Marking of Lime and Limestone Products[2]

C 51 Terminology Relating to Lime and Limestone (As Used by the Industry)[2]

C 110 Test Methods for Physical Testing of Quicklime, Hydrated Lime, and Limestone[2]

E 11 Specification for Wire-Cloth Sieves for Testing Purposes[3]

3. Terminology

3.1 *Definitions*—Unless otherwise specified, for definitions of terms used in this standard, refer to Terminology C 51.

4. General Requirements

4.1 Quicklime shall be slaked and aged in accordance with the printed directions of the manufacturer. The resulting lime putty shall be stored until cool.

4.2 Lime putty prepared in accordance with Appendix X1.4.2 and adjusted to standard consistency in accordance with Test Methods C 110, shall show no pops or pits when tested in accordance with Test Methods C 110.

4.3 Lime putty prepared as above shall have a plasticity figure of not less than 200.

5. Chemical Composition

5.1 The quicklime shall conform to the following requirements as to chemical composition, calculated on a nonvolatile basis:

	Calcium Lime	Magnesium Lime
Calcium oxide, min, %	75	
Magnesium oxide, min, %		20
Calcium and magnesium oxide, min, %	95	95
Silica, alumina, and oxide of iron, max, %	5	5
Carbon dioxide, max, %:		
If sample is taken at place of manufacture	3	3
If sample is taken at any other place	10	10

6. Residue

6.1 The quicklime shall contain no more than 15 weight % of residue.

7. Sampling, Inspection, etc.

7.1 The sampling, inspection, rejection, retesting, packing, and marking shall be conducted in accordance with Methods C 50.

8. Test Methods

8.1 Conformance to chemical requirements shall be determined in accordance with Test Methods C 25.

8.2 Conformance to plasticity and residue requirements shall be determined in accordance with Test Methods C 110.

9. Keywords

9.1 building (structural); calcium oxide; dolomitic lime; high calcium lime; lime putty; magnesium oxide; plasticity; quicklime; residue; slaking

---

[1] This specification is under the jurisdiction of ASTM Committee C-7 on Lime and is the direct responsibility of Subcommittee C07.02 on Structural Lime.

Current edition approved Nov. 30, 1979. Published January 1980. Originally published as C 5 – 13 T. Last previous edition C 5 – 59 (1974).

[2] *Annual Book of ASTM Standards*, Vol 04.01.

[3] *Annual Book of ASTM Standards*, Vol 14.02.

APPENDIX

(Nonmandatory Information)

X1. SLAKING AND PREPARATION OF LIME PUTTY

X1.1 Introduction

X1.1.1 Quicklime can never be used as such for structural purposes; it must always be slaked first. Since the method of slaking is an important factor in determining the quality of the finished product, the following directions for the preparation of lime putty are given, not as a part of the specification, but as information for the further protection of the purchaser.

X1.1.2 Different kinds of lime vary considerably in the way in which they behave with water. A little supervision over the operation of slaking will amply pay for itself by ensuring the production of the greatest possible quantity and the best possible quality of putty. To find out how to slake a new lot of lime, it is safest to try a little of it and see how it works. Since different lots of the same brand of lime vary somewhat, and since the weather conditions at the time have a decided influence, it is wise to try a sample from each lot used, whether familiar with the brand or not.

X1.2 Classification of Limes

X1.2.1 In a bucket, put two or three lumps of lime about the size of one's fist, or, in the case of granular lime, an equivalent amount. Add sufficient water to just barely cover the lime, and note how long it takes for slaking to begin. Slaking has begun when pieces split off from the lumps or when the lumps crumble. Water of the same temperature should be used for test and field practice.

X1.2.2 If slaking begins in less than 5 min, the lime is quick slaking; from 5 to 30 min, medium slaking; over 30 min, slow slaking.

X1.3 Directions for Slaking

X1.3.1 Slake quicklime in accordance with the printed directions of the manufacturer. When such directions are not provided, proceed as follows:

X1.3.2 For quick-slaking lime, always add the lime to the water, not the water to the lime. Have sufficient water at first to cover all the lime completely. Have a plentiful supply of water available for immediate use—a hose throwing a good stream, if possible. Watch the lime constantly. At the slightest appearance of escaping steam, hoe thoroughly and quickly, and add enough water to stop the steaming. Do not be afraid of using too much water with this kind of lime.

X1.3.3 For medium–slaking lime, add the water to the lime. Add enough water so that the lime is about half submerged. Hoe occasionally if steam starts to escape. Add a little water now and then if necessary to prevent the putty from becoming dry and crumbly. Be careful not to add any more water than required, and not too much at a time.

X1.3.4 For slow–slaking lime, add enough water to the lime to moisten it thoroughly. Let it stand until the reaction has started. Cautiously add more water, a little at a time, taking care that the mass is not cooled by the fresh water. Do not hoe until the slaking is practically complete. If the weather is very cold, it is preferable to use hot water, but if this is not available, the mortar box may be covered in some way to retain the heat.

X1.4 Preparation of Putty for Use

X1.4.1 After slaking, prepare putty for use as follows:

X1.4.2 *White Coat*—After slaking and aging finishing quicklime in accordance with the printed directions of the manufacturer, store the putty until cool. If no printed directions are provided by the manufacturer, prepare the putty for use as follows: After the action has ceased, run off the putty through a No. 10 (2.00-mm) sieve conforming to Specification E 11, and store for a minimum of 2 weeks.

X1.4.3 *Base Coats*—After the action has ceased, run off the putty through a No. 8 (2.36-mm) sieve conforming to Specification E 11. Add sand up to equal parts by weight, all of the hair or other fibers required, and store for a minimum of 2 weeks.

X1.4.4 *Masons' Mortar*—After the action has ceased, add part or all of the sand required, and store for a minimum of 24 h.

Designation: C 11 – 98

Standard Terminology Relating to Gypsum and Related Building Materials and Systems[1]

This standard is issued under the fixed designation C 11; the number immediately following the designation indicates the year of original adoption or, in the case of revision, the year of last revision. A number in parentheses indicates the year of last reapproval. A superscript epsilon (ε) indicates an editorial change since the last revision or reapproval.

1. Scope

1.1 This standard contains terminology relating to gypsum and related building materials and systems. The terms are generically defined. More specific and expanded definitions may appear in appropriate standards.

2. Referenced Documents

2.1 *ASTM Standards:*

C 22/C 22M Standard Specification for Gypsum[2]

C 28 Specification for Gypsum Plasters[2]

C 35 Specification for Inorganic Aggregates for Use in Gypsum Plaster[2]

C 36 Specification for Gypsum Wallboard[2]

C 37 Specification for Gypsum Lath[2]

C 52 Specification for Gypsum Partition Tile or Block[3]

C 59 Specification for Gypsum Casting and Molding Plaster[2]

C 61 Specification for Gypsum Keene's Cement[2]

C 79 Specification for Gypsum Sheathing Board[2]

C 317 Specification for Gypsum Concrete[2]

C 318 Specification for Gypsum Formboard[3]

C 472 Test Methods for Physical Testing of Gypsum, Gypsum Plasters and Gypsum Concrete[2]

C 473 Test Methods for Physical Testing of Gypsum Board Products and Gypsum Lath[2]

C 475 Specification for Joint Compound and Joint Tape for Finishing Gypsum Board[2]

C 557 Specification for Adhesives for Fastening Gypsum Wallboard to Wood Framing[2]

C 587 Specification for Gypsum Veneer Plaster[2]

C 588 Specification for Gypsum Base for Veneer Plasters[2]

E 84 Test Method for Surface Burning Characteristics of Building Materials[4]

E 96 Test Methods for Water Vapor Transmission of Materials[5]

E 119 Test Methods for Fire Tests of Building Construction and Materials[4]

E 337 Test Method for Measuring Humidity With a Psychrometer (The Measurement of Wet- and Dry-Bulb Temperatures)[6]

3. Terminology

accelerator, *n*—a material that will shorten the setting time of gypsum plasters.

accessories, *n*—linear formed metal, metal and paper, or plastic members fabricated for the purpose of forming corners, edges, control joints, or decorative effects in conjunction with gypsum board and plaster assemblies.

adhesive, *n*—a substance capable of holding materials together by surface attachment.

admixture, *n*—a material other than water, aggregate, or inorganic cementitious material that is used as an ingredient in gypsum plaster or cement plaster and is added to the batch immediately before or during job mixing.

aggregate, *n*—an inert granular material which may be added to gypsum plasters **(C 35).**

all purpose compound, *n*—a compound formulated and manufactured to serve as a taping or finishing compound, or both.

anhydrite, *n*—the mineral consisting primarily of anhydrous calcium sulfate, $CaSO_4$.

arris (of an arch), *n*—the outside corner or angle formed by the meeting of a wall surface with the curved surface of an arch (see Fig. 1).

base coat, *n*—the first layer or layers of plaster applied over a lath or other substrate. The first application is normally called a scratch coat and the second application is referred to as a brown coat.

base coat floating, *n*—the finishing act of spreading, compacting, and smoothing of the base coat plaster to a reasonably true plane.

bedding coat, *n*—that coat of plaster to receive aggregate or other decorative material of any size, impinged or embedded into its surface, before it sets.

bond plaster, *n*—a calcined gypsum plaster specially formulated for application over rough monolithic concrete as a bonding coat for a subsequent gypsum plaster layer **(C 28).**

bridging, *n*—*in framing*, sections of wood or metal pieces used between framing members to stiffen, give lateral

[1] This terminology is under the jurisdiction of ASTM Committee C-11 on Gypsum and Related Building Materials and Systems and are the direct responsibility of Subcommittee C11.91 on Terminology and Editorial.

Current edition approved Oct. 10, 1998. Published March 1999. Originally published as C 11 – 16 T. Last previous edition C 11 – 95a.

[2] *Annual Book of ASTM Standards*, Vol 04.01.

[3] Discontinued: See *1982 Annual Book of ASTM Standards*, Vol 04.01.

[4] *Annual Book of ASTM Standards*, Vol 04.07.

[5] *Annual Book of ASTM Standards*, Vol 04.06.

[6] *Annual Book of ASTM Standards*, Vol 11.03.

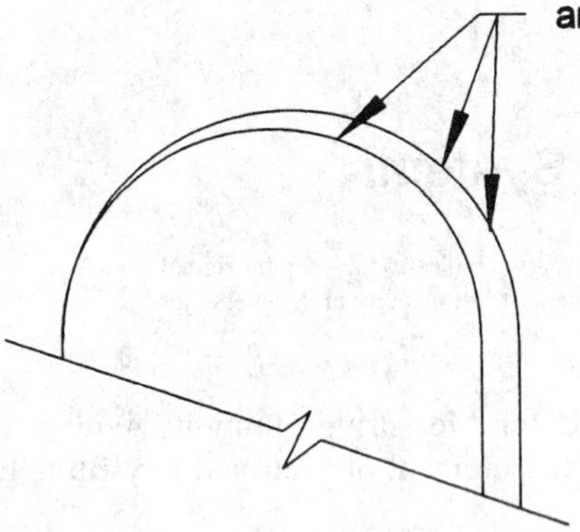

FIG. 1 Arris (of an Arch)

support, and minimize rotation.

brown coat, *n*—the second coat in three-coat gypsum plaster application.

building construction joint, *n*—a designed division of a building that allows movement of all component parts of the building, in any plane, which may be caused by thermal, seismic, wind loading or any other force. The construction of the separation is accomplished by one of the following methods: (*1*) manufactured devices suitable for this application, or (*2*) by field fabrication of suitable materials.

calcined gypsum, *n*—a dry powder; primarily calcium sulfate hemihydrate, resulting from calcination of gypsum; cementitious base for production of most gypsum plasters: also called plaster of paris; sometimes called stucco.

calcium sulfate, *n*—the chemical compound $CaSO_4$.

cementitious material, *n*—a material that, when mixed with water, with or without aggregate, provides the plasticity and the cohesive and adhesive properties necessary for placement, and the formation of a rigid mass.

combined water, *n*—the water chemically held, as water of crystallization, by the calcium sulfate dihydrate or hemihydrate crystal.

compressive strength, *n*—the maximum load sustained by a standard specimen of a material when subjected to a crushing force.

consistency, *n*—a property of a material determined by the complete flow force relation **(C 557).**

consistency (normal), *n*—the number of millilitres of water per 100 g of gypsum plaster or gypsum concrete required to produce a mortar or a slurry of specified fluidity **(C 472).**

control (expansion-contraction) joint, *n*—a designed separation in the system materials that allows for movement caused by expansion or contraction of the system. The construction of the separation is accomplished by one of the following methods: (1) manufactured devices suitable for this application, or (2) by field fabrication of suitable materials.

core (of gypsum board), *n*—the hardened material filling the space between the face and back papers consisting substantially of rehydrated gypsum with additives.

cored tile or block, *n*—see **gypsum tile or block** **(C 52).**

corner reinforcement, exterior, *n*—a preformed section of wire or expanded sheet steel, for the reinforcement of exterior stucco external corners (arrises).

cure (portland cement plaster or stucco), *v*—(*1*) to provide conditions conducive to the hydration process of portland cement plaster or stucco.

(*2*) to maintain proper temperature and a sufficient quantity of water within the plaster to ensure cement hydration.

dash-bond coat, *n*—a thick slurry of portland cement, sand, and water, machine-sprayed or hand-dashed with a paddle or stiff-bristled broom to any acceptable surface, to provide improved adhesion and a key for the subsequent portland cement plaster or stucco coat.

density, *n*—the weight per unit volume of a material **(C 472).**

edge (of gypsum board), *n*—the paper-bound edge as manufactured.

embedding compound—see **taping compound.**

end (of gypsum board), *n*—the end perpendicular to the paper-bound edge. The gypsum core is always exposed.

expansion joint, n—see **control (expansion-contraction) joint**.

face, *n*—the surface designed to be left exposed to view or to receive decoration or additional finishes.

featured edge, *n*—an edge configuration of the paper bound edge of gypsum board that provides special design or performance.

fineness modulus, *n*—an empirical factor obtained by adding total percentages of a sample of aggregate retained on each of a specified series of sieves and dividing by 100. The sieve sizes used are: No. 100 (150 μm), No. 50 (300 μm), No. 30 (600 μm), No. 16 (1.18 mm), No. 8 (2.36 mm), No. 4 (4.75 mm),⅜ in. (9.5 mm), ¾ in. (19.0 mm), 1½in. (38.1 mm) and larger, increasing in the ratio of 2 to 1.

finish coat, *n*—the final layer of plaster applied over a basecoat or other substrate.

finish coat floating, *n*—the finishing act of spreading, compacting, and smoothing the finish coat plaster or stucco to a specified surface texture.

finishing compound, *n*—(*sometimes called topping compound*) a compound specifically formulated and manufactured for use over taping or all purpose compounds to provide a smooth and level surface for the application of decoration.

fire-resistance classification, *n*—a standard rating of fire-resistance and protective characteristics of a building construction or assembly **(E 119).**

flame spread classification, *n*—a standard rating of relative surface burning characteristics of a building material as compared to a standard material **(E 84).**

flexural strength, *n*—the maximum load sustained by a standard specimen of a sheet material when subjected to a bending force.

framing member, *n*—stud, plate, track, joist, furring, and other support to which a gypsum panel product, or metal plaster base is attached.

free water, *n*—all water contained by gypsum board or plaster in excess of that chemically held as water of crystallization.

gauging plaster, *n*—a calcined gypsum plaster used with lime for a finish coat **(C 28).**

grout, *n*—gypsum or portland cement plaster used to fill crevices or to fill hollow metal frames.

gypsum, *n*—the mineral consisting primarily of fully hydrated calcium sulfate, $CaSO_4 \cdot 2H_2O$ or calcium sulfate dihydrate **(C 22).**

gypsum backing board, *n*—a ¼ in, to ⅝ in. gypsum board for use as a backing for gypsum wallboard, acoustical tile, or other dry cladding.

water resistant gypsum backing board—a gypsum board designed for use on walls primarily as a base for the application of ceramic, or plastic tile.

gypsum base for veneer plasters, *n*—a gypsum board used as the base for application of a gypsum veneer plaster **(C 588).**

gypsum board, *n*—the generic name for a family of sheet products consisting of a noncombustible core primarily of gypsum with paper surfacing.

gypsum concrete, *n*—a calcined gypsum mixed with wood chips, or aggregate, or both, used primarily for poured roof decks **(C 317).**

gypsum core board, *n*—a ¾ in. (19.0 mm) to 1 in. (25.4 mm) gypsum board consisting of a single board or factory laminated multiple boards, used as a gypsum stud or core in semisolid or solid gypsum board partitions.

gypsum formboard[7], *n*—a gypsum panel product used as the permanent form for poured gypsum roof decks **(C 318).**

gypsum lath, *n*—a gypsum board used as the base for application of gypsum plaster **(C 37).**

perforated gypsum lath[7]—a gypsum lath having perforations to provide mechanical keying of the basecoat plaster.

foil-backed gypsum lath—the same as plain gypsum lath except that in addition, the back surface shall be covered with a continuous sheet of pure bright finished aluminum foil.

type X lath—a gypsum lath specially manufactured to provide specific fire-resistant characteristics.

gypsum molding plaster, *n*—a calcined gypsum plaster used primarily for plaster casts or molds, sometimes used as a gauging plaster **(C 59, C 28).**

gypsum neat plaster, *n*—a calcined gypsum plaster without aggregate; common usage is for gypsum plaster used for base coats **(C 28).**

gypsum panel products, *n*—the general name for a family of sheet products consisting essentially of gypsum.

gypsum plaster, *n*—the generic name for a family of powdered cementitious products consisting primarily of calcined gypsum with additives to modify physical characteristics, and having the ability, when mixed with water, to produce a plastic mortar or slurry which can be formed to the desired shape by various methods and will subsequently set to a hard, rigid mass **(C 28).**

gypsum sheathing, *n*—a gypsum board used as a backing for exterior surface materials, manufactured with water-repellant paper and may be manufactured with a water-resistant core **(C 79).**

gypsum tile or block[7], *n*—a cast gypsum building unit **(C 52).**

gypsum wallboard, *n*—a gypsum board used primarily as an interior surfacing for building structures **(C 36).**

foil-backed gypsum wallboard—a gypsum wallboard with the back surface covered with a continuous sheet of pure bright finished aluminum foil **(C 36).**

type X gypsum wallboard—a gypsum wallboard specially manufactured to provide specific fire-resistant characteristics **(C 36).**

hemihydrate, *n*—the dry powder, calcium sulfate hemihydrate, resulting from calcination of $CaSO_4 \cdot 2H_2O$, calcium sulfate dihydrate. See **calcined gypsum.**

joining, *n*—the juncture of two separate plaster applications of the same coat, usually within a single surface plane.

joint compound, *n*—a compound used for taping or finishing gypsum board, or both.

joint reinforcing metal, *n*—strips of expanded metal, woven or welded wire mesh used to reinforce corners and other areas of plaster and lath.

joint tape, *n*—a type of paper, metal, fabric, glass mesh, or other material, commonly used with a cementitious compound, to reinforce the joints between adjacent gypsum boards **(C 475).**

Keene's cement, *n*—an anhydrous gypsum plaster characterized by a low mixing water requirement and special setting properties, primarily used with lime to produce hard, dense finish coats **(C 61).**

key, *n*—the grip or mechanical bond of one coat of plaster to another coat, or to a plaster base. It may be accomplished physically by the penetration of wet mortar or crystals into paper fibers, perforations, scoring irregularities, or by the embedment of the lath.

lath—see **gypsum lath.**

load-bearing partition, *n*—a partition designed to support a portion of the building structure.

machine direction, *n*—the direction parallel to the paper-bound edge of the gypsum board.

masonry cement, *n*—a hydraulic cement for use in mortars for masonry construction, containing one or more of the following materials: portland cement, portland blast furnace slag cement, portland-pozzolan cement, natural cement, slag cement, or hydraulic lime; and in addition usually containing one or more materials such as hydrated lime, limestone, chalk, calcareous shell, talc, slag, or clay as prepared for this purpose.

mechanical bonds, *n*—the attachment created when plaster penetrates, into or through, the substrate, or envelops irregularities in the surface of the substrate.

mortar, *n*—a mixture of gypsum plaster with aggregate or hydrate lime, or both, and water to produce a trowelable fluidity.

neat gypsum plaster—see **gypsum neat plaster.**

perm, *n*—a unit of measurement of water vapor permeance; a metric perm, or 1 g/24 h·m^2·mm Hg. U.S. unit, 1 grain/h·ft^2·in. Hg **(E 96).**

permeability, *n*—the property of a porous material that permits a fluid (or gas) to pass through it; in construction, commonly refers to water vapor permeability of a sheet material or assembly and is defined as water vapor permeance per unit thickness. Metric unit of measurement, metric perms per centimetre of thickness. See **water vapor**

[7] This product is no longer manufactured, and the specifications for it have been discontinued but the definition is retained for reference purposes.

transmission, perm, permeance **(E 96).**

permeance (water vapor), *n*—the ratio of the rate of water vapor transmission (WVT) through a material or assembly between its two parallel surfaces to the vapor pressure differential between the surfaces. Metric unit of measurement is the metric perm, 1 g/24 h·m^2·mm Hg; U.S. unit, 1 grain/h·ft^2·in. Hg. See **water vapor transmission, permeability, perm** **(E 96).**

plaster—see **gypsum plaster, gypsum neat plaster**.

plaster bond, *n*—the state of adherence between plaster coats or between plaster and a plaster base, produced by adhesive or mechanical interlock of plaster with base or special supplementary materials.

plastic cement, *n*—a hydraulic cement to which one or more plasticizing agents (but not more than 12 % by volume) have been added during intergrinding or blending to increase the workability and molding qualities of the resultant cement paste, mortar, or plaster.

plasticity, *n*—that property of freshly mixed cement paste, mortar, or plaster which determines its workability and molding qualities.

portland cement, *n*—a hydraulic cement produced by pulverizing clinker consisting essentially of hydraulic calcium silicates, and usually containing one or more forms of calcium sulfate as an interground addition.

portland cement plaster, *n*—a plaster mix in which portland cement or combinations of portland and masonry cements or portland cement and lime are the principal cementitious materials mixed with aggregate.

purity, *n*—the percentage of $CaSO_4 \cdot ½H_2O$ in the calcined gypsum portion of a gypsum plaster or gypsum concrete, as defined by Specification C 28, for Gypsum Plasters. The percentage of $CaSO_4 \cdot 2H_2O$ in the gypsum or the gypsum portion of fully hydrated, dry, set gypsum plaster **(C 471, C 28).**

ready-mixed plaster, *n*—a calcined gypsum plaster with aggregate added during manufacture **(C 28).**

relative humidity, *n*—the ratio of actual water vapor pressure to the saturation water vapor pressure at the same temperature, expressed as a percentage **(E 337).**

retarder, *n*—a material that will lengthen the setting time of gypsum plaster.

round edge, *n*—a rounded, paper-bound edge formation on gypsum board, commonly used for gypsum lath **(C 37, C 473).**

saddle-tie, *n, for furring*—a single or double strand of wire used to attach furring members to framing members of wall or ceiling assemblies. See Fig. 2.

saddle-tie, *v, for furring*—to attach furring members to framing members of wall or ceiling assemblies using a single or double strand of wire. See Fig. 2.

saddle-tie, *v, for wire hangers*—to attach wire hangers to main runners. See Fig. 3.

set, *n*—the hardening and hydration of a gypsum plaster. See **setting time** **(C 472).**

setting time, *n*—the elapsed time required for a gypsum plaster to attain a specified hardness and strength after mixing with water **(C 472).**

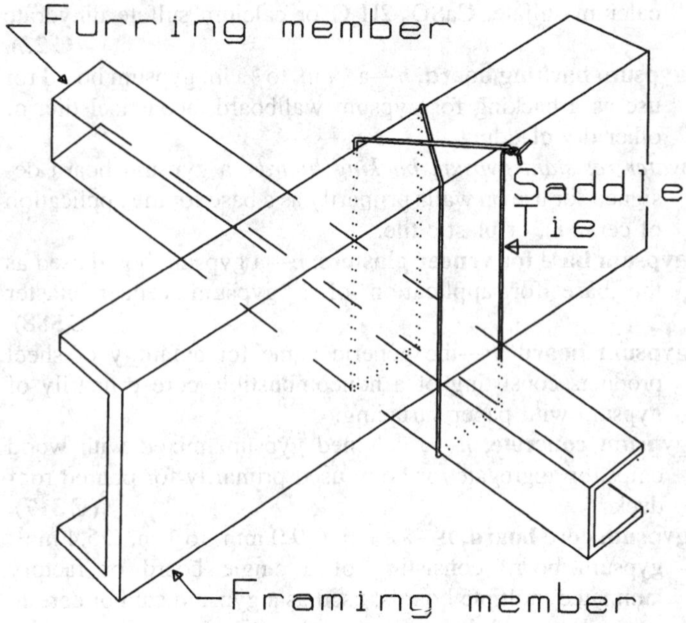

FIG. 2 Saddle-Tie

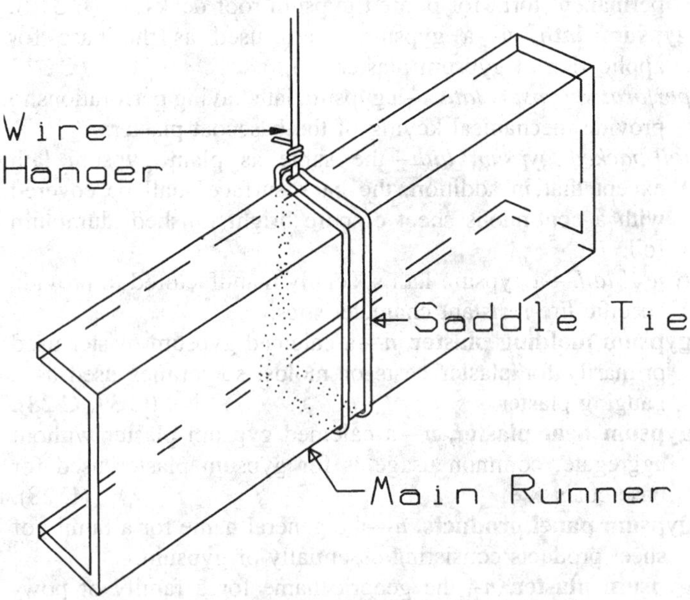

FIG. 3 Saddle-Tie

stucco, *n*—a portland cement-aggregate plaster mix designed for use on exterior surfaces. See **portland cement plaster**.

synthetic gypsum, *n*—a chemical product, consisting primarily of calcium sulfate dihydrate ($CaSO_4 \cdot 2H_2O$) resulting primarily from an industrial process.

tapered edge, *n*—an edge formation of gypsum board which provides a shallow depression at the paper-bound edge to receive joint reinforcement **(C 473, C 36).**

taping compound, *n*—*(sometimes called embedding compound)* a compound specifically formulated and manufactured for use in embedding of joint reinforcing tape at gypsum board joints.

topping compound—see **finishing compound**

veneer plaster, *n*—a calcined gypsum plaster specially manufactured to provide high strength, hardness, and abrasion resistance when applied in thin coats over a gypsum base for veneer plasters **(C 587).**

wallboard—see **gypsum wallboard** **(C 36).**

water absorption, *n*—the amount of water absorbed by a material under specified test conditions commonly expressed as weight percent of the test specimen.

water-repellent paper, *n*—gypsum board paper surfacing which has been formulated or treated to resist water penetration.

water-resistant core, *n*—a gypsum board specially formulated to resist water penetration.

water vapor transmission (WVT), *n*—the rate of water vapor flow, under steady specified conditions, through a unit area of a material, between its two parallel surfaces and normal to the surfaces. Metric unit of measurement is 1 g/24 h·m^2. See **permeability,permeance, perm** **(E 96).**

wood-fibered plaster, *n*—a calcined gypsum plaster containing shredded or ground wood fiber added during manufacture **(C 28).**

Designation: C 22/C22M – 96$^{\epsilon 1}$

Standard Specification for Gypsum[1]

This standard is issued under the fixed designation C 22/C22M; the number immediately following the designation indicates the year of original adoption or, in the case of revision, the year of last revision. A number in parentheses indicates the year of last reapproval. A superscript epsilon (ε) indicates an editorial change since the last revision or reapproval.

ϵ^1 NOTE—The formula in paragraph 4.1 was editorially corrected to match the formula in 1.1.

1. Scope *

1.1 This specification covers gypsum, calcium sulfate combined with two molecules of water in crystalline form and having the approximate chemical formula $CaSO_4 \cdot 2H_2O$.

1.2 The values stated in either inch-pound units or SI (metric) are to be regarded separately as the standard. Within the text, the SI units are shown in brackets. The values stated in each system shall be used independently of the other. Values from the two systems shall not be combined.

2. Referenced Documents

2.1 *ASTM Standards:*

C 11 Terminology Relating to Gypsum and Related Building Materials and Systems[2]

C 471 Test Methods for Chemical Analysis of Gypsum and Gypsum Products[2]

C 472 Test Methods for Physical Testing of Gypsum, Gypsum Plasters and Gypsum Concrete[2]

3. Terminology

3.1 *Definitions*—Definitions shall be in accordance with Terminology C 11.

4. Chemical Composition

4.1 Gypsum shall contain not less than 70.0 weight percent $CaSO_4 \cdot 2H_2O$.

4.2 The chemical composition, within the limits prescribed by 4.1, shall be specified in the purchase order or contract.

5. Physical Properties

5.1 *Size*—If necessary, gypsum shall be crushed or ground, or both, to the size specified in the purchase order.

6. Sampling

6.1 *Bulk*—When gypsum is shipped in bulk, samples of approximately 5 lb (2 kg) each shall be taken at regular intervals during the complete loading or unloading of the carrier, in such a way as to accumulate not less than 200 lb (90 kg) of material and to represent correctly the percentages of both the coarse particles and the fines in the shipment. This material shall be broken to pass a 1-in. (25.0-mm) sieve, thoroughly mixed, and reduced by quartering to provide not less than a 1-lb (0.5-kg) sample for the laboratory.

6.2 *Packages*—When gypsum is shipped in packages, not less than 1 % of the packages shall be sampled. Samples shall be taken from both the surface and the center of the packages. These samples shall then be broken, mixed, and quartered in accordance with 6.1.

6.3 *Laboratory Samples*—Each laboratory sample shall be placed immediately in an airtight container and shipped to the laboratory for testing.

7. Test Methods

7.1 The chemical composition of gypsum shall be determined in accordance with Test Methods C 471.

7.2 The physical properties of gypsum shall be determined in accordance with Test Methods C 472.

8. Inspection

8.1 Inspection of the gypsum shall be agreed upon between the purchaser and the supplier as part of the purchase agreement.

9. Rejection

9.1 Gypsum that fails to conform to the requirements of this specification may be rejected. Rejection shall be reported to the producer or supplier promptly and in writing. The notice of rejection shall contain a statement documenting how the gypsum has failed to conform to the requirements of this specification.

10. Certification

10.1 When specified in the purchase agreement, a manufacturer's or supplier's report shall be furnished at the time of shipment certifying that the product is in compliance with this specification.

11. Packaging and Package Marking

11.1 Gypsum shall be shipped either in packages or in bulk.

11.2 When shipped in packages for resale, the following

---

[1] This specification is under the jurisdiction of ASTM Committee C-11 on Gypsum and Related Building Materials and Systems and is the direct responsibility of Subcommittee C11.01 on Specifications and Test Methods for Gypsum Products.

Current edition approved May 10, 1996. Published July 1996. Originally published as C 22 – 19.Last previous edition C 22 – 95a.

[2] *Annual Book of ASTM Standards*, Vol 04.01.

***A Summary of Changes section appears at the end of this standard.**

information (Note) shall be legibly marked on each package or on a tag of suitable size attached thereto:

11.2.1 Name of manufacturer,

11.2.2 Description of material, and

11.2.3 Net or gross weights, or both, of the package.

11.3 When shipped in bulk, a card containing the required information in accordance with 11.2 shall be conspicuously placed in the carrier.

NOTE 1—State law may require additional information.

12. Keywords

12.1 calcium sulfate; gypsum

SUMMARY OF CHANGES

This section identifies the location of changes to this specification that have been incorporated since the last issue. Committee C-11 has identified those changes that affect the technical interpretation or use of the specification.

(1) Section 5.1 was revised.

Designation: C 25 – 98

Standard Test Methods for Chemical Analysis of Limestone, Quicklime, and Hydrated Lime[1]

This standard is issued under the fixed designation C 25; the number immediately following the designation indicates the year of original adoption or, in the case of revision, the year of last revision. A number in parentheses indicates the year of last reapproval. A superscript epsilon (ε) indicates an editorial change since the last revision or reapproval.

This standard has been approved for use by agencies of the Department of Defense.

1. Scope

1.1 These test methods cover the chemical analysis of high-calcium and dolomitic limestone, quicklime, and hydrated lime. These test methods are classified as either standard (preferred) or alternative (optional).

1.2 The standard test methods are those that employ classical gravimetric or volumetric analytical procedures and are typically those required for referee analyses where chemical specification requirements are an essential part of contractual agreement between buyer and seller.

1.3 Alternative or optional test methods are provided for those who wish to use procedures shorter or more convenient than the standard methods for the routine determinations of certain constituents. Optional test methods may sometimes be preferred to the standard test methods, but frequently the use of modern and expensive instrumentation is indicated which may not be accessible to everyone. Therefore, the use of these test methods must be left to the discretion of each laboratory.

1.4 The analytical procedures appear in the following order:

	Section
Aluminum Oxide	15
Available Lime Index	28
Calcium and Magnesium Oxide:	
Alternative EDTA Titration Method	31
Calcium Carbonate Equivalent	33
Calcium Oxide:	
Gravimetric Method	16
Volumetric Method	17
Carbon Dioxide by Standard Method	22
Combined Oxides of Iron and Aluminum	12
Ferrous Iron	X5
Free Calcium Oxide	X6
Free Moisture in Hydrated Lime	21
Free Moisture in Limestone	20
Free Silica	29
Insoluble Matter Including Silicon Dioxide:	
Standard Method	8
Optional Perchloric Acid Method	9
Insoluble Matter Other Than Silicon Dioxide	11
Loss on Ignition	19
Magnesium Oxide	18
Manganese:	
Bismuthate Method	X4
Periodate (Photometric) Method	27
pH Determination of Alkaline Earth Solutions	34
Phosphorus:	
Titrimetric Method	X3
Molybdovanadate Method	26
Silicon Dioxide	10
Strontium Oxide	X2
Sulfur Trioxide	23
Total Carbon:	
Direct Combustion-Thermal Conductivity Cell Method	32
Total Iron:	
Standard Method, Potassium Dichromate Titration	13
Potassium Permanganate Titration Method	X1
Ortho-Phenanthroline, Photometric Method	14
Total Sulfur:	
Sodium Carbonate Fusion	24
Combustion-Iodate Titration Method	25
Unhydrated Oxides	30

1.5 *This standard does not purport to address all of the safety concerns, if any, associated with its use. It is the responsibility of the user of this standard to establish appropriate safety and health practices and determine the applicability of regulatory limitations prior to use.* For specific precautionary statements, see Notes 11, 13, 27, 47, X2.1, and X5.1.

2. Referenced Documents

2.1 *ASTM Standards:*

C 50 Methods of Sampling, Inspection, Packing and Marking of Lime and Limestone Products[2]

C 51 Terminology Relating to Lime and Limestone (as Used by the Industry)[2]

C 911 Specification for Quicklime, Hydrated Lime, and Limestone for Chemical Uses[2]

D 1193 Specification for Reagent Water[3]

E 29 Practice for Using Significant Digits in Test Data to

[1] These test methods are under the jurisdiction of ASTM Committee C-7 on Lime and are the direct responsibility of Subcommittee C07.05 on Chemical Tests.

Current edition approved July 10, 1998. Published September 1998. Originally published as C 25 – 19T. Last previous edition C 25 – 96a.

[2] *Annual Book of ASTM Standards,* Vol 04.01.

[3] *Annual Book of ASTM Standards,* Vol 11.01.

Determine Conformance with Specifications[4]

E 50 Practices for Apparatus, Reagents, and Safety Precautions for Chemical Analysis of Metals[5]

E 70 Test Method for pH of Aqueous Solutions with the Glass Electrode[6]

E 173 Practice for Conducting Interlaboratory Studies of Methods for Chemical Analysis of Metals[5]

E 200 Practice for Preparation, Standardization, and Storage of Standard and Reagent Solutions for Chemical Analysis[6]

E 691 Practice for Conducting an Interlaboratory Study to Determine the Precision of a Test Method[4]

E 832 Specification for Laboratory Filter Papers[4]

3. Terminology

3.1 *Definitions*—Unless otherwise specified, for definitions of terms used in these test methods refer to Terminology C 51.

4. Significance and Use

4.1 These test methods provide accurate and reliable analytical procedures to determine the chemical constituents of limestone, quicklime, and hydrated lime (Note 1). The percentages of specific constituents which determine a material's quality or fitness for use are of significance depending upon the purpose or end use of the material. Results obtained may be used in relation to specification requirements.

4.2 Because quicklime and hydrated lime quickly absorb water and carbon dioxide from the air, precision and bias are extremely dependent upon precautions taken during sample preparation and analysis to minimize excessive exposure to ambient conditions.

NOTE 1—These test methods can be applied to other calcareous materials if provisions are made to compensate for known interferences.

5. General Apparatus and Materials and Reagents

5.1 *General Apparatus and Materials:*

5.1.1 *Balance*—The balance shall be of an analytical type with a capacity not to exceed 200 g. It may be of conventional design or it may be a constant-load, direct-reading type. It shall be capable of reproducing weighings within 0.0002 g with an accuracy of ±0.0002 g. Rapid weighing devices that may be provided such as a chain, damper, or heavy riders shall not increase the basic inaccuracy by more than 0.0001 g at any reading and with any load within the rated capacity of the balance.

5.1.2 *Weights*—Weights used for analysis shall conform to Class S-1 requirements of the National Institute of Standards and Technology as described in NIST Circular 547[7]. They shall be checked at least once a year or when questioned, and adjusted to within allowable tolerances for Class S-1 weights. All new sets of weights purchased shall have the weights of 1 g and larger made of stainless steel or other corrosion-resistant alloy not requiring protective coating and shall meet the density requirements for Class S.

5.1.3 *Glassware and Laboratory Containers*—Standard volumetric flasks, burets, pipets, dispensers, etc., shall be carefully selected precision grade or better and shall be calibrated, if necessary, to meet the requirements of each operation. Standard-type interchangeable ground glass or TFE-fluorocarbon joints are recommended for all volumetric glassware. Polyethylene containers are recommended for all aqueous solutions of alkalies and for standard solutions where the presence of dissolved silica or alkali from the glass would be objectionable.

5.1.4 *Desiccators*—Desiccators shall be provided with a good desiccant such as anhydrous magnesium perchlorate, activated alumina, sulfuric acid, or phosphoric anhydride. Anhydrous calcium sulfate may also be used provided it has been treated with a color-changing indicator to show when the desiccant has lost its effectiveness. Calcium chloride and silica gel are not satisfactory desiccants for this type of analysis.

5.1.5 *Filter Paper*—Filter paper shall conform to the requirements of Specification E 832, Type II (quantitative). Class E shall be used for coarse and gelatinous precipitates. When medium-textured paper is required, Class F filter paper shall be used. When a retentive paper is needed, Class G shall be used.

Recommendations: Class	Filter Pore Size (microns)	Filter Speed
E	20 to 25	fast speed
F	8	medium speed
G	2.5	slow speed

5.1.6 *Crucibles*—Platinum crucibles and tight fitting lids should preferably be made of pure unalloyed platinum and be of 25 to 35-mL capacity. Where alloyed platinum is used for greater stiffness or to obviate sticking of fused material to crucible or lid, the alloyed platinum should not decrease in weight by more than 0.2 mg when heated at 1200°C for 1 h.

5.1.7 *Muffle Furnace*—The electric muffle furnace should be capable of continuous operation up to 1000°C and be capable of intermittent operation at higher temperatures if required. It should have an indicating pyrometer accurate to ±25°C.

5.2 *Reagents:*

5.2.1 *Purity of Reagents*—Reagent grade chemicals shall be used in all tests. Unless otherwise indicated, it is intended that all reagents shall conform to the specifications of the Committee on Analytical Reagents of the American Chemical Society[8] where such specifications are available. Other grades may be used provided it is first ascertained that the reagent is of sufficiently high purity to permit its use without lessening the accuracy of the determination. In addition to this, it is desirable in many cases for the analyst to ensure the accuracy of his results by running blanks or checking against a comparable

[4] *Annual Book of ASTM Standards*, Vol 14.02.

[5] *Annual Book of ASTM Standards*, Vol 03.05.

[6] *Annual Book of ASTM Standards*, Vol 15.05.

[7] Available from National Institute of Standards and Technology, Gaithersburg, MD 20899.

[8] *Reagent Chemicals, American Chemical Society Specifications*, American Chemical Society, Washington, DC. For suggestions on the testing of reagents not listed by the American Chemical Society, see *Analar Standards for Laboratory Chemicals*, BDH Ltd., Poole, Dorset, U.K., and the *United States Pharmacopeia and National Formulary*, U.S. Pharmacopeia Convention, Inc. (USPC), Rockville, MD.

sample of known composition.

5.2.2 *Purity of Water*—Unless otherwise indicated, references to water are understood to mean distilled water or other water of equivalent purity. Water conforming to Specification D 1193 meets these requirements.

5.2.3 *Concentration of Reagents:*

5.2.3.1 *Concentrated Acids and Ammonium Hydroxide*—When acids and ammonium hydroxide are specified by name or chemical formula only, it shall be understood that concentrated reagents approximating the following specific gravities or concentrations are intended:

Acetic acid ($HC_2H_3O_2$)	99.5 %
Hydrochloric acid (HCl)	sp gr 1.19
Hydrofluoric acid (HF)	48 %
Nitric acid (HNO_3)	sp gr 1.42
Perchloric acid ($HClO_4$)	70 %
Phosphoric acid (H_3PO_4)	85 %
Sulfuric acid (H_2SO_4)	sp gr 1.84
Ammonium hydroxide (NH_4OH)	sp gr 0.90

5.2.3.2 *Dilute Reagents*—The concentration of dilute acids and NH_4OH except when standardized, are specified as a ratio stating the number of measured volumes of the concentrated reagent to be diluted with a given number of measured volumes of water. In conformance with international practice, new and revised methods will use the "plus" designation instead of the ratio (:) symbol as the specified designation of dilution; for example, H_2SO_4 (5 + 95) means 5 volumes of concentrated H_2SO_4 (sp gr 1.84) diluted with 95 volumes of water.

5.2.3.3 *Standard Solutions*—Concentrations of standard solutions shall be expressed as normalities (*N*) or as equivalents in grams per millilitre of the component to be determined, for example: 0.1 *N* $K_2Cr_2O_7$ solution (1 mL = 0.004 g Fe_2O_3). The average of at least three determinations shall be used for all standardizations. The standardization used to determine the strength of the standard solutions is described in the text under each of the appropriate procedures.

6. General Procedures

6.1 *Sampling*—Samples of lime and limestone for chemical analysis shall be taken and prepared in accordance with the requirements of Methods C 50 applicable to the material to be tested.

6.2 *Tared or Weighed Crucibles*—The tare weight of crucibles shall be determined by preheating the empty crucible to constant weight at the same temperature and under the same conditions as shall be used for the final ignition of a residue and cooling in a desiccator for the same period of time used for the crucible containing the residue.

6.3 *Constancy of Weight of Ignited Residue*—To definitely establish the constancy of weight of the ignited residue, the residue and container shall be ignited at the specified temperature and time, cooled to room temperature in a desiccator, and weighed. The residue and container shall then be reheated for at least 30 min at the same temperature, cooled in a desiccator for the same period of time, and reweighed. Additional ignition periods may be required until two consecutive weights do not differ by more than 0.2 mg, at which time it shall be considered that constant weight has been attained. For ignition loss, each reheating period shall be 5 min.

6.4 *Calculation:*

6.4.1 The calculations included in the individual procedures sometimes assume that the exact weight specified has been used. Accurately weighed samples which are approximately but not exactly equal to the weight specified may be used provided appropriate corrections are made in the calculation. Unless otherwise stated, weights of all samples and residues should be recorded to the nearest 0.0001 g.

6.4.2 In all mathematical operations on a set of observed values, the equivalent of two more places of figures than in the single observed values shall be retained. For example, if observed values are read or determined to the nearest 0.1 mg, carry numbers to the nearest 0.001 mg in calculation.

6.5 *Rounding Figures*—Rounding figures to the nearest significant place required in the report should be done after the calculations are completed, in order to keep the final results free from calculation errors. The rounding procedure should follow the principle outlined in Practice E 29.

7. Performance Requirements for Test Methods

7.1 *Referee Analyses*—The reference test methods that appear in Sections 8 through 32, or any other test methods qualified in accordance with 7.3, are required for referee analysis in those cases where conformance to the requirements of a chemical specification are questioned. In these cases a limestone, quicklime, or hydrated lime shall not be rejected for failure to conform to chemical requirements unless all sample preparation and analysis of any one constituent is made entirely by reference test methods prescribed in the appropriate sections of this test method or by other qualified test methods. Exception can be made when specific test methods are prescribed in the standard specification for the limestone, quicklime, or hydrated lime in question. The test methods actually used for the analysis shall be designated.

7.1.1 When there is a question regarding acceptance, referee analyses shall be made in duplicate. If the two results do not agree within the permissible variation given in Table 1, the determination including sample preparation shall be repeated

TABLE 1 Maximum Permissible Variations in Results[A]

(Column 1) Constituent	(Column 2) Maximum Difference Between Duplicates	(Column 3) Maximum Difference of the Average of Duplicates from SRM Certificate Values[B]
Al as Al_2O_3	0.20	±0.30
Ca as CaO	0.20	±0.30
Mg as MgO	0.20	±0.30
C (lime and hydrated lime)	0.20	±0.30
C (limestone)	0.60	±0.60
Fe as Fe_2O_3	0.10	±0.15
Si as SiO_2	0.15	±0.30
Mn	0.05	±0.10
P	0.02	±0.05
Sr as SrO	0.05	[C]
S	0.03	±0.05

[A] For demonstrating the performance of rapid test methods the SRM closest in overall composition to the limestone shall be used (Table 2). In the case of quicklime or hydrated lime, the SRM closest in overall composition, after heating at 1000°C for 1 h, to the product composition shall be used, except for C and S determinations (Note 3).

[B] Interelement corrections may be used for any standardization provided improved accuracy can be demonstrated.

[C] No SRM currently available.

in duplicate until the results agree within the permissible variation. When the results agree within the permissible variation, their average shall be accepted as the correct value. For the purpose of comparing results, the percentages shall be calculated to one more significant figure than reported as indicated in the test methods. When a blank determination is specified, one shall be made with each individual analysis or with each group of two or more samples analyzed on the same day for a given constituent.

7.1.2 Test results from Referee methods intended for use as a basis for product acceptance or rejection, or for manufacturer's certification, can be used only after demonstration of precise and accurate analyses by meeting the requirements of 7.1.3, or except when demonstrated under 7.3.2.1. Such demonstrations may be made concurrently with analysis of the limestone, quicklime, or hydrated lime product being tested. The demonstration is required only for those constituents being used as a basis for acceptance, rejection, or certification of a limestone, quicklime, or hydrated lime, but may be made for any constituent of limestone, quicklime, or hydrated lime product for which a standard exists. Such demonstrations must be made annually.

7.1.3 Demonstrations shall be made by analysis of each constituent of concern in a SRM limestone, quicklime, or hydrated lime (Notes 2 and 3). Duplicate samples shall be run on different days. The same test methods to be used for analysis of the limestone, quicklime, or hydrated lime being tested shall be used for analysis of the SRM. If the duplicate results do not agree within the permissible variation given in Table 1, the determinations shall be repeated, following identification and correction of problems or errors, until a set of duplicate results do agree within the permissible variation.

NOTE 2—The term SRM refers to approved Standard Reference Materials listed in Table 2.

NOTE 3—There are no SRMs that are quicklime or hydrated lime as supplied. When analyzing a quicklime or hydrated lime the SRM in carbonate form needs to be converted to closely resemble the matrix of the product being tested. To accomplish this conversion, heat the chosen SRM for 1 h at 1000°C, immediately prior to analysis and protect it from hydration or carbonation with sealed containers and desiccation during cooling. Carbon and sulfur may be driven off during heating, rendering the converted SRM unsuitable as a standard for carbon and sulfur determinations. For carbon and sulfur determinations use the appropriate SRM in its normal matrix.

7.1.4 The average of the results of acceptable duplicate determinations for each constituent may differ from the SRM certificate value by no more than the value shown in Column 3 of Table 1. When no SRM certificate value is given, a generally accepted accuracy standard for that constituent has not been identified. In such cases, only the differences between duplicate values as specified in 7.1.3 shall apply and notification of this exception shall be reported.

7.1.5 In questions concerning the acceptance or rejection of a limestone, quicklime, or hydrated lime product, upon request data shall be made available to all parties involved demonstrating that precise and accurate results were obtained with SRM samples by the same analyst making the acceptance determination.

7.2 *Optional Analyses*—The alternative test methods, as opposed to reference methods, provide procedures that are, in some cases, shorter or more convenient to use for routine determination of some constituents (Note 4). In some instances longer, more complex procedures have been retained as alternative test methods to permit comparison of results by different procedures or for use when unusual materials are being examined, or when unusual preparation for analysis is required. Results from alternative test methods may be used as a basis for acceptance or rejection.

NOTE 4—It is not intended that the use of reference test methods be confined to referee analysis. A reference test method may be used in preference to an alternative test method when so desired. A reference test method must be used where an alternative test method is not provided.

7.2.1 Duplicate analyses and blank determinations are left to the discretion of the analyst when using the alternative test methods. The final results should include the number of determinations performed and whether or not they were corrected for blank values.

TABLE 2 Approved SRM List

(SRM)	Al as % Al203	Ca as % CaO	Mg as % MgO	Fe as % Fe203	Si as % SiO2	% Mn	% P	Sr as % SrO	% S	Ti as % TiO2	K as % K2O	Na as % Na_2O	% L.O.I.
ECRM-752-1[A]	0.12	55.4	0.15	0.045	0.70	0.008	NC[B]	0.019	0.007	0.009	0.02	NC	43.4
IPT 35	0.24	53.8	0.70	0.14	1.98	0.009	0.003	0.04	NC	0.013	0.10	0.004	43.0
IPT 44	0.33	50.5	2.93	0.30	2.69	0.012	0.006	0.04	NC	0.019	0.12	0.002	42.9
IRSID DO 1-1[C]	0.55	52.69	0.60	1.04	1.99	0.022	0.022	NC	0.040	0.030	NC	NC	NC
NIST 1C	1.30	50.3	0.42	0.55	6.84	0.019	0.017	0.030	NC	0.07	0.28	0.020	39.9
NIST 88B	0.34	29.95	21.0	0.277	1.13	0.012	0.002	0.0076	NC	(0.016)[D]	0.103	0.029	(46.98)
BCS 368	0.17	30.8	20.9	0.23	0.92	0.05	NC	0.0089	NC	<0.01	NC	(<0.01)	46.7
IRSID 702-1	0.40	30.05	20.51	0.629	2.22	0.098	0.024	NC	0.027	0.022	NC	NC	NC
GBW 07214	0.017	54.95	0.67	0.071	0.38	0.007	0.0011	NC	0.020	NC	NC	NC	43.57
GBW 07215	0.50	51.56	2.67	0.292	1.17	0.014	0.0013	NC	0.196	NC	NC	NC	43.22
GBW 07216	0.027	36.55	16.59	0.226	0.092	0.022	0.0018	NC	0.014	NC	NC	NC	46.23
GBW 07217	0.295	30.60	20.73	0.376	0.96	0.048	0.0012	NC	0.174	NC	NC	NC	46.30
GBW 03106	0.64	50.38	2.28	0.29	4.38	0.0055	0.006	NC	0.006	0.034	0.14	0.070	41.58
GBW 03108	0.88	47.49	3.63	1.97	3.84	0.15	0.017	NC	0.036	0.14	0.23	0.024	41.52
IPT 48	0.17	31.0	21.2	0.17	0.45	0.011	0.0096	0.009	NC	0.006	0.026	0.013	46.9

[A] This SRM is still available, but its name has been changed from BCS 393 to ECRM 752-1.

[B] NC = not certified.

[C] This SRM has been found to be unavailable commercially. The use of private stock, though, is still allowed.

[D] ( ) = not certified, data for information only.

7.3 *Performance Requirements for Alternative Test Methods:*

7.3.1 *Definition and Scope*—When analytical data obtained in accordance with this section is required, any test method may be used that meets the requirements of 7.3.2. A test method is considered to consist of the specific procedures, reagents, supplies, equipment, instrument, etc. selected and used in a consistent manner by a specific laboratory.

7.3.1.1 If more than one instrument is used for the same analysis, use of each instrument shall constitute a separate test method and each must be qualified separately.

7.3.2 *Qualification of a Test Method*—Prior to use each test method (see 7.3.1) must be qualified for each material that will be tested. Qualification data or, if applicable, requalification data shall be made available.

7.3.2.1 Using the test method chosen, make single determinations for each constituent under consideration on the SRM which in overall composition most closely resembles the limestone, quicklime, or hydrated lime to be tested (Note 2). Complete two rounds of tests on nonconsecutive days repeating all steps of sample preparations. Calculate the differences between values and the averages of values from the two rounds of tests. Blank determinations are not required, if it has been determined that blank values do not affect the validity of the data. Blank or interference-corrected data must be so designated.

7.3.2.2 The differences between duplicates obtained for any single constituent shall not exceed the limits shown in Column 2 of Table 1.

7.3.2.3 For each constituent the average of the duplicates obtained shall be compared to the SRM Certificate value and shall not differ from the certified value by more than the value in Column 3 of Table 1. The qualification testing shall be conducted with newly prepared specimens.

7.3.2.4 The standardization, if applicable, used for qualification and analysis of each constituent shall be determined by valid curve-fitting procedures (Note 5). Restandardization shall be performed as frequently as required to ensure that the accuracy and precision in Table 1 are maintained.

NOTE 5—An actual drawing of a curve is not required, if such a curve is not needed for the method in use. A point-to-point, saw-tooth curve that is artificially made to fit a set of data points does not constitute a valid curve-fitting procedure.

7.3.3 *Partial Results*—Test methods that provide acceptable results for some constituents, but not for others, may be used only for those components for which acceptable results are obtained.

7.3.4 *Report of Results*—Chemical analyses obtained by qualified alternative test methods shall be indicated as having been obtained by alternative methods and the type of test method used shall be designated.

7.3.5 *Rejection of Material*—See 7.1 and 7.2.

7.3.6 *Requalification of a Test Method:*

7.3.6.1 Requalification of a test method, as defined in 7.3.2, shall be required annually.

7.3.6.2 Requalification also shall be required upon receipt of substantial evidence that the test method may not be providing data in accordance with Table 1. Such requalification may be limited to those constituents indicated to be in error and shall be carried out prior to further use of the method for analysis of those constituents.

7.3.6.3 Substantial evidence that a test method may not be providing data in accordance with Table 1 shall be considered to have been received when a laboratory is informed that analysis of the same material by Reference Test Methods run in accordance with 7.1.1, a certified value of an approved SRM, or an accepted value of a known secondary standard differs from the value obtained by the test method in question by more than twice the value of Column 2 of Table 1 for one or more constituents. When indirect test methods are involved, as when a value is obtained by difference, corrections shall be made for minor constituents in order to put the analyses on a comparable basis prior to determining the differences (Note 6). For any constituents affected, a test method also shall be requalified after any substantial repair or replacement of one or more critical components of an instrument essential to the test method.

NOTE 6—Instrumental analyses can usually detect only the element sought. Therefore, to avoid controversy, the actual procedure used for the elemental analysis should be noted when differences with reference procedures exist. For example, Combined Oxides of Iron and Aluminum by Wet Test should be compared to the sum of Fe_2O_3 and Al_2O_3 obtained instrumentally.

7.3.6.4 If an instrument or piece of equipment is replaced even by one of identical make and model, or is significantly modified, a previously qualified test method using such new or modified instrument or equipment shall be considered a new method and must be qualified in accordance with 7.3.2.

7.4 *Precision and Bias*—Different analytical test methods are subject to individual limits of precision and bias. It is the responsibility of the user to demonstrate that the test methods used at least meet the requirements shown in Table 1.

8. Insoluble Matter Including Silicon Dioxide (Standard Method)

8.1 *Scope*—This test method is based on a double evaporation to dryness of the hydrochloric acid solution of the limestone or lime sample to convert silicon dioxide (SiO_2) to the insoluble form. The acid-insoluble residue of a typical limestone consists of free silica and a mixture of minerals such as clay, mica, feldspar, tourmaline, barytes, garnet, zircon, rutile, etc.

8.2 *Summary of Test Method*—After dissolution in hydrochloric acid, the silica is dehydrated by a double evaporation to dryness. After each dehydration, the dry salts are redissolved with dilute hydrochloric acid, the solution is filtered, and the siliceous residue and other insoluble matter separated. The two papers containing the residues are combined, ignited, and weighed.

8.3 *Procedure:*

8.3.1 Weigh 0.5 g of quicklime or hydrated lime, or 1.0 g of limestone ground to pass a No. 50 (250-μm) sieve (Note 7). If the sample is a limestone or hydrated lime, ignite in a covered platinum crucible in an electric muffle (Note 8) at 950°C for 15 min or longer to effect complete decomposition. Transfer to an evaporating dish, preferably of platinum (Note 9), containing about 10 mL of water, mix to a thin slurry, add 5 to 10 mL of

HCl, and digest with the aid of gentle heat and agitation until solution is complete (Note 10).

NOTE 7—Due to the rapidity with which quicklime and hydrated lime absorb water and carbon dioxide from the air, samples must be protected in tightly stoppered containers at all times. Samples for analysis are to be weighed quickly and the sample container re-stoppered immediately after the sample has been removed.

NOTE 8—Ignition of the sample in an electric muffle is far superior to flame ignition. However, if an electric muffle is not available, flame ignition and the blast lamp may be used.

NOTE 9—If a platinum dish is not available, porcelain may be used. A glass container positively must *not* be used.

NOTE 10—Alternatively, the loss on ignition (LOI) can be determined first, using 0.5 g of sample. The insoluble matter including silicon dioxide can then be assayed using the ignited product that remains in the LOI crucible.

8.3.2 Evaporate the solution to dryness on a steam bath. When dry or nearly so, cover the dish and place it in an air bath or drying oven or on a metal triangle resting on a hot plate. Heat for 1 h at 100°C, remove the dish from the heat, and allow the dish and contents to cool slightly.

8.3.3 Drench the cooled mass with 20 mL (1 + 1) HCl and place on the water bath for 10 min. Filter the mixture containing the insoluble residue through a retentive filter of suitable size. Wash filter thoroughly with warm, diluted (5 + 95) HCl and then twice with hot water. Reserve the paper and residue.

8.3.4 Evaporate the filtrate to dryness, dehydrate and extract the residue with HCl as before, but this time heat the acidified solution for 1 to 2 min. Filter through a second and smaller piece of retentive filter paper and wash as before. Retain the filtrate for iron, aluminum, calcium, and magnesium determinations; combine the two wet papers containing the separated residues and transfer to a weighed platinum crucible.

8.3.5 Char carefully without allowing the paper to inflame, and then ignite at 1000°C for 30 min in an electric muffle furnace (Note 8). Cool in a desiccator and weigh. The increase in weight represents the insoluble matter including SiO_2.

8.4 *Calculation*—Calculate the percentage of insoluble matter including silicon dioxide to the nearest 0.01 % as follows:

$$\text{Insoluble matter including } SiO_2 = (A/B) \times 100 \quad (1)$$

where:

A = mass of ignited residue, g, and
B = original mass of sample, g.

8.5 *Precision and Bias*—This test method was originally approved for publication before the inclusion of precision and bias statements within standards was mandated. The user is cautioned to verify by the use of reference materials, if available, that the precision and bias of this test method are adequate for the contemplated use.

9. Insoluble Matter Including Silicon Dioxide (Optional Perchloric Acid Method)

9.1 *Scope*—In this test method the insoluble matter including silicon dioxide is determined gravimetrically as in the standard method except that perchloric acid is used to dehydrate the silica. The procedure is more rapid than in the standard method because only a single dehydration is necessary. Fuming perchloric acid is a very powerful dehydrating agent, and silicic acid can usually be completely converted to the insoluble silicon dioxide in less than 20 min. This test method has been determined by other agencies such as the Association of Official Agricultural Chemists (AOAC) to be comparable to the standard hydrochloric acid method.

9.2 *Summary of Test Method*—The sample is decomposed without prior ignition by a mixture of nitric (HNO_3) and perchloric ($HClO_4$) acids and evaporated to fumes of $HClO_4$. The fuming perchloric acid is refluxed at this temperature for a short period of time to completely dehydrate the silica. The residue of silica and insoluble matter is filtered and washed free of acids and salts. The filter paper containing the residue is burned off, the resultant ash is ignited at high temperature until the ash is white, and then is weighed.

9.3 *Procedure:*

NOTE 11: **Precaution**—Perchloric acid ($HClO_4$) is an extremely reactive liquid. When using $HClO_4$, there are precautions to be followed which, if unheeded, may lead to serious explosions. Contact of the hot concentrated acid with organic matter must be absolutely avoided. Any organic matter in the sample must first be destroyed by the addition of nitric acid (HNO_3) to the sample prior to fuming with $HClO_4$. All evaporations involving $HClO_4$ must be done in a well-ventilated hood made of nonporous and inorganic material, preferably Type 316L stainless steel. Facilities should be provided for washdown procedures that should be performed regularly and thoroughly. These precautions on perchloric acid use are fully discussed in Practices E 50.

9.3.1 Weigh 0.5 g of quicklime or hydrated lime, or 1 g of limestone ground to pass a No. 50 (250-μm) sieve. Transfer the sample to a 250-mL beaker, wet carefully with a few millilitres of water, and dissolve cautiously with 10 mL of concentrated nitric acid. Add 20 mL of perchloric acid and boil until dense white fumes appear. If the solution darkens at this point, add several millilitres of HNO_3 until the solution clears. Heat again to fumes.

9.3.2 With the beaker covered, boil gently for 15 min to completely dehydrate the silica. Never allow contents to become solid or go to dryness, otherwise the separation of silica will be incomplete. If this happens, add more $HClO_4$ and repeat the dehydration.

9.3.3 Cool, add 50 mL of water, heat to boiling, and filter immediately using medium textured paper. Wash paper and residue thoroughly (at least 15 times) with hot water. Test with pH paper until washings are free of acid (Note 12). Reserve the filtrate for iron, aluminum, calcium, and magnesium determinations.

NOTE 12—The filter paper and silica residue must be washed free of perchlorate salts to prevent small explosions from occurring in the crucible when the filter paper is charred and ignited.

9.3.4 Place the filter paper and contents in a weighed platinum or porcelain crucible and heat gently with a low flame until paper chars without inflaming, or alternatively char in an electric muffle at 300 to 400°C. Slowly raise the temperature until the carbon has been burned and the ash is white. Finally, ignite at 1000°C for 30 min. Cool in a desiccator and weigh as insoluble matter including SiO_2.

9.4 *Calculation*—Calculate the percentage of insoluble matter including silicon dioxide to the nearest 0.01 % as follows:

$$\text{Insoluble matter including } SiO_2, \% = (A/B) \times 100 \quad (2)$$

where:
A = mass of ignited residue, g, and
B = original mass of sample, g.

9.5 *Precision and Bias:*

9.5.1 Four laboratories cooperated in testing on four limestone samples and three laboratories cooperated in testing on an additional eight limestone samples thereby obtaining the precision data summarized in Table 3.

9.5.2 The user is cautioned to verify by the use of reference materials, if available, that the bias of this test method is adequate for the contemplated use.

10. Silicon Dioxide

10.1 *Scope*—For control purposes or routine determinations, a separate analysis of SiO_2 may not be necessary. However, for certain applications in process industries, the amount of silica derived from the lime or limestone could be important. To satisfy situations such as this, silicon dioxide may be determined by volatilizing the SiO_2 from the insoluble residue with hydrofluoric acid and the percent SiO_2 determined by the difference in mass obtained.

10.2 *Procedure:*

10.2.1 To the ignited residue in the platinum crucible (8.3.5 or 9.3.4), add 5 mL of water, 5 mL of hydrofluoric acid (HF), and 1 or 2 drops of H_2SO_4.

NOTE 13: **Precaution**—All acids should be handled with care, but extra precaution is required with hydrofluoric acid. This is a very dangerous acid, harmful to eyes and skin; rubber gloves and goggles should be worn when using this acid. It does its work silently and leaves a festering sore that is slow to heal. Any acid that touches the skin should be immediately washed off with copious quantities of water. A physician should be notified immediately if any acid is sprayed into the eyes or if prolonged contact with the skin occurs.

10.2.2 Evaporate to dryness on a hot plate and heat in an electric muffle at 1000°C (Note 8) for 2 or 3 min. Cool in a desiccator and weigh. The difference between this mass and the mass of insoluble matter including silicon dioxide is the mass of SiO_2.

10.3 *Calculation*—Calculate the percent of silicon dioxide to the nearest 0.01 % as follows:

$$SiO_2, \% = ([A - B]/C) \times 100 \quad (3)$$

where:
A = mass of ignited residue, g (insoluble matter including SiO_2),
B = mass of ignited residue less SiO_2, g, and
C = original mass of sample, g.

10.4 *Precision and Bias:*

10.4.1 Three laboratories cooperated in testing on four limestone samples and two laboratories cooperated in testing on an additional eight limestone samples thereby obtaining the precision data summarized in Table 3.

10.4.2 The user is cautioned to verify by the use of reference materials, if available, that the bias of this test method is adequate for the contemplated use.

11. Insoluble Matter

11.1 *Scope*—The difference between the mass of insoluble matter (including silicon dioxide) and silicon dioxide represents the mass of insoluble matter other than silicon dioxide. The insoluble matter contains the remnants of any clay, siliceous minerals, or other refractory material present in limestone. The elemental components are mainly iron and aluminum which should be removed and added to the main filtrate from the SiO_2 separation. If the insoluble matter including silica is reported as such and no hydrofluoric acid treatment is indicated, then there is no need to make a recovery of the metals and the insoluble residue may be discarded.

11.2 *Procedure*—The insoluble matter left in the crucible after the silica is volatilized with HF may be dissolved by fusing the residue with 2 to 3 g of sodium carbonate (Na_2CO_3) (Note 14). Cool the melt and dissolve it in diluted HCl. Add the solution to the filtrate from the dehydration and separation of insoluble matter including silicon dioxide (8.3.4 or 9.3.3).

NOTE 14—Fusion with pyrosulfate is to be avoided because this will introduce undesirable sulfates into the solution.

11.3 An alternative fusion can also be made using either lithium metaborate or lithium tetraborate as opposed to using sodium carbonate.

11.4 *Calculation*—Calculate the percentage of insoluble matter other than silicon dioxide to the nearest 0.01 % as follows:

$$\text{Insoluble matter other than } SiO_2, \% = A - B \quad (4)$$

TABLE 3 Precision Summary of Classical Test Methods

Section	Test Method	Average,[A] % Found	Range,[A] % Found	Repeatability (R_1, E 173)	Reproducibility (R_2, E 173)
8	Insol + SiO_2 (Standard)	...	...	...	...
9	Insol + SiO_2 (Optional)	1.405	0.09–6.40	0.184	0.351
10	SiO_2	1.177	0.03–5.36	0.128	0.146
11	Insoluble Matter	0.242	0.02–0.93	0.169	0.204
12	Combined Oxides	0.459	0.22–1.21	0.181	0.282
13	Fe_2O_3	0.180	0.05–0.36	0.064	0.183
15	Al_2O_3	0.268	0.10–0.88	0.165	0.223
16	CaO (Gravimetric)	54.46	53.4–55.1	0.558	1.020
17	CaO (Volumetric)	30.57	30.4–30.7	0.371	1.132
17	CaO (Volumetric)	53.82	49.6–55.3	0.187	0.298
18	MgO (Gravimetric)	0.817	0.19–2.28	0.158	0.210
18	MgO (Gravimetric)	21.34	21.1–21.5	0.652	1.716
19	Loss on Ignition	43.73	43.6–43.9	0.158	0.463

[A] Average and range of the limestones tested.

where:
A = insoluble matter including SiO_2, %, and
B = SiO_2, %.

11.5 *Precision and Bias:*

11.5.1 Three laboratories cooperated in testing on four limestone samples and two laboratories cooperated in testing on an additional eight limestone samples thereby obtaining the precision data summarized in Table 3.

11.5.2 The user is cautioned to verify by the use of reference materials, if available, that the bias of this test method is adequate for the contemplated use.

12. Combined Oxides (Iron, Aluminum, Phosphorus, Titanium, Manganese)

12.1 *Scope*—The combined oxides describe a group of metals that form precipitates with ammonium hydroxide which may then be ignited to their respective oxides. Historically, it has been the practice to report the combined oxides present in limestone samples as a group because it was not always easy or desirable to determine each metal oxide separately. The group of metal oxides consists primarily of the oxides of iron and aluminum, with minor amounts of titanium dioxide (TiO_2), phosphorus pentoxide (P_2O_5), and manganese oxide (Mn_3O_4) also present. Where separate determinations are preferred, the combined oxides are usually weighed first, iron oxide is then assayed separately, and aluminum oxide is finally determined by calculating the difference between the percent combined oxides and the percent Fe_2O_3. The other metal oxides are generally assumed to be present in trace amounts and are often disregarded. When necessary, these metals may be analyzed separately and appropriate corrections made in the Al_2O_3 analysis.

12.2 *Summary of Test Method*—In this test method, aluminum, iron, titanium, and phosphorus are precipitated from the filtrate after SiO_2 removal, by means of ammonium hydroxide. With care, little if any manganese will be precipitated. The precipitate is ignited and weighed as the combined metal oxides.

12.3 *Special Solution:*

12.3.1 *Methyl Red Solution* (0.2 %)—Dissolve 2 g of methyl red indicator with 1 L of 95 % ethyl alcohol.

12.4 *Procedure:*

12.4.1 To the acid solution from the determination of SiO_2 (8.3.4 or 9.3.3), add hydrochloric acid (HCl) if necessary to ensure a total of 10 to 15 mL of HCl.

NOTE 15—Sufficient hydrochloric acid must be present before the solution is rendered ammoniacal to prevent the precipitation of magnesium.

12.4.2 If a platinum evaporating dish has been used for the dehydration of SiO_2, or a fusion made in the platinum crucible containing the HF-insoluble residue, iron may have been partially reduced. The iron must then be oxidized by adding 1 mL of saturated bromine water to the filtrate. Boil the filtrate to eliminate the excess bromine completely before adding methyl red indicator.

12.4.3 Dilute with water to a volume of 200 to 250 mL, add a few drops of methyl red solution, and heat just to boiling. Add NH_4OH (1 + 1) (Note 16) until the color of the solution becomes distinctly yellow, then add 1 drop in excess (Note 17). Heat the solution containing the precipitate to boiling and boil for 50 to 60 s. Remove from heat and allow the precipitate to settle (not more than 5 min). Filter using medium-textured paper and wash the precipitate two or three times without delay with a hot, 2 % solution of ammonium chloride (NH_4Cl) (Note 18).

NOTE 16—The NH_4OH used to precipitate the hydroxides must be free of any dissolved carbon dioxide (CO_2).

NOTE 17—At the neutral point, it usually takes 1 drop of NH_4OH (1 + 1) to change the color of the solution from red to orange and another drop to change the color from orange to yellow. If the color fades during the precipitation or while heating, add more of the indicator. The boiling should not be prolonged as the precipitate may peptize and be difficult to retain on the filter. The solution should be distinctly yellow when it is ready to filter. If it is not, restore the yellow color with more NH_4OH (1 + 1).

NOTE 18—Two drops of methyl red indicator solution should be added to the NH_4Cl solution in the wash bottle followed by NH_4OH (1 + 1) added dropwise until the color just changes to yellow. If the color reverts to red at any time due to heating, it should be brought back to yellow by the addition of a drop of NH_4OH (1 + 1).

12.4.4 Set aside the filtrate and dissolve any precipitate from the paper with 40 mL hot (1 + 3) HCl, pouring the hot acid through the paper into the beaker in which the precipitation was made. Wash the filter paper thoroughly with hot HCl (1 + 19) followed by hot water and reserve the paper. Boil the solution and precipitate the hydroxides with NH_4OH as before. The precipitate is filtered through a fresh piece of medium textured filter paper and washed four or five times (Note 19) with a hot 2 % solution of NH_4Cl. Combine filtrates for Ca and calcium magnesium analysis.

NOTE 19—If perchloric acid has been used, the final precipitate should be washed at least eight times to remove all traces of perchlorate salts (Note 11).

12.4.5 Place the moist precipitate and the two filter papers in a weighed platinum crucible (Note 9), heat slowly until the papers are charred, and finally ignite to constant weight at 1050 to 1100°C. Cool in a desiccator and weigh.

12.5 *Calculation*—Calculate the percentage of ammonium hydroxide group (combined oxides) to the nearest 0.01 % as follows:

$$\text{Combined oxides, \%} = (A/C) \times 100 \tag{5}$$

where:
A = mass of the combined oxides, g, and
C = original mass of sample, g.

12.6 *Precision and Bias:*

12.6.1 Four laboratories cooperated in testing on four limestone samples and three laboratories cooperated in testing on an additional seven limestone samples thereby obtaining the precision data summarized in Table 3.

12.6.2 The user is cautioned to verify by the use of test reference materials, if available, that the bias of this test method is adequate for the contemplated use.

13. Total Iron, Standard Method

13.1 *Scope*—Iron in limestone is usually present as pyrite (FeS_2) with occasional occurrences of other discrete iron minerals. The amount present varies according to the location and geological history of the deposit. During lime calcination, most if not all of the iron minerals present in the limestone ore will be converted to iron oxide or sulfate.

13.2 *Summary of Test Method*—In this test method, the total Fe_2O_3 content of the sample is determined from the ignited combined oxides by fusing the oxides with potassium pyrosulfate and leaching the melt with sulfuric acid. The iron is reduced to the ferrous state with stannous chloride and titrated with a standard solution of potassium dichromate ($K_2Cr_2O_7$).

13.3 *Special Solutions:*

13.3.1 *Stannous Chloride Solution* (50 g/L)—Dissolve 5 g of $SnCl_2 \cdot 2H_2O$ in 10 mL of HCl and dilute to 100 mL with water. Add several pieces of mossy tin metal to the bottle to preserve the $SnCl_2$ solution.

13.3.2 *Sodium Diphenylamine Sulfonic Acid Indicator* (2 g/L)—Dissolve 0.20 g sodium diphenylamine sulfonate in 100 mL of water. Store in a dark-colored bottle.

13.3.3 *Mercuric Chloride Solution* (5 %)—Dissolve 5 g of $HgCl_2$ in 100 mL of water.

13.3.4 *Potassium Dichromate, Standard Solution* (0.05 *N*)—Dry pure crystals of $K_2Cr_2O_7$ at 110°C, then pulverize and dry at 180°C to constant weight. Dissolve 2.4518 g of pulverized $K_2Cr_2O_7$ in water and dilute to 1 L. This is a primary standard, 1 mL = 0.0040 g Fe_2O_3.

13.4 *Procedure:*

13.4.1 To the combined oxides of iron and aluminum (Note 20) in the platinum crucible, add 3 to 4 g of potassium pyrosulfate ($K_2S_2O_7$). Fuse at low heat until the oxides form a clear melt in the crucible. Cool, break up the button by gently tapping the crucible on the bench, and wash fragments into a small beaker with hot H_2SO_4 (5 + 95). Add 5 mL of H_2SO_4 (sp gr 1.82) to the contents in the beaker, and heat to dissolve the fused mass. Evaporate the solution to fumes of sulfuric acid and fume strongly for about 10 min. Cool, add 20 mL of water, and warm to dissolve the salts. There may be traces of silica appearing at this point, which for most routine work can be ignored. If the analyst prefers to determine it, however, the precipitate can be filtered, washed, and ignited. The recovered SiO_2 can then be added to the mass of SiO_2 previously found and its mass deducted from the gross mass of iron and aluminum reported (Note 20).

NOTE 20—When the iron is present in small quantities, it is not always desirable to determine it in the ignited oxides from the 0.5-g sample. Under these conditions, the alternative procedure should be used with a larger sample weight.

NOTE 21—The recovered SiO_2 is usually small, but could be as much as 1 to 2 mg, even after two evaporations.

13.4.2 To the sulfuric acid solution, add 10 mL HCl (1 + 1) and heat to near boiling. Add dropwise stannous chloride solution (Note 22) until the yellow color of the ferric iron just disappears. Add 2 or 3 drops of $SnCl_2$ in excess.

NOTE 22—If the stannous chloride has little effect and more than 5 to 10 mL are required, it has probably become oxidized to stannic chloride and a fresh supply should be obtained.

13.4.3 Cool the mixture and add approximately 100 mL of cold water. Add 10 mL of mercuric chloride solution, stir, and allow to stand for 3 to 5 min.

NOTE 23—A slight, white, silky precipitate should form. If the precipitate appears gray or black, it indicates too much $SnCl_2$ was added and the analysis must be repeated.

13.4.4 Add 5 mL of H_3PO_4 and 3 drops of sodium diphenylamine sulfonate indicator.

13.4.5 Titrate with standard 0.05 *N* $K_2Cr_2O_7$ solution adding the solution slowly while stirring constantly. The end point is indicated by a change in color from green to deep blue-violet.

13.5 *Calculation:*

$$Fe_2O_3, \% = (A/C) \times B \times 100 \tag{6}$$

where:

A = $K_2Cr_2O_7$ used in titration, mL,
B = 0.004 (Fe_2O_3 equivalent of $K_2Cr_2O_7$), and
C = sample, g.

13.6 *Precision and Bias:*

13.6.1 Four laboratories cooperated in testing on four limestone samples and three laboratories cooperated in testing on an additional seven limestone samples thereby obtaining the precision data summarized in Table 3.

13.6.2 The user is cautioned to verify by the use of reference materials, if available, that the bias of this test method is adequate for the contemplated use.

14. Total Iron by Ortho-Phenanthroline Photometric Method

14.1 *Scope*—When the iron oxide content is very low, less than 0.1 %, and an accurate analysis at this low level is required, it is preferable to determine iron using procedures that have better sensitivity than the titrimetric methods. For an accurate determination of minute amounts of iron, the ortho phenanthroline method has proved invaluable.[9] In general, the method consists of reducing the iron to the ferrous state and then adding a slight excess of 1, 10 phenanthroline, which forms a complex with ferrous iron, giving an orange-pink color. The color intensity is proportional to the iron content of the solution.

14.2 *Summary of Test Method*—The bulk of the iron in the sample is dissolved with HCl, the silica dehydrated and separated by filtration, and the insoluble matter including SiO_2, ignited in a platinum crucible and treated with HF and H_2SO_4 to expel the SiO_2 and recover the small amount of iron that may not have dissolved with HCl. The acidified solution is transferred to a volumetric flask and diluted to volume. The iron is reduced with hydroxylamine hydrochloride and the color of the ferrous complex is developed with 1,10 phenanthroline and compared against a set of iron standards similarly treated.

14.3 *Special Solutions:*

14.3.1 *Hydroxylamine Hydrochloride* (10 g/100)—Dissolve 10 g of hydroxylamine hydrochloride in 100 mL of water. Prepare fresh every week.

14.3.2 *Ammonium Acetate* (20 g/100)—Dissolve 200 g in 1 L of water.

14.3.3 *1,10 (Ortho) Phenanthroline* (0.1 g/100)—Dissolve 1.0 g in 1 L of hot water.

[9] Sandel, E. B., *Colorimetric Determination of Traces of Metals,* 3rd Ed., Interscience Publications, 1959.

14.3.4 *Iron Standard Solution* (1 mL = 1.0 mg Fe_2O_3)—Dissolve 0.7000 g of pure iron wire by heating gently in 20 mL of HCl (1 + 1) and dilute to 1 L in a volumetric flask.

14.3.4.1 *Iron Work Standard Solution* (1 mL = 0.01 mg Fe_2O_3)—Transfer 10 mL of the iron standard solution to a 1 L volumetric flask and dilute to volume with water.

14.3.5 *Preparation of Calibration Curve*—To each of six 50 mL volumetric flasks, add, respectively, 1, 2, 4, 6, 8, and 10 mL of working iron standard solution. When diluted to volume, each mL of the prepared standard solutions will contain, respectively 0.2, 0.4, 0.8, 1.2, 1.6, and 2.0 micrograms Fe_2O_3.

14.3.5.1 Add to each flask in the following sequence, mixing after each addition, 1 mL of hydroxylamine hydrochloride solution, 5 mL of ammonium acetate, and 5 mL of 1,10 phenanthroline. Roll a small piece of congo red paper into a ball and insert it into the volumetric flask. Add NH_4OH (1 + 1) until the congo red indicator turns bright red, then add 1 drop of NH_4OH (1 + 1) in excess. Dilute to 50 mL, mix, and let stand for 15 to 20 min. Determine the absorbance of the solution in a spectrophotometer at a wavelength setting of 510 nm using water in the reference cell. Prepare a calibration curve by plotting the absorbance versus the concentration of Fe_2O_3 in μg/mL of solution.

14.4 *Procedure:*

14.4.1 Weigh 1 g of the properly prepared sample in 10 mL HCl (1 + 1) and evaporate rapidly to dryness. Add 50 mL of HCl (1 + 4) and heat to dissolve the salts. Filter the insoluble matter including SiO_2 through a retentive paper and wash several times with hot water. Reserve the residue. Heat the filtrate to boiling.

14.4.2 Place the paper containing the insoluble matter from the evaporated HCl solution in a platinum crucible. Char the paper at low heat without inflaming, then ignite at higher heat until the carbon has been completely burned off. Cool, add 1 mL H_2SO_4 and 10 to 15 mL HF and evaporate to fumes of sulfuric acid. Cool, dilute the contents of the crucible with water, and warm to dissolve salts. Transfer the acidified solution to the main solution containing the bulk of the iron.

14.4.3 Transfer the combined solutions to a 100 mL volumetric flask and dilute to volume. Pipet the aliquot containing 0.02 to 0.10 mg Fe_2O_3 into a 50 mL volumetric flask. Dilute to about 25 mL and add in the following sequence, mixing well after each addition: 1 mL hydroxylamine hydrochloride, 5 mL ammonium acetate, and 5 mL of 1,10 phenanthroline. Roll a small piece of congo red paper into a ball and insert into the volumetric flask. Add NH_4OH (1 + 1) until the congo red indicator turns a bright red, then add one drop of NH_4OH (1 + 1) in excess. Dilute to 50 mL, mix and let stand for 15 to 20 min. Determine the absorbance of the solution in a spectrophotometer at a wavelength setting of 510 nm using water in the reference cell. Compare against a set of standards similarly treated.

14.5 *Calculation:*

14.5.1 Calculate the % Fe_2O_3 as follows:

$$\%\ Fe_2O_3 = \frac{C \times D}{W \times 10^4} \quad (7)$$

where:

C = concentration of Fe_2O_3 in sample solution, μg/mL (determined from calibration curve),
D = dilution factor, and
W = sample mass, g.

14.6 *Precision and Bias:*

14.6.1 The number of laboratories, materials, and determinations in this study does not meet the minimum requirements for determining precision prescribed in Practice E 691:

	Test Methods C 25	Practice E 691 Minimum
Laboratories	2	6
Materials	5	4
Determinations	4	2

14.6.2 The following precision statements are provisional. Within five years, additional data will be obtained and processed which does meet the requirements of Practice E 691.

14.6.2.1 Precision, characterized by repeatability, *Sr* and *r*, and reproducibility, *SR* and *R*, has been determined for the following test method and materials to be:

Precision Statement for Test Method: Material	% Fe_2O_3 Average	*Sr*	*SR*	Color *r*	*R*
S-1143	0.0358	0.0058	0.0201	0.0163	0.0564
S-1145	0.0480	0.0053	0.0214	0.0148	0.0599
S-1141	0.1688	0.0466	0.0640	0.1306	0.1792
S-1142	0.2025	0.0141	0.0631	0.0396	0.1767
S-1144	0.9252	0.0562	0.2096	0.1574	0.5870

15. Aluminum Oxide

15.1 *Scope*—Aluminum oxide, for the purpose of this test method, is considered to be the difference between the combined oxides and Fe_2O_3. When phosphorus or titanium are determined, their oxides must also be deducted.

15.2 *Procedure*—Subtract the percent Fe_2O_3 obtained in accordance with Sections 13 and 14 from the percent combined oxides (Section 12). Report the remainder as percent Al_2O_3. In special cases where P_2O_5 and TiO_2 need to be reported, a correction for these oxides must be made.

15.3 *Calculation*—Calculate the percent Al_2O_3 as follows:

$$Al_2O_3, \% = A - B \quad (8)$$

where:

A = combined oxides (Al_2O_3 + Fe_2O_3), %, and
B = Fe_2O_3, %.

15.4 *Precision and Bias:*

15.4.1 Four laboratories cooperated in testing on four limestone samples and three laboratories cooperated in testing on an additional seven limestone samples thereby obtaining the precision data summarized in Table 3.

15.4.2 The user is cautioned to verify by the use of reference materials, if available, that the bias of this test method is adequate for the contemplated use.

16. Calcium Oxide by Gravimetric Method

16.1 *Scope*—Calcium is separated from magnesium by means of a double precipitation as the oxalate after the determination of the ammonium hydroxide group. The precipitate is converted to CaO by ignition and weighed. The gravimetric method should be used when a recovery of aluminum is indicated or when a determination of strontium by gravimetric analysis is required.

16.2 *Summary of Test Method*—Calcium is precipitated with ammonium oxalate $(NH_4)_2C_2O_4$, filtered, ignited to the oxide, and redissolved with HCl. Any of the NH_4OH group of metals that escaped precipitation before may be recovered at this point by the addition of a small amount of NH_4OH and boiling. Any precipitate that separates out is assumed to be $Al(OH)_3$ and after ignition to Al_2O_3 this amount is added to the mass of Al_2O_3 calculated in 16.2. Calcium is precipitated a second time as the oxalate, filtered, washed, ignited, and weighed as CaO.

16.3 *Special Solutions:*

16.3.1 *Ammonium Oxalate Solution* (saturated)—Dissolve 45 g of ammonium oxalate $(NH_4C_2O_4)$ in 1 L of hot water. When cooled to room temperature the supersaturated solution will partially crystallize out and the supernatant solution will then be saturated with ammonium oxalate.

16.3.2 *Ammonium Oxalate Wash Solution* (1 g/L)—Dissolve 1 g of $(NH_4)_2C_2O_4$ in 1 L of water.

16.4 *Procedure:*

16.4.1 Add 30 mL of HCl (1 + 1) and 20 mL of 10 % oxalic acid to the combined filtrates from the iron and aluminum hydroxide precipitation and heat the solution to boiling. To the boiling solution, add ammonium hydroxide (1 + 3) slowly until a precipitate begins to form. At this point, add the ammonium hydroxide still more slowly (dropwise, with a pipet) while stirring continuously until the methyl red just turns yellow. Add 25 mL of hot saturated ammonium oxalate solution while stirring. Remove from the heat and let stand until the precipitate has settled and the supernatant liquid is clear. Allow to cool for a minimum of 1 h, and filter using a retentive paper. Wash the paper and precipitate with five 10-mL portions of cold, neutral 0.1 % solution of $(NH_4)_2C_2O_4$ (Note 24). Reserve filtrate for the magnesium determination.

NOTE 24—Hot solutions should be avoided when washing the CaC_2O_4 precipitate. One litre of hot water will dissolve 5 mg of CaO. One litre of cold 0.1 % $(NH_4)_2C_2O_4$ solution will dissolve only 0.1 mg of CaO.

16.4.2 Place the wet filter and precipitate in a platinum crucible, and char the paper without inflaming at low heat. Increase the heat to burn off all the carbon and ignite at 1000°C for about 10 min. Cool, dissolve the ignited oxide in 50 mL of dilute HCl (1 + 4), and dilute to about 100 mL with water. Add a few drops of methyl red indicator and neutralize with NH_4OH till the color of indicator changes to yellow. Heat just to boiling. If a small amount of $Al(OH)_3$ separates, filter it, wash with a hot 2 % solution of NH_4Cl, ignite, weigh, and add this to the mass of Al_2O_3 determined in 15.2.

16.4.3 Heat the filtrate to boiling and add slowly, while stirring, 35 mL of saturated $(NH_4)_2C_2O_4$ solution. Digest, filter, and wash as in 16.4.1. Combine the filtrate and washing with the ones reserved from the first precipitation, and retain for the determination of MgO. Place the filter in a tared platinum crucible with cover and carefully char the paper without inflaming. Increase the heat to burn off the carbon and ignite the calcium oxide in the covered platinum crucible at 1000°C. Cool in a desiccator and weigh as CaO. Repeat the ignition to constant weight avoiding any hydration or carbonation of the lime.

16.5 *Calculation*—Calculate the percent calcium oxide (CaO) as follows:

$$CaO, \% = (M/W) \times 100 \tag{9}$$

where:
M = mass of CaO, g, and
W = mass of sample, g.

16.6 *Precision and Bias:*

16.6.1 Two laboratories cooperated in testing on four limestone samples and obtained the precision data summarized in Table 3.

16.6.2 The user is cautioned to verify by the use of reference materials, if available, that the bias of this test method is adequate for the contemplated use.

17. Calcium Oxide by Volumetric Method

17.1 *Scope*—This volumetric test method is used mostly for ordinary control work in the plant laboratory, but it is capable of giving exact results, especially with those products that are free of interfering elements. Traces of strontium, barium, magnesium, or oxalate that may be present will also be titrated and calculated as calcium on an equivalence, not weight, basis.

17.2 *Summary of Test Method*—In this test method, the calcium oxalate precipitate is dissolved with dilute sulfuric acid and the liberated oxalic acid is titrated with standard potassium permanganate. The calcium equivalent of the oxalic acid is determined and the grams of CaO calculated.

17.3 *Special Solutions:*

17.3.1 *Potassium Permanganate, Standard Solution* (0.175 *N*):

17.3.1.1 Dissolve 5.64 g of potassium permanganate $(KMnO_4)$ in 1 L of water and boil gently for 20 to 30 min. Dilute again to 1 L, cover and allow to age for several days. Filter through purified asbestos or a wad of glass wool, and standardize against the National Institute of Standards and Technology's standard sample 40C of sodium oxalate $(Na_2C_2O_4)$ or equivalent as follows:

17.3.1.2 Transfer 0.5 g of the standard sodium oxalate dried at 105°C to a 400-mL beaker. Add 250 mL of diluted H_2SO_4 (5 + 95) freshly boiled for 10 to 15 min and cooled to 27 ± 3°C. Stir until the oxalate has dissolved. Add 40 to 42 mL of the standard $KMnO_4$ solution at the rate of 25 to 35 mL/min, while stirring slowly. Let stand until the pink color disappears (about 60 s) (Note X1.2).

17.3.1.3 Heat the contents of the beaker to 60°C and complete the titration at this temperature by adding $KMnO_4$ solution until a slight pink color persists for 30 s. Add the last 0.5 to 1 mL dropwise, allowing each drop to become decolorized before the next one is added.

17.3.1.4 Determine the exact normality of the $KMnO_4$ solution from the following:

$$N = W/V \times 0.06701 \tag{10}$$

where:
N = normality of $KMnO_4$ solution,
W = mass of standard sodium oxalate,

V = $KMnO_4$ used to titrate sodium oxalate, mL, and
0.06701 = sodium oxalate equivalent to 1 mL of 1 N $KMnO_4$ solution, g.

17.3.1.5 Determine the CaO equivalent of the $KMnO_4$ solution as follows:

$$F = N \times 0.02804 \quad (11)$$

where:
F = CaO equivalent of the $KMnO_4$ solution in grams CaO/mL,
N = normality of $KMnO_4$ solution, and
0.02804 = CaO equivalent to 1 mL of 1 N $KMnO_4$ solution, g.

17.4 *Procedure:*

17.4.1 Add 30 mL of HCl (1 + 1) and 20 mL of 10 % oxalic acid to the combined filtrates from the iron and aluminum hydroxide precipitation and heat the solution to boiling. To the boiling solution, add ammonium hydroxide (1 + 3) slowly until a precipitate begins to form. At this point, add the ammonium hydroxide still more slowly (dropwise, with a pipet) while stirring continuously until the methyl red just turns yellow. Add 25 mL of hot saturated ammonium oxalate while stirring. Remove from the heat and let stand until the precipitate has settled and the supernatant liquid is clear. Allow to cool and filter at the end of 1 h. Wash the paper with cold water, limiting the total washings to 125 mL (Note 25). Retain the filtrate for magnesium.

NOTE 25—A Gooch crucible may be used instead of filter paper to filter the CaC_2O_4 precipitate.

17.4.2 With a jet of hot water, wash the precipitate from the paper into the beaker in which the precipitation was made. Fold the paper and leave it adhering to the upper portion of the beaker. Add to the contents of the beaker 250 mL of hot, diluted H_2SO_4 (1 + 19) and heat to 80 to 90°C.

17.4.3 Titrate with 0.175 N $KMnO_4$ solution until the pink end point is obtained. Drop the folded filter paper that contained the original precipitate into the liquid and macerate it with a stirring rod; the pink color of the solution will be discharged (Note 26). Finish the titration by adding the $KMnO_4$ standard solution dropwise until the end point is again obtained.

NOTE 26—There will always be some fine particles of precipitate imbedded in the pores of the filter paper which are dissolved by the acid in solution. The filter paper is not introduced at the beginning of the titration to avoid introduction of traces of organic matter due to the action of the hot sulfuric acid on the paper; these would consume $KMnO_4$ and give high results for CaO.

17.5 *Calculation*—Calculate the percentage of CaO in the sample using the CaO equivalent from 17.3.1.5 as follows:

$$\text{CaO, \%} = (V \times F)/W \times 100 \quad (12)$$

where:
V = $KMnO_4$ solution used in titration, mL,
F = CaO equivalent of $KMnO_4$, and
W = original mass of sample, g.

17.6 *Precision and Bias:*

17.6.1 Two laboratories cooperated in testing on twelve limestone samples and obtained the precision data summarized in Table 3.

17.6.2 The user is cautioned to verify by the use of reference materials, if available, that the bias of this test method is adequate for the contemplated use.

18. Magnesium Oxide

18.1 *Scope*—Magnesium oxide in lime and limestone may vary from a few tenths to 2 % for high-calcium limestone to as much as 22 % for dolomitic limestone. The pyrophosphate gravimetric method has been used successfully throughout the industry to determine magnesium within this wide range.

18.2 *Summary of Test Method*—In this test method, magnesium is doubly precipitated as magnesium ammonium phosphate from the filtrate after removal of calcium. The precipitate is ignited and weighed as magnesium pyrophosphate ($Mg_2P_2O_7$). The MgO equivalent is then calculated.

18.3 *Special Solutions:*

18.3.1 *Ammonium Phosphate, Dibasic Solution* (250 g/L)—Dissolve 250 g of dibasic ammonium phosphate ($(NH_4)_2HPO_4$) in 1 L of water.

18.3.2 *Ammonium Hydroxide Wash Solution* (5 + 95)—Dilute 50 mL of NH_4OH with 950 mL of water and add 1 or 2 mL of HNO_3.

18.4 *Procedure:*

18.4.1 Add 2 drops of methyl red indicator to the combined filtrates from the determination of calcium, acidify with HCl, and concentrate to about 250 mL. Add to this solution about 10 mL of the $(NH_4)_2HPO_4$ solution, 250 g/L, and cool the solution to room temperature. Add NH_4OH slowly while stirring constantly until the solution is alkaline or the crystalline magnesium ammonium phosphate begins to form; then add about 15 to 20 mL of NH_4OH in excess and continue stirring for several more minutes. Allow the beaker and precipitate to stand in a cool place overnight. Filter and wash with cold dilute ammonium hydroxide wash solution (5 + 95).

18.4.2 Dissolve the precipitate with hot diluted HCl (1 + 9) and wash the filter paper well with hot diluted HCl (1 + 99). Dilute the solution to 100 mL, cool to room temperature, and add 1 mL of the 20 % solution of $(NH_4)_2HPO_4$. Precipitate the magnesium ammonium phosphate as before and allow to stand for about 2 h in a cool place.

18.4.3 Filter the precipitate on paper or in a tared Gooch crucible, washing with diluted NH_4OH (5 + 95). If filtered through a Gooch, place directly in a muffle at 400°C and raise heat to 1100°C. If filtration was through paper, place paper and precipitate in a weighed platinum or porcelain crucible. Slowly char the paper without inflaming and carefully burn off the resulting carbon (Note 27). Ignite at 1100°C for ½ h, cool in desiccator, and weigh as $Mg_2P_2O_7$ (Note 28).

NOTE 27: **Caution**—Extreme caution should be exercised during this ignition. Reduction of the phosphate precipitate can result if carbon is in contact with it at high temperatures. There is also a danger of occluding carbon in the precipitate if ignition is too rapid.

NOTE 28—For research purposes or in the most exacting types of work, the manganese content of the pyrophosphate residue should be determined and deducted as $Mn_2P_2O_7$.

18.5 *Calculation*—Calculate the percentage of MgO to the nearest 0.01 % as follows:

$$\text{MgO, \%} = A \times 36.2/B \quad (13)$$

where:

A = $Mg_2P_2O_7$, g,
B = sample, g, and
36.2 = molecular ratio of 2MgO to $Mg_2P_2O_7 \times 100$.

18.6 *Precision and Bias:*

18.6.1 Four laboratories cooperated in testing on three limestone samples and three laboratories cooperated in testing on an additional nine limestone samples thereby obtaining the precision data summarized in Table 3.

18.6.2 The user is cautioned to verify by the use of reference materials, if available, that the bias of this test method is adequate for the contemplated use.

19. Loss on Ignition

19.1 *Scope*—Loss on ignition (LOI) is the loss in weight expressed as percent of the initial "as received" sample weight obtained after ignition of the sample at 1000°C to constant weight. The loss in weight is due to a release of free moisture, chemically combined "lattice" or "hydroxy" water, CO_2, SO_2, and volatile pyrolytic products of any organic material that may be present.

19.2 *Summary of Test Method*—The tared crucible containing the weighed sample is ignited to constant weight. The loss in weight is the LOI of the sample.

19.3 *Procedure*—Transfer approximately 1 g of the sample prepared to pass a 100-mesh (149-μm) U.S. standard sieve to a tare-weighed porcelain or platinum crucible. Cover with a lid and weigh accurately to within 0.1 mg. When testing quicklime, the crucible cover is not required. Also, quicklime may be placed directly into a muffle at 1000°C avoiding preignition. Pre-ignite in a muffle furnace at approximately 400°C for 30 min. Then increase muffle temperature to 1000°C ±20°C, and maintain at this temperature for a minimum of 20 min or until constant mass is obtained. The difference between the original mass of the sample and the final mass represents the loss on ignition.

19.4 *Calculation*—Calculate LOI as follows:

$$\text{LOI, \%} = (A - B)/C \times 100 \tag{14}$$

where:

A = mass of crucible + sample, g,
B = mass of crucible plus sample after ignition, g, and
C = mass of sample, g.

19.5 *Precision and Bias:*

19.5.1 Fifteen laboratories cooperated in testing on three samples of high calcium limestone to obtain the precision data for percent LOI given in 19.5.2 and 19.5.3.

19.5.2 The repeatability (Practice E 691 [r]) was found to be 0.158 % LOI.

19.5.3 The reproducibility (Practice E 691 [R]) was found to be 0.463 % LOI.

19.5.4 The user is cautioned to verify by the use of reference materials, if available, that the bias of this test method is adequate for the contemplated use.

20. Free Moisture in Limestone

20.1 *Scope*—For the purpose of this test method, the conventional definition of "hygroscopic moisture" or "free water" (also known as "free-moisture") is accepted; that is, the amount of water and any other volatile matter than can be expelled from a sample of the material by drying to constant weight at a temperature slightly above the boiling point of water.

20.2 *Summary of Test Method*—The sample in a container is heated in a drying oven at 115 to 120°C constant weight. The loss in weight represents the free moisture.

20.3 *Special Apparatus:*

20.3.1 *Bottle,* weighing, low-form, glass-stoppered, or wide-form, large porcelain crucible.

20.4 *Procedure*—Weigh 1 g of the prepared sample in the stoppered weighing bottle. Remove the stopper and heat in a drying oven at 115 to 120°C for 2 h. Quickly stopper, cool in a desiccator, and weigh, lifting the stopper momentarily just before weighing. The use of a similar weighing bottle as a counterpoise carried through all the operations is a desirable procedure unless a single pan balance is used. The loss in weight represents "free moisture" loss at 120°C.

20.5 *Calculation*—Calculate the percent "free moisture" as follows:

$$\text{Free-moisture, \%} = (A - B)/C \times 100 \tag{15}$$

where:

A = mass of crucible and sample before heating, g,
B = mass of crucible and sample after heating at 120°C, g, and
C = original mass of sample, g.

20.6 *Precision and Bias*—The precision and bias of this test method have not been determined.

21. Free Moisture in Hydrated Lime

21.1 *Scope*—The free moisture in hydrated lime is that water that is released from the sample at a temperature of 115 to 120°C. This distinguishes it from the hydroxyl water that is chemically bound to the lime and which cannot be liberated except at higher temperatures.

21.2 *Summary of Test Method*—Free moisture in hydrated lime is determined by aspirating a slow stream of CO_2-free air over the sample in a container placed inside a 115 to 120°C oven. The loss in weight of the sample is equal to the free moisture of the hydrated lime.

21.3 *Special Apparatus:*

21.3.1 *Sample Flask E,* illustrated in Fig. 1, consists of a 50-mL flat-bottom, glass-stoppered flask, supplied with a ground glass joint and solid ground glass stopper.

21.3.1.1 The flask shall be fitted with an interchangeable hollow ground-glass stopper, equipped with two glass entry tubes for conducting the dry air over the sample.

21.3.2 *Purifying Train* (Fig. 1), located outside the oven *F* for conducting the dry air over the samples, shall consist of a series of scrubbers and absorption bulbs to remove CO_2 and moisture from the air. The apparatus are arranged in the following order starting from the air source:

21.3.2.1 *Soda-Lime Tower A,* at the air inlet to remove CO_2 from the air.

21.3.2.2 *Bottle B,* containing lime water to show when the soda lime is exhausted.

21.3.2.3 *Fleming Jar C,* containing sulfuric acid to remove water from the air.

21.3.2.4 *Absorption Bulb D,* filled with Anhydrone (magnesium perchlorate) to complete the drying of the air.

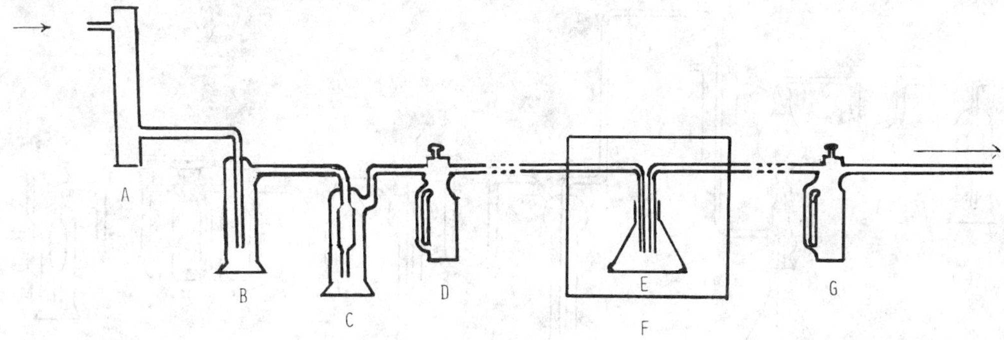

A Soda-Lime Tower at inlet to remove CO_2.
B Bottle containing lime water to show when soda lime tower is exhausted.
C Fleming jar containing sulfuric acid to remove water from air.
D Absorption bulb filled with Anhydrone (Magnesium Perchlorate) to complete drying of air.
E 50-mL sample flask.
F Drying oven operating at 110°C.
G Absorption bulb filled with Anhydrone to prevent moisture backup into sample.

FIG. 1 Apparatus for Free Moisture in Hydrated Lime

21.3.2.5 *Sample Flask E.*

21.3.2.6 *Drying Oven F.*

21.3.2.7 *Absorption Bulb G,* also filled with Anhydrone and located on the exit side of the sample bulb as a protective barrier against atmospheric moisture.

21.4 *Procedure*—Weigh 2.5 to 3 g of the prepared sample, and using glazed paper folded in the shape of a funnel, transfer it rapidly into the previously weighed bottle and immediately restopper it. Insert the bottle into the 120°C oven and quickly exchange stoppers. Connect the sample bottle to the purifying train by means of flexible tubing and pass a slow current of dry CO_2-free air through the apparatus for 2 h. Disconnect the sample bottle from the train, remove it from the oven with another quick exchange of stoppers, and place it in a desiccator to cool. When cool, remove it to the balance case for several minutes before weighing it, and just before weighing, lift the stopper slightly for an instant to relieve any vacuum that may exist in the bottle. The loss in weight of the sample represents "free moisture" loss as 120°C. Use a bottle similar to the one containing the sample as a counterpoise in all weighings unless a single-pan balance is used.

21.5 *Calculation*—Calculate the percent "free moisture" in the sample as follows:

$$\text{Free moisture, \%} = (A - B)/C \times 100 \quad (16)$$

where:

A = mass of sample flask + sample, g,
B = mass of sample flask after drying, g, and
C = mass of sample, g.

21.6 *Precision and Bias*—The precision and bias of this test method have not been determined.

22. Carbon Dioxide by Standard Method

22.1 *Scope*—Carbon dioxide in limestone is sometimes determined to verify the presence of carbonates other than calcium or magnesium. These may include carbonates of iron, manganese, and occasionally traces of other substances. Samples of lime and hydrated lime are analyzed for CO_2 to check for the presence of carbonates, most of which are there as uncalcined limestone.

22.2 *Summary of Test Method*—The sample is decomposed with HCl and the liberated CO_2 is passed through a series of scrubbers to remove water and sulfides. The CO_2 is absorbed with Ascarite, a special sodium hydroxide absorbent, and the gain in weight of the absorbtion tube is determined and calculated as percent CO_2.

22.3 *Special Apparatus:*

22.3.1 The apparatus illustrated in Fig. 2 consists of the following:

22.3.1.1 *Purifying Jar A,* Fleming, containing sulfuric acid.

22.3.1.2 *Drying Tube B,* U-shaped with side arms and glass stoppers. Side arms are shaped to hold rubber tubing. Contains Anhydrone on left side and Ascarite on right side.

22.3.1.3 *Erlenmeyer Flask C,* 250-mL, 24/40 ground-glass joint.

22.3.1.4 *Separatory Funnel D,* with ground-glass stopper and interchangeable hollow ground-glass joint. A delivery tube bent at the end extends into the sample flask about ½ in. from the bottom. Used to introduce acid into flask.

22.3.1.5 *Condenser E.*

22.3.1.6 *Gas-Washing Bottle F,* 250-mL, with fritted disk containing distilled water to retain most of the acid volatilized from the alkalimeter.

22.3.1.7 *U-Tube G,* containing mossy zinc to remove the last traces of HCl.

22.3.1.8 *Gas-Washing Bottle H,* 250-mL, with fritted disk, containing concentrated H_2SO_4 and trap *I,* to remove any SO_3 mist that may have been carried over.

22.3.1.9 *Absorption Bulb J,* containing Anhydrone to remove last traces of water vapor.

22.3.1.10 *CO_2 Absorption Bulb,* containing Ascarite filled as follows: On the bottom of the bulb, place a layer of glass wool extending above the bottom outlet and on top of this a layer of Anhydrone about ⅜ in. thick; immediately above this is placed another layer of glass wool, and Ascarite is then added to almost fill the bulb. A top layer of Anhydrone about ⅜ in. thick is placed on top of the Ascarite and topped off with a covering of glass wool.

22.3.1.11 *U-Guard Tube L,* filled with Anhydrone in left

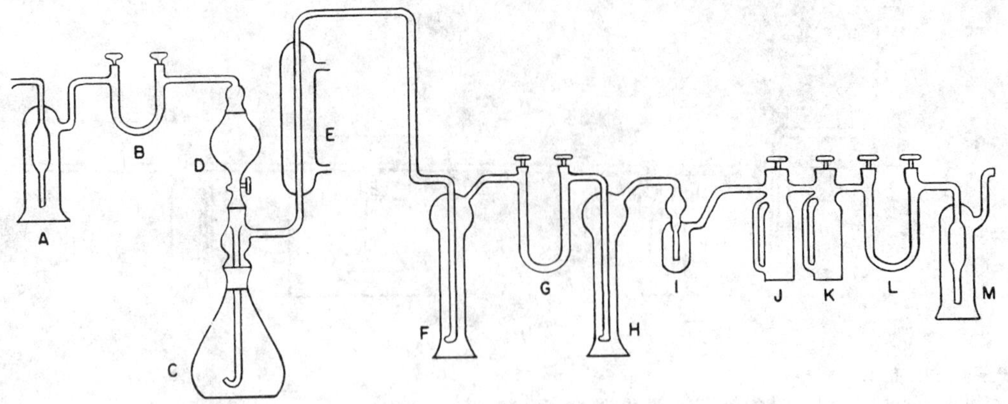

A Purifying jar, Fleming, containing concentrated H_2SO_4.
B Drying tube, U-shaped, Ascarite in right side, Anhydrone in left side.
C Erlenmeyer flask, 250-mL, 24/40 glass joint.
D Separatory funnel.
E Condenser.
F Gas-washing bottle, 250-mL with fritted disk, containing water to retain most of the acid volatilized from the alkalimeter.
G U-tube containing mossy zinc to remove the last traces of HCl.
H Gas-washing bottle, 250-mL with fritted disk, containing concentrated H_2SO_4.
I Trap.
J Absorption bulb containing Anydrone.
K CO_2 absorption bulb containing Ascarite.
L U-guard tube with Anhydrone in left side and Ascarite in right side.
M Purifying jar, Fleming, containing concentrated H_2SO_4.

FIG. 2 Apparatus for Carbon Dioxide by Standard Method

side and Ascarite in right side.

22.3.1.12 *Purifying Jar M,* Fleming, containing H_2SO_4.

22.4 *Procedure:*

22.4.1 Weigh an indicated amount of prepared sample, 0.5 g for limestone and 5 g for lime or hydrated lime, and transfer to the 250-mL Erlenmeyer flask. Connect the sample flask to apparatus as shown in the diagram (Fig. 2). Purge the system free of carbon dioxide by passing a current of CO_2-free air through the apparatus for 10 to 15 min.

22.4.2 Weigh the absorption bulb and attach it to the train. Remove the glass stopper from separatory funnel, place 50 mL of dilute HCl (1 + 1) in the separatory funnel (*D*) and replace the stopper with the interchangeable hollow ground-glass joint through which passes a tube for admitting purified air. Open the stopcock of the separatory funnel and admit air through the top of the funnel to force the hydrochloric acid into the Erlenmeyer flask (*C*).

22.4.3 Start cold water circulating through the condenser (*E*) and, with CO_2-free air passing at a moderate rate through the absorption train, place a small hot plate or gas burner under the sample flask and boil for about 2 min. Remove the hot plate and continue the flow of purified air at about three bubbles per second for 30 min to sweep the apparatus free of CO_2. Close the absorption bulb, disconnect it from the train and weigh, opening the stopper momentarily to equalize the pressure. Use a second absorption bulb as counterpoise in all weighings unless a single pan balance is used.

22.5 *Calculation*—Calculate the percent CO_2 as follows:

$$CO_2, \% = (A - B)/C \times 100 \quad (17)$$

where:

A = mass of absorption bulb + CO_2, g,
B = mass of absorption bulb before the run, g, and
C = mass of sample, g.

22.6 *Precision and Bias*—The precision and bias of this test method have not been determined.

23. Sulfur Trioxide

23.1 *Scope*—This test method will determine sulfur compounds, mostly present as sulfates in lime and limestone, that are soluble in dilute HCl. Iron pyrites and other sulfides will not be included because they will either be volatilized as H_2S or not react at all with the acid.

23.2 *Summary of Test Method*—In this test method, sulfate is precipitated from an acid solution of the lime or limestone with barium chloride ($BaCl_2$) and the SO_3 equivalent is calculated.

23.3 *Special Solution:*

23.3.1 *Barium Chloride Solution* (100 g/L)—Dissolve 100 g of barium chloride ($BaCl_2 \cdot 2H_2O$) in 1 L of water.

23.4 *Procedure*—Select and weigh the prepared sample into a 250-mL beaker containing 50 mL of cold water in accordance with the following:

Expected SO_3 Range, %	Sample Weight, g
0.001 to 0.500	10.00
0.500 to 2.50	5.00
2.50 to 12.5	2.00

Stir until all lumps are broken and the lighter particles are in suspension. Add 50 mL of diluted HCl (1 + 1) and heat until the reaction has stopped and decomposition is complete. Digest for several minutes at a temperature just below boiling. Add a few drops of methyl red indicator and render the solution alkaline (yellow color) with NH_4OH (1 + 1). Heat the solution to boiling. Filter through a medium-textured paper and wash the residue thoroughly with hot water. Dilute the filtrate to 250 mL, add 5 mL of HCl (1 + 1), heat to boiling, and add slowly 10 mL of hot $BaCl_2$ solution. Continue the boiling until the precipitate is well formed, stir well, and allow to stand overnight at room temperature. Take care to keep the volume of solution between 225 and 250 mL, and add water for this purpose if necessary. Filter through a retentive paper and wash the precipitate with hot water. Place the paper and contents in a weighed platinum crucible, and slowly char the paper without flaming. Burn off all the carbon, ignite in a muffle at 1000°C, cool in a desiccator, and weigh.

23.5 *Calculation*—Calculate the percentage of SO_3 to the nearest 0.001 as follows:

$$SO_3, \% = A \times 0.343/W \times 100 \quad (18)$$

where:

A = mass of $BaSO_4$, g,
W = mass of sample, g, and
0.343 = molecular ratio of SO_3 to $BaSO_4$.

23.6 *Precision and Bias:*

23.6.1 Six laboratories cooperated in testing on four samples of limestone and lime materials covering the range from 0.04 to 5.15 % SO_3 to obtain the precision data given in 23.6.2 and 23.6.3.

23.6.2 The repeatability (Practice E 173 R_1) was found to be (0.135 % SO_3 per weight in grams of sample analyzed).

23.6.3 The reproducibility (Practice E 173 R_2) was found to be (0.271 % SO_3 per weight in grams of sample analyzed).

23.6.4 The user is cautioned to verify by the use of reference materials, if available, that the bias of this test method is adequate for the contemplated use.

24. Total Sulfur by Sodium Carbonate Fusion

24.1 *Scope*—Sulfur in limestone is chiefly, if not wholly, present as sulfide, usually as pyrite. If the total sulfur obtained in the following test method is in excess of that present as soluble sulfate, the difference can be assumed to be present as iron disulfide.

24.2 *Summary of Test Method*—The sample is fused with sodium carbonate and the ignited mass is leached in water and dissolved with HCl. The solution is made ammoniacal and the hydroxide precipitate is filtered. The sulfur in the filtrate is precipitated with a 10 % solution of barium chloride. The precipitate is ignited and weighed as barium sulfate ($BaSO_4$) and the SO_3 equivalent is calculated.

24.3 *Procedure:*

24.3.1 Select and weigh the prepared sample into a porcelain crucible in accordance with the following:

Expected S Range, %	Sample Weight, g	Na_2CO_3 Weight, g
0.001 to 0.20	10.00	5.00
0.200 to 1.00	5.00	2.50
1.00 to 5.00	2.00	1.00

Add the indicated amount of Na_2CO_3 and mix well. Heat in a muffle at 600°C for 15 min. Increase the heat 50°C every 15 min until 1000°C is reached and maintain at this temperature for 15 min. (Note 29). Cool, place the crucible and cover in a 400-mL beaker, and cover with hot water. Add 10 mL bromine water (Note 30) and then add sufficient HCl (1 + 1) to make the solution slightly acid to methyl red. Boil until solution is complete and all bromine has been expelled. Remove the crucible and wash with hot water.

NOTE 29—Since not enough flux is used to produce more than a sintering, the air entering the crucible after the bulk of the carbon dioxide has been released effects very speedy oxidation in the porous mass.

NOTE 30—It has been found that 10 mL of 30 % hydrogen peroxide (H_2O_2) may be substituted for the bromine water to accomplish oxidation without affecting the analytical result.

24.3.2 Add a few drops of methyl red indicator and render the solution alkaline with NH_4OH (1 + 1). Heat the solution to boiling, filter using a retentive paper and wash with hot water. To the filtrate add 5 mL of HCl (1 + 1), adjust the volume to about 250 mL, and bring the solution to boiling. To the boiling solution, add 10 mL of hot $BaCl_2$ solution, slowly and with stirring. Allow to stand overnight. Filter through a retentive paper and wash the precipitate with hot water. Place paper and contents in a weighed platinum crucible and slowly char the paper without flaming. Burn off the carbon and ignite in a muffle at 1000°C for 1 h. Cool in a desiccator and weigh as $BaSO_4$.

24.4 *Calculation*—Calculate the percentage of sulfur to the nearest 0.001 as follows:

$$S, \% = A \times 13.73/W \quad (19)$$

where:

A = mass of $BaSO_4$, g,
W = sample, g, and
13.73 = mass of molecular ratio of S to $BaSO_4$ × 100.

24.5 *Precision and Bias:*

24.5.1 Six laboratories cooperated in testing on four samples of limestone and lime materials covering the range from 0.021 to 2.15 % sulfur to obtain the precision data given in 24.5.2 and 24.5.3.

24.5.2 The repeatability (Practice E 173 R_1) was found to be (0.065 % S per weight in grams of sample analyzed).

24.5.3 The reproducibility (Practice E 173 R_2) was found to be (0.094 % S per weight in grams of sample analyzed).

24.5.4 The user is cautioned to verify by the use of reference materials, if available, that the bias of this test method is adequate for the contemplated use.

25. Total Sulfur by the Combustion-Iodate Titration Method

25.1 *Scope*—This test method covers the determination of sulfur in concentration from 0.005 to 1 %. At the combustion temperature of approximately 1650°C, complete combustion of the sulfur in the sample will take place regardless of sulfur form or sample matrix.

25.2 *Summary of Test Method*—A major portion of the sulfur in various types of lime and limestone samples is converted to oxides of sulfur, primarily sulfur dioxide (SO_2), by combustion in a stream of oxygen at the elevated temperature of a high-frequency induction furnace. During the combustion, the SO_2 is absorbed in an acidified starch-iodine solution and titrated with potassium iodate. The latter is standardized against limestone standard samples of known sulfur content to compensate for characteristics of a given apparatus and for day-to-day variation in the percentage of sulfur recovered as SO_2. Compensation is also made for the blank due to accelerators and crucibles.

25.3 *Apparatus*[10].

25.3.1 *Induction Furnace*—The induction furnace shall be supplied with a rheostat used to control the power input to the reduction coil that will avoid heating some types of samples too rapidly during the early stages of combustion. The train of the induction furnace shall include an oxygen purifier, described in 25.3.3.

25.3.2 *Automatic Titrator*—This apparatus shall consist of an absorption and titration vessel of appropriate volume and contain an inlet bubbler tube for the sulfur gases with a float valve to prevent backflow of liquid when the sample is starting to consume oxygen. The vessel must be shaped to effect complete absorption of SO_2 in a small volume of solution. The titrator comes equipped with a buret that should be approximately 10 mL in capacity marked with 200 divisions. The automatic titrator utilizes a photoelectric cell to activate a titrator inlet valve that allows the titration to proceed without the presence of an operator.

25.3.3 *Oxygen Purifiers*—Reagent-grade oxygen from a commercial tank is passed through a suitable two-stage reduction valve to provide an even and adequate flow of oxygen through a purifying train consisting of a sulfuric acid tower, an absorption bulb containing 20 to 30-mesh inert base impregnated with NaOH, and another absorption bulb containing anhydrous magnesium perchlorate Mg (ClO_4). A flowmeter precedes the induction furnace assembly.

25.3.4 *Combustion Crucibles*—The crucibles for use with the induction furnace must be of adequate thickness to retain the molten slag and have a sulfur blank as low and consistent as possible. The crucibles for use in the induction furnace must have adequate capacity and may be provided with suitable covers.

25.3.5 *Glass Accelerator Scoop.*

25.3.6 *Starch Dispenser*—A plastic bottle with a device for dispensing a few millilitres of starch solution at a time.

25.3.7 *Timer,* having a 0 to 15-min range in ¼-min intervals. Turns off the furnace at end of preset time and automatically resets.

25.3.8 *Loading Funnel*—Three-legged funnel that fits over the crucible and simplifies addition of sample.

25.4 *Reagents:*[11]

25.4.1 *Copper (Low-Sulfur) Ring Accelerator.*

25.4.2 *Iron (Low-Sulfur) Accelerator*—Iron chips. (For samples containing very low percentages of sulfur, the use of iron powder is recommended because of its low blank.)

25.4.3 *Tin Metal (Low-Sulfur) Accelerator,* granular.

25.4.4 *Potassium Iodate* (KIO_3) *Crystal.*

25.4.5 *Potassium Iodide* (KI) *Crystal.*

25.4.6 *Starch,* soluble.

25.5 *Special Solutions:*

25.5.1 *Potassium Iodate,* standard solutions.

25.5.1.1 KIO_3 *Standard Solution A*—Dissolve 0.2227 g KIO_3 in 900 mL of water containing 1 g sodium hydroxide (NaOH) and dilute to 1 L. For a 0.500-g sample, the buret reads directly in percent sulfur.

25.5.1.2 *Starch-Iodide Solution*—Transfer 2 g of soluble starch (for example, Arrowroot) to a 50-mL beaker, add a few millilitres of water, and stir into a smooth paste. Slowly add starch to 500 mL of distilled water while stirring. Add 4 g of NaOH and continue stirring the solution until the appearance changes from cloudy to translucent. Add 6 g of potassium iodide (KI), stir until the KI is dissolved, and dilute to 1 L.

NOTE 31—Discard any starch solution that imparts a red tinge to the blue color when titrating.

25.6 *Calibration*—This test method and instrument should be standardized by using a limestone sample of known sulfur content as determined by the Total Sulfur Method by Sodium Carbonate Fusion, Section 24. The Leco instrument, in addition, may be standardized daily by running limestone reference materials whose sulfur content, as determined by the Total Sulfur Method, ranges from 0.02 to 0.05 %. The limestone standards are run to determine the day-to-day variations in the test method and to verify that the electronics in the Leco are working properly.

25.6.1 It has been found through round robin studies that the practice of pre-igniting samples at 1000°C causes erratic recovery of sulfur. This practice should not be used.

25.7 *Procedure:*

25.7.1 Allow 15 min for the electronics in the furnace assembly and titrator to warm up.

25.7.2 Set the grid-current tap switch to low, medium, or high position. Determine the position on a test run, with the sample and accelerators that will give a complete combustion at approximately 400 mA as indicated on the plate current ammeter.

25.7.3 Set the automatic timer to the estimated time required to evolve the sulfur in the sample completely, as follows:

Sample	Time, Min
Quicklime	8
Hydrated Lime	10
Limestone	12

25.7.4 Weigh the sample and brush carefully into the combustion crucible using the loading funnel. The correct sample weight is determined by the estimated sulfur content of the sample as follows:

Sulfur %	Sample Weight (g)
0.001	0.500

25.7.5 The choice of accelerators is left to the discretion of the user, as each furnace will burn differently in accordance with type and amount used. Generally, the more accelerator used, the greater the furnace temperature. Tin metal, iron chip, iron powder, and copper ring have been found to be suitable materials. Porous covers should be used to prevent splattering of the hot flux and damage to the combustion tube. Do not

[10] The apparatus describes commercially available units manufactured and sold by the Leco Corp., St. Joseph, MI. Although the description of the apparatus is directed toward this commercially available equipment, it does not restrict the use of other equivalent equipment which may be available or may be constructed, as long as it follows the general principles outlined under the summary of the test method.

[11] All the reagents listed are available from Leco Corp., St. Joseph, MI. Other reagents may be substituted provided they are of the same purity and consistency.

re-use crucibles or covers.

25.7.6 Run a blank determination before each series (of sulfur determinations) using a crucible that contains all the accelerators but no sample.

25.7.7 Place the crucible and sample on the pedestal and lift into position in the combustion tube.

25.7.8 With the oxygen flow at 1 L/min, close stopcock on bottom of titration vessel, and add the HCl to the middle of the bell-shaped portion of the titration vessel. Always fill to the same level.

25.7.8.1 Add one measure of starch solution to the titration vessel. Fill the iodate buret.

25.7.9 Turn the double throw switch on the titrator to the end-point position (down). Slowly rotate the end-point control in a clockwise direction until it has added KIO_3 in the amount to give a solid medium blue color. After the indicator light (for solenoid valve) has stopped blinking, place the switch in the neutral position and fill the KIO_3 buret again. Turn the switch to the titrate position.

25.7.10 Turn on the power of the high-frequency furnace. The temperature will rise in the crucible as indicated by the plate current ammeter on the induction furnace that must indicate a reading of at least 400 mA before complete combustion of sulfur can take place.

25.7.11 As sulfur dioxide is given off, the unit will begin titrating automatically. The titration is finished when the indicator light stops blinking for a period of time, or the iodate in the buret stops falling over a period of time.

25.7.12 Inspect the crucible for a proper burn. A rough, bumpy surface or appearance of non-combustion indicates that the furnace temperature was too low. Sticking of the porous cover to the crucible indicates that the furnace temperature may have been too hot. Both conditions indicate poor sulfur recovery and may be helped by a slight change in accelerator amounts.

25.8 *Calculation:*

25.8.1 Calculation of Furnace Factor (*F*)

$$F = \frac{R}{(A - B)/(W \times 2)} \quad (20)$$

where:

A = buret reading as % Sulfur (S),
B = buret reading for Blank determination,
R = % Sulfur (by Sodium Carbonate Fusion Method) of the reference material, and
W = weight of sample, g.

25.8.2 Calculate the percentage of sulfur in the sample by using furnace factor *F*.

$$\% S = F \times \frac{A - B}{W \times 2} \quad (21)$$

where:

A = buret reading as % Sulfur (S),
B = buret reading for Blank determination,
F = furnace factor, and
W = weight of sample, g.

25.9 *Precision and Bias:*

25.9.1 Nine laboratories cooperated in testing on three samples of high-calcium limestone to obtain the precision data for % sulfur given in 25.9.2 and 25.9.3.

25.9.2 The repeatability (Practice E 691 [r]) was found to be 0.0070 % sulfur.

25.9.3 The reproducibility (Practice E 691 [R]) was found to be 0.0120 % sulfur.

25.9.4 The user is cautioned to verify by the use of reference materials, if available, that the bias of this test method is adequate for the contemplated use.

26. Phosphorus by Molybdovanadate Method

26.1 *Scope*—This method is suitable for the determination of small amounts of phosphorous in lime and limestone samples. The procedure is based on the fact that phosphorous in its ortho form will combine with ammonium molybdovanadate to yield a yellow color that can be measured spectrophotometrically. Total phosphate is determined after a strong oxidation decomposition with perchloric acid.

26.2 *Summary of Test Method*—The sample is decomposed with perchloric acid, the solution filtered, SiO_2 expelled, and the insoluble residue fused with Na_2CO_3. Ammonium molybdovanadate which is then added reacts with the phosphorous in solution to form the heteropoly phosphomolybdovanadate complex. The absorbance of the solution is measured with a photometer at 430 nm and compared against standards similarly treated.

26.3 *Special Solutions:*

26.3.1 *Phosphorous Standard Stock Solution* (0.5 mg P/mL)—Weigh 1.0983 g of potassium dihyrogen phosphate, KH_2PO_4, into a 250-mL beaker and dampen with about 5 to 10 mL of water. Add 10 mL HNO_3 and 25 mL $HClO_4$ (Note 32), heat on a hot plate, and evaporate to heavy fumes of $HClO_4$. Cover and boil until the solution is colorless or slightly yellow (10 to 15 min). Cool the solution, transfer to a 500-mL volumetric flask, dilute to volume, and mix. Store in a borosilicate or plastic bottle with a screw cap.

26.3.2 *Phosphorous Working Standard* (0.05 mg P/mL)—Dilute 50 mL of stock solution 26.3.1 to 500 mL with distilled water. Store in a Pyrex or plastic bottle with screw cap.

26.3.4 *Ammonium Molybdovanadate Solution:*

26.3.4.1 Dissolve 1.25 g of ammonium metavanadate in 400 mL of 1 + 1 nitric acid in a 1 L volumetric flask.

26.3.4.2 Dissolve 50 g of ammonium molybdate in 400 mL of distilled water.

26.3.4.3 Pour solution from 26.3.4.2 into solution 26.3.4.1, mix, and dilute to volume with distilled water.

26.4 *Preparation of Standard Curve:*

26.4.1 To each of seven individual 50 mL volumetric flasks, add with a buret 0, 1, 2, 4, 6, 10, and 14 mL of phosphorous working standard solution corresponding to 0, 0.05, 0.10, 0.20, 0.30, 0.50, and 0.70 mg of phosphorous, respectively.

26.4.2 Add 1 mL of perchloric acid and dilute to about 20 mL with water. Add 10 mL of the molybdovanadate solution, swirling the contents of the flask during the addition. Dilute to volume with distilled water, mix well, and allow to stand for 10 min. The prepared standard solutions will contain 0 (blank), 1.0, 2.0, 4.0, 6.0, 10.0, and 14.0 micrograms (μg) P/mL.

26.4.3 Determine the absorbance of each standard solution in the spectrophotometer at a wavelength of 430 nm using the blank standard as the reference solution. Prepare a calibration curve by plotting absorbance versus concentration of phosphorous in μg/mL.

26.5 *Procedure:*

26.5.1 Weigh 5.0 g of prepared sample into a 250-mL beaker and dampen with about 5 to 10 mL of water. Add 10 mL HNO_3 and 25 mL $HClO_4$ (Note 32), heat on hot plate, and evaporate to heavy fumes of $HClO_4$. Cover and boil until solution is colorless or slightly yellow (10 to 15 min). Cool slightly and add 50 mL H_2O. Filter through retentive filter paper and wash thoroughly with hot H_2O (Note 33). Reserve filtrate.

NOTE 32—If a special perchloric acid hood is not available, the use of perchloric may be omitted. Instead, add 10 mL HNO_3 and evaporate to dryness. Take up salts with 5 mL HCl, break up residue, and again evaporate to dryness. Boil up with 1 mL of HNO_3 and 20 mL of water, and filter, etc.

NOTE 33—The filter paper and silica residue must be washed free of perchlorate salts to prevent small explosions from occurring in the crucible when the filter paper is charred and ignited.

26.5.2 Place paper and contents in a platinum crucible and heat gently with a low flame until paper chars. Ignite at a higher temperature until ash is white. Cool, add 2 drops of H_2SO_4 (1 + 1) and 10 to 15 mL HF. Evaporate cautiously to dryness and heat in a muffle at 1000°C for 2 min (Note 34). Fuse residue with 0.5 g of Na_2CO_3, cool, add 5 mL H_2O and 1 mL $HClO_4$ and warm to dissolve. Combine the filtrates and transfer to a 100 mL volumetric flask. Dilute to volume and mix.

NOTE 34—The treatment of the residue with HF and H_2SO_4 may be omitted if the SiO_2 in the sample is low, <3 %.

26.5.3 Pipet 25 mL of solution into a 50 mL-volumetric flask and add 10 mL ammonium molydovanadate solution, dilute to volume and mix well. Allow to stand 30 min and measure the absorbance at 430 nm using the blank standard solution in the reference cell. Compare against a set of standards similarly treated.

26.6 *Calculation:*

26.6.1 Calculate % P_2O_5 as follows:

$$\% \; P_2O_5 = \frac{C \times D \times 2.2913}{W \times 10^4} \tag{22}$$

where:

C = concentration of P in sample solution, μg/mL determined from calibration curve,

D = dilution factor, and

W = g of sample.

26.7 *Precision and Bias:*

26.7.1 The number of laboratories, materials, and determinations in this study does not meet the minimum requirements for determining precision prescribed in Practice E 691:

	Test Methods C 25	Practice E 691 Minimum
Laboratories	3	6
Materials	5	4
Determinations	4	2

26.7.2 The following precision statements are provisional. Within five years, additional data will be obtained and processed which does meet the requirements of Practice E 691.

26.7.2.1 Precision, characterized by repeatability, *Sr* and *r,* and reproducibility, *SR* and *R,* has been determined for the following test method and materials to be:

Precision Statement for Test Method: Material	% P_2O_5 Average	*Sr*	*SR*	Color *r*	*R*
S-1145	0.0031	0.0010	0.0017	0.0029	0.0046
S-1142	0.0091	0.0019	0.0031	0.0053	0.0087
S-1141	0.0221	0.0014	0.0043	0.0040	0.0122
S-1144	0.0657	0.0063	0.0144	0.0175	0.0404
S-1143	0.1353	0.0077	0.0147	0.0215	0.0413

27. Manganese by the Periodate (Photometric) Method

27.1 *Scope*—In this method, periodate is the oxidizing agent used to convert manganous into permanganate ion whose color can then be read in a photometer at a wavelength of 545 nm. This method is capable of determining small amounts of Mn as low as 10 ppm.

27.2 *Summary of Test Method*—The same sample solution prepared in the determinations of phosphorous by molybdovanadate (26.5.1 to 26.5.2) can be used for the determination of manganese by periodate. The acid solution is oxidized to permanganate by potassium periodate. Photometric measurement is made at 545 nm.

27.3 *Special Solutions:*

27.3.1 *Manganese Standard Solution* (1 mL to 50 μg Mn)—Dissolve 0.0500 g pure manganese (Mn) metal in 20 mL H_2SO_4 (1 + 9) and dilute to 1 L with water.

27.3.2 *Nitric-Phosphoric Acid Mixture*—Add 800 mL HNO_3 and 200 mL H_3PO_4 to 400 mL of H_2O and dilute to 2 L.

27.4 *Preparation of Calibration Curve:*

27.4.1 Transfer 0, 1, 2, 3, 5 and 10 mL of manganese standard solution to six 150 mL beakers, respectively. Add 25 mL of acid mixture to each and heat but do not boil. While heating, add potassium periodate (KIO_3) crystals, a few milligrams at a time (about 0.3 g total), until the permanganic color is fully developed. Keep solution near boiling for 10 min after color develops; the total heating period should last about 30 min.

27.4.2 Allow to cool, transfer to a 50-mL volumetric flask, dilute to volume and mix. Read absorbance at 545 nm using the "0" standard (blank) in the reference cell and construct a calibration curve by plotting absorbance versus concentration of Mn in μg/mL.

27.5 *Procedure:*

27.5.1 Weigh 2.0 g of prepared sample into a 250-mL beaker and dampen with about 5 to 10 mL of water. Add 10 mL HNO_3 and 20 mL $HClO_4$ (Note 35), heat on hot plate, and evaporate to heavy fumes of $HClO_4$. Cover and boil until solution is colorless or slightly yellow (10 to 15 min). Cool slightly and add 50 mL H_2O. Filter through retentive filter paper and wash thoroughly with hot H_2O (Note 33). Reserve filtrate.

NOTE 35—If a special perchloric hood is not available, omit the use of $HClO_4$. Instead, evaporate to dryness twice with nitric acid and finally boil with 10 mL HNO_3 and 50 mL H_2O.

27.5.2 Transfer paper and contents to a platinum crucible and heat until paper chars. Ignite at a higher temperature until ash is white. Cool, expel SiO_2 with HF and H_2SO_4, evaporate to dryness (Note 34) and fuse residue with Na_2CO_3. Cool, add 10 mL H_2O and 2 mL HNO_3 and warm to dissolve. Combine solution with filtrate reserved in 27.5.1 and transfer to a 100 mL volumetric flask. Dilute to volume and mix.

27.5.3 Transfer an aliquot containing <500 μg Mn to a 150 mL beaker. Add 25 mL of acid mixture, heat to near boiling and develop the permanganate color by small additions of KIO_3 as directed in 27.4.1. Cool the solution, transfer to a 50 mL volumetric flask, dilute to volume and mix.

27.5.4 Record the absorbance at 545 nm using the blank standard solution in reference cell as in preparation of standard curve, and compare against a set of standards similarly treated.

27.6 *Calculation:*

27.6.1 Calculate the percent Mn as follows:

$$\% \text{ Mn} = \frac{C \times D}{W \times 10^4} \tag{23}$$

where:
C = concentration of Mn in sample solution μg/mL determined from calibration curve,
D = dilution factor, and
W = g of sample.

27.7 *Precision and Bias:*

27.7.1 The number of laboratories, materials, and determinations in this study does not meet the minimum requirements for determining precision prescribed in Practice E 691:

	Test Methods C 25	Practice E 691 Minimum
Laboratories	4	6
Materials	5	4
Determinations	4	2

27.7.2 The following precision statements are provisional. Within five years, additional data will be obtained and processed which does meet the requirements of Practice E 691.

27.7.2.1 Precision, characterized by repeatability, *Sr* and *r*, and reproducibility, *SR* and *R*, has been determined for the following test method and materials to be:

Precision Statement for Test Method:		% Mn		Color	
Material	Average	*Sr*	*SR*	*r*	*R*
S-1145	0.0011	0.0004	0.0005	0.0011	0.0014
S-1143	0.0025	0.0005	0.0007	0.0014	0.0020
S-1142	0.0147	0.0010	0.0010	0.0028	0.0028
S-1141	0.0271	0.0012	0.0024	0.0034	0.0066
S-1144	0.1096	0.0072	0.0108	0.0200	0.0304

28. Available Lime Index

28.1 *Scope*—The available lime index of high-calcium quicklime and hydrated lime designates those constituents that enter into the reaction under the conditions of this specified test method, otherwise known as the "rapid sugar test method." The interpretation of results obtained by this test method shall be restricted by this definition.

28.2 *Summary of Test Method*—The sample is slaked and dispersed with water. The lime is solubilized by reaction with sugar to form calcium sucrate which is then determined by titration against standard acid using phenolphthalein as the indicator.

28.3 *Special Solutions:*

28.3.1 *Hydrochloric Acid, Standard* (1.000 *N*)—Prepare a solution by diluting 83 mL of HCl to 1 L with CO_2-free water. Standardization of sock solution should be performed on a regular basis at a minimum of once per month. For precision and bias information on standardization with Na_2CO_3 or Tris-(Hydroxymethyl) Amino-Methane see Practice E 200.

28.3.2 *Standardization of HCl with Na_2CO_3:*

28.3.2.1 Transfer approximately 20 g of primary standard anhydrous sodium carbonate (Na_2CO_3) to a platinum dish or crucible, and dry at 250°C for 4 h. Cool in a desiccator.

28.3.2.2 Weigh accurately 4.4 g to the nearest 0.1 mg of the dried Na_2CO_3 and transfer to a 500-mL flask. Add 50 mL of CO_2-free water, swirl to dissolve the carbonate, and add two drops of a 0.1 % solution of methyl red in alcohol. Titrate with the HCl solution to the first appearance of a red color, and boil the solution carefully, until the color is discharged (Note 32). Cool to room temperature, and continue the titration, alternating the addition of HCl solution and the boiling and cooling to the first appearance of a faint red color that is not discharged on further heating.

28.3.2.3 *Calculation*—Calculate normality as follows:

$$A = (B \times 18.87)/C \tag{24}$$

where:
A = normality of the HCl solution,
B = Na_2CO_3 used, g, and
C = HCl solution consumed, mL.

NOTE 37—This titration can also be performed potentiometrically with the aid of a glass electrode and a calomel electrode.

28.3.3 *Standardization of HCl with TRIS (THAM)—[Tris-(Hydroxymethyl) Amino-Methane]:*

28.3.3.1 Transfer an appropriate amount of primary standard tris-(hydroxymethyl) amino-methane to suitable dish or crucible and dry in a vacuum at 70°C for 24 h (refer to Practice E 200). As an alternative, Tris can also be dried at 105°C (±5°C) for 2 h in a regular laboratory drying oven. Cool in a desiccator to room temperature.

28.3.3.2 *Preparation of Mixed Indicator*—Mix 100 mg of Bromocresol Green with 2 mL 0.1 *N* NaOH and dilute with CO_2-free water to 100 mL. Dissolve 100 mg of Alizarin Red S in 100 mL CO_2-free H_2O. Mix equal portions of Bromocresol Green and Alizarin Red S solutions to form mixed indicator.

28.3.3.3 Weigh approximately 8 g of TRIS to the nearest 0.1 mg and dissolve in 50 mL of CO_2-free water. Add 6 drops of mixed indicator (Note 33) and titrate to a bright yellow end-point with acid.

NOTE 38—This titration can also be performed potentiometrically using a suitable pH meter to a pH of 4.70.

28.3.3.4 *Calculation*—Calculate normality as follows:

$$A = B/(0.121136 \times C) \tag{25}$$

where:
A = normality of the HCl solution,
B = Tris-(Hydroxymethyl) Amino-Methane used, g,
C = HCl solution consumed, mL.

28.3.4 *Phenolphthalein Indicator (4 %)*—Dissolve 4 g of dry phenolphthalein in 100 mL of 95 % alcohol.

28.3.5 *Sucrose Solution*—(40 g of pure cane sugar in solid

form may be used per sample in place of a sugar solution.) Prepare a 40 % solution (w/v) using pure cane sugar and CO_2-free water in a large beaker and stir until dissolved. Add several drops of phenolphthalein indicator solution. Add 0.1 *N* NaOH solution dropwise with stirring until a faint pink color persists. Stock solution of sugar may be made for convenience; however, it should not be stored for more than two days. As an alternative the acidity of each lot of sugar can be determined, and a correction applied to the titration.

28.4 *Procedure:*

28.4.1 *Procedure for Quicklime:*

28.4.1.1 The sample as received at the laboratory shall be thoroughly mixed and a representative sample with minimum weight of 100 g shall be taken and pulverized to pass a No. 50 sieve for analysis. Weigh rapidly 2.804 g of the finely pulverized sample, brush carefully into a 500-mL Erlenmeyer flask containing about 40 mL of CO_2-free water (Note 39), and immediately stopper the flask.

NOTE 39—Water should not be added to the sample because, especially with quicklime, there is a tendency for the material to cake and form lumps difficult to completely dissolve in the sugar solution later. On the other hand, if the lime is added to a little water, a better dispersion of the fine particles occurs, leading to a more rapid dissolution of the sample. It is possible that in the case of quicklime, some slaking action occurs to facilitate the dispersion and solution.

28.4.1.2 Remove the stopper. Place the flask on a hot plate and immediately add 50 mL of boiling CO_2-free water to the flask. Swirl the flask and boil actively 1 min for complete slaking. Remove from the hot plate, stopper the flask loosely, and place in a cold-water bath to cool to room temperature.

28.4.1.3 Add 100 mL of the neutralized sugar solution (or approximately 40 g of pure cane sugar). Stopper the flask, swirl, and let stand for 15 min to react. (Reaction time should not be less than 10 min nor more than 20 min.) Swirl at 5-min intervals during reaction. Remove stopper, add 4 to 5 drops of 4 % phenolphthalein indicator solution and wash down the stopper and sides of the flask with CO_2-free water.

28.4.1.4 When titrating (Note 40), first add about 90 % of the acid requirement from a 100-mL buret. Finish the titration, more carefully at approximately one drop per second, to the first disappearance of the pink color, which persists for 3 s. Note the endpoint and ignore any further return of color.

NOTE 40—A mechanical stirrer may be used during the titration if desired. Put a clean magnetic stirrer bar into the flask and place the flask on the magnetic stirrer. Adjust to stir as rapidly as possible without incurring any loss by spattering. Unless the operator is familiar with previous analyses of the lime under test, and in cases where the available lime content varies to extremes, it is good practice to run a preliminary test by slow titration to determine the proper amount of acid required to neutralize the sample.

28.4.2 *Procedure for Hydrated Lime*—The procedure for determining $Ca(OH)_2$ is the same as for CaO with the exception that cold CO_2-free water is used and the boiling and cooling steps are omitted.

28.5 *Calculation:*

28.5.1 Calculate for CaO as follows:

$$\text{Available lime (CaO), \%} = N \times V \times 2.804/W \quad (26)$$

where:

N = normality of acid solution,
V = standard HCl used (1.000 *N*), mL,
W = weight of sample, g, and
2.804 = CaO, g, equivalent to 1 mL of standard acid × 100, or 1 mL of standard HCl = 1 % CaO if exactly 2.804 g of sample is used.

28.5.2 Calculate for $Ca(OH)_2$ as follows:

$$\text{Available lime [Ca(OH)}_2\text{], \%} = N \times V \times 3.704/W \quad (27)$$

where:

N = normality of acid solution,
V = standard HCl (1.000 *N*), mL,
W = weight of sample, g, and
3.704 = $Ca(OH)_2$, g, equivalent to 1 mL of standard acid × 100 or 1 mL of standard HCl = 1.32 % $Ca(OH)_2$ when exactly 2.804 g of sample is used.

28.6 *Precision and Bias:*

28.6.1 Twenty-four laboratories cooperated in the testing of three high calcium quicklimes and one high calcium hydrated lime thereby obtaining the repeatability (*r*) and reproducibility (*R*) (Practice E 691) data contained in Table 4.

28.6.2 The user is cautioned that the repeatability and reproducibility are considered adequate for the purposes of the test method. However, higher levels of MgO and nonreactive CaO increase the value of the repeatability (within lab) slightly and the reproducibility value (between labs) noticeably.

28.6.3 Due to the lack of a recognized industry standard, the bias of this test method has not been determined.

29. Free Silica

29.1 *Scope*—Free silica is usually present in the acid-insoluble residue of lime and limestone samples. This method is applicable to the determination of free silica when it exceeds 0.05 %.

29.2 *Summary of Test Method*—After dissolution of a large sample of lime or limestone, the insoluble matter including SiO_2 is separated, ashed, and the oxides fused with pyrosulfate. The silicic acid liberated from the clay minerals in the insoluble matter is dissolved in a hot solution of sodium hydroxide but the free silica is unaffected.

29.3 *Procedure:*

29.3.1 Weigh 5.0 g of the prepared sample in a 400-mL beaker, add 25 mL of HCl (1 + 1), and heat to a boil. Using a retentive paper, filter the insoluble matter including SiO_2 and wash several times with hot water. Discard the filtrate.

29.3.2 Place the paper containing the insoluble matter in a platinum crucible and char the paper without inflaming at low heat. Increase the heat to burn off the carbon, but do not exceed 600 ± 50°C. Cool, add approximately 10 g of fused and powdered $KHSO_4$ or $K_2S_2O_7$ in the platinum crucible, and blend thoroughly with a small spatula. Fuse thoroughly over a gas burner, first by gradual heating to prevent loss of SO_3. When the fusion becomes quietly molten, finish the fusion at a

TABLE 4 Precision Data (Results in % Available CaO)

Material	Average	*r*	*R*
Hydrated lime	71.967	0.367	0.963
Magnesian quicklime	88.495	0.479	1.784
Shaft kiln quicklime	94.393	0.398	1.405
Rotary kiln quicklime	94.438	0.337	1.092

dull red heat not over 800°C.

NOTE 41—Do not continue heating to a point where salts begin to freeze on top of the melt and on the sides of the crucible because of the difficulty of subsequent solution.

29.3.3 Cool the crucible and its contents and dissolve the melt by heating with 150 to 200 mL of water in a 400-mL beaker. To the warm solution, cautiously add approximately 12 g of NaOH pellets a few at a time to dissolve the precipitated silicic acid (Note 41). Digest on a hot plate for 30 min at 80 to 90°C.

NOTE 42—Free silica is not transformed to silicic acid by the fusion and is not affected by the caustic treatment.

29.3.4 Filter quickly, using a retentive paper; thoroughly scrub and wash the beaker with hot water. Wash the filter paper and its contents ten times with hot water; five times with hot dilute HCl (1 + 1) to dissolve iron and other contaminants of the free silica, and finally five times with hot water or until free of acid. Transfer the paper and residue to a tared platinum crucible, ignite at 1000°C to constant weight, cool in a dessicator, and weigh as free silica.

29.3.5 To the ignited and weighed residue add 2 drops of H_2SO_4 (1 + 1) and approximately 10 mL of HF, evaporate to dryness, ignite and weigh (Note 38). If the mass of the residue exceeds 0.001 g, the determination may have been improperly accomplished. Replicate determinations should then be made to verify the validity of the results, or the residue examined by X-ray diffraction to determine whether refractory or highly insoluble minerals are present.

NOTE 43—Occasionally the free silica is contaminated by compounds not decomposed during fusion or dissolved in the subsequent treatments. The residue then is treated with HF and sulfuric and SiO_2 expelled.

29.4 *Calculation*—Calculate the percent free silica as follows:

$$\text{Free silica, \%} = (A - B)/W \times 100 \tag{28}$$

where:
A = crucible + insoluble residue, g,
B = crucible minus SiO_2, g, and
W = sample, g.

29.5 *Precision and Bias*—The precision and bias of this test method have not been determined.

30. Unhydrated Oxides Calculated on As-Received Basis

30.1 *Scope*—From the analytical determinations made in accordance with the preceding sections, it is possible to calculate combined water, $CaCO_3$, and $CaSO_4$ in samples of lime and hydrated lime. Unhydrated oxides of MgO in hydrated lime can also be calculated.

30.2 *Summary of Test Method*—Determine the percent of free water, LOI, CO_2, SO_3, CaO, and MgO in accordance with the preceding sections. Calculate combined H_2O, calcium carbonate, calcium sulfate, and unhydrated magnesium oxide using the following procedures.

30.3 *Procedure:*

30.3.1 Calculate percentage of combined water in quicklime from the loss on ignition and CO_2 determinations, as follows:

$$\text{Combined } H_2O, \% = \% \text{ LOI} - \% (CO_2 + \text{FM}) \tag{29}$$

30.3.2 Calculate the percentage of combined water in hydrated lime from the LOI, CO_2, and free moisture (FM) determinations as follows:

$$\text{Combined } H_2O, \% = \% \text{ LOI} - \% (CO_2 + \text{FM}) \tag{30}$$

30.3.3 Calculate the percent unhydrated oxides on the as-received basis as follows (Notes 44, 45, 46):

NOTE 44—The calculations involved in determining the percentage of unhydrated oxides on the as-received basis are illustrated in Table 5.

NOTE 45—This method for calculating the percentage of hydrated oxides is intended primarily for use with Type S hydrated lime.

NOTE 46—It is recognized that the results from this method of calculation may not be in strict accord with the actual composition of the material. Experience indicates, however, that these results provide an index to the performance of the material in practice. The value obtained by this calculation shall be reported to the nearest 0.5 %.

30.3.3.1 Calculate the CaO equivalents of the CO_2 and SO_3 as follows:

$$CO_2 \times 1.275 = \text{equivalent CaO } (A) \text{ as } CaCO_3 \tag{31}$$

$$SO_3 \times 0.700 = \text{equivalent CaO } (B) \text{ as } CaSO_4 \tag{32}$$

30.3.3.2 Subtract the sum of these CaO equivalents from the total CaO:

$$\text{Total CaO} - (A + B) = \text{hydrated CaO } (C) \tag{33}$$

30.3.3.3 Calculate the H_2O equivalent of the remaining CaO as follows:

$$\text{Hydrated CaO } (C) \times 0.3213 = \text{equivalent } H_2O\ (D) \text{ as } Ca(OH)_2 \tag{34}$$

30.3.3.4 Subtract this value for H_2O (D) from the combined water as calculated in 30.3.1 and 30.3.2.

$$\text{Combined water} - H_2O\ (D) = \text{equivalent } H_2O\ (E) \text{ as } Mg(OH)_2 \tag{35}$$

30.3.3.5 Calculate the MgO equivalent of the remaining water (E) as follows:

$$H_2O\ (E) \times 2.238 = \text{equivalent MgO } (F) \text{ as } Mg(OH)_2 \tag{36}$$

30.3.3.6 Subtract this MgO equivalent (F) from the total MgO as determined from preceding sections to obtain the percentage of unhydrated MgO on the as-received basis.

$$\text{Total MgO} - \text{MgO equivalent } (F) = \text{unhydrated MgO} \tag{37}$$

TABLE 5 Example of Calculation, Unhydrated Oxides

Compound	Values Determined From Chemical Analysis, %	Residual Values	Factors	Calculated Values, %
CO_2	0.40		× 1.275	= equivalent CaO = 0.51
SO_3	0.48		× 0.700	= equivalent CaO = 0.34
CaO	42.79 − (0.51 + 0.34 = 0.85)	= 41.94	× 0.3213	= equivalent H_2O = 13.48
H_2O	25.09 − 13.48	= 11.61	× 2.238	= equivalent MgO = 25.98
MgO	30.68 − 25.98			= unhydrated MgO = 4.70

31. Calcium and Magnesium Oxide (Alternative EDTA Titration Method)

31.1 *Scope*—This test method is a rapid EDTA complexometric method for determining calcium and magnesium in lime and limestone products. Ordinarily the EDTA procedure is designed to follow routine separations, that is, single dehydration of silica and a single precipitation with NH_4OH of the combined oxides of iron and aluminum. For expediency, the assays can be run directly without prior separation of the combined oxides of iron and aluminum by using the complexing action of EDTA at appropriate pH levels.

31.2 *Summary of Test Method:*

31.2.1 In this test method, calcium and magnesium are determined by EDTA titration after separation of silica and the NH_4OH group during a routine analysis of lime and limestone. The assays may also be made after a direct HCl decomposition followed by removal of the silica and insoluble.

31.2.2 If interfering elements are present in large enough quantities to cause problems, the interferences may be suppressed by the addition of complexing or masking agents such as triethanolamine or cyanide.

31.2.3 For the determination of calcium, the solution is adjusted to a pH of 12 to 12.5 with KOH solution and titrated with EDTA to a blue end point using hydroxy naphthol blue as the indicator. Both CaO and MgO are then titrated from a solution buffered to pH 10 with $NH_3 \cdot NH_4Cl$ using Calmagite as the indicator. MgO is calculated by subtracting the EDTA equivalent to CaO present from the EDTA equivalent to CaO + MgO.

31.3 *Reagents:*

31.3.1 *EDTA Solution* (0.4 %)—Dissolve 4 g of disodium dihydrogen ethylenediamine tetraacetic acid in water and dilute to 1 L.

31.3.2 *Potassium Hydroxide, Standard Solution* (1 *N*)—Dissolve 56 g of KOH in 1 L of distilled water.

31.3.3 *Ammonia Buffer* (pH 10.5)—Dissolve 67.5 g of NH_4Cl in 300 mL of distilled water, add 570 mL of NH_4OH, and dilute to 1 L.

31.3.4 *Hydroxy Naphthol Blue* (calcium indicator)—Disodium salt of 1-(2-naphthol azo-3,6 disulfonic acid)2 naphthol-4-sulfonic acid on suitable carrier.

31.3.5 *Calmagite* (magnesium + calcium indicator)—1-(hydroxyl-4-methyl-2 phenylazo)-2 naphthol-4 sulfonic acid on a suitable carrier.

NOTE 47—Both the hydroxy naphthol blue and Calmagite indicators are manufactured by Mallinckrodt Chemical Works, St. Louis, MO. Each indicator is a diluted mixture of dye plus an inert carrier.

31.3.6 *Hydrochloric Acid* (1 + 1).

31.3.7 *Hydrochloric Acid* (1 + 9).

31.3.8 *Triethanolamine* (1 + 2).

31.3.9 *Potassium Cyanide Solution* (20 g/L)—Dissolve 2 g of KCN in 100 mL of water.

31.3.10 *Calcium Standard Solution* (1.00 mg CaO/mL)—Weigh 1.785 g of $CaCO_3$, primary standard grade. Dissolve in HCl (1 + 9) and dilute to 1 L with distilled water.

31.3.11 *Magnesium Standard Solution* (1.00 mg MgO/mL)—Dissolve 0.603 g of magnesium metal turnings in HCl and dilute to 1 L with distilled water.

31.4 *Apparatus:*

31.4.1 *Magnetic Stirrer* with light.

31.4.2 *Stirring Bar,* TFE-fluorocarbon-covered.

31.5 *Standardization:*

31.5.1 *Calcium*—Pipet 10 mL of standard CaO solution into an Erlenmeyer flask and add 100 mL of distilled water. To prevent precipitation of calcium, add about 10 mL of EDTA titrating solution. Adjust to pH 12 to 12.5 with about 15 mL of 1 *N* KOH solution and stir. Add 0.2 to 0.3 g of hydroxy naphthol blue indicator (Note 48) and complete titration to a clear blue end point (Note 49). Titrate three or more aliquots and use the average to calculate the titer of the CaO solution.

$$\text{CaO titer, mg/mL} = 10 \text{ mg CaO standard/mL EDTA titration} \quad (38)$$

NOTE 48—The amount of indicator is usually a matter of individual preference, but if Mallinckrodt reagents are used, the recommended dosages are considered the proper amounts to add for easy end point detection.

NOTE 49—The use of a magnetic stirrer with light may be very helpful in detecting the color change. The end point is the first definite blue color that remains stable for at least 30 s.

31.5.2 *Magnesium*—Pipet 10 mL of the 1.00-mg/mL standard MgO solution into an Erlenmeyer flask and add 100 mL of distilled water. Adjust to pH 10 with 10 mL of the ammonium chloride-ammonium hydroxide buffer solution, and add 0.3 to 0.4 g of Calmagite indicator (Note 48). Titrate with disodium EDTA, noting the color change from a red to deep blue end point (Note 49). Titrate three or more aliquots and use the average to calculate the magnesium titer.

$$\text{MgO titer, mg/mL} = 10 \text{ mg MgO standard/mL EDTA titration} \quad (39)$$

31.6 *Procedure:*

31.6.1 If using limestone, dry the sample at 110°C to constant weight and cool to room temperature.

31.6.2 Weigh 0.5 g of the dry sample into a 250-mL beaker, add 10 mL of HCl (1 + 1), and carefully evaporate to dryness on hot plate. Dissolve residue in 25 mL of HCl (1 + 9), dilute to about 100 mL with water, and digest at low heat for 15 min. Cool, transfer to a 250-mL volumetric flask, dilute to volume, mix, and let settle or filter using a medium textured paper.

31.6.3 Alternatively, if the standard procedure has been used to determine the silica and insoluble and iron and aluminum, the filtrate remaining after the precipitation of the iron and aluminum with ammonia can be used for the Ca and Mg EDTA determination. Continue the analysis as follows: Acidify the filtrate with HCl, transfer to a 250-mL volumetric flask, dilute to volume with distilled water, and mix.

31.6.4 *Calcium Oxide:*

31.6.4.1 Pipet a 20-mL aliquot from the 250-mL volumetric flask and transfer to an Erlenmeyer flask. Dilute to 150 mL with water, adjust the pH to 12 with about 30 mL of 1 *N* KOH solution, and stir.

NOTE 50—Precipitation of calcium hydroxide can be prevented, if necessary, by adding one half to one third of the estimated volume of disodium EDTA solution before the addition of the potassium hydroxide. Presence of a large precipitate can cause loss of sharpness in the end point.

31.6.4.2 If the sample is known to contain significant quantities (>1 %) of iron, manganese, and heavy metals (Note

51), add 2 to 3 drops of 2 % potassium cyanide solution, or 10 mL of triethanolamine.

NOTE 51—If silica and phosphate or excessive amounts (>5 %) of iron and aluminum are present, they should first be removed by a double ammonia precipitation as in the standard procedure.

NOTE 52: **Precaution**—Cyanides should be used with utmost care because they can be very toxic compounds. They should be kept basic and prescribed disposal procedures used. Waste cyanide solutions must not be poured down laboratory sinks. Such solutions can be rendered relatively harmless by making them strongly basic, pH 12 or higher, with sodium hydroxide, adding calcium hypochlorite, and letting the solution stand for 24 h.

31.6.4.3 Add 0.2 to 0.3 g of hydroxy naphthol blue indicator (Note 43) and titrate as in 36.5.1 to a clear blue end point (Note 49).

31.6.4.4 *Calculation:*

$$\text{CaO, \%} = \text{CaO titer} \times \text{mL EDTA} \times 1.25/\text{weight of sample, g} \quad (40)$$

31.6.5 *Magnesium Oxide:*

31.6.5.1 Pipet a 20-mL aliquot of the acid solution from the 250-mL volumetric flask and transfer to an Erlenmeyer flask. Dilute with 100 mL of water, add about 20 mL of ammonia buffer to pH 10, and stir.

31.6.5.2 Add 2 to 3 drops of 2 % potassium cyanide solution, or 10 mL of triethanolamine and 1 drop of triethanolamine (**Caution**—See Note 52) (Note 51). Add EDTA standard solution equivalent to calcium titration; then add approximately 0.4 g of Calmagite indicator (Note 43).

31.6.5.3 Titrate to a blue end point with the 0.4 % EDTA solution (Note 49).

31.6.5.4 The titration determines both the calcium and magnesium present in solution. Subtraction of the EDTA titration obtained for calcium from the total titration gives the titration value for magnesium.

31.6.5.5 *Calculation:*

$$\text{mL EDTA solution equivalent to MgO} = (\text{mL EDTA solution used in CaO + MgO titration}) - (\text{mL EDTA solution used in CaO titration}) \quad (41)$$

$$\text{MgO, \%} = \text{mL EDTA equivalent to MgO} \times \text{MgO EDTA titer} \times 1.25/\text{weight of sample, g} \quad (42)$$

31.7 *Precision and Bias:*[12]

31.7.1 The precision of this test method was tested by ten laboratories using three limestone and one dolomite reference samples. The results shown in Tables 6 and 7 are summarized as follows:

31.7.1.1 The overall precision (1 sigma) between laboratories (reproducibility) for CaO is ±0.31 absolute units.

31.7.1.2 The overall precision (1 sigma) between laboratories for MgO is ±0.28 absolute units.

31.7.1.3 The overall precision (1 sigma) within laboratories (repeatability) for CaO is ±0.24 absolute units.

31.7.1.4 The overall precision (1 sigma) within laboratories for MgO is ±0.22 absolute units.

[12] Supporting data for this test method have been filed at ASTM Headquarters. Request RR: C-7-1000.

TABLE 6 Precision of Calcium Oxide

Sample Tested	CaO Found	Repeatability 1 sigma	Reproducibility 1 sigma
A-1a	53.48	0.31	0.27
A-2b	30.98	0.28	0.41
L-1	53.92	0.29	0.26
P-1a	53.88	0.31	0.28
Pooled		0.30	0.31

TABLE 7 Precision of Magnesium Oxide

Sample Tested	MgO Found	Repeatability Within Laboratories	Reproducibility Between Laboratories
A-1a	1.58	0.19	0.18
A-2b	21.58	0.24	0.41
L-1	0.58	0.21	0.22
P-1a	1.45	0.23	0.24
Pooled	...	0.22	0.28

32. Total Carbon by the Direct Combustion-Thermal Conductivity Cell Method

32.1 This test method covers the determination of carbon in lime and hydrated lime samples having a carbon concentration in the range from 0.005 to 5 %.

32.2 *Summary of Test Method*—All the carbon in the sample is converted to CO_2 by combustion in a stream of purified oxygen using an induction furnace. Under some conditions, CO is formed and a catalyst furnace is used to convert it to CO_2. The products of combustion are swept into the carbon analyzer where the CO_2 is selectively absorbed by a molecular sieve. The CO_2 is later released and swept by a fresh stream of oxygen past a thermal conductivity cell which senses the amount of CO_2 present. The signal from the sensor is amplified and electronically converted to % C which is displayed on a digital readout panel.

32.3 *Apparatus:*

32.3.1 *Combustion Furnace and Carbon Determinator*—This unit is commercially available.[13]

32.3.2 *Crucibles*—Use crucibles recommended by the manufacturer of the instrument.

32.3.3 *Oxygen Cylinder with Two-Stage Regulator.*

32.3.4 *Purifying Train,* consisting of a sulfuric acid tower, an absorption bulb containing Ascarite and another absorption bulb containing Anhydrone. A flowmeter precedes the induction furnace assembly.

32.3.5 *Catalyst Furnace,* to convert CO to CO_2.

32.3.6 *Combustion Tube with Built-In Jet.*

32.4 *Reagents:*

32.4.1 *Iron Chip Accelerator,* low-carbon.

32.4.2 *Copper-Tin Accelerator.*

32.5 *Preparation of Apparatus:*

32.5.1 Assemble the apparatus. Start the flow of oxygen at 1500 mL/min and the carrier gas flow at the rate recommended by the manufacturer of the apparatus.

[13] A commercially available unit is manufactured and sold by the Leco Corp., St. Joseph, MI. For the sake of convenience, the description of the apparatus and the test method are directed toward this commercially available equipment. However, this does not restrict the use of other equivalent equipment which may be available or which may be constructed as long as it follows the general principles outlined under the summary of the test method.

32.5.2 Test the furnace and the analyzer to ensure the absence of leaks and make the required electrical power connections. Prepare the induction furnace and analyzer for calibration and sample analysis according to the manufacturer's manual of instructions.

32.5.3 After the instrument has been prepared for calibration, check the CO_2 collection time which should have an 80 to 90-s time duration. The collection time can be checked by operating without a sample. If it is outside of this range, make the necessary adjustment of the collect relay as provided for in the manual of instructions.

32.6 *Calibration:*

32.6.1 Select NIST steel calibration standards containing approximately 0.02 to 1 % carbon (Note 53).

NOTE 53—The accuracy of this test method is dependent to a large extent on the accuracy of the methods used to certify the carbon concentration in the calibration standards as well as upon their homogeneity. Tests made on NIST steel standards have shown that they are sufficiently homogeneous to permit the use of samples as small as 20 mg.

32.6.2 Condition the analyzer if more than 2 h have elapsed since the last sample was run and determine the blank reading as follows:

32.6.2.1 Load into a crucible 1.5 g of iron chip accelerator and 1 g of tin-coated copper.

32.6.2.2 Proceed as directed in 32.7.1 to 32.7.4.

32.6.2.3 Repeat 32.6.2.1 and 32.6.2.2 a sufficient number of times to establish that low and consistent blank readings are obtained.

32.6.2.4 The normal blank reading should be 0.007 to 0.009. If it is out of this range, determine the cause, correct it, and repeat the steps as directed in 32.6.2.1 through 32.6.2.2.

32.6.3 *Calibration Curve:*

32.6.3.1 The calibration of the analyzer should be done using carbon standards that bracket the percent carbon in the samples estimated from previous tests, using the following guidelines:

% Carbon in Sample	High-Carbon Standard, %	Low-Carbon Standard, %
0.5–1	1.0	0.4
0.2–0.5	0.6	0.2
0.1–0.2	0.2	0.1
0.02–0.1	0.1	0.01

32.6.3.2 Treat each standard as directed in 32.7.1 to 32.7.4.

32.6.3.3 Select two standards, one slightly higher and the other slightly lower than the carbon concentration of the sample.

32.6.3.4 Run the high standard first and observe the reading. The carbon reading should be within ±1 % of the standard carbon value plus the blank (Note 54). If it is within this range, proceed to 32.6.3.7.

NOTE 54—For example, with a blank reading of 0.009, a carbon standard which has a standardized value of 0.428 % should read 0.433 to 0.441.

32.6.3.5 If the reading is out of this range, run a duplicate sample.

32.6.3.6 If the second sample has approximately the same value as the first sample, adjust the slope control (Note 55) to bring the reading to within the tolerance of the standard.

NOTE 55—Not all manufacturers may have this particular feature in their instrument, in which case the manual of instructions supplied with each instrument should be adhered to very closely and the instructions followed.

32.6.3.7 Run a carbon standard that has a carbon value lower than the sample as directed in 32.6.3.1 to 32.6.3.3.

32.6.3.8 The reading should be within +1 % of the carbon value plus the blank (Note 55).

32.6.3.9 If the reading is within tolerance range, proceed to sample analysis, 32.7.1 to 32.7.4.

32.6.3.10 If the reading is out of the tolerance range, run a duplicate sample. If the second sample has the same out of tolerance value as the first sample, repeat 32.6.3.4 to 32.6.3.9, readjusting the slope control to bring the high standard within the tolerance of the standard.

32.6.3.11 If the standards are still out of line, refer to the manufacturer's manual for instruction, determine the cause, correct it, and repeat the standardization procedure, 32.6.3.2 to 32.6.3.8.

32.7 *Procedure:*

32.7.1 Stabilize the furnace and analyzer as directed in 32.5.1 to 32.5.3.

32.7.2 Weigh duplicate samples to the nearest 1 mg, using the following guidelines:

Carbon Content, %	Sample Weight, g
0 to 0.1	1.000
0.1 to 2	0.500
2 to 5	0.250

32.7.3 Transfer the sample to a combustion crucible, add 1.5 g of iron chip accelerator and 1 g of the tin-coated copper accelerator. Place the crucible on the furnace pedestal, raise the pedestal into position, and lock the system. Start the flow of oxygen at 1500 mL/min, and flush the system for 30 s.

32.7.4 Start the cycle timer which energizes the furnace and starts the programmer, after having set it to provide a combustion cycle of 1 min. Using the variable transformer, manually control the plate current within the range from 350 to 450 mA. When the cycle is complete, remove and discard the crucible. Record the reading.

32.8 *Calculation*—Subtract the value found for the blank in 32.6.2.3 from the reading found in 32.7.4 and record the net reading. Calculate the percentage of carbon as follows:

$$\text{Carbon, \%} = A/B \tag{43}$$

where:

A = net carbon content, %, and
B = sample mass, g.

32.9 *Precision and Bias*—The precision and bias of this test method have not yet been determined.

33. Calcium Carbonate Equivalent

33.1 *Scope*—The calcium carbonate equivalence (CCE) test is used to determine the neutralizing capability of a calcareous material and to report this value in terms of percent calcium carbonate equivalents (% $CaCO_3$).

33.2 *Significance and Use*—Calcareous materials such as crushed limestone, hydrated lime and pulverized slags (from the production of steel) have been used extensively as soil modifiers or agricultural liming materials. A measure of their neutralizing capability can be determined through the use of this method of test. Not all neutralizing components of acalcareous material may be beneficial, therefore, the chemical analysis is suggested.

33.3 *Apparatus:*

33.3.1 *pH Metre,* conforming to Sections 4 and 5 of Test Method E 70.

33.3.2 *Mechanical Stirrer.*

33.3.3 *Sieve,* No. 60 (250-μm).

33.4 *Reagents:*

33.4.1 *Bromocresol Green Indicator Solution* (0.1 g/100 mL)—Dissolve 0.1 g (3′3″5′5″ tetrabromo-m-cresol-sulphonphthalein sodium salt) in 100 mL water.

33.4.2 *Hydrochloric Acid,* Standard (1.0 *N*)—Prepare a 1.0 *N* solution of hydrochloric acid (HCl) and standardize. Refer to available lime index test method for procedure on standardization of hydrochloric acid. For further information on preparation and standardization of solutions see Practice E 200.

33.4.3 *Hydrogen Peroxide* (30 %)—Concentrated hydrogen peroxide (H_2O_2).

33.4.4 *Phenolphthalein Indicator Solution* (1 g/100 mL)—Dissolve 1.0 g phenolphthalein in 100 mL of alcohol.

33.4.5 *Sodium Hydroxide Standard Solution* (0.5 V̇AN)—Dissolve 20 g of sodium hydroxide (NaOH) in 150 mL of carbon dioxide—free water. Cool the solution to 25°C and dilute to 1 L. Prepare free from carbonates and store in a polyethylene bottle provided with a siphon tube and with guard tubes containing soda-lime, or other suitable device, to prevent absorption of carbon dioxide (CO_2) from the air:

33.4.5.1 Standardize against 1.0 *N* hydrochloric acid with three 50 mL aliquots using phenolphthalein solution as indicator. Take the average of the three results as the normality of the Sodium Hydroxide Standard Solution.

$$\text{Normality (NaOH)} = \frac{V_2 N_2}{V_1} \qquad (44)$$

where:

V_1 = volume of NaOH Standard Solution,
V_2 = volume of HCl used to neutralize NaOH, and
N_2 = normality of HCl Standard Solution.

33.5 *Preparation of Sample:*

33.5.1 Weigh and dry the limestone and slag samples to constant weight at a temperature of 110 ± 5°C. Record the as-received and dry weights in order to calculate moisture content. Use lime samples as received. Grind the sample to pass a No. 60 (250-μm) sieve, mix thoroughly, and store in airtight containers.

33.6 *Procedure:*

33.6.1 *Indicator Titration Method*—Using Table 8 as a guide (see Note 56), weigh the material, to the nearest 0.1 mg, and place into a 500-mL Erlenmeyer flask; add 100 mL of 1.0 *N* HCl and boil gently for 5 min. Cool, and titrate the excess acid with 0.5 *N* NaOH solution using phenolphthalein as indicator. Calculate CCE as described in 33.6.3.

33.6.2 *Potentiometric Titration Method* (applicable to agricultural liming materials containing large amounts of ferrous iron or coloring matter, but not to silicate materials)—Proceed as in 33.6.1 through "Cool . . . ". Transfer to a 400-mL beaker and insert glass and calomel electrodes of a pH metre and a mechanical stirrer. Stir at a moderate speed to avoid splash. Deliver from a buret the 0.5 *N* NaOH solution rapidly to pH 5, then dropwise until the solution attains pH 7 and remains constant for 1 min while stirring. Report as $CaCO_3$ equivalent as described in 33.6.3.

33.6.3 Calculation of Percent Calcium Carbonate Equivalence:

33.6.3.1 *Dry Basis:*

$$\% \, CaCO_{3(dry)} = \frac{5.0045\,(V_1N_1 - V_2N_2)}{W} \qquad (45)$$

where:

V_1 = HCl solution used, mL,
N_1 = normality of HCl solution,
V_2 = NaOH solution required for titration of excess acid, mL,
N_2 = normality of NaOH solution, and
W = sample weight, g.

33.6.3.2 *As-received basis (limestone only):*

$$\% \, CaCO_{3(as\ received)} = [1 - (A - B/A)] \times \% \, CaCO_{3(dry)} \qquad (46)$$

where:

TABLE 8 Recommended Sample Weight[A]

Sample Type	Formula	Purity, %	Theoretical C.C.E.	Sample Weight, g
Magnesium oxide	MgO	97	248.33	1.9
Dolomitic lime	CaO MgO	97	207.69	2.3
Calcium oxide	CaO	97	178.48	2.6
Dolomitic Type N hydrate	Ca(OH)2 MgO	98	175.01	2.6
Dolomitic Type S hydrate	Ca(OH)2 Mg(OH)2	98	151.21	3.0
Calcium hydroxide	Ca(OH)2	98	135.11	3.4
Dolomitic limestone	CaCO3 MgCO3	97	108.56	4.3
Calcium carbonate	CaCO3	97	100.00	4.6

[A] For those materials not listed in Table 8: To determine the amount of 1.0 *N* HCl required for neutralization, a preliminary titration using 2 g of sample (weighed to the nearest 0.1 mg) should be performed to a yellow end point using Bromocresol green indicator. Using the following formula, determine the sample weight need for testing and proceed to 33.6.1.

$$\text{Wr, grams (g)} = \frac{\text{Wu} * 100}{\text{Vu} + 10}$$

where:

Wr = sample weight required for testing, g,
Wu = sample weight used in preliminary titration, g, and
Vu = volume of 1.0 *N* HCl used in preliminary titration, mL.

A = as received sample weight from 33.5.1, and
B = dry sample weight from 33.5.1.

34. pH of Alkaline Earth Solutions

34.1 *Scope*—This method covers the electrometric pH analysis of solutions made from a variety of lime, hydrate, and limestone-bearing materials.

34.2 *Summary of Test Method*—In this test method a sample of material is mixed with CO_2-free deionized water so that the pH of the solution is determined electrometrically. To ensure that an accurate measurement of pH is obtained for samples with a pH greater than 11.5, it is critical that all samples and buffer solutions be at 25.0°C when measured.

34.3 *Special Apparatus:*

34.3.1 *pH Metre*—A commercially available pH metre capable of calibration with a minimum of two different buffers (Note 56).

NOTE 56—It is necessary that the pH metre be equipped with a combination pH electrode as well as a temperature compensation probe. It is strongly recommended that a combination pH and temperature compensation probe be used.

34.3.2 *Constant Temperature Water Bath*—The water bath should be large enough to hold all samples and buffer solutions necessary to perform this method.

34.4 *Special Solutions:*

34.4.1 *Potassium Chloride Solution (0.2 M)*—Dry reagent grade KCl crystals at 110°C for 1 h. Dissolve 14.91 g of the dry KCl in CO_2-free deionized water and dilute to 1 L.

34.4.2 *Sodium Hydroxide Solution (0.2 M)*—Dissolve 8.00 g of NaOH in CO_2-free deionized water and dilute to 1 L. Standardize this solution with potassium acid phthalate using phenolphthalein as the indicator.

34.4.3 *Hydroxide-Chloride Buffer Solution (pH 13.0 at 25.0°C)*—Dilute 25.0 mL of the KCl solution and 66.0 mL of the NaOH solution to 100 mL with CO_2-free deionized water.

34.4.4 *Phenolphthalein Indicator*—Dissolve 4 g of dry phenolphthalein in 100 mL of 95 % alcohol.

33.4.5 Commercially available pH buffer solutions, pH 7 and pH 10.

34.5 *Standardization of Sodium Hydroxide Solution:*

34.5.1 *Initial Standardization*—Transfer 50.0 mL of the sodium hydroxide solution to be standardized to a well rinsed 50-mL Class A buret by means of a pipette. Accurately weigh out 8.0 ± 0.1 g of reagent grade potassium hydrogen phthalate which has been dried in an oven at 110°C for 1 h. Record the weight to the nearest 0.1 mg, then quantitatively transfer the weighed material into a 400-mL beaker. Add 185 mL of CO_2-free deionized water which is at a temperature of 26.0 ± 2.0°C. Stir this mixture gently until all of the crystals dissolve. Add five drops of the phenolphthalein indicator solution and proceed to titrate the prepared standard with the sodium hydroxide solution to be standardized. Titrate to an endpoint of light pink which will persist for at least 6 s.

34.5.2 *Calculation of Sodium Hydroxide Normality*—The normality of the sodium hydroxide solution can be calculated from the following equation:

$$\text{Normality of NaOH} = \frac{\text{Wt. of potassium hydrogen phthalate}}{\text{mL of NaOH} \times \text{factor of } 0.204228} \quad (47)$$

This preliminary standardization may reveal the necessity for either dilution with CO_2-free deionized water or the 0.2 molarity specified.

34.5.3 *Correction if Base is "too weak"*—If the molarity (normality) is determined to be too weak, it can be adjusted by the addition of additional sodium hydroxide. An example of this calculation follows. If 1.0 L of 0.1979 N solution needs to be strengthened to 0.2000 N, calculate how many grams of sodium hydroxide need to be added as follows:

$$\begin{aligned} &1.0\text{ L} \times 0.1979\text{ M} = 0.1979\text{ mole} \\ &0.2\text{ M} - 0.1979\text{ M} = 0.0021\text{ M} \\ &\text{NaOH} = 39.997\text{ g/mole} \\ &39.997\text{ g/mole} \times 0.0021\text{ mole} = 0.0830\text{ g needed} \end{aligned} \quad (48)$$

After the addition of additional sodium hydroxide is completed, recheck the standardization as specified in section 34.4.

34.5.4 *Correction if Base is "too strong"*—If the molarity (normality) is determined to be too strong, it can be adjusted by the addition of CO_2-free deionized water. An example of this calculation follows. If 1.0 L of 0.2019 N solution needs to be weakened to 0.2000 N, calculate how many millilitres of CO_2-free deionized water need to be added as follows:

$$\begin{aligned} &1.0\text{ L} \times 0.2019\text{ M} = 0.2019\text{ mole} \\ &0.2019 - 0.200 = 0.0019\text{ L or } 1.9\text{ mL of } CO_2\text{-free deionized water needed} \end{aligned} \quad (49)$$

After the addition of additional CO_2-free deionized water is completed, recheck the standardization as specified in section 34.5.

34.6 *Calibration of pH Metre*—Due to the wide variety of commercially available pH equipment, a detailed calibration procedure cannot be incorporated into this test method. Follow the manufacturer's instructions for calibration of your particular equipment.

34.6.1 Using the appropriate combination of the commercially available pH 7 and pH 10.0 buffers, and the Hydroxide-Chloride buffer (pH 13.0), choose the pair of buffers which best brackets the pH of the sample being tested. Standardize your pH metre in accordance with the manufacturer's instructions. For samples with pH less than 11.5, the temperature of the samples and buffers should be kept at 25.0 ± 3.0°C (Note 58).

NOTE 57—For alkaline solutions in this pH range the temperature does not cause errors to the extent that it does with samples with a pH in excess of 11.5, therefore the temperature is not as critical.

For samples with a pH greater than 11.5, it is imperative that all solutions be at 25.0 ± 0.1°C while they are being measured. If all buffers and test solutions are not within this range, due to the high pH being measured, large errors in the measured pH will occur. This can easily be accomplished by the use of a constant temperature water bath (Note 59).

NOTE 58—In strongly alkaline solutions, especially above pH 12, the concentration of hydroxide ions is so high that the small amount of hydrogen ions produced by the ionization of water is sufficient to change the pH appreciably with temperature.

34.7 *Procedure—Materials with pH Less than 11.5:*

34.7.1 Weigh 10.0 g of the prepared sample and record the weight to the nearest 0.1 mg. Quantitatively transfer the weighed material into a 250-mL Erlenmeyer flask containing 200 mL of CO_2-free deionized water. Add a magnetic stir bar, immediately stopper the flask with a rubber stopper, and stir for 30 min.

34.7.2 Let the solution stand for at least 30 min to allow the suspended material to settle out from the suspension, as well as to allow the sample temperature to stabilize at 25.0 ± 3.0°C. A constant temperature water bath may be necessary to obtain the desired temperature range.

34.7.3 Once the temperature of the sample is stabilized at 25.0 ± 3.0°C, using a 50-mL transfer pipette, transfer 50 mL of the settle solution to a 100-mL beaker containing a magnetic stir bar.

34.7.4 Slowly stir the sample solution using a magnetic stirrer. Using a previously standardized pH metre, standardized in accordance with section 34.6, quickly measure the pH of the sample. Report the pH to four significant figures.

34.8 *Procedure—Materials with pH of 11.5 and greater:*

34.8.1 Weigh 10.0 g of the prepared sample and record the weight to the nearest 0.1 mg. Quantitatively transfer the weighed material into a 250-mL Erlenmeyer flask containing 200 mL of CO_2-free deionized water. Add a magnetic stir bar, immediately stopper the flask with a rubber stopper, and stir for 30 min.

34.8.2 Place the sample solution into a constant temperature water bath set at 25.0°C. Let the solution stand for at least 30 min to allow the suspended material to settle out from the suspension as well as to allow the sample temperature to stabilize at 25.0 ± 0.1°C.

34.8.3 Once the temperature of the sample is stabilized at 25.0 ± 0.1°C, using a 50-mL transfer pipette, transfer 50 mL of the settled solution to a 100-mL beaker containing a magnetic stir bar.

34.8.4 Slowly stir the sample solution using a magnetic stirrer. Using a previously standardized pH metre, standardized in accordance with section 34.6, quickly measure the pH of the sample solution (Note 58). It is imperative that the temperature of the test solution remain at 25.0 ± 0.1°C during the pH measurement. If needed, the pH measurement should be carried out in a constant temperature water bath. Report the pH to four significant figures.

34.9 *Precision and Bias:*

34.9.1 The number of laboratories, materials, and determinations in this study does meet the minimum requirements for determinating precision prescribed in Practice E 691:

	Test Methods C 25	Practice E 691 Minimum
Laboratories	7	6
Materials	12	4
Determinations:	3	2

34.9.2 *Precision Statement for Test Method: pH*—Precision, characterized by repeatability, *Sr* and *r,* and reproducibility, *SR* and *R,* has been determined for the following materials to be:

Materials	Average	*Sr*	*SR*	*r*	*R*
Sample A kiln dust	12.3817	0.0069	0.2791	0.0194	0.7814
Sample B waste lime	12.2998	0.0297	0.4004	0.0831	1.1210
Sample C dolo hydrate	12.4073	0.0123	0.2408	0.0345	0.6741
Sample D dolo lime	12.3942	0.0177	0.2785	0.0495	0.7799
Sample E −325 stone	11.2199	0.0274	0.3355	0.0766	0.9393
Sample F cyclone dust	12.4090	0.0164	0.2336	0.0459	0.6541
Sample G bottom ash	12.1948	0.0120	0.4049	0.0336	1.1338
Sample H fly ash	9.6776	0.0557	0.3821	0.1559	1.0699
Sample I fly ash	11.3193	0.0143	0.2191	0.0402	0.6136
Sample J hydrate	12.4111	0.0244	0.2994	0.0684	0.8384
Sample K h. hyd. lime	12.5477	0.0069	0.1875	0.0193	0.5249
Sample L quicklime	12.4331	0.0260	0.1767	0.0728	0.4949

35. Keywords

35.1 alternative method; aluminum; available lime index; calcium; carbon; carbon dioxide; chemical analysis; free moisture, lime; free moisture, limestone; free silica; hydrated lime; instrumental analysis; iron; limestone; loss on ignition, (LOI); magnesium; manganese; methods, classical; methods, historical; oxides; oxides, unhydrated; phosphorus; quicklime; reference method; silicon dioxide; Standard Reference Material (SRM); strontium; sulfur

APPENDIXES

(Nonmandatory Information)

HISTORICAL ASTM METHODS

INTRODUCTION

The following test methods have been published as information only and are a collection of classical methods that had historically been widely used by the industry. Because they are presently of limited use, they have been deleted from the main body of reference test methods. Since some laboratories may still wish to perform these analyses, they have been placed in these appendixes. The precision and bias testing has not been performed for these methods. Therefore, users are advised to develop their own laboratory precision and to check against standard reference materials. Subcommittee C07.05 encourages comments and evaluation of these methods.

X1. TOTAL IRON BY THE POTASSIUM PERMANGANATE TITRATION METHOD

X1.1 *Scope*—This test method permits a more accurate analysis of the trace amounts of iron that may be present because a larger sample is used.

X1.2 *Summary of Test Method*—The sample is dissolved in HCl and the SiO_2 is dehydrated and separated from the soluble salts by filtration. The small amount of iron left in the insoluble residue is dissolved with sulfuric acid after SiO_2 is expelled with HF, and the recovered iron is combined with the bulk of the iron in the main filtrate. Iron is reduced with $SnCl_2$ and titrated with $KMnO_4$ solution.

X1.3 *Special Solutions:*

X1.3.1 *Manganese Sulfate Solution* (70 g/L)—Dissolve 70 g of crystalline manganese sulfate ($MnSO_4$) in 500 mL of water. Add 140 mL H_3PO_4 and 130 mL H_2SO_4. Dilute to 1 L.

X1.3.2 *Potassium Permanganate, Standard Solution* (0.05 *N*):

X1.3.2.1 Dissolve 1.60 g of potassium permanganate ($KMnO_4$) in 1 L of water and boil gently in a flask for 20 to 30 min. Cool, dilute again to 1 L, stopper, and allow to stand for several days in the dark. Filter through purified asbestos or a wad of glass fiber and standardize against the National Institute of Standards and Technology's standard sample 40C, sodium oxalate ($Na_2C_2O_4$), or equivalent in accordance with X1.3.2.2 through X1.3.2.4.

NOTE X1.1—Alternatively, potassium dichromate may be used in place of potassium permanganate as in the procedure of 13.4.

X1.3.2.2 Transfer 0.15 g of the standard sodium oxalate, dried at 105°C, to a 400-mL beaker. Add 250 mL of diluted H_2SO_4 (5 + 95), which has been previously boiled for 10 min and cooled to 27 ± 3°C. Stir until the oxalate has dissolved. Add 40 to 42 mL of the standard $KMnO_4$ solution at the rate of 25 to 30 mL/min, while stirring slowly. Let stand until the pink color disappears (about 60 s).

NOTE X1.2—If the pink color should persist because of the addition of too much $KMnO_4$ solution, discard and begin again, adding a few millilitres less of the $KMnO_4$ solution initially.

X1.3.2.3 Heat the contents of the beaker to 60°C and complete the titration at this temperature by adding $KMnO_4$ solution until a slight pink color persists for 30 s. Add the last 0.5 to 1 mL dropwise, allowing each drop to become decolorized before the next one is added.

X1.3.2.4 One millilitre of standard $KMnO_4$ solution will be approximately equivalent to 0.004 g of ferric oxide (Fe_2O_3). Determine the exact Fe_2O_3 equivalent of the solution from the following:

$$N = (W/V) \times 14.925 \tag{X1.1}$$

where:

N	=	normality of $KMnO_4$ solution,
W	=	sodium oxalate, g,
V	=	$KMnO_4$ solution used to titrate the sodium oxalate solution, mL, and
14.925	=	sodium oxalate equivalent to 1 mL of 1 *N* $KMnO_4$ solution, g.

Fe_2O_3 equivalent of $KMnO_4 = N \times 0.07984$, where 0.07984 = g Fe_2O_3 equivalent to 1 mL of 1 *N* $KMnO_4$ solution.

X1.4 *Procedure:*

X1.4.1 Weigh 2 to 5 g (depending upon the amount of iron present) of the properly prepared sample into a beaker and add 25 mL of HCl (1 + 1). After the reaction ceases, evaporate the solution to dryness. Add 50 mL of HCl (1 + 4), heat to dissolve the salts, filter the insoluble matter including SiO_2 using a medium-textured paper, then wash the paper and contents several times with hot water. Retain the filter paper. Heat the solution to boiling, add a few drops of methyl red indicator, and precipitate the iron by adding enough NH_4OH to change the color of the indicator from red to distinct yellow. Allow the precipitate to settle slightly, filter, and wash using hot water. Dissolve contents of the paper by placing paper and contents in 50 mL of hot HCl (1 + 3). Filter into a clean beaker and wash. Reserve filtrate.

X1.4.2 Place the paper containing the insoluble matter from the evaporated HCl solution in a platinum crucible. Char the paper without inflaming, at low heat, increase the heat to ash the paper, and ignite at 1000°C for 15 min. Cool, add 1 mL of H_2SO_4 and 10 to 15 mL of HF, and evaporate to fumes of sulfur trioxide. Cool, dilute the contents of the crucible with water, and warm to dissolve salts. Transfer the acidified solution to the main solution containing the bulk of the iron.

X1.4.3 Heat the solution to near boiling. Reduce the iron by adding $SnCl_2$ dropwise from a pipet, while swirling the beaker constantly, until the yellow color of the ferric iron just disappears (Note 23). Add 2 or 3 drops of $SnCl_2$ in excess, but no more. Cool the beaker quickly in running water, then add all at once 10 mL of $HgCl_2$ solution. Stir and allow to stand for 3 to 5 min (Note 24). Transfer with washing to a 600-mL of 1-L beaker containing 300 mL of cold water and 25 mL of the $MnSO_4$ solution. Titrate with standard 0.05 *N* $KMnO_4$ solution, added very slowly while stirring constantly, until a permanent pink end point is obtained. A blank should also be determined.

X1.5 *Calculation*—Calculate the percent iron oxide (Fe_2O_3) to the nearest 0.01 % as follows:

$$Fe_2O_3, \% = (A \times B)/C \times 100 \quad \text{(X1.2)}$$

where:
A = 0.05 *N* $KMnO_4$ solution used in titration, mL,
B = Fe_2O_3 equivalent of $KMnO_4$, and
C = sample weighed, g.

X1.6 *Precision and Bias*—The precision and bias of this test method have not been determined.

X2. STRONTIUM OXIDE BY THE GRAVIMETRIC SULFATE (ETHER) METHOD

X2.1 *Scope*—There may be certain applications that require a separate determination of strontium content or an exact determination of calcium. This test method is designed to determine the SrO present which will allow a correction for the percent CaO as obtained in previous sections.

X2.2 *Summary of Test Method*—The alkaline earth oxides are converted to nitrates. The different solubilities of calcium and strontium nitrate in absolute alcohol and ether are used to separate the two metals from each other. After filtration, the strontium nitrate precipitate is dissolved with water and the strontium precipitated with H_2SO_4 filtered, ignited, and weighed as strontium sulfate ($SrSO_4$).

X2.3 *Procedure:*

X2.3.1 Transfer the weighed oxides obtained in the gravimetric determination of calcium oxide (Section 16) to a small flask of 20-mL capacity and dissolve in HNO_3. Evaporate to dryness and heat at 150 to 160°C. Treat the thoroughly dried nitrates with as little (rarely over 2 mL) of a mixture of equal parts of absolute alcohol and ether as may be needed to dissolve the calcium salt, hastening solution by occasional, gentle agitation. After standing overnight in the stoppered flask, collect the insoluble matter on the smallest possible filter and wash with more of the above mixture of alcohol and ether.

NOTE X2.1: **Precaution**—Ethers tend to form peroxides by air oxidation on standing which are highly explosive. Peroxide containing ethers should not be distilled unless the peroxide is first destroyed by a suitable agent such as sodium sulfate or ferrous sulfate and should not be allowed to stand indefinitely in the laboratory.

X2.3.2 Let the filter air dry, then pass a few millilitres of hot water through the filter to dissolve the strontium nitrate. A few tenths of a milligram of residue may remain, consisting of material other than lime or alkaline earth, whose mass should therefore be deducted from the mass of the lime determined by the gravimetric method (Section 16).

NOTE X2.2—This correction is unnecessary if it can be shown that the impurities are derived from the glass of the small flask in which the nitrates of calcium and strontium were evaporated.

X2.3.3 To the solution of strontium nitrate, in a small beaker add a few drops of H_2SO_4 and a quantity of alcohol equal to the volume of the nitrate solution. The precipitate of strontium sulfate is allowed to develop and settle at room temperature for a period of 12 h.

X2.3.4 Filter the $SrSO_4$ on a small filter paper, and wash with 50 % alcohol. Ignite at low temperature, cool, moisten with dilute H_2SO_4 (1 + 1), evaporate to dryness, again ignite, and weigh as strontium sulfate ($SrSO_4$). The sulfate may be tested spectroscopically for traces of calcium and barium if it is considered necessary.

X2.4 *Calculation:*

X2.4.1 Calculate the mass of strontium oxide as follows:

$$SrO, \% = (A/B) \times 56.41 \quad \text{(X2.1)}$$

where:
A = mass of $SrSO_4$,
B = original mass of sample, g, and
56.41 = molecular ratio of SrO to $SrSO_4 \times 100$.

X2.4.2 Deduct the percent SrO from the percent CaO determined in Section 17.

X2.5 *Precision and Bias*—This test method was originally approved for publication before the inclusion of precision and bias statements within standards was mandated. The user is cautioned to verify by the use of reference materials, if available, that the precision and bias of this test method are adequate for the contemplated use.

X3. PHOSPHORUS BY THE TITRIMETRIC METHOD

X3.1 *Scope*—This is a titrimetric test method for determining phosphorus in samples of limestone, quicklime, or lime hydrate. A fairly large sample weight is required because this test method is not sensitive enough for an accurate analysis at the low levels of phosphorus frequently encountered in limestone rock.

X3.2 *Summary of Test Method*—The sample is dissolved in dilute nitric acid, the solution filtered, the residue fused with a little sodium carbonate and its nitric acid solution added to the other. An HF-HNO_3 treatment is included if the rock contains more than 3 % SiO_2. Ammonium molybdate is added to the nitric acid solution to precipitate phosphorus as ammonium phosphomolybdate. The phosphomolybdate is dissolved in an excess of a standard solution of sodium hydroxide which is then back-titrated with standard acid in the presence of phenolphthalein indicator.

X3.3 *Special Solutions:*

X3.3.1 *Ammonium Molybdate Solution:*

X3.3.1.1 *Solution A*—Add 400 mL of nitric acid to 600 mL of water. Mix and cool.

X3.3.1.2 *Solution B*—Dissolve 118 g of 85 % molybdic acid with 400 mL of water and add 80 mL of NH_4OH. Cool.

X3.3.1.3 Add Solution B to Solution A slowly with constant stirring, let solution stand for 24 h. Filter immediately before using.

X3.3.2 *Potassium Nitrate Solution* (10 g/L)—Dissolve 10 g of potassium nitrate (KNO_3) in water freshly boiled to expel CO_2 and cooled. Dilute to 1 L.

X3.3.3 *Sodium Hydroxide, Standard Solution* (0.3 *N*)—Dissolve 12 g of sodium hydroxide (NaOH) in 1 L of water that has been freshly boiled to expel CO_2, and cooled. Add 10 mL of a freshly filtered, saturated solution of barium hydroxide $Ba(OH)_2$ (Note X3.1). Shake the solution frequently for several hours and filter it. Store in a stoppered flask protected from the CO_2 of the air by means of a guard tube packed with Ascarite. Standardize against standard acid phthalate such as NIST Standard Sample No. 84 or benzoic acid, NIST Standard Sample No. 39 furnished by the National Institute of Standards and Technology[6], as follows:

NOTE X3.1—$Ba(OH)_2$ precipitates the carbonate as insoluble $BaCO_3$.

X3.3.3.1 Lightly crush 5 to 6 g of standard acid potassium phthalate to a fineness of approximately 10 mesh and dry for 2 h at 100°C. Place in a small glass-stoppered container and cool in a desiccator.

X3.3.3.2 Accurately weigh about 1 g of the dried standard and transfer it to a 500-mL Erlenmeyer flask.

X3.3.3.3 Add 50 mL of CO_2-free distilled water and swirl gently to dissolve. Add 3 drops of phenolphthalein indicator and titrate with the sodium hydroxide to the first pink color.

X3.3.3.4 Calculate the normality of NaOH as follows:

$$N = W/V \times 204.228 \quad \text{(X3.1)}$$

where:

N = normality of NaOH solution,
W = grams of potassium acid phthalate,
V = millilitres of NaOH used, and
204.228 = gram equivalent weight of potassium acid phthalate.

X3.3.3.5 Calculate the phosphorus pentoxide (P_2O_5) equivalent (Note X3.2) of the NaOH solution in grams per millilitre as follows:

$$E = N \times 0.003086 \quad \text{(X3.2)}$$

where:

E = P_2O_5 equivalent of the NaOH solution, g/mL,
N = normality of the NaOH solution, and
0.003086 = P_2O_5 equivalent of 1 *N* NaOH solution, g/mL.

NOTE X3.2—Phosphorus reacts to form ammonium phosphomolybdate $2(NH_4)_3PO_4 \cdot 12MoO_3$. The precipitate reacts with the NaOH solution thus:

2(NH4)3PO4 · 12MoO3 + 46NaOH = 2(NH4)2HPO4
+ (NH4)2MoO4 + 23Na2MoO4 + 22H2O

The equivalent weight of P_2O_5 (0.003086) is obtained by dividing the molecular weight of P_2O_5 (141.96) by 46 (for 46 moles of NaOH in the equation) and by 1000 (number of millilitres in 1 L). As the actual composition of the precipitate is influenced by the conditions under which the precipitation is made, it is essential that all the details of the procedure be followed closely as prescribed.

X3.3.4 *Sulfuric Acid, Standard* (0.15 *N*)—Dilute 4.0 mL of H_2SO_4 to 1 L with water that has been freshly boiled and cooled. Standardize against the standard NaOH solution. Determine the ratio in strength of the standard H_2SO_4 solution to the standard NaOH solution by dividing the volume of NaOH solution by the volume of H_2SO_4 solution used in titration.

X3.4 *Procedure:*

X3.4.1 Dissolve 10 g of the prepared sample by adding 80 mL of dilute HNO_3 (1 + 1) and heating on hot plate. Add 350 mg of ferric nitrate salt ($Fe(NO_3)_3 \cdot 9\ H_2O$) (Note X3.3) and 15 g of ammonium chloride (NH_4Cl), make ammoniacal, then add 5 mL of NH_4OH (1 + 1) in excess. Heat to boiling, filter on medium-textured paper and wash with diluted NH_4OH (5 + 95). Discard the filtrate.

NOTE X3.3—Small amounts of phosphorus in the sample can be concentrated very easily this way. The iron combines with the phosphate to form $FePO_4$ that is very insoluble in NH_4OH and affords the analyst the opportunity to separate it from major matrix elements.

X3.4.2 Place the precipitating beaker under the funnel and punch a small hole in the tip of the filter cone by pushing a platinum wire through it. Pour 25 mL of hot HNO_3 (1 + 4) through the paper and wash free of iron with hot 5 % HNO_3. Discard the paper.

X3.4.3 Evaporate the nitric acid of solution to dryness and to the dry salts add 30 mL of aqua regia mix (160 mL of HNO_3 + 40 mL HCl (1 + 1). Boil on hot plate until reddish-brown fumes cease to come off and for several minutes thereafter. Add 100 mL of water and boil.

X3.4.4 Filter into a 500-mL Erlenmeyer flask and wash with hot water. Ignite the residue, cool, add 1 mL of HNO_3 and 10 mL of HF and evaporate to dryness (Note X3.4). Evaporate twice with nitric acid to expel all the HF. Fuse residue with a pinch of Na_2CO_3 (0.5 g), dissolve in dilute nitric acid, and add to the filtrate in the Erlenmeyer flask.

NOTE X3.4—The treatment of the residue with HF and HNO_3 may be omitted if the SiO_2 in the sample is low, <3 %.

X3.4.5 Neutralize the nitric acid solution with NH_4OH until iron hydroxide just begins to form. Reacidify with nitric acid slowly until precipitate just dissolves, then add 1 mL of HNO_3 for every 100 mL of solution. Heat the solution to 40°C and add 40 mL of ammonium molybdate ($(NH_4)_2MoO_4$) (filtered). Stopper the flask, shake for several minutes, and allow to settle overnight (12 h).

X3.4.6 Filter through a retentive paper. Wash the flask and precipitate five times with 15-mL portions of a 1 % solution of KNO_3 and then wash the paper ten times with 10-mL portions of the same solution. The paper should be carefully washed each time from the rim downward and then allowed to drain completely before washing with the next portion of wash solution. (The wash solution must be free of CO_2.)

X3.4.7 Return the paper and precipitate to the flask, add about 75 mL of CO_2-free water, and add enough standard sodium hydroxide solution to decompose the precipitate and leave approximately 2 mL in excess. Record the millilitres of NaOH solution added.

X3.4.8 Stopper the flask and shake to disintegrate the paper. Wash the stopper and sides of the flask with CO_2-free water, add 6 drops of phenolphthalein indicator solution, and titrate the excess sodium hydroxide with standard H_2SO_4 until the pink color of indicator just disappears. Record the millilitres of standard acid.

X3.5 *Calculation:*

NOTE X3.5—Titanium and vanadium interfere in this determination but should not be encountered in significant amounts in this class of material.

X3.5.1 Calculate the percent P_2O_5 as follows:

$$P_2O_5, \% = \{E(V_1 - V_2R)\}/W \times 100 \quad (X3.3)$$

where:

E = P_2O_5 equivalent of the NaOH solution, g/mL,
V_1 = NaOH solution used, mL,
V_2 = H_2SO_4 solution used, mL,
R = ratio in strength of the NaOH solution to the H_2SO_4 solution, and
W = mass of sample, g.

X3.6 *Precision and Bias*—The precision and bias of this method have not been determined.

X4. MANGANESE BY THE BISMUTHATE METHOD

X4.1 *Scope*—This is the standard titrimetric method that may be used to determine manganese in lime and limestone samples when the concentration of manganese is at least 0.05 %.

X4.2 *Summary of Test Method*—After dissolution of the sample, bivalent manganese is oxidized to permanganic acid by the use of sodium bismuthate ($NaBiO_3$) in nitric acid solution, filtering, washing with nitric acid, adding a measured excess of ferrous sulfate and then determining this excess by titration with a standard solution of permanganate.

X4.3 *Apparatus:*

X4.3.1 *Filtering Crucible*—An asbestos Gooch crucible or a fritted glass crucible of fine porosity.

X4.4 *Reagents:*

X4.4.1 *Sodium Bismuthate*—The sodium bismuthate shall contain enough active oxygen to correspond to at least 75 % $NaBiO_3$.

X4.4.2 *Ferrous Ammonium Sulfate Solution* (12 g/L)—Dissolve 12 g of $Fe(NH_4)_2(SO_4)_2 \cdot 6H_2O$ in 1 L of cool H_2SO_4 (5 + 95).

X4.4.3 *Standardization of Ferrous Ammonium Sulfate Solution*—Measure into a 250-mL Erlenmeyer flask, 50 mL of cold nitric acid (3 + 97). Add about 0.5 g of sodium bismuthate ($NaBiO_3$), agitate, dilute with 50 mL of HNO_3 (3 + 97) and filter through asbestos or a fritted glass crucible. To the filtrate add 50 mL of the ferrous ammonium sulfate solution and titrate with standard $KMnO_4$ solution to pink color. Determine the ratio (R) of the standard $KMnO_4$ solution to the standard ferrous ammonium sulfate solution by dividing the volume of $KMnO_4$ solution by the volume of $Fe(NH_4)_2(SO_4)_2$ solution used in the titration.

X4.4.4 *Nitric Acid* (3 + 97).

X4.4.5 *Potassium Permanganate Standard Solution* (0.05 N)—The same solution as prepared in the Total Iron, Standard Method, Section 13.

X4.5 *Procedure:*

X4.5.1 Dissolve 10 g of the sample in 100 mL of diluted HNO_3 (1 + 1), filter, and wash the residue with hot water. Ignite the residue in platinum, fuse with a little Na_2CO_3, and add the nitric acid solution of the melt to the main filtrate.

X4.5.2 Dilute the solution to 150 mL and add 0.5 g of sodium bismuthate. Heat for a few minutes or until the pink color has disappeared and MnO_2 has precipitated. If MnO_2 does not precipitate, add more bismuthate.

X4.5.3 Clear the solution by slowly adding ferrous ammonium sulfate solution (12 g/L) until the MnO_2 dissolves. Boil to expel brown fumes, cool to 10 to 15°C, add 0.5 to 1.5 g of $NaBiO_3$ (or an amount equal to at least 26 times the weight of manganese present), and agitate vigorously for 1 min. Add 50 mL of cold HNO_3 (3 + 97) and filter through a Gooch crucible or fritted glass crucible of fine porosity. Wash with 50 mL of cold HNO_3 (3 + 97) or until the washings run colorless.

X4.5.4 Add enough ferrous ammonium sulfate solution (12 g/L) from a buret to discharge completely the permanganate color and then add 1.0 to 2.0 mL in excess. Record the buret reading. Titrate the excess ferrous with 0.05 N $KMnO_4$ solution to a faint pink end point.

X4.6 *Calculation*—Calculate the percentage of manganese as follows:

$$Mn, \% = (V_1R - V_2)N/W \times 1.0988 \quad (X4.1)$$

where:

V_1 = $Fe(NH_4)_2(SO_4)_2$ solution used, mL,
R = ratio in strength of $KMnO_4$ solution to the $Fe(NH_4)_2(SO_4)_2$ solution,
V_2 = $KMnO_4$ solution used, mL,
N = normality of $KMnO_4$ solution,
W = mass of sample, g, and
1.0988 = manganese equivalent of $KMnO_4 \times 100$.

X4.7 *Precision and Bias*—The precision and bias of this test method have not been determined.

X5. FERROUS IRON BY TITRATION WITH POTASSIUM PERMANGANATE

X5.1 *Scope*—The determination of ferrous iron in carbonate rocks is difficult if not impossible in the presence of carbonaceous matter. Nevertheless, acceptable results are sometimes obtainable if there is not much of such matter, and if it does not give a colored solution with acid. But even then, it is only the soluble iron, existing chiefly if not wholly as carbonate, that is determinable.

X5.2 *Summary of Test Method*—The limestone sample is decomposed with dilute sulfuric acid in a steam or CO_2 atmosphere and the soluble ferrous iron is titrated with $KMnO_4$. In the absence of carbonaceous matter, it is possible to determine ferrous iron in the insoluble residue. Metallic iron, if introduced by grinding media, will also be determined as ferrous iron, and precautions should be taken to avoid this contaminate.

X5.3 *Special Solutions:*

X5.3.1 Potassium Permanganate, *Standard Solution* (0.05 *N*)—Prepare as in Total Iron, Standard Method (Section 13).

X5.3.2 *Diphenylamine Sulfonate Indicator Solution*—Dissolve 0.2 g of sodium diphenylamine sulfonate in 100 mL of water.

X5.3.3 *Potassium Dichromate, Standard Solution* (0.028 *N*)—Weigh 2.728 g of dry National Institute of Standards and Technology potassium dichromate ($K_2Cr_2O_7$) or other pure, dry $K_2Cr_2O_7$ and dilute to 2 L in a volumetric flask. The solution contains the equivalent of 0.002 g FeO/mL.

X5.4 *Procedure:*

X5.4.1 *In Absence of Carbonaceous Matter:*

X5.4.1.1 Weigh several grams of prepared sample into a flask of about 250-mL capacity, add 25 mL of water, and boil water until air is expelled. While still boiling, add diluted H_2SO_4 (1 + 1) a little at a time until effervescence ceases, then add a few millilitres in excess (Note X5.2). Calcium sulfate precipitates, but the iron will remain in solution. Remove the flame and tightly insert a stopper through which passes a small stopcock funnel. When the flask is cool, or nearly so, pour cold water into the funnel and cautiously open the cock so the water may be drawn into the flask. Keep filling the funnel as fast as it empties until the solution in the flask amounts to 100 to 150 mL (Note X5.3). Immediately titrate the iron solution in the flask with 0.05 *N* $KMnO_4$ solution (Note X5.4). The ferrous iron thus found is mostly, if not altogether, that existing as carbonate.

NOTE X5.1: **Warning**—The contents of the flask may erupt violently unless this addition is made with care.

NOTE X5.2—With a carbonate that is wholly decomposable without the aid of heat, solution may be accomplished in a flask filled with CO_2, using cold acid. This mode of acid decomposition lessens the danger of attacking silicates if these are present.

NOTE X5.3—Such precaution to exclude air is hardly necessary, in most cases; however, since in the presence of H_2SO_4 the oxidation of ferrous iron is exceedingly slow. It is ordinarily quite sufficient to equalize the internal and external pressures by opening the cock, removing the stopper, and pouring in cold water.

NOTE X5.4—If MnO_2 is absent, the potassium dichromate method of titration may be substituted and HCl used instead of H_2SO_4. In this case, there is no separation of an insoluble calcium salt which renders easier the subsequent determination of iron in the insoluble residue.

X5.4.2 *Ferrous Iron in the Insoluble Residue in Absence of Carbonaceous Matter:*

X5.4.2.1 Filter the titrated solution (X5.4.1.1) through a medium-sized, hardened filter. Transfer the residue completely to the paper and wash with hot water. Rinse the contents of the paper into a large platinum crucible (80 to 100-mL capacity) with a snugly fitting cover of transparent polypropylene and having a plastic tube for the introduction of CO_2 and a small plastic funnel for the addition of acid and the escape of vapors and steam.

X5.4.2.2 Add 18 mL of H_2SO_4 (1 + 1) and air-free water until the crucible is at least half full. Drop in a small coil of platinum wire to prevent bumping, fit the cover tightly, and pass CO_2 through for at least 10 min as the mixture is gradually brought to boiling. Stop the stream of gas as soon as steam escapes, immediately add 7 mL of HF through the funnel, and boil gently for 10 to 15 min, or until decomposition is complete.

X5.4.2.3 Prepare an 800-mL beaker containing 500 mL of air-free water, 12 g of boric acid, and 20 mL of H_2SO_4 (1 + 3). Quickly transfer the crucible cover and contents to the 800-mL beaker. With the aid of a stirring rod, remove the crucible and cover from the beaker and rinse it off with water. Add 3 to 5 drops of sodium diphenylamine sulfonate indicator and titrate with the $K_2Cr_2O_7$ standard solution to a purple color that persists for 20 to 30 s.

NOTE X5.5—Potassium dichromate is preferred because it reacts much less rapidly on organic matter than permanganate.

X5.4.3 *Ferrous Iron in Presence of Insoluble Carbonaceous Matter:*

X5.4.3.1 Weigh 2 to 5 g of prepared sample into a flask of about 250-mL capacity, pass in a stream of CO_2 through a delivery tube and introduce dilute H_2SO_4 (1 + 1) till effervescence ceases, then add a few more millilitres in excess. Heat if necessary to decompose the limestone (Note X5.6) and filter quickly (Note X5.7) through asbestos or sintered glass fiber (in an atmosphere of carbon dioxide if much iron is present). Wash the filter and the residue a few times with water and titrate the filtrate at once with 0.05 *N* $KMnO_4$ solutions. Discard the residue (Note X5.8).

NOTE X5.6—Heat is not needed with limestone if active agitation is used; but with dolomites, heat is necessary.

NOTE X5.7—It is important to allow the acid to act no longer than is necessary and to filter quickly.

NOTE X5.8—A determination of the ferrous iron in the insoluble matter is usually not worth attempting because of the admixed organic matter.

X5.5 *Calculation*—Calculate the percent FeO as follows:

$$\text{FeO, \%} = V \times N \times 7.184/W \quad \text{(X5.1)}$$

where:

V = titrant ($KMnO_4$ or $K_2Cr_2O_7$), mL,
N = normality of $KMnO_4$ or $K_2Cr_2O_7$ solution,
W = weight of sample, g, and

7.184 = equivalent weight of FeO × 100.

X5.6 *Precision and Bias*—The precision and bias of this test method have not been determined.

X6. TEST METHOD FOR FREE CALCIUM OXIDE IN HIGH-CALCIUM HYDRATED LIME[14]

X6.1 Scope

X6.1.1 This test method covers the determination of the amount of free calcium oxide (CaO) in high-calcium hydrated lime capable of being hydrated by steam at atmospheric pressure.

X6.2 Summary of Test Method

X6.2.1 This test method is based on the principle of gain in weight of CaO when it is hydrated to form calcium hydroxide ($Ca(OH)_2$).

X6.2.2 The sample is dried, subjected to steam at atmospheric pressure, again dried, and the gain in weight calculated to CaO.

X6.3 Apparatus

X6.3.1 *Drying Oven,* thermostatically controlled, CO_2-free atmosphere.

X6.3.2 *Analytical Balance.*

X6.3.3 *Steam Bath,* as shown in Fig. X6.1.

X6.4 Procedure

X6.4.1 Weigh a clean, dry, 10-mL Erlenmeyer flask on an analytical balance.

NOTE X6.1—Once started, complete the test without interruption.

X6.4.2 Add to the flask, by means of a widestem funnel, 3 to 5 g of the lime to be tested. Reweigh to get the exact sample weight.

NOTE X6.2—The flask should be kept stoppered at all times, except when weighing, drying, or steaming.

X6.4.3 Place the flask in the drying oven that has been previously heated to 120°C, and maintain at this temperature for 30 min. Remove, stopper, cool in a desiccator, and weigh.

NOTE X6.3—Oven atmosphere can be kept CO_2 free by placing therein a container of soda lime, quicklime, or other suitable CO_2-absorbing medium.

X6.4.4 Place the flask and the sample in a vigorously boiling steam bath and steam for 30 min. Protect the flask in the steam bath by a drip shield during the steaming period. Remove the flask to the oven, dry for 30 min at 120°C, stopper, cool in a desiccator, and weigh.

X6.4.5 Record all weighings to the fourth decimal place.

X6.5 Calculation

X6.5.1 Calculate the percentage of water and CaO as follows:

$$\text{Free water, \%} = [(W_1 - W_2)/S] \times 100$$

$$\textit{Free CaO, \%} = \frac{(W_3 - W_2) \times 3.114}{S} \times 100 \quad \text{(X6.1)}$$

where:

W_1 = weight of sample and flask before drying,
W_2 = weight of sample and flask after drying,
W_3 = weight of sample and flask after steaming and drying, and
S = weight of sample.

[14] This test method was taken from the Appendix of Specification C 911.

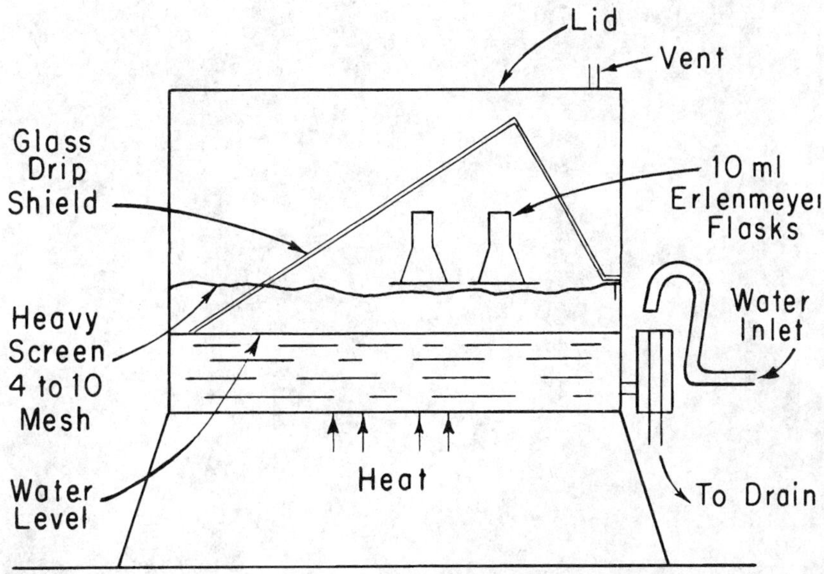

FIG. X6.1 Steam Chamber

Standard Specification for Gypsum Plasters[1]

This standard is issued under the fixed designation C 28; the number immediately following the designation indicates the year of original adoption or, in the case of revision, the year of last revision. A number in parentheses indicates the year of last reapproval. A superscript epsilon (ε) indicates an editorial change since the last revision or reapproval.

This standard has been approved for use by agencies of the Department of Defense. Consult the DoD Index of Specifications and Standards for the specific year of issue which has been adopted by the Department of Defense.

[ε1] NOTE—Scope section corrected editorially and editorial changes made throughout in June 1997.

1. Scope

1.1 This specification covers four gypsum plasters; gypsum mill-mixed plaster, gypsum neat plaster, gypsum wood fibered plaster and gypsum gauging plaster.

1.2 The values stated in inch-pound units are to be regarded as the standard. The SI (metric) values given in parenthesis are approximate and are provided for information purposes only.

1.3 The text of this standard references notes and footnotes which provide explanatory material. These notes and footnotes (excluding those in tables and figures) shall not be considered as requirements of the standard.

1.4 The following safety hazards caveat pertains only to the test methods described in the appendix to this specification. *This standard does not purport to address all of the safety concerns, if any, associated with its use. It is the responsibility of the user of this standard to establish appropriate safety and health practices and determine the applicability of regulatory limitations prior to use.*

2. Referenced Documents

2.1 *ASTM Standards:*

C 11 Terminology Relating to Gypsum and Related Building Materials and Systems[2]

C 22 Specification for Gypsum[2]

C 35 Specification for Inorganic Aggregates for Use in Gypsum Plaster[2]

C 471M Test Methods for Chemical Analysis of Gypsum and Gypsum Products[2]

C 472 Test Methods for Physical Testing of Gypsum, Gypsum Plasters, and Gypsum Concrete[2]

C 778 Specification for Standard Sand[2]

E 11 Specification for Wire-Cloth Sieves for Testing Purposes[3]

3. Terminology

3.1 *Definitions*—Definitions shall be in accordance with Terminology C 11.

3.2 *Definitions of Terms Specific to This Standard:*

3.2.1 *gypsum gauging plaster for finish coat, n*—a calcined gypsum plaster designed to be mixed with lime putty for the finish coat.

3.2.2 *gypsum neat plaster, n*—calcined gypsum mixed at the mill with other ingredients to control working quality and setting time.

3.2.2.1 *Discussion:* Neat plaster is either fibered or unfibered. The addition of aggregate is required on the job.

3.2.3 *gypsum mill-mixed plaster, n*—calcined gypsum plaster, mixed at the mill with a mineral aggregate, designed to function as a base coat to receive various finish coats.

3.2.3.1 *Discussion:* Other materials are not prohibited from being added to control setting time and other desirable working properties.

3.2.4 *gypsum wood-fibered plaster, n*—A calcined gypsum plaster in which wood fiber is used as an aggregate.

4. Materials

4.1 *Calcined Gypsum*—Calcined gypsum shall be manufactured from gypsum meeting the requirements of Specification C 22.

4.2 *Vermiculite Aggregate*, Specification C 35.

4.3 *Perlite Aggregate*, Specification C 35.

4.4 *Sand Aggregate*, Specification C 35.

4.5 *Ottawa Sand*, Specification C 778 for 20–30 sand.

4.6 *Wood Fiber*, non staining wood fiber.

5. Composition

5.1 When tested in accordance with Test Methods C 471M, calcined gypsum shall contain no less than 66.0 weight % $CaSO_4 \cdot 1/2H_2O$.

5.2 *Aggregate*—plasters shall contain a volume of aggregate no greater than that shown in Table 1.

NOTE 1—An approximation of the aggregate content may be made by using the test methods in the appendix. A more accurate method for determining sand in plaster is described in Test Methods C 471M. Test Methods C 471M requires that the original sand and plaster components be available separately.

6. Mechanical Properties

6.1 *Compressive Strength*—When tested in accordance with Test Methods C 472 the compressive strength shall be as specified in Table 1.

6.2 *Setting Time*—When tested in accordance with Test Methods C 472 using the Vicat method, the setting time shall be as specified in Table 1.

---

[1] This specification is under the jurisdiction of ASTM Committee C-11 on Gypsum and Related Building Materials and Systems and is the direct responsibility of Subcommittee C11.01 on Specifications and Test Methods for Gypsum Products.

Current edition approved Dec. 10, 1996. Published February 1997. Originally published as C 28 – 20 T. Last previous edition C 28 – 92.

[2] *Annual Book of ASTM Standards*, Vol 04.01.

[3] *Annual Book of ASTM Standards*, Vol 14.02.

TABLE 1 Composition and Physical Properties

Product	Aggregate Type	Composition, Aggregate per 100 lb of Plaster, not more than		Setting Time[A]	Compressive Strength[B] not less than	
		ft³	m³	hours	psi	MPa
Mill-Mixed Plaster over Lath Base	Vermiculite	2	0.057	1½ to 8	450	3.1
	Perlite	2	0.057	1½ to 8	600	4.1
	Sand	2.5	0.071	1½ to 8	700	4.8
Mill-Mixed Plaster over Masonry	Vermiculite	3	0.085	1½ to 8	325	2.2
	Perlite	3	0.085	1½ to 8	400	2.8
	Sand	3	0.085	1½ to 8	400	2.8
Neat Plaster	none	. . .	. . .	2 to 16[C]	750[D]	5.2
Wood Fibered Plaster	none	. . .	. . .	1½ to 8	1200	8.3
Gauging Plaster	none	. . .	. . .	[E]	1200	8.3

[A] Setting time shall be not less than the shorter time shown nor more than the longer time shown.
[B] Compressive strengths are as tested at the setting times specified in Test Methods C 472.
[C] When mixed with 3 parts by weight of Ottawa sand.
[D] When mixed with 2 parts by weight of Ottawa sand.
[E] When not retarded, the setting time shall be not less than 20 nor more than 40 minutes and, when retarded, not less than 40 minutes.

NOTE 2—*Setting Time*—Attention is directed to conditions affecting job set of gypsum plasters that are usually beyond the control of the producer. Materials added at the site of application such as water and aggregates affect job sets. In addition, the various bases with different absorptive values may affect job performance. The plaster may be accelerated to provide a setting time of not more than 4 h. Setting times greater than 2 h may result in lower compressive strengths.

6.3 *Fineness of Gauging Plaster for Finish Coat*—When tested in accordance with Test Method C 472 all shall pass a No. 14 (1.4 mm) sieve and not less than 60 % shall pass a No. 100 (150 μm) sieve.[4]

7. Sampling

7.1 At least 1 % of the packages, but not less than 5 packages, shall be sampled. Packages to be sampled shall be selected at random. Samples shall be taken both from the outer portion and the center of each package. The materials so obtained shall be thoroughly mixed to provide a composite sample of not less than 7 lb (3.2 kg). This composite sample shall be placed immediately in a clean, dry, airtight container for delivery to the laboratory.

8. Packaging and Package Marking

8.1 Gypsum plasters shall be dry and free of lumps, and shall be shipped in packages.

8.2 When shipped for resale, the following information shall be legibly marked on each package or on a tag of suitable size attached to the package:

8.2.1 Name of manufacturer or seller,

8.2.2 Brand, and

8.2.3 Net weight of the package.

9. Inspection

9.1 Inspection of the gypsum plaster shall be agreed upon between the purchaser and the supplier as part of the purchase agreement.

10. Rejection

10.1 Rejection of gypsum plaster that fails to conform to the requirements of this specification shall be reported to the producer or supplier promptly and in writing. The notice of rejection shall contain a statement documenting requirements of this specification.

11. Certification

11.1 When specified in the purchase agreement, a manufacturer's or supplier's report shall be furnished at the time of shipment certifying that the product is in compliance with this specification.

12. Keywords

12.1 aggregate; gauging plaster; gypsum plaster; mill-mixed plaster; neat plaster; plaster; wood fibered plaster

[4] Detailed requirements for these sieves are given in Specification E 11.

APPENDIX

(Nonmandatory Information)

X1. DETERMINATION OF AGGREGATE CONTENT OF MILL-MIXED PLASTER

INTRODUCTION

The determination of the aggregate content of mill-mixed plaster is normally not precise. Most gypsum plasters contain natural impurities that cannot be separated from the aggregate by either mechanical or chemical means. A better estimate can be made by obtaining the approximate purity of the gypsum plaster used in the product from the manufacturer.

An accurate method for determining sand in set plaster is contained in Test Methods C 471M. Separate specimens of the sand and gypsum plaster used in the mill-mixed plaster are required for Test Methods C 471M.

X1.1 Scope

X1.1.1 This appendix covers mechanical and chemical test methods for separating the aggregate from gypsum plaster.

X1.2 Mechanical Separation

X1.2.1 *Significance and Use:*

X1.2.1.1 This test method is used to estimate the weight or volume of aggregate contained in mill-mixed gypsum plasters by mechanical separation.

X1.2.2 *Interferences:*

X1.2.2.1 It is assumed by this test method that the aggregate meets Specification C 35 and that only a minor portion of the aggregate passes through a 100-mesh (150 µm) sieve. To the extent that a greater portion passes the 100 mesh sieve, the results will show lower values for volume of the aggregate, and higher values for the weight of the gypsum plaster, than actually in the sample.

X1.2.3 *Apparatus:*

X1.2.3.1 *Sieve*—Specification E 11, 100 mesh (150 µm).

X1.2.3.2 *Balance*—with a capacity of not less than 100 g and a precision of at least 0.1 g.

X1.2.4 *Procedure:*

X1.2.4.1 Weigh 100 g of the sample to ±0.1 g.

X1.2.4.2 Sieve the 100 g specimen through the 100 mesh (150 µm) sieve. Examine the material on the sieve to be sure small lumps or agglomerates of plaster are not retained. The retained portion is assumed to be aggregate. If the weight ratio of the aggregate and plaster is to be determined, weigh each portion to the nearest 0.1 g. If the volume ratio of aggregate to the weight of plaster is to be determined, measure the volume of aggregate retained on the sieve to the nearest mL and weigh the material passing through the sieve to the nearest 0.1 g.

X1.2.5 *Calculation and Report:*

X1.2.5.1 *Weight Ratio*—Calculate and report the weight ratio as pounds of aggregate/100 lb (45 kg) of plaster per Eq 1.

$$W_R = \frac{w_a}{w_p} \times 100 \qquad (1)$$

where:

W_R = weight ratio, lbs of aggregate/100 lbs (45 kg) of plaster,
w_a = weight of aggregate, g, and,
w_p = weight of plaster, g.

X1.2.5.2 *Volume Ratio*—Calculate and report the volume ratio as cubic feet of aggregate/100 lbs (45 kg) of plaster per Eq 2.

$$V_R = \frac{v_a}{w_p} \times 1.60 \qquad (2)$$

where:

V_R = Volume ratio, cubic feet of aggregate per 100 lbs of plaster,
v_a = volume of aggregate, mL, and
w_p = weight of plaster, g.

X1.2.6 *Precision and Bias:*

X1.2.6.1 It is not practical to specify the precision or the bias of this test method because of the unknown variabilities in the particle sizes of the aggregates.

X1.3 Chemical Separation

X1.3.1 *Significance and Use:*

X1.3.1.1 This test method is used to estimate the weight or volume of aggregate contained in mill-mixed gypsum plasters by chemical separation.

X1.3.2 *Interferences:*

X1.3.2.1 It is assumed in this test method that all of the material insoluble in ammonium acetate is aggregate and the soluble material is plaster. In many cases there is a significant amount of insoluble material in the plaster. To the extent that this is true it will produce erroneous calculated amounts of aggregate and plaster. If possible, obtain the approximate purity of the plaster from the manufacturer. This information will greatly increase the accuracy of the results.

X1.3.2.2 In the volume calculation it is assumed that the volume of the insoluble material in the plaster is insignificant compared to the volume of the aggregate. This is usually the case.

X1.3.3 *Apparatus:*

X1.3.3.1 *Analytical Balance*—with a precision of 0.001 g.

X1.3.3.2 *Drying Oven*—capable of maintaining a temperature of 100 to 110°C.

X1.3.4 *Reagents:*

X1.3.4.1 *Ammonium Acetate Solution* (250 g/L)—dissolve 250 g of ammonium acetate ($NH_4C_2H_3O_2$) in water and dilute to 1 L.

X1.3.4.2 *Ammonium Hydroxide* (1+59)—Mix one volume of concentrated (NH_4OH) (sp gr 0.90) with 59 volumes of water.

X1.3.5 *Procedure:*

X1.3.5.1 Accurately weigh 40 ± 0.05 g of the sample into a 1-L beaker. Add 600 to 700 mL of ($NH_4C_2H_3O_2$) solution that is slightly alkaline to litmus paper. If acidic, add a few mL of (NH_4OH) (1+59) to the stock ($NH_4C_2H_3O_2$) solution to render it slightly alkaline prior to the addition of the test sample.

X1.3.5.2 Warm the suspension to a temperature of 70 ± 5°C and stir continuously for 20 to 30 min. Filter the warm suspension through a small Buchner funnel or Gooch crucible in which filter paper has previously been formed, the funnel and mat having been dried at 110°C to a constant weight within 0.01 g. Refilter the first 100 mL of the filtrate. Wash the aggregate remaining in the beaker onto the filter with an additional 100 mL of warm ($NH_4C_2H_3O_2$) solution. Wash the beaker and residue with 200 to 300 mL of water, dry the funnel and aggregate at 100°C to constant weight.

X1.3.6 *Calculation and Report:*

X1.3.6.1 *Weight Ratio*—Calculate and report the weight ratio as pounds of aggregate/100 lbs (45 kg) of plaster per Eq 3.

$$W_R = \frac{w_1 - \dfrac{w_1 - w_2}{p}}{\dfrac{w_1 - w_2}{p}} \times 100 \quad (3)$$

where:

W_R = weight ratio, lbs of aggregate/100 lbs (45 kg) of plaster,
w_1 = Weight of original specimen, g,
w_2 = weight of insoluble residue retained on filter, g, and
p = Purity of gypsum plaster expressed as a decimal (that is, 85 % purity = 0.85). If purity is unknown use 0.9 (90 %) as an estimate.

X1.3.6.2 *Volume Ratio*—Calculate the volume ratio as cubic feet of aggregate/100 lbs (45 kg) of plaster per Eq 4.

$$V_R = \frac{v_1}{\dfrac{w_1 - w_2}{p}} \times 1.6 \quad (4)$$

where:

V_R = volume ratio, cubic feet of aggregate/100 lbs (45 kg) of plaster,
v_1 = volume of insoluble residue retained on filter, mL,
w_1 = weight of original specimen, g,
w_2 = weight of insoluble residue retained on filter, g, and
p = purity of gypsum plaster expressed as a decimal (that is, 85 % purity = 0.85). If purity is unknown use 0.9 (90 %) as an estimate.

X1.3.7 *Precision and Bias:*

X1.3.7.1 It is not practical to specify the precision or the bias of this test method because of the unknown variabilities in the purity of the plaster.

Standard Specification for Inorganic Aggregates for Use in Gypsum Plaster[1]

This standard is issued under the fixed designation C 35; the number immediately following the designation indicates the year of original adoption or, in the case of revision, the year of last revision. A number in parentheses indicates the year of last reapproval. A superscript epsilon (ϵ) indicates an editorial change since the last revision or reapproval.

This specification has been approved for use by agencies of the Department of Defense. Consult the DoD Index of Specifications and Standards for the specific year of issue which has been adopted by the Department of Defense.

1. Scope*

1.1 This specification covers perlite, vermiculite, natural and manufactured sand for use as gypsum plaster aggregates.

1.2 The values stated in inch-pound units are to be regarded as the standard. The values given in parentheses are provided for information purposes only.

1.3 *This standard does not purport to address all of the safety concerns, if any, associated with its use. It is the responsibility of the user of this standard to establish appropriate safety and health practices and determine the applicability of regulatory limitations prior to use.*

2. Referenced Documents

2.1 *ASTM Standards:*

C 11 Terminology Relating to Gypsum and Related Building Materials and Systems[2]

C 29/C 29M Test Method for Unit Weight and Voids in Aggregate[3]

C 40 Test Method for Organic Impurities in Fine Aggregates for Concrete[3]

C 136 Test Method for Sieve Analysis of Fine and Coarse Aggregates[3]

C 471 Test Methods for Chemical Analysis of Gypsum and Gypsum Products[2]

D 75 Practice for Sampling Aggregates[4]

E 11 Specification for Wire-Cloth Sieves for Testing Purposes[5]

3. Terminology

3.1 Definitions shall be in accordance with Terminology C 11.

3.2 *Descriptions of Terms Specific to This Standard:*

3.2.1 *manufactured sand, n*—the fine material resulting from the crushing and classification by screening, or otherwise, of rock, gravel or blast furnace slag.

3.2.2 *natural sand, n*—the fine granular material resulting from the natural disintegration of rock or from the crushing of friable sandstone.

3.2.3 *perlite aggregate, n*—a siliceous volcanic glass expanded by heat.

3.2.4 *vermiculite aggregate, n*—a micaceous mineral expanded by heat.

4. Mechanical Requirements

4.1 *Grading*—Shall be as follows when tested in accordance with Section 7.

4.1.1 Perlite and vermiculite shall be graded within the limits specified in Table 1.

4.1.2 *Natural and Manufactured Sand*—Shall be graded with not more than 50 % retained between any two consecutive sieves shown in Table 1, nor more than 25 % between the No. 50 (300 μm) and No. 100 (150 μm) sieves.

4.2 *Density:*

4.2.1 Perlite shall have a density of not less than 6 nor more than 12 lb/ft^3 (96 to 192 kg/m^3) when tested in accordance with Test Method C 29/C 29M using the shoveling procedure.

4.2.2 Vermiculite shall have a density of not less than 6 nor more than 10 lb/ft^3 (96 to 160 kg/m^3) when tested in accordance with Test Method C 29/C 29M using the shoveling procedure.

5. Chemical Composition

5.1 Water-soluble impurities in sand shall not exceed 0.15 weight % and sodium ion content shall not exceed 0.02 weight % when tested in accordance with Section 7.

5.2 Sand shall develop a color no darker than the standard, when subjected to the colorimetric test for organic impurities in accordance with Test Method C 40, unless it is established by adequate tests that the impurities causing the color are not harmful in plaster.

6. Sampling

6.1 *Apparatus*—Suitable thief tube.

6.2 *Bulk Aggregate*—Obtain specimens of sand in accordance with Practice D 75.

6.3 *Bagged Aggregate*—Obtain from one bag randomly selected from each 100 bags but not less than 6 bags from the shipment. For smaller shipments, not less than 6 % of the number of bags shall constitute the sample.

6.3.1 Secure specimens by inserting a suitable thief tube the full distance between diagonally opposite corners of the bag with the bag lying in a horizontal position. Combine the

---

[1] This specification is under the jurisdiction of ASTM Committee C-11 on Ceiling and Walls and is the direct responsibility of Subcommittee C11.02 on Specifications and Test Methods for Accessories and Related Products.

Current edition approved July 15, 1995. Published October 1995. Originally published as C 35-21 T. Last previous edition C 35 – 89a.

[2] *Annual Book of ASTM Standards*, Vol 04.01.

[3] *Annual Book of ASTM Standards*, Vol 04.02.

[4] *Annual Book of ASTM Standards*, Vol 04.03.

[5] *Annual Book of ASTM Standards*, Vol 14.02.

*** A Summary of Changes section appears at the end of this specification.**

portions to produce a composite specimen having a volume not less than 1 ft^3 (0.028 m^3). Prepare and test not less than one composite specimen separately for each 2000 bags of aggregate.

6.4 Reduce specimens obtained in accordance with 6.2 or 6.3 by quartering or riffling to obtain specimens of proper size for individual tests.

7. Test Methods

7.1 *Apparatus:*

7.1.1 *Sieve*—Specification E 11, Nos. 4, 8, 16, 30, 50, 100, 200 (4.75, 2.36, 1.18 mm, 600 µm, 300 µm, 150 µm, 15 µm).

7.1.2 *Graduated 250 mL Cylinder.*

7.1.3 *250 mL Beaker.*

7.1.4 *Tared Evaporating Dish.*

7.1.5 *Fast Paper Filter.*

7.1.6 *100 mL Volumetric Flask.*

7.1.7 *Deionized Water.*

7.1.8 *Spatula.*

7.1.9 *Weighing Scale.*

7.1.10 *Oven.*

7.1.11 *Desiccator.*

7.1.12 *Mechanical Shaker.*

7.1.13 *500 mL Graduated Cylinder.*

7.2 *Sieve Analysis for Sand*—Use Test Method C 136.

7.3 *Sieve Analysis for Perlite or Vermiculite*—Use Test Method C 136 except as modified as follows.

7.3.1 Use a specimen of 500 mL bulk volume.

7.3.1.1 Measure the volume of each sieve fraction in a 250 mL graduated cylinder. Pour the specimen loosely into the graduate without tamping or shaking. Level the surface with a spatula and read the volume to within ±2 mL.

7.3.2 Express the volumes of the individual sieve fractions as percentages of the sum of the volumes of all fractions.

7.3.3 Calculate the cumulative percentages on each designated sieve by summing the individual percentages of all fractions larger than that sieve.

7.4 *Density of Lightweight Aggregate*—Test Method C 29/C 29M; Use shoveling procedure.

7.5 *Water Soluble Impurities of Sand:*

7.5.1 *Significance and Use*—Impurities can affect the setting properties, strength and bond performance of gypsum plaster. These procedures determine the water soluble impurities of the sand being evaluated.

7.5.2 *Apparatus:*

7.5.2.1 *Analytical Balance,* with a precision of 0.0001 g.

7.5.2.2 *Balance,* capable of weighing at least 10 g to a precision of 0.1 g.

7.5.2.3 *Desiccator,* containing calcium chloride or equivalent desiccant.

7.5.2.4 *Drying Oven,* capable of being controlled to 250°F (121°C).

7.5.2.5 *Filter Paper*—A "fast" filter paper shall be used.

NOTE 1—S & S Black Ribbon or Whitman No. 41 have been found suitable for this purpose.[6]

7.5.2.6 *Atomic Absorption or Ion Chromatography Equipment.*

7.5.2.7 *Mechanical Shaker.*

[6] Available from VWR Scientific.

TABLE 1 Grading Requirements

Sieve Size	Percentage Retained on Each Sieve, Cumulative					
	Perlite, by volume		Vermiculite, by volume		Sand, by weight	
	Max	Min	Max	Min	Max	Min
No. 4 (4.75-mm)	0	...	0	...	0	...
No. 8 (2.36-mm)	5	0	10	0	5	0
No. 16 (1.18-mm)	60	5	75	40	30	5
No. 30 (600-µm)	95	45	95	65	65	30
No. 50 (300-µm)	98	75	98	75	95	65
No. 100 (150-µm)	100	85	100	90	100	90
No. 200 (15-µm)					100	95

7.5.3 *Procedure:* (Water Soluble Impurities.)

7.5.3.1 Weigh approximately 10 g of air-dried sand and record the weight to 0.1 g. Transfer to a 250-mL beaker. Add 100 mL of distilled water. Heat to boiling and allow to simmer on a hot plate for 5 min. Filter through a fast paper into a tared evaporating dish. Wash with hot distilled water until the volume of filtrate is about 125 mL. Evaporate to dryness in an oven without allowing the temperature of the dish to rise above 250°F (121°C). Cool residue in a dessicator, weigh to 0.001 g and calculate the percentage of residue on the initial weight of sand.

7.5.3.2 *Report*—Report percentage as water soluble impurities.

7.5.3.3 *Precision and Bias*—No estimate of the precision or bias, or both, is available at this time.

7.5.4 *Procedure*—(Sodium Ion Content.)

7.5.4.1 Weigh approximately 5 g of sand and record the weight to 0.01 g. Transfer to 100-mL volumetric flask. Add approximately 80 mL of deionized water to the flask. Shake on a mechanical shaker for 4 h. Remove and bring to a full volume with deionized water and allow to settle. Analyze for Na ion according to general atomic absorption in accordance with Test Methods C 471 or ion chromatography procedures.[7]

NOTE 2—To minimize contamination, care must be exercised to be sure all equipment is clean.

7.5.4.2 *Calculation and Report*—Calculate and report the percent Na as follows:

$$\mu g/g = \frac{\text{Curve concentration} \times \text{any dilution factor} \times 100 \text{ mL final volume}}{\text{initial weight of sample}}$$

$$\frac{\mu g/g}{10\,000} = \%\ \text{metal ion}\ (Na^+)$$

7.5.4.3 *Precision and Bias*—No estimate of the precision or bias is available at this time.

7.5.5 *Organic Impurities*—Test Method C 40.

8. Inspection

8.1 Inspection of the material shall be agreed upon between the purchaser and the producer or supplier as part of the purchase agreement.

9. Rejection and Rehearing

9.1 Rejection of material shall be reported to the producer

[7] Ion Chromatographs supplied by Dionex Inc., 1228 Titan Way, Sunnyvale, CA 94086 have been found suitable for this purpose.

or supplier within 10 working days from the receipt of the shipment by the purchaser. Rejection of the material shall be based either on the specific cause of failure to conform to the requirements of this specification or the failure of the material to meet the minimum weight or approximate volume as stated on the shipment.

9.2 Failure to meet the minimum weight or approximate volume shall be determined as follows:

9.2.1 *Individual Packages:*

9.2.1.1 The contents, on a weight basis, are 5 % less than that indicated on the packages, or

9.2.1.2 The contents, on a volume basis, are 10 % less than that indicated on the package.

9.2.2 *The Entire Shipment:*

9.2.2.1 On a weight basis when the average contents of two packages for each 100 but not less than six packages selected at random, in any one shipment, are less than that indicated on the package. The net weight of the contents shall be determined by weighing the package or packages and deducting the weight of the container.

9.2.2.2 On a volume basis when the average contents of two packages for each 100 but not less than six packages selected at random, in any one shipment, differ by more than 5 % more or 10 % less from that indicated on the package. The volume of the contents in the package shall be calculated by determining the weight of the contents of the package and then obtaining the weight per unit of volume of the aggregate, from an average weight package of the samples selected in 6.3.2, by the shoveling procedure given in Test Method C 29/C 29M, and then dividing the weight of the contents of the bag by the unit of volume of the aggregate.

10. Certification

10.1 When specified in the purchase order, the producer's or supplier's certification shall be furnished to the purchaser stating that the material was prepared or manufactured in accordance with this specification and has been found to meet the specified requirements.

11. Packaging and Package Marking

11.1 When lightweight aggregates covered by this specification are delivered in packages, the name of the type of aggregate, and minimum weight and approximate volume of the contents shall be plainly indicated thereon.

11.2 When delivered in bulk quantities, the bill of lading shall show the quantity by weight or unit of volume, and the name of the supplier.

11.2.1 Bulk materials shall be delivered in clean transport vessels, free of contaminants.

12. Keywords

12.1 aggregates; gypsum plaster; inorganic; manufactured sand; natural sand; perlite; vermiculite

SUMMARY OF CHANGES

This section identifies the location of changes to this specification that have been incorporated since the last issue. Committee C-11 has highlighted those changes that affect the technical interpretation or use of this specification.

(1) The scope was revised.

(2) Sections 3 through 8 were editorially rewritten and put into mandatory language.

(3) Paragraphs 11.1 and 11.2 were changed to reflect approved C-11 language.

(4) Keywords were added.

(5) Paragraph 7.5.1 was revised.

(6) Paragraph 7.5.3.1 was revised.

(7) Paragraph 7.5.4.1 was revised.

(8) Paragraphs 7.6.4 and 7.6.5 were renumbered as 7.5.5 and 7.5.6.

(9) Paragraph 5.2 was added.

(10) Paragraph 7.5.7 was added.

Designation: C 36 – 97

Standard Specification for Gypsum Wallboard[1]

This standard is issued under the fixed designation C 36; the number immediately following the designation indicates the year of original adoption or, in the case of revision, the year of last revision. A number in parentheses indicates the year of last reapproval. A superscript epsilon (ε) indicates an editorial change since the last revision or reapproval.

This specification has been approved for use by agencies of the Department of Defense. Consult the DoD Index of Specifications and Standards for the specific year of issue which has been adopted by the Department of Defense.

1. Scope*

1.1 This specification covers gypsum wallboard that is designed to be used for walls, ceilings, or partitions and affords a surface suitable to receive decoration.

1.2 The values stated in inch-pound units are to be regarded as the standard. The SI (metric) values given in parentheses are approximate and are provided for information only.

2. Referenced Documents

2.1 *ASTM Standards:*

C 11 Terminology Related to Gypsum and Related Building Materials and Systems[2]

C 473 Test Methods for Physical Testing of Gypsum Board Products and Gypsum Lath[2]

C 645 Specification for Non-Load Bearing (Axial) Steel Studs, Runners (Tracks), and Rigid Furring Channels for Screw Application of Gypsum Board[2]

E 84 Test Method for Surface Burning Characteristics of Building Materials[3]

E 96 Test Methods for Water Vapor Transmission of Materials[4]

E 119 Test Methods for Fire Tests of Building Construction and Materials[3]

3. Terminology

3.1 Definitions used in this standard shall be in accordance with Terminology C 11.

4. Materials and Manufacture

4.1 Gypsum wallboard shall consist of a noncombustible core, essentially gypsum, surfaced with paper bonded to the core.

4.1.1 Aluminum foil shall be bonded to the back surface of foil-backed gypsum wallboard.

4.2 Gypsum wallboard, type X (special fire-resistant) designates gypsum wallboard complying with this specification that provides not less than 1-h fire-resistance for boards ⅝ in. (15.9 mm) thick or ¾-h fire-resistance for boards ½ in. (12.7 mm) thick, applied parallel with and on each side of load bearing 2 by 4 wood studs spaced 16 in. (406 mm) on centers with 6d coated nails, 1⅞ in. (48 mm) long, 0.0915 in. (2.3 mm) diameter shank, ¼ in. (6.4 mm) diameter heads, spaced 7 in. (178 mm) on centers with wallboard joints staggered 16 in. (406 mm) on each side of the partition and tested in accordance with Test Methods E 119.

4.3 Gypsum wallboard shall have a flame spread index of not more than 25 when tested in accordance with Test Method E 84.

NOTE 1—Consult producers for independent test data on assembly details and fire resistance classifications for other types of construction. See official fire test reports for assembly particulars, materials, and classifications.

5. Physical Properties

5.1 Specimens shall be tested in accordance with Test Methods C 473.

5.1.1 Specimens shall be taken from the samples obtained in accordance with 8.2.

5.1.2 *Flexural Strength*—The specimens shall be tested face up and face down. The average breaking load shall be not less than the following:

	Method A		Method B	
Thickness, in. (mm)	Load, lbf (N) Bearing Edges Across Fiber of Surfacing	Load, lbf (N) Bearing Edges Parallel to Fiber of Surfacing	Load, lbf (N) Bearing Edges Across Fiber of Surfacing	Load, lbf (N) Bearing Edges Parallel to Fiber of Surfacing
¼ (6.4)	50 (222)	20 (89)	46 (205)	16 (71)
⁵⁄₁₆ (7.9)	65 (289)	25 (111)	62 (276)	21 (93)
⅜ (9.5)	80 (356)	30 (133)	77 (343)	26 (116)
½ (12.7)	110 (489)	40 (178)	107 (476)	36 (160)
⅝ (15.9)	150 (667)	50 (222)	147 (654)	46 (205)
¾ (19.0)	170 (756)	60 (267)	167 (743)	56 (249)

5.1.3 *Humidified Deflection*—The specimens shall have an average deflection of not more than the following:

Thickness, in. (mm)	Humidified Deflection, Eighths of an inch (mm)
¼ (6.4)	not applicable
⁵⁄₁₆ (7.9)	not applicable
⅜ (9.5)	15 (48)
½ (12.7)	10 (32)
⅝ (15.9)	5 (16)
¾ (19.0)	5 (16)

5.1.4 *Core, End, and Edge Hardness*—The specimens shall have an average hardness of not less than 15 lbf (67 N)

[1] This specification is under the jurisdiction of ASTM Committee C-11 on Gypsum and Related Building Materials and Systems and is the direct responsibility of Subcommittee C 11.01 on Specifications and Test Methods for Gypsum Products.

Current edition approved April 10, 1997. Published June 1997. Originally published as C 36 – 21 T. Last previous edition C 36 – 95b.

[2] *Annual Book of ASTM Standards*, Vol 04.01.

[3] *Annual Book of ASTM Standards*, Vol 04.07.

[4] *Annual Book of ASTM Standards*, Vol 04.06.

*** A Summary of Changes section appears at the end of this specification.**

when tested by Method A and 11 lbf (49 N) when tested by Method B.

5.1.5 *Nail Pull Resistance*—The specimen shall have an average nail-pull resistance of not less than the following:

Thickness, in. (mm)	Method A Nail Pull Resistance, lbf (N)	Method B Nail Pull Resistance, lbf (N)
¼ (6.4)	40 (178)	36 (160)
5/16 (7.9)	50 (222)	46 (205)
⅜ (9.5)	60 (267)	56 (249)
½ (12.7)	80 (356)	77 (343)
⅝ (15.9)	90 (400)	87 (387)
¾ (19.0)	100 (445)	97 (432)

5.2 *Foil-Backed Gypsum Wallboard:*

5.2.1 Foil-backed gypsum wallboard shall meet all of the requirements for gypsum wallboard. In addition, aluminum foil shall be bonded to the back surface.

5.2.2 When tested in accordance with Test Methods E 96 (Desiccant Method), the permeance of foil-backed gypsum wallboard shall be not more than 0.30 perm (17 ng/Pa·s·m^2) for the condition of 50 % relative humidity on the face of the board, and 0 % relative humidity on the foil-covered back side of the board.

6. Dimensions and Permissible Variations

6.1 Specimens shall be taken from the samples obtained in accordance with 8.2.

6.2 Thickness, width, length, and end squareness shall be determined in accordance with Test Methods C 473.

6.2.1 *Thickness*—The nominal thickness shall be ¼, 5/16, ⅜, ½, ⅝, or ¾ in. (6.4, 7.9, 9.5, 12.7, 15.9 or 19.0 mm), with permissible variations in the nominal thickness of ±1/64 in. (0.4 mm) with permissible local variations of ±1/32 in. (0.8 mm) from the nominal thickness.

6.2.2 *Width*—The nominal width shall be up to 48 in. (1220 mm), with widths up to 54 in. (1370 mm) permitted, with a permissible variation of 3/32 in. (2.4 mm) under the specified width.

6.2.3 *Length*—The nominal length and permissible variation shall be as follows:

Thickness in. (mm)	Length ft (mm)	Variation in. (mm)
¼ (6.4)	4 to 12 (1220 to 3660)	±¼ (6.4)
5/16 (7.9)	4 to 14 (1220 to 4270)	±¼ (6.4)
⅜ (9.5)	4 to 16 (1220 to 4880)	±¼ (6.4)
½ (12.7)	4 to 16 (1220 to 4880)	±¼ (6.4)
⅝ (15.9)	4 to 16 (1220 to 4880)	±¼ (6.4)
¾ (19.0)	4 to 16 (1220 to 4880)	±¼ (6.4)

6.2.4 *Tapered Edge Depth*—The average thickness of the edge of recessed or tapered edge shall be not less than 0.020 in. (0.51 mm) but not more than 0.090 in. (2.29 mm) less than the average thickness of the gypsum wallboard.

6.2.5 *End Squareness*—Corners shall be square with a permissible variation of ±⅛ in. (±3.2 mm) in the full width of the board.

6.3 *Edges and Ends*—The edges and ends shall be straight and either square, recessed, beveled, featured, tapered, or featured and tapered.

7. Workmanship, Finish, and Appearance

7.1 The surfaces of gypsum wallboard shall be true and free from imperfections that would render the wallboard unfit for use with or without decoration.

8. Sampling

8.1 When required by the purchase agreement, samples of gypsum wallboard shall be taken at the place of manufacture or at the destination. If the samples are taken other than at the place of manufacture, such samples shall be taken within 24 h of the receipt of the material unless otherwise specified in the purchase agreement.

8.2 At least 0.25 % of the number of gypsum wallboards in a shipment, but not less than three boards, shall be so selected as to be representative of the shipment and shall constitute a sample for purpose of tests by the purchaser or user.

9. Inspection

9.1 Inspection of the gypsum wallboard shall be agreed upon between the purchaser and the supplier as part of the purchase agreement.

10. Rejection

10.1 Rejection of gypsum wallboard that fails to conform to the requirements of this specification shall be reported to the producer or supplier promptly and in writing. The notice of rejection shall contain a statement documenting how the gypsum wallboard has failed to conform to the requirements of this specification.

11. Certification

11.1 When specified in the purchase agreement, a producer's or supplier's report shall be furnished at the time of shipment certifying that the product is in compliance with this specification.

12. Packaging and Package Marking

12.1 Unless otherwise required by the purchase agreement, each board or package shall have legibly marked thereon the following: the thickness; the name of the producer or supplier; the brand name (if any); and the ASTM designation for the product.

13. Shipping, Handling and Storage

13.1 Gypsum wallboard shall be shipped so as to be kept dry.

13.2 Gypsum wallboard shall be stored so as to be kept dry. Where necessary to store gypsum wallboard outside, it shall be stacked off the ground, supported on a level platform, and fully protected from weather and direct sunlight exposure.

13.3 Gypsum wallboard shall be neatly stacked flat with care taken to prevent sagging or damage to edges, ends, and surfaces.

14. Keywords

13.1 ceiling; foil-backed; gypsum; gypsum wallboard; gypsum wallboard, type X; partitions; wall

APPENDIX

(Nonmandatory Information)

This Appendix gives general information and also suggestions for inclusions to be made elsewhere by the specifier. It is not part of this specification.

X1. ALTERNATE DEFINITION FOR TYPE X

X1.1 Gypsum wallboard, type X (special fire-resistant) designates gypsum wallboard providing a greater fire resistance than regular gypsum wallboard of the same thickness. Type X (special fire-resistant) gypsum wallboard, when tested in accordance with Test Methods E 119, shall provide the following minimum fire resistance ratings for the assemblies described:

X1.1.1 1 h for a ⅝ in. (15.9 mm) thickness applied to a partition in a single layer application on each side of 3⅝ in. (92 mm) deep non-loadbearing galvanized steel studs complying with Specification C 645 spaced 24 in. (610 mm) on center. The ⅝ in. (15.9 mm) thick gypsum wallboard 48 in. (1220 mm) wide shall be attached using 1 in. (25 mm) long drywall screws spaced 8 in. (203 mm) on center along the edges and ends, and 12 in. (305 mm) along intermediate studs. All joints shall be oriented parallel to and located over studs and staggered on opposite sides of the assembly. All joints shall be filled with joint compound, covered with joint tape and covered with an additional coat of joint compound. All screw heads shall be covered with joint compound; and

X1.1.2 2 h for a ½ in. (12.7 mm) thickness applied to a partition in a double layer application on each side of 2½ in. (64 mm) deep non-loadbearing galvanized steel studs complying with Specification C 645 spaced 24 in. (610 mm) on center. The base layer 48 in. (1220 mm) wide shall be attached using 1 in. (25 mm) long drywall screws spaced 12 in. (305 mm) on center along board edges, ends and along intermediate studs. Joints shall be oriented parallel to and located over studs and staggered on opposite sides of the assembly. The face layer 48 in. (1220 mm) wide shall be attached using 1⅝ in. (41 mm) long drywall screws spaced 12 in. (305 mm) along board edges, ends and along intermediate studs. Joints shall be oriented parallel to and located over studs, offset 24 in. (610 mm) from the base layer joints, and staggered on opposite sides of the assembly. All joints in the face layer shall be filled with joint compound, covered with joint tape and covered with an additional coat of joint compound. All screw heads in the face layer shall be covered with two coats of joint compound.

SUMMARY OF CHANGES

This section identifies the location of changes to this specification that have been incorporated since the last issue. Committee C-11 has highlighted those changes that affect the technical interpretation or use of this specification.

(1) The tapered edge depth range in 6.2.4 was revised from 0.015 (0.38 mm) - 0.075 in. (1.90 mm) to 0.020 (0.51 mm) - 0.090 in. (2.29 mm).

Designation: C 37 – 95

Standard Specification for Gypsum Lath[1]

This standard is issued under the fixed designation C 37; the number immediately following the designation indicates the year of original adoption or, in the case of revision, the year of last revision. A number in parentheses indicates the year of last reapproval. A superscript epsilon (ε) indicates an editorial change since the last revision or reapproval.

This specification has been approved for use by agencies of the Department of Defense. Consult the DoD Index of Specifications and Standards for the specific year of issue which has been adopted by the Department of Defense.

1. Scope*

1.1 This specification covers plain and foil-backed types of gypsum lath, designed to be used as a base for the application of gypsum plaster (see Specification C 28).

1.2 The values stated in inch-pound units are to be regarded as the standard.

2. Referenced Documents

2.1 *ASTM Standards:*

C 28 Specification for Gypsum Plasters[2]

C 473 Test Methods for Physical Testing of Gypsum Board Products and Gypsum Lath[2]

E 96 Test Methods for Water Vapor Transmission of Materials[3]

3. Materials and Manufacture

3.1 Plain gypsum lath shall consist of a noncombustible core, essentially gypsum, surfaced with paper bonded to the core.

3.2 Foil-backed gypsum lath shall be plain gypsum lath with the back surface covered with aluminum foil.

NOTE—Foil-backed gypsum lath shall be installed with the foil surface against the framing members. The foil surface is not adapted to receive or retain plaster.

3.3 Type X (special fire-resistant) designates gypsum lath complying with this specification that provides at least 1 h fire-resistant classification rating for lath ⅜ in. (9.5 mm) thick, applied at right angles to each side of 2 ft by 4 ft wood studs 16 in. (406 mm) on center and nailed with blue lath nails 5 in. (127 mm) on center over which is applied ½ in. (12.7 mm) 1:2 gypsum-sand plaster.

4. Physical Properties

4.1 *Flexural Strength*—Specimens shall be tested in accordance with Test Methods C 473. When tested face up and when tested face down the specimens shall have an average breaking load of not less than the following:

	Method A		Method B	
Thickness, in. (mm)	Load, lbf(N) Bearing Edges Across Fiber of Surfacing	Load, lbf(N) Bearing Edges Parallel to Fiber of Surfacing	Load, lbf(N) Bearing Edges Across Fiber of Surfacing	Load, lbf(N) Bearing Edges Parallel to Fiber of Surfacing
⅜ (6.4)	60 (267)	25 (111)	56 (249)	21 (93)
½ (12.7)	100 (445)	35 (156)	97 (431)	31 (138)

4.2 *Humidified Deflection*—Specimens taken from plain gypsum lath shall have an average deflection of no more than the following when tested in accordance with Test Methods C 473:

Thickness, in. (mm)	Humidified Deflection, in. (mm)
⅜ (9.5)	15/8 (48)
½ (12.7)	10/8 (32)

4.3 *Core, End, and Edge Hardness*—Specimens taken from the gypsum lath shall have an average hardness of not less than 15 lbf (67 N) when tested by Method A and 11 lbf (49 N) when tested by Method B for the core, ends, or edges when tested in accordance with Test Methods C 473.

4.4 *Nail-Pull Resistance*—Specimens taken from the gypsum lath shall have an average nail-pull resistance of not less than the following when tested in accordance with Test Methods C 473:

Thickness, in. (mm)	Method A Nail Pull Resistance, lbf (N)	Method B Nail Pull Resistance, lbf (N)
⅜ (9.5)	50 (222)	46 (205)
½ (12.7)	70 (312)	66 (294)

4.5 *Permeance*—Specimens taken from foilbacked gypsum lath shall have a permeance not to exceed 0.30 perm for the condition of 50 % relative humidity on Side I (the face of the lath) and 0 % relative humidity on Side II (the foil-covered back of the lath) when tested in accordance with the Desiccant Procedure of Test Methods E 96.

5. Dimensions and Permissible Variations

5.1 *Thickness*—The nominal thickness of gypsum lath shall be ⅜ or ½ in. (9.8 or 12.7 mm), with permissible variations in the nominal thickness of ±1⁄32 in. (±0.8 mm), and with permissible local variations ±1⁄16 in. (±1.6 mm) from the nominal thickness.

[1] This specification is under the jurisdiction of ASTM Committee C-11 on Gypsum and Related Building Materials and Systems and is the direct responsibility of Subcommittee C11.01 on Specifications and Test Methods for Gypsum Products.

Current edition approved Dec. 10, 1995. Published February 1996. Originally published as C 37 – 21 T. Last previous edition C 37 – 92.

[2] *Annual Book of ASTM Standards*, Vol 04.01.

[3] *Annual Book of ASTM Standards*, Vol 04.06.

* A Summary of Changes Section appears at the end of this specification.

5.2 *Width*—The nominal width of lath shall be up to 48 in. (1220 mm), with permissible variations of 3⁄16 in. (4.8 mm) under and 1⁄8 in. (3.2 mm) over the specified width.

5.3 *Length*—The nominal length of gypsum lath shall be 32, 36, 48 and up to 144 in. (813, 914, 1219 mm and 3.7 m), with permissible variations of 1⁄4 in. (6.4 mm) under and 1⁄8 in. (3.2 mm) over the specified length.

5.4 *Edges and Ends*—Edges and ends of gypsum lath shall be straight and solid. Edges shall be either round or square. The corners shall be square with a permissible variation of ±1⁄4 in. (± 6.4 mm) in the full width of the lath.

6. Workmanship, Finish, and Appearance

6.1 The surfaces of gypsum lath shall be such that they will readily receive and retain gypsum plasters. Gypsum lath shall be free of any cracks and imperfections that would render it unfit for its intended use.

6.2 Gypsum lath having burred or broken corners shall be considered acceptable provided that the burred or broken portion is not more than 1 in. (25.4 mm) in either dimension.

7. Sampling

7.1 At least 0.25 % of the number of gypsum lath in a shipment, but not less than three laths, shall be so selected as to be representative of the shipment and shall constitute a sample for purpose of tests by the purchaser or user.

8. Inspection

8.1 Inspection of the gypsum lath shall be agreed upon between the purchaser and the supplier as part of the purchase agreement.

9. Rejection

9.1 Rejection of gypsum lath that fails to conform to the requirements of this specification shall be reported to the producer or supplier promptly and in writing. The notice of rejection shall contain a statement documenting how the gypsum lath has failed to conform to the requirements of this specification.

10. Certification

10.1 When specified in the purchase agreement, a manufacturer's or supplier's report shall be furnished at the time of shipment certifying that the product is in compliance with this specification.

11. Packaging and Package Marking

11.1 Gypsum lath shall be shipped so as to be kept dry and free of moisture.

11.2 When shipped for resale, the name of the manufacturer or supplier and the brand shall be legibly marked on each lath or package.

12 Keywords

12.1 gypsum board; gypsum lath

SUMMARY OF CHANGES

This section identifies the location of changes to this specification that have been incorporated since the last issue. Committee C-11 has highlighted those changes that affect the technical interpretation or use of this specification.

(1) Paragraph 9.1 was revised to change permissive language into mandatory language.

(2) A keywords section was added.

Designation: C 50 – 94

Standard Practice for Sampling, Inspection, Packing, and Marking of Lime and Limestone Products[1]

This standard is issued under the fixed designation C 50; the number immediately following the designation indicates the year of original adoption or, in the case of revision, the year of last revision. A number in parentheses indicates the year of last reapproval. A superscript epsilon (ε) indicates an editorial change since the last revision or reapproval.

These methods have been approved for use by agencies of the Department of Defense. Consult the DoD Index of Specifications and Standards for the specific year of issue which has been adopted by the Department of Defense.

1. Scope

1.1 This practice covers sampling, inspection, rejection, retesting, packing, and marking of lime and lime products and of limestone as it may be used in the chemical, agricultural, and process industries.

NOTE—For sampling of limestone for use as a highway material, see Practice D 75.

1.2 The values stated in inch-pound units are to be regarded as the standard.

1.3 *This standard does not purport to address all of the safety concerns, if any, associated with its use. It is the responsibility of the user of this standard to establish appropriate safety and health practices and determine the applicability of regulatory limitations prior to use.* Specific precautions are given in S4.1.4.1.

2. Referenced Documents

2.1 *ASTM Standards:*

C 702 Practice for Reducing Samples of Aggregate to Testing Size[2]

D 75 Practice for Sampling Aggregates[3]

E 11 Specification for Wire-Cloth Sieves for Testing Purposes[4]

SAMPLING

3. General

3.1 Samples of lime and lime products shall be taken at the place of manufacture or at the destination, as may be agreed upon by the seller and the purchaser. If the samples are taken elsewhere than at the place of manufacture of the material, such samples shall be taken within 24 h of the receipt of the material. When inspection and sampling are to be conducted at the point of manufacture, the manufacturer shall afford the inspector representing the purchaser all reasonable facilities for inspection and sampling of finished material, which shall be so conducted as not to interfere unnecessarily with the operation of the plant.

3.2 Samples may be taken from conveyors delivering to bins or piles, from filled bins by means of sampling tubes, from filled bins at the point of discharge, from piles by means of sampling tubes or shovels, from packaged materials, or from loaded cars.

3.3 Sampling and preparation shall be conducted as expeditiously as possible in order to avoid undue exposure of the material to the air. Samples shall not be taken from broken packages.

3.4 Samples shall be taken in triplicate. Samples of quicklime and of hydrated lime shall be immediately sealed in airtight, moistureproof containers. Each of the triplicate samples shall weigh not less than 5 lb (2.3 kg), provided screen size does not form a part of the specification. If screen size forms a part of the specification, the required weight of triplicate samples shall be as prescribed in Table 1.

3.5 One sample shall be immediately delivered to the consignee; one sample shall be delivered to the consignor, if requested, within 10 days from the date of sampling; the third sample shall be retained with the seal unbroken until such time as the material has been accepted or a referee test called for or 90 days elapsed time, whichever is less.

4. Lump or Granular Material in Bulk

4.1 *Piles, Railcars, Truckloads, or Barges*—Material shall be selected so that it will represent an average of all parts of the pile, railcar, truckload, or barge and shall not contain a disproportionate share of the top or bottom layers. This material shall comprise at least 10 shovelsful taken from different parts of the pile, railcar, truckload, or barge from holes dug approximately 1 ft (304 mm) below the surface. The total quantity so taken shall weigh not less than 50 lb (23 kg) for a unit of material, with a proportionate quantity for multiples of this shipping unit, provided that at no time shall the total amount of material so taken weigh less than 50 lb (23 kg). Triplicate 5-lb (2.3-kg) samples shall be taken by mixing and quartering of the total sample, with intermediate crushing if necessary, if screen size has not been specified; if screen size has been specified, triplicate samples as prescribed in Table 1 shall be taken.

4.2 *At Conveyors to or from Bins, Piles, Railcars, Trucks, or Barges*—Material shall be taken at regular intervals during the entire time of movement of the mass being sampled. The quantity to be taken at each interval and the number of such intervals shall be so proportioned that the final quantity of material taken shall conform to the requirements prescribed in 4.1.

---

[1] This practice is under the jurisdiction of ASTM Committee C-7 on Lime and is the direct responsibility of Subcommittee C07.06 on Physical Tests.

Current edition approved Sept. 15, 1994. Published November 1994. Originally published as C 50 – 22 T. Last previous edition C 50 – 86 (1991).

[2] *Annual Book of ASTM Standards*, Vol 04.02.

[3] *Annual Book of ASTM Standards*, Vol 04.03.

[4] *Annual Book of ASTM Standards*, Vol 14.02.

TABLE 1 Weight of Triplicate Samples

Screen Size of Products, max, in. (mm)	Sample Size, min, lb (kg)
¾ (19.0)	5 (2.3)
1 (25)	10 (4.5)
1½ (37.5)	30 (13.6)
2 (50)[A]	75 (34)

[A] If a screen size of larger than 2 in. (50 mm) forms a part of the specification, the minimum sample size shall be at least 100 times the weight of the largest particle.

4.3 These bulk sampling methods shall not be applied to any limestone product that has not been crushed or screened, or both, to minus 3½ in. (88.9 mm). Otherwise erratic, and non-representative samples may result. The minimum mass of the sample taken should conform to the provisions of Table 1 of Practice D 75.

5. Lump or Granular Material in Packages

5.1 At least 1 % of the packages shall be sampled, but in no case shall less than 5 packages be sampled. Individual packages shall be taken from various parts of the unit being sampled. Each package so taken shall be opened and not less than 1 lb (450 g) removed therefrom and placed in a container. The material removed shall be thoroughly mixed and quartered. Triplicate 5-lb (2.3-kg) samples shall then be taken if screen size has not been specified, or not less than amounts prescribed in Table 1 if screen size has been specified.

6. Powdered Material in Bulk

6.1 *Bins, Piles, or Cars*—Either a sampling tube shall be used that takes a core not less than 1 in. (25 mm) in diameter and that is of sufficient length to permit the taking of a sample from the top to the bottom of the mass being sampled, or samples shall be taken from holes dug into at least ten separate parts of the mass. The samples so taken shall be thoroughly mixed and quartered, and the triplicate samples for testing shall be taken from the material so selected and prepared. A minimum quantity of 50 lb (23 kg) shall be taken for a unit of 30 tons (27 metric tons) of material, with a proportionate quantity for multiples of this unit, provided that at no time shall less than 50 lb (23 kg) be taken as the original portion. The representative sample shall be mixed and quartered and triplicate samples of not less than 5 lb (2.3 kg) shall be taken for testing.

6.2 *At Conveyors to or from Bins, Piles, or Cars, at Packing Machines, etc.*—Sampling at conveyors, packing machines, etc., shall conform to the provisions of 4.2.

7. Powdered Material in Packages

7.1 At least 1 % of the packages shall be sampled, but in no case shall fewer than 5 packages be sampled. Individual packages shall be taken from various parts of the unit being sampled.

7.2 The method of package sampling may be carried out by either of the following two methods, as may be agreed by the contracting parties.

7.2.1 A 1-lb (450-g) sample shall be taken from the top of each package.

7.2.2 A sampling tube shall be used that takes a core not less than 1 in. (25 mm) in diameter. The tube shall be inserted into the package being sampled so that it will take a core of material from substantially the entire length of the package.

7.3 Samples removed from individual packages shall be thoroughly mixed and quartered, and triplicate samples of not less than 5 lb (2.3 kg) each shall be taken for testing.

8. Treatment of Sample at Laboratory

8.1 The sample as received at the laboratory shall be thoroughly mixed, quartered, and a representative sample taken and pulverized to pass a No. 50 (300-μm) sieve[5] for analysis (see Method C 702). The remaining unpulverized portion shall be resealed for further possible tests.

REJECTION

9. Rejection

9.1 Rejection of material based on failure to pass the tests prescribed in the specifications shall be reported to the manufacturer within 1 week after tests have been completed, and the cause for rejection shall be stated.

9.2 The samples that represent rejected material shall be kept in airtight, moistureproof containers for at least 1 week from the date of the original test report.

RETESTING

10. Retesting

10.1 Either of the contracting parties may make claim for a retest within 1 week of the date of the original test report. The expense of the retest shall be borne by the party demanding such retest.

10.2 Should the contracting parties be unable to reach a mutually satisfactory agreement based upon the results of the original test, the third sample of the material shall be delivered unopened to a mutually satisfactory referee laboratory for test and the results of this referee test shall be binding on both parties.

PACKING

11. Packing

11.1 Lime and limestone products may be shipped in bulk or in containers agreed upon between the manufacturer and the purchaser. The most common units for hydrated lime are paper bags holding 50 lb (23 kg), 25 lb (11.5 kg), 10 lb (4.5 kg), or 5 lb (2.3 kg).

MARKING

12. Package Marking

12.1 Unless otherwise agreed upon by the manufacturer and the purchaser, each package shall have legibly marked thereon the name of the product, the net weight of its contents, the name of the manufacturer, the place of manufacture, and the brand name, if any.

12.2 In addition to the above information, the following may be marked on each package of shipment: "The contents meet the requirements of ASTM Practice C 50."

[5] Detailed requirements for this sieve are given in Specification E 11.

GOVERNMENT/MILITARY PROCUREMENT

These requirements apply only to Federal/Military procurement, not domestic sales or transfers.

S1. Referenced Documents

S1.1 *Federal Specification:*
UU-S-48 Sacks, Shipping, Paper[6]

S1.2 *Federal Standard:*
Fed. Std. No. 123 Marking for Shipment (Civil Agencies)[6]

S1.3 *Military Standards:*
MIL-STD-129 Marking for Shipment and Storage[6]
MIL-STD-147 Palletized Unit Loads for 40″ by 48-in. Pallets[6]

S1.4 *Code of Federal Regulations:*
29 CFR 1910 Occupational Safety and Health Standards[6]

S1.5 *National Motor Freight Traffic Association, Inc., Agent:*
National Motor Freight Classification[7]

S1.6 *Uniform Classification Committee, Agent:*
Uniform Freight Classification[8]

S2. Ordering Information

S2.1 Orders for material under this specification shall include the following information:

S2.1.1 ASTM designation and date of issue,

S2.1.2 Name, form, and type of material,

S2.1.3 Quantity required,

S2.1.4 Sampling and inspection if other than as specified (see S3.2),

S2.1.5 Level of packing if specified, and if other than as specified (see S4.1 and S4.1.1), and

S2.1.6 When palletization is required (S4.1.3).

S3. Quality Assurance

S3.1 *Responsibility for Inspection*—Unless otherwise specified in the contract or purchase order, the supplier is responsible for the performance of all inspection requirements as specified herein. Except as otherwise specified in the contract or order, the supplier may use his own or any other facilities suitable for the performance of the inspection requirements specified herein, unless disapproved by the purchaser. The purchaser reserves the right to perform any of the inspections set forth in the specification where such inspections are deemed necessary to assure that supplies and services conform to prescribed requirements.

S3.2 *Sampling and Inspection*—Unless otherwise specified (see 14.1.4), sampling and inspection shall be as specified in this specification.

S3.3 *Test Replication*—Each test shall be conducted in duplicate. Failure of either test of any analysis shall be cause for rejection.

S3.4 *Inspection of Preparation for Delivery*—The packaging, packing, and marking shall be inspected to verify conformance to the requirements of Section S4.

S4. Preparation for Delivery

S4.1 When specified (see S2.1.5), the material shall be prepared for shipment as specified for level A, B, or C.

S4.1.1 *Level A or B*—Unless otherwise specified (see S2.1.5), the material shall be packed in 50 or 100-lb (22.7 or 45.4-kg) quantities as specified in Federal Specification UU-S-48.

S4.1.2 *Level C*—The material shall be packed in 50 or 100-lb (22.7 or 45.4-kg) quantities in weather-resistant containers constructed to assure safe delivery of contents. The containers shall comply with Uniform Freight Classification rules or National Motor Freight Classification rules, as applicable. The containers shall be sufficiently strong to withstand transportation and to permit storage in tiers at least 10 ft high without damage to containers or contents.

S4.1.3 *Palletization*—When specified (see S2.1.6), palletization shall be in accordance with MIL-STD-147.

S4.1.4 *Marking*—Marking shall be in accordance with Fed. Std. No. 123 or MIL-STD-129.

S4.1.4.1 *Precautionary Marking for Calcium Oxide and Quicklime*—Each container shall be marked essentially as follows:

WARNING! CAUSES SEVERE BURNS TO SKIN AND EYES

Avoid contact with skin, eyes, and clothing.
Keep container tightly closed and dry.
When handling, wear goggles or face shield.
Comply with OSHA 29 CFR 1910.1000 for air contaminants.

In case of accident get first aid or call physician immediately!

---

[6] Available from Standardization Documents Order Desk, Bldg. 4 Section D, 700 Robbins Ave., Philadelphia, PA 19111-5094, Attn: NPODS.

[7] Available from American Trucking Association, Inc., 2200 Mill Rd., Alexandria, VA 22314.

[8] Available from Uniform Classification Committee, 222 South Riverside Plaza, Chicago, IL 60606.

Designation: C 51 – 98

Standard Terminology Relating to Lime and Limestone (as used by the Industry)[1]

This standard is issued under the fixed designation C 51; the number immediately following the designation indicates the year of original adoption or, in the case of revision, the year of last revision. A number in parentheses indicates the year of last reapproval. A superscript epsilon (ε) indicates an editorial change since the last revision or reapproval.

These definitions have been approved for use by agencies of the Department of Defense. Consult the DoD Index of Specifications and Standards for the specific year of issue which has been adopted by the Department of Defense.

INTRODUCTION

Where appropriate, the various terms defined below should be prefixed with one or other of the adjectives "high-calcium," "magnesian," or "dolomitic." (Examples: dolomitic quicklime; high-calcium hydraulic hydrated lime; magnesian or dolomitic limestone.)

The composition of a limestone should be given in terms of a percentage of the carbonates present. In limestone of interest to the lime industry, it is usually assumed that the material consists almost entirely of carbonates. Where this assumption is not valid, the percentage of noncarbonate material should be determined, and the composition expressed in terms of the carbonate material present.

This terminology is as commonly used by the industry. For specific application of lime or a limestone product, see the appropriate ASTM specification.

agricultural lime—either ground quicklime or hydrated lime whose calcium and magnesium content is capable of neutralizing soil acidity.

agricultural limestone—ground or pulverized limestone whose calcium and magnesium content is capable of neutralizing soil acidity.

air-slaked lime—the product containing various proportions of the oxides, hydroxides, and carbonates of calcium and magnesium which results from the exposure of quicklime to the air in sufficient quantity to show physical signs of hydration (difficult to determine visually in pulverized quicklime).

alkaline earth solutions (AES)—an aqueous solution of the oxide or hydroxide of an element of group IIa in the periodic table, such as calcium or magnesium. These solutions may be strongly alkaline. See **pH**.

available lime—those constituents of a lime which enter into a desired reaction under the conditions of a specific method or process.

building or construction lime—a lime whose chemical and physical characteristics and method of processing make it suitable for the ordinary or special construction uses of the product.

calcareous—originating from predominately calcium carbonate or one of its derivative forms.

calcia—the chemical compound calcium oxide (CaO).

chemical lime—a quicklime or hydrated lime whose chemical and physical characteristics and method of processing make it suitable for one or more of the many and varied chemical and industrial uses of the product.

DISCUSSION—The chemical forms of calcium oxide (CaO), calcium hydroxide ($Ca(OH)_2$), magnesium oxide (MgO), or magnesium hydroxide ($Mg(OH)_2$) alone or in combination may be produced either primarily or as a by-product of materials other than limestone, for example, $Ca(OH)_2$ formed by acetylene generation from calcium carbide (CaC_2), water treatment sludges, etc.

dead burned dolomite—dolomitic limestone that has been heated with or without additives to a temperature sufficiently high and for a long enough time to decompose the carbonate structure so as to form calcium oxide and periclase in a matrix that provides resistance to subsequent hydration and recombination with carbon dioxide.

dolomitic—indicates the presence of 35 to 46 % magnesium carbonate ($MgCO_3$) in the limestone from which the material was formed.

dolomitic limestone—see **limestone.**

di-hydrated or double hydrated lime—dolomitic lime which has been hydrated under greater than atmospheric pressure and contains less than 8 % unhydrated oxides.

finishing hydrated lime—hydrated lime suitable for use in the finish coat of plaster.

finishing quicklime—quicklime suitable (after slaking to a lime putty) for use in the finish coat of plaster.

fluxing lime—a term referring to quicklime used as an agent in the manufacture of steel or glass.

fluxstone—a term referring to limestone (high-calcium, magnesian, or dolomitic) used as an agent in the manufacture of iron and steel or glass.

high-calcium—indicates the presence of 0 to 5 % magnesium carbonate ($MgCO_3$) in the limestone from which the material was formed.

high-calcium limestone—see **limestone.**

hydrated lime—a dry powder obtained by treating quicklime with water enough to satisfy its chemical affinity for water under the conditions of its hydration. It consists essentially

[1] This terminology is under the jurisdiction of ASTM Committee C-7 on Lime and is the direct responsibility of Subcommittee C07.08 on Nomenclature.

Current edition approved Feb. 10, 1998. Published June 1998. Originally published as C 51 – 22. Last previous edition C 51 – 95.

of calcium hydroxide or a mixture of calcium hydroxide and magnesium oxide or magnesium hydroxide, or both.

DISCUSSION—The chemical forms of calcium oxide (CaO), calcium hydroxide ($Ca(OH)_2$), magnesium oxide (MgO), or magnesium hydroxide ($Mg(OH)_2$) alone or in combination may be produced either primarily or as a by-product of materials other than limestone, for example, $Ca(OH)_2$ formed by acetylene generation from calcium carbide (CaC_2), water treatment sludges, etc.

hydraulic hydrated lime—the hydrated dry cementitious product obtained by calcining a limestone containing silica and alumina to a temperature short of incipient fusion so as to form sufficient free lime (CaO) to permit hydration, and at the same time, leaving unhydrated sufficient calcium silicates to give a dry powder meeting hydraulic property requirements.

lime—a general term which includes the various chemical and physical forms of quicklime, hydrated lime, and hydraulic lime. It may be high-calcium, magnesian, or dolomitic.

DISCUSSION—The chemical forms of calcium oxide (CaO), calcium hydroxide ($Ca(OH)_2$), magnesium oxide (MgO), or magnesium hydroxide ($Mg(OH)_2$) alone or in combination may be produced either primarily or as a by-product of materials other than limestone, for example, $Ca(OH)_2$ formed by acetylene generation from calcium carbide (CaC_2), water treatment sludges, etc.

lime mortar—a lime putty mixed with an aggregate, suitable for masonry purposes.

lime putty—the product obtained by slaking quicklime with water according to the directions of the manufacturer or by mixing hydrated lime and water to a desired consistency.

limestone—an initially sedimentary rock consisting chiefly of calcium carbonate or of the carbonates of calcium and magnesium. Limestone may be of high calcium, magnesian, or dolomitic.

(*1*) *dolomitic limestone*—limestone containing from 35 to 46 % magnesium carbonate ($MgCO_3$).

(*2*) *magnesium limestone*—a limestone containing from 5 to 35 % $MgCO_3$.

(*3*) *high-calcium limestone*—a limestone containing from 0 to 5 % $MgCO_3$.

liming material—a general term which includes the various chemical and physical forms of lime, limestone, mollusk shells, marl, and slag whose calcium and magnesium compounds are capable of neutralizing acidity.

magnesia—the chemical compound magnesium oxide (MgO).

magnesian—indicates the presence of 5 to 35 % magnesium carbonate ($MgCO_3$) in the limestone from which the material was formed.

magnesian limestone—see **limestone.**

masons hydrated lime—hydrated lime suitable for use for masonary purposes.

masons quicklime—quicklime suitable (after slaking to a lime putty) for use for masonry purposes.

milk of lime—a suspension of hydrated lime (or slaked quicklime) in water in such proportions as to resemble milk in appearance.

DISCUSSION—The chemical forms of calcium oxide (CaO), calciuim hydroxide ($Ca(OH)_2$), magnesium oxide (MgO), or magnesium hydroxide ($Mg(OH)_2$) alone or in combination may be produced either primarily or as a by-product of materials other than limestone, for example, $Ca(OH)_2$ formed by acetylene generation from calcium carbide (CaC_2), water treatment sludges, etc.

mono-hydrated lime—dolomitic lime which has been hydrated at atmospheric pressure and contains more than 8 % unhydrated oxides.

pH—the negative logarithm of the hydrogen ion concentration, which can be greatly affected by temperature, particularly under alkaline conditions. It is therefore important to measure alkaline earth solutions (AES) at a specific recommended temperature of 25°C.

quicklime—a calcined limestone, the major part of which is calcium oxide or calcium oxide in association with magnesium oxide, capable of slaking with water.

DISCUSSION—The chemical forms of calcium oxide (CaO), calcium hydroxide ($Ca(OH)_2$), magnesium oxide (MgO), or magnesium hydroxide ($Mg(OH)_2$) alone or in combination may be produced either primarily or as a by-product of materials other than limestone, for example, $Ca(OH)_2$ formed by acetylene generation from calcium carbide (CaC_2), water treatment sludges, etc.

quicklime sizes—the different sizes depending upon the type of limestone, kind of kiln used, or treatment subsequent to calcining. The sizes commonly recognized are as follows:

(*1*) *large lump*—8 in. (203 mm) and smaller,

(*2*) *pebble or crushed*—2½ in. (64 mm) and smaller,

(*3*) *ground, screened or granular*—¼ in. (6.4 mm) and smaller, and

(*4*) *pulverized*—substantially all passing a No. 20 (850-μm) sieve.

refractory lime—lime (usually of a dolomitic type) that has been extremely hard burned so that it will possess little or no tendency for conversion of the oxides to hydroxides.

run-of-kiln quicklime—quicklime as drawn or discharged from a kiln.

spray lime—a hydrated lime of such fineness that at least 95 % of the particles will pass a No. 325 (45-μm) sieve.

white wash—a combination of hydrated lime (or slaked quicklime), water, and other materials to be used as a paintlike coating.

Designation: C 59 – 95

Standard Specification for
Gypsum Casting Plaster and Gypsum Molding Plaster[1]

This standard is issued under the fixed designation C 59; the number immediately following the designation indicates the year of original adoption or, in the case of revision, the year of last revision. A number in parentheses indicates the year of last reapproval. A superscript epsilon (ε) indicates an editorial change since the last revision or reapproval.

This specification has been approved for use by agencies of the Department of Defense. Consult the DoD Index of Specifications and Standards for the specific year of issue which has been adopted by the Department of Defense.

1. Scope*

1.1 This specification covers gypsum casting plaster and gypsum molding plaster materials consisting essentially of calcined gypsum.

1.2 The values stated in inch-pound units are to be regarded as the standard. The SI equivalents of inch-pound units shown are approximate.

1.3 The text of this specification references footnotes that provide explanatory material. These footnotes shall not be considered as requirements of the specification.

2. Referenced Documents

2.1 *ASTM Standards:*

C 11 Terminology Relating to Gypsum and Related Building Materials and Systems[2]

C 471 Test Methods for Chemical Analysis of Gypsum and Gypsum Products[2]

C 472 Test Methods for Physical Testing of Gypsum, Gypsum Plasters and Gypsum Concrete[2]

E 11 Specification for Wire-Cloth Sieves for Testing Purposes[3]

3. Terminology

3.1 Definitions of terms shall be in accordance with Terminology C 11.

3.2 *Description of Terms Specific to This Standard:*

3.2.1 *gypsum casting plaster, n*—a calcined gypsum plaster used primarily to form objects, such as lamp bases, art ware, and novelties, etc.

3.2.2 *gypsum molding plaster, n*—a calcined gypsum plaster used primarily for plaster casts or molds, sometimes used as a gauging plaster.

4. Composition

4.1 Gypsum casting plaster and gypsum molding plaster shall have a purity of not less than 85 % $CaSO_4 \cdot 2H_2O$ when tested in accordance with Test Methods C 471.

5. Physical Properties

5.1 *Setting Time*—Gypsum casting plaster and gypsum molding plaster shall set in not less than 10 nor more than 50 min when tested in accordance with Test Methods C 472 using the procedure for gypsum concrete and all gypsum plasters, except gypsum neat plaster.

5.2 *Compressive Strength*—Gypsum casting plaster and gypsum molding plaster shall have a compressive strength of not less than 1800 psi (12.4 MPa) when tested in accordance with Test Methods C 472.

5.3 *Fineness*—Gypsum casting plaster and gypsum molding plaster shall all pass a No. 30 (600-μm) sieve and not less than 90 % shall pass a No. 100 (150-μm) sieve[4], when tested in accordance with Test Methods C 472.

6. Sampling

6.1 Not less than 1 % of the packages, but not less than 5 packages, shall be sampled. Packages to be sampled shall be selected at random. Samples shall be taken both from the outer portion and the center of each package. The materials so obtained shall be thoroughly mixed to provide a composite sample of not less than 7 lb (3.2 kg). This composite sample shall be placed immediately in a clean, dry, airtight container for delivery to the laboratory.

7. Inspection

7.1 Inspection of the gypsum casting and molding plaster shall be agreed upon by the purchaser and the supplier as part of the purchase agreement.

8. Rejection

8.1 Rejection of gypsum casting plaster and gypsum molding plaster that fails to conform to the requirements of this specification shall be reported to the producer or supplier promptly and in writing. The notice of rejection shall contain a statement documenting how the gypsum casting and molding plaster has failed to conform to the requirements of this specification.

9. Certification

9.1 When specified in the purchase agreement, a producer's or supplier's report shall be furnished at the time of shipment certifying that the product is in compliance with this specification.

---

[1] This specification is under the jurisdiction of ASTM Committee C-11 on Gypsum and Related Building Materials and Systems and is the direct responsibility of Subcommittee C11.01 on Specifications and Test Methods for Gypsum Products.

Current edition approved Dec. 10, 1995. Published February 1996. Originally published as C 59 – 26. Last previous edition C 59 – 91.

[2] *Annual Book of ASTM Standards*, Vol 04.01.

[3] *Annual Book ot ASTM Standards*, Vol 14.02.

[4] Detailed requirements for these sieves are given in ASTM Specification E 11.

*** A Summary of Changes section appears at the end of this specification.**

10. Packaging and Package Marking

10.1 Gypsum casting plaster and gypsum molding plasters shall be dry and free of lumps and shall be shipped in packages.

10.2 When shipped, the following information shall be legibly marked on each package or on a tag of suitable size attached thereto:

10.2.1 Name of manufacturer or seller,

10.2.2 Brand,

10.2.3 Description, and

10.2.4 Net weight of package. Gross weight shall be shown where required.

11. Keywords

11.1 casting plaster; gypsum; gypsum plaster; molding plaster

SUMMARY OF CHANGES

This section identifies the location of changes to this standard that have been incorporated since the last issue. Committee C-11 has highlighted those changes that affect the technical interpretation or use of this standard.

(1) Sections 1.3, 3 and 11 were added.
(2) Sections 1.1 and 6.1 were revised.
(3) The note in section 4.1 was deleted.

Designation: C 61 – 95

Standard Specification for Gypsum Keene's Cement[1]

This standard is issued under the fixed designation C 61; the number immediately following the designation indicates the year of original adoption or, in the case of revision, the year of last revision. A number in parentheses indicates the year of last reapproval. A superscript epsilon (ε) indicates an editorial change since the last revision or reapproval.

This specification has been approved for use by agencies of the Department of Defense. Consult the DoD Index of Specifications and Standards for the specific year of issue which has been adopted by the Department of Defense.

1. Scope*

1.1 This specification covers various grades of anhydrous calcined gypsum known as gypsum Keene's cement, designed for use in the base and finish coats of gypsum plaster.

NOTE 1—The setting time is accelerated by the addition of other materials.

NOTE 2—Keene's cements are generally available as quick-setting and standard-setting types. However, various grades of gypsum Keene's cement of different fineness and setting time are available. Grades intended for use in Scagiola castings and other special purposes should conform to the requirements of this specification in all respects except fineness and setting time.

1.2 The values stated in inch-pound units are to be regarded as the standard. The SI units in parentheses are for information purposes only.

1.3 The text of this specification references notes and footnotes that provide explanatory material. These notes and footnotes shall not be considered as requirements of the specification.

1.4 The following safety hazards caveat pertains only to the test methods portion, Section 7, of this specification: *This standard does not purport to address all of the safety concerns, if any, associated with its use. It is the responsibility of the user of this standard to establish appropriate safety and health practices and determine the applicability of regulatory limitations prior to use.*

2. Referenced Documents

2.1 *ASTM Standards:*

C 11 Terminology Relating to Gypsum and Related Building Materials and Systems[2]

C 471 Test Methods for Chemical Analysis of Gypsum and Gypsum Products[2]

C 472 Test Methods for Physical Testing of Gypsum, Gypsum Plasters, and Gypsum Concrete[2]

E 11 Specification for Wire-Cloth Sieves for Testing Purposes[2]

3. Terminology

3.1 *Definitions*—Definitions of terms used in this specification shall be as defined in Terminology C 11.

4. Chemical Composition

4.1 *Combined Water*—The cement shall show a combined water content not more than 2 %.

5. Physical Properties

5.1 *Setting Time*—The cement shall set in not less than 20 min and not more than 6 h.

5.2 *Compressive Strength*—Gypsum Keene's cement shall have a compressive strength of not less than 2500 psi (17 MPa).

5.3 *Fineness*—The cement shall all pass a No. 14 (1.40-mm) sieve; not less than 98 % shall pass a No. 40 (425-μm) sieve, and not less than 80 % shall pass a No. 100 (150-μm) sieve.[3]

6. Sampling

6.1 Randomly select at least 1 % of the packages, but not less than 5 packages. Take samples from both the outer portion and the center of each package. Mix the samples so obtained to provide a composite sample of at least 15 lb (6.8 kg). Immediately place the sample in a clean, dry, airtight container for delivery to the laboratory.

7. Test Methods

7.1 Determine the chemical analysis of gypsum Keene's cement in accordance with Test Methods C 471.

7.2 Determine the physical properties of gypsum Keene's cement in accordance with Test Methods C 472, except that in the determination of the setting time and compressive strength, the consistency used shall be such that a 35-g conical plunger, described in Test Methods C 472, shall give a penetration of 30 ± 2 mm, 20 s after release, the measurement being made 5 min after the addition of the gypsum Keene's cement to the gauging water.

7.2.1 Determine the setting time by the Vicat method.

7.2.2 Soak standard Keene's cement in water for 20 min with occasional stirring before filling the mold in the compressive strength determination.

[1] This specification is under the jurisdiction of ASTM Committee C-11 on Gypsum and Related Building Materials and Systems and is the direct responsibility of Subcommittee C11.01 on Specifications and Test Methods for Gypsum Products.

Current edition approved Oct. 10, 1995. Published December 1995. Originally published as C 61 – 26 T. Last previous edition C 61 – 94.

[2] *Annual Book of ASTM Standards*, Vol 04.01.

[3] Detailed requirements for these sieves are given in ASTM Specification E 11.

*** A Summary of Changes section appears at the end of this specification.**

8. Inspection

8.1 Inspection of the gypsum Keene's cement shall be agreed upon between the producer or supplier and the purchaser as part of the purchase agreement.

9. Rejection

9.1 Rejection of gypsum Keene's cement that fails to conform to the requirements of this specification shall be reported to the producer or supplier promptly and in writing. The written notice of rejection shall contain a statement documenting how the gypsum Keene's cement has failed to conform to the requirements of this specification.

10. Certification

10.1 When specified in the purchase agreement, a producer's or supplier's report shall be furnished at the time of shipment certifying that the product is in compliance with this specification.

11. Packaging and Marking

11.1 Gypsum Keene's cement shall be dry, free of lumps, and shipped in packages.

11.2 When shipped for resale, the following information shall be legibly marked on each package or on a tag of suitable size attached thereto:

11.2.1 Name of producer or supplier,

11.2.2 Brand,

11.2.3 Description, and

11.2.4 Net and gross weights of package.

12. Keywords

12.1 anhydrous gypsum; gypsum; gypsum Keene's cement; gypsum plaster; Keene's cement; plaster

APPENDIX

(Nonmandatory Information)

X1. FIELD TEST FOR GYPSUM KEENE'S CEMENT

X1.1 Scope

X1.1.1 This method is a simple field test for the identification of gypsum Keene's cement. It does not indicate the quality of the cement, which shall be determined by the laboratory tests in the body of this specification.

X1.2 Significance and Use

X1.2.1 This method is of value to users in determining whether the material at hand is gypsum Keene's cement or some other material.

X1.3 Procedure

X1.3.1 Take a cupful of the material, mix with water to the consistency of a thick paste, and transfer to a plate or piece of glass. Let stand until fairly firm and definite signs of set having begun are manifest. This will be some time less than 2 h, depending on climatic conditions.

X1.3.2 Divide the sample, allowing one half to remain undisturbed. Take the other half, add a little water, remix, and then allow it to "set up" again on the plate or glass.

X1.4 Interpretation of Results

X1.4.1 If the material is gypsum Keene's cement, the remixed portion will, within a few hours, become as hard and strong as the portion that was not remixed.

X1.5 Precision and Bias

X1.5.1 It is not possible to state either the precision or bias of this test method because the method only determines whether a material is or is not gypsum Keene's cement and there are no numerical results generated.

SUMMARY OF CHANGES

This section identifies the location of changes to this specification that have been incorporated since the last issue. Committee C-11 has highlighted those changes that affect the technical interpretation of this specification.

(1) Paragraph 1.3 was added.

Standard Specification for Treated Core and Nontreated Core Gypsum Sheathing Board[1]

This standard is issued under the fixed designation C 79/C 79M; the number immediately following the designation indicates the year of original adoption or, in the case of revision, the year of last revision. A number in parentheses indicates the year of last reapproval. A superscript epsilon (ϵ) indicates an editorial change since the last revision or reapproval.

1. Scope

1.1 This specification covers treated core and nontreated core gypsum sheathing, which is designed to be used as a sheathing on buildings.

1.2 The values stated in either inch-pound or SI (metric) units are to be regarded separately as the standard. Within the text, the SI units are shown in brackets. The values stated in each system shall be used independently of the other. Values from the two systems shall not be combined.

2. Referenced Documents

2.1 *ASTM Standards:*

C 11 Terminology Relating to Gypsum and Related Building Materials and Systems[2]

C 473 Test Methods for Physical Testing of Gypsum Board Products and Gypsum Lath[2]

C 645 Specification for Nonstructural Steel Framing Members[2]

C 1264/C 1264M Specification for Sampling, Inspection, Rejection, Certification, Packaging, Marking, Shipping, Handling, and Storage of Gypsum Board[2]

E 119 Test Methods for Fire Tests of Building Construction and Materials[3]

3. Terminology

3.1 Definitions used in this specification shall be in accordance with Terminology C 11.

4. Materials and Manufacture

4.1 Gypsum sheathing shall consist of a noncombustible core, essentially gypsum, surfaced on both the face and the back with water-repellent paper bonded to the core.

4.1.1 Treated core gypsum sheathing shall have a water-resistant material incorporated into the core.

4.1.2 Nontreated core gypsum sheathing shall not be required to have a water-resistant material incorporated in the core.

4.2 *Gypsum Sheathing, Type X (Special Fire-Resistant),* designates gypsum sheathing complying with this specification that provides not less than 1-h fire-resistance for boards ⅝ in. [15.9 mm] thick or ¾-h fire-resistance for boards ½ in. [12.7 mm] thick, applied parallel with and on each side of load bearing 2 by 4 wood studs spaced 16 in. [406 mm] on centers with 6d coated nails, 1⅞ in. [48 mm] long, 0.0915-in. [2.3-mm] diameter shank, ¼-in. [6.4-mm] diameter heads, spaced 7 in. [178 mm] on center with gypsum sheathing joints staggered 16 in. [406 mm] on each side of the partition and tested in accordance with Test Methods E 119.

NOTE—Consult producers for independent test data on assembly details and fire resistance classifications for other types of construction. See official fire test reports for assembly particulars, materials, and classifications.

5. Physical Properties

5.1 Specimens shall be taken from the samples obtained in accordance with 8.1.

5.1.1 Specimens shall be tested in accordance with Test Methods C 473.

5.1.2 *Flexural Strength*—The specimens shall be tested face up and face down. The average breaking load shall be not less than the following:

Thickness in. [mm]	Method A Bearing Edges Perp to Panel Length lbf [N]	Method A Bearing Edges Par to Panel Length lbf [N]	Method B Bearing Edges Perp to Panel Length lbf [N]	Method B Bearing Edges Par to Panel Length lbf [N]
⅜ [9.5]	80 [356]	30 [133]	77 [343]	26 [116]
4/10 [10.2]	88 [391]	32 [142]	85 [378]	28 [125]
½ [12.7]	110 [489]	40 [178]	107 [476]	36 [160]
⅝ [15.9]	150 [667]	50 [222]	147 [654]	46 [205]

5.1.3 *Humidified Deflection*—The specimens shall have an average deflection of not more than the following:

Thickness in. [mm]	Deflection eighths of an inch [mm]
⅜ [9.5]	15 [48]
4/10 [10.2]	12 [38]
½ [12.7]	10 [32]
⅝ [15.9]	5 [16]

5.1.4 *Core, End, and Edge Hardness*—The specimens shall have an average hardness of not less than 15 lbf [67 N] when tested by Method A and 11 lbf [49 N] when tested by Method B.

5.1.5 *Nail Pull Resistance*—The specimens shall have an average nail-pull resistance of not less than the following:

Thickness in. [mm]	Method A lbf [N]	Method B lbf [N]
⅜ [9.5]	60 [267]	56 [249]
4/10 [10.2]	70 [312]	67 [298]
½ [12.7]	80 [356]	77 [343]
⅝ [15.9]	90 [400]	87 [387]

5.1.6 *Water Resistance of Treated Core Gypsum Sheathing*—The specimens shall have an average water absorption of not more than 10 weight % after 2-h immersion.

5.1.7 *Surface Water Absorption of Gypsum Sheathing*—The specimens shall have an average surface water absorp-

[1] This specification is under the jurisdiction of ASTM Committee C-11 on Gypsum and Related Building Materials and Systems and is the direct responsibility of Subcommittee C11.01 on Specifications and Test Methods for Gypsum Products.

Current edition approved April 10, 1997. Published June 1997. Originally published as C 79 – 30 T. Last previous edition C 79 – 95.

[2] *Annual Book of ASTM Standards,* Vol 04.01.

[3] *Annual Book of ASTM Standards,* Vol 04.07.

tion of not more than 1.6 g after 2 h of elapsed time.

6. Dimensions and Permissible Variations

6.1 Specimens shall be taken from the samples obtained in accordance with 8.1.

6.2 Thickness, width, length, and end squareness shall be determined in accordance with Test Methods C 473.

6.2.1 *Thickness*—The nominal thickness shall be 3/8, 4/10, 1/2, or 5/8 in. [9.5, 10.2, 12.7, or 15.9 mm] with permissible variations in the thickness of ±1/32 in. [±0.8 mm], and with permissible local variations of ±1/16 in. [±1.6 mm] from the nominal thickness.

6.2.2 *Width*—The nominal width shall be 24 or 48 in. [610 or 1220 mm], with a permissible variation of ±1/8 in. [±3 mm] from the specified width.

6.2.3 *Length*—The nominal length shall be from 6 to 12 ft [1830 to 3660 mm] with a permissible variation of ±1/4 in. [±6 mm] from the specified length.

6.2.4 *End Squareness*—Shall be square with a permissible variation of ±1/8 in. [±3 mm] in the full width of the board.

6.3 *Edges and Ends*—Shall be straight and solid. Edges shall be either square or V-tongue and groove.

7. Finish and Appearance

7.1 Gypsum sheathing shall be free of any cracks and imperfections that would render it unfit for its intended use.

8. Sampling, Inspection, Rejection, Certification, Packaging, Marking, Shipping, Handling, and Storage

8.1 Sampling, inspection, rejection, certification, packaging, marking, shipping, handling, and storage of gypsum sheathing shall be in accordance with Specification C 1264/C 1264M.

8.2 Treated core gypsum sheathing shall be marked legibly on the face or the back as treated core.

9. Keywords

9.1 gypsum; gypsum board; gypsum sheathing; sheathing; Type X; water resistant core

APPENDIX

(Nonmandatory Information)

X1. INTRODUCTION

This appendix gives general information and also suggestions for inclusions to be made elsewhere by the specifier. They are not part of this specification.

X2. ALTERNATE DEFINITION FOR TYPE X

X2.1 Gypsum sheathing, Type X (special fire-resistant) designates gypsum sheathing providing a greater fire resistance than regular gypsum sheathing of the same thickness. Type X (special fire-resistant) gypsum sheathing, when tested in accordance with Test Methods E 119, shall provide the following minimum fire resistance for the assemblies described below:

X2.1.1 One hour for a 5/8-in. [15.9-mm] thickness applied to a partition in a single layer application on each side of 3 5/8-in. [92-mm] deep non-loadbearing galvanized steel studs complying with Specification C 645, spaced 24 in. [610 mm] on center. The 5/8-in. [15.9-mm] thick gypsum sheathing 48-in. [1220-mm] wide shall be attached using 1-in. [25-mm] long drywall screws spaced 8 in. [203 mm] on center along the edges and ends, and 12 in. [305 mm] along intermediate studs. All joints shall be oriented parallel to and located over studs and staggered on opposite sides of the assembly.

X2.1.2 Two hours for a 1/2-in. [12.7-mm] thickness applied to a partition in a double layer application on each side of 2 1/2-in. [64-mm] deep non-loadbearing galvanized steel studs complying with Specification C 645, spaced 24 in. [610 mm] on center. The 48-in. [1220-mm] wide base layer shall be attached using 1-in. [25-mm] long drywall screws spaced 12 in. [305 mm] on center along board edges, ends, and along intermediate studs. Joints shall be oriented parallel to and located over studs and staggered on opposite sides of the assembly. The 48-in. [1220-mm] wide face layer shall be attached using 1 5/8-in. [41-mm] long drywall screws spaced 12 in. [305 mm] along board edges, ends, and along intermediate studs. Joints shall be oriented parallel to and located over studs, offset 24 in. [610 mm] from the base layer joints and staggered on opposite sides of the assembly.

Summary of Changes

Committee C-11 has identified the location of selected changes to this standard since the last issue (C 79) that may impact the use of this standard.

(*1*) The title was revised to include "treated core and nontreated core."

(*2*) SI units were placed in brackets.

(*3*) References to C 1264/C 1264M were added to 2.1 and 8.1.

(*4*) The new marking requirement was added to 8.2.

Standard Specification for Masonry Cement[1]

This standard is issued under the fixed designation C 91; the number immediately following the designation indicates the year of original adoption or, in the case of revision, the year of last revision. A number in parentheses indicates the year of last reapproval. A superscript epsilon (ε) indicates an editorial change since the last revision or reapproval.

This specification has been approved for use by agencies of the Department of Defense. Consult the DoD Index of Specifications and Standards for the specific year of issue which has been adopted by the Department of Defense.

1. Scope

1.1 This specification covers three types of masonry cement for use where mortar for masonry is required.

1.2 The values stated in SI units are to be regarded as the standard. Values in SI units shall be obtained by measurement in SI units or by appropriate conversion of measurements made in other units, using the Rules for Conversion and Rounding given in IEEE/ASTM SI 10.

1.3 The text of this standard refers to notes and footnotes which provide explanatory material. These notes and footnotes (excluding those in tables and figures) shall not be considered as requirements of the standard.

1.4 The following safety hazards caveat pertains only to Sections 19 and 20 of this specification. *This standard does not purport to address all of the safety concerns, if any, associated with its use. It is the responsibility of the user of this standard to establish appropriate safety and health practices and determine the applicability of regulatory limitations prior to use.*

2. Referenced Documents

2.1 *ASTM Standards:*

C 109/C 109M Test Method for Compressive Strength of Hydraulic-Cement Mortars (Using 2-in. or 50-mm Cube Specimens)[2]

C 114 Test Methods for Chemical Analysis of Hydraulic Cement[2]

C 151 Test Method for Autoclave Expansion of Portland Cement[2]

C 183 Practice for Sampling and the Amount of Testing of Hydraulic Cement[2]

C 185 Test Method for Air Content of Hydraulic Cement Mortar[2]

C 187 Test Method for Normal Consistency of Hydraulic Cement[2]

C 188 Test Method for Density of Hydraulic Cement[2]

C 219 Terminology Relating to Hydraulic Cement[2]

C 230 Specification for Flow Table for Use in Tests of Hydraulic Cement[2]

C 266 Test Method for Time of Setting of Hydraulic Cement Paste by Gillmore Needles[2]

C 270 Specification for Mortar for Unit Masonry[3]

C 305 Practice for Mechanical Mixing of Hydraulic Cement Pastes and Mortars of Plastic Consistency[2]

C 430 Test Method for Fineness of Hydraulic Cement by the 45-μm (No. 325) Sieve[2]

C 511 Specification for Moist Cabinets, Moist Rooms, and Water Storage Tanks Used in the Testing of Hydraulic Cements and Concretes[2]

C 778 Specification for Standard Sand[2]

IEEE/ASTM SI 10 Standard for Use of the International System of Units (SI): The Modern Metric System[4]

3. Terminology

3.1 *Definitions:*

3.1.1 *masonry cement*—a hydraulic cement, primarily used in masonry and plastering construction, consisting of a mixture of portland or blended hydraulic cement and plasticizing materials (such as limestone, hydrated or hydraulic lime) together with other materials introduced to enhance one or more properties such as setting time, workability, water retention, and durability.

3.1.2 Other terms used in this standard are defined in Terminology C 219.

4. Classification

4.1 *Type N*—For use in preparation of Specification C 270 Type N mortar without further addition of cements or hydrated lime, and for use in preparation of Specification C 270 Type S or Type M mortar when cement is added in accordance with the requirements of C 270.

4.2 *Type S*—For use in preparation of Specification C 270 Type S mortar without further addition of cements or hydrated lime.

4.3 *Type M*—For use in preparation of Specification C 270 Type M mortar without further addition of cements or hydrated lime.

5. Physical Properties

5.1 Masonry cement shall conform to the applicable requirements prescribed in Table 1 for its classification.

6. Staining

6.1 This requirement shall apply only when the purchaser requires that the cement shall be nonstaining to limestone. Nonstaining cement shall contain not more than 0.03 % of

[1] This specification is under the jurisdiction of ASTM Committee C-1 on Cement and is the direct responsibility of Subcommittee C01.11 on Masonry Cement.

Current edition approved Feb. 10, 1998. Published June 1998. Originally published as C 91 – 32 T. Last previous edition C 91 – 97.

[2] *Annual Book of ASTM Standards*, Vol 04.01.

[3] *Annual Book of ASTM Standards*, Vol 04.05.

[4] *Annual Book of ASTM Standards*, Vol 14.02.

TABLE 1 Physical Requirements

Masonry Cement Type	N	S	M
Fineness, residue on a 45-μm (No. 325) sieve, max, %	24	24	24
Autoclave expansion, max, %	1.0	1.0	1.0
Time of setting, Gillmore method:			
Initial set, minutes, not less than	120	90	90
Final set, minutes, not more than	1440	1440	1440
Compressive strength (average of 3 cubes): The compressive strength of mortar cubes, composed of 1 part cement and 3 parts blended sand (half graded standard sand, and half standard 20–30 sand) by volume, prepared and tested in accordance with this specification shall be equal to or higher than the values specified for the ages indicated below:			
7 days, MPa (psi)	3.4 (500)	9.0 (1300)	12.4 (1800)
28 days, MPa (psi)	6.2 (900)	14.5 (2100)	20.0 (2900)
Air content of mortar, prepared and tested in accordance with requirements of this specification:			
Min, volume %	8	8	8
Max, volume %	21	19	19
Water retention value, min, % of original flow	70	70	70

water-soluble alkali when determined in accordance with Section 14.

NOTE 1—The amount and nature of the staining material in limestones seems to vary with the stone. The alkali in any cement may, therefore, induce markedly different staining on different stone, even though the stone may have come apparently from the same source.

NOTE 2—Staining should not be confused with efflorescence. Nonstaining cements will not prevent efflorescence.

7. Sampling

7.1 The masonry cement shall be sampled in accordance with Practice C 183.

8. Temperature and Humidity

8.1 The temperature and relative humidity of the air in the vicinity of the mixing slab and dry materials, molds, base plates, and mixing bowl shall conform to the requirements of Test Method C 109/C 109M.

8.2 The moist cabinet or moist room shall conform to the requirements of Specification C 511.

9. Fineness

9.1 Determine the residue on the 45-μm (No. 325) sieve in accordance with Test Method C 430.

10. Normal Consistency

10.1 Determine normal consistency by the Vicat apparatus in accordance with Test Method C 187.

11. Autoclave Expansion

11.1 Determine autoclave expansion in accordance with Test Method C 151. After molding, store the bars in the moist cabinet or room for 48 h ± 30 min before removal from the molds for measurement and testing in the autoclave. Calculate the difference in length of the test specimen before and after autoclaving to the nearest 0.01 % of the effective gage length and report as the autoclave expansion of the masonry cement.

12. Time of Setting

12.1 Determine the time of setting by the Gillmore needle method in accordance with Test Method C 266.

13. Density

13.1 Determine the density of the masonry cement in accordance with Test Method C 188, using kerosine as the liquid. Use the density so determined in the calculation of the air content of the mortars.

14. Staining Test (Determination of Water-Soluble Alkali)

14.1 *Procedure*—Determine water-soluble alkali in accordance with Test Methods C 114.

14.2 *Calculation*—Calculate the percent of water-soluble alkali expressed as Na_2O to the nearest 0.01 % in accordance with Test Methods C 114.

15. Apparatus for Mortar Tests

15.1 *Weights and Weighing Devices, Seives, Glass Graduates, Specimen (Cube) Molds, Tamper, Trowel, and Testing Machine*, as described in Test Method C 109/C 109 M.

15.2 *Flow Table*, conforming to the requirements prescribed in Specification C 230.

15.3 *Mixing Apparatus*, conforming to the requirements prescribed in Practice C 305.

15.4 *Measure, Straightedge, Tamper, Tapping Stick, and Spoon*, conforming to the requirements given in Test Method C 185.

15.5 *Specimen (Cube) Molds*—Molds shall be prepared in accordance with Test Method C 109/C 109 M.

16. Blended Sand

16.1 The sand shall be a blend of equal parts by weight of graded standard sand and standard 20–30 sand conforming to Specification C 778.

17. Preparation of Mortar

17.1 *Proportions for Mortar*—Mortar for air entrainment, compressive strength, and water retention tests shall be proportioned to contain 1620 g of sand and a mass of cement, in grams, equal to 15 times the net mass in kilograms printed on the bag. The sand shall consist of 810 g of graded standard sand and 810 g of 20–30 standard sand (Note 3). The quantity of water, measured in millilitres shall be such as to produce a flow of 110 ± 5 as determined by Test Method C 109/C 109M using 25 drops of the flow table.

NOTE 3—For example, if the net mass printed on the bag is 32 kg, the mortar contains 480 g of cement. The specified mortar proportions are approximately the 1:3 nominal proportions by volume, commonly specified for construction, as indicated in the following calculation:

If a bag of masonry cement contained 28 L of cement, and 28 L of loose damp sand contained 36 kg of dry sand,

$$A = 1620 \times (C/B) \quad (1)$$

where:

A = number of grams of cement to be used in the mortar with 1620 g of sand,
B = 3 × 36 = 108 kg, the mass of dry sand in 84 (or 3 × 28) L of loose damp sand, and
C = mass per bag (kg).

17.2 *Mixing of Mortars*—Mix the mortar in accordance with Practice C 305.

18. Air Entrainment

18.1 *Procedure*—If the mortar has the correct flow, use a separate portion of the mortar for the determination of entrained air. Determine the mass of 400 mL of mortar in accordance with Test Method C 185.

18.2 *Calculation*—Calculate the air content of the mortar and report it to the nearest 1 % as follows:

$$D = (W_1 + W_2 + V_w)/[(W_1/S_1) + (W_2/S_2) + V_w] \quad (2)$$
$$A = 100 - (W_m/4D)$$

where:

D = density of air-free mortar, g/cm^3,
W_1 = mass of cement, g,
W_2 = mass of sand, g,
V_w = millilitres-grams of water used,
S_1 = density of cement, g/cm^3
S_2 = density of standard sand, 2.65 g/cm^3,
A = volume percent of entrained air, and
W_m = mass of 400 mL of mortar, g.

19. Compressive Strength

19.1 *Test Specimens:*

19.1.1 *Molding*—Immediately after determining the flow and mass of 400 mL of mortar, return all of the mortar to the mixing bowl and remix for 15 s at the medium speed. Then mold the test specimens in accordance with Test Method C 109/C 109 M, except that the elapsed time for mixing mortar, determining flow, determining air entrainment, and starting the molding of cubes shall be within 8 min.

19.1.2 *Storage*—Immediately after molding, store all test specimens in the molds on plane plates in a moist cabinet or moist room for 48 to 52 h in such a manner that the upper surfaces shall be exposed to the moist air. Then remove the cubes from the molds, and place them in the moist cabinet or moist room for 5 days in such a manner as to allow free circulation of air around at least five faces of the specimens. At the age of 7 days, immerse the cubes for the 28-day tests in saturated lime water in storage tanks of noncorrodible materials.

19.2 *Procedure:*

19.2.1 Test the cube specimens immediately after their removal from the moist cabinet or moist room for 7-day specimens, and immediately after their removal from storage water for all other specimens. If more than one specimen at a time is removed from the moist cabinet or moist room for 7-day tests, cover these cubes with a damp cloth until the time of testing.

19.2.2 The remainder of the testing procedure shall conform to Test Method C 109/C 109 M.

20. Water Retention

20.1 *Apparatus:*

20.1.1 *Water-Retention Apparatus*—For the water-retention test, an apparatus essentially the same as that shown in Fig. 1 shall be used. This apparatus consists of a water aspirator or other source of vacuum controlled by a mercury relief column and connected by way of a three-way stopcock to a funnel upon which rests a perforated dish. The perforated dish shall be made of metal not attacked by masonry mortar. The metal in the base of the dish shall have a thickness of 1.7 to 1.9 mm and shall conform to the requirements given in Fig. 1. The stopcock bore shall have a 4.0 ± 0.5 mm in diameter, and the connecting glass tubing shall have a minimum inside diameter of 4 mm. A mercury manometer, connected as shown in Fig. 1, indicates the vacuum. The contact surfaces of the funnel and perforated dish shall be plane and are permitted to be lapped to ensure intimate contact. An air-tight seal shall be maintained between the funnel and the dish during a test. This shall be accomplished by either of the following procedures: (*1*) a synthetic (grease-resistant) rubber gasket shall be permanently sealed to the top of the funnel using petrolatum or light grease to ensure a seal between the gasket and dish, or (*2*) the top of the funnel shall be lightly coated with petrolatum or light grease to ensure a seal between the funnel and dish. Care shall be taken to ensure that none of the holes in the perforated dish are clogged. Hardened, very smooth, not rapid filter paper shall be used. It shall be 150 mm in diameter and be placed so as to completely cover the perforations in the dish.

20.1.2 *Straightedge*—A steel straightedge not less than 200 mm (8 in.) long and not less than 2 mm (1/16 in.) nor more than 3 mm (1/8 in.) in thickness.

20.1.3 *Other Apparatus*—Other apparatus required for the water retention test shall conform to the applicable requirements of Section 15.

20.2 *Procedure:*

20.2.1 Adjust the mercury relief column to maintain a vacuum of 51 ± 3 mm as indicated by the manometer. Seat the perforated dish on the greased gasket or greased rim of the funnel. Place a wetted filter paper in the bottom of the dish. Turn the stopcock to apply the vacuum to the funnel and check the apparatus for leaks and to determine that the required vacuum is obtained. Then turn the stopcock to shut off the vacuum from the funnel.

20.2.2 Mix the mortar to a flow of 110 ± 5 % in accordance with Practice C 305. Immediately after making the flow test, return the mortar on the flow table to the mixing bowl and remix the entire batch for 15 s at medium speed. Immediately after remixing the mortar, fill the perforated dish with the mortar to slightly above the rim. Tamp the mortar 15 times with the tamper. Apply ten of the tamping strokes at approximately uniform spacing adjacent to the rim of the dish and with the long axis of the tamping face held at right angles to the radius of the dish. Apply the remaining five tamping strokes at random points distributed

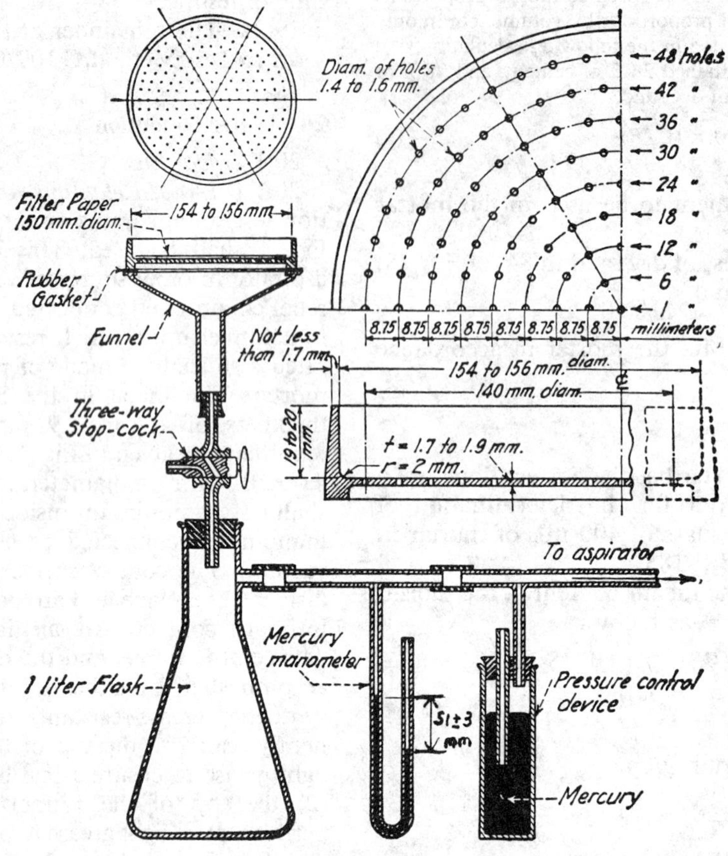

NOTE—The gasket is to be synthetic rubber. The stopcock and the bore of the tubing should measure at least 4 mm. A check valve or water trap, or both, is suggested for the connection to the aspirator.

FIG. 1 Apparatus Assembly for the Water-Retention Test

over the central area of the dish. The tamping pressure shall be just sufficient to ensure filling of the dish. On completion of the tamping, the top of the mortar should extend slightly above the rim of the dish. Smooth off the mortar by drawing the flat side of the straightedge (with the leading edge slightly raised) across the top of the dish. Then cut off the mortar to a plane surface flush with the rim of the dish by drawing the straightedge with a sawing motion across the top of the dish in two cutting strokes, starting each cut from near the center of the dish. If the mortar is pulled away from the side of the dish during the process of drawing the straightedge across the dish, gently press the mortar back into contact with the side of the dish using the tamper.

20.2.3 Turn the stopcock to apply the vacuum to the funnel. The time elapsed from the start of mixing the cement and water to the time of applying the vacuum shall not exceed 8 min. After suction for 60 s, quickly turn the stopcock to expose the funnel to atmospheric pressure. Immediately slide the perforated dish off from the funnel, touch it momentarily on a damp cloth to remove droplets of water, and set the dish on the table. Then, using the bowl scraper (rubber scraper as used in Practice C 305), plow and mix the mortar in the dish for 15 s. Upon completion of mixing, place the mortar in the flow mold and determine the flow. The entire operation shall be carried out without interruption and as quickly as possible, and shall be completed within an elapsed time of 11 min after the start of mixing the cement and water for the first flow determination. Both flow determinations shall be made in accordance with Test Method C 109/C 109M.

20.3 *Calculation*—Calculate the water retention value for the mortar as follows:

$$\text{Water retention value} = (A/B) \times 100 \quad (3)$$

where:

A = flow after suction, and
B = flow immediately after mixing.

21. Storage

21.1 The cement shall be stored in such a manner as to permit easy access for the proper inspection and identification of each shipment, and in a suitable weathertight building that will protect the cement from dampness and minimize warehouse set.

22. Inspection

22.1 Every facility shall be provided to the purchaser for the necessary inspection and sampling.

22.2 All packages shall be in good condition at the time of inspection.

23. Rejection

23.1 At the option of the purchaser, the cement shall be rejected if it fails to meet any of the requirements of this specification.

23.2 At the option of the purchaser, packages more than 2 % below the mass marked thereon shall be rejected. At the option of the purchaser, the entire shipment represented shall be rejected if the average mass of packages in any shipment as shown by weighing fifty packages taken at random is less than that marked on the packages.

23.3 At the option of the purchaser, cement remaining in storage prior to shipment for a period greater than six months after testing shall be retested and, at the option of the purchaser shall be rejected if it fails to meet any of the requirements of this specification.

24. Manufacturer's Certification

24.1 Upon request of the purchaser in the contract or order, a manufacturer's report shall be furnished at the time of shipment stating the results of the tests made on samples of the material taken during production or transfer and certifying that the applicable requirements of this specification have been met.

25. Packaging and Package Marking

25.1 When masonry cement is delivered in packages, the brand, name of the manufacturer, type of masonry cement, and net mass of the package in kilograms (see Note 4) shall be indicated plainly thereon. Similar information shall be provided in the shipping documents accompanying the shipment of masonry cement in bulk.

NOTE 4—To facilitate the change to SI units, a standard SI package size of 32 kg for Type N, 34 kg for Type S, and 36 kg for Type M will provide convenient mass increments reasonably similar to the traditional 70-, 75-, and 80-lb packages.

26 Keywords

26.1 masonry; masonry cement; mortar

Standard Test Method for Compressive Strength of Hydraulic Cement Mortars (Using 2-in. or [50-mm] Cube Specimens)[1]

This standard is issued under the fixed designation C 109/C 109M; the number immediately following the designation indicates the year of original adoption or, in the case of revision, the year of last revision. A number in parentheses indicates the year of last reapproval. A superscript epsilon (ϵ) indicates an editorial change since the last revision or reapproval.

This standard has been approved for use by agencies of the Department of Defense.

1. Scope

1.1 This test method covers determination of the compressive strength of hydraulic cement mortars, using 2-in. or [50-mm] cube specimens.

NOTE 1—Test Method C 349 provides an alternative procedure for this determination (not to be used for acceptance tests).

1.2 This test method covers the application of the test using either inch-pound or SI units. The values stated in either system shall be regarded separately as standard. Within the text, the SI units are shown in brackets. The values stated in each system are not exact equivalents; therefore, each system shall be used independently of the other. Combining values from the two systems may result in nonconformance with the specification.

1.3 Values in SI units shall be obtained by measurement in SI units or by appropriate conversion, using the Rules for Conversion and Rounding given in Standard IEEE/ASTM SI 10, of measurements made in other units.

1.4 *This standard does not purport to address all of the safety concerns, if any, associated with its use. It is the responsibility of the user of this standard to establish appropriate safety and health practices and determine the applicability of regulatory limitations prior to use.*

2. Referenced Documents

2.1 *ASTM Standards:*

C 230 Specification for Flow Table for Use in Tests of Hydraulic Cement[2]

C 305 Practice for Mechanical Mixing of Hydraulic Cement Pastes and Mortars of Plastic Consistency[2]

C 349 Test Method for Compressive Strength of Hydraulic Cement Mortars (Using Portions of Prisms Broken in Flexure)[2]

C 511 Specification for Moist Cabinets, Moist Rooms and Water Storage Tanks Used in the Testing of Hydraulic Cements and Concretes[2]

C 670 Practice for Preparing Precision and Bias Statements for Test Methods for Construction Materials[3]

C 778 Specification for Standard Sand[2]

C 1005 Specification for Weights and Weighing Devices for Use in Physical Testing of Hydraulic Cements[2]

IEEE/ASTM SI 10 Standard for Use of the International System of Units (SI): The Modern Metric System[4]

3. Summary of Test Method

3.1 The mortar used consists of 1 part cement and 2.75 parts of sand proportioned by mass. Portland or air-entraining portland cements are mixed at specified water/cement ratios. Water content for other cements is that sufficient to obtain a flow of 110 ± 5 in 25 drops of the flow table. Two-inch or [50-mm] test cubes are compacted by tamping in two layers. The cubes are cured one day in the molds and stripped and immersed in lime water until tested.

4. Significance and Use

4.1 This test method provides a means of determining the compressive strength of hydraulic cement and other mortars and results may be used to determine compliance with specifications. Further, this test method is referenced by numerous other specifications and test methods. Caution must be exercised in using the results of this test method to predict the strength of concretes.

5. Apparatus

5.1 *Weights and Weighing Devices*, shall conform to the requirements of Specification C 1005. The weighing device shall be evaluated for precision and bias at a total load of 2000 g.

5.2 *Glass Graduates*, of suitable capacities (preferably large enough to measure the mixing water in a single operation) to deliver the indicated volume at 20°C. The permissible variation shall be ±2 mL. These graduates shall be subdivided to at least 5 mL, except that the graduation lines may be omitted for the lowest 10 mL for a 250-mL graduate and for the lowest 25 mL of a 500-mL graduate. The main graduation lines shall be circles and shall be numbered. The least graduations shall extend at least one seventh of the way around, and intermediate

[1] This test method is under the jurisdiction of ASTM Committee C-1 on Cement and is the direct responsibility of Subcommittee C01.27 on Strength.

Current edition approved Jan. 10, 1999. Published May 1999. Originally published as C 109 – 34 T. Last previous edition C 109 – 98.

[2] *Annual Book of ASTM Standards*, Vol 04.01.

[3] *Annual Book of ASTM Standards*, Vol 04.02.

[4] *Annual Book of ASTM Standards*, Vol 14.02.

graduations shall extend at least one fifth of the way around.

5.3 *Specimen Molds*, for the 2-in. or [50-mm] cube specimens shall be tight fitting. The molds shall have not more than three cube compartments and shall be separable into not more than two parts. The parts of the molds when assembled shall be positively held together. The molds shall be made of hard metal not attacked by the cement mortar. For new molds the Rockwell hardness number of the metal shall be not less than 55 HRB. The sides of the molds shall be sufficiently rigid to prevent spreading or warping. The interior faces of the molds shall be plane surfaces and shall conform to the tolerances of Table 1.

5.4 *Mixer, Bowl and Paddle*, an electrically driven mechanical mixer of the type equipped with paddle and mixing bowl, as specified in Practice C 305.

5.5 *Flow Table and Flow Mold*, conforming to the requirements of Specification C 230.

5.6 *Tamper*, a nonabsorptive, nonabrasive, nonbrittle material such as a rubber compound having a Shore A durometer hardness of 80 ± 10 or seasoned oak wood rendered nonabsorptive by immersion for 15 min in paraffin at approximately 392°F or [200°C], shall have a cross section of about ½by 1 in. or [13 by 25 mm] and a convenient length of about 5 to 6 in. or [120 to 150 mm]. The tamping face shall be flat and at right angles to the length of the tamper.

5.7 *Trowel*, having a steel blade 4 to 6 in. [100 to 150 mm] in length, with straight edges.

5.8 *Moist Cabinet or Room,* conforming to the requirements of Specification C 511.

5.9 *Testing Machine*, either the hydraulic or the screw type, with sufficient opening between the upper bearing surface and the lower bearing surface of the machine to permit the use of verifying apparatus. The load applied to the test specimen shall be indicated with an accuracy of ±1.0 %. If the load applied by the compression machine is registered on a dial, the dial shall be provided with a graduated scale that can be read to at least the nearest 0.1 % of the full scale load (Note 2). The dial shall be readable within 1 % of the indicated load at any given load level within the loading range. In no case shall the loading range of a dial be considered to include loads below the value that is 100 times the smallest change of load that can be read on the scale. The scale shall be provided with a graduation line equal to zero and so numbered. The dial pointer shall be of sufficient length to reach the graduation marks; the width of the end of the pointer shall not exceed the clear distance between the smallest graduations. Each dial shall be equipped with a zero adjustment that is easily accessible from the outside of the dial case, and with a suitable device that at all times until reset, will indicate to within 1 % accuracy the maximum load applied to the specimen.

5.9.1 If the testing machine load is indicated in digital form, the numerical display must be large enough to be easily read. The numerical increment must be equal to or less than 0.10 % of the full scale load of a given loading range. In no case shall the verified loading range include loads less than the minimum numerical increment multiplied by 100. The accuracy of the indicated load must be within 1.0 % for any value displayed within the verified loading range. Provision must be made for adjusting to indicate true zero at zero load. There shall be provided a maximum load indicator that at all times until reset will indicate within 1 % system accuracy the maximum load applied to the specimen.

NOTE 2—As close as can be read is considered 1⁄50 in. or [0.5 mm] along the arc described by the end of the pointer. Also, one half of the scale interval is about as close as can reasonably be read when the spacing on the load indicating mechanism is between 1⁄25 in. or [1 mm] and 1⁄16 in. or [1.6 mm]. When the spacing is between 1⁄16 in. or [1.6 mm] and ⅛ in. or [3.2 mm], one third of the scale interval can be read with reasonable certainty. When the spacing is ⅛ in. or [3.2 mm] or more, one fourth of the scale interval can be read with reasonable certainty.

5.9.2 The upper bearing shall be a spherically seated, hardened metal block firmly attached at the center of the upper head of the machine. The center of the sphere shall lie at the center of the surface of the block in contact with the specimen. The block shall be closely held in its spherical seat, but shall be free to tilt in any direction. The diagonal or diameter (Note 3) of the bearing surface shall be only slightly greater than the diagonal of the face of the 2-in. or [50-mm] cube in order to facilitate accurate centering of the specimen. A hardened metal bearing block shall be used beneath the specimen to minimize wear of the lower platen of the machine. The bearing block surfaces intended for contact with the specimen shall have a Rockwell hardness number not less than 60 HRC. These surfaces shall not depart from plane surfaces by more than 0.0005 in. or [0.013 mm] when the blocks are new and shall be maintained within a permissible variation of 0.001 in. or [0.025 mm].

NOTE 3—A diameter of 3⅛ in. or [79.4 mm], is satisfactory, provided that the lower bearing block has a diameter slightly greater than the diagonal of the face of the 2-in. or [50-mm] cube but not more than 2.9 in. or [74 mm], and is centered with respect to the upper bearing block and held in position by suitable means.

6. Materials

6.1 *Graded Standard Sand*:

6.1.1 The sand (Note 4) used for making test specimens

TABLE 1 Permissible Variations of Specimen Molds

Parameter	2-in. Cube Molds		[50-mm] Cube Molds	
	New	In Use	New	In Use
Planeness of sides	<0.001 in.	<0.002 in.	[<0.025 mm]	[<0.05 mm]
Distance between opposite sides	2 in. ± 0.005	2 in. ± 0.02	[50 mm ± 0.13 mm]	[50 mm ± 0.50 mm]
Height of each compartment	2 in. + 0.01 in. to − 0.005 in.	2 in. + 0.01 in. to − 0.015 in.	[50 mm + 0.25 mm to − 0.13 mm]	[50 mm + 0.25 mm to − 0.38 mm]
Angle between adjacent faces[A]	90 ± 0.5°	90 ± 0.5°	90 ± 0.5°	90 ± 0.5°

[A] Measured at points slightly removed from the intersection. Measured separately for each compartment between all the interior faces and the adjacent face and between interior faces and top and bottom planes of the mold.

shall be natural silica sand conforming to the requirements for graded standard sand in Specification C 778.

NOTE 4—*Segregation of Graded Sand*—The graded standard sand should be handled in such a manner as to prevent segregation, since variations in the grading of the sand cause variations in the consistency of the mortar. In emptying bins or sacks, care should be exercised to prevent the formation of mounds of sand or craters in the sand, down the slopes of which the coarser particles will roll. Bins should be of sufficient size to permit these precautions. Devices for drawing the sand from bins by gravity should not be used.

7. Temperature and Humidity

7.1 *Temperature*—The temperature of the air in the vicinity of the mixing slab, the dry materials, molds, base plates, and mixing bowl, shall be maintained between 68 and 81.5°F or [20 and 27.5°C]. The temperature of the mixing water, moist closet or moist room, and water in the storage tank shall be set at 73.5 ± 3.5°F or [23 ± 2°C] and shall not vary from this temperature by more than± 3°F or [±1.7°C].

7.2 *Humidity*—The relative humidity of the laboratory shall be not less than 50 %. The moist closet or moist room shall conform to the requirements of Specification C 511.

8. Test Specimens

8.1 Make two or three specimens from a batch of mortar for each period of test or test age.

9. Preparation of Specimen Molds

9.1 Apply a thin coating of release agent to the interior faces of the mold and non-absorptive base plates. Apply oils and greases using an impregnated cloth or other suitable means. Wipe the mold faces and the base plate with a cloth as necessary to remove any excess release agent and to achieve a thin, even coating on the interior surfaces. When using an aerosol lubricant, spray the release agent directly onto the mold faces and base plate from a distance of 6 to 8 in. or [150 to 200 mm] to achieve complete coverage. After spraying, wipe the surface with a cloth as necessary to remove any excess aerosol lubricant. The residue coating should be just sufficient to allow a distinct finger print to remain following light finger pressure (Note 5).

9.2 Seal the surfaces where the halves of the mold join by applying a coating of light cup grease such as petrolatum. The amount should be sufficient to extrude slightly when the two halves are tightened together. Remove any excess grease with a cloth.

9.3 After placing the mold on its base plate (and attaching, if clamp-type) carefully remove with a dry cloth any excess oil or grease from the surface of the mold and the base plate to which watertight sealant is to be applied. As a sealant, use paraffin, microcrystalline wax, or a mixture of three parts paraffin to five parts rosin by mass. Liquify the sealant by heating between 230 and 248°F or [110 and 120°C]. Effect a watertight seal by applying the liquefied sealant at the outside contact lines between the mold and its base plate.

NOTE 5—Because aerosol lubricants evaporate, molds should be checked for a sufficient coating of lubricant immediately prior to use. If an extended period of time has elapsed since treatment, retreatment may be necessary.

NOTE 6—*Watertight Molds*—The mixture of paraffin and rosin specified for sealing the joints between molds and base plates may be found difficult to remove when molds are being cleaned. Use of straight paraffin is permissible if a watertight joint is secured, but due to the low strength of paraffin it should be used only when the mold is not held to the base plate by the paraffin alone. A watertight joint may be secured with paraffin alone by slightly warming the mold and base plate before brushing the joint. Molds so treated should be allowed to return to the specified temperature before use.

10. Procedure

10.1 *Composition of Mortars*:

10.1.1 The proportions of materials for the standard mortar shall be one part of cement to 2.75 parts of graded standard sand by weight. Use a water-cement ratio of 0.485 for all portland cements and 0.460 for all air-entraining portland cements. The amount of mixing water for other than portland and air-entraining portland cements shall be such as to produce a flow of 110 ± 5 as determined in accordance with 10.3 and shall be expressed as weight percent of cement.

10.1.2 The quantities of materials to be mixed at one time in the batch of mortar for making six and nine test specimens shall be as follows:

	Number of Specimens	
	6	9
Cement, g	500	740
Sand, g	1375	2035
Water, mL		
Portland (0.485)	242	359
Air-entraining portland (0.460)	230	340
Other (to flow of 110 ± 5)	...	...

10.2 *Preparation of Mortar*:

10.2.1 Mechanically mix in accordance with the procedure given in Practice C 305.

10.3 *Determination of Flow*:

10.3.1 Carefully wipe the flow-table top clean and dry, and place the flow mold at the center. Place a layer of mortar about 1 in. or [25 mm] in thickness in the mold and tamp 20 times with the tamper. The tamping pressure shall be just sufficient to ensure uniform filling of the mold. Then fill the mold with mortar and tamp as specified for the first layer. Cut off the mortar to a plane surface, flush with the top of the mold, by drawing the straight edge of a trowel (held nearly perpendicular to the mold) with a sawing motion across the top of the mold. Wipe the table top clean and dry, being especially careful to remove any water from around the edge of the flow mold. Lift the mold away from the mortar 1 min after completing the mixing operation. Immediately, drop the table through a height of ½ in. or [13 mm] 25 times in 15 s. Using the calipers, determine the flow by measuring the diameters of the mortar along the lines scribed in the table top, adding the four readings. The total of the four readings from the calipers equals the percent increase of the original diameter of the mortar.

10.3.2 For portland and air-entraining portland cements, merely record the flow.

10.3.3 In the case of cements other than portland or air-entraining portland cements, make trial mortars with varying percentages of water until the specified flow is obtained. Make each trial with fresh mortar.

10.4 *Molding Test Specimens*:

10.4.1 Immediately following completion of the flow test,

return the mortar from the flow table to the mixing bowl. Quickly scrape the bowl sides and transfer into the batch the mortar that may have collected on the side of the bowl and then remix the entire batch 15 s at medium speed. Upon completion of mixing, the mixing paddle shall be shaken to remove excess mortar into the mixing bowl.

10.4.2 When a duplicate batch is to be made immediately for additional specimens, the flow test may be omitted and the mortar allowed to stand in the mixing bowl 90 s without covering. During the last 15 s of this interval, quickly scrape the bowl sides and transfer into the batch the mortar that may have collected on the side of the bowl. Then remix for 15 s at medium speed.

10.4.3 Start molding the specimens within a total elapsed time of not more than 2 min and 30 s after completion of the original mixing of the mortar batch. Place a layer of mortar about 1 in. or [25 mm] (approximately one half of the depth of the mold) in all of the cube compartments. Tamp the mortar in each cube compartment 32 times in about 10 s in 4 rounds, each round to be at right angles to the other and consisting of eight adjoining strokes over the surface of the specimen, as illustrated in Fig. 1. The tamping pressure shall be just sufficient to ensure uniform filling of the molds. The 4 rounds of tamping (32 strokes) of the mortar shall be completed in one cube before going to the next. When the tamping of the first layer in all of the cube compartments is completed, fill the compartments with the remaining mortar and then tamp as specified for the first layer. During tamping of the second layer bring in the mortar forced out onto the tops of the molds after each round of tamping by means of the gloved fingers and the tamper upon completion of each round and before starting the next round of tamping. On completion of the tamping, the tops of all cubes should extend slightly above the tops of the molds. Bring in the mortar that has been forced out onto the tops of the molds with a trowel and smooth off the cubes by drawing the flat side of the trowel (with the leading edge slightly raised) once across the top of each cube at right angles to the length of the mold. Then, for the purpose of leveling the mortar and making the mortar that protrudes above the top of the mold of more uniform thickness, draw the flat side of the trowel (with the leading edge slightly raised) lightly once along the length of the mold. Cut off the mortar to a plane surface flush with the top of the mold by drawing the straight edge of the trowel (held nearly perpendicular to the mold) with a sawing motion over the length of the mold.

10.5 *Storage of Test Specimens*—Immediately upon completion of molding, place the test specimens in the moist closet or moist room. Keep all test specimens, immediately after molding, in the molds on the base plates in the moist closet or moist room from 20 to 72 h with their upper surfaces exposed to the moist air but protected from dripping water. If the specimens are removed from the molds before 24 h, keep them on the shelves of the moist closet or moist room until they are 24-h old, and then immerse the specimens, except those for the 24-h test, in saturated lime water in storage tanks constructed of noncorroding materials. Keep the storage water clean by changing as required.

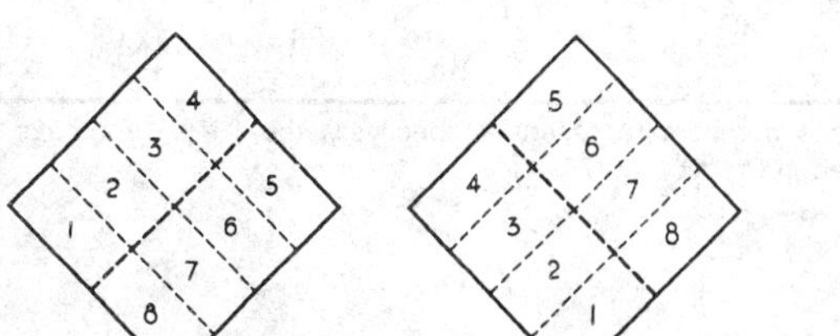

FIG. 1 Order of Tamping in Molding of Test Specimens

10.6 *Determination of Compressive Strength*:

10.6.1 Test the specimens immediately after their removal from the moist closet in the case of 24-h specimens, and from storage water in the case of all other specimens. All test specimens for a given test age shall be broken within the permissible tolerance prescribed as follows:

Test Age	Permissible Tolerance
24 h	±½ h
3 days	±1 h
7 days	±3 h
28 days	±12 h

If more than one specimen at a time is removed from the moist closet for the 24-h tests, keep these specimens covered with a damp cloth until time of testing. If more than one specimen at a time is removed from the storage water for testing, keep these specimens in water at a temperature of 73.5± 3.5°F or [23 ± 2°C] and of sufficient depth to completely immerse each specimen until time of testing.

10.6.2 Wipe each specimen to a surface-dry condition, and remove any loose sand grains or incrustations from the faces that will be in contact with the bearing blocks of the testing machine. Check these faces by applying a straightedge (Note 7). If there is appreciable curvature, grind the face or faces to plane surfaces or discard the specimen. A periodic check of the cross-sectional area of the specimens should be made.

NOTE 7—*Specimen Faces*—Results much lower than the true strength will be obtained by loading faces of the cube specimen that are not truly plane surfaces. Therefore, it is essential that specimen molds be kept scrupulously clean, as otherwise, large irregularities in the surfaces will occur. Instruments for cleaning molds should always be softer than the metal in the molds to prevent wear. In case grinding specimen faces is necessary, it can be accomplished best by rubbing the specimen on a sheet of fine emery paper or cloth glued to a plane surface, using only a moderate pressure. Such grinding is tedious for more than a few thousandths of an inch (hundredths of a millimetre); where more than this is found necessary, it is recommended that the specimen be discarded.

10.6.3 Apply the load to specimen faces that were in contact with the true plane surfaces of the mold. Carefully place the specimen in the testing machine below the center of the upper bearing block. Prior to the testing of each cube, it shall be ascertained that the spherically seated block is free to tilt. Use no cushioning or bedding materials. Bring the spherically seated block into uniform contact with the surface of the specimen. Apply the load rate at a relative rate of movement between the upper and lower platens corresponding to a loading on the specimen with the range of 200 to 400 lbs/s [900 to 1800 N/S]. Obtain this designated rate of movement of the platen during the first half of the anticipated maximum load and make no adjustment in the rate of movement of the platen in the latter half of the loading especially while the cube is yielding before failure.

NOTE 8—It is advisable to apply only a very light coating of a good quality, light mineral oil to the spherical seat of the upper platen.

11. Calculation

11.1 Record the total maximum load indicated by the testing machine, and calculate the compressive strength as follows:

$$fm = P/A \quad (1)$$

where:

fm = compressive strength in psi or [MPa],
P = total maximum load in lbf or [N], and
A = area of loaded surface in^2 or [mm^2].

Either 2-in. or [50-mm] cube specimens may be used for the determination of compressive strength, whether inch-pound or SI units are used. However, consistent units for load and area must be used to calculate strength in the units selected. If the cross-sectional area of a specimen varies more than 1.5 % from the nominal, use the actual area for the calculation of the compressive strength. The compressive strength of all acceptable test specimens (see Section 12) made from the same sample and tested at the same period shall be averaged and reported to the nearest 10 psi [0.1 MPa].

12. Report

12.1 Report the flow to the nearest 1 % and the water used to the nearest 0.1 %. Average compressive strength of all specimens from the same sample shall be reported to the nearest 10 psi [0.1 MPa].

13. Faulty Specimens and Retests

13.1 In determining the compressive strength, do not consider specimens that are manifestly faulty.

13.2 The maximum permissible range between specimens from the same mortar batch, at the same test age is 8.7 % of the average when three cubes represent a test age and 7.6 % when two cubes represent a test age (Note 9).

NOTE 9—The probability of exceeding these ranges is 1 in 100 when the within-batch coefficient of variation is 2.1 %. The 2.1 %is an average for laboratories participating in the portland cement and masonry cement reference sample programs of the Cement and Concrete Reference Laboratory.

13.3 If the range of three specimens exceeds the maximum in 13.2, discard the result which differs most from the average and check the range of the remaining two specimens. Make a retest of the sample if less than two specimens remain after disgarding faulty specimens or disgarding tests that fail to comply with the maximum permissible range of two specimens.

NOTE 10—Reliable strength results depend upon careful observance of all of the specified requirements and procedures. Erratic results at a given test period indicate that some of the requirements and procedures have not been carefully observed; for example, those covering the testing of the specimens as prescribed in 10.6.2 and 10.6.3. Improper centering of specimens resulting in oblique fractures or lateral movement of one of the heads of the testing machine during loading will cause lower strength results.

14. Precision and Bias

14.1 *Precision*—The precision statements for this test method are listed in Table 2 and are based on results from the Cement and Concrete Reference Laboratory Reference Sample Program. They are developed from data where a test result is the average of compressive strength tests of three cubes molded from a single batch of mortar and tested at the same age. A significant change in precision will not be noted when a test result is the average of two cubes rather than three.

14.2 These precision statements are applicable to mortars made with cements mixed, and tested at the ages as noted. The appropriate limits are likely, somewhat larger for tests at younger ages and slightly smaller for tests at older ages.

14.3 *Bias*—The procedure in this test method has no bias because the value of compressive strength is defined in terms of the test method.

15. Keywords

15.1 compressive strength; hydraulic cement mortar; hydraulic cement strength; mortar strength; strength

TABLE 2 Precision

		Test Age, Days	Coefficient of Variation 1s %[A]	Acceptable Range of Test Results d2s %[A]
Portland Cements				
Constant water-cement ratio:				
Single-lab		3	4.0	11.3
		7	3.6	10.2
	Av		3.8	10.7
Multi-lab		3	6.8	19.2
		7	6.4	18.1
	Av		6.6	18.7
Blended Cements				
Constant flow mortar:				
Single-lab		3	4.0	11.3
		7	3.8	10.7
		28	3.4	9.6
	Av		3.8	10.7
Multi-lab		3	7.8	22.1
		7	7.6	21.5
		28	7.4	20.9
	Av		7.6	21.5
Masonry Cements				
Constant flow mortar:				
Single-lab		7	7.9	22.3
		28	7.5	21.2
	Av		7.7	21.8
Multi-lab		7	11.8	33.4
		28	12.0	33.9
	Av		11.9	33.7

[A] These numbers represent, respectively, the (1s %) and (d2s %) limits as described in Practice C 670.

The American Society for Testing and Materials takes no position respecting the validity of any patent rights asserted in connection with any item mentioned in this standard. Users of this standard are expressly advised that determination of the validity of any such patent rights, and the risk of infringement of such rights, are entirely their own responsibility.

This standard is subject to revision at any time by the responsible technical committee and must be reviewed every five years and if not revised, either reapproved or withdrawn. Your comments are invited either for revision of this standard or for additional standards and should be addressed to ASTM Headquarters. Your comments will receive careful consideration at a meeting of the responsible technical committee, which you may attend. If you feel that your comments have not received a fair hearing you should make your views known to the ASTM Committee on Standards, 100 Barr Harbor Drive, West Conshohocken, PA 19428.

Standard Test Methods for Physical Testing of Quicklime, Hydrated Lime, and Limestone[1]

This standard is issued under the fixed designation C 110; the number immediately following the designation indicates the year of original adoption or, in the case of revision, the year of last revision. A number in parentheses indicates the year of last reapproval. A superscript epsilon (ϵ) indicates an editorial change since the last revision or reapproval.

This standard has been approved for use by agencies of the Department of Defense.

1. Scope

1.1 These test methods cover physical testing of quicklime and hydrated lime, and of limestone not otherwise covered in ASTM standards.[2]

NOTE 1—Quicklime and hydrated lime have a high affinity for moisture and carbon dioxide. Caution should be taken to protect both hydrated and quicklime during sampling, storage, and testing (see Practice C 50).

1.2 The test procedures appear in the following order:

	Section
Air Entrainment	13
Apparent Loose Density of Hydrated Lime, Pulverized Quicklime, and Limestone	17
Apparent Packed Density of Hydrated Lime, Pulverized Quicklime, and Limestone	18
Autoclave Expansion of Hydrated Lime	8
Dry Brightness of Pulverized Limestone	16
Dry Screening by Air Jet Sieve	20
Fineness of Pulverized Quicklime and Hydrated Lime by Air Permeability	19
Limestone Grindability	21
Particle Size of Pulverized Limestone	15
pH of Pulverized Limestone	14
Plasticity of Lime Putty	7
Popping and Pitting of Hydrated Lime	9
Residue and Sieve Analysis	5
Settling Rate of Hydrated Lime	11
Slaking Rate of Quicklime	12
Specific Gravity of Hydrated Lime Products	22
Standard Consistency of Lime Putty	6
Water Retention of Hydrated Lime	10
Wet Sieve Analysis of Agricultural Liming Materials	23

1.3 The values stated in SI units are to be regarded as the standard.

1.4 *This standard does not purport to address all of the safety concerns, if any, associated with its use. It is the responsibility of the user of this standard to establish appropriate safety and health practices and determine the applicability of regulatory limitations prior to use.*

2. Referenced Documents

2.1 *ASTM Standards:*

C 25 Test Methods for Chemical Analysis of Limestone, Quicklime, and Hydrated Lime[3]
C 28 Specification for Gypsum Plasters[3]
C 50 Practice for Sampling, Inspection, Packing, and Marking of Lime and Limestone Products[3]
C 51 Terminology Relating to Lime and Limestone (as used by the Industry)[3]
C 91 Specification for Masonry Cement[3]
C 109 Test Method for Compressive Strength of Hydraulic Cement Mortars (Using 2-in. or 50-mm Cube Specimens)[3]
C 117 Test Method for Material Finer than 75–Micrometre (No. 200) Sieve in Mineral Aggregates by Washing[4]
C 136 Test Method for Sieve Anlaysis of Fine and Coarse Aggrgates[4]
C 150 Specification for Portland Cement[3]
C 185 Test Method for Air Content of Hydraulic Cement Mortar[3]
C 188 Test Method for Density of Hydraulic Cement[3]
C 204 Test Method for Fineness of Hydraulic Cement by Air Permeability Apparatus[3]
C 207 Specification for Hydrated Lime for Masonry Purposes[3]
C 230 Specification for Flow Table for Use in Tests of Hydraulic Cement[3]
C 305 Practice for Mechanical Mixing of Hydraulic Cement Pastes and Mortars of Plastic Consistency[3]
C 430 Test Method for Fineness of Hydraulic Cement by the 45-μm (No. 325) Sieve[3]
C 472 Test Methods for Physical Testing of Gypsum, Gypsum Plasters and Gypsum Concrete[3]
C 670 Practice for Preparing Precision and Bias Statements for Test Methods for Construction Materials[4]
C 702 Practice for Reducing Field Samples of Aggregate to Testing Size[4]
C 778 Specification for Standard Sand[3]
C 1005 Specification for Weights and Weighing Devices for Use in the Physical Testing of Hydraulic Cements[3]

[1] These test methods are under the jurisdiction of ASTM Committee C-7 on Lime and are the direct responsibility of Subcommittee C07.06 on Physical Tests.

Current edition approved Nov. 10, 1998. Published May 1999. Originally published as C 110 – 34 T. Last previous edition C 110 – 96a.

[2] For tests on limestone as aggregate, see Vol 04.02 of the *Annual Book of ASTM Standards*. For tests on limestone as building stone, see Vol 04.05 of the *Annual Book of ASTM Standards*.

[3] *Annual Book of ASTM Standards*, Vol 04.01.

[4] *Annual Book of ASTM Standards*, Vol 04.02.

D 75 Practice for Sampling Aggregates[5]
E 11 Specification for Wire-Cloth and Sieves for Testing Purposes[6]
E 29 Practice for Using Significant Digits in Test Data to Determine Conformance with Specifications[6]
E 691 Practice for Conducting an Interlaboratory Study to Determine the Precision of a Test Method[6]

3. Terminology

3.1 *Definitions*—Unless otherwise specified, for definitions of terms used in this standard see Terminology C 51.

4. General Procedures

4.1 *Sampling*—Samples of lime and limestone for chemical analysis shall be taken and prepared in accordance with the requirements of Practice C 50 applicable to the material to be tested.

4.2 *Tared or Weighed Crucibles*—The tare weight of crucibles shall be determined by preheating the empty crucible to constant weight at the same temperature and under the same conditions as shall be used for the final ignition of a residue and cooling in a desiccator for the same period of time used for the crucible containing the residue.

4.3 *Constancy of Weight of Ignited Residue*—To definitely establish the constancy of weight of the ignited residue, the residue and container shall be ignited at the specified temperature and time, cooled to room temperature in a desiccator, and weighed. The residue and container shall then be reheated for at least 30 min at the same temperature, cooled in a desiccator for the same period of time and reweighed. Additional ignition periods may be required until two consecutive weights do not differ by more than 0.2 mg, at which time it shall be considered that constant weight has been attained. For ignition loss, each reheating period shall be 5 min.

4.4 *Calculation*:

4.4.1 The calculations included in the individual procedures sometimes assume that the exact weight specified has been used. Accurately weighed samples which are approximately but not exactly equal to the weight specified may be used provided appropriate corrections are made in the calculation. Unless otherwise stated, weights of all samples and residues should be recorded to the nearest 0.0001 g.

4.4.2 In all mathematical operations on a set of observed values, the equivalent of two more places of figures than in the single observed values shall be retained. For example, if observed values are read or determined to the nearest 0.1 mg, carry numbers to the nearest 0.001 mg in calculation.

4.5 *Rounding Figures*—Rounding of figures to the nearest significant place required in the report should be done after the calculations are completed, in order to keep the final results free from calculation errors. The rounding procedure should follow the principle outlined in Practice E 29.

5. Residue and Sieve Analysis

5.1 *Significance and Use*:

5.1.1 This test method determines the residue obtained from slaking quicklime. Residue, in this case, is largely unreacted material such as uncalcined limestone or dolomite, overburned quicklime, or gross impurities, or a combination of these.

5.2 *Apparatus*:

5.2.1 The sieves used shall conform to the requirements of Specification E 11. Preferably the sieves should have a 4-in. depth.

5.2.2 If sieve calibrations are required, follow the method as outlined in Test Method C 430.

5.2.3 *Spray Nozzle* conforming to the requirements of Test Method C 430.[7]

5.2.4 *Pressure Gage* shall be 3-in. (75-mm) minimum diameter, and shall be graduated in 1-psi (6.9 kPa) increments, and shall have a maximum capacity of 30-psi (207 kPa). The accuracy at 10 psi (69 kPa) shall be ±0.25 psi (±1.7 kPa).[7]

5.2.5 Attach a pressure gage to the water faucet and a rubber tubing to the output side of the pressure gage. On the other end of the rubber tubing attach the spray nozzle (see 5.2.3).

5.3 *Residue of Quicklime*:

5.3.1 Select a representative 2.5-kg (5-lb) sample of the quicklime. Break lime selected for this test so as to entirely pass a 25.0-mm (1-in.) square mesh screen. Test the pulverized lime as received. Place the sample in a box of wood or of some material of similarly low thermal conductivity, and an experienced operator should slake it with sufficient water at 21 to 27°C (70 to 80°F) to produce the maximum quantity of lime putty, carefully avoiding "burning" or "drowning" the lime. Allow it to stand for 1 h and then wash through an 850-μm (No. 20) sieve by a stream of water having a moderate pressure. Do not rub any material through the sieve. Continue the washing until the residue on the screen appears to consist entirely of coarse particles, but in no case continue the washing for more than 30 min. Dry the residue to constant weight at a temperature of 100 to 107°C (212 to 225°F) and calculate the percentage residue, based on the original weight of the sample.

5.4 *Sieve Analysis of Hydrated Lime*:

5.4.1 Select a 100-g sample of the hydrated lime as received and place on a 600-μm (No. 30) sieve, which is nested above a 75-μm (No. 200) sieve. Wash the material through the sieves by means of a stream of water from the nozzle attached to a rubber tubing (see 5.2.5) after adjusting the water pressure to 10 psi (69 kPa) ±0.25 psi (±1.7 kPa). Carefully wash the sample through the sieves without allowing any splashing over the sides of the sieve. After the sample is washed through the top sieve, separate the two sieves and continue washing through the 75-μm (No. 200) sieve until the water coming through the sieve is clear, that is, no particles can be seen in a beaker of the rinse water, but in no case continue the washing longer than 30 min. Take care not to let water accumulate on the 75-μm (No. 200) sieve, because the openings will become clogged and the operation cannot be completed in 30 min.

5.4.2 Calculate the percentage residue retained on each sieve, based on the original weight of the sample. The weight of the material retained on the 600-μm (No. 30) sieve shall be

[5] *Annual Book of ASTM Standards*, Vol 04.03.
[6] *Annual Book of ASTM Standards*, Vol 14.02.

[7] A wet washing spray attachment, Soiltest Model CL-364, or equivalent has been found suitable for this purpose. Available from Soiltest, Inc., 86 Albrecht Drive, P.O. Box 8004, Lake Bluff, IL 60044-8004.

added to the weight of the material retained on the 75-µm sieve to obtain the correct weight of the material retained on the 75-µm sieve.

5.5 *Sieve Analysis of Limestone and Dry Quicklime*:

5.5.1 Select the desired sieves and nest them with the coarsest sieves on top. Weigh a 100-g sample of the material to be tested and place it on the top sieve. Conduct the sieving operation by means of a lateral and vertical motion of the sieve accompanied by a jarring action to keep the sample moving continuously over the surface of the sieve. Continue sieving until not more than 1 % of the residue passes any sieve during 1 min. If mechanical sieving is used, the device shall be such as to impart the type of agitation described in the hand sieving operation. Continue the shaking for a period of 15 min.

5.5.2 Weigh the residue retained on each sieve to the nearest 0.1 g. Report the results of the sieve analysis as follows: (*1*) total percentages passing each sieve, (*2*) total percentages retained on each sieve, or (*3*) percentages retained between consecutive sieves, depending upon the form of the specification for the use of the material under test.

5.6 *Precision and Bias*:

5.6.1 No precision data are available due to the limited use of these test methods. Therefore, users are advised to develop their own laboratory precision. No statement is being made about the bias of these test methods.

6. Standard Consistency of Lime Putty

6.1 *Significance and Use*:

6.1.1 In order to measure certain physical properties of a lime putty, such as plasticity, it is necessary to have a uniform or standard consistency (viscosity), since the property measurement is affected by the consistency level.

6.2 *Apparatus*:

6.2.1 *Modified Vicat Apparatus*—The apparatus, constructed as shown in Fig. 1, shall consist of a bracket, *A*, bearing a movable brass rod, *B*, 6.3 mm in diameter and of suitable length to fit the Vicat bracket. A plunger, *C*, 12.5 mm in diameter, made of aluminum tubing, shall be attached to the lower end of the rod. The total weight of the rod with plunger shall be 30 g. The lower end of the plunger shall be closed without shoulders or curvature and the tube may be loaded with shot to the specified weight. The total weight required may also be obtained by means of a weight, *D*, screwed into the rod. The rod can be held in any position by means of a screw, *E*, and has a mark midway between the ends which moves under a scale, *F*, graduated in millimetres, attached to the bracket, *A*.

6.2.2 *Mold*—The conical ring mold shall be made of a noncorroding, nonabsorbent material, and shall have an inside diameter of 70 mm at the base and 60 mm at the top, and a height of 40 mm.

6.2.3 *Base Plate*—The base plate for supporting the ring mold shall be of plate glass and about 100 mm square.

6.2.4 *Mechanical Mixers*.[8]

6.3 *Standard Consistency Determination*:

6.3.1 *Mechanical Mixing Procedure Using the Vac-U-Mixer*—To a measured amount of water contained in an

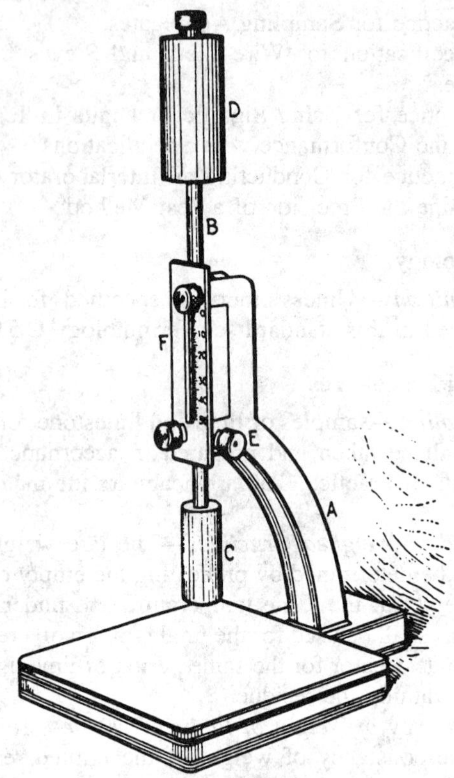

FIG. 1 Modified Vicat Apparatus

800-cm^3 Vac-U-Mix bowl, add 300 g of hydrated lime and hand mix for 10 s with a stiff spatula (Note 2). Cover putty to prevent evaporation of water. After the applicable soaking period, 30 min maximum for Type S, special hydrated lime, and not less than 16 h nor more than 24 h for Type N, normal hydrated lime, insert the paddle assembly and mix the putty for 30 s with the mechanical mixer. Remove the paddle assembly and scrape down any putty adhering to it and to the sides of the mixing bowl. Remix for 30 s and determine the consistency as prescribed in 6.3. If the penetration is less than 15 mm, return all of the material to the mixer bowl, add additional water, and remix for 15 s. If the penetration is greater than 25 mm, repeat the test.

NOTE 2—Most lime hydrates will require 250 to 300 mL of water to produce a putty of proper consistency for this test if 300 g of lime are used.

6.3.2 *Mechanical Mixing Procedure Using the Hobart N-50 Mixer*—To a measured amount of water contained in the N-50 mixing bowl, add 600 g of hydrated lime and hand mix for 10 s with a stiff spatula (Note 3). Cover putty to prevent evaporation of water. After the applicable soaking period, 30 min maximum for Type S, special hydrated lime, and not less than 16 h nor more than 24 h for Type N, normal hydrated lime, insert the paddle assembly and mix the putty for 1 min at a slow speed. Stop the mixer and scrape down the paddle and the sides of the mixing bowl. Remix for 4 min at a slow speed. Determine the consistency as prescribed in 6.3.3. If the penetration is less than 15 mm, return all of the material to the mixing bowl, add additional water, and remix for 15 s. If the penetration is more than 25 mm, repeat the test.

NOTE 3—Most lime hydrates will require 500 to 600 mL of water to produce a putty of proper consistency for this test if 600 g of lime are used.

[8] A Vac-U-Mixer or an N-50 Hobart Mixer, or equivalent, has been found suitable for this purpose.

6.3.3 *Consistency Determination*—To determine consistency, place the mold with its larger end resting on the glass base plate and fill with the lime putty. Then strike off the putty flush with the top of the mold. Center the lime putty, confined in the ring mold resting on the plate, under the rod of the modified Vicat apparatus (Fig. 1). Bring the plunger end, *C*, in contact with the surface of the lime putty and take an initial reading. Release the rod and take the final reading 30 s after the plunger is released. The lime putty is of standard consistency when a penetration of 20 ± 5 mm is obtained. Record both the total amount of water required to bring the putty to standard consistency and the actual penetration. Proceed with the plasticity determination in accordance with 7.3.

6.4 *Precision and Bias*:

6.4.1 The precision and bias of this test method has not been determined.

7. Plasticity of Lime Putty

7.1 *Significance and Use*:

7.1.1 This test method provides a measure of the degree of stiffening of lime putty of standard consistency as water is withdrawn from it by a standard suction base plate.

7.1.2 Plasticity is an important property when applying mixtures containing lime putty to porous or absorptive surfaces such as in plastering, stuccoing, and masonry construction.

7.2 *Apparatus*:

7.2.1 Determine the plasticity of lime putty using the plasticimeter shown in Fig. 2.[9]

7.2.2 *Cleaning and Care of Base Plates*—In making the plasticity determinations, much of the success attainable depends upon the condition of the base plates. In the case of porcelain plates which are reused, improper cleaning results in clogging of the pores with reduction in the rate of absorption. After a plate has been used, wipe the excess lime off and immerse the plate in clear water for not less than 2 h, after which transfer it without drying to a dilute solution of hydrochloric acid (HCl, 1 + 9) where it shall be kept immersed for another 2 h. Then transfer to a receptacle containing running water for at least 1 h. The plate is then free of acid. After the removal of excess water, place the plate in an oven overnight at a temperature of between 100 and 110°C (212 and 230°F) for drying. Before using, cool the plate to room temperature.

7.2.3 *Absorption of Plasticimeter Base Plates*:

7.2.3.1 *Total Absorption*—Plasticimeter base plates when immersed in water at room temperature for a period of 24 h shall absorb not less than 40 g of water. Before making the determination, dry the porcelain plates overnight in an oven at temperatures of between 100 and 110°C (212 and 230°F) and permit to cool to room temperature. Dry the plaster plates overnight over calcium chloride at room temperature. After immersion and before weighing, wipe off the excess water with a damp cloth.

7.2.3.2 *Rate of Absorption (Note 4)*—When tested over an area 70 mm (2¾ in.) in diameter, the water absorbed shall be in accordance with the following:

Time, min	Water Absorbed, mL
1	8 to 14
2	5 to 7½
3	4 to 6½
4	4 to 6
5	3½ to 5½

NOTE 4—A convenient apparatus for determining the rate of absorption consists of a buret sealed onto an inverted glass funnel from which the stem has been removed. The diameter of the larger end of the funnel shall be ground so as to be 70 mm (2¾ in.) in internal diameter. The funnel may be attached to the plate on which the measurement is being made by melted paraffin. The paraffin should not be too hot. A little experience will indicate when it is of the proper consistency.

Constants of the Machine:

Absorption of Porcelain Base Plate—minimum of 40 g in 24 h. For rate of absorption of base plates see 7.2.3.2.

Dimension of Base Plate—25 mm (1 in.) in thickness by 100 mm (4 in.) in diameter.

Dimensions of Disk—0.8 mm (1⁄32 in.) in thickness by 76 mm (3 in.) in diameter.

Speed of Vertical Shaft—1 revolution in 6 min, 40 s.

Torque on Disk when Bob Reading is 100—1.41 N·m.

FIG. 2 Emley Plasticimeter

7.3 *Plasticity Determination*:

7.3.1 Lubricate a ring mold such as is described in 6.2.2 with a thin film of water, place on a porcelain base plate (see 7.2.2 and 7.2.3) or a disposable plaster base plate (see 7.2.3), fill with the paste which has been adjusted to standard consistency as described in 6.3.3, and strike off level. Remove the mold by raising it vertically without distorting the paste. Center the base plate and paste in the instrument and turn the carriage up by hand until the surface of the paste is in contact with the disk and the distance between the disk and the top of the base plate is 32 mm (1¼ in.). Throw the carriage into gear and start the motor. It is essential that the motor be started exactly 120 s after the first portion of the paste has been placed in the mold. Record the time when the first portion of paste is placed in the mold as zero time; the motor is therefore started at 2 min. Take care to protect the specimen from drafts during the test.

7.3.2 Record the scale reading at 1-min intervals until the

[9] Emley Plasticimeters are no longer manufactured. If test is required, contact the chairman of Subcommittee C07.06 through ASTM Headquarters, 100 Barr Harbor Drive, West Conshohocken, PA 19428.

test is completed. Consider the test complete when: (*1*) the scale reading reaches 100, (*2*) any reading is less than the one before, or (*3*) the scale reading remains constant for three consecutive readings (2 min) and the specimen has visibly ruptured or broken loose from the base plate. Note the time and the scale reading at the end of the test.

7.4 *Calculation*:

7.4.1 Calculate the plasticity figure as follows:

$$P = \sqrt{F^2 + (10T)^2} \quad (1)$$

where

P = plasticity figure,
F = scale reading at the end of the test, and
T = time in minutes from the time when the first portion of paste was put in the mold to the end of the test.

7.5 *Precision and Bias*:

7.5.1 There are as yet insufficient analyzed data to permit preparation of a precision and bias statement for this test method. When data are collected and analyzed, precision and bias statements will be proposed.

8. Autoclave Expansion of Hydrated Lime

8.1 *Significance and Use*:

8.1.1 Expansion of pressed tablets of hydrated lime generally indicates the presence of unhydrated oxides of magnesium and calcium. The relation of the degree of expansion in this test method to field performance has not been determined.

8.2 *Apparatus*:

8.2.1 *Mold and Press*—A steel mold capable of producing a press tablet at least 0.032 m (1.25 in.) in diameter and 0.006 m (0.25 in.) thick, and able to sustain at least 88.9 kN (20 000 lbf) pressure from a suitable press. It should be provided with a release jig also.

8.2.2 *Autoclave*, capable of holding 1034 kPa (150 psi) for 2 h.

8.2.3 *Micrometer*, dial-type, capable of measuring 2.54 µm (0.0001 in.).

8.2.4 *Microscope*, with graduated lens for measuring 0.10 mm.

8.3 *Procedure*:

8.3.1 Weigh out 15 g of hydrated sample, place in the mold, and press into a tablet. Press to 33.4 kN (7500 lbf) for 10 s, then increase pressure to 88.9 kN (20 000 lbf) or more. Hold for 10 s before releasing. Press tablet from mold with jig and draw three diameter lines across the surface of the tablet using a lead pencil. Draw two diameter lines normal to each other and draw the third bisecting the 90° angles of the other two. Measure the diameters with a dial micrometer and place the tablet on the autoclave rack. Use aluminum foil to protect the tablets from water dripping. Autoclave at 862 to 1034 kPa (125 to 150 psi) for 2 h. Begin timing when the pressure reaches 345 kPa (50 psi). After the autoclaving interval, allow the autoclave to cool, remove the tablet, and remeasure the diameters. Calculate the average percent expansion of the tablet from the before and after measurements.

8.4 *Expansion of Hydrated Lime-Portland Cement-Aggregate*:

8.4.1 *Materials*:

8.4.1.1 *Standard Cement*—Type I or Type II portland cement.

8.4.1.2 *Standard Aggregate*—Pulverized limestone, minus 212-µm (No. 70) sieve, having less than 0.5 % silicon dioxide (SiO_2).

8.4.2 *Procedure*:

8.4.2.1 *Test Tablet*—Make up a pressed tablet in accordance with the procedure outlined in 8.3.1 using the following mixture for the sample:

Standard portland cement	14 g
Hydrated lime	8 g
Standard aggregate (pulverized limestone)	72 g

Blend the mix until homogeneous.

8.4.2.2 *Standard Tablet*—Make up a pressed tablet in accordance with the procedure outlined in 8.3.1 using the following mixture for the sample:

Standard portland cement	7 g
Standard aggregate (pulverized limestone)	16 g

Blend the mix until homogeneous.

8.4.2.3 Autoclave and calculate expansions of the test tablet and the standard tablet in accordance with 8.3.1.

8.4.2.4 Determine the autoclave expansion of hydrated lime for masonry purposes by subtracting the average percent expansion of the standard tablet from the sample tablet.

8.5 *Precision and Bias*:

8.5.1 No precision data are available due to the limited use of this test method. Therefore, users are advised to develop their own laboratory precision. No statement is being made about the bias of this test method.

9. Popping and Pitting of Hydrated Lime

9.1 *Significance and Use*:

9.1.1 Pops and pits are caused by the hydration and expansion of coarse particles of unhydrated lime or lime-impurity reaction products present in the hydrated lime. The level of popping and pitting in the sample is indicative of the potential for the appearance of surface defects in plastering applications.

9.2 *Gauging Plaster*:

9.2.1 The gauging plaster used for the popping and pitting test shall conform to the Test Methods section of Specification C 28 and shall have a setting time of not more than 1 h when tested in accordance with Test Methods C 472. Test the gauging plaster without lime in the manner described in 9.3 to ensure its freedom from pops and pits. If any pops or pits are found, provide another lot of gauging plaster that is free of pops and pits when subjected to this test.

9.3 *Procedure*:

9.3.1 Mix 100 g of hydrated lime with sufficient water to bring to such a consistency as to give a penetration of 20 ± 5 mm when tested in accordance with 6.3.3. Mix into this putty, 25 g of gauging plaster (9.2.1), adding more water as required to maintain workable consistency. Spread on a glass plate to make a pat at least 150 by 200 mm (6 by 8 in.) by approximately 3 mm (⅛ in.) in thickness. Trowel to a smooth finish. Allow to stand overnight.

9.3.2 Place the specimen and plate on a rack in the steam bath so that water is not in contact with the specimen to be tested. Provide a sloping cover above the specimen to prevent condensed steam from dripping onto the surface of the specimen. Raise the temperature of the water in the steam bath to boiling and maintain at boiling for 5 h. Remove the specimens from the bath and examine for pops and pits.

9.3.3 The pitting potential of hydrated lime can be determined in conjunction with autoclave expansion as in 8.3.1. However, it is not necessary to measure diameter, if only the pitting potential is to be determined. After following the procedure for expansion in 8.3.1, examine the pressed tablet under the measuring microscope, and count and measure the pits in millimetres.

10. Water Retention of Hydrated Lime

10.1 *Significance and Use*:

10.1.1 This test method measures the ability of the hydrated lime in a plastic mix with sand to retain water, and hence retain consistency of the mix, when subjected to an applied suction. This ability, measured as a percent of the original consistency, is indicative of the workability to be expected in a masonry containing the lime.

10.2 *Proportioning and Mixing*:

10.2.1 *Apparatus*—The apparatus used shall conform to Practice C 305.

10.2.2 *Proportions*—The mortar tested shall be composed of 500 g of lime and 1500 g of standard sand conforming to 13.2.4. If hydrated lime putty is used, use that weight of putty that is equivalent to 500 g of dry hydrated lime.

10.2.3 *Mechanical Mixing*:

10.2.3.1 Place the dry paddle and the dry bowl in the mixing position in the mixer.

10.2.3.2 Place a measured quantity of water in the bowl.

10.2.3.3 Add the lime to the water, then start the mixer and mix at slow speed (140 ± 5 r/min) for 30 s.

10.2.4 Add the entire quantity of sand slowly over a 30-s period while mixing at slow speed.

10.2.5 Stop the mixer, change to medium speed (285 ± 10 rpm) and mix for 30 s.

10.2.6 Stop the mixer and let the mortar stand for 1½ min. (During the first 15 s of this interval, quickly scrape down into the batch any mortar that may have collected on the side of the bowl, then for the remainder of this interval cover the bowl with the lid.)

10.2.7 Finish the mixing for 1 min at medium speed.

10.2.8 In any case requiring a remixing interval, any mortar adhering to the side of the bowl shall be quickly scraped down into the batch prior to remixing.

10.3 *Consistency*:

10.3.1 *Apparatus*—The flow table and mold used for the measurement of consistency of the mortar shall conform to Specification C 230.

10.3.2 *Procedure*—Carefully wipe dry the flow table top and place the flow mold at the center. Immediately after completing the mixing operation, fill the mold with mortar gently pressed into place by the finger tips to ensure uniform filling free of voids. Smooth off the mortar level with the top of the mold by aid of a trowel, and remove the mold. Immediately drop the table through a height of 13 mm (½ in.), 25 times in 15 s. The flow is the resulting increase in diameter of the mortar mass, expressed as the percentage of the original diameter. The mortar may be adjusted, if the flow is below 100 %, by additions of water until the flow is within the range from 100 to 115 %. Make each adjustment by returning the mortar to the original mixing bowl, add water, and then mix at medium speed (285± 10 r/min) for 30 s. If the flow of the original mortar is greater than 115 %, prepare a new batch.

10.4 *Water Retention Test*:

10.4.1 *Apparatus*—The equipment used to determine water retention shall conform to the same apparatus used for the water retention test in Specification C 91 (see Fig. 3).

10.4.2 *Procedure*:

10.4.2.1 Adjust the mercury relief column to maintain a vacuum of 50.8 mm as measured on the manometer. Seat the perforated dish on the greased gasket of the funnel. Place a wetted filter paper in the bottom of the dish. Turn the stopcock to apply the vacuum to the funnel and check the apparatus for leaks and to determine that the required vacuum is obtained. Then turn the stopcock to shut off the vacuum from the funnel.

10.4.2.2 Immediately after the final consistency test (10.3.2), return all of the mortar to the bowl and remix the entire batch for 15 s at medium speed. Immediately after remixing the mortar, fill the perforated dish with the mortar to slightly above the rim. Tamp the mortar 15 times with the tamper. Apply ten of the tamping strokes at approximately uniform spacing adjacent to the rim of the dish and with the long axis of the tamping face held at right angles, to the radius of the dish. Apply the remaining five tamping strokes at random points distributed over the central area of the dish. The tamping pressure shall be just sufficient to ensure filling of the dish. On completion of the tamping, the top of the mortar should extend slightly above the rim of the dish. Smooth off the mortar by drawing the flat side of the straightedge (with the leading edge slightly raised) across the top of the dish. Then cut off the mortar to a plane surface flush with the rim of the dish by drawing the straightedge with a sawing motion across the top of the dish in two cutting strokes, starting each cut near the center of the dish. If the mortar is pulled away from the side of the dish during the process of cutting off the excess mortar, gently press the mortar back into contact with the side of the dish using the tamper.

10.4.2.3 Turn the stopcock to apply vacuum to the funnel. After suction for 60 s, quickly turn the stopcock to expose the funnel to atmospheric pressure. Immediately slide the perforated dish off the funnel, touch it momentarily on a damp cloth to remove droplets of water, and set the dish on the table. Then using the bowl scraper (rubber scraper as specified in Practice C 305), plow and mix the mortar in the dish for 15 s. Upon completion of mixing, place the mortar in the flow mold and determine the flow. Carry out the entire operation without interruption and as quickly as possible. Not more than 30 min should be required for completion, starting from the completion of the mixing of the mortar for the first flow determination.

10.4.3 *Calculation*:

10.4.3.1 Calculate the water retention value for the mortar as follows:

$$\text{Water retention value} = (A/B) \times 100 \tag{2}$$

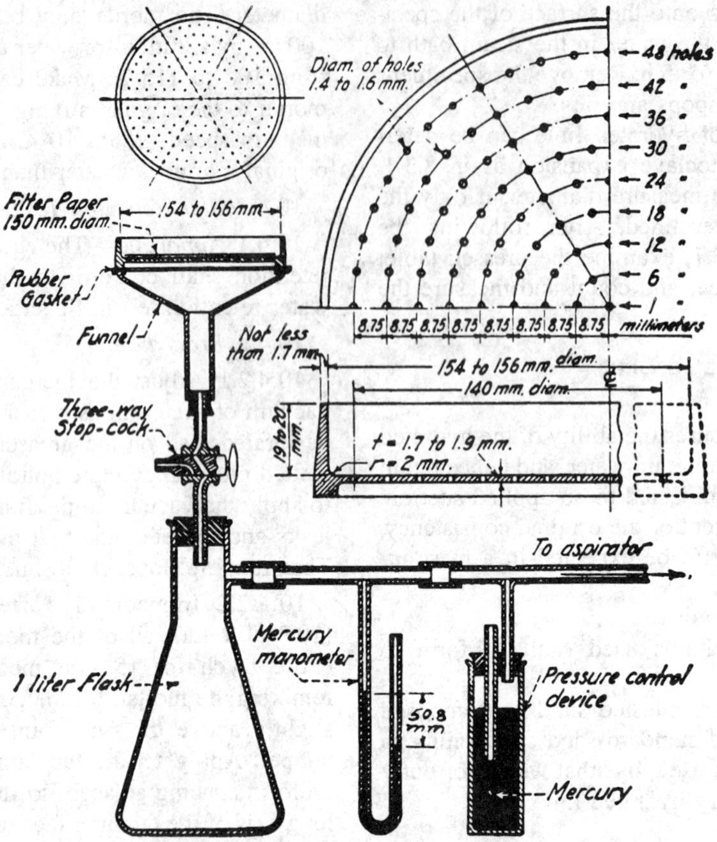

FIG. 3 Apparatus Assembly for the Water Retention Test

where:

A = flow after suction, and

B = flow immediately after mixing.

10.5 *Precision and Bias*:

10.5.1 No precision data are available due to the limited use of this test method. Therefore, users are advised to develop their own laboratory precision.

11. Settling Rate of Hydrated Lime

11.1 *Significance and Use*:

11.1.1 This test method provides a measure of the rate of settling of a hydrated lime slurry, a form in which this material is frequently used. In some applications a slow settling slurry is desirable; in others, fast settling is preferred.

11.2 *Procedure*:

11.2.1 Place 10.0 g of lime hydrate in a 100-mL glass-stoppered graduated cylinder (internal diameter about 24 mm). Wet with 50 mL of carbon dioxide (CO_2) free distilled water at 23 ± 1.7°C (73.4± 3°F) and mix thoroughly by alternately inverting and righting the cylinder slowly for a period of 2 min. Allow the graduate and contents to stand at 23 ± 1.7°C for 30 min and then dilute to the 100-mL mark with CO_2-free distilled water at 23 ± 1.7°C. Mix contents again thoroughly as before and allow to stand undisturbed at 23 ± 1.7°C for 24 h.

11.3 *Report*:

11.3.1 Report the sedimentation height in millilitres after ¼, ½, ¾, 1, 2, 4, and 24 h, reading the bottom of the meniscus.

NOTE 5—Slight variations in results of this test method on a sample run in different laboratories or by different operators are permissible. The test is not an absolute one, but is designed to distinguish between fast and slow settling hydrates.

11.4 *Precision and Bias*:

11.4.1 No precision data are available due to the limited use of this test method. Therefore, users are advised to develop their own laboratory precision.

12. Slaking Rate of Quicklime

12.1 Significance and Use:

12.1.1 The temperature rise in 30 s is a measure of the reactivity of the softer-burned portion of the quicklime. Total slaking time provides a measure of the overall degree of reactivity of the material. Total temperature rise is largely dependent on the available lime content of the sample.

12.1.2 These slaking parameters provide an indication of the performance of the quicklime to be expected in industrial slaking systems. Slaking characteristics have an effect on lime slurry properties such as settling characteristics, viscosity, particle size, and reaction rate.

12.2 *Apparatus*:

12.2.1 *Mechanical Stirrer* ,[10] speed 400 ± 50 r/min, fitted with a special stirring rod.

12.2.2 *Modified Dewar Flask* ,[11] 665-mL, fitted with special rubber gasket covers.

12.2.3 *Thermometer*, dial-type, 0 to 100°C range in 1°C increments or thermocouple with a response time equivalent to

[10] A Fisher 14-498, or equivalent, has been found suitable for this purpose.

[11] A Fisher 10-197, or equivalent, has been found suitable for this purpose.

or faster than the dial thermometer.

12.2.4 *Torsion Balance*.

12.2.5 *Sieve*, 203-mm (8-in.), 3.35-mm (No. 6), conforming to Specification E 11.

12.2.6 An apparatus essentially the same as that illustrated in Fig. 4 and Fig. 5 shall be used. The apparatus consists of a covered reaction container fitted with a mechanical stirrer and thermometer. The quicklime charge shall be stirred with a mechanical stirrer fitted with a stainless steel rod, the end of which is formed into a loop to follow the contour of the reaction container. The vacuum reaction flask shall be provided with a cover consisting of two circular pieces of gasket rubber sheet, approximately 3 mm (1/8 in.) thick. The first piece is provided with a single radial slot that slides over the stirring rod and the thermometer. The second piece (top) has a similar slot plus a hole to provide for the dial thermometer. When the two cover pieces are in place, the slot on the lower piece is at right angles to the slot on the upper piece with the thermometer stem extending through the lower slot. The apparatus may be assembled by any convenient supporting equipment.

12.3 *Procedure*:

12.3.1 Prepare the sample of quicklime (as rapidly as possible to prevent sample deterioration) so that a majority of the material passes a 3.35 mm (No. 6) sieve. Place the sample in an airtight container and allow to come to room temperature before testing. The slaking rate of lime is significantly affected by the particle size of the sample and must be as close to a 3.35 mm (No. 6) sieve as possible. It is not necessary that 100 % of the sample pass a 3.35 mm (No. 6) sieve, but all of the sample, including the plus 3.35 mm (plus No. 6) fraction, must be used in the test.

12.3.2 *Slaking Rate*—Adjust the temperature of about 500 mL of distilled water in accordance with the schedule given in Table 1, and add the specified amount to the Dewar flask. Set the agitator revolving at 400 ± 50 r/min. The temperature of the water in the flask must be ±0.5°C of the desired temperature. Quarter and weigh out the required amount of the prepared quicklime sample. Add the quicklime to the water without delay and simultaneously begin timing. Put the covers in place immediately. Take a reading at each 30 s interval.

12.3.3 Continue readings until less than 0.5°C temperature change is noted in each of three consecutive readings. The total active slaking time will then be the time at which the first of the

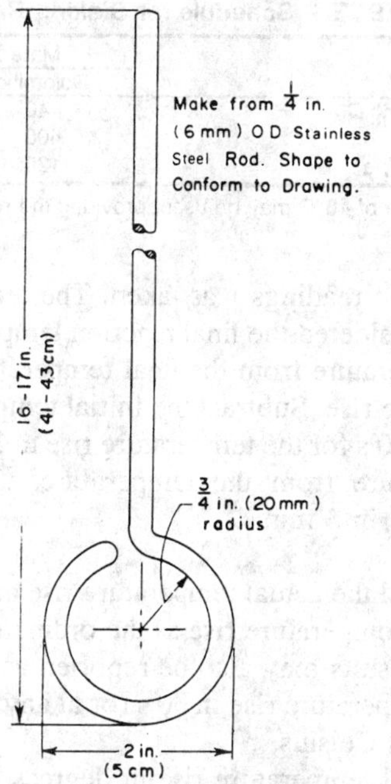

FIG. 5 Stirring Rod Detail

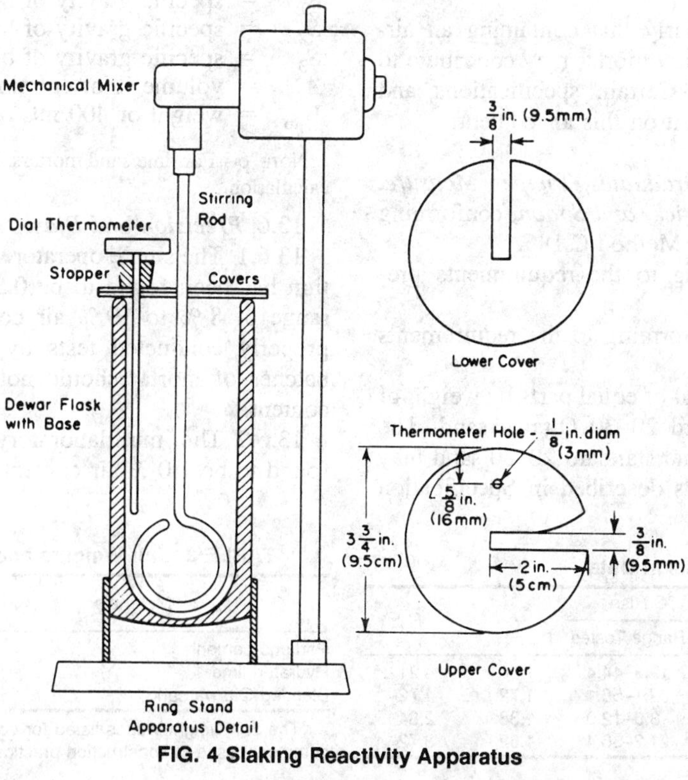

FIG. 4 Slaking Reactivity Apparatus

TABLE 1 Schedule for Slaking Rate

	Material to Be Tested	
	Dolomitic	High Calcium
Temperature of water,° C	40	25[A]
Quantity of water, mL	400	400
Quantity of quicklime, g	120	100

[A]Initial temperature of 40°C may be used, provided the report of results states the initial temperature.

three consecutive readings was taken. The temperature at this time will be considered the final reaction temperature. Subtract the initial temperature from the final temperature to obtain the total temperature rise. Subtract the initial temperature from the temperature at 30 s for the temperature rise in 30 s. Subtract the initial temperature from the temperature at 3 min for the temperature rise in 3 min.

12.4 *Report*:

12.4.1 Record the actual temperature rise and plot a suitable curve showing temperature rise as the ordinate and time as the abscissa. The results may also be reported as:

12.4.1.1 Temperature rise in 30 s (or at any other designated time) in degrees Celsius,

12.4.1.2 Total temperature rise in degrees Celsius, and

12.4.1.3 Total active slaking time in minutes.

12.5 *Precision and Bias*:

12.5.1 Twelve laboratories cooperated in the testing of five high calcium quicklimes and four dolomitic quicklimes thereby obtaining the repeatability (*r*) and reproducibility (*R*) (Practice E 691) data contained in Table 2.

12.5.2 Due to the lack of a recognized industry standard, the bias of this test method has not been determined. The variety of reporting options also complicates obtaining a suitable bias statement.

13. Air Entrainment

13.1 *Significance and Use*:

13.1.1 Hydrated lime, particularly that containing an air-entraining additive, used in masonry mortar may contribute to the air content of the mortar. Certain specifications and applications of mortar place a limit on this air content.

13.2 *Apparatus*:

13.2.1 *Scales, Sieves, Glass Graduates, Tamper, Measure, Straightedge, Spatula, Tapping Stick, and Spoon*, conforming to the requirements given in Test Method C 185.

13.2.2 *Flow Table*, conforming to the requirements prescribed in Specification C 230.

13.2.3 *Mixing Apparatus*, conforming to the requirements as prescribed in Practice C 305.

13.2.4 The sand shall be a blend of equal parts by weight of graded Ottawa sand and standard 20–30 Ottawa sand. The fineness of graded Ottawa sand and standard 20–30 sand may be checked by using the methods described in Specification C 778.

13.3 *Preparation of Mortar*:

13.3.1 *Proportions for Mortar*—Portland cement-hydrated lime mortar for measurement of air entrainment shall be proportioned to conform, in batch size, to the unit weights by volume of cementitious material and aggregate as shown in Table 3. The portland cement shall conform to Specification C 150, and the hydrated lime to Specification C 207. The quantity of water, measured in millilitres, shall be such as to produce a flow of 110 ± 5 % as determined by the flow table. Proportions for the generally used batch sizes based on Table 2 material unit weight shall contain the weights as prescribed in Table 4.

13.3.2 *Mixing of Mortars*—Mix the mortar in accordance with the procedure for mixing pastes in Practice C 305.

13.3.3 *Determination of Flow*—Determine the flow in accordance with the Procedure section of Test Method C 109.

13.4 *Procedure*:

13.4.1 If the mortar has the correct flow, use a separate portion of the mortar for the determination of entrained air. Determine the weight of 400 mL of mortar in accordance with Test Method C 185.

13.5 *Calculation*:

13.5.1 Calculate the air content of the mortar and report it to the nearest 0.1 % as follows:

$$D = (W_1 + W_2 + W_3 + V_w)/[(W_1/S_1) + (W_2/S_2) + (W_3/S_3) + V_w]$$
$$A = 100 - (W_m/4D) \quad (3)$$

where:

D = density of air-free mortar,
W_1 = weight of portland cement, g,
W_2 = weight of hydrated lime, g,
W_3 = weight of blended Ottawa sand, g,
V_w = water used, mL,
S_1 = specific gravity of portland cement,
S_2 = specific gravity of hydrated lime,
S_3 = specific gravity of blended Ottawa sand,
A = volume % of entrained air, and
W_m = weight of 400 mL of mortar, g.

NOTE 6—For lime/sand mortars, W_1 and S_1 should be dropped from the calculation.

13.6 *Precision and Bias*:

13.6.1 The single operator within laboratory standard deviation has been found to be 0.56 % air content throughout the range of 8 % to 19 % air content. Therefore results of two properly conducted tests by the same operator on similar batches of mortar should not differ by more than 1.6 % air content.

13.6.2 The multilaboratory standard deviation has been found to be 1.0 % air content throughout the range of 8 % to

TABLE 2 Precision Data

Material	Labs	Results in °C Rise		r	R
		Time	Range Tested		
High Calcium	12	30 s	12.3–44.4	1.56	4.21
High Calcium	11	3 min	32.1–56.1	1.72	4.72
Dolomitic	10	30 s	3.6–12.0	1.38	2.84
Dolomitic	9	3 min	21.2–36.4	1.62	3.72

TABLE 3 Unit Weights and Apparent Specific Gravities

Materials	Unit weight, kg/m³(lb/ft³)[A]	Specific Gravity
Portland cement	1 504 (94)	3.15
Hydrated lime	800 (50)	2.30
Blended Ottawa sand	1 280 (80)	2.65

[A]The unit weight values listed for cementitious materials are assumed values commonly used in construction practice.

TABLE 4 Weight of Materials for Mortar Batch

Mortar Type (C 270)	Proportions by Volume	Portland Cement, g	Hydrated Lime, g	Blended Ottawa Sand, g
M	1:¼:3	564	75	1 440
S	1:½:4½	376	100	1 440
N	1:1:6	282	150	1 440
O	1:2:9	188	200	1 440
Lime/Sand	N/A	0	500	1 500

19 % air content. Therefore, results of two different laboratories on similar batches of mortar should not differ from each other by more than 2.8 % air content (see Test Method C 185).

14. pH of Pulverized Limestone

14.1 Refer to the "pH of Alkaline Earth Solutions" section in Test Methods C 25 for the appropriate test method.

15. Particle Size of Pulverized Limestone

15.1 *Significance and Use*:

15.1.1 Particle size of pulverized limestone, as the word is used in these methods, is the percent distribution of the equivalent spherical diameter of the individual particles expressed in micrometres, using the principle of sedimentation and Stokes' law for particle size determination. It is intended for use with pulverized limestones with not more than 0.5 % residue on a 45-μm (No. 325) sieve.

15.2 *Apparatus*:

15.2.1 *Soil Hydrometer*, ASTM 152H.[12]

15.2.2 *Sedimentation Cylinder*, ASTM, 1000-mL capacity.

15.2.3 *Rubber Stopper*, Size 12.

15.2.4 *Thermometer*, 0 to 105°C.

15.2.5 *Stop Watch.*

15.2.6 *Regular Clock or Watch.*

15.2.7 *Mixer* .[13]

15.2.8 *Water Bath.*

15.2.9 *Balance* .[14]

15.2.10 *Watch Glass.*

15.2.11 *Graph Paper* ,[15] 3 cycles × 70 divisions.

15.2.12 *Sieve*, 45-μm (No. 325), stainless steel cloth, brass frame, 8-in. diameter.

15.2.13 *Sieve*, 500-mesh, stainless steel cloth, brass frame, 4-in. diameter, 5-in. tall frame.

15.3 *Reagents*:

15.3.1 *Particle-Dispersing Agent* [16] (30 mL of 25 % solution is diluted up to 400 mL with distilled water).

15.4 *Procedure*:

15.4.1 Determine meniscus correction by inserting the hydrometer in the sedimentation cylinder filled to mark with distilled water. Record the reading at the top of the meniscus and at the bottom of the meniscus. The difference between the two readings is the meniscus correction. For example, in Fig. 6, the correction for the hydrometer used is 1.2. This reading is *added* to each R to obtain R_r.

15.4.2 Calibrate the hydrometer by adding 30 mL of the particle-dispersing solution to the sedimentation cylinder, then bringing up to the mark with distilled water at 27°C. Mix thoroughly and take a hydrometer reading (read at the top of the meniscus). Repeat after cooling the cylinder to 17°C and adjusting the meniscus so it is on the mark. Assume a straight-line relationship and draw a line that gives the *composite correction factor.* This factor is the difference between the reading and zero. These are the corrections entered in Table 5 and should be determined for each hydrometer. Four factors are compensated for in the correction factor: (*1*) *Temperature:* Hydrometers and cylinders are calibrated at 20°C; variations from this temperature produce inaccuracy in the hydrometer reading; (*2*) *Specific gravity:* Addition of dispersant changes the specific gravity of the solution; (*3*) *Meniscus correction:* Hydrometers are graduated to read at the bottom of the meniscus but opaque calcium carbonate solutions require readings at the top of the meniscus; and (*4*) *Hydrometers:* In spite of the supposed similarity in volume of the hydrometers (ASTM 152H), variations of as much as 1.0-scale divisions between two similar hydrometers have been noted. The correction factor brings all four into line with one another. It is not necessary to repeat this calibration unless changing to a different hydrometer.

15.4.3 Weigh 40 g of sample.

15.4.4 Add approximately 300 mL of distilled water to the mixer, 30 mL of the particle-dispersing solution, followed by 40 g of unknown sample. Cover. Agitate for exactly 2 min at high speed.

15.4.5 Transfer the slurry quantitatively to the 1000-mL sedimentation cylinder. Make up to approximately 3.2 mm (⅛ in.) above the mark since it must be read from the top (as the bottom of the meniscus is not visible) and this will approximate the 1000-mL calibration of the cylinder. Cylinder temperature can be adjusted to 20°C by running cool water on the outside of the cylinder and stirring with a thermometer until 20°C is reached. Cap with the rubber stopper. Mix well by inverting the cylinder 15 or more times. Remove the stopper and put the cylinder in a water bath that has been previously adjusted to as close to 20°C as is possible. Start the stop watch and note the time on the clock. At exactly 4½ min after start, carefully insert the hydrometer to the approximate point where the reading is to be made. Take the reading at exactly 5 min. Record the reading and temperature (Note 7). Remove the hydrometer and wash clean of any slurry. Cover the cylinder with the watch glass.

NOTE 7—Temperature must be taken inside the cylinder and not in the water bath.

15.4.6 Take additional readings at 15, 30, 60, 120, *or* 180 min; 300 *or* 360 min; and 1200 *or* 1440 min after the start.

15.4.7 Take a 25-g sample and run a 500-mesh wet-sieve test. The opening of the 500-mesh sieve is approximately 25 μm. From this result calculate the percent finer than 25 μm. Do not discard the plus 500-mesh but use this with the 45-μm (No. 325) sieve to obtain the percent finer than 44 μm. The opening

[12] Available from Taylor Instrument Co., Catalog No. 22297.

[13] The Hamilton Beach Model No. 210 mixer, or equivalent, has been found suitable for this purpose.

[14] The O'Haus CG 311 balance, or equivalent, has been found suitable.

[15] Dietzgen No. 340-L310 or Keuffel and Esser No. 359-71G graph paper, or equivalent, has been found suitable.

[16] Daxad 30, a particle dispersing agent, has been found suitable, and is available from the Dewey and Almy Div. of W. R. Grace Co.

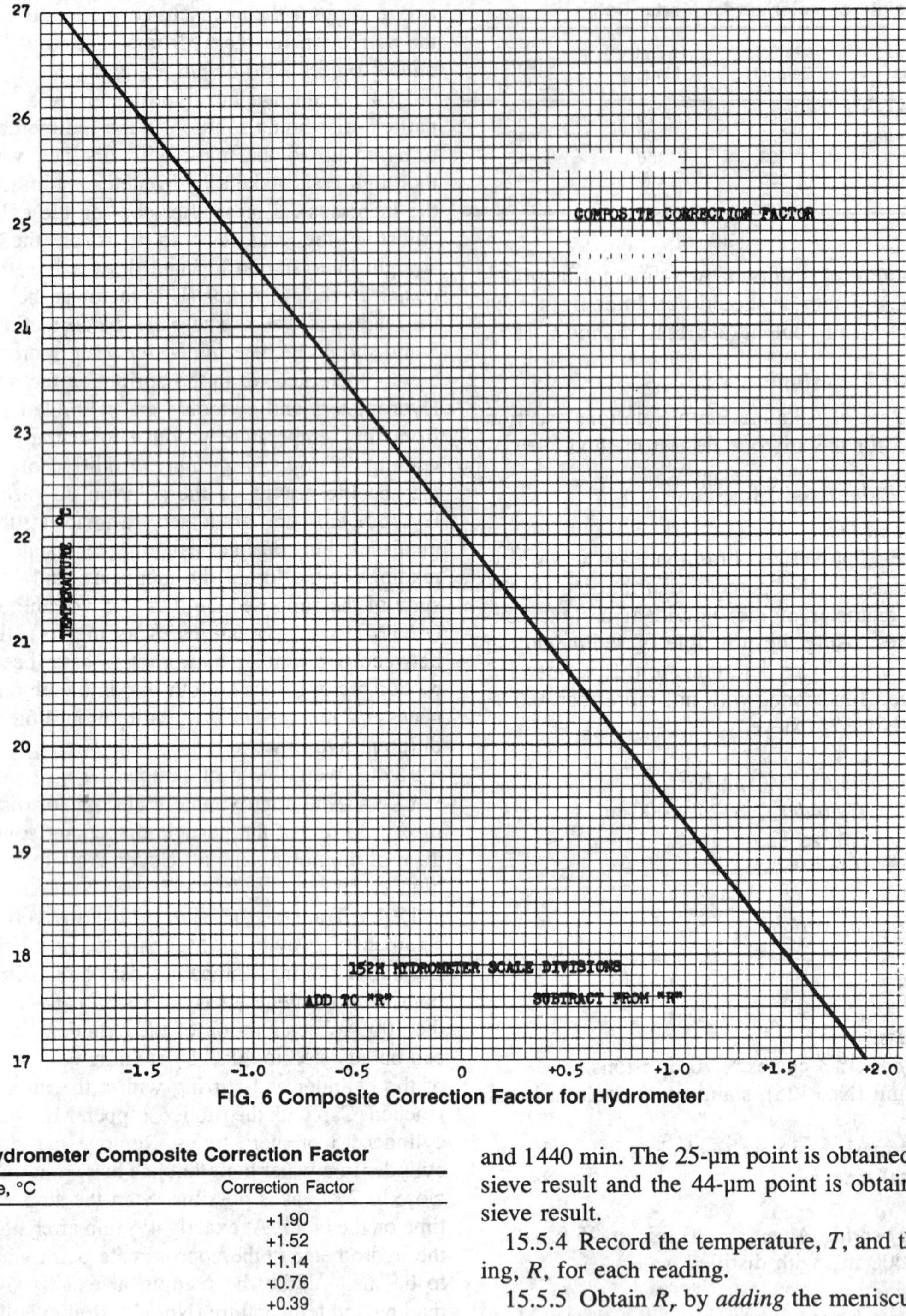

FIG. 6 Composite Correction Factor for Hydrometer

TABLE 5 Hydrometer Composite Correction Factor

Temperature, °C	Correction Factor
17	+1.90
18	+1.52
19	+1.14
20	+0.76
21	+0.39
22	0.00
23	−0.38
24	−0.76
25	−1.14
26	−1.52
27	−1.90

of the 45-μm sieve is 44 μm.

15.5 *Calculation*:

15.5.1 Arrange the data on a sample sheet.

15.5.2 Record the date and clock readings as readings are taken.

15.5.3 Readings are usually taken at 5, 15, 30, 60, 180, 360, and 1440 min. The 25-μm point is obtained from the 500-mesh sieve result and the 44-μm point is obtained from the 45-μm sieve result.

15.5.4 Record the temperature, *T*, and the hydrometer reading, *R*, for each reading.

15.5.5 Obtain R_r by *adding* the meniscus correction to each *R* value.

15.5.6 Obtain R_c, the corrected hydrometer reading, from Fig. 6. This value can be different for each hydrometer and must be individually determined.

15.5.7 Obtain *L* from Table 6 using R_r values.

15.5.8 $\sqrt{L/T}$ is found from Fig. 7 and the values for *L* and *T* (time). For times not in Fig. 7, calculate the $\sqrt{L/T}$ since the values for *L* and *T* (in minutes) are known.

15.5.9 Find *D* at 20°C in terms of $\sqrt{L/T}$ using Table 7.

15.5.10 . To correct *D* for temperature, use Table 8 and find ΔD in terms of $\sqrt{L/T}$. Multiply by ΔT (ΔT is the difference in temperature between 20°C and the actual temperature of the

TABLE 6 Effective Depth, *L*, for Hydrometer 152H

R_r	L, cm	R_r	L, cm
0	16.3	31	11.2
1	16.1	32	11.1
2	16.0	33	10.9
3	15.8	34	10.7
4	15.6	35	10.6
5	15.5	36	10.4
6	15.3	37	10.2
7	15.2	38	10.1
8	15.0	39	9.9
9	14.8	40	9.7
10	14.7	41	9.6
11	14.5	42	9.4
12	14.3	43	9.2
13	14.2	44	9.1
14	14.0	45	8.9
15	13.8	46	8.8
16	13.7	47	8.6
17	13.5	48	8.4
18	13.3	49	8.3
19	13.2	50	8.1
20	13.0	51	7.9
21	12.9	52	7.8
22	12.7	53	7.6
23	12.5	54	7.4
24	12.4	55	7.3
25	12.2	56	7.1
26	12.0	57	7.0
27	11.9	58	6.8
28	11.7	59	6.6
29	11.5	60	6.5
30	11.4		

test). This will give a value to be subtracted from the *D* found in 15.5.9 if the temperature is above 20°C. If the temperature is below 20° C, this correction should be added.

15.5.11 Find *P* by using Table 8 and the value for R_c.

15.5.12 The values of D_c are now plotted against the values of *P*.

15.6 *Precision and Bias*:

15.6.1 The precision and bias of this test method has not been determined.

16. Dry Brightness of Pulverized Limestone

16.1 *Summary of Test Method*:

16.1.1 A sample of the dry material is compressed and its reflectance measured on a reflectometer that has previously been standardized.

16.2 *Significance and Use*:

16.2.1 This test method provides a measure of the reflectance, or whiteness, or both of ground calcium carbonate products by comparison with a standard, using green and blue filters.

16.3 *Apparatus*:

16.3.1 *Reflectometer* :[17]

16.3.2 *Dry Powder Press* (See Fig. 8)[18]—Instructions, as supplied by the manufacturer, for preparation of the sample and use of the powder press shall be explicitly followed.

16.3.3 *White Porcelain Standard Plaque*, to be used as secondary standard.

16.4 *Reagent*:

16.4.1 *Barium Sulfate* ($BaSO_4$)—Use Eastman Kodak[19] Chemical No. 6091, white reflectance standard only.

16.5 *Calibration and Standardization*:

16.5.1 *Zero Scale Calibration (bottom of scale standardization)*:

16.5.1.1 Place the black glass provided with the instrument over the specimen port, so that the shiny side is towards the opening. The glass should be positioned so that no light escapes from the black glass-opening interface.

16.5.1.2 The processor is then adjusted to read *zero reflectance.*

16.5.2 *Standardizing of the White Standard (standardization of the upper part of the scale)*:

16.5.2.1 A primary standard pellet (barium sulfate) which is free from surface flaws should be positioned over the specimen port so that no light can escape at the pellet-opening interface.

16.5.2.2 The Eastman Kodak barium sulfate reflectance standard is provided with reflectance values at various wavelengths. Since some variation is possible between lots of $BaSO_4$, the values used to standardize the reflectometer must be calculated. A normal *Y* value will be between 99.0 and 98.5, depending on the lot number.

16.5.2.3 After this has been accomplished, a reading of the white standard plaque can be taken and the values of *X*, *Y*, and *Z* recorded. This plaque can then be used as a secondary standard for future standardizations. This reduces the necessity of making a barium sulfate pellet for every test series.

16.6 *Procedure*:

16.6.1 The reflectometer must be given ample warm-up time prior to the sample readings.

16.6.2 The reflectometer must first be standardized; this consists of standardization of the bottom of the scale and standardization of the upper part of the scale.

16.6.3 Sample pellets should then be pressed (Note 8) following manufacturer instructions explicitly (Note 9).

NOTE 8—Ground products with more than 0.5 % residue on a 45-μm (No. 325) screen will require special care in preparing the sample cup. The coarser the product, the harder to obtain a compact, smooth surface.

NOTE 9—Some reflectometers and spectrophotometers can measure reflectance with the powder sample in a horizontal position, thus eliminating the necessity to prepare a sample pellet. Also, coated ground limestones are difficult to pelletize. Loose powder samples should be smoothed in a convenient sized container until the surface is level and free from cracks and other surface defects.

16.6.4 After the reflectometer has been standardized, the sample pellets are centered beneath the opening and positioned so that no light escapes from the pellet-opening interface.

16.6.5 The samples are then read for *X*, *Y*, *Z*, *L*, *a*, and *b* values. These values are recorded.

16.6.6 To determine if the values of the reflectometer have drifted, the white standard (either the barium sulfate pellet or the porcelain plaque) is placed over the specimen port and read. Values should be the same as those placed in the processor during the standardization procedure.

[17] The Hunter D25A-Tristimulus colorimeter, or equivalent, has been found suitable for this purpose.

[18] The Bausch and Lomb or Carl Zeiss Dry Powder Press Assembly, or equivalent, has been found satisfactory for this purpose.

[19] Registered trademark.

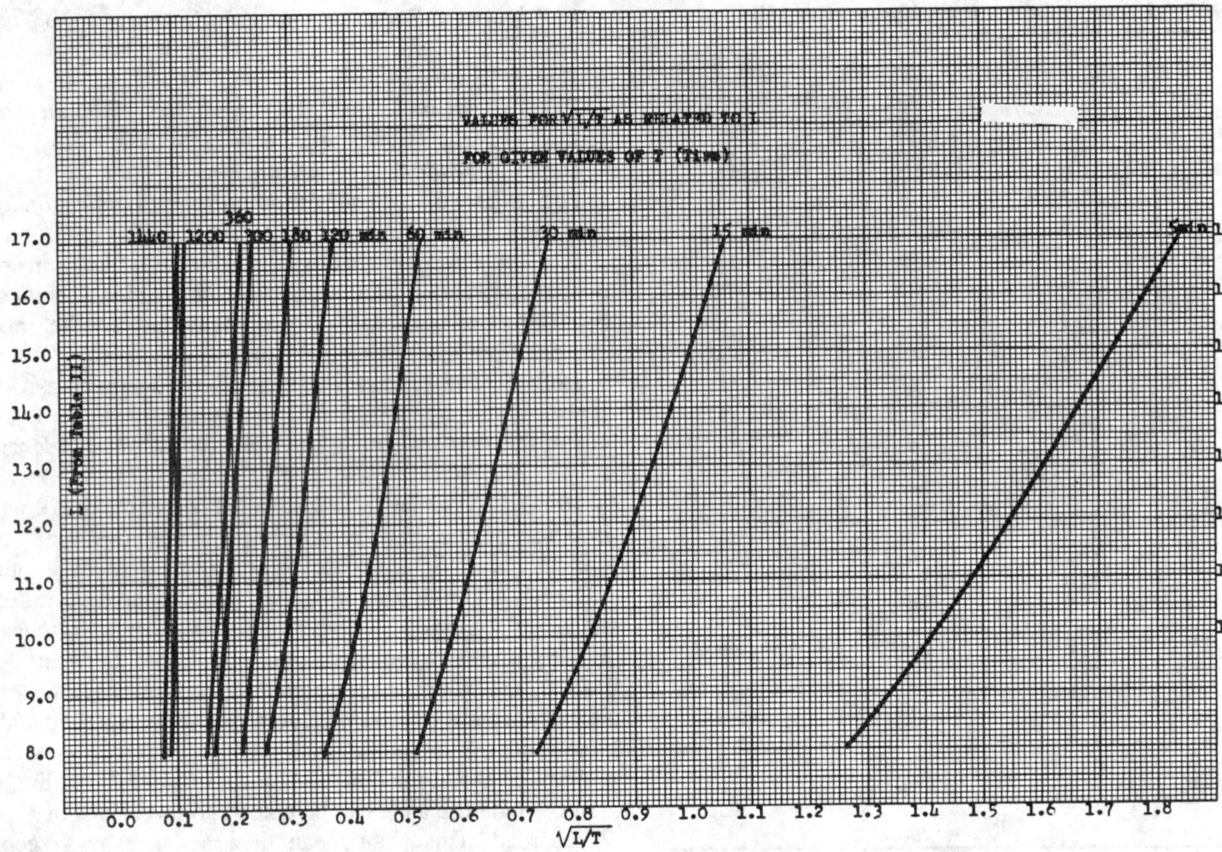

FIG. 7 Values for $\sqrt{L/T}$ as Related to *L* for Given Values of *T* (Time)

TABLE 7 *D* as Related to $\sqrt{L/T}$ at 20°C

$\sqrt{L/T}$	*D*, µm	$\sqrt{L/T}$	*D*, µm
0.05	0.7	1.05	14.0
0.10	1.4	1.10	14.7
0.15	2.0	1.15	15.4
0.20	2.7	1.20	16.0
0.25	3.4	1.25	16.7
0.30	4.0	1.30	17.4
0.35	4.7	1.35	18.0
0.40	5.4	1.40	18.7
0.45	6.0	1.45	19.4
0.50	6.7	1.50	20.1
0.55	7.4	1.55	20.8
0.60	8.0	1.60	21.4
0.65	8.7	1.65	22.1
0.70	9.4	1.70	22.8
0.75	10.0	1.75	23.4
0.80	10.7	1.80	24.1
0.85	11.4	1.85	24.8
0.90	12.0		
0.95	12.7		
1.00	13.4		

TABLE 8 ΔD from $\sqrt{L/T}$

$\sqrt{L/T}$	ΔD
2.0	0.32
1.9	0.31
1.8	0.29
1.7	0.27
1.6	0.26
1.5	0.24
1.4	0.23
1.3	0.21
1.2	0.19
1.1	0.18
1.0	0.16
0.9	0.15
0.8	0.13
0.7	0.11
0.6	0.10
0.5	0.08
0.4	0.07
0.3	0.05
0.2	0.03
0.1	0.02

16.7 *Report:*

16.7.1 The *Y* value is recorded as the dry brightness of that specific limestone.

16.8 *Precision and Bias:*

16.8.1 The same instrument, operator, and standard should reproduce ±0.2 %. Different instrument (Note 10), operators, and standard should agree ±1.0 %.

NOTE 10—It is recognized that there are various manufacturers of reflectometers, and testing has been undertaken to relate *X*, *Y*, and *Z* tristimulus color values from one instrument to another. If results of this comparison testing are desired, please contact the Pulverized Limestone Association.

TABLE 9 Values for P as Related to R_c, Using $a = 0.988$ and $W = 40$

NOTE 1—Calculate to nearest 0.1 of R_c. For a reading of 24.7, take reading of 24.5 which is 60.0 and add 2 × 0.4 or 60.8%.

R_c	P	R_c	P	R_c	P	R_c	P
0.0	0.0	11.5	28.5	23.0	57.0	34.5	85.0
0.5	1.5	12.0	30.0	23.5	58.0	35.0	86.5
1.0	2.5	12.5	31.0	24.0	59.0	35.5	87.5
1.5	3.5	13.0	32.0	24.5	60.0	36.0	89.0
2.0	5.0	13.5	33.5	25.0	62.0	36.5	90.0
2.5	6.0	14.0	35.0	25.5	63.0	37.0	91.5
3.0	7.5	14.5	36.0	26.0	64.0	37.5	92.5
3.5	8.5	15.0	37.5	26.5	65.0	38.0	94.0
4.0	10.0	15.5	38.5	27.0	66.5	38.5	95.0
4.5	11.0	16.0	39.5	27.5	67.5	39.0	96.5
5.0	12.5	16.5	40.5	28.0	69.0	39.5	97.5
5.5	13.5	17.0	42.0	28.5	70.0	40.0	99.0
6.0	15.0	17.5	43.0	29.0	71.5	40.5	100.0
6.5	16.0	18.0	44.5	29.5	72.5		
7.0	17.5	18.5	45.5	30.0	74.0		
7.5	18.5	19.0	47.0	30.5	75.0		
8.0	20.0	19.5	48.0	31.0	76.5		
8.5	21.0	20.0	49.5	31.5	77.5		
9.0	22.5	20.5	50.5	32.0	79.0		
9.5	23.5	21.0	52.0	32.5	80.0		
10.0	25.0	21.5	53.0	33.0	81.5		
10.5	26.0	22.0	54.5	33.5	82.5		
11.0	27.5	22.5	56.5	34.0	84.0		

FIG. 8 Dry-Powder Press

17. Apparent Loose Density of Hydrated Lime, Pulverized Quicklime, and Limestone

17.1 *Significance and Use*:

17.1.1 This test method determines the loose or unsettled density of hydrated lime, pulverized quicklime, and limestone. It provides for an approximate measure of the maximum volume occupied by a given weight of hydrated lime, pulverized quicklime, or limestone.

17.2 *Apparatus*:

17.2.1 *Flour Sifter*—A 114 to 127 mm (4½ to 5 in.) kitchen-type flour sifter of either the squeeze handle type or the hand crank type. It shall be able to hold at least 300 g of hydrated lime or 500 g of limestone or quicklime. The wire mesh openings should be between 0.8 and 1.5 mm.

17.2.2 *Density Cup*, 400 mL cylindrical cup as described in the Apparatus Section of Test Method C 185.

17.2.3 *Balance*, suitable for weighing at least 800 g accurately to 0.1 g.

17.2.4 *Clock or Watch*.

17.2.5 *Straight Edge*.

17.3 *Procedure*:

17.3.1 Weigh the empty density cup to the nearest 0.1 g on a balance. Place the tared cup on a solid table with a suitable mat inserted underneath the cup to collect excess sample spilling over the cup. Fill the flour sifter with more than enough material to fill the density cup. Start the clock and the sifting device to facilitate the flow of powder into the cup. Overflow the cup until there is a cone of excess material.

17.3.2 After 3 min, carefully remove the excess powder by passing the edge of a spatula blade parallel with, and in contact with, the top of the cup. Move the spatula smoothly and keep it level at all times to prevent packing or pulling the sample out of the cup.

17.3.3 After the cup is level, lightly tap it with the edge of the spatula to settle the powder. Wipe the outside of the cup with a lintless cloth or paper towel. Avoid spilling the sample while transferring the cup to the balance for weighing.

17.3.4 Weigh the cup and sample to the nearest 0.1 g and determine the weight of the sample by difference.

17.3.5 The loose density of the material is calculated and reported as grams per cubic centimetre, or as pounds per cubic foot.

17.4 *Calculation*:

17.4.1 Calculate the loose density as follows:

$$D = W/V \tag{4}$$

where:

D = loose density,
W = weight of sample, g, and
V = volume of cup, cm^3.

17.4.2 For reporting as pounds per cubic foot, multiply grams per cubic centimetre by 62.43.

17.5 *Precision and Bias*:

17.5.1 *Single Operator Precision*—The single operator standard deviation has been found to be 0.4 lb/ft^3(Note 10). Therefore, results of two properly conducted tests by the same operator on the same material should not differ by more than 1.13 lb/ft^3.

17.5.2 *Multilaboratory Precision*—The multilaboratory standard deviation has been found to be 1.26 lb/ft^3(Note 11). Therefore, results of two properly conducted tests from two different laboratories on samples of the same material should not differ by more than 3.6 lb/ft^3.

NOTE 11—These numbers represent, respectively, the (1s) and (d2s) limits as described in Practice C 670.

17.5.3 The above precision statements are based on a multilaboratory testing program for determination of the loose bulk density of hydrated lime. No statement is made regarding the precision of this method as it relates to other materials. Due to a lack of a recognized industry standard, the bias of this test method has not been determined.

18. Apparent Packed Density of Hydrated Lime, Pulverized Quicklime, and Limestone

18.1 Significance and Use:

18.1.1 This test method determines the packed or settled density of hydrated lime, pulverized quicklime, and limestone. It provides for determining the minimum volume occupied by a given weight of hydrated lime, pulverized quicklime or limestone.

18.2 *Apparatus*:

18.2.1 *Graduated Cylinder*, 100 mL capacity.

18.2.2 *Balance*, accurate to 0.1 g.

18.3 *Procedure*:

18.3.1 Weigh to within 0.1 g a 25 g sample of powdered material and transfer it to the graduated cylinder.

18.3.2 Allow powder to settle by gently tapping the cylinder on a desk top cushioned with a thick magazine or writing tablet so that compaction occurs without fluffing.

18.3.3 Record the volume of the lime after each 100 taps and continue tapping until compaction volume change is less than 0.5 mL/100 taps.

18.3.4 Calculate the density in grams per cubic centimetre or in pounds per cubic foot to the nearest pound.

18.4 *Calculation*:

18.4.1 Calculate the packed density as follows:

$$D = W/V \quad (5)$$

where:
D = packed density,
W = weight of sample, g, and
V = final volume of sample, cm^3.

18.4.2 For reporting as pounds per cubic foot, multiply grams per cubic centimetre by 62.43.

18.5 *Precision and Bias*:

18.5.1 *Single Operator Precision*—The single operator standard deviation has been found to be 0.5 lb/ft^3(Note 11). Therefore, results of two properly conducted tests by the same operator on the same material should not differ by more than 1.4 lb/ft^3.

18.5.2 *Multilaboratory Precision*—The multilaboratory standard deviation has been found to be 1.7 lb/ft^3(Note 11). Therefore, results of two properly conducted tests from two different laboratories on samples of the same material should not differ by more than 4.8 lb/ft^3.

18.5.3 The above precision statements are based on a multilaboratory testing program for determination of the packed bulk density of hydrated lime. No statement is made regarding the precision of this method as it relates to other materials. Due to a lack of a recognized industry standard, the bias of this test method has not been determined.

19. Fineness of Pulverized Quicklime and Hydrated Lime by Air Permeability

19.1 *Significance and Use*:

19.1.1 This test method covers the determination of fineness of pulverized quicklime and hydrated lime using the Blaine air permeability apparatus described in Test Method C 204. Fineness in terms of surface area shall be expressed as total surface area in square centimetres per gram, or square metres per kilogram.

19.1.2 This test method provides, in general, relative rather than absolute fineness values. For the complete description of the apparatus and the procedures for use, refer to Test Method C 204.

19.2 *Precision and Bias*:

19.2.1 Although precision for the test method for fineness of portland cement by air permeability apparatus has been reported in Test Method C 204, the precision of this test method has not been determined for pulverized lime and hydrated lime. When sufficient data has been obtained and analyzed, a statement of precision will be provided. In the meantime users of this test method are advised to develop their own.

20. Dry Screening of Hydrated Lime, Pulverized Quicklime, and Limestone by Air Jet Sieving

20.1 *Significance and Use*:

20.1.1 This test method uses a rotating slit nozzle to supply a stream of air directed at the backside of a test sieve, keeping the screen from "blinding." The aerated material is then pulled back through the sieve by a vacuum source.

20.1.2 The advantages of dry screening by air jet sieving are twofold. The material being tested is less likely to "blind" the screen because of the recurring counterflow of an air stream to the back of the sieve. Also, dry screening avoids the error introduced by the interaction of the test material with soluble liquid media.

20.1.3 This test method is suitable for screening material from a nominal 300 μm (50 mesh) in size to 20 μm (635 mesh).

NOTE 12—Blinding of the sieves can occur at various sizes depending on the materials being sieved. Experience has shown 45 μm (325 mesh) to be the lower limit with some hydrates. Other hydrates and pulverized quicklime may be sieved to 32 μm (450 mesh). Limestone can be sieved to 20 μm (635 mesh).

20.2 *Apparatus*:

20.2.1 *An Enclosed Device*, capable of creating a vacuum on the backside of a sieve causing a rotating slit nozzle to supply an air stream perpendicular to the bottom of the sieve. The purpose is to suspend all material on the sieve by the air stream on a rotating basis.[20]

20.2.2 *Balance*, suitable for weighing accurately to 0.01 g.

NOTE 13—Selection of balance with regard to accuracy is dependent on the sample size chosen and residue retained and must be consistent with the accuracy required. Therefore, a balance weighing accurately to 0.001 g may be desired.

20.2.3 *Brush*, soft bristle.

20.2.4 *Sieve Cover*—A hard plastic transparent cover used to create a vacuum on the sieve.

20.2.5 *Test Sieves*—The sieves should be constructed using a woven wire, either brass or stainless steel, mounted on a substantial frame. Electroformed sieves are not recommended because of increased blinding and cleaning problems, making them impractical to use under most conditions. The sieves shall be approximately 8 in. in diameter and conform to Specification E 11. A flexible collar must be used to ensure an air tight

[20] The apparatus describes commercially available units sold by the Alpine American Corporation of Natick, Massachusetts. Although the description of the apparatus is directed toward this commercially available equipment, it does not restrict the use of other equivalent equipment which may be available or may be constructed, as long as it follows the general principles outlined under the summary of this test method.

fit between the sieve and the device.

20.3 *Procedure*:

20.3.1 After placing the appropriate sieve into position, weigh (to the nearest 0.01 g) a sample of the test material and place it on the sieve.

NOTE 14—The amount of sample and duration of sieving are dependent upon the type of material and gradation and therefore should be adapted to individual conditions. Generally, the larger the sample size, the more representative of the material tested and the less significant are errors of technique, therefore, the results are the more exact. Sample weights can vary from 20 g for material finer than 40 μm up to 50 g for larger, heavier materials.

20.3.2 Place cover on sieve, set timer to 6 min and start vacuum (maintain vacuum according to manufacturer's recommendation). Any material clinging to the cover or edge of the sieve can be removed by light tapping with a mallet or similar device (see Note 14). If agglomerations form, they can be broken apart with a soft bristle brush.

NOTE 15—Static electrical charges can often develop on the cover (if it is made of plastic) causing it to hold a heavy film of the material being sieved. If tapping will not loosen the material, a static face sheet[21] may be used to wipe the cover surface before starting the test.

20.3.3 After screening, clean the sieve with a fine bristle brush being careful not to damage the mesh and then weigh the residue to the nearest 0.01 g.

20.4 *Calculation*:

20.4.1 Calculate percent passing as follows:

$$[(S - R)/S] \times 100 = \text{percent passing} \quad (6)$$

where:

S = sample weight, g, and
R = weight of sieve residue, g.

20.5 *Precision and Bias*:

20.5.1 There are as yet insufficient analyzed data to permit preparation of a precision and bias statement for this test method.

21. Limestone Grindability Determination by the Laboratory Ball Mill Method

21.1 *Scope*:

21.1.1 This test method is used to determine the relative grindability or ease of pulverization of limestones of differing hardness and to report this as a grindability index.

21.1.2 This test method is applicable to all types of limestone.

21.2 *Summary of Test Method*:

21.2.1 Limestone of a specified size range is wet ground in a ball mill therein receiving a specified amount of grinding energy. The amount of minus 75-μm (200-mesh) limestone produced is measured by wet sieving and reported as the percent passing 75-μm (200-mesh) after 5000 revolutions. This is the grindability index.

21.3 *Significance and Use*:

21.3.1 This test method is useful for comparison and acceptance testing of limestone for applications where fine ground limestone is desired.

21.4 *Apparatus*:

21.4.1 *Jar Mill* ,[22] operated at 110 ± 10 r/min.

21.4.2 *Mill Jar* ,[23] ceramic 14 cm (5½in.) diameter by 17.2 cm (6¾in.) high.

21.4.3 *Grinding Media*, 160 ± 1 g total, consisting of seven ceramic 21 by 21 mm (13⁄16 by 13⁄16 in.) cylindrical grinding media[24] (about 23 g each).

21.4.4 The sieves used shall conform to the requirements of Specification E 11.

21.4.5 Weights and weighing devices, shall conform to the requirements of Specification C 1005.

21.4.6 *Drying Oven*, capable of maintaining 100°C.

21.4.7 *A Chipmunk Crusher* ,[25] capable of breaking large rocks to less than 6.35 mm (¼ in.).

21.4.8 *Riffle Sample Splitter* ,[26] open pan, 12.7 mm (½ in.) chute width.

21.4.9 *Stopwatch.*

21.5 *Reagents and Materials*:

21.5.1 *Milling Solution*, a 0.1 % solution of acrylate based dispersant.[27] The dispersant chosen should not increase the solubility of limestone in water.[28]

21.6 *Sampling*:

21.6.1 Sample in accordance with Practice D 75.

21.6.2 Reduce the sample in accordance with Practice C 702 and prepare by sieving out the material that passes a 850-μm (No. 20) sieve[29] and is retained on a 425-μm (No. 40) sieve.[29]

21.7 *Procedure*:

21.7.1 Weigh seven grinding media, make adjustments (by substitutions or filing) to bring total weight to 160 g ± 1 g.

21.7.2 If the jar mill has provision for automatic shut-off, set it for 5000 revolutions, otherwise determine the mill r/min by counting the revolutions in an accurately timed period (using stopwatch) and then calculate the exact time required for 5000 revolutions.

21.7.3 Weigh out 20 ± 0.01 g of dried 20 by 40 mesh limestone. Record actual weight as $W1$.

21.7.4 Add 180 mL of milling solution to clean and empty mill jar.

21.7.5 Add the seven grinding media and quantitatively transfer the limestone sample to the mill jar and secure the top.

21.7.6 Place the mill jar on the mill rollers and operate the mill for the exact time required to make 5000 revolutions.

21.7.7 Quantitatively transfer the limestone slurry from the jar mill by rinsing the entire contents onto a coarse sieve (for

[21] Commercially available.

[22] U.S. Stoneware No. 753-RM rotating mill, or equivalent, has been found satisfactory for this purpose.

[23] U.S. Stoneware mill jar No. 774-B-00, or equivalent, has been found satisfactory for this purpose.

[24] U.S. Stoneware burundum 13⁄16in. grinding media, or equivalent, has been found satisfactory for this purpose.

[25] Bico-Braun Crusher (jaw, chipmunk) Model No. 241-34, or equivalent, has been found satisfactory for this purpose.

[26] Curtis Matheson Scientific Model No. 095-851, or equivalent, has been found satisfactory for this purpose.

[27] Alcosperse 149 from Alco Chemical Corp., or equivalent, has been found satisfactory for this purpose.

[28] Distilled or deionized water should be used for milling solution or solubility tests.

[29] U.S. Standard Sieves 6, 20, 40 and 200 mesh sizes; 20.3 cm (8 in.) diameter by 5 cm (2 in.) depth with stainless steel wire cloth.

example, 3.35 mm (No. 6)) and an underlying 75-μm (No. 200) sieve. Rinse the media and coarse sieve and separate the sieves.

21.7.8 Wet sieve the sample remaining on the 75-μm (No. 200) sieve to remove the finer material.

21.7.9 Dry and weigh the residue from the 75-μm (No. 200) sieve and record as $W2$ (to the nearest 0.01 g).

21.8 *Calculation*:

21.8.1 Calculate the grindability index (*GI*) as follows:

$$GI = (W1 - W2)/W1 \times 100 \quad (7)$$

21.9 *Precision and Bias*:

21.9.1 The precision and bias of this test method has not been determined at this time.

22. Specific Gravity of Hydrated Lime Products

22.1 *Significance and Use:*

22.1.1 This test method covers the determination of the specific gravity of hydrated lime. The specific gravity of hydrated lime is needed for calculations of air content (see Section 13) and Blaine Surface Area.

22.1.2 The specific gravity of hydrated lime is defined as the mass of a unit volume of the solids.

22.2 *Apparatus:*

22.2.1 *Le Chatelier Flask*—The standard flask, which is circular in cross section with shape and dimensions conforming essentially to Fig. 1 of Test Method C 188 (Note 16). The requirements in regard to tolerance, inscription and length, spacing, and uniformity of graduation shall be rigidly observed. There shall be a space of at least 10 mm between the highest graduation mark and the lowest point of grinding for the glass stopper.

22.2.1.1 The material of construction shall be best quality glass, transparent and free of striae. The glass shall be chemically resistant and shall have small thermal hysteresis. The flasks shall be thoroughly annealed before being graduated. They shall be of sufficient thickness to ensure reasonable resistance to breakage.

22.2.1.2 The neck shall be graduated from 0 to 1 mL and from 18 to 24 mL in 0.1 mL graduations. The error of any indicated capacity shall not be greater than 0.05 mL.

22.2.1.3 Each flask shall bear a permanent identification number and the stopper, if not interchangeable ground, shall bear the same number. Interchangeable ground-glass parts shall be marked on both members with the standard-taper symbol followed by the size designation. The standard temperature shall be indicated, and the unit of capacity shall be shown by the letters "mL" placed above the highest graduation mark.

NOTE 16—The design is intended to ensure complete drainage of the flask when emptied, and stability of standing on a level surface, as well as accuracy and precision of reading.

22.3 *Reagents:*

22.3.1 *Ethyl Alcohol (Abolute)*, free of water, shall be used in the density determination. The ethanol used should not have been denatured with methanol.

22.4 *Procedure:*

22.4.1 Determine the specific gravity of hydrated lime on the material as received, unless otherwise specified.

22.4.2 Fill the flask (Note 17) with denatured ethyl alcohol to a point on the stem between the 0 and 1-mL mark. Dry the inside of the flask above the level of the liquid, if necessary, after pouring. Record the first reading after the flask has been immersed in the water bath (Note 18) in accordance with 22.3.4.

NOTE 17—It is advisable to use a rubber pad on the table top when filling or rolling the flask.

NOTE 18—Before the hydrated lime has been added to the flask, a loose fitting lead-ring weight around the stem of the flask will be helpful in holding the flask in an upright position in the water bath, or the flask may be held in the water bath by a buret clamp.

22.4.3 Introduce a quantity of hydrated lime, weighed to the nearest 0.05 g, (about 50 g for hydrated lime) in small increments at the same temperature as the liquid (Note 17). Take care to avoid splashing and see that the hydrated lime does not adhere to the inside of the flask above the liquid. A vibrating apparatus may be used to accelerate the introduction of the hydrated lime into the flask and to prevent the lime from sticking to the neck. After all the hydrated lime has been introduced, place the stopper in the flask and roll the flask in an inclined position (Note 17), or gently whirl it in a horizontal circle as to free the lime from air until no further air bubbles rise to the surface of the liquid. If a proper amount of hydrated lime has been added, the level of the liquid will be in its final position at some point of the upper series of graduations. Take the final reading after the flask has been immersed in the water bath in accordance with 22.3.4.

22.4.4 Immerse the flask in a constant-temperature water bath for sufficient periods of time in order to avoid flask temperature variations greater than 0.2°C between the intial and final readings.

22.5 *Calculation:*

22.5.1 The difference between the first and the final readings represents the volume of liquid displaced by the mass of hydrated lime used in the test.

22.5.2 Calculate the hydrated lime density, *p*, as follows:

$$p(\text{Mg/m}^3) = p(\text{g/cm}^3) = \frac{\text{mass of hydrated lime (g)}}{\text{displaced volume (cm}^3\text{)}} \quad (8)$$

NOTE 19—The displaced volume in milliliters is numerically equal to the displaced volume in cubic centimeters.

22.6 *Precision and Bias:*

22.6.1 There are as yet insufficient analyzed data to permit preparation of a precision and bias statement for this test method. When data is collected and analyzed, precision and bias statements will be proposed.

23. Wet Sieve Analysis of Agricultural Liming Materials

23.1 *Scope*:

23.1.1 This test method covers the determination of gradation of an Agricultural Liming material by wash. Material finer than 75 micormetres (No. 200) sieve will be removed from the aggregate during the test, prior to dry sieving of the coarser fraction.

23.1.2 The results of this procedure can be used to determine compliance to a particle size distribution relative to applicable specifications for agricultural liming materials.

23.1.3 This method can **not** be used for the sieve analysis of quicklime or materials containing quicklime, because of the exothermic reaction with water.

23.2 *Significance and Use*:

23.2.1 Material finer than the 75 micrometres (No. 200) sieve can be separated from larger particles more efficiently and completely by wet sieving than through the use of dry sieving. In sieving fine aggregate, the finer fractions can adhere to the coarser fractions through a normal dry sieving operation. Further, due to electrostatic charges, fine material can also adhere to the coarser sieves. By removing the finer fractions through the wash test, these problems are avoided. Therefore, when an accurate determination of an agricultural liming material is desired, this test method should be used.

23.2.2 The results of this test method are included in the calculation of gradation, and the total amount of material finer than 75 micrometres by washing, plus that obtained by dry sieving the same sample, is reported with the results. Usually the additional amount of material finer than 75 micrometres obtained in the subsequent dry sieving process is a small amount. If it is large, the efficiency of the washing operation should be checked. It could, also, be an indication of degradation of the material.

23.3 *Apparatus and Materials*:

23.3.1 *Balance*—A balance or scale readable and accurate to 0.1 g or 0.1 % of the test load, whichever is greater, at any point within the range of use.

23.3.2 *Sieves*—The sieves shall be mounted on substantial frames, either full height or extended, and constructed in a manner that will prevent loss of material during sieving. The sieves shall be of equal diameter and shape to allow for nesting. The sieves shall conform to Specification E 11.

23.3.3 *Sieve Pan*—A pan of the same diameter as the sieves.

23.3.4 *Container*—A pan or vessel of a size sufficient to contain the sample covered with water and to permit vigorous agitation without loss of any part of the sample or water.

23.3.5 *Oven*—An oven of sufficient size, capable of maintaining a uniform temperature of 110° ± 5°C.

NOTE 20—The use of a mechanical apparatus to perform the washing operation is not precluded, provided the results are consistent with those obtained using manual operations. The use of some mechanical washing equipment with some samples may cause degradation of the sample.

23.4 *Sampling*:

23.4.1 Since sampling is equally as important as testing, it is critical that samples be obtained that are representative of the materials being investigated. Therefore, sampling must be conducted in as conscientious a manner as possible. Incremental sampling should be employed, combining the increments into one bulk sample.

23.4.2 Use the procedures described in Practice C 50, or Practice D 75.

23.4.3 To reduce the bulk sample to the amount needed for testing, the use of mechanical splitters or the process of quartering must be used. At no time can a random lot be drawn from the bulk sample for testing.

23.5 *Procedure*:

23.5.1 Dry the test sample to constant mass at a temperature of 110 ± 5°C. Determine the mass to the nearest 0.1 % of the mass of the test sample. The amount of sample required for this test shall be between 200 to 400 g. The test sample shall be the end result of the reduction. Reduction to an exact predetermined mass shall not be permitted.

23.5.2 After drying and determining the mass, place the test sample in the container and add sufficient water to cover it. Agitate the sample with sufficient vigor to result in complete separation from the coarser particles of all particles finer than the 75 micrometre sieve from the coarser particles, and to bring the fine material into suspension. Immediately pour the wash water containing the suspended and dissolved solids over the nested sieves (see Note 21), arranged with the coarser sieve on top. Take care to avoid, as much as feasible, the decantation of coarser particles of the sample.

NOTE 21—It is recommended that a #16 mesh sieve or one of approximate size be used to nest on top of the # 200 mesh sieve for protection from the sample.

23.5.3 Add a second charge of water to the sample in the container, agitate, and decant as before. Repeat this operation until the wash water is clear (see Note 22).

NOTE 22—If mechanical washing equipment is used, the charging of water, agitating, and decanting may be a continuous operation.

23.5.4 Return all material retained on the nested sieves by flushing to the washed sample. Dry the washed material to constant mass at a temperature of 110 ± 5° C and determine the mass to the nearest 0.1 g. Calculate the amount of material passing a 75 micrometre sieve by washing according to 23.6.1.

23.5.5 Select the sieves of a suitable size as desired or required by specification and those sieves necessary to regulate the amount of material between the various screens. Nest the screens in order of decreasing size of opening, from top to bottom, on a sieve pan. Be sure to nest the 75 micrometre sieve just above the pan. Place the dried sample on the top sieve. Shake the nested sieves briskly or by use of a mechanical device for a sufficient period to assure that the sample has been adequately screened (see Note 23).

NOTE 23—To be adequately screened, no more than 1 % of the residue on any individual sieve will pass that sieve during 1 min. of continuous sieving. For a more detailed explanation refer to C 136-84a, paragraph 7.4.

23.5.6 The amount of material on an eight inch sieve after shaking is limited to approximately 200 g, so that all particles during a test have access to the screen medium (see Note 24). If the amount remaining on a sieve is greater than 200 g, the sieve analysis must be repeated. Place an intermediate sieve between the overloaded sieve and the one nested above.

NOTE 24—A twelve inch screen can be used, but the amount of material retained on a sieve is limited to 6 kg/m^2.

23.5.7 Record the weight of material from each size increment and the sieve pan by weighing on a balance to the nearest 0.1 g (see Note 25). Total the retained masses of all the size increments and the sieve pan from the dry sieve test. If this amount differs from the dry mass of the sample after washing by more than 0.3 %, the results should not be used for acceptance purposes.

NOTE 25—A soft bristle can be used to help remove material from sieves, but at no time can material be forced through a sieve.

23.6 *Calculation*:

23.6.1 Calculate the amount of a material passing a 75 micrometre sieve by washing as follows:

$$A = [(B - C) / B] \times 100 \quad (9)$$

where:

A = percentage of material finer than a 75 micrometre sieve by washing,
B = original dry mass of sample, g, and
C = dry mass of sample after washing, g.

23.6.2 Calculation of amount of material passing individual sieves follows:

23.6.2.1 Add weight of material finer than a 75 micrometre sieve by washing (B – C, in 23.6.1) to the weight of material finer than a 75 micrometre sieve obtained in the dry screening, if washed.

23.6.2.2 Calculate the percent retained on each sieve and the pan as follows:

$$A = (B / C) \times 100 \quad (10)$$

where:

A = percentage of material retained on each sieve (pan),
B = dry mass of sample, g, retained on each sieve (pan), and
C = original dry mass of sample, g.

23.6.2.3 Calculate the cumulative percent retained for each screen by adding the percentage retained on that sieve and all sieves of larger diameter opening.

23.6.2.4 For each screen, subtract the cumulative percent retained from 100 to obtain the percent passing.

23.7 *Precision and Bias*:

23.7.1 The precision and bias of this method has not been determined by this subcommittee at this time. For precision and bias data from similar methods refer to Test Method C 117 and Test Method C 136.

24. Keywords

24.1 apparent density; dry brightness; dry screening; fineness; lime putty; limestone grindability; particle size; slaking rate

Designation: C 114 – 99

Standard Test Methods for Chemical Analysis of Hydraulic Cement[1]

This standard is issued under the fixed designation C 114; the number immediately following the designation indicates the year of original adoption or, in the case of revision, the year of last revision. A number in parentheses indicates the year of last reapproval. A superscript epsilon (ε) indicates an editorial change since the last revision or reapproval.

1. Scope

1.1 These test methods cover the chemical analyses of hydraulic cements. Any test methods of demonstrated acceptable precision and bias may be used for analysis of hydraulic cements, including analyses for referee and certification purposes, as explained in Section 3. Specific chemical test methods are provided for ease of reference for those desiring to use them. Thcy are grouped as Reference Test Methods and Alternative Test Methods. The reference test methods are long accepted wet chemical test methods which provide a reasonably well-integrated basic scheme of analysis for hydraulic cements. The alternative test methods generally provide individual determination of specific components and may be used alone or as alternates and determinations within the basic scheme at the option of the analyst and as indicated in the individual method.

1.2 *Contents:*

Section	Subject
2	Referenced Documents
3	Number of Determinations and Permissible Variations
3.1	Referee Analyses
3.2	Optional Analyses
3.3	Performance Requirements for Rapid Test Methods
3.4	Precision and Bias
4	General
4.1	Interferences and Limitations
4.2	Apparatus and Materials
4.3	Reagents
4.4	Sample Preparation
4.5	General Procedures
4.6	Recommended Order for Reporting Analyses
	Reference Test Methods
5	Insoluble Residue
6	Silicon Dioxide
6.2	Cements with Insoluble Residue Less Than 1 %
6.3	Cements with Insoluble Residue Greater Than 1 %
7	Ammonium Hydroxide Group
8	Ferric Oxide
9	Phosphorus Pentoxide
10	Titanium Dioxide
11	Zinc Oxide
12	Aluminum Oxide
13	Calcium Oxide
14	Magnesium Oxide
15	Sulfur
15.1	Sulfur Trioxide
15.2	Sulfide
16	Loss On Ignition
16.1	Portland Cement
16.2	Portland Blast-Furnace Slag Cement and Slag Cement
17	Sodium and Potassium Oxides
17.1	Total Alkalis
17.2	Water-Soluble Alkalis
18	Manganic Oxide
19	Chloride
20	Chloroform-Soluble Organic Substances
	Alternative Test Methods
21	Calcium Oxide
22	Magnesium Oxide
23	Loss on Ignition
23.1	Portland Blast-Furnace Slag Cement and Slag Cement
24	Titanium Dioxide
25	Phosphorus Pentoxide
26	Manganic Oxide
27	Free Calcium Oxide

Appendices	Title
X1	Example of Determination of Equivalence Point for the Chloride Determination
X2	CO_2 Determinations in Hydraulic Cements

1.3 The values stated in SI units are to be regarded as the standard.

1.4 *This standard does not purport to address all of the safety concerns, if any, associated with its use. It is the responsibility of the user of this standard to establish appropriate safety and health practices and determine the applicability of regulatory limitations prior to use.* See 6.3.2.1 and Note 43 for specific caution statements.

2. Referenced Documents

2.1 *ASTM Standards:*

C 25 Standard Test Methods for Chemical Analysis of Limestone, Quicklime, and Hydrated Lime[2]

C 115 Test Method for Fineness of Portland Cement by the Turbidimeter[2]

C 150 Specification for Portland Cement[2]

C 183 Practice for Sampling and the Amount of Testing of Hydraulic Cement[2]

C 595 Specification for Blended Hydraulic Cements[2]

[1] These test methods are under the jurisdiction of ASTM Committee C-1 on Cement and are the direct responsibility of Subcommittee C01.23 on Compositional Analysis.

Current edition approved Jan. 10, 1999. Published February 1999. Originally published as C 114 – 34 T. Last previous edition C 114 – 98.

[2] *Annual Book of ASTM Standards,* Vol 04.01.

D 1193 Specification for Reagent Water[3]

E 29 Practice for Using Significant Digits in Test Data to Determine Conformance with Specifications[4]

E 275 Practice for Describing and Measuring Performance of Ultraviolet, Visible, and Near Infrared Spectrophotometers[5]

E 350 Test Methods for Chemical Analysis of Carbon Steel, Low-Alloy Steel, Silicon Electrical Steel, Ingot Iron, and Wrought Iron[6]

E 617 Specification for Laboratory Weights and Precision Mass Standards[4]

E 832 Specification for Laboratory Filter Papers[4]

3. Number of Determinations and Permissible Variations

3.1 *Referee Analyses*—The reference test methods that follow in Sections 5 through 20, or other test methods qualified according to 3.3, are required for referee analysis in those cases where conformance to chemical specification requirements are questioned. In these cases, a cement shall not be rejected for failure to conform to chemical requirements unless all determinations of constituents involved and all necessary separations prior to the determination of any one constituent are made entirely by reference test methods prescribed in the appropriate sections of this test method or by other qualified test methods, except when specific test methods are prescribed in the standard specification for the cement in question. The test methods actually used for the analysis shall be designated.

3.1.1 Referee analyses, when there is a question regarding acceptance, shall be made in duplicate and the analyses shall be made on different days. If the two results do not agree within the permissible variation given in Table 1, the determination shall be repeated until two or three results agree within the permissible variation. When two or three results do agree within the permissible variation, their average shall be accepted as the correct value. When an average of either two or three results can be calculated, the calculation shall be based on the three results. For the purpose of comparing analyses and calculating the average of acceptable results, the percentages shall be calculated to the nearest 0.01 (or 0.001 in the case of chloroform-soluble organic substances), although some of the average values are reported to 0.1 as indicated in the test methods. When a blank determination is specified, one shall be made with each individual analysis or with each group of two or more samples analyzed on the same day for a given component.

3.1.2 Referee analyses or analyses intended for use as a basis for acceptance or rejection of a cement or for manufacturer's certification shall be made only after demonstration of precise and accurate analyses by the test methods in use by meeting the requirements of 3.1.3, except when demonstrated under 3.3.2.1. Such demonstration may be made concurrently with analysis of the cement being tested and must have been made within the preceding two years. The demonstration is required only for those constituents being used as a basis for acceptance, rejection, or certification of a cement, but may be made for any constituent of cement for which a standard exists.

3.1.3 Initial qualification of the operator/analyst shall be demonstrated by analysis of each constituent of concern in at least one NIST SRM cement (Note 1) no matter what test

TABLE 1 Maximum Permissible Variations in Results[A]

(Column 1) Component	(Column 2) Maximum Difference Between Duplicates[D]	(Column 3) Maximum Difference of the Average of Duplicates from SRM Certificate Values[B,C,D]
SiO_2 (silicon dioxide)	0.16	±0.2
Al_2O_3 (aluminum oxide)	0.20	±0.2
Fe_2O_3 (ferric oxide)	0.10	±0.10
CaO (calcium oxide)	0.20	±0.3
MgO (magnesium oxide)	0.16	±0.2
SO_3 (sulfur trioxide)	0.10	±0.1
LOI (loss on ignition)	0.10	±0.10
Na_2O (sodium oxide)	0.03	±0.05
K_2O (potassium oxide)	0.03	±0.05
TiO_2 (titanium dioxide)	0.02	±0.03
P_2O_5 (phosphorus pentoxide)	0.03	±0.03
ZnO (zinc oxide)	0.03	±0.03
Mn_2O_3 (manganic oxide)	0.03	±0.03
S (sulfide sulfur)	0.01	[F]
Cl (chloride)	0.02	[F]
IR (insoluble residue)	0.10	[F]
Cx (free calcium oxide)	0.20	[F]
CO_2 (carbon dioxide)	0.12	[F,G]
Alk_{sol} (water-soluble alkali)[E]	0.75/*w*	[F]
Chl_{sol} (chloroform-soluble organic substances)	0.004	[F]

[A] When seven SRM cements are required, as for demonstrating the performance of rapid test methods, at least six of the seven shall be within the prescribed limits and the seventh shall differ by no more than twice that value. When more than seven SRM's are used, as for demonstrating the performance of rapid test methods, at least 77 % shall be within the prescribed limits, and the remainder by no more than twice the value. When a lesser number of SRM cements are required, all of the values shall be within the prescribed limits.

[B] Interelement corrections may be used for any oxide standardization provided improved accuracy can be demonstrated when the correction is applied to all seven SRM cements.

[C] Where an SRM certificate value includes a subscript number, that subscript number shall be treated as a valid significant figure.

[D] Where no value appears in Column 3, SRM certificate values do not exist. In such cases, only the requirement for differences between duplicates shall apply.

[E] w = weight, in grams, of samples used for the test.

[F] Not applicable. No certificate value given.

[G] Demonstrate performance by analysis, in duplicate, of at least one Portland cement. Prepare three standards, each in duplicate: Standard A shall be selected Portland cement; Standard B shall be Standard A containing 2.00 % Certified $CaCO_3$ (such as NIST 915a); Standard C shall be Standard A containing 5.00 % Certified $CaCO_3$. Weigh and prepare two separate specimens of each standard. Assign the CO_2 content of Standard A as the average of the two values determined, provided they agree within the required limit of Column 2. Assign CO_2 values to Standards B and C as follows: Multiply the Certified $CaCO_3$ value (Y) for CO_2 (from the certificate value) by the mass fraction of Certified $CaCO_3$ added to that standard (percentage added divided by 100); multiply the value determined for Standard A by the mass fraction of Standard A in each of the other standards (that is, 0.98 and 0.95 for Standards B and C, respectively); add the two values for Standard A and for Standard B, respectively; call these values B and C.

Example:

$B = 0.98A + 0.02Y.$

$C = 0.95A + 0.05Y.$

Where for Certified $CaCO_3$, if Y = 39.9 %

$B = 0.98A + 0.80$ % by mass.

$C = 0.95A + 2.00$ % by mass.

Maximum difference between the duplicate CO_2 values for Standards B and C, respectively, shall be 0.17 and 0.24 % by mass. Averages of the duplicate values for Standards B and C shall differ from their assigned values (B and C) by no more than 10 % of those respective assigned values.

[3] *Annual Book of ASTM Standards,* Vol 11.01.

[4] *Annual Book of ASTM Standards,* Vol 14.02.

[5] *Annual Book of ASTM Standards,* Vol 03.06.

[6] *Annual Book of ASTM Standards,* Vol 03.05.

method is used (for example, gravimetric, instrumental). Duplicate samples shall be run on different days. The same test methods to be used for analysis of cement being tested shall be used for analysis of the NIST SRM cement. If the duplicate results do not agree within the permissible variation given in Table 1, the determinations shall be repeated, following identification and correction of problems or errors, until a set of duplicate results do agree within the permissible variation.

NOTE 1—The term SRM samples refers to NIST Portland-Cement Chemical Standard Reference Materials.

3.1.4 The average of the results of acceptable duplicate determinations for each constituent may differ from the SRM certificate value by no more than the value shown in column 2 of Table 1 after correction for minor components when needed. When no SRM certificate value is given, a generally accepted accuracy standard for that constituent does not exist. In such cases, only the differences between duplicate values as specified in 3.1.3 shall apply.

3.1.5 Data demonstrating that precise and accurate results were obtained with NIST SRM cements by the same analyst making the acceptance determination shall be made available on request to all parties concerned when there is a question of acceptance of a cement.

3.2 *Optional Analyses*—The alternative test methods provide, in some cases, procedures that are shorter or more convenient to use for routine determination of certain constituents than are the reference test methods (Note 2). Longer, more complex procedures, in some instances, have been retained as alternative test methods to permit comparison of results by different procedures or for use when unusual materials are being examined, where unusual interferences may be suspected, or when unusual preparation for analysis is required. Test results from alternative test methods may be used as a basis for acceptance or rejection when it is clear that a cement does or does not meet the specification requirement. Any change in test method procedures from those procedures listed in Sections 5 through 27 requires method qualification in accordance with 3.3.

NOTE 2—It is not intended that the use of reference test methods be confined to referee analysis. A reference test method may be used in preference to an alternative test method when so desired. A reference test method must be used where an alternative test method is not provided.

3.2.1 Duplicate analyses and blank determinations are not required when using the alternative test methods. If, however, a blank determination is desired for an alternative test method, one may be used and it need not have been obtained concurrently with the analysis. The final results, when corrected for blank values, should, in either case, be so designated.

3.3 *Performance Requirements for Rapid Test Methods:*[7]

3.3.1 *Definition and Scope*—Where analytical data obtained in accordance with this test method are required, any test method may be used that meets the requirements of 3.3.2. A test method is considered to consist of the specific procedures, reagents, supplies, equipment, instrument, etc. selected and used in a consistent manner by a specific laboratory. See Note 3 for examples of procedures.

NOTE 3—Examples of test methods used successfully by their authors for analysis of hydraulic cement are given in the list of references. Included are test methods using atomic absorption X-ray spectrometry, and spectrophotometry-EDTA.

3.3.1.1 If more than one instrument, even though substantially identical, is used in a specific laboratory for the same analyses, use of each instrument shall constitute a separate test method and each must be qualified separately.

3.3.2 *Qualification of a Test Method*—Prior to use for analysis of hydraulic cement, each test method (see 3.3.1) must be qualified individually for such analysis. Qualification data, or if applicable, requalification data, shall be made available pursuant to the Manufacturer's Certification Section of the appropriate hydraulic cement specification.

3.3.2.1 Using the test method chosen, make single determinations for each oxide under consideration on at least any seven of the SRM samples (Note 1). Complete two rounds of tests on different days repeating all steps of sample preparations. Calculate the differences between values and averages of the values from the two rounds of tests.

3.3.2.2 When seven SRM's are used in the qualification procedure, at least six of the seven differences between duplicates obtained of any single component shall not exceed the limits shown in Column 2 of Table 1 and the remaining differences by no more than twice that value. When more than seven SRM's are used, the values for at least 77 % of the samples shall be within the prescribed limits, while the values for the remainder shall differ by no more than twice that value.

3.3.2.3 For each component and each SRM, the average obtained shall be compared to the certified concentrations. Where a certificate value includes a subscript number, that subscript shall be assumed to be a significant number. When seven SRM's are used in the qualification procedure, at least six of the seven averages for each component (oxide) shall not differ from the certified concentrations by more than the value shown in Column 3 of Table 1, and the remaining average by more than twice that value. When more than seven SRM's are used in the qualification procedure, at least 77 % of the averages for each component (oxide) shall not differ from the certified concentrations by more than the value shown in Column 3 of Table 1, and the remaining average(s) by more than twice that value. The standardization, if needed, used for qualification and for analysis of each constituent shall be determined by valid curve-fitting procedures. The qualification testing shall be conducted with newly prepared specimens.

NOTE 4—An actual drawing of a curve is not required if such curve is not needed for the method in use. A point-to-point, saw-tooth curve that is artificially made to fit a set of data points does not constitute a valid curve-fitting procedure.

3.3.3 *Partial Results*—Test Methods that provide acceptable results for some components but not for others may be used only for those components for which acceptable results are obtained.

3.3.4 *Report of Results*—Chemical analyses obtained by qualified rapid test methods and reported pursuant to the

[7] Gebhardt, R. F., "Rapid Methods for Chemical Analysis of Hydraulic Cement," ASTM STP 985, 1988.

Manufacturer's Certification Section of the appropriate hydraulic cement specification shall be indicated as having been obtained by rapid methods and the type of test method used shall be designated.

3.3.5 *Rejection of Material*—See 3.1 and 3.2.

3.3.6 *Requalification of a Test Method:*

3.3.6.1 Requalification of a test method shall be required upon receipt of substantial evidence that the test method may not be providing data in accordance with Table 1 for one or more constituents. Such requalification may be limited to those constituents indicated to be in error and shall be carried out prior to further use of the method for analysis of those constituents.

3.3.6.2 Substantial evidence that a test method may not be providing data in accordance with Table 1 shall be considered to have been received when a laboratory is informed that analysis of the same material by Reference Test Methods run in accordance with 3.1.1, the final average of a CCRL sample, a certificate value of an NIST SRM, or an accepted value of a known secondary standard differs from the value obtained by the test method in question by more than twice the value shown in Column 2 of Table 1 for one or more constituents. When indirect test methods are involved, as when a value is obtained by difference, corrections shall be made for minor constituents in order to put analyses on a comparable basis prior to determining the differences. (See Note 5.) For any constituents affected, a test method also shall be requalified after any substantial repair or replacement of one or more critical components of an instrument essential to the test method.

NOTE 5—Instrumental analyses can usually detect only the element sought. Therefore, to avoid controversy, the actual procedure used for the elemental analyses should be noted when actual differences with reference procedures can exist. For example, P_2O_5 and TiO_2 are included with Al_2O_3 in the usual wet test method and sulfide sulfur is included in most instrumental procedures with SO_3.

3.3.6.3 If an instrument or piece of equipment is replaced, even if by one of identical make or model, or is significantly modified, a previously qualified test method using such new or modified instrument or equipment shall be considered a new method and must be qualified in accordance with 3.3.2.

3.4 *Precision and Bias*—Different analytical test methods are subject to individual limits of precision and bias. It is the responsibility of the user to demonstrate that the test methods used at least meet the limits of precision and bias shown in Table 1.

4. General

4.1 *Interferences and Limitations:*

4.1.1 These test methods were developed primarily for the analysis of portland cements. However, except for limitations noted in the procedure for specific constituents, the reference test methods provide for accurate analyses of other hydraulic cements that are completely decomposed by hydrochloric acid, or where a preliminary sodium carbonate fusion is made to ensure complete solubility. Some of the alternative test methods may not always provide accurate results because of interferences from elements which are not removed during the procedure.

4.1.2 When using a test method that determines total sulfur, such as most instrumental test methods, sulfide sulfur will be determined with sulfate and included as such. In most hydraulic cements, the difference resulting from such inclusion will be insignificant, less than 0.05 weight %. In some cases, notably slags and slag-containing cements but sometimes other cements as well, significant levels of sulfide may be present. In such cases, especially if there is a question of meeting or not meeting a specification limit or when the most accurate results are desired, analytical test methods shall be chosen so that sulfate and sulfide can be reported separately.

4.2 *Apparatus and Materials:*

4.2.1 *Balance*—The analytical balance used in the chemical determinations shall conform to the following requirements:

4.2.1.1 The balance shall have a capacity of not more than 200 g. It may be of conventional design, either with or without "quick-weighing" devices, or it may be a constant-load, direct-reading type. It shall be capable of reproducing results within 0.0002 g with an accuracy of ±0.0002 g. Direct-reading balances shall have a sensitivity not exceeding 0.0001 g (Note 6). Conventional two-pan balances shall have a maximum sensibility reciprocal of 0.0003 g. Any rapid weighing device that may be provided, such as a chain, damped motion, or heavy riders, shall not increase the basic inaccuracy by more than 0.0001 g at any reading and with any load within the rated capacity of the balance.

NOTE 6—The sensitivity of a direct-reading balance is the weight required to change the reading one graduation. The sensibility reciprocal for a conventional balance is defined as the change in weight required on either pan to change the position of equilibrium one division on the pointer scale at capacity or at any lesser load.

4.2.2 *Weights*—Weights used for analysis shall conform to Types I or II, Grades S or O, Classes 1, 2, or 3 as described in Specification E 617. They shall be checked at least once a year, or when questioned, and adjusted at least to within allowable tolerances for Class 3 weights (Note 7). For this purpose each laboratory shall also maintain, or have available for use, a reference set of standard weights from 50 g to 10 mg, which shall conform at least to Class 3 requirements and be calibrated at intervals not exceeding five years by the National Institute of Standards and Technology (NIST). After initial calibration, recalibration by the NIST may be waived provided it can be shown by documented data obtained within the time interval specified that a weight comparison between summations of smaller weights and a single larger weight nominally equal to that summation, establishes that the allowable tolerances have not been exceeded. All new sets of weights purchased shall have the weights of 1 g and larger made of stainless steel or other corrosion-resisting alloy not requiring protective coating, and shall meet the density requirements for Grades S or O.

NOTE 7—The scientific supply houses do not presently list weights as meeting Specification E 617. They list weights as meeting NIST or OIML standards. The situation with regard to weights is in a state of flux because of the trend toward internationalization. Hopefully this will soon be resolved.

NIST Classes S and S-1 and OIML Class F_1 weights meet the requirements of this standard.

4.2.3 *Glassware and Laboratory Containers*—Standard volumetric flasks, burets, and pipets should be of precision grade or better. Standard-taper, interchangeable, ground-glass joints are recommended for all volumetric glassware and distilling apparatus, when available. Wherever applicable, the use of special types of glassware, such as colored glass for the protection of solutions against light, alkali-resistant glass, and high-silica glass having exceptional resistance to thermal shock is recommended. Polyethylene containers are recommended for all aqueous solutions of alkalies and for standard solutions where the presence of dissolved silica or alkali from the glass would be objectionable. Such containers shall be made of high-density polyethylene having a wall thickness of at least 1 mm.

4.2.4 *Desiccators*—Desiccators shall be provided with a good desiccant, such as magnesium perchlorate, activated alumina, or sulfuric acid. Anhydrous calcium sulfate may also be used provided it has been treated with a color-change indicator to show when it has lost its effectiveness. Calcium chloride is not a satisfactory desiccant for this type of analysis.

4.2.5 *Filter Paper*—Filter paper shall conform to the requirements of Specification E 832, Type II, Quantitative. When coarse-textured paper is required, Class E paper shall be used, when medium-textured paper is required, Class F paper shall be used, and when retentive paper is required, Class G shall be used.

4.2.6 *Crucibles*—Platinum crucibles for ordinary chemical analysis should preferably be made of pure unalloyed platinum and be of 15 to 30-mL capacity. Where alloyed platinum is used for greater stiffness or to obviate sticking of crucible and lid, the alloyed platinum should not decrease in weight by more than 0.2 mg when heated at 1200°C for 1 h.

4.2.7 *Muffle Furnace*—The muffle furnace shall be capable of operation at the temperatures required and shall have an indicating pyrometer accurate within ±25°C, as corrected, if necessary, by calibration. More than one furnace may be used provided each is used within its proper operating temperature range.

4.3 *Reagents:*

4.3.1 *Purity of Reagents*—Reagent grade chemicals shall be used in all tests. Unless otherwise indicated, it is intended that all reagents shall conform to the specifications of the Committee on Analytical Reagents of the American Chemical Society, where such specifications are available.[8] Other grades may be used, provided it is first ascertained that the reagent is of sufficiently high purity to permit its use without lessening the accuracy of the determination.

4.3.2 Unless otherwise indicated, references to water shall mean water conforming to the numerical limits for Type II reagent water described in Specification D 1193.

4.3.3 *Concentration of Reagents:*

4.3.3.1 *Prepackaged Reagents*—Commercial prepackaged standard solutions or diluted prepackaged concentrations of a reagent may be used whenever that reagent is called for in the procedures provided that the purity and concentrations are as specified. Verify purity and concentration of such reagents by suitable tests.

4.3.3.2 *Concentrated Acids and Ammonium Hydroxide*—When acids and ammonium hydroxide are specified by name or chemical formula only, it shall be understood that concentrated reagents of the following specific gravities or concentrations by weight are intended:

Acetic acid ($HC_2H_3O_2$)	99.5 %
Hydrochloric acid (HCl)	sp gr 1.19
Hydrofluoric acid (HF)	48 %
Nitric acid (HNO_3)	sp gr 1.42
Phosphoric acid (H_3PO_4)	85 %
Sulfuric acid (H_2SO_4)	sp gr 1.84
Ammonium hydroxide (NH_4OH)	sp gr 0.90

4.3.3.3 The desired specific gravities or concentrations of all other concentrated acids shall be stated whenever they are specified.

4.3.4 *Diluted Acids and Ammonium Hydroxide*—Concentrations of diluted acids and ammonium hydroxide, except when standardized, are specified as a ratio stating the number of volumes of the concentrated reagent to be added to a given number of volumes of water, for example: HCl (1+99) means 1 volume of concentrated HCl (sp gr 1.19) added to 99 volumes of water.

4.3.5 *Standard Solutions*—Concentrations of standard solutions shall be expressed as normalities (*N*) or as equivalents in grams per millilitre of the component to be determined, for example: 0.1 *N* $Na_2S_2O_3$ solution or $K_2Cr_2O_7$ (1 mL = 0.004 g Fe_2O_3). The average of at least three determinations shall be used for all standardizations. When a material is used as a primary standard, reference has generally been made to the standard furnished by the National Bureau of Standards. However, when primary standard grade materials are otherwise available they may be used or the purity of a salt may be determined by suitable tests.

4.3.6 *Nonstandardized Solutions*—Concentrations of nonstandardized solutions prepared by dissolving a given weight of the solid reagent in a solvent shall be specified in grams of the reagent per litre of solution, and it shall be understood that water is the solvent unless otherwise specified, for example: NaOH solution (10 g/L) means 10 g of NaOH dissolved in water and diluted with water to 1 L. Other nonstandardized solutions may be specified by name only, and the concentration of such solutions will be governed by the instructions for their preparation.

4.3.7 *Indicator Solutions:*

4.3.7.1 *Methyl Red*—Prepare the solution on the basis of 2 g of methyl red/L of 95 % ethyl alcohol.

4.3.7.2 *Phenolphthalein*—Prepare the solution on the basis of 1 g of phenolphthalein/L of 95 % ethyl alcohol.

4.4 *Sample Preparation:*

4.4.1 Before testing, pass representative portions of each sample through a No. 20 (850-μm) sieve, or any other sieve having approximately 20 openings/1 in., in order to mix the sample, break up lumps, and remove foreign materials. Discard the foreign materials and hardened lumps that do not break up on sieving or brushing.

4.4.2 By means of a sample splitter or by quartering, the representative sample shall be reduced to a laboratory sample

---

[8] *Reagent Chemicals, American Chemical Society Specifications,* American Chemical Society, Washington, DC. For suggestions on the testing of reagents not listed by the American Chemical Society, see *Analar Standards for Laboratory Chemicals,* BDH Ltd., Poole, Dorset, U.K., and the *United States Pharmacopeia and National Formulary,* U.S. Pharmacopeia Convention, Inc. (USPC), Rockville, MD.

of at least 50 g. Where larger quantities are required for additional determinations such as water-soluble alkali, chloride, duplicate testing, etc., prepare a sample of at least 100 g.

4.4.3 Pass the laboratory sample through a U.S. No. 100 sieve (sieve opening of 150 μm). Further grind the sieve residue so that it also passes the No. 100 sieve. Homogenize the entire sample by again passing it through the sieve.

4.4.4 Transfer the sample to a clean, dry, glass container with an airtight lid and further mix the sample thoroughly.

4.4.5 Expedite the above procedure so that the sample is exposed to the atmosphere for a minimum time.

4.5 *General Procedures:*

4.5.1 *Weighing*—The calculations included in the individual test methods assume that the exact weight specified has been used. Accurately weighed samples, that are approximately but not exactly equal to the weight specified, may be used provided appropriate corrections are made in the calculations. Unless otherwise stated, weights of all samples and residues should be recorded to the nearest 0.0001 g.

4.5.2 *Tared or Weighed Crucibles*—The tare weight of crucibles shall be determined by preheating the empty crucible to constant weight at the same temperature and under the same conditions as shall be used for the final ignition of a residue and cooling in a desiccator for the same period of time used for the crucible containing the residue.

4.5.3 *Constancy of Weight of Ignited Residues*—To definitely establish the constancy of weight of an ignited residue for referee purposes, the residue shall be ignited at the specified temperature and for the specified time, cooled to room temperature in a desiccator, and weighed. The residue shall then be reheated for at least 30 min, cooled to room temperature in a desiccator, and reweighed. If the two weights do not differ by more than 0.2 mg, constant weight is considered to have been attained. If the difference in weights is greater than 0.2 mg, additional ignition periods are required until two consecutive weights agree within the specified limits. For ignition loss, each reheating period shall be 5 min.

4.5.4 *Volatilization of Platinum*—The possibility of volatilization of platinum or alloying constituents from the crucibles must be considered. On reheating, if the crucible and residue lose the same weight (within 0.2 mg) as the crucible containing the blank, constant weight can be assumed. Crucibles of the same size, composition, and history shall be used for both the sample and the blank.

4.5.5 *Calculation*—In all operations on a set of observed values such as multiplying or dividing, where possible, retain the equivalent of two more places of figures than in the single observed values. For example, if observed values are read or determined to the nearest 0.1 mg, carry numbers to the nearest 0.001 mg in calculation.

4.5.6 *Rounding Figures*—Rounding of figures to the number of significant places required in the report should be done after calculations are completed, in order to keep the final results substantially free of calculation errors. The rounding procedure should follow the principle outlined in Practice E 29.[9]

[9] See also the *ASTM Manual on Presentation of Data and Control Chart Analysis, STP 15D*, 1976.

NOTE 8—The rounding procedure referred to in 4.5.6, in effect, drops all digits beyond the number of places to be retained if the next figure is less than 5. If it is more than 5, or equal to 5 and subsequent places contain a digit other than 0, then the last retained digit is increased by one. When the next digit is equal to 5 and all other subsequent digits are 0, the last digit to be retained is unchanged when it is even and increased by one when it is odd. For example 3.96 (50) remains 3.96 but 3.95 (50) becomes 3.96.

4.6 *Recommended Order for Reporting Analyses*—The following order is recommended for reporting the results of chemical analysis of portland cement:

Major Components:
- SiO_2 (silicon dioxide)
- Al_2O_3 (aluminum oxide)
- Fe_2O_3 (ferric oxide)
- CaO (calcium oxide)
- MgO (magnesium oxide)
- SO_3 (sulfur trioxide)
- Loss on ignition

Minor Components:
- Na_2O (sodium oxide)
- K_2O (potassium oxide)
- TiO_2 (titanium dioxide)
- P_2O_5 (phosphorus pentoxide)
- ZnO (zinc oxide)
- Mn_2O_3 (manganic oxide)
- Sulfide sulfur

Separate Determinations:
- Insoluble residue
- Free calcium oxide
- CO_2 (Carbon Dioxide)
- Water-soluble alkali
- Chloroform—soluble organic substances

REFERENCE TEST METHODS

5. Insoluble Residue (*Reference Test Method*)

5.1 *Summary of Test Method:*

5.1.1 In this test method, insoluble residue of a cement is determined by digestion of the sample in hydrochloric acid followed, after filtration, by further digestion in sodium hydroxide. The resulting residue is ignited and weighed (Note 9).

NOTE 9—This test method, or any other test method designed for the estimation of an acid-insoluble substance in any type of cement, is empirical because the amount obtained depends on the reagents and the time and temperature of digestion. If the amount is large, there may be a little variation in duplicate determinations. The procedure should be followed closely in order to reduce the variation to a minimum.

5.1.2 When this test method is used on blended cement, the decomposition in acid is considered to be complete when the portland-cement clinker is decomposed completely. An ammonium nitrate solution is used in the final washing to prevent finely-ground insoluble material from passing through the filter paper.

5.2 *Reagents:*

5.2.1 *Ammonium Nitrate Solution* (20 g NH_4NO_3/L).

5.2.2 *Sodium Hydroxide Solution* (10 g NaOH/L).

5.3 *Procedure:*

5.3.1 To 1 g of the sample (Note 10) add 25 mL of cold water. Disperse the cement in the water and while swirling the mixture, quickly add 5 mL of HCl. If necessary, warm the solution gently, and grind the material with the flattened end of a glass rod for a few minutes until it is evident that decomposition of the cement is complete (Note 11). Dilute the solution to 50 mL with hot water (nearly boiling) and heat the covered mixture rapidly to near boiling by means of a high-temperature hot plate. Then digest the covered mixture for 15 min at a temperature just below boiling (Note 12). Filter the solution through a medium-textured paper into a 400-mL beaker, wash the beaker, paper, and residue thoroughly with hot water, and reserve the filtrate for the sulfur trioxide determination, if desired (Note 13). Transfer the filter paper and contents to the original beaker, add 100 mL of hot (near boiling) NaOH solution (10 g/L), and digest at a temperature just below boiling for 15 min. During the digestion, occasionally stir the mixture and macerate the filter paper. Acidify the solution with HCl using methyl red as the indicator and add an excess of 4 or 5 drops of HCl. Filter through medium-textured paper and wash the residue at least 14 times with hot NH_4NO_3 solution (20 g/L) making certain to wash the entire filter paper and contents during each washing. Ignite the residue in a weighed platinum crucible at 900 to 1000°C, cool in a desiccator, and weigh.

NOTE 10—If sulfur trioxide is to be determined by turbidimetry it is permissible to determine the insoluble residue on a 0.5-g sample. In this event, the percentage of insoluble residue should be calculated to the nearest 0.01 by multiplying the weight of residue obtained by 200. However, the cement should not be rejected for failure to meet the insoluble residue requirement unless a 1-g sample has been used.

NOTE 11—If a sample of portland cement contains an appreciable amount of manganic oxide, there may be brown compounds of manganese which dissolve slowly in cold diluted HCl but rapidly in hot HCl in the specified strength. In all cases, dilute the solution as soon as decomposition is complete.

NOTE 12—In order to keep the solutions closer to the boiling temperature, it is recommended that these digestions be carried out on an electric hot plate rather than in a steam bath.

NOTE 13—Continue with the sulfur trioxide determination (15.1.2.1 through 15.1.3) by diluting to 250 or 200 mL as required by the appropriate section.

5.3.2 *Blank*—Make a blank determination, following the same procedure and using the same amounts of reagents, and correct the results obtained in the analysis accordingly.

5.4 *Calculation*—Calculate the percentage of the insoluble residue to the nearest 0.01 by multiplying the weight in grams of the residue (corrected for the blank) by 100.

6. Silicon Dioxide (*Reference Test Method*)

6.1 *Selection of Test Method*—For cements other than portland and for which the insoluble residue is unknown, determine the insoluble residue in accordance with Section 5 of these test methods. For portland cements and other cements having an insoluble residue less than 1 %, proceed in accordance with 6.2. For cements having an insoluble residue greater than 1 % proceed in accordance with 6.3.

6.2 *Silicon Dioxide in Portland Cements and Cements with Low Insoluble Residue:*

6.2.1 *Summary of Test Method*—In this test method silicon dioxide (SiO_2) is determined gravimetrically. Ammonium chloride is added and the solution is not evaporated to dryness. This test method was developed primarily for hydraulic cements that are almost completely decomposed by hydrochloric acid and should not be used for hydraulic cements that contain large amounts of acid-insoluble material and require a preliminary sodium carbonate fusion. For such cements, or if prescribed in the standard specification for the cement being analyzed, the more lengthy procedure in 6.3 shall be used.

6.2.2 *Reagent*—Ammonium chloride (NH_4Cl).

6.2.3 *Procedure:*

6.2.3.1 Mix thoroughly 0.5 g of the sample and about 0.5 g of NH_4Cl in a 50-mL beaker, cover the beaker with a watch glass, and add cautiously 5 mL of HCl, allowing the acid to run down the lip of the covered beaker. After the chemical action has subsided, lift the cover, add 1 or 2 drops of HNO_3, stir the mixture with a glass rod, replace the cover, and set the beaker on a steam bath for 30 min (Note 14). During this time of digestion, stir the contents occasionally and break up any remaining lumps to facilitate the complete decomposition of the cement. Fit a medium-textured filter paper to a funnel, transfer the jelly-like mass of silicic acid to the filter as completely as possible without dilution, and allow the solution to drain through. Scrub the beaker with a policeman and rinse the beaker and policeman with hot HCl (1+99). Wash the filter two or three times with hot HCl (1+99) and then with ten or twelve small portions of hot water, allowing each portion to drain through completely. Reserve the filtrate and washings for the determination of the ammonium hydroxide group (Note 15).

NOTE 14—A hot plate may be used instead of a steam bath if the heat is so regulated as to approximate that of a steam bath.

Under conditions where water boils at a lower temperature than at sea level: such as at higher elevations, 30 min may not be sufficient to recover all of the silica. In such cases, increase the time of digestion as necessary to get complete recovery of the silica. In no case should this time exceed 60 min.

NOTE 15—Determine the ammonium hydroxide group in accordance with the procedure described in 7.1 through 7.3.

6.2.3.2 Transfer the filter paper and residue to a weighed platinum crucible, dry, and ignite, at first slowly until the carbon of the paper is completely consumed without inflaming, and finally at 1100 to 1200°C for 1 h. Cool in a desiccator and weigh. Reignite to constant weight. Treat the SiO_2 thus obtained, which will contain small amounts of impurities, in the crucible with 1 or 2 mL of water, 2 drops of H_2SO_4 (1+1), and about 10 mL of HF, and evaporate cautiously to dryness. Finally, heat the small residue at 1050 to 1100°C for 5 min, cool in a desiccator, and weigh. The difference between this weight and the weight previously obtained represents the weight of SiO_2. Consider the weighed residue remaining after the volatilization of SiO_2 as combined aluminum and ferric oxides and add it to the result obtained in the determination of the ammonium hydroxide group.

6.2.3.3 If the HF residue exceeds 0.0020 g, the silica determination shall be repeated, steps should be taken to ensure complete decomposition of the sample before a silica separation is attempted, and the balance of the analysis (ammonium hydroxide group, CaO, and MgO) determined on the new silica filtrate provided the new silica determination has a HF residue of 0.0020 g or less except as provided in 6.2.3.4 and 6.2.3.5.

6.2.3.4 If two or three repeated determinations of a sample of portland cement consistently show HF residues higher than 0.0020 g, this is evidence that contamination has occurred in sampling or the cement has not been burned properly during manufacture. In such a case, do not fuse the large HF residue with pyrosulfate for subsequent addition to the filtrate from the silica separation. Instead, report the value obtained for the HF residue. Do not ignite the ammonium hydroxide group in the crucible containing this abnormally large HF residue.

6.2.3.5 In the analysis of cements other than portland, it may not always be possible to obtain HF residues under 0.0020 g. In such cases, add 0.5 g of sodium or potassium pyrosulfate ($Na_2S_2O_7$ or $K_2S_2O_7$) to the crucible and heat below red heat until the small residue of impurities is dissolved in the melt (Note 16). Cool, dissolve the fused mass in water, and add it to the filtrate and washings reserved for the determination of the ammonium hydroxide group.

NOTE 16—A supply of nonspattering pyrosulfate may be prepared by heating some pyrosulfate in a platinum vessel below red heat until the foaming and spattering cease, cooling, and crushing the fused mass.

6.2.3.6 *Blank*—Make a blank determination, following the same procedure and using the same amounts of reagents, and correct the results obtained in the analysis accordingly.

6.2.4 *Calculation*—Calculate the percentage of SiO_2 to the nearest 0.1 multiplying the mass in grams of SiO_2 by 200 (100 divided by the mass (see 6.2.3.1) or equivalent mass (see 6.3.2.1) of the sample used (0.5 g)).

6.3 *Silicon Dioxide in Cements with Insoluble Residue Greater Than 1 %:*

6.3.1 *Summary of Test Method*—This test method is based on the sodium carbonate fusion followed by double evaporation to dryness of the hydrochloric acid solution of the fusion product to convert silicon dioxide (SiO_2) to the insoluble form. The solution is filtered and the insoluble siliceous residue is ignited and weighed. Silicon dioxide is volatilized by hydrofluoric acid and the loss of weight is reported as pure SiO_2.

6.3.2 *Procedure:*

6.3.2.1 Weigh a quantity of the ignited sample equivalent to 0.5 g of the as-received sample calculated as follows:

$$W = [(0.5\ (100.00 - I)]/100 \quad (1)$$

where:

W = weight of ignited sample, g, and
I = loss of ignition, %.

The ignited material from the loss on ignition determination may be used for the sample. Thoroughly mix the sample with 4 to 6 g of Na_2CO_3 by grinding in an agate mortar. Place a thin layer of Na_2CO_3 on the bottom of a platinum crucible of 20 to 30-mL capacity, add the cement-Na_2CO_3 mixture, and cover the mixture with a thin layer of Na_2CO_3. Place the covered crucible over a moderately low flame and increase the flame gradually to a maximum (approximately 1100°C) and maintain this temperature until the mass is quiescent (about 45 min). Remove the burner, lay aside the cover of the crucible, grasp the crucible with tongs, and slowly rotate the crucible so that the molten contents spread over the sides and solidify as a thin shell on the interior. Set the crucible and cover aside to cool. Rinse off the outside of the crucible and place the crucible on its side in a 300-mL casserole about one third full of water. Warm the casserole and stir until the cake in the crucible disintegrates and can be removed easily. By means of a glass rod, lift the crucible out of the liquid, rinsing it thoroughly with water. Rinse the cover and crucible with HCl (1+3); then add the rinse to the casserole. Very slowly and cautiously add 20 mL of HCl (sp gr 1.19) to the covered casserole. Remove the cover and rinse. If any gritty particles are present, the fusion is incomplete and the test must be repeated, using a new sample. **Caution:** Subsequent steps of the test method must be followed exactly for accurate results.

6.3.2.2 Evaporate the solution to dryness on a steam bath (there is no longer a gelatinous appearance). Without heating the residue any further, treat it with 5 to 10 mL of HCl, wait at least 2 min, and then add an equal amount of water. Cover the dish and digest for 10 min on the steam bath or a hot plate. Dilute the solution with an equal volume of hot water, immediately filter through medium-textured paper and wash the separated SiO_2 thoroughly with hot HCl (1+99), then with hot water. Reserve the residue.

6.3.2.3 Again evaporate the filtrate to dryness, and bake the residue in an oven for 1 h at 105 to 110°C. Cool, add 10 to 15 mL of HCl (1+1), and digest on the steam bath or hot plate for 10 min. Dilute with an equal volume of water, filter immediately on a fresh filter paper, and wash the small SiO_2 residue thoroughly as described in 6.3.2.2. Stir the filtrate and washings and reserve for the determination of the ammonium hydroxide group in accordance with 7.1 through 7.3.

6.3.2.4 Continue the determination of silicon dioxide in accordance with 6.2.3.2.

7. Ammonium Hydroxide Group (*Reference Test Method*)

7.1 *Summary of Test Method*—In this test method aluminum, iron, titanium, and phosphorus are precipitated from the filtrate, after SiO_2 removal, by means of ammonium hydroxide. With care, little if any manganese will be precipitated. The precipitate is ignited and weighed as the oxides.

7.2 *Procedure:*

7.2.1 To the filtrate reserved in accordance with 6.2.3.1 (Note 17) which should have a volume of about 200 mL, add HCl if necessary to ensure a total of 10 to 15 mL of the acid. Add a few drops of methyl red indicator and heat to boiling. Then treat with NH_4OH (1+1) (Note 18), dropwise until the color of the solution becomes distinctly yellow, and add one drop in excess (Note 19). Heat the solution containing the precipitate to boiling and boil for 50 to 60 s. In the event difficulty from bumping is experienced while boiling the ammoniacal solution, a digestion period of 10 min on a steam bath, or on a hot plate having the approximate temperature of a steam bath, may be substituted for the 50 to 60-s boiling period. Allow the precipitate to settle (not more than 5 min) and filter using medium-textured paper (Note 20). Wash, with hot ammonium nitrate (NH_4NO_3, 20 g/L) (Note 21), twice for a small precipitate to about four times for a large one.

NOTE 17—If a platinum evaporating dish has been used for the dehydration of SiO_2, iron may have been partially reduced. At this stage, add about 3 mL of saturated bromine water to the filtrate and boil the filtrate to eliminate the excess bromine before adding the methyl red indicator. If difficulty from bumping is experienced during the boiling, the following alternate techniques may be helpful: (*1*) a piece of filter paper, approximately 1 cm^2 in area, positioned where the bottom and side of the beaker merge and held down by the end of a stirring rod may solve the difficulty, and (*2*) use of 400-mL beakers supported inside a cast aluminum cup has also been found effective.

NOTE 18—The NH_4OH used to precipitate the hydroxides must be free of contamination with carbon dioxide (CO_2).

NOTE 19—It usually takes 1 drop of NH_4OH (1+1) to change the color of the solution from red to orange and another drop to change the color from orange to yellow. If desired, the addition of the indicator may be delayed until ferric hydroxide ($Fe(OH)_3$) is precipitated without aluminum hydroxide ($Al(OH)_3$) being completely precipitated. In such a case, the color changes may be better observed. However, if the content of Fe_2O_3 is unusually great, it may be necessary to occasionally let the precipitate settle slightly so that the color of the supernatant liquid can be observed. If the color fades during the precipitation, add more of the indicator. Observation of the color where a drop of the indicator strikes the solution may be an aid in the control of the acidity. The boiling should not be prolonged as the color may reverse and the precipitate may be difficult to retain on the filter. The solution should be distinctly yellow when it is ready to filter. If it is not, restore the yellow color with more NH_4OH (1+1) or repeat the precipitation.

NOTE 20—To avoid drying of the precipitate with resultant slow filtration, channeling, or poor washing, the filter paper should be kept nearly full during the filtration and should be washed without delay.

NOTE 21—Two drops of methyl red indicator solution should be added to the NH_4NO_3 solution in the wash bottle, followed by NH_4OH (1+1) added dropwise until the color just changes to yellow. If the color reverts to red at any time due to heating, it should be brought back to yellow by the addition of a drop of NH_4OH (1+1).

7.2.2 Set aside the filtrate and transfer the precipitate and filter paper to the same beaker in which the first precipitation was effected. Dissolve the precipitate with hot HCl (1+2). Stir to thoroughly macerate the paper and then dilute the solution to about 100 mL. Reprecipitate the hydroxides as described in 7.2.1. If difficulty from bumping is experienced while boiling the acid solution containing the filter paper, it may be obviated by diluting the hot 1+2 solution of the mixed oxides with 100 mL of boiling water and thus eliminate the need for boiling. Filter the solution and wash the precipitate with about four 10-mL portions of hot NH_4NO_3 solution (20 g/L) (Note 21). Combine the filtrate and washings with the filtrate set aside and reserve for the determination of CaO in accordance with 13.3.1.

7.2.3 Place the precipitate in a weighed platinum crucible, heat slowly until the papers are charred, and finally ignite to constant weight at 1050 to 1100°C taking care to prevent reduction, and weigh as the ammonium hydroxide group.

7.2.4 *Blank*—Make a blank determination, following the same procedure and using the same amounts of reagents, and correct the results obtained in the analysis accordingly.

7.3 *Calculation*—Calculate the percentage of ammonium hydroxide group to the nearest 0.01 by multiplying the weight in grams of ammonium hydroxide group by 200 (100 divided by the weight of sample used (0.5 g)).

8. Ferric Oxide (*Reference Test Method*)

8.1 *Summary of Test Method*—In this test method, the Fe_2O_3 content of the cement is determined on a separate portion of the cement by reducing the iron to the ferrous state with stannous chloride ($SnCl_2$) and titrating with a standard solution of potassium dichromate ($K_2Cr_2O_7$). This determination is not affected by any titanium or vanadium that may be present in the cement.

8.2 *Reagents:*

8.2.1 *Barium Diphenylamine Sulfonate Indicator Solution*—Dissolve 0.3 g of barium diphenylamine sulfonate in 100 mL of water.

8.2.2 *Potassium Dichromate, Standard Solution* (1 mL = 0.004 g Fe_2O_3)—Pulverize and dry primary standard potassium dichromate ($K_2Cr_2O_7$) reagent, the current lot of NBS 136, at 180 to 200°C to constant weight. Weigh accurately an amount of dried reagent equal to 2.45700 g times the number of litres of solution to be prepared. Dissolve in water and dilute to exactly the required volume in a single volumetric flask of the proper size. This solution is a primary standard and requires no further standardization.

NOTE 22—Where large quantities of standard solution are required, it may be desirable for certain laboratories to use commercially-produced primary standard potassium dichromate for most determinations. Such a material may be used provided that the first solution made from the container is checked, as follows: Using a standard solution of NBS 136, prepared as described in 8.2.2, analyze, in duplicate, samples of a NBS SRM cement (see Note 1), by the procedure given in 8.3.1.3 and 8.3.1.4. Repeat using a similar solution prepared from the commercial primary standard dichromate. The average percentages of Fe_2O_3 found by each method should not differ by more than 0.06 %.

8.2.3 *Stannous Chloride Solution*—Dissolve 5 g of stannous chloride ($SnCl_2 \cdot 2H_2O$) in 10 mL of HCl and dilute to 100 mL. Add scraps of iron-free granulated tin and boil until the solution is clear. Keep the solution in a closed dropping bottle containing metallic tin.

8.3 *Procedure*—For cements other than portland and for which the insoluble residue is unknown, determine the insoluble residue in accordance with the appropriate sections of these test methods. When insoluble residue is known, proceed in accordance with 8.3.1 or 8.3.2 as is appropriate for the cement being analyzed.

8.3.1 For portland cements and cements having insoluble residue lower than 1 %, weigh 1 g of the sample into a 500-mL Phillips beaker or other suitable container. Add 40 mL of cold water and, while the beaker is being swirled, add 10 mL of HCl. If necessary, heat the solution and grind the cement with the flattened end of a glass rod until it is evident that the cement is completely decomposed. Continue the analysis in accordance with 8.3.3.

8.3.2 For cements with insoluble residue greater than 1 %, weigh a 0.500 g sample, blend with 1 g $LiBO_2$ using a mortar and pestle, and transfer to a previously fired 8-mL carbon crucible that has 0.1 g $LiBO_2$ sprinkled in the bottom (Note 23). Cover with 0.1 g $LiBO_2$ that was used to chemically wash the mortar and pestle (Note 24). Place the uncovered crucible in a furnace set at 1100°C for 15 min. Remove the crucible from the furnace and check for complete fusion (Note 25). If the fusion is incomplete, return the crucible to the furnace for another 30 min. Again, check for complete fusion. If the fusion is still incomplete, discard the sample and repeat the fusion procedure using 0.250 g sample or a smaller quantity with the same amount of $LiBO_2$. When the fusion is complete, gently swirl the melt and pour into a 150-mL glass beaker containing 10 mL concentrated HCl and 50 mL water. Stir continuously until the fusion product is dissolved, usually 10 min or less

(Note 26). If a stirring bar is used, remove and rinse the bar. Continue the analysis in accordance with 8.3.3.

NOTE 23—The firing loosens the carbon on the surface, reducing the possibility of the fusion product sticking to the crucible.

NOTE 24—A chemical wash is a dry rinse of the equipment in which the blending was done so that any sample adhering to this equipment will be loosened and transferred to the crucible.

NOTE 25—When fusion is incomplete, the sample may not be completely melted or there may be particles on top of the bead. Usually, if the bead forms a small smooth spherical ball when taken from the furnace and before it is swirled, the sample is completely fused.

NOTE 26—There are usually some carbon particles that are in suspension, undissolved in the solution, but they will not interfere with the completion of the analysis.

8.3.3 Heat the solution to boiling and treat it with the $SnCl_2$ solution, added dropwise while stirring and boiling, until the solution is decolorized. Add 1 drop in excess and cool the solution to room temperature by placing the beaker in a pan of cool water. After cooling and without delay, rinse the inside of the vessel with water, and add all at once 10 mL of a cool, saturated mercuric chloride ($HgCl_2$) solution. Stir the solution vigorously for 1 min by swirling the beaker and add 10 mL of H_3PO_4 (1+1) and 2 drops of barium diphenylamine sulfonate indicator. Add sufficient water so that the volume after titration will be between 75 and 100 mL. Titrate with the standard $K_2Cr_2O_7$ solution. The end point shall be taken as the point at which a single drop causes an intense purple coloration that remains unchanged on further addition of standard $K_2Cr_2O_7$ solution.

8.3.4 *Blank*—Make a blank determination following the same procedure and using the same amounts of reagents. Record the volume of $K_2Cr_2O_7$ solution required to establish the end point as described in 8.3.3. As some iron must be present to obtain the normal end point, if no definite purple color is obtained after the addition of 4 drops of the standard $K_2Cr_2O_7$ solution, record the blank as zero.

8.4 *Calculation:*

8.4.1 Calculate the percentage of Fe_2O_3 to the nearest 0.01 % (to be reported to the nearest 0.1) as follows:

$$Fe_2O_3, \% = E(V - B) \times 100/W \quad (2)$$

where:

E = Fe_2O_3 equivalent of the $K_2Cr_2O_7$ solution, g/mL,
V = millilitres of $K_2Cr_2O_7$ solution required by the sample determination,
B = millilitres of $K_2Cr_2O_7$ solution required by the blank determination, and
W = mass of sample within 0.1 mg.

9. Phosphorus Pentoxide (*Reference Test Method*)

9.1 *Summary of Test Method*—This colorimetric test method is applicable to the determination of P_2O_5 in portland cement. Under the conditions of the test, no constituent normally present in portland cement will interfere.

9.2 *Apparatus:*

9.2.1 *Spectrophotometer* (Note 27):

9.2.1.1 The instrument shall be equipped to measure absorbance of solutions at a spectral wavelength of 725 nm.

9.2.1.2 Wavelength measurements shall be repeatable within ±1 nm or less.

9.2.1.3 In the absorbance range from 0.1 to 1.0, the absorbance measurements shall be repeatable within ±1 % or less.

9.2.1.4 To establish that the spectrophotometer will permit a satisfactory degree of accuracy, qualify the instrument in accordance with 3.3.2 using the procedure in 9.4.1 through 9.4.9.

NOTE 27—For the measurement of the performance of the spectrophotometer, refer to Practice E 275.

9.3 *Reagents:*

9.3.1 *Ammonium Molybdate Solution*—Into a 1-L volumetric flask introduce 500.0 mL of 10.6 *N* H_2SO_4 (9.3.7). Dissolve 25.0 g of ammonium molybdate ($(NH_4)_6MO_7O_{24} \cdot 4H_2O$) in about 250 mL of warm water and transfer to the flask containing the H_2SO_4, while swirling the flask. Cool, dilute to 1 L with water, and store in a plastic bottle.

9.3.2 *Ascorbic Acid Powder*—For ease in dissolving, the finest mesh available should be used.

9.3.3 *Hydrochloric Acid, Standard* (6.5 ± 0.1 *N*)—Dilute 540 mL of concentrated HCl (sp gr 1.19) to 1 L with water. Standardize against standard NaOH solution (9.3.6) using phenolphthalein as indicator. Determine the exact normality and adjust to 6.5 ± 0.1 *N* by dilution with water. Restandardize to ensure that the proper normality has been achieved.

9.3.4 *Phosphate, Standard Solution A*—Dissolve 0.1917 g of oven-dried potassium dihydrogen phosphate (KH_2PO_4) in water and dilute to 1 L in a volumetric flask.

9.3.5 *Phosphate, Standard Solution B*—Dilute 50.0 mL of phosphate solution A to 500 mL with water.

9.3.6 *Sodium Hydroxide, Standard Solution* (1 *N*)—Dissolve 40.0 g of sodium hydroxide (NaOH) in water, add 10 mL of a freshly filtered saturated solution of barium hydroxide ($Ba(OH)_2$), and dilute to 1 L with water that has been recently boiled and cooled. Shake the solution from time to time during a several-hour period, and filter into a plastic bottle. Keep the bottle tightly closed to protect the solution from CO_2 in the air. Standardize against acid potassium phthalate or benzoic acid acidimetric standards furnished by the National Bureau of Standards (standard samples 84f and 350), using the test methods in the certificates accompanying the standard samples. Determine the exact normality of the solution.

9.3.7 *Sulfuric Acid, Standard* (10.6 ± 0.1 *N*)—To a 1-L volumetric flask cooled in water add about 600 mL of water and then, slowly, with *caution,* 300 mL of concentrated H_2SO_4 (sp gr 1.84). After cooling to room temperature, dilute to 1 L with water. Standardize against the standard NaOH solution (9.3.6) using phenolphthalein as indicator. Determine the normality and adjust to 10.6 ± 0.1 *N* by dilution with water. Restandardize to ensure that the proper normality has been achieved.

9.4 *Procedure:*

9.4.1 Prepare a series of phosphate solutions to cover the range from 0 to 0.5 % P_2O_5. Prepare each solution by adding a suitable volume of standard phosphate solution B and 25.0 mL of the 6.5 *N* hydrochloric acid to a 250-mL volumetric flask (Note 28). Dilute to the mark with water.

NOTE 28—One millilitre of standard phosphate solution B/250 mL of solution is equivalent to 0.004 % P_2O_5 for a 0.25-g cement sample. Aliquots of 0, 12.5, 25, 50, 74, 100, and 125 mL are equivalent to P_2O_5

contents in the sample of 0, 0.05, 0.10, 0.20, 0.30, 0.40, and 0.50 %.

9.4.2 Prepare a blank by adding 25.0 mL of the standard HCl to a 250-mL volumetric flask and diluting to 250 mL with water.

9.4.3 Develop colors in the series of phosphate solutions, and in the blank, in accordance with 9.4.6 through 9.4.8.

9.4.4 Plot the net absorbance (absorbance of standard minus that of the blank) values obtained as ordinates and the corresponding P_2O_5 concentrations as abscissas. Draw a smooth curve through the points.

NOTE 29—A suitable paper for plotting the calibration curve is a 10 by 15-in. (254 by 381-mm) linear cross section paper having 20 by 20 divisions to the inch. The percentage of P_2O_5 can then be plotted on the long dimension using five divisions equal to 0.01 % P_2O_5. A scale of one division equal to 0.005 absorbance units is suitable as the ordinate (short dimension of the paper). Scales other than this may be used but under no circumstances should a scale division less than 1/20 in. (1.3 mm) be used for 0.005 units of absorbance or for 0.005 % P_2O_5. A separate calibration curve should be made for each spectrophotometer used, and the calibration curve checked against standard phosphate solution whenever a new batch of ammonium molybdate reagent is used.

9.4.5 Transfer 0.250 g of the sample to a 250-mL beaker and moisten with 10 mL of cold water to prevent lumping. Add 25.0 mL of the standard HCl and digest with the aid of gentle heat and agitation until solution is complete. Filter into a 250-mL volumetric flask and wash the paper and the separated silica thoroughly with hot water. Allow the solution to cool and then dilute with water to 250 mL.

9.4.6 Transfer a 50.0-mL aliquot (Note 30) of the sample solution to a 250-mL beaker, add 5.0 mL of ammonium molybdate solution and 0.1 g of ascorbic acid powder. Mix the contents of the beaker by swirling until the ascorbic acid has dissolved completely. Heat the solution to vigorous boiling and then boil, uncovered, for 1.5 ± 0.5 min. Cool to room temperature and transfer to a 50-mL volumetric flask. Rinse the beaker with one small portion of water and add the rinse water to the flask. Dilute to 50 mL with water.

NOTE 30—The range of the test can be extended by taking a smaller aliquot of the sample solution. In such instances the decrease in the aliquot volume must be made up by the blank solution (9.4.5) to maintain the proper acidity of the final solution. Thus, if a 25-mL aliquot of the sample solution is taken (instead of the usual 50 mL), a 25-mL aliquot of the blank solution should be added before proceeding with the test. The result of the test must then be calculated accordingly.

9.4.7 Measure the absorbance of the solution against water as the reference at 725.0 nm.

9.4.8 Develop on a 50.0-mL aliquot of the blank solution prepared in 9.4.2 in the same manner as was used in 9.4.6 for the sample solution. Measure the absorbance in accordance with 9.4.7 and subtract this absorbance value from that obtained for the sample solution in 9.4.6 in order to obtain the net absorbance for the sample solution.

9.4.9 Using the net absorbance value found in 9.4.8, record the percentage of P_2O_5 in the cement sample as indicated by the calibration curve. Report the percentage of P_2O_5 to the nearest 0.01.

10. Titanium Dioxide (*Reference Test Method*)

10.1 *Summary of Test Method*—In this test method titanium dioxide (TiO_2) in portland cement is determined colorimetrically using Tiron reagent. Under the conditions of the test iron is the only constituent of portland cement causing a very slight interference equivalent to 0.01 % for each 1 % of Fe_2O_3 present in the sample.

10.2 *Apparatus:*

10.2.1 *Spectrophotometer* (Note 31):

10.2.1.1 The instrument shall be equipped to measure absorbance of solutions at a spectral wavelength of 410 nm.

10.2.1.2 Wavelength measurements shall be repeatable within ±1 nm or less.

10.2.1.3 In the absorbance range from 0.1 to 1.0, the absorbance measurements shall be repeatable within ±1 % or less.

10.2.1.4 To establish that the spectrophotometer will permit a satisfactory degree of accuracy, qualify the instrument in accordance with 3.3.2 using the procedure in 10.4.1 through 10.4.6 of this test method.

NOTE 31—For the measurement of the performance of the spectrophotometer, refer to Practice E 275.

10.3 *Reagents:*

10.3.1 *Buffer* (pH 4.7)—68 g of $NaC_2H_3O_2 \cdot 3H_2O$, plus 380 mL of water, plus 100 mL of 5.0 *N* CH_3COOH.

10.3.2 *Ethylenedinitrilo Tetraacetic Acid Disodium Salt, Dihydrate* (0.2 *M* EDTA)—Dissolve 37.5 g of EDTA in 350 mL of warm water, and filter. Add 0.25 g of $FeCl_3 \cdot 6H_2O$ and dilute to 500 mL.

10.3.3 *Hydrochloric Acid* (1+6).

10.3.4 *Hydrochloric Acid, Standard* (6.5 *N*)—Dilute 540 mL of concentrated HCl (sp gr 1.19) to 1 L with water.

10.3.5 *Ammonium Hydroxide* (NH_4OH, 1+1).

10.3.6 *Potassium Pyrosulfate* ($K_2S_2O_7$).

10.3.7 *Titanium Dioxide, Stock Solution A*—Fuse slowly in a platinum crucible over a very small flame 0.0314 g of NBS SRM 154b (TiO_2 = 99.74 %) or its replacements with about 2 or 3 g of $K_2S_2O_7$. Allow to cool, and place the crucible in a beaker containing 125 mL of H_2SO_4 (1+1). Heat and stir until the melt is completely dissolved. Cool, transfer to a 250-mL volumetric flask, and dilute the solution to volume.

10.3.7.1 *Titanium Dioxide, Dilute Standard Solution B* (1 mL = 0.0125 mg TiO_2)—Pipet 50 mL of stock TiO_2 solution into a 500-mL volumetric flask, and dilute to volume. One millilitre of this solution is equal to 0.0125 mg of TiO_2, which is equivalent to 0.05 % TiO_2 when used as outlined in 10.4.4 through 10.4.6.

10.3.8 *Sulfuric Acid* (1+1).

10.3.9 *Tiron* (disodium-1,2-dihydroxybenzene-3,5 disulfonate).

10.4 *Procedure:*

10.4.1 Prepare a series of TiO_2 solutions to cover the range from 0 to 1.0 % TiO_2. Prepare each solution in a 50-mL volumetric flask.

NOTE 32—One millilitre of dilute TiO_2 standard solution B per 50 mL (10.3.7.1) is equivalent to 0.05 % TiO_2 for a 0.2500-g cement sample. Aliquots of 0, 5, 10, 15, and 20 mL of dilute TiO_2 standard solution are equivalent to TiO_2 contents in the sample of 0, 0.25, 0.50, 0.75, and 1.0 %. Dilute each to 25 mL with water.

10.4.2 Develop color in accordance with 10.4.4 starting

with second sentence. Measure absorbance in accordance with 10.4.5.

10.4.3 Plot absorbance values obtained as ordinates and the corresponding TiO_2 concentrations as abscissas. Draw a smooth curve through the points.

NOTE 33—A suitable paper for plotting the calibration curve is a linear cross section paper having 10 × 10 divisions to 1 cm. A scale division equivalent to 0.002 absorbance and 0.002 % TiO_2 should be used. A separate calibration curve should be made for each spectrophotometer used.

10.4.4 Transfer a 25.0-mL aliquot of the sample solution prepared in 9.4.5 into a 50-mL volumetric flask (Note 34). Add 5 mL tiron and 5 mL EDTA, mix, and then add NH_4OH (1+1) dropwise, mixing thoroughly after each drop, until the color changes through yellow to green, blue, or ruby red. Then; just restore the yellow color with HCl (1+6) added dropwise and mixing after each drop. Add 5 mL buffer, dilute to volume and mix.

10.4.5 Measure the absorbance of the solution against water as the reference at 410 nm.

NOTE 34—The range of the test can be extended by taking a smaller aliquot. The results of the test must then be calculated accordingly.

10.4.6 Using the absorbance value determined in 10.4.5, record the percentage of TiO_2 in the cement sample as indicated by the calibration curve to the nearest 0.01. Correct for the iron present in the sample to obtain the true TiO_2 as follows: True TiO_2 = measured % TiO_2 − (0.01 × % Fe_2O_3). Report the percent of TiO_2 to the nearest 0.01.

11. Zinc Oxide (*Reference Test Method*)[10]

11.1 Any test method may be used that meets the requirements of Section 3.3 and Table 1.

12. Aluminum Oxide (*Reference Test Method*)

NOTE 35—In the reference test method, Al_2O_3 is calculated from the ammonium hydroxide group by subtracting the separately determined constituents that usually are present in significant amounts in the ammonium hydroxide precipitate. These are Fe_2O_3, TiO_2 and P_2O_5. Most instrumental test methods for Al_2O_3 analysis give Al_2O_3 alone if standardized and calibrated properly.

12.1 *Calculation:*

12.1.1 Calculate the percentage of Al_2O_3 by deducting the percentage of the sum of the Fe_2O_3, TiO_2, and P_2O_5 from the percentage of ammonium hydroxide group. All determinations shall be by referee test methods described in the appropriate sections herein. All percentages shall be calculated to the nearest 0.01 %. Report the Al_2O_3 to the nearest 0.1 %. For nonreferee analyses, the percentages of Fe_2O_3, TiO_2, and P_2O_5 can be determined by any procedure for which qualification has been shown.

13. Calcium Oxide (*Reference Test Method*)

13.1 *Summary of Test Method:*

13.1.1 In this test method, manganese is removed from the filtrate after the determination of SiO_2 and the ammonium hydroxide group. Calcium is then precipitated as the oxalate. After filtering, the oxalate is redissolved and titrated with potassium permanganate ($KMnO_4$).

NOTE 36—For referee analysis or for the most accurate determinations, removal of manganese in accordance with 13.3.2 must be made. For less accurate determinations, and when only insignificant amounts of manganese oxides are believed present, 13.3.2 may be omitted.

13.1.2 Strontium, usually present in portland cement as a minor constituent, is precipitated with calcium as the oxalate and is subsequently titrated and calculated as CaO. If the SrO content is known and correction of CaO for SrO is desired as, for example, for research purposes or to compare results with SRM certificate values, the CaO obtained by this method may be corrected for SrO. In determining conformance of a cement to specifications, the correction of CaO for SrO should not be made.

13.2 *Reagents:*

13.2.1 *Ammonium Oxalate Solution* (50 g/L).

13.2.2 *Potassium Permanganate, Standard Solution* (0.18 *N*)—Prepare a solution of potassium permanganate ($KMnO_4$) containing 5.69 g/L. Let this solution stand at room temperature for at least 1 week, or boil and cool to room temperature. Siphon off the clear solution without disturbing the sediment on the bottom of the bottle; then filter the siphoned solution through a bed of glass wool in a funnel or through a suitable sintered glass filter. Do not filter through materials containing organic matter. Store in a dark bottle, preferably one that has been painted black on the outside. Standardize the solution against 0.7000 to 0.8000 g of primary standard sodium oxalate, according to the directions furnished with the sodium oxalate and record the temperature at which the standardization was made (Note 37).

13.2.2.1 Calculate the CaO equivalent of the solution as follows:

1 mL of 1 *N* $KMnO_4$ solution is equivalent to 0.06701 g of pure sodium oxalate.

NormalityofKmnO$_4$

$$= \frac{\text{weight of sodium oxalate} \times \text{fraction of its purity}}{\text{mL of } KMnO_4 \text{ solution} \times 0.06701} \quad (3)$$

1 mL of 1 *N* $KMnO_4$ solution is equivalent to 0.02804 g of CaO.

$$F = \frac{\text{normality of } KMnO_4 \text{ solution} \times 0.02804 \times 100}{0.5}$$

where F = CaO equivalent of the $KMnO_4$ solution in % CaO/mL based on a 0.5-g sample of cement.

NOTE 37—Because of the instability of the $KMnO_4$ solution, it is recommended that it be restandardized at least bimonthly.

13.3 *Procedure:*

13.3.1 Acidify the combined filtrates obtained in the precipitations of the ammonium hydroxide group (7.2.2). Neutralize with HCl to the methyl red end point, make just acid, and add 6 drops of HCl in excess.

13.3.2 *Removal of Manganese*—Evaporate to a volume of about 100 mL. Add 40 mL of saturated bromine water to the hot solution and immediately add NH_4OH until the solution is

[10] The 1988 revision of these test methods deleted the colorimetric method for determination of ZnO using an extraction with CCl_4. Those interested in this test method should refer to the *1987 Annual Book of ASTM Standards,* Volume 04.01.

distinctly alkaline. Addition of 10 mL of NH_4OH is generally sufficient. A piece of filter paper, about 1 cm^2 in area, placed in the heel of the beaker and held down by the end of a stirring rod aids in preventing bumping and initiating precipitation of hydrated manganese oxides (MnO). Boil the solution for 5 min or more, making certain that the solution is distinctly alkaline at all times. Allow the precipitate to settle, filter using medium-textured paper, and wash with hot water. If a precipitate does not appear immediately, allow a settling period of up to 1 h before filtration. Discard any manganese dioxide that may have been precipitated. Acidify the filtrate with HCl using litmus paper as an indicator, and boil until all the bromine is expelled (Note 38).

13.3.3 Add 5 mL of HCl, dilute to 200 mL, and add a few drops of methyl red indicator and 30 mL of warm ammonium oxalate solution (50 g/L) (Note 39). Heat the solution to 70 to 80°C, and add NH_4OH (1+1) dropwise, while stirring until the color changes from red to yellow (Note 40). Allow the solution to stand without further heating for 60 ± 5 min (no longer), with occasional stirring during the first 30 min.

13.3.4 Filter, using retentive paper, and wash the precipitate 8 to 10 times with hot water, the total amount of water used in rinsing the beaker and washing not to exceed 75 mL. During this washing, water from the wash bottle should be directed around the inside of the filter paper to wash the precipitate down, then a jet of water should be gently directed towards the center of the paper in order to agitate and thoroughly wash the precipitate. Acidify the filtrate with HCl and reserve for the determination of MgO.

13.3.5 Place the beaker in which the precipitation was made under the funnel, pierce the apex of the filter paper with the stirring rod, place the rod in the beaker, and wash the precipitate into the beaker by using a jet of hot water. Drop about 10 drops of H_2SO_4 (1+1) around the top edge of the filter paper. Wash the paper five more times with hot water. Dilute to 200 mL, and add 10 mL of H_2SO_4 (1+1). Heat the solution to a temperature just below boiling, and titrate it immediately with the 0.18 *N* $KMnO_4$ solution (Note 41). Continue the titration slowly until the pink color persists for at least 10 s. Add the filter paper that contained the original precipitate and macerate it. If the pink color disappears continue the titration until it again persists for at least 10 s.

NOTE 38—Potassium iodide starch paper may be used to indicate the complete volatilization of the excess bromine. Expose a strip of moistened paper to the fumes from the boiling solution. The paper should remain colorless. If it turns blue bromine is still present.

NOTE 39—If the ammonium oxalate solution is not perfectly clear, it should be filtered before use.

NOTE 40—This neutralization must be made slowly, otherwise precipitated calcium oxalate may have a tendency to run through the filter paper. When a number of these determinations are being made simultaneously, the following technique will assist in ensuring slow neutralization. Add two or three drops of NH_4OH to the first beaker while stirring, then 2 or 3 drops to the second, and so on, returning to the first beaker to add 2 or 3 more drops, etc., until the indicator color has changed in each beaker.

NOTE 41—The temperature of the 0.18 *N* $KMnO_4$ solution at time of use should not vary from its standardization temperature by more than 10°F (5.5°C). Larger deviations could cause serious error in the determination of CaO.

13.3.6 *Blank*—Make a blank determination, following the same procedure and using the same amounts of reagents (Note 42), and record the millilitres of $KMnO_4$ solution required to establish the end point.

NOTE 42—When the amount of calcium oxalate is very small, its oxidation by $KMnO_4$ is slow to start. Before the titration, add a little $MnSO_4$ to the solution to catalyze the reaction.

13.4 *Calculation:*

13.4.1 Calculate the percentage of CaO to the nearest 0.1 as follows:

$$\text{CaO, \%} = E(V - B) \tag{4}$$

where:

E = CaO equivalent of the $KMnO_4$ solution in % CaO/mL based on a 0.5-g sample,

V = millilitres of $KMnO_4$ solution required by the sample, and

B = millititres of $KMnO_4$ solution required by the blank.

13.4.2 If desired calculate the percentage of CaO corrected for SrO as follows:

$$\text{CaO}_c\ \% = \text{CaO}_i\ \% - 0.54\ \text{SrO}\ \% \tag{5}$$

where:

CaO_c = CaO corrected for SrO, and

CaO_i = initial CaO as determined in 13.4.1

0.54 = $\frac{56.08}{103.62}$ = molecular weight ratio $\frac{\text{CaO}}{\text{SrO}}$

14. Magnesium Oxide (*Reference Test Method*)

14.1 *Summary of Test Method*—In this test method, magnesium is precipitated as magnesium ammonium phosphate from the filtrate after removal of calcium. The precipitate is ignited and weighed as magnesium pyrophosphate ($Mg_2P_2O_7$). The MgO equivalent is then calculated.

14.2 *Reagent*—Ammonium phosphate, dibasic (100 g/L) $(NH_4)_2HPO_4$.

14.3 *Procedure:*

14.3.1 Acidify the filtrate from the determination of CaO (13.3.4) with HCl and evaporate by boiling to about 250 mL. Cool the solution to room temperature, add 10 mL of ammonium phosphate, dibasic, $(NH_4)_2HPO_4$ (100 g/L), and 30 mL of NH_4OH. Stir the solution vigorously during the addition of NH_4OH and then for 10 to 15 min longer. Let the solution stand for at least 8 h in a cool atmosphere and filter. Wash the residue five or six times with NH_4OH (1+20) and ignite in a weighed platinum or porcelain crucible, at first slowly until the filter paper is charred and then burn off (see Note 43), and finally at 1100°C for 30 to 45 min. Weigh the residue as magnesium pyrophosphate ($Mg_2P_2O_7$).

14.3.2 *Blank*—Make a blank determination following the same procedure and using the same amounts of reagents, and correct the results obtained in the analysis accordingly.

14.4 *Calculation:*

14.4.1 Calculate the percentage of MgO to the nearest 0.1 as follows:

$$\text{MgO, \%} = W \times 72.4 \tag{6}$$

where:

W = grams of $Mg_2P_2O_7$, and

72.4 = molecular ratio of $2MgO$ to $Mg_2P_2O_7$ (0.362) divided by the weight of sample used (0.5 g) and multiplied by 100.

NOTE 43: **Caution**—Extreme caution should be exercised during this ignition. Reduction of the phosphate precipitate can result if carbon is in contact with it at high temperatures. There is also danger of occluding carbon in the precipitate if ignition is too rapid.

15. Sulfur (See Note 44)

15.1 *Sulfur Trioxide: (Reference Test Method):*

15.1.1 *Summary of Test Method*—In this test method, sulfate is precipitated from an acid solution of the cement with barium chloride ($BaCl_2$). The precipitate is ignited and weighed as barium sulfate ($BaSO_4$) and the SO_3 equivalent is calculated.

15.1.2 *Procedure:*

15.1.2.1 To 1 g of the sample add 25 mL of cold water and, while the mixture is stirred vigorously, add 5 mL of HCl (Note 45). If necessary, heat the solution and grind the material with the flattened end of a glass rod until it is evident that decomposition of the cement is complete (Note 46). Dilute the solution to 50 mL and digest for 15 min at a temperature just below boiling. Filter through a medium-textured paper and wash the residue thoroughly with hot water. Dilute the filtrate to 250 mL and heat to boiling. Add slowly, dropwise, 10 mL of hot $BaCl_2$ (100 g/L) and continue the boiling until the precipitate is well formed. Digest the solution for 12 to 24 h at a temperature just below boiling (Note 47). Take care to keep the volume of solution between 225 and 260 mL and add water for this purpose if necessary. Filter through a retentive paper, wash the precipitate thoroughly with hot water, place the paper and contents in a weighed platinum crucible, and slowly char and consume the paper without inflaming. Ignite at 800 to 900°C, cool in a desiccator, and weigh.

NOTE 44—When an instrumental test method is used for sulfur or when comparing results of classical wet and instrumental test methods, consult 4.1.2 of these test methods.

NOTE 45—The acid filtrate obtained in the determination of the insoluble residue (5.3.1) may be used for the determination of SO_3 instead of using a separate sample.

NOTE 46—A brown residue due to compounds of manganese may be disregarded (see Note 11).

NOTE 47—If a rapid determination is desired, the time of digestion may be reduced to as little as 3 h. However, the cement may be rejected for failure to meet the specification requirement only on the basis of results obtained when using 12 to 24-h digestion times.

15.1.2.2 *Blank*—Make a blank determination following the same procedure and using the same amounts of reagents, and correct the results obtained in the analysis accordingly.

15.1.3 *Calculation*—Calculate the percentage of SO_3 to the nearest 0.01 as follows:

$$SO_3, \% = W \times 34.3 \quad (7)$$

where:

W = grams of $BaSO_4$, and
34.3 = molecular ratio of SO_3 to $BaSO_4$ (0.343) multiplied by 100.

15.2 *Sulfide: (Reference Test Method)*

15.2.1 *Summary of Test Method*—In this test method sulfide sulfur is determined by evolution as hydrogen sulfide (H_2S) from an acid solution of the cement into a solution of ammoniacal zinc sulfate ($ZnSO_4$) or cadmium chloride ($CdCl_2$). The sulfide sulfur is then titrated with a standard solution of potassium iodate (KIO_3). Sulfites, thiosulfates, and other compounds intermediate between sulfides and sulfates are assumed to be absent. If such compounds are present, they may cause an error in the determination.

15.2.2 *Apparatus:*

15.2.2.1 *Gas-Generating Flask*—Connect a dry 500-mL boiling flask with a long-stem separatory funnel and a small connecting bulb by means of a rubber stopper. Bend the stem of the funnel so that it will not interfere with the connecting bulb, adjust the stem so that the lower end is close to the bottom of the flask, and connect the opening of the funnel with a source of compressed air. Connect the bulb with an L-shaped glass tube and a straight glass tube about 200 mm in length. Insert the straight glass tube in a tall-form, 400-mL beaker. A three-neck distilling flask with a long glass tubing in the middle opening, placed between the source of compressed air and the funnel, is a convenient aid in the regulation of the airflow. Rubber used in the apparatus shall be pure gum grade, low in sulfur, and shall be cleaned with warm HCl.

15.2.3 *Reagents:*

15.2.3.1 *Ammoniacal Cadmium Chloride Solution*—Dissolve 15 g of cadmium chloride ($CdCl_2 \cdot 2H_2O$) in 150 mL of water and 350 mL of NH_4OH. Filter the solution after allowing it to stand at least 24 h.

15.2.3.2 *Ammoniacal Zinc Sulfate Solution*—Dissolve 50 g of zinc sulfate ($ZnSO_4 \cdot 7H_2O$) in 150 mL of water and 350 mL of NH_4OH. Filter the solution after allowing it to stand at least 24 h.

15.2.3.3 *Potassium Iodate, Standard Solution* (0.03 *N*)—Prepare a solution of potassium iodate (KIO_3) and potassium iodide (KI) as follows: Dry KIO_3 at 180°C to constant weight. Weigh 1.0701 g of the KIO_3 and 12 g of KI. Dissolve and dilute to 1 L in a volumetric flask. This is a primary standard and requires no standardization (Note 48). One millilitre of this solution is equivalent to 0.0004809 g of sulfur.

NOTE 48—The solution is very stable, but may not maintain its titer indefinitely. Whenever such a solution is over 1 year old it should be discarded or its concentration checked by standardization.

15.2.3.4 *Stannous Chloride Solution*—To 10 g of stannous chloride ($SnCl_2 \cdot 2H_2O$) in a small flask, add 7 mL of HCl (1+1), warm the mixture gently until the salt is dissolved, cool the solution, and add 95 mL of water. This solution should be prepared as needed, as the salt tends to hydrolyze.

15.2.3.5 *Starch Solution*—To 100 mL of boiling water, add a cool suspension of 1 g of soluble starch in 5 mL of water and cool. Add a cool solution of 1 g of sodium hydroxide (NaOH) in 10 mL of water, then 3 g of potassium iodide (KI), and mix thoroughly.

15.2.4 *Procedure:*

15.2.4.1 Place 15 mL of the ammoniacal $ZnSO_4$ or $CdCl_2$ solution (Note 49) and 285 mL of water in a beaker. Put 5 g of the sample (Note 50) and 10 mL of water in the flask and shake the flask gently to wet and disperse the cement completely. This step and the addition of $SnCl_2$ should be performed rapidly to prevent the setting of the cement. Connect the flask with the funnel and bulb. Add 25 mL of the $SnCl_2$ solution through the funnel and shake the flask. Add 100 mL of HCl (1+3) through the funnel and shake the flask. During these shakings keep the funnel closed and the delivery tube in the ammoniacal $ZnSO_4$ or $CdCl_2$ solution. Connect the funnel with the source of compressed air, open the funnel, start a slow stream of air, and heat the flask and contents slowly to boiling. Continue the boiling gently for 5 or 6 min. Cut off the heat, and continue the passage of air for 3 or 4 min. Disconnect the delivery tube and leave it in the solution for use as a stirrer. Cool the solution to 20 to 30°C (Note 51), add 2 mL of the starch solution and 40 mL of HCl (1+1) and titrate immediately with the 0.03 N KIO_3 solution until a persistent blue color is obtained (Note 52).

NOTE 49—In general, the $ZnSO_4$ is preferable to the $CdCl_2$ solution because $ZnSO_4$ is more soluble in NH_2OH than is $CdCl_2$. The $CdCl_2$ solution may be used when there is doubt as to the presence of a tracc of sulfide sulfur, as the yellow cadmium sulfide (CdS) facilitates the detection of a trace.

NOTE 50—If the content of sulfur exceeds 0.20 or 0.25 %, a smaller sample should be used so that the titration with the KIO_3 solution will not exceed 25 mL.

NOTE 51—The cooling is important as the end point is indistinct in a warm solution.

NOTE 52—If the content of sulfur is appreciable but not approximately known in advance, the result may be low due to the loss of H_2S during a slow titration. In such a case the determination should be repeated with the titration carried out more rapidly.

15.2.4.2 Make a blank determination, following the same procedure and using the same amounts of reagents. Record the volume of KIO_3 solution necessary to establish the end point as described in 15.2.4.1.

15.2.5 *Calculation*—Calculate the percentage of sulfide sulfur (see 15.2.1) as follows:

$$\text{Sulfide, \%} = E(V - B) \times 20 \quad (8)$$

where:

E = sulfide equivalent of the KIO_3 solution, g/mL,
V = millilitres of KIO_3 solution required by the sample,
B = millilitres of KIO_3 solution required by the blank, and
20 = 100 divided by the weight of sample used (5 g).

16. Loss on Ignition (*Reference Test Methods*)

16.1 *Portland Cement:*

16.1.1 *Summary of Test Method*—In this test method, the cement is ignited in a muffle furnace at a controlled temperature. The loss is assumed to represent the total moisture and CO_2 in the cement. This procedure is not suitable for the determination of the loss on ignition of portland blast-furnace slag cement and of slag cement. A test method suitable for such cements is described in 16.2.1 through 16.2.3.

16.1.2 *Procedure*—Weigh 1 g of the sample in a tared platinum crucible. Cover and ignite the crucible and its contents to constant weight in a muffle furnace at a temperature of 950 ± 50°C. Allow a minimum of 15 min for the initial heating period and at least 5 min for all subsequent periods.

16.1.3 *Calculation*—Calculate the percentage of loss on ignition to the nearest 0.1 by multiplying the loss of weight in grams by 100.

16.2 *Portland Blast-Furnace Slag Cement and Slag Cement:*

16.2.1 *Summary of Test Method*—Since it is desired that the reported loss on ignition represent moisture and CO_2, this test method provides a correction for the gain in weight due to oxidation of sulfides usually present in portland blast-furnace slag cement and slag cement by determining the increase in SO_3 content during ignition. An optional test method providing for a correction based on the decrease in sulfide sulfur during ignition is given in 23.1.1 through 23.1.3.1.

16.2.2 *Procedure:*

16.2.2.1 Weigh 1 g of cement into a tared platinum crucible and ignite in a muffle furnace at a temperature of 950 ± 50°C for 15 min. Cool to room temperature in a desiccator and weigh. Without checking for constant weight, carefully transfer the ignited material to a 400-mL beaker. Break up any lumps in the ignited cement with the flattened end of a glass rod.

16.2.2.2 Determine the SO_3 content by the test method given in 15.1.1.1 through 15.1.3.1 (Note 53). Also determine the SO_3 content of a portion of the same cement that has not been ignited, using the same procedure.

NOTE 53—Some of the acid used for dissolving the sample may first be warmed in the platinum crucible to dissolve any adhering material.

16.2.3 *Calculation*—Calculate the percentage loss of weight occurring during ignition and add 0.8 times the difference between the percentages of SO_3 in the ignited sample and the original cement (Note 54). Report the corrected percentage as loss on ignition.

NOTE 54—If a gain in weight is obtained during ignition, subtract the percentage gain from the correction for SO_3.

17. Sodium and Potassium Oxides (*Reference Test Methods*)

17.1 *Total Alkalies:*

17.1.1 *Summary of Test Method*—This test method[11] covers the determination of sodium oxide (Na_2O) and potassium oxide (K_2O) by flame photometry or atomic absorption.

NOTE 55—This test method is suitable for hydraulic cements that are completely decomposed by hydrochloric acid and should not be used for determination of total alkalies in hydraulic cements that contain large amounts of acid-insoluble material, for example, pozzolan cements. It may be used to determine acid-soluble alkalies for such cements. An alternate test method of sample dissolution for such cements is in preparation.

17.1.2 *Apparatus:*

17.1.2.1 *Instrument*—Any type flame photometer or atomic absorption unit may be used provided it can be demonstrated that the required degree of accuracy and precision is as indicated in 17.1.3.

NOTE 56—After such accuracy is established, for a specific instrument, further tests of instrument accuracy are not required except when it must

[11] The 1963 revision of these test methods deleted the classical (J. L. Smith) gravimetric method for the determination of Na_2O and K_2O in cements. Those interested in this method should refer to the *1961 Book of ASTM Standards,* Part 4.

The 1983 revision of these test methods deleted the details of the flame photometric procedure for the determination of Na_2O and K_2O. Those interested in this method should refer to the *1982 Annual Book of ASTM Standards,* Part 13.

be demonstrated that the instrument gives results within the prescribed degree of accuracy by a single series of tests using the designated standard samples.

NOTE 57—For normal laboratory testing, it is recommended that the accuracy of the instrument be routinely checked by the use of either a National Institute of Standards and Technology cement or cement of known alkali content.

17.1.2.2 The instrument shall consist at least of an atomizer and burner; suitable pressure-regulating devices and gages for fuel and oxidant gas; an optical system, capable of preventing excessive interference from wavelengths of light other than that being measured; and a photosensitive indicating device.

17.1.3 *Initial Qualification of Instrument*—Qualify the instrument in accordance with 3.3.2 to establish that an instrument provides the desired degree of precision and accuracy.

17.1.4 *Reagents and Materials:*

17.1.4.1 *Laboratory Containers*—All glassware shall be made of borosilicate glass and all polyethylene shall comply with the requirements of 4.2.3.

17.1.4.2 *Calcium Carbonate*—The calcium carbonate ($CaCO_3$) used in the preparation of the calcium chloride stock solution (17.1.5.1) shall contain not more than 0.020 % total alkalies as sulfate.

NOTE 58—Materials sold as a primary standard or ACS "low alkali" grade normally meet this requirement. However, the purchaser should assure himself that the actual material used conforms with this requirement.

17.1.4.3 *Potassium Chloride* (KCl).

17.1.4.4 *Sodium Chloride* (NaCl).

17.1.4.5 Commercially available solutions may be used in place of those specified in 17.1.5.

17.1.5 *Preparation of Solutions:*

17.1.5.1 *Calcium Chloride Stock Solution*—Add 300 mL of water to 112.5 g of $CaCO_3$ in a 1500-mL beaker. While stirring, slowly add 500 mL of HCl. Cool the solution to room temperature, filter into a 1-L volumetric flask, dilute to 1 L, and mix thoroughly. This solution contains the equivalent of 63 000 ppm (6.30 %) CaO.

17.1.5.2 *Sodium-Potassium Chloride Stock Solution*—Dissolve 1.8858 g of sodium chloride (NaCl) and 1.583 g of potassium chloride (KCl) (both dried at 105 to 110°C for several hours prior to weighing) in water. Dilute to 1 L in a volumetric flask and mix thoroughly. This solution contains the equivalent of 1000 ppm (0.10 %) each of Na_2O and K_2O. Separate solutions of Na_2O and of K_2O may be used provided that the same concentration solutions are used for calibration for cement analysis as were used for the calibration when qualifying the instrument in accordance with 17.1.3.

17.1.5.3 *Standard Solutions*—Prepare the standard solutions prescribed for the instrument and method used. Measure the required volume of NaCl-KCl stock solutions in calibrated pipets or burets. The calcium chloride stock solutions, if needed, may be measured in appropriate graduated cylinders. If the instrument being used requires an internal standard, measure the internal standard solution with a pipet or buret. Place each solution in a volumetric flask, dilute to the indicated volume, and mix thoroughly.

17.1.5.4 If more dilute solutions are required by the method in use, pipet the required aliquot to the proper sized volumetric flask, add any necessary internal standard, dilute to the mark, and mix thoroughly.

17.1.6 *Calibration of Apparatus:*

NOTE 59—No attempt is made in this section to describe in detail the steps for putting the instrument into operation since this will vary considerably with different instruments. The manufacturer's instructions should be consulted for special techniques or precautions to be employed in the operation, maintenance, or cleaning of the apparatus.

17.1.6.1 Turn on the instrument and allow it to warm up in accordance with the manufacturer's instructions. (A minimum of 30 min is required for most instruments.) Adjust the fuel and oxidant gas pressures as required by the instrument being used. Light and adjust the burner for optimum operation. Make any other adjustments that may be necessary to establish the proper operating conditions for the instrument.

17.1.7 *Procedure:*

17.1.7.1 *Solution of the Cement*—Prepare the solution of the cement in accordance with the procedure specified by the instrument manufacturer. If no procedure is specified, or if desired, proceed as specified in 17.1.7.1.1 or 17.1.7.1.2 (Note 59).

NOTE 60—The presence of SiO_2 in solution affects the accuracy of some flame photometers. In cases where an instrument fails to provide results within the prescribed degree of accuracy outlined in 3.3.2.1 through 3.3.3 tests should be made on solutions from which the SiO_2 has been removed. For this removal proceed as in 17.1.7.1.2.

17.1.7.1.1 Place 1.000 ± 0.001 g of the cement in a 150-mL beaker and disperse with 20 mL of water using a swirling motion of the beaker. While still swirling add 5.0 mL of HCl all at once. Dilute immediately to 50 mL with water. Break up any lumps of cement remaining undispersed with a flat-end stirring rod. Digest on a steam bath or hot plate for 15 min, then filter through a medium-textured filter paper into a 100-mL volumetric flask. Wash beaker and paper thoroughly with hot-water, cool contents of the flask to room temperature, dilute to 100 mL, and mix the solution thoroughly. Continue as given in 17.1.7.2.

17.1.7.1.2 Place 1.000 ± 0.001 g of cement into a platinum evaporating dish and disperse with 10 mL of water using a swirling motion. While still swirling, add 5.0 mL of HCl all at once. Break up any lumps with a flat-end stirring rod and evaporate to dryness on a steam bath. Make certain that the gelatinous appearance is no longer evident. Treat the residue with 2.5 mL of HCl and about 20 mL of water. Digest on a steam bath for 5 to 10 min and filter immediately through a 9-cm medium-textured filter paper into a 100-mL volumetric flask. Wash thoroughly with repeated small amounts of hot water until the total volume of solution is 80 to 95 mL. Cool to room temperature, dilute to the mark, and mix thoroughly.

When it has been demonstrated that the removal of SiO_2 is necessary to obtain the required accuracy described in 3.3.2.1 through 3.3.3 for a specific flame photometer, SiO_2 must always be removed when making analyses that are used as the basis for rejection of a cement for failure to comply with specifications or where specification compliance may be in question. Where there is no question as to specification compliance, analyses may be made by such instruments without SiO_2 removal provided the deviations from certificate values obtained by the tests prescribed in 3.3.2.1 through 3.3.3 are not more than twice the indicated limits.

17.1.7.2 If the test method in use requires more dilute solutions, an internal standard, or both, carry out the same dilutions as in 17.1.5.4 as needed. The standard and the sample solutions to be analyzed must be prepared in the same way and to the same dilution as the solutions of standard cements analyzed for the qualification of the instrument.

17.1.7.3 *Procedure for Na_2O* (Note 62)—Warm up and adjust the instrument for the determination of Na_2O as described in 17.1.6.1. Immediately following the adjustment and without changing any instrumental settings, atomize the cement solution and note the scale reading (Note 61). Select the standard solutions which immediately bracket the cement solution in Na_2O content and observe their readings. Their values should agree with the values previously established during calibration of the apparatus. If not, recalibrate the apparatus for that constituent. Finally, alternate the use of the unknown solution and the bracketing standard solutions until readings of the unknown agree within one division on the transmission or meter scale, or within 0.01 weight percent for instruments with digital readout, and readings for the standards similarly agree with the calibration values. Record the average of the last two readings obtained for the unknown solution.

NOTE 61—The order in determining Na_2O or K_2O is optional. In all cases, however, the determination should immediately follow the adjustment of the instrument for that particular constituent.

17.1.7.4 If the reading exceeds the scale maximum, either transfer a 50-mL aliquot of the solution prepared in 17.1.7.1 to a 100-mL volumetric flask or, if desired, prepare a new solution by using 0.500 g of cement and 2.5 mL of HCl (instead of 5.0 mL) in the initial addition of acid. In the event silica has to be removed from the 0.5-g sample of cement, treat the dehydrated material with 1.25 mL of HCl and about 20 mL of water, then digest, filter, and wash. In either case, add 5.0 mL of calcium chloride stock solution (17.1.5.1) before diluting to mark with water. Dilute to the mark. Proceed as in 17.1.5.4 if more dilute solutions are required by the test method in use. Determine the alkali content of this solution as described in (17.1.7.3) and multiply by a factor of 2 the percentage of alkali oxide.

17.1.7.5 *Procedure for K_2O*—Repeat the procedure described in 17.1.7.3 except that the instrument shall be adjusted for the determination of K_2O. For instruments that read both Na_2O and K_2O simultaneously, determine K_2O at the same time as determining Na_2O.

17.1.8 *Calculation and Report*—From the recorded averages for Na_2O and K_2O in the unknown sample, report each oxide to the nearest 0.01 %.

17.2 *Water-Soluble Alkalies:*

NOTE 62—The determination of water-soluble alkali should not be considered as a substitute for the determination of total alkali according to 17.1.2.1 to 17.1.8. Moreover, it is not to be assumed that in this method all water-soluble alkali in the cement will be dissolved. Strict adherence to the procedure described is essential where there is a specified limit on the content of water-soluble alkali or where several lots of cement are compared on the basis of water-soluble alkali.

17.2.1 *Procedure:*

17.2.1.1 Weigh 25.0 g of sample into a 500-mL Erlenmeyer flask and add 250 mL of water. Stopper the flask with a rubber stopper and shake continuously for 10 min at room temperature. Filter through a Büchner funnel which contains a well-seated retentive, dry filter paper, into a 500-mL filtering flask, using a weak vacuum. Do not wash.

17.2.1.2 Transfer a 50-mL aliquot (Note 63) of the filtrate to a 100-mL volumetric flask and acidify with 0.5 mL of concentrated HCl (sp gr 1.19). Add 9.0 mL of stock $CaCl_2$ solution (63 000 ppm CaO), described in 17.1.5.1, to the 100-mL flask, and dilute the solution to 100 mL. If the test method in use requires more dilute solutions, an internal standard, or both, carry out the same dilutions as in 17.1.5.4, as needed. Determine the Na_2O and K_2O contents of this solution as described in 17.1.7.3 and 17.1.7.5. Record the parts per million of each alkali in the solution in the 100-mL flask.

NOTE 63—The aliquot of the filtrate taken for the analysis should be based on the expected water-soluble alkali content. If the expected level of either K_2O or Na_2O is more than 0.08 weight % of cement, or if the water soluble alkali level is unknown, a 50-mL aliquot as given in 17.2.1.2 should be used to make up the initial test solution. If either the Na_2O or K_2O exceeds 0.16 %, place a 50-mL aliquot of the solution from 17.2.1.2 in a 100-mL volumetric flask, add 5 mL of $CaCl_2$ stock solution, and dilute to 100 mL. When the level of either K_2O or Na_2O is less than 0.08 %, take a 100-mL aliquot from the original filtrate (obtained by 17.2.1.1), add 1 mL of HCl, and evaporate on a hot plate in a 250-mL beaker to about 70 mL. Add 8 mL of stock $CaCl_2$ solution and transfer the sample to a 100-mL volumetric flask, rinsing the beaker with a small portion of distilled water. Cool the solution to room temperature and dilute to 100 mL.

17.2.2 *Calculations*—Calculate the percentage of the water-soluble alkali, expressed as Na_2O, to the nearest 0.01 as follows:

$$\text{Total water-soluble alkali, as } Na_2O = A + E \quad (9)$$

$$A = B/(V \times 10)$$

$$C = D/(V \times 10)$$

$$E = C \times 0.658$$

where:

A = percentage of water-soluble sodium oxide (Na_2O),
V = millilitres of original filtrate in the 100-mL flask,
B = parts per million of Na_2O in the solution in the 100-mL flask,
C = percent of water-soluble potassium oxide (K_2O),
D = parts per million of K_2O in the 100-mL flask,
E = percentage Na_2O equivalent to K_2O determined, and
0.658 = molecular ratio of Na_2O to K_2O.

18. Manganic Oxide (*Reference Method*)

18.1 *Summary of Method*—In this procedure, manganic oxide is determined volumetrically by titration with sodium arsenite solution after oxidizing the manganese in the cement with sodium metabismuthate ($NaBiO_3$).

18.2 *Reagents:*

18.2.1 *Sodium Arsenite, Standard Solution* (1 mL = 0.0003 g Mn_2O_3)—Dissolve in 100 mL of water 3.0 g of sodium carbonate (Na_2CO_3) and then 0.90 g of arsenic trioxide (As_2O_3), heating the mixture until the solution is as complete as possible. If the solution is not clear or contains a residue, filter the solution. Cool it to room temperature, transfer to a volumetric flask, and dilute to 1 L.

18.2.1.1 Dissolve 0.58 g of potassium permanganate ($KMnO_4$) in 1 L of water and standardize it against about 0.03 g of sodium oxalate ($Na_2C_2O_4$) oxidimetric standard furnished by the National Bureau of Standards (Standard Sample No. 40 or its replacement) according to the directions furnished with the sodium oxalate. Put 30.0 mL of the $KMnO_4$ solution in a 250-mL Erlenmeyer flask. Add 60 mL of HNO_3 (1+4) and 10 mL of sodium nitrite ($NaNO_2$, 50 g/L) to the flask. Boil the solution until the HNO_2 is completely expelled. Cool the solution, add $NaBiO_3$, and finish by titrating with the standard sodium arsenite ($NaAsO_2$) solution as described in 18.3.2. Calculate the manganic oxide (Mn_2O_3) equivalent of the $NaAsO_2$ solution, g/mL, as follows:

$$E = (A \times 7.08)/BC \tag{10}$$

where:

E = Mn_2O_3 equivalent of the $NaAsO_2$ solution, g/mL,
A = grams of $Na_2C_2O_4$ used,
B = millilitres of $KMnO_4$ solution required by the $Na_2C_2O_4$,
C = millilitres of $NaAsO_2$ solution required by 30.0 mL of $KMnO_4$ solution, and
7.08 = molecular ratio of Mn_2O_3 to 5 $Na_2C_2O_4$ (0.236) multiplied by 30.0 (millilitres of $KMnO_4$ solution).

18.2.2 *Sodium Metabismuthate* ($NaBiO_3$).

18.2.3 *Sodium Nitrite Solution* (50 g $NaNO_2$/L).

18.3 *Procedure:*

18.3.1 Weigh 1.0 to 3.0 g of the sample (Note 64) into a 250-mL beaker and treat it with 5 to 10 mL of water and then with 60 to 75 mL of HNO_3 (1+4). Boil the mixture until the solution is as complete as possible. Add 10 mL of $NaNO_2$ solution (50 g/L) to the solution and boil it until the nitrous acid is completely expelled (Note 65), taking care not to allow the volume of the solution to become so small as to cause the precipitation of gelatinous SiO_2. There may be some separated SiO_2, which may be ignored, but if there is still a red or brown residue, use more $NaNO_2$ solution (50 g/L) to effect a complete decomposition, and then boil again to expel the nitrous acid. Filter the solution through a medium-textured paper into a 250-mL Erlenmeyer flask and wash the filter paper with water.

NOTE 64—The amount of cement taken for analysis depends on the content of manganese, varying from 1 g for about 1 % of Mn_2O_3 to 3 g for 0.25 % or less of Mn_2O_3.

NOTE 65—When $NaNO_2$ is added, the expulsion of HNO_2 by boiling must be complete. If any HNO_2 remains in the solution, it will react with the added $NaBiO_3$ and decrease its oxidizing value. If there is any manganese in the cement, the first small quantity of $NaBiO_3$ should bring out a purple color.

18.3.2 The solution should have a volume of 100 to 125 mL. Cool it to room temperature. To the solution add a total of 0.5 g of $NaBiO_3$ in small quantities, while shaking intermittently. After the addition is completed, shake the solution occasionally for 5 min and then add to it 50 mL of cool HNO_3 (1+33) which has been previously boiled to expel nitrous acid. Filter the solution through a pad of ignited asbestos in a Gooch crucible or a carbon or fritted-glass filter with the aid of suction. Wash the residue four times with the cool HNO_3 (1+33). Titrate the filtrate immediately with the standard solution of $NaAsO_2$. The end point is reached when a yellow color is obtained free of brown or purple tints and does not change upon further addition of $NaAsO_2$ solution.

18.3.3 *Blank*—Make a blank determination, following the same procedure and using the same amounts of reagents, and correct the results obtained in the analysis accordingly.

18.4 Calculate the percentage of Mn_2O_3 to the nearest 0.01 as follows:

$$Mn_2O_3, \% = (EV/S) \times 100 \tag{11}$$

where:

E = Mn_2O_3 equivalent of the $NaAsO_2$ solution, g/mL,
V = millilitres of $NaAsO_2$ solution required by the sample, and
S = grams of sample used.

19. Chloride (*Reference Test Method*)

19.1 *Summary of Test Method*—In this test method total chloride content of portland cement is determined by the potentiometric titration of chloride with silver nitrate. The procedure is also applicable to hardened concrete, clinker, and portland cement raw mix. Under the conditions of the test, no constituent normally present in these materials will interfere.

NOTE 66—Species that form insoluble silver salts or stable silver complexes in acid solution interfere with potentiometric measurements. Thus, iodides and bromides interfere while fluorides will not. Sulfide salts in concentrations typical of these materials should not interfere because they are decomposed by acid treatment.

19.2 *Apparatus:*

19.2.1 *Chloride, Silver/Sulfide Ion Selective Electrode,* or a silver billet electrode coated with silver chloride (Note 67), with an appropriate reference electrode.

19.2.2 *Potentiometer,* with millivolt scale readable to 1 mV or better. A digital read-out is preferred but not required.

19.2.3 *Buret,* Class A, 10-mL capacity with 0.05-mL divisions. A buret of the potentiometric type, having a displaced delivery tip, is convenient, but not required.

NOTE 67—Suitable electrodes are available from Orion, Beckman Instruments, and Leeds and Northrup. Carefully following the manufacturer's instructions, add filling solution to the electrodes. The silver billet electrodes must be coated electrolytically with a thin, even layer of silver chloride. To coat the electrode, dip the clean silver billet of the electrode into a saturated solution of potassium chloride (about 40 g/L) in water and pass an electric current through the electrode from a 1½ to 6-V dry cell with the silver billet electrode connected to the positive terminal of the battery. A carbon rod from an all-dry cell or other suitable electrode is connected to the negative terminal and immersed in the solution to complete the electrical circuit. When the silver chloride coating wears off, it is necessary to rejuvenate the electrode by repeating the above procedure. All of the old silver chloride should first be removed from the silver billet by rubbing it gently with fine emery paper followed by water rinsing of the billet.

19.3 *Reagents:*

19.3.1 *Sodium Chloride* (NaCl), primary standard grade.

19.3.2 *Silver Nitrate* ($AgNO_3$), reagent grade.

19.3.3 *Potassium Chloride* (KCl), reagent grade (required for silver billet electrode only).

19.3.4 *Reagent Water* conforming to the requirements of Specification D 1193 for Type III reagent water.

19.4 *Preparation of Solutions:*

19.4.1 *Sodium Chloride, Standard Solution* (0.05 *N* NaCl)—Dry sodium chloride (NaCl) at 105 to 110°C to a constant weight. Weigh 2.9222 g of dried reagent. Dissolve in water and dilute to exactly 1 L in a volumetric flask and mix thoroughly. This solution is the standard and requires no further standardization.

19.4.2 *Silver Nitrate, Standard Solution* (0.05 *N* $AgNO_3$)—Dissolve 8.4938 g of silver nitrate ($AgNO_3$) in water. Dilute to 1 L in a volumetric flask and mix thoroughly. Standardize against 5.00 mL of standard 0.05 *N* sodium chloride solution diluted to 150 mL with water following the titration test method given in 19.5.4 beginning with the second sentence. The exact normality shall be calculated from the average of three determinations as follows:

$$N = 0.25/V \tag{12}$$

where:

N = normality of $AgNO_3$ solution,
0.25 = milliequivalents NaCl (5.0 mL × 0.05 *N*), and
V = volume of $AgNO_3$ solution, mL.

Commercially available standard solutions may be used provided the normality is checked according to the standardization procedure.

19.4.3 *Methyl Orange Indicator*—Prepare a solution containing 2 g of methyl orange per litre of 95 % ethyl alcohol.

19.5 *Procedure:*

19.5.1 Weigh a 5.0-g sample of the cement or a 10.0-g sample of concrete into a 250-mL beaker (Note 68). Disperse the sample with 75-mL of water. Without delay slowly add 25 mL of dilute (1+1) nitric acid, breaking up any lumps with a glass rod. If the smell of hydrogen sulfide is strongly evident at this point, add 3 mL of hydrogen peroxide (30 % solution) (Note 69). Add 3 drops of methyl orange indicator and stir. Cover the beaker with a watch glass and allow to stand for 1 to 2 min. If a yellow to yellow-orange color appears on top of the settled solids, the solution is not sufficiently acidic. Add additional dilute nitric acid (1+1) dropwise while stirring until a faint pink or red color persists. Then add 10 drops in excess. Heat the covered beaker rapidly to boiling. Do not allow to boil for more than a few seconds. Remove from the hot plate (Note 70).

NOTE 68—Use a 5-g sample for cement, and 10 g for concrete and other materials having an expected chloride content of less than about 0.15 % Cl. Use proportionally smaller samples for materials with higher chloride concentrations. Use cement and other powdered materials as is without grinding. Coarse samples require grinding to pass a 20-mesh sieve. If a sample is too fine, excessive silica gel may form during digestion with nitric acid, thereby slowing subsequent filtration.

NOTE 69—Slags and slag cements contain sulfide sulfur in concentrations that can interfere with the determination.

NOTE 70—It is important to keep the beaker covered during heating and digestion to prevent the loss of chloride by volatilization. Excessive amounts of acid should not be used since this results in early removal of the silver chloride coating from the silver billet electrode. A slurry that is only slightly acidic is sufficient.

19.5.2 Wash a 9-cm coarse-textured filter paper with four 25-mL increments of water using suction filtering provided by a 250-mL or 500-mL Büchner funnel and filtration flask. Discard the washings and rinse the flask once with a small portion of water. Reassemble the suction apparatus and filter the sample solution. Rinse the beaker and the filter paper twice with small portions of water. Transfer the filtrate from the flask to a 250-mL beaker and rinse the flask once with water. The original beaker may be used (Note 71). Cool the filtrate to room temperature. The volume should not exceed 175 mL.

NOTE 71—It is not necessary to clean all the slurry residue from the sides of the beaker nor is it necessary that the filter remove all of the fine material. The titration may take place in a solution containing a small amount of solid matter.

19.5.3 For instruments equipped with dial readout it is necessary to establish an approximate "equivalence point" by immersing the electrodes in a beaker of water and adjusting the instrument to read about 20 mV lower than mid-scale. Record the approximate millivoltmeter reading. Remove the beaker and wipe the electrodes with absorbent paper.

19.5.4 To the cooled sample (Note 72) beaker from 19.5.2, carefully pipet 2.00 mL of standard 0.05 *N* NaCl solution. Place the beaker on a magnetic stirrer and add a TFE-fluorocarbon-coated magnetic stirring bar. Immerse the electrodes into the solution taking care that the stirring bar does not strike the electrodes; begin stirring gently. Place the delivery tip of the 10-mL buret, filled to the mark with standard 0.05 *N* silver nitrate solution, in (preferably) or above the solution (Note 73).

NOTE 72—It is advisable to maintain constant temperature during measurement, for the solubility relationship of silver chloride varies markedly with temperature at low concentrations.

NOTE 73—If the tip of the buret is out of the solution, any adhering droplet should be rinsed onto the beaker with a few millilitres of water following each titration increment.

19.5.5 Gradually titrate, record the amount of standard 0.05 *N* silver nitrate solution required to bring the millivoltmeter reading to −60.0 mV of the equivalence point determined in the water.

19.5.6 Continue the titration with 0.20-mL increments. Record the buret reading and the corresponding millivoltmeter reading in columns 1 and 2 of a four-column recording form like that shown in Appendix X1. Allow sufficient time between each addition for the electrodes to reach equilibrium with the sample solution. Experience has shown that acceptable readings are obtained when the minimum scale reading does not change within a 5-s period (usually within 2 min).

19.5.7 As the equivalence point is approached, the equal additions of $AgNO_3$ solution will cause larger and larger changes in the millivoltmeter readings. Past the equivalence point the change per increment will again decrease. Continue to titrate until three readings past the approximate equivalence point have been recorded.

19.5.8 Calculate the difference in millivolt readings between successive additions of titrant and enter the values in Column 3 of the recording form. Calculate the difference between consecutive values in Column 3 and enter the results in Column 4. The equivalence point of the titration will be within the maximum Δ mV interval recorded in Column 3. The precise equivalence point can be interpolated from the data listed in Column 4 as shown in the Appendix X1.

19.5.9 *Blank*—Make a blank determination using 75 mL of water in place of the sample, following the same procedure starting with the third sentence of 19.5.1 without delay. Correct the results obtained in the analysis accordingly (Note 74) by subtracting the blank.

19.6 *Calculation*—Calculate the percent chloride to the nearest 0.001 % as follows:

$$\text{Cl, \%} = \frac{3.545\,[(V_1 - V_2)N - 0.10]}{W} \tag{13}$$

where:

V_1 = millilitres of 0.05 N $AgNO_3$ solution used for sample titration (equivalence point),
V_2 = millilitres of 0.05 N $AgNO_3$ solution used for blank titration (equivalence point),
N = exact normality of 0.05 N $AgNO_3$ solution,
0.10 = milliequivalents of NaCl added (2.0 mL × 0.05 N), and
W = weight of sample, g.

NOTE 74—For nonreferee analysis the blank may be omitted.

20. Chloroform-Soluble Organic Substances (*Reference Test Method*)

20.1 *Summary of Test Method*—This test method[12] was specially designed for the determination of Vinsol resin and tallow in portland cement, although mineral oil, common rosin, calcium stearate, and other fatty acid compounds, and probably some other substances, if present, will be included in the determination. Extreme care is necessary in the entire procedure. The test method may be applied to types of cement other than portland cement, although if the cement contains a large amount of acid-insoluble matter, the emulsions may separate slowly, and less vigorous shaking, more chloroform, and more washing may be necessary.

20.2 *Reagents:*

20.2.1 *Chloroform*—If the blank determination as described in 20.3.5 exceeds 0.0015 g, the chloroform should be distilled before use. Chloroform recovered in the procedure may be slightly acid but can be reused for the portions to be shaken with the aqueous acid solution of the sample in the 1-L funnel. Chloroform used for washing the filter and transferring the extract should be fresh or distilled from fresh chloroform.

20.2.2 *Stannous Chloride* ($SnCl_2$).

20.3 *Procedure:*

20.3.1 Place 40 g of cement in a 1-L Squibb separatory funnel (Note 75) and mix it with 520 mL of water added in two approximately equal portions. Shake vigorously immediately after the addition of the first portion to effect complete dispersion. Then add the second portion and shake again. At once add rapidly 185 mL of HCl in which 10 g of $SnCl_2$ (Note 76) have been dissolved, rapidly insert the stopper in the funnel, invert, and shake with a swirling motion for a few seconds to loosen and disperse all the cement, taking care to avoid the development of great internal pressure due to unnecessarily violent shaking. Release internal pressure immediately by opening and closing the stopcock. Repeat the shaking and release the pressure until the decomposition of the cement is complete. If necessary, break up persistent lumps with a long glass rod. Cool to room temperature rapidly by allowing tap water to run on the flask.

NOTE 75—The use of grease to lubricate the stopcocks and glass stoppers of the separatory funnels should be avoided. Wetting the stopcocks with water before using will assist in their easy operation.

NOTE 76—The purpose of the $SnCl_2$ is to prevent the oxidation of sulfide sulfur to elemental sulfur, which is soluble in chloroform.

20.3.2 Add 75 mL of chloroform to the solution, stopper the funnel, shake it vigorously for 5 min, and allow the water and chloroform to stand 15 min to separate. Draw off the lower chloroform layer into a 125-mL Squibb separatory funnel, including the scum (Note 77) and a few millilitres of the aqueous layer, making certain that all the scum is transferred. Keep the amount of the aqueous layer transferred to an absolute minimum, since excessive water in the 125-mL funnel may result in incomplete extraction of the scum and may cause an emulsion which does not separate readily. Shake the funnel vigorously to ensure the complete extraction of the scum. Allow the chloroform to separate, and draw it into a 250-mL Squibb separatory funnel which contains 50 mL of water and a few drops of HCl, making sure to keep the scum behind in the 125-mL funnel. Shake the 250-mL funnel, and draw the chloroform into another 250-mL funnel that contains 50 mL of water and a few drops of HCl. Shake this funnel as in the case of the first 250-mL funnel. When the chloroform separates, draw it into a standard-taper flat-bottom boiling flask (Note 78), taking care not to allow any water to enter the flask.

NOTE 77—There is usually a dark colored scum at the liquid interface. It may contain chloroform-soluble organic substance after shaking in the funnel, where the proportion of water to chloroform is great. It may be concentrated and confined to a small volume by gently twirling the funnel after the scum has been drawn into the narrower part of the funnel.

NOTE 78—The liquid is later distilled. No cork or rubber stoppers should be used. A 250 or 300-mL soil analysis flask, fitted with a condenser tube by means of a ground joint, is satisfactory. The tube may be bent near the neck and the remaining part fitted with a water-cooling jacket. Chloroform thus recovered may be reused as described in 20.2.1.

20.3.3 Add 25 mL of chloroform to the solution in the original 1-L separatory funnel, and carry out the operations as described in 20.3.2, retaining the original wash water in the 250-mL funnels. Repeat, using another 25-mL portion of chloroform.

20.3.4 Distill the combined chloroform extracts in the boiling flask until their volume is reduced to 10 to 15 mL. Filter the remaining liquid into a weighed 100-mL glass beaker or platinum dish (Note 79) through a small medium-textured filter paper that has been washed with fresh chloroform. Rinse the flask and wash the paper with several small portions of fresh chloroform. Evaporate the extracts at a low temperature (not over 63°C) to dryness (Note 80) and heat it in an oven at 57 to 63°C for 3 min. Pass dry air into the vessel for 15 s, cool, and weigh. Repeat the heating and weighing until two successive weighings do not differ by more than 0.0010 g. The higher of the last two weights shall be taken as the true weight.

[12] The 1965 revision of these test methods deleted the methoxyl test method for determining Vinsol resin. Those interested in this test method should refer to the *1966 Book of ASTM Standards*, Part 9.

NOTE 79—A platinum dish is preferable, as it quickly attains the temperature of the balance. If a glass beaker is used, it should be allowed to stand in the case of the balance for at least 20 min before weighing.

NOTE 80—Care should be taken in drying the extract, as many of the chloroform-soluble organic substances are somewhat volatile when heated for a long time at even moderate temperatures. With protection from the accumulation of dust, the solution may be evaporated at room temperature overnight.

When a quick evaporation is desired, the solution may be evaporated on a hot plate at low heat under a stream of dry air through a glass tube (about 10 mm in inside diameter) until it is about 3 mm in depth. Then remove the vessel from the hot plate and continue a slow stream of dry air until the residue appears dry. Then continue with a more rapid stream of dry air for 5 min at room temperature before placing the vessel in the oven at 57 to 63°C. After each 3-min heating period in the oven, pass dry air into the vessel for about 15 s before weighing. The air may be dried by passing it through a cheap desiccant, such as calcium chloride or sulfuric acid, followed by a desiccant of high efficiency, such as magnesium perchlorate or anhydrous calcium sulfate, with care taken to avoid the carrying of dust from the desiccant by the air. Instead of using compressed air, which is often contaminated with oil, dirt, and moisture, one can place the chloroform solution under a bell glass and induce a stream of air through the desiccants by means of an aspirator or vacuum pump.

When Vinsol resin is known to be the only substance present, the residue is more stable and may be heated at 100 to 105°C, instead of 57 to 63°C, in order to expel all possible traces of chloroform.

20.3.5 *Blank*—Make a blank determination. Ignite a 40-g sample of the cement at 950 to 1000°C for 1 h (Note 81) and regrind. Treat this ignited sample by the same procedure and using the same reagents as in the analysis and correct the results accordingly.

NOTE 81—Care should be taken to completely burn off the organic substance. A 100-mL flat platinum dish, in which the sample is well spread out, and a muffle furnace are advised for this purpose. If such a furnace is not available, a large high-temperature burner of the Meker type may be used. Thorough stirring of the sample should be done frequently—every 5 min when a burner is used.

20.4 *Calculation*—Calculate the percentage of chloroform-soluble organic substances to the nearest 0.001 by multiplying the weight in grams of residue (Note 82) by 2.5 (100 divided by the weight of the sample used (40 g)).

NOTE 82—If the organic substance in the cement is tallow, the residue is the fatty acids resulting from the hydrolysis of the tallow in the hot acid solution, and its weight should be multiplied by 1.05 to give the weight of the original glycerides in the tallow. If the original substance is calcium stearate, the residue is stearic acid, and its weight multiplied by 1.07 gives the weight of calcium stearate.

ALTERNATIVE TEST METHODS

21. Calcium Oxide (*Alternative Test Method*)

21.1 *Summary of Test Method:*

21.1.1 This test method covers the gravimetric determination of CaO after removal of SiO_2 and the ammonium hydroxide groups and double precipitation of calcium as the oxalate. The precipitate is converted to CaO by ignition and is weighed.

21.1.2 Strontium, usually present in portland cement as a minor constituent, is precipitated with calcium as the oxalate and is subsequently calculated as CaO. If the SrO content is known and correction of CaO for SrO is desired as, for example, for research purposes or to compare results with SRM certificate values, the CaO obtained by this test method may be corrected by subtracting percent SrO. In determining conformance of a cement to specifications the correction of CaO for SrO should not be made.

21.2 *Procedure* (Note 83):

21.2.1 Acidify the combined filtrates obtained in the determination of the ammonium hydroxide group (7.1 through 7.3) and, if necessary, evaporate to a volume of about 200 mL. Add 5 mL of HCl, a few drops of methyl red indicator solution, and 30 mL of warm ammonium oxalate solution (50 g/L) (Note 39). Heat the solution to 70 to 80°C and add NH_4OH (1+1) dropwise with stirring until the color changes from red to yellow (see Note 40). Allow the solution to stand without further heating for 1 h (not longer), with occasional stirring during the first 30 min. Filter using a retentive paper and wash moderately with cold ammonium oxalate solution (1 g/L). Reserve the filtrate and washings.

NOTE 83—When analyses are being made for determining conformity to specifications and there is a possibility that sufficient manganese will be present to cause the percentage of magnesium determined by alternate test methods to exceed the specification limit, manganese may be removed as directed in 13.3.2 before CaO is determined by this alternative test method.

21.2.2 Transfer the precipitate and filter paper to the beaker in which the precipitation was made. Dissolve the oxalate in 50 mL of hot HCl (1+4) and macerate the filter paper. Dilute to 200 mL with water, add a few drops of methyl red indicator and 20 mL of ammonium oxalate solution, heat the solution nearly to boiling, and precipitate calcium oxalate again by neutralizing the acid solution with NH_4OH as described in 13.3.1. Allow the solution to stand 1 to 2 h (standing for 2 h at this point does no harm), filter, and wash as before. Combine the filtrate with that already obtained and reserve for the determination of MgO (14.3.1).

21.2.3 Dry the precipitate in a weighed covered platinum crucible. Char the paper without inflaming, burn the carbon at as low a temperature as possible, and, finally, heat with the crucible tightly covered in an electric furnace or over a blast lamp at a temperature of 1100 to 1200°C. Cool in a desiccator and weigh as CaO. Repeat the ignition to constant weight.

21.2.4 *Blank*—Make a blank determination, following the same procedure and using the same amounts of reagents, and correct the results obtained in the analysis accordingly.

21.3 *Calculation:*

21.3.1 Calculate the percentage of CaO to the nearest 0.1 by multiplying the weight in grams of CaO by 200 (100 divided by the weight of sample used (0.5 g)).

21.3.2 Correct the percent CaO for SrO, if desired, by subtracting the percent SrO.

22. Magnesium Oxide (*Alternative Test Method*)

22.1 *Summary of Test Method*—This alternative test method is a volumetric procedure suitable for use when the determinations of silicon dioxide (SiO_2), aluminum oxide (Al_2O_3), ferric oxide (Fe_2O_3), and calcium oxide (CaO) are omitted.

22.2 *Rapid Volumetric Test Method (Titration of Magnesium Oxyquinolate):*

22.3 *Reagents:*

22.3.1 *Ammonium Nitrate Solution* (20 g NH_4NO_3/L).

22.3.2 *Ammonium Oxalate Solution* (50 g/L).

22.3.3 *Hydroxyquinoline Solution*—Dissolve 25 g of 8-hydroxyquinoline in 60 mL of acetic acid. When the solution is complete, dilute to 2 L with cold water. One millilitre of this solution is equivalent to 0.0016 g of MgO.

22.3.4 *Potassium Bromate-Potassium Bromide, Standard Solution* (0.2 *N*)—Dissolve 20 g of potassium bromide (KBr) and 5.57 g of potassium bromate ($KBrO_3$) in 200 mL of water and dilute to 1 L. Obtain the ratio of the strength of this solution to that of the 0.1 *N* $Na_2S_2O_3$ solution (22.2.6) as follows: To 200 mL of water in a 500-mL Erlenmeyer flask add 25.0 mL of the 0.2 *N* $KBrO_3$-KBr solution, measured from a pipet or buret. Add 20 mL of HCl, stir, and add immediately 10 mL of potassium iodide (KI) (250 g/L). Mix well and titrate at once with the $Na_2S_2O_3$ solution until nearly colorless. Add 2 mL of starch solution and titrate to the disappearance of the blue color. Calculate the ratio in strength of the $KBrO_3$-KBr solution to the $Na_2S_2O_3$ solution by dividing the volume of $Na_2S_2O_3$ solution by the volume of $KBrO_3$-KBr solution used in the titration.

22.3.5 *Potassium Iodide Solution* (250 g KI/L).

22.3.6 *Sodium Thiosulfate, Standard Solution* (0.1 *N*)—Dissolve 25 g of sodium thiosulfate ($Na_2S_2O_3 \cdot 5H_2O$) in 200 mL of water, add 0.1 g of sodium carbonate (Na_2CO_3), and dilute to 1 L. Let stand at least 1 week. Standardize this solution directly against primary standard potassium dichromate ($K_2Cr_2O_7$). One millilitre of 0.10 *N* $Na_2S_2O_3$ solution is equivalent to 0.000504 g of MgO.

22.3.7 *Starch Solution*—To 500 mL of boiling water add a cold suspension of 5 g of soluble starch in 25 mL of water, cool to room temperature, add a cool solution of 5 g of sodium hydroxide (NaOH) in 50 mL of water, add 15 g of KI, and mix thoroughly.

22.4 *Procedure:*

22.4.1 Disperse 0.5 g (Note 84) of the sample of cement in a 400-mL beaker with 10 mL of water, using a swirling motion. While still swirling, add 10 mL of HCl all at once. Dilute immediately to 100 mL. Heat gently and grind any coarse particles with the flattened end of a glass rod until decomposition is complete, add 2 or 3 drops of HNO_3 and heat to boiling (Note 85).

NOTE 84—If SiO_2, ammonium hydroxide group, and CaO are separated and determined in accordance with the appropriate sections for either the reference or alternative test methods, the remaining filtrate may be used for the determination of MgO as described in 22.4.1, starting with the third from the last sentence of 22.4.2, "Add 5 mL of HCl. . .".

NOTE 85—In the case of cements containing blast-furnace slag or a significant quantity of sulfide sulfur, add 12 drops of HNO_3 and boil for 20 min to oxidize iron and remove sulfide.

22.4.2 Add 3 drops of methyl red indicator to the solution and then add NH_4OH until the solution is distinctly yellow. Heat this solution to boiling and boil for 50 to 60 s. In the event difficulty from bumping is experienced while boiling the ammoniacal solution, a digestion period of 10 min on a steam bath, or a hot plate having the approximate temperature of a steam bath, may be substituted for the 50 to 60-s boiling period. Remove from the burner, steam bath, or hot plate and allow to stand until the precipitate has settled. Using medium-textured paper, filter the solution without delay, wash the precipitate twice with hot NH_4NO_3 (20 g/L), and reserve the filtrate. Transfer the precipitate with the filter paper to the beaker and dissolve in 10 mL of HCl (1+1). Macerate the filter paper. Dilute to about 100 mL and heat to boiling. Reprecipitate, filter, and wash the hydroxides as above. Combine this filtrate and washings with those from the first precipitation taking care that the volume does not exceed 300 mL (Note 86). Add 5 mL of HCl, a few drops of methyl red indicator solution and 30 mL of warm ammonium oxalate solution (50 g/L). Heat the solution to 70 to 80°C and add NH_4OH (1+1) dropwise, while stirring, until the color changes from red to yellow (see Note 40). Allow the solution to stand without further heating for 15 min on a steam bath.

NOTE 86—In the case of cements containing blast-furnace slag, or which are believed to contain a significant quantity of manganese, acidify with HCl, evaporate to about 100 mL, and remove the manganese, using the procedure described in 13.3.1.

22.4.3 Add 10 to 25 mL of the 8-hydroxyquinoline reagent (Note 87) and then 4 mL of NH_4OH/100 mL of solution. Stir the solution on a mechanical stirring machine for 15 min and set aside until the precipitate has settled (Note 88). Filter the solution using medium-textured paper and wash the precipitate with hot NH_4OH (1+40). Dissolve the precipitate in 50 to 75 mL of hot HCl (1+9) in a 500-mL Erlenmeyer flask. Dilute the resulting solution to 200 mL and add 15 mL of HCl. Cool the solution to 25°C and add 10 to 35 mL of the 0.2 *N* $KBrO_3$-KBr solution (Note 89) from a pipet or buret. Stir the solution and allow to stand for about 30 s to ensure complete bromination. Add 10 mL of KI (250 g/L). Stir the resulting solution well and then titrate with the 0.1 *N* $Na_2S_2O_3$ solution until the color of the iodine becomes faintly yellow. At this point add 2 mL of the starch solution and titrate the solution to the disappearance of the blue color.

NOTE 87—An excess of the 8-hydroxyquinoline reagent is needed to avoid a low result for MgO, but too great an excess will yield high results. The following guide should be used to determine the amount of reagent added:

Approximate Content of MgO, %	Approximate Amount of Reagent Required, mL
0 to 1.5	10
1.5 to 3.0	15
3.0 to 4.5	20
4.5 to 6.0	25

NOTE 88—The precipitate should be filtered within an hour. Prolonged standing may cause high results.

NOTE 89—The amount of the standard $KBrO_3$–KBr solution used should be as follows:

Approximate Content of MgO, %	Amounts of Standard $KBrO_3$–KBr Solution, mL
0 to 1	10
1 to 2	15
2 to 3	20
3 to 4	25
4 to 5	30
5 to 6	35

22.4.4 *Blank*—Make a blank determination, following the same procedure and using the same amounts of reagents, and correct the results obtained in the analysis accordingly.

22.5 *Calculation*—Calculate the percentage of MgO to the nearest 0.1 as follows: (Note 90)

$$MgO, \% = E(V_1R - V_2) \times 200 \quad (14)$$

where:

E = MgO equivalent of the $Na_2S_2O_3$ solution, g/mL,
V_1 = millilitres of $KBrO_3$–KBr solution used,
R = ratio in strength of the $KBrO_3$–KBr solution to the $Na_2S_2O_3$ solution,
V_2 = millilitres of $Na_2S_2O_3$ solution used, and
200 = 100 divided by the weight of sample used (0.5 g).

NOTE 90—V_1R represents the volume of $Na_2S_2O_3$ solution equivalent to the volume of $KBrO_3$–KBr solution used. V_2 represents the amount of $Na_2S_2O_3$ required by the excess $KBrO_3$–KBr which is not reduced by magnesium oxyquinolate.

23. Loss on Ignition

23.1 *Portland Blast-Furnace Slag Cement and Slag Cement (Alternative Test Method):*

23.1.1 *Summary of Test Method*—This test method covers a correction for the gain in weight due to oxidation of sulfides usually present in such cement by determining the decrease in the sulfide sulfur content during ignition. It gives essentially the same result as the reference test method (16.2.1 through 16.2.3) which provides for applying a correction based on the increase in SO_3 content.

23.1.2 *Procedure:*

23.1.2.1 Weigh 1 g of cement in a tared platinum crucible, cover, and ignite in a muffle furnace at a temperature of 950 ± 50°C for 15 min. Cool to room temperature in a desiccator and weigh. After weighing carefully transfer the ignited material to a 500-mL boiling flask. Break up any lumps in the ignited cement with the flattened end of a glass rod.

23.1.2.2 Determine the sulfide sulfur content of the ignited sample using the procedure described in 15.2.1 through 15.2.5. Using the same procedure, also determine the sulfide sulfur content of a portion of the cement that has not been ignited.

23.1.3 *Calculation*—Calculate the percentage loss of weight occurring during ignition (23.1.2.1) and add twice the difference between the percentages of sulfide sulfur in the original sample and ignited sample as determined in 23.1.2.2. Report this value as the loss on ignition.

NOTE 91—If a gain of weight is obtained during the ignition, subtract the percentage of gain from the correction for sulfide oxidation.

24. Titanium Dioxide (*Alternative Test Method*)

24.1 *Summary of Test Method*—In this test method, titanium dioxide (TiO_2) is determined colorimetrically by comparing the color intensity of the peroxidized solution of the titanium in the sample with the color intensity of a peroxidized standard solution of titanic sulfate.

24.2 *Interferences*—Interfering elements in the peroxide method for TiO_2 are vanadium, molybdenum, and chromium. In very small quantities the interference of the last two is negligible. However, vanadium in very small quantities causes interference and, as some cements contain this element, the Na_2CO_3 fusion (24.5.4) and extraction with water are necessary.

24.3 *Apparatus:*

24.3.1 *Colorimeter*—The apparatus shall consist of a colorimeter of the Kennicott or Duboscq type, or other colorimeter or spectrophotometer designed to measure light transmittancy, and suitable for measurements at wavelengths between 400 and 450 nm.

24.4 *Reagents:*

24.4.1 *Ammonium Chloride* (NH_4Cl).

24.4.2 *Ammonium Nitrate* (20 g NH_4NO_3/L).

24.4.3 *Ferrous Sulfate Solution* (1 mL = 0.005 g Fe_2O_3)—Dissolve 17.4 g of ferrous sulfate ($FeSO_4 \cdot 7H_2O$) in water containing 50 mL of H_2SO_4 and dilute to 1 L. One millilitre is equivalent to 1 % of Fe_2O_3 in 0.5 g of sample.

24.4.4 *Hydrogen Peroxide* (30 %)—Concentrated hydrogen peroxide (H_2O_2).

24.4.5 *Sodium Carbonate* (20 g Na_2CO_3/L).

24.4.6 *Sodium or Potassium Pyrosulfate* ($Na_2S_2O_7$ or $K_2S_2O_7$).

24.4.7 *Titanic Sulfate, Standard Solution* (1 mL = 0.0002 g TiO_2)—Use standard TiO_2 furnished by the National Bureau of Standards (Standard Sample 154 or its replacements). Dry for 2 h at 105 to 110°C. Transfer a weighed amount, from 0.20 to 0.21 g of the TiO_2 to a 125-mL Phillips beaker. Add 5 g of ammonium sulfate ($(NH_4)_2SO_4$) and 10 mL of H_2SO_4 to the beaker and insert a short-stem glass funnel in the mouth of the beaker. Heat the mixture cautiously to incipient boiling while rotating the flask over a free flame. Continue the heating until complete solution has been effected and no unattacked material remains on the wall of the flask (Note 92). Cool and rapidly pour the solution into 200 mL of cold water while stirring vigorously. Rinse the flask and funnel with H_2SO_4 (1+19), stir, and let the solution and washings stand for at least 24 h. Filter into a 1-L volumetric flask, wash the filter thoroughly with H_2SO_4 (1+19), dilute to the mark with H_2SO_4 (1+19), and mix.

NOTE 92—There may be a small residue, but it should not contain more than a trace of TiO_2 if the operations have been properly performed.

24.4.8 Calculate the TiO_2 equivalent of the titanic sulfate solution, g/mL, as follows:

$$E = AB/1000 \quad (15)$$

where:

E = TiO_2 equivalent of the $Ti(SO_4)_2$ solution, g/mL,
A = grams of standard TiO_2 used (corrected for loss on drying),
B = percentage of TiO_2 in the standard TiO_2 as certified by the National Institute of Standards and Technology, divided by 100, and
1000 = number of millilitres in the volumetric flask.

24.5 *Procedure:*

24.5.1 Mix thoroughly 0.5 g of the sample of cement and about 0.5 g of NH_4Cl in a 50-mL beaker, cover the beaker with a watch glass, and add cautiously 5 mL of HCl, allowing the acid to run down the lip of the covered beaker. After the chemical action has subsided, lift the cover, stir the mixture with a glass rod, replace the cover, and set the beaker on a steam bath for 30 min (Note 93). During this time of digestion, stir the contents occasionally and break up any remaining lumps to facilitate the complete decomposition of the cement. Fit a medium-textured filter paper to a funnel and transfer the precipitate to the filter. Scrub the beaker with a rubber policeman and rinse the beaker and policeman. Wash the filter two or three times with hot HCl (1+99) and then with ten or twelve small portions of hot water, allowing each portion to drain through completely.

NOTE 93—A hot plate may be used instead of a steam bath if the heat is so regulated as to approximate that of a steam bath.

24.5.2 Transfer the filter and residue to a platinum crucible (Note 94), dry, and ignite slowly until the carbon of the paper is completely consumed without inflaming. Treat the SiO_2 thus obtained with 0.5 to 1 mL of water, about 10 mL of HF, and 1 drop of H_2SO_4, and evaporate cautiously to dryness (Note 95).

NOTE 94—When it is desired to shorten the procedure for purposes other than referee analysis, usually with little sacrifice of accuracy, the procedure given in 24.5.2 may be omitted.

NOTE 95—When a determination of SiO_2 is desired in addition to one of TiO_2, the SiO_2 may be obtained and treated with HF as directed in 6.2.3.1 through 6.2.4.

24.5.3 Heat the filtrate to boiling and add NH_4OH until the solution becomes distinctly alkaline, as indicated by an ammoniacal odor. Add a small amount of filter paper pulp to the solution and boil for 50 to 60 s. Allow the precipitate to settle, filter through a medium-textured paper, and wash twice with hot NH_4NO_3 solution (20 g/L). Place the precipitate in the platinum crucible in which the SiO_2 has been treated with HF and ignite slowly until the carbon of the paper is consumed.

NOTE 96—When a determination of ammonium hydroxide group is desired in addition to one of TiO_2, the precipitation and ignition may be made as described in 7.2.1 through 7.2.4. However, the crucible must contain the residue from the treatment of the SiO_2 with HF unless circumstances permit its omission as indicated in Note 95.

24.5.4 Add 5 g of Na_2CO_3 to the crucible and fuse for 10 to 15 min (see 24.2.1). Cool, separate the melt from the crucible, and transfer to a small beaker. Wash the crucible with hot water, using a policeman. Digest the melt and washings until the melt is completely disintegrated, then filter through a 9-cm medium-textured filter paper and wash a few times with Na_2CO_3 (20 g/L). Discard the filtrate. Place the precipitate in the platinum crucible and ignite slowly until the carbon of the paper is consumed.

24.5.5 Add 3 g of $Na_2S_2O_7$ or $K_2S_2O_7$ to the crucible and heat below red heat until the residue is dissolved in the melt (Note 97). Cool and dissolve the fused mass in water containing 2.5 mL of H_2SO_4. If necessary, reduce the volume of the solution (Note 98), filter into a 100-mL volumetric flask through a 7-cm medium-textured filter paper, and wash with hot water. Add 5 mL of H_3PO_4, and cool the solution to room temperature. Add H_2O_2 (1.0 mL of 30 % strength or its equivalent) (Note 99), dilute to the mark with water, and mix thoroughly.

NOTE 97—Start the heating with caution because pyrosulfates (also known as fused bisulfates) as received often foam and spatter in the beginning due to an excess of H_2SO_4. Avoid an unnecessarily high temperature or unnecessarily prolonged heating, as fused pyrosulfates may attack platinum. A supply of nonspattering pyrosulfates may be prepared by heating some pyrosulfate in a platinum vessel to eliminate the excess H_2SO_4 and crushing the cool fused mass.

NOTE 98—If the solution is evaporated to too small a volume and allowed to cool, there may be a precipitate of sulfates difficult to redissolve. In case of over-evaporation, do not permit the contents to cool, but add hot water and digest on a steam bath or hot plate until the precipitate is redissolved, with the possible exception of a small amount of SiO_2.

NOTE 99—Hydrogen peroxide deteriorates on standing. Its strength may be determined by adding a measured volume of the solution to 200 mL of cold water and 10 mL of H_2SO_4 (1+1) and titrating with a standard solution of potassium permanganate ($KMnO_4$) prepared in accordance with 13.2.2. If the standard solution contains 0.0056357 g of $KMnO_4$/mL, 49.5 mL of it will be required by 0.50 mL of H_2O_2 (30 %).

24.5.6 Prepare from the standard $Ti(SO_4)_2$ solution a suitable reference standard solution or a series of reference standard solutions in 100-mL volumetric flasks, depending upon the type of colorimeter to be used. To each solution add 3 g of $Na_2S_2O_7$ or $K_2S_2O_7$ dissolved in water, an amount of $FeSO_4$ solution equivalent to the Fe_2O_3 content in 0.5 g of the cement under test, 2.5 mL of H_2SO_4, and 5 mL of H_3PO_4 (Note 100). When the solution is at room temperature, add H_2O_2 (1.0 mL of 30 % strength or its equivalent), dilute to the mark with water, and mix thoroughly (Note 101).

NOTE 100—The color imparted to the solution by $Fe_2(SO_4)_2$ is partly offset by the bleaching effect of H_2SO_4, H_3PO_4, and alkali salts on ferric and peritanic ions. The directions should be followed closely for the highest degree of precision. However, when it is desired to shorten this procedure for purposes other than referee analysis, the addition of pyrosulfate, $FeSO_4$ solution and H_3PO_4 to the color comparison solutions may be omitted provided the Fe_2O_3 of the sample cement is less than 5 %. This usually leads to little sacrifice to accuracy.

NOTE 101—The color develops rapidly and is stable for a sufficient period of time, but if the peroxidized solution is allowed to stand a long time, bubbles of oxygen may appear and interfere with color comparison. When the contents of a tube are first mixed, there may be fine bubbles which should be allowed to clear up before the comparison is made. Comparison between the standard and unknown solution should be made not less than 30 min after addition of H_2O_2.

24.5.7 Compare the color, light transmittancy, or absorbance of the unknown solution with the reference standard solution. The technique of comparing colored solutions or measuring transmittancy or absorbance depends on the type of apparatus (see 24.5.8 to 24.5.10) and should be in accordance with standard practice appropriate to the particular type used or with instructions supplied by the manufacturer of the equipment. If the peroxidized solution of cement is compared with a single standard peroxidized solution, bear in mind that a single peroxidized solution cannot be used for the whole range in TiO_2 content that may be encountered. The difference in volume or depth for the two liquids should not exceed 50 % of the smaller value. All solutions should contain the prescribed concentrations of H_2SO_4, H_3PO_4, $Fe_2(SO_4)_3$, and persulfate except under the circumstances indicated in Note 101.

24.5.8 *Colorimeter of the Kennicott Type*—By means of a plunger in a reservoir of standard peroxidized solution, adjust the amount of solution through which light passes until it gives the same color intensity as the peroxidized solution of the sample.

24.5.9 *Colorimeter of the Duboscq Type*—Lower or raise the plungers in the cups until the two solutions give the same color intensity when viewed vertically. The color matching may be done either visually or photoelectrically.

24.5.10 *Colorimeter Designed to Measure Light Transmittancy*—The measurement should be made between

400 to 450 nm and may be made either visually or photoelectrically. In most colorimeters of this type, the instrument is calibrated with standard solutions and a calibration curve showing the relation of light transmittancy or absorbance to TiO_2 content is prepared in advance of the analysis of the sample for TiO_2.

24.5.11 *Blank*—Make a blank determination, following the same procedure and using the same amounts of reagent, and correct the results obtained in the analysis accordingly.

24.6 *Calculation*—Calculate the percentage of TiO_2 to the nearest 0.01. When a colorimeter designed to measure light transmittancy is used, read the percentage of TiO_2 from a calibration curve showing the relation of light intensity to TiO_2 content. When the peroxidized solution of the sample is compared with a single reference standard solution, calculate the percentage of TiO_2 as follows (Note 102):

24.6.1 *For Colorimeters of the Kennicott Type:*

$$TiO_2, \% = (100\ VE/S) \times (D/C) \quad (16)$$

24.6.2 *For Colorimeters of the Duboscq Type:*

$$TiO_2, \% = (100\ VE/S) \times (F/G) \quad (17)$$

where:

V = millilitres of standard $Ti(SO_4)_2$ solution in the peroxidized standard solution,
E = TiO_2 equivalent of the standard $Ti(SO_4)_2$ solution, g/mL,
S = grams of sample used,
C = total volume of the peroxidized reference standard solution, mL,
D = volume of peroxidized reference standard solution that matches the peroxidized solution of the sample, mL,
F = depth of peroxidized reference standard solution through which light passes, and
G = depth of peroxidized solution of the sample through which light passes.

NOTE 102—The difference between D and C or between F and G should not exceed 50 % of the smaller value.

25. Phosphorus Pentoxide (*Alternative Test Method*)

25.1 *Summary of Test Method*—In this test method, phosphorus is determined volumetrically by precipitation of the phosphorus as ammonium phosphomolybdate and titration with NaOH and H_2SO_4.

25.2 *Reagents:*

25.2.1 *Ammonium Molybdate Solution*—Prepare the solution in accordance with 9.3.1.

25.2.2 *Ammonium Nitrate* (NH_4NO_3).

25.2.3 *Potassium Nitrate Solution* (10 g/L)—Dissolve 10 g of potassium nitrate (KNO_3) in water freshly boiled to expel CO_2 and cooled, and dilute to 1 L.

25.2.4 *Sodium Hydroxide, Standard Solution* (0.3 *N*)—Dissolve 12 g of sodium hydroxide (NaOH) in 1 L of water that has been freshly boiled to expel CO_2, and cooled. Add 10 mL of a freshly filtered, saturated solution of barium hydroxide ($Ba(OH)_2$). Shake the solution frequently for several hours, and filter it. Protect it from contamination by CO_2 in the air. Standardize the solution against standard acid potassium phthalate (Standard Sample No. 84) or benzoic acid (Standard Sample No. 39) furnished by the National Institute of Standards and Technology, according to the directions furnished with the standard. Calculate the phosphorus pentoxide (P_2O_5) equivalent (Note 103) of the solution, g/mL, as follows:

$$E = N \times 0.003086 \quad (18)$$

where:

E = P_2O_5 equivalent of the NaOH solution, g/mL,
N = normality of the NaOH solution, and
0.003086 = P_2O_5 equivalent of 1 *N* NaOH solution, g/mL.

NOTE 103—The value of the solution is based on the assumption that the phosphorus in cement is precipitated as ammonium phosphomolybdate ($2(NH_4)_3PO_4 \cdot 12MoO_3$) and that the precipitate reacts with the NaOH solution thus:

$$2(NH_4)_3PO_4 \cdot 12MoO_3 + 46NaOH = 2(NH_4)_2HPO_4 + (NH_4)_2MoO_4 + 23Na_2MoO_4 + 22H_2O \quad (19)$$

The number of 0.003086 is obtained by dividing the molecular weight of P_2O_5 (141.96) by 46 (for 46 NaOH in the equation) and by 1000 (number of millilitres in 1 L).

As the actual composition of the precipitate is influenced by the conditions under which the precipitation is made, it is essential that all the details of the procedure are followed closely as prescribed.

25.2.5 *Sodium Nitrite* (50 g $NaNO_2$/L).

25.2.6 *Sulfuric Acid, Standard Solution* (0.15 *N*)—Dilute 4.0 mL of H_2SO_4 to 1 L with water that has been freshly boiled and cooled. Standardize against the standard NaOH solution. Determine the ratio in strength of the standard H_2SO_4 solution to the standard NaOH solution by dividing the volume of NaOH solution by the volume of H_2SO_4 solution used in the titration.

25.3 *Procedure:*

25.3.1 Weigh 1 to 3 g of the sample (Note 104) and 10 g of NH_4NO_3 into a 150-mL beaker. Mix the contents, add 10 mL of HNO_3, and stir quickly, using the flattened end of a glass rod to crush lumps of cement, until the cement is completely decomposed and the thick gel of silica (SiO_2) is broken up. Cover the beaker with a watch glass, place it on a water bath or a hot plate at approximately 100°C for 15 to 20 min, and stir the contents occasionally during the heating. Add 20 mL of hot water to the beaker and stir the contents. If the cement contains an appreciable amount of manganese, as shown by the presence of a red or brown residue, add a few millilitres of $NaNO_2$ (50 g/L) to dissolve this residue. Boil the contents of the beaker until all nitrous fumes are completely expelled. This procedure should not take more than 5 min, and water should be added to replace any lost by evaporation. Filter, using medium-textured paper, into a 400-mL beaker under suction and with a platinum cone to support the filter paper. Wash the residue of SiO_2 with hot water until the volume of filtrate and washings is about 150 mL.

NOTE 104—The amounts of sample and reagents used depend on the content of phosphorus in the cement. The minimum requirements are sufficient if the cement contains 0.5 % P_2O_5 or more. The maximum amounts are required if the content of P_2O_5 is 0.1 % or less.

25.3.2 Heat the solution to 69 to 71°C, remove it from the heat source, and immediately add 50 to 100 mL of the ammonium molybdate solution. Stir the solution vigorously for 5 min, wash down the sides of the beaker with cool KNO_3

solution (10 g/L), cover the beaker with a watch glass, and allow to stand 2 h. Using suction, filter the precipitate (Note 105), decanting the solution with as little disturbance to the precipitate as possible. Stir the precipitate in the beaker with a stream of the cool KNO_3 solution, decant the liquid, then wash the precipitate onto the filter. Scrub the stirring rod and beaker with a policeman and wash the contents onto the filter. Wash and precipitate until it is acid-free (Note 106), allowing each portion of wash solution to be sucked completely through before adding the next.

NOTE 105—The filter may be a small medium-textured filter paper supported by a platinum cone, or a small Hirsch funnel may be used with filter paper cut to fit and a thin mat of paper pulp or acid-washed asbestos pulp. The filtration should be carried out with care to avoid any loss of the precipitate. The filter should fit well, and the suction should be started before filtration and maintained until the end of the washing.

NOTE 106—About ten washings are usually required. Test the tenth washing with one drop of neutral phenolphthalein indicator and half a drop of the standard NaOH solution. If a definite pink color lasts at least 5 min, the precipitate is considered to be acid-free; otherwise, continue the washing.

25.3.3 Transfer the filter and precipitate to the beaker in which the precipitation took place, using small damp pieces of paper to wipe out the funnel and to pick up portions of the precipitate that may remain on it. Add 20 mL of cool CO_2-free water to the beaker, and break up the filter by stirring rapidly with the policeman that was used to scrub the beaker. Add an excess of the 0.3 *N* NaOH solution, stir the contents until all trace of yellow has disappeared, wash down the policeman and sides of the beaker with 50 mL of cool, CO_2-free water, and add 2 drops of neutral phenolphthalein indicator solution. Treat the solution with a measured quantity of the 0.15 *N* H_2SO_4 solution, sufficient to destroy completely the pink color. Complete the titration with the NaOH solution until there is a definite faint pink color that lasts at least 5 min.

25.3.4 *Blank*—Make a blank determination, following the same procedure and using the same amounts of reagents, and correct the results obtained in the analysis accordingly.

25.4 *Calculation*—Calculate the percentage of P_2O_5 to the nearest 0.01 as follows:

$$P_2O_5, \% = [E(V_1 - V_2R)/S] \times 100 \quad (20)$$

where:

E = P_2O_5 equivalent of the NaOH solution, g/mL,
V_1 = millilitres of NaOH solution used,
V_2 = millilitres of H_2SO_4 solution used,
R = ratio in strength of the H_2SO_4 solution to the NaOH solution, and
S = grams of sample used.

26. Manganic Oxide (*Alternative Test Method*)

26.1 *Summary of Test Method*—In this test method manganic oxide is determined volumetrically by titration with potassium permanganate solution.

26.2 *Reagents:*

26.2.1 *Potassium Permanganate, Standard Solution* (0.18 *N*)—Prepare a solution of potassium permanganate ($KMnO_4$) and standardize as described in 13.2.2, except that the manganic oxide (Mn_2O_3) equivalent of the solution is calculated instead of the calcium oxide (CaO) equivalent. Calculate the Mn_2O_3 equivalent of the solution as follows:

$$E = (B \times 0.3534)/A \quad (21)$$

where:

E = Mn_2O_3 equivalent of the $KMnO_4$ solution, g/mL,
B = grams of $Na_2C_2O_4$ used,
A = millilitres of $KMnO_4$ solution required by the $Na_2C_2O_4$, and
0.3534 = mole ratio of 3 Mn_2O_3 to 10 $Na_2C_2O_4$.

26.2.2 *Zinc Oxide* (ZnO), powder.

26.3 *Procedure:*

26.3.1 Place 2 g of the sample in a 250-mL beaker and add about 50 mL of water to the cement. Stir the mixture until it is in suspension and then add about 15 mL of HCl. Heat the mixture gently until the solution is as complete as possible. Add 5 mL of HNO_3 and 50 mL of water to the solution and boil it until most of the chlorine has been expelled. If necessary, add hot water to maintain the solution at a volume of about 100 mL. Stop the boiling and add ZnO powder to the solution until the acid is neutralized. Add an excess of 3 to 5 g of ZnO powder to the solution and boil it for a few minutes.

26.3.2 Without filtering, and while keeping the solution hot (90 to 100°C) by intermittent or continuous heating, titrate the solution with the 0.18 *N* $KMnO_4$ solution until a drop of it gives a permanent pink color (Note 107). When the end point is approached, add the standard solution dropwise. After each drop, stir the solution, allow the precipitate to settle a little, and observe the color of the stratum of the solution by looking through the side of the beaker.

NOTE 107—In the case of a cement in which the approximate content of Mn_2O_3 is unknown, a preliminary determination may be made with rapid titration, 0.5 to 1 mL of the standard solution being added at a time, and without an attempt to keep the solution close to the boiling point.

26.3.3 *Blank*—Make a blank determination, following the same procedure and using the same amounts of reagents, and correct the results obtained in the analysis accordingly.

26.4 *Calculation*—Calculate the percentage of Mn_2O_3 to the nearest 0.01 as follows:

$$Mn_2O_3, \% = EV \times 50 \quad (22)$$

where:

E = Mn_2O_3 equivalent of the $KMnO_4$ solution, g/mL,
V = millilitres of $KMnO_4$ solution used, and
50 = 100 divided by the weight of sample used (2 g).

27. Free Calcium Oxide (*Alternative Test Methods*)

27.1 *Summary of Test Methods*—These are rapid test methods for the determination of free calcium oxide in fresh clinker. When applied to cement or aged clinker, the possibility of the presence of calcium hydroxide should be kept in mind since these methods do not distinguish before free CaO and free $Ca(OH)_2$. Two test methods are provided. Alternate Test Method A is a modified Franke procedure in which uncombined lime is titrated with dilute perchloric acid after solution in an ethylacetoacetate-isobutylalcohol solvent. Alternate Test Method B is an ammonium acetate titration of the alcohol–glycerin solution of uncombined lime with $Sr(NO_3)_2$ as an accelerator.

27.2 *Modified Franke Test Method (Alternative Test Method A):*

27.2.1 *Apparatus:*

27.2.1.1 *Refluxing Assembly,* consisting of a flask with flat-bottom short neck Erlenmeyer flask with 250-mL capacity. The water-cooled refluxing condenser should have a minimum length of 300 mm. The flask and reflux condenser shall be connected with standard tapered ground glass joints. The reflux condenser shall be fitted with an absorption tube containing a desiccant, such as indicating silica gel, and a material for the removal of CO_2, such as Ascarite. The absorption tube shall be inserted with a rubber stopper in the upper end of the reflux column.

27.2.1.2 *Buret,* having a 10-mL capacity and graduated in units not more than 0.05 mL.

27.2.1.3 *Vacuum Filtration Assembly,* consisting of a Gooch crucible size No. 3, 25-mL capacity in which is placed a suitablc filter paper, 21-mm size, a Walter crucible holder, a 500-mL vacuum flask, and vacuum source. The crucible is half filled with compressed filter pulp.

27.2.1.4 *Glass Boiling Beads:*

27.2.2 *Solutions Required:*

27.2.2.1 *Ethyl Acetoacetate–Isobutyl Alcohol Solvent*—3 parts of volume of ethyl acetoacetate and 20 parts by volume of isobutyl alcohol.

27.2.2.2 *Thymol Blue Indicator*—Dissolve 0.1 g of thymol blue indicator powder in 100 mL of isobutyl alcohol.

27.2.2.3 *Perchloric Acid, Standard Solution* (0.2 *N*)—Dilute 22 mL of 70 to 72 % perchloric acid to 1 L with isobutyl alcohol. Standardize this solution as follows: Ignite 0.1000 g of primary standard calcium carbonate in a platinum crucible at 900 to 1000°C. Cool the crucible and contents in a desiccator and weigh to the nearest 0.0001 g to constant weight. Perform the weighings quickly to prevent absorption of water and CO_2. Immediately transfer the CaO without grinding to a clean, dry Erlenmeyer flask and reweigh the empty crucible to the nearest 0.0001 g to determine the amount of CaO added. Then follow procedure beginning with "Add 70 mL of the ethyl acetoacetate–isobutyl alcohol. . ." in 27.2.3.1. Calculate the CaO equivalents of the standard perchloric acid solution in grams per millilitre by dividing the weight of CaO used by the volume of perchloric acid required for the titration.

27.2.3 *Procedure:*

27.2.3.1 Weigh 1.0000 g of ground sample (Note 108) and transfer it into a clean, dry 250-mL Erlenmeyer flask. Add four to five glass boiling beads. Add 70 mL of prepared ethyl acetoacetate–isobutyl alcohol solvent. Agitate the flask to disperse the sample.

NOTE 108—Thorough grinding of the sample is essential for proper exposure of the free lime grains that often are occluded in crystals of tricalcium silicate in the cement. However, exposure of the sample to the air must be kept at a minimum to prevent carbonation of the free lime. In particular, direct breathing into the sample must be avoided. The sample should be sufficiently fine to easily pass a No. 200 (75-μm) sieve but actual sieving is not recommended. If the sample is not to be immediately tested, it must be kept in an airtight container to avoid unnecessary exposure to the atmosphere.

27.2.3.2 Attach the flask to a reflux condenser and bring the material to a boil. Reflux for 15 min.

27.2.3.3 Remove flask from condenser, stopper, and cool rapidly to room temperature.

27.2.3.4 Filter the sample and solution using the vacuum assembly. Wash the flask and residue with small increments (10 to 15 mL) of isobutyl alcohol until a total of 50 mL has been used for wash solution.

27.2.3.5 Add 12 drops of the thymol blue indicator to the filtrate and immediately titrate with 0.2 *N* perchloride acid to the first distinct color change.

27.2.4 *Calculation*—Calculate the percent free calcium oxide to the nearest 0.1 % as follows:

$$\text{free CaO, \%} = \frac{EV \times 100}{W} \tag{23}$$

where:

E = CaO equivalent of the perchloric acid, g/mL,
V = millilitres of perchloric acid solution required by sample, and
W = weight of the sample, g.

27.3 *Rapid $Sr(NO_3)_2$ Method (Alternative Test Method B):*

27.3.1 *Reagents:*

27.3.1.1 *Ammonium Acetate, Standard Solution* (1 mL = 0.005 g CaO)—Prepare a standard solution of ammonium acetate ($NH_4C_2H_3O_2$) by dissolving 16 g of desiccated ammonium acetate in 1 L of ethanol in a dry, clean, stoppered bottle. Standardize this solution by the same procedure as described in 27.3.2.1, except use the following in place of the sample: ignite to constant weight approximately 0.1 g of calcium carbonate ($CaCO_3$) in a platinum crucible at 900 to 1000°C, cool the contents in a desiccator, and weigh to the nearest 0.0001 g. Perform the weighings quickly to prevent absorption of water and CO_2. Immediately transfer the CaO without grinding to a 250-mL boiling flask (containing glycerin–ethanol solvent and $Sr(NO_3)_2$), and reweigh the empty crucible to determine the weight of CaO to the nearest 0.0001 g. Continue as described in 27.3.2.1 and 27.3.2.2. Calculate the CaO equivalent of the ammonium acetate in grams per millilitre by dividing the weight of CaO used by the volume of solution required.

27.3.1.2 *Phenolphthalein Indicator*—Dissolve 1.0 g of phenolphthalein in 100 mL of ethanol (Formula 2B) (Note 109).

27.3.1.3 *Glycerin-Ethanol Solvent* (1+2)—Mix 1 volume of glycerin with 2 volumes of ethanol (Formula 2B). To each litre of this solution, add 2.0 mL of phenolphthalein indicator solution.

NOTE 109—Ethanol denatured in accordance with Formula 2B (99.5 % ethanol and 0.5 % benzol) is preferred but may be replaced by isopropyl alcohol, A.R.

27.3.1.4 *Strontium Nitrate* ($Sr(NO_3)_2$), reagent grade.

27.3.2 *Procedure:*

27.3.2.1 Transfer 60 mL of the glycerin–ethanol solvent into a clean, dry, 250-mL standard-taper flat-bottom boiling flask. Add 2 g of anhydrous strontium nitrate ($Sr(NO_3)_2$), and adjust the solvent to slightly alkaline with a dropwise addition of a freshly prepared dilute solution of NaOH in ethanol until a faint pink color is formed. Weigh 1.000 g of the finely ground sample (Note 109) into the flask, add encapsulated stirring bar, and immediately attach a water-cooled condenser (with a standard 24/40 glass joint). Boil the solution in the flask on a magnetic stirrer hot plate for 20 min with mild stirring.

27.3.2.2 Remove the condenser and filter the contents of the flask on a small polypropylene Büchner funnel under vacuum, using a 250-mL filtering flask with side tube. Bring the filtrate to a boil and immediately titrate with standard ammonium acetate solution to a colorless end point.

27.3.3 *Calculation*—Calculate the percent free CaO to the nearest 0.1 % as follows:

$$\text{free CaO, \%} = EV \times 100 \quad (24)$$

where:

E = CaO equivalent of the ammonium acetate solution, g/mL, and

V = millilitres of ammonium acetate solution required by the sample.

28. Keywords

28.1 chemical analysis; compositional analysis; hydraulic cements

APPENDIX

(Nonmandatory Information)

X1. EXAMPLE OF DETERMINATION OF EQUIVALENCE POINT FOR THE CHLORIDE DETERMINATION

(Column 1) $AgNO_3$, mL	(Column 2) Potential, mV	(Column 3) Δ mV[A]	(Column 4) Δ^2 mV[B]
1.60	125.3		
		5.8	
1.80	119.5		1.4
		7.2	
2.00	112.3		1.3
		8.5	
2.20	103.8		1.3
		9.8	
2.40	94.0		0.6
		9.2	
2.60	84.8		2.3
		6.9	
2.80	77.9		0.8
		6.1	
3.00	71.8		1.3
		4.8	
3.20	67.0		

The equivalence point is in the maximum Δ mV interval (Column 3) and thus between 2.20 and 2.40 mL. The exact equivalence point in this 0.20 increment is calculated from the Δ^2 mV (Column 4) data as follows:

$$E = 2.20 + \frac{1.3}{1.3 + 0.6} \times 0.20 = 2.337 \text{ mL. Round to } 2.34.$$

[A] Differences between successive readings in Column 2.

[B] Differences between successive Δ readings in Column 3 "second differentials".

X2. CO_2 DETERMINATIONS IN HYDRAULIC CEMENTS

X2.1 Scope

X2.1.1 This appendix contains information about methods for determination of carbon dioxide (CO_2) in hydraulic cement. The methods listed received a favorable evaluation by Task Group C01.23.04.

X2.1.2 Section X2.2 lists the analytical methods that received a favorable evaluation, briefly describes each method, suggests analytical techniques or cautions that may be useful, and indicates limitations to some of the methods.

X2.1.3 The methods listed as X2.2.1, X2.2.4, X2.2.5, and X2.2.6 determine total carbon calculated as CO_2. For that reason, they are not appropriate for determination of carbon dioxide in fly ash, limestones containing carbon in the form of graphite or kerogen, in other carbon bearing materials, or in blended cements produced from these materials.

X2.1.4 The methods listed in X2.2.2 and X2.2.3 can determine actual CO_2 directly rather than by calculation from total carbon. They are suggested for analysis of blended cements and blended cement ingredients which are likely to contain non-carbonate carbon.

X2.1.5 The split loss on ignition method in X2.2.1 can give misleading results when used with materials containing $CaOH_2$ (calcium hydroxide). This can occur with aged cement, cement made from aged clinker, or high free lime clinker, in addition to cements with a lime or hydrated lime ingredient.

X2.2 Analytical Methods

X2.2.1 *Split Loss on Ignition*—This procedure is comparable to the analytical method described in Test Methods C 114, Section 16, with the following modifications:

1. A crucible of known mass and containing a sample of known mass is initially heated at 500°C for 2 h.
2. After being cooled to room temperature in a desiccator, and its mass determined, the crucible with sample is then heated at 950°C for 2 h.
3. Finally, the crucible with sample is cooled and its mass is determined as per Step No. 2.
4. The difference in residue masses after the respective heat treatments is assumed to be carbon dioxide.

TGA results indicated that $CaOH_2$ can lose a significant portion of its mass above 500°C. Thus, the Split Loss on Ignition procedure should not be used when situations described in X2.1.5 exist.

X2.2.2 *Thermogravimetric Analysis (TGA)*—This method involves the determination of sample mass loss at various temperatures. The heating of a sample through a temperature range allows for mass loss differentiation based on mineral form (for example, $CaCO_3$, $MgCO_3$, $CaOH_2$, etc.). Specific operational information is provided by the equipment manufacturers. If free carbon is present, an inert atmosphere (for example, nitrogen) should be used for sample analysis.

TABLE X2.1 Cooperative Test Series No. 1

	Single Determinations			
	Carbon Dioxide Unknown[A]	Determinations Base Cement[B]	Added CO_2, c% Determined[C]	Added CO_2, % Known[D]
Split LOI	2.40	0.45	1.97	2.00
	2.52	0.56	1.99	
	2.41	0.36	2.07	
	2.39	0.32	2.09	
	2.41	0.36	2.07	
	2.28	0.27	2.02	
Test Methods E 350	2.00	0.02	1.98	
	2.00	0.02	1.98	
Induction Furnace/IR	2.46	0.40	2.07	
	2.53	0.48	2.08	
	2.38	0.40	2.00	
	2.42	0.48	1.97	
Average			2.02	
Standard Deviation			0.05	

[A] The Unknown was prepared by blending/grinding a mixture of 5.00 % NIST SRM 1C Argillaceous Limestone and 95.00 % CCRL Portland Cement Reference Sample No. 85. According to the Certificate of Analysis, the SRM 1C had a loss on ignition of 39.9 %. For the purpose of the cooperative test series, the loss on ignition was assumed to be CO_2 only.

[B] The Base Cement was CCRL Portland Cement Reference Sample No. 85.

[C] The Determined % Added CO_2 was obtained by subtracting the Base Cement % CO_2 from the Unknown % CO_2.

[D] The addition of 5.00 % NIST SRM 1C (with a loss on ignition value of 39.9 %) would provide 2.00 % Added CO_2. (Again, it was assumed that the SRM 1C loss on ignition was only carbon dioxide.)

TABLE X2.2 Cooperative Test Series No. 2

	Averages from Three Determinations			
	Carbon Dioxide Unknown[A]	Determinations Base Cement[B]	Added CO_2, % Determined[C]	Added CO_2 % Known[D]
Split LOI	2.00	0.41	1.59	1.60
	1.65	0.32	1.33	
	2.02	0.46	1.56	
	1.91	0.35	1.56	
	2.10	0.43	1.67	
	1.91	0.41	1.50	
	1.98	0.46	1.52	
XRFA	1.68[E]	0.00[E]	1.68[E]	
Induction Furnace/IR	2.23	0.28	1.95	
	1.96	0.28	1.68	
	1.95	0.40	1.55	
TGA	1.77	0.20	1.57	
	1.87	0.25	1.62	
Average			1.60	
Standard Deviation			0.14	

[A] The Unknown was prepared by blending/grinding a mixture of 4.00 % NIST SRM 1C Argillaceous Limestone and 96.00 % CCRL Portland Cement Reference Sample No. 85. According to the Certificate of Analysis, the SRM 1C had a loss on ignition of 39.9 %. For the purpose of the cooperative test series, the loss on ignition was assumed to be CO_2 only.

[B] The Base Cement was CCRL Portland Cement Reference Sample No. 85.

[C] The Determined % Added CO_2 was obtained by subtracting the Base Cement % CO_2 from the Unknown % CO_2.

[D] The addition of 4.00 % NIST SRM 1C (with a loss on ignition value of 39.9 %) would provide 1.60 % Added CO_2. (Again, it was assumed that the SRM 1C loss on ignition was only carbon dioxide.)

[E] The XRF instrument was calibrated using standards composed of the Base Cement (that is, CCRL No. 85) and NIST SRM 1C. It was assumed that the Base Cement contained 0 % CO_2.

X2.2.3 *Test Methods C 25, Section 22*—"Standard Test Methods for Chemical Analysis of Limestone, Quicklime, and Hydrated Lime." This method involves decomposition of the sample with HCl. The liberated CO_2 is then passed through a series of scrubbers to remove water and sulfides. The CO_2 is absorbed with special Sodium Hydroxide Absorbent (Ascarite). The gain in mass of the absorption tube is determined and calculated as percent CO_2. Calcium carbonate, for instance, can be calculated by multiplying the determined CO_2 content by a conversion factor (for example, $CO_2 \times 2.2742 = CaCO_3$).

X2.2.4 *X-ray Fluorescence Spectroscopy*—In this method the sample is ground to a fine particle size, pressed into a flat pellet and irradiated with the chosen instrument. Carbon content is determined by comparing the collected carbon emissions to calibration standards.

X2.2.5 *Combustion by Induction Furnace/IR*—This method involves volatilization by induction furnace and detection by infrared absorption. Suitable calibration standards (for example, calcium carbonate and synthetic carbon) are available from some instrument manufacturers. NIST cement SRM's with known additions of NIST argillaceous limestone (or other suitable standards) should also be considered to check instrument calibration.

X2.2.6 *"Test Methods E 350*—This method, Total Carbon by the Combustion Gravimetric Method from Test Methods for Chemical Analysis of Carbon Steel, Low-Alloy Steel, Silicon Electrical Steel, Ingot Iron, and Wrought Iron," is suitable for the determination of carbon in concentrations from 0.05 to 1.80 % (as carbon dioxide 0.18 to 6.60 %). The test method involves burning the sample in a stream of oxygen; the carbon dioxide in the evolved gases is then collected in a suitable absorbent and its mass determined. Time of analysis is less than 10 min.

X2.3 Cooperative Test Results

X2.3.1 Tables X2.1 and X2.2 list results from two series of cooperative tests using several of the analytical methods evaluated by the Task Group. In all, five of the six methods receiving favorable ratings were included. The sixth method, Test Methods C 25, was specifically not tested in these cooperative series, but was rated favorably because of long history of use with related materials.

X2.3.2 Table X2.1 includes results of single determinations using three of the different analytical methods. The methods used were Split Loss on Ignition, Test Methods E 350, and Combustion by Induction Furnace with Infra-Red Detection.

X2.3.3 Table X2.2 includes results based on average of three determinations. Results from four of the different analytical methods are included. Methods used were Split Loss on Ignition, X-Ray Fluorescence Analysis, Induction furnace with Infra-Red Detection, and Thermo-Gravimetric Analysis. Test Methods E 350 was not used in this series of tests.

For additional useful information on details of cement test methods, reference may be made to the "Manual of Cement Testing," which appears in Vol 04.01 of the *Annual Book of ASTM Standards.*

Standard Test Method for Fineness of Portland Cement by the Turbidimeter[1]

This standard is issued under the fixed designation C 115; the number immediately following the designation indicates the year of original adoption or, in the case of revision, the year of last revision. A number in parentheses indicates the year of last reapproval. A superscript epsilon (ϵ) indicates an editorial change since the last revision or reapproval.

1. Scope

1.1 This test method covers determination of the fineness of portland cement as represented by a calculated measure of specific surface, expressed as square centimetres of total surface area per gram, or square metres of total surface area per kilogram, of cement, using the Wagner turbidimeter.[2]

1.2 *This standard does not purport to address all of the safety concerns, if any, associated with its use. It is the responsibility of the user of this standard to establish appropriate safety and health practices and determine the applicability of regulatory limitations prior to use.*

1.3 The values stated in SI units are to regarded as the standard. The inch-pound equivalents may be approximate.

2. Referenced Documents

2.1 *ASTM Standards:*

C 114 Test Methods for Chemical Analysis of Hydraulic Cement[3]

C 430 Test Method for Fineness of Hydraulic Cement by the 45-μm (No. 325) Sieve[3]

C 670 Practice for Preparing Precision and Bias Statements for Test Methods for Construction Materials[4]

3. Significance and Use

3.1 The purpose of this test method is to determine whether or not the hydraulic cement under test meets the Wagner turbidimetric fineness requirements of the applicable hydraulic cement specification for which the test is being made. Fineness of the cement component is only one of the many characteristics that influence the strength capabilities of concrete.

4. Apparatus

4.1 *Nature of Apparatus*—The Wagner turbidimeter consists essentially of a source of light maintained at constant intensity and adjusted so that approximately parallel rays of light pass through a suspension of the cement to be tested and impinge upon the sensitive plate of a photoelectric cell. The current generated in the cell is measured by means of a microammeter and the indicated reading is a measure of the turbidity of the suspension. General considerations indicate that turbidity is in turn a measure of the surface area of the suspended sample of cement. The apparatus shall consist specifically of the parts described in 4.2 through 4.7 and shall be constructed in accordance with the detailed design and dimensional requirements shown in Fig. 1 and Table 1, except that the case may be either of wood or of metal.

4.2 *Turbidimeter,* mounted in a suitable wood or metal case including the following features:

4.2.1 *Source of Light*—The source of light (Fig. 1) shall consist of a concentrated-filament electric lamp of between 3 and 6 cd operated by a source of constant emf. The lamp shall be mounted rigidly in the socket. A clean, bright parabolic metallic reflector shall be rigidly mounted behind the lamp, focused so that approximately parallel rays of light will pass through the sedimentation tank and impinge upon the photoelectric cell. The light intensity shall be regulated by two rheostats of approximately 6 and 30 Ω, respectively, and they shall possess such characteristics that uniform changes in light intensity may be obtained over the full range of resistance. The rheostats shall be mounted in parallel with each other and in series with the lamp.

4.2.2 *Heat-Absorbing Device*—The light shall pass through a suitable heat absorbing device before entering the sedimentation tank in order that radiant heat from the beam shall be absorbed, the device being either (*1*) a water cell or (*2*) a special heat-absorbing glass filter. The water cell shall be made from 76-mm (3-in.) outside diameter seamless brass tubing, 3-mm (1/8-in.) thick wall, 102-mm (4 in.) in length with glass windows sealed in the ends. The cell shall contain a hole for filling with distilled water. The hole shall be sealed with a metal plug. The cell, when mounted on the movable shelf, may have the plug in either the top or bottom position. The heat-absorbing device shall be so arranged that essentially all rays of light entering the sedimentation tank shall first pass through the heat-absorbing device.

4.2.3 *Retarding Filter*—A light-retarding glass or other device shall be provided that will reduce the intensity of light from that corresponding to 100 μA to a reading of 20 to 30 μA. The light intensity shall be uniformly retarded over the entire area of that portion of the cell which is exposed to light during a test. The retarding filter shall be mounted in a carrier on the shield and shall be capable of being swung out of the light path by means of a handle.

4.2.4 *Sedimentation Tank*—The sedimentation tank shall be either (*1*) constructed of 5 to 6-mm (3/16 to 1/4-in.) plate glass or borosilicate glass cemented or sealed together to form a rectangular tank, or (*2*) a molded glass tank having walls approximately 5-mm (3/16 in.) thick with plane surfaces. The inside dimensions of the rectangular tank shall

---

[1] This test method is under the jurisdiction of ASTM Committee C-1 on Cement and is the direct responsibility of Subcommittee C01.25 on Fineness.

Current edition approved Dec. 10, 1996. Published February 1997. Originally published as C 115 – 34 T. Last previous edition C 115 – 96.

[2] This turbidimeter was developed by L. A. Wagner, Research Associate of the Cement Reference Laboratory, National Institute of Standards and Technology, Washington, DC. A description of the apparatus and the original mathematical derivations of formulas used are given in the paper: Wagner, L. A., "A Rapid Method for the Determination of the Specific Surface of Portland Cement," *Proceedings*, ASTM, ASTEA, Vol 33, Part II, 1933, p. 553.

[3] *Annual Book of ASTM Standards*, Vol 04.01.

[4] *Annual Book of ASTM Standards*, Vol 04.02.

TABLE 1 Turbidimeter Apparatus Dimensions (see Fig. 1)

Letter	mm	in.	Letter	mm	in.
A	445	$17\frac{1}{2}$	T	51	2
B	438	$17\frac{1}{4}$	U	22.2	$\frac{7}{8}$
C	381	15	V	34.9	$1\frac{3}{8}$
D	105	$4\frac{1}{8}$	W	13	$\frac{1}{2}$
E	3.0	$\frac{1}{8}$	X	38	$1\frac{1}{2}$
F	28.6	$1\frac{1}{8}$	Y	15.9	$\frac{5}{8}$
G	85.7	$3\frac{3}{8}$	Z	38 ± 0.76	$1\frac{1}{2}$ ± 0.03
H	33.3	$1\frac{5}{16}$	AA	51 ± 2.5	2 ± 0.10
I	102	4	AB	267	$10\frac{1}{2}$
J	39.7	$1\frac{9}{16}$	AC	...	$\frac{5}{8}$[A]
K	55.6	$2\frac{3}{16}$	AD	...	11[A]
L	65.1	$2\frac{9}{16}$	AE	39.7	$1\frac{9}{16}$
M	28.6	$1\frac{1}{8}$	AF	55.6	$2\frac{3}{16}$
N	3.0	$\frac{1}{8}$	AG	66.7	$2\frac{5}{8}$
O	51	2	AH	203	8
P	51	2	AI	23.6	0.93
Q	3.0	$\frac{1}{8}$	AJ	1.38[A]	$3\frac{1}{2}$[A]
R	61.1	$2\frac{13}{32}$	AK	cm[A]	in.[A]
S	69.8	$2\frac{3}{4}$			

[A] These are pure numbers such as parts of a thread designation and numbers of links per unit; consequently, they do not correspond with the column titles.

be 51 mm by 38 mm (2 by $1\frac{1}{2}$ in.) by 203 mm (8 in.) in height. The permissible variation on the inside dimensions of the tank shall be ±2.5 mm (0.1 in.) in length and ±0.76 mm (0.03 in.) in width. The 51-mm faces of the tank shall be equidistant within 0.25 mm (0.1 in.) at all points. A mark shall be placed on the side of the tank to indicate a volumetric content of 335 mL, which is the level to which the tank will be filled in a test. A tank filled to the mark with clear kerosine and placed in the turbidimeter light beam shall yield uniform microammeter readings, within ±0.1 μA, for the entire usable portion of the tank.

4.2.5 *Photoelectric Cell*—The means of measuring the light intensity shall be a sensitive photoelectric cell[5] connected directly to a microammeter. A hood with a horizontal slot 13 mm ($\frac{1}{2}$ in.) in height by 35 mm ($1\frac{3}{8}$ in.) in width shall be mounted over the photoelectric cell. The front of the hood shall be 25 ± 1 mm (1 ± $\frac{1}{16}$ in.) in front of the face of the cell. The face of the photocell shall be parallel to the tank faces within 0.5 mm (0.02 in.).

4.2.6 *Shield*—A metallic shield having a slot 16 mm ($\frac{5}{8}$ in.) in height by 38 mm ($1\frac{1}{2}$ in.) in width, as indicated in Fig. 1, shall be placed between the heat absorbing device and the sedimentation tank.

4.2.7 *Elevating Device*—The source of the light, the heat-absorbing device, the photoelectric cell, the retarding filter, and the shield shall be mounted on a movable shelf which may be raised or lowered by two connected lead screws, and which may be readily and accurately adjusted so that the turbidity of the suspension may be determined at any desired depth. The center of the light source, the heat absorbing device, the photocell, the center of the slots of the metal shield, and the hood shall be on a straight line which is parallel to the shelf. The sedimentation tank shall be mounted on a base which is independent of the rest of the apparatus so that the tank shall be free from vibration caused by moving the shelf. Care shall be taken that the shelf shall be level at all points of elevation and that the tank shall be normal to the shelf. The distance between the tank and the edges of the opening in the shelf shall vary not more than $\frac{1}{64}$

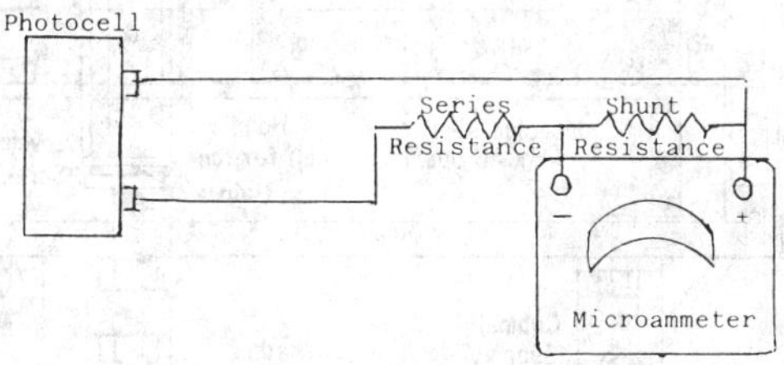

Microammeter Internal Resistance = 90 Ω
Shunt Equivalent Resistance:

$$470\ \Omega \text{ in parallel with } 100\ \Omega = \frac{470 \times 100}{470 + 100} = 82\ \Omega$$

Meter & Shunt Equivalent + Resistance:

$$90\ \Omega \text{ in parallel with } 82\ \Omega = \frac{90 \times 82}{90 + 82} = 43\ \Omega$$

Series Resistance = 90 − 43 = 47 Ω

FIG. 1 Illustrated Example of D'Arsonval Meter Circuit for I_r Determination

in. (0.4 mm) between the "30–50" and "0" positions. The level of the light beam with reference to the surface of the suspension shall be indicated by a pointer which will travel along a scale mounted on the cabinet. The zero of the scale shall indicate that position at which the center lines of the slots for the light beam are at the same elevation as the surface of the liquid in the tank when filled to the 335-mL level. The lines on the scale to be marked 7.5, 10, 15, 20, 25, and 30–50, shall be located at distances from the zero mark equal to suspension depth values, *h*, in Table 2. The scale, when compared with a standard scale accurate to within 0.1 mm at all points, shall not show a deviation at any point greater than 0.25 mm and shall indicate the positions at which the pointer should be located when turbidity readings for these values of *h* are taken. The interior of the turbidimeter cabinet and the exterior surfaces of the shelf, the parabolic reflector, the heat absorbing device, the shield, and the photoelectric cell hood shall be painted with a dull flat black paint.

NOTE 1—The requirement of the 0 to 50 markings on the scale shall apply only to new Wagner Turbidimeters and not to equipment in use

[5] Weston Photronic type, Model 594YY is acceptable.

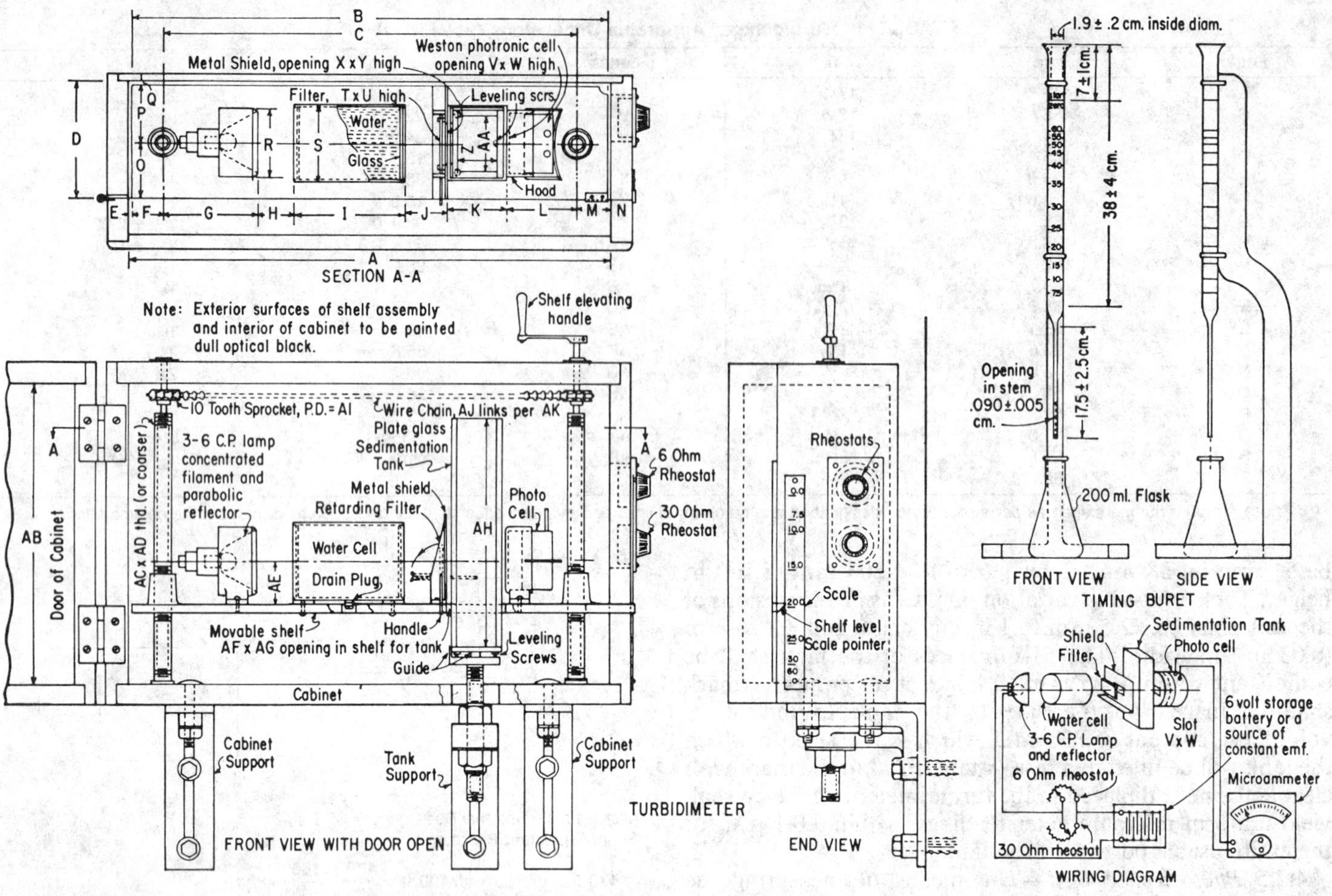

FIG. 2 Dimensional Details of Turbidimeter Fineness Test Apparatus (see Table 1)

which meets the other requirements of this method.

4.3 *Microammeters:*

4.3.1 *D'Arsonval-Type Microammeters* shall have a range from 0 to 50 µA and shall be readable to 0.1 µA. New microammeters shall be accurate to ±0.5 % of full scale value at any part of the scale value at any part of the scale at 77°F (25°C). For microammeters, in use, the accuracy shall be the same as for new instruments except that the accuracy at 40 and 50 µA shall be ±1 % of full scale. The internal resistance of the microammeter shall be between 50 and 150 Ω. The microammeter shall not be mounted upon a working surface containing or consisting of iron or steel, or near other magnetic influence.

4.3.2 *Digital Microammeter:*

NOTE 2—A meter with a range of 199.9 µA is satisfactory for use and enables the operator to read the theoretical I_0 directly without supplementary devices. The high internal resistance of the digital microammeter does not affect the linearity of readings at the light intensity levels encountered in a Wagner turbidimetric determination of fineness.

4.4 *Source of Current*—A 6-V automobile starting and lighting storage battery or a source of constant emf shall be used for supplying current to the lamp.

4.5 *Sieve*—The sieve shall conform to the requirements of Test Method C 430.

4.6 *Stirring Apparatus*—The stirring apparatus shall consist of either (*1*) a cylindrical brush, 19 mm (¾ in.) in diameter and about 45 mm (1¾ in.) in length, with an end approximately fitting the contour of the bottom of a 22-mm (⅞-in.) diameter test tube, or (*2*) any other stirring device that will be equally efficient in dispersion as measured by specific surface determinations on a standard sample. The stirring apparatus shall rotate at a speed of approximately 3500 r/min.

4.7 *Timing Buret*—The time of settling for the different-sized particles shall be obtained by use of a buret from which kerosine is allowed to flow. The buret shall consist of a glass tube having a capillary tube fused into the lower end. The upper end of the large tube shall be flared to serve as a funnel for introducing kerosine into the tube. The buret shall conform to the limiting dimensions given in Table 3. The graduation lines on the buret shall be complete circles. A filter made of No. 325 (45-µm) wire cloth shall be used with the timing buret and a cover shall be placed over the top of the buret when it is not in use.

4.8 *Weights and Weighing Devices*, shall conform to the requirements of Methods C 114.

5. Materials

5.1 *Suspending Liquid*—Clear white kerosine shall be used with the turbidimeter apparatus. The kerosine shall not be reused.

TABLE 2 Values of h, d, and h/d^2 to be Used in Calibration of the Turbidimeter Apparatus

Particle Diameter, d, µm	Depth of Suspension, h, cm	h/d^2
50	15	0.00600
45	15	0.00741
40	15	0.00938
35	15	0.01224
30	15	0.01667
25	13.1	0.0210
20	10	0.0250
15	6.6	0.0293
10	3.3	0.0330
7.5	2.1	0.0373

TABLE 3 Buret Dimensions[A]

	Dimension, cm	Permissible Variation, cm
Length of large tube	38	±4
Inside diameter of large tube	1.9	±0.2
Length of capillary	17.5	±2.5
Diameter of capillary	0.09	±0.005
Top of buret to zero line	7	±1

[A] Since glass tubing of desired dimensions is not always obtainable, the wide permissible variations listed above allow selection of dimensions to produce a buret having a duration of flow which will permit calibration as described in 6.1.1.

6. Test Specimen or Sample

6.1 *Size of Test Sample*—Select the size of the sample of cement for test so that the initial microammeter reading is between 12 and 20 µA.

NOTE 3—The following approximations will be helpful in many instances in selecting the size of sample: 0.25 g for normal fineness cements and 0.20 g for high fineness cements.

7. Calibration

7.1 *Calibration of Turbidimeter*—Calibrate the turbidimeter apparatus in accordance with the following procedure:

7.1.1 *Calibration of Buret Scale:*

7.1.1.1 For calibration of the buret scale use a kerosine having a known viscosity and density for the temperature at which the calibration is to be made. Density and viscosity of the kerosine should be determined. Calculate the times of flow from the buret that correspond to the times of settling for the different sized particles, from the following equation:

$$t = [1{,}837{,}000\eta/(\rho_1 - \rho_2)] \times (h/d^2) \quad (1)$$

where:

t = time of settling, or time of flow, s,
η = viscosity of kerosine at the temperature of calibration, P,
ρ_1 = density of cement particles, Mg/m³ (g/cm³) = 3.15 for portland cement (Note 4),
ρ_2 = density of kerosine, Mg/m³ at the temperature of calibration,
h = depth of suspension to level of light, cm, and
d = diameter of particle, µm.

Values of h/d^2 are given in Table 2.

7.1.1.2 Fill the buret with kerosine at the calibrating temperature, start a timing clock at the instant the kerosine in the buret drains past the zero line, and mark on the buret the levels reached by the draining kerosine for each of the time intervals, t, calculated as described above. At these marks, etch permanent lines and numbers on the buret indicating the corresponding diameters (Note 5). The construction and the graduation of the buret shall be such that at the temperature of calibration the time required for the kerosine to pass the permanent lines of the buret agrees with the calculated time of settling within 1 percent, except that the permissible variation shall be not less than 1 s.

NOTE 4—The density of portland cement does not vary greatly and in this work it is considered constant at 3.15. A variation of 0.15 from this value when substituted in Stokes' law gives a variation of 2.5 % in the diameter of the particle measured.

NOTE 5—By using the calibrated buret the apparatus may be used within the normal range of room temperatures without further correction, the change in rate of flow of the kerosine from the buret automatically compensating for change in viscosity of the suspension due to temperature. The temperature of the kerosine in the buret and that of the suspension should be kept the same within 0.5°C (1°F). This condition will ordinarily exist if the supply of kerosine is kept in the same room as the apparatus.

Care must be taken to ascertain that only clean kerosine is used in the buret, and, in addition, the capillary should be examined frequently to make sure that no small pieces of lint or other foreign material have become lodged in it.

7.1.2 *Calibration of No. 325 (45-µm) Sieve*—Calibration shall be made in accordance with Method C 430, basing the percentage sieve correction on the difference between the test residue obtained and the assigned residue value indicated by the electroformed sheet sieve fineness specified for the standard sample, expressed as a percentage of the test residue.

7.1.3 *Determination of the Proper Light Intensity, I_r:*

7.1.3.1 Fill the sedimentation tank to the mark with clear kerosine.

7.1.3.2 For a turbidimeter equipped with a digital-type microammeter with a range of 199.9 µA, place the sedimentation tank containing clear kerosine against the guides of the tank support in the turbidimeter. Set the light intensity so that the microammeter reads 100.0 µA. Readjust the light intensity until equilibrium is established at this reading. Place the light retarding filter in the path of the light beam. When equilibrium is established, record this meter reading as I_r. This meter reading becomes the reference value for the intensity of the light. Meter readings are also taken at the beginning and end of each test through the light retarding filter alone to check the requirements of 8.3.8.

7.1.3.3 For a turbidimeter equipped with a D'Arsonval-type microammeter, place the sedimentation tank containing clear kerosine against the guides of the turbidimeter. Set the light intensity so that the microammeter reads 50.0 µA with the light retarding filter removed from the light path. Readjust the light intensity until equilibrium is established at this reading. Connect resistors or resistor combinations as indicated in Fig. 1 so that the meter reading is reduced to an equilibrium value between 23.0 and 25.0 µA. Increase the light intensity until the equilibrium light reading obtained between 23.0 and 25.0 µA is exactly doubled. Readjust the light intensity until the equilibrium is established at this doubled meter reading level. Place the light-retarding filter in the path of the light beam. Remove the resistors from the meter circuit. When equilibrium is established, record this meter reading as I_r. This becomes the reference value for the intensity of the light. Adjust the rheostats at the beginning of each test to give this meter reading through the filter plus the tank of clear kerosine. This meter reading corresponds to a meter reading of 100.0 μA through the tank containing clear kerosine with the light filter removed from the light path. Meter readings are also taken at the beginning and at the end of each test through the light retarding filter alone to check the requirements of 8.3.8.

NOTE 6—A half-watt carbon resistor preferably with 5 % tolerance limits can be used to obtain the desired meter readings. The shunt resistor or resistor combination as shown in Fig. 1 shall have a resistance value between 85 and 100 % of the value of the internal resistance of the meter as printed on the card in the meter cover. The series resistor shall have a resistance value equal to the difference between the internal resistance of the meter and the calculated resistance of the meter and shunt combination.

NOTE 7—A less exact but still acceptable procedure for determining a meter setting that corresponds to a current of 100.0 μA through the meter with the light filter removed from the light path can be achieved by eliminating the series resistor in the meter circuit. The internal resistance of the photocell is sufficiently large to cause only a minor error of approximately 0.5 % in the desired setting when the series resistor is omitted from the circuit.

7.1.3.4 *Example*—If the internal resistance of the meter is 90 Ω, and the meter is shunted with a 470-Ω resistor in parallel with a 100 Ω resistor, the combined resistance of the meter and shunts would be reduced to 43 Ω. This would reduce the meter reading from 50.0 μA to a calculated value of 23.9 μA.

7.1.4 *Determination of the Value of K in the Specific Surface Equation*—Using the regular procedure in 8.1, 8.2, and 8.3; and with the light adjusted in accordance with the appropriate procedure in 7.1.3, make a specific surface determination on the current lot of National Institute of Standards and Technology's Reference Material No. 114.[6] Determine the value of K in the specific surface equation in 10.1.1 by dividing the assigned specific surface of the standard sample by the value obtained for the expression in brackets [ ]. Repeat this calibration procedure twice more. Use the average of the three results (Note 8) obtained for the value of K during calibration in specific surface equation in 10.1.1.

7.1.5 *Frequency of Calibration*—The turbidimeter apparatus shall have been calibrated within 6 months prior to testing a cement sample.

NOTE 8—The three individual values of K determined during calibration should not vary from the average value by more than 1.5 %.

8. Procedure

8.1 *Sieve Determination*—Make a No. 325 (45-μm) sieve determination on a 1-g sample of cement in accordance with the procedure stated in Test Method C 430. Record the percentage (corrected) of cement passing the sieve as r.

8.2 *Preparation of the Suspension:*

8.2.1 *Dispersion of the Sample*—Place the cement sample in a test tube together with 10 to 15 mL of kerosine and 5 drops of dispersing agent—oleic acid or aged linseed oil (Note 9). Stir the mixture for 1 min, using the apparatus described in 4.6, and then transfer to the sedimentation tank. Wash the stirring apparatus and the tube with clear kerosine, which is then added together with additional clear kerosine until the total volume of the suspension in the tank is 335 mL (Note 10).

NOTE 9—The oleic acid, if used, shall be of USP grade. The aged linseed oil, if used, shall have a density in the range from 0.948 to 0.953 Mg/m^3 (g/cm^3). These dispersing agents have slight or no effect on the viscosity of the kerosine. Each of these dispersing agents changes with age or with exposure to heat and light—hence should be kept in well-stoppered, dark-glass bottles. Dropping bottles should be emptied, thoroughly cleaned, dried, and refilled periodically. Where conditions are such that poor dispersion is obtained with oleic acid or when foaming of the suspension occurs (possibly due to traces of water in the kerosine) aged linseed oil shall be used as the dispersing agent. The same dispersing agent shall be used for both the instrument calibration and the fineness determination.

NOTE 10—Since it is important that the level of the liquid in the tank coincide with the zero mark on the scale of the instrument (see 4.2.7), it is correspondingly important that the volume of the suspension be close to 335 mL, as specified. This may be accomplished by the use of a flask, calibrated to deliver 335 mL, from which all the kerosine used in making up a suspension is taken; or a point gage may be used, consisting essentially of a flat plate approximately 50 by 63 mm (2 by 2½ in.), to the center of which is attached vertically a metal pin of such length that, when the plate is placed in position on top of the tank, the point of the pin is at the 335-mL level. Either of these methods should ensure the position of the liquid level within a few tenths of a millimetre.

8.2.2 *Agitation of the Suspension*—Just before placing the tank in the path of the light beam, agitate the contents to effect a uniformity of the suspension. Cover the tank with a ground-glass cover and oscillate 180° about a horizontal axis through the center of the tank, turning the tank upside down and back again to its original upright position approximately once each second for about 1 min (Note 11). Take care to prevent loss of kerosine (Note 12). The suspension is then ready to place in its proper position in the path of the light beam.

NOTE 11—The procedure described for oscillating the suspension shall be carefully followed. The procedure used in shaking a cocktail shall not be employed.

NOTE 12—The lid to the tank shall fit so that no kerosine drains down the outside faces of the tank. The faces of the glass tank shall be clean when the suspension is added, and care shall be taken to avoid the necessity of cleaning the faces until the determination is complete.

8.3 *Operation of Turbidimeter*—Operate the turbidimeter apparatus in accordance with the following procedure:

8.3.1 With retarding filter and sedimentation tank containing approximately 100 mL of clear kerosine in the light path, adjust the light to the intensity, I_r, as determined in 7.1.3 by taking repeated readings at 1-min intervals until an unchanging value indicates that the lamp and the photoelectric cell are in equilibrium. Remove the tank, check and record the intensity of the lamp.

NOTE 13—To protect the microammeter, the lamp shall be turned on only with a suspension or the retarding filter in position to reduce the light intensity to a value within the range of the microammeter. A freshly charged storage battery should be momentarily short-circuited to reduce the voltage to constant value. If the microammeter indication continues to fluctuate, look for loose connections in the lamp and photoelectric cell circuits, see that the rheostat contacts are clean, and make certain that the lamp, socket, and reflector are rigidly fixed in their mounting.

8.3.2 Weigh the sample of cement in the appropriate amount and record the weight to the nearest 0.0002 g. Prepare a suspension in accordance with 8.2.

[6] Obtainable at the National Institute of Standards and Technology, (NIST), Gaithersburgh, MD 20899.

8.3.3 Place the shelf level pointer at the 30–50 µm position.

8.3.4 Fill the buret to the predetermined height with kerosine from the same lot and at the same temperature as the kerosine used in the suspension, and start oscillation of the tank containing the suspension in accordance with 8.2.2. Continue the oscillation until the kerosine drains to the zero line on the buret; then stop the agitation. Immediately place the tank in position in the path of the light beam.

8.3.5 Immediately remove the retarding filter from the light path and close the cabinet door.

8.3.6 Read the microammeter to the nearest 0.1 µA at the instant the kerosine in the buret drains past marks 50, 45, 40, 35, and 30.

8.3.7 Raise the shelf successively to the marks 25, 20, 15, 10, and 7.5 on the pointer scale, reading the microammeter at each position as the kerosine drains past the corresponding mark on the buret.

8.3.8 Lower the shelf to the 30–50 µm position. Replace the filter in the path of the light beam, remove the tank, and check the intensity of the lamp. If the microammeter indication has shifted more than 0.3 µA from the initial setting through the filter alone, the test must be repeated.

9. Data Recording

9.1 The form shown in Table 4 is suggested for the recording of turbidimeter data and for the calculation of specific surface.

10. Calculation

10.1 *Calculation of Specific Surface:*

TABLE 4 Illustrative Form for Recording Turbidimeter Test Data and Calculation of Specific Surface

NOTE—Values shown are for purpose of illustration only.

Sample identification	sample X
Passing 45-µm (No. 325) sieve, *r*, corrected, *T*	90.4
Calibration factor, *K*	33.9
Meter reading through filter alone, µA:	
Before test	17.5
After test	17.5
Weight of sample tested, g	0.25

Particle Size, µm	*I*, µA	log *I*	
50	17.3	. . .	1.238^A
45	17.4	1.241	
40	17.6	1.246	
35	17.9	1.253	
30	18.4	1.265	
25	19.1	1.281	
20	20.1	1.303	
15	21.6	1.334	
10	23.9	1.378	
7.5	25.6	. . .	1.408^A
	0.75 × 1.408 =	1.056	
		1.500	
	Sum =	12.857	
	9.5 × 1.238 =	11.761	
	Difference =	1.096	

$S = (33.9 \times 90.4 \times 0.762)/1.096 = 2131\ cm^2/g$

(Round to 2130 cm^2/g or 213 m^2/kg)

^A For convenience in calculation, log I_{50} and log $I_{7.5}$ are recorded in a separate column.

10.1.1 Calculate the turbimetric specific surface as follows:

$$S = K\left[\frac{r(2 - \log I_{50})}{1.5 + 0.75 \log I_{7.5} + \log I_{10} + \log I_{15} \ldots + \log I_{45} - 9.5 \log I_{50}}\right] \quad (2)$$

where:

S = specific surface of the sample, m^2/kg,

K = calibration factor as determined in 7.1.4

r = corrected weight percent of sample passing the 45-µm (No. 325) sieve (Note 14), and $I_{7.5}, I_{10}, I_{15} \cdots, I_{50}$ = microammeter readings, µA, that correspond to the particle diameters 7.5, 10, 15 · · ·, 50 µm.

NOTE 14—The value of *K* as determined in 7.1.3 applies only to a material having the density of portland cement (approximately 3.15 Mg/m^3 or g/cm^3). For any other material, the corresponding value must be calculated for this factor which, in the derivation of the formula, varies inversely as the density of the particles (g/cm^3).

10.1.2 To calculate the specific surface values in square metres per kilogram, multiply the surface area in cm^2/g by the factor of 0.1.

10.1.3 Round values in cm^2/g to the nearest 10 units (in m^2/kg to the nearest unit).

10.2 *Specific Surface From First Turbidity Reading I_{50}*—This test method may be used for successive determinations from the same plant, provided that the same size test samples are used and that there is no great change in fineness, color, or other properties of the cement. Under these conditions the specific surface of a sample may be calculated from the first turbidity reading, I_{50}, by using the following equation:

$$S = c(2 - \log I_{50}) \quad (3)$$

where:

S = specific surface of the sample, m^2/kg,

I_{50} = microammeter reading corresponding to particle diameter of 50 µm, and

c = transmittancy constant for the particular plant.

NOTE 15—The transmittancy constant, *c*, may be evaluated by substituting in the above equation known values of *S* and $2 - \log I_{50}$ as determined from a complete turbidimeter test. The average of not less than five values of *c*, as determined from complete tests, should be used in the above equation.

10.3 If fineness of a sample as determined in accordance with 10.2 fails to conform with the specification requirements for fineness, a retest shall be made using the complete procedure in 8.3, 9.1, and 10.1.

11. Precision and Bias

11.1 *Single-Operator Precision*—The single-operator coefficient of variation has been found to be 2.6 % (Note 16). Therefore, results of two properly conducted tests by the same operator, on the same sample should not differ by more than 7.3 % (Note 13) of their average.

11.2 *Multilaboratory Precision*—The multilaboratory coefficient of variation has been found to be 3.8 % (Note 16). Therefore, results of two different laboratories on identical samples of a material should not differ from each other by more than 10.6 % (Note 16) of their average.

NOTE 16—These numbers represent, respectively, the (D1S %) and (D2S %) limits as described in Practice C 670.

11.3 Since there is no accepted reference material suitable for determining any bias that may be associated with this test method, no statement on bias is being made.

12. Keywords

12.1 fineness; portland cement; turbidimeter; Wagner

APPENDIX

X1. CALCULATION OF PARTICLE SIZE DISTRIBUTION

X1.1 For research or comparative purposes particle-size distributions to a minimum of 7.5 μm may be of value.

X1.2 Calculate the particle size distribution as follows: The form shown in Table X1.1 is suggested for the calculation of particle size distribution.

TABLE X1.1 Illustrative Form for Calculation of Particle Size Distribution from Recorded Turbimetric Test Data

Particle Size, μm	I, μA	log$_I$	Difference	Average Particle Size, d_1	d_1 × Difference	Weight, %	
						Fractional	Cumulative
50	17.3	1.238	0.003	47.5	.143	2.34	90.4
45	17.4	1.241	0.005	42.5	.213	3.49	88.1
40	17.6	1.246	0.007	37.5	.263	4.31	84.6
35	17.9	1.253	0.012	32.5	.390	6.39	80.3
30	18.4	1.265	0.016	27.5	.440	7.21	73.9
25	19.1	1.281	0.022	22.5	.495	8.11	66.7
20	20.1	1.303	0.031	17.5	.543	8.89	58.6
15	21.6	1.334	0.044	12.5	.550	9.01	49.7
10	23.9	1.378	0.030	8.75	.263	4.31	40.7
7.5	25.6	1.408	0.592	3.75	2.220	36.36	36.4
(0)	(100.0)	(2.000)			. . .		
					5.520		

$F = r/\text{Sum}(d_1 \times \text{diff}) = 90.4/5.520 = 16.377$

$\%\ \text{fraction} = F(d_1 \times \text{diff})$

Designation: C 141 –97

Standard Specification for Hydraulic Hydrated Lime for Structural Purposes[1]

This standard is issued under the fixed designation C 141; the number immediately following the designation indicates the year of original adoption or, in the case of revision, the year of last revision. A number in parentheses indicates the year of last reapproval. A superscript epsilon (ε) indicates an editorial change since the last revision or reapproval.

1. Scope

1.1 This specification covers hydraulic hydrated lime for structural purposes.

1.2 Hydraulic hydrated lime may be used in the scratch or brown coat of plaster, stucco, mortar, or in portland-cement concrete either as blend, amendment, or admixture.

1.3 The values stated in inch-pound units are to be regarded as the standard.

1.4 The following precautionary caveat pertains only to the test method portion, Section 10 of this specification: *This standard does not purport to address all of the safety concerns, if any, associated with its use. It is the responsibility of the user of this standard to establish appropriate safety and health practices and determine the applicability of regulatory limitations prior to use.*

2. Referenced Documents

2.1 *ASTM Standards:*

C 25 Test Methods for Chemical Analysis of Limestone, Quicklime, and Hydrated Lime[2]

C 109 Test Method for Compressive Strength of Hydraulic Cement Mortars (Using 2-in. or 50-mm Cube Specimens)[2]

C 150 Specification for Portland Cement[2]

C 151 Test Method for Autoclave Expansion of Portland Cement[2]

C 184 Test Method for Fineness of Hydraulic Cement by the 150-μm (No. 100) and 75- μm (No. 200) Sieves[2]

C 187 Test Method for Normal Consistency of Hydraulic Cement[2]

C 230 Specification for Flow Table for Use in Tests of Hydraulic Cement[2]

C 266 Test Method for Time of Setting of Hydraulic-Cement Paste by Gillmore Needles[2]

C 305 Practice for Mechanical Mixing of Hydraulic Cement Pastes and Mortars of Plastic Consistency[2]

C 778 Specification for Standard Sand[2]

E 11 Specification for Wire-Cloth Sieves for Testing Purposes[3]

3. Terminology

3.1 *Definitions*:

3.1.1 *hydraulic hydrated lime*—the hydrated dry cementitious product obtained by calcining a limestone containing silica and alumina, or a synthetic mixture of similar composition, to a temperature short of incipient fusion so as to form sufficient free lime (CaO) to permit hydration and at the same time leaving unhydrated sufficient calcium silicates to give the dry powder, meeting the requirements herein prescribed, its hydraulic properties.

NOTE 1—The purchaser may increase the hydraulicity by the addition of pulverized portland cement clinker, or a pulverized pozzolan, either natural or artificial.

3.1.1.1 *high calcium hydraulic hydrated lime*—a lime that contains not more than 5 % magnesium oxide (of the nonvolatile portion).

3.1.1.2 *magnesium hydraulic hydrated lime*—a lime containing more than 5 % magnesium oxide (of the nonvolatile portion).

4. Chemical Composition

4.1 The hydraulic hydrated lime shall conform to the following requirements as to chemical composition, calculated to the nonvolatile basis:

	Min	Max
Calcium and magnesium oxides (CaO and MgO calculated to the nonvolatile basis), %	65	75
Silica (SiO_2 calculated to the nonvolatile basis), %	16	26
Iron and aluminum oxides (Fe_2O_3 and Al_2O_3 caculated to the nonvolatile basis), %	...	12
Carbon dioxide (CO_2 on an as received basis), %	...	8

5. Fineness

5.1 The sample shall leave a residue of not more than 0.5 % on a No. 30 (600-μm) sieve and not more than 10 % on a No. 200 (75-μm) sieve when tested as described in 10.2.

6. Time of Setting

6.1 The neat lime paste mixed to normal consistency shall not develop an initial set in less than 2 h as determined by the Gillmore needle method (see 10.4). Final set shall be attained within 24 h.

7. Soundness

7.1 The neat bars when made, stored, and autoclaved as described in 10.5 shall not have an expansion of more than 1.0 %.

8. Compressive Strength

8.1 The average compressive strength of not less than three 2-in. (50-mm) cubes, made, stored, and tested in accordance with 10.6, shall be not less than 250 psi (1.7 MPa) at the age of 7 days. The average strength attained at 28 days shall be not less than 500 psi (3.4 MPa).

[1] This specification is under the jurisdiction of ASTM Committee C-7 on Lime and is the direct responsibility of Subcommittee C07.10 on Hydraulic Lime.

Current edition approved Nov. 10, 1997. Published June 1998. Originally published as C 141 – 38 T. Last previous edition C 141 – 96.

[2] *Annual Book of ASTM Standards*, Vol 04.01.

[3] *Annual Book of ASTM Standards*, Vol 14.02.

9. Sampling

9.1 Each sample selected for purpose of tests shall weigh at least 5 lb (2.3 kg) and shall represent not more than 50 000 lb (22 680 kg). If only one sample is taken it shall weigh at least 10 lb (4.5 kg).

9.2 The sample shall be given a preliminary sieving by being passed through a No. 20 (850-μm) sieve in order to thoroughly mix the sample, break up lumps, and remove foreign materials.

9.3 Samples shall be shipped and stored in airtight, moistureproof containers.

10. Test Methods:

10.1 *Chemical Analysis:*

10.1.1 Analyses for chemical composition shall be carried out in accordance with Test Methods C 25.

10.2 *Fineness:*

10.2.1 *Apparatus*—The No. 30 (600-μm) and No. 200 (75-μm) sieves shall conform to Specification E 11.

10.2.2 *Procedure*—Place a 100-g sample of the material as received on a No. 30 (600-μm) sieve nested above a No. 200 (75-μm) sieve. Wash the material by means of a stream of water from a faucet (Note 2). Continue the washing until the water coming through the sieve is clear, but in no case shall the washing be continued for more than 15 min (Note 3). Dry the residue on both sieves to constant weight at a temperature between 212 and 248°F (100 and 120°C). Calculate the weights of material retained as a percentage of the original sample, the weight of the material retained on the No. 30 sieve being added to the weight of the material retained on the No. 200 sieve to obtain the correct weight of the material retained on the No. 200 sieve. For additional information, also see Test Method C 184.

NOTE 2—A piece of rubber tubing attached to a water faucet may be used for the washing. The velocity of the water, which may be increased by pinching the tubing, shall not be sufficient to cause any splashing of the sample over the sides of the sieve.

NOTE 3—Care shall be taken not to let water accumulate on the No. 200 (75-μm) sieve, because the openings will become clogged and the operation cannot be completed in 15 min.

10.3. *Normal Consistency:*

10.3.1 Determine normal consistency by the Vicat apparatus in accordance with Test Method C 187.

10.4 *Time of Setting:*

10.4.1 Determine time of setting by the Gillmore needle method in accordance with Test Method C 266.

10.5 *Autoclave Expansion:*

10.5.1 Determine autoclave expansion in accordance with Test Method C 151, with the following modifications:

10.5.1.1 The material used in preparing the test specimens shall be a blend of 75 weight % of the hydraulic lime to be tested with 25 weight % of a portland cement conforming to Type I of Specification C 150.

10.5.1.2 Dry-mix the hydraulic lime and portland cement before water is added, and prepare the paste in accordance with Test Method C 151.

10.5.1.3 Remove the bars from the mold when they are 48 ± 2 h old, measure, and return them to the moist closet. When 7 days old, 5 days after removing from the mold, autoclave them in the usual manner. For the purpose of this test, consider the difference in length of the test specimens at the 48-h period and after autoclaving as the autoclave expansion.

10.6 *Compressive Strength:*

10.6.1 *Apparatus:*

10.6.1.1 *Scales and Weights, Sieves, Glass Graduates, Specimen Molds, Tamper, Trowel, and Testing Machine*, in accordance with Test Method C 109.

10.6.1.2 *Flow Table*, conforming to the requirements of Specification C 230.

10.6.1.3 *Mixing Apparatus*, conforming to the requirements of Practice C 305.

10.6.2 *Standard Sand*—Use 20–30 sand meeting the requirements of Specification C 778.

10.6.3 *Proportions for Standard Mortar*—The mortar shall consist of 1 part of hydraulic hydrated lime to 3 parts of standard Ottawa sand, by weight. The quantity of water measured in millilitres shall be such as to produce a flow of 100 to 115 as determined by the flow table.

10.6.4 *Preparation of Standard Mortar*—Mix the mortar in accordance with Section 6 of Practice C 305, using 500 g of hydraulic lime and 1500 g of standard Ottawa sand with the necessary water to give the desired flow.

10.6.5 *Determination of Flow*—Determine the flow in accordance with Section 8.3 of Test Method C 109.

10.6.6 *Molding of Test Specimens:*

10.6.6.1 Prepare molds in accordance with Section 7 of Test Method C 109.

10.6.6.2 The temperature of the air in the vicinity of the mixing slab and of the dry materials, molds, base plates, and mixing bowl, shall be maintained between 68 and 81.5°F (20 and 27.5°C). The temperature of the mixing water, moist closet or moist room, and water in the storage tank shall not vary from 73.4°F (23°C) by more than ±3°F (±1.7°C). The moist closet or moist room shall be so constructed as to provide storage facilities for test specimens at a relative humidity of not less than 90 %.

10.6.6.3 Mold test specimens in accordance with 8.4 of Test Method C 109.

10.6.7 *Storage of Test Specimens*—Keep all test specimens, immediately after molding, in the molds on plane plates in a damp closet, maintained at a relative humidity of 90 % or more, for from 48 to 52 h in such a manner that the upper surfaces are exposed to the moist air. Then remove the specimens from the molds and place in the air of the laboratory for 5 days in such a manner as to allow free circulation of air around at least 5 faces of the specimens. At the age of 7 days immerse the specimens for the 28-day tests in saturated lime water in storage tanks of noncorrodible materials.

10.6.8 *Testing of Specimens:*

10.6.8.1 Test the specimens for 7-day and 28-day strengths after removal from storage as described in 10.6.7.

10.6.8.2 The remainder of the testing procedure shall be in accordance with 8.6 and Section 9 of Test Method C 109.

10.6.9 *Faulty Specimens*—Specimens that are manifestly faulty or that give strengths differing by more than 15 % from the average value of all test specimens made from the same sample and tested at the same period, shall not be considered in determining the compressive strength.

11. Storage

11.1 The hydraulic hydrated lime shall be stored in such a manner as to permit easy access for proper inspection and identification of each shipment, and in a suitable weathertight building that will protect the hydraulic hydrated lime from damage.

12. Inspection

12.1 Every facility shall be provided the purchaser for sampling and inspection either at the place of manufacture or at the destination, as may be specified by the purchaser.

13. Rejection

13.1 *Product*—The hydraulic hydrated lime may be rejected if it fails to meet any of the requirements of this specification. However, hydraulic hydrated lime failing to meet the 7-day requirements may be held awaiting the result of the 28-day tests before rejection. At least 10 days from the time of sampling shall be allowed for completion of the 7-day test, and at least 31 days shall be allowed for completion of the 28-day test.

13.2 *Packaging*—Packages varying more than 10 % from the weight marked thereon may be rejected, and if the average weight of packages in any shipment, as shown by weighing 50 packages taken at random, is less than the weight marked thereon, the entire shipment may be rejected.

14. Packaging and Marking

14.1 The hydraulic hydrated lime shall be delivered in suitable containers with the brand and name of the manufacturer and the net weight plainly marked thereon. All packages shall be in good condition at the time of inspection.

15. Keywords

15.1 compressive strength; high calcium hydraulic lime; hydraulic hydrated lime; hydraulic lime; magnesium hydraulic lime; mortar; portland cement concrete; plaster; stucco

Designation: C 150 – 98

Standard Specification for Portland Cement[1]

This standard is issued under the fixed designation C 150; the number immediately following the designation indicates the year of original adoption or, in the case of revision, the year of last revision. A number in parentheses indicates the year of last reapproval. A superscript epsilon (ε) indicates an editorial change since the last revision or reapproval.

This standard has been approved for use by agencies of the Department of Defense.

1. Scope

1.1 This specification covers eight types of portland cement, as follows (see Note 1):

1.1.1 *Type I*—For use when the special properties specified for any other type are not required.

1.1.2 *Type IA*—Air-entraining cement for the same uses as Type I, where air-entrainment is desired.

1.1.3 *Type II*—For general use, more especially when moderate sulfate resistance or moderate heat of hydration is desired.

1.1.4 *Type IIA*—Air-entraining cement for the same uses as Type II, where air-entrainment is desired.

1.1.5 *Type III*—For use when high early strength is desired.

1.1.6 *Type IIIA*—Air-entraining cement for the same use as Type III, where air-entrainment is desired.

1.1.7 *Type IV*—For use when a low heat of hydration is desired.

1.1.8 *Type V*—For use when high sulfate resistance is desired.

1.2 When both SI and inch-pound units are present, the SI units are the standard. The inch-pound units are approximations listed for information only.

2. Referenced Documents

2.1 *ASTM Standards:*

C 33 Specification for Concrete Aggregates[2]

C 109/C 109M Test Method for Compressive Strength of Hydraulic Cement Mortars (Using 2-in. or 50-mm Cube Specimens)[3]

C 114 Test Methods for Chemical Analysis of Hydraulic Cement[3]

C 115 Test Method for Fineness of Portland Cement by the Turbidimeter[3]

C 151 Test Method for Autoclave Expansion of Portland Cement[3]

C 183 Practice for Sampling and the Amount of Testing of Hydraulic Cement[3]

C 185 Test Method for Air Content of Hydraulic Cement Mortar[3]

C 186 Test Method for Heat of Hydration of Hydraulic Cement[3]

C 191 Test Method for Time of Setting of Hydraulic Cement by Vicat Needle[3]

C 204 Test Method for Fineness of Hydraulic Cement by Air Permeability Apparatus[3]

C 226 Specification for Air-Entraining Additions for Use in the Manufacture of Air-Entraining Portland Cement[3]

C 266 Test Method for Time of Setting of Hydraulic Cement Paste by Gillmore Needles[3]

C 451 Test Method for Early Stiffening of Hydraulic Cement (Paste Method)[3]

C 452 Test Method for Potential Expansion of Portland Cement Mortars Exposed to Sulfate[3]

C 465 Specification for Processing Additions for Use in the Manufacture of Hydraulic Cements[3]

C 563 Test Method for Optimum SO_3 in Hydraulic Cement Using 24-h Compressive Strength[3]

C 1038 Test Method for Expansion of Portland Cement Mortar Bars Stored in Water[3]

E 29 Practice for Using Significant Digits in Test Data to Determine Conformance with Specifications[4]

3. Terminology

3.1 *Definitions:*

3.1.1 *portland cement*—a hydraulic cement produced by pulverizing clinker consisting essentially of hydraulic calcium silicates, usually containing one or more of the forms of calcium sulfate as an interground addition.

3.1.2 *air-entraining portland cement*—a hydraulic cement produced by pulverizing clinker consisting essentially of hydraulic calcium silicates, usually containing one or more of the forms of calcium sulfate as an interground addition, and with which there has been interground an air-entraining addition.

4. Ordering Information

4.1 Orders for material under this specification shall include the following:

4.1.1 This specification number and date,

[1] This specification is under the jurisdiction of ASTM Committee C-1 on Cement and is the direct responsibility of Subcommittee C01.10 on Portland Cement.

Current edition approved Nov. 10, 1998. Published May 1999. Originally published as C 150 – 40 T. Last previous edition C 150 – 97a.

[2] *Annual Book of ASTM Standards*, Vol 04.02.

[3] *Annual Book of ASTM Standards*, Vol 04.01.

[4] *Annual Book of ASTM Standards*, Vol 14.02.

TABLE 1 Standard Chemical Requirements

Cement Type^{A}	I and IA	II and IIA	III and IIIA	IV	V
Silicon dioxide (SiO_2), min, %	...	20.0^{B,C}	...	...	...
Aluminum oxide (Al_2O_3), max, %	...	6.0	...	...	...
Ferric oxide (Fe_2O_3), max, %	...	6.0^{B,C}	...	6.5	...
Magnesium oxide (MgO), max, %	6.0	6.0	6.0	6.0	6.0
Sulfur trioxide (SO_3),^{D} max, %					
When (C_3A)^{E} is 8 % or less	3.0	3.0	3.5	2.3	2.3
When (C_3A)^{E} is more than 8 %	3.5	^{F}	4.5	^{F}	^{F}
Loss on ignition, max, %	3.0	3.0	3.0	2.5	3.0
Insoluble residue, max, %	0.75	0.75	0.75	0.75	0.75
Tricalcium silicate (C_3S),^{E} max, %	...	...	...	35^{B}	...
Dicalcium silicate (C_2S),^{E} min, %	...	...	...	40^{B}	...
Tricalcium aluminate (C_3A)^{E} max, %	...	8	15	7^{B}	5^{C}
Tetracalcium aluminoferrite plus twice the tricalcium aluminate^{E} ($C_4AF + 2(C_3A)$), or solid solution ($C_4AF + C_2F$), as applicable, max, %	...	...	...	...	25^{C}

^{A}See Note 1.

^{B}Does not apply when the heat of hydration limit in Table 3 is specified.

^{C}Does not apply when the sulfate resistance limit in Table 3 is specified.

^{D}There are cases where optimum SO_3(using Test Method C 563) for a particular cement is close to or in excess of the limit in this specification. In such cases where properties of a cement can be improved by exceeding the SO_3 limits stated in this table, it is permissible to exceed the values in the table, provided it has been demonstrated by Test Method C 1038 that the cement with the increased SO_3 will not develop expansion in water exceeding 0.020 % at 14 days. When the manufacturer supplies cement under this provision, he shall, upon request, supply supporting data to the purchaser.

^{E}All values calculated as described in this note shall be rounded according to Practice E 29. When evaluating conformance to a specification, round values to the same number of places as the corresponding table entry before making comparisons. The expressing of chemical limitations by means of calculated assumed compounds does not necessarily mean that the oxides are actually or entirely present as such compounds.

When expressing compounds, C = CaO, S = SiO_2, A = Al_2O_3, F = Fe_2O_3. For example, $C_3A = 3CaO \cdot Al_2O_3$.

Titanium dioxide and phosphorus pentoxide (TiO_2 and P_2O_5) shall not be included with the Al_2O_3 content. See Note 2.

When the ratio of percentages of aluminum oxide to ferric oxide is 0.64 or more, the percentages of tricalcium silicate, dicalcium silicate, tricalcium aluminate, and tetracalcium aluminoferrite shall be calculated from the chemical analysis as follows:

Tricalcium silicate = $(4.071 \times \% CaO) - (7.600 \times \% SiO_2) - (6.718 \times \% Al_2O_3) - (1.430 \times \% Fe_2O_3) - (2.852 \times \% SO_3)$

Dicalcium silicate = $(2.867 \times \% SiO_2) - (0.7544 \times \% C_3S)$

Tricalcium aluminate = $(2.650 \times \% Al_2O_3) - (1.692 \times \% Fe_2O_3)$

Tetracalcium aluminoferrite = $3.043 \times \% Fe_2O_3$

When the alumina-ferric oxide ratio is less than 0.64, a calcium aluminoferrite solid solution (expressed as ss($C_4AF + C_2F$)) is formed. Contents of this solid solution and of tricalcium silicate shall be calculated by the following formulas:

ss($C_4AF + C_2F$) = $(2.100 \times \% Al_2O_3) + (1.702 \times \% Fe_2O_3)$

Tricalcium silicate = $(4.071 \times \% CaO) - (7.600 \times \% SiO_2) - (4.479 \times \% Al_2O_3) - (2.859 \times \% Fe_2O_3) - (2.852 \times \% SO_3)$.

No tricalcium aluminate will be present in cements of this composition. Dicalcium silicate shall be calculated as previously shown.

^{F}Not applicable.

4.1.2 Type or types allowable. If no type is specified, Type I shall be supplied,

4.1.3 Any optional chemical requirements from Table 2, if desired,

4.1.4 Type of setting-time test required, Vicat or Gillmore. If not specified, the Vicat shall be used,

TABLE 2 Optional Chemical Requirements^{A}

Cement Type	I and IA	II and IIA	III and IIIA	IV	V	Remarks
Tricalcium aluminate (C_3A),^{B} max, %	...	...	8	...	...	for moderate sulfate resistance
Tricalcium aluminate (C_3A),^{B} max, %	...	...	5	...	...	for high sulfate resistance
Sum of tricalcium silicate and tricalcium aluminate,^{B} max, %	...	58^{C}	...	...	...	for moderate heat of hydration
Equivalent Alkalies ($Na_2O + 0.658K_2O$), max, %	0.60^{D}	0.60^{D}	0.60^{D}	0.60^{D}	0.60^{D}	low-alkali cement

^{A}These optional requirements apply only when specifically requested. Verify availability before ordering. See note in Section 4.

^{B}All values calculated as described in this note shall be rounded according to Practice E 29. When evaluating conformance to a specification, round values to the same number of places as the corresponding table entry before making comparisons. The expressing of chemical limitations by means of calculated assumed compounds does not necessarily mean that the oxides are actually or entirely present as such compounds.

When expressing compounds, C = CaO, S = SiO_2, A = Al_2O_3, F = Fe_2O_3. For example, $C_3A = 3CaO \cdot Al_2O_3$.

Titanium dioxide and phosphorus pentoxide (TiO_2 and P_2O_5) shall not be included with the Al_2O_3 content. See Note 2.

When the ratio of percentages of aluminum oxide to ferric oxide is 0.64 or more, the percentages of tricalcium silicate, dicalcium silicate, tricalcium aluminate, and tetracalcium aluminoferrite shall be calculated from the chemical analysis as follows:

Tricalcium silicate = $(4.071 \times \% CaO) - (7.600 \times \% SiO_2) - (6.718 \times \% Al_2O_3) - (1.430 \times \% Fe_2O_3) - (2.852 \times \% SO_3)$

Dicalcium silicate = $(2.867 \times \% SiO_2) - (0.7544 \times \% C_3S)$

Tricalcium aluminate = $(2.650 \times \% Al_2O_3) - (1.692 \times \% Fe_2O_3)$

Tetracalcium aluminoferrite = $3.043 \times \% Fe_2O_3$

When the alumina-ferric oxide ratio is less than 0.64, a calcium aluminoferrite solid solution (expressed as ss ($C_4AF + C_2F$)) is formed. Contents of this solid solution and of tricalcium silicate shall be calculated by the following formulas:

ss($C_4AF + C_2F$) = $(2.100 \times \% Al_2O_3) + (1.702 \times \% Fe_2O_3)$

Tricalcium silicate = $(4.071 \times \% CaO) - (7.600 \times \% SiO_2) - (4.479 \times \% Al_2O_3) - (2.859 \times \% Fe_2O_3) - (2.852 \times \% SO_3)$.

No tricalcium aluminate will be present in cements of this composition. Dicalcium silicate shall be calculated as previously shown.

^{C}The optional limit for heat of hydration in Table 3 shall not be requested when this optional limit is requested.

^{D}Specify this limit when the cement is to be used in concrete with aggregates that are potentially reactive and no other provisions have been made to protect the concrete from deleteriously reactive aggregates. Refer to Specification C 33 for information on potential reactivity of aggregates.

4.1.5 Any optional physical requirements from Table 3, if desired.

NOTE 1—Cement conforming to the requirements for all types are not carried in stock in some areas. In advance of specifying the use of cement other than Type I, determine whether the proposed type of cement is, or can be made, available.

5. Additions

5.1 The cement covered by this specification shall contain no addition except as follows:

5.1.1 Water or calcium sulfate, or both, may be added in amounts such that the limits shown in Table 1 for sulfur trioxide and loss-on-ignition shall not be exceeded.

5.1.2 Processing additions used in the manufacture of the cement shall have been shown to meet the requirements of Specification C 465 in the amounts used or greater.

5.1.3 Air-entraining portland cement shall contain an interground addition conforming to the requirements of Specification C 226.

6. Chemical Composition

6.1 Portland cement of each of the eight types shown in Section 1 shall conform to the respective standard chemical requirements prescribed in Table 1. In addition, optional chemical requirements are shown in Table 2.

NOTE 2—When comparing oxide analyses and calculated compounds from different sources or from different historic times, be aware that they may not have been reported on exactly the same basis. Chemical data obtained by Reference and Alternate Test Methods of Test Methods C 114 (wet chemistry) may include titania and phosphorus as alumina unless proper correction has been made (see Test Methods C 114), while data obtained by rapid instrumental methods usually do not. This can result in small differences in the calculated compounds. Such differences are usually within the precision of the analytical methods, even when the methods are properly qualified under the requirements of Test Methods C 114.

7. Physical Properties

7.1 Portland cement of each of the eight types shown in Section 1 shall conform to the respective standard physical requirements prescribed in Table 4. In addition, optional physical requirements are shown in Table 3.

8. Sampling

8.1 When the purchaser desires that the cement be sampled and tested to verify compliance with this specification, perform sampling and testing in accordance with Practice C 183.

8.2 Practice C 183 is not designed for manufacturing quality control and is not required for manufacturer's certification.

9. Test Methods

9.1 Determine the applicable properties enumerated in this specification in accordance with the following test methods:

9.1.1 *Air Content of Mortar*—Test Method C 185.

9.1.2 *Chemical Analysis*—Test Methods C 114.

9.1.3 *Strength*—Test Method C 109.

9.1.4 *False Set*—Test Method C 451.

9.1.5 *Fineness by Air Permeability*—Test Method C 204.

9.1.6 *Fineness by Turbidimeter*—Test Method C 115.

9.1.7 *Heat of Hydration*—Test Method C 186.

9.1.8 *Autoclave Expansion*—Test Method C 151.

9.1.9 *Time of Setting by Gillmore Needles*—Test Method C 266.

9.1.10 *Time of Setting by Vicat Needles*—Test Method C 191.

9.1.11 *Sulfate Resistance*—Test Method C 452 (sulfate expansion).

9.1.12 *Calcium Sulfate (expansion of) Mortar*—Test Method C 1038.

9.1.13 *Optimum SO_3*—Test Method C 563.

10. Inspection

10.1 Inspection of the material shall be made as agreed upon between the purchaser and the seller as part of the purchase contract.

11. Rejection

11.1 The cement shall be rejected if it fails to meet any of the requirements of this specification.

11.2 At the option of the purchaser, retest, before using, cement remaining in bulk storage for more than 6 months or cement in bags in local storage in the custody of a vendor for more than 3 months after completion of tests and reject the cement if it fails to conform to any of the requirements of this specification. Cement so rejected shall be the responsibility of the owner of record at the time of resampling for retest.

11.3 Packages shall identify the mass contained as net weight. At the option of the purchaser, packages more than 2 %

TABLE 3 Optional Physical Requirements[A]

Cement Type[A]	I	IA	II	IIA	III	IIIA	IV	V
False set, final penetration, min, %	50	50	50	50	50	50	50	50
Heat of hydration:								
7 days, max, kJ/kg (cal/g)	...	...	290 (70)[C]	290 (70)[C]	...	...	250 (60)[D]	...
28 days, max, kJ/kg (cal/g)	...	...	...	...	...	...	290 (70)[D]	...
Strength, not less than the values shown:								
Compressive strength, MPa (psi)								
28 days	28.0 (4060)	22.0 (3190)	28.0 (4060) 22.0[C] (3190)[C]	22.0 (3190) 18.0[C] (2610)[C]	...	...	...	...
Sulfate resistance,[E] 14 days, max, % expansion	...	...	...[F]	...[F]	...	...	...	0.040

[A]These optional requirements apply only when specifically requested. Verify availability before ordering. See Note in Section 4.

[C]The optional limit for the sum of the tricalcium silicate and tricalcium aluminate in Table 2 shall not be requested when this optional limit is requested. These strength requirements apply when either heat of hydration or the sum of tricalcium silicate and tricalcium aluminate requirements are requested.

[D]When the heat of hydration limit is specified, it shall be instead of the limits of C3S, C2S, C3A, SiO_2, and Fe_2O_3 listed in Table 1.

[E]When the sulfate resistance is specified, it shall be instead of the limits of C_3A, $C_4AF + 2\ C_3A$, SiO_2, and Fe_2O_3 listed in Table 1.

[F]Cement meeting the high sulfate resistance limit for Type V are deemed to meet the moderate sulfate resistance requirement of Type II.

TABLE 4 Standard Physical Requirements

Cement Type[A]	I	IA	II	IIA	III	IIIA	IV	V
Air content of mortar,[B] volume %:								
max	12	22	12	22	12	22	12	12
min	...	16	...	16	...	16	...	...
Fineness,[C] specific surface, m²/kg (alternative methods):								
Turbidimeter test, min	160	160	160	160	...	...	160	160
Air permeability test, min	280	280	280	280	...	...	280	280
Autoclave expansion, max, %	0.80	0.80	0.80	0.80	0.80	0.80	0.80	0.80
Strength, not less than the values shown for the ages indicated as follows:[D]								
Compressive strength, MPa (psi):								
1 day	...	...	...	...	12.0 (1740)	10.0 (1450)	...	...
3 days	12.0 (1740)	10.0 (1450)	10.0 (1450) 7.0[E] (1020)[E]	8.0 (1160) 6.0[E] (870)[E]	24.0 (3480)	19.0 (2760)	...	8.0 (1160)
7 days	19.0 (2760)	16.0 (2320)	17.0 (2470) 12.0[E] (1740)[E]	14.0 (2030) 9.0[E] (1310)[E]	...	...	7.0 (1020)	15.0 (2180)
28 days	...	...	...	...	...	...	17.0 (2470)	21.0 (3050)
Time of setting (alternative methods):[F]								
Gillmore test:								
Initial set, min, not less than	60	60	60	60	60	60	60	60
Final set, min, not more than	600	600	600	600	600	600	600	600
Vicat test:[G]								
Time of setting, min, not less than	45	45	45	45	45	45	45	45
Time of setting, min, not more than	375	375	375	375	375	375	375	375

[A] See Note 1.
[B] Compliance with the requirements of this specification does not necessarily ensure that the desired air content will be obtained in concrete.
[C] The testing laboratory shall select the fineness method to be used. However, when the sample fails to meet the requirements of the air-permeability test, the turbidimeter test shall be used, and the requirements in this table for the turbidimetric method shall govern.
[D] The strength at any specified test age shall be not less than that attained at any previous specified test age.
[E] When the optional heat of hydration or the chemical limit on the sum of the tricalcium silicate and tricalcium aluminate is specified.
[F] The time-of-setting test required shall be specified by the purchaser. In case he does not so specify, the requirements of the Vicat test only shall govern.
[G] The time of setting is that described as initial setting time in Test Method C 191.

below the mass marked thereon shall be rejected and if the average mass of packages in any shipment, as shown by determining the mass of 50 packages selected at random, is less than that marked on the packages, the entire shipment shall be rejected.

12. Manufacturer's Statement

12.1 At the request of the purchaser, the manufacturer shall state in writing the nature, amount, and identity of any air-entraining addition and of any processing addition used, and also, if requested, shall supply test data showing compliance of such air-entraining addition with Specification C 226 and of such processing addition with Specification C 465.

13. Packaging and Package Marking

13.1 When the cement is delivered in packages, the words" Portland Cement," the type of cement, the name and brand of the manufacturer, and the mass of the cement contained therein shall be plainly marked on each package. When the cement is an air-entraining type, the words" air-entraining" shall be plainly marked on each package. Similar information shall be provided in the shipping documents accompanying the shipment of packaged or bulk cement. All packages shall be in good condition at the time of inspection.

NOTE 3—With the change to SI units, it is desirable to establish a standard SI package for portland cements. To that end 40 kg (88.18 lb) provides a convenient, even-numbered mass reasonably similar to the traditional 94-lb (42.6384-kg) package.

14. Storage

14.1 The cement shall be stored in such a manner as to permit easy access for proper inspection and identification of each shipment, and in a suitable weather-tight building that will protect the cement from dampness and minimize warehouse set.

15. Manufacturer's Certification

15.1 Upon request of the purchaser in the contract or order, a manufacturer's report shall be furnished at the time of shipment stating the results of tests made on samples of the material taken during production or transfer and certifying that the cement conforms to applicable requirements of this specification.

16. Keywords

16.1 hydraulic cement; portland cement; specification

The American Society for Testing and Materials takes no position respecting the validity of any patent rights asserted in connection with any item mentioned in this standard. Users of this standard are expressly advised that determination of the validity of any such patent rights, and the risk of infringement of such rights, are entirely their own responsibility.

This standard is subject to revision at any time by the responsible technical committee and must be reviewed every five years and if not revised, either reapproved or withdrawn. Your comments are invited either for revision of this standard or for additional standards and should be addressed to ASTM Headquarters. Your comments will receive careful consideration at a meeting of the responsible technical committee, which you may attend. If you feel that your comments have not received a fair hearing you should make your views known to the ASTM Committee on Standards, 100 Barr Harbor Drive, West Conshohocken, PA 19428.

Designation: C 151 – 98a

Standard Test Method for Autoclave Expansion of Portland Cement[1]

This standard is issued under the fixed designation C 151; the number immediately following the designation indicates the year of original adoption or, in the case of revision, the year of last revision. A number in parentheses indicates the year of last reapproval. A superscript epsilon (ε) indicates an editorial change since the last revision or reapproval.

1. Scope

1.1 This test method covers determination of the autoclave expansion of portland cement by means of a test on a neat cement specimen.

1.2 The values stated in inch-pound units are to be regarded as the standard.

1.3 *This standard does not purport to address all of the safety concerns, if any, associated with its use. It is the responsibility of the user of this standard to establish appropriate safety and health practices and determine the applicability of regulatory limitations prior to use.*For specific precaution statements, see the section on Safety Precautions.

2. Referenced Documents

2.1 *ASTM Standards:*

C 187 Test Method for Normal Consistency of Hydraulic Cement[2]

C 305 Practice for Mechanical Mixing of Hydraulic Cement Pastes and Mortars of Plastic Consistency[2]

C 490 Practice for Use of Apparatus for the Determination of Length Change of Hardened Cement Paste, Mortar, and Concrete[2]

C 511 Specification for Moist Cabinets, Moist Rooms, and Water Storage Tanks Used in the Testing of Hydraulic Cements and Concretes[2]

C 1005 Specification for Weights and Weighing Devices for Use in the Physical Testing of Hydraulic Cements[2]

3. Significance and Use

3.1 The autoclave expansion test provides an index of potential delayed expansion caused by the hydration of CaO, or MgO, or both, when present in portland cement.[3]

4. Apparatus

4.1 *Weighing Devices and Weights*, for determining the mass of materials conforming to the requirements of Specification C 1005.

4.2 *Glass Graduates*, 200 or 250-mL capacity, and conforming to the requirements of Practice C 490.

4.3 *Molds*, 1 by 1-in. (25.4 by 25.4-mm) cross section, conforming to the requirements of Practice C 490.

4.4 *Flat Trowel*, having a straight-edged steel blade 4 to 6 in. (100 to 150 mm) in length.

4.5 *Autoclave*, consisting of a high-pressure steam vessel provided with a thermometer well. The autoclave shall be equipped with an automatic pressure control and a rupture disk with a bursting pressure of 350 psi (2.4 MPa) ± 5 %. In locations where the use of a rupture disk is not permitted, the autoclave shall be equipped with a safety valve. In addition, the autoclave shall be equipped with a vent valve to allow the escape of air during the early part of the heating period and to release any steam pressure remaining at the end of the cooling period. The pressure gage shall have a nominal capacity of 600 psi (4.1 MPa), a dial with a nominal diameter of 4½in. (114 mm) and shall be graduated from 0 to 600 psi (0 to 4.1 MPa) with scale divisions not exceeding 5 psi (0.03 MPa). The error in the gage shall not exceed ±3 psi (±0.02 MPa) at the operating pressure of 295 psi (2 MPa). The capacity of the heating unit shall be such that with maximum load (water plus specimens) the pressure of the saturated steam in the autoclave may be raised to a gage pressure of 295 psi in 45 to 75 min from the time the heat is turned on. The automatic pressure control shall be capable of maintaining the gage pressure at 295± 10 psi (2 ± 0.07 MPa) for at least 3 h. A gage pressure of 295 ± 10 psi corresponds to a temperature of 420 ± 3°F (216± 2°C). The autoclave shall be designed to permit the gage pressure to drop from 295 psi to less than 10 psi in 1½h after the heat supply has been shut off.

4.5.1 *Rupture Disk*—The rupture disk shall be made of a material having a tensile strength that is relatively insensitive to temperature in the range 68 to 420°F (20 to 216°C) and that is electrochemically compatible with the pipe leading to it and to its holder.[4]

4.6 *Length Comparator*—The comparator used for measuring length change of specimens shall conform to the requirements of Practice C 490.

5. Temperature and Humidity

5.1 *Molding Room*—Maintain the temperature of the molding room, dry materials and mixing water, and the relative

---

[1] This test method is under the jurisdiction of ASTM Committee C-1 on Cement and is the direct responsibility of Subcommittee C01.31 on Volume Change.

Current edition approved Oct. 10, 1998. Published February 1999. Originally published as C 151 – 40 T. Last previous edition C 151 – 98.

[2] *Annual Book of ASTM Standards*, Vol 04.01.

[3] Gonnerman, H. F., Lerch, W. and Whiteside, T. M. , "Investigations of the Hydration Expansion Characteristics of Portland Cements," *Portland Cement Association Research Department Bulletin 45*, 1953, pp. 1-168.

[4] A list of suppliers of autoclave rupture disks and gaskets is maintained at ASTM Headquarters.

humidity of the molding room within the limits of Practice C 490.

5.2 *Moist Storage Facilities*—Maintain the temperature and humidity of the moist storage facilities to the requirements of Specification C 511.

6. Safety Precautions

6.1 The pressure gage shall have a capacity of 600 psi (4.1 MPa). A gage with too small or too large a capacity may be a hazard, since for pressure above the specified maximum working pressure, with a smaller capacity gage, the pressure may be off scale, and with a larger capacity gage, the arc of movement may be too small to invite attention. The operator shall be sure the gage hand has not passed the maximum graduation on the scale.

6.2 Test the pressure gage for proper operation. Always use a thermometer together with the pressure gage, so as to provide a means of detecting any failure of the pressure gage to operate properly, and to indicate any unusual condition.

6.3 Maintain the automatic control in proper working order at all times.

6.4 Set the safety valve to relieve the pressure at about 6 to 10 % above the maximum of 305 psi (2.1 MPa) specified in this test method, that is, at about 330 psi (2.3 MPa). Unless the manufacturer has given specific instructions as to maintenance of the safety valve, test the valve twice each year.[5] Test with a gage testing device, or by adjusting the automatic controls to allow the autoclave to reach a pressure of about 330 psi, at which pressure the safety valve will either open or be adjusted to open. Direct the safety valve discharge away from the operator.

NOTE 1—Unexpected combinations of conditions may occur. For example, in one case the automatic control had failed, the safety valve had become stuck, and the gage hand, which at first glance appeared to be at about zero, had passed the maximum graduation and had come to stop on the wrong side of the pin. This condition of the gage was finally detected and the pressure, then of an unknown magnitude, was released before failure could occur in the apparatus.

6.5 Wear heavy leather work gloves to prevent burning of the hands when removing the top of the autoclave at the end of the test. Direct the vent valve away from the operator. When removing the autoclave lid, tilt it so that any steam escaping from beneath the lid will be discharged away from the operator. Care shall be taken to avoid scalding by any liquid that may have been used in the autoclave well.

6.6 The operator shall be made aware that for many autoclave pressure gages the return of the gage hand to the initial rest or starting point does not necessarily indicate zero pressure within the autoclave; there may then still remain an appreciable dangerous pressure.

7. Number of Test Specimens

7.1 Make at least one test specimen.

8. Preparation of Specimen Molds

8.1 Prepare the specimen molds in accordance with the requirements of Practice C 490 except that molds need not be sealed.

9. Preparation of Test Specimens

9.1 *Mixing Cement Paste*—Prepare the standard batch consisting of 650 g of cement and sufficient water to give a paste of normal consistency in accordance with the procedure described in Test Method C 187. Mix this batch in accordance with the procedure described in Practice C 305. Both a time of setting specimen and an autoclave bar may be made from the same batch. If the paste to be used for time of setting is the paste sample already used to determine normal consistency, then the autoclave bar may be prepared immediately from the remainder of the batch.

9.2 *Molding Specimens*—Immediately following preparation of the time of setting specimen or completion of mixing, mold the test specimen in two approximately equal layers, each layer being compacted with the thumbs or forefingers by pressing the paste into the corners, around the gage studs, and along the surface of the mold until a homogeneous specimen is obtained. Compact the top layer, cut off the paste flush with the top of the mold with a thin-edged trowel, and smooth the surface with a few strokes of the flat trowel. During the operations of mixing and molding, protect the hands with rubber gloves.

9.3 *Storage of Test Specimens*—After filling the mold, place it in the moist closet or moist room. Store the specimens in the molds in the moist enclosure for at least 20 h; if removed from the molds before 24 h, they shall be kept in the moist closet or moist room until time of test.

10. Procedure

10.1 At 24 h ± 30 min after molding, remove the specimens from the moist atmosphere, immediately obtain a length comparator reading for each specimen, and place in the autoclave at room temperature in a rack so that all sides of the specimen will be exposed to saturated steam. The autoclave shall contain enough water, at an initial temperature of 68 to 82°F (20 to 28°C), to maintain an atmosphere of saturated steam vapor during the entire test. Ordinarily 7 to 10 % of the volume of the autoclave should be occupied by the water.

10.2 To permit air to escape from the autoclave during the early portion of the heating period, leave the vent valve open until steam begins to escape. (**Caution**—See the section on Safety Precautions). Close the valve and raise the temperature of the autoclave at a rate that will bring the gage pressure of the steam to 295 psi (2 MPa) in 45 to 75 min from the time the heat is turned on. Maintain the 295 ± 10 psi (2 ± 0.07 MPa) pressure for 3 h. At the end of the 3-h period, shut off the heat supply and cool the autoclave at such a rate that the pressure will be less than 10 psi at the end of 1½ h. At the end of the 1½-h period, slowly release any remaining pressure by partially opening the vent valve until atmospheric pressure is attained. Then open the autoclave and place the test specimen in water at a temperature above 194°F (90°C). Cool the water surrounding the bars at a uniform rate by adding cold water so that the temperature of the water will be lowered to 74°F (23°C) in 15 min. Maintain the water surrounding the specimens at 74°F for an additional 15 min; then, surface-dry the

[5] A list of facilities for testing autoclave safety valves is maintained at ASTM Headquarters.

specimens and obtain a length comparator reading for each specimen.

11. Calculation

11.1 Calculate the change in length of the test specimen by subtracting the length comparator reading before autoclaving from that after autoclaving, and report as percent of effective gage length to the nearest 0.01 %. Report the percentage of increase in length as the autoclave expansion. Indicate a decrease in length by a minus sign prefixed to the percent value.

12. Precision and Bias

12.1 *Precision*—The single-operator (within-laboratory) standard deviation has been found to be 0.024 % throughout the range of 0.11 % to 0.94 % expansions. Therefore, results of two properly conducted tests by the same operator for expansions of similar batches should not differ from each other by more than 0.07 % expansion. The multi-laboratory (between laboratory) standard deviation has been found to be 0.030 % throughout the range of 0.11 % to 0.94 % expansions. Therefore, results of two properly conducted tests from two different laboratories for expansions of similar batches should not differ from each other by more than 0.09 % expansion.

12.2 *Bias*—Since there is no accepted reference material suitable for determining bias, bias has not been determined.

13. Keywords

13.1 autoclave; autoclave expansion; expansion; portland cement; soundness

For additional useful information on details of cement test methods, reference may be made to the "Manual of Cement Testing," which appears in the *Annual Book of ASTM Standards,* Vol 04.01.

Designation: C 183 – 97

Standard Practice for Sampling and the Amount of Testing of Hydraulic Cement[1]

This standard is issued under the fixed designation C 183; the number immediately following the designation indicates the year of original adoption or, in the case of revision, the year of last revision. A number in parentheses indicates the year of last reapproval. A superscript epsilon (ε) indicates an editorial change since the last revision or reapproval.

1. Scope

1.1 This practice covers procedures for sampling and for the amount of testing of hydraulic cement after it has been manufactured and is ready to be offered for sale.

1.2 The values stated in SI units are to be regarded as the standard. The inch-pound units in parentheses are for information purposes only.

1.2.1 A ton as used in this practice is 907 kg (2000 lb).

1.2.2 Values in SI units shall be obtained by measurement in SI units or by appropriate conversion, using the Rules for Conversion and Rounding given in Standard IEEE/ASTM SI 10, of measurement made in other units.

1.3 *This standard does not purport to address all of the safety concerns, if any, associated with its use. It is the responsibility of the user of this standard to establish appropriate safety and health practices and determine the applicability of regulatory limitations prior to use.*

2. Referenced Documents

2.1 *ASTM Standards:*

C 91 Specification for Masonry Cement[2]

C 109 Test Method for Compressive Strength of Hydraulic Cement Mortars (Using 2-in. or 50-mm Cube Specimens)[2]

C 114 Test Methods for Chemical Analysis of Hydraulic Cement[2]

C 115 Test Method for Fineness of Portland Cement by the Turbidimeter[2]

C 150 Specification for Portland Cement[2]

C 151 Test Method for Autoclave Expansion of Portland Cement[2]

C 157 Test Method for Length Change of Hardened Hydraulic-Cement Mortar and Concrete[3]

C 185 Test Method for Air Content of Hydraulic Cement Mortar[2]

C 186 Test Method for Heat of Hydration of Hydraulic Cement[2]

C 191 Test Method for Time of Setting of Hydraulic Cement by Vicat Needle[2]

C 204 Test Method for Fineness of Hydraulic Cement by Air Permeability Apparatus[2]

C 227 Test Method for Potential Alkali Reactivity of Cement-Aggregate Combinations (Mortar-Bar Method)[3]

C 265 Test Method for Calcium Sulfate in Hydrated Portland Cement Mortar[2]

C 266 Test Method for Time of Setting of Hydraulic-Cement Paste by Gillmore Needles[2]

C 451 Test Method for Early Stiffening of Hydraulic Cement (Paste Method)[2]

C 452 Test Method for Potential Expansion of Hydraulic Cement Mortars Exposed to Sulfate[2]

C 563 Test Method for Optimum SO_3 in Hydraulic Cement Using 24 h Compressive Strength[2]

C 595 Specification for Blended Hydraulic Cements[2]

C 845 Specification for Expansive Hydraulic Cement[2]

C 1012 Test Method for Length Change of Hydraulic-Cement Mortars Exposed to a Sulfate Solution[2]

E 11 Specification for Wire-Cloth Sieves for Testing Purposes[4]

IEEE/ASTM SI 10 Standard for Use of the International System of Units (SI): The Modern Metric System[4]

2.2 *ACI Standard:*

225.1R Guide to the Selection and Use of Hydraulic Cements[5]

3. Terminology

3.1 *Definitions of Terms Specific to This Standard:*

3.1.1 *lot* (of cement)—specific quantity of cement offered for inspection at any one time. A lot may be one or more storage bins filled consecutively. A lot may also be the contents of one or more transport units representing cement drawn from the same storage bin.

3.1.2 *reduced testing rate*—test program that provides for the testing of only two samples from any given lot of samples obtained and prepared for testing at the normal rate as described herein. The program utilizes probability factors and is so designed that when results from the two samples fulfill the requirements of the program it may be said with 95 % confidence that less than 5 % of the samples would be outside the specification limits.

4. Significance and Use

4.1 The sampling procedures described are intended for use in the procurement of samples of hydraulic cement after it has been manufactured and is ready to be offered for sale. They are not intended as sampling procedures for quality control purposes during manufacturing. The testing procedures outlined cover the amount of testing to be done and provide guidance for reporting on conformance or non-

---

[1] This practice is under the jurisdiction of ASTM Committee C-1 on Cement, and is the direct responsibility of Subcommittee C01.95 on Methods of Test.

Current edition approved Oct. 10, 1995. Published December 1995. Originally published as C 183 – 44. Last previous edition C 183 – 95.

[2] *Annual Book of ASTM Standards*, Vol 04.01.

[3] *Annual Book of ASTM Standards*, Vol 04.02.

[4] *Annual Book of ASTM Standards*, Vol 14.02.

[5] Detailed requirements for this sieve are given in Specification E 11.

conformance of cements with requirements of purchase specifications.

4.2 This practice is referenced as the procedure for sampling masonry cement (Specification C 91), portland cement (Specification C 150), blended hydraulic cement (Specification C 595), and expansive hydraulic cement (Specification C 845).

4.3 Most building codes and construction specifications require that hydraulic cement to be used in the work meet the applicable requirements of the relevant purchase specifications, such as Specifications C 91, C 150, C 595, or C 845. If the code or specification requires sampling of the manufactured cement, the provisions given in 4.4 are applicable. Not much cement is sold on the basis of such sampling and testing. A useful discussion of sampling and testing cement is contained in ACI 225.1R.

4.4 The procedures covered in this practice should be done by or for purchasers of hydraulic cement who are using a code or specification that requires sampling and testing to determine if the samples conform to the relevant acceptance specifications. The testing is done using specified methods to determine whether the samples yield test results that conform to the specification, and the tests serve as a basis for acceptance or rejection of the lot of material sampled.

4.5 It is neither intended nor required that all cements be tested using all the test methods referenced in Section 2.

5. Kinds and Size of Samples and by Whom Taken

5.1 A cement sample secured from a conveyor, from bulk storage, or from a bulk shipment in one operation, shall be termed a "grab sample." A sample obtained during a 10-min interval using an automatic sampling device that continuously samples a cement stream may also be termed a grab sample. Grab samples taken at prescribed intervals over a period of time may be combined to form a "composite sample" representative of the cement produced during that period of time.

5.2 All samples, whether grab or composite, shall have a mass of at least 5 kg (10 lb).

5.3 The purchaser may designate a representative to supervise the sampling, packing, and shipping of samples when it is so specified in the purchase contract.

5.4 Package the samples in moisture-proof, airtight containers numbered consecutively in the order in which the samples are taken. The purchase contract shall state who will pay for the costs of sampling, packaging, shipping and testing the samples.

NOTE 1—Polyvinyl chloride sample containers, upon occasion, have been found to affect the air-entraining potential of a cement sample. The same problem might be experienced with containers made from other plastics.

6. Testing-Time Requirements for the Completion of Tests

6.1 When tests of hydraulic cement are made at a laboratory other than that of the cement manufacturer, the cement sampling schedule, sample transportation time, and sample testing schedule must be coordinated among the purchaser, the manufacturer, and the testing laboratory so that the tests results will be available when required.

6.2 The manufacturer of the cement shall make the cement available to be sampled for testing early enough before the time the test results are needed so that at least the applicable time intervals listed in 6.3 exist.

6.3 When this has been done, the testing laboratory shall provide test results not later than the indicated number of days after sampling:

Test Methods	Time Interval, days
C 109 (1-day results), C 114, C 115, C 151, C 185, C 191, C 204, C 451, C 265, C 266, C 563	8
C 109 (3-day results)	10
C 109 and C 186 (7-day results)	14
C 227, C 452, and C 1012 (14-day results)	21
C 109 and C 186 (28-day results)	35
C 157 (34-day results)	41
C 227 (56-day results)	63
C 227 (91-day results)	98

7. Sampling

7.1 The cement may be sampled by any of the applicable methods described in this section.

7.1.1 *From the Conveyor Delivering to Bulk Storage*—Take one grab sample, having a mass of at least 5 kg (10 lb), at approximately 6-h intervals.

7.1.2 *Transfer Sampling*—Sample cement in storage while the cement is being transferred from one bin to another. Take one grab sample from the transfer stream for each 360 Mg (400 tons) of cement, or fraction thereof, but take no less than two grab samples and combine them to produce a composite sample.

7.1.3 *Other Sampling Methods*—When neither of the above sampling methods is applicable, samples may, when authorized by the purchaser, be taken by one of the following methods:

7.1.3.1 *From Bulk Storage at Points of Discharge*—Withdraw cement from the discharge openings in a steady stream until sampling is completed. Estimate the quantity of cement in Mg to be withdrawn from one discharge opening as $0.055\ d^3 \times 0.2$, where d is the depth in metres of cement above the discharge opening. If a high circular silo is being sampled, take all samples from one opening. If the quantity of the cement in the bin exceeds 1100 Mg (1200 tons) when low rectangular bins are being sampled, discharge openings employed in the sampling shall be such that for no opening shall the number of samples represent more than one half the contents of the bin or more than 1800 Mg (2000 tons). In sampling bulk storage at points of discharge, while the cement is flowing through the openings, take samples at such intervals so that at least two grab samples shall be secured for each 360 Mg (400 tons) in the bin or silo.

7.1.3.2 *From Bulk Storage and Bulk Shipment by Means of a Slotted Tube Sampler*—When the depth of the cement to be sampled does not exceed 2.1 m (7 ft), obtain samples using a slotted tube sampler similar to that shown in Fig. 1. It shall be between 1.5 and 1.8 m (5 and 6 ft) long and

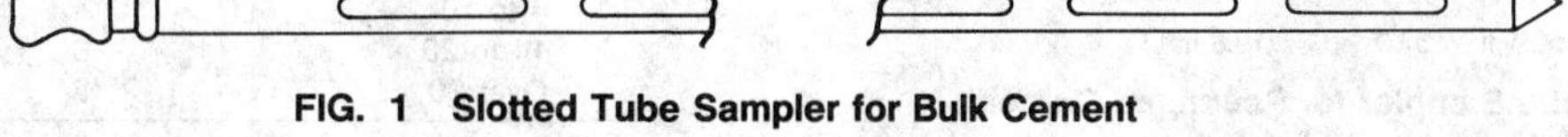

FIG. 1 Slotted Tube Sampler for Bulk Cement

approximately 35 mm (1 3/8 in.) in outside diameter and consist of two polished brass telescopic tubes with registering slots that are opened or closed by rotation of the inner tube, the outer tube being provided with a sharp point to facilitate penetration. Take samples from well-distributed points and various depths of the cement so that the samples taken will represent the cement involved.

7.1.3.3 *From Packaged Cement by Means of Tube Sampler*—Insert the sampler, shown in Fig. 2, diagonally into the valve of the bag and place the thumb over the air hole. Then withdraw the sampler. Take one sample from a bag in each 4.5 Mg (5 tons) or fraction thereof.

7.1.3.4 *From Bulk Shipment of Car or Truck:*

(a) Single Shipment—If only one car or truck is being loaded and the loading is continuous and all from the same source, take a 5-kg (10-lb) sample. If not continuous or unknown, combine five or more portions from different points in the load to form the test sample.

(b) Multiple Shipments—When the shipment consists of several cars or trucks loaded from the same source and on the same day, sample the shipment at the rate of one sample for each 90 Mg (100 tons) of cement or fraction thereof, but take not less than two samples. Consider cement represented by such samples as a lot, and test the samples in accordance with the procedure outlined in the section on Amount of Testing.

7.2 *Protection of Samples*—As samples are taken, place them directly in moisture-proof airtight containers to avoid moisture absorption and aeration of the sample. If the samples are placed in cans, fill the can completely and immediately seal. Use moisture-proof multiple-wall paper bags or plastic bags if they are strong enough to avoid breakage, and if they can be sealed immediately after filling in such a manner as to eliminate excess air in the sample and avoid moisture absorption and aeration of the sample. Samples shall be treated as described in the section on Preparation of Sample.

8. Preparation of Sample

8.1 Before testing, pass each sample through an 850-µm (No. 20) sieve,[6] or any other sieve having approximately the same size openings, in order to mix the sample, break up lumps, and remove foreign material. Discard the foreign materials and hardened lumps that do not break up on sieving or brushing. Store the cement in airtight moisture-proof containers to prevent aeration or absorption of moisture prior to test.

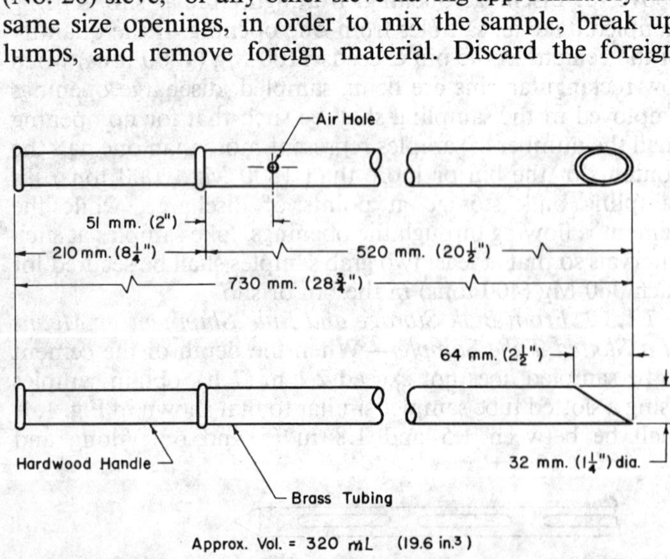

FIG. 2 Tube Sampler for Packaged Cement

9. Amount of Testing

9.1 *General*—When required, the purchaser shall specify the amount of testing for heat of hydration (Test Method C 186), alkali reactivity (Test Method C 227), and sulfate resistance (Test Method C 1012). Make all other tests on individual grab or composite samples chosen as specified herein under Selection of Samples for Testing. Do only those tests required by the applicable specification.

9.2 *Normal Testing*—Determine the number of samples to be tested in accordance with Table 1. The normal testing rate shall be used under the following conditions:

9.2.1 Before the quality history has been established,

9.2.2 When no samples from a particular mill have been tested within a year,

9.2.3 When the quality history is based entirely on data more than two years old, and

9.2.4 When it is deemed necessary to recalculate the critical limit because of indicated lack of control as shown by the control chart of the range.

NOTE 2—Random grab samples taken at inappropriate times, such as immediately following the repair or adjustment of manufacturing equipment, or from inappropriate places, such as from the top surface of the material in a car, will not suitably reflect the properties of a cement, and therefore should not be used as the basis for acceptance or rejection of a lot of cement.

9.3 *Reduced Testing*—After the quality history has been established, test at the reduced testing rate. If the results of these tests are within the critical range, make additional tests (total equal to the number of tests at the normal rate as shown in Table 1).

NOTE 3—When the quality history indicates that the results for a given requirement will probably be within the critical range, and substantial delay in completion of the tests would result from making additional tests (for example, compressive strength), it may be desirable to make the tests at the normal rate, rather than the reduced testing rate.

9.4 *Selection of Samples for Testing*—Take samples to be tested from each lot by some random method. The following method is suggested: Place a group of consecutively numbered markers equal to the number of samples in a container and mix, then draw one marker at a time from the container until the number drawn is equal to the number of samples to be tested at the normal rate. If the testing is to be done at the reduced rate, mix the drawn markers and draw two to select the numbers of the samples to be tested.

9.5 *Establishing a Quality History and Control Charts:*

9.5.1 *Quality History*—The quality history shall represent cement from the same source as the cement to be tested, and

TABLE 1 Number of Samples for Test

Lot Size—Number of Samples	Number of Tests	
	Normal Rate	Reduced Rate
2	2	2
3	3	2
4 to 10	4	2
11 to 20	6	2
Over 20	8	2

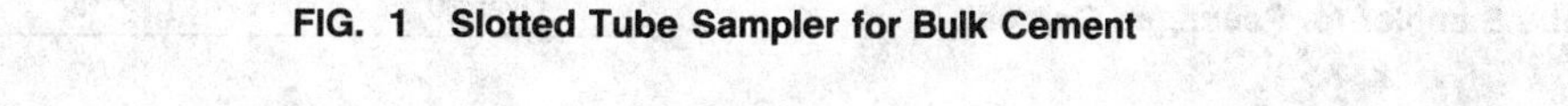

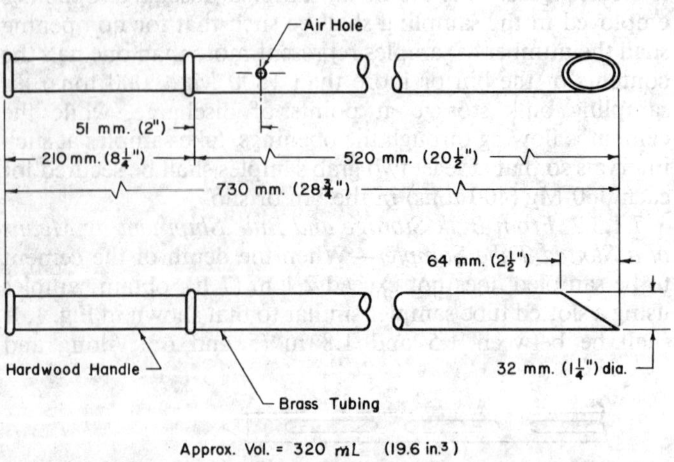

shall be based on data not more than 2 years old. There shall be available test results for not less than 40 test samples representing not less than 7 lots of cement. The test samples shall conform to the applicable provisions of this practice. A pair shall be two test samples from the same lot, in numerical sequence. Several pairs from the same lot may be used where available. The number of paired samples representing a large lot may be reduced as follows: From the consecutively numbered group of tested samples representing the entire lot, select a subgroup by some random method. List the numbers identifying the subgroup in numerical sequence, and pair in the order of listing. Compute the range (difference between the test results of a pair) for each pair of test results. Total the ranges and divide their sum by the total number of ranges used to obtain the average range, $\bar{r}$. Compute the average range, $\bar{r}$, for each included physical and chemical property limited by specification requirements.

9.5.2 *Critical Limit*—Calculate the critical limit, *C*, for each included physical and chemical property limited by a specification requirement. First, multiply the average range, $\bar{r}$, by the probability factor, 2.49, this will yield a number that for convenience is called *d*. If the requirement has a maximum specification limit, obtain *C* by subtracting *d* from the specification limit and, if a minimum, add *d* to the specification limit. Maintain quality history charts.

NOTE 4—Improved estimates of the range $\bar{r}$, and consequently of *C*, will result if the test results are not rounded. For example, the test result of 21.78 % for SiO_2 is preferred to the rounded value of 21.8 %. For the fineness, the calculated value of 3243 is preferred to the rounded value of 3240.

9.5.3 *Control Chart of the Range*—Maintain a control chart of the range to indicate when the critical limit needs to be recomputed. Multiply the average range. $\bar{r}$, as obtained in 9.5.2, by the probability factor 3.267 to obtain the upper control limit for the range between each consecutive pair of test results. The horizontal scale of the chart will be successive groups of two, and the vertical scale will be the range. Where the range chart indicates lack of control (points beyond the upper control limit), the critical limit, *C*, may need to be recalculated. Consider the occurrence of two consecutive points beyond the upper control limit for the range, or the occurrence of three points beyond the upper control limit in any series of five consecutive points cause to recalculate the critical limit. Where it becomes necessary to recalculate the critical limit, discontinue reduced testing until a new quality history has been established.

NOTE 5—Examples of the calculation of $\bar{r}$, *d*, and quality history and control charts are shown in Table 2 and Figs. 3 and 4. The specification limits used in these examples are hypothetical.

9.6 *Reporting for Normal Testing*—When the testing is done at the normal testing rate, report the cement as complying with the specification if it meets the specification requirements, and report it as failing to meet the specification requirements if it does not meet each of the requirements as specified.

9.7 *Reporting for Reduced Testing*—When the testing is done at the reduced testing rate, report the cement as complying with the specification if the average of the test results is further from the specified limit than the critical limit. If the average of the results for one or more requirements are between the critical limit and the specification limit, test additional samples (total equal to the number of tests at the normal rate) for that requirement, and if on completion of the additional tests, all of the results meet the specified requirements report the cement as complying with the specification. Report the cement as failing to meet the specification requirements if any test result does not conform to the respective requirements.

9.8 When a cement is reported as failing to meet the specification requirements, state in the report which requirement the cement failed and the applicable limit.

TABLE 2 Test Data, Type I Low-Alkali Cement[A]

Lot No.	Sample No.	Alkalies %	Range %	7-day Strength Average of 3 Specimens		Range	
				MPa	(psi)	MPa	(psi)
88	1	0.58		35.5	(5150)		
	13	0.61	0.03	37.0	(5358)	1.44	(208)
	17	0.57		32.2	(4675)		
	21	0.55	0.02	33.1	(4800)	0.86	(125)
91	1	0.55		32.0	(4633)		
	5	0.55	0.00	33.9	(4917)	1.95	(283)
	13	0.57		34.3	(4975)		
	21	0.54	0.03	35.2	(5108)	0.92	(133)
98	5	0.55		33.8	(4896)		
	13	0.56	0.01	34.2	(4957)	0.42	(61)
	17	0.56		35.4	(5133)		
	21	0.56	0.00	36.3	(5267)	0.92	(133)
106	5	0.42		35.6	(5158)		
	13	0.45	0.03	34.1	(4950)	1.44	(208)
	17	0.47		33.3	(4832)		
	21	0.39	0.08	32.6	(4728)	0.72	(104)
107	4	0.47		34.1	(4938)		
	8	0.46	0.01	34.8	(5042)	0.72	(104)
	12	0.40		32.3	(4683)		
	20	0.41	0.01	33.7	(4892)	1.44	(208)
111	4	0.45		36.1	(5233)		
	8	0.44	0.01	36.9	(5350)	0.80	(117)
	12	0.41		35.6	(5163)		
	20	0.40	0.01	36.2	(5246)	0.57	(83)
112	3	0.45		36.8	(5333)		
	7	0.48	0.03	34.2	(4958)	2.59	(375)
	15	0.48		34.5	(4996)		
	19	0.49	0.01	35.3	(5113)	0.80	(117)
113	2	0.49		34.0	(4937)		
	15	0.46	0.03	33.1	(4803)	0.92	(133)
	20	0.47		34.4	(4994)		
	24	0.49	0.02	34.0	(4925)	0.48	(69)
120	1	0.46		32.5	(4717)		
	6	0.46	0.00	33.2	(4814)	0.67	(98)
	11	0.46		32.2	(4675)		
	21	0.46	0.00	33.2	(4808)	0.92	(133)
123	6	0.46		36.6	(5304)		
	11	0.45	0.01	36.3	(5267)	0.26	(38)
	21	0.44		35.3	(5117)		
	26	0.44	0.00	35.8	(5196)	0.55	(79)
Total	40		0.34			19.39	2811

Calculation of Critical Limit and Control Limit

	Alkalies	Strength MPa	Strength psi
Specification limit	0.60	30.0	4350
$\bar{r}$	0.017	0.969	141
d = 2.49 $\bar{r}$	0.042	2.413	350
Critical limit	(0.60 − 0.042)	(30 + 2.4)	(4350 + 350)
	0.558	32.4	4700
3.267 $\bar{r}$	0.0555	3.17	459
Control limit	0.056	3.2	459

10. Noncompliance and Retest

10.1 If any test result fails to meet the specification requirement, the lot of cement shall not be reported as not

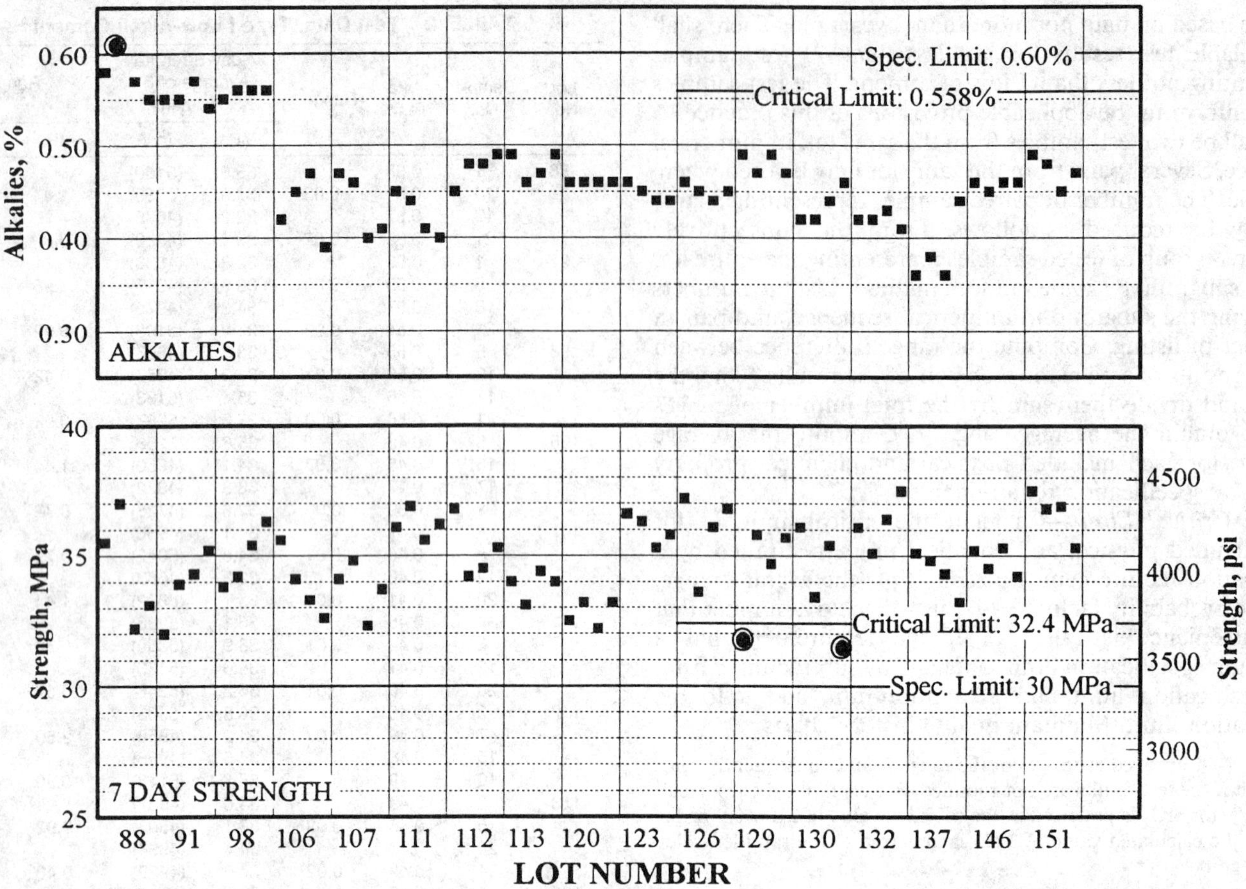

FIG. 3 Quality History Chart

complying with the specification unless noncompliance is confirmed by retest as described in 10.2.

10.2 A retest is considered to be an additional test of a certain property that is made when the initial test of that property produces a result not complying with the specification requirements. A retest may consist of either a single determination or a set of replicate determinations.

10.3 Retests shall be conducted in accordance with the provisions, if given, of the applicable specification. If no provisions are given, the following procedure shall be used:

10.3.1 Make the retest on a portion of the same sample as was used for the initial test. Use referee methods whenever they are provided for determination of the property requiring retest and in such case use only the results obtained by referee methods. The retest shall consist of the same number of determinations required for the initial test, or, if a within-laboratory precision statement is given which is based on a specified number of replicates (that is, duplicate or triplicate determinations), the number of replicates used as the basis of such precision statement. If two or more determinations are required, the value reported shall be the average of all results that are within the limits of precision of the method at the 95 % confidence level, as stated in the applicable specification or as generally recognized.

11. Keywords

11.1 hydraulic cement; sampling; testing

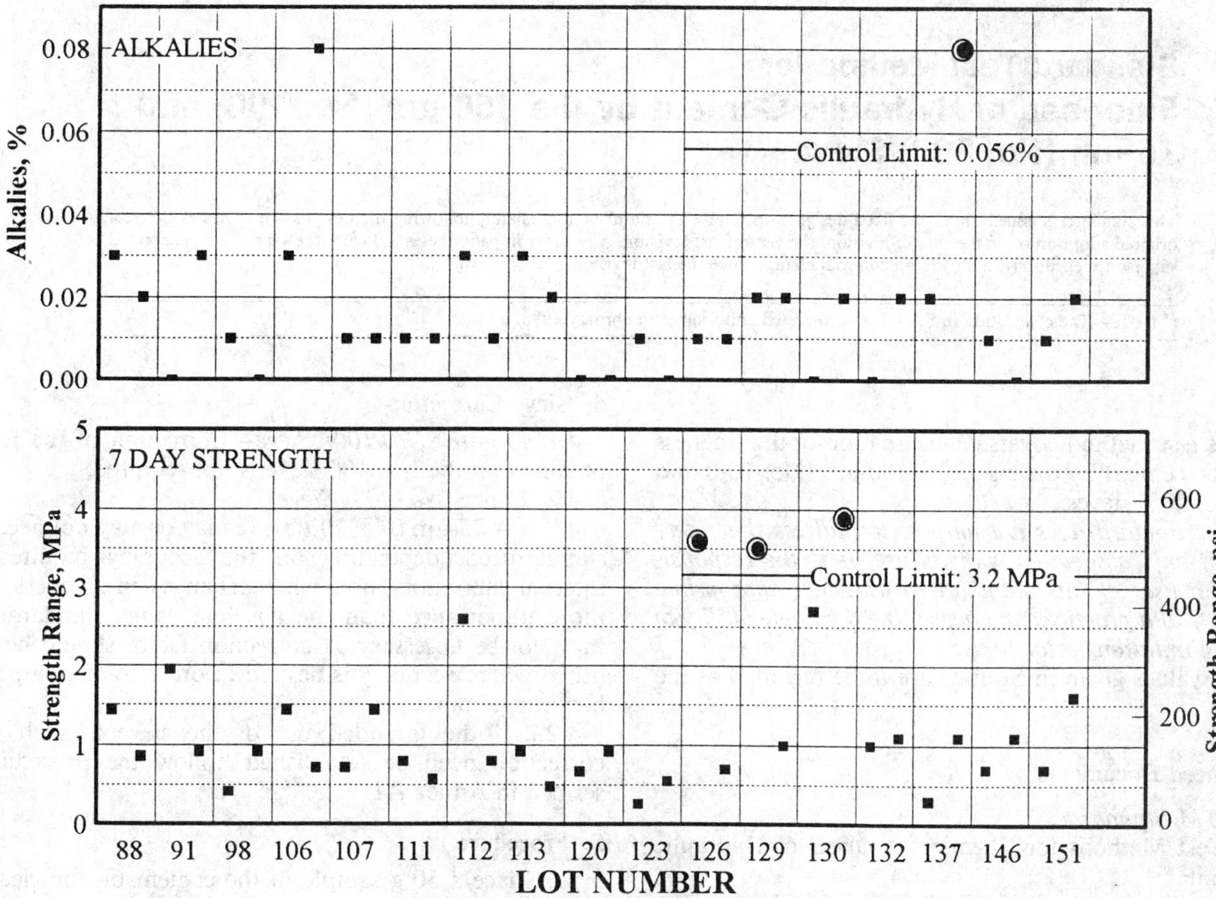

FIG. 4 Control Chart for Range

The American Society for Testing and Materials takes no position respecting the validity of any patent rights asserted in connection with any item mentioned in this standard. Users of this standard are expressly advised that determination of the validity of any such patent rights, and the risk of infringement of such rights, are entirely their own responsibility.

This standard is subject to revision at any time by the responsible technical committee and must be reviewed every five years and if not revised, either reapproved or withdrawn. Your comments are invited either for revision of this standard or for additional standards and should be addressed to ASTM Headquarters. Your comments will receive careful consideration at a meeting of the responsible technical committee, which you may attend. If you feel that your comments have not received a fair hearing you should make your views known to the ASTM Committee on Standards, 100 Barr Harbor Drive, West Conshohocken, PA 19428.

Standard Test Method for Fineness of Hydraulic Cement by the 150-μm (No. 100) and 75-μm (No. 200) Sieves[1]

This standard is issued under the fixed designation C 184; the number immediately following the designation indicates the year of original adoption or, in the case of revision, the year of last revision. A number in parentheses indicates the year of last reapproval. A superscript epsilon (ε) indicates an editorial change since the last revision or reapproval.

$^{\epsilon 1}$ NOTE—The calculation in 6.1.1.1 was corrected editorially in February 1996.

1. Scope

1.1 This test method covers determination of the fineness of hydraulic cement by means of the 150-μm (No. 100) and 75-μm (No. 200) sieves.

1.2 *This standard does not purport to address the safety concerns, if any, associated with its use. It is the responsibility of the user of this standard to establish appropriate safety and health practices and determine the applicability of regulatory limitations prior to use.*

1.3 The values given in SI units are to be regarded as the standard.

2. Referenced Documents

2.1 *ASTM Standards:*

C 114 Test Methods for Chemical Analysis of Hydraulic Cement[2]

E 11 Specification for Wire-Cloth Sieves for Testing Purposes[3]

3. Apparatus

3.1 *Sieves*—Standard 150-μm (No. 100) or 75-μm (No. 200) sieves conforming to the requirements of Specification E 11. Wire cloth for standard sieves for cement shall be woven (not twilled) from brass, bronze, or other suitable wire, and mounted without distortion on a frame about 50 mm (2 in.) below the top of the frame. The joint between the cloth and frame shall be smoothly filled with solder to prevent lodging of the cement. The sieve frame shall be circular, 203 mm (8 in.) in diameter and provided with a pan and cover.

3.2 *Balance*—The analytical balance used in fineness determinations shall conform to the requirements of 4.2.1 of Test Methods C 114.

3.3 *Weights*—The weights used in fineness determinations shall conform to the requirements of 4.2.2 of Test Methods C 114.

3.4 *Brush*—A bristle brush will be required for use in cleaning the 150-μm (No. 100) or 75-μm (No. 200) sieve. A 25 or 38-mm (1 or 1½-in.) brush with a 250-mm (10-in.) handle is a convenient size.

4. Sieve Correction

4.1 *150-μm (No. 100) Sieve*—Correction of the residues obtained on the No. 100 sieve is not required.

4.2 *75-μm (No. 200) Sieve:*

4.2.1 A 75-μm (No. 200) sieve may or may not need to be standardized, depending on the accuracy required. For internal laboratory use, where changes in fineness are of more importance than the absolute value, standardization may not be necessary. A correction factor should be established where accuracy is desired in order to compare results between laboratories.

4.2.2 If the intended use of the sieve is such that a correction need be established, follow the procedure described in Annex A1.

5. Procedure

5.1 Place a 50-g sample of the cement on the clean, dry 150-μm (No. 100) or 75-μm (No. 200) sieve with the pan attached. Do not use washers, shot, slugs, or coins on the sieve. While holding the sieve and uncovered pan in both hands, sieve with a gentle wrist motion until most of the fine material has passed through and the residue looks fairly clean. This operation usually requires only 3 or 4 min. When the residue appears clean place the cover on the sieve and remove the pan. Then, with the sieve and cover held firmly in one hand, gently tap the side of the sieve with the handle of the brush used for cleaning the sieve. Dust adhering to the sieve will thus be dislodged and the underside of the sieve may then be swept clean. Empty the pan and thoroughly wipe it out with a cloth or waste, replace the sieve in the pan, and carefully remove the cover. Return any coarser material that has been caught in the cover during the tapping to the sieve.

5.2 Continue the sieving without the cover as described in 5.1 for 5 or 10 min, depending on the condition of the cement. The gentle wrist motion involves no danger of spilling the residue, which shall be kept well spread out on the sieve. Continuously rotate the sieve throughout the sieving. This open sieving may usually be continued safely for 9 min or more, but take care that it is not continued too long. Then replace the cover and clean, following the same process as described in 5.1. If the cement is in proper condition, there should now be no appreciable dust remaining in the residue nor adhering to the sieve or pan.

5.3 Make 1-min tests as follows: Hold the sieve, with the pan and cover attached, in one hand in a slightly inclined position and move it forward and backward in the plane of inclination; at the same time gently strike the side about 150

[1] This test method is under the jurisdiction of ASTM Committee C-1 on Cement and is the direct responsibility of Subcommittee C01.25 on Fineness.

Current edition approved Sept. 15, 1994. Published November 1994. Originally published as C 184 – 44. Last previous edition C 184 – 90.

[2] *Annual Book of ASTM Standards*, Vol 04.01.

[3] *Annual Book of ASTM Standards*, Vol 14.02.

times/min against the palm of the other hand on the upstroke. Perform the sieving over a white paper. Return any material escaping from the sieve or pan and collecting on the paper to the sieve. After every 25 strokes, turn the sieve about one sixth of a revolution, in the same direction (see Note). Continue the sieving operation until not more than 0.05 g of the material passes through in 1 min of continuous sieving. Then transfer the residue on the sieve to the balance pan, taking care to brush the sieve cloth thoroughly from both sides to ensure the removal of all the residue from the sieve.

NOTE—Sieve covers may be marked with three straight lines, intersecting at 60°, through the center. If one of the lines is marked with an arrowhead and the habit of starting with this point under the right hand has been formed, one can easily follow the progress of the 1-min tests. Be certain to take a full 10 s for the 25-stroke period. This will mean that the 150 strokes, plus turning the sieve, will take more than 1 min total. It is important that a full minute of sieving takes place.

6. Calculation

6.1 Calculate the fineness of the samples as follows:

6.1.1 *When No Correction is Used:*

$$F = 100 - [(R_s \times 100)/W] \quad (1)$$

where:

F = fineness of cement expressed as the percentage passing the 150-µm (No. 100) or the 75-µm (No. 200) sieve,
R_s = residue from sample retained on the 150-µm (No. 100) or the 75-µm (No. 200) sieve, g, and
W = weight of sample, g.

6.1.1.1 *Example—150-µm (No. 100) Sieve:*

Residue from sample being tested = 0.82 g
Sample weight = 50.00 g
Amount passing 150-µm sieve, % = 100 − [(0.82 × 100)/50]
= 100 − 1.64 = 98.36 % (round to 98.4 %)

6.1.2 *Using Correction*—The correction determined by the procedure in the Annex is a point correction valid in the same general fineness area as the sample used in the standardization procedure:

$$R_c = [(R_s \times 100)/W] + C \quad (2)$$

$$F = 100 - R_c \quad (3)$$

where:

F = fineness of cement expressed as the percentage passing the 75-µm (No. 200) sieve,
R_s = residue from sample retained on the 75-µm (No. 200) sieve, g,
C = sieve correction, %, which may be either plus or minus, determined by the procedures given in the Annex,
R_c = corrected residue, %, and
W = weight of sample, g.

6.1.2.1 *Example—75-µm (No. 200) Sieve:*

Residue from sample being tested = 0.82 g
Sieve correction applied to percent of residue = −0.2 %
Sample weight = 50.00 g
Corrected residue = [(0.82 × 100)/50] −0.2 = 1.64 − 0.2 = 1.44 %
Amount passing 75-µm sieve, % = 100 − 1.44 = 98.56 %
(round to 98.6 %)

7. Report

7.1 Report the fineness as the amount passing the number sieve on which the determination was made.

8. Mechanical Sieving

8.1 Mechanical sieving devices may be used, but the cement shall not be rejected if it meets the fineness requirement when tested by the hand method described in Section 5.

9. Precision and Bias

9.1 No precision data are available due to the limited use of this test method. Therefore, users are advised to develop their own laboratory precision.

9.2 Since there is no accepted reference material suitable for determining any bias that may be associated with this test method, no statement on bias is being made.

10. Keywords

10.1 fineness; hydraulic cement; sieve

ANNEX

(Mandatory Information)

A1. METHOD FOR DETERMINATION OF SIEVE CORRECTION

A1.1 To determine the percentage of a test material passing through a nominal 75-µm (No. 200) sieve, two additional sieves are needed: a 90-µm (No. 170) and a 63-µm (No. 230) sieve. Using NIST SRM No. 1004 (No. 140 to No. 400 sieves), Glass Bead Standard, determine the effective openings of each of the three sieves.[4]

NOTE A1—NIST SRM No. 1004 tends to blind the screen during use and the beads lodged in the meshes are difficult to remove. Tapping the frame of the sieve and gentle brushing of the cloth from the underside will aid in recovering all the beads and keep the bead loss down to approximately 50 mg.

A1.2 Select a test sample of calibration. It is important that the sample for calibration be of the same typical fineness as the range at which the sieve is to be used. Determine the percent residue of the test material on each of the three sieves, in accordance with Section 5. Plot the average percent residues *versus* the effective openings determined above. The percent residue at the nominal opening of the 75-µm (No. 200) sieve can be read off the plot and the difference

[4] Available from the National Institute of Standards and Technology, Gaithersburg, MD 20899.

between the actual obtained residue and the nominal percent residue at 75-µm may be used algebraically as a correction for that sieve only in the same general fineness area.

A1.3 *Example—Determination of Sieve Correction Factor for 75-µm (No. 200) Sieve*

A1.3.1 Using the SRM 1004, the following was determined:

(*a*) Effective opening of nominal sieves being corrected:

75-µm (No. 200)	72 µm
90-µm (No. 170)	90 µm
63-µm (No. 230)	65 µm

(*b*) Test sample residue, % (to be plotted on graph paper):

at 90 µm	1.1 %
at 72 µm	2.3 %
at 65 µm	2.8 %

(*c*) From the above plot:

Residue at 75-µm = 2.1 %

A1.3.2 Since the sieve is woven too tighly and retains more than it should if it were a 75-µm opening, the correction, *C*, to be applied to the residue in percent is −0.2 %.

For additional useful information on details of cement test methods, reference may be made to the "Manual of Cement Testing," *Annual Book of ASTM Standards,* **Vol 04.01.**

times/min against the palm of the other hand on the upstroke. Perform the sieving over a white paper. Return any material escaping from the sieve or pan and collecting on the paper to the sieve. After every 25 strokes, turn the sieve about one sixth of a revolution, in the same direction (see Note). Continue the sieving operation until not more than 0.05 g of the material passes through in 1 min of continuous sieving. Then transfer the residue on the sieve to the balance pan, taking care to brush the sieve cloth thoroughly from both sides to ensure the removal of all the residue from the sieve.

NOTE—Sieve covers may be marked with three straight lines, intersecting at 60°, through the center. If one of the lines is marked with an arrowhead and the habit of starting with this point under the right hand has been formed, one can easily follow the progress of the 1-min tests. Be certain to take a full 10 s for the 25-stroke period. This will mean that the 150 strokes, plus turning the sieve, will take more than 1 min total. It is important that a full minute of sieving takes place.

6. Calculation

6.1 Calculate the fineness of the samples as follows:

6.1.1 *When No Correction is Used:*

$$F = 100 - [(R_s \times 100)/W] \tag{1}$$

where:

F = fineness of cement expressed as the percentage passing the 150-µm (No. 100) or the 75-µm (No. 200) sieve,

R_s = residue from sample retained on the 150-µm (No. 100) or the 75-µm (No. 200) sieve, g, and

W = weight of sample, g.

6.1.1.1 *Example—150-µm (No. 100) Sieve:*

Residue from sample being tested = 0.82 g
Sample weight = 50.00 g
Amount passing 150-µm sieve, % = 100 − [(0.82 × 100)/50]
= 100 − 1.64 = 98.36 % (round to 98.4 %)

6.1.2 *Using Correction*—The correction determined by the procedure in the Annex is a point correction valid in the same general fineness area as the sample used in the standardization procedure:

$$R_c = [(R_s \times 100)/W] + C \tag{2}$$

$$F = 100 - R_c \tag{3}$$

where:

F = fineness of cement expressed as the percentage passing the 75-µm (No. 200) sieve,

R_s = residue from sample retained on the 75-µm (No. 200) sieve, g,

C = sieve correction, %, which may be either plus or minus, determined by the procedures given in the Annex,

R_c = corrected residue, %, and

W = weight of sample, g.

6.1.2.1 *Example—75-µm (No. 200) Sieve:*

Residue from sample being tested = 0.82 g
Sieve correction applied to percent of residue = −0.2 %
Sample weight = 50.00 g
Corrected residue = [(0.82 × 100)/50] −0.2 = 1.64 − 0.2 = 1.44 %
Amount passing 75-µm sieve, % = 100 − 1.44 = 98.56 %
(round to 98.6 %)

7. Report

7.1 Report the fineness as the amount passing the number sieve on which the determination was made.

8. Mechanical Sieving

8.1 Mechanical sieving devices may be used, but the cement shall not be rejected if it meets the fineness requirement when tested by the hand method described in Section 5.

9. Precision and Bias

9.1 No precision data are available due to the limited use of this test method. Therefore, users are advised to develop their own laboratory precision.

9.2 Since there is no accepted reference material suitable for determining any bias that may be associated with this test method, no statement on bias is being made.

10. Keywords

10.1 fineness; hydraulic cement; sieve

ANNEX

(Mandatory Information)

A1. METHOD FOR DETERMINATION OF SIEVE CORRECTION

A1.1 To determine the percentage of a test material passing through a nominal 75-µm (No. 200) sieve, two additional sieves are needed: a 90-µm (No. 170) and a 63-µm (No. 230) sieve. Using NIST SRM No. 1004 (No. 140 to No. 400 sieves), Glass Bead Standard, determine the effective openings of each of the three sieves.[4]

NOTE A1—NIST SRM No. 1004 tends to blind the screen during use and the beads lodged in the meshes are difficult to remove. Tapping the frame of the sieve and gentle brushing of the cloth from the underside will aid in recovering all the beads and keep the bead loss down to approximately 50 mg.

A1.2 Select a test sample of calibration. It is important that the sample for calibration be of the same typical fineness as the range at which the sieve is to be used. Determine the percent residue of the test material on each of the three sieves, in accordance with Section 5. Plot the average percent residues *versus* the effective openings determined above. The percent residue at the nominal opening of the 75-µm (No. 200) sieve can be read off the plot and the difference

[4] Available from the National Institute of Standards and Technology, Gaithersburg, MD 20899.

between the actual obtained residue and the nominal percent residue at 75-μm may be used algebraically as a correction for that sieve only in the same general fineness area.

A1.3 *Example—Determination of Sieve Correction Factor for 75-μm (No. 200) Sieve*

A1.3.1 Using the SRM 1004, the following was determined:

(*a*) Effective opening of nominal sieves being corrected:

75-μm (No. 200)	72 μm
90-μm (No. 170)	90 μm
63-μm (No. 230)	65 μm

(*b*) Test sample residue, % (to be plotted on graph paper):

at 90 μm	1.1 %
at 72 μm	2.3 %
at 65 μm	2.8 %

(*c*) From the above plot:

Residue at 75-μm = 2.1 %

A1.3.2 Since the sieve is woven too tighly and retains more than it should if it were a 75-μm opening, the correction, *C*, to be applied to the residue in percent is −0.2 %.

For additional useful information on details of cement test methods, reference may be made to the "Manual of Cement Testing," *Annual Book of ASTM Standards,* **Vol 04.01.**

The American Society for Testing and Materials takes no position respecting the validity of any patent rights asserted in connection with any item mentioned in this standard. Users of this standard are expressly advised that determination of the validity of any such patent rights, and the risk of infringement of such rights, are entirely their own responsibility.

This standard is subject to revision at any time by the responsible technical committee and must be reviewed every five years and if not revised, either reapproved or withdrawn. Your comments are invited either for revision of this standard or for additional standards and should be addressed to ASTM Headquarters. Your comments will receive careful consideration at a meeting of the responsible technical committee, which you may attend. If you feel that your comments have not received a fair hearing you should make your views known to the ASTM Committee on Standards, 100 Barr Harbor Drive, West Conshohocken, PA 19428.

Designation: C 185 – 99

American Association State Highway and Transportation Officials Standard
AASHTO No.: T137

Standard Test Method for Air Content of Hydraulic Cement Mortar[1]

This standard is issued under the fixed designation C 185; the number immediately following the designation indicates the year of original adoption or, in the case of revision, the year of last revision. A number in parentheses indicates the year of last reapproval. A superscript epsilon (ϵ) indicates an editorial change since the last revision or reapproval.

This standard has been approved for use by agencies of the Department of Defense.

1. Scope

1.1 This test method covers the determination of the air content of hydraulic cement mortar under the conditions hereinafter specified.

1.2 The values stated in SI units are to be regarded as the standard.

1.3 Values in SI shall be obtained by measurement in SI units or by appropriate conversion, using the Rules for Conversion and Rounding in IEEE/ASTM SI 10, of measurements made in other units.

1.4 *This standard does not purport to address all of the safety concerns, if any, associated with its use. It is the responsibility of the user of this standard to establish appropriate safety and health practices and determine the applicability of regulatory limitations prior to use.*

2. Referenced Documents

2.1 *ASTM Standards:*

C 91 Specification for Masonry Cement[2]

C 109 Test Method for Compressive Strength of Hydraulic-Cement Mortars (Using 2-in. or [50-mm] Cube Specimens)[2]

C 183 Practice for Sampling and the Amount of Testing of Hydraulic Cement[2]

C 230 Specification for Flow Table for Use in Tests of Hydraulic Cement[2]

C 305 Practice for Mechanical Mixing of Hydraulic Cement Pastes and Mortars of Plastic Consistency[2]

C 511 Specification for Moist Cabinets, Moist Rooms, and Water Storage Tanks Used in the Testing of Hydraulic Cements and Concretes[2]

C 778 Specification for Standard Sand[2]

C 1005 Specification for Weights and Weighing Devices for Use in the Physical Testing of Hydraulic Cements[2]

E 438 Specification for Glasses in Laboratory Apparatus[3]

E 694 Specification for Laboratory Glass Volumetric Apparatus[3]

IEEE/ASTM SI 10 Standard for Use of the International System of Units (SI): The Modern Metric System[3]

3. Summary of Test Method

3.1 Prepare a mortar with standard sand and the cement to be tested, using a water content sufficient to give a required flow. Compact the mortar into a measure of known volume and determine mass. Calculate the air content from the measured density of the mortar, the known densities of the constituents, and the mixture proportions.

4. Significance and Use

4.1 The purpose of this test method is to determine whether or not the hydraulic cement under test meets the air-entraining or non-air-entraining requirements of the applicable hydraulic cement specification for which the test is being made. The air content of concrete is influenced by many factors other than the potential of the cement for air entrainment.

5. Apparatus

5.1 *Flow Table, Flow Mold, and Caliper*, shall conform to Specification C 230.

5.2 *Measure*—A cylindrical measure having an inside diameter of 76 ± 2 mm and a depth (approximately 88 mm) adjusted by standardization with water to contain 400 ± 1 mL at 23.0 ± 2.0°C (Note 1). For the purposes of this test, the capacity of the measure in millilitres is the mass of the water content of the measure, in grams, divided by 0.9976, no correction in mass being made for the buoyant effect of air. The measure shall have a uniform wall thickness. The thickness of the wall and bottom shall not be less than 2.9 mm. The total mass of the empty measure shall not be more than 900 g. The measure shall be made of a metal not attacked by the cement mortar.

NOTE 1—Calibrate the 400-mL measure by filling with distilled water at 23.0± 2.0°C to a point where the meniscus extends appreciably above the top of the measure, placing a clean piece of plate glass on the top of the measure, and allowing the excess water to be squeezed out. The absence of air bubbles as seen through the glass ensures that the measure is completely full. Care shall be taken that the excess water is wiped from the sides of the container before weighing.

5.3 *Mixer, Bowl, and Paddle*, shall conform to the apparatus section of Practice C 305.

5.4 *Straightedge*—A steel straightedge not less than 200

---

[1] This test method is under the jurisdiction of ASTM Committee C-1 on Cement and is the direct responsibility of Subcommittee C01.21 on Air Entrainment.

Current edition approved Jan. 10, 1999. Published May 1999. Originally published as C 185 – 44 T. Last previous edition C 185 – 97.

[2] *Annual Book of ASTM Standards*, Vol 04.01.

[3] *Annual Book of ASTM Standards*, Vol 14.02.

mm long and not less than 1.5 mm nor more than 3.5 mm in thickness.

5.5 *Weights and Weighing Devices*, shall conform to Specification C 1005. Evaluate the weighing device for precision and accuracy at a total load of 2 kg.

5.6 *Glass Graduates*—Glass graduates of 250-mL capacity, conforming to the requirements of Specifications E 438 and E 694.

5.7 *Tamper*, conforming to the requirements of Test Method C 109. The tamping face of the tamper shall be flat and at right angles to the length of the tamper.

5.8 *Tapping Stick*, a piece of hard wood having a diameter of approximately 16 mm and a length of approximately 152 mm.

5.9 *Spoon*—A metal restaurant-type serving spoon not less than 230 mm in length and with a bowl approximately 100 mm in length.

6. Temperature and Humidity

6.1 Maintain the temperature of the room and dry materials at 23 ± 3°C.

6.2 Condition the mixing water and the measure, if it is being calibrated at 23 ± 2°C.

6.3 Maintain the relative humidity of the laboratory at not less than 50 %.

7. Standard Sand

7.1 Use sand conforming to the requirements of Specification C 778 for 20–30 sand.

8. Sampling

8.1 Sample the cement in accordance with Practice C 183.

9. Procedure

9.1 *Batch*—Proportion the standard mortar using 350 g cement to 1400 g 20–30 standard sand and sufficient water to give a flow of 87½ ± 7½ % when determined in accordance with 9.3.

NOTE 2—Masonry cements require different sand, sample mass, and flow. Refer to Specification C 91.

9.2 *Mixing of Mortar*—Mix the mortar in accordance with Practice C 305.

9.3 *Flow Determination*—Carefully wipe dry the flow-table top and place the flow mold at the center of it. Using the spoon, place a layer of mortar about 25 mm in thickness in the mold and tamp 20 times with the tamper. The tamping pressure shall be just sufficient to ensure uniform filling of the mold. Overfill the top of the mold approximately 20 mm with mortar and tamp as specified for the first layer. Then cut off the mortar to a plane surface, flush with the top of the mold, by drawing the straightedge with a sawing motion across the top of the mold. Wipe the flow table top clean and dry, being especially careful to remove any water from around the edge of the mold. Lift the mold away from the mortar 1 min after completing the mixing operation. Immediately drop the table 10 times in accordance with Specification C 230. The flow is the resulting increase in average diameter of the mortar mass, as determined with the calipers, measured on at least four diameters at approximately equispaced intervals, expressed as a percentage of the original diameter. Make trial mortars with varying percentages of water until the specified flow is obtained. Make each trial with fresh mortar.

9.4 *Mass per 400 mL of Mortar*—When the quantity of mixing water has been found that produces a flow of 87½ ± 7½ %, immediately determine the mass per 400 mL of mortar, using the mortar remaining in the mixing bowl after the flow has been determined. In the determination of the mass per 400 mL, do not use the portion of the mortar used in the flow determination. Using the spoon, place the mortar gently into the 400–ml measure in three equal layers. Tamp each layer 20 times around the inner surface of the measure. For the final layer of mortar, overfill the 400–ml measure approximately 20 mm. The position of the tamper shall be that: the broad side of the tamper is parallel to the radius and is perpendicular to the inner surface of the measure. Each layer is tamped in one complete revolution (rotation) with only sufficient pressure to adequately fill the measure and eliminate voids within the mortar. After the measure has been filled and tamped in the above prescribed manner, tap the sides of the measure lightly with the side of the tapping stick, one each at five different points at approximately equal spacing around the outside of the measure, in order to preclude entrapment of extraneous air (Note 3). No obvious space shall be left between the mortar and the inner surface of the measure as a result of the tamping operation. Then cut the mortar off to a plane surface, flush with the top of the measure, by drawing the straightedge with a sawing motion across the top of the measure, making two passes over the entire surface, the second pass being made at right angles to the first. If, in the striking-off operation, loose sand grains cause the straightedge to ride above the top surface of the measure, these grains shall be removed, and the operation repeated. Complete the entire operation of filling and striking off the measure within 1½ min. Wipe off all mortar and water adhering to the outside of the measure. Determine the mass of the measure and its contents. Subtract the mass of the container, and record the mass of the mortar in grams.

NOTE 3—This operation may be facilitated by placing the measure on a steady flat-surfaced support of lesser diameter than the measure while filling and wiping.

10. Calculation

10.1 Calculate the air content of the mortar from the following formula which is based on the batch proportions given in 9.1, taking the specific gravity of portland cement as 3.15 and of 20–30 standard sand as 2.65. When the hydraulic cement is other than portland, the appropriate value for its specific gravity shall be substituted for the value of 3.15 and the formula rederived accordingly:

$$\text{Air content, volume \%} = 100 - W[(182.7 + P)/(2000 + 4P)] \quad (1)$$

where:

W = mass of 400 mL of mortar, g, and

P = percentage of mixing water, based on mass of cement used.

NOTE 4—This formula is derived as follows:

$$\text{Air content, volume \%} = 100[1 - (W_a/W_c)] \quad (2)$$

where:

W_a = actual mass per unit of volume as determined by this test method.
= W/400 g/mL, where W is the mass in grams of the specified 400 mL of mortar (9.4),

W_c = theoretical mass per unit of volume, calculated on an air-free basis as follows and using the values for quantities of materials and specific gravities as given in 9.1 and 10.1,

$$= \frac{350 + 1400 + 350 \times P \times 0.01}{\frac{350}{3.15} + \frac{1400}{2.65} + \frac{350 \times P \times 0.01}{1}}$$

$$= \frac{(5 + 0.01P)}{(1.827 + 0.01P)}, \text{ and}$$

P = percentage of mixing water, based on mass of cement.

Substituting for W_a and W_c we have:

$$\text{Air content, volume \%} = 100\left(1 - \frac{W}{400} \times \frac{1.827 + 0.01P}{5 + 0.01P}\right)$$

$$\text{Air content, volume \%} = 100 - \frac{W}{4} \times \frac{(1.827 + 0.01P)}{(5 + 0.01P)}$$

$$\text{Air content, volume \%} = 100 - 2.5W\frac{(182.7 + P)}{(5000 + 10P)}$$

$$\text{Air content, volume \%} = 100 - W\frac{(182.7 + P)}{(2000 + 4P)}$$

10.2 Make only one determination of air content on a batch.

NOTE 5—Difficulty has occasionally been experienced with this test method by some persons. These difficulties usually have taken the form of values for air content that are abnormally high and that may be greater than the specified maxima in specifications for hydraulic cement. The air-entraining potential of the sand may be reduced by washing in the laboratory with potable water by decantation until a clear, colorless supernatant liquid is obtained. In case of dispute, such freshly washed sand shall be used.

11. Report

11.1 When this test method is used for specific action in compliance testing, report the value for the air content to the nearest 1 %.

12. Precision and Bias

12.1 The single-operator, within-laboratory, standard deviation has been found to be 0.56 % air content throughout the range of 8 to 19 % air. Therefore, results of two properly conducted tests by the same operator on similar batches should not differ by more than 1.6 % air.

12.2 The multilaboratory standard deviation has been found to be 1.0 % air content throughout the range from 8 to 19 % air. Therefore, results of two different laboratories on similar batches should not differ from each other by more than 2.8 % air.

13. Keywords

13.1 air content; hydraulic cement mortar

For additional useful information on details of cement tests methods, references may be made to the "Manual of Cement Testing," which appears in the *Annual Book of ASTM Standards,* Vol 04.01.

Designation: C 186 – 98

Standard Test Method for Heat of Hydration of Hydraulic Cement[1]

This standard is issued under the fixed designation C 186; the number immediately following the designation indicates the year of original adoption or, in the case of revision, the year of last revision. A number in parentheses indicates the year of last reapproval. A superscript epsilon (ε) indicates an editorial change since the last revision or reapproval.

1. Scope

1.1 This test method covers the determination of the heat of hydration of a hydraulic cement by measuring the heat of solution of the dry cement and the heat of solution of a separate portion of the cement that has been partially hydrated for 7 and for 28 days, the difference between these values being the heat of hydration for the respective hydrating period.

1.2 The results of this test method may be inaccurate if some of the components of the hydraulic cement are insoluble in the nitric acid/hydrofluoric acid solution.

1.3 The values stated in SI units are to be regarded as the standard. The values given in parentheses are for information only.

1.4 Values in SI units shall be obtained by measurement in SI units or by appropriate conversion, using the Rules for Conversion and Rounding given in Standard IEEE/ASTM SI 10, or measurements made in other units.

1.5 *This standard does not purport to address all of the safety concerns, if any, associated with its use. It is the responsibility of the user of this standard to establish appropriate safety and health practices and determine the applicability or regulatory limitations prior to use.*

2. Referenced Documents

2.1 *ASTM Standards:*

C 109 Test Method for Compressive Strength of Hydraulic Cement Mortars (Using 2-in. or 50-mm Cube Specimens)[2]

C 114 Test Methods for Chemical Analysis of Hydraulic Cement[2]

C 670 Practice for Preparing Precision and Bias Statements for Test Methods for Construction Materials[3]

C 1005 Specification for Weights and Weighing Devices for Use in the Physical Testing of Hydraulic Cements[2]

E 11 Specification for Wire-Cloth Sieves for Testing Purposes[4]

IEEE/ASTM SI 10 Standard for Use of the International System of Units (SI): The Modern Metric System[4]

3. Significance and Use

3.1 The purpose of this test is to determine if the hydraulic cement under test meets the heat of hydration requirement of the applicable hydraulic cement specification.

3.2 This test may also be used for research purposes when it is desired to determine the heat of hydration of hydraulic cement at any age.

NOTE 1—When tests are performed for research purposes, useful additional information can be obtained by determining fineness, chemical and compound compositions.

3.3 Determination of the heat of hydration of hydraulic cements provides information that is helpful for calculating temperature rise in mass concrete.

4. Apparatus

4.1 *Calorimetric Apparatus:*

4.1.1 *Calorimeter*—The calorimeter, such as that illustrated in Fig. 1 shall consist of a 0.5-L (1-pt), wide-mouth vacuum jar, with cork stopper, or other suitable non-reactive stopper held in a suitably insulated container (see 4.1.2) to keep the vacuum jar in position and to protect the jar from undue temperature fluctuations. The vacuum jar shall be coated on the interior with a material resistant to hydrofluoric acid, such as a baked phenolic resin, a baked vinyl chloride acetate resin, or beeswax. The acid-resistant coating shall be intact and free of cracks at all times; it shall be examined frequently and renewed whenever necessary. As another means of protecting the vacuum jar, a plastic liner of suitable size may be used instead of coating the interior of the jar. The contents of the vacuum jar shall not change more than 0.001°C/min per degree difference from room temperature when filled with 425 g of the acid specified in 6.2, stoppered, and allowed to stand unstirred for 30 min. The temperature for this check shall approximate the starting temperatures to be used in making the determination.

4.1.2 *Insulated Container*—The container shall have an insulating layer of a material such as non-reactive foam, cotton, or fiber-glass, which shall be at least 25 mm (1 in.) in thickness and shall encase the sides and bottom of the vacuum jar, but shall be so arranged as to permit easy removal of the jar.

4.1.3 *Differential and Reference Thermometers*—The adjustable differential thermometer shall be of the Beckmann-type, graduated at least to 0.01°C, and shall have a range of approximately 6°C. The thermometer shall be so adjusted that the upper limit of the scale approximates room temperature. The portion of the thermometer that will rest inside the calorimeter shall be protected with a coating resistant to hydrofluoric acid (see 4.1.1). The thermometer shall be equipped with a suitable reading lens. The Beckmann thermometer zero must be determined by immersion in a liquid and comparison with the reference thermometer

[1] This test method is under the jurisdiction of ASTM Committee C-1 on Cement and is the direct responsibility of Subcommittee C01.26 on Heat of Hydration.

Current edition approved Jan. 10, 1998. Published July 1998. Originally published as C 186 – 44 T. Last previous edition C 186 – 97.

[2] *Annual Book of ASTM Standards*, Vol 04.01.

[3] *Annual Book of ASTM Standards*, Vol 04.02.

[4] *Annual Book of ASTM Standards*, Vol 14.02.

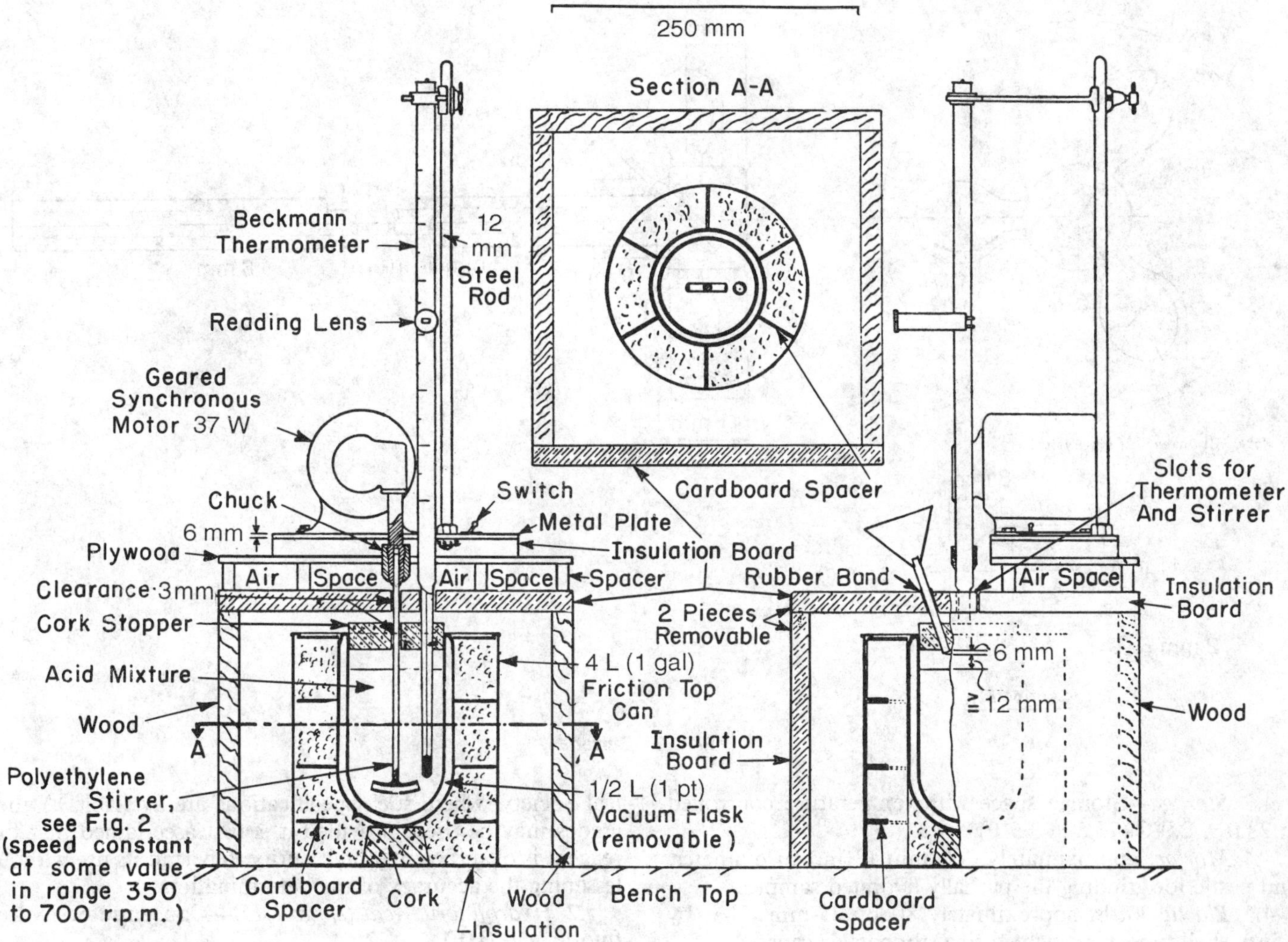

Metric Units	Inch-Pound Equivalents
250 mm	10 in.
12 mm	1/2 in.
37 W	1/20 hp

Metric Units	Inch-Pound Equivalents
6 mm	1/4 in.
3 mm	1/8 in.
0.47 litre	1 pt

FIG. 1 Calorimeter

described. An accurate reference thermometer of the appropriate range and having 0.1°C divisions shall be placed in the proximity of the calorimetric apparatus and shall be used for room temperature readings and for establishing the Beckmann thermometer zero.

4.1.4 *Funnel*—The funnel through which the sample is introduced into the calorimeter shall be of glass or plastic and shall have a stem approximately 75 mm (3 in.) in length and with an inside diameter of not less than 6 mm (1/4 in.).

4.1.5 *Stirring Assembly*—The stirrer shall be a three-bladed polyethylene propeller having the dimensions shown in Fig. 2, and shall extend as closely as possible to the bottom of the calorimeter. The motor shall be of the constant-speed type, at least 37 W (1/20 hp), and shall be equipped with a geared speed reducer so that one speed, in the range of 350 to 700 r/min, can be maintained constant.

NOTE 2—The stirrer shown in Fig. 2 may be readily made from a commercially available three-bladed polyethylene propeller having a propeller diameter of 34 mm (1 3/8 in.), shaft diameter of 6 mm (1/4 in.), and a shaft length of approximately 455 mm (18 in.). The function of the stirrer is two-fold: to maintain uniform temperature throughout the liquid and to supply sufficient agitation to keep the solid in suspension in the acid mixture. Since a stirrer capable of keeping the solid in suspension generates considerable heat in the calorimeter, it is important that the stirrer speed, and hence the rate of heat generation, be maintained constant. Because such constancy is difficult to achieve with other types of motors, a synchronous motor with a geared speed reducer is recommended.

4.2 *Mixer*—A moderate-speed mechanical mixer, such as a milk-shake type stirrer, capable of intimately mixing the cement and water to a uniform paste.

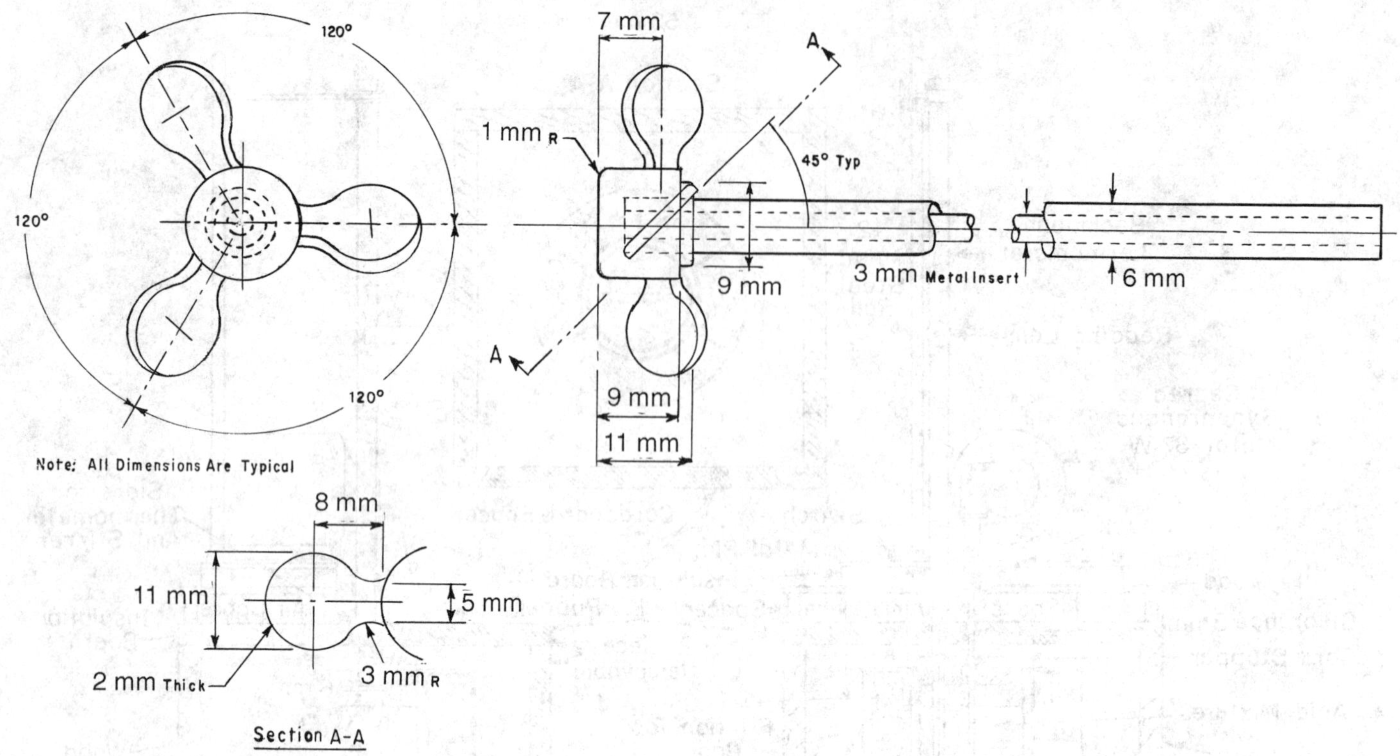

FIG. 2 Stirrer

4.3 *Storage*—Storage space with temperature controlled at 23.0 ± 2.0°C (73.5 ± 3.5°F).

4.4 *Mortar*, approximately 200 mm (8 in.) in diameter, and pestle for grinding the partially hydrated samples.

4.5 *Plastic Vials*, approximately 80 by 25-mm (3⁵⁄₃₂ by 1-in.), shell-type, with tight-fitting stoppers or caps.

4.6 *Drying Oven*, maintained at 100 to 110°C.

4.7 *Weighing Bottles*, approximately 40 mm high and 25 mm wide, with matching stoppers.

4.8 *Stop Watch or Clock Timer.*

4.9 *Sieves*, 150-μm (No. 100) and 850-μm (No. 20), conforming to Specification E 11.

4.10 *Crucibles*, platinum, 30-mL capacity, with covers, for loss on ignition determination.

4.11 *Muffle Furnace*, or suitable burners capable of maintaining a temperature of 900 to 950°C.

4.12 *Analytical Balance and Analytical Weights*, conforming to the requirements prescribed in Test Methods C 114 for weighing out calorimetric samples and for loss on ignition weighings.

4.13 *Weights and Weighing Devices*, conforming to the requirements of Specification C 1005. The weighing device shall be evaluated at a total load of 1000 g.

5. Reagents and Materials

5.1 *Purity of Reagents*—Reagent grade chemicals shall be used in all tests. Unless otherwise indicated, it is intended that all reagents shall conform to the specifications of the Committee on Analytical Reagents of the American Chemical Society, where such specifications are available.[5] Other grades may be used, provided it is first ascertained that the reagent is of sufficiently high purity to permit its use without lessening the accuracy of the determination.

5.2 *Hydrofluoric Acid (sp gr 1.15)*—Concentrated hydrofluoric acid (HF).

5.3 *Nitric Acid (2.00 N)*—The 2.00 *N* HNO_3, for use in the calorimeter, shall be prepared and standardized in large quantities. Optionally, the dilute HNO_3 may be made up with 127 mL of concentrated HNO_3 (sp gr 1.42) per litre of solution, provided that heat capacity determinations are made with each batch of diluted HNO_3 so prepared.

5.4 *Wax*—Paraffin wax, or other suitable wax, for sealing vials.

5.5 *Zinc Oxide (ZnO)*—The ZnO shall be heated at 900 to 950°C for 1 h, then cooled in a desiccator, ground to pass a 150-μm (No. 100) sieve, and stored. Immediately prior to a heat capacity determination, 7 g of the ZnO so prepared shall be heated for not more than 5 min at 900 to 950°C, cooled to room temperature in a desiccator, and weighed accurately for introduction into the calorimeter.

NOTE 3—The rate of solution of the ZnO varies with the preliminary treatment. The procedure described results in a product which dissolves at about the same rate as the dry cement.

[5] *Reagent Chemicals, American Chemical Society Specifications*, American Chemical Society, Washington, DC. For suggestions on the testing of reagents not listed by the American Chemical Society, see *Analar Standards for Laboratory Chemicals*, BDH Ltd., Poole, Dorset, U.K., and the *United States Pharmacopeia and National Formulary*, U.S. Pharmacopeial Convention, Inc. (USPC), Rockville, MD.

6. Determination of Heat Capacity of Apparatus

6.1 To determine the heat capacity of the system (that is, the number of joules or calories required to raise the temperature of the calorimeter and contents 1°C), measure the corrected temperature rise obtained by dissolving 7 g of ignited ZnO in the specified acid mixture (see 6.2 to 6.7).

6.2 Transfer approximately 400 g of the 2.00 *N* HNO_3, which has been cooled to the temperature indicated by the lower range of the Beckmann thermometer (ordinarily about 4 to 5°C below room temperature), into the vacuum jar, add 8.0 mL of HF (sp gr 1.15), weigh, and add sufficient additional 2.00 *N* HNO_3 to bring the total weight of the solution to 425.0 g. Then, assemble the calorimeter and start the stirring motor. Take care that the stirrer blades or shaft do not touch the thermometer, the sides or bottom of the jar, or the cork stopper. The lower end of the funnel stem shall extend approximately 6 mm (¼ in.) below the lower surface of the stopper and at least 12 mm (½ in.) above the level of the liquid. The upper end of the bulb of the Beckmann thermometer shall be at least 38 mm (1½ in.) below the surface of the liquid. Place it at the same depth in all determinations. After an initial stirring period of at least 20 min to allow the temperature of the system to become uniform, record the temperature of the room to the nearest 0.1°C, the temperature of the acid to the nearest 0.001°C, record the time, and then immediately introduce the prepared ZnO through the funnel at a uniform rate (see Note 4). Complete the introduction of the ZnO in not less than 1 or more than 2 min. Brush any ZnO clinging to the funnel stem into the acid mixture by means of a small "camel's-hair" brush.

NOTE 4—The temperature of the sample shall be identical with that of the room when the sample is introduced into the calorimeter.

6.3 Read the temperature, to the nearest 0.001°C, at 20 min and again at 40 min after beginning the introduction of the sample. The first 20-min period is the uncorrected temperature rise, which covers the solution period. The second 20-min period is the rating period, and the temperature difference between the 20 and 40-min readings is the correction to be added to or subtracted from the uncorrected temperature rise, according to whether the calorimeter temperature rises or falls during the rating period.

6.4 Calculate the corrected temperature rise as follows:

$$R_o = \theta_{20} - \theta_o$$
$$R = R_o - (\theta_{40} - \theta_{20}) \quad (1)$$

where:

R_o = observed temperature rise, °C,
θ_{20} = calorimeter temperature at the end of the solution period,
θ_o = calorimeter temperature when sample was introduced,
R = corrected temperature rise, °C, and
θ_{40} = calorimeter temperature at the end of the rating period.

6.5 Calculate the heat capacity of the calorimeter and contents as follows (see Note 5):

$$C = \frac{W[1072 + 0.4(30 - t) + 0.5(T - t)]}{R} \quad (2)$$

where:

C = heat capacity, kJ/°C,
W = mass of ZnO, g,
t = final temperature of the calorimeter, °C (θ_{20} plus temperature, °C, at which the Beckmann thermometer reading is zero),
T = temperature of the ZnO (room temperature), °C, when introduced into the calorimeter, and
R = corrected temperature rise, °C.

NOTE 5—The heat of solution of ZnO is 1072 kJ/kg (256.1 cal/g) at 30°C. This value increases 0.4 kJ/kg (0.1 cal/g) for each degree decrease in temperature below 30°C. The heat capacity of ZnO is 0.5 kJ/kg·K (0.12 cal/g·°C). The heat required to bring the ZnO to the final temperature of the calorimeter must be included in the effective heat of solution.

6.6 If more than a trace of ZnO is found adhering to the tip of the funnel or to the stopper when the calorimeter is opened, reject the test.

6.7 Redetermine the heat capacity at the following times:

6.7.1 When the Beckmann thermometer is reset,

6.7.2 When a new coating is applied to thermometer, stirrer, or flask,

6.7.3 When a new thermometer, stirrer, or flask is put in service,

6.7.4 When a new batch of acid is used, and

6.7.5 At other times when, according to the judgment of the operator, the need is indicated.

7. Sampling and Test Specimens

7.1 *Preparation of Cement Paste*—Store the cement and the mixing water in a constant-temperature room at 23.0 ± 2.0 °C (73.5 ± 3.5°F) until the materials are at ambient temperature before preparation of the paste. Mix 150 g of cement and 60 mL of distilled water by means of a spatula, and then vigorously stir the mixture with a mechanical stirrer for 5 min. Place approximately equal representative portions of the paste in four or more plastic vials, filling the vials to within about 13 mm (½ in.) of the top. Immediately after filling the vials, close them with tight-fitting stoppers or caps. (If there is any doubt regarding the tightness of the seal, the sealed ends of the vials should be dipped in molten paraffin wax.) Store the vials in an upright position in a water bath at 23 ± 2.0°C until the time of test.

7.2 *Preparation of Partially Hydrated Sample for Heat of Solution Test*—At the specified age of test or age of interest, remove a vial of the partially hydrated sample from storage within the test time tolerances of Test Method C 109, and, during a 20-min initial stirring period of the calorimeter, break the plastic away from the sample and rapidly crush the entire sample with a mortar and pestle so that all the material will pass through a 850-µm (No. 20) sieve; then quickly place the sample in a well-stoppered weighing bottle. Take care, particularly with the 7-day partially hydrated sample, to expose the sample to the air as little as possible, and thus minimize the action of CO_2 or the loss of moisture from the sample.

8. Procedure

8.1 *Calorimetric Procedure, Dry Cement*—Determine the heat of solution of the dry cement sample according to the procedure described for the heat capacity determination (Section 6), but use a 3-g sample (weighed to the nearest

0.001 g) of the dry cement instead of the prepared ZnO (see Note 4). (Exercise care in securing a uniform and representative sample.) Calculate and report the results on the ignited mass basis (8.3). Perform the heat of solution test on the dry cement just prior to the test on the corresponding 7-day partially hydrated sample.

8.2 *Calorimetric Procedure, Partially Hydrated Sample*—For the heat of solution of the partially hydrated sample follow the same procedure as for the dry cement described in 8.1, but use a 4.18 ± 0.05-g calorimetric sample of the partially hydrated cement, weighed to the nearest 0.001 g (see Note 4). Calculate the results on the ignited basis.

8.3 *Loss on Ignition:*

8.3.1 *Portland Cement*—Immediately before and after the calorimetric sample is being weighed out, weigh a sample of similar amount into a platinum crucible for determination of loss on ignition, the value to be used being the average of the two determinations. Ignite the dry cement at 950 ± 50°C for at least 1½ h or to constant mass. Immediately place the crucible containing the sample in a desiccator and allow to cool to room temperature; then quickly weigh the crucible. When determining the loss on ignition of the hydrated cement, first dry the weighed sample in an oven at 100 to 110°C for 1 h; then place the sample in a muffle furnace at 950 ± 50°C overnight, or bring to constant mass. Reduce the mass of the cement sample that was introduced into the calorimeter to the ignited mass basis for use in the final calculations as follows:

$$W_i = (A/B)W \tag{3}$$

where:
W_i = mass of calorimetric sample, on ignited basis, g,
A = mass of ignited sample, g,
B = mass of sample before ignition, g, and
W = mass of calorimetric sample, g.

8.3.2 *Blended Hydraulic Cements*—In addition to the procedures described in 8.3.1, determine the loss on ignition by the reference method given in Test Methods C 114 for portland blast-furnace slag cement and slag cement.

8.3.2.1 Determine the SO_3 content by the reference method given in Test Methods C 114 (see Note 6). Also determine the SO_3 content of a portion of the same cement that has not been ignited, using the same procedure.

8.3.2.2 Calculate the percentage of mass gain from sulfide sulfur as follows:

$$G = 0.8\,(S_1 - S_2) \tag{4}$$

where:
G = percent mass gain in ignited sample,
S_1 = SO_3 determined on ignited sample, and
S_2 = SO_3 determined on unignited sample.
0.8 = molecular weight ratio of 4(0)/SO_3

NOTE 6—Some of the acid used for dissolving the sample may first be warmed in the platinum crucible to dissolve any adhering material.

8.3.2.3 Calculate the mass of the dry calorimetric sample on the ignited basis as follows:

$$W_i = \frac{\left(A - \frac{BG}{100}\right) W}{B} \tag{5}$$

where:
W_i = mass of dry calorimetric sample, on ignited basis, g,
A = mass of ignited dry sample, g,
B = mass of dry sample before ignition, g,
G = percentage mass gain from sulfide sulfur, and
W = mass of dry calorimetric sample, g.

Calculate the mass of the partially hydrated calorimetric sample on the ignited basis as follows: (Note 7)

$$W_i = \frac{A\,W\left(1 - \frac{G}{100}\right)}{B} \tag{6}$$

where:
W_i = mass of calorimetric sample, on ignited basis, g,
A = mass of partially hydrated sample after ignition, g,
B = mass of partially hydrated sample before ignition, g,
G = percentage mass gain from sulfide sulfur, and
W = mass of partially hydrated calorimetric sample, g.

NOTE 7—An assumption is made in the calculation that the same percentage of sulfide sulfur is present prior to ignition in the partially hydrated sample as was determined in the cement. Tests have confirmed that the assumption is reasonably correct and will not alter the precision of the test method.

9. Calculation

9.1 *Heat of Solution of Dry Cement*—Calculate the corrected temperature rise as described in 6.3 and 6.4. Also, correct the heat of solution value if the final calorimeter temperature of the heat of solution test is different from the temperature of the calorimetric sample when introduced. Thus, for the dry cement, which has a specific heat of approximately 0.8 kJ/kg·K (0.2 cal/g·°C), if the final calorimeter temperature exceeds the temperature of the cement sample at the time it was introduced, add a correction of 0.8 kJ/kg·K (0.2 cal/g·°C) difference in those temperatures when calculating the heat of solution. Calculate the heat of solution of the dry cement as follows:

$$H_1 = (RC/W_i) - 0.8(T - t_d) \tag{7}$$

where:
H_1 = heat of solution of dry cement, kJ/kg,
R = corrected temperature rise, °C,
C = heat capacity, kJ/°C,
W_i = mass of sample on ignited basis, g,
T = room temperature, when sample is introduced, °C, and
t_d = final calorimeter temperature at end of determination on dry cement, °C.

9.2 *Heat of Solution of Partially Hydrated Sample*—Calculate the heat of solution of the partially hydrated sample in the same way as for the dry cement (9.1), except make additional corrections, as follows:

9.2.1 Since an increase of 1°C in the temperature at which the heat of solution test occurs causes a decrease of approximately 1.3 kJ/kg (0.3 cal/g) in the heat of solution, if the temperature of the heat of solution test of the partially hydrated sample exceeds the temperature of the dry cement determination, a correction of 1.3 kJ/kg·K (°C) difference in temperature shall be added to the heat of solution value obtained for the partially hydrated sample.

9.2.2 Also, correct the heat of solution value if the final calorimeter temperature of the solution test is different from the temperature of the calorimetric sample when introduced. Thus, for the partially hydrated sample, which has a specific heat of approximately 1.7 kJ/kg (0.4 cal/g) of ignited cement, if the final calorimeter temperature exceeds the temperature of the sample at the time it was introduced, add a correction of 1.7 kJ/kg·K (°C) difference in those temperatures when calculating the heat of solution.

9.2.3 Calculate the heat of solution of the partially hydrated sample as follows:

$$H_2 = (RC/W_i) - 1.7(T - t_h) - 1.3(t_d - t_h) \quad (8)$$

where:

H_2 = heat of solution of partially hydrated sample, kJ/kg,
R, C, W_i, and T = the same definition as in 9.1 except that they relate to the partially hydrated sample,
t_d = the same numerical value as in 9.1, and
t_h = final calorimeter temperature at end of determination on partially hydrated sample, °C.

9.3 *Heat of Hydration*—A final calorimeter temperature of 25°C shall be considered as the basis to which the heat of hydration shall be referred, and the effects of variation in that temperature should be kept in mind when considering test results. An increase in the final temperature raises the heat of hydration approximately 0.4 kJ/kg·K (0.1 cal/g·°C) of ignited cement. For example, if the final temperature is 27°C, 0.8 kJ/kg (0.2 cal/g) should be subtracted from the observed heat of hydration in order to refer the results to 25°C. In borderline cases, proper correction should be made for the effects of final calorimeter temperature. Calculate the heat of hydration of the cement to the nearest kilojoule, as follows:

$$H = H_1 - H_2 - 0.4(t_h - 25.0) \quad (9)$$

where:

H = heat of hydration of ignited cement, kJ/kg,
H_1 = heat of solution of dry cement (9.1),
H_2 = heat of solution of partially hydrated sample (9.2), and
t_h = the same numerical value as in 9.2.3

NOTE 8—To convert cal/g to kJ/kg multiply by 4.184 in accordance with Standard IEEE/ASTM SI 10.

10. Retests

10.1 In case of failure to meet the 28-day requirement for heat of hydration, a reserve sample of paste may be tested at a later age and a correction of 2.1 kJ/kg (0.5 cal/g) per day of excess age added to bring the retested heat of solution to a 28-day basis. The period over which this correction may be made shall be limited to 4 days. In case of failure to meet the 7-day requirement, a complete retest including mixing of the paste should be made.

11. Report

11.1 Report the following information:

11.1.1 Sample identification, which may include the source and type of hydraulic cement and sampling date, and

11.1.2 The heat of hydration results at each of the test ages required by the applicable specification.

12. Precision and Bias

12.1 *Precision:*

12.1.1 *Single-Operator Precision*—The single-operator standard deviations have been found to be 12.2 kJ/kg(1s) (2.91 cal/g)(1s) and 14.8 kJ/kg(1s) (3.54 cal/g)(1s) for the determinations of heat of solution and heat of hydration, respectively. Therefore, results of two properly conducted tests by the same operator on samples of the same cement should not differ from each other by more than 34 kJ/kg (8 cal/g) in the determination of heat of solution or 42 kJ/kg (10 cal/g) in the determination of heat of hydration.[6]

12.1.2 *Multilaboratory Precision*—The multilaboratory standard deviations have been found to be 18.5 kJ/kg(1s) (4.42 cal/g)(1s) and 16.9 kJ/kg(1s) (4.03 cal/g)(1s) for the determinations of heat of solution and heat of hydration respectively. Therefore, results of two properly conducted tests from two different laboratories on samples of the same cement should not differ from each other by more than 52 kJ/kg (13 cal/g) in the determination of heat of solution or 48 kJ/kg (11 cal/g) in the determination of heat of hydration.[6]

12.2 *Bias*—Since there is no accepted reference material, no statement on bias is being made.

13. Keywords

13.1 blended cement; heat of hydration; heat of solution; hydraulic cements; portland cement

[6] These numbers represent, respectively the (1s) and (d2s) limits as described in Practice C 670.

Standard Test Method for Normal Consistency of Hydraulic Cement[1]

This standard is issued under the fixed designation C 187; the number immediately following the designation indicates the year of original adoption or, in the case of revision, the year of last revision. A number in parentheses indicates the year of last reapproval. A superscript epsilon (ε) indicates an editorial change since the last revision or reapproval.

1. Scope

1.1 This test method covers the determination of the normal consistency of hydraulic cement.

1.2 The values stated in SI units are to be regarded as the standard. The values given in parentheses are for information only.

1.3 *This standard does not purport to address all of the safety concerns, if any, associated with its use. It is the responsibility of the user of this standard to establish appropriate safety and health practices and determine the applicability of regulatory limitations prior to use.*

2. Referenced Documents

2.1 *ASTM Standards:*

C 305 Practice for Mechanical Mixing of Hydraulic Cement Pastes and Mortars of Plastic Consistency[2]

C 490 Practice for Use of Apparatus for the Determination of Length Change of Hardened Cement Paste, Mortar, and Concrete[2]

C 1005 Specification for Weights and Weighing Devices for Use in the Physical Testing of Hydraulic Cements[2]

D 1193 Specification for Reagent Water[3]

E 177 Practice for Use of the Terms Precision and Bias in ASTM Test Methods[4]

3. Significance and Use

3.1 This test method is intended to be used to determine the amount of water required to prepare hydraulic cement pastes for testing.

4. Apparatus

4.1 *Weights and Weighing Devices*—The weights and weighing devices shall conform to the requirements of Specification C 1005. The weighing device shall be evaluated for precision and accuracy at a total load of 1000 g.

4.2 *Glass Graduates,* 200 or 250-mL capacity, and conforming to the requirements of Practice C 490.

4.3 *Vicat Apparatus*—The Vicat apparatus shall consist of a frame *A* (Fig. 1) bearing a movable rod *B,* weighing 300 g, one end *C,* the plunger end, being 10 mm in diameter for a distance of at least 50 mm, and the other end having a removable needle *D,* 1 mm in diameter and 50 mm in length. The rod *B* is reversible, and can be held in any desired position by a set screw *E,* and has an adjustable indicator *F,* which moves over a scale (graduated in millimetres) attached to the frame *A.* The paste is held in a rigid conical ring *G,* resting on a plane nonabsorptive square base plate *H,* about 100 mm on each side. The rod *B* shall be made of stainless steel having a hardness of not less than 35 HRC (Note), and shall be straight with the plunger end which is perpendicular to the rod axis. The ring shall be made of a noncorroding, nonabsorbent material, and shall have an inside diameter of 70 mm at the base and 60 mm at the top, and a height of 40 mm. In addition to the above, the Vicat apparatus shall conform to the following requirements:

Weight of movable rod	300 ± 0.5 g (0.661 lb ± 8 grains) in
Diameter of plunger end of rod	10 ± 0.05 mm (0.394 ± 0.002 in.)
Diameter of needle	1 ± 0.05 mm (0.039 ± 0.002 in.)
Inside diameter of ring at bottom	70 ± 3 mm (2.75 ± 0.12 in.)
Inside diameter of ring at top	60 ± 3 mm (2.36 ± 0.12 in.)
Height of ring	40 ± 1 mm (1.57 ± 0.04 in.)
Graduated scale	The graduated scale, when compared with a standard scale accurate to within 0.1 mm at all points, shall not show a deviation at any point greater than 0.25 mm.

NOTE—The requirement that the rod be made of stainless steel shall apply only to new Vicat apparatus or replacement rods and not to equipment in use which meets the other requirements of this test method.

5. Temperature and Humidity

5.1 The temperature of the air in the vicinity of the mixing slab, the dry cement, molds, and base plates shall be maintained between 20 and 27.5°C (68 and 81.5°F). The temperature of the mixing water shall not vary from 23.0°C (73.5°F) by more than ±2.0°C (3.5°F).

5.2 The relative humidity of the laboratory shall be not less than 50 %.

6. Procedure

6.1 *Preparation of Cement Paste*—Mix 650 g of cement with a measured quantity of water following the procedure prescribed in the Procedure for Mixing Pastes of Practice C 305. The water shall conform to the numerical limits of Specification D 1193 for Type III or Type IV grade of reagent water.

6.2 *Molding Test Specimen*—Quickly form the cement paste, prepared as described in 6.1, into the approximate shape of a ball with gloved hands. Then toss six times through a free path of about 150 mm (6 in.) from one hand to another so as to produce a nearly spherical mass that may be easily inserted into

---

[1] This test method is under the jurisdiction of ASTM Committee C-1 on Cement and is the direct responsibility of Subcommittee C01.22 on Workability.

Current edition approved July 10, 1998. Published December 1998. Originally published as C 187 – 44. Last previous edition C 187 – 86 (1991).

[2] *Annual Book of ASTM Standards,* Vol 04.01.

[3] *Annual Book of ASTM Standards,* Vol 11.01.

[4] *Annual Book of ASTM Standards,* Vol 14.02.

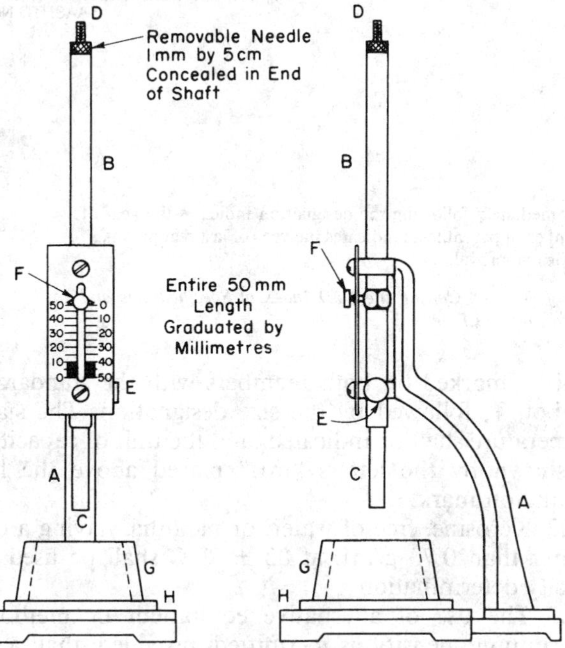

FIG. 1 Vicat Apparatus

the Vicat ring with a minimum amount of additional manipulation. Press the ball, resting in the palm of one hand, into the larger end of the conical ring *G,* Fig. 1, held in the other hand, completely filling the ring with paste. Remove the excess at the larger end by a single movement of the palm of the hand. Place the ring on its larger end on the base plate *H,* and slice off the excess paste at the smaller end at the top of the ring by a single oblique stroke of a sharp-edged trowel held at a slight angle with the top of the ring, and smooth the top, if necessary, with a few light touches of the pointed end of the trowel. During these operations of cutting and smoothing, take care not to compress the paste.

6.3 *Consistency Determination*—Center the paste confined in the ring, resting on the plate, under the rod *B,* Fig. 1, the plunger end *C* of which shall be brought in contact with the surface of the paste, and tighten the set-screw *E.* Then set the movable indicator *F* to the upper zero mark of the scale, or take an initial reading, and release the rod immediately. This must not exceed 30 s after completion of mixing. The apparatus shall be free of all vibrations during the test. The paste shall be of normal consistency when the rod settles to a point 10 ± 1 mm below the original surface in 30 s after being released. Make trial pastes with varying percentages of water until the normal consistency is obtained. Make each trial with fresh cement.

7. Calculation

7.1 Calculate the amount of water required for normal consistency to the nearest 0.1 % and report it to the nearest 0.5 % of the weight of the dry cement.

8. Precision and Bias

8.1 The single operator-instrument precision has been found to be 0.25(1S), and the multilaboratory precision has been found to be 0.35(1S) as defined in Practice E 177; therefore, the results of two properly conducted tests by the same operator in a laboratory should agree within 0.7 percentage point, and test results between two laboratories should agree with 1.0 percentage point 95 % of the time.

9. Keywords

9.1 consistency; normal consistency; Vicat needles

For additional useful information on details of cement test methods, reference may be made to the "Manual of Cement Testing," which appears in the *Annual Book of ASTM Standards,* Vol 04.01.

American Association State Highway and Transportation Officials Standard AASHTO No.: T 133

Standard Test Method for Density of Hydraulic Cement[1]

This standard is issued under the fixed designation C 188; the number immediately following the designation indicates the year of original adoption or, in the case of revision, the year of last revision. A number in parentheses indicates the year of last reapproval. A superscript epsilon (ε) indicates an editorial change since the last revision or reapproval.

This standard has been approved for use by agencies of the Department of Defense. Consult the DoD Index of Specifications and Standards for the specific year of issue which has been adopted by the Department of Defense.

1. Scope

1.1 This test method covers determination of the density of hydraulic cement. Its particular usefulness is in connection with the design and control of concrete mixtures.

1.2 The density of hydraulic cement is defined as the mass of a unit volume of the solids.

1.3 The values stated in SI units are to be regarded as the standard.

1.4 *This standard does not purport to address all of the safety concerns, if any, associated with its use. It is the responsibility of the user of this standard to establish appropriate safety and health practices and determine the applicability of regulatory limitations prior to use.*

2. Referenced Documents

2.1 *ASTM Standards:*

C 114 Test Methods for Chemical Analysis of Hydraulic Cement[2]

C 670 Practice for Preparing Precision and Bias Statements for Test Methods for Construction Materials[3]

3. Apparatus

3.1 *Le Chatelier flask*—The standard flask which is circular in cross section with shape and dimensions conforming essentially to Fig. 1 (Note 1). The requirements in regard to tolerance, inscription and length, spacing, and uniformity of graduation will be rigidly observed. There shall be a space of at least 10 mm between the highest graduation mark and the lowest point of grinding for the glass stopper.

3.1.1 The material of construction shall be best quality glass, transparent and free of striae. The glass shall be chemically resistant and shall have small thermal hysteresis. The flasks shall be thoroughly annealed before being graduated. They shall be of sufficient thickness to ensure reasonable resistance to breakage.

3.1.2 The neck shall be graduated from 0 to 1 mL and from 18 to 24 mL in 0.1-mL graduations. The error of any indicated capacity shall not be greater than 0.05 mL.

3.1.3 Each flask shall bear a permanent identification number and the stopper, if not interchangeably ground, shall bear the same number. Interchangeable ground-glass parts shall be marked on both members with the standard-taper symbol, ₮, followed by the size designation. The standard temperature shall be indicated, and the unit of capacity shall be shown by the letters "mL" placed above the highest graduation mark.

3.2 Kerosine, free of water, or naphtha, having a density greater than 0.73 g/mL at 23 ± 2° C shall be used in the density determination.

3.3 The use of alternative equipment or methods for determining density is permitted provided that a single operator can obtain results within ± 0.03 Mg/m^3 of the results obtained using the flask method.

NOTE 1—The design is intended to ensure complete drainage of the flask when emptied, and stability of standing on a level surface, as well as accuracy and precision of reading.

4. Procedure

4.1 Determine the density of cement on the material as received, unless otherwise specified. If the density determination on a loss-free sample is required, first ignite the sample as described in the test for loss on ignition in section 16.1 of Test Methods C 114.

4.2 Fill the flask (Note 2) with either of the liquids specified in 3.2 to a point on the stem between the 0 and the 1-mL mark. Dry the inside of the flask above the level of the liquid, if necessary, after pouring. Record the first reading after the flask has been immersed in the water bath (Note 3) in accordance with 4.4.

NOTE 2—It is advisable to use a rubber pad on the table top when filling or rolling the flask.

NOTE 3—Before the cement has been added to the flask, a loose-fitting, lead-ring weight around the stem of the flask will be helpful in holding the flask in an upright position in the water bath, or the flask may be held in the water bath by a buret clamp.

4.3 Introduce a quantity of cement, weighed to the nearest 0.05 g, (about 64 g for portland cement) in small increments at the same temperature as the liquid (Note 2). Take care to avoid splashing and see that the cement does not adhere to the inside of the flask above the liquid. A vibrating apparatus may be used to accelerate the introduction of the cement into the flask and to prevent the cement from sticking to the neck. After all the cement has been introduced, place the stopper in the flask and roll the flask in an inclined position (Note 2), or gently whirl it in a horizontal circle, so as to free the cement from air until no further air bubbles rise to the surface of the liquid. If a proper amount of cement has been added, the level of the liquid will be in its final position at some point of the upper series of graduations. Take the final

---

[1] This test method is under the jurisdiction of ASTM Committee C-1 on Cement, and is the direct responsibility of Subcommittee C01.25 on Fineness.

Current edition approved Sept. 10, 1995. Published November 1995. Originally published as C 188 – 44. Last previous edition C 188 – 89(1995)$^{\epsilon 1}$.

[2] *Annual Book of ASTM Standards,* Vol 04.01.

[3] *Annual Book of ASTM Standards,* Vol 04.02.

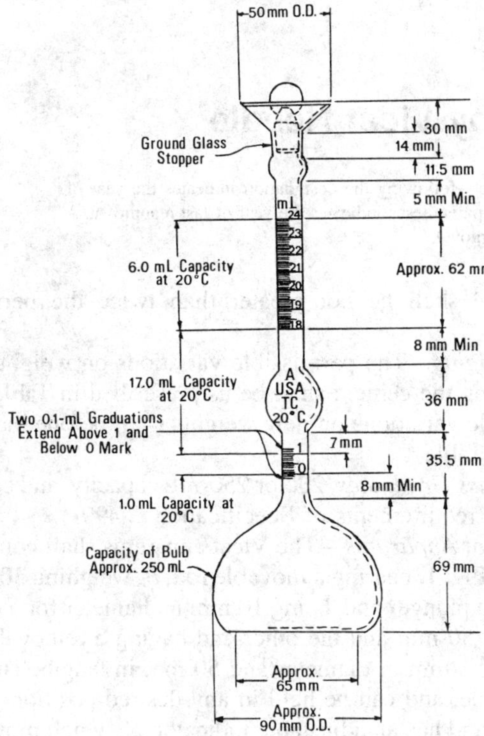

NOTE—Variations of a few millimetres in such dimensions as total height of flask, diameter of base, etc., are to be expected and will not be considered sufficient cause for rejection. The dimensions of the flask shown in Fig. 1 apply only to new flasks and not to flasks in use which meet the other requirements of this test method.

FIG. 1 Le Chatelier Flask for Density Test

reading after the flask has been immersed in the water bath in accordance with 4.4.

4.4 Immerse the flask in a constant-temperature water bath for sufficient periods of time in order to avoid flask-temperature variations greater than 0.2°C between the initial and the final readings.

5. Calculation

5.1 The difference between the first and the final readings represents the volume of liquid displaced by the mass of cement used in the test.

5.2 Calculate the cement density, ρ, as follows:

$$\rho(\text{Mg/m}^3) = \rho(\text{g/cm}^3) = \text{mass of cement, g/displaced volume, cm}^3$$

NOTE 4—The displaced volume in millilitres is numerically equal to the displaced volume in cubic centimetres.

NOTE 5—Density in megagrams per cubic metre (Mg/m^3) is numerically equal to grams per cubic centimetre (g/cm^3). Calculate the cement density, ρ, to three decimal places and round to the nearest 0.01 Mg/m^3.

NOTE 6—In connection with proportioning and control of concrete mixtures, density may be more usefully expressed as specific gravity, the latter being a dimensionless number. Calculate the specific gravity as follows: Sp gr = cement density/water density at 4°C (at 4°C the density of water is 1 Mg/m^3(1g/cm^3)).

6. Precision and Bias

6.1 The single-operator standard deviation for portland cements has been found to be 0.012.[4] Therefore, the results of two properly conducted tests by the same operator on the same material should not differ by more than 0.03.[4]

6.2 The multilaboratory standard deviation for portland cements has been found to be 0.037.[4] Therefore, the results of two properly conducted tests from two different laboratories on samples of the same cement should not differ by more than 0.10.[4]

6.3 Since there is no accepted reference material suitable for determining any bias that might be associated with this test method, no statement on bias is being made.

7. Keywords

7.1 density; hydraulic cement; specific gravity

[4] These numbers represent the 1s and d2s limits described in Practice C 670.

American Association State
Highway and Transportation Officials Standard
AASHTO No.: T131

Standard Test Method for Time of Setting of Hydraulic Cement by Vicat Needle[1]

This standard is issued under the fixed designation C 191; the number immediately following the designation indicates the year of original adoption or, in the case of revision, the year of last revision. A number in parentheses indicates the year of last reapproval. A superscript epsilon (ε) indicates an editorial change since the last revision or reapproval.

1. Scope

1.1 This method covers determination of the time of setting of hydraulic cement by means of the Vicat needle.

NOTE 1—For the method for determining time of setting by Gillmore needles, see Test Method C 266.

1.2 The values stated in SI units are to be regarded as the standard. Values in parentheses are for information only.

1.3 *This standard does not purport to address all of the safety concerns, if any, associated with its use. It is the responsibility of the user of this standard to establish appropriate safety and health practices and determine the applicability of regulatory limitations prior to use.* See Note 2 for a specific warning statement.

NOTE 2—**Warning:** Fresh hydraulic cementitious mixtures are caustic and may cause chemical burns to skin and tissue upon prolonged exposure. The use of gloves, protective clothing, and eye protection is recommended. Wash contact area with copious amounts of water after contact. Wash eyes for a minimum of 15 min. Avoid exposure of the body to clothing saturated with the liquid phase of the unhardened material. Remove contaminated clothing immediately after exposure.

2. Referenced Documents

2.1 *ASTM Standards:*

C 187 Test Method for Normal Consistency of Hydraulic Cement[2]

C 266 Test Method for Time of Setting of Hydraulic-Cement Paste by Gillmore Needles[2]

C 305 Practice for Mechanical Mixing of Hydraulic Cement Pastes and Mortars of Plastic Consistency[2]

C 490 Practice for Use of Apparatus for the Determination of Length Change of Hardened Cement Paste, Mortar, and Concrete[2]

3. Apparatus

3.1 *Scales*—The scales shall conform to the following requirements: On scales in use the permissible variation at a load of 9.8 N shall be ± 0.01 N. The permissible variation on new scales shall be one half of this value. The sensibility reciprocal[3] shall be not greater than twice the permissible variation.

3.2 *Weights*—The permissible variations on weights in use in weighing the cement shall be as prescribed in Table 1. The permissible variations on new weights shall be one half of the values in Table 1.

3.3 *Glass Graduates*, 200 or 250-mL capacity, and conforming to the requirements of Specification C 490.

3.4 *Vicat Apparatus*—The Vicat apparatus shall consist of a frame, *A*, Fig. 1, bearing a movable rod, *B*, weighing 300 g, one end, *C*, the plunger end, being 10 mm in diameter for a distance of at least 50 mm and the other end having a removable steel needle, *D*, 1 mm in diameter and 50 mm in length. The rod *B* is reversible, and can be held in any desired position by a set screw, *E*, and has an adjustable indicator, *F*, which moves over a scale (graduated in millimetres) attached to the frame, *A*. The paste is held in a conical ring, *G*, resting on a plate of similar planeness, corrosivity, and absorptivity to that of glass, *H*, about 100 mm square. The ring shall be made of a noncorroding, nonabsorbing material, and shall have an inside diameter of 70 mm at the base and 60 mm at the top and a height of 40 mm. In addition to the above, the Vicat apparatus shall conform to the following requirements:

Weight of plunger	300 ± 0.5 g (0.661 lb ± 8 grains)
Diameter of larger end of plunger	10 ± 0.05 mm (0.394 ± 0.002 in.)
Diameter of needle	1 ± 0.05 mm (0.039 ± 0.002 in.)
Inside diameter of ring at bottom	70 ± 3 mm (2.75± 0.12 in.)
Inside diameter of ring at top	60 ± 3 mm (2.36± 0.12 in.)
Height of ring	40 ± 1 mm (1.57 ± 0.04 in.)
Graduated scale	The graduated scale, when compared with a scale accurate to within 0.1 mm at all points, shall not show any point greater than 0.25 mm.

4. Temperature and Humidity

4.1 The temperature of the air in the vicinity of the mixing slab, the dry cement, molds, and base plates shall be maintained between 20 and 27.5°C (68 and 81.5°F). The temperature of the mixing water and of the moist closet or moist room shall not vary from 23°C (73.4°F) by more than ± 1.7°C (3°F).

[1] This method is under the jurisdiction of ASTM Committee C-1 on Cement and is the direct responsibility of Subcommittee C01.30 on Time of Set.

Current edition approved Jan. 10, 1999. Published May 1999. Originally published as C 191 – 44. Last previous edition C 191 – 92.

[2] *Annual Book of ASTM Standards*, Vol 04.01.

[3] Generally, defined, the sensibility reciprocal is the change in load required to change the position of rest of the indicating element or elements of a nonautomatic-indicating scale a definite amount at any load. For more complete definition, see "Specifications, Tolerances, and Regulations for Commercial Weighing and Measuring Devices," *Handbook H44*, National Bureau of Standards, September 1949, pp. 92 and 93.

TABLE 1 Permissible Variations on Weights

Weight, g	Permissible Variations on Weights in Use, plus or minus, g
500	0.18
300	0.15
250	0.13
200	0.10
100	0.07
50	0.04
20	0.02
10	0.02
5	0.01
2	0.01
1	0.01

4.2 The relative humidity of the laboratory shall be not less than 50 %. The moist closet or moist room shall be so constructed as to provide storage facilities for test specimens at a relative humidity of not less than 90 %.

5. Preparation of Cement Paste[4]

5.1 Mix 650 g of cement with the percentage of mixing water required for normal consistency following the procedure described in Practice C 305. Distilled water is preferable and shall be used for all referee or cooperative tests.

5.2 The test specimen used for the determination of normal consistency may be used for the additional determination of time of setting by Vicat needle following the procedure described in 6.2.

[4] See Test Method C 187.

6. Procedure

6.1 *Molding Test Specimen*—Quickly form the cement paste, prepared as described in the section on preparation of cement paste, into a ball with the gloved hands and toss six times from one hand to the other, maintaining the hands about 6 in. (152 mm) apart. Press the ball, resting in the palm of the hand, into the larger end of the conical ring, *G*, Fig. 1, held in the other hand, completely filling the ring with paste. Remove the excess at the larger end by a single movement of the palm of the hand. Place the ring on its larger end on a plate of similar planeness, corrosivity, and absorptivity to that of glass, *H*, and slice off the excess paste at the smaller end at the top of the ring by a single oblique stroke of a sharpedged trowel held at a slight angle with the top of the ring. Smooth the top of the specimen, if necessary, with one or two light touches of the pointed end of the trowel. During the operation of cutting and smoothing, take care not to compress the paste. Immediately after molding, place the test specimen in the moist closet or moist room and allow it to remain there except when determinations of time of setting are being made. The specimen shall remain in the conical mold, supported by the plate of similar planeness, corrosivity, and absorptivity to that of glass, *H*, throughout the test period. A time of set specimen and an autoclave bar may be made from the same batch.

6.2 *Time of Setting Determination*— Allow the time of setting specimen to remain in the moist cabinet for 30 min after molding without being disturbed. Determine the penetration of the 1-mm needle at this time and every 15 min thereafter (every 10 min for Type III cements) until a penetration of 25 mm or less is obtained. For the penetration test, lower the needle *D* of

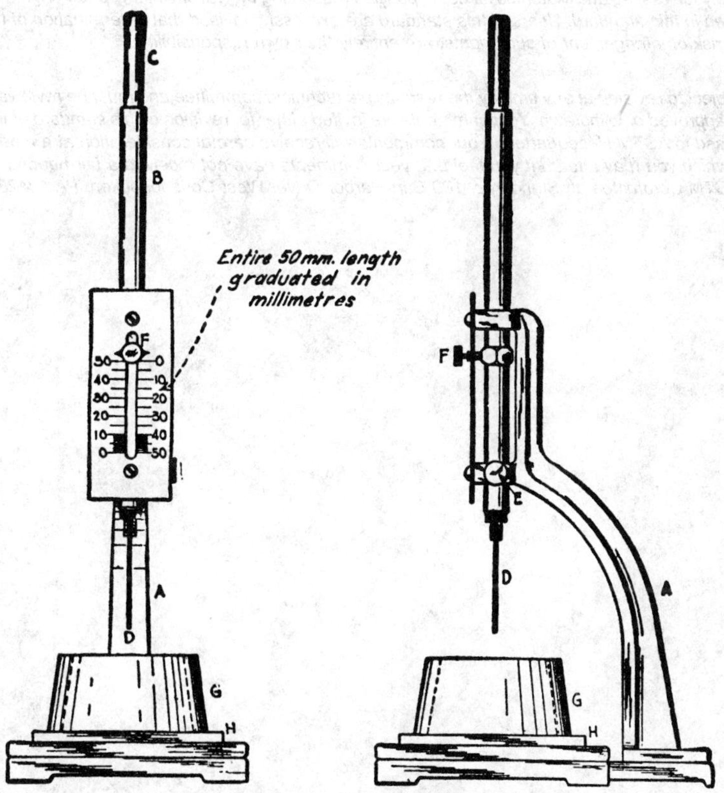

FIG. 1 Vicat Apparatus

the rod *B* until it rests on the surface of the cement paste. Tighten the set screw, *E*, and set the indicator, *F*, at the upper end of the scale, or take an initial reading. Release the rod quickly by releasing the set screw, *E*, and allow the needle to settle for 30 s; then take the reading to determine the penetration. (If the paste is obviously quite soft on the early readings, the fall of the rod may be retarded to avoid bending the 1-mm needle, but the rod shall be released only by the set screw when actual determinations for the setting time are made.) No penetration test shall be made closer than ¼ in. (6.4 mm) from any previous penetration and no penetration test shall be made closer than ⅜ in. (9.5 mm) from the inside of the mold. Record the results of all penetration tests and, by interpolation, determine the time when a penetration of 25 mm is obtained. This is the initial setting time. The final setting time is when the needle does not sink visibly into the paste.

6.3 *Precautions*—All the apparatus shall be free from vibration during the penetration test. Take care to keep the 1-mm needle straight, and the needle must be kept clean as the collection of cement on the sides of the needle may retard the penetration, while cement on the point may increase the penetration. The time of setting is affected not only by the percentage and the temperature of the water used and the amount of kneading the paste received, but also by the temperature and humidity of the air, and its determination is therefore only approximate.

7. Precision and Bias

7.1 *Precision*:

7.1.1 The single-operator (within-laboratory) standard deviation has been found to be 12 min for the initial time of setting, throughout the range of 49 to 202 min, and 20 min for the final time of settings throughout the range of 185 to 312 min. Therefore, results of two properly conducted tests by the same operator on Vicat initial time of setting of similar paste should not differ from each other by more than 34 min and on Vicat final time of setting of similar pastes should not differ from each other by more than 56 min.

7.1.2 The multilaboratory (between-laboratory) standard deviation has been found to be 16 min for the initial time of setting throughout the range of 49 to 207 min, and 43 min for the final time of setting throughout the range of 185 to 312 min. Therefore, results of two properly conducted tests from two different laboratories on Vicat initial time of setting of similar pastes should not differ from each other by more than 45 min, and on Vicat final time of setting of similar pastes should not differ from each other by more than 122 min.

7.2 *Bias*—Since there are no accepted reference materials suitable for determining the bias for the procedure in this test method, no statement on bias is presented.

For additional useful information on details of cement test methods, reference may be made to the "Manual of Cement Testing," which appears in the *Annual Book of ASTM Standards*, Vol 04.01.

Designation: C 204 – 96a

American Association State
Highway and Transportation Officials Standard
AASHTO No.: T 153

Standard Test Method for Fineness of Hydraulic Cement by Air Permeability Apparatus[1]

This standard is issued under the fixed designation C 204; the number immediately following the designation indicates the year of original adoption or, in the case of revision, the year of last revision. A number in parentheses indicates the year of last reapproval. A superscript epsilon (ε) indicates an editorial change since the last revision or reapproval.

1. Scope

1.1 This test method covers determination of the fineness of hydraulic cement, using the Blaine air permeability apparatus, in terms of the specific surface expressed as total surface area in square centimetres per gram, or square metres per kilogram, of cement. Although the test method may be, and has been, used for the determination of the measures of fineness of various other materials, it should be understood that, in general, relative rather than absolute fineness values are obtained.

1.1.1 This test method is known to work well for portland cements. However, the user should exercise judgement in determining its suitability with regard to fineness measurements of cements with densities, or porosities that differ from those assigned to Standard Reference Material No. 114.

1.2 The values stated in SI units are to be regarded as the standard.

1.3 *This standard does not purport to address all of the safety concerns, if any, associated with its use. It is the responsibility of the user of this standard to establish appropriate safety and health practices and determine the applicability of regulatory limitations prior to use.*

2. Referenced Documents

2.1 *ASTM Standard:*

C 670 Practice for Preparing Precision and Bias Statements for Test Methods for Construction Materials[2]

2.2 *Other Documents:*

UU-P-236 Federal Specification for Paper; Filtering[3]

No. 114 National Institute of Standards and Technology Standard Reference Material[4]

3. Apparatus

3.1 *Nature of Apparatus*—The Blaine air permeability apparatus consists essentially of a means of drawing a definite quantity of air through a prepared bed of cement of definite porosity. The number and size of the pores in a prepared bed of definite porosity is a function of the size of the particles and determines the rate of airflow through the bed. The apparatus, illustrated in Fig. 1, shall consist specifically of the parts described in 3.2 to 3.8.

3.2 *Permeability Cell*—The permeability cell shall consist of a rigid cylinder 12.70 ± 0.10 mm in inside diameter, constructed of austenitic stainless steel. The interior of the cell shall have a finish of 0.81 μm (32 μin.). The top of the cell shall be at right angles to the principal axis of the cell. The lower portion of the cell must be able to form an airtight fit with the upper end of the manometer, so that there is no air leakage between the contacting surfaces. A ledge ½ to 1 mm in width shall be an integral part of the cell or be firmly fixed in the cell 55 ± 10 mm from the top of the cell for support of the perforated metal disk. The top of the permeability cell shall be fitted with a protruding collar to facilitate the removal of the cell from the manometer.

NOTE 1—AISI Type 303 stainless steel (UNS designation S30300) has been found to be suitable for the construction of the permeability cell and the plunger.

3.3 *Disk*—The disk shall be constructed of noncorroding metal and shall be 0.9 ± 0.1 mm in thickness, perforated with 30 to 40 holes 1 mm in diameter equally distributed over its area. The disk shall fit the inside of the cell snugly. The center portion of one side of the disk shall be marked or inscribed in a legible manner so as to permit the operator always to place that side downwards when inserting it into the cell. The marking or inscription shall not extend into any of the holes, nor touch their peripheries, nor extend into that area of the disk that rests on the cell ledge.

3.4 *Plunger*—The plunger shall be constructed of austenitic stainless steel and shall fit into the cell with a clearance of not more than 0.1 mm. The bottom of the plunger shall sharply meet the lateral surfaces and shall be at right angles to the principal axis. An air vent shall be provided by means of a flat 3.0 ± 0.3 mm wide on one side of the plunger. The top of the plunger shall be provided with a collar such that when the plunger is placed in the cell and the collar brought in contact with the top of the cell, the distance between the bottom of the plunger and the top of the perforated disk shall be 15 ± 1 mm.

3.5 *Filter Paper*—The filter paper shall be medium retentive, corresponding to Type 1, Grade B, as prescribed in Federal Specification for Paper; Filtering (UU-P-236). The filter paper disks shall be circular, with smooth edges, and shall have the same diameter (Note 2) as the inside of the cell.

NOTE 2—Filter paper disks that are too small may leave part of the sample adhering to the inner wall of the cell above the top disk. When too large in diameter, the disks have a tendency to buckle and cause erratic results.

3.6 *Manometer*—The U-tube manometer shall be constructed according to the design indicated in Fig. 1, using nominal 9-mm outside diameter, standard-wall, glass tubing.

---

[1] This test method is under the jurisdiction of ASTM Committee C-1 on Cement and is the direct responsibility of Subcommittee C01.25 on Fineness.

Current edition approved Dec. 10, 1996. Published February 1997. Originally published as C 204 – 46 T. Last previous edition C 204 – 96.

[2] *Annual Book of ASTM Standards*, Vol 04.02.

[3] Available from Standardization Documents Order Desk, Bldg. 4, Section D, 700 Robbins Ave., Phila., PA 19111-5094, Attn: NPODS.

[4] Available from National Institute of Standards and Technology, Gaithersburg, MD 20899.

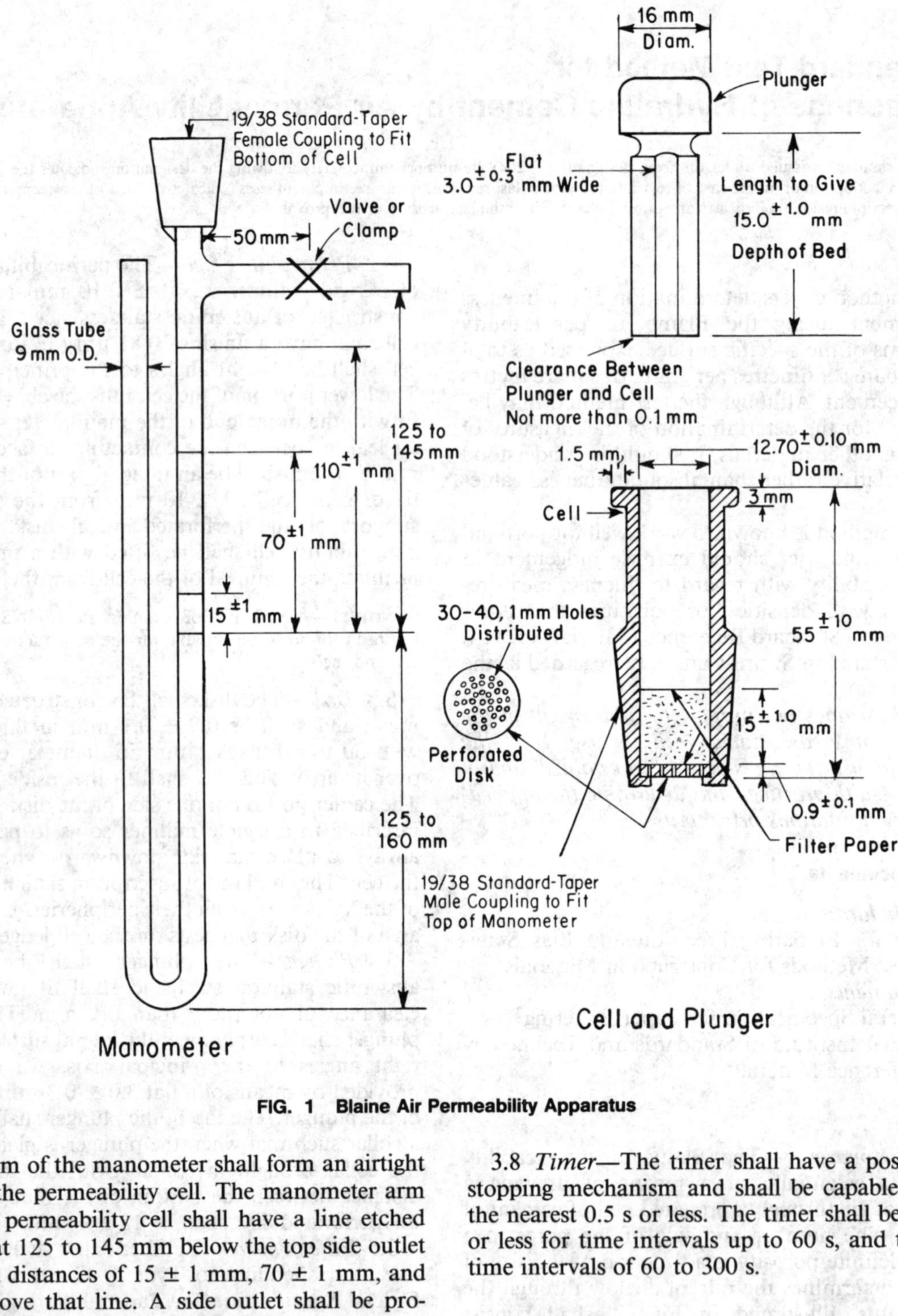

FIG. 1 Blaine Air Permeability Apparatus

The top of one arm of the manometer shall form an airtight connection with the permeability cell. The manometer arm connected to the permeability cell shall have a line etched around the tube at 125 to 145 mm below the top side outlet and also others at distances of 15 ± 1 mm, 70 ± 1 mm, and 110 ± 1 mm above that line. A side outlet shall be provided at 250 to 305 mm above the bottom of the manometer for use in the evacuation of the manometer arm connected to the permeability cell. A positive airtight valve or clamp shall be provided on the side outlet not more than 50 mm from the manometer arm. The manometer shall be mounted firmly and in such a manner that the arms are vertical.

3.7 *Manometer Liquid*—The manometer shall be filled to the midpoint with a nonvolatile, nonhygroscopic liquid of low viscosity and density, such as dibutyl phthalate (dibutyl 1,2-benzene-dicarboxylate) or a light grade of mineral oil.

3.8 *Timer*—The timer shall have a positive starting and stopping mechanism and shall be capable of being read to the nearest 0.5 s or less. The timer shall be accurate to 0.5 s or less for time intervals up to 60 s, and to 1 % or less for time intervals of 60 to 300 s.

4. Calibration of Apparatus

4.1 *Sample*—The calibration of the air permeability apparatus shall be made using the current lot of NIST Standard Reference Material No. 114. The sample shall be at room temperature when tested.

4.2 *Bulk Volume of Compacted Bed of Powder*—Determine the bulk volume of the compacted bed of powder by the mercury displacement method as follows:

4.2.1 Place two filter paper disks in the permeability cell, pressing down the edges, using a rod having a diameter

TABLE 1 Density of Mercury, Viscosity of Air (η), and $\sqrt{\eta}$ at Given Temperatures

Room Temperature, °C	Density of Mercury, Mg/m³	Viscosity of Air, η μPa·s	$\sqrt{\eta}$
16	13.56	17.88	4.23
18	13.55	17.98	4.24
20	13.55	18.08	4.25
22	13.54	18.18	4.26
24	13.54	18.28	4.28
26	13.53	18.37	4.29
28	13.53	18.47	4.30
30	13.52	18.57	4.31
32	13.52	18.67	4.32
34	13.51	18.76	4.33

slightly smaller than that of the cell, until the filter disks are flat on the perforated metal disk; then fill the cell with mercury, ACS reagent grade or better, removing any air bubbles adhering to the wall of the cell. Use tongs when handling the cell. If the cell is made of material that will amalgamate with mercury, the interior of the cell shall be protected by a very thin film of oil just prior to adding the mercury. Level the mercury with the top of the cell by lightly pressing a small glass plate against the mercury surface until the glass is flush to the surface of the mercury and rim of the cell, being sure that no bubble or void exists between the mercury surface and the glass plate. Remove the mercury from the cell, weigh, and record the weight of the mercury. Remove one of the filter disks from the cell. Using a trial quantity of 2.80 g of cement (Note 3) compress the cement (Note 4) in accordance with 4.5 with one filter disk above and one below the sample. Into the unfilled space at the top of the cell, add mercury, remove entrapped air, and level off the top as before. Remove the mercury from the cell, weigh, and record the weight of the mercury.

4.2.2 Calculate the bulk volume occupied by the cement to the nearest 0.005 cm³ as follows:

$$V = (W_A - W_B)/D \quad (1)$$

where:

V = bulk volume of cement, cm³,

W_A = grams of mercury required to fill the cell, no cement being in the cell,

W_B = grams of mercury required to fill the portion of the cell not occupied by the prepared bed of cement in the cell, and

D = density of mercury at the temperature of test, Mg/m³ (see Table 1).

4.2.3 Make at least two determinations of bulk volume of cement, using separate compactions for each determination. The bulk volume value used for subsequent calculations shall be the average of two values agreeing within ±0.005 cm³. Note the temperature in the vicinity of the cell and record at the beginning and end of the determination.

NOTE 3—It is not necessary to use the standard sample for the bulk volume determination.

NOTE 4—The prepared bed of cement shall be firm. If too loose or if the cement cannot be compressed to the desired volume, adjust the trial quantity of cement used.

4.3 *Preparation of Sample*—Enclose the contents of a vial of the standard cement sample in a jar, approximately 120 cm³ (4 oz), and shake vigorously for 2 min to fluff the cement and break up lumps or agglomerates. Allow the jar to stand unopened for a further 2 min, then remove the lid and stir gently to distribute throughout the sample the fine fraction that has settled on the surface after fluffing.

4.4 *Weight of Sample*—The weight of the standard sample used for the calibration test shall be that required to produce a bed of cement having a porosity of 0.500 ± 0.005, and shall be calculated as follows:

$$W = \rho V(1 - \epsilon) \quad (2)$$

where:

W = grams of sample required,

ρ = density of test sample (for portland cement a value of 3.15 Mg/m³ or 3.15 g/cm³ shall be used),

V = bulk volume of bed of cement, cm³, as determined in accordance with 4.2, and

ϵ = desired porosity of bed of cement (0.500 ± 0.005) (Note 5).

NOTE 5—The porosity is the ratio of volume of voids in a bed of cement to the total or bulk volume of the bed, V.

4.5 *Preparation of Bed of Cement*—Seat the perforated disk on the ledge in the permeability cell, inscribed or marked face down. Place a filter paper disk on the metal disk and press the edges down with a rod having a diameter slightly smaller than that of the cell. Weigh to the nearest 0.001 g the quantity of cement determined in accordance with 4.4 and place in the cell. Tap the side of the cell lightly in order to level the bed of cement. Place a filter paper disk on top of the cement and compress the cement with the plunger until the plunger collar is in contact with the top of the cell. Slowly withdraw the plunger a short distance, rotate about 90°, repress, and then slowly withdraw. Use of fresh paper filter disks is required for each determination.

4.6 *Permeability Test:*

4.6.1 Attach the permeability cell to the manometer tube, making certain that an airtight connection is obtained (Note 6) and taking care not to jar or disturb the prepared bed of cement.

4.6.2 Slowly evacuate the air in the one arm of the manometer U-tube until the liquid reaches the top mark, and then close the valve tightly. Start the timer when the bottom of the meniscus of the manometer liquid reaches the second (next to the top) mark and stop when the bottom of the meniscus of liquid reaches the third (next to the bottom) mark. Note the time interval measured and record in seconds. Note the temperature of test and record in degrees Celsius.

4.6.3 In the calibration of the instrument, make at least three determinations of the time of flow on each of three separately prepared beds of the standard sample (Note 7). The calibration shall be made by the same operator who makes the fineness determination.

NOTE 6—A little stopcock grease should be applied to the standard taper connection. The efficiency of the connection can be determined by attaching the cell to the manometer, stoppering it, partially evacuating the one arm of the manometer, then closing the valve. Any continuous drop in pressure indicates a leak in the system.

NOTE 7—The sample may be refluffed and reused for preparation of the test bed, provided that it is kept dry and all tests are made within 4 h of the opening of the sample.

4.7 *Recalibration*—The apparatus shall be recalibrated (Note 8):

4.7.1 At periodic intervals, the duration of which shall not exceed 2½ years, to correct for possible wear on the plunger or permeability cell, or upon receipt of evidence that the test is not providing data in accordance with the precision and bias statement in Section 8.

4.7.2 If any loss in the manometer fluid occurs, recalibrate starting with 4.5, or

4.7.3 If a change is made in the type or quality of the filter paper used for the tests.

NOTE 8—It is suggested that a secondary sample be prepared and used as a fineness standard for the check determinations of the instrument between regular calibrations with the standard cement sample.

5. Procedure

5.1 *Temperature of Cement*—The cement sample shall be at room temperature when tested.

5.2 *Size of Test Sample*—The weight of sample used for the test shall be the same as that used in the calibration test on the standard sample, with these exceptions: When determining the fineness of Type III or other types of fine-ground portland cement whose bulk for this weight is so great that ordinary thumb pressure will not cause the plunger collar to contact the top of the cell, the weight of the sample shall be that required to produce a test bed having a porosity of 0.530 ± 0.005. When determining the fineness of materials other than portland cement, or if for a portland cement sample one of the required porosities cannot be attained, the weight of the sample shall be adjusted so that a firm, hard bed is produced by the compacting process. In no case, however, shall more than thumb pressure be used to secure the proper bed, nor shall such thumb pressure be used that the plunger "rebounds" from the cell top when pressure is removed.

5.3 *Preparation of Bed of Cement*—Prepare the test bed of cement in accordance with the method described in 4.5.

5.4 *Permeability Tests*—Make the permeability tests in accordance with the method described in 4.6, except that only one time-of-flow determination need be made on each bed.

6. Calculation

6.1 Calculate the specific surface values in accordance with the following equations:

$$S = \frac{S_s\sqrt{T}}{\sqrt{T_s}} \tag{3}$$

$$S = \frac{S_s\sqrt{\eta_s}\sqrt{T}}{\sqrt{T_s}\sqrt{\eta}} \tag{4}$$

$$S = \frac{S_s(b-\epsilon_s)\sqrt{\epsilon^3}\sqrt{T}}{\sqrt{\epsilon_s{}^3}\sqrt{T_s}(b-\epsilon)} \tag{5}$$

$$S = \frac{S_s(b-\epsilon_s)\sqrt{\epsilon^3}\sqrt{\eta_s}\sqrt{T}}{\sqrt{\epsilon_s{}^3}\sqrt{T_s}\sqrt{\eta}(b-\epsilon)} \tag{6}$$

$$S = \frac{S_s\rho_s(b_s-\epsilon_s)\sqrt{\epsilon^3}\sqrt{T}}{\rho(b-\epsilon)\sqrt{\epsilon_s{}^3}\sqrt{T_s}} \tag{7}$$

$$S = \frac{S_s\rho_s(b_s-\epsilon_s)\sqrt{\eta_s}\sqrt{\epsilon^3}\sqrt{T}}{\rho(b-\epsilon)\sqrt{\epsilon_s{}^3}\sqrt{T_s}\sqrt{\eta}} \tag{8}$$

where:

S = specific surface of the test sample, m^2/kg,

TABLE 2 Values for Porosity of Cement Bed

Porosity of Bed, ϵ	$\sqrt{\epsilon^3}$
0.495	0.348
0.496	0.349
0.497	0.350
0.498	0.351
0.499	0.352
0.500	0.354
0.501	0.355
0.502	0.356
0.503	0.357
0.504	0.358
0.505	0.359
0.506	0.360
0.507	0.361
0.508	0.362
0.509	0.363
0.510	0.364
0.525	0.380
0.526	0.381
0.527	0.383
0.528	0.384
0.529	0.385
0.530	0.386
0.531	0.387
0.532	0.388
0.533	0.389
0.534	0.390
0.535	0.391

S_s = specific surface of the standard sample used in calibration of the apparatus, m^2/kg,

T = measured time interval, s, of manometer drop for test sample (Note 9),

T_s = measured time interval, s, of manometer drop for standard sample used in calibration of the apparatus (Note 9),

η = viscosity of air, micro pascal seconds (µPa·s), at the temperature of test of the test sample (Note 9),

η_s = viscosity of air, micro pascal seconds (µPa·s), at the temperature of test of the standard sample used in calibration of the apparatus (Note 9),

ϵ = porosity of prepared bed of test sample (Note 9),

ϵ_s = porosity of prepared bed of standard sample used in calibration of apparatus (Note 9),

ρ = density of test sample (for portland cement a value of 3.15 Mg/m^3 or 3.15 g/cm^3 shall be used),

ρ_s = density of standard sample used in calibration of apparatus (assumed to be 3.15 Mg/m^3 or 3.15 g/cm^3),

b = a constant specifically appropriate for the test sample (for hydraulic cement a value of 0.9 shall be used), and

b_s = 0.9, the appropriate constant for the standard sample.

NOTE 9—Values for $\sqrt{\eta}$, $\sqrt{\epsilon 3}$, and $\sqrt{T}$ may be taken from Tables 1, 2, and 3, respectively.

6.1.1 Equations 3 and 4 shall be used in calculations of fineness of portland cements compacted to the same porosity as the standard fineness sample. Equation 3 is used if the temperature of test of the test sample is within ±3°C of the temperature of calibration test, and Eq 4 is used if the temperature of test of the test sample is outside of this range.

TABLE 3 **Time of Airflow**

T = time of airflow in seconds; $\sqrt{T}$ = the factor for use in the equations

T	$\sqrt{T}$	T	$\sqrt{T}$	T	$\sqrt{T}$	T	$\sqrt{T}$	T	$\sqrt{T}$	T	$\sqrt{T}$
26	5.10	51	7.14	76	8.72	101	10.05	151	12.29	201	14.18
26½	5.15	51½	7.18	76½	8.75	102	10.10	152	12.33	202	14.21
27	5.20	52	7.21	77	8.77	103	10.15	153	12.37	203	14.25
27½	5.24	52½	7.25	77½	8.80	104	10.20	154	12.41	204	14.28
28	5.29	53	7.28	78	8.83	105	10.25	155	12.45	205	14.32
28½	5.34	53½	7.31	78½	8.86	106	10.30	156	12.49	206	14.35
29	5.39	54	7.35	79	8.89	107	10.34	157	12.53	207	14.39
29½	5.43	54½	7.38	79½	8.92	108	10.39	158	12.57	208	14.42
30	5.48	55	7.42	80	8.94	109	10.44	159	12.61	209	14.46
30½	5.52	55½	7.45	80½	8.97	110	10.49	160	12.65	210	14.49
31	5.57	56	7.48	81	9.00	111	10.54	161	12.69	211	14.53
31½	5.61	56½	7.52	81½	9.03	112	10.58	162	12.73	212	14.56
32	5.66	57	7.55	82	9.06	113	10.63	163	12.77	213	14.59
32½	5.70	57½	7.58	82½	9.08	114	10.68	164	12.81	214	14.63
33	5.74	58	7.62	83	9.11	115	10.72	165	12.85	215	14.66
33½	5.79	58½	7.65	83½	9.14	116	10.77	166	12.88	216	14.70
34	5.83	59	7.68	84	9.17	117	10.82	167	12.92	217	14.73
34½	5.87	59½	7.71	84½	9.19	118	10.86	168	12.96	218	14.76
35	5.92	60	7.75	85	9.22	119	10.91	169	13.00	219	14.80
35½	5.96	60½	7.78	85½	9.25	120	10.95	170	13.04	220	14.83
36	6.00	61	7.81	86	9.27	121	11.00	171	13.08	222	14.90
36½	6.04	61½	7.84	86½	9.30	122	11.05	172	13.11	224	14.97
37	6.08	62	7.87	87	9.33	123	11.09	173	13.15	226	15.03
37½	6.12	62½	7.91	87½	9.35	124	11.14	174	13.19	228	15.10
38	6.16	63	7.94	88	9.38	125	11.18	175	13.23	230	15.17
38½	6.20	63½	7.97	88½	9.41	126	11.22	176	13.27	232	15.23
39	6.24	64	8.00	89	9.43	127	11.27	177	13.30	234	15.30
39½	6.28	64½	8.03	89½	9.46	128	11.31	178	13.34	236	15.36
40	6.32	65	8.06	90	9.49	129	11.36	179	13.38	238	15.43
40½	6.36	65½	8.09	90½	9.51	130	11.40	180	13.42	240	15.49
41	6.40	66	8.12	91	9.54	131	11.45	181	13.45	242	15.56
41½	6.44	66½	8.15	91½	9.57	132	11.49	182	13.49	244	15.62
42	6.48	67	8.19	92	9.59	133	11.53	183	13.53	246	15.68
42½	6.52	67½	8.22	92½	9.62	134	11.58	184	13.56	248	15.75
43	6.56	68	8.25	93	9.64	135	11.62	185	13.60	250	15.81
43½	6.60	68½	8.28	93½	9.67	136	11.66	186	13.64	252	15.87
44	6.63	69	8.31	94	9.70	137	11.70	187	13.67	254	15.94
44½	6.67	69½	8.34	94½	9.72	138	11.75	188	13.71	256	16.00
45	6.71	70	8.37	95	9.75	139	11.79	189	13.75	258	16.06
45½	6.75	70½	8.40	95½	9.77	140	11.83	190	13.78	260	16.12
46	6.78	71	8.43	96	9.80	141	11.87	191	13.82	262	16.19
46½	6.82	71½	8.46	96½	9.82	142	11.92	192	13.86	264	16.25
47	6.86	72	8.49	97	9.85	143	11.96	193	13.89	266	16.31
47½	6.89	72½	8.51	97½	9.87	144	12.00	194	13.93	268	16.37
48	6.93	73	8.54	98	9.90	145	12.04	195	13.96	270	16.43
48½	6.96	73½	8.57	98½	9.92	146	12.08	196	14.00	272	16.49
49	7.00	74	8.60	99	9.95	147	12.12	197	14.04	274	16.55
49½	7.04	74½	8.63	99½	9.97	148	12.17	198	14.07	276	16.61
50	7.07	75	8.66	100	10.00	149	12.21	199	14.11	278	16.67
50½	7.11	75½	8.69	100½	10.02	150	12.25	200	14.14	280	16.73

6.1.2 Equations 5 and 6 shall be used in calculation of fineness of portland cements compacted to some porosity other than that of the standard fineness sample used in the calibration test. Equation 5 is used if the temperature of test of the test sample is within ±3°C of the temperature of calibration test of the standard fineness sample, and Eq 6 is used if the temperature of test of the test sample is outside of this range.

6.1.3 Equations 7 and 8 shall be used in calculation of fineness of materials other than hydraulic cement. Equation 7 shall be used when the temperature of test of the test sample is within ±3°C of the temperature of calibration test, and Eq 8 is used if the temperature of test of the test sample is outside of this range.

6.1.4 It is recommended that values of b be determined on no less than three samples of the material in question. Test each sample at a minimum of four different porosities over a porosity range of at least 0.06. Correlation coefficients should exceed 0.9970 for the correlation of $\sqrt{\epsilon^3 T}$ versus ϵ on each sample tested (see Appendix X1).

6.2 To calculate the specific surface values in square metres per kilogram, multiply the surface area in cm^2/g by the factor of 0.1.

6.3 Round values in cm^2/g to the nearest 10 units (in m^2/kg to the nearest unit). Example: 3447 cm^2/g is rounded to 3450 cm^2/g or 345 m^2/kg.

7. Report

7.1 For portland cements and portland cement-based materials, report results on a single determination on a single bed.

7.2 For very high fineness materials with long time intervals, report the average fineness value of two permeability tests, provided that the two agree within 2 % of each other. If they do not agree, discard the values and repeat the test (Note 10) until two values so agreeing are obtained.

NOTE 10—Lack of agreement indicates a need for checks of procedure and apparatus. See also the "Manual of Cement Testing."[2]

8. Precision and Bias

8.1 *Single-Operator Precision*—The single-operator coefficient of variation for portland cements has been found to be 1.2 % (Note 11). Therefore, results of two properly conducted tests, by the same operator, on the same sample, should not differ by more than 3.4 % (Note 11) of their average.

8.2 *Multilaboratory Precision*—The multilaboratory coefficient of variation for portland cements has been found to be 2.1 % (Note 11). Therefore, results of two different laboratories on identical samples of a material should not differ from each other by more than 6.0 % (Note 11) of their average.

NOTE 11—These numbers represent, respectively, the 1s % and d2s % limits as described in Practice C 670.

8.3 Since there is no accepted reference material suitable for determining any bias that may be associated with Test Method C 204, no statement is being made.

9. Keywords

9.1 air permeability; apparatus; fineness

APPENDIX

(Nonmandatory Information)

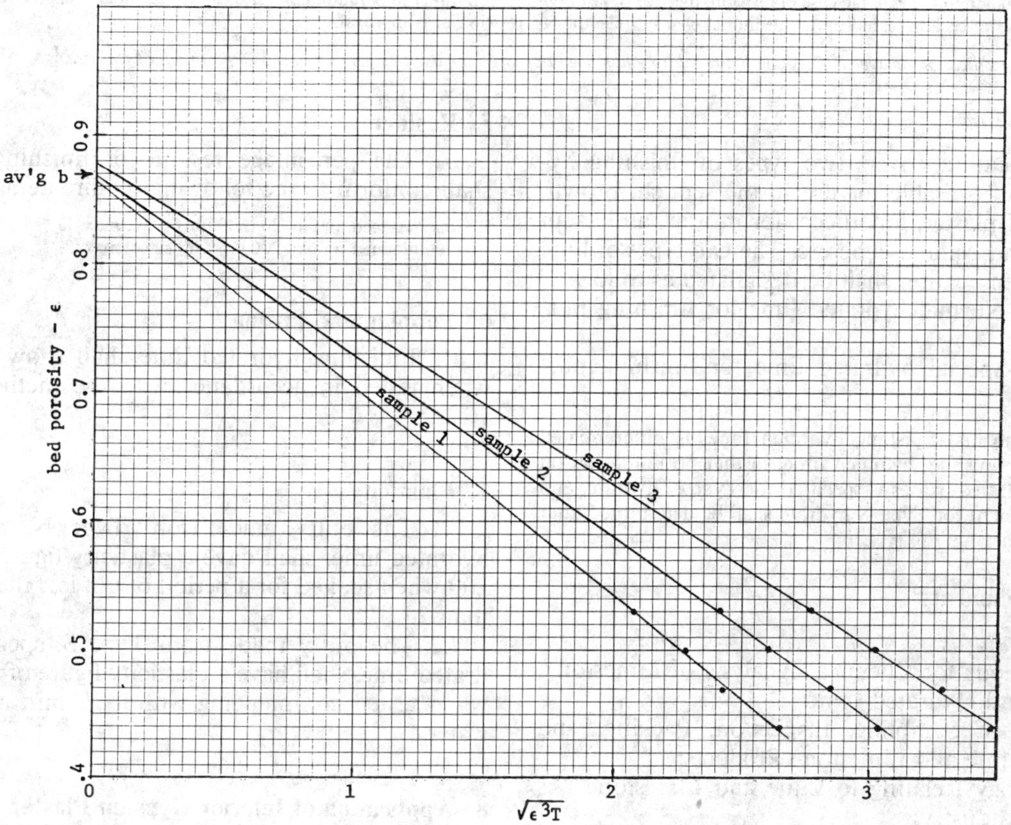

Material: Silica flour
ρ = density of test sample = 2.65 Mg/m³
V = bulk volume of sample bed = 1.887 cm³
ϵ = desired porosity of test
W = grams of sample required = $\rho V(1 - \epsilon)$
T = measured test time interval, seconds

Computed values of b by linear regression:
Sample 1 b = 0.863 (correlation coefficient = 0.9980)
Sample 2 b = 0.869 (correlation coefficient = 0.9993)
Sample 3 b = 0.879 (correlation coefficient = 0.9973)
Average b = 0.870

ϵ	W	T	$\sqrt{\epsilon^3 T}$
		Sample 1	
0.530	2.350	29.0	2.078
0.500	2.500	42.0	2.291
0.470	2.650	57.5	2.443
0.440	2.800	82.5	2.651
		Sample 2	
0.530	2.350	39.0	2.410
0.500	2.500	55.5	2.634
0.470	2.650	79.0	2.864
0.440	2.800	108.5	3.040
		Sample 3	
0.530	2.350	51.5	2.769
0.500	2.500	73.0	3.021
0.470	2.650	104.0	3.286
0.440	2.800	141.5	3.472

FIG. X1.1 Illustrative Method for the Determination of the Value for the Constant *b* (for use in fineness calculations of materials other than portland cement)

Designation: C 206 – 84 (Reapproved 1997)

Standard Specification for Finishing Hydrated Lime[1]

This standard is issued under the fixed designation C 206; the number immediately following the designation indicates the year of original adoption or, in the case of revision, the year of last revision. A number in parentheses indicates the year of last reapproval. A superscript epsilon (ε) indicates an editorial change since the last revision or reapproval.

This specification for type S finishing hydrated lime has been approved for use by agencies of the Department of Defense.

1. Scope

1.1 This specification covers two types of finishing hydrated lime that are suitable for use in the scratch, brown, and finish coats of plaster, for stucco, for mortar, and as an addition to portland-cement concrete. The two types of lime sold under this specification shall be designated as follows:

1.1.1 *Type N*—Normal hydrated lime for finishing purposes, and

1.1.2 *Type S*—Special hydrated lime for finishing purposes.

NOTE—Type N, normal finishing hydrated lime, is differentiated from Type S, special finishing hydrated lime, in that no limitation on the amount of unhydrated oxides is specified for Type N hydrate, and the plasticity requirement for Type N hydrate shall be determined after soaking for 16 to 24 h.

2. Referenced Documents

2.1 *ASTM Standards:*

C 25 Test Methods for Chemical Analysis of Limestone, Quicklime, and Hydrated Lime[2]

C 50 Methods of Sampling, Inspection, Packing, and Marking of Lime and Limestone Products[2]

C 51 Terminology Relating to Lime and Limestone (As Used by the Industry)[2]

C 110 Test Methods for Physical Testing of Quicklime, Hydrated Lime, and Limestone[2]

C 842 Specification for Application of Interior Gypsum Plaster[2]

3. Terminology

3.1 *Definitions*—For definitions of terms relating to hydrated lime, refer to Terminology C 51.

4. Chemical Composition

4.1 Hydrated lime for finishing purposes shall conform to the following requirements as to chemical composition:

	Type N	Type S
Calcium and magnesium oxides (nonvolatile basis), min, %	95	95
Carbon dioxide (as-received basis), max, %		
If sample is taken at the place of manufacture	5	5
If sample is taken at any other place	7	7
Unhydrated oxides (as-received basis), max, %	. . .	8

5. Residue

5.1 The percentage residue of finishing hydrated lime shall conform to the following requirements:

Residue retained on No. 30 (600-μm) sieve, max, %	0.5
Residue retained on No. 200 (75-μm) sieve, max, %	15

6. Popping and Pitting

6.1 Finishing hydrated lime shall show no pops or pits when tested in accordance with the method prescribed in 10.1.2.

7. Plasticity

7.1 The putty made from Type N, normal finishing hydrated lime, shall have a plasticity figure of not less than 200 when soaked for a period of not less than 16 h nor more than 24 h.

7.2 The putty made from Type S, special finishing hydrated lime, shall have a plasticity figure of not less than 200 when tested commencing within 30 min after mixing with water.

8. Application of Interior Gypsum Plaster

8.1 For recommended application procedures refer to Specification C 842.

9. Sampling, Inspection, etc.

9.1 The sampling, inspection, rejection, retesting, packing, and marking shall be conducted in accordance with Methods C 50.

10. Test Methods

10.1 The properties enumerated in this specification shall be determined in accordance with the following methods:

10.1.1 *Chemical Analysis*—Test Methods C 25.

10.1.2 *Physical Tests*—Test Methods C 110.

11. Package Marking

11.1 Type N hydrated lime, in bags, conforming to this specification shall be soaked for a minimum of 16 h prior to use.

12. Keywords

12.1 finishing lime; masonry; plaster; plasticity; popping and pitting; residue; Type N; Type S; unhydrated oxides

---

[1] This specification is under the jurisdiction of ASTM Committee C-7 on Lime and is the direct responsibility of Subcommittee C07.02 on Structural Lime.

Current edition approved May 25, 1984. Published July 1984. Originally published as C 206 – 46. Replaces C 6 – 49 (1974). Last previous edition C 206 – 84.

[2] *Annual Book of ASTM Standards*, Vol 04.01.

Designation: C 207 – 91 (Reapproved 1997)

Standard Specification for Hydrated Lime for Masonry Purposes[1]

This standard is issued under the fixed designation C 207; the number immediately following the designation indicates the year of original adoption or, in the case of revision, the year of last revision. A number in parentheses indicates the year of last reapproval. A superscript epsilon (ε) indicates an editorial change since the last revision or reapproval.

This specification has been approved for use by agencies of the Department of Defense. Consult the DoD Index of Specifications and Standards for the specific year of issue which has been adopted by the Department of Defense.

1. Scope

1.1 This specification covers four types of hydrated lime. Types N and S are suitable for use in mortar, in scratch and brown coats of cement plaster, for stucco, and for addition to portland-cement concrete. Types NA and SA are air-entrained hydrated limes that are suitable for use in any of the above uses where the inherent properties of lime and air-entrainment are desired. The four types of lime sold under this specification shall be designated as follows:

1.1.1 *Type N*—Normal hydrated lime for masonry purposes.

1.1.2 *Type S*—Special hydrated lime for masonry purposes.

1.1.3 *Type NA*—Normal air-entraining hydrated lime for masonry purposes.

1.1.4 *Type SA*—Special air-entraining hydrated lime for masonry purposes.

NOTE 1—Type S, special hydrated lime, and Type SA, special air-entraining hydrated lime, are differentiated from Type N, normal hydrated lime, and Type NA, normal air-entraining hydrated lime, principally by their ability to develop high, early plasticity and higher water retentivity, and by a limitation on their unhydrated oxide content.

NOTE 2—For normal (Type N) and special (Type S) finishing hydrated lime, refer to Specification C 206.

NOTE 3—Some building codes prohibit the use of air-entraining materials in mortar, because of the accompanying reduction in bond and compressive strength. Where increased freeze-thaw resistance is important, air-entraining may be beneficial. Air-entraining lime should not be used as a finishing lime.

2. Referenced Documents

2.1 *ASTM Standards:*

C 25 Test Methods for Chemical Analysis of Limestone, Quicklime, and Hydrated Lime[2]

C 50 Practice for Sampling, Inspection, Packing, and Marking of Lime and Limestone Products[2]

C 51 Terminology Relating to Lime and Limestone (As Used by the Industry)[2]

C 110 Test Methods for Physical Testing of Quicklime, Hydrated Lime, and Limestone[2]

C 206 Specification for Finishing Hydrated Lime[2]

C 226 Specification for Air-Entraining Additions for Use in the Manufacture of Air-Entraining Portland Cement[2]

[1] This specification is under the jurisdiction of ASTM Committee C-7 on Lime and is the direct responsibility of Subcommittee C07.02 on Structural Lime.

Current edition approved Sept. 15, 1991. Published November 1991. Originally published as C 207 – 46 T. Last previous edition C 207 – 76 (1988)$^{\epsilon 1}$.

[2] *Annual Book of ASTM Standards*, Vol 04.01.

3. Terminology

3.1 *Definitions:*

3.1.1 *hydrated lime*—The hydrated lime covered by Types N or S in this specification shall contain no additives for the purpose of entraining air. The air content of cement-lime mortars made with Types N or S shall not exceed 7 % as determined in accordance with the requirements of Test Methods C 110. Types NA and SA shall contain an air-entraining additive as specified by Section 4 of this specification. The air content of cement-lime mortars made with Types NA or SA shall have a minimum of 7 % and a maximum of 14 % when tested in accordance with the requirements of Test Methods C 110.

NOTE 4—Air-entraining cement should not be used in running the air-content test.

3.1.2 For definitions of terms relating to hydrated lime, refer to Terminology C 51.

4. Additions

4.1 Types NA and SA hydrated lime covered by this specification shall contain additives for the purpose of entraining air, and such additives shall conform to the requirements of Specification C 226.

5. Manufacturer's Statement

5.1 At the request of the purchaser, the manufacturer shall state in writing the nature, amount, and identity of the air-entraining agent used and of any processing addition that may have been used, and also, if requested, shall supply test data showing compliance of such air-entraining addition with the provisions of Specification C 226.

6. Chemical Composition

6.1 Hydrated lime for masonry purposes shall conform to the following requirements as to chemical composition:

	Hydrate Types:			
	N	NA	S	SA
Calcium and magnesium oxides (nonvolatile basis), min, %	95	95	95	95
Carbon dioxide (as-received basis), max, %:				
If sample is taken at place of manufacture	5	5	5	5
If sample is taken at any other place	7	7	7	7
Unhydrated oxides (as-received basis), max, %	...	...	8	8

7. Residue, Popping, and Pitting

7.1 The four types of hydrated lime for masonry purposes

shall conform to one of the following requirements:

7.1.1 The residue retained on a No. 30 (600-μm) sieve shall not be more than 0.5 %, or

7.1.2 If the residue retained on a No. 30 (600-μm) sieve is over 0.5 %, the lime shall show no pops or pits when tested in accordance with the method prescribed in 11.1.2.

8. Plasticity

8.1 The putty made from Type S, special hydrate, or Type SA, special air-entraining hydrate, shall have a plasticity figure of not less than 200 when tested commencing within 30 min after mixing with water.

9 Water Retention

9.1 Hydrated lime mortar made with Type N (normal hydrated lime) or Type NA (normal air-entraining hydrated lime), after suction for 60 s, shall have a water retention value of not less than 75 % when tested in a standard mortar made from the dry hydrate or from putty made from the hydrate which has been soaked for a period of 16 to 24 h.

9.2 Hydrated lime mortar made with Type S (special hydrated lime) or Type SA (special air-entraining hydrated lime), after suction for 60s, shall have a water retention value of not less than 85 % when tested in a standard mortar made from the dry hydrate.

10. Sampling and Inspection

10.1 The sampling, inspection, rejection, retesting, packing, and marking shall be conducted in accordance with Methods C 50.

11. Test Methods

11.1 Determine the properties enumerated in this specification in accordance with the following methods:

11.1.1 *Chemical Analysis*—Test Methods C 25.

11.1.2 *Physical Tests*—Test Methods C 110.

12. Special Package Marking

12.1 When Types NA or SA air-entraining hydrated lime are delivered in packages, the name and brand of the manufacturer, the type under this specification, and the words "AIR-ENTRAINING" shall be plainly indicated thereon or in case of bulk shipments, so indicated on shipping notices.

13. Keywords

13.1 hydrated lime; masonry; plasticity; popping and pitting; residue; stucco; Type N; Type NA; Type S; Type SA; unhydrated oxides; water retention; air entraining

Designation: C 219 – 99

Standard Terminology Relating to Hydraulic Cement[1]

This standard is issued under the fixed designation C 219; the number immediately following the designation indicates the year of original adoption or, in the case of revision, the year of last revision. A number in parentheses indicates the year of last reapproval. A superscript epsilon (ϵ) indicates an editorial change since the last revision or reapproval.

1. Scope

1.1 This terminology defines terms relating to hydraulic cements, their components, characteristics, properties, and the testing thereof. Some terms may have wider application than just to hydraulic cement.

1.2 See individual standards for terms applicable primarily therein, including meanings that may be more restrictive than those given here, and for explanations and descriptions of terms as they apply to those standards.

1.3 *This standard does not purport to address all of the safety concerns, if any, associated with its use. It is the responsibility of the user of this standard to establish appropriate safety and health practices and determine the applicability of regulatory limitations prior to use.*

2. Referenced Documents

2.1 *ASTM Standards:*

C 11 Terminology Relating to Gypsum and Related Building Materials and Systems[2]

C 51 Terminology Relating to Lime and Limestone (as used by the industry)[2]

C 125 Terminology Relating to Concrete and Concrete Aggregates[3]

C 294 Descriptive Nomenclature for Constituents of Natural Mineral Aggregates[3]

2.2 *ACI Standard:*

ACI 116R Cement and Concrete Terminology[4]

2.3 *API Standard:*

API 10A Specification for Oil-Well Cements[5]

3. Significance and Use

3.1 In definitions of cements, ingredients are cited only when they are inherent to the definition, for example portland-pozzolan cement. For ingredients and their quantity limits, if any, that are permitted or prohibited by a specification for a particular cement, see the applicable specification for that cement.

3.2 In definitions of materials including cements, the method of production is included only if it is inherent to the definition.

3.3 Related terms may be found in other terminology documents such as Terminology C 11, Terminology C 51, Terminology C 125, ACI 116R, and API 10A .

4. Terminology

addition, *n*—a material that is interground or blended in limited amounts into a hydraulic cement during manufacture either as a "processing addition" to aid in manufacturing and handling the cement or as a "functional addition" to modify the use properties of the finished product.

air content, *n*—of freshly mixed mortar the volume of air (and other gases) in mortar, expressed as a percentage of total volume of mortar.

air-entraining hydraulic cement, *n*— hydraulic cement containing an air entraining addition in such amount as to cause the product to entrain air in mortar within specified limits.

alkali equivalent, n—deprecated term; see **equivalent alkalies.**

aluminous cement, n—deprecated term.

anhydrite, n—see **calcium sulfate.**

blast-furnace slag, *n*—the nonmetallic product, consisting essentially of silicates and aluminosilicates of calcium and other bases, that is developed in a molten condition simultaneously with iron in a blast furnace.

blended hydraulic cement, *n*—a hydraulic cement consisting of two or more inorganic constituents (at least one of which is not portland cement or portland cement clinker) which separately or in combination contribute to the strength-gaining properties of the cement, (made with or without other constituents, processing additions and functional additions, by intergrinding or other blending).

calcium sulfate, *n*—*as applied to cement manufacturing*, a product composed essentially of calcium sulfate in any hydration state or states: anhydrite ($CaSO_4$), gypsum ($CaSO_4 \cdot 2H_2O$), and hemihydrate ($CaSO_4 \cdot \frac{1}{2}H_2O$).

cementitious material, hydraulic, *n*—an inorganic material or a mixture of inorganic materials which sets and develops strength by chemical reaction with water by formation of hydrates, and which is capable of doing so underwater.

early stiffening, *n*—rapid loss of plasticity or rapid development of rigidity in freshly mixed hydraulic cement paste, mortar, or concrete.

false set, n—with little evolution of heat, which can be

[1] This terminology is under the jurisdiction of ASTM Committee C-1 on Cement and is the direct responsibility of Subcommittee C01.91 on Editorial and Terminology.

Current edition approved Feb. 10, 1999. Published May 1999. Originally published as C 219 – 48 T. Last previous edition C 219 – 98.

[2] *Annual Book of ASTM Standards*, Vol 04.01.

[3] *Annual Book of ASTM Standards*, Vol 04.02.

[4] American Concrete Institute, PO Box 9094, Farmington Hills, MI 48333.

[5] American Petroleum Institute, 1220 L. Street, NW Washington, DC 2005.

dispelled by further mixing without the addition of water.

flash set, n—with evolution of considerable heat, which cannot be dispelled by further mixing without the addition of water.

equivalent alkalies, Na_2Oeq, *n—in hydraulic cement*, total combined sodium and potassium oxides, expressed as a percentage, calculated from the analysis, by means of the formula ($Na_2Oeq = \% Na_2O + 0.658 \times \% K_2O$).

DISCUSSION—This term is used to indicate the levels of alkalies in a material when only Na_2O and K_2O are considered. The value, 0.658, is the molecular ratio of Na_2O to K_2O. Alkali equivalent, soda equivalent, and sodium equivalent are deprecated terms.

expansive hydraulic cement, *n*—a hydraulic cement that forms a paste when mixed with water, and increases in volume a controlled amount during the early hardening period occurring after setting.

DISCUSSION—Expansive hydraulic cements are used to compensate for volume decrease due to shrinkage or to induce tensile stress in concrete reinforcement.

false set—see **early stiffening**

flash set—see **early stiffening**

fog room, *n*—a moist room in which the humidity is controlled by atomization of water.

gage length, *n—in cement testing*, the nominal length between the innermost ends of metal studs that are molded into a test specimen with the axis of the stud in each end of the test specimen coincident with the longitudinal axis of the test specimen and with each other.

granulated blast-furnace slag, *n*— the glassy granular material formed when molten blast-furnace slag is rapidly chilled, as by immersion in water.

gypsum, *n*—see **calcium sulfate**.

hydration, *n*—the chemical reaction between hydraulic cement and water forming new compounds most of which have strength-producing properties.

hydraulic cement, *n*—a cement that sets and hardens by chemical interaction with water and that is capable of doing so under water.

hydraulic mortar, *n*—deprecated term; see mortar.

length change, *n—in cement testing*, an increase or decrease in linear dimension due to causes other than applied load, usually measured along the longitudinal axis of a test specimen and expressed as a percentage of a gage length.

lot, *n*—a quantity of a single material that is considered as a unit, such as that manufactured during a single production run, offered at one time for sale, or offered at one time for inspection, or the contents of one or more transport containers drawn from one or more bins of material from a single production run, sequentially packaged from one or more bins, or a group of samples secured from one of the above.

masonry cement, *n*—a hydraulic cement manufactured for use in mortars for masonry construction or in plasters, or both, which contains a plasticizing material and, possibly, other performance-enhancing addition(s).

moist room, *n*—an enclosed room for storage and curing of paste, mortar, and concrete specimens in which temperature and high relative humidity can be controlled within specified limits.

mortar, *n*—a mixture of finely divided hydraulic cementitious material, fine aggregate, and water in either the unhardened or hardened state; hydraulic mortar.

natural cement, *n*—a hydraulic cement produced by calcining a naturally occurring argillaceous limestone at a temperature below the sintering point and then grinding to a fine powder.

DISCUSSION—Argillaceous limestone is described in Standard C 294.

normal consistency, *n*—a degree of plasticity of a hydraulic-cement paste that is appropriate for testing as measured by a stipulated method.

DISCUSSION—The result of tests for normal consistency is reported as the mass of water required to achieve this plasticity divided by the mass of hydraulic cement, expressed as a percentage.

paste, cement, *n*—an unhardened or hardened mixture of finely divided hydraulic cementitious material and water.

portland blast-furnace slag cement, *n*—a hydraulic cement consisting of an intimately interground mixture of portland-cement clinker and granulated blast-furnace slag or an intimate and uniform blend of portland cement and fine granulated blast-furnace slag in which the amount of the slag constituent is within specified limits.

portland cement, *n*—a hydraulic cement produced by pulverizing portland-cement clinker, and usually containing calcium sulfate.

portland-cement clinker, *n*—a clinker, partially fused by pyroprocessing, consisting predominantly of crystalline hydraulic calcium silicates.

portland-pozzolan cement, *n*—a hydraulic cement consisting of an intimate and uniform blend of portland cement or portland blast-furnace slag cement and fine pozzolan produced by intergrinding portland cement clinker and pozzolan, by blending portland cement or portland blast-furnace slag cement and finely divided pozzolan, or a combination of intergrinding and blending, in which the amount of the pozzolan constituent is within specified limits.

pozzolan, *n*—a siliceous or siliceous and aluminous material, which in itself possesses little or no cementitious value but which will, in finely divided form and in the presence of moisture, chemically react with calcium hydroxide at ordinary temperatures to form cementitious hydrates.

slag cement, *n*—a hydraulic cement consisting mostly of an intimate and uniform blend of ground, granulated blast-furnace slag and, one or both of portland cement or hydrated lime, in which the amount of the slag constituent is within specified limits.

5. Keywords

5.1 cement terminology; definitions; hydraulic cement terminology; terminology; terms

Designation: C 226 – 96

Standard Specification for Air-Entraining Additions for Use in the Manufacture of Air-Entraining Hydraulic Cement[1]

This standard is issued under the fixed designation C 226; the number immediately following the designation indicates the year of original adoption or, in the case of revision, the year of last revision. A number in parentheses indicates the year of last reapproval. A superscript epsilon (ε) indicates an editorial change since the last revision or reapproval.

This specification has been approved for use by agencies of the Department of Defense. Consult the DoD Index of Specifications and Standards for the specific year of issue which has been adopted by the Department of Defense.

1. Scope

1.1 This specification covers the requirements and methods for establishing the suitability of a material for use as an air-entraining addition to be interground with the clinker in the manufacture of air-entraining hydraulic cement conforming to Specifications C 150, C 595, and C 1157.

1.2 The values stated in SI units are to be regarded as the standard.

2. Referenced Documents

2.1 *ASTM Standards:*

C 33 Specification for Concrete Aggregates[2]

C 39 Test Method for Compressive Strength of Cylindrical Concrete Specimens[2]

C 109 Test Method for Compressive Strength of Hydraulic-Cement Mortars (Using 2-in. or 50-mm Cube Specimens)[3]

C 114 Test Methods for Chemical Analysis of Hydraulic Cement[3]

C 115 Test Method for Fineness of Portland Cement by the Turbidimeter[3]

C 138 Test Method for Unit Weight, Yield, and Air Content (Gravimetric) of Concrete[2]

C 143 Test Method for Slump of Hydraulic Cement Concrete[2]

C 150 Specification for Portland Cement[3]

C 151 Test Method for Autoclave Expansion of Portland Cement[3]

C 175 Specifications for Air-Entraining Portland Cement[4]

C 185 Test Method for Air Content of Hydraulic Cement Mortar[3]

C 187 Test Method for Normal Consistency of Hydraulic Cement[3]

C 191 Test Method for Time of Setting of Hydraulic Cement by Vicat Needle[3]

C 192 Practice for Making and Curing Concrete Test Specimens in the Laboratory[2]

C 204 Test Method for Fineness of Portland Cement by Air Permeability Apparatus[3]

C 231 Test Method for Air Content of Freshly Mixed Concrete by the Pressure Method[2]

C 293 Test Method for Flexural Strength of Concrete (Using Simple Beam with Center-Point Loading)[2]

C 595/C 595M Specification for Blended Hydraulic Cements[3]

C 596 Test Method for Drying Shrinkage of Mortar Containing Portland Cement[3]

C 666 Test Method for Resistance of Concrete to Rapid Freezing and Thawing[2]

C 1157 Performance Specification for Blended Hydraulic Cement[3]

2.2 *ACI Standards:*[5]

ACI 211.1-77 Recommended Practice for Selecting Proportions for Normal and Heavyweight Concrete

3. General Requirements

3.1 Air-entraining additions shall conform to the respective requirements prescribed in this specification.

3.2 The trade name, source, and character of the material, and means for the quantitative identification of the proposed addition in the finished cement, shall be furnished by the maker or seller of the addition, and that information shall form a part of the record of tests of the addition.

3.3 Air-entraining additions shall be evaluated by testing cements ground with and without the additions. The cements ground without the additions shall be referred to in this specification as "control" cements.

3.4 An air-entraining addition under this specification, when interground with hydraulic cement, shall produce a cement that complies with the appropriate Specification C 150, C 595, or C 1157 and when evaluated by the results of tests made according to the procedures herein described, shall also comply with the following requirements as to the effect of the addition on the properties of the cement:

3.4.1 The time of setting of cement containing the addition shall not vary from the time of setting of the respective "control" cement by more than 50 %.

3.4.2 The percentage autoclave expansion for cement containing the addition shall not exceed the percentage autoclave expansion for the corresponding "control" cement by more than 0.1.

3.4.3 The compressive strength of standard mortar cubes made with cement containing the addition shall be not less

---

[1] This specification is under the jurisdiction of ASTM Commitee C-1 on Cement and is the direct responsibility of Subcommittee C01.20 on Additions.

Current edition approved July 10, 1996. Published September 1996. Originally published as C 226 – 50 T. Last previous edition C 226 – 95.

[2] *Annual Book of ASTM Standards,* Vol 04.02.

[3] *Annual Book of ASTM Standards,* Vol 04.01.

[4] Discontinued, see 1949 *Annual Book of ASTM Standards,* Part 3.

[5] Available from the American Concrete Institute, P.O. Box 19150, Detroit, MI 48219.

than 80 % of the compressive strength of similar cubes made with the corresponding "control" cement.

3.4.4 The percentage length change of air-stored mortar bars made with cement containing the addition, based on an initial measurement at the age of 7 days, and expressed as a percentage change in length, shall be not more than 0.01 greater than that of similar mortar bars made with the corresponding "control" cement and similarly tested.

3.4.5 The percentage of air entrained in the concrete made with cement containing the addition shall exceed by at least 2.5 the percentage air in similar concrete prepared with the corresponding "control" cement. (See 4.1.1 for the limitation of air-entraining properties of the "control" cement.)

3.4.6 The compressive strength of the concrete made with cement containing the addition shall be not less than 80 % of the compressive strength of similar concrete made with the corresponding "control" cement.

3.4.7 The flexural strength of concrete made with cement containing the addition shall be not less than 85 % of the flexural strength of corresponding concrete made with the "control" cement.

3.4.8 In the freezing and thawing test, the durability factor of the concrete made with the cement containing the proposed addition shall be not less than 80 % of the durability factor of similar concrete made with the corresponding "control" cement and containing the reference addition as specified in 4.3. (See 11.1.3 for the method of calculating the durability factor.)

4. Materials

4.1 *Cements:*

4.1.1 In cases where it is desired that the proposed air-entraining addition be accepted for general use, tests shall be made on six lots of cement ground at cement plants, using commercial grinding equipment. From each of three different samples of clinkers, two lots of cement shall be ground, representing respectively: a Type I portland cement containing not less than 9 % tricalcium aluminate (C_3A), calculated as specified in Table 1 of Specification C 150 and a Type II and a Type III portland cement conforming to Specification C 150. One lot, the "control" shall be ground without the proposed air-entraining addition; the proposed addition shall be interground with the other lot, using the addition in such amounts as to produce the air/entrainment required in Specification C 150. Not more than two of the three clinkers shall be produced by or ground at the same mill.

4.1.2 In cases where it is desired that the proposed air-entraining addition be limited in use to specific types of cement in specific cement manufacturing plants, either or both less in number than required in 4.1.1, the tests and test procedure shall be as specified in 4.1.1, except that the number of cements to be tested shall be limited to those under specific consideration.

4.1.3 The two companion cements made from any one clinker shall be ground to the same fineness (within 50 cm^2/g) as measured by the turbidimeter test (7.1.3) or (within 100 cm^2/g) as measured by the air permeability apparatus, and the sulfur trioxide (SO_3) content expressed as a percentage of the cement weight and reported to the nearest 0.01 %, shall differ by no more than 0.24 for all types of cement. Each "control" cement shall comply with all of the requirements applicable to that type of cement, as prescribed in Specification C 150.

4.1.4 The percentage of each of the following constituents shall be determined for each lot of cement tested: silicon dioxide (SiO_2), aluminum oxide (Al_2O_3), iron oxide (Fe_2O_3), calcium oxide (CaO), magnesium oxide (MgO), SO_3, ignition loss, insoluble residue, sodium oxide (Na_2O), and potassium oxide (K_2O). There shall also be calculated the potential percentages of the following compounds: tricalcium silicate, dicalcium silicate, tricalcium aluminate, and tetracalcium aluminoferrite. Determinations for the percentage of addition shall be made on the cements containing the addition, using the method proposed therefore by the maker or seller of the addition.

4.2 *Aggregates:*

4.2.1 The fine and coarse aggregates used in the tests shall conform to the requirements of Specification C 33, except that the grading of the aggregates shall conform to the requirements given in Table 1.

4.2.2 The coarse aggregate shall be carefully separated on the 25.0-mm (1-in.), 19.0-mm (¾-in.), 12.5-mm (½-in.), 9.5-mm (⅜-in.), and 4.75-mm (No. 4) sieves, and then recombined, using equal quantities by weight of each of the resulting four sizes.

4.2.3 The fine and coarse aggregates used in the tests of any two companion cements (that is, a cement containing the addition and the corresponding "control" cement) shall each come from a single lot of such aggregate.

4.3 *Reference Addition:*

4.3.1 The reference addition used in the concrete mixture specified in Section 8, from which specimens will be made for tests for resistance to freezing and thawing as specified in 10.2.3, shall be any one of the four materials (Vinsol resin, Darex, N-TAIR, or Airalon) that have been declared acceptable by ASTM under the former Specifications C 175 – 48 T.

4.3.2 The reference addition to be used will be designated by the person or agency for whom the testing will be performed. If no reference addition is designated, the material known commercially as "Vinsol resin" shall be used. The Vinsol resin used shall be neutralized with 15 percent by mass of sodium hydroxide (NaOH). The air contents of the concrete containing the reference addition and the concrete containing the proposed addition shall agree within 0.5 percentage points.

4.3.3 The reference addition, used as an admixture, is intended for use with control cements only in concrete for freezing-and-thawing tests to establish the durability factor by means of which the concretes containing the addition under test may be evaluated.

4.3.4 *Preparation of Standard Reference Solutions:*

4.3.4.1 Place 50 g (total solids in the case of solution or pastes) of the designated reference addition in 500 mL of freshly distilled water in a 1000-mL flask and mix thoroughly until the solids are completely dissolved or the paste or solution is uniformly diluted. After surface foam has been dissipated, dilute to 1000 mL and mix thoroughly.

4.3.4.2 In the case of Vinsol resin, the neutralized solution shall be made as follows: Dissolve 7.50 g of cp NaOH in 100 mL of distilled water. Add a few drops of this solution to 300

TABLE 1 Grading Requirements of Aggregates

Sieve	Percentage Passing
Fine Aggregate	
4.75-mm (No. 4)	100
1.18-mm (No.16)	65 to 75
300-μm (No. 50)	15 to 20
150-μm (No. 100)	2 to 5
Coarse Aggregate	
25.0-mm (1-in.)	100
19.0-mm (¾ in.)	75
12.5-mm (½-in.)	50
9.5-mm (⅜-in.)	25
4.75-mm (No.4)	0

to 350 mL of distilled water contained in a 600-mL beaker. Add 50.00 g of dry, unneutralized Vinsol resin in pulverized form to the beaker and stir until all of the resin is wetted and well dispersed. Then add all of the NaOH solution to this suspension and stir until all of the resin is in solution. Transfer to a measuring flask, dilute to 1000 mL and mix thoroughly. From this stock standard solution prepare a dilute standard solution by diluting 100 mL of the stock solution to 1000 mL.

5. Sampling Cement

5.1 Samples of the plant-ground cements shall be obtained during grinding. Prior to the commencement of the sampling of a given lot of cement, the mill shall have run for approximately 4 h to establish equilibrium. Notes shall be kept as to the rate and continuity of the feed of the addition, the form in which the addition is used, strength of the solution, and the mill temperature. Fineness of the grinding should be checked during the grinding.

5.2 The quantity of sample shall be not less than 272 kg (600 lb) for the cement containing the proposed addition and for the corresponding control cement.

5.3 As the cement samples are secured, they shall be placed in metal drums provided with gasket-fitted lids. The drums shall be tightly closed at the end of the sampling period. Prior to use, the samples of a given lot of cement shall be well blended to form a uniform, representative composite.

6. Test Methods

6.1 Determine the properties enumerated in this specification in accordance with the methods prescribed in Sections 7 to 11.

7. Tests on Cement

7.1 Test the cement samples in accordance with the following methods:

7.1.1 *Chemical Analysis of Cement*—Test Methods C 114.

7.1.2 *Determination of Addition in the Finished Cement*—Determine the percentage of the addition in the finished cement by the method furnished by the manufacturer or seller of the proposed addition. The method shall be adequate for the qualitative and quantitative determination of the addition in the finished cement, and shall be fully described in the report of the tests on the addition.

7.1.3 *Fineness of Cement*—Test Method C 115 or C 204.

7.1.4 *Normal Consistency*—Test Method C 187.

7.1.5 *Time of Setting*—Determine the time of setting with the Vicat needle, in accordance with Test Method C 191.

7.1.6 *Autoclave Expansion*—Test Method C 151.

7.1.7 *Air Content of Mortar*—Test Method C 185.

7.1.8 *Compressive Strength of Mortar*—Test Method C 109.

7.1.9 *Length Change of Mortar*—For each lot of cement, determine the length change of mortar bars according to Test Method C 596 except, after demolding, cure the specimens in saturated lime water for 6 days (7 days total age) and measure the length of each specimen at 7 days of age. Then, place the specimens in air storage for the remainder of the test period. Obtain a length comparator reading for each specimen at 28 days, 180 days, and 365 days of total age. Report the length change for each bar as a percentage of its length at 7 days of age.

8. Concrete Mixtures

8.1 *Preparation and Weighing*—Prepare all materials used in preparing the concrete mixtures, and make all weighings, as prescribed in Practice C 192. Report the amount of mixing water on the basis of saturated surface-dry aggregates.

8.2 *Proportions*—Design one concrete mixture, having an actual cement content of 307 ± 3 kg/m^3 (517 ± 5 lb/yd^3), and use this mixture in all of the concrete tests specified herein. Adjust the water content of the mixture to provide concrete having a consistency equal to a 64 ± 13-mm (2½ ± ½-in.) slump. Adjust the ratio of fine aggregate to total aggregate to the optimum for concrete to be consolidated by hand-rodding (suggested trial values[6] for the percentage of fine aggregate in the total aggregate, by absolute volume, are shown in Table 2).

8.3 *Mixing of Concrete*—Mix the concrete in accordance with Practice C 192 except as follows: Hand mixing shall not be permitted. The rated capacity of the mixer shall be not more than twice the size of the batch used. Use a preliminary (buttering) batch of the same proportions as the test batch to coat the mixer and discard it just prior to receiving the batch of the test concrete, which shall be mixed continuously for 2 min following the addition of the mixing water. Do not scrape the mixer after buttering. At the end of the mixing period dump the concrete without scraping from the mixer into a metal wheelbarrow or pan, and then turn with a shovel before making the slump, unit weight, and air determinations, and also before being placed in the test specimen molds.

9. Tests on Plastic Concrete

9.1 Test samples of the plastic concrete in accordance with the following methods:

9.1.1 *Consistency*—Test Method C 143.

9.1.2 *Unit Weight and Air Content*—Make a determination of the unit weight, yield, and air content for each batch, in accordance with Test Method C 138, except that the air

[6] Values for concrete without entrained air are based on the ACI Standard Recommended Practice for Selecting Proportions for Normal and Heavyweight Concrete (ACI 211.1-77).

TABLE 2 Fine Aggregate in Total Aggregate, percent

Coarse Aggregate, Maximum Size 25 mm (1 in.)	Concrete Without Entrained Air	Concrete With Entrained Air
Angular	45	41
Rounded	40	36

content of the concrete may be determined in accordance with Test Method C 231.

10. Test Specimens of Hardened Concrete

10.1 *Number of Specimens*—Make six test specimens from each condition of concrete to be compared for each test and age. Make test specimens representing each test and each condition of concrete from at least three separate batches. An equal number of specimens for each variable should be made on any given day. When it is impossible to make at least one specimen for each variable on a given day, complete the mixing of the entire series of specimens in as few days as possible and repeat one of the mixes each day as a standard of comparison.

10.2 *Types of Specimens:*

10.2.1 *Compressive Strength*—Compressive strength test specimens shall be 152- by 305-mm (6- by 12-in.) cylinders, and shall be made and cured as prescribed in Practice C 192.

10.2.2 *Flexural Strength*—Make and cure flexural strength test specimens as specified in Practice C 192, except that the size of the beams shall be 76 by 101 by 406 mm (3 by 4 by 16 in.).

10.2.3 *Resistance to Freezing and Thawing*—Make and cure test specimens for resistance to freezing and thawing as specified in Practice C 192, except that the size of the beams shall be 76 by 101 by 406 mm (3 by 4 by 16 in.).

11. Tests on Hardened Concrete

11.1 Test specimens of hardened concrete, as specified in Section 10, in accordance with the following methods, and at the ages herein specified.

11.1.1 *Compressive Strength*—Use Test Method C 39, and test specimens at ages of 3, 7, and 28 days, and 3 months, except that for Type III cement a test shall be made also at 1 day.

11.1.2 *Flexural Strength*—Test Method C 293. Test flexural strength specimens at the ages of 3, 7, and 28 days, 3 months, and 1 year, except that for Type III cement a test shall be made also at 1 day.

11.1.3 *Resistance to Freezing and Thawing*—Use Test Method C 666, except that the specimens shall be tested at the end of the 28-day curing period.

12. Report

12.1 The report covering the results of the evaluation of a material proposed as an air-entraining addition under this specification shall include the following information:

12.1.1 Trade name, source, and character of the material and means for the quantitative identification of the proposed addition in the finished cement—all as furnished by the manufacturer or seller of the addition.

12.1.2 Identification of the cements as to their type. General geographical location of mills where the test lots of cements were ground. (Company and brand names need not be given.)

12.1.3 Detailed results of all analyses and tests prescribed in this specification, as well as pertinent information required.

12.1.4 Comparison of test results to determine compliance with the requirements prescribed in 3.4.

12.1.5 Name and location of the laboratory or laboratories that made the tests covered by the report.

12.1.6 A statement indicating whether the evaluation was made to establish acceptability of the addition for general use (4.1.1), or whether the evaluation was made to establish acceptability for use with specific types of cement in specific cement mills (4.1.2). In the latter case, the type or types of cement and the brand names and location of mills that are covered by the tests shall be given.

13. Keywords

13.1 addition; air entraining; hydraulic cement

Standard Specification for Flow Table for Use in Tests of Hydraulic Cement[1]

This standard is issued under the fixed designation C 230; the number immediately following the designation indicates the year of original adoption or, in the case of revision, the year of last revision. A number in parentheses indicates the year of last reapproval. A superscript epsilon (ε) indicates an editorial change since the last revision or reapproval.

1. Scope

1.1 This specification coves requirements for the flow table and accessory apparatus (Note 1) used in making flow tests for consistency of mortars in tests of hydraulic cement.

NOTE 1—To help clarify the design of the flow table and accessory apparatus see the drawing in Fig. 1. This drawing is for informational purposes only.

1.2 The values stated in either inch-pound units or SI units shall be regarded separately as standard. The values stated are not exact equivalents; therefore, each system must be used independently of the other. It is permissible to use an inch-pound caliper and mold with a SI flow table or a SI caliper and mold with an inch-pound flow table. It is not permissible to mix a SI mold with an inch-pound caliper or an inch-pound mold with a SI caliper.

2. Flow Table and Frame

2.1 The flow table apparatus shall consist of an integrally cast rigid iron frame and a circular rigid table top 10 ± 0.1 in. [255± 2.5 mm] in diameter, with a shaft attached perpendicular to the table top by means of a screw thread. The table top and shaft with contact shoulder shall be mounted on a frame in such a manner that it can be raised and dropped vertically through the specified height of 0.500 ± 0.005 in. [12.7± 0.13 mm] for new tables and of 0.500 ± 0.015 in. [12.7 ± 0.38 mm] for tables in use, by means of a rotated cam. The table top shall have a fine machined plane surface, free of blowholes and surface defects. The top shall be scribed with eight equidistant lines 2⅝ in. [68 mm] long, extending from the outside circumference toward the center of the table. Each line shall end with a scribed arc, ¼ in. [6 mm] long, whose center point is the center of the table top with a radius of 2⅜ in. [59.5 mm]. The scribe lines shall be made with a 60° tool to a depth of 0.01 in. [0.25 mm]. The table top shall be of cast brass or bronze having a Rockwell hardness number not less than 25 HRB with an edge thickness of 0.3 in. [7.5 mm], and shall have six integral radial stiffening ribs. The table top and attached shaft shall weigh 9 ± 0.1 lb [4.08± 0.05 kg] and the weight shall be symmetrical around the center of the shaft.

2.2 The cam and vertical shaft shall be of medium carbon machinery steel, hardened on the end of the shaft contacting the cam and the tip of the cam contacting the shaft. The shaft shall be straight and the difference between the diameter of the shaft and the diameter of the bore of the frame shall be not less than 0.002 in. [0.05 mm] and not more than 0.003 in. [0.08 mm] for new tables and shall be maintained at 0.002 to 0.010 in. [0.05 to 0.25 mm] for tables in use. The end of the shaft shall not fall upon the cam at the end of the drop, but shall make contact with the cam not less than 120° from the point of drop. The face of the cam shall be a smooth spiraled curve of uniformly increasing radius from ½ to 1¼ in. [13 to 32 mm] in 360° and there shall be no appreciable jar as the shaft comes into contact with the cam. The cam shall be so located and the contact faces of the cam and shaft shall be such that the table does not rotate more than one revolution in 25 drops. The surfaces of the frame and of the table that come into contact at the end of the drop shall be maintained smooth, plane, and horizontal and parallel with the upper surface of the table and shall make continuous contact over a full 360°.

2.3 The supporting frame of the flow table shall be integrally cast of fine-grained, high-grade cast iron. The frame casting shall have three integral stiffening ribs extending the full height of the frame and located 120° apart. The top of the frame shall be chilled to a depth of approximately ¼ in. [6 mm], and the face shall be ground and lapped square with the bore to give 360° contact with the shaft shoulder. The underside of the base of the frame shall be ground to secure a complete contact with the steel plate beneath.

2.4 The flow table shall be driven by a motor (Note 2), connected to the cam shaft through an enclosed worm gear speed reducer and flexible coupling. The speed of the cam shaft shall be approximately 100 r/min. The motor drive mechanism shall not be fastened or mounted on the table base plate or frame.

NOTE 2—A ½₀-hp [40-W] motor has been found adequate.

2.5 The performance of a flow table shall be considered satisfactory if, in calibration tests, the table gives a flow value that does not differ by more than 5 percentage points from flow

[1] This specification is under the jurisdiction of ASTM Committee C-1 on Cement and is the direct responsibility of Subcommittee C01.22 on Workability.

Current edition approved Aug. 10, 1998. Published March, 1999. Originally published as C 230 – 49 T. Last previous edition C 230 – 97.

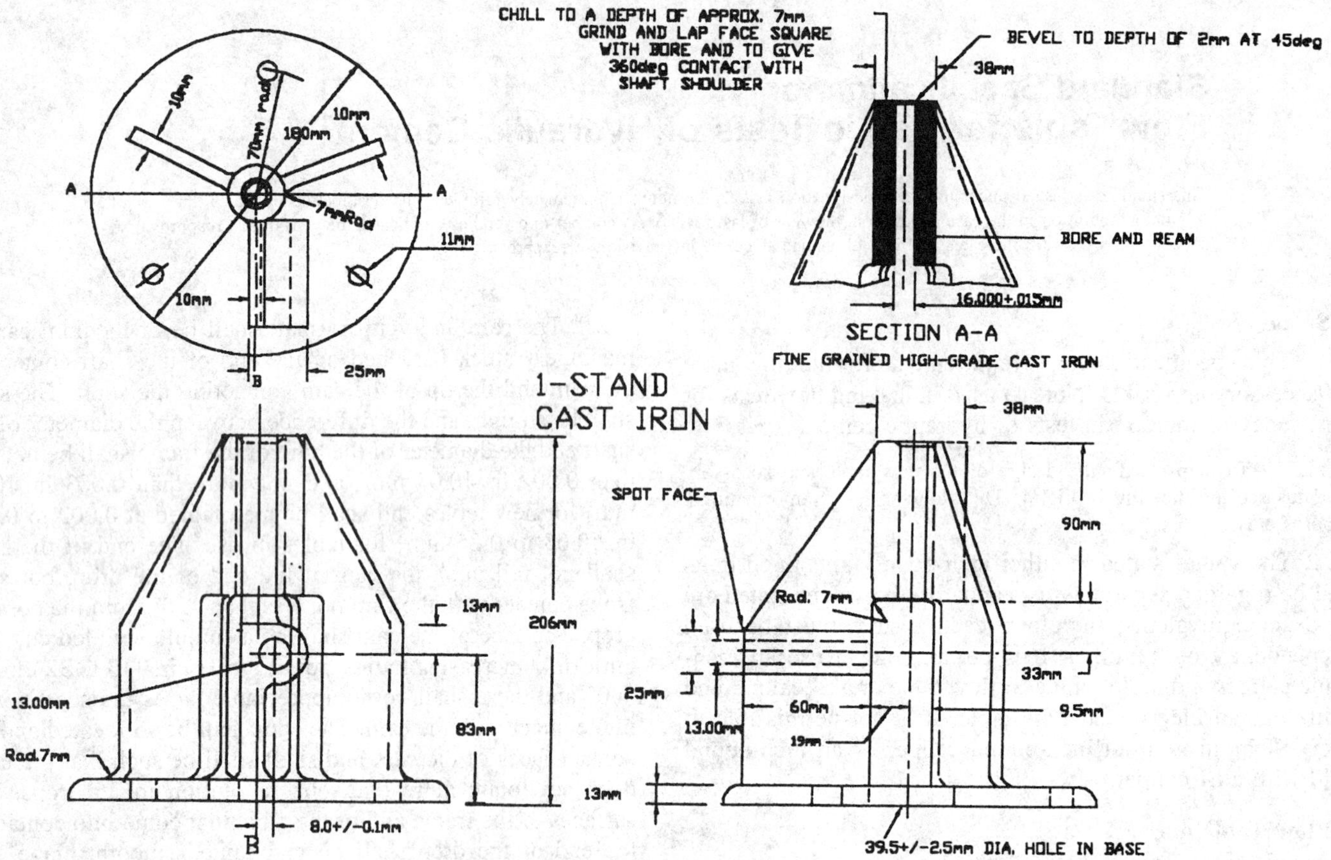

FIG. 1 Flow Table and Accessory Apparatus (Partial)

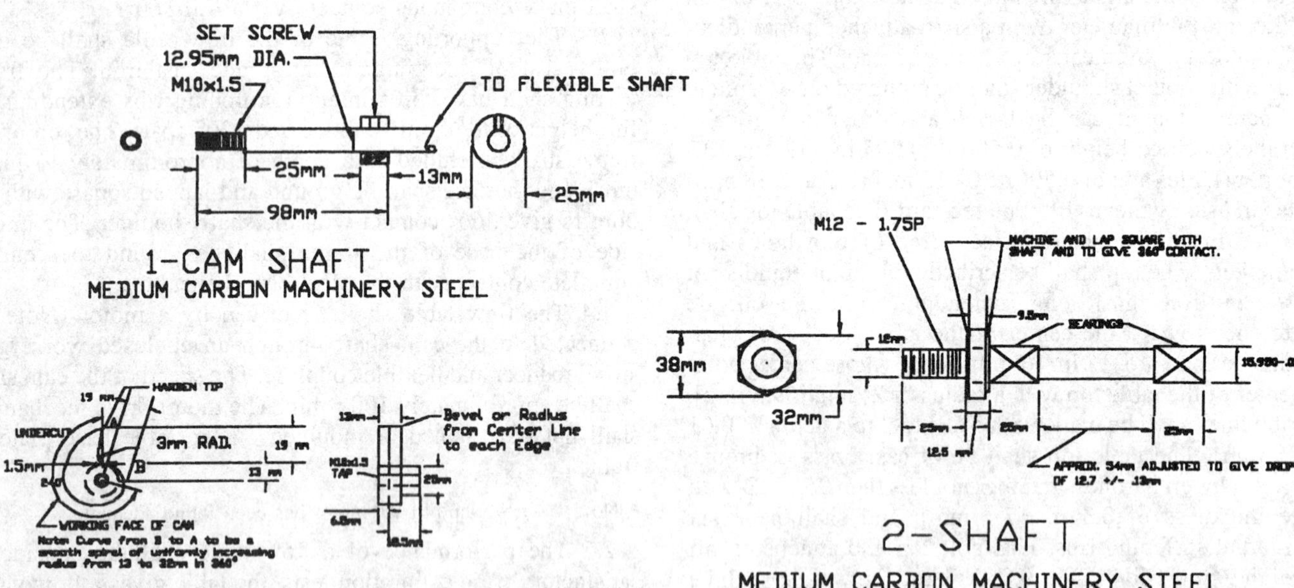

FIG. 1 Flow Table and Accessory Apparatus (Partial) *(continued)*

values obtained with a suitable calibration material.[2]

3. Flow Table Mounting

3.1 The flow table frame shall be tightly bolted to a cast iron or steel plate at least 1 in. [25 mm] thick and 10 in. [250 mm] square. The top surface of this plate shall be machined to a smooth plane surface. The plate shall be anchored to the top of a concrete pedestal by four ½-in. [13-mm] bolts that pass through the plate and are imbedded at least 6 in. [150 mm] in the pedestal. The pedestal shall be cast inverted on the base plate. A positive contact between the base plate and the

[2] Such a material can be obtained from the Cement and Concrete Reference Laboratory at the National Institute of Standards and Technology, Gaithersburg, MD 20899.

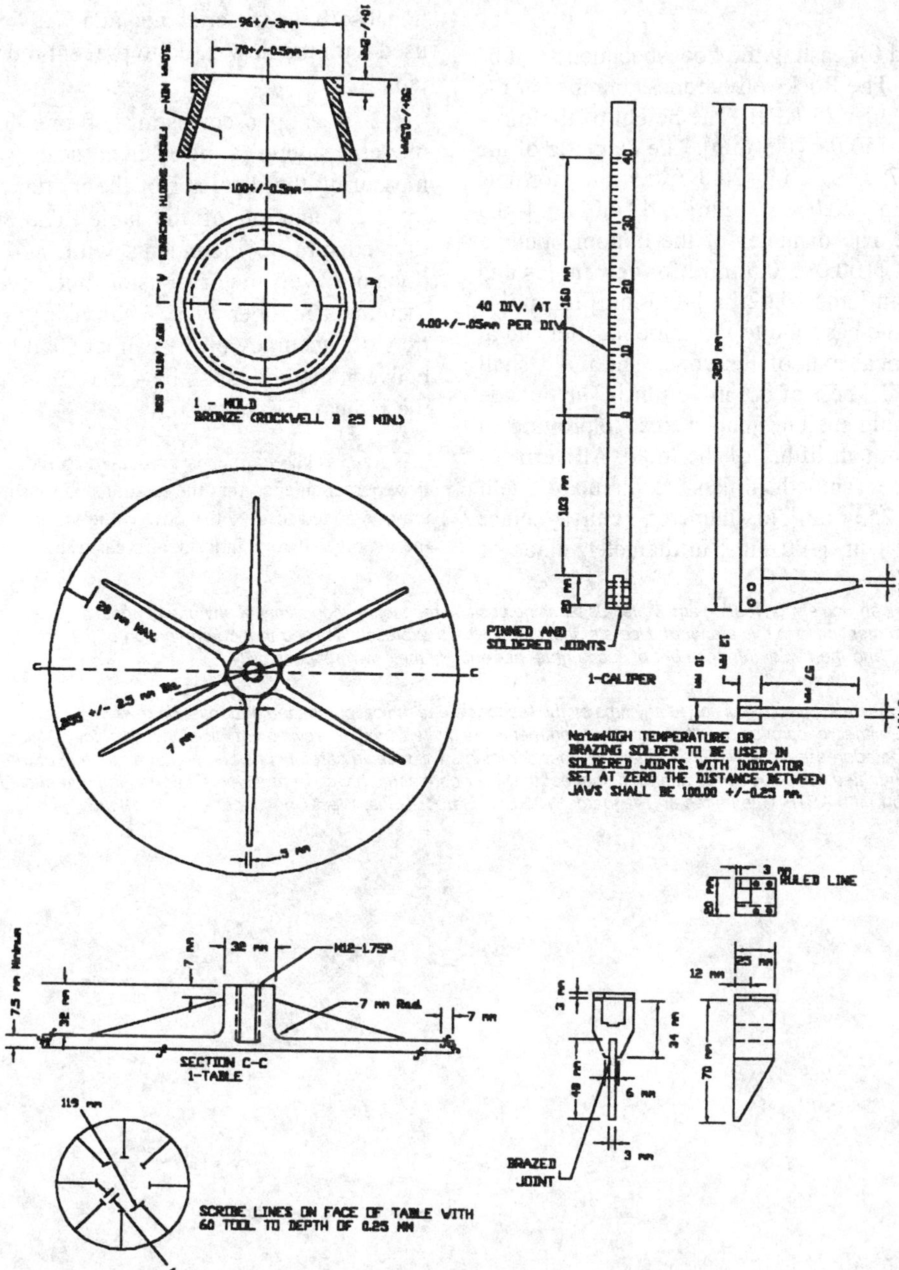

FIG. 1 Flow Table and Accessory Apparatus (Partial) *(continued)*

pedestal shall be obtained at all points. No nuts or other such leveling devices shall be used between the plate and the pedestal. Leveling shall be effected by suitable means under the base of the pedestal.

3.2 The pedestal shall be 10 to 11 in. [250 to 275 mm] square at the top, and 15 to 16 in. [375 to 400 mm] square at the bottom, 25 to 30 in. [625 to 750 mm] in height, and shall be of monolithic construction, cast from concrete weighing at least 140 lb/ft^3 [2240 kg/m^3]. A stable gasket cork padding, ½ in. [13 mm] thick and the same size as the pedestal bottom or four pieces of padding ½ in. [13 mm] thick and approximately 4 in. [100 mm] square, shall be inserted under the pedestal or the four corners, respectively. The flow table shall be checked frequently for levelness of the table top, stability of the pedestal, and tightness of the bolts and nuts in the table base and the pedestal table. (A torque of 20 lb·ft [27 N·m] is recommended when tightening those fastenings.)

3.3 The table top, after the frame has been mounted on the pedestal, shall be level along two diameters at right angles to each other, in both the raised and lowered positions.

4. Flow Table Lubrication

4.1 The vertical shaft of the table shall be kept clean and shall be lightly lubricated with a light oil (SAE-10). Oil shall not be present between the contact faces of the table top and the supporting frame. Oil on the cam face will lessen wear and promote smoothness of operation. The table should be raised and permitted to drop a dozen or more times just prior to use if it has not been operated for some time.

5. Mold and Caliper

5.1 The conical mold for casting the flow specimen shall be of cast bronze or brass. The Rockwell hardness number of the metal shall be not less than 25 HRB. The height of the mold shall be 2.00 ± 0.02 in. [50.0 ± 0.5 mm]. The diameter of the top opening shall be 2.75± 0.02 in. [70.0 ± 0.5 mm] for new molds and 2.75 + 0.05 in. [70.0 + 1.3 mm] and –0.02 in. [–0.5 mm] for molds in use. The diameter of the bottom opening shall be 4.00 ± 0.02 in. [100.0 ± 0.5 mm] for new molds and 4.00 + 0.05 in. [+ 1.3 mm] and –0.02 in. [–0.5 mm] for molds in use. The surfaces of the base and top shall be parallel and at right angles to the vertical axis of the cone. The mold shall have a minimum wall thickness of 0.2 in. [5 mm]. The outside of the top edge of the mold shall be shaped so as to provide an integral collar for convenient lifting of the mold. All surfaces shall be machined to a smooth finish. A circular shield approximately 10 in. [255 mm] in diameter, with a center opening approximately 4 in. [100 mm] in diameter, made of nonabsorbing material not attacked by the cement, shall be used with the flow mold to prevent mortar from spilling on the table top.

5.2 A caliper consisting of one fixed jaw and one jaw movable along a permanent scale, shall be provided for measuring the diameter of the mortar after it has been spread by the operation of the table. The scale shall be machine divided into 40 increments with 0.16 in. [4.0 mm] between divisions with major division lines every 5 divisions and the increment number every 10 divisions (Note 3). The construction and accuracy of the caliper shall be such that the distance between the jaws shall be 4 ± 0.01 in. [100 ± 0.25 mm] when the indicator is set at zero.

NOTE 3—The caliper is graduated to indicate one fourth of the actual flow percentage, so that the readings of four measurements on the caliper may be added to give the flow value without the necessity of calculating the average of four individual measurements of the total flow.

Standard Test Method for Bleeding of Cement Pastes and Mortars[1]

This standard is issued under the fixed designation C 243; the number immediately following the designation indicates the year of original adoption or, in the case of revision, the year of last revision. A number in parentheses indicates the year of last reapproval. A superscript epsilon (ε) indicates an editorial change since the last revision or reapproval.

1. Scope

1.1 This test method covers the determination of the bleeding rate and bleeding capacity of cement paste and mortar.

1.2 The values stated in SI units are to be regarded as the standard. The values given in parentheses are for information only.

1.3 *This standard does not purport to address all of the safety concerns, if any, associated with its use. It is the responsibility of the user of this standard to establish appropriate safety and health practices and determine the applicability of regulatory limitations prior to use.* For specific hazard statements, see Section 5.

2. Referenced Documents

2.1 *ASTM Standards:*

C 109 Test Method for Compressive Strength of Hydraulic Cement Mortars (Using 2-in. or 50-mm Cube Specimens)[2]

C 230 Specification for Flow Table for Use in Tests of Hydraulic Cement[2]

C 305 Practice for Mechanical Mixing of Hydraulic Cement Pastes and Mortars of Plastic Consistency[2]

C 778 Specification for Standard Sand[2]

C 1005 Specification for Weights and Weighing Devices for Use in the Physical Testing of Hydraulic Cements[2]

3. Significance and Use

3.1 This test method is intended primarily for use by those interested in research into the relative bleeding characteristics of cement pastes and mortars.

4. Apparatus

4.1 *Weights and Weighing Devices*—The weights and weighing devices shall conform to the requirements of Specification C 1005. The weighing device shall have a capacity of at least 5000 g and shall be evaluated for precision and accuracy at a total load of 5000 g.

4.2 *Tamper*—The tamper shall be made of a nonabsorptive, nonabrasive material such as medium-hard rubber or seasoned oak wood rendered nonabsorptive by immersion for 15 min in paraffin at approximately 200°C (392°F), and shall have a cross section of 13 by 25 mm (½ by 1 in.) and a convenient length (127 or 152 mm (5 or 6 in.)). The tamping face of the tamper shall be flat and at right angles to the length of the tamper.

4.3 *Liquid Displacement Bleeding Apparatus*[3]—The bleeding apparatus shall consist of a container for the paste or mortar to be tested, a collecting ring (Note 1), and a measuring buret with funnel, as illustrated in Fig. 1. The container shall be made of noncorroding metal, 127 mm (5 in.) in diameter and 102 mm (4 in.) in height. An outer ring 152 mm (6 in.) in diameter shall be soldered to the perimeter of the cup 12.7 mm (0.5 in.) below the top rim of the cup and extended 38 mm (1.5 in.) above the rim. The collecting ring shall be 76 mm (3.0 in.) in inside diameter, flanged as shown in Fig. 1, to direct the rising water into the funnel, and bevelled on the lower rim as shown in the lower left insert of Fig. 1 to trap only the rising water within the area of the 76-mm diameter ring. This ring shall be made of noncorroding metal and shall be centered and supported at the proper depth by brackets as shown in Fig. 1. The container and ring shall be of sufficient thickness and rigidity to maintain their shape under normal handling. The funnel and buret assembly shall be made of glass and so joined that there will be no rough edges to inhibit the rise of the water into the buret. The outside diameter of the funnel shall be 73.7 mm (2.9 in.). The rim of the funnel shall be ground on a plane parallel to the axis of the funnel and buret, but the resulting edge shall be rounded. The buret shall be of 25-mL capacity and not over 460 mm in over-all length. The constricted capillary, stopcock, and aspirator may be connected to the top of the buret through a suitable rubber stopper. The buret assembly shall be supported by a ring stand or similar equipment so that it can be conveniently raised or lowered over the collecting ring and held rigidly in place during the test period.

NOTE 1—At times, water bubbles rise up the outside wall of the collecting ring and enter the buret. It has been found that a band of inert epoxy put around the outside of the collecting ring blocks these bubbles and prevents them entering the buret.

4.4 *Mechanical Mixer*—A mechanical mixer shall be used that conforms to the requirements of Practice C 305.

4.5 *Flow Table*—The flow table shall conform to the requirements of Specification C 230.

4.6 *Flow Mold*—The flow mold shall conform to the requirements of Specification C 230.

[1] This test method is under the jurisdiction of ASTM Committee C-1 on Cement and is the direct responsibility of Subcommittee C01.22 on Workability.

Current edition approved Nov. 10, 1995. Published January 1996. Originally published as C 243 – 50 T. Last previous edition C 243 – 85 (1989).

[2] *Annual Book of ASTM Standards*, Vol 04.01.

[3] The application of the liquid-displacement principle to the collection and continuous measurement of bleeding water is due to R. C. Valore, Jr. of the National Institute of Standards and Technology, Washington, DC, who, with the collaboration of colleagues J. E. Bowling and R. L. Blaine, developed the idea into the present apparatus. See the paper by these authors on "The Direct and Continuous Measurement of Bleeding in Portland Cement-Water Mixtures," *Proceedings*, ASTM, Vol 49, 1949, p. 891.

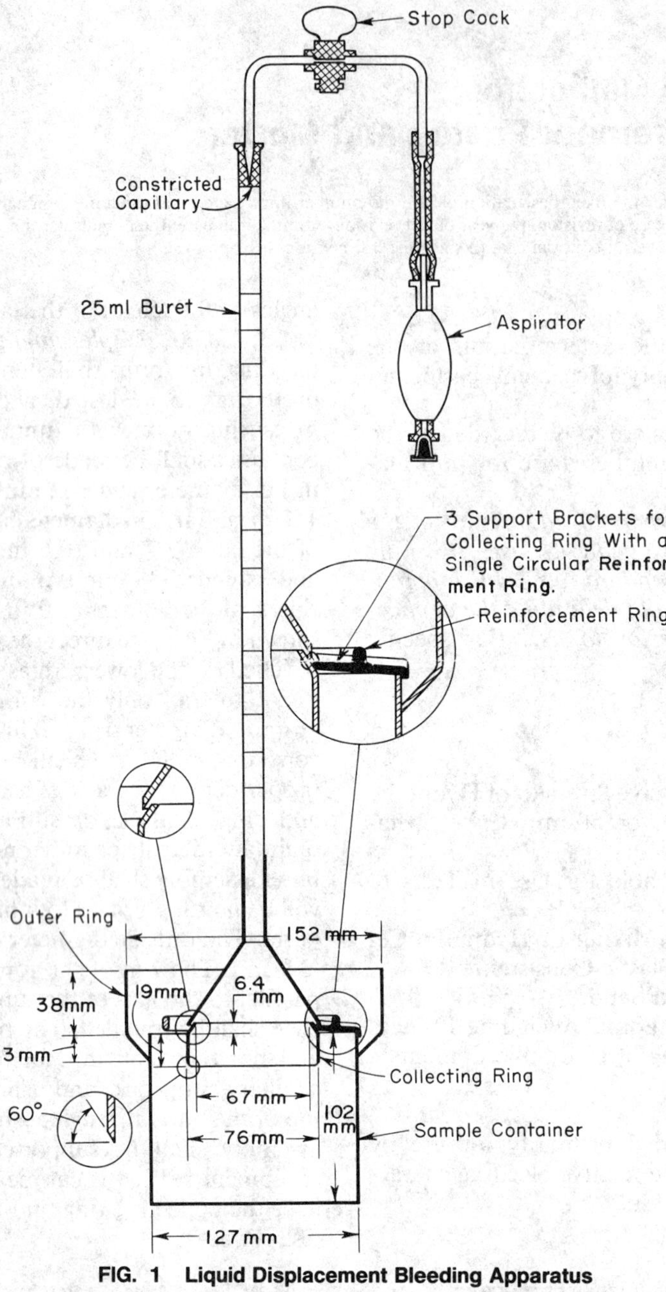

FIG. 1 Liquid Displacement Bleeding Apparatus

5. Hazards

5.1 Many chlorinated hydrocarbons are considered to be toxic substances which can be absorbed through the skin, but the major exposure is from inhalation of vapors. Chronic poisoning can occur from long continued exposure to small amounts. Acute poisoning can follow a single heavy exposure. Avoid skin contact and work with chlorinated hydrocarbons only in a properly vented fume hood.

5.2 Place the bleeding apparatus in a location subject to a minimum vibration and set on a sponge rubber mat or equivalent.

6. Temperature and Humidity

6.1 The temperature of the room and the temperature of the cement, sand, and water shall be 23 ± 1.7°C (73.4 ± 3°F).

6.2 The relative humidity of the laboratory shall be not less than 50 %.

7. Tests on Pastes

7.1 *Batch*—The paste mix shall consist of 2300 g of cement and 1035 g of water.

7.2 *Mixing*—When mechanically mixing pastes, the materials for the batch shall be introduced in the following manner:

7.2.1 Place all the mixing water in the bowl.

7.2.2 Add the cement to the water and mix for 3 min at slow speed (140 ± 5 rpm).

7.2.3 Let the paste stand for 3 min. During the first 15 s of this interval, quickly scrape down into the batch any paste that may have collected on the side of the bowl; then for the remainder of this period, cover with a lid or damp cloth.

7.2.4 Mix for 3 min at slow speed.

7.3 *Procedure*—Immediately following completion of mixing, carefully pour the paste into the sample container and screed with a steel straightedge to make certain that the paste height is level with the rim of the 127-mm (5-in.) diameter container. Screeding should be completed within 15 s. Take the completion of screeding as the starting time for the test. Place the collecting ring in the center of the paste surface and embed it to a depth of approximately 19.1 mm (¾ in.) using the three support brackets as a guide. Cover the surface of the paste with 1,1,1-trichloroethane to a depth of 31.8 mm (1¼ in.) (Note 2). Lower the funnel-buret assembly to within 6.4 mm (¼ in.) of and directly over the collecting ring. Draw 1,1,1-trichloroethane up into the buret by means of an aspirator to the "zero" mark of the buret and close the stopcock. Four minutes after the starting time, take the first reading; take readings thereafter at 2-min intervals until 30 min after the starting time, and then at 10-min intervals until cessation of bleeding (Note 3).

NOTE 2—Approximately 500 mL of 1,1,1-trichloroethane are required per test. This may be reclaimed by siphoning off the liquid at the end of the test and placing it in a separatory funnel. The 1,1,1-trichloroethane may be tinted with an oil-soluble dye for ease in reading the level in the buret. Attention is drawn to the need for using 1,1,1-trichloroethane in a well ventilated room or under a ventilation hood to avoid any cumulative toxic effects.

NOTE 3—To compensate for density changes as increasing quantities of water gather in the buret, zero the meniscus immediately before taking each reading.

8. Tests on Mortars

8.1 *Batch*—The mortar mix shall consist of 930 g of cement and 2325 g of graded sand as specified in Specification C 778. The amount of mixing water, measured in grams, shall be such as to produce a flow of between 105 and 110 as determined in accordance with 8.3 and shall be expressed as weight percent of the cement (Note 4).

8.2 *Mixing*—When mixing mortars, the materials for the batch shall be introduced in the following manner:

8.2.1 Place all the mixing water in the bowl.

8.2.2 Add the cement to the water, then start the mixer and mix for 30 s at slow speed (140 ± 5 rpm).

8.2.3 Add the entire quantity of sand slowly over a 30-s period, while mixing at slow speed.

8.2.4 Mix at slow speed for an additional 2 min.

8.2.5 Stop the mixer and let the mortar stand for 3 min. During the first 15 s of this interval, quickly scrape down into the batch any mortar that may have collected on the side of the bowl; then for the remainder of the interval, cover the bowl with a lid or damp cloth.

8.2.6 Finish by mixing for 3 min at slow speed.

NOTE 4—As a guide for the initial trial mortar, the weight percent of water of the cement to produce the specified flow will be about 50 to 52 % for portland cement containing air-entraining material, and about 52 to 54 % for portland cement not containing air-entraining material.

8.3 *Determination of Flow*—Determine flow of the mortar mix in accordance with the procedure described in the Determination of Flow Section of Test Method C 109, with the following exceptions:

8.3.1 The table shall be dropped through a height of 13 mm (½ in.) ten times in 6 s.

8.3.2 When a batch having the proper flow has been obtained, discard the portion used in making the flow determination and do not use it in the bleeding determination.

8.4 *Procedure*—Immediately following the flow test, remix for 30 s the mortar remaining in the mixing bowl. Within a total elapsed time of not more than 2 min and 30 s after completion of the original mixing of the mortar batch, begin placing the mortar in the sample container. Place the mortar in three layers, tamp each layer 40 times with the tamper, and strike off the surface level with the edge of the 127-mm (5-in.) diameter container by screeding with a steel straightedge. Conduct the remainder of the test in the same manner as that for the paste (7.3), using as a starting time the completion of striking off the surface of the mortar in the container.

9. Calculation

9.1 *Bleeding Rate*—Calculate the initial rate of bleeding as follows, using values taken during the first 30 min of the test when the bleeding should be proceeding at a uniform rate (Note 5):

$$R_B = V_1/At$$

where:

R_B = bleeding rate, mL/cm²,

V_1 = volume of bleeding water, mL, measured during time interval, t,

A = area covered by the 76.2-mm (3.0-in) diameter collecting ring, cm², and

t = time during which bleeding occurs at a uniform rate, s.

9.2 *Bleeding Capacity*—Calculate the bleeding capacity as follows:

$$C_B = V_2/V_3$$

where:

C_B = bleeding capacity, mL/mL,

V_2 = total volume of bleeding water, mL, and

V_3 = volume of the cylinder of paste or mortar directly under the 76.2-mm (3.0-in) diameter collecting ring, mL.

10. Report

10.1 Calculate and report bleeding rate and bleeding capacity values to the nearest three significant digits.

NOTE 5—Since many cements will not bleed at a constant rate for the entire 30-min period, it may be necessary to plot the bleeding values against time to determine the correct initial slope of the curve, that is, initial bleeding rate.

11. Precision and Bias

11.1 It is not practicable to specify the precision of the procedure in this test method.

12. Keywords

12.1 bleeding capacity; bleeding rate; cement mortar; cement paste

For additional useful information on details of cement test methods, reference may be made to the "Manual of Cement Testing," ***Annual Book of ASTM Standards,*** **Vol 04.01.**

Designation: C 265 – 99

Standard Test Method for Calcium Sulfate in Hydrated Portland Cement Mortar[1]

This standard is issued under the fixed designation C 265; the number immediately following the designation indicates the year of original adoption or, in the case of revision, the year of last revision. A number in parentheses indicates the year of last reapproval. A superscript epsilon (ε) indicates an editorial change since the last revision or reapproval.

1. Scope

1.1 This test method covers the measurement of water-soluble (water-extractable) SO_3 in hardened hydraulic cement mortar. This measurement is assumed to represent unreacted, free calcium sulfate remaining in the mortar.

1.2 The values stated in SI units are to be regarded as the standard. The values given in parentheses are for information only.

1.3 Values in SI units were obtained by measurement in SI units or by appropriate conversion using the Rules for Conversion and Rounding given in Practice E 380 of measurements made in other units.

1.4 *This standard does not purport to address all of the safety concerns, if any, associated with its use. It is the responsibility of the user of this standard to establish appropriate safety and health practices and determine the applicability of regulatory limitations prior to use.*

2. Referenced Documents

2.1 *ASTM Standards:*

C 109/C 109M Test Method for Compressive Strength of Hydraulic Cement Mortars (Using 2-in. or 50-mm Cube Specimens)[2]

C 114 Test Methods for Chemical Analysis of Hydraulic Cement[2]

C 150 Specification for Portland Cement[2]

C 219 Terminology Relating to Hydraulic Cement[2]

C 305 Practice for Mechanical Mixing of Hydraulic Cement Pastes and Mortars of Plastic Consistency[2]

C 595M Specification for Blended Hydraulic Cements [Metric][2]

C 778 Specification for Standard Sand[2]

C 1157 Performance Specification for Blended Hydraulic Cement [Metric][2]

D 1193 Specification for Reagent Water[3]

E 11 Specification for Wire-Cloth Sieves for Testing Purposes[4]

IEEE/ASTM SI-10 Standard for Use of the International System of Units (SI): The Modern Metric System[4]

3. Terminology

3.1 *Definitions*—The terms used in this test method are defined in accordance with Terminology C 219.

4. Significance and Use

4.1 This test method is intended for use by manufacturers of hydraulic cement and those interested in research on a suitable method for determining whether calcium sulfate has or has not been used in an amount considered to be optimum. Also, for any such cement having an above optimum SO_3 content, this test method establishes whether that excess exceeds the limit allowed in Specification C 595. This test method also can provide useful information on other hydraulic cements, such as those specified in Specifications C 150 and C 1157.

5. Apparatus

5.1 *Sieve*—A 2.36-mm (No.8) sieve conforming to Specification E 11.

5.2 *Mixer, Bowl, and Paddle*—An electrically driven mechanical mixer of the type equipped with a paddle and bowl, as specified in the Apparatus section of Practice C 305.

5.3 *Polyethylene Containers*—Polyethylene bags of 1-L (1-qt) capacity or approximately 360-mm (14-in.) sheet material, made using polyethylene at least 0.10-mm (0.004-in.) in thickness. Bags, if used, must be watertight.

5.4 *Mortar and Pestle*—A mortar of 1.5-L (1½-qt) size, and a pestle, both of which shall be iron or porcelain.

5.5 *Water Bath*—A water bath thermostatically controlled at 23 ± 0.15°C (73.4 ± 0.3°F).

6. Reagents and Materials

6.1 *Mixing Water*—Reagent water conforming to the numerical limits of Type II of Specification D 1193.

6.2 *Graded Sand*—Graded sand conforming to the requirements for graded sand in Specification C 778.

7. Temperature and Humidity

7.1 Maintain the temperature of the molding room at 24 ± 4°C (75.2 ± 7.2°F). Adjust the temperature of the dry materials and the mixing water, prior to mixing, so that the temperature of the mortar, immediately upon completion of mixing is 23 ± 1°C (73.4 ± 1.8°F).

---

[1] This test method is under the jurisdiction of ASTM Committee C-1 on Cement and is the direct responsibility of Subcommittee C01.28 on Sulfate Content.

Current edition approved Jan. 10, 1999. Published May 1999. Originally published as C 265 – 51 T. Last previous edition C 265 – 91.

[2] *Annual Book of ASTM Standards*, Vol 04.01.

[3] *Annual Book of ASTM Standards*, Vol 11.01.

[4] *Annual Book of ASTM Standards*, Vol 14.02.

7.2 Maintain the humidity of the laboratory so that it is not less than 50 %.

8. Preparation of Test Mortar

8.1 Proportion the mortar in accordance with the instructions for a six-cube batch given in the section on Composition of Mortars of Test Method C 109, except use a water-cement ratio of 0.5, by mass. Mix mechanically in accordance with Procedure for Mixing Mortars of Practice C 305.

8.2 Split the batch immediately after mixing then place approximately 450 g of mortar in each of the two polyethylene containers. Obtain the temperature of each portion to ensure that it is within the range from 23 ± 1°C (73.4 ± 1.8°F). If bags are used, twist the neck of each and seal with a rubber band; double the neck over and seal with a second band. If sheets are used to contain the samples, gather the corners and edges to make a bag and seal in similar manner.

9. Storage of Test Mortar

9.1 Immediately immerse both containers in the water bath maintained at (23 ± 0.15°C) 73.4 ± 0.3°F .

10. Preparation of Cement Extract

10.1 At 24¼h after the instant the cement and water were mixed together, remove the containers successively from the water bath, pulverizing each sample in turn with the pestle in the mortar as rapidly as possible. Immediately upon removal from the water bath treat each sample individually as follows: First pulverize approximately one third of the specimen until most of it passes the 2.36-mm (No. 8) sieve. Return the coarse residue to the mortar and grind another one third of the specimen as before. Repeat the process until all of the specimen is pulverized to pass the 2.36-mm (No. 8). Place 400 g of the pulverized material in a 400-mL beaker and add 100 mL of water at a temperature of 23 ± 0.6°C (73.4 ± 1°F). Quickly mix to uniform consistency with a spatula; then stir mechanically for 2 min. The type and speed of the stirrer shall be such that all particles of the slurry will be suspended, with no settling. Excessive stirring which might raise the temperature of the slurry appreciably shall be avoided.

10.2 Filter the slurry on a dry No. 2A Büchner funnel with the aid of suction using a dry 9-cm, medium, ashless paper. Complete filtration within a period of 2 min. Filter again, whether the filtrate is turbid or not, using a new dry filter paper without suction.

11. Analysis of the Extract

11.1 Transfer 50.0 mL of the clear extract by means of a 50-mL pipet to a 400-mL beaker, dilute with distilled water to about 250 mL, and add 5 mL of concentrated hydrochloric acid (HCl, sp gr 1.19). Heat to boiling, and proceed in accordance with the sulfur trioxide determination section of Test Methods C 114. Perform a blank on the reagents in use in the analysis.

11.1.1 After the $BaCl_2$ solution is added in accordance with the sulfur trioxide determination section of Test Methods C 114, boil the solution vigorously for 15 min.

11.2 When the amount of filtrate is insufficient to provide a 50-mL aliquot, use 25 mL of the extract and a factor twice that given in the section on Calculation.

12. Calculation and Report

12.1 Calculate the SO_3 content to the nearest 0.01 g/L of solution, as follows:

$$SO_3, g/L = W \times 6.86 \quad (1)$$

where:

W = grams of barium sulfate ($BaSO_4$) precipitated.

12.2 Report the average of the two tests.

13. Precision and Bias

13.1 *Precision*—The single-operator (within-laboratory) standard deviation has been found to be 0.077 g/L SO_3 throughout the range of 0.20 to 0.85 g/L SO_3. Therefore, results of two properly conducted tests by the same operator on water-extractable SO_3 in hardened hydraulic cement mortars of the same material should not differ from each other by more than 0.22 g/L SO_3. The multi-laboratory (between-laboratory) standard deviation has been found to be 0.137g/L SO_3 throughout the range of 0.20 to 0.85 g/L SO_3. Therefore, results of two properly conducted tests on the same material, from two different laboratories on water-extractable SO_3 in hardened hydraulic cement mortars should not differ from each other by more than 0.39 g/L SO_3.

13.2 *Bias*—No justifiable statement can be made on the bias of the procedure in this test method for measuring the water-extractable SO_3 in hardened hydraulic cement mortars, because no reference samples are available.

14. Keywords

14.1 calcium sulfate; hydraulic cement; sulfate content

For additional useful information on details of cement test methods, reference may be made to the "Manual of Cement Testing," ***Annual Book of ASTM Standards,*** **Vol 04.01.**

Designation: C 266 – 99

Standard Test Method for Time of Setting of Hydraulic-Cement Paste by Gillmore Needles[1]

This standard is issued under the fixed designation C 266; the number immediately following the designation indicates the year of original adoption or, in the case of revision, the year of last revision. A number in parentheses indicates the year of last reapproval. A superscript epsilon (ε) indicates an editorial change since the last revision or reapproval.

1. Scope

1.1 This test method covers the determination of the time of setting of hydraulic-cement paste by means of the Gillmore needles.

1.2 The values stated in SI units are to be regarded as the standard. The values given in parentheses are for information only.

1.3 *This standard does not purport to address all of the safety concerns, if any, associated with its use. It is the responsibility of the user of this standard to establish appropriate safety and health practices and determine the applicability of regulatory limitations prior to use.* See Note 1 for a specific warning statement.

NOTE 1—**Warning:** Fresh hydraulic cementitious mixtures are caustic and may cause chemical burns to skin and tissue upon prolonged exposure. The use of gloves, protective clothing, and eye protection is recommended. Wash contact area with copious amounts of water after contact. Wash eyes for a minimum of 15 min. Avoid exposure of the body to clothing saturated with the liquid phase of the unhardened material. Remove contaminated clothing immediately after exposure.

2. Referenced Documents

2.1 *ASTM Standards:*

C 151 Test Method for Autoclave Expansion of Portland Cement[2]

C 183 Practice for Sampling and the Amount of Testing of Hydraulic Cement[2]

C 187 Test Method for Normal Consistency of Hydraulic Cement[2]

C 305 Practice for Mechanical Mixing of Hydraulic Cement Pastes and Mortars of Plastic Consistency[2]

C 490 Practice for Use of Apparatus for the Determination of Length Change of Hardened Cement Paste, Mortar, and Concrete[2]

C 511 Specification for Moist Cabinets, Moist Rooms, and Water Storage Tanks Used in the Testing of Hydraulic Cements and Concretes[2]

C 670 Practice for Preparing Precision and Bias Statements for Test Methods for Construction Materials[3]

C 1005 Specification for Reference Masses and Devices for Determining Mass for Use in the Physical Testing of Hydraulic Cements[2]

D 1193 Specification for Reagent Water[4]

3. Summary of Test Method

3.1 Sufficient water is added to the cement that is being tested to produce a paste of normal consistency. A specimen is molded from this paste and is tested for setting times by means of the Gillmore initial and final needles. The initial setting time is the time required for the test specimen to bear the initial Gillmore needle without appreciable indentation, while the time required for the test specimen to bear the final Gillmore needle without appreciable indentation is the final setting time.

4. Significance and Use

4.1 The purpose of this test method is to establish whether a cement complies with a specification limit on setting time. The time of setting is affected not only by the percentage and temperature of the water used, and the amount of kneading the paste received, but also by the temperature and humidity of the air.

5. Apparatus

5.1 *Trowel*, having a steel blade 100 to 150 mm (4 to 6 in.) in length, with a straight edge.

5.2 *Mixer, Bowl, Paddle, and Scraper*, conforming to the requirements of Practice C 305.

5.3 *Glass Graduates*, conforming to the requirements of Specification C 490.

5.4 *Mass Determining Devices*, conforming to the requirements of Specification C 1005.

5.5 *Plane Non-absorptive Plates*, 102 mm ± 3 mm (4 in. ± 0.125 in.) square.

5.6 *Gillmore Needles*, conforming to the following requirements:

5.6.1 The initial setting-time needle shall have a weight of 113.4 ± 0.5 g (0.250 ± 0.001 lb) and a tip diameter of 2.12± 0.05 mm (0.084± 0.002 in.).

5.6.2 The final setting time needle shall have a weight of

---

[1] This test method is under the jurisdiction of ASTM Committee C-1 on Cement and is the direct responsibility of Subcommittee C01.30 on Time of Set.

Current edition approved Jan. 10, 1999. Published May 1999. Originally published as C 266 – 51 T. Last previous edition C 266 – 89 (1995)ε1.

[2] *Annual Book of ASTM Standards*, Vol 04.01.

[3] *Annual Book of ASTM Standards*, Vol 04.02.

[4] *Annual Book of ASTM Standards*, Vol 11.01.

453.6 ± 0.5 g (1.000 ± 0.001 lb) and a tip diameter of 1.06± 0.05 mm (0.042± 0.002 in.).

5.6.3 The needle tips shall be cylindrical for a distance of 4.8 ± 0.5 mm (0.189 ± 0.020 in.). The needle ends shall be plane and at right angles to the axis of the rod and shall be maintained in a clean condition. The Gillmore needles should preferably be mounted as shown in Fig. 1 (b).

6. Reagents

6.1 *Mixing Water*—Potable water is satisfactory for routine tests. For all referee and cooperative tests, reagent water conforming to the requirements of Specification D 1193 for Type III or Type IV grade of reagent water shall be used.

7. Sampling

7.1 When the test is part of acceptance testing, sample the cement in accordance with Practice C 183.

8. Conditioning

8.1 The temperature of the room, dry materials, paddle,

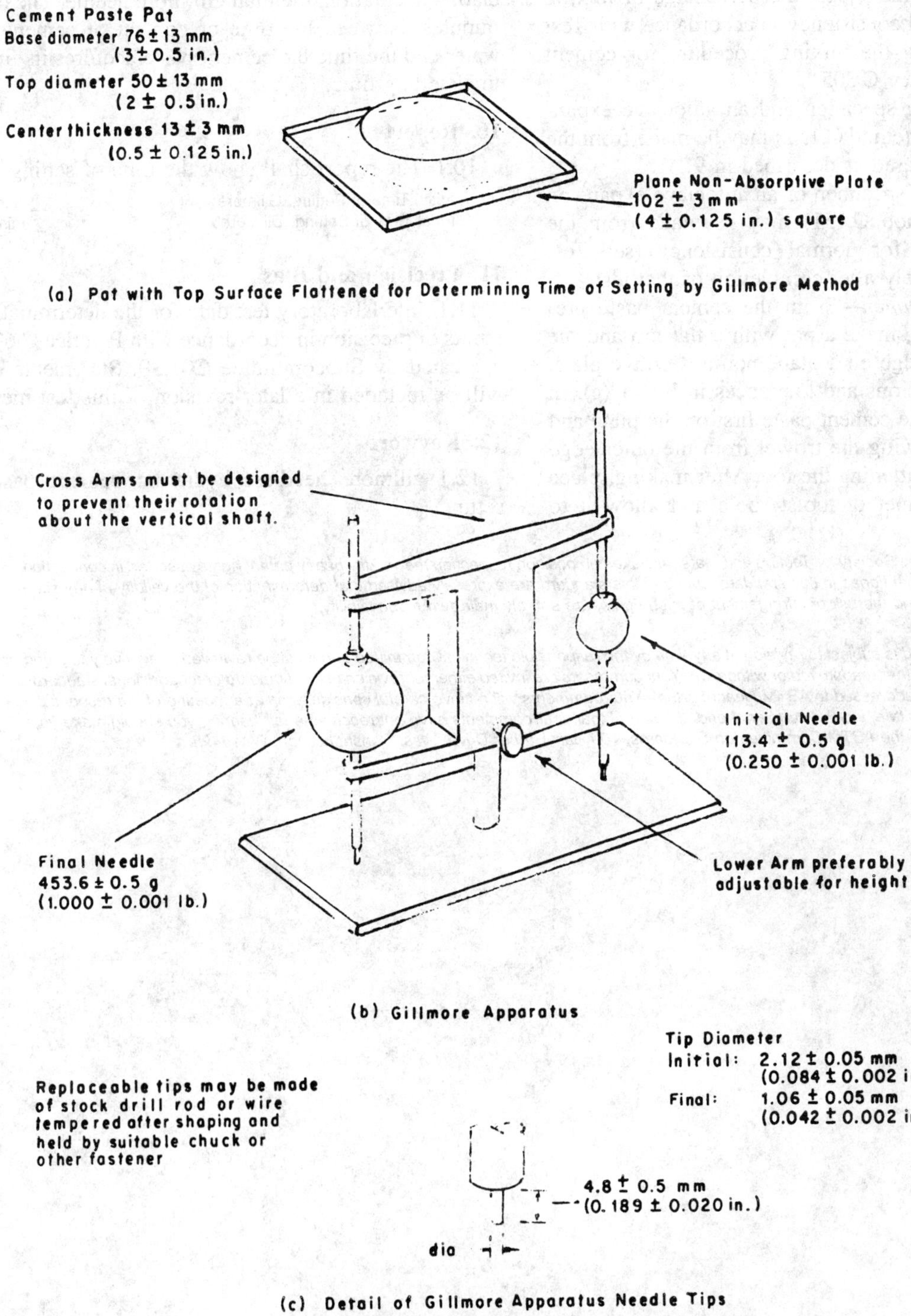

FIG. 1 Gillmore Apparatus and Test Specimen

bowl, and plane non-absorptive plates shall be maintained between 20 and 27.5°C (68 and 81.5°F). The temperature of the mixing water shall not vary from 23°C (73.4°F) by more than ± 1.7°C (±3°F).

8.2 The relative humidity of the laboratory shall be not less than 50%.

8.3 The moist cabinet or moist room shall conform to the requirements of Specification C 511.

9. Procedure

9.1 *Preparation of Cement Paste*:

9.1.1 Mix 650 g of cement with the percentage of mixing water required for normal consistency in accordance with Test Method C 187, following the mixing procedure for cement pastes described in Practice C 305.

9.1.2 A time of setting specimen and an autoclave expansion specimen (see Test Method C 151) may be made from the same batch of paste prepared as described in 9.1.1.

9.1.3 A time of setting specimen or an autoclave expansion specimen (see Test Method C 151) may be made from the remainder of the batch for normal consistency (see Test Method C 187) immediately after completion of that test.

9.2 *Molding Test Specimen*—From the cement paste prepared as described in 9.1, make a pat with a flat top and the sides tapering to a thin edge on a plane non-absorptive plate, conforming to the dimensions and tolerances in Fig. 1 (a). In molding the pat, flatten the cement paste first on the plate and then form the pat by drawing the trowel from the outer edge toward the center, then flattening the top. After making, place the pat in the moist cabinet or moist room and allow it to remain there except when the determinations of time of setting are being made.

9.3 *Time of Setting Determination*—In determining the time of setting, hold the needles in a vertical position and apply lightly to the surface of the pat. Consider the cement paste to have acquired its initial set when the pat will bear, without appreciable indentation, the initial Gillmore needle. The difference, in minutes, between the time of contact of cement and mixing water and the time the cement paste acquires its initial set is the initial setting time. Consider the cement paste to have acquired its final set when the pat will bear, without appreciable indentation, the final Gillmore needle. The difference, in minutes, between the time of contact of cement and mixing water and the time the cement paste acquires its final set is the final setting time.

10. Report

10.1 The report shall show the time of setting as follows:

Initial Time of Setting, Gillmore __ min
Final Time of Setting, Gillmore __min

11. Precision and Bias

11.1 Interlaboratory test data for the determination of statements of precision in accordance with Practice C 670 are being evaluated by Subcommittee C01.30. Statements of precision will be included in a later revision of this test method.

12. Keywords

12.1 gillmore needles; hydraulic cement paste; time of setting

American Association State
Highway and Transportation Officials Standard
AASHTO No.: T 162

Standard Practice for Mechanical Mixing of Hydraulic Cement Pastes and Mortars of Plastic Consistency[1]

This standard is issued under the fixed designation C 305; the number immediately following the designation indicates the year of original adoption or, in the case of revision, the year of last revision. A number in parentheses indicates the year of last reapproval. A superscript epsilon (ε) indicates an editorial change since the last revision or reapproval.

1. Scope

1.1 This practice covers the mechanical mixing of hydraulic cement pastes and mortars of plastic consistency.

1.2 *This standard does not purport to address all of the safety concerns, if any, associated with its use. It is the responsibility of the user of this standard to establish appropriate safety and health practices and determine the applicability of regulatory limitations prior to use.*

2. Referenced Documents

2.1 *ASTM Standards:*

C 778 Specification for Standard Sand[2]

3. Significance and Use

3.1 This practice is intended for use in the mechanical mixing of pastes and mortars for the testing of hydraulic cements.

4. Apparatus

4.1 *Mixer*—The mixer shall be an electrically driven mechanical mixer of the epicyclic type, which imparts both a planetary and a revolving motion to the mixer paddle. The mixer shall have a minimum of two speeds, controlled by definite mechanical means. (Rheostat adjustment of speed will not be acceptable.) The first, or slow speed shall revolve the paddle at a rate of 140 ± 5 r/min, with a planetary motion of approximately 62 r/min. The second speed shall revolve the paddle at a rate of 285 ± 10 r/min, with a planetary motion of approximately 125 r/min. The electric motor shall be at least 124 W (1/6 hp).[3] The mixer shall be equipped with a clearance adjustment bracket such as the one shown in Fig. 1 (Note 1) to provide that the clearance between the lower end of the paddle and the bottom of the bowl is not greater than 2.5 mm but not less than 0.8 mm (Note 2) when the bowl is in the mixing position.

NOTE 1—When the bracket is in the proper position beneath the motor housing, the lugs are to the front and facing upward and the heads of the adjustment screws are to the rear and facing downward in the path of the sliding frame that holds the bowl. It is intended that the bracket be fastened at the front housing connection by inserting replacement screws on an appropriate size upward through the opening in each lug and into the existing threaded holes in the bottom of the motor housing. The original stops for the sliding frame are to be filed down if they prevent the frame from coming in contact with the adjustment screws.

NOTE 2—This is the approximate diameter of a grain of 20-30 sand as described in Specification C 778.

4.2 *Paddle*—The paddle shall be readily removable, made of stainless steel, and shall conform to the basic design shown in Fig. 2.[2] The dimensions of the paddle shall be such that when in the mixing position the paddle outline conforms to the contour of the bowl used with the mixer, and the clearance between corresponding points on the edge of the paddle and the side of the bowl in the position of closest approach shall be approximately 4.0 mm but not less than 0.8 mm.

4.3 *Mixing Bowl*—The removable mixing bowl shall have a nominal capacity of 4.73 L, shall be of the general shape and comply with the limiting dimensions shown in Fig. 3,[3] and shall be made of stainless steel. The bowl shall be so equipped that it will be positively held in the mixing apparatus in a fixed position during the mixing procedure. There shall be provided a lid, made of a nonabsorbing material not attacked by the cement.

4.4 *Scraper*—The scraper shall consist of a semirigid rubber blade attached to a handle about 150 mm long. The blade shall be about 75 mm long, 50 mm wide, and tapered to a thin edge about 2 mm thick.

NOTE 3—A kitchen tool known as a plate and bowl scraper conforms to these requirements.

4.5 *Supplementary Apparatus*—The balances, weights, glass graduates, and any other supplementary apparatus used in measuring and preparing the mortar materials prior to mixing shall conform to the respective requirements for such apparatus as specified in the method for the particular test for which the mortar is being prepared.

5. Temperature and Humidity

5.1 The temperature of the room shall be maintained between 20 and 27.5°C (68 and 81.5°F), and the temperature of the dry materials, paddle, and bowl shall be within the above range at the time of test. The temperature of the mixing water shall not vary from 23°C (73.4°F) by more than ±1.7°C (3°F).

5.2 The relative humidity of the laboratory shall be not less than 50 %.

6. Materials, Proportioning, and Consistency

6.1 The materials and their proportions and quantities shall conform to the requirements contained in the partic-

[1] This practice is under the jurisdiction of ASTM Committee C-1 on Cement and is the direct responsibility of Subcommittee C01.22 on Workability.

Current edition approved March 15, 1994. Published May 1994. Originally published as C 305 – 53 T. Last previous edition C 305 – 91.

[2] *Annual Book of ASTM Standards*, Vol 4.01.

[3] The Model N-50 Mixer, manufactured by the Hobart Corporation, Troy, OH, used in conjunction with paddle, special stainless steel "B" flat beater, part No. 274688 and the mixing bowl part No. 78575-2, is considered to conform to these requirements.

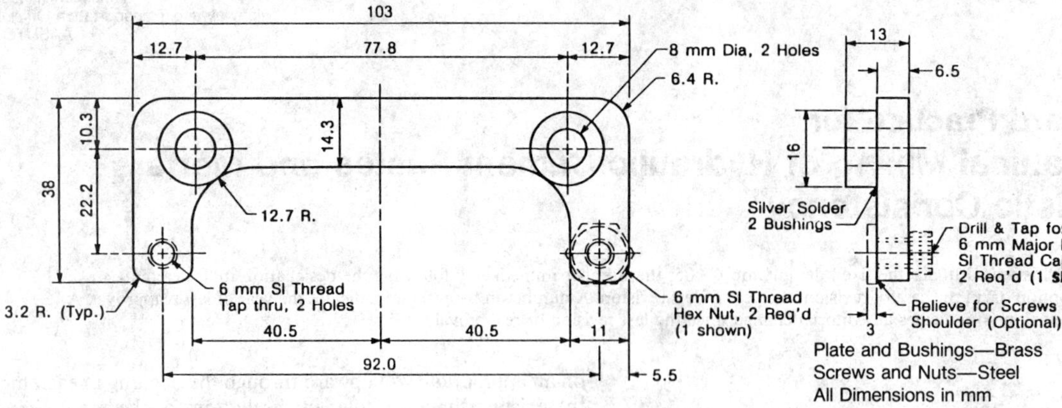

FIG. 1 Clearance Adjustment Bracket

ular method for which the paste or mortar is being prepared.

7. Procedure for Mixing Pastes

7.1 Place the dry paddle and the dry bowl in the mixing position in the mixer. Then introduce the materials for a batch into the bowl and mix in the following manner:

7.1.1 Place all the mixing water in the bowl.

7.1.2 Add the cement to the water and allow 30 s for the absorption of the water.

7.1.3 Start the mixer and mix at slow speed (140 ± 5 r/min) for 30 s.

7.1.4 Stop the mixer for 15 s and during this time scrape down into the batch any paste that may have collected on sides of the bowl.

7.1.5 Start the mixer at medium speed (285 ± 10 r/min) and mix for 1 min.

8. Procedure for Mixing Mortars

8.1 Place the dry paddle and the dry bowl in the mixing position in the mixer. Then introduce the materials for a batch into the bowl and mix in the following manner:

8.1.1 Place all the mixing water in the bowl.

8.1.2 Add the cement to the water; then start the mixer and mix at the slow speed (140 ± 5 r/min) for 30 s.

8.1.3 Add the entire quantity of sand slowly over a 30-s period, while mixing at slow speed.

8.1.4 Stop the mixer, change to medium speed (285 ± 10 r/min), and mix for 30 s.

8.1.5 Stop the mixer and let the mortar stand for 1½ min. During the first 15 s of this interval, quickly scrape down into the batch any mortar that may have collected on the side of the bowl; then for the remainder of this interval, cover the bowl with the lid.

8.1.6 Finish by mixing for 1 min at medium speed (285 ± 10 r/min).

8.1.7 In any case requiring a remixing interval, any mortar adhering to the side of the bowl shall be quickly scraped down into the batch with the scraper prior to remixing.

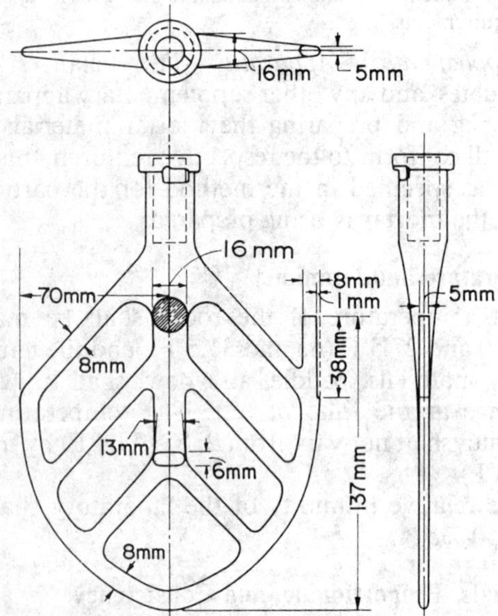

FIG. 2 Paddle

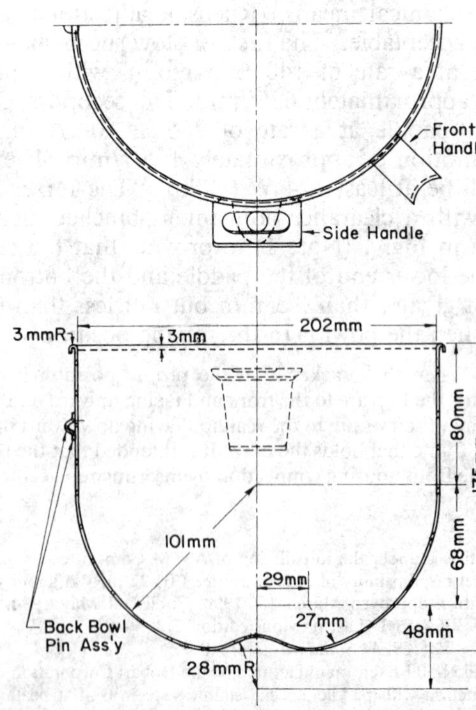

FIG. 3 Mixing Bowl

NOTE 4—**Caution**—The clearances between the paddle and the bowl specified in this practice are suitable when using mortar made with standard sand as described in Specification C 778. To permit the mixer to operate freely and to avoid serious damage to the paddle and bowl when coarser aggregates are used, it may be necessary to set the clearance adjustment bracket to provide greater clearances than those specified in 4.1.

For additional useful information on details of cement test methods, reference may be made to the "Manual of Cement Testing," which appears in the *Annual Book of ASTM Standards*, Vol 04.01.

Designation: C 317 – 98

Standard Specification for Gypsum Concrete[1]

This standard is issued under the fixed designation C 317; the number immediately following the designation indicates the year of original adoption or, in the case of revision, the year of last revision. A number in parentheses indicates the year of last reapproval. A superscript epsilon (ϵ) indicates an editorial change since the last revision or reapproval.

This standard has been approved for use by agencies of the Department of Defense.

1. Scope

1.1 This specification covers mill-mixed gypsum concrete, consisting essentially of calcined gypsum and suitable aggregate, requiring the addition of water only at the job site. Gypsum concrete is intended for use in the construction of poured-in-place roof decks or slabs. Two classes, based on the compressive strength and density, are covered.

1.2 The values stated in inch-pound units are to be regarded as the standard. The SI (metric) values given in parentheses are approximate and are provided for information only.

2. Referenced Documents

2.1 *ASTM Standards:*

C 11 Terminology Relating to Gypsum and Related Building Materials and Systems[2]

C 28 Specification for Gypsum Plasters[2]

C 35 Specification for Inorganic Aggregates for Use in Gypsum Plaster[2]

C 472 Test Methods for Physical Testing of Gypsum, Gypsum Plasters and Gypsum Concrete[2]

3. Terminology

3.1 Definitions of terms shall be in accordance with Terminology C 11.

4. Materials and Manufacture

4.1 Gypsum concrete shall consist of calcined gypsum to which may be added aggregates, wood chips, or wood shavings in proportion to meet the applicable requirements of this specification.

4.2 *Calcined Gypsum*—Specification C 28.

4.3 *Aggregates*—Specification C 35.

4.4 *Wood Chips or Wood Shavings,* shall be dry wood, uniform and clean in appearance, passing a 1-in. (25-mm) sieve, not more than 1⁄16-in. (1.6-mm) thick.

5. Physical Properties

5.1 *Setting Time*—Gypsum concrete shall set in not less than 20 min and not more than 90 min.

5.2 *Compressive Strength*—Gypsum concrete shall have the following compressive strength for the respective classes:

Class	Compressive Strength, min psi (MPa)
A	500 (3.5)
B	1000 (6.9)

5.3 *Density*—Gypsum concrete, Class A, shall have a density of not more than 60 lb/ft^3 (960 kg/m^3).

6. Sampling

6.1 Randomly select at least 1 % of the packages, but not less than five packages. Take samples from both the outer portion and the center of each package. Mix the samples so obtained to provide a composite sample of not less than 15 lb (6.8 kg). Immediately place the sample in a clean, dry, airtight container for delivery to the laboratory.

7. Test Methods

7.1 Determine the physical properties of gypsum concrete in accordance with Test Methods C 472.

7.1.1 Determine the setting time by the Vicat method.

8. Inspection

8.1 Inspection of the gypsum concrete shall be agreed upon between the purchaser and the supplier as part of the purchase agreement.

9. Rejection

9.1 Rejection of gypsum concrete that fails to conform to the requirements of this specification shall be reported to the producer or supplier promptly and in writing. The notice of rejection shall contain a statement documenting how the gypsum concrete has failed to conform to the requirements of this specification.

10. Certification

10.1 When specified in the purchase agreement, a producer's or supplier's report shall be furnished at the time of shipment certifying that the product is in compliance with this specification.

11. Packaging and Marking

11.1 Gypsum concrete shall be shipped dry and free of lumps.

11.2 When shipped for resale, the following information

[1] This specification is under the jurisdiction of ASTM Committee C-11 on Gypsum and Related Building Materials and is the direct responsibility of Subcommittee C11.01 on Specifications and Test Methods for Gypsum Products.

Current edition approved Apr. 10, 1998. Published December 1998. Originally published as C 317 – 53 T. Last previous edition C 317 – 93.

[2] *Annual Book of ASTM Standards,* Vol 04.01.

shall be legibly marked on each package or on a tag of suitable size attached thereto:

11.2.1 Name of the manufacturer or the seller,

11.2.2 Brand,

11.2.3 Description, and

11.2.4 Net and gross weight of the package.

12. Keywords

12.1 aggregates; calcined gypsum; gypsum; gypsum concrete

Designation: C 318/C 318M – 96

Standard Specification for Gypsum Formboard[1]

This standard is issued under the fixed designation C 318/C 318M; the number immediately following the designation indicates the year of original adoption or, in the case of revision, the year of last revision. A number in parentheses indicates the year of last reapproval. A superscript epsilon (ε) indicates an editorial change since the last revision or reapproval.

1. Scope*

1.1 This specification covers the minimum requirements for gypsum formboard designed for use as a permanent form for poured-in-place reinforced gypsum concrete roof decks.

1.2 The values stated in either inch-pound units or SI (metric) are to be regarded separately as standard. Within the text, the SI units are shown in brackets. The values stated in each system shall be used independent of the other. Values from the two systems shall not be combined.

1.3 The text of this standard references notes which provide explanatory material. These notes shall not be considered requirements of the standard.

2. Referenced Documents

2.1 *ASTM Standards:*

C 11 Terminology Relating to Gypsum and Related Building Materials and Systems[2]

C 473 Test Methods for Physical Testing of Gypsum Board Products and Gypsum Lath[2]

C 1264 Specification for Sampling, Inspection, Rejection, Certification, Packaging, Marking, Shipping, Handling and Storage of Gypsum Board[2]

G 21 Practice for Determining Resistance of Synthetic Polymeric Materials to Fungi[3]

3. Terminology

3.1 *Definitions*—Definitions of terms used in this standard shall be in accordance with Terminology C 11.

4. Materials and Manufacture

4.1 Gypsum formboard shall consist of a noncombustible core, essentially gypsum, containing not more than 15 % by weight of fiber, either mineral or organic, synthetic or natural.

4.2 The face surface shall be specifically treated to resist fungus growth.

5. Physical Properties

5.1 Specimens shall be tested in accordance with Test Methods C 473.

NOTE—Since this product is laid between subpurlins without mechanical attachment, a nail pull requirement is not specified for this gypsum panel product.

5.1.1 Specimens shall be taken from the samples obtained in accordance with Specification C 1264.

5.1.2 *Flexural Strength*—The specimens shall be tested face up and face down. The average breaking load shall be not less than the following:

	Bearing Edges Across Fiber of Surfacing, lbf [N]	Bearing Edges Parallel Fiber of Surfacing, Face, lbf [N]	Bearing Edges Parallel Fiber of Surfacing, Back, lbf [N]
Method A	160 [712]	60 [267]	40 [178]
Method B	157 [698]	56 [249]	36 [160]

5.1.3 *Humidified Deflection*—The specimens shall have an average deflection of not more than the following:

Thickness, in. (mm)	Humidified Deflection, Eighths of an in. (mm)
½ [12.7]	10 [32]

5.1.4 *Core, End, and Edge Hardness*—The specimens shall have an average hardness of not less than 15 lbf [67 N] when tested by Method A and 11 lbf [49 N] when tested by Method B.

5.1.5 *Resistance to Fungi*—The specimens shall obtain a rating not more than 1 (one) when judged in accordance with Practice G 21, paragraph 9.3, (Observation for Visible Effects).

6. Dimensions and Permissible Variations

6.1 Specimens shall be taken from the samples obtained in accordance with Section 8.

6.2 Thickness, width, length, and end squareness shall be determined in accordance with Test Methods C 473.

6.2.1 *Thickness*—The nominal thickness shall be ½ in. [12.7 mm] with permissible variations in the nominal thickness of ±1⁄64 in. [0.4 mm] with permissible local variations of ±1⁄32 in. [0.8 mm] from the nominal thickness.

6.2.2 *Width*—The nominal width shall be 32 in. [813 mm] with a permissible variation of ⅛ in. [3 mm] over the specified width, and ¼ in. [6 mm] under the specified width.

6.2.3 *Length*—The nominal length of gypsum formboard shall conform to the purlin spacing of multiples thereof, with permissible variations of ±¼ in. [6 mm] from the nominal length.

6.2.4 *End Squareness*—Corners shall be square with a permissible variation of ±⅛ in. [±3 mm] in the full width of the board.

6.3 *Edges and Ends*—The edges and ends shall be square (90 ± 10°), straight, and solid.

[1] This specification is under the jurisdiction of ASTM Committee C-11 on Gypsum and Related Building Materials and Systems and is the direct responsibility of Subcommittee C11.01 Specifications and Test Methods for Gypsum Products.

Current edition approved May 10, 1996. Published July 1996. Originally published as C 318 – 53 T. Last previous edition C 318 – 78. Discontinued 1985 and reinstated 1996.

[2] *Annual Book of ASTM Standards,* Vol 04.01.

[3] *Annual Book of ASTM Standards,* Vol 14.02.

*** A Summary of Changes section appears at the end of this standard.**

7. Finish and Appearance

7.1 The surfaces of gypsum formboard shall be true and free from imperfections that would render it unfit for use with or without decoration.

8. Sampling, Inspection, Rejection, Certification, Packaging and Package Marking, Shipping, Handling and Storage

8.1 The sampling, inspection, rejection, certification, packaging and package marking, shipping, handling and storage shall be conducted in accordance with Specification C 1264.

9. Keywords

9.1 gypsum; gypsum concrete; gypsum formboard; gypsum panel product; roof decks

SUMMARY OF CHANGES

This document is based on Specification C 318 withdrawn in 1985. The need has arisen to reinstate the product specification into ASTM. This document is based on the 1985 version with updates required to conform to current *ASTM Form and Style* and test methods.

(*1*) One new requirement to evaluate the resistance to fungi is introduced.

Designation: C 348 – 97

Standard Test Method for Flexural Strength of Hydraulic-Cement Mortars[1]

This standard is issued under the fixed designation C 348; the number immediately following the designation indicates the year of original adoption or, in the case of revision, the year of last revision. A number in parentheses indicates the year of last reapproval. A superscript epsilon (ε) indicates an editorial change since the last revision or reapproval.

1. Scope

1.1 This test method covers the determination of the flexural strength of hydraulic-cement mortars.

1.2 The values stated in SI units are to be regarded as the standard. The values shown in parentheses are for information only.

1.3 Values in SI units shall be obtained by measurement in SI units or by appropriate conversion, using the Rules for Conversion and rounding given in Standard IEEE/ASTM SI 10, of measurements made in other units.

1.4 *This standard does not purport to address all of the safety concerns, if any, associated with its use. It is the responsibility of the user of this standard to establish appropriate safety and health practices and determine the applicability of regulatory limitations prior to use.*

2. Referenced Documents

2.1 *ASTM Standards:*

C 109/C 109M Test Method for Compressive Strength of Hydraulic Cement Mortars (Using 2-in. or 50-mm Cube Specimens)[2]

C 230 Specification for Flow Table for Use in Tests of Hydraulic Cement[2]

C 305 Practice for Mechanical Mixing of Hydraulic Cement Pastes and Mortars of Plastic Consistency[2]

C 349 Test Method for Compressive Strength of Hydraulic Cement Mortars (Using Portions of Prisms Broken in Flexure)[2]

C 670 Practice for Preparing Precision and Bias Statements for Test Methods for Construction Materials[3]

C 778 Specification for Standard Sand[2]

IEEE/ASTM SI 10 Standard for Use of the International System of Units (SI): The Modern Metric System[4]

3. Summary of Test Method

3.1 The test mortar used consists of 1 part cement and 2.75 parts of sand proportion by mass. Portland or air-entraining portland cements are mixed at specified water-cement ratios. Water content of other cements is that sufficient to obtain a flow of 110 ± 5 % with 25 drops of the flow table. Test prisms, 40 by 40 by 160-mm are molded by tamping in two layers. Prisms are cured one day in the molds and stripped until tested by center point loading.

4. Significance and Use

4.1 This test method provides a means for determining the flexural strength of hydraulic cement mortars. Portions of the mortar prisms tested in flexure according to this test method may be used for the determination of compressive strength in accordance with Test Method C 349.

4.2 The values are determined from this test method for research or reference purposes only and are not used for determining compliance with specification requirements.

5. Apparatus

5.1 *Weights, Weighing Devices and Glass Graduates*, shall conform to the Apparatus Section of Test Method C 109/C 109M.

5.2 *Mixer, Bowl and Paddle*, an electrically driven mechanical mixer of the type equipped with paddle and mixing bowl, as specified in the Apparatus Section of Practice C 305.

5.3 *Flow Table and Flow Mold*, shall conform to Specification C 230.

5.4 *Specimen Molds*—Molds for the 40 by 40 by 160-mm prism specimens shall be triple-gang molds and shall be so designed that the specimens will be molded with their longitudinal axes in a horizontal position. The molds shall be made of a hard metal, not attacked by cement mortar, and with a Rockwell hardness of not less than HRB 55. The parts of the molds shall be matchmarked and, when assembled, shall be tight-fitting and positively held together. The sides of the molds shall be sufficiently rigid to prevent spreading or warping. The interior faces of the molds shall be plane surfaces with a permissible variation, in any 50-mm line on a surface, of 0.03 mm for new molds and 0.05 mm for molds in use. The distance between opposite sides shall be 40 ± 0.13 mm for new molds and 40 ± 0.3 mm for molds in use. The height of the molds shall be 40 mm with permissible variations of +0.25 and −0.15 mm for new molds, and +0.25 and −0.40 mm for molds in use. The inside length of the molds shall be 160 ± 2.5 mm. The angle between adjacent interior faces and top and bottom planes of the mold shall be 90 ± 0.5°, measured at points slightly removed from the intersections of the faces. The base plate shall be approximately 10 mm in thickness and shall have a plane surface 200 by 180 mm with a permissible variation in any 50-mm line on the surface of 0.03 mm.

5.5 *Tamper*—The tamper (see Fig. 1) shall be made of nonabsorptive, nonabrasive material, such as a rubber compound having a Shore A durometer hardness of 80 ± 10 or seasoned oak wood rendered nonabsorptive by immersion for 15 min in paraffin at approximately 200°C (392°F). The face of the tamper shall be 22 by 85 mm.

[1] This test method is under the jurisdiction of ASTM Committee C-1 on Cement and is the direct responsibility of Subcommittee C01.27 on Strength.

Current edition approved May 10, 1997. Published June 1998. Originally published as C 348 – 54 T. Last previous edition C 348 – 95.

[2] *Annual Book of ASTM Standards*, Vol 04.01.

[3] *Annual Book of ASTM Standards*, Vol 04.02.

[4] *Annual Book of ASTM Standards*, Vol 14.02.

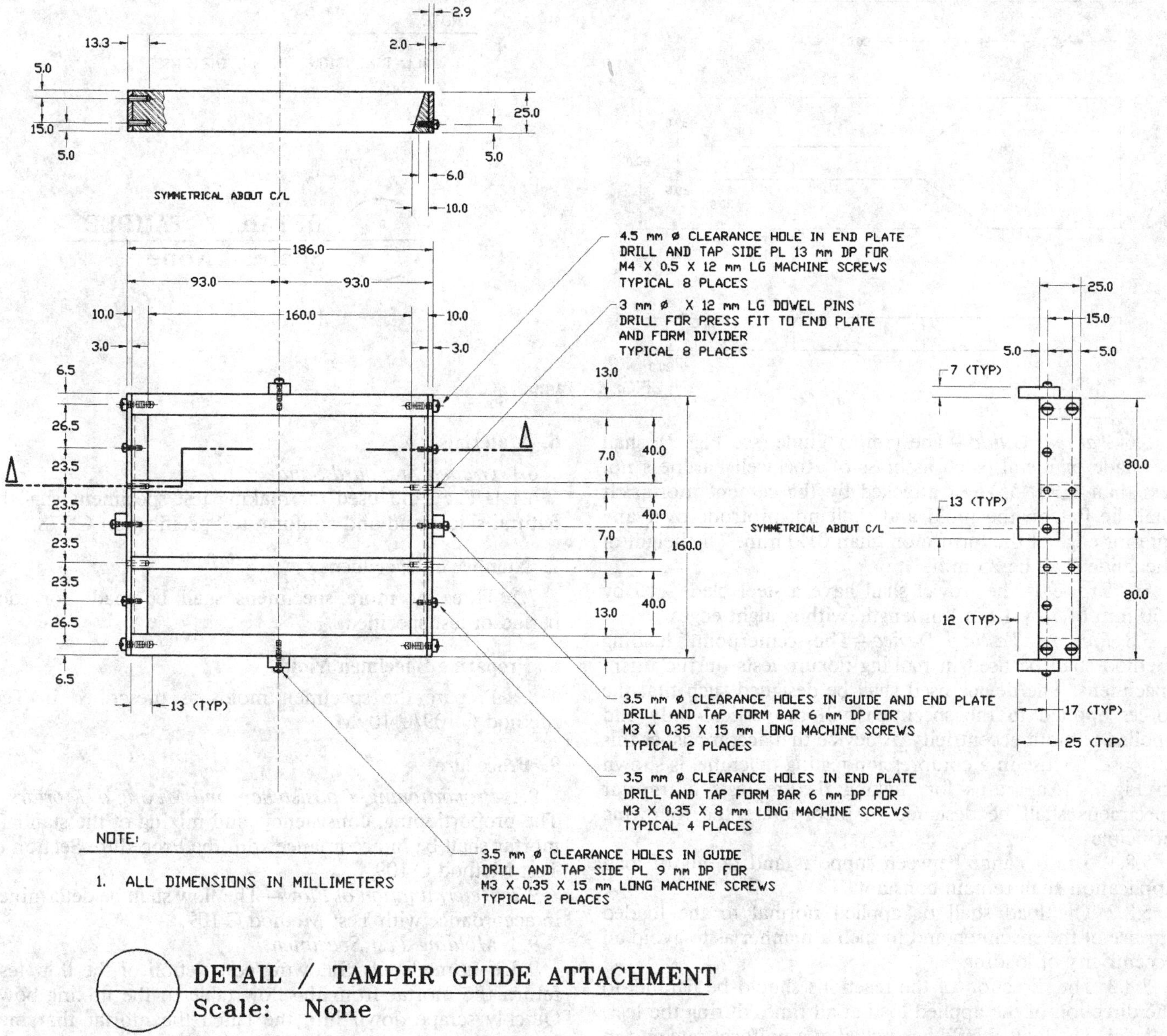

FIG. 2 Tamper Guide Attachment

horizontal position about 25 mm above the mortar level and then thrust directly downward with sufficient force to squeeze out a small amount of mortar from under the tamping surface. Fill the molds with mortar which shall be uniformly distributed and tamped in the same manner as the bottom layer. Then remove the tamper guide and smooth off the specimens by drawing the flat side of the trowel (with the leading edge slightly raised) once along the length of the molds. Cut the mortar off flush with the top of the molds by the straight edge of the trowel (held nearly perpendicular to the molds) with a sawing motion over the length of the molds. Following the cutting operation repair tears or cracks in the top surfaces and then make the surfaces of the specimens plane by two or three light longitudinal strokes of the trowel held with the leading edge slightly raised.

9.4 *Storage of Test Specimens*—Store the test specimens in accordance with Test Method C 109/C 109M.

9.5 *Determination of Flexural Strength:*

9.5.1 Test the specimens immediately after their removal from the moist closet in the case of 24-h specimens and from the storage water in the case of all other specimens. All test specimens for a designated test age shall be broken within the permissible tolerance prescribed below:

Test Age	Permissible Tolerance, h
24 h	±½
3 days	±1
7 days	±3
28 days	±12

When the portions of prisms are tested as modified cubes in accordance with Test Method C 349, test the prisms early enough so that the modified cubes are also broken within the above tolerances. If more than one specimen at a time is removed from the moist closet for the 24-h test, cover these specimens with waterproof plastic until the time of test. For

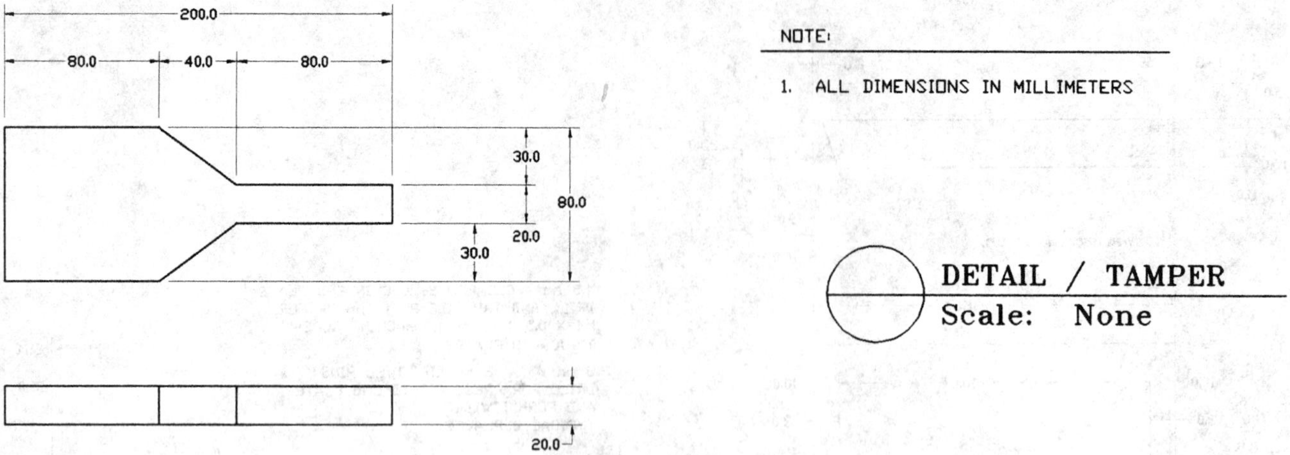

FIG. 1 Tamper

5.6 *Tamper Guide*—The tamper guide (see Fig. 2), shall be made of metal (such as brass of Rockwell hardness not less than HRB 55) not attacked by the cement mortar. It shall lie flat on the mold and shall not protrude over any interior edge of the form more than 0.40 mm. The height of the guide shall be 25 mm.

5.7 *Trowel*—The trowel shall have a steel blade 115 by 250 mm (4.5 by 10 in.) in length, with straight edges.

5.8 *Flexure Testing Device*—The centerpoint loading method shall be used in making flexure tests on the prism specimens. The device used shall be designed such that the forces applied to the specimen will be vertical only and applied without eccentricity. A device that accomplishes this purpose, for use in a compression testing machine, is shown in Fig. 3. Apparatus for making flexure tests of mortar specimens shall be designed to incorporate the following principles:

5.8.1 The distance between supports and points of load application shall remain constant.

5.8.2 The load shall be applied normal to the loaded surface of the specimen and in such a manner as to avoid all eccentricity of loading.

5.8.3 The direction of the reactions should be parallel to the direction of the applied load at all times during the test.

5.8.4 The load should be applied at a uniform rate and in a manner to avoid shock.

5.9 *Compression Testing Machine*—The compression testing machine used with the flexure testing device, as shown in Fig. 3, shall be of the hydraulic type and conform to the requirements prescribed in Test Method C 109/C 109M.

NOTE 1—Most hydraulic compression machines designed for breaking 50-mm (2-in.) cubes have a relatively small diameter lower bearing surface directly centered below the upper spherically seated head, on which close-fitting pedestals of appropriate heights are set for breaking 50-mm cubes and 50-mm (2-in.) and 50 by 100-mm (2 by 4-in.) or 75 by 150-mm (3 by 6-in.) cylinders. The base plate of the flexure testing apparatus shown in Fig. 3 is designed to rest on the low pedestal intended for tests on 75 by 150-mm (3 by 6-in.).

NOTE 2—In the absence of self-centering arrangements on machines with large lower bearing surfaces, the center of this surface directly below the center of the upper spherically seated head shall be accurately located. A circle or concentric circles of appropriate diameters shall be scribed on the lower bearing surface around this point. A cylindrical pedestal of appropriate diameter and height shall be obtained. End faces of the pedestal must be plane and parallel and at 90° to the axis of the cylinder. The upper face shall have a diameter of 77.5 mm (3.05 in.).

6. Materials

6.1 *Graded Standard Sand:*

6.1.1 The sand used for making test specimens shall be natural silica sand and conform to Specification C 778.

7. Number of Specimens

7.1 Three or more specimens shall be made for each period of test specified.

8. Preparing Specimen Molds

8.1 Prepare the specimen molds as prescribed in Test Method C 109/C 109M.

9. Procedure

9.1 *Proportioning, Consistency, and Mixing of Mortars*—The proportioning, consistency, and mixing of the standard mortar shall be in accordance with the Procedure Section of Test Method C 109.

9.2 *Determination of Flow*—The flow shall be determined in accordance with Test Method C 109.

9.3 *Molding Test Specimens:*

9.3.1 Immediately following completion of the flow test, return the mortar from the flow table to the mixing bowl. Quickly scrape down into the batch the mortar that may have collected on the side of the bowl and then remix the entire batch 15 s at medium speed. Upon completion of mixing, the mixing paddle shall be shaken to remove excess mortar into the mixing bowl.

9.3.2 When a duplicate batch is to be made immediately for additional specimens, the flow test may be omitted and the mortar allowed to remain in the mixing bowl for 90 s without covering. During the last 15 s of this interval, quickly scrape down into the batch the mortar that may have collected on the side of the bowl. Then remix for 15 s at medium speed.

9.3.3 Start molding the specimens within a total elapsed time of not more than 2 min and 30 s after completion of the original mixing of the mortar batch.

9.3.4 Evenly distribute a layer of mortar about 20 mm in thickness in each of the three molds with the tamper guide in place. Then compact the mortar in each mold by twelve strokes of the tamper, applied in three rounds of four strokes each, as shown in Fig. 4. Complete the twelve strokes in about 15 s. For each stroke hold the tamper face in

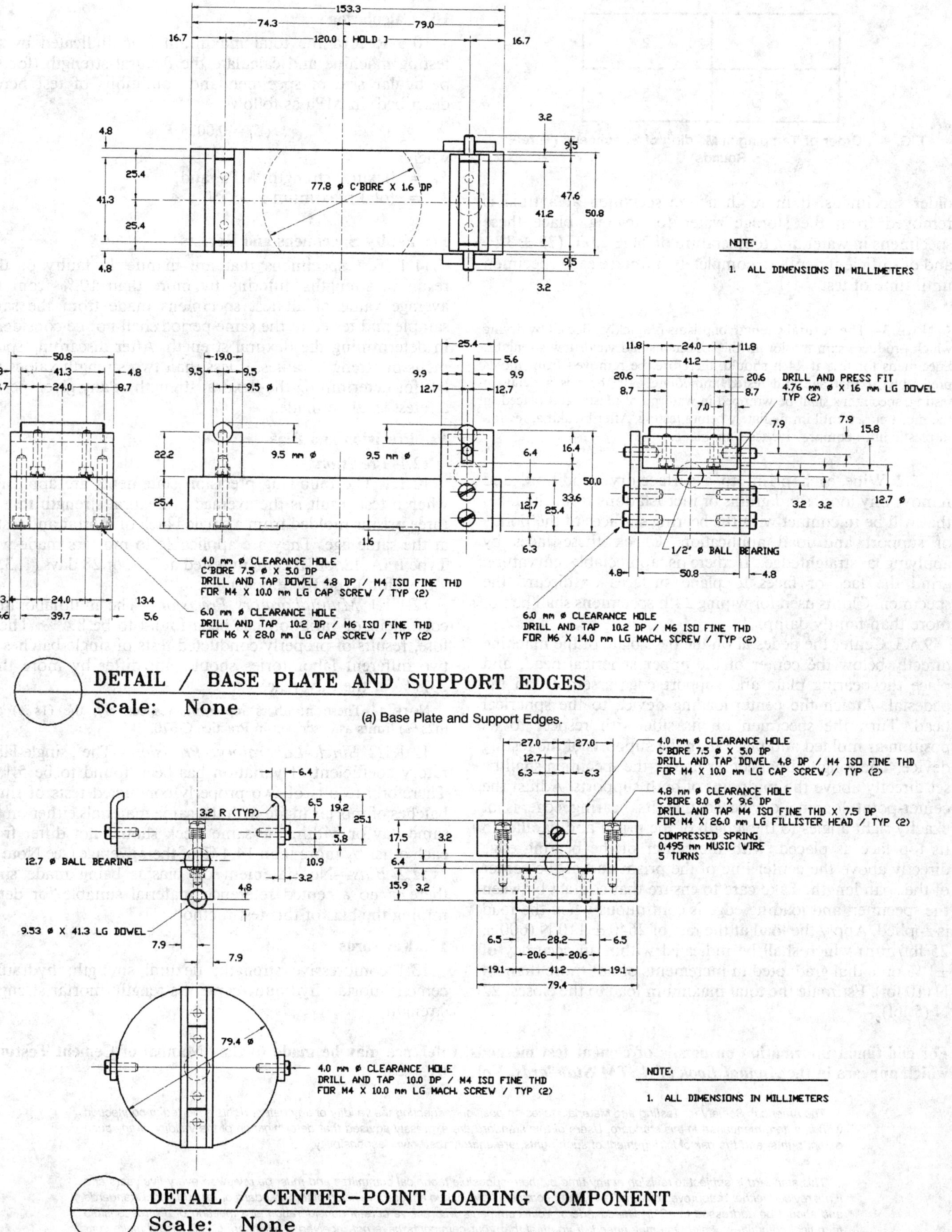

(a) Base Plate and Support Edges.

(b) Center-Point Loading Component.

FIG. 3 Special Components for Adapting Compression Testing Machine for Testing 40 by 40 by 160-mm Prisms

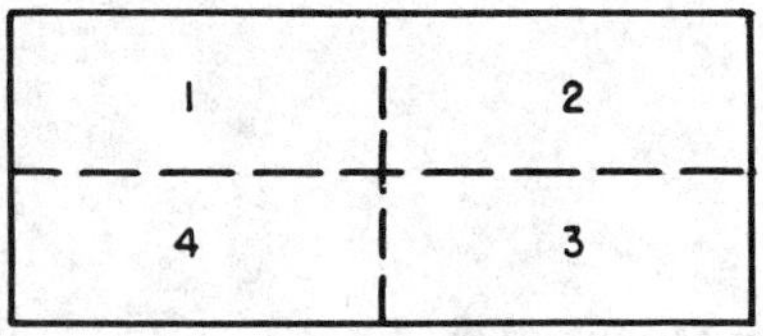

FIG. 4 Order of Tamping in Molding of Specimens (Three Rounds)

older specimens, if more than one specimen at a time is removed from the storage water for testing, place these specimens in water at a temperature of 23 ± 2.0°C (73 ± 3°F) and of sufficient depth to completely immerse each specimen until time of test.

NOTE 3—The flexural strength of prisms is quickly affected by drying which produces skin tension in the specimens and yields low strengths. Specimens for test at 24 h should, therefore, be removed from molds preferably at that age and tested immediately. If there is a delay in testing, specimens shall be wrapped in waterproof plastic and placed in the moist closet until immediately before testing. After breaking, rewrap in plastic until compressive tests are made.

9.5.2 Wipe each prism to a surface-dry condition, and remove any loose sand grains or incrustations from the faces that will be in contact with the bearing surfaces of the points of support and load application. Check these faces by applying a straightedge. If there is appreciable curvature, grind the face or faces to plane surfaces or discard the specimen. Cloths used for wiping 24-h specimens shall be not more than faintly damp.

9.5.3 Center the pedestal on the base plate of the machine directly below the center of the upper spherical head, and place the bearing plate and support edge assembly on the pedestal. Attach the center-loading device to the spherical head. Turn the specimen on its side with respect to its position as molded and place it on the supports of the testing device. The longitudinal center line of the specimen shall be set directly above the midpoint of both supports. Adjust the center-point loading device so that its bearing edge is at exactly right angles to the length of the prism and parallel to its top face as placed, with the center of the bearing edge directly above the center line of the prism and at the center of the span length. Take care to ensure that contact between the specimen and loading edge is continuous when the load is applied. Apply the load at the rate of 2640 ± 110 N (600 ± 25 lbf/min) which shall be indicated within an accuracy of ±1 % on a dial graduated in increments of not more than 44 N (10 lbf). Estimate the total maximum load to the closest 22 N (5 lbf).

10. Calculation

10.1 Record the total maximum load indicated by the testing machine and calculate the flexural strength (for the particular size of specimen and conditions of test herein described) in MPa as follows:

$$S_f = 0.0028\,P$$

where:

S_f = flexural strength, MPa, and
P = total maximum load, N.

11. Faulty Specimens and Retests

11.1 Test specimens that are manifestly faulty or that result in strengths differing by more than 10 % from the average value of all test specimens made from the same sample and tested at the same period shall not be considered in determining the flexural strength. After discarding specimens or strength values, if less than two strength values are left for determining the flexural strength at any given period a retest shall be made.

12. Precision and Bias

12.1 *Precision:*

12.1.1 The following precision statements are applicable when a test result is the average of flexural strength tests of three prisms molded from a single batch of mortar and tested at the same age. They are applicable to mortars made with Type I, IA, IS, or III cement tested at 3, 7, or 28 days (1, 3, or 7 days for Type III cement).

12.1.1.1 *Multilaboratory Precision*—The multilaboratory coefficient of variation has been found to be 8.4 %. Therefore, results of properly conducted tests of single batches by two different laboratories should not differ by more than 23.8 % of their average.

NOTE 4—These numbers represent, respectively, the (1s %) and (d2s %) limits as described in Practice C 670.

12.1.1.2 *Single-Laboratory Precision*—The single-laboratory coefficient of variation has been found to be 5.1 %. Therefore, results of two properly conducted tests of single batches of mortar made with the same materials either on the same day or within the same week should not differ from each other by more than 14.4 % of their average (see Note 4).

12.2 *Bias*—No statement on bias is being made since there is no accepted reference material suitable for determining the bias of this test method.

13. Keywords

13.1 compressive strength; flexural strength; hydraulic-cement mortar; hydraulic-cement strength; mortar strength; strength

For additional information on details of cement test methods, reference may be made to the "Manual of Cement Testing," which appears in the *Annual Book of ASTM Standards,* Vol 04.01.

Designation: C 349 – 97

Standard Test Method for Compressive Strength of Hydraulic-Cement Mortars (Using Portions of Prisms Broken in Flexure)[1]

This standard is issued under the fixed designation C 349; the number immediately following the designation indicates the year of original adoption or, in the case of revision, the year of last revision. A number in parentheses indicates the year of last reapproval. A superscript epsilon (ε) indicates an editorial change since the last revision or reapproval.

1. Scope

1.1 This test method covers the determination of the compressive strength of hydraulic-cement mortars, using for the test specimens portions of prisms made and broken in flexure in accordance with Test Method C 348.

1.2 The values stated in SI units are to be regarded as the standard. The values in parentheses are for information only.

1.3 Values in SI units shall be obtained by measurement in SI units or by appropriate conversion, using the Rules for Conversion and rounding given in Standard IEEE/ASTM SI 10, of measurements made in other units.

1.4 *This standard does not purport to address all of the safety concerns, if any, associated with its use. It is the responsibility of the user of this standard to establish appropriate safety and health practices and determine the applicability of regulatory limitations prior to use.*

2. Referenced Documents

2.1 *ASTM Standards:*

C 109/C 109M Test Method for Compressive Strength of Hydraulic Cement Mortars (Using 2-in. or 50-mm Cube Specimens)[2]

C 348 Test Method for Flexural Strength of Hydraulic Cement Mortars[2]

C 670 Practice for Preparing Precision and Bias Statements for Test Methods for Construction Materials[3]

IEEE/ASTM SI 10 Standard for Use of the International System of Units (SI): The Modern Metric System[4]

3. Significance and Use

3.1 This method provides a means of obtaining compressive strength values from the same specimens previously used for flexural strength determinations by Test Method C 348. The compressive strength values are for reference purposes, and not as substitutes for values (using 50-mm (2-in.) cubes) obtained by Test Method C 109/C 109M for cement acceptance.[2]

4. Apparatus

4.1 *Bearing Plates*—The bearing plates shall be not less than 25 mm in thickness and shall be made of a hard metal. The surfaces of the bearing plates that are placed in contact with the specimens shall be 40.0 ± 0.1-mm by 50.8-mm rectangles, with the 50.8-mm dimension at right angles to the longitudinal axis of the prism. The bearing plate surfaces in contact with the specimen shall have a Rockwell hardness of not less than 60 HRC. These surfaces shall not depart from plane surfaces by more than 0.01 mm when the plates are new and shall be maintained within a permissible variation of 0.03 mm.

4.2 *Device for Aligning Bearing Plates*—A device for aligning the bearing plates to ensure the proper location of the upper plate with reference to the lower plate is shown in Fig. 1. If the upper bearing plate does not have free movement within the aligning plates, these shall be separated from the bottom plate by insertion of 0.05-mm shims on each side of the bottom bearing plate, and the two 6-mm pins in the upper bearing plate filed down just sufficiently to permit free up and down movement in the slots at the top of the aligning plates.

4.3 *Testing Machine*—The testing machine shall conform to the requirements prescribed in Test Method C 109, except that the testing machine shall be of the hydraulic type.

5. Test Specimens

5.1 Both portions from each prism broken in flexure shall be used for compression testing, except that the broken portions of prisms selected for the compression test shall have a length of not less than 65 mm and shall be free of cracks, chipped surfaces, or other obvious defects.

6. Procedure

6.1 *Determination of Compressive Strength*—During the interval between flexure tests of the prisms and testing the broken portions as modified cubes, cover the specimens for the 24-h test with waterproof plastic and completely immerse all other specimens in water at a temperature of 23 ± 2.0°C (73 ± 3°F) until the time of testing. Wipe the specimen to a surface-dry condition, and remove any sand grains or incrustations from the faces that will be in contact with the bearing plates of the testing apparatus. Check these faces by application of a straightedge. If there is appreciable curvature, grind the face or faces to plane surfaces or discard the specimen (Note 1). Center the pedestal usually provided for breaking 50 by 100-mm (2 by 4-in.) cylinders on the base bearing block of the machine, and center the bearing plate assembly on top of this pedestal. If the testing machine has no provisions for automatic accurate centering of a pedestal exactly below the center of its upper spherical bearing head, a hardened steel cylinder of suitable diameter and height and with parallel plane end faces may be used, provided that the bearing plate assembly is centered accurately below the

---

[1] This test method is under the jurisdiction of ASTM Committee C-1 on Cement and is the direct responsibility of Subcommittee C01.27 on Strength.

Current edition approved May 10, 1997. Published June 1998. Originally published as C 349 – 54 T. Last previous edition C 349 – 95.

[2] *Annual Book of ASTM Standards*, Vol 04.01.

[3] *Annual Book of ASTM Standards*, Vol 04.02.

[4] *Annual Book of ASTM Standards*, Vol 14.02.

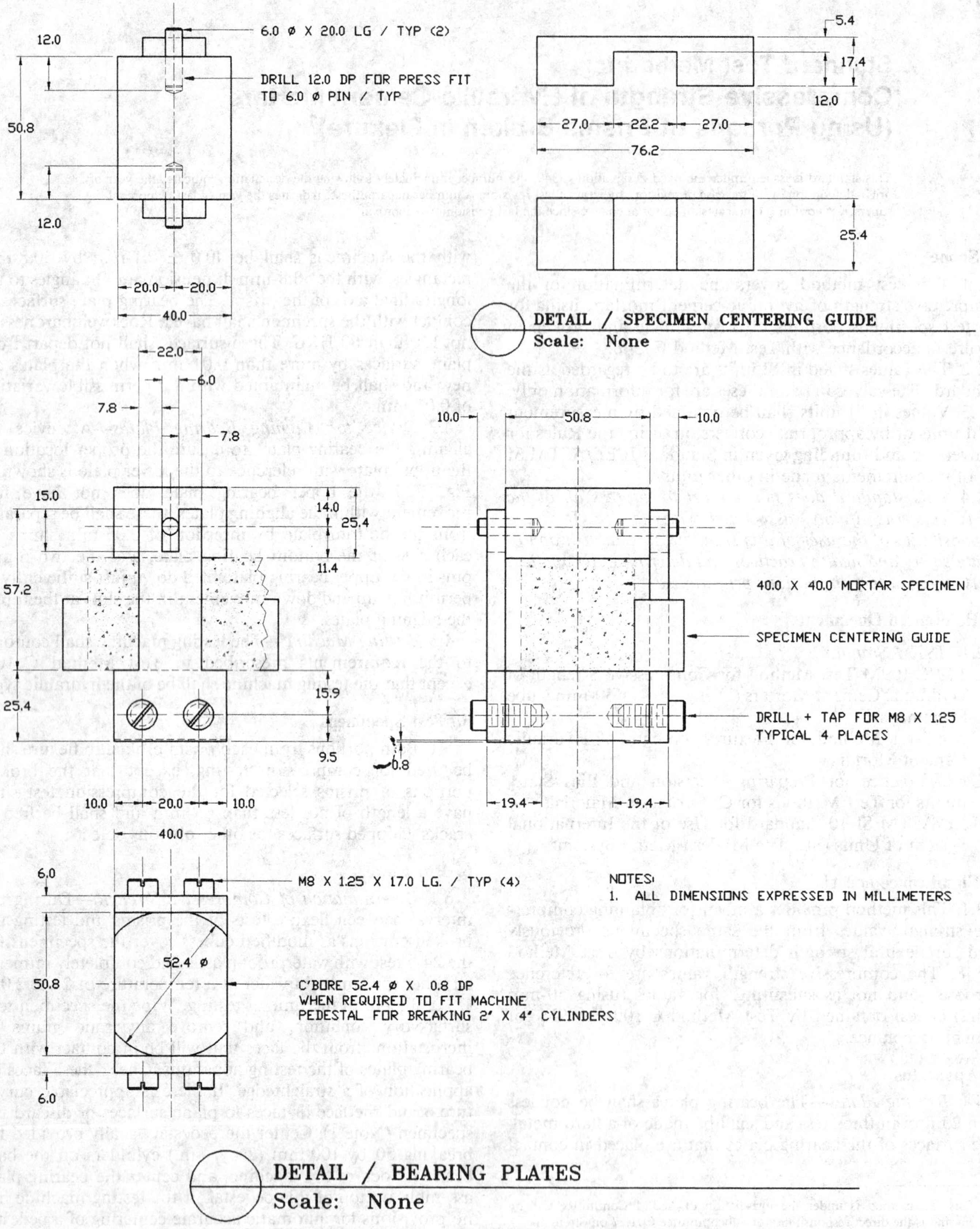

SI Equivalents

in.	1/32	1/4	5/16	3/8	0.462	1/2	9/16	5/8	3/4	7/8	1	1 1/16	1 3/4	2	2 1/16	2 1/2	3
mm	0.8	6.4	7.9	9.5	11.7	13	14.3	16	19	22.2	25	27	44.5	51	52.4	64	76

FIG. 1 Bearing Plates and Bearing Plate Aligning Device for Testing 40 by 40 by 160-mm Mortar Prisms as Modified Cubes

center of the upper bearing head after the specimen has been placed in proper position in the bearing plate assembly. Apply the specimen aligning guide to the outside of one of the aligning plates of the bearing plate assembly, with the lugs at each end resting on or slightly above the edge of the bearing face of the bottom plate. Turn the specimen on its side with respect to its position as molded and place it in the device with the bottom as molded in contact with the aligning lugs, holding the aligning guide against the aligning plate firmly with one hand. Then remove the aligning guide without disturbing the position of the specimen and apply the load in accordance with the Procedure section of Test Method C 109/C 109M. The testing of the broken portions as modified cubes shall follow breaking in flexure within 10 min for 24-h specimens and within 30 min for all other specimens.

NOTE 1: *Modified Cube Faces*—Results much lower than the true strength will be obtained by loading faces of the modified cube that are not truly plane surfaces. It is essential, therefore, that molds be kept scrupulously clean, as otherwise large irregularities in the surfaces will occur. Instruments for cleaning molds should always be softer than the metal in the molds to prevent wear. In case grinding of modified cube faces is necessary, it can be accomplished best by rubbing the specimen on a sheet of fine emery paper or cloth glued to a plane surface, using only moderate pressure. Since such grinding is tedious for more than a few thousandths of an inch or hundredth of a millimetre, it is recommended that where more than this is found necessary, the specimen be discarded.

7. Calculation

7.1 Record the total maximum load indicated by the testing machine and calculate the compressive strength in megapascals to the nearest 0.1 MPa, as follows:

$$S_c = 0.00062\ P$$

where:

S_c = compressive strength, MPa, and
P = total maximum load, N.

8. Faulty Specimens and Retests

8.1 Specimens that are manifestly faulty or that result in strengths differing by more than 10 % from the average value of all test specimens made from the same sample and tested at the same period shall not be considered in determining the compressive strength. After discarding strength values, if less than two strength values are left for determining the compressive strength at any given period a retest shall be made.

NOTE 2—Reliable strength results depend upon careful observance of all of the specified requirements and procedures. Erratic results at a given test period indicate that some of the requirements and procedures have not been carefully observed; for example, those covering the testing of the modified cubes, as prescribed in Sections 5 and 6. Specimens exhibiting oblique fractures on breaking, due to improper centering in the compression machine or to lateral movement of one of the testing machine heads during loading, will often indicate lower strengths than specimens showing a normal pyramidal fracture.

9. Precision and Bias

9.1 *Precision*—The following precision statements are applicable when a test result is the average of compressive strength tests of six modified cubes (using portions of prisms broken in flexure) molded from a single batch of mortar and tested at the same age. They are applicable to mortars made with Type I, IA, IS, or III cement tested at 3, 7, or 28 days (1, 3, or 7 days for Type III cement).

9.1.1 *Multilaboratory Precision*—The multilaboratory coefficient of variation has been found to be 6.3 %. Therefore, results of properly conducted tests of single batches by two different laboratories should not differ by more than 17.8 % of their average (Note 3).

NOTE 3—These numbers represent, respectively, the (1s%) and (d2s%) limits as described in Practice C 670.

9.1.2 *Single-Laboratory Precision*—The single-laboratory coefficient of variation has been found to be 3.5 %. Therefore, results of two properly conducted tests of single batches of mortar made with the same materials either on the same day or within the same week should not differ from each other by more than 9.9 % of their average (Note 3).

9.2 *Bias*—The procedure in this test method has no bias because the value of compressive strength is defined in terms of the test method.

10. Keywords

10.1 compressive strength; hydraulic-cement mortar; hydraulic-cement strength; mortar strength; strength

For additional useful information on details of cement test methods, reference may be made to the "Manual of Cement Testing," which appears in the *Annual Book of ASTM Standards*, Vol 04.01.

Standard Test Method for Early Stiffening of Portland Cement (Mortar Method)[1]

This standard is issued under the fixed designation C 359; the number immediately following the designation indicates the year of original adoption or, in the case of revision, the year of last revision. A number in parentheses indicates the year of last reapproval. A superscript epsilon (ε) indicates an editorial change since the last revision or reapproval.

1. Scope

1.1 This test method covers the determination of early stiffening in portland-cement mortar.

1.2 The values stated in SI units are the standard. The values in parentheses are for information only.

1.3 *This standard does not purport to address all of the safety concerns, if any, associated with its use. It is the responsibility of the user of this standard to establish appropriate safety and health practices and determine the applicability of regulatory limitations prior to use.* See Note 1 for a specific warning statement.

NOTE 1—**Warning:** Fresh hydraulic cementitious mixtures are caustic and may cause chemical burns to skin and tissue upon prolonged exposure. The use of gloves, protective clothing, and eye protection is recommended. Wash contact area with copious amounts of water after contact. Wash eyes for a minimum of 15 min. Avoid exposure of the body to clothing saturated with the liquid phase of the unhardened material. Remove contaminated clothing immediately after exposure.

2. Referenced Documents

2.1 *ASTM Standards:*

C 150 Specification for Portland Cement[2]

C 183 Practice for Sampling and the Amount of Testing of Hydraulic Cement[2]

C 185 Test Method for Air Content of Hydraulic Cement Mortar[2]

C 187 Test Method for Normal Consistency of Hydraulic Cement[2]

C 305 Practice for Mechanical Mixing of Hydraulic Cement Pastes and Mortars of Plastic Consistency[2]

C 490 Practice for Use of Apparatus for the Determination of Length Change of Hardened Cement Paste, Mortar, and Concrete[2]

C 670 Practice for Preparing Precision and Bias Statements for Test Methods for Construction Materials[3]

C 778 Specification for Standard Sand[2]

C 1005 Specification for Reference Masses and Devices for Determining Mass for Use in the Physical Testing of Hydraulic Cements[2]

D 1193 Specification for Reagent Water[4]

E 1 Specification for ASTM Thermometers[5]

3. Terminology

3.1 *Definitions:*

3.1.1 *early stiffening*—the early development of stiffness in the working characteristics of a portland-cement paste, mortar, or concrete. Varieties include false set and flash set.

3.1.2 *false set*—the early development of stiffness in the working characteristics of a portland-cement paste, mortar, or concrete without the evolution of much heat, which stiffness can be dispelled and plasticity regained by further mixing without addition of water; also known as "grab set,"" premature stiffening," "hesitation set," and "rubber set."

3.1.3 *flash set*—the early development of stiffness in the working characteristics of a portland-cement paste, mortar, or concrete, usually with the evolution of considerable heat, which stiffness cannot be dispelled nor can the plasticity be regained by further mixing without addition of water; also known as "quick set."

4. Summary of Test Method

4.1 A mortar is prepared with the cement to be tested, using specified quantities of cement, standard sand, and water. Using the modified Vicat apparatus, measurements of penetration are made at stipulated intervals after the beginning of the mixing procedure. Upon completion of the first series of penetration measurements, the mortar is returned to the mixer to be remixed. Following the remix procedure, an additional penetration, termed the remix penetration, is determined. The report is a tabulation of the penetration measurements.

5. Significance and Use

5.1 The purpose of this test method is to determine the degree to which a cement mortar develops early stiffening. It is intended primarily for use by those interested in research on methods for determining the potential false set of portland cement.

5.2 When used for estimating the relative tendency of a cement to manifest early stiffening, a judgment may be made by comparing the behavior in the penetration series (see 10.4.1 and 10.4.2) and the remix procedure (see 10.4.3) to differentiate a relatively less serious and less persistent tendency to

[1] This test method is under the jurisdiction of ASTM Committee C-1 on Cement and is the direct responsibility of Subcommittee C01.30 on Time of Set.

Current edition approved Jan. 10, 1999. Published May 1999. Originally published as C 359 – 55 T. Last previous edition C 359 – 89 (1995)$^{\epsilon 1}$.

[2] *Annual Book of ASTM Standards*, Vol 04.01.

[3] *Annual Book of ASTM Standards*, Vol 04.02.

[4] *Annual Book of ASTM Standards*, Vol 11.01.

[5] *Annual Book of ASTM Standards*, Vol 14.03.

early stiffening from one that is more persistent and, consequently, more serious.

5.3 Severe false setting in a cement may cause difficulty from a placing and handling standpoint, but it is not likely to cause difficulties where concrete is mixed for a longer time than usual, as usually occurs in transit mixing, or where it is remixed prior to placing or transporting, in concrete pumping operations. It is most likely noticed where concrete is mixed for a short period of time in stationary mixers and transported to the forms in nonagitating equipment, as on some paving jobs.

5.4 Cements with severe false setting usually require slightly more mixing water to produce the same consistency, which may result in slightly lower strengths and increased drying shrinkage.

5.5 Flash set of a severity sufficient to cause difficulties from a placing and handling standpoint usually will cause the cement to fail the requirements for time of setting in Specification C 150.

6. Apparatus

6.1 *Vicat Apparatus*, conforming to the requirements of Test Method C 187, with the following modifications:

6.1.1 The 1-mm needle shall be replaced by a weight, such that the total weight of the 10-mm plunger, indicator, and added weight shall be 400 ± 0.5 g.

6.2 *Spoon*, conforming to the requirements of Test Method C 185.

6.3 *Mixer, Bowl, Paddle, and Scraper*, conforming to the requirements of Practice C 305.

6.4 *Glass Graduates*, conforming to the requirements of Specification C 490.

6.5 *Masses and Mass Determining Devices* , conforming to the requirements of Specification C 1005.

6.6 *Thermometer*, ASTM No. 1C or 1F, conforming to the requirements of Specification E 1.

6.7 *Trowel*, having a steel blade 100 to 150 mm (4 to 6 in.) in length, with a straightedge.

6.8 *Clock Timer*.

6.9 *Containers*, approximately 50 by 50 by 150 mm (2 by 2 by 6 in.) inside dimensions (Fig. 1). These containers, which may be made of any suitable materials such as sheet metal or plastic, shall be rigid, watertight, and at least 50 mm (2 in.) deep.

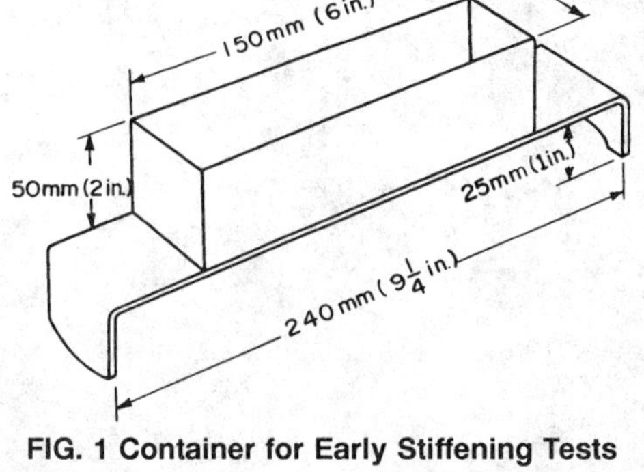

FIG. 1 Container for Early Stiffening Tests

7. Reagents and Materials

7.1 *Standard Sand, 20-30 and Graded*, conforming to the requirements of Specification C 778.

7.2 *Mixing Water*—Potable water is satisfactory for routine tests. For all referee and cooperative tests, reagent water conforming to the requirements of Specification D 1193 for Type III or Type IV grades of reagent water shall be used.

8. Sampling

8.1 Sample the cement in accordance with Methods C 183.

9. Conditioning

9.1 The temperature of the room, dry materials, paddle, bowl, and containers shall be maintained between 20 and 27.5°C (68 and 81.5°F). The temperature of the mixing water shall not vary from 23°C (73.4°F) by more than ±1.7°C (±3°F).

9.2 The relative humidity of the laboratory shall not be less than 50 %.

10. Procedure

10.1 *Batch*—Mix at one time 600 g of cement, 300 g of graded standard sand, 300 g of 20-30 standard sand, and 180 mL of water for all cements except Types III and IIIA, for which the amount of water shall be 192 mL.

10.2 *Mixing of Mortar*—The mixing shall be done in the mechanical mixer as follows:

10.2.1 Place the sand and cement in the dry bowl, and mix the dry materials for a few seconds with the spoon.

10.2.2 Place the bowl in the mixer, set the paddle in place, and mix the dry materials for 10 s at a slow speed (140 ± 5 r/min).

10.2.3 With the mixer operating at a slow speed (140± 5 r/min), add the entire quantity of mixing water within 5 s. Stop the mixer, quickly change to a medium speed (285 ± 10 r/min), and continue the mixing for 1 min, timing from the first addition of water.

10.2.4 Stop the mixer, scrape the sides of the mixing bowl with the rubber scraper, and quickly place the thermometer in the mortar. Allow it to stand undisturbed for the remainder of a 45-s interval from the time of stopping the mixer.

10.2.5 Read the temperature, remove the thermometer, start the mixer, and mix for 15 s at a medium speed (285 ± 10 r/min). If the mortar temperature is not in the range from 23 ± 1.7°C (73.4± 3°F), discard the batch and adjust the temperature of the water or sand, or both, to give the required temperature.

10.3 *Filling Container*:

10.3.1 Immediately after completion of the mixing, remove the bowl from the mixer and with a spoon, uniformly distribute a portion of the mortar into the container until the container is heaping full. Quickly and gently place each spoonful of mortar in the container. When removing the mortar from the bowl, do not remove the material pushed up on the side of the bowl by the paddle. After the container has been filled, reassemble the mixer, cover the bowl with a lid, and retain the remaining mortar for a remix test to be performed later. To compact the mortar in the container, lift the container approximately 80 mm (3 in.) from the table with both hands and rap it twice against the surface of the table.

10.3.2 With the leading edge slightly raised, strike off the mortar with one stroke of the trowel along the length of the container. Then remove the excess mortar by means of a sawing motion with the straightedge of the trowel along the length of the container in a direction opposite to that used in striking off. Then, smooth the surface of the mortar with a single stroke of the trowel.

10.4 *Penetration Tests*:

10.4.1 After filling the container, immediately place the 10-mm plunger of the Vicat apparatus, Fig. 1 of Test Method C 187, in contact with the surface of the mortar at the midpoint of the container on the longitudinal center line. Set the movable indicator at zero. Release the plunger 3 min after the beginning of the wet mixing and record, as the initial penetration, the depth in millimetres to which the plunger has settled below the surface 10 s after being released. Generally, the plunger will settle to the bottom of the container, and the initial penetration will, accordingly, be recorded as 50 + mm.

10.4.2 Immediately withdraw and clean the plunger. In a similar manner, determine, after moving the Vicat apparatus to the desired location, the penetrations at intervals of 5, 8, and 11 min after the beginning of mixing. Do not move the filled container until these measurements are completed. Make all penetrations along the longitudinal center line of the container. Obtain 5 and 8-min penetrations at a distance of approximately 40 mm (1½ in.) from each end of the container, respectively, and determine the 11-min penetration at a point approximately midway between the points at which the initial and 5-min penetrations were determined.

10.4.3 At the completion of the measurement of the 11-min penetration, immediately return the mortar in the container to the bowl. Start the mixer, raise the bowl into mixing position, and remix the contents of the bowl at medium speed (285 ± 10 r/min) for 1 min. Fill a clean container as outlined in 10.3.1 and 10.3.2, and determine the penetration 45 s after completion of mixing.

11. Report

11.1 The report shall show the depths of the penetrations as follows:

Initial penetration	___mm
5-min penetration	___mm
8-min penetration	___mm
11-min penetration	___mm
Remix penetration	___mm

12. Precision and Bias

12.1 Interlaboratory test data for the determination of statements of precision and bias in accordance with Practice C 670 are being evaluated by Subcommittee C01.30. Statements of precision and bias will be included in a later revision of this test method.

NOTE 2—For additional useful information on details of cement test methods, reference may be made to the "Manual of Cement Testing," which appears in the *Annual Book of ASTM Standards*, Vol 04.01.

13. Keywords

13.1 early stiffening; false set; flash set; portland-cement mortar

Designation: C 400 - 98

Standard Test Methods for Quicklime and Hydrated Lime for Neutralization of Waste Acid[1]

This standard is issued under the fixed designation C 400; the number immediately following the designation indicates the year of original adoption or, in the case of revision, the year of last revision. A number in parentheses indicates the year of last reapproval. A superscript epsilon (ϵ) indicates an editorial change since the last revision or reapproval.

1. Scope

1.1 This test method covers the testing of quicklime and hydrated lime for use in the treatment of waste acid solutions and includes the characterization of the liming material and of the acid waste.

NOTE 1—When agreed upon by the manufacturer and the purchaser, this method may be used directly to evaluate the lime requirement for a plant waste acid and in the preparation of the lime slurry; the pH, the time, temperature, and other conditions of treatment may be adjusted to conform to plant practice. Otherwise the test shall be performed as described in this method.

NOTE 2—Under some conditions of test the lime requirement may vary substantially from that indicated by ultimate chemical analysis or by the use of a determination of available alkalinity by titration to a stoichiometric end point.

1.2 *This standard does not purport to address all of the safety concerns, if any, associated with its use. It is the responsibility of the user of this standard to establish appropriate safety and health practices and determine the applicability of regulatory limitations prior to use.*

2. Referenced Documents

2.1 *ASTM Standards:*

C 50 Methods of Sampling, Inspection, Packing, and Marking of Lime and Limestone Products[2]

E 70 Test Method for pH of Aqueous Solutions with the Glass Electrode[3]

3. Terminology

3.1 *Definitions:*

3.1.1 *neutralization coefficient*—the number of parts of lime required to neutralize one million parts by weight of sulfuric acid solution (1.5 %) to a pH of 4.4 in 30 min (see Note 1).

3.1.2 *pH*—a measure of the hydrogen ion concentration of a sample and representing the logarithm of the reciprocal (negative logarithm) of the activity of hydrogen ions, calculated as follows:

$$pH = \log \frac{1}{(H)^+} = -\log(H^+)$$

where:

H^+ = activity of hydrogen ions.

4. Significance and Use

4.1 Since all limes and waste acid solutions are different, this test method evaluates the relative quantity of a given lime needed for the neutralization of a specific quantity of acid.

4.2 Liming materials have a specific basicity factor or measure of alkalinity which can be used for comparing their relative neutralizing power. It avoids dependence on chemical analysis and is determined as grams of calcium oxide equivalents per kilogram of liming material.

4.3 Likewise, specific waste acids have an acid value that can be expressed as grams of hydrochloric acid equivalent per litre of acid waste. This value is related on a stoichiometric basis to lime neutralization requirements of a liquid acid waste.

5. Apparatus

5.1 *pH Measuring Apparatus,* conforming to the requirements of Section 5, Apparatus, and Section 6, Reagents and Materials, of Test Method E 70. A time-pH recording device or attachment is desirable but not required.

5.2 *Analytical Balance,* capable of reproducing results within 0.005 g.

5.3 *Stirrer,* motorized, capable of rotating at speeds from 120 to 600 r/min.

5.4 *Miscellaneous Equipment*—Chemically resistant glassware,[4] metal stands for support of the equipment, etc.

6. Purity of Reagents

6.1 Reagent grade chemicals shall be used in all tests. Unless otherwise indicated, it is intended that all reagents shall conform to the specifications of the Committee on Analytical Reagents of the American Chemical Society, where such specifications are available.[5] Other grades may be used, provided it is first ascertained that the reagent is of sufficiently

[1] These test methods are under the jurisdiction of ASTM Committee C-7 on Lime and are the direct responsibility of Subcommittee C07.05 on Chemical Tests.

Current edition approved May 10, 1998. Published December 1998. Originally published as C 400 - 57 T. Last previous edition C 400 - 93.

[2] *Annual Book of ASTM Standards,* Vol 04.01.

[3] *Annual Book of ASTM Standards,* Vol 15.05.

[4] Borosilicate glass has been found satisfactory for this purpose.

[5] "Reagent Chemicals, American Chemical Society Specifications," Am. Chemical Soc., Washington, DC. For suggestions on the testing of reagents not listed by the American Chemical Society, see "Reagent Chemicals and Standards," by Joseph Rosin, D. Van Nostrand Co., Inc., New York, NY, and the "United States Pharmacopeia."

high purity to permit its use without lessening the accuracy of the determination.

6.2 Unless otherwise indicated, references to water shall be understood to mean distilled water.

7. Test Method A—for Determining Lime Requirement for the neutralization of a standard sulfuric acid or waste acid solution within a time limit of thirty minutes.

7.1 *Reagents:*

7.1.1 *Sulfuric Acid, Standard Solution* (1.5 ± 0.05 %)—Thoroughly mix 15.8 g of concentrated sulfuric acid (95 % H_2SO_4) with water, dilute to 1 L, analyze to within 0.005 % H_2SO_4, and store in a suitable stoppered bottle.

7.2 *Preparation of Sample:*

7.2.1 Take samples of quicklime and hydrated lime in accordance with Methods C 50.

7.2.2 *Quicklime*—Weigh a sample of quicklime (2.50 ± 0.01 g) ground to pass a No. 50 mesh sieve. Add to a 500-mL tall-form beaker containing 50 mL of boiling water and continue to boil for 2 min. Remove from hot plate and continue stirring with a magnetic stirrer until the slurry reaches room temperature (at least 10 min).

7.2.3 *Hydrated Lime*—Mix a sample of hydrated lime (2.50 ± 0.01 g) with 50 mL of water in a 500-mL tall-form beaker for at least 10 min.

7.3 *Procedure:*

7.3.1 Place the pH electrodes within the beaker about ½ in. (13 mm) above the surface of the lime slurry sample. Set the motor stirrer to the highest speed at which splattering or immersion of the electrodes does not occur. Measure 250 g of H_2SO_4 (1.5 ± 0.05 %) into a separate 500-mL beaker. Add this acid to the sample in a single motion and record the pH as a function of time for a period of 30 min (see Note 1).

7.3.2 Test a second slurry sample using more or less than 250 g of H_2SO_4 (1.5 ± 0.05 %), as indicated by the pH value observed after 30 min reaction in the first test. Repeat the procedure at least four times or until sufficient data are available to give a family of curves that indicate pH at 30 min as a function of grams of lime per 1000 g of H_2SO_4 (1.5 %), with particular emphasis in the region of pH equals 4.4.

7.4 *Results:*

7.4.1 *Neutralization Rate Curves:*

7.4.1.1 Plot pH versus time in minutes for each of the lime-acid ratios used in the initial series of tests. This family of curves serves as the basis for calculation of the lime requirements.

7.4.2 *Lime Requirement Curve:*

7.4.2.1 Plot the pH in 30 min, as determined from the neutralization rate curves described in Section 10, against grams of lime used per 1000 g of H_2SO_4 (1.5 %).

7.4.3 *Neutralization Coefficient:*

7.4.3.1 Interpolate the quantity of lime required to achieve a pH of 4.4 in 30 min from the lime requirement curve (Section 11), correct for the actual acid concentration, and calculate as parts of lime required per million parts of H_2SO_4 (1.5 %) as follows:

$$N = (1.5/Z) \times X \times 1000$$

where:

N = neutralization coefficient (see 3.1.1),
X = grams of lime per 1000 g of H_2SO_4 (1.5 %), and
Z = concentration of the H_2SO_4 to the nearest 0.005 %, as determined by analysis.

7.5 *Precision and Bias:*

7.5.1 The precision and bias of this test method has not been determined at this time. Replicate samples are recommended.

8. Test Method B—For Determining Acid Value-Basicity Factor Relationship

8.1 *Procedure:*

8.1.1 *Basicity Factor*—Weigh 2.804 g ± 1 mg of the sample that has been prepared to pass a No. 100 (150-μm) sieve and transfer to a 500-mL Erlenmeyer flask containing approximately 20 mL of cold water. (For quicklime samples, heat to boiling to ensure complete slaking). Add from a buret 100 mL of 1.0 *N* hydrochloric acid and stopper with a two-hole rubber stopper (Note 3). The quantity of acid may be increased or decreased depending on the equivalent strength of the material being tested. Place the flask on a hot plate and boil for 15 min. (Glass beads may be added to prevent bumping.) Remove the flask from the hot plate and cool in water. Add several drops of phenolphthalein indicator solution and titrate the excess acid with 0.5 *N* sodium hydroxide solution. Calculate the basicity factor as follows:

$$\text{Basicity factor} = [(V_1N_1 - V_2N_2)/W] \times 0.02804$$

where:

V_1 = HCl (mL) required for titration of the sample,
N_1 = normality of HCl,
V_2 = NaOH solution (mL) required for titration of the excess acid,
N_2 = normality of NaOH solution, and
W = sample (g).

NOTE 3—A water-cooled condenser with a tapered, ground-glass joint and a magnetic stir plate may be used during the boiling step.

8.1.2 *Acid Value*—Pipet 10 mL of the sample of acidic waste into a 250-mL Erlenmeyer flask. This quantity may be increased or decreased depending on the strength of the acidic waste. Add an excess of 25 to 30 mL of 0.5 *N* sodium hydroxide solution and boil for 3 min. Filter through a small Büchner funnel with the aid of vacuum. Wash the residue on the funnel with water. Add several drops of phenolphthalein indicator solution to the filtrate and titrate the excess sodium hydroxide with 1.0 *N* hydrochloric acid solution. Calculate acid value as follows:

$$\text{Acid value} = [(V_1N_1 - V_2N_2)/V] \times 0.03646$$

where:

V_1 = NaOH solution (mL) required for titration of the sample,
N_1 = normality of NaOH solution,
V_2 = HCl (mL) required for titration of the excess NaOH solution,
N_2 = normality of HCl, and
V = sample (mL).

8.2.3 *Relationship of Acid Value to Basicity Factor*—

Calculate the acid value-basicity factor relationship as follows:

$$\frac{\text{Acid value}}{\text{Basicity factor}} \times 0.7691$$

= lime or limestone product (kg) per litre of Acidic Waste.

NOTE 4—When agreed upon between the manufacturer and the purchaser, Test Method A may also be used.

8.2 *Precision and Bias:*

8.2.1 The precision and bias of this test method has not been determined at this time. Replicate samples are recommended.

9. Keywords

9.1 acid value; alkalinity; basicity; lime requirement; liming material; neutralization; neutralization coefficient; pH; waste acid

Designation: C 430 – 96

Standard Test Method for Fineness of Hydraulic Cement by the 45-μm (No. 325) Sieve[1]

This standard is issued under the fixed designation C 430; the number immediately following the designation indicates the year of original adoption or, in the case of revision, the year of last revision. A number in parentheses indicates the year of last reapproval. A superscript epsilon (ϵ) indicates an editorial change since the last revision or reapproval.

1. Scope

1.1 This test method covers the determination of the fineness of hydraulic cement by means of the 45-μm (No. 325) sieve.

1.2 The values stated in SI units are to be regarded as the standard. The inch-pound equivalents of SI units may be approximate.

1.3 *This standard does not purport to address all of the safety concerns, if any, associated with its use. It is the responsibility of the user of this standard to establish appropriate safety and health practices and determine the applicability of regulatory limitations prior to use.*

2. Referenced Documents

2.1 *ASTM Standards:*

E 11 Specification for Wire-Cloth Sieves for Testing Purposes[2]

E 161 Specification for Precision Electroformed Sieves (Square Opening Series)[2]

E 177 Practice for Use of the Terms Precision and Bias in ASTM Test Methods[2]

3. Apparatus

3.1 *Sieve:*

3.1.1 *Sieve Frame*—The sieve frame shall be of metal not subject to corrosion by water and shall be circular, either 51 ± 6 mm (2.0 ± 0.25 in.) when woven-wire cloth is mounted in the frame or 76 ± 6 mm (3.0 ± 0.25 in.) in diameter when an electroformed sheet is mounted in the frame. The depth of the sieve from the top of the frame to the cloth or sheet shall be 76 ± 6 mm (3.0 ± 0.25 in.). The frame shall have either side walls of 89 ± 6 mm (3.5 ± 0.25 in.) in total height, or legs at least 12 mm (0.5 in.) in length, sufficient to allow air circulation beneath the sieve cloth or electroformed sheet.

3.1.2 *Sieve Cloth or Electroformed Sheet*—The sieve frame shall be fitted with either a 45-μm (No. 325) stainless steel AISI Type 304 woven-wire sieve cloth, conforming to the requirements of Specification E 11 or a 45-μm electroformed reinforced nickel sieve sheet conforming to the requirements of Specification E 161 with the exception that the number of openings shall be 71 ± 2 per linear centimetre (180 ± 5 per linear inch).

3.1.3 *Cloth or Sheet Mounting:*

3.1.3.1 *Woven-Wire Cloth Mounting*—Type 304 stainless steel woven-type cloth shall be mounted in the frame without distortion, looseness, or wrinkling. For a sieve fabricated by soldering the cloth to the frame, the joint shall be made smooth to prevent the cement from catching in the joints between the sieve cloth and the frame. Two-piece sieves shall clamp tightly on the cloth to prevent the cement from catching in the joints between the sieve cloth and the frame.

3.1.3.2 *Electroformed Sieve Sheet Mounting*—Electroformed reinforced nickel sieve sheet shall be mounted in the frame without distortion, looseness, or wrinkling. The joint between the sieve cloth and the frame shall be made smooth with a material impervious to water.

3.2 *Spray Nozzle*—The spray nozzle (Fig. 1) shall be constructed of metal not subject to corrosion by water and shall be 17.5 mm (0.69 in.) in inside diameter with a central hole drilled in line with the longitudinal axis, an intermediate row of eight holes drilled 6 mm (0.23 in.) center-to-center at an angle of 5° from the longitudinal axis, and an outer row of eight holes drilled 11 mm (0.44 in.) center-to-center at an angle of 10° from the longitudinal axis. All holes shall be 0.5 mm (0.02 in.) in diameter. The spray nozzle shall have been checked within six months prior to testing a cement sample to ensure that the flow rate is between 1500 and 3000 g/min at 69 ± 3 kPa (10 ± 0.4 psi).

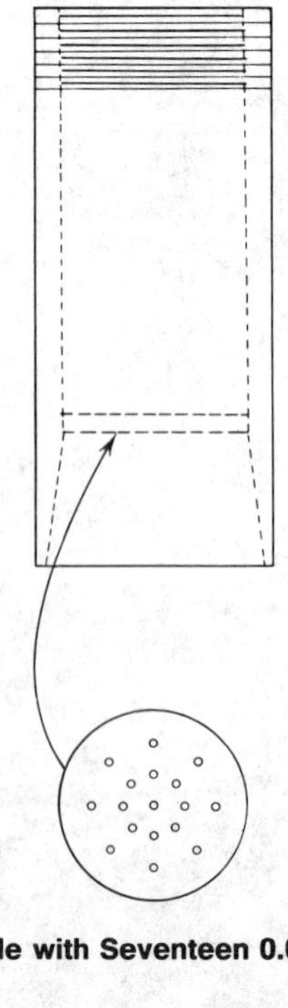

FIG. 1 Spray Nozzle with Seventeen 0.02-in. (0.51-mm) Holes

[1] This test method is under the jurisdiction of ASTM Committee C-1 on Cement and is the direct responsibility of Subcommittee C01.25 on Fineness.

Current edition approved Dec. 10, 1996. Published February 1997. Originally published as C 430 – 59 T. Last previous edition C 430 – 95.

[2] *Annual Book of ASTM Standards*, Vol 14.02.

3.3 *Pressure Gage*—The pressure gage shall be 76 mm (3.0 in.) minimum diameter, and shall be graduated in 7-kPa (1-psi) increments, and shall have a maximum capacity of 207 kPa (30 psi). The accuracy at 69 kPa (10 psi) shall be ±2 kPa (±0.25 psi).

4. Calibration of 45-μm (No. 325) Sieve

4.1 Place 1.000 g of the current lot of National Institute of Standards and Technology standard sample No. 114 on the clean, dry, 45-μm (No. 325) sieve and proceed as in Section 5. The sieve correction factor is the difference between the test residue obtained and the assigned residue value indicated by the electroformed sheet sieve fineness specified for the standard sample, expressed as a percentage of the test residue.

NOTE 1—It should be observed that the sieve correction as specified is a factor to be multiplied by the residue obtained and that the amount to be added to or subtracted from the residue in any given test is therefore proportional to the amount of residue.

Example of Determination of Sieve Correction Factor:

Residue on 45-μm (No. 325) Sieve, sample No. 114		12.2 %
Residue for a 1-g sample		0.122 g
Residue on sieve being calibrated		0.093 g
Difference	=	+0.029 g
Correction factor = +0.029/0.093 × 100 = +31.18	=	+31.2 %

5. Procedure

5.1 Place a 1.000 g sample of the cement on the clean, dry 45-μm (No. 325) sieve. Wet the sample thoroughly with a gentle stream of water. Remove the sieve from under the nozzle and adjust the pressure on the spray nozzle to 69 ± 4 kPa (10 ± 0.5 psi). Return the sieve to its position under the nozzle and wash for 1 min, moving the sieve with a circular motion in a horizontal plane at the rate of one motion per second in the spray. The bottom of the spray nozzle should extend below the top of the sieve frame about 12 mm (0.5 in.). Immediately upon removing the sieve from the spray, rinse once with about 50 cm^3 of distilled or deionized water, using caution not to lose any of the residue, and then blot the lower surface gently upon a damp cloth. Dry the sieve and residue in an oven or over a hot plate (see Note 2), supporting the sieve in such a manner that air may pass freely beneath it. Cool the sieve; then brush the residue from the sieve and weigh on an analytical balance capable of reproducing results within 0.0005 g.

NOTE 2—Care should be taken not to heat the sieve hot enough to soften the solder.

6. Cleaning of 45-μm (No. 325) Sieves

6.1 *Frequency of Cleaning and Calibration*—Sieves fitted with woven wire sieve cloth shall be cleaned after no more than five determinations. Sieves fitted with an electroformed reinforced sieve sheet having 71 openings per linear centimetre shall be cleaned after no more than three determinations. Both types of sieves shall be recalibrated after no more than 100 determinations.

6.2 *Acceptable Cleaning Procedures*—One option for cleaning is to place the sieve in a low-power (150 W maximum power input) ultrasonic bath containing an appropriate laboratory cleaning solution. The bath is to be operated for sufficient time (approximately 10 to 15 min at room temperature) to remove particles lodged in the openings. Be apprised that electroformed sieve sheets containing more than 71 openings per linear centimetre may well be damaged by ultrasonic cleaning. An option for cleaning which does not require an ultrasonic bath can also be employed. Immerse the sieve in a bath of appropriate laboratory cleaning solution heated to just below boiling point. Cover with a watch glass to reduce evaporation. Continue this soaking for a time sufficient to loosen lodged particles with a rinse following the bath. Overnight soaking in similar but unheated cleaning solutions is also acceptable, provided a rinse following the bath is able to wash away lodged particles. Cleaning or rinsing with dilute hydrochloric or acetic acid solutions is to be avoided. Appropriate cleaning solutions are restricted to soap or detergent-type solutions.

7. Calculation

7.1 Calculate the fineness of the cement to the nearest 0.1 % as follows:

$$R_c = R_s \times (100 + C) \quad (1)$$

$$F = 100 - R_c \quad (2)$$

where:

F = fineness of the cement expressed as the corrected percentage passing the 45-μm (No. 325) sieve,
R_c = corrected residue, %,
R_s = residue from the sample retained on the 45-μm (No. 325) sieve, g, and
C = sieve correction factor (determined as prescribed in Section 4) which may be either plus or minus.

Example:

Sieve correction factor, C	=	+31.2 %
Residue from sample being tested, R_s	=	0.088 g
Corrected residue, R_c	=	0.088 × (100 + 31.2)
	=	11.5 %
Corrected amount passing, F	=	100 − 11.5 % = 88.5 %

8. Precision and Bias

8.1 *Normal Fineness Product*—The multilaboratory precision has been found to be ±0.75 % (IS) as defined in Practice E 177, therefore, results of properly conducted tests from two different laboratories on identical samples of cement should agree 95 % of the time within ±2.1 %.

8.2 *High Fineness Product*—The multilaboratory precision has been found to be ±0.50 % (IS) as defined in Practice E 177; therefore, results of properly conducted tests from two different laboratories on identical samples of cement should agree 95 % of the time within ±1.4 %.

NOTE 3—The use of outside threads instead of inside threads as shown in this figure is permissible.

8.3 Since there is no accepted reference material suitable for determining the bias for the procedure in this test method, no statement on bias is being made.

9. Keywords

9.1 fineness of hydraulic cement

Designation: C 442 – 98

Standard Specification for Gypsum Backing Board and Coreboard[1]

This standard is issued under the fixed designation C 442; the number immediately following the designation indicates the year of original adoption or, in the case of revision, the year of last revision. A number in parentheses indicates the year of last reapproval. A superscript epsilon (ε) indicates an editorial change since the last revision or reapproval.

This specification has been approved for use by agencies of the Department of Defense.

1. Scope

1.1 This specification covers gypsum board designed to be used as a base in multilayer systems or as a gypsum stud or core in semi-solid or solid gypsum board partitions.

1.2 The values stated in inch-pound units are to be regarded as the standard. The (SI) metric values given in parentheses are approximate and are provided for information purposes only.

NOTE 1—The terms *backing board* and *coreboard* refer to different end uses of the gypsum board specified in this specification. The term *gypsum backing board* in this specification shall include gypsum coreboard unless otherwise stated.

2. Referenced Documents

2.1 *ASTM Standards:*

C 473 Test Methods for Physical Testing of Gypsum Panel Products[2]

C 645 Specification for Nonstructural Steel Framing Members[2]

C 1264/C 1264M Specification for Sampling, Inspection, Rejection, Certification, Packaging, Marking, Shipping, Handling, and Storage[2]

E 84 Test Method for Surface Burning Characteristics of Building Materials[3]

E 96 Test Methods for Water Vapor Transmission of Materials[4]

E 119 Test Methods for Fire Tests of Building Construction and Materials[3]

3. Terminology

3.1 *Definitions of Terms Specific to This Standard:*

3.1.1 *gypsum backing board*—a ¼ to ⅝-in. (6.4 to 16-mm) gypsum board used as a backing for gypsum wallboard, acoustical tile, or other dry cladding.

3.1.2 *gypsum coreboard*—a ¾ to 1-in. (19 to 25.4-mm) gypsum board used as a gypsum stud or core in semisolid or solid gypsum board partitions.

4. Materials and Manufacture

4.1 Gypsum backing board shall consist of a noncombustible core, essentially gypsum, surfaced with paper bonded to the core.

4.2 Gypsum coreboard may be a single board or may be composed of two factory laminated boards to provide up to 1-in. (25.4-mm) total nominal thickness.

4.3 Foil backed backing board shall have the back surface covered with aluminum foil.

5. Physical Properties

5.1 *Foil Backed Backing Board*—Foil-backed gypsum backing board shall have a maximum permeance of 0.30 perms when tested in accordance with Test Methods E 96 for the condition of 50 % relative humidity on Side I and 0 % relative humidity on Side II, the foil covered side of the board.

5.2 *Flame Spread*—Gypsum backing board shall have a maximum flame spread rating not to exceed 25 when tested in accordance with Test Method E 84.

5.3 *Type X Gypsum Backing Board*—Gypsum backing board, Type X (special fire-resistant) designates gypsum backing board complying with this specification that provides not less than 1-h fire-resistance for boards ⅝-in. (15.9 mm) thick or ¾-h fire-resistance for boards ½-in. (12.7 mm) thick, applied parallel with and on each side of load bearing 2 by 4 wood studs spaced 16 in. (406 mm) on center with 6d coated nails, 1⅞-in. (48-mm) long, 0.0915-in. (2.3-mm) diameter shank, ¼-in. (6.4-mm) diameter heads, spaced 7 in. (178 mm) on center with gypsum backing board joints staggered 16 in. (406 mm) on each side of the partition and tested in accordance with Test Methods E 119.

NOTE 2—Consult backing board manufacturer for independent test data on assembly particulars, materials, and ratings for other types of construction.

5.4 *Flexural Strength*—Specimens shall be tested in accordance with Test Methods C 473. When tested face up and when tested face down the specimens shall have an average breaking load of not less than the following:

	Method A		Method B	
Thickness, in. (mm)	Load, lbf (N) Bearing Edges Across Fiber of Surfacing	Load, lbf (N) Bearing Edges Parallel to Fiber of Surfacing	Load, lbf (N) Bearing Edges Across Fiber of Surfacing	Load, lbf (N) Bearing Edges Parallel to Fiber of Surfacing
¼ (6.4)	50 (222)	20 (89)	46 (205)	16 (71)
⅜ (9.5)	80 (356)	30 (133)	77 (343)	26 (116)
½ (12.7)	110 (489)	40 (178)	107 (476)	36 (160)
⅝ (15.9)	140 (622)	50 (222)	137 (609)	46 (205)
¾ (19)	170 (756)	60 (267)	167 (743)	56 (249)
1 (25.4)	230 (1023)	80 (356)	228 (1014)	77 (343)

[1] This specification is under the jurisdiction of ASTM Committee C-11 on Gypsum and Related Building Materials and Systems and is the direct responsibility of Subcommittee C11.01 on Specifications and Test Methods for Gypsum Products.

Current edition approved April 10, 1998. Published December 1998. Originally published as C 442 – 59 T. Last previous edition C 442 – 97.

[2] *Annual Book of ASTM Standards,* Vol 04.01.

[3] *Annual Book of ASTM Standards,* Vol 04.07.

[4] *Annual Book of ASTM Standards,* Vol 04.06.

5.5 *Humidified Deflection*—Specimens taken from gypsum backing board shall have an average deflection of no more than the following when tested in accordance with Test Methods C 473:

Thickness, in. (mm)	Humidified Deflection, in. (mm)
¼ (6.4)	not applicable
⅜ (9.5)	1⁵⁄₈ (48)
½ (12.7)	1⁰⁄₈ (32)
⅝ (15.9)	⅝ (16)

5.6 *Core, End, and Edge Hardness*—Specimens taken from the gypsum backing board shall have an average hardness of not less than 15 lbf (67 N) when tested by Method A and 11 lbf (49 N) when tested by Method B for the core, ends, or edges when tested in accordance with Test Methods C 473.

5.7 *Nail-Pull Resistance*—Specimens taken from the gypsum backing board shall have an average nail-pull resistance of not less than the following when tested in accordance with Test Methods C 473:

Thickness, in. (mm)	Test Method A Nail Pull Resistance, lbf (N)	Test Method B Nail Pull Resistance, lbf (N)
¼ (6.4)	40 (178)	36 (160)
⅜ (9.5)	60 (267)	56 (249)
½ (12.7)	80 (356)	77 (343)
⅝ (15.9)	90 (400)	87 (387)
¾ (19)	Note 3	Note 3
1 (25.4)	Note 3	Note 3

NOTE 3—Testing is not required for thicknesses greater than ⅝ in. (15.9 mm).

6. Dimensions and Permissible Variations

6.1 *Thickness*—The nominal thickness of gypsum backing board shall be from ¼ to 1 in. (6.4 to 25.4 mm), with permissible variations in the nominal thickness of ±1⁄32 in. (±0.8 mm) and with permissible local variation of ±1⁄16 in. (±1.6 mm) from the nominal thickness.

6.2 *Width*—The nominal width of gypsum backing board shall be from 16 to 48 in. (406 to 1220 mm) with permissible variations of ±0 to −⅛ in. (+0 to −3.2 mm) from the specified width.

6.3 *Length*—The nominal length of gypsum backing board shall be from 4 to 16 ft (1220 to 4877 mm) with permissible variations of ±¼ in. (±6.4 mm) from the specified length.

6.4 *Edges and Ends*—The edges and ends of gypsum backing shall be straight and solid. The corners shall be square with a permissible variation of ±⅛ in. (±3.2 mm) in the full width of the board. Edges shall be square, round, V-tongue and groove, or featured.

7. Workmanship, Finish, and Appearance

7.1 The surfaces of the gypsum backing board shall be suitable for the application of facing materials specified by the manufacturer of the backing board.

8. Sampling, Inspection, Rejection, Certification, Packaging, Marking, Shipping, Handling, and Storage

8.1 Sampling, inspection, rejection, certification, packaging, marking, shipping, handling, and storage of gypsum backing board and coreboard shall be in accordance with Specification C 1264/C 1264M.

APPENDIX

(Nonmandatory Information)

This Appendix gives general information and also suggestions for inclusions to be made elsewhere by the specifier. They are not part of this specification.

X1. Alternate Definition for Type X Gypsum Backing Board

X1.1 Gypsum backing board, Type X (special fire-resistant) designates gypsum sheathing board providing a greater fire resistance than regular gypsum backing board of the same thickness. Type X (special fire-resistant) gypsum backing board, when tested in accordance with Test Methods E 119, shall provide the following minimum fire resistance for the assemblies described:

X1.1.1 One hour for a ⅝-in. (15.9-mm) thickness applied to a partition in a single layer application on each side of 3⅝-in. (92-mm) deep non-load bearing galvanized steel studs complying with Specification C 645, spaced 24 in. (610 mm) on center. The ⅝-in. (15.9-mm) thick gypsum backing board, 48-in. (1220-mm) wide shall be attached using 1-in. (25-mm) long drywall screws spaced 8 in. (203 mm) on center along the edges and ends, and 12 in. (305 mm) along intermediate studs. All joints shall be oriented parallel to and located over studs and staggered on opposite sides of the assembly. All joints shall be filled with joint compound, covered with joint tape and covered with an additional coat of joint compound. All screw heads shall be covered with joint compound, and

X1.1.2 Two hours for a ½-in. (12.7-mm) thickness applied to a partition in a double layer application on each side of 2½-in. (64-mm) deep non-load bearing galvanized steel studs complying with Specification C 645, spaced 24 in. (610 mm) on center. The 48-in. (1220-mm) wide base layer shall be attached using 1-in. (25-mm) long drywall screws spaced 12 in. (305 mm) on center along board edges, ends, and along intermediate studs. Joints shall be oriented parallel to and located over studs and staggered on opposite sides of the assembly. The 48-in. (1220-mm) wide face layer shall be attached using 1⅝-in. (41-mm) long drywall screws spaced 12 in. (305 mm) along board edges, ends and along intermediate studs. Joints shall be oriented parallel to and located over studs, offset 24 in. (610 mm) from the base layer joints, and staggered on opposite sides of the assembly. All joints in the face layer shall be filled with joint compound, covered with joint tape and covered with an additional coat of joint compound. All screw heads in the face layer shall be covered with two coats of joint compound.

Standard Test Method for Early Stiffening of Hydraulic Cement (Paste Method)[1]

This standard is issued under the fixed designation C 451; the number immediately following the designation indicates the year of original adoption or, in the case of revision, the year of last revision. A number in parentheses indicates the year of last reapproval. A superscript epsilon (ε) indicates an editorial change since the last revision or reapproval.

1. Scope

1.1 This test method covers the determination of early stiffening in hydraulic-cement paste.

1.2 The values stated in SI units are to be regarded as the standard. The values given in parentheses are for information only.

1.3 *This standard does not purport to address all of the safety concerns, if any, associated with its use. It is the responsibility of the user of this standard to establish appropriate safety and health practices and determine the applicability of regulatory limitations prior to use.* See Note 1 for a specific warning statement.

NOTE 1—**Warning:** Fresh hydraulic cementitious mixtures are caustic and may cause chemical burns to skin and tissue upon prolonged exposure. The use of gloves, protective clothing, and eye protection is recommended. Wash contact area with copious amounts of water after contact. Wash eyes for a minimum of 15 min. Avoid exposure of the body to clothing saturated with the liquid phase of the unhardened material. Remove contaminated clothing immediately after exposure.

2. Referenced Documents

2.1 *ASTM Standards:*

C 150 Specification for Portland Cement[2]

C 183 Practice for Sampling and the Amount of Testing of Hydraulic Cement[2]

C 187 Test Method for Normal Consistency of Hydraulic Cement[2]

C 305 Practice for Mechanical Mixing of Hydraulic Cement Pastes and Mortars of Plastic Consistency[2]

C 490 Practice for Use of Apparatus for the Determination of Length Change of Hardened Cement Paste, Mortar, and Concrete[2]

C 670 Practice for Preparing Precision and Bias Statements for Test Methods for Construction Materials[3]

C 1005 Specification for Reference Masses and Devices for Determining Mass for Use in the Physical Testing of Hydraulic Cements[2]

D 1193 Specification for Reagent Water[4]

3. Terminology

3.1 *Definitions:*

3.1.1 *early stiffening, n*—the early development of stiffness in the working characteristics of a portland-cement paste, mortar, or concrete. Varieties include false set and flash set.

3.1.2 *false set, n*—the early development of stiffness in the working characteristics of a portland-cement paste, mortar, or concrete without the evolution of much heat, which stiffness can be dispelled and plasticity regained by further mixing without addition of water; also known as "grab set," "premature stiffening,"" hesitation set," and "rubber set."

3.1.3 *flash set, n*—the early development of stiffness in the working characteristics of a portland-cement paste, mortar, or concrete, usually with the evolution of considerable heat, which stiffness cannot be dispelled nor can the plasticity be regained by further mixing without addition of water; also known as "quick set."

4. Summary of Test Method

4.1 A paste is prepared with the cement to be tested, using sufficient water to give a required initial penetration as measured by the Vicat apparatus at a stipulated time after completion of mixing. A second penetration, termed the final penetration, is measured at a later stipulated time. The ratio of final penetration to initial penetration is calculated as a percentage.

5. Significance and Use

5.1 The purpose of this test method is to determine the degree to which a cement paste develops early stiffening or to establish whether or not a cement complies with a specification limit on early stiffening.

5.2 When used for establishing compliance with a specification limit, the specification requirement is customarily stated in terms of the minimum allowable final penetration, in percent, calculated in accordance with Section 11. When used for estimating the relative tendency of a cement to manifest early stiffening, additional information of value may be obtained if the remix procedure described in Section 10 is employed. Under some conditions, a judgment may be made by comparing the behavior in the initial test and in the remix procedure to differentiate a relatively less serious and less persistent tendency to early stiffening from one that is more persistent and, consequently, more serious.

5.3 Severe false setting in a cement may cause difficulty from a placing and handling standpoint, but it is not likely to

---

[1] This test method is under the jurisdiction of ASTM Committee C-1 on Cement and is the direct responsibility of Subcommittee C01.30 on Time of Set.

Current edition approved Jan. 10, 1999. Published May 1999. Originally published as C 451 – 60 T. Last previous edition C 451 – 95.

[2] *Annual Book of ASTM Standards*, Vol 04.01.

[3] *Annual Book of ASTM Standards*, Vol 04.02.

[4] *Annual Book of ASTM Standards*, Vol 11.01.

cause difficulties where concrete is mixed for a longer time than usual, as usually occurs in transit mixing, or where it is remixed prior to placing or transporting, as occurs in concrete pumping operations. It is most likely to be noticeable where concrete is mixed for a short period of time in stationary mixers and transported to the forms in nonagitating equipment, such as on some paving jobs.

5.4 Cements with severe false setting usually require slightly more water to produce the same consistency, which may be expected to result in slightly lower strengths and increased drying shrinkage.

5.5 Flash set of a severity sufficient to cause difficulties from a placing and handling standpoint usually will cause the cement to fail the requirements for time of setting in Specification C 150.

6. Apparatus

6.1 *Vicat Apparatus*, conforming to the requirements of Test Method C 187.

6.2 *Trowel*, having a steel blade 100 to 150 mm (4 to 6 in.) in length, with straightedge.

6.3 *Mixer, Bowl, Paddle, and Scraper*, conforming to the requirements of Practice C 305.

6.4 *Glass Graduates*, conforming to the requirements of Practice C 490.

6.5 *Masses and Mass Determining Devices* conforming to the requirements of Specification C 1005.

7. Reagent

7.1 *Mixing Water*—Potable water is satisfactory for routine tests. For all referee and cooperative tests, reagent water conforming to the requirements of Specification D 1193 for Type III or Type IV grades of reagent water shall be used.

8. Sampling

8.1 When the test is part of acceptance testing, sample the cement in accordance with Practice C 183.

9. Conditioning

9.1 The temperature of the room, dry materials, paddle, bowl, conical ring, and base plate shall be maintained between 20 and 27.5°C (68 and 81.5°F). The temperature of the mixing water shall not vary from 23°C (73.4°F) by more than ±1.7°C (±3°F).

9.2 The relative humidity of the laboratory shall not be less than 50 %.

10. Procedure

10.1 *Preparation of Cement Paste*—Mix 500 g of cement with sufficient water to produce a paste with an initial penetration of 32 ±4 mm using the following procedure.

10.1.1 Place the dry paddle and the dry bowl in the mixing position in the mixer.

10.1.2 Introduce the materials for a batch into the bowl and mix in the following manner:

10.1.2.1 Place all the mixing water in the bowl.

10.1.2.2 Add the cement to the water and allow 30 s for the water to absorb.

10.1.2.3 Start the mixer and mix at a slow speed (140± 5 r/min) for 30 s.

10.1.2.4 Stop the mixer for 15 s, and during this time scrape down into the batch any paste that may have collected on the sides of the bowl.

10.1.2.5 Start the mixer at a medium speed (285 ±10 r/min) and mix for 2½ min.

10.2 *Molding Test Specimens*—Quickly form the cement paste into a ball with gloved hands. Press the ball, resting in the palm of one hand, into the larger end of the conical ring G, Fig. 1 of Test Method C 187, held in the other hand, completely filling the ring with paste. Remove the excess at the larger end by a single movement of the palm of the hand. Place the ring on its larger end on a glass plate, H, and slice off the excess paste at the smaller end at the top of the ring by a single oblique stroke of a sharp-edged trowel held at a slight angle with the top of the ring. Smooth the top of the specimen, if necessary, with one or two light touches of the pointed end of the trowel. During the operation of cutting and smoothing, take care not to compress the paste.

10.3 *Determination of Initial Penetration*—Set the paste confined in the ring resting on the glass plate, H, under the rod, B, Fig. 1 of Test Method C 187, about one third of the diameter from the edge, and the plunger end, C, shall be brought in contact with the surface of the paste and the tightened setscrew, E. Then set the movable indicator, F, to the upper zero mark of the scale, and release the rod exactly 20 s after completion of the mixing. The apparatus shall be free of all vibrations during the test. Consider the paste to have proper consistency when the rod settles to a point 32 ±4 mm below the original surface in 30 s after being released. Make trial pastes with varying percentages of water until this consistency is obtained. This consistency is the initial penetration. During the 30-s interval for the initial penetration, return the excess paste to the bowl and cover the bowl and mixing paddle with a lid.

10.4 *Determination of Final Penetration*—After completion of the initial reading, remove the plunger from the paste, clean it, and reset the ring and plate in a new position. This operation shall be performed with as little disturbance as possible to the paste confined in the Vicat ring. Then bring the plunger again in contact with the surface of the paste, tighten the setscrew, and set the movable indicator to the upper zero mark of the scale. Release the plunger a second time 5 min after completion of the mixing period, and determine the final penetration 30 s after the plunger is released.

10.5 *Determination of Remix Penetration*—If the penetrations determined by the foregoing procedure show the cement to be stiffening rapidly, information as to the nature of the stiffening may be obtained by testing as follows:

10.5.1 After completing the measurement of the 5-min penetration, immediately return the paste in the ring to the bowl.

10.5.2 Start the mixer, raise the bowl into mixing position, and remix the contents of the bowl at a medium speed (285 ±10 r/min) for 1 min.

10.5.3 Fill the ring and determine the penetration following the procedures specified in 10.2 and 10.3.

11. Calculation

11.1 Calculate the percent final penetration, based on the ratio of final penetration to initial penetration, as follows:

$$P, \% = \frac{B}{A} \times 100 \quad (1)$$

where:

P = percent final penetration,
A = initial penetration, mm, and
B = final penetration, mm.

12. Report

12.1 Report the results of the test as follows:

Initial penetration	___ mm
Final penetration	___ mm
Percent final penetration	___ %
Remix penetration	___ mm

13. Precision and Bias

13.1 Interlaboratory test data for the determination of statements of precision and bias in accordance with Practice C 670 are being evaluated by Subcommittee C01.30. Statements of precision and bias will be included in a later revision of this test method.

14. Keywords

14.1 early stiffening; false set; flash set; portland-cement paste.

The American Society for Testing and Materials takes no position respecting the validity of any patent rights asserted in connection with any item mentioned in this standard. Users of this standard are expressly advised that determination of the validity of any such patent rights, and the risk of infringement of such rights, are entirely their own responsibility.

This standard is subject to revision at any time by the responsible technical committee and must be reviewed every five years and if not revised, either reapproved or withdrawn. Your comments are invited either for revision of this standard or for additional standards and should be addressed to ASTM Headquarters. Your comments will receive careful consideration at a meeting of the responsible technical committee, which you may attend. If you feel that your comments have not received a fair hearing you should make your views known to the ASTM Committee on Standards, 100 Barr Harbor Drive, West Conshohocken, PA 19428.

Designation: C 452 – 95

Standard Test Method for Potential Expansion of Portland-Cement Mortars Exposed to Sulfate[1]

This standard is issued under the fixed designation C 452; the number immediately following the designation indicates the year of original adoption or, in the case of revision, the year of last revision. A number in parentheses indicates the year of last reapproval. A superscript epsilon (ε) indicates an editorial change since the last revision or reapproval.

1. Scope

1.1 This test method, which is applicable only to portland cements, covers the determination of the expansion of mortar bars made from a mixture of portland cement and gypsum in such proportions that the mixture has a sulfur trioxide (SO_3) content of 7.0 mass %.

1.2 The values stated in SI (Practice E 380) units are to be regarded as the standard.

1.3 *This standard does not purport to address all of the safety concerns, if any, associated with its use. It is the responsibility of the user of this standard to establish appropriate safety and health practices and determine the applicability of regulatory limitations prior to use.*

2. Referenced Documents

2.1 *ASTM Standards:*

C 109/C 109M Test Method for Compressive Strength of Hydraulic Cement Mortars (Using 50-mm (2-in.) Cube Specimens)[2]

C 150 Specification for Portland Cement[2]

C 230 Specification for Flow Table for Use in Tests of Hydraulic Cement[2]

C 305 Practice for Mechanical Mixing of Hydraulic Cement Pastes and Mortars of Plastic Consistency[2]

C 471 Test Methods for Chemical Analysis of Gypsum and Gypsum Products[2]

C 490 Practice for Use of Apparatus for the Determination of Length Change of Hardened Cement Paste, Mortar, and Concrete[2]

C 511 Specification for Moist Cabinets, Moist Rooms, and Water Storage Tanks Used in the Testing of Hydraulic Cements and Concretes[2]

C 778 Specification for Standard Sand[2]

C 1005 Specification for Weights and Weighing Devices for Use in the Physical Testing of Hydraulic Cements[2]

D 1193 Specification for Reagent Water[3]

E 380 Practice for Use of the International System of Units (SI) (The Modernized Metric System)[4]

3. Significance and Use

3.1 This test method is used primarily by those interested in research on methods for determining the potential sulfate resistance of portland cement. This test method is also used to establish that a sulfate-resisting portland cement meets the performance requirements of Specification C 150.

4. Apparatus

4.1 *Weights and Weighing Devices*, conforming to the requirements of Specification C 1005.

4.2 *Flow Table*, conforming to the requirements of Specification C 230.

4.3 *Mixer, Bowl, and Paddle*, conforming to the requirements of Practice C 305.

4.4 *Trowel and Tamper*, conforming to the requirements of Test Method C 109.

4.5 *Glass Graduates, Molds, and Length Comparator*, conforming to the requirements of Practice C 490.

5. Temperature and Humidity

5.1 *Molding Room, Dry Materials, and Mixing Water*—The temperature of the molding room, dry materials, and mixing water shall be maintained between 20 and 27.5°C (68 and 81.5°F) and the relative humidity of the molding room shall not be less than 50 %.

5.2 *Moist Cabinet or Room*, conforming to the requirements of Specification C 511.

6. Materials

6.1 The sand used for making the test mortar shall conform to Specification C 778.

6.2 The gypsum[5] used for addition to the portland cement shall be high grade natural gypsum with 100 % passing the 150-µm (No. 100) sieve, at least 94 % passing the 75-µm (No. 200) sieve, and at least 90 % passing the 45-µm (No. 325) sieve. Calculate the percentage of cement and gypsum required to provide a mixture containing 7.0 mass % SO_3 as follows:

$$\text{Cement, \%} = [(g - 7.0)/(g - c)] \times 100 \quad (1)$$

$$\text{Gypsum, \%} = [(7.0 - c)/(g - c)] \times 100 \quad (2)$$

where:

c = SO_3 content of the portland cement, %,
g = SO_3 content of the gypsum, %, and
7.0 = SO_3 content of the cement-gypsum mixture, %.

6.3 If the SO_3 content of the gypsum is unknown, the gypsum shall be analyzed for SO_3 content using Test

[1] This test method is under the jurisdiction of ASTM Committee C-1 on Cement and is the direct responsibility of Subcommittee C01.29 on Sulfate Resistance.

Current edition approved June 15, 1995. Published August 1995. Originally published as C 452 – 60. Last previous edition C 452 – 89.

[2] *Annual Book of ASTM Standards*, Vol 04.01.

[3] *Annual Book of ASTM Standards*, Vol 11.01.

[4] *Annual Book of ASTM Standards*, Vol 14.02.

[5] Terra Alba No. 1, available from the U.S. Gypsum Co., Southard, OK plant, meets these requirements.

Methods C 471. The SO_3 content shall be determined to the nearest 0.1 %.

6.4 *Purity of Water*—Unless otherwise indicated, references to water shall be understood to mean reagent water conforming to Type IV of Specification D 1193.

7. Number and Dimensions of Test Specimens

7.1 Six 25 by 25 by 285-mm (or 1 by 1 by 11¼-in.) test specimens, three from each of two batches, shall be made for each cement.

7.2 In routine tests, 25 by 25 by 160-mm (or 1 by 1 by 6¼-in.) specimens may be used, but in case of dispute, results obtained with 25 by 25 by 285-mm (or 1 by 1 by 11¼-in.) specimens shall govern.

8. Preparing Specimen Molds

8.1 The molds shall be prepared in accordance with Specification C 490.

9. Proportioning, Consistency, and Mixing of Mortar

9.1 The quantities of dry materials required for each batch shall be 400 g (cement plus gypsum) and 1100 g of sand. The amount of mixing water shall be 194 mL for all non-air-entraining portland cements and 184 mL for all air-entraining portland cements.

9.2 Mix the mortar according to the procedure for mixing mortars of Practice C 305, except after placing the mixing water in the bowl, add the gypsum and mix at the slow speed for 15 s, then stop the mixer, add the cement, and continue as prescribed in Practice C 305.

10. Procedure

10.1 *Molding Test Specimens:*

10.1.1 Immediately upon completion of the flow test, return the mortar to the bowl, scrape down into the batch any mortar on the side of the bowl, and remix at the medium speed (285 ± 10 r/min) for 15 s. Remove the paddle and the bowl from the mixer and shake the excess mortar from the paddle into the bowl. Fill the mold in two layers, each layer being compacted with the tamper. Work the mortar into the corners, around the gage studs, and along the surfaces of the mold with the tamper until a homogeneous specimen is obtained. After the top layer has been compacted, cut the mortar off flush with the top of the mold and smooth the surface with a few strokes of the trowel.

10.2 *Storage of Test Specimens:*

10.2.1 *Initial Storage*—Cure the specimens in the molds in the moist closet at 23 ± 1.7°C (73.4 ± 3°F) for 22 to 23 h. Then remove them from the molds, properly identify, and place in water at 23 ± 1.7°C (73.4 ± 3°F) for at least 30 min prior to making the initial length measurement.

10.2.2 *Subsequent Storage*—After the bars have been removed from the molds and measured, store them horizontally in water at 23 ± 1.7°C (73.4 ± 3°F). Store the specimens with at least 6-mm (¼-in.) clearance on all sides except for the necessary supports. Cover the specimens with at least 13 mm (½ in.) of water. The ratio of the volume of water to a volume of the bars shall not exceed 5 to 1 to prevent excessive leaching. Replenish the water with fresh water every 7 days for the first 28 days and every 28 days thereafter.

11. Length Measurement

11.1 Measure the specimens for length by means of the length comparator. Remove them from the water storage, one at a time, and wipe with a damp cloth before measuring.

11.2 Make the first reading at the age of 24 h ± 15 min from the time the cement and water are mixed together. Measure the specimen again at the age of 14 days.

NOTE—Additional information of value may be obtained by returning the specimen to water storage after the 14-day test and making additional measurements at later ages.

12. Calculation

12.1 Calculate the difference in length of the specimen at 24 h and at 14 days to the nearest 0.001 % of the effective gage length and report as the expansion of the specimen at that period. All specimens remaining after 14 days must comprise a set having at least three specimens with a permissible range depending on the number of the remaining specimens, as follows:

No. of Specimens	Maximum Permissible Range, %
3	0.010
4	0.011
5	0.012
6	0.012

13. Report

13.1 Report the average of the specimens comprising the set to the nearest 0.001 %.

14. Precision and Bias

14.1 *Precision:*

14.1.1 The single-operator standard deviation has been found to be: 0.003 % for expansions between 0.01 and 0.04 %. Therefore, results of two properly conducted tests by the same operator on the same material should not differ from each other by more than 0.009 %.

14.1.2 The multilaboratory standard deviation has been found to be 0.005 % for expansions between 0.01 and 0.04 %. Therefore, results of two properly conducted tests on the same material in two different laboratories should not differ from each other by more than 0.014 %.

14.1.3 The precision of this test method has been evaluated by cooperative testing.[6]

14.2 *Bias*—Since there is no accepted reference material suitable for determining the bias for this test method, no statement on bias is being made.

[6] A research report is available from ASTM Headquarters. Request RR: 001-1004.

For additional useful information on details of cement test methods, reference may be made to the "Manual of Cement Testing," which appears in the *Annual Book of ASTM Standards*, **Vol 04.01.**

Designation: C 465 – 99

Standard Specification for Processing Additions for Use in the Manufacture of Hydraulic Cements[1]

This standard is issued under the fixed designation C 465; the number immediately following the designation indicates the year of original adoption or, in the case of revision, the year of last revision. A number in parentheses indicates the year of last reapproval. A superscript epsilon (ε) indicates an editorial change since the last revision or reapproval.

1. Scope

1.1 This specification pertains to the criteria and tests to be used for determining whether a processing addition, when used in the recommended amount at the option of the cement producer in the manufacture of hydraulic cements, meets the requirements as prescribed by definition in Specifications C 150, C 1157, C 845, and C 595. The materials listed in the following former ASTM Specifications shall be considered as meeting the requirements of this specification:

C 150 – 62, for Portland Cement[2]
C 205 – 58 T, for Portland Blast-Furnace Slag Cement[3]
C 340 – 58 T, for Portland Pozzolan Cement[3]
C 358 – 58, for Slag Cement[3]

1.2 The following safety hazards caveat pertains only to the test methods described in this specification. *This standard does not purport to address all of the safety concerns, if any, associated with its use. It is the responsibility of the user of this standard to establish appropriate safety and health practices and determine the applicability of regulatory limitations prior to use.*

2. Referenced Documents

2.1 *ASTM Standards:*

C 33 Specification for Concrete Aggregates[4]
C 39 Test Method for Compressive Strength of Cylindrical Concrete Specimens[4]
C 78 Test Method for Flexural Strength of Concrete (Using Simple Beam with Third-Point Loading)[4]
C 109/C 109M Test Method for Compressive Strength of Hydraulic Cement Mortars (Using 2-in. or [50-mm] Cube Specimens)[5]
C 114 Test Methods for Chemical Analysis of Hydraulic Cement[5]
C 115 Test Method for Fineness of Portland Cement by the Turbidimeter[5]
C 138 Test Method for Unit Weight, Yield, and Air Content (Gravimetric) of Concrete[4]
C 143 Test Method for Slump of Hydraulic Cement Concrete[4]
C 150 Specification for Portland Cement[5]
C 151 Test Method for Autoclave Expansion of Portland Cement[5]
C 173 Test Method for Air Content of Freshly Mixed Concrete by the Volumetric Method[4]
C 185 Test Method for Air Content of Hydraulic Cement Mortar[5]
C 187 Test Method for Normal Consistency of Hydraulic Cement[5]
C 191 Test Method for Time of Setting of Hydraulic Cement by Vicat Needle[5]
C 192 Practice for Making and Curing Concrete Test Specimens in the Laboratory[4]
C 204 Test Method for Fineness of Hydraulic Cement by Air Permeability Apparatus[5]
C 205 – 58T Specification for Portland Blast–Furnace Slag Cement[3]
C 226 Specification for Air-Entraining Additions for Use in the Manufacture of Air-Entraining Portland Cement[5]
C 231 Test Method for Air Content of Freshly Mixed Concrete by the Pressure Method[4]
C 293 Test Method for Flexural Strength of Concrete (Using Simple Beam with Center-Point Loading)[4]
C 340 – 58T Specification for Portland Pozzolan Cement[3]
C 358 – 58 Specification for Slag Cement[3]
C 595 Specification for Blended Hydraulic Cements[5]
C 596 Test Method for Drying Shrinkage of Mortar Containing Portland Cement[5]
C 617 Practice for Capping Cylindrical Concrete Specimens[4]
C 845 Specification for Expansive Hydraulic Cement[5]
C 1157 Performance Specification for Blended Hydraulic Cement[5]
D 891 Test Methods for Specific Gravity, Apparent, of Liquid Industrial Chemicals[6]
E 203 Test Method for Water Using Karl Fischer Reagent[6]

3. Materials

3.1 *Cements:*

---

[1] This specification is under the jurisdiction of ASTM Committee C-1 on Cement and is the direct responsibility of Subcommittee C01.20 on Additions.

Current edition approved Feb. 10, 1999. Published May 1999. Originally published as C 465 – 61 T. Last previous edition C 465 – 96.

[2] See *1961 Book of ASTM Standards*, Part 4.

[3] Discontinued, see *1958 Book of ASTM Standards*, Part 4.

[4] *Annual Book of ASTM Standards*, Vol 04.02.

[5] *Annual Book of ASTM Standards*, Vol 04.01.

[6] *Annual Book of ASTM Standards*, Vol 15.05.

3.1.1 In cases where it is desired that the proposed addition be accepted for general use in portland cement, tests shall be made on cements prepared from at least five different clinkers. As a minimum, these clinkers shall represent two Type I cements containing not less than 9.0 % C_3A, one Type II cement, and two Type III cements, all conforming to Specification C 150.

3.1.2 In cases where it is also desired that the proposed addition be used in blended cements, the test and test procedures shall be as specified with a control and an addition for cement conforming to the appropriate Specification C 595 or C 1157.

3.1.3 Processing additions which have been shown to meet the requirements of this specification may also be used in cements conforming to Specification C 845. Testing of the addition with these special cements, where desired, shall be done using the tests and test procedures as specified with a control cement and a cement containing the addition, both conforming to Specification C 845.

3.1.4 In cases where it is desired that the proposed addition be limited in use to specific types of cement less in number than required in 3.1.1, the tests and test procedures shall be as specified, and at least two pairs of cements shall be prepared from two clinkers from different plants for each type under specific consideration.

3.1.5 In cases where it is desired that the proposed addition be limited in use to a single plant, the tests and test procedures shall be as specified and at least two pairs of cements shall be prepared from clinker representing each type under specific consideration.

3.1.6 The two companion cements to be made from any one clinker shall be ground to the same fineness within 7 m^2/kg when tested in accordance with Test Method C 115 or within 13 m^2/kg when tested in accordance with Test Method C 204, and the SO_3 content, expressed as a percentage of the cement mass and reported to the nearest 0.01 %, shall differ by not more than 0.3, so as to afford comparable samples for indicating the effect of the addition on the cement. Each control cement shall comply with all requirements in the specification applicable to that type of cement, and shall not contain the proposed addition when tested by the method furnished by the producer or seller of the addition.

3.1.7 The percentage of each of the following shall be determined for each lot of cement tested: silicon dioxide (SiO_2), aluminum oxide (Al_2O_3), ferric oxide (Fe_2O_3), calcium oxide (CaO), magnesium oxide (MgO), sulfur trioxide (SO_3), ignition loss, insoluble residue, sodium oxide (Na_2O), and potassium oxide (K_2O). There shall also be calculated the potential percentages of the following compounds: tricalcium silicate, dicalcium silicate, tricalcium aluminate, and tetracalcium aluminoferrite. Determinations for the percentage of the addition shall be made, both on the control cements and on those with which the addition was interground, using the method proposed therefore by the sponsor.

3.2 *Aggregates*—The fine and coarse aggregates shall comply with Specification C 33; the coarse aggregate shall comply with the grading requirements for Size No. 57 or Size No. 67. A sufficient quantity from a single lot of coarse aggregate and from a single lot of fine aggregate shall be provided to complete all tests. To prevent the segregation of particle sizes in the fine aggregate, a single lot of sand sufficient for all tests shall either (*1*) be separated on the 4.75-mm (No. 4), 1.18-mm (No. 16), 300 µm (No. 50), and 150 µm (No. 100) sieves and then be recombined in the required quantity for each batch; or (*2*) be blended while in a damp condition, and maintained in that condition for the duration of the tests. Under option (*2*), lots of appropriate size for single mortar and concrete batches shall be carefully split or quartered from the entire batch.

4. General Requirements

4.1 Processing additions shall conform to the respective requirements in this specification.

4.2 The trade name, source, character of the material, and means for the quantitative determination of the addition in the finished cement shall be furnished by the sponsor, manufacturer, or supplier of the addition, and the information shall form a part of the record of tests of the addition. If the processing addition is a liquid, the specific gravity and percent water content shall also be part of the record.

4.2.1 The specific gravity, run in accordance with 7.1.1 shall be within ±0.05 units of the value reported in 4.2.

4.3 Processing additions shall be evaluated by comparing cements containing the addition to otherwise identical cements from the same source without the addition, or containing a processing addition which has been shown to comply with this specification using control cements without any additions, hereinafter designated the "control" cement.

4.4 The amount of the processing addition to be interground with the cement for evaluation purposes shall be determined by the sponsor of the addition.

4.4.1 The amount of the addition in the cement containing the addition and showing compliance with the requirements of this specification shall be determined quantitatively by means of the quantitative determination required by 4.2.

4.4.2 The amount of addition, so determined, shall be used to state the amount of addition that shows compliance with this specification.

4.4.3 When tests on cements containing the addition show compliance with the requirements of this specification, the addition in cement may be used in any amount up to the maximum amount showing compliance.

4.5 The cement produced for evaluation purposes with the processing addition shall comply with the appropriate Specifications C 150, C 845, C 1157M, or C 595, except that it contains the addition under test. The effect of the addition on the properties of the cement shall also be within the following limits:

4.5.1 The percentage of water by mass of cement required for normal consistency of cement containing the addition shall not exceed that required by the corresponding control cement by more than 1.0. For those cements not limited to a fixed water requirement, the percentage of water by mass of cement required for standard consistency of the mortar used for strength determinations as described in 4.5.4 shall not be increased by more than 2.0 by the addition over that required for the control cement.

4.5.2 The time of setting of cement containing the addition

shall not vary from the time of setting of the corresponding control cement by more than 1 h or 50 %, whichever is the lesser.

4.5.3 The autoclave expansion of cement containing the addition, expressed as a percentage change in length, shall be not more than 0.10 greater than that of the corresponding control cement.

4.5.4 The compressive strength of mortar cubes made with cement containing the addition, in accordance with Test Method C 109/C 109M, and tested at 1, 3, 7, and 28 days for all types, shall be compared with strengths obtained with the control cement at similar ages. The grand average of these individual strength percentages shall be not less than 95 % of the control cement values. It is required that cubes for companion cements be made and tested on the same days, with storage of specimens side by side in the same section of the moist cabinet during the 24-h curing period. Retesting of companion cements on the same, or a following, day is required in order to provide six, rather than three, test specimens for each cement and age of test.

4.5.5 The ultimate drying shrinkage (percent) of mortar made with cement containing the addition shall not be more than 0.025 greater than that of similar mortar made with the corresponding control cement when tested in accordance with Test Method C 596.

4.5.6 The compressive strength of the concrete made with cement containing the addition shall be compared with strengths obtained with the control cement at similar ages. The grand average of these individual strength percentages shall be not less than 90 % of the values for the control cement.

4.5.7 The flexural strength of concrete made with cement containing the addition shall be compared with strengths obtained with the control cement at similar ages. The grand average of these individual strength percentages shall be not less than 90 % of the values for the control cement.

4.5.8 The amount of air-entraining addition required to produce 19 ± 3 % air in the mortar test made in accordance with Test Method C 185, with the cement containing the addition under test, shall be not greater than 120 % of the amount required to produce, within± 1 %, the air content obtained with the control cement. The air-entraining addition used shall meet the requirements of Specification C 226.

4.6 Processing additions which provide maximum effects as grinding aids or pack set inhibitors may increase cement flowability to a point where mill retention time is reduced sufficiently to affect significantly the particle size distribution of the resulting cement and its physical-chemical properties. Since mill retention times are controllable by mechanical means in full-scale grinding mills, the true physical-chemical effects of the test additive may, in instances where full-scale tests have shown mill retention time reductions to have significant effects on the properties of the resulting cement, be determined for acceptance purposes by making supplementary laboratory or pilot-mill grinds.

4.7 In the event that the effect of the addition on the properties of cement are determined on the basis of laboratory or pilot mill grinds, this fact shall be entered in the report specified in Section 13, and the specific tests shall be indicated.

5. Sampling Cement

5.1 Samples of the plant-ground cement shall be taken from the product stream during grinding. Prior to the start of sampling a given lot of cement, the mill shall have run for 4 h or long enough to have reached equilibrium under the general conditions that are to govern during the sampling period.

NOTE 1—Records should be kept as to the rate and continuity of feed of the addition, the form in which the addition is used, strength of solution, magnitude of circulating loads, mill discharge temperature, and feed rate of clinker and gypsum. Product fineness should be determined during the grinding immediately subsequent to sampling.

5.2 As the cement samples are taken, they shall be placed in sealable containers which shall be sealed immediately at the end of the sampling period. Prior to use, the samples of a given lot of cement shall be thoroughly blended to form a uniform, representative composite.

6. Test Methods

6.1 Determine the properties enumerated in this specification in accordance with the test methods prescribed in Sections 7-12.

7. Tests on Liquid Processing Additions

7.1 Test the liquid processing addition, in the form normally marketed, in accordance with the following test methods:

7.1.1 *Specific Gravity*—Use Test Methods D 891, Test Method A or B.

7.1.2 *Water Content*—Use Test Method E 203.

8. Tests on Cement

8.1 Test cement in accordance with the following standards:

8.1.1 *Chemical Analysis of Cement*—Test Methods C 114.

8.1.2 *Compound Composition*—Specification C 150.

8.1.3 *Fineness of Cement*—Test Method C 115 or C 204.

8.1.4 *Normal Consistency*—Test Method C 187.

8.1.5 *Time of Setting (Vicat)*—Test Method C 191.

8.1.6 *Autoclave Expansion*—Test Method C 151.

8.1.7 *Air Content of Mortar*—Test Method C 185.

8.1.8 *Compressive Strength of Mortar*— Test Method C 109/C 109M.

8.1.9 *Drying Shrinkage of Mortar*—Test Method C 596.

9. Concrete Mixtures

9.1 *Preparation and Weighing*—Prepare all materials used in making the concrete mixtures and make all weighings as prescribed in Practice C 192. Report the amount of mixing water used in each batch on the basis of saturated, surface-dry aggregates.

9.2 *Proportions*—Design one basic concrete mixture having an actual cement content of 307 ± 3 kg/m^3 (517± 5 lb/yd^3), and use in all concrete tests herein specified. Adjust the water content of mixtures to provide concrete having a consistency equal to a 64 ± 13-mm (2½ ± ½-in.) slump in each case. Adjust the ratio of fine to coarse aggregate to the optimum for concrete to be consolidated by hand rodding. Recommended trial values for the percentage of fine aggregate in the total aggregate, by absolute volume, are as follows:

Coarse Aggregate, Maximum 25.0 mm (1 in.)	Concrete Without Entrained Air
Angular	45
Rounded	40

9.3 *Mixing of Concrete*—Mix the concrete in accordance with Practice C 192 except as follows: Hand mixing will not be permitted. The rated capacity of the machine mixer shall not be more than twice the size of the batch used.

10. Tests on Freshly Mixed Concrete

10.1 Test samples of the freshly mixed concrete for slump in accordance with Test Method C 143; unit weight in accordance with Test Method C 138; and air content in accordance with Test Method C 138, C 231, or C 173.

11. Test Specimens of Hardened Concrete

11.1 *Number of Specimens*—At least three specimens shall be made for each test condition. For each cement containing an addition and its companion control cement, make three rounds of concrete mixed on different days. One round of mixes on a given day shall include both the cement containing the addition and its companion control cement. From each round, make at least one test specimen for each test condition. If necessary, to obtain enough concrete for all test specimens to be made in any one round, it may be necessary to make more than one concrete batch for each round.

11.2 *Types of Specimens*:

11.2.1 *Compressive Strength*—Compressive strength test specimens shall be cylinders made and cured as prescribed in Practice C 192. Cylinders shall be capped as prescribed in Practice C 617.

11.2.2 *Flexural Strength*—Flexural strength test specimens shall be beams made and cured as prescribed in Practice C 192.

12. Test on Hardened Concrete

12.1 Test the specimens on hardened concrete, as specified in Section 11, in accordance with the following methods and at the specified ages:

12.1.1 *Compressive Strength*—Test specimens in accordance with Test Method C 39 at ages 3, 7, 28 days, and 3 months, except also test Type III cement at 24 h.

12.1.2 *Flexural Strength*—Test specimens in accordance with Test Method C 293, or Test Method C 78, ages at 3, 7, 28 days, and 3 months, except also test Type III cement at 24 h. By either method of test, turn the specimen on its side with respect to its position as molded and center it on the bearing blocks.

13. Report

13.1 The report covering the results of the evaluation of a material proposed for use as a processing addition in the manufacture of portland cement under this specification shall include the following information:

13.1.1 Trade name, source and character of the material, and the amount recommended for use, together with means for determination of the proposed addition in the finished cement, all as furnished by the sponsor, manufacturer, or seller of the addition,

13.1.2 If the proposed processing addition is a liquid, the specific gravity and percent water content by mass,

13.1.3 Detailed results of all analyses and tests prescribed by this specification, and the amount of the addition used, as well as other pertinent information required,

13.1.4 Comparison of test results to determine compliance with the requirements prescribed in 4.4,

13.1.5 Name and location of the laboratory or laboratories that made the tests covered by the report,

13.1.6 Include as an appendix to the report, letters of certification from the various cement manufacturers stating the name of the addition, the amount used, and the type of cement in which used, and

13.1.7 The highest amount of addition that has demonstrated compliance with the specification in a particular type of cement (see 4.4.3).

14. Manufacturer's Certification of Liquid Processing Aids

14.1 Upon request of the purchaser in the contract or order, a manufacturer's report shall be furnished at the time of shipment stating the water content and specific gravity of the samples of the material taken during production or transfer and certifying that the specific gravity value is within the specified limits of the originally qualified product and that the product meets the requirements of this Specification.

15. Keywords

15.1 additions; hydraulic cements; processing

Designation: C 471M – 96
METRIC

Standard Test Methods for Chemical Analysis of Gypsum and Gypsum Products [Metric][1]

This standard is issued under the fixed designation C 471M; the number immediately following the designation indicates the year of original adoption or, in the case of revision, the year of last revision. A number in parentheses indicates the year of last reapproval. A superscript epsilon (ε) indicates an editorial change since the last revision or reapproval.

1. Scope*

1.1 These test methods cover the chemical analysis of gypsum and gypsum products, including gypsum ready-mixed plaster, gypsum wood-fibered plaster and gypsum concrete.

NOTE 1—Gypsum reference standard materials, prepared by Domtar, Inc are available through Brammer Standards Company, Inc.

1.2 The test methods appear in the following order:

	Sections
Alternative Procedure for Analysis for Calcium Sulfate by Ammonium Acetate Method	17 to 22
Alternative Procedure for Analysis for Sodium Chloride by the Coulometric Method	23 to 29
Complete Procedure	5 to 16
Determination of Sand in Set Plaster	30 to 36
Optional Procedure for Analysis for Sodium by Flame Photometry	47 to 54
Optional Procedure for Analysis for Sodium by the Atomic Absorption Method	40 to 46
Preparation of Sample	4
Wood-Fiber Content in Wood-Fiber Gypsum Plaster	37 to 39

1.3 The values stated in SI units are to be regarded as the standard.

1.4 *This standard does not purport to address all of the safety concerns, if any, associated with its use. It is the responsibility of the user of this standard to establish appropriate safety and health practices and determine the applicability of regulatory limitations prior to use.*

2. Referenced Documents

2.1 *ASTM Standards:*

C 11 Terminology Relating to Gypsum and Related Gypsum Building Materials and Systems[2]
C 22/C 22M Specification for Gypsum[2]
C 28 Specification for Gypsum Plasters[2]
C 59 Specification for Gypsum Casting and Molding Plaster[2]
C 61 Specification for Gypsum Keene's Cement[2]
C 317 Specification for Gypsum Concrete[2]
C 842 Specification for Application of Interior Gypsum Plaster[2]
D 1193 Specification for Reagent Water[3]
D 1428 Methods of Test for Sodium and Potassium in Water and Water-Formed Deposits by Flame Photometry[3]
D 2013 Method of Preparing Coal Samples for Analysis[4]
E 11 Specification for Wire-Cloth Sieves for Testing Purposes[5]

3. Terminology

3.1 *Definitions*—Definitions shall be in accordance with Terminology C 11.

3.2 *Definitions of Terms Specific to This Standard:*

3.2.1 *sample as received, n*—a representative portion of raw gypsum or gypsum product in the state received by the testing laboratory, including aggregates, impurities and water content.

3.2.2 *dried sample, n*—a sample dried of free water.

3.2.3 *riffle, n*—a hand feed sample divider device that divides the sample into two parts of approximately the same weight. **(D 2013)**

4. Preparation of Sample

4.1 *General Procedures*—Details of sample preparation will vary according to the type of material being tested.

4.1.1 *Sample As Received*—Use a sufficient amount of sample such that, after sieving, not less than 50 g of sample will remain for testing. Weigh the entire sample immediately after opening the container in which the material was received. This will become the weight of the sample as received.

4.1.2 *Drying*—Dry the sample in accordance with Section 7. This will be the weight of the dried sample.

4.1.3 *Crushing and Grinding*—Crush and grind the sample by hand with a mortar and pestle or by mechanical crushing and grinding equipment to pass a 250-μm (No. 60) sieve. Take care, particularly with mechanical equipment, not to expose the sample to temperatures of more than 52°C. Clean the equipment thoroughly between samples. Thoroughly remix the ground sample and store it in an airtight container to avoid contamination.

4.1.4 *Rehydrating*—Thoroughly blend and rehydrate samples which contain calcium sulfate in forms other than $CaSO_4 \cdot 2H_2O$ and natural anhydrite. Place the sample in distilled water and keep it wet for not less than 48 h. Dry the hydrated sample in an oven at 45 ± 3°C to constant weight and recrush or grind it in accordance with 4.1.3.

---

[1] These test methods are under the jurisdiction of ASTM Committee C-11 on Gypsum and Related Building Materials and Systems and are the direct responsibility of Subcommittee C11.01 on Specifications and Test Methods for Gypsum Products.

Current edition approved Nov. 10, 1996. Published January 1997. Originally published as C 471 – 61. Last previous edition C 471 – 95.

[2] *Annual Book of ASTM Standards*, Vol 04.01.

[3] *Annual Book of ASTM Standards*, Vol 11.01.

[4] *Annual Book of ASTM Standards*, Vol 05.05.

[5] *Annual Book of ASTM Standards*, Vol 14.02.

*** A Summary of Changes section appears at the end of these test methods.**

4.1.5 *Sample Reduction*—Thoroughly mix and reduce large samples as required by quartering or by the use of a riffle to obtain a specimen of approximately 50 g.

4.2 *Gypsum* (C 22/C 22M)—Gypsum samples will be received in the form of rocks, powder or both. If necessary reduce and crush the entire dried sample in accordance with 4.1.3 and 4.1.5.

4.3 *Gypsum Plaster,* (C 28).

4.3.1 *Gypsum Ready-Mixed Plaster or Gypsum Wood-Fibered Plaster*—Screen the dried sample through a 150-µm (No. 100) sieve[6] and discard the residue retained on the sieve. Reweigh the remaining sample and calculate the percentage of the dried sample. Reduce the sample in accordance with 4.1.5. Thoroughly blend and rehydrate the specimen in accordance with 4.1.4.

4.3.2 *Gypsum Neat Plaster or Gypsum Gauging Plaster*—Reduce the dried sample in accordance with 4.1.5. Thoroughly blend and rehydrate the specimen in accordance with 4.1.4.

4.4 *Gypsum Casting and Molding Plaster,* (C 59)—Reduce the dried sample in accordance with 4.1.5. Thoroughly blend and rehydrate the specimen in accordance with 4.1.4.

4.5 *Gypsum Keene's Cement,* (C 61)—Reduce the dried sample in accordance with 4.1.5. Blend in no more than 1 % molding plaster or K_2SO_4 and rehydrate the specimen in accordance with 4.1.4.

4.6 *Gypsum Concrete,* (C 317)—Screen the dried sample through a 150-µm (No. 100) sieve[6] and discard the residue retained on the sieve. Reweigh the remaining sample and calculate the percentage of the dried sample. Reduce the sample in accordance with 4.1.5. Thoroughly blend and rehydrate the specimen in accordance with 4.1.4.

4.7 *Gypsum Board*—Cut or break the dried sample into small pieces. Using a mortar and pestle, strike the pieces of the sample to loosen the paper face. Remove the pieces of paper by hand as they are separated from the core of the gypsum board. Carefully scrape any remaining powder from the paper. When all the paper has been removed from the pieces of the sample, reduce the sample in accordance with 4.1.5. Thoroughly blend and rehydrate the specimen in accordance with 4.1.4.

COMPLETE PROCEDURE

5. Reagents

5.1 *Purity of Reagents*—Use reagent grade chemicals in all tests. Unless otherwise indicated, use reagents that conform to the specifications of the Committee on Analytical Reagents of the American Chemical Society, where such specifications are available.[7] If it is necessary to use other grades first ascertain that the reagent is of sufficiently high purity to permit its use without lessening the accuracy of the determination.

5.1.1 *Ammonium Chloride* (NH_4Cl).

5.1.2 *Ammonium Hydroxide* (sp gr 0.90)—Concentrated ammonium hydroxide (NH_4OH).

5.1.3 *Ammonium Nitrate* (25 g/L)—Dissolve 25 g of ammonium nitrate (NH_4NO_3) in water and dilute to 1 L.

5.1.4 *Ammonium Oxalate* ($(NH_4)_2C_2O_4$).

5.1.5 *Barium Chloride* (100 g/L)—Dissolve 100 g of barium chloride ($BaCl_2 \cdot 2H_2O$) in water and dilute to 1 L.

5.1.6 *Calcium Chloride* ($CaCl_2$)—Anhydrous Calcium Chloride with a combined water of not more than 5 %.

5.1.7 *Hydrochloric Acid* (sp gr 1.19)—Concentrated hydrochloric acid (HCl).

5.1.8 *Hydrochloric Acid* (1+4)—Mix 1 volume of HCl (sp gr 1.19) with 4 volumes of water.

5.1.9 *Hydrochloric Acid* (1+5)—Mix 1 volume of HCl (sp gr 1.19) with 5 volumes of water.

5.1.10 *Nitric Acid* (sp gr 1.42)—Concentrated nitric acid (HNO_3).

5.1.11 *Potassium Chromate Solution* (100 g/L)—Dissolve 5 g of potassium chromate (K_2CrO_4) in 50 mL of water, mix, add 10 drops of 0.05 N silver nitrate ($AgNO_3$) solution, allow to stand for 5 min, and filter.

5.1.12 *Potassium Permanganate* (5.6339 g/L)—Dissolve 5.6339 g of potassium permanganate ($KMnO_4$) in water and dilute to 1 L.

5.1.13 *Silver Nitrate, Standard Solution* (0.05 N)—Prepare and standardize a 0.05 N silver nitrate ($AgNO_3$) solution.

5.1.14 *Sodium Ammonium Phosphate* ($NaNH_4HPO_4$).

5.1.15 *Sulfuric Acid* (sp gr 1.84)—Concentrated sulfuric acid (H_2SO_4).

5.1.16 *Sulfuric Acid* (1+6)—Carefully mix 1 volume of H_2SO_4 (sp gr 1.84) with 6 volumes of water.

5.1.17 *Nitric Acid* (0.1 N)—Mix 1.4 mL of HNO_3 (sp gr 1.42) with 200 mL of water.

5.1.18 *Phenolphthalein Indicator Solution*—Dissolve 0.25 g of phenolphthalein in 30 mL of methanol and dilute to 50 mL with water.

5.1.19 *Sodium Hydroxide Solution* (0.1 N)—Dissolve 1 g of sodium hydroxide (NaOH) in 250 mL of water.

5.1.20 *Water*—Reagent water shall be in accordance with Specification D 1193, type II. Specification D 1193 gives the following values for type II grade water.

Electrical conductivity, max, µS/cm at 298 K (25-C)	1.0
Electrical resistivity, min, MΩ·cm at 298 K (25-C)	1.0
Total organic carbon (TOC), max, µg/L	50.0
Sodium, max, µg/L	5.0
Chlorides max, µg/L	5.0
Total silica, max, µg/L	5.0

6. Apparatus

6.1 *Analytical Balance*—Capable of weighing not less than 1 g at a precision of 0.0001 g.

6.2 *Balance*—Capable of weighing not less than 100 g at a precision of 0.001 g.

6.3 *Drying Oven*—A mechanical convection oven set at 45 ± 3°C.

6.4 *Desiccator*—Capable of being tightly sealed and containing calcium chloride or equivalent desiccant.

6.5 *Calcining Oven or Furnace*—Capable of achieving and maintaining temperatures to not less than 1000°C.

6.6 *Weighing Bottles*—Borosilicate glass or ceramic containers with lids that can be sealed tightly.

[6] Detailed requirements for this sieve are given in Specification E 11.

[7] *Reagent Chemicals, American Chemical Society Specifications,* American Chemical Society, Washington, DC. For suggestions on the testing of reagents not listed by the American Chemical Society, see *Analar Standards for Laboratory Chemicals,* BDH Ltd., Poole, Dorset, U.K., and the *United States Pharmacopeia and National Formulary,* U.S. Pharmaceutical Convention, Inc. (USPC), Rockville, MD.

6.7 *Hot Plate*—A controllable hot plate capable of heating casseroles to approximately 120°C.

6.8 *Porcelain Casseroles*—With a capacity of 50 to 100 mL.

6.9 *Filtering Funnels.*

6.10 *Filter Paper*—Ashless filter paper Whatman #42 or equivalent.[8]

6.11 *Porcelain Crucibles.*

6.12 *Mortar and Pestle.*

6.13 *Mechanical Jaw Crusher*—Capable of crushing gypsum rocks up to 50 mm diameter.

6.14 *Mechanical Grinder*—Burr mill or equivalent capable of grinding the granular output of the jaw crusher specified in 6.13.

7. Free Water

7.1 *Significance and Use*—The free water analysis determines the amount of free water contained in the sample as opposed to chemically combined water, and prepares the sample for further analysis.

7.2 *Procedure:*

7.2.1 Weigh a sample of the material as received of not less than 50 g to a precision of 0.001 g and spread it out in a thin layer in a suitable vessel. Place in an oven and dry at 45 ± 3°C for 2 h, then cool in a desiccator and weigh again. The loss of weight corresponds to the free water.

7.2.2 Retain the sample in a sealed container or in the desiccator for further analysis.

7.3 *Calculation and Report*—Calculate and report loss in weight as a percentage of the sample as received or of the dried sample as required.

7.4 *Precision and Bias*—Neither the precision nor the bias for the free water analysis has been determined.

8. Combined Water

8.1 *Significance and Use*—The combined water analysis determines the percent of chemically combined water and is used to calculate the purity of gypsum or the amount of gypsum or gypsum plaster in gypsum products.

8.2 *Interferences*—Some organic materials may partially decompose and give high results. If hydrated compounds other than gypsum are present they may decompose and give high results. Take care that the oven used does not exceed the maximum temperature required, or some carbonates, if present, may decompose and give high results.

8.3 *Procedure:*

8.3.1 For each sample, place three weighing bottles with lids in the preheated calcining oven or furnace and heat for 2 h at 215 to 230°C. Place in the desiccator and allow to cool to room temperature. Weigh the bottles and lids to the nearest 0.0001 g and record the tare weights.

8.3.2 Weigh out three specimens of approximately 1 g each of the sample as prepared in Section 4 and dried in Section 7 to a precision of 0.0001 g in the previously tared weighing bottles and record the total weight with lids.

8.3.3 Place the specimens in the calcining oven with the lids placed loosely on each bottle or crucible for 2 h or until constant weight has been obtained.

8.3.4 Place the lids tightly on the weighing bottles, remove from the oven, and place in the desiccator to cool to room temperature.

8.3.5 Weigh each specimen to a precision of 0.0001 g and record the weights.

8.3.6 Retain the residues for carbon dioxide analysis.

8.4 *Calculation and Report*—Calculate and report the average loss in weight of the three specimens as a percentage of the sample as received or of the dried sample, as required, to the nearest 0.001 g and record the tare weights.

8.5 *Precision and Bias*—Neither the precision nor the bias for the combined water analysis has been determined.

9. Carbon Dioxide

9.1 *Summary of Test Method*—The sample is decomposed with HCl and the liberated CO_2 is passed through a series of scrubbers to remove water and sulfides. The CO_2 is absorbed with Ascarite, a special sodium hydroxide absorbent,[9] and the gain in weight of the absorption tube is determined and calculated as percent CO_2.

9.2 *Significance and Use*—The carbon dioxide analysis is useful in estimating carbonates and organic carbon for chemical balance.

9.3 *Special Reagents:*

9.3.1 *Magnesium Perchlorate Desiccant*[10]—for drying.

9.3.2 *Sodium Hydroxide Absorbent*[9]—a coarse sodium hydroxide coated silica.

9.4 *Special Apparatus*—The apparatus illustrated in Fig. 1 consists of the following:

9.4.1 *Purifying Jar A*, Fleming, containing sulfuric acid.

9.4.2 *Drying Tube B*, U-shaped with side arms and glass-stoppers. Side arms are shaped to hold rubber tubing. Contains Anhydrone on left side and Ascarite on right side.

9.4.3 *Erlenmeyer Flask C*, 250 mL, 24/20 ground-glass joint.

9.4.4 *Separatory Funnel D*, with ground glass stopper and interchangeable hollow ground-glass joint. A delivery tube bent at the end extends into the sample flask approximately 15 mm from the bottom and is used to introduce acid into flask.

9.4.5 *Condenser E.*

9.4.6 *Gas-Washing Bottle F*, 250 mL, with fritted disk containing distilled water to retain most of the acid volatilized from the alkalimeter.

9.4.7 *U-Tube G*, containing mossy zinc to remove the last traces of HCl.

9.4.8 *Gas Washing Bottle H*, 250 mL, with fritted disk, containing concentrated H_2SO_4 and trap *I*, to remove any SO_3 mist that is carried over.

9.4.9 *Absorption Bulb J*, containing Anhydrone to remove last traces of water vapor.

9.4.10 *CO_2 Absorption Bulb*, containing Ascarite filled as follows: On the bottom of the bulb, place a layer of glass wool extending above the bottom outlet and on top of this a layer of Anhydrone approximately 10 mm thick; immedi-

[8] Whatman No. 42 or an equivalent filter paper has been found suitable for this purpose.

[9] Ascarite, manufactured by Arthur H. Thomas has been found satisfactory for this purpose.

[10] Anhydrone, manufactured by J. T. Baker Inc. has been found satisfactory for this purpose.

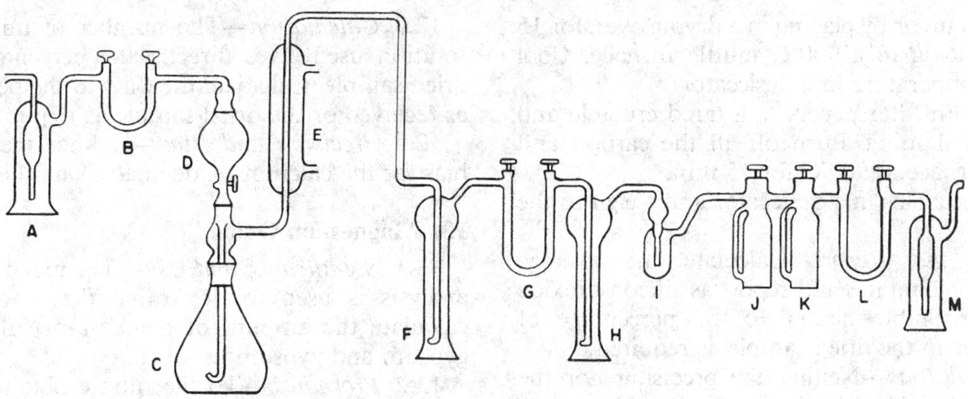

FIG. 1 Apparatus for Carbon Dioxide Analysis

ately above this place another layer of glass wool, then add Ascarite to almost fill the bulb. Place a top layer of Anhydrone approximately 10 mm thick on top of the Ascarite and top it off with a covering of glass wool.

9.4.11 *U-Guard Tube L*, filled with Anhydrone in left side and Ascarite in right side.

9.4.12 *Purifying Jar M*, Fleming, containing H_2SO_4.

9.5 *Procedure:*

9.5.1 After drying as described in Section 8 place the residue obtained in the 250 mL Erlenmeyer flask (*C*). Connect the flask to the apparatus as shown in Fig. 1. Purge the system free of carbon dioxide by passing a current of CO_2-free air through the apparatus for 10 to 15 min.

9.5.2 Weigh the absorption bulb to 0.0001 g and attach it to the train. Remove the glass stopper from the separatory funnel, place 50 mL of dilute HCl (1+1) in the separatory funnel (*D*) and replace the stopper with the interchangeable hollow ground-glass joint through which passes a tube for admitting purified air. Open the stopcock of the separatory funnel and admit air through the top of the funnel to force the hydrochloric acid into the Erlenmeyer flask (*C*).

9.5.3 Start cold water circulating through the condenser (*E*) and, with CO_2-free air passing at a moderate rate through the absorption train, place a small hot plate or gas burner under the sample flask and boil for approximately 2 min. Remove the hot plate and continue the flow of purified air at approximately three bubbles per second for 10 min to sweep the apparatus free of CO_2. Close the absorption bulb, disconnect it from the train and weigh, opening the stopper momentarily to equalize the pressure. Use a second absorption bulb as counterpoise in all weighings unless a single pan balance is used.

9.6 *Calculation*—Calculate the percent CO_2 to the dried sample as follows:

$$\text{Percent } CO_2 = ((A - B)/C \times 100)(1 - D) \quad (1)$$

where:

A = mass of absorption bulb + CO_2 g,
B = mass of absorption bulb before the run, g,
C = mass of specimen, g, and
D = percent combined water as determined in Section 8 as a decimal.

Calculate the percent CO_2 to the sample as received as follows:

$$\text{Percent } CO_2 = E(1 - F) \quad (2)$$

where:

E = result of Eq 1, and
F = percent free water as determined in Section 7 as a decimal.

9.7 *Precision and Bias*—Neither the precision nor the bias for the carbon dioxide analysis has been determined.

10. Silicon Dioxide and Other Acid Insoluble Matter

10.1 *Summary of Test Method*—The gypsum and other acid soluble components of the sample are dissolved in dilute hydrochloric acid (HCl). The residue is weighed and calculated as silicon dioxide and other acid insoluble matter.

10.2 *Significance and Use*—The silicon dioxide and other acid insoluble matter analysis determines and is used to report the percentage of one of the inert impurities in gypsum and gypsum products.

10.3 *Procedure*—Perform in triplicate.

10.3.1 Weigh approximately 1 g of the specimen prepared in Section 4 to the nearest 0.0001 g.

10.3.2 Place the specimen in a porcelain casserole. Add approximately 50 mL of 1+5 hydrochloric acid. Evaporate slowly and carefully to apparent dryness on a hot plate. The evaporation should take approximately 20 min. Make a blank determination with one casserole in parallel. Cool to room temperature.

10.3.3 Add enough hydrochloric acid (sp gr 1.19) to wet the solid residue. Add 20 mL of water, boil and filter through filter paper.[8] Wash the filter paper thoroughly using not less than 100 mL of room temperature water to render the precipitate chloride free. The most effective washing technique is to use many small quantities of wash water rather than fill the funnel to the brim two or three times. Test the filtrate for chloride by collecting a small amount and adding a few drops of 0.1 normal silver nitrate ($AgNO_3$) solution. A white precipitate indicates more washing is needed. Discard this test solution.

10.3.4 Place all the filtrate back in the same casserole. Evaporate to dryness and heat to 120°C for 1 h and cool. To the cooled casserole add enough HCl (sp gr 1.19) to wet the solid residue. Add 50 mL of water and boil.

10.3.5 Wash the second contents of the casserole through another filter[8] paper. Thoroughly wash the residue in the filter paper until chloride free as in 10.3.3. Retain the filtrate for the iron and aluminum oxide analysis.

10.3.6 Dry sufficient crucibles by placing in a cold muffle

furnace during warm up or by placing in a drying oven for 15 to 20 min, then placing in a 900°C muffle furnace. Cool crucibles to room temperature in a desiccator.

10.3.7 Transfer both filter papers to a tared crucible and char slowly without flaming. Burn off all the carbon and ignite in a muffle furnace at 900°C for 15 min.

10.3.8 Cool the crucibles in a desiccator and weigh to the nearest 0.0001 g.

10.4 *Calculation and Report*—Calculate the average weight of the three precipitates and report as silicon dioxide (SiO_2) and other insoluble matter to the percentage of sample as received or to the dried sample as required.

10.5 *Precision and Bias*—Neither the precision nor the bias for the silicon dioxide and other acid insoluble matter has been determined.

11. Iron and Aluminum Oxides

11.1 *Significance and Use*—The iron and aluminum oxides (Fe_2O_3 + Al_2O_3) analysis is used to determine the quantity of these metal oxides in gypsum or gypsum products.

11.2 *Procedure*—To the filtrate, obtained as described in Section 10, add a few drops of nitric acid (HNO_3), and boil to ensure oxidation of the iron. Add 2 g of ammonium chloride (NH_4Cl) previously dissolved in water. Make alkaline with ammonium hydroxide (NH_4OH). Digest hot for a few minutes until the precipitate coagulates. Filter, wash, ignite the precipitate at 1000°C for 30 min or to constant weight in a muffle furnace and weigh as Fe_2O_3 + Al_2O_3. Save the filtrate for the CaO analysis.

NOTE 2—The addition of a pinch of ashless filter paper pulp will aid in the filtration of the precipitate.

11.3 *Calculation*—Calculate Fe_2O_3 + Al_2O_3 to the percentage of sample as received or the dried sample as required. This precipitate may be further treated to separate the two oxides, but this is generally unnecessary.

11.4 *Precision and Bias*—Neither the precision nor the bias for the iron and aluminum oxides analysis has been determined.

12. Calcium Oxide

12.1 *Significance and Use*—The calcium oxide (CaO) analysis is used to determine the amount of CaO and calculate the amount of calcium carbonate ($CaCO_3$) in gypsum and gypsum products.

12.2 *Procedure:*

12.2.1 To the filtrate obtained as described in Section 11 add 5 g of ammonium oxalate ($(NH_4)_2C_2O_4$) dissolved in water. Digest hot for 30 min, making sure that the solution is always alkaline with NH_4OH. Filter, wash, and ignite the precipitate at 1000°C for 2 h to constant weight in a platinum crucible in a muffle furnace.

12.2.2 *Alternative Method*—To the filtrate obtained as described in Section 11 add 5 g of $(NH_4)_2C_2O_4$ dissolved in water. Digest hot for 30 min, making sure that the solution is always alkaline with NH_4OH. Filter and wash, transfer the precipitate to a beaker, and wash the filter paper with hot H_2SO_4 (1+6), catching the washing in the same beaker. Heat gently to complete solution, adding more H_2SO_4 if necessary. While still warm, titrate with potassium permangonate ($KMnO_4$) solution (5.6339 g/L) until the pink color persists.

12.3 *Calculation*—The number of milliliters of $KMnO_4$ solution used gives directly the percentage of lime in the dried sample. Calculate the CaO to the percentage of sample as received or the dried sample as required.

12.4 *Precision and Bias*—Neither the precision nor the bias for the calcium oxide analysis has been determined.

13. Magnesium Oxide

13.1 *Significance and Use*—The magnesium oxide (MgO) analysis is used to determine the amount of MgO and calculate the amount of magnesium carbonate $MgCO_3$ in gypsum and gypsum products.

13.2 *Procedure*—To the filtrate obtained as described in 12.2.1 or 12.2.2, add enough water to give a total volume of approximately 600 mL. Cool, and add 10 mL of NH_4OH and 5 g of sodium ammonium phosphate ($NaNH_4HPO_4$) dissolved in water. Stir vigorously until a precipitate begins to form. Let stand overnight. Filter, using a Gooch crucible, and wash with NH_4NO_3 solution. Ignite at 1000°C for 2 h to constant weight in a muffle furnace.

13.3 *Calculation*—Multiply this weight by 0.36207 to find the weight of magnesium oxide (MgO). Calculate the MgO to the percentage of sample as received or to the dried sample as required.

13.4 *Precision and Bias*—Neither the precision nor the bias for the magnesium oxide analysis has been determined.

14. Sulfur Trioxide

14.1 *Summary of Test Method*—In this test method, sulfate is precipitated from an acid solution of the gypsum with barium chloride ($BaCl_2$). The precipitate is filtered and weighed as barium sulfate ($BaSO_4$) and the sulfur trioxide (SO_3) equivalent is calculated.

14.2 *Significance and Use*—The specification for gypsum and some gypsum products specifies the amount of calcium sulfate ($CaSO_4$) required, either in the dihydrate ($CaSO_4 \cdot 2H_2O$) or hemihydrate ($CaSO_4 \cdot \frac{1}{2}H_2O$) form. This procedure assumes that an insignificant amount of sulfate other than calcium sulfate is present. This test method is used to determine compliance to the gypsum and gypsum product specifications. It is also commonly used in quality control work.

14.3 *Interference*—This test method has been developed for natural gypsum and for impurities generally found associated with natural gypsum. Synthetic gypsum may have an additional number of interfering elements and compounds, consequently, this procedure may not give accurate results and is not recommended. This test method has a number of interferences that theoretically affect the results. Co-precipitation and occlusion can be problems if the solution is either too acidic or too basic. Calculations using SO_3 analysis are most accurate on samples that are known to be completely hydrated or completely dehydrated.

14.4 *Procedure:*

14.4.1 Having properly selected and prepared the samples as specified in Section 4, weigh a representative specimen of approximately 0.5 g, to the nearest 0.0001 g.

14.4.2 Place the weighed sample into a 400-mL beaker. Add 50 mL of HCl (1+5). Boil and disperse with the flattened end of a glass rod while stirring until the sample is completely broken down. Add approximately 100 mL boiling water and continue boiling for 15 min, with this step to be extended as required, so the combined boiling time is not less than 1 h.

14.4.3 Using filter paper[8], filter into a clean 600-mL flask and rinse the 400-mL beaker thoroughly with hot distilled water. Carefully wash the sides of the 400-mL beaker while wiping the insides with a rubber-tipped glass rod making sure all splatters and insoluble are washed into the filter paper. Dry and burn off the filter paper leaving the residue that can be dried and weighed for insoluble matter should this test method not be otherwise conducted.

14.4.4 Dilute the filtrate to 400 to 500 mL. Add 1 to 2 drops of 0.1 % methyl red indicator. Prepare a 400 to 500-mL sample of 0.05 to 0.1 N HCl. Add 1 to 2 drops of 0.1 % methyl red indicator. Compare the color of this solution to the color of the filtrate. Dilute the filtrate or add HCl (1+5) solution as necessary to match the pH of the 0.05 to 0.1 N HCl solution.

14.4.5 Boil the filtrate solution and add 20 mL of near-boiling 10 % barium chloride solution, preferably with the help of a pipette, drop by drop while stirring. The barium chloride solution should be prepared not less than one day before use. Continue boiling the solution for 10 to 15 min and digest hot for 3 h or until the precipitate settles.

14.4.6 Filter[8] and wash with hot water to render the precipitate chloride free. One hundred twenty five to 150 mL of distilled water should be adequate. The filtrate can be tested for chloride by collecting a small amount and adding a few drops of 0.1 N $AgNO_3$ solution. A white precipitate indicates more washing is needed. Alternately, filtering crucibles[11] may be used for quick filtering if the particular crucibles to be used are tested prior to use by refiltering the filtrate from the crucibles with filter paper, and no more than 2 mg is collected on the filter paper.

14.4.7 Ignite the precipitate and paper in a tared crucible, and slowly char the paper without inflaming. Burn off all the carbon and ignite in a muffle furnace at 800 to 900°C or using bright red heat over a Bunsen burner for 15 to 20 min. Dry the filtering crucibles by placing in a cold muffle furnace during warm-up or in a drying oven prior to igniting in a muffle furnace at 800 to 900°C for 15 to 20 min.

NOTE 3—Thoroughly cleans crucibles before each use and heat in a furnace at 800 to 900°C and cool in a desiccator before taring.

14.4.8 Cool all crucibles in a desiccator and weigh to the nearest 0.0001 g.

14.5 *Calculation*—Multiply the weight of the precipitate by 0.343 to determine the weight of sulfur trioxide (SO_3). Calculate the SO_3 to the percentage of sample as received or to the dried sample as required.

14.6 *Precision and Bias*—Neither the precision nor the bias for the sulfur trioxide analysis has been determined.

15. Chlorides

15.1 *Significance and Use*—Small amounts of chloride in gypsum or gypsum products can have a detrimental effect on their use. This procedure is used to measure the amount of chlorides present and report it as sodium chloride.

15.2 *Procedure:*

[11] Gooch or Coors filtering crucibles have been found suitable for this purpose.

15.2.1 Weigh approximately 20.0 g of sample as prepared in Section 4 to 0.001 g and transfer to a 400-mL beaker. Add 150 mL of water, stir, and heat to just below the boiling point. Cover with a watch glass and maintain at just below boiling (not less than 80°C) for 1 h with occasional stirring. Filter with suction on a Buchner funnel fitted with a medium filter paper. Wash the residue with four 20-mL portions of hot water.

15.2.2 Add 2 drops of phenolphthalein indicator solution to the filtrate. If the filtrate fails to turn pink, add 0.1 N NaOH solution dropwise with stirring until a faint pink color develops. Add 0.1 N HNO_3 dropwise until the pink color just disappears.

15.2.3 If the chloride content is very low, transfer the entire filtrate quantitatively to a 400-mL beaker and proceed as described in 15.2.4. If larger amounts of chloride are expected, transfer the filtrate quantitatively to a 250-mL volumetric flask, cool to room temperature, and dilute to 250 mL. Take a suitable aliquot, transfer to a 400-mL beaker, and dilute to a volume of 100 to 250 mL.

15.2.4 Place the beaker containing the sample on a white surface, add 0.5 mL (10 drops) of K_2CrO_4 solution and titrate with $AgNO_3$ solution using a micro buret having a 10-mL capacity and graduated in divisions of 0.02 mL. Titrate until a faint but definite orange color is visible.

15.2.5 Perform a blank titration using the same volume of water as the sample volume and the same amount of K_2CrO_4 solution. Titrate to the same color as obtained with the sample.

15.3 *Calculation*—Subtract the volume of $AgNO_3$ solution used for the blank titration from the volume used for the sample to give the net titration. A 1-mL net titration is equivalent to 0.002923 g of sodium chloride (NaCl). Calculate the NaCl as a percentage of the sample as received or the dried sample as required.

15.4 *Precision and Bias*—Neither the precision nor the bias for the chloride analysis has been determined.

16. Report

16.1 Report the results obtained in the analysis as follows:

	Percent
Free water	...
Combined water	...
Carbon dioxide (CO_2)	...
Silicon dioxide (SiO_2) and insoluble matter	...
Iron and aluminum oxides ($Fe_2O_3 + Al_2O_3$)	...
Lime (CaO)	...
Magnesium oxide (MgO)	...
Sulfur trioxide (SO_3)	...
Sodium chloride (NaCl)	...
Total	100.00±

NOTE 4—Since it is frequently advisable to recalculate the results obtained in the chemical analysis in order that they may be more enlightening, the following is submitted for consideration:

(*1*) Multiply the percentage of combined water by 4.778 to obtain purity or percentage gypsum. To calculate the percentage of $CaSO_4 \cdot \frac{1}{2}H_2O$ in plasters, multiply the percentage of gypsum by 0.8430.

(*2*) Multiply the percentage of combined water by 2.222 to obtain the amount of SO_3 combined as gypsum.

(*3*) Subtract the result obtained in (*2*) from the total SO_3 found by analysis to obtain the excess SO_3.

(*4*) Multiply the excess SO_3 by 1.700 to obtain the percentage anhydrite, $CaSO_4$.

(*5*) Multiply the percentage of gypsum found in (*1*) by 0.3257 to obtain the percentage of CaO combined as gypsum.

(*6*) Multiply the percentage of anhydrite found in (*4*) by 0.4119 to obtain the percentage of CaO combined as anhydrite.

(*7*) Add (*5*) and (*6*) together. Then subtract this result from the total CaO percentage found by analysis.

(*8*) Multiply the excess CaO percentage by 1.785 to obtain the percentage of calcium carbonate.

(*9*) Multiply the percentage of MgO by 2.091 to obtain the percentage of magnesium carbonate.

NOTE 5—Having made the calculations in Note 4, the results may be reported as follows:

	Percent
Gypsum ($CaSO_4 \cdot 2H_2O$)	...
Anhydrite ($CaSO_4$ natural and manufactured) (Note 4)	...
Silicon dioxide and insoluble (SiO_2 + Ins.)	...
Iron and aluminum oxide (R_2O_3)	...
Calcium carbonate ($CaCO_3$)	...
Magnesium carbonate ($MgCO_3$)	...
Sodium chloride (NaCl)	...
Total	100.00±

NOTE 6—The presence of the different forms of $CaSO_4$ may be determined by a microscopic examination. A paper titled "Gypsum Analysis with the Polarizing Microscope" containing suggested methods can be found in ASTM STP 861.[12]

ALTERNATIVE PROCEDURE FOR ANALYSIS FOR CALCIUM SULFATE BY THE AMMONIUM ACETATE METHOD[13]

17. Significance and Use

17.1 This test method covers the determination of calcium sulfate in gypsum and gypsum products by extraction with ammonium acetate solution, and may be used as an alternative method.

18. Reagents and Materials

18.1 *Ammonium Acetate Solution*—Dissolve 454 g of ammonium acetate in 2 L of water. Add sufficient NH_4OH to make the solution distinctly ammoniacal, using phenolphthalein as the indicator.

18.2 *Ammonium Hydroxide Wash Solution*—Dilute 100 mL of concentrated ammonium hydroxide (NH_4OH, sp gr 0.90) to 1 L with water.

18.3 *Filter Aid*—Diatomaceous silica, analytical grade.

18.4 *Phenolphthalein Indicator Solution.*

19. Procedure Using Gooch Crucible

19.1 Weigh rapidly approximately 4 g of the well-mixed sample and transfer to a 600-mL beaker. Make all weighings to 0.001 g, except weigh the crucibles and their contents to 0.0001 g.

19.2 Without delay, weigh approximately 1 g of the well-mixed sample in a tared weighing bottle having a ground-glass stopper. Dry the sample and weighing bottle to constant weight at 45°C. Stopper weighing bottles immediately upon removal from the oven in order to prevent absorption of moisture from the air upon cooling.

19.3 If the percentage by weight of combined water held by the calcium sulfate is desired, heat the sample and weighing bottle to constant weight at 220°C.

19.4 To the contents of the 600-mL beaker (19.1), add 350 mL of the ammonium acetate solution, and stir the mixture thoroughly to loosen all of the solid matter from the bottom of the beaker. Add 0.2000 g of redried diatomaceous silica to the mixture, heat the beaker and contents to 70°C on a steam bath, and maintain at that temperature for 30 min, while stirring frequently. During heating, keep the solvent ammoniacal by additions of NH_4OH and phenolphthalein, if indicated. Meanwhile, heat a supply of the ammonium acetate solution to 70°C, keeping it also distinctly ammoniacal. Filter the mixture, with suction, through a tared Gooch crucible, stirring frequently during filtration to keep the diatomaceous earth suspended in the liquid. Wash the Gooch crucible containing the residue with five 10-mL portions of the warm acetate solution, draining thoroughly after each washing. Wash in the same manner with eight 10-mL portions of the NH_4OH wash solution. Take care to wash the upper walls of the Gooch crucible. Drain the crucible dry with suction, place in an oven at 70°C, and dry to constant weight (Note 7). Allow the crucible to cool in a desiccator before weighing.

NOTE 7—Avoid overheating in all oven drying of ammonium acetate residues; that is, place crucibles well away from the heating elements. This is of particular importance for samples high in impurities, as these impurities often have water of hydration that is lost on local overheating.

20. Procedure Using Tared Filter Papers

NOTE 8—This procedure is suggested where several samples are to be analyzed at once. It has been found that gravity filtration on six samples will proceed as rapidly as it is possible to handle the samples.

20.1 Dry a quarter-folded, 110-mm quantitative filter paper overnight at 70°C in a wideform, glass-stoppered, 30 by 60-mm weighing bottle. After drying, cool the weighing bottle and paper in a desiccator, and weigh.

20.2 Treat the sample exactly as described in 19.1, 19.2, and 19.4 prior to the filtration. Filter the mixture by gravity through a 70-mm glass funnel, stirring frequently during filtration to keep the diatomaceous silica suspended in the liquid. Wash the filter paper and residue with five 10-mL portions of warm acetate solution, draining thoroughly after each washing. Wash in the same manner with eight 10-mL portions of the NH_4OH wash solution. After final draining, replace the paper and residue in the weighing bottle, and dry at 70°C to constant weight. Cool the weighing bottle, paper, and residue in a freshly prepared desiccator before weighing; this is essential, due to the hygroscopic character of paper.

21. Calculation

21.1 Calculate the percentage of loss in weight at 45°C (free water) as follows:

$$\text{Loss in weight at 45°C, \%} = [(A - B)/C] \times 100 \quad (3)$$

where:

A = original weight of sample and weighing bottle,
B = weight of sample and weighing bottle dried to constant weight at 45°C, and
C = original weight of sample.

Calculate the weight of the 4-g sample (19.1), corrected for loss on heating to constant weight at 45°C.

21.2 Calculate the percentage of combined water as follows:

[12] Green, George W., "Gypsum Analysis with the Polarizing Microscope," *The Chemistry and Technology of Gypsum, ASTM STP 861*, ASTM, 1984, pp. 22–47.

[13] This procedure was developed by L. S. Wells and W. F. Clarke, National Bureau of Standards, and modified by B. E. Kester, United States Gypsum Co.

$$\text{Combined water, \%} = [(B - D)/(B - E)] \times 100 \quad (4)$$

where:

B = weight of sample and weighing bottle dried to constant weight at 45°C,

D = weight of sample and weighing bottle dried to constant weight at 220°C, and

E = weight of weighing bottle.

21.3 Calculate the percentage of $CaSO_4 \cdot XH_2O$ on the basis of the sample dried to constant weight at 45°C as follows:

$$CaSO_4 \cdot XH_2O, \% = [F - (G - H)/F] \times 100 \quad (5)$$

where:

F = weight of sample, corrected for loss on heating to constant weight at 45°C,

G = weight of dried crucible and contents (19.4) or weight of weighing bottle and contents (20.2), and

H = weight of crucible plus diatomaceous silica used as filter aid (19.4), or weight of weighing bottle, diatomaceous silica used as a filter aid and the weight of filter paper (20.2).

22. Precision and Bias

22.1 Neither the precision nor the bias for the analysis of calcium sulfate by the ammonium acetate method has been determined.

ALTERNATIVE PROCEDURE FOR ANALYSIS FOR SODIUM CHLORIDE BY THE COULOMETRIC METHOD[14]

23. Significance and Use

23.1 This test method covers the determination of sodium chloride in gypsum and gypsum products by the coulometric method, and may be used as an alternative method.

24. Interferences

24.1 The presence of sulfide, sulfhydryl, or other silver reactive substances will lead to high results. Such interfering substances may be removed by alkaline oxidation with hydrogen peroxide.

25. Apparatus

25.1 *Chloride Meter:*

25.1.1 The instrument shall be equipped to measure the concentration of dissolved chloride in aqueous solutions by the coulometric method.[15]

25.1.2 The instrument shall be capable of measuring chloride concentrations in the range from 10 to 260 mg/L with a repeatability of ±1 mg/L.

26. Reagents

26.1 *Acid Buffer Solution*—Dissolve 100 mL of 99.5 % acetic acid ($HC_2H_3O_2$) and 6.5 mL of concentrated nitric acid (sp gr 1.42) in approximately 200 mL of water and dilute to 500 mL.

26.2 *Diluted Standard Solution* (100 mg Cl/L)—Dilute 5.00 mL of stock standard solution to 500 mL.

26.3 *Gelatin Solution*—Add 2.5 g of gelatin and 0.5 g of thymol blue to 250 mL of water and dissolve by stirring continuously while bringing to a boil. With the solution just boiling, continue stirring until all the thymol blue is dissolved. Add 0.5 g of thymol, cool, and dilute the solution to 500 mL.

NOTE 9—The gelatin solution holds the precipitated silver chloride (AgCl) in suspension and also indicates the presence of the acid buffer. The solution will keep for 3 months at room temperature or longer if refrigerated. Warm the refrigerated solution to room temperature before use.

26.4 *Stock Standard Solution* (10 g Cl/L)—Dissolve 8.240 g of dried sodium chloride (NaCl) in water and dilute to 500 mL.

27. Procedure

27.1 Weigh 20.0 g of the well-mixed sample and transfer to a 150-mL beaker.

27.2 Add 50 mL of water, boil, allow the solid material to settle, and filter off the solution. Add an additional 50 mL of water to the solids, boil, and pour the contents of the beaker into the filter. Wash the residue with 100 mL of hot water, adding the washing to the filtrate. Cool and dilute with water to 250 mL.

27.3 Switch on the chloride meter and allow a period of 25 min before use. Set the counter to zero.

27.4 Place a magnetic stirring bar in the test beaker, add 10 mL of diluted standard solution, 3 mL of acid buffer solution, and 5 drops of gelatin solution. Place the test beaker on the platform and lower the electrodes into the solution. Press the "start" button until the pilot light is extinguished. The counter will begin to register after a few seconds. Do not remove the electrodes from the sample until the pilot light comes on. Read the chloride content from the counter. The reading should be 100 ± 1 mg Cl/L. If this reading is not obtained, refer to the manufacturer's instruction manual. Reset the counter to zero.

27.5 Repeat the procedure used in 27.4, using 10 mL of the sample solution instead of the diluted standard solution. Read the result as milligrams of chlorine per litre. When all tests are completed, lower the electrodes into reagent water.

28. Calculation

28.1 Calculate the amount of NaCl as a percentage of the sample as received or dried sample as follows:

$$\text{NaCl, \%} = 0.00206 \times A \quad (6)$$

where:

A = chloride meter reading, mg Cl/L.

29. Precision and Bias

29.1 Neither the precision nor the bias for the sodium chloride analysis by the coulometric method has been determined.

[14] This procedure was developed by Westroc Industries Limited.

[15] The EEL Chloride Meter, available from Fisher Laboratory Supplies Co., has been found satisfactory. Other instruments available for the coulometric determination of chloride are the Aminco Chloride Titrator, available from American Instrument Co. Ltd.; the Buchler Chloridometer, available from Buchler Instruments Division of Nuclear Chicago Corp.; and the Fiske/Marius Chlor-O-Counter available from Johns Scientific.

DETERMINATION OF SAND IN SET PLASTER

30. Significance and Use

30.1 This test method is used for determining the sand content of samples of aggregated plaster taken from job sites to determine compliance with Specification C 842.

31. Summary of Test Method

31.1 This test method for the determination of the sand content of set gypsum plaster requires for accurate results the following determinations:

31.1.1 Determination of the percentage of insoluble matter in the sand used with the plaster,

31.1.2 Determination of the percentage of insoluble matter in the gypsum neat plaster, and

31.1.3 Determination of the percentage of insoluble matter in the sanded calcined plaster.

NOTE 10—If samples of the original gypsum neat plaster and the sand are not available, an approximation of the insoluble matter may be obtained by the use of this method on plaster and sand from the same sources as those from which the plaster to be analyzed was originally prepared.

32. Sampling

32.1 Where plaster to be tested is part of a two-coat or three-coat plastering operation, take the sample for analysis from that portion of the entire plaster sheet that comprises the single coat being tested. Separate succeeding coats of plaster by use of a stiff putty knife or similar implement. Not less than 500 g shall be taken as a sample, the sample preferably being obtained from different sections of the wall or ceiling under examination.

33. Reagents

33.1 *Ammonium Acetate* (250 g/L)—Dissolve 250 g of ammonium acetate ($NH_4C_2H_3O_2$) in water and dilute to 1 L.

33.2 *Ammonium Hydroxide* (1+59)—Mix 1 volume of concentrated ammonium hydroxide (NH_4OH) (sp gr 0.90) with 59 volumes of water.

34. Procedure

34.1 In a clean porcelain mortar, grind the set plaster sample to the size of the largest sand particles present, or smaller, so that approximately 100 % of the sample will pass a 2.36-mm (No. 8) sieve. Fine grinding makes solution of the gypsum faster. Place approximately 200 g of the ground sample in a porcelain casserole or evaporating dish, and calcine on a sand bath. Stir the sample continuously with a thermometer during the heating, and adjust the rate of heating so that 20 to 30 min will be required to raise the temperature of the sample to 160 ± 5°C. Cool the sample to room temperature in a desiccator.

34.2 After cooling, weigh accurately 20 ± 0.05 g of the calcined sample into a 600-mL beaker. Add 300 to 350 mL of $NH_4C_2H_3O_2$ solution, which should be slightly alkaline to litmus paper. If acidic, add a few millilitres of NH_4OH (1+59) to the stock $NH_4C_2H_3O_2$ solution to render it slightly alkaline prior to the addition to the test sample.

34.3 Warm the suspension to a temperature of 70 ± 5°C and stir continuously for 20 to 30 min. Filter the warm suspension with the aid of suction through a small Büchner funnel or Gooch crucible in which filter paper has previously been placed. Refilter the first 100 mL of the filtrate. Wash the sand remaining in the beaker onto the filter with an additional 100 mL of warm ammonium acetate solution. Wash the beaker and residue with 200 to 300 mL of water, dry the funnel and sand at 100°C to constant weight. The weight of the residue is the weight of insoluble matter.

34.4 *Insoluble Matter in Sand*—Determine the weight of insoluble matter in sand as described in 34.1 to 34.3, except that no grinding of the sample is necessary.

34.5 *Insoluble Matter in Gypsum Neat Plaster*—Determine the weight of insoluble matter in the gypsum neat plaster as described in 34.1 to 34.3, except that only a 5-g sample is required and no grinding of the sample is necessary.

35. Calculation

35.1 *Insoluble Matter in Plaster*—Multiply by 5 the weight of the insoluble matter obtained as described in 34.3 to obtain the percentage of insoluble matter in sanded plaster.

35.2 *Insoluble Matter in Sand*—Multiply by 5 the weight of the insoluble matter in sand obtained as described in 34.4 to obtain the percentage of insoluble matter in sand.

35.3 *Insoluble Matter in Gypsum Neat Plaster*—Multiply by 20 the weight of the insoluble matter in gypsum neat plaster obtained as described in 34.5 to obtain the percentage of insoluble matter in gypsum neat plaster.

35.4 Calculate the percentage of sand in the sanded plaster as follows:

$$X = [(C - B)/(A - B)] \times 100 \tag{7}$$

where:

X = % of sand in sanded plaster,
A = % of insoluble matter in the sand,
B = % of insoluble matter in the gypsum neat plaster, and
C = % of insoluble matter in the sanded plaster.

35.5 To express the results as a ratio of the parts of sand per part of plaster by weight, the following equation may be used:

$$\text{Ratio of sand to plaster} = X/(100 - X) \tag{8}$$

NOTE 11—The results obtained by the above procedure indicate the amount of sand originally mixed with the gypsum neat plaster before it had been gaged with water or set.

36. Precision and Bias

36.1 Neither the precision nor the bias for the analysis of sand in set plaster has been determined.

WOOD-FIBER CONTENT IN WOOD-FIBER GYPSUM PLASTER

37. Significance and Use

37.1 This test method is used to determine the weight of wood fiber in wood-fibered plaster.

38. Procedure

38.1 Place a 100-g sample of wood-fiber plaster, prepared as described in Section 4 on a 600-μm (No. 30) sieve[6] nested over a 150-μm (No. 100) sieve.[6] Wash the plaster on the 600-μm sieve with a stream of cold water, removing the 600-μm sieve when the fiber on it is practically or entirely free of plaster. Next, wash the material on the 150-μm sieve until the bulk of the plaster has been washed through the sieve and the residue is mainly fiber. Transfer the material retained on the 150-μm sieve to a 300-mL, vitreous enamel, lipped pan, adding the charge on the 600-μm sieve if the fiber contains any adhering particles of plaster. Elutriate the material in the pan (purify by washing and straining, effecting as clean a separation of fiber from plaster as is feasible), catching the elutriated fibers on a 150-μm sieve. To avoid loss of the fine particles of fiber, it may be necessary to make the transfer from the pan to the 150-μm sieve by several stages of washing, stirring the charge, and quickly pouring upon the sieve the fiber flotations, repeating the elutriation procedure several times. Examine the fiber collected on the 150-μm sieve and repeat the elutriation if it seems desirable.

38.2 Dry the sieves (or sieve, as the case may be) and the residue contained therein overnight in an oven maintained at a temperature of 45°C. Carefully invert the sieves, or sieve, over a piece of white paper, and transfer the residual material to the paper by brushing the bottom of the inverted sieve. Examine the transferred material visually, noting whether the separation of fibers from plaster has been complete. Then transfer the material to a weighed platinum crucible and dry to constant weight at a temperature of 45°C. If the previous visual examination of the charge on the white paper showed that the fiber was practically free of particles of plaster, report as the percentage of fiber the weight of the fiber dried at 45°C, divided by 100. If, on the other hand, the visual examination revealed the presence of an appreciable quantity of plaster associated with the fiber, carefully ignite the contents of the crucible to constant weight. In this case, report as the percentage of fiber the loss on ignition, divided by 100.

39. Precision and Bias

39.1 Neither the precision nor the bias for the analysis of wood-fiber content in wood-fiber gypsum plaster has been determined.

OPTIONAL PROCEDURE FOR ANALYSIS FOR SODIUM BY THE ATOMIC ABSORPTION METHOD

40. Significance and Use

40.1 This test method covers the determination of sodium in gypsum and gypsum products by the atomic absorption method, and may be used as an optional procedure.

41. Interferences

41.1 Sodium is partially ionized in the air-acetylene flame. The effects of ionization may be significantly overcome by the addition of 1 to 2 g/L of another alkali to blanks, standards, and samples. Alternatively, the air-hydrogen flame may be used, as it produces less ionization and less visible emission than the air-acetylene flame.

42. Apparatus

42.1 *Atomic Absorption Spectrophotometer:*

42.1.1 The instrument shall be equipped to measure the concentration of dissolved sodium in aqueous solutions using either the air-acetylene or air-hydrogen flame.

42.1.2 The instrument shall be capable of measuring sodium concentrations within the optimum analytical range of 0.1 to 0.5 absorbance units while providing a coefficient of variation of approximately 0.5 to 2 %.

43. Reagents

43.1 *Solvent*—Use deionized water to prepare all solutions. If an alkali is to be included for the purpose of suppressing sodium ionization, it is most convenient to add it to the solvent at the start. In this way a constant concentration of alkali in blank, standards, and sample solution is ensured.

43.2 *Stock Standard Solution* (1.000 g Na/L)—Dissolve 2.5418 g of dried sodium chloride (NaCl) in water and dilute to 1 L with water.

43.3 *Dilute Standard Solutions*—Prepare dilute standard solutions bracketing the absorbance range of the dilute sample solution, using the stock standard solution. (Solutions having a concentration less than approximately 0.500 g/L are unstable for periods of more than one day.)

44. Procedure

44.1 Take 18 g of the well-mixed sample and transfer to a 150-mL beaker.

44.2 Add 50 mL of water, boil, allow the solids to settle, and decant the supernatant liquid into a filter. Add an additional 50 mL of water to the solids, boil, and pour the contents of the beaker into the filter. Wash the residue with 100 mL of hot water, adding the washing to the filtrate. Cool the filtrate to room temperature and dilute to 500 mL in a volumetric flask to make the stock sample solution. Take 10 mL of the solution and make up to 500 mL in a second volumetric flask, to make the dilute sample solution.

44.3 Determine the absorbance readings on the dilute standard solutions and the solvent blank at a wavelength of 589.0 to 589.6 nm, following the manufacturer's instruction manual. Subtract the absorbance value for the blank from the absorbance values for the dilute standard solutions and prepare a curve relating sodium concentration in milligrams per litre to absorbance values.

NOTE 12—If the absorbance of the dilute sample solution is known to lie within the linear range, that is, the sodium concentration is below approximately 1 mg/L only one standard and the solvent blank are needed to prepare the curve.

44.4 Determine absorbance readings on the dilute sample solution using the same technique, followed with the dilute standard solutions. The concentration of sodium in the dilute sample solution in milligrams per litre is then found by consulting the standard curve.

45. Calculation

45.1 Calculate the amount of NaCl as a percentage of the sample received or the dried sample as required as follows:

$$\text{Sodium calculated as \% NaCl} = A \times 6.3553/S \tag{9}$$

where:

A = concentration of dilute sample solution, mg/L, and
S = weight of sample, g.

46. Precision and Bias

46.1 Neither the precision nor the bias for the analysis of sodium by the atomic absorption method has been determined.

OPTIONAL PROCEDURE FOR ANALYSIS FOR SODIUM BY FLAME PHOTOMETRY

47. Significance and Use

47.1 This test method covers the determination of sodium in gypsum and gypsum products by flame photometry, and may be used as an optional procedure. This test method is based on Methods D 1428.

48. Interferences

48.1 Radiation interferences caused by elements other than that being determined are the greatest contributor to error in flame photometry. Some effects are positive and others negative. Of the elements encountered in this analysis, the greatest effect is that of one alkali metal on another. The foreign-element effects cannot be entirely compensated for without employing calibration standards closely duplicating the composition of the sample. However, the effects may be minimized by operating at the lowest practical sodium concentration range or by removal of the interfering elements. For example, aluminum has a depressing effect on alkali-metal emission, which may be of serious consequence. Aluminum should be removed from the extraction liquid prior to flame photometry if its concentration has been found, by preliminary tests, to exceed that of the sodium.

48.2 Self-absorption causes the curve of intensity versus concentration to decrease its slope at higher concentrations, tending to reduce accuracy. Bracketing the unknown by known standard solutions tends to minimize this interference.

49. Apparatus and Materials

49.1 *Flame Photometer*—The instrument shall consist of an atomizer and burner; suitable pressure-regulating devices and gages for fuel and air or oxygen; an optical system, consisting of suitable light-dispensing or filtering devices capable of preventing excessive interference from light of wavelengths other than that being measured; and a photosensitive indicating device.

49.2 *Supply of Fuel and Air or Oxygen*—The supplies of fuel and air or oxygen shall be maintained at pressures somewhat higher than the controlled operating pressure of the instrument.

50. Reagents

50.1 Prepare the following stock solutions from reagents that have been dried to constant weight at 105°C. Store the stock solutions in polyethylene or equally alkali-metal-free containers.

50.2 *Sodium Chloride Solution* (2.5418 g/L)—Dissolve 2.5418 g of sodium chloride (NaCl) in water and dilute to 1 L with water. This stock standard solution contains 1.000 g/L of sodium ion.

50.3 *Dilute Standard Solutions*—Prepare dilute standard solutions from stock standard solution, bracketing the expected sodium concentration range of the diluted sample extraction liquid. For example, if the expected range of the sample extraction liquid is between 0 and 0.010 g/L, prepare eleven equally spaced standards in tenths of the maximum.

51. Calibration of Instrument

51.1 Select the proper photocell; the blue-sensitive phototube having a range from 320 to 620 nm is required for sodium determination. Open the slit width to approximately one fourth of the maximum opening, set the instrument to the maximum sensitivity range, and balance the meter to obtain electrical equilibrium.

51.2 Feed fuel and air or oxygen to the burner and ignite the emitted mixture. Adjust fuel and air or oxygen pressures and follow the procedures for warm-up time prescribed by the instrument manufacturer. Set the scale-reading dial at approximately 95 % of full scale, introduce a solution containing the maximum amount of sodium in the range to be covered, and allow the emitted light to strike the photocell.

51.3 Select the proper filter, if a filter-type instrument is used. For instruments employing spectral dispersing devices, turn the wavelength dial back and forth slowly and carefully in the vicinity of 589 nm. The galvanometer can be made to pass through a maximum deflection. This wavelength setting produces maximum sensitivity, and should not be disturbed during the test.

51.4 Continue to atomize the maximum standard of the range to be covered, and set the scale-reading of the dial at exactly full scale (100 or 1000) (Note 13). Adjust the gain so as to balance the galvanometer needle.

NOTE 13—For instruments equipped with a variable slit, carry out the procedure described above with the slit width between fully closed and one fourth open for the preliminary test. In determining the proper slit width for optimum instrument performance, consideration must be given to the fact that the intensity of the emission line is approximately proportional to the slit width, whereas continuous background intensity increases as the square of the slit width. A decrease in slit width results in decreased illumination of the phototube for a given concentration and is compensated for by increasing the gain of the instrument. The most favorable operating conditions are obtained with the smallest slit width that does not result in instability of the galvanometer needle when it is set to give full-scale reading with the maximum standard in the range to be covered. When altering the slit width, determine the background by atomizing a zero standard, and check for sensitivity setting and instrument stability with the maximum standard in the range to be covered. Lower ranges require wider slit widths. Determine and record the optimum slit width for each range and element to be covered. Use these values in all subsequent tests.

51.5 Determine the emission intensity of all standards.

51.6 Plot emission intensity (scale reading) versus concentration on linear graph paper. For the lower ranges, the curve thus prepared approximates a straight line but may not intersect zero because of background intensity. At higher ranges, the curves show a decrease in slope with increasing concentration. Record on graphs all data in regard to slit width, fuel pressure, and air or oxygen pressure.

52. Procedure

52.1 Weigh 25 g of the well-mixed sample and transfer to a 150-mL beaker.

52.2 Add 50 mL of water, boil, allow the solid material to settle, and decant the supernatant liquid into a filter. Add an additional 50 mL of water to the solids, boil and pour the contents of the beaker into the filter. Wash the residue with 75 mL of hot water, adding the washing to the filtrate. Cool and dilute with water to 200 mL in a volumetric flask to make the stock sample solution.

52.3 Take 5 mL of the stock sample solution and make up to 100 mL in a volumetric flask to make the dilute sample solution.

NOTE 14—If the concentration of sodium in the sample is found to be greater than the maximum standard, further dilute sample solution with water to bring the concentration within the range. If the concentration of sodium in the sample is less than one tenth of the value of the maximum standard, prepare a new dilute sample solution from the stock sample solution to bring the concentration within the range.

52.4 Turn the instrument on and feed fuel and air or oxygen to the burner. Ignite the gas mixture. For instruments with an adjustment slit, set the width to the value determined as outlined in Note 13.

52.5 Place the scale-reading dial at maximum. Atomize the sample and allow its emitted light to strike the photocell. Set the wavelength to 589 nm as described and adjust the gain to balance the galvanometer. Determine the emission intensity of the sample.

52.6 Refer to the standard curve prepared above and read off the concentration of sodium ion in the dilute sample solution in milligrams per litre.

53. Calculation

53.1 Calculate the amount of NaCl as follows:

$$\text{Sodium, calculated as NaCl, } \% = A \times 0.04067 \quad (10)$$

where:

A = concentration of sodium, mg/L read off the standard curve.

54. Precision and Bias

54.1 Neither the precision nor bias of the analysis for sodium by flame photometry has been determined.

55. Keywords

55.1 ammonium acetate method; atomic absorption; chemical analysis; coulometric method; flame photometry; gypsum; gypsum concrete; gypsum board; gypsum products; plaster; sand in set plaster; wood-fiber plaster

SUMMARY OF CHANGES

This section identifies the location of changes to these test methods that have been incorporated since the last issue. Committee C-11 has highlighted those changes that affect the technical interpretation or use of these test methods.

(1) Section 5.1.20 was revised.

Standard Test Methods for Physical Testing of Gypsum, Gypsum Plasters and Gypsum Concrete[1]

This standard is issued under the fixed designation C 472; the number immediately following the designation indicates the year of original adoption or, in the case of revision, the year of last revision. A number in parentheses indicates the year of last reapproval. A superscript epsilon (ε) indicates an editorial change since the last revision or reapproval.

This standard has been approved for use by agencies of the Department of Defense.

1. Scope

1.1 These test methods cover the physical testing of gypsum, gypsum plasters, and gypsum concrete.

1.2 The test methods appear in the following sections:

	Sections
Precautions for Physical Tests	4
Reagents and Materials	5
Free Water	6
Fineness	7
Normal Consistency of Gypsum Plaster	8
Normal Consistency of Gypsum Concrete	9
Setting Time	10
Setting Time (Temperature Rise Method)	11
Compressive Strength	12
Density	13

1.3 The values regarded as the standard are either in inch-pound units or SI (metric). The values stated first shall be regarded as the standard. Values following in parentheses are approximate and are provided for information purposes only.

1.4 The text of this standard references notes and footnotes which provide explanatory material. These notes and footnotes (excluding those in tables and figures) are not requirements of the standard.

1.5 *This standard does not purport to address all of the safety concerns, if any, associated with its use. It is the responsibility of the user of this standard to establish appropriate safety and health practices and determine the applicability of regulatory limitations prior to use.* For a specific precautionary statement, see X1.4.1.1.

2. Referenced Documents

2.1 *ASTM Standards:*

C 11 Terminology Relating to Gypsum and Related Building Materials and Systems[2]

C 778 Specification for Standard Sand[2]

E 11 Specification for Wire-Cloth Sieves for Testing Purposes[3]

3. Terminology

3.1 *Definitions:*

3.1.1 For useful definitions refer to Terminology C 11.

4. Precautions for Physical Tests

4.1 Gypsum products are greatly affected by small amounts of impurities introduced by careless laboratory manipulation. In order to obtain accurate results, it is absolutely essential to observe the following precautions:

4.1.1 Keep all apparatus thoroughly clean. Remove all traces of set plaster.

NOTE 1—For mixing pastes and mortars, a 500-ml rubber dental bowl is a convenience.

NOTE 2—Use care when drying gypsum, gypsum plasters, or gypsum concrete. Exceeding the specified drying temperatures may calcine the specimens, which will cause inaccurate test results.

5. Reagents and Materials

5.1 *Distilled or Deionized Water*—free of chlorides and sulfates at a temperature of 21 ±1°C (70 ±2°F).

5.2 *Standard Sand*— Specification C 778, 20–30 sand.

6. Free Water

6.1 *Significance and Use*—This test method determines the free water contained in gypsum, gypsum plasters, and gypsum concrete samples, and prepares the sample for subsequent testing.

6.2 *Apparatus*:

6.2.1 *Balance*, capable of weighing not less than 500 g at a precision of 0.1 g.

6.2.2 *Drying Oven*, set at 45 ± 3°C.

6.2.3 *Desiccator*, containing calcium chloride or equivalent desiccant.

6.3 *Procedure*:

6.3.1 Weigh a sample of not less than 500 g of the material as received to the nearest 0.1 g and spread it to a thin layer in a suitable vessel. Place in an oven (6.2.2) and dry for 2 h; then cool in a desiccator (6.2.3) and weigh again.

[1] These test methods are under the jurisdiction of ASTM Committee C-11 on Gypsum and Related Building Materials and Systems and are the direct responsibility of Subcommittee C11.01 on Specifications and Test Methods for Gypsum Products.

Current edition approved Oct. 10, 1998. Published May 1999. Originally published as C 472–61. Last previous edition C 472–93.

[2] *Annual Book of ASTM Standards*, Vol 04.01.

[3] *Annual Book of ASTM Standards*, Vol 14.02.

6.3.2 Retain the dried sample in an airtight container until used for the fineness test (Section 7).

6.4 *Report*—Report the loss in weight as a percentage of the original weight of the sample as received to a precision of 0.1%.

6.5 *Precision and Bias*—The precision and bias of the free water test method have not been determined.

7. Fineness

7.1 *Significance and Use*—This test method covers a procedure for determining the fineness of gypsum and gypsum plasters and is used to determine compliance with gypsum and gypsum plaster specifications. The degree of correlation between the results of this test method and service performance has not been determined.

7.2 *Apparatus*:

7.2.1 *Sieves*, as required by a particular gypsum or gypsum plaster specification, in accordance with Specification E 11.

7.2.2 *Balance*, capable of weighing not less than 1000 g at a precision of 0.1 g.

7.2.3 *Mechanical Sieving Machine*

7.3 *Procedure*:

7.3.1 Determine fineness by sieving a known weight of the dried sample through sieves of the specified sizes (Note 3 and Note 4). The size of the sample to be used in determining fineness depends upon the particle size of the material. If the material will pass a 6.3-mm (¼-in.) sieve, a 100-g sample will be sufficient; if the largest particles are more than 25 mm (1 in.) in diameter, use not less than a 1000-g sample. With these limitations the size of sample to be used is left to the discretion of the operator. Shake the sample through each sieve with as little abrasion as possible (Note 5). Weigh the amount of material retained on each sieve and calculate the fineness, expressed as a percentage of the weight of the original sample.

7.3.2 If a mechanical sieving machine is used, make a series of tests at one minute intervals, with each type of material and sieve size combination to be used. When not more than 0.5 g passes each sieve in a one minute interval, use that time as the standard sieving time for that test procedure on that machine.

NOTE 3—For suggested method of sieving gypsum through a 45-µm (No. 325) sieve, see Appendix X1.

NOTE 4—The sizes of the sieves to be used are given in the specifications of ASTM covering the particular product in question.

NOTE 5—When sieving by hand through a 150-µm (No. 100) sieve, use a lateral motion, and tap the side of the sieve with the palm of the hand. Continue without brushing until not more than 0.5 g passes through during 1 min of sieving. If the sieve openings become clogged, transfer the retained material temporarily to another vessel, invert the sieve over a sheet of paper on the table and tap it sharply against the table. Then transfer all the retained material back into the sieve and continue sieving.

7.4 *Report*—Report the percent by weight retained on each sieve or passing particular sieves in accordance with the material specification to the nearest full percentage point.

7.5 *Precision and Bias*—The precision and bias of the fineness test method have not been determined.

8. Normal Consistency of Gypsum Plaster

8.1 *Significance and Use*—This test method is used to determine the volume of water required for mixing gypsum plaster when performing the setting time and compressive strength tests.

8.2 *Apparatus*:

8.2.1 *Modified Vicat Apparatus*—The modified Vicat apparatus (Fig. 1)[4] shall consist of a bracket, *A*, bearing a movable brass rod, *B*, 6.3 mm in diameter and of suitable length to fit the Vicat Bracket. A conical plunger made of aluminum with an apex angle of 53° 08 min and a height of 45 mm shall be attached to the lower end of the rod. The total weight of the rod and plunger shall be 35 g. A means shall be provided for adding a weight, *G*, to the rod, increasing the total weight to 50 g. The rod shall be capable of being held in any position by a screw, *E*. The rod shall have a mark, *D*, midway between the ends which moves under a scale, *F*, graduated in millimetres, attached to the bracket, *A*.

8.2.2 *Mold*—The conical ring mold shall be made of a noncorroding, nonabsorbent material, and shall have an inside diameter of 60 mm at the base and 70 mm at the top and a height of 40 mm.

8.2.3 *Base Plate*—The base plate for supporting the ring mold shall be of plate glass and about 100 mm square.

8.2.4 *Balance*, capable of weighing not less than 500 g at a precision of 0.1 g.

[4] This method is described by Kuntze, R. A., "An Improved Method for the Normal Consistency of Gypsum Plasters," *ASTM Bulletin No. 246*, ASTM, May 1960, p. 35.

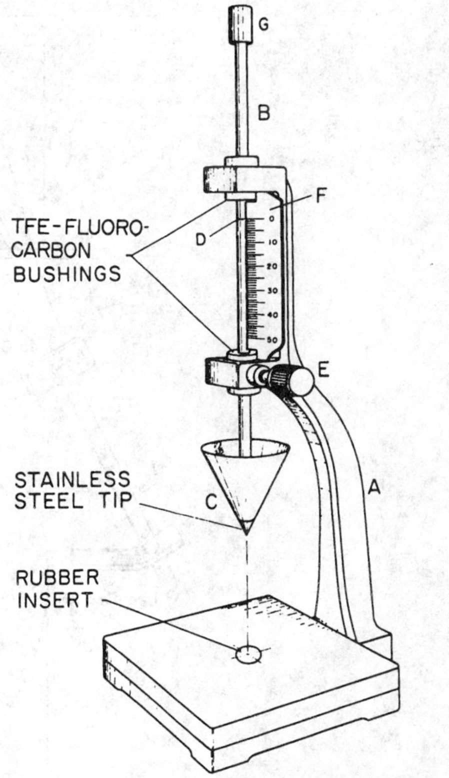

A—Hinged support bracket
B—Plunger support rod
C—Conical plunger
D—Graduation mark
E—Lock screw
F—Modified scale
G—Weight

FIG. 1 Modified Vicat Apparatus (Conical Plunger Method)

8.2.5 *Graduated Cylinder*, shall be made to deliver the indicated volume at 20°C (68°F) and shall be subdivided to not less than 1 mL.

8.3 *Procedure*:

8.3.1 Clean the plunger, mold, and base plate of the modified Vicat apparatus. Apply a thin coat of petroleum jelly or other suitable lubricant to the upper surface of the base plate in order to prevent leaks during the test.

8.3.2 Sift a weighed quantity of the sample (200 to 300 g as required to fill the mold) into a known volume of water (See 5.1). If the plaster is unretarded, add to the mixing water 0.2 g of sodium citrate per 100 g of sample. After allowing the sample to soak for 2 min, stir the mixture for 1 min to an even fluidity. Pour this sample into the ring mold, work slightly to remove air bubbles, and then strike off flush with the top of the mold. Wet the plunger of the modified Vicat apparatus and lower it to the surface of the sample at approximately the center of the mold. Read the scale and release the plunger immediately. After the rod has settled, read the scale again. Make three determinations on each mix, care being taken to have the mold completely filled and the plunger clean and wet.

8.3.3 Test gypsum molding plaster and gypsum gauging plaster with different amounts of water until a penetration of 30 ± 2 mm is obtained. Use the rod and plunger for this determination weighing 35 g.

8.3.4 Test all gypsum mixtures containing aggregates with different amounts of water until a penetration of 20 ± 2 mm is obtained. Add weight to the rod and plunger to bring the total weight to 50 g for these determinations.

8.3.5 Mix gypsum neat plaster with standard sand (see 5.2) in the ratio of 200 g of sand to 100 g of plaster before testing for normal consistency.

8.4 *Report*:

Report the normal consistency as the average number of millilitres of water required to be added to 100 g of the dry mixture to the nearest whole millilitre.

8.5 *Precision and Bias*—The precision and bias of the test method for normal consistency of gypsum plaster have not been determined.

9. Normal Consistency of Gypsum Concrete

9.1 *Significance and Use*—This test method is used to determine the volume of water required for mixing gypsum concrete when performing the setting time and compressive strength tests.

9.2 *Apparatus*:

9.2.1 *Consistometer* (Fig. 2)—The consistometer consists of a conical vessel made of noncorroding, nonabsorbent material, and having an inside diameter of 9 in. (229 mm) at the top and 1¾in. (44.5 mm) at the bottom, and a height of 5½ in. (139.7 mm). It shall be provided with a sliding gate at the bottom and supported so that the bottom is 4 in. (102 mm) above the base plate. The base plate shall be of plate glass, free of scratches and about 18 in. (457 mm) square.

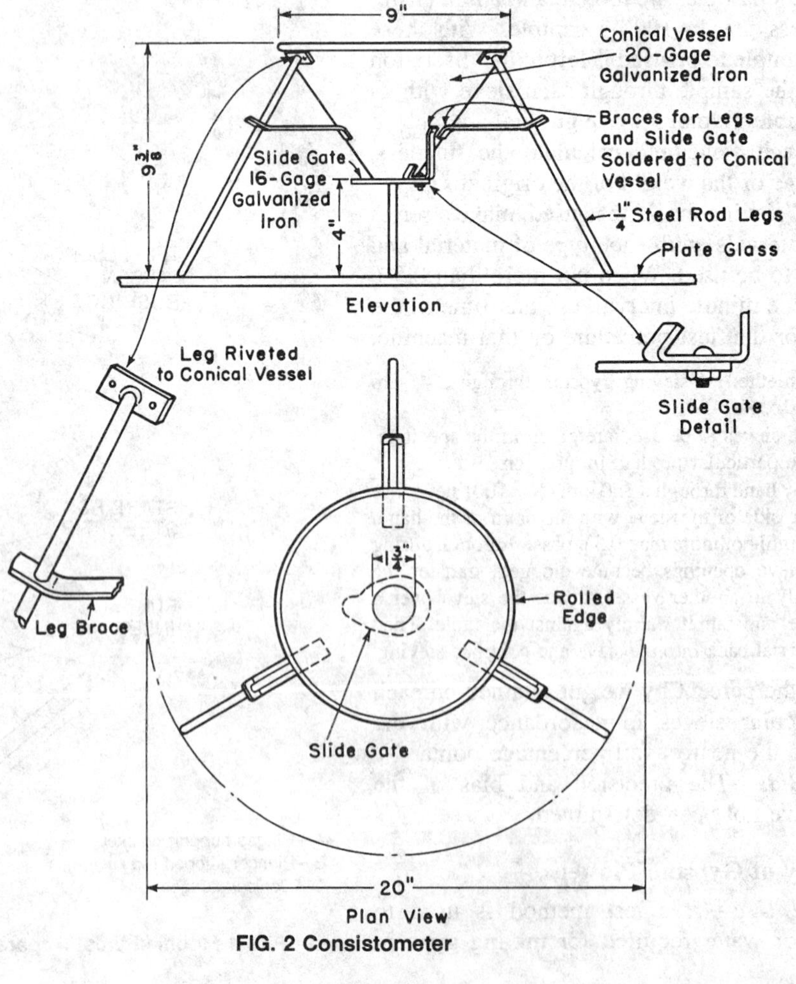

FIG. 2 Consistometer

9.2.2 *Balance*, capable of weighing not less than 2000 g at a precision of 1 g.

9.3 *Procedure*:

9.3.1 Clean and dry the consistometer and the base plate and close the sliding gate.

9.3.2 Sift 2000 g of the sample into a known volume of water (see 5.1) to which 1.0 g of sodium citrate has previously been added. After allowing the sample to soak for 1 min, stir the mixture for 3 min to an even fluidity. Pour the mixture into the consistometer until level with the top. Then rapidly and completely open the sliding gate, allowing the mixture to run out onto the base plate. When the sliding gate is opened, take care to avoid jarring the consistometer.

9.3.3 Measure the resulting patty along its major and minor axes and determine the average diameter.

9.3.4 Test gypsum concrete with different amounts of water until a patty diameter of 380 ± 13 mm (15 ± ½in.) is obtained.

9.4 *Report*—Report the normal consistency as the number of millilitres of water to the nearest whole millilitre required to be added to 100 g of the gypsum concrete.

9.5 *Precision and Bias*—The precision and bias of the test method for normal consistency of gypsum concrete have not been determined.

10. Setting Time

10.1 *Significance and Use*—This test method is used to determine the setting time of gypsum plaster and gypsum concrete, and is used to determine compliance with gypsum plaster and gypsum concrete specifications. Since variable job conditions influence the setting time, the degree of correlation between this test method and service performance has not been determined.

10.1.1 This test method does not determine the time of complete hydration of gypsum plaster or gypsum concrete. To determine the time of complete hydration see Setting Time, (Temperature Rise Method,) Section 11.

10.2 *Apparatus*:

10.2.1 *Vicat Apparatus*—The Vicat apparatus (Fig. 3) shall consist of a frame, *A*, bearing a movable rod, *B*, weighing 300 g, one end, *C*, the plunger end, being 10 mm in diameter for a distance of not less than 50 mm, the other end having a removable needle, *D*, 1 mm in diameter and 50 mm in length. The rod, *B*, shall be reversible, shall be able to be held in any position by a screw, *E*, and shall have an adjustable indicator, *F*, that moves over a scale (graduated in millimetres) attached to the frame, *A*. The paste shall be held in a rigid conical ring, *G*, resting on a glass plate, *H*, about 100 mm square. The ring shall be made of a noncorroding, nonabsorbent material and shall have an inside diameter of 70 mm at the base and 60 mm at the top, and a height of 40 mm.

10.2.2 In addition, the Vicat apparatus shall conform to the following requirements:

Weight of plunger	300 ± 0.5 g
Diameter of larger end of plunger	10 ± 0.05 mm
Diameter of needle	1 ± 0.05 mm
Inside diameter of ring at bottom	70 ± 3 mm
Inside diameter of ring at top	60 ± 3 mm
Height of ring	40 ± 1 mm
Graduated scale	The graduated scale, when compared with a standard scale accurate to within 0.1 mm at all points, shall not show a deviation at any point greater than 0.25 mm.

10.2.3 *Balance*, capable of weighing not less than 500 g at a precision of 0.1 g.

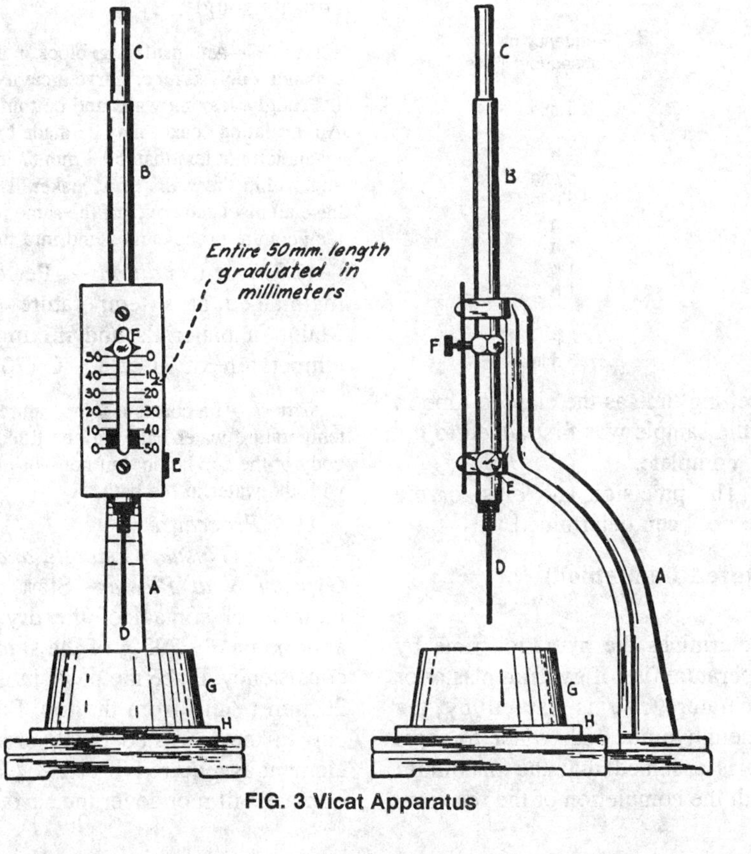

FIG. 3 Vicat Apparatus

10.2.4 *Stopwatch or Clock*, for determining setting time.

10.2.5 *Temperature and Humidity Controlled Cabinet*, capable of maintaining a temperature of 21 ± 1°C (70 ± 2°F) and a relative humidity of not less than 85 %.

10.3 *Procedure*:

10.3.1 *Gypsum Concrete and All Gypsum Plasters, Except Gypsum Neat Plaster*—Start the timing of the test approximately at the moment of contact of the dry material with the water (see 5.1). Mix enough paste to fill the conical ring. For the quantity of water and directions for mixing, see Section 8, but do not add retarder. Place the conical ring on the base plate and fill the ring with the paste level with the top of the ring. Position the end of the needle so that it just touches the top surface of the paste and tighten the screw. Release the screw so that the needle falls freely into the paste. After each penetration, wipe the needle clean, and move the base plate and ring slightly so that the needle will not strike the same place twice. Test the sample at intervals according to the table in 10.3.3. Setting time is complete when the needle no longer penetrates to the bottom of the paste. When setting times are expected to be longer than ½h, store the test specimens in the cabinet described in 10.2.5.

10.3.2 *Gypsum Neat Plaster*—Test gypsum neat plaster for setting time as mixed with three parts by weight of standard sand (see 5.2). Mix dry a 100-g sample of the gypsum neat plaster and 300 g of the sand, and then add sufficient water to produce a normal consistency. For the quantity of water (see 5.1) and mixing instructions, see Section 8, but do not use retarder. Stir for 1 min, to an even, lump-free consistency. Place the mortar in the conical ring and test for setting time as described in 10.3.1

10.3.3 *Frequency of Testing*—Test the materials according to the following schedule:

Kind of Material	First Test	Frequency of Subsequent Tests
Molding plaster	15 min	5 min
Keene's cement:		
Standard	15 min	1 h
Quick set	15 min	5 min
Gypsum concrete	15 min	5 min
Ready mixed plaster	1½ h	1 h
Neat plaster	2 h	1 h
Wood-fibered plaster	1½ h	1 h
Bond plaster	1½ h	1 h
Gauging plaster:		
Slow set	40 min	2 h
Quick set	15 min	5 min

10.4 *Report*—Report the setting time as the elapsed time in minutes from the time when the sample was first added to the water to the time when set is complete.

10.5 *Precision and Bias*—The precision and bias for the setting time test method have not been determined.

11. Setting Time (Temperature Rise Method)

11.1 *Significance and Use*:

11.1.1 This test method determines the hydration time by measuring the maximum temperature rise of gypsum plaster or gypsum concrete. The temperature rise is caused by the reaction of calcium sulfate hemihydrate with water to form calcium sulfate dihydrate. It is assumed that the maximum temperature rise coincides with the completion of the reaction. This test method is not used to determine compliance with specifications for gypsum plaster or gypsum concrete unless it is specified in the purchase contract.

11.1.2 The setting time determined by this test method is usually longer than the setting time determined in Section 10.

11.1.3 The degree of correlation between this test method and service performance has not been determined.

11.2 *Apparatus*:

11.2.1 *Potentiometer*—A single- or multiple-channel recording potentiometer or thermistor bridge shall be used to record the temperature change of the sample under test. The chart speed shall be not less than 1 in. (25 mm)/h. Imprints recording the temperature shall not be longer than 1 min apart for each sample.

11.2.2 *Temperature Sensors*—Temperature changes shall be indicated by either thermocouples or thermistors which shall be movable or in a fixed position. Temperature sensing elements shall be of such capacity and sensitivity that, when connected to the recording potentiometer, a temperature change of 0.5°C (1°F) in the sample shall be recorded on the chart.

11.2.3 *Sample Cups*—Cups with a capacity of 175 to 275 mL (6 to 9 oz) shall be used. The cup containing the mixture under test shall be placed inside a matching cup held in an insulated block or beaker, as specified in 11.2.4. The temperature sensor, in this case, shall be positioned 25 to 33 % of the distance up from the bottom and between the inner and outer cup. Alternately, the cup containing the test mixture shall be positioned over a spring-loaded sensor to ensure close contact with the bottom of the cup.

11.2.4 *Insulated Block or Beaker*, with walls and bottom having a thermal resistance of not less than R6 (1.1 ($m^2 \cdot K$)/ W) (6 ($h \cdot ft^2 \cdot °F$)/Btu), and a centered cavity into which the sample cup fits snugly.

NOTE 6—An insulating block may be constructed from a block of expanded polystyrene. Carve a cavity in the block so that the sample cup fits snugly, leaving walls and bottom not less than 40 mm (1.5 in.) thick. An insulating beaker may be made by stuffing glass fiber or mineral wool insulation not less than 50.8 mm (2 in.) thick around the sample container and within the walls of a beaker. The insulated block or beaker should have an insulated cover of the same insulating material through which the temperature probe can extend into the test material.

11.3 *Test Conditions*— Perform tests in a room or cabinet maintained at a temperature of 21 ± 1.1°C (70 ± 2°F). Maintain materials and mixing water used for the test at a temperature of 21 ± 1.1°C (70 ± 2°F).

NOTE 7—If a constant-temperature cabinet is not available, a constant-temperature water bath may be fitted with a cover which will admit the body of the cup holder but not its rim, so that the cup holder is in contact with the water in the bath.

11.4 *Procedure*:

11.4.1 *Gypsum Concrete and All Gypsum Plasters, Except Gypsum Neat Plaster*—Start the timing of the test at the moment of contact of the dry material with the water. Mix approximately 200 g of the sample to make a paste of normal consistency. Place the paste in a clean dry sample cup to about 20 mm (¾in.) from the top. Place the filled cup in an empty cup in the insulated block or beaker and adjust the sensing element as required in 11.2.3. Cover the cup with a watch glass or plastic film or cover the surface of the paste with mineral oil to a depth of approximately 5 mm (¼ in.).

11.4.2 *Gypsum Neat Plaster*—Prepare the plaster in accordance with 10.3.2 and test in accordance with 11.4.1.

11.5 *Report*—Report the setting time as the elapsed time in minutes from the time when the sample was first added to the water to the time when maximum temperature rise is attained.

11.6 *Precision and Bias*—The precision and bias for the setting time, temperature rise method, have not been determined.

12. Compressive Strength

12.1 *Significance and Use*—This test method is used to determine the compressive strength of gypsum plaster and gypsum concrete and is used to determine compliance with applicable specifications. The correlation between this test method and service performance has not been determined.

12.2 *Apparatus*:

12.2.1 *Specimen Molds*—Molds for making test specimens shall be 2-in. (50.8 mm) split cube molds made of noncorrodible material and shall be sufficiently rigid to prevent spreading during molding. The molds shall have not more than three cube compartments and shall be separable into not more than two parts. When assembled, the parts of the molds shall be held firmly together, and dimensions shall conform to the following requirements: Interior faces shall be plane surfaces with a maximum variation of 0.001 in. (0.03 mm) for new molds and 0.002 in. (0.05 mm) for old molds; distance between opposite faces, and height of the molds, measured separately for each cube compartment, shall be 2 ± 0.005 in. (50.8 ± 0.13 mm) for new molds or 2 ± 0.020 in. (50.8 ± 0.51 mm) for old molds. Angle between adjacent interior faces and between interior faces and top and bottom planes of the mold shall be 90 ± 0.5°, measured at points slightly removed from the intersection of the faces.

12.2.2 *Testing Machine*, either the hydraulic or the screw type, with sufficient opening between the upper bearing surface and the lower bearing surface of the machine for the use of verifying apparatus. The load applied to the test specimen shall be indicated with an accuracy of ±1.0 %. The load indicating device shall have a means of indicating maximum load. The upper bearing shall be a spherically seated, hardened metal block firmly attached at the center of the upper head of the machine. The center of the sphere shall lie at the center of the surface of the block in contact with the specimen. The block shall be free to tilt in any direction. A hardened metal bearing block shall be used beneath the specimen to minimize wear of the lower platen of the machine. The lower bearing block shall have concentric circles at ¼-in. (6-mm) intervals centered on the block to facilitate accurate centering of the test specimen. The bearing block surfaces intended for contact with the specimen shall have a Rockwell hardness number not less than 60 HRC. These surfaces shall not depart from plane surfaces by more than 0.0005 in. (0.013 mm) when the blocks are new and 0.001 in. (0.025 mm) after use.

12.2.3 *Drying Oven*, a mechanical convection oven maintained at a temperature of 45 ± 3°C and a relative humidity not more than 50 %.

12.2.4 *Desiccator*, containing magnesium perchlorate or calcium chloride that is capable of being tightly sealed and of such a size as to hold not less than one set of 6 cubes.

12.3 *Test Specimens* :

12.3.1 Mix sufficient sample at normal consistency to produce not less than 1000 mL (34 fluid oz) of mixed mortar and cast into six specimen molds (see 12.2.1). Premix neat gypsum plaster dry with two parts by weight of standard sand (see 5.2). For the quality and quantity of water, see 5.1 and Section 8, but do not add retarder. Place the required amount of water in a clean 2-L (2-qt) mixing bowl.

12.3.2 For all gypsum plasters except gypsum concrete, add the required amount of dried plaster and allow to soak for 2 min. Mix vigorously (about 150 complete circular strokes per minute) for 1 min with a metal spoon, stiff-bladed spatula or mechanical mixer (see Note 8) to produce a mortar of uniform consistency. For gypsum concrete, soak for 1 min, and stir vigorously (about 150 complete circular strokes per minute) with a large spoon or mechanical mixer for 3 min.

NOTE 8—The Hobart N-50 mixer has been found satisfactory for this purpose.

12.3.3 If setting times as determined in accordance with Section 10 are more than the maximum limits shown in Table 1, discard the cubes and adjust the setting time by adding freshly ground gypsum accelerator to the plaster, or add molding plaster to Keene's cement. Do not use more gypsum accelerator or molding plaster than 1 % of the dry weight of plaster or Keene's cement.

12.3.4 Coat the molds with a thin film of mineral oil or other mold release agent and place on an oiled glass or metal plate. Place a layer of mortar about 25 mm (1 in.) in depth in each mold and puddle ten times across the mold between each pair of opposite faces with a 25 mm (1 in.) wide spatula to remove air bubbles. Fill the molds to a point slightly above the tops of the molds, by the same filling and puddling procedure used for the first layer. Also fill the conical mold for the Vicat apparatus described in 10.2.1 and 10.2.2 and determine the setting time (see 10.3). As soon as the mortar or paste has set, cut off the excess to a plane surface flush with the top of the mold, using a stiff broad knife or similar implement.

12.3.5 Place the filled molds in moist air (90 to 100 % relative humidity). Remove the cubes from the molds at any time after they are thoroughly hardened, but retain in the moist air for not less than a total of 16 h. Place the cubes in an oven as specified in 12.2.3. Dry the cubes at 45 ± 3°C to constant

TABLE 1 Setting-Time Limits for Mortar

Kind of Material	Setting Time, min	
	min	max
Molding plaster	20	140
Keene's cement:		
Standard	40	120
Quick set	20	40
Gypsum concrete	20	40
Ready-mixed plaster	90	120
Neat plaster (with 2 parts sand)	120	150
Wood-fibered plaster	90	120
Bond plaster	120	150
Gauging plaster:		
Slow set	40	120
Quick set	20	40
Veneer plaster	30	90

weight as determined by weighing once each day, but do not dry longer than 7 days. Then place the cubes in a desiccator as described in 12.2.4 for a minimum of 16 h before testing. Test the cubes immediately on removal from the desiccator.

12.4 *Procedure*—As soon as the cube specimens have been dried (12.3.5), determine their compressive strengths. (If density is to be determined, do so at this time. (See Section 13)). Position the cubes in the testing machine so that the load is applied on surfaces formed by faces of the molds, not on top and bottom. Apply the load continuously and without shock, at a constant rate within the range 15 to 40 psi/s (100 to 280 kPa/s). Do not exceed 40 psi/s (275 kPa/s) after 50 % of the maximum load has been reached.

12.5 *Report*—Report the average compressive strength as the compressive strength of the material in psi (Mpa) rounded to the nearest whole psi (0.1 Mpa) except that if the strengths of one or two of the cubes vary more than 15 % from the average of the six, discard them and report the compressive strength as the average of the remaining specimens. In case the compressive strengths of three or more cubes vary more than 15 % from the average, discard the results and repeat the test.

12.6 *Precision and Bias*—The precision and bias for the compressive strength test method have not been determined.

13. Density

13.1 *Significance and Use*—This test method is used to determine the density of cast gypsum concrete and is used to determine compliance with the specification for gypsum concrete. The correlation between this test method and service performance has not been determined.

13.2 *Apparatus*:

13.2.1 *Balance*, capable of weighing not less than 1500 g to the nearest gram.

13.3 *Procedure*—Determine the density of gypsum concrete by weighing the six cubes to the nearest gram after drying and cooling, as described in 12.3.5, and before determining the compressive strength.

13.4 *Calculation*— Calculate the density by multiplying the total weight of the six cubes by 0.0794.

13.5 *Report*—Report the density in pounds per cubic foot to the nearest 1 lb/ft^3(10 kg/m^3).

13.6 *Precision and Bias*—The precision and bias for the density test method have not been determined.

14. Keywords

14.1 compressive strength; density; free water; fineness; gypsum; gypsum concrete; gypsum plasters; normal consistency; setting time

APPENDIX

(Nonmandatory Information)

X1. ALCOHOL WASH METHOD OF SIEVING GYPSUM AND GYPSUM PRODUCTS

X1.1 Scope

X1.1.1 This appendix covers a sieve test for fine gypsum and gypsum products using an alcohol wash method.

X1.2 Significance and Use

X1.2.1 It is impractical to sieve dry gypsum through a fine sieve, and water cannot be used as a washing agent without introducing errors due to hydration or solution. This method is used when a purchase order specifies fineness through fine sieves.

X1.3 Apparatus

X1.3.1 Sieves, complying with Specification E 11.

X1.3.2 Mechanical Sieving Machine.

X1.3.3 Vessel, with a diameter not less than 50 mm greater than the sieves, and a depth not less than 75 mm greater than the height of the sieves.

X1.3.4 Balance, capable of reading to a precision of 0.01 g.

X1.4 Reagents

X1.4.1 Isoprophyl alcohol, 99%. The alcohol may be reused after decanting or filtering, provided its strength does not fall below 95%

X1.4.1.1 **Caution**—When using isoprophyl alcohol, keep fire away, provide good ventilation, and avoid excessive inhalation of vapor.

X1.5 Calibration and Standardization

X1.5.1 See 7.3.2 for method of calibrating a mechanical sieving machine.

X1.6 Procedure

X1.6.1 Place the alcohol in a vessel (X1.3.3) to a depth of 50 mm more than the height of the sieve. Weigh approximately 50 g of the sample to a precision of 0.01 g and place on a 45-μm or coarser sieve up to and including the 150 μm (No. 100).

X1.6.2 Lower the sieve into the alcohol until the specimen is wetted throughout. Lift the sieve out of the alcohol, with a swirling motion, permitting the alcohol to drain through the specimen back into the vessel. Repeat this operation not less than eight times, until the alcohol passes freely through the sieve and the specimen is essentially free from fines. Wash the residue retained on the sieve with about 100 mL of clear alcohol, and then blot the bottom of the sieve with a soft, dry, lint-free cloth. Dry the sieve containing the residue at 45 ± 3°C.

X1.6.3 Shake the sieve on a mechanical shaker for 2 min ± 2 s. If desired, to permit simultaneous determinations of coarser fractions, transfer the residue after drying and before shaking to a series of coarser sieves up to and including the 150 μm nested in sequential order above the 45-μm sieve. Weigh the material retained on each sieve to the nearest 0.01 g.

X1.7 Calculation

X1.7.1 Calculate the percent of the original specimen retained on or through each sieve as follows:

$$\%\ retained = (B/A) \cdot 100$$

$$\%\ through = \frac{A-B}{A} \cdot 100$$

where:

A = Weight of original specimen, and
B = weight of specimen retained on sieve.

X1.8 Report

X1.8.1 Report the percent retained or through each sieve to the nearest 0.1 percent.

X1.9 Precision and Bias

X1.9.1 The precision and bias of the alcohol wash method of sieving gypsum and gypsum products have not been determined.

SUMMARY OF CHANGES

Committee C-11 has identified the location of selected changes to this standard since the last issue (C 473–93) that may impact the use of this standard.

(1) Nonmandatory terms such as may, can, and others listed by BOCA were eliminated throughout.
(2) Drying temperatures for plaster products were all changed to be consistent with Test Methods C 471M.
(3) Most volume and weight measurements were converted to metric units.
(4) The numbering system was revised to be more consistent with other standards.

Designation: C 473 – 97$^{\epsilon 1}$

Standard Test Methods for Physical Testing of Gypsum Panel Products[1]

This standard is issued under the fixed designation C 473; the number immediately following the designation indicates the year of original adoption or, in the case of revision, the year of last revision. A number in parentheses indicates the year of last reapproval. A superscript epsilon (ε) indicates an editorial change since the last revision or reapproval.

ε1 NOTE—6.1 was corrected editorially in July 1997.

1. Scope

1.1 These test methods cover the physical testing of gypsum panel products.

1.2 The test methods appear in the following order:

	Section
Flexural Strength (Method A)	7
Core, End and Edge Hardness (Method A)	8
Nail Pull Resistance (Method A)	9
Flexural Strength (Method B)	11
Core, End and Edge Hardness (Method B)	12
Nail Pull Resistance (Method B)	13
Humidified Deflection	14
End Squareness	15
Nominal Thickness	16
Recessed- or Tapered Edge Depth	17
Width	18
Length	19
Water Resistance of Core-Treated Water-Repellent Gypsum Panel Products	20
Surface Water Resistance of Gypsum Panel Products with Water-Repellent Surfaces	21

1.3 The text of these test methods references notes and footnotes that provide explanatory material. These notes and footnotes, excluding those in tables and figures, shall not be considered as requirements of these test methods.

1.4 *This standard does not purport to address all of the safety concerns, if any, associated with its use. It is the responsibility of the user of this standard to establish appropriate safety and health practices and determine the applicability of regulatory limitations prior to use.*

2. Referenced Documents

2.1 *ASTM Standards:*

C 11 Terminology Relating to Gypsum and Related Building Materials and Systems[2]

E 4 Practices for Force Verification of Testing Machines[3]

3. Terminology

3.1 Definitions used in these test methods are in accordance with Terminology C 11.

3.2 *Descriptions of Terms Specific to This Standard:*

3.2.1 *edge, n*—the paper-bound edge, or long edge, as manufactured.

3.2.2 *ends, n*—the end perpendicular to the paper-bound edge or long edge.

4. Sampling

4.1 For use in these test methods, a sample shall consist of not less than 0.25 % of the number of gypsum panel products in a shipment, but not less than three gypsum panel products. Test specimens shall be taken from the sample lot as indicated in each method.

5. Conditioning

5.1 Condition specimens to constant weight at a temperature of 85 ± 15°F (29.5 ± 8.5°C) and relative humidity of 50 ± 2 % (Note 1). Test specimens individually after constant weight is reached and within 10 min upon removal from the conditioning chamber.

NOTE 1—Laboratory test results indicated that conditioning temperatures may range from 70 to 100°F (21 to 38°C) without significantly affecting the results. However, relative humidity is critical. This is the reason for testing specimens within 10 min upon removal from the conditioning chamber.

Method A—Constant Rate of Loading

INTRODUCTION

The following procedures are applicable to so-called "shot machines" that utilize lead or steel shot that applies a load to the specimen at a constant stress rate or any other device that applies the load at a constant stress rate.

6. Apparatus—Method A

6.1 *Testing Machine*—Apply the test load at a uniform rate of 1 lbf/s (4.45 N/s), with a permissible variation of ±10 % in the rate. Apply the load with a device independent of manual control of the rate of loading during an individual sample test (see Fig. 1). The deformation of the specimen controls the cross head speed.

7. Flexural Strength (Method A)

7.1 *Summary of Test Method*—Flexural properties of gypsum panel products are evaluated by supporting the specimen near the ends and applying a transverse load midway between the supports.

7.2 *Significance and Use*—This test method covers a procedure for evaluating the ability of gypsum panel products to withstand flexural stresses during handling or use of the material. This test method is used to determine compliance with gypsum panel product specifications. The degree of correlation between this test and service performance has not been fully determined.

7.3 *Apparatus*—See 6.1.

7.3.1 *Specimen Supports*—All specimen bearing supports and load surfaces shall be true, shall engage the full width of the test specimen, shall be of such strength they will not deform under the test load, and shall be rounded to a radius of 0.125 in. (3.2 mm). The supports shall be spaced 14 ±

---

[1] These test methods are under the jurisdiction of ASTM Committee C-11 on Gypsum and Related Building Materials and Systems and are the direct responsibility of Subcommittee C11.01 on Specifications and Test Methods for Gypsum Products.

Current edition approved April 10, 1997. Published June 1997. Originally published as C 473 – 61. Last previous edition C 473 – 95.

[2] *Annual Book of ASTM Standards*, Vol 04.01.

[3] *Annual Book of ASTM Standards*, Vol 03.01 and 04.02.

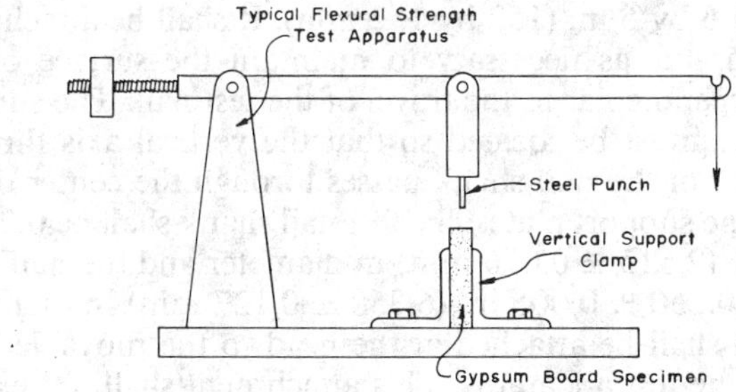

FIG. 1 Apparatus Schematic for Core, End, and Edge Hardness Test

0.016 in. (356 ± 0.4 mm) on centers measured at the points of surface contact with the specimen. The supports shall be attached to a plate that shall be attached rigidly to the test apparatus. The supports shall be parallel to each other and perpendicular to the long dimension of the test specimen. The bearing surface of the end supports shall be on the same plane to ensure contact along the entire bearing surfaces. Supports shall be not less than 12 in. (305 mm) in length. When a load is being applied, the loading bar shall be in contact with the full width of the specimen. The load-bearing surface shall be parallel to the bearing surface of the specimen supports. The specimen and supports shall be located so that the vertical axis through the center of the load-bearing surface passes through the specimen at one half of the specimen width and one half of the distance between the supports.

7.4 *Sampling*—Sampling shall be in accordance with Section 4.

7.5 *Test Specimens:*

7.5.1 Cut four specimens, each 12 in. (305 mm) by approximately 16 in. (406 mm) from each gypsum panel product in the sample, two having the 16-in. dimension parallel to the edge and two having the 16-in. dimension perpendicular to the edge. Cut the specimens not less than 4 in. (102 mm) away from the ends and edges of the gypsum panel product except where the product is 24-in. or less in width, where it is impossible to do so.

7.5.2 Condition specimens in accordance with Section 5.

7.6 *Procedure:*

7.6.1 Use the apparatus described in 7.3, supporting each specimen centrally on fixed parallel bearings, 14 in. (356 mm) on centers, and apply the load on a similar bearing midway between the supports. Position perforated gypsum lath so the load is not applied over perforations.

7.6.2 For specimens with the long dimension parallel to the edge, test one specimen from each gypsum panel product face up and the other face down. For specimens with the long dimension perpendicular to the edge, test one specimen from each gypsum panel product face up and the other face down.

7.6.3 For gypsum panel products with surfacing material, the load is determined at the time of transverse failure of the surfacing material and core through the thickness of the specimen. The specimen shall not show a break in the bond between the surfacing and the core before transverse failure of the gypsum panel product. For gypsum panel products without surfacing material, the load is determined at the time of transverse failure through the thickness of the specimen.

7.7 *Report*—Calculate and report the average breaking load in pound-force or newtons for each test condition, rounded to the nearest 1 lbf (N). The test conditions are: (*1*) parallel, face up; (*2*) parallel, face down; (*3*) perpendicular, face up; and, (*4*) perpendicular, face down.

7.8 *Precision and Bias*—The precision and bias of this test method has not been determined.

8. Core, End, and Edge Hardness (Method A)

8.1 *Summary of Test Method*—The core, end, and edge hardness of gypsum panel products is evaluated by determining the force required to push a steel punch into the area of test.

8.2 *Significance and Use*—This test method covers a procedure for evaluating the ability of the gypsum panel product core, ends, and edges to resist crushing during handling or use of the material. This test method is used to determine compliance with gypsum panel product specifications. The degree of correlation between these test methods and service performance has not been determined.

8.3 *Apparatus*—See 6.1. Clamp or secure the specimen to the base of the test apparatus shown in Fig. 1 with the face of the specimen perpendicular to the base of the test apparatus and parallel to the movement of the steel punch. Secure the steel punch to the moving head of the test apparatus, positioned so that its center axis is parallel with the line of travel and so that the punch is inserted into the core equidistant from the surfaces of the specimen. Use a steel punch 0.099 ± 0.003 in. (2.515 ± 0.076 mm) in diameter and more than ½ in. (12.7 mm) long with the bearing surface machined to a 90° angle in reference to the punch center line (see Fig. 1 for test setup schematic).

8.4 *Sampling*—Sampling shall be in accordance with Section 4.

8.5 *Test Specimens:*

8.5.1 *Specimens for Core Hardness*—Using a table or band saw, cut three specimens from the center of each of the three gypsum panel product samples approximately 12 by not less than 3 in. (305 by 76 mm), with the 12-in. dimension perpendicular to the edges of the products. The smaller, not less than 3 in., dimension will vary depending on the specific punch length and opening in the testing machines. Place the sample in the tester, with the loading arm in a horizontal position and with the punch resting on the specimen. Saw cuts along the 12-in. dimension shall be smooth and square with the smaller, not less than 3-in., dimension, as well as with the specimen surface.

8.5.2 *Specimens for End Hardness*—Cut specimens the same size and method as described in 8.5.1 from each opposite mill-cut end of the gypsum panel product, with the 12 in. (305 mm) dimension perpendicular to the edges, and run tests only on the mill-cut end of the specimens.

8.5.3 *Specimens for Edge Hardness*—Cut specimens the same size and method as described in 8.5.1 from the opposite edges of the gypsum panel product, with the 12 in. (305 mm) dimension of the specimens parallel with the edges. For gypsum panel products with surfacing material, cut a ¼ to ½-in. (6 to 13-mm) strip of surfacing material from the edge of the test specimens exposing the core for test. Remove as little of the core as possible when removing the surfacing material. Prepare the first test area 2 ± ½ in. (51 ± 13 mm) from one edge of the specimen, with the next two test areas approximately 4 in. (102 mm) apart.

8.5.4 Condition specimens in accordance with Section 5.

8.6 *Procedure*—Using the apparatus in 8.3, clamp or secure the specimen in a fixed vertical position on its 12 in. (305 mm) dimension edge (see Fig. 1 for test setup schematic. If necessary, modify the flexural strength test equipment to conduct the core, end, and edge hardness test). Place this test surface not less than 1 in. (25 mm) above the grip of the specimen holder. Conduct three tests spaced approximately 4 in. (102 mm) apart on each specimen, with the first test area 2 ± ½ in. (51 ± 13 mm) from one edge of the specimen. Position the steel punch over the test area and apply the load. Report as the core, end, or edge hardness the load required to push the steel punch a distance of ½ in. (13 mm) into the core of the specimen.

8.6.1 *Procedure for Core Hardness*—Test either of the 12-in. (305-mm) saw cut edges.

8.6.2 *Procedure for End Hardness*—Test the 12-in. (305-mm) mill-cut end.

8.6.3 *Procedure for Edge Hardness*—Test areas along the 12-in. (305-mm) edge. For products with surfacing material, test where the surfacing material has been removed.

8.7 *Report*—Report the core, end, and edge hardness as the average load in pounds-force or newtons rounded to the nearest 1 lbf (N) for the two specimens from all of the panels, a total of six test results. Discard any single test value that varies from the average by more than 15 % and repeat the test on the same sample to obtain a new value.

8.8 *Precision and Bias*—The precision and bias of this test method has not been determined.

9. Nail Pull Resistance (Method A)

9.1 *Summary of Test Method*—The ability of gypsum panel products to resist nail pull-through is evaluated by determining the load required to push a standard nail head through the product.

9.2 *Significance and Use*—This test method covers a procedure for evaluating the ability of gypsum panel products to resist nail pull-through by determining the load required to force a standard nail head through the product. This test method is used to determine compliance with gypsum panel product specifications. The degree of correlation between these test methods and service performance has not been determined.

9.3 *Apparatus*—See 6.1. Figure 2 shows a schematic of a nail pull resistance apparatus adaptable to a typical flexural strength apparatus used commonly in the gypsum industry. The specimen support plate shall have a 3-in. (76-mm) diameter hole in the center. The support plate shall be not less than 6 by 6 in. (152 by 152 mm). It shall be attached to the apparatus as necessary to maintain the surface of the plate perpendicular to the travel of the test nail. The support plate also must be located so that the vertical axis through the center of the nail shank passes through the center of the hole in the support plate. The test nail shank shall be 0.099 ± 0.003 in. (2.515 ± 0.076 mm) in diameter and the nail head shall be 0.250 ± 0.005 in. (6.350 ± 0.127 mm) in diameter. The nail shall be attached at the head to the movable cross head of the test apparatus. This attachment shall not exceed the diameter of the nail head and shall be placed directly over the head. It also shall be attached so that the vertical axis through the nail shank is parallel with the travel.

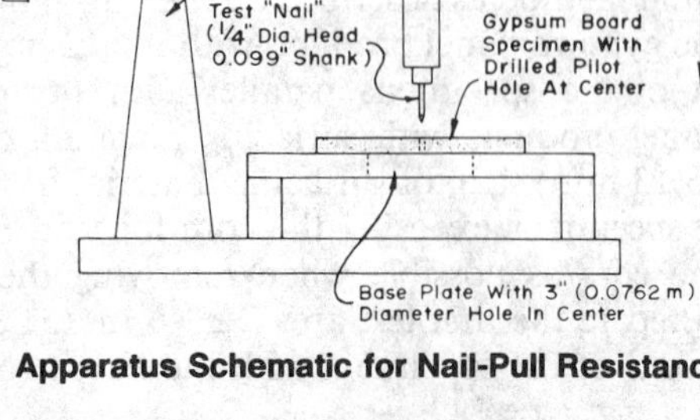

FIG. 2 Apparatus Schematic for Nail-Pull Resistance Test

9.4 *Sampling*—Sampling shall be in accordance with Section 4.

9.5 *Test Specimens:*

9.5.1 Cut five specimens 6 by 6 in. (152 by 152 mm), from a strip cut across the width of each 48-in. (1220-mm) gypsum panel product starting not less than 4 in. (102 mm) in from the edge. Cut a total of five specimens from more than one full width strip from gypsum panel products less than 38 in. (965 mm) wide. Do not cut the specimen closer than 3 in. (76 mm) from the product edges.

9.5.2 Drill a 7⁄64-in. (2.778-mm) diameter pilot hole through the thickness of the gypsum panel product in the center of each test specimen and perpendicular to the surface.

9.5.3 Condition specimens in accordance with Section 5.

9.6 *Procedure*—Using the apparatus in 9.3, center the specimen face up on the base plate and manually lower the test nail, inserting the nail shank into the pilot hole, until the nail head is flush with the surface of the specimen. Apply the load until the maximum load is achieved. Normally, this will occur as the nail head breaks through the gypsum panel product surface.

9.7 *Report*—Calculate the average maximum load in pounds-force or newtons, rounded to the nearest 1 lbf (N). Discard any single test that varies from the average by more than 15 %. If more than 15 % of the test values are discarded, repeat the test.

9.8 *Precision and Bias*—The precision and bias of this test method has not been determined.

Method B—Constant Rate of Cross Head Speed

INTRODUCTION

The following procedures are applicable to so-called universal test machines. These instruments often use electronic load cells and data acquisition equipment to gather information about the specimen being tested. The instruments generally in use move the cross head at a constant speed, as opposed to having the deformation on the specimen control cross head speed, which would be a true constant rate of strain instrument.

10. Apparatus

10.1 *Testing Machine*, of any type that is of sufficient capacity and that is capable of providing a rate of loading at a strain rate of 1.0 in./min ± 5 s (25 mm/min ± 5 s). The load platform of the test equipment shall be of sufficient size to accept the specimen holder and load applying apparatus.

The testing machine shall be power operated and shall apply the load continuously and without shock.

10.2 The test apparatus shall be verified according to Practices E 4, the three sections titled "Basis of Verification," "Corrections," and "Time Interval Between Verifications."

10.3 *Load-Indicating Equipment*, capable of reading to 1-lbf (4.45-N) increments, shall be provided for the test apparatus. A means shall be provided to indicate the maximum load attained during the test.

11. Flexural Strength (Method B)

11.1 *Summary of Test Method*—Flexural properties of gypsum panel products are evaluated by supporting the specimen near the ends and applying a transverse load midway between the supports.

11.2 *Significance and Use*—This test method covers a procedure for evaluating the ability of gypsum panel products to withstand flexural stresses during handling or use of the material. This test method is used to determine compliance with gypsum panel product specifications. The degree of correlation between this test and service performance has not been fully determined.

11.3 *Apparatus*—See Section 10.

11.3.1 *Specimen Supports*—All bearing supports and load surfaces shall be true, shall engage the full width of the test specimen, shall be of such strength they will not deform under the test load, and shall be rounded to a radius of 0.125 in. (3.2 mm). The specimen supports shall be spaced 14 ± 0.016 in. (356 ± 0.41 mm) on centers measured at the points of surface contact with the specimen. The supports shall be attached to a plate that shall be rigidly attached to the test apparatus. Test specimen supports shall be parallel to each other and perpendicular to the long dimension of the test specimen. The bearing surface of the end supports shall be on the same plane to ensure contact along the entire bearing surfaces. Supports shall be not less than 12 in. (305 mm) in length. When a load is being applied, the loading bar shall be in contact with the full width of the specimen. The load-bearing surface shall be parallel to the bearing surface of the specimen supports. The specimen and specimen supports shall be located so that the vertical axis through the center of the load-bearing surface passes through the specimen at one half of the specimen width and one half of the distance between the supports (see Fig. 3).

11.4 *Sampling*—Sampling shall be in accordance with Section 4.

11.5 *Test Specimen:*

11.5.1 Cut four specimens, each 12 in. (305 mm) by approximately 16 in. (406 mm) from each gypsum panel product in the sample, two having the 16-in. dimension parallel to the edge and two having the 16-in. dimension perpendicular to the edge. Cut the specimens not less than 4 in. (102 mm) away from the ends and edges of the gypsum panel product except where the product is 24-in. or less in width, where it is impossible to do so.

11.5.2 Condition specimens in accordance with Section 5.

11.6 *Procedure:*

11.6.1 Use the apparatus in 11.3, supporting each specimen centrally on the fixed parallel supports, 14-in. (356-mm) on centers, and apply the load on a similar bearing midway between the supports. Position perforated gypsum lath so the load is not applied over perforations.

11.6.2 For specimens with the long dimension parallel to the edge, test one specimen from each gypsum panel product face up and the other face down. For specimens with the long dimension perpendicular to the edge, test one specimen from each gypsum panel product face up and the other face down.

11.6.3 For gypsum panel products with surfacing material, the load is determined at the time of transverse failure of the surfacing material and core through the thickness of the specimen. The specimen shall not show a break in the bond between the surfacing and the core before transverse failure of the gypsum panel product. For gypsum panel products without surfacing material, the load is determined at the time of transverse failure through the thickness of the specimen.

11.7 *Report*—Calculate and report the average breaking load in pounds-force or newtons for each test condition, rounded to the nearest 1 lbf (N). The test conditions are: (*1*) parallel, face up; (*2*) parallel, face down; (*3*) perpendicular, face up; and, (*4*) perpendicular, face down.

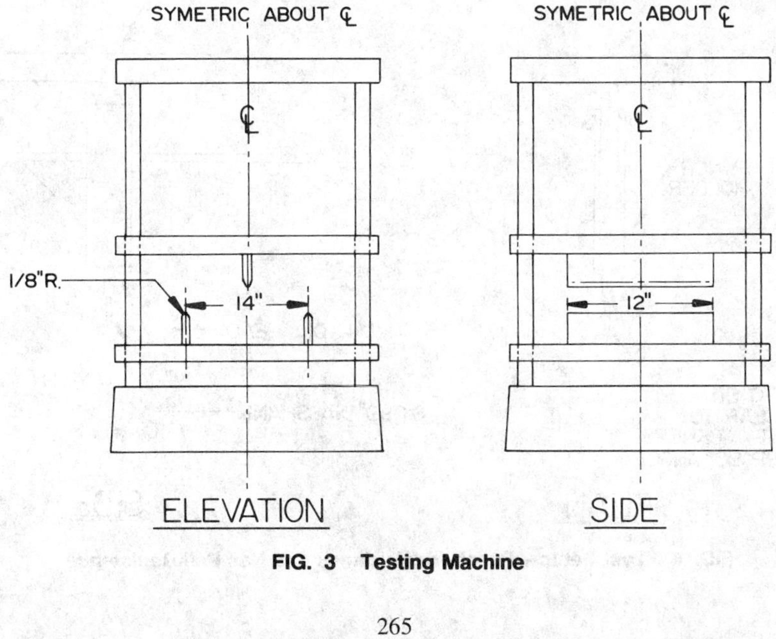

FIG. 3 Testing Machine

11.8 *Precision and Bias*—The precision and bias for this test method has not been determined.

12. Core, End, and Edge Hardness (Method B)

12.1 *Summary of Test Method*—The core, end, and edge hardness of gypsum panel products is evaluated by determining the force required to push a steel punch into the area of test.

12.2 *Significance and Use*—This test method covers a procedure for evaluating the ability of the gypsum panel product core, ends, and edges to resist crushing during handling or use of the material. This test method is used to determine compliance with gypsum panel product specifications. The degree of correlation between this test method and service performance has not been fully determined.

12.3 *Apparatus*—See Section 10.

12.3.1 A means of securing the specimen to the base of the test apparatus shall be provided. The face of the specimen shall be perpendicular to the base of the test apparatus and parallel to the movement of the steel punch. The steel punch shall be secured to the moving head of the test apparatus. It shall be positioned so that its center axis is parallel with the line of travel. It also shall be positioned so that the punch is inserted into the core equidistant from the surfaces of the specimen. The steel punch shall be 0.099 ± 0.003 in. (2.515 ± 0.076 mm) in diameter and more than ½ in. (12.7 mm) long with the bearing surface machined to a 90° angle in reference to the punch center line.

NOTE 2—See Figs. 4 and 5 for examples of test setup configurations. Other configurations may be used as long as the load is applied to the specimen as specified.

12.4 *Sampling*—Sampling shall be in accordance with Section 4.

12.5 *Test Specimens:*

12.5.1 *Specimens for Core Hardness*—Using a table or band saw, cut a specimen not less than 12 by 3 in. (305 by 76 mm) from the center of each gypsum panel product. The 12-in. (305-mm) dimension shall be perpendicular to the edges of the gypsum panel product. Cuts along the 12 in. (305 mm) dimension shall be smooth and perpendicular to the face of the specimen.

12.5.2 *Specimens for End Hardness*—Cut a specimen not less than 12 by 3 in. (305 by 76 mm) from each mill-cut end of each gypsum panel product. Cut the 12-in. (305-mm) dimension perpendicular to the edges of the gypsum panel product.

12.5.3 *Specimens for Edge Hardness*—Cut a specimen not less than 12 by 3 in. (305 by 76 mm) from each edge of each gypsum panel product. Cut the 12-in. (305-mm) dimension perpendicular to the edge of the gypsum panel product. For gypsum panel products with surfacing material, cut a ¼ to ½-in. (6 to 13-mm) strip of surfacing material from the edge of the test specimens exposing the core for test. Remove as little of the core as possible when removing the surfacing material. Prepare the first test area 2 ± ½ in. (51 ± 13 mm) from one edge of the specimen, with the next two test areas spaced approximately 4 in. (102 mm) apart.

12.5.4 Condition specimens in accordance with Section 5.

12.6 *Procedure*—Using the apparatus in 12.3, clamp or secure the specimen in a fixed vertical position on its 12 in. (305 mm) dimension edge. If necessary, modify the flexural strength test equipment to conduct the core, end, and edge hardness test. Place this test surface not less than 1 in. (25 mm) above the grip of the specimen holder. After the specimen is in place, zero the load indicator. Conduct three tests spaced approximately 4 in. (102 mm) apart on each specimen, with the first test area 2 ± ½ in. (51 ± 13 mm) from one edge of the specimen. Position the steel punch over the test area and apply the load at a strain rate of 1 in./min ± 5 s (25 mm/min ± 5 s). Report as the core, end, or edge hardness the load required to push the steel punch a distance of ½ in. (13 mm) into the core of the specimen.

12.6.1 *Procedure for Core Hardness*—Test either of the 12-in. (305-mm) saw cut edges.

12.6.2 *Procedure for End Hardness*—Test the 12-in. (305-mm) mill-cut end.

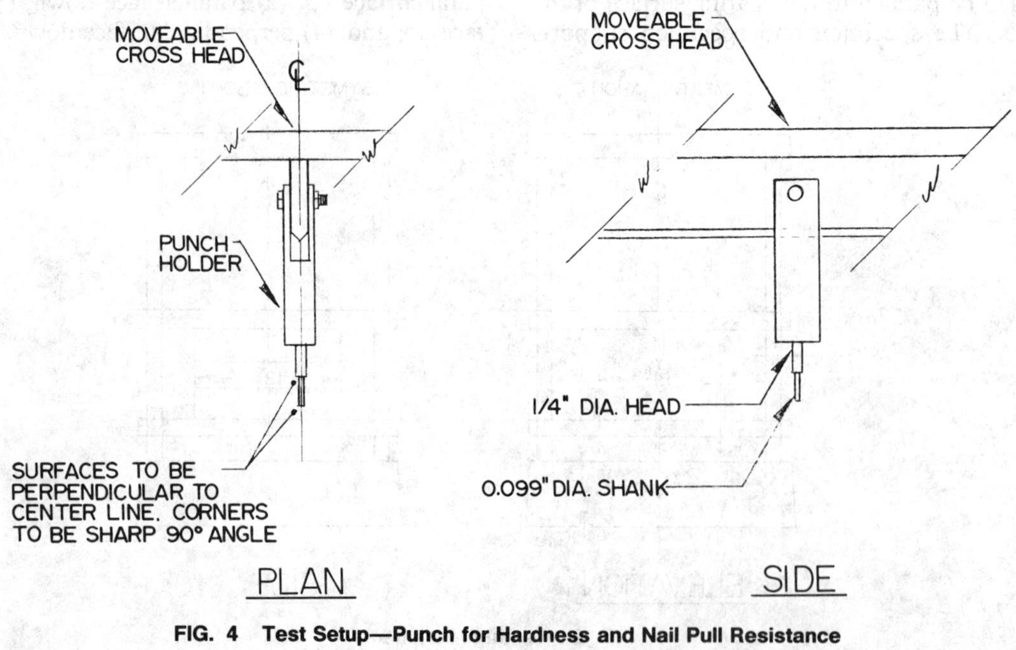

FIG. 4 Test Setup—Punch for Hardness and Nail Pull Resistance

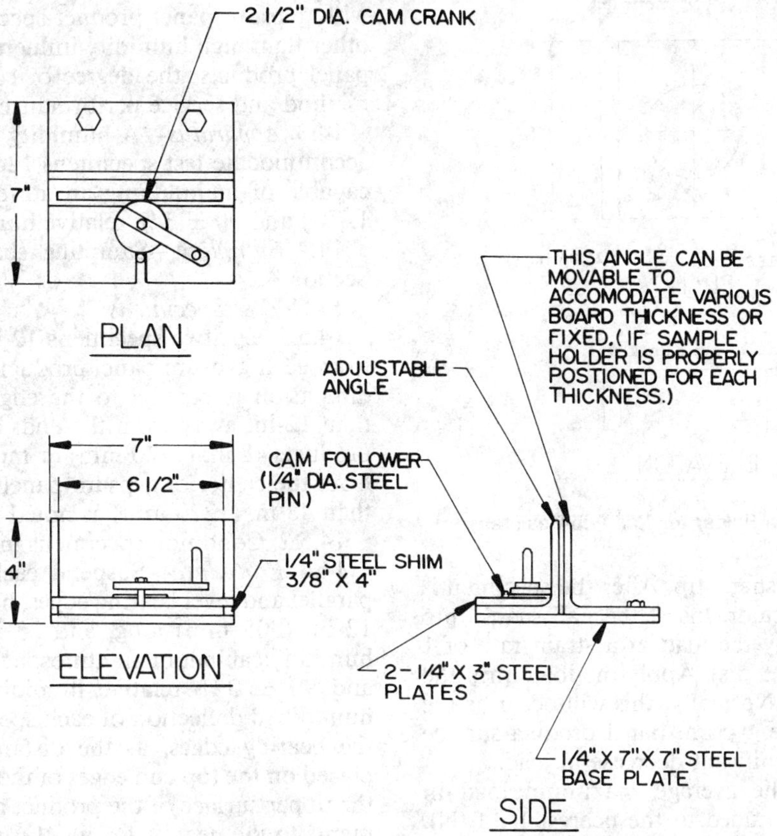

FIG. 5 Test Setup—Sample Holder for Hardness Test

12.6.3 *Procedure for Edge Hardness*—Test areas along the 12-in. (305-mm) edge. For products with surfacing material, test where the surfacing material has been removed.

12.7 *Report*—Report the core, end, and edge hardness as the average load in pounds-force or newtons rounded to the nearest 1 lbf (N) for the two specimens from all of the panels, a total of six test results. Discard any single test value that varies from the average by more than 15 % and repeat the test on the same sample to obtain a new value.

12.8 *Precision and Bias*—The precision and bias of this test method has not been determined.

13. Nail Pull Resistance (Method B)

13.1 *Summary of Test Method*—The ability of gypsum panel products to resist nail pull-through is evaluated by determining the load required to push a standard nail head through the product.

13.2 *Significance and Use*—This test method covers a procedure for evaluating the ability of gypsum panel products to resist nail pull-through by determining the load required to force a standard nail head through the product. This test method is used to determine compliance with gypsum panel product specifications. The degree of correlation between these test methods and service performance has not been determined.

13.3 *Apparatus*—See Section 10.

13.3.1 *Specimen Support Plate*, shall be true and have a 3-in. (76-mm) diameter hole in the center. The support plate shall be not less than 6 by 6 in. (152 by 152 mm). It shall be attached to the apparatus as necessary to maintain the surface of the plate perpendicular to the travel of the test nail. The support plate also must be located so that the vertical axis through the center of the nail shank passes through the center of the hole in the support plate. The test nail shank shall be 0.099 ± 0.003 in. (2.515 ± 0.076 mm) in diameter and the nail head shall be 0.250 ± 0.005 in. (6.350 ± 0.127 mm) in diameter. The nail shall be attached at the head to the movable cross head of the test apparatus. This attachment shall not exceed the diameter of the nail head and shall be placed directly over the head. It shall also be attached so that the vertical axis through the nail shank is parallel with the travel. See Figs. 4 and 6 for examples of the test set-up configuration.

13.4 *Sampling*—Sampling shall be in accordance with Section 4.

13.5 *Test Specimens:*

13.5.1 Cut five specimens 6 by 6 in. (152 by 152 mm) from a strip cut across the width of each 48-in. (1220-mm) gypsum panel product starting not less than 4 in. (102 mm) in from the edge. Cut a total of five specimens from more than one full width strip from gypsum panel products less than 38 in. (965 mm) wide. Do not cut the specimen closer than 3 in. (76 mm) from the product edges.

13.5.2 Drill a 7⁄64-in. (2.778-mm) diameter pilot hole through the thickness of the gypsum panel product in the center of each test specimen and perpendicular to the surface.

13.5.3 Condition specimens in accordance with Section 5.

13.6 *Procedure*—Using the apparatus in 13.3, place the specimen on the specimen support plate so that the drilled

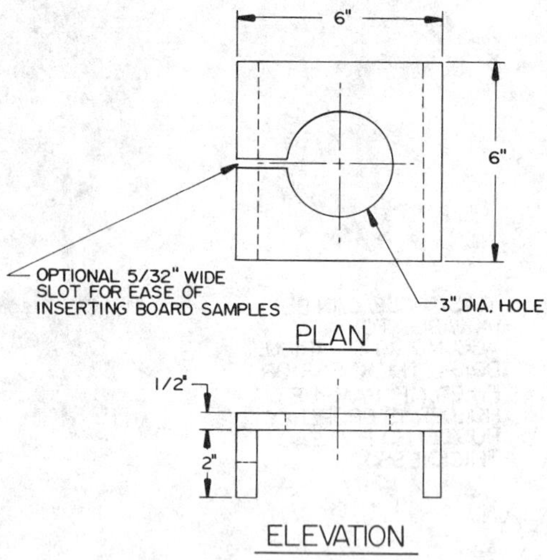

FIG. 6 Test Setup—Sample Holder for Nail Pull Resistance

hole is aligned with the nail shank tip. After the specimen is in place, zero the load indicator. Insert the nail shank into the specimen hole and apply the load at a strain rate of 1 in./min ± 5 s (25 mm/min ± 5 s). Apply the load until the maximum load is achieved. Normally, this will occur as the nail head breaks through the gypsum panel product surface and is followed by a very significant decrease in load.

13.7 *Report*—Calculate the average maximum load in pounds-force or newtons, rounded to the nearest 1 lbf (N). Discard any single test that varies from the average by more than 15 %. If more than 15 % of the test values are discarded, repeat the test.

13.8 *Precision and Bias*—The precision and bias of this test method has not been determined.

14. Humidified Deflection

14.1 *Summary of Test Method*—The humidified deflection of gypsum panel products is evaluated by supporting a specimen that has been cut with the long dimension perpendicular to the machine direction, supported horizontally, and subjected to high humidity for 48 h.

14.2 *Significance and Use*—This test method covers a procedure for evaluating the deflection of gypsum panel products when horizontally suspended and subjected to high humidity. This test method is used to determine compliance with gypsum panel product specifications. Since job factors other than high humidity influence the deflection of gypsum panel products, the degree of correlation between this test method and service performance has not been determined.

14.3 *Apparatus*—A humidity cabinet of sufficient size to accommodate test specimens 12 by 24 in. (305 by 610 mm) capable of maintaining an atmosphere of 90 ± 3°F (32 ± 1.7°C) and 90 ± 3 % relative humidity for 48 h.

14.4 *Sampling*—Sampling shall be in accordance with Section 4.

14.5 *Test Specimens:*

14.5.1 Cut two specimens 12 by 24 in. (305 by 610 mm) from each gypsum panel product in the sample so the 12-in. dimension is parallel to the edges. Cut specimens not less than 12-in. away from the ends and edges of gypsum panel products 48 in. (1220 mm) or more in width. Cut specimens from the center of gypsum panel products that are narrower than 48 in. (1220 mm) as noted in Fig. 7.

14.5.2 Condition specimens in accordance with Section 5.

14.6 *Procedure*—Suspend each specimen face down on parallel and level bearing edges, having ⅛-in. (3.2-mm) radii, 12 in. (305 mm) long, and 23 in. (584 mm) apart in the humidity cabinet in an atmosphere of 90 ± 3°F (32 ± 1.7°C) and 90 ± 3 % relative humidity for 48 h. Measure the humidified deflection of each specimen, while suspended on the bearing edges, as the distance between a straightedge placed on the top end edges of the gypsum panel product and the upper surface of the product at its center. Take measurements to the nearest 1/16 in. (1.6 mm).

14.7 *Report*—Report the humidified deflection as the average of the specimens to the nearest ⅛ in. (3 mm).

14.8 *Precision and Bias*—The precision and bias of this test method has not been determined.

15. End Squareness

15.1 *Summary of Test Method*—The end squareness of gypsum panel products is evaluated by drawing a line perpendicular to the edge of a full width specimen to its opposite edge and measuring the length of the side of the triangle formed.

15.2 *Significance and Use*—This test method covers a procedure for determining the end squareness of gypsum panel products. This test method is used to determine compliance with gypsum panel product specifications. The

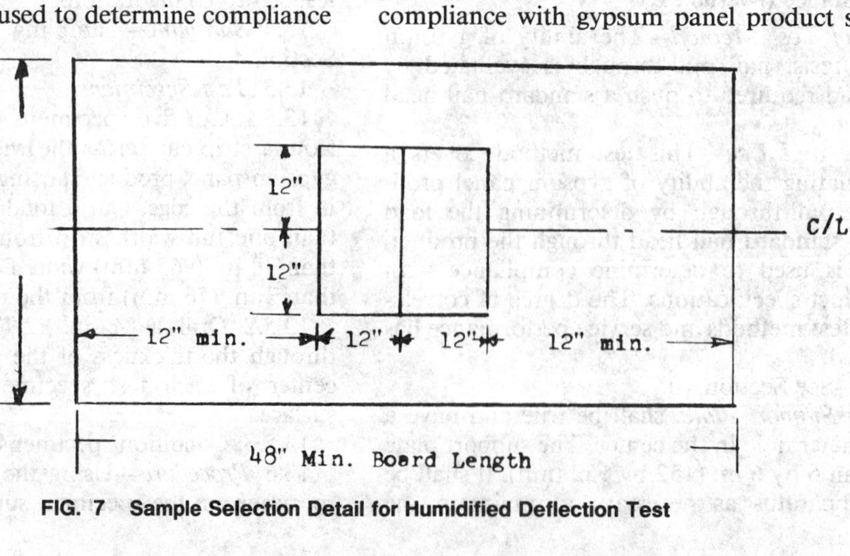

FIG. 7 Sample Selection Detail for Humidified Deflection Test

degree of correlation between this test method and service performance has not been determined.

15.3 *Apparatus*—A T-shaped or L-shaped square for drawing or testing right angles. The long edge of the square shall be not less than 48 in.

15.4 *Sampling*—Sampling shall be in accordance with Section 4.

15.5 *Test Specimens*—Use full-size gypsum panel products with undamaged ends for test specimens.

15.6 *Procedure*—Place one edge of the square flush against the edge of the gypsum panel product with the apex of the square at one corner of the product. Determine the distance *b*, to the nearest 1/16 in. (1.6 mm) at the specified distance *a*. Distance is measured from the apex of the square (see Fig. 8).

15.7 *Report*—Report the average of distance *b* to the nearest 1/8 in. (3.2 mm).

15.8 *Precision and Bias*—The precision and bias of this test method has not been determined.

16. Thickness

16.1 *Summary of Test Method*—The nominal thickness of gypsum panel products is evaluated by measuring the thickness of the specimen with a micrometer.

16.2 *Significance and Use*—This test method covers a procedure for determining the thickness of gypsum panel products, and is used to determine compliance with gypsum panel product specifications. The degree of correlation between this test method and service performance has not been determined.

16.3 *Apparatus*—A micrometer gage for determining the thickness measurement to 0.001 in. (0.025 mm). The micrometer shall be a 1-in. (25.4-mm) hand micrometer with 1/4-in. (6.35-mm) diameter anvils graduated to 0.01 in. (0.25 mm).

16.4 *Sampling*—Sampling shall be in accordance with Section 4.

16.5 *Test Specimens:*

16.5.1 Use full-size gypsum panel products for test specimens.

16.5.2 Mark the product to be tested, score on both sides, and break at 12 in. (305 mm) from each end, discarding the two ends prior to testing. Retain the remaining part of each gypsum panel product as a test specimen.

16.6 *Procedure*—Mark measurement points at approximately equal points across the width at each end of the gypsum panel product excluding the tapered, beveled, or recessed edges if applicable. Engage the micrometer firmly, but do not indent the surfacing material of the gypsum panel product. Measure the thickness of the gypsum panel product with the micrometer not less than 1/2 in. (13 mm) from cut ends. Take a minimum of not less than seven readings at each end of the 54-in. (1370-mm) wide gypsum panel product, six readings at each end of the 48-in. (1220-mm) wide gypsum panel product, five readings at each end of the 32-in. (813-mm) wide gypsum panel product, four readings at each end of the 24-in. (610-mm) wide gypsum panel product, or three readings at each end of the 16-in. (406-mm) wide gypsum panel product.

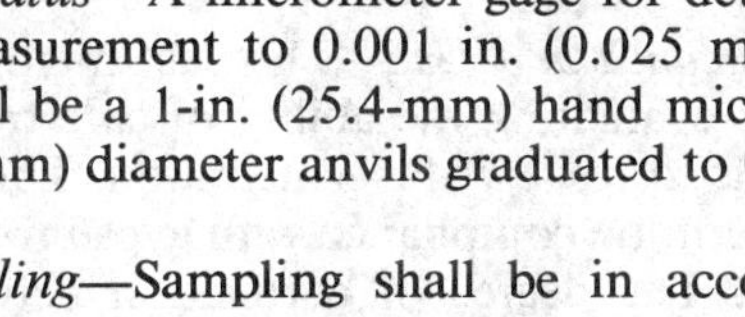

FIG. 8 End Squareness Determination

16.7 *Report*—Calculate the nominal thickness of each specimen as the average of the individual micrometer reading taken on each gypsum panel product. Calculate and report the sample thickness as the average measured thickness of all the specimens in the sample to the nearest 1/64 in. (0.4 mm) and local variations in any one specimen to the nearest 1/32 in. (0.8 mm).

16.8 *Precision and Bias*—The precision and bias of this test method has not been determined.

17. Recessed- or Tapered-Edge Depth

17.1 *Summary of Test Method*—The recessed- or tapered-edge depth of gypsum panel products is evaluated by comparing the thickness near the edge of the specimen to the thickness in the field of the specimen.

17.2 *Significance and Use*—This test method provides a procedure for determining the depth of the recess or taper on the edge of gypsum panel products board and is used to determine compliance with gypsum panel product specifications. The degree of correlation between this test method and service performance has not been determined.

17.3 *Apparatus*—Use the micrometer gage described in 16.3. Engage the micrometer firmly but do not indent the surfacing material. Take all readings to the nearest 0.001 in. (0.025 mm) of specimen thickness.

17.4 *Sampling*—Sampling shall be in accordance with Section 4.

17.5 *Test Specimens:*

17.5.1 Use full-size gypsum panel products for test specimens.

17.5.2 Mark the individual products to be tested, score on both sides, and break at 12 in. (305 mm) from each end, discarding the two ends prior to testing. Retain the remaining part of each gypsum panel product as a test specimen.

17.6 *Procedure:*

17.6.1 At one corner of the specimen, determine thickness 1/2 in. (12.7 mm) from the cut end and 6 in. (152 mm) from the edge.

17.6.2 At the same corner, also determine the thickness 6 in. (152 mm) from the cut end and 3/8 in. (9.5 mm) from the edge.

17.6.3 Subtract the specimen thickness 6 in. (152 mm) from the cut end and 3/8 in. (9.5 mm) from the edge, as in 17.6.2, from the specimen thickness at 1/2 in. (12.7 mm) from the cut end and 6 in. (152 mm) from the edge, as in 17.6.1, to determine the individual recessed or taper-edge depth at that specimen corner.

17.6.4 Repeat the procedure in 17.6.1, 17.6.2, and 17.6.3 on the remaining three corners of this specimen and on the remaining two specimens of the sample.

17.6.5 Determine the average recessed or tapered-edge depth from all the individual recessed or tapered-edge depths, as in 17.5.

17.6.6 Determine the variation of any individual recessed or tapered-edge depth from the average as determined in 17.6.5 using the subtraction method.

17.6.7 Reject or discard any individual recessed- or tapered-edge depth calculation, as in 17.6.3, if the variation from average, as in 17.6.6, exceeds 0.050 in. (1.27 mm). To constitute a valid test do not discard more than three individual calculations.

17.7 *Report*—Report the average recessed- or tapered-edge depth as in 17.6.5 after any adjustments for exclusions of data as in 17.6.6 or 17.7.7 to the nearest 0.001 in. (0.025 mm).

17.8 *Precision and Bias*—The precision and bias of this test method has not been determined.

18. Width

18.1 *Summary of Test Method*—The width of gypsum panel products is evaluated by measuring the edge-to-edge dimension of a full size specimen.

18.2 *Significance and Use*—This test method provides a procedure for determining the width of gypsum panel products and is used to determine compliance with gypsum panel products specifications. The degree of correlation between this test method and service performance has not been determined.

18.3 *Apparatus*—Any accurate measuring device capable of measuring the width of the gypsum panel product to the nearest 1/32 in. (1 mm).

18.4 *Sampling*—Sampling shall be in accordance with Section 4.

18.5 *Test Specimens*—Use full-size gypsum panel products for test specimens.

18.6 *Procedure:*

18.6.1 Measure the width of an individual gypsum panel product by placing the measuring device perpendicular to the edges of the gypsum panel product when measurements are taken. The width of the specimen is the distance between extremities of the edges, so in cases where the edge or edges are beveled (slope in or out) with the face or back of the specimen, or are round, determine the overall distance. One exception is tongue and groove edge specimens, the width of which is the distance between the rip of the tongue and the bottom of the groove.

18.6.2 The width of an individual gypsum panel product is the maximum of two readings taken not less than 48 in. (1220 mm) apart and about 6 in. (152 mm) away from both ends of the specimen. Cut off 6 in. (152 mm) from both ends before taking measurements at ends of tongue and groove-edge specimens.

18.7 *Report*—Report the width as the average width of not less than three specimens to the nearest 1/32 in. (1 mm).

18.8 *Precision and Bias*—The precision and bias of this test method has not been determined.

19. Length

19.1 *Summary of Test Method*—The length of gypsum panel products is evaluated by measuring the end-to-end dimension of a full size specimen.

19.2 *Significance and Use*—This test method provides a procedure for determining the length gypsum panel products and is used to determine compliance with gypsum panel products specifications. The degree of correlation between this test method and service performance has not been determined.

19.3 *Apparatus*—Any accurate measuring device capable of measuring the width of the gypsum panel product to the nearest 1/16 in. (2 mm).

19.4 *Sampling*—Sampling shall be in accordance with Section 4.

19.5 *Test Specimens*—Use full-size gypsum panel products for test specimens.

19.6 *Procedure:*

19.6.1 Measure the length of an individual gypsum panel product by placing the measuring device parallel with the edges when measurements are taken. The length of the specimen is the distance between extremities of the ends, so in cases where the ends are beveled, determine the overall distance.

19.6.2 The length of a specimen is the maximum of two readings taken about 3 in. (76 mm) away from both edges of the specimen.

19.7 *Report*—Report the length as the average length of not less than three specimens to the nearest 1/8 in. (3 mm).

19.8 *Precision and Bias*—The precision and bias of this test method has not been determined.

20. Water Resistance of Core-Treated Water-Repellent Gypsum Panel Products

20.1 *Summary of Test Method*—The water resistance of core-treated water-repellent gypsum panel products is evaluated by immersing the specimen in water for the specified period of time.

20.2 *Significance and Use*—This test method provides a procedure for evaluating the ability of core-treated gypsum panel products to resist water penetration. This test method is used to determine compliance with gypsum panel product specifications. The degree of correlation between this test method and service performance has not been determined.

20.3 *Apparatus:*

20.3.1 *Water Bath*, not less than 16 by 16 by 3 in. (406 by 406 by 12.7 mm) capable of maintaining the water at a constant temperature of 70 ± 5°F (21 ± 0.5°C).

20.3.2 *Glass Rods*, 0.25 in. (6 mm) in diameter, or other supports capable of keeping the specimens off the bottom of the water bath.

20.4 *Sampling*—Sampling shall be in accordance with Section 4.

20.5 *Test Specimens:*

20.5.1 Prepare a 12 by 12-in. (305 by 305-mm) test specimen cut from the center of the gypsum panel product being tested by scoring and breaking. Cut specimens approximately centered between both edges and not less than 6 in. (152 mm) from either end. Do not treat the edges of the specimen nor damage any surfacing material.

20.5.2 Condition the specimens in accordance with Section 5, except at a temperature of 70 ± 5°F (21 ± 2°C), and then weigh to within 0.5 g.

20.6 *Procedure:*

20.6.1 Position the specimen so that it rests flatwise on glass rods or other supports in the water bath with a head of 1 in. (25 mm) of water over the top of the specimen. If the specimen floats, place sufficient weight on the specimen to keep it in contact with the supports under the 1-in. (25-mm) head of water.

20.6.2 After 2 h have elapsed, remove the specimen from the bath, wipe excess water from the surfaces and edges of the specimen and weigh immediately to within 0.5 g.

20.7 *Report*—Calculate the gain in weight for each specimen as a percentage of the original weight as determined in 20.5.2 to the nearest 1 % and report the average percentage as the water absorption of the gypsum panel product.

20.8 *Precision and Bias*—The precision and bias of this test method has not been determined.

21. Surface Water Resistance of Gypsum Panel Products with Water-Repellent Surfaces

21.1 *Summary of Test Method*—The surface water resistance of gypsum panel products with water-repellent surfaces is evaluated by damming a specified area on the treated surface, filling that area with water, and allowing it to remain undisturbed for the specified time period.

21.2 *Significance and Use*—This test method provides a procedure for evaluating the ability of water-repellent surfacing of gypsum panel products to resist water penetration. This test method is used to determine compliance with gypsum panel product specifications. The degree of correlation between this test method and service performance has not been determined.

21.3 *Apparatus*—A Cobb Tester conditioned to a temperature of 70 ± 5°F (21 ± 0.5°C).

21.4 *Sampling*—Sampling shall be in accordance with Section 4.

21.5 *Test Specimens:*

21.5.1 Cut three specimens 5 in. (127 mm) square from the field of each sample piece of gypsum panel product.

21.5.2 Condition the specimens in accordance with Section 5, except at a temperature of 70 ± 5°F (21 ± 2°C), to constant weight within 0.1 % and then weigh to within 0.05 g.

21.6 *Procedure:*

21.6.1 Place the conditioned test specimens in the Cobb Tester with the side to be exposed to water facing up. Fill the Cobb ring with tap water adjusted to a temperature of 70 ± 1°F (21 ± 0.5°C), resulting in a 1 in. (25.4 mm) head of water above the test area of the specimen.

21.6.2 After 2 h have elapsed, pour the water out of the Cobb ring and remove the specimen from the Cobb Tester. Immediately remove excess water from the specimen by blotting and weigh to the nearest 0.05 g.

21.7 *Report*—Calculate average difference between the dry and wet weights of the test specimens to the nearest 0.1 g and report this as the surface water absorption of the gypsum panel product.

21.8 *Precision and Bias*—The precision and bias of this test method has not been determined.

22. Keywords

22.1 absorption testing; core hardness testing; dimension measurements; edge hardness testing; end hardness testing; flexural strength; gypsum panel products; humidified sag; nail pull testing

APPENDIX

(Nonmandatory Information)

X1. INTRODUCTION

X1.1 ASTM specifications for gypsum panel products do not contain a specification for weight. The procedure, however, is referenced in other industry standards, and is therefore retained as an appendix item.

X2. WEIGHT

X2.1 *Test Specimens*—A sample shall consist of not less than three full size pieces of the gypsum panel product.

X2.2 *Procedure*—Measure the length and width of each specimen in accordance with Sections 17 and 18. Weight the specimens on a suitable scale and record the weight to the nearest ¼ lb (0.1 kg).

X2.3 *Calculation*—Calculate the weight in accordance with the following:

$$W = K \times P/L \times B$$

where:

W = weight per calculated area, lb/1000 ft^2 (kg/100 m^2),
L = length of specimen, in. (mm),
B = width of specimen, in. (mm),
P = weight of specimen, lb (kg), and
K = 1.44×10^5 in.^{2} (9.29×10^7 mm^2).

X2.4 *Report*—Report the weight, as the average calculated weight, in lb/1000 ft^2 (kg/100 m^2) of the three gypsum specimens.

SUMMARY OF CHANGES

Committee C-11 has identified the location of selected changes to this standard since the last issue (C 473-95) that may impact the use of this standard.

(1) The terms "gypsum board" and "gypsum lath" were replaced by "gypsum panel products" throughout this standard, including the title.

(2) Terminology was revised specifically to paper to include all types of surfacing material, as well as gypsum panel products without any kind of surfacing material.

(3) Editorial revisions were made throughout bringing consistency to the various sections in Methods A and B, and to be in compliance with ASTM format.

(4) Allowances for testing 54-in. wide products were given where needed.

Designation: C 474 – 97

Standard Test Methods for Joint Treatment Materials for Gypsum Board Construction[1]

This standard is issued under the fixed designation C 474; the number immediately following the designation indicates the year of original adoption or, in the case of revision, the year of last revision. A number in parentheses indicates the year of last reapproval. A superscript epsilon (ε) indicates an editorial change since the last revision or reapproval.

1. Scope

1.1 These test methods cover the physical testing of joint compound, paper joint tape, and an assembly of joint compound and paper joint tape.

1.1.1 Joint treatment materials are specified in Specification C 475.

1.2 The test methods appear in the following order:

	Section
Joint Compound Tests:	
Viscosity	5
Shrinkage	6
Check Cracking	7
Putrefaction	8
Paper Joint Tape Tests:	
Tensile Strength	9
Dimensional Stability	10
Width	11
Thickness	12
Assemblages of Paper Joint Tape and Joint Compound:	
Bond of Paper Joint Tape to Joint Compound	13
Cracking of Joint Compound at Tape Edges	14

1.3 The values stated in inch-pound units are to be regarded as the standard. The SI (metric) values given in parentheses are approximate and are provided for information purposes only.

1.4 The text of this standard references notes and footnotes which provide explanatory material. These notes and footnotes (excluding those in tables and figures) shall not be considered as requirements of the standard.

1.5 *This standard does not purport to address all of the safety concerns, if any, associated with its use. It is the responsibility of the user of this standard to establish appropriate safety and health practices and determine the applicability of regulatory limitations prior to use.*

2. Referenced Documents

2.1 *ASTM Standards:*

C 11 Terminology Relating to Gypsum and Related Building Materials and Systems[2]

C 36 Specification for Gypsum Wallboard[2]

C 472 Test Methods for Physical Testing of Gypsum, Gypsum Plasters, and Gypsum Concrete[2]

C 475 Specification for Joint Compound and Joint Tape for Finishing Gypsum Board[2]

D 685 Practice for Conditioning Paper and Paper Products for Testing[3]

D 828 Test Method for Tensile Properties of Paper and Paperboard Using Constant Rate-of-Elongation Apparatus[3]

E 100 Specification for ASTM Hydrometers[4]

2.2 *TAPPI Standard:*

T 410 Grammage of Paper and Paperboard (Weight Per Unit Area)[5]

3. Terminology

3.1 *Definitions*—For definitions of terms relating to gypsum, see Terminology C 11.

3.2 *Descriptions of Terms Specific to this Standard:*

3.2.1 *bond, n—in joint systems,* the quality of adhesion between the paper joint tape and joint compound.

3.2.1.1 *Discussion*—A 0 % bond means that no paper fiber is adhering to the joint compound. A 100 % bond means that there is cohesive failure of the paper joint tape.

3.2.2 *check cracking, n—in joint systems,* short, narrow cracks randomly oriented in the surface of the dried joint compound.

3.2.3 *joint compound, powder, n*—A drying-type or setting-type cementitious material to be mixed with water.

3.2.4 *joint compound, ready-mix, n*—A drying-type cementitious material that is factory mixed in ready-to-use form.

4. Specimen Preparation

4.1 *Joint Compound, Powder:*

4.1.1 Mix 300 g of joint compound, powder, with approximately 150 to 160 mL of water.

4.1.2 Allow the specimen to stand for 30 min (Note 1), remix and adjust the temperature to 77 ± 2°F (25 ± 1°C) by placing the container holding the specimen in warm or cool water.

NOTE 1—Allow setting type compounds to stand for one half of their setting times, as determined by Test Methods C 472 but not more than 30 min, prior to remixing.

4.1.3 Measure the viscosity in accordance with Section 5.

4.1.4 If the measured viscosity is not between 480 and 520

[1] These test methods are under the jurisdiction of ASTM Committee C-11 on Gypsum and Related Building Materials and Systems and are the direct responsibility of Subcommittee C11.02 on Specifications and Test Methods for Accessories and Related Products.

Current edition approved April 10, 1997. Published June 1997. Originally published as C 474 – 61. Last previous edition C 474 – 96.

[2] *Annual Book of ASTM Standards*, Vol 04.01.

[3] *Annual Book of ASTM Standards*, Vol 15.09.

[4] *Annual Book of ASTM Standards*, Vol 14.03.

[5] Available from Technical Association of the Pulp and Paper Industry, Technology Park, P.O. Box 105113, Atlanta, GA 30348.

Brabender units, repeat 4.1.1 through 4.1.3 with an increase or decrease in water as necessary.

4.1.5 Record the volume of water used to adjust the viscosity to 500 ± 20 Brabender units. (Note 2)

NOTE 2—Making note of the volumes of water, in millilitres per 100 g of material, used to adjust the viscosity, will facilitate specimen preparation in other tests.

4.2 *Joint Compound, Ready-Mix:*

4.2.1 Remix joint compound, ready-mix, to reincorporate any separated ingredients. Adjust the temperature to 77 ± 2°F (25 ± 1°C) by placing the container holding the specimen in warm or cool water.

4.2.2 Measure the viscosity in accordance with Section 5.

4.2.3 If the viscosity is more than 520 Brabender units, add water to achieve a viscosity of 500 ± 20 Brabender units. (Note 2)

4.2.3.1 If the original sample viscosity is less than 480 Brabender units, test as received.

JOINT COMPOUND

5. Joint Compound Viscosity

5.1 *Significance and Use:*

5.1.1 This test method provides a procedure for measuring joint compound viscosity.

5.2 *Sampling:*

5.2.1 Sampling shall be in accordance with Specification C 475.

5.3 *Specimen Preparation:*

5.3.1 Prepare specimens in accordance with Section 4.

5.4 *Apparatus:*

5.4.1 *Viscosity Specimen Container*, metal or plastic with an open top having an inside diameter of 2½ to 3 in. (64 to 76 mm) and a height of 2½ to 3 in. (64 to 76 mm).

5.4.2 *Viscometer*[6], adjusted to operate at 78 ± 1 r/min, and with a 250 cm-g sensitivity cartridge.

5.4.3 *Viscometer Pin*, having dimensions as follows:

	in. (mm)
Shaft diameter	0.187 ± 0.015 (4.75 ± 0.38)
Pin diameter	0.094 ± 0.015 (2.39 ± 0.38)
Immersion depth (from bottom of spindle)	1.625 ± 0.015 (41.3 ± 0.38)
Length of pin projecting from shaft	0.750 ± 0.015 (19.1 ± 0.38)
Upper pin from bottom of shaft	0.313 ± 0.015 (7.95 ± 0.38)
Lower pin from bottom of shaft	0.125 ± 0.015 (3.28 ± 0.38)

5.5 *Procedure:*

5.5.1 Fill the viscosity container with the mixed specimen until level with the top of the container.

5.5.1.1 Remove all air bubbles by puddling the sample container with a spatula and sharply rapping the bottom of the container on a hard flat surface.

5.5.2 Lock the filled container in the center of the viscometer spindle platform. Raise the platform until the level of the specimen reaches the mark on the viscometer pin and lock the platform in place.

5.5.3 Start the viscometer. Read the viscosity after the pen starts to trace a straight line (usually within 1 min). If the tracing remains inconsistent, estimate the average viscosity reading.

5.6 *Report:*

5.6.1 Report the viscosity of the joint compound specimen in Brabender units.

5.7 *Precision and Bias:*

5.7.1 Precision and bias of this test method have not been determined.

6. Shrinkage

6.1 *Significance and Use:*

6.1.1 This test is used to measure the amount of shrinkage in joint compound. The degree of correlation between this test and service performance has not been determined.

6.2 *Sampling:*

6.2.1 Sampling shall be in accordance with Specification C 475.

6.3 *Specimen Preparation:*

6.3.1 Specimen preparation shall be in accordance with Section 4.

6.4 *Apparatus:*

6.4.1 *Plastic or Rubber Film*, approximately 5 by 5 in. (130 by 130 mm). Any thin, flexible film that peels clean from a partially dried patty may be used.[7]

6.4.2 *Balance*, having a sensitivity of 10 mg (Figs. 1 and 2).

6.4.3 *Beaker, Ring Stand, and Wire Cradle* (see Fig. 1).

6.4.4 *Forced Air Drying Oven*, capable of being maintained at 90 to 120°F (32 to 49°C).

6.4.5 *Spatula*, having a blade approximately 4 in. by ½ in. (100 by 13 mm).

6.4.6 *Steel-Reinforced Broad Knife*, a 5 to 8 in. (130 by 200 mm) drywall broad knife reinforced by a steel bar, 1 in. (25.4 mm) wide by ⅛ in. (3.2 mm) thick, by the knife width, attached to the back of the knife blade ¼ in. (6 mm) from the edge.

6.4.7 *Hydrometer*, having a range of 0.7 to 0.8 sp gr, in accordance with Specification E 100.

6.4.8 *Volumetric Container*, a container which has a volume between 25 and 300 cm³.

6.5 *Reagents and Materials:*

6.5.1 *Displacement Fluids.*

6.5.1.1 *Minerals Spirits*, odorless.

6.5.1.2 *Kerosine.*

6.6 *Preparation of Apparatus:*

6.6.1 *Support Plates*—Cover three plastic or glass plates with plastic or rubber film.

6.6.2 Obtain and record the tare weight of each assembly.

6.7 *Calibration:*

6.7.1 *Volumetric Container*—Determine the container volume in cubic centimetres and its tare weight in grams.

6.7.2 *Mineral Spirits, Kerosine*—Using the hydrometer, determine the specific gravity and record the result as M.

6.8 *Determination of Wet Volume:*

6.8.1 Prepare a specimen to determine the wet specific gravity by weighing the specimen in the volumetric container.

6.8.1.1 Adjust the temperature to 77 ± 2°F (25 ± 1°C) by placing the container holding the specimen in warm or cool water.

[6] The Brabender "Visco-Corder" Model VC-3, manufactured by C.W. Brabender Instruments Inc., South Hackensack, NJ has been found satisfactory.

[7] Rubber dental dam dusted with talc, polyethylene, or PTFE films have been found satisfactory for this use.

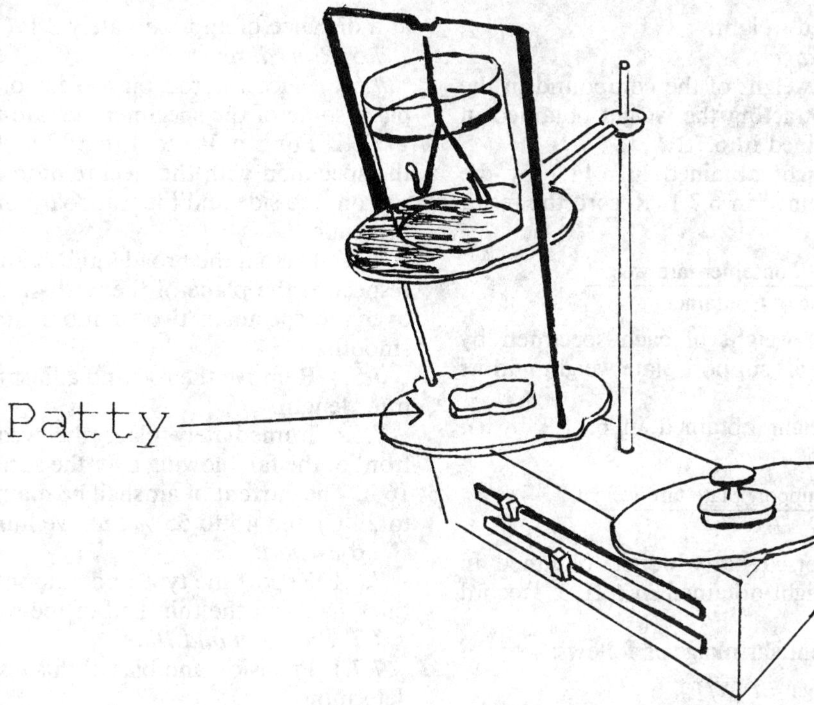

FIG. 1 Wire Cradle in Kerosine

6.8.1.2 Puddle the specimen with the spatula to remove entrapped air bubbles.

6.8.1.3 Fill the container completely and strike off the surface flush with the top using the steel-reinforced broad knife.

6.8.1.4 Weigh the filled container to the nearest 0.01 g. Record the weight of the filled container.

6.9 *Preparation of Specimen to Determine Dry Volume:*

6.9.1 Place approximately 30 g of specimen onto each prepared support plate.

6.9.1.1 Spread the specimen into an elongated patty 3/16 to 1/4 in. (4.8 to 6.4 mm) thick with a spatula.

6.9.1.2 Remove any specimen remaining on the spatula and add to the patty.

NOTE 3—The patty may be scored across its narrowest width to facilitate breaking the patty after it is dry.

6.9.1.3 Weigh and record the total weight of each patty, film and plate.

6.9.1.4 When testing setting type joint compounds, place the patties in the drying oven one h after the setting time has been reached as determined by Test Methods C 472.

6.9.1.5 When testing drying type joint compounds, place the patties in the drying oven immediately after weighing.

6.9.2 Dry patties at a temperature between 100° and 120°F (32° to 49°C) for 16 to 24 h.

6.9.3 Strip off the plastic or rubber film, and continue to dry until constant weight is reached.

6.10 *Determination of Dry Volume:*

6.10.1 Immerse each patty in a beaker of the displacement fluid, such that they do not touch the sides of the beaker, until constant weight is reached.

NOTE 4—Each patty may be broken into two or three pieces to fit in the beaker without touching the sides.

6.10.2 Remove each patty from the displacement fluid. Using a cloth moistened in the same fluid, lightly blot off the excess fluid from the surface of the patty.

6.10.3 With the wire cradle suspended in the beaker of displacement fluid, weigh each patty in air on the pan of the balance (Fig. 1), and record as its air weight.

6.10.4 Next, weigh each patty in the wire cradle, making sure that the patty is completely immersed in the liquid and that it does not touch the sides of the beaker (Fig. 2). Record

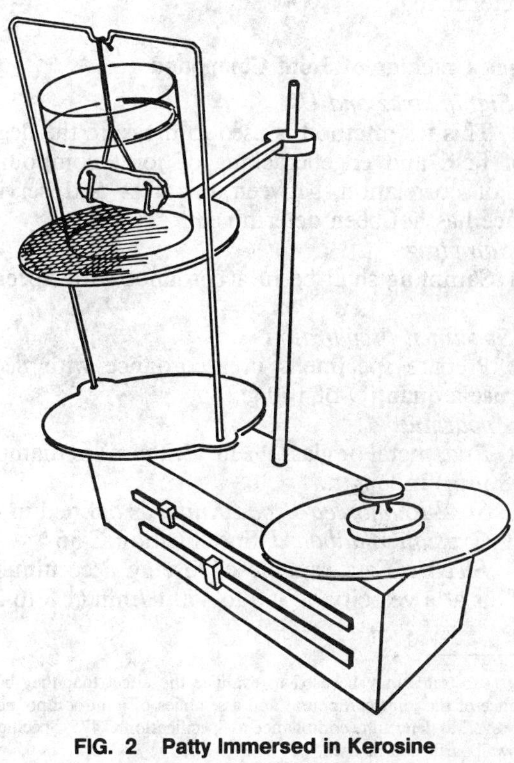

FIG. 2 Patty Immersed in Kerosine

these weights as the immersed weight.

6.11 *Calculation of Shrinkage:*

6.11.1 Determine the net weight of the compound in the volumetric container by subtracting the weight obtained in 6.8.1.4 from the weight obtained in 6.7.1.

6.11.2 Divide the net weight obtained in 6.11.1 by the volume of the container obtained in 6.7.1. Record the result as G, grams per millilitre.

$$G = \frac{\text{Total weight} - \text{Container tare weight}}{\text{Volume of Container}}$$

6.11.3 Determine the net weight of each specimen by subtracting the tare weight of its support plate weight and its weight obtained in 6.9.1.3.

6.11.4 Divide the net weight obtained in 6.11.3 by G. Record as V, volume of patty.

$$V = \frac{\text{Dry weight} - \text{Support plate tare weight}}{G}$$

6.11.5 Subtract the immersed patty weight obtained in 6.10.4 from the air patty weight obtained in 6.10.3. Record as D, the weight difference.

6.11.6 Calculate the percent shrinkage as follows:

$$\%\ \text{Shrinkage} = D(X/V),$$

where:

D = the difference in air and immersed weight,
X = 100/displacement fluids specific gravity, and
V = weight of wet patty/specific gravity of the compound.

6.12 *Report:*

6.12.1 Take the average of the three patties tested. If there is a difference between the percent shrinkage of the three patties of more than 1.5 %, completely retest an additional three specimens and take the average of the six.

6.13 *Precision and Bias:*

6.13.1 Precision and bias of this test method have not been determined.

7. Check Cracking of Joint Compound

7.1 *Significance and Use*

7.1.1 This test method is used to measure the degree and type of field and edge cracking of joint compound. The degree of correlation between this test and service performance has not been determined.

7.2 *Sampling:*

7.2.1 Sampling shall be in accordance with Specification C 475.

7.3 *Specimen Preparation:*

7.3.1 Prepare specimens in accordance with Section 4, except use a quantity of 100 g.

7.4 *Apparatus:*

7.4.1 *Rod,* metal or glass, 1/8 in. (3.2 mm) in diameter by 7 in. (178 mm) in length.

7.4.2 *Steel-Reinforced Broad Knife,* as defined in 6.4.6.

7.4.3 *Gypsum Wallboard,* Specification C 36.[8]

7.4.4 *Electric Fan,* capable of forcing a continuous current of air at a velocity of 350 to 450 ft/min (1.8 to 2.3 m/s) at a distance of approximately 3 ft (1 m).

7.5 *Procedure:*

7.5.1 Place the rod on a piece of gypsum wallboard and place some of the specimen next to the rod.

7.5.2 Form a 3½ to 4 in. (90 to 100 mm) wide wedge of the specimen with the steel-reinforced broad knife, with the rod on one side and the wallboard on the other side to guide the knife.

7.5.2.1 Hold the broad knife at an angle less than 45° with respect to the plane of the wallboard. Draw the broad knife over the specimen two or more times to leave the surface smooth.

7.5.3 Remove the rod and adjust the wedge to a 5 in. (130 mm) length.

7.5.4 Immediately place the wedge-shaped specimen in front of the fan blowing over the surface of the wedge for 8 to 16 h. The current of air shall be maintained at 70 to 85°F (21 to 29°C) and 45 to 55 % relative humidity.

7.6 *Report:*

7.6.1 Report the type and amount of cracking in both the thick half and the thin half of the wedge.

7.7 *Precision and Bias:*

7.7.1 Precision and bias of this test method have not been determined.

8. Putrefaction

8.1 *Significance and Use:*

8.1.1 This test method is used to evaluate the tendency of the joint compound to putrefy. The degree of correlation between this test and service performance has not been determined.

8.2 *Sampling:*

8.2.1 Sampling shall be in accordance with Specification C 475.

8.3 *Apparatus:*

8.3.1 *Humidity Cabinet,* a chamber capable of maintaining 85 to 95°F (29 to 35°C) and 85 to 95 % relative humidity.

8.3.2 *Glass Container,* capable of being sterilized in an autoclave, having a minimum volume of 250 mL.[9]

8.3.3 *Cover,* made of glass or aluminum foil large enough to cover the glass container and capable of being sterilized in an autoclave.

8.3.4 *Autoclave,* capable of maintaining 260°F (126°C) at 21 psi (145 kPa) steam pressure for not less than 15 min.

8.4 *Preparation of Apparatus:*

8.4.1 Sterilize the glass container and cover in an autoclave at 21 psi (144.8 kPa) and 260°F (126°C) for 15 min or more.

8.5 *Procedure:*

8.5.1 *Joint Compound, Powder*—Mix 50 g of joint compound with water, as determined in 4.1 in the glass container. Allow it to soak 30 min and then remix. Place the cover over the glass container and place in the humidity cabinet.

8.5.2 *Joint Compound, Ready-mix*—Select an unopened container that has not exceeded the producer's specified shelf life.

[8] Other substrates may be used to evaluate the effect that they have on the performance of the joint compound and assemblies of joint compound and joint tape. However, to determine compliance to Specification C 475, Specification C 36 gypsum wallboard shall be used.

[9] Deep Petri dishes and 250 mL beakers have been found suitable for this purpose.

8.5.2.1 Open the container. If the material in the container has separated, mix thoroughly.

8.5.2.2 Remove 100 g of joint compound, ready-mix, from the container.

8.5.2.3 Put the specimen in the glass container and cap with the cover.

8.5.2.4 Place in the humidity cabinet.

8.6 *Interpretation of Results:*

8.6.1 Observe daily for putrefaction.

8.7 *Report:*

8.7.1 Report the number of days required to produce putrefaction.

8.8 *Precision and Bias:*

8.8.1 Precision and bias of this test method have not been determined.

PAPER JOINT TAPE

9. Tensile Strength

9.1 *Significance and Use:*

9.1.1 This test method is used to evaluate the cross-directional strength of paper tape used to reinforce the joints created by gypsum wallboard construction. The degree of correlation between this test and service performance has not been determined.

9.2 *Sampling:*

9.2.1 Sampling shall be in accordance with Specification C 475.

9.2.2 A sample for the purpose of testing shall consist of not less than 12 ft (3.6 m) of tape from each roll.

9.2.3 Take ten specimens in the cross direction at not less than 1 ft (300 mm) intervals. Cut each specimen to 1 in. (25 mm) by roll width.

9.3 *Apparatus:*

9.3.1 The apparatus shall be in accordance with Test Method D 828 except that the distance of the jaw spacing shall be reduced to ½ ± 1/64 in. (12.7 ± 0.4 mm); the rate of loading shall be 5 ± 1.5 lbf/s (22.4 ± 6.7 N/s).

9.4 *Procedure:*

9.4.1 Condition the specimens for a minimum of 24 h at 72 ± 4°F (22 ± 2°C) and 50 ± 2 % relative humidity.

9.4.2 Test in accordance with Test Method D 828 with equipment as set forth in 9.3.

9.5 *Interpretation of Results:*

9.5.1 Accept or reject results of the individual test specimens in accordance with Test Method D 828.

9.5.2 If any results are rejected, test additional specimens so that there are at least ten test results for the evaluation of any unit of tape.

9.6 *Report:*

9.6.1 Report test results as pounds-force per inch (newtons per millimetre) of width of specimen.

9.7 *Precision and Bias:*

9.7.1 Precision and bias of this test method have not been determined.

10. Dimensional Stability

10.1 *Significance and Use:*

10.1.1 This test method is used to determine the length-wise and crosswise expansion characteristics of the paper joint tape used to reinforce the joints created in gypsum wallboard construction. The degree of correlation between this test and service performance has not been determined.

10.2 *Sampling:*

10.2.1 Sampling shall be in accordance with Specification C 475.

10.3 *Specimen Preparation:*

10.3.1 Cut not less than three specimens of tape 10 to 16 in. (250 to 400 mm) long from the roll to be tested.

10.3.2 Condition the specimens for a minimum of 16 h at 72 ± 4°F (22 ± 2°C) and 50 ± 2 % relative humidity.

10.4 *Apparatus:*

10.4.1 *Steel Rule, Caliper Rule, Cathetometer* or other measuring device capable of resolving 0.005 in. over a minimum span of 10 in. A metric rule must be capable of resolving 0.10 mm over a minimum span of 250 mm.

10.4.2 *Sharp Knife, 4 to 5× Magnifying Glass, Pencil.*

10.4.3 *Glass Container* at least 2½ in. (63.5 mm) deep and 6 in. (152.4 mm) in diameter.

10.5 *Reagents and Materials:*

10.5.1 *Water*, distilled or deionized.

10.6 *Procedure:*

10.6.1 Place the conditioned specimens on a flat surface.

10.6.1.1 At about ½ in. (10 mm) from each end of the paper tape, cut two ½ in. (10 mm) long reference marks for the length-wise measurement crosswise of the tape with a sharp knife.

10.6.1.2 Place the rule on the tape so that one edge is centered lengthwise.

10.6.2 Move the rule so that starting mark (A) coincides with the reference mark at one end of the tape. Record the value at A. Take the reading at the other reference mark. Read to the nearest 0.005 in. (0.10 mm). (See Fig. 3).

NOTE 5—Measuring from the 1 in. (25 mm) mark and subtracting this value may be a useful method to obtain an accurate measurement. Reading the rule with a magnifying glass of 4 to 5× is recommended.

10.6.3 Place the rule across the width of the tape.

10.6.3.1 Align the starting mark (B) of the rule with one edge of the tape. Record the value at B.

10.6.3.2 Take the reading at the opposite edge of the tape. Mark the location of this crosswise measurement by drawing a 1½ in. (40 mm) long pencil mark across the tape without marring the edges of the tape. (See Fig. 4.)

10.6.4 Roll up the tape and submerge it in the container full of water at 72 ± 4°F (22 ± 2°C).

10.6.5 After 30 min, remove the tape from the water and

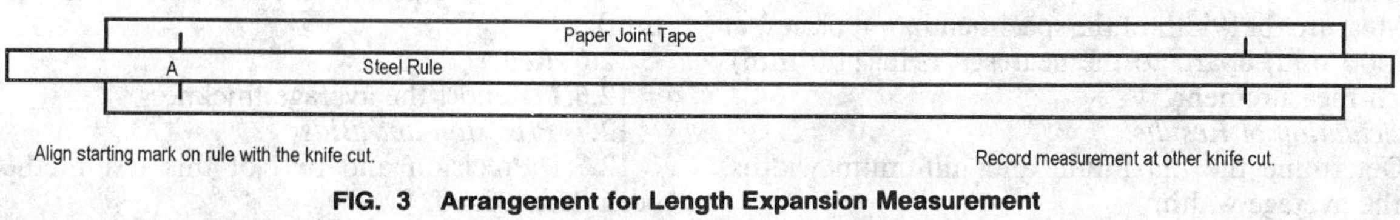

FIG. 3 Arrangement for Length Expansion Measurement

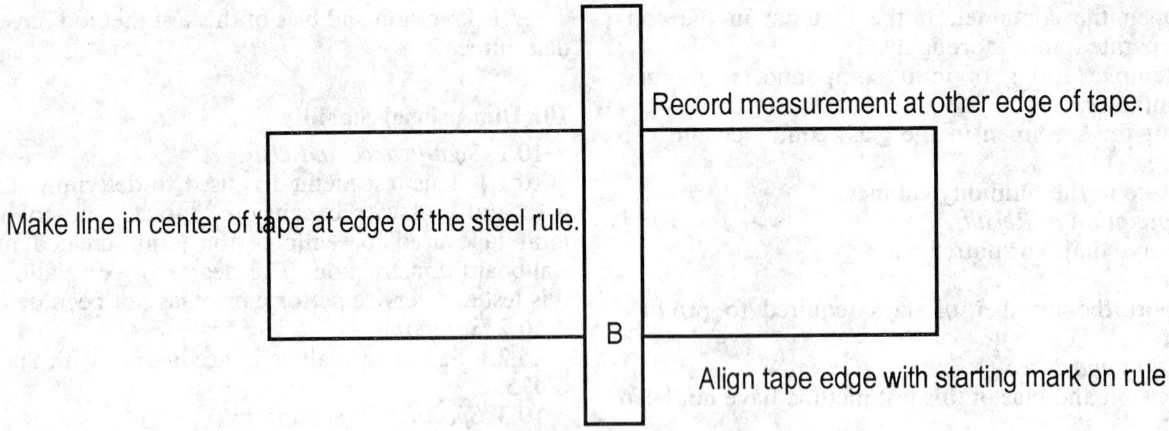

FIG. 4 Arrangement for Width Expansion Measurement

roll it out on the flat surface. Repeat the lengthwise and crosswise measurements.

10.7 *Calculation of Results:*

10.7.1 Subtract A from the reading obtained in 10.6.2 and record as the dry length of the tape at this point.

10.7.2 Subtract B from the reading obtained in 10.6.3.2 and record as the dry width of the tape at this point.

10.7.3 Determine the amount of expansion by subtracting the original measurement from the final measurement. Divide the expansion in inches (millimetres) by the original reading and multiply by 100 to obtain the percentage expansion.

$$\% \text{ Expansion} = \frac{(\text{Wet Measurement} - \text{Dry Measurement})}{\text{Dry Measurement}} \times 100$$

10.8 *Report:*

10.8.1 Report test results as percent expansion in the lengthwise and crosswise directions.

10.9 *Precision and Bias:*

10.9.1 Precision and bias of this test method have not been determined.

11. Width

11.1 *Significance and Use:*

11.1.1 This test method is used to determine the average variation in width of the paper joint tape. The degree of correlation between this test and service performance has not been determined.

11.2 *Sampling:*

11.2.1 Sampling shall be in accordance with Specification C 475.

11.3 *Specimen Preparation:*

11.3.1 Cut a specimen of paper joint tape not less than 12 ft (3.6 m) long from the roll to be tested.

11.4 *Apparatus:*

11.4.1 *Steel Rule, Caliper Rule,* or other measuring device capable of resolving 1/32 in. A metric device must be capable of resolving 1.0 mm.

11.5 *Procedure:*

11.5.1 Measure the width of the specimen in ten places, at least 1 ft (300 mm) apart, to the nearest 1/32 in. (1.0 mm). Record each measurement.

11.6 *Calculation of Results:*

11.6.1 Determine the maximum and minimum widths. Calculate the average width.

11.6.2 Determine the difference between the maximum and minimum width by subtraction.

11.7 *Report:*

11.7.1 Report the average width, and the difference between the maximum and minimum width.

11.8 *Precision and Bias:*

11.8.1 Precision and bias of this test method have not been determined.

12. Thickness

12.1 *Significance and Use:*

12.1.1 This test method is used to determine the average thickness of paper joint tape. The degree of correlation between this test and service performance has not been determined.

12.2 *Sampling:*

12.2.1 Sampling shall be in accordance with Specification C 475.

12.3 *Specimen Preparation:*

12.3.1 Cut a specimen of tape not less than 12 ft (3.6 m) long from the roll to be tested.

12.3.2 Condition the specimen for a minimum of 24 h as described in Practice D 685.

12.4 *Apparatus:*

12.4.1 *Paper Micrometer* having circular faces of 0.25 to 0.33 in.^{2} (160 to 213 mm^2) in area. Faces shall be under steady pressure of 7 to 9 psi (48 to 62 kPa).

NOTE 6—For details see TAPPI T 410, except waive Section 7.

12.5 *Procedure:*

12.5.1 Measure the thickness of the specimen in 10 places, between the edges, at least 1 ft (300 mm) apart. Skived areas on the tape edges shall not be included in the area measured for thickness.

NOTE 7—Skiving is not present on all tapes.

12.5.2 Record the thickness to the nearest 0.001 in. (0.01 mm).

12.6 *Report:*

12.6.1 Report the average thickness.

12.7 *Precision and Bias:*

12.7.1 Precision and bias of this test method have not been determined.

ASSEMBLAGES OF PAPER JOINT TAPE AND JOINT COMPOUND

13. Bond of Paper Joint Tape to Joint Compound

13.1 *Significance and Use:*

13.1.1 This test method is used to determine the bonding properties of the paper joint tape. The degree of correlation between this test and service performance has not been determined.

13.2 *Sampling:*

13.2.1 Sampling shall be in accordance with Specification C 475.

13.3 *Specimen Preparation:*

13.3.1 Select specimens of paper joint tape in accordance with 9.2.2.

13.3.2 Cut specimens of paper joint tape into at least 12 in. (300 mm) specimens for use in test.

13.3.3 Prepare specimens of joint compound in accordance with Section 4, except use a quantity of 200 g.

13.4 *Apparatus:*

13.4.1 *Feeler Gage Strips* (two required for each test), each 12 in. (300 mm) long, ½ in. (13 mm) wide, 0.025 in. (0.64 mm) thick with a small hole drilled in one end.

13.4.2 *Steel-Reinforced Broad Knife,* in accordance with 6.4.6.

13.4.3 *Gypsum Wallboard,* Specification C 36[8], two pieces, 6 by 14 in. (150 by 350 mm) with the 14-in. (350-mm) length in the machine direction of the paper.

13.4.4 *Overlay Transparency Grid*—A transparent photo copy of 10 by 10 divisions/in. graph paper. An area 2 by 5 in. (51 by 127 mm) enclosing 1000 square divisions is outlined.

13.5 *Procedure:*

13.5.1 Place two feeler gage strips parallel to each other about 4 in. (102 mm) apart and fasten to the face of the gypsum wallboard with a thumb tack through the hole in the end.

13.5.2 Using the steel reinforced broad knife, apply an amount of joint compound sufficient to cover the area between the feeler gage strips. Spread the specimen evenly between the feeler gage strips leaving the specimen slightly thicker than the strips.

13.5.3 Center a 12-in. (305-mm) length of the paper tape in the specimen. Press one end of the tape into the specimen and hold it in place.

13.5.4 Embed the tape by applying two or three pressure strokes with the steel-reinforced broad knife. Wipe away from the end being held so the excess joint compound is squeezed out.

NOTE 8—The thickness of the joint compound plus the tape is about 0.025 in. (0.64 mm).

13.5.4.1 Carefully remove the feeler gages before drying.

13.5.5 Allow the test assembly to dry to constant weight in an atmosphere of 75 ± 5°F (24 ± 2°C) and 50 ± 5 % relative humidity.

13.5.6 When the test assembly is dry, use a sharp knife to make a cut across and perpendicular to the tape 3½ in. (90 mm) from one end. Make a second cut 5 in. (130 mm) from and parallel to the first cut. Make two diagonal cuts across the tape connecting the opposite corners of the 5-in. (130 mm) section. With the tip of the knife, peel back the tabs formed by the "X" cuts and pull up sharply.

13.5.6.1 Make a second test by repeating 13.5.6 below the first test.

13.5.7 Using a sharp pencil, lightly outline the areas where fiber remains attached to the compound. Align the overlay transparency grid so that the grid outline matches the 2 by 5-in. (51 by 127-mm) sides of the tape bond area.

13.6 *Calculation of Results:*

13.6.1 Using the overlay transparency grid, count the number of squares that are more than half bare of fiber separated from the tape and outlined by pencil.

13.6.2 Subtract this number from 1000 and divide by 10 to determine the percent bond. Record the average of the two tests.

13.7 *Report:*

13.7.1 Report the average percent bond failure.

13.8 *Precision and Bias:*

13.8.1 Precision and bias of this test method have not been determined.

14. Cracking of Joint Compound at Tape Edges

14.1 *Significance and Use:*

14.1.1 This test method is used to determine the adhesion of the joint compound to the edges of the paper joint tape. The degree of correlation between this test and service performance has not been determined.

14.2 *Sampling:*

14.2.1 Sampling shall be in accordance with Specification C 475.

14.3 *Specimen Preparation:*

14.3.1 Select specimens of paper joint tape in accordance with 9.2.2.

14.3.2 Cut specimens of paper joint tape into not less than 12 in. (305 mm) specimens.

14.3.3 Prepare specimens of joint compound in accordance with Section 4, except use a quantity of 200 g.

14.4 *Apparatus:*

14.4.1 *Feeler Gage Strips,* two required for each test, each 12-in. (305 mm) long, ½ in. (13 mm) wide, 0.040 in. (1 mm) thick.

14.4.2 *Steel-Reinforced Broad Knife,* see 13.4.2.

14.4.3 *Gypsum Wallboard,* Specification C 36[8], approximately 6 by 12 in. (152 by 305 mm).

14.4.4 *Electric Fan,* in accordance with 7.4.4.

14.5 *Procedure:*

14.5.1 Place the two feeler gage strips on the face of the wallboard about 4 in. (102 mm) apart.

14.5.2 Fill the space between the feeler gage strips with joint compound specimen and level to the thickness of the feeler gage strips.

14.5.3 After leveling to the correct thickness, center a specimen of paper joint tape between the feeler gage strips.

14.5.3.1 Using the steel-reinforced broad knife at about a 45° angle, press the tape specimen firmly into contact with the joint compound.

14.5.3.2 Use a final flat stroke of the knife to smooth the joint compound on both sides of the tape. Do not press so hard that the joint compound will be squeezed out from under the tape, but hard enough so that the tape is in full contact with the joint compound.

14.5.4 Remove the feeler gage strips.

14.5.5 Place this assembly in front of the electric fan

blowing a current of air over the surface of the assembly. Maintain a current of air at 100 ± 5°F (38 ± 2°C) and 25 ± 5 % relative humidity.

14.5.5.1 Alternatively, in very dry air (under 20 % relative humidity) at temperatures between 75 and 90°F (24 to 32°C) just allow the electric fan to blow a current of air over the surface.

14.5.6 After 1 h, examine the assembly for cracks along the edges of the tape with a magnifying glass of 4 to 5×.

14.6 *Calculation of Results:*

14.6.1 Measure the total length of any cracking at the edge of the paper joint tape and joint compound.

14.6.2 Divide the total length of cracked edges by the total length of the edges exposed and multiply 100 to obtain percent of cracked edge.

14.7 *Report:*

14.7.1 Report the percentage of cracking along the edges of the tape.

14.8 *Precision and Bias:*

14.8.1 Precision and bias of this test method have not been determined.

15. Keywords

15.1 bond of tape; cracking; joint compound; joint tape; joint treatment; putrefaction; shrinkage; tensile strength; topping compound; viscosity

Designation: C 475 – 94

Standard Specification for Joint Compound and Joint Tape for Finishing Gypsum Board[1]

This standard is issued under the fixed designation C 475; the number immediately following the designation indicates the year of original adoption or, in the case of revision, the year of last revision. A number in parentheses indicates the year of last reapproval. A superscript epsilon (ε) indicates an editorial change since the last revision or reapproval.

This specification has been approved for use by agencies of the Department of Defense. Consult DoD Index of Specifications and Standards for the specific year of issue which has been adopted by the Department of Defense.

1. Scope

1.1 This specification covers all-purpose, taping and finishing joint compounds, and paper joint tape.

1.2 This specification also covers the minimum requirements for an assembly of taping and all-purpose compound and joint tape.

1.3 The values stated in inch-pound units are to be regarded as the standard.

2. Referenced Documents

2.1 *ASTM Standards:*

C 11 Terminology Relating to Gypsum and Related Building Materials and Systems[2]

C 474 Test Methods for Joint Treatment Materials for Gypsum Board Construction[2]

3. Terminology

3.1 *Definitions*—Definitions shall be in accordance with Terminology C 11.

3.2 *Description of Terms Specific to This Standard:*

3.2.1 *joint-tape, n*—a strip of paper reinforcing material designed to be embedded in the joint compound.

4. Physical Properties

4.1 *General*—The physical properties shall be determined in accordance with Test Methods C 474.

4.2 *Joint Compounds.*

4.2.1 *Check Cracking*—No cracks in the thinner half of the wedge and no deep fissure cracks in the thicker half of the wedge.

4.2.2 *Putrefaction*—No putrefaction in less than 4 days.

4.2.3 *Shrinkage*—Not more than 35 % by volume.

4.3 *Joint Tape:*

4.3.1 *Width*—Shall be not less than 1⅞ in. (47.6 mm) nor more than 2¼ in. (57.2 mm). The width of an individual roll shall vary not more than $\pm 1/32$ in. (0.8 mm).

4.3.2 *Thickness*—Shall be not more than 0.012 in. (0.30 mm).

4.3.3 *Tensile Strength*—Shall be not less than 30 lbf/in. (524 N/mm) in the cross direction.

4.3.4 *Dimensional Stability*—Shall expand not more than 0.4 % lengthwise and not more than 2.5 % crosswise.

4.4 *Assembly of Taping or All-Purpose Compound and Joint Tape:*

4.4.1 *Bond* between the joint tape and the joint compound shall be not less than 90 %.

4.4.2 *Edge Cracking* shall be not more than 10 % along the edges of the tape specimen.

5. Sampling

5.1 *Sampling of Joint Compound*—At least 1 %, but not less than one package or container shall be sampled and shall be so selected as to be representative of the shipment. The material so obtained shall be thoroughly blended and divided. Provide not less than one 6-lb (2.72-kg) sample for the test.

5.2 *Sampling of Joint Tape*—At least one roll for every ten cases or a fraction thereof, but not more than ten rolls per shipment shall constitute a sample.

6. Inspection

6.1 Inspection of the joint compound and joint tape shall be agreed upon between the purchaser and the supplier as part of the purchase agreement.

7. Rejection

7.1 Rejection of joint compound and joint tape that fails to conform to requirements of this specification shall be reported to the producer or supplier promptly and in writing. The notice of rejection shall contain a statement documenting how the joint compound or joint tape has failed to conform to the requirements of this specification.

8. Certification

8.1 When specified in the purchase agreement, a producer's or supplier's report shall be furnished at the time of shipment certifying that the product is in compliance with this specification.

9. Packaging and Package Marking

9.1 Joint compound and joint tape shall be marked on each item or package with the name, brand or trademark of the manufacturer or seller, size, and volume or net weight of contents.

9.2 Joint compound packaged in 3-½ through 6-½ gal (U.S. liquid) (13 through 25 L) straight-sided, open-head pails not less than 10 in. (250 mm) nor more than 19 in. (480 mm) in overall height and having a full diameter unrestricted opening at the top, shall be labeled with a cautionary label meeting the requirements of 9.2.1 through 9.2.5.

---

[1] This specification is under the jurisdiction of ASTM Committee C-11 on Gypsum and Related Building Materials and Systems and is the direct responsibility of Subcommittee C11.02 on Specifications and Test Methods for Accessories and Related Products.

Current edition approved Aug. 15, 1994. Published October 1994. Originally published as C 475 – 61 T. Last previous edition C 475 – 93.

[2] *Annual Book of ASTM Standards*, Vol 04.01.

9.2.1 The cautionary label shall be placed on the side of the container between the upper rim and the container base.

9.2.2 The cautionary label shall be not less than 10 in.^{2} (65 000 mm^2) in area with no single dimension less than 1¾ in. (45 mm) and with a perimeter of not less than 12¾in. (325 mm).

9.2.2.1 Alternatively, two cautionary labels shall be permitted to be used. Each of the two labels shall be not less than 6 in.^{2} (40 000 mm) in area with no single dimension less than 1¾ in. (45 mm) and with a perimeter of not less than 9 in. (225 mm). At least one of the labels shall be in English and the letter sizes of both labels shall conform to 9.2.4.1 and 9.2.4.3.

9.2.3 The cautionary label shall have a border or another form of contrast which delineates it from any other information on the container.

9.2.4 The cautionary label shall be in a noticeable layout and shall contain a signal word panel, a pictorial and a message as specified in 9.2.4.1 through 9.2.4.3 and shall be printed in colors that contrast with the background of the label.

9.2.4.1 The signal word panel shall contain the word "WARNING" in upper case letters, preceded by a safety alert symbol consisting of an exclamation mark inside an equilateral triangle with the apex of the triangle oriented upward. The word "WARNING" shall be on one line and shall be in letters not less than 5/16 in. (7.9 mm) high. Where two cautionary labels are used they shall conform to 9.2.2.1 and the word "WARNING" shall be in letters not less than ¼ in. (6.4 mm) high.

9.2.4.2 the pictorial shall depict a child looking or falling into a bucket. The pictorial shall have an overall height of not less than 1½ in. (38 mm). Additional cautionary symbols shall be allowed.

9.2.4.3 The message shall contain the words "Children can fall into bucket and drown" followed by the words "Keep children away from bucket with even a small amount of liquid"; additional cautionary statements shall be allowed. The lettering of the cautionary message shall be in sans serif letters in combinations of upper and lower case letters. The upper case letters shall be not less than ⅛ in. (3.2 mm) high. Where two cautionary labels are used they shall conform to 9.2.2.1 and the upper case letters shall be not less than 5/64 in. (2 mm) high.

9.2.5 The label shall be not easily removed, other than by the use of tools or solvents. If removable labels are used for container recycling purposes, the label shall bear the words "DO NOT REMOVE THIS LABEL."

10 Keywords

10.1 all-purpose compound; bond; check cracking; dimensional stability; edge cracking; finishing compound; joint compound; joint tape; putrefaction; shrinkage; taping compound; tensile strength; topping compound

Designation: C 490 – 97

Standard Practice for Use of Apparatus for the Determination of Length Change of Hardened Cement Paste, Mortar, and Concrete[1]

This standard is issued under the fixed designation C 490; the number immediately following the designation indicates the year of original adoption or, in the case of revision, the year of last revision. A number in parentheses indicates the year of last reapproval. A superscript epsilon (ε) indicates an editorial change since the last revision or reapproval.

1. Scope

1.1 This practice covers the requirements for the apparatus and equipment used to prepare specimens for the determination of length change in hardened cement paste, mortar, and concrete, the apparatus and equipment used for the determination of these length changes, and the procedures for its use.

1.2 Methods for the preparation and curing of test specimens, conditions of testing and curing, and detailed procedures for calculating and reporting test results are contained in applicable test methods.

1.3 The values stated in inch-pound units are to be regarded as the standard.

1.4 Values in SI units shall be obtained by measurement in SI units or by appropriate conversion, using the Rules for Conversion and Rounding given in IEEE/ASTM SI 10, of measurement made in other units.

2. Referenced Documents

2.1 *ASTM Standards:*

C 511 Specification for Moist Cabinets, Moist Rooms, and Water Storage Tanks Used in the Testing of Hydraulic Cements and Concretes[2]

C 1005 Specification for Weights and Weighing Devices for Use in the Physical Testing of Hydraulic Cement[2]

IEEE/ASTM SI 10 Standard for Use of the International System of Units (SI): The Modern Metric System

3. Terminology

3.1 *length change*—an increase or decrease in the linear dimension of a test specimen, measured along the longitudinal axis, due to causes other than applied load.

4. Significance and Use

4.1 This practice is intended to provide standard requirements for apparatus common to many test methods used in connection with cement and concrete and standardized procedures for its use. The detailed requirements as to materials, mixtures, specimens, conditioning of specimens, number of specimens, ages at which measurements are to be made, interpretation of results, and precision and bias are left to be dealt with in specific test methods.

5. Apparatus

5.1 *Weights and Weighing Devices*, shall conform to the requirements of Specification C 1005.

5.2 *Glass Graduates*, of suitable capacities (large enough to measure the mixing water for paste and mortar mixtures in a single operation) shall be made to deliver the indicated volume at 20°C (68°F).

5.2.1 The permissible variation for graduates of 100 to 150-mL capacities shall be ±1.0 mL, for graduates of 200 to 300-mL capacities shall be ±2.0 mL, and for all larger graduates shall be ±0.5 % of the rated capacity.

5.2.2 These graduates shall be subdivided to at least 5 mL with the following exceptions:

5.2.2.1 The graduation lines may be omitted for the lowest 15 mL for a 150-mL graduate,

5.2.2.2 The graduation lines may be omitted for the lowest 25 mL for a 250-mL graduate, and

5.2.2.3 The graduation lines may be omitted for the lowest 50 mL for a 500-mL graduate.

5.2.3 The main graduation lines shall extend at least three quarters of the way around the graduate and shall be numbered.

5.3 *Molds*, shall have either one or two compartments and shall be constructed as shown in Figs. 1 or 2. Molds for test specimens used in determining the length change of cement pastes and mortars shall provide for 1 by 1 by 11¼-in. prisms having a 10-in. gage length, or for 25 by 25 by 285-mm prisms having a 250-mm gage length. Molds for test specimens used in the length change of concretes shall provide for prisms of the desired cross section having a 10-in. or 250-mm gage length. In some routine tests, 1 by 1 by 6¼-in. specimens with a gage length of 5 in. or 25 by 25 by 106-mm specimens with a gage length of 125 mm are permitted, but in case of dispute, results obtained with specimens of 10-in. (250-mm) gage length shall govern.

5.3.1 The gage length shall be considered as the nominal length between the innermost ends of the gage studs. The parts of the molds shall be tight fitting and firmly held together when assembled, and their surfaces shall be smooth and free of pits. The molds shall be made of steel or other hard metal not readily attacked by the cement paste, mortar, or concrete. The sides of the molds shall be sufficiently rigid to prevent spreading or warping. For the molds shown in Fig. 1, the tolerance on dimension A is ±0.03 in. For the molds shown in Fig. 2, the tolerance on dimension A is ±0.7 mm.

5.3.2 Each end plate of the mold shall be equipped to hold properly in place, during the setting period, one of the gage studs shown in Figs. 1 or 2. The gage studs shall be of

[1] This practice is under the jurisdiction of ASTM Committee C-1 on Cement and is the direct responsibility of Subcommittee C01.95 on Methods of Test.

Current edition approved July 10, 1997. Published June 1998. Originally published as C 490 – 62. Last previous edition C 490 – 96.

[2] *Annual Book of ASTM Standards*, Vol 04.01.

[3] *Annual Book of ASTM Standards*, Vol 14.02.

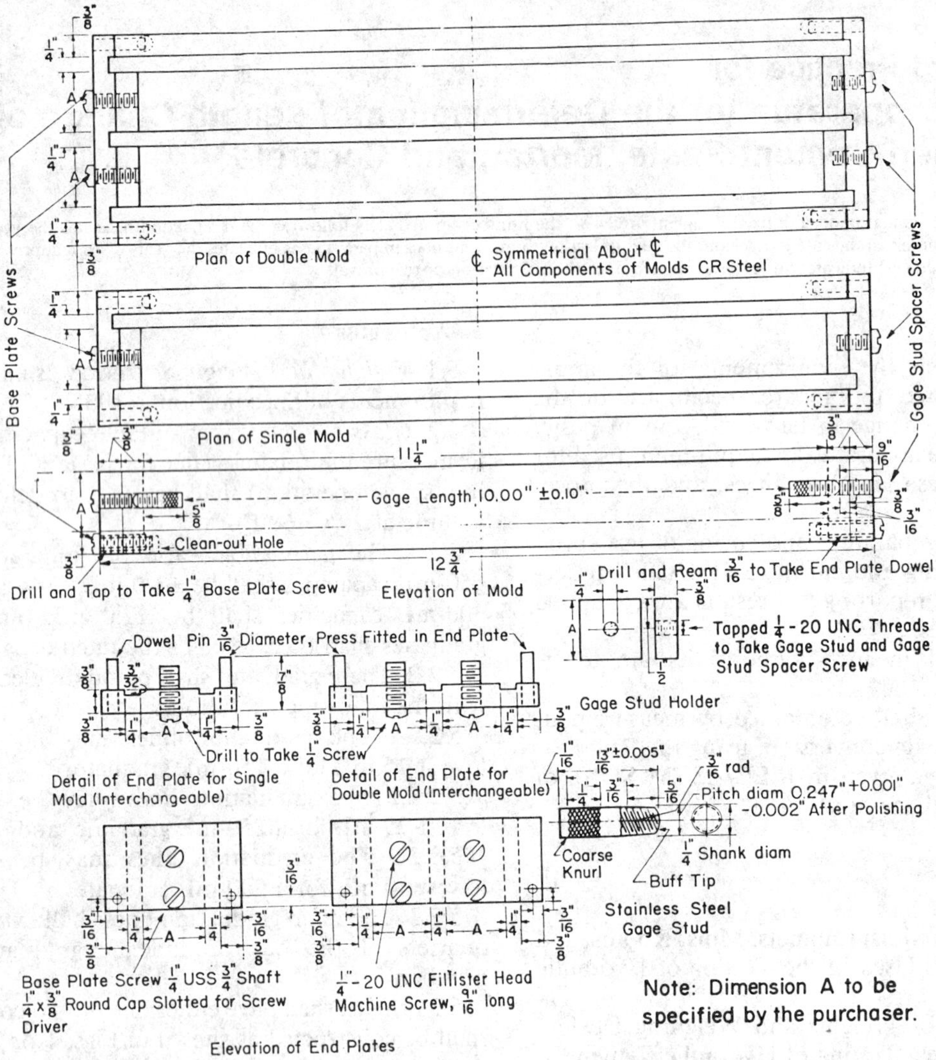

FIG. 1 Molds (Inch-Pound Units)

American Iron and Steel Institute (AISI)[4] Type 316 stainless steel or other corrosion-resistant metal of similar hardness. Gage studs of Invar or similar metal shall be used when specimens are tested at widely different temperatures. To prevent restraint of the gage studs before demolding of the specimen, the device for holding the gage studs in position shall be so arranged that, if necessary, it can be partially or completely released after the compaction of the paste or mortar into place in the mold. The gage studs shall be set so that their principal axes coincide with the principal axis of the test specimen. For the molds shown in Fig. 1, gage studs shall extend into the specimen 0.625 ± 0.025 in. and the distance between the inner ends of the gage studs shall be 10.00 ± 0.10 in. and 10 in. shall be considered the gage length for calculating length change. For the molds shown in Fig. 2, gage studs shall extend into the specimen 17.5 ± 0.5 mm and the distance between the inner ends of the gage studs shall be 250.0 ± 2.5 mm and 250 mm shall be considered the gage length for calculating length change.

5.4 *Length Comparator*, for determining length change of specimens, shall be designed to accommodate the size of specimen employed and to provide or permit a positive means of contact with the gage studs and the convenient and rapid obtaining of comparator readings (Note 1).

5.4.1 The comparator for determining length changes of specimens produced in the molds shown in Fig. 1 shall provide a dial micrometer or other measuring device graduated to read in 0.0001-in. units, accurate within 0.0001 in. in any 0.0010-in. range, and within 0.0002 in. in any 0.0100-in. range, and sufficient range (at least 0.3 in.) in the measuring device to allow for small variations in the actual length of various specimens. The terminals of the comparator shall be plane, polished and heat-treated. They shall be fitted with collars held in place with set screws. The collars shall extend 0.062 ± 0.003 in. beyond the plane face of the terminal and have an inside diameter 0.02 in. greater than the average diameter of the portion of the gage studs that must fit into the collars.

5.4.2 The comparator for determining length changes of specimens produced in the molds shown in Fig. 2 shall

[4] Details on this material are available from the American Iron and Steel Institute, 1133 15th St. N.W., Washington, DC 20005.

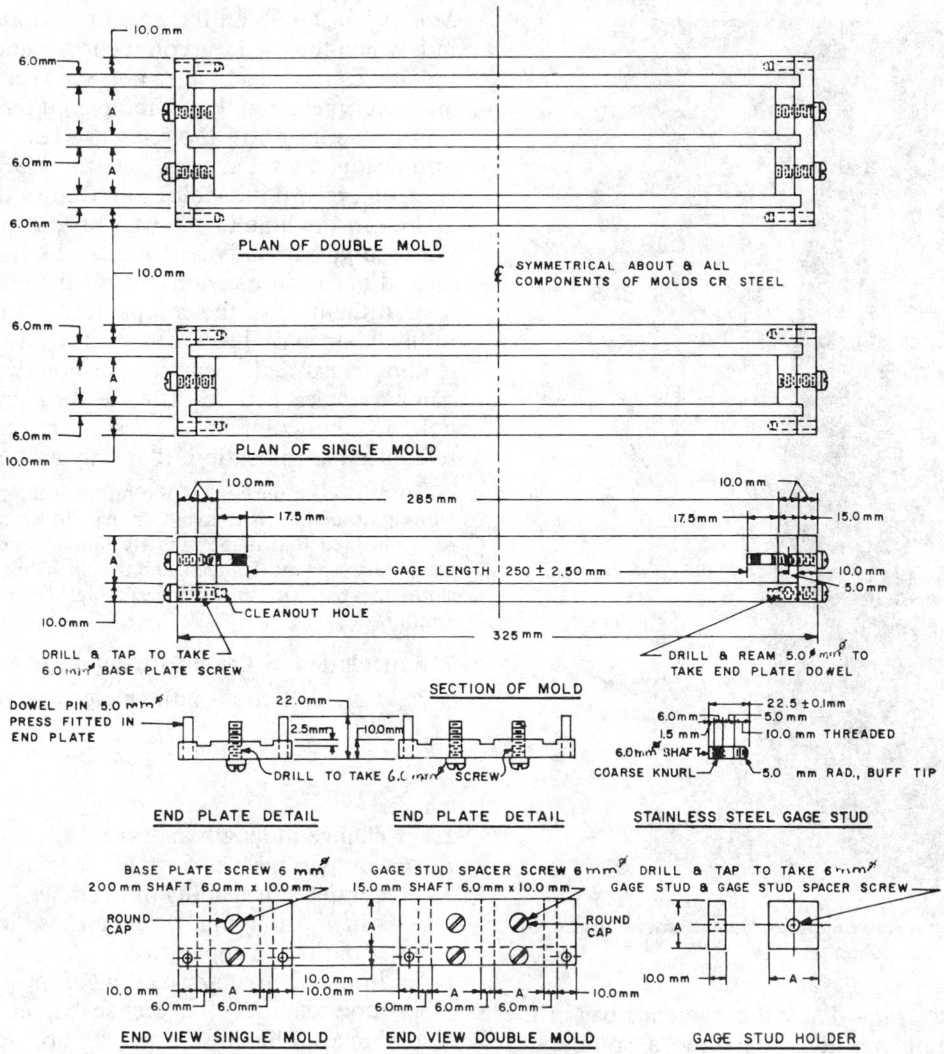

FIG. 2 Molds (SI Units)

provide a dial micrometer or other measuring device graduated to read in 0.001 or 0.002-mm units, accurate within 0.002 mm in any 0.020-mm range, and within 0.004 mm in any 0.200-mm range, and sufficient range (at least 8.0 mm) in the measuring device to allow for small variations in the actual length of various specimens. The terminals of the comparator shall be plane, polished and heat-treated. They shall be fitted with collars held in place with set screws. The collars shall extend 1.5 ± 0.1 mm beyond the plane face of the terminal and have an inside diameter 0.5 mm greater than the average diameter of the gage studs that must fit into the collars.

NOTE 1—One type of instrument that has been found satisfactory for use with small prisms is shown in Fig. 3. A horizontal comparator should be used with prisms with a cross section greater than 9 in.^{2} or 58 cm^2.

5.4.3 The design shall provide a means for checking the measuring device against a reference bar at regular intervals. The reference bar shall have an overall length of 11⅝ ± 1⁄16 in. or 6⅝ ± 1⁄16 in. (300 ± 1.5 mm or 170 ± 1.5 mm), whichever is appropriate for the specimen in use. The bar shall be of a steel alloy having a coefficient of thermal expansion not greater than two millionths per degree Celsius. Each end of the reference bar shall be fitted with heat treated, hardened, and polished tips machined to the same shape as the contact end of the gage studs used in test specimens. Except for the tips, which are attached after heat treatment, no part of the reference bar shall be heat treated. The central 4 in. (100 mm) of the length of the reference bar shall be covered by a rubber tube with a wall at least ⅛ in. (3 mm) thick to minimize the effect of temperature change during handling. The reference bar shall be provided near one end with a positioning mark.

NOTE 2—Alloys such as invar lose their low coefficient of thermal expansion properties when heat treated.

6. Procedure

6.1 *Preparation of Molds*—Prior to the molding of specimens, the outside joints of the mold and the contact lines of the mold and base plate shall be sealed to prevent loss of mixing water from a freshly molded specimen. Thinly cover the interior surfaces of the mold with mineral oil. After this operation, set the gage studs, taking care to keep them clean, and free of oil, grease, and foreign matter.

FIG. 3 Type of Suitable Apparatus for Measurement of Length Changes

6.2 *Use of Reference Bar*—Place the reference bar in the instrument in the same position each time a comparator reading is taken. Check the dial gage setting of the measuring device by use of the reference bar at least at the beginning and end of the readings made within a half day when the apparatus is kept in a room maintained at constant temperature. Check it more often when kept in a room where the temperature is not constant.

NOTE 3—The equation given in the section on calculation of length change contemplates that a comparator reading for the reference bar will be recorded each time the reference bar is used and a difference calculated for each test specimen reading. Alternatively, the dial gage setting can be reset, if necessary, to its original setting with the reference bar in place each time the reference bar is read. Doing so simplifies the calculation of length change by canceling the comparator reading of the reference bar from the values of L_x and L_i. If this procedure is used, care should be taken to ensure that the dial set screw is tightened adequately each time the dial is reset.

6.3 *Obtaining Comparator Readings*—Rotate specimens slowly in the measuring instrument while the comparator reading is being taken. Record the minimum reading of the dial if the rotation causes a change in the dial reading. Place specimens in the instrument with the same end up each time a comparator reading is taken.

6.3.1 *Obtaining Comparator Readings of Specimens Stored Moist*—Clean the hole in the base of the comparator into which the gage stud on the lower end of the bar fits (this hole tends to collect water and sand and should be cleaned after every reading). Read and record the comparator indication of the length of the reference bar. Take one bar out of immersion, blot the pins, put the bar in the comparator, read, and record the indication. Return the bar to immersion and clean the hole in the base of the comparator. Take out the second bar and treat it in a like manner. Return the second bar to immersion, record the reading, and clean the hole in the base of the comparator. Continue the procedure until all bars have been read, returned to immersion, and the readings recorded, cleaning the hole in the bottom of the comparator each time. After reading the last bar, clean the hole in the comparator base and read and record the reference-bar indication. Blot only around the pins (Note 4).

NOTE 4—The purpose of the minimal blotting of the pins and no blotting of the bars is to avoid drying and shrinkage of the bars. It has been observed that if the pins are blotted, and the bar placed in the comparator and the dial read, and the bar is then wiped gently with a dry cloth, the bar will shrink measurably. Therefore, drying should be minimized.

7. Calculation of Length Change

7.1 Calculate the length change at any age as follows:

$$L = \frac{(L_x - L_i)}{G} \times 100$$

where:

L = change in length at x age, %,
L_x = comparator reading of specimen at x age minus comparator reading of reference bar at x age; in inches when using Fig. 1 apparatus, in millimetres when using Fig. 2 apparatus,
L_i = initial comparator reading of specimen minus comparator reading of reference bar at that same time; in inches when using Fig. 1 apparatus, in millimetres when using Fig. 2 apparatus, and
G = nominal gage length, 10 when using Fig. 1 apparatus, 250 when using Fig. 2 apparatus.

7.2 Calculate length change values for each specimen to the nearest 0.001 % and report averages to the nearest 0.01 %.

8. Temperature, Humidity, and Time

8.1 *Molding Room*—The temperature of the molding room and dry materials shall be maintained between 20 and 27.5°C (68 and 81.5°F). The relative humidity shall be not less than 50 %. The temperature of the mixing water shall be 23.0 ± 2.0°C (73.5 ± 3.5°F).

8.2 *Moist Storage Facility*—The temperature and humidity of the air in the moist storage facility shall conform to the requirements of Specification C 511.

8.3 *Time*—Comparator readings shall be taken at specified time intervals or ages. All intervals and ages shall be met within ±2 %.

9. Keywords

9.1 cement paste; comparator; concrete; length change apparatus; molds; mortar

Standard Specification for Moist Cabinets, Moist Rooms, and Water Storage Tanks Used in the Testing of Hydraulic Cements and Concretes[1]

This standard is issued under the fixed designation C 511; the number immediately following the designation indicates the year of original adoption or, in the case of revision, the year of last revision. A number in parentheses indicates the year of last reapproval. A superscript epsilon (ε) indicates an editorial change since the last revision or reapproval.

1. Scope

1.1 This specification covers requirements for moist cabinets, moist rooms, and water storage tanks used for storage of paste, mortar, and concrete test specimens.

1.2 Values in SI units shall be obtained by measurement in SI units or by appropriate conversion, using the Rules for Conversion and rounding given in Standard IEEE/ASTM SI 10, of measurements made in other units.

1.3 *This standard does not purport to address all of the safety concerns, if any, associated with its use. It is the responsibility of the user of this standard to establish appropriate safety and health practices and determine the applicability of regulatory limitations prior to use.*

2. Referenced Documents

2.1 *ASTM Standards:*

C 911 Specification for Quicklime, Hydrated Lime, and Limestone for Chemical Uses[2]

E 77 Test Methods for the Inspection and Verification of Thermometers[3]

IEEE/ASTM SI 10 Standard for Use of the International System of Units (SI): The Modern Metric System[4]

3. Terminology

3.1 *Definitions:*

3.1.1 *moist cabinet, n*—a compartmented storage facility of moderate dimensions with controlled temperature and relative humidity.

3.1.2 *moist room, n*—a "walk-in" storage facility with controlled temperature and relative humidity, commonly called a fog room when the prescribed relative humidity is achieved by the atomization of water.

4. Requirements

4.1 *General*—The atmosphere in a moist cabinet or room shall have a temperature of 23.0 ± 2.0°C and a relative humidity of not less than 95 %. The moisture in the atmosphere shall be saturated to the degree needed to ensure that the exposed surfaces of all specimens in storage will both look moist and feel moist at all times. All storage units shall be equipped with recording thermometers. The recording thermometer shall be calibrated at least every six months or whenever there is a question of accuracy. Perform the verification of the recording thermometer by comparing the temperature reading of the recording thermometer with the temperature reading of a reference thermometer during the normal operation of the moist cabinet or room. The thermometer used as the reference thermometer must be accurate and readable to 0.5°C. A copy of the certificate or report which verifies the accuracy shall be available in the laboratory (Note 1). Position the reference thermometer in a readable position in air as near as is practical to the recording temperature probe. Keep the door to the moist cabinet or room closed for at least 5 min prior to taking readings. Record the temperatures of both the recording thermometer and the reference thermometer. When taking these readings, the reference thermometer shall remain in the moist cabinet or room and be read immediately upon opening the door. If the difference between the two temperatures is greater than 1°C, the recording thermometer shall be adjusted to within 0.5°C of the reference thermometer. The use of humidity recording devices is optional. Shelves on which fresh specimens are placed shall be level.

NOTE 1—The ice-point method describe in Test Method E 77 may be used to ensure that no damage to the reference thermometer has occurred during shipping.

4.1.1 The air in a moist storage unit must be nearly saturated with moisture in order to provide specified storage conditions. In many cases, saturation is below optimum during periods when specimens are being placed in or removed from storage. Measurements of relative humidity should not be made at such obviously inopportune times.

4.1.2 The recordings from the recording thermometers shall be audited in order to ascertain the adequacy of the mechanisms used to control the moist room air temperature.

4.1.3 The air temperature inside the storage facility (moist cabinet or moist room) shall be controlled with provisions made for heating or cooling, or both, as may be necessary. This shall be accomplished in one of two ways:

4.1.3.1 Thermostatically control the air temperature within

---

[1] This specification is under the jurisdiction of ASTM Committee C-1 on Cement and is the direct responsibility of Subcommittee C01.95 on Coordination of Standards.

Current edition approved Nov. 10, 19987. Published March 1999. Originally published as C 511 – 68. Last previous edition C 511 – 97.

[2] *Annual Book of ASTM Standards*, Vol 04.01.

[3] *Annual Book of ASTM Standards*, Vol 14.03.

[4] *Annual Book of ASTM Standards*, Vol 14.02.

the storage facility when surrounding space is not conditioned. In this case the sensing element for the controls shall be located inside the storage facility.

4.1.3.2 Thermostatically control the space surrounding the storage facility and manually control the temperature within the storage facility.

4.1.4 In either of the preceding cases, the laboratory shall demonstrate the ability of the controls to maintain the required temperature in the storage facility over an extended period of time. Charts from the recording thermometer that indicate that the temperatures are within the temperature limits specified in 4.1 shall be required as evidence of this ability.

4.2 *Moist Cabinets*—A moist cabinet shall be constructed of durable materials and the doors shall be tight-fitting. The specified relative humidity shall be maintained by the use of one or more fog sprays, water sprays, or curtains of water on the inner walls that are so directed that the discharge will collect in a pool at or near the bottom of the moist storage section.

4.3 *Moist Rooms*:

4.3.1 *General*—The walls of a moist room shall be constructed of durable materials, and all openings shall be provided with tight-fitting doors or windows (Note 2). The specified relative humidity may be maintained in any convenient and suitable manner (Note 3).

NOTE 2—Well insulated walls will substantially help maintain necessary conditions.

NOTE 3—A fog spray found suitable for this purpose is shown in Fig. 1.

4.3.2 *Moist Rooms Used in Cement Testing*—Durable shelving that is properly shielded to prevent droplets of water from

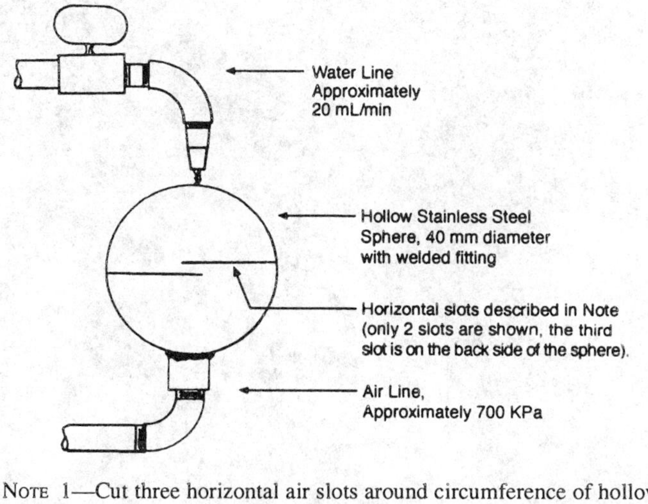

NOTE 1—Cut three horizontal air slots around circumference of hollow sphere using a 0.20 mm thick diamond lapidary saw covering 120° to 150° each and spaced approximately 5 mm apart. Air passing through these slots strikes the water (which is flowing over the outer surface of the sphere) to produce a spray.

FIG. 1 Example of a Fog Spray for Maintaining Relative Humidity in Moist Rooms (Full Scale)

falling on the surfaces of freshly molded specimens shall be available within each room.

4.3.3 *Moist Rooms Used in Concrete Testing*—Atmospheric conditions within each room shall be such that test specimens in storage shall have free water maintained on their entire surface area at all times. Specimens shall not be exposed to dripping or running water.

4.4 *Water Storage Tanks*:

4.4.1 *General*—Tanks shall be constructed of noncorroding materials. Provision for automatic control of water temperature at 23.0 ± 2.0°C shall be made where a tank is located in a room not having temperature controlled within that range and in any other instance where difficulty in maintaining temperatures within the specified range is encountered. Each tank located in a space not controlled at 23.0 ± 2.0°C shall be equipped with a recording thermometer with its sensing element in the storage water. Water storage recording thermometers shall be checked for accuracy at least every six months by comparing their output with that of a reference thermometer placed in the water adjacent to the probe of the recording thermometer. Record the temperatures of both the recording thermometers and the reference thermometer. Adjustments shall be made to the recording thermometer if the difference in the observed readings exceed 1°C. The adjusted reading of the recording thermometer shall be within 0.5°C of the reference thermometer. The requirements for the reference thermometer are listed earlier in the *General* Section of this specification. The water in a storage tank shall be saturated with calcium hydroxide to prevent leaching of calcium hydroxide from the specimens (Note 4). Water not saturated with calcium hydroxide (high-calcium hydrated lime) may affect test results due to leaching of lime from the test specimens and shall not be used in storage tanks. To maintain saturation with calcium hydroxide, excess calcium hydroxide shall be present. For the purposes of lime saturation to prevent leaching, lime means calcium hydroxide only (hydrated lime, such as Type CH, meeting the requirements of Specification C 911), not calcium carbonate. The water in the storage tank shall be thoroughly stirred at intervals not to exceed one month to help replace calcium ions that have depleted. Tanks shall be cleaned and refilled with water containing 3 g/L of calcium hydroxide at intervals not to exceed 24 months (Note 5).

NOTE 4—pH is not a reliable indicator of lime saturation in storage tank water since severe reductions in dissolved calcium ions can occur before pH values are significantly reduced.

NOTE 5—The 3 g/L level is intended to provide a quantity of calcium hydroxide approximately two times that required for initial saturation.

4.4.2 Do not use continuously running fresh water or demineralized water in storage tanks because it may effect test results due to excessive leaching. A closed system, circulating the saturated lime water between or among storage tanks, may be used.

5. Keywords

5.1 cement paste; concrete; moist cabinets; moist rooms; mortar; water storage tanks

The American Society for Testing and Materials takes no position respecting the validity of any patent rights asserted in connection with any item mentioned in this standard. Users of this standard are expressly advised that determination of the validity of any such patent rights, and the risk of infringement of such rights, are entirely their own responsibility.

This standard is subject to revision at any time by the responsible technical committee and must be reviewed every five years and if not revised, either reapproved or withdrawn. Your comments are invited either for revision of this standard or for additional standards and should be addressed to ASTM Headquarters. Your comments will receive careful consideration at a meeting of the responsible technical committee, which you may attend. If you feel that your comments have not received a fair hearing you should make your views known to the ASTM Committee on Standards, 100 Barr Harbor Drive, West Conshohocken, PA 19428.

Designation: C 514 – 96

Standard Specification for Nails for the Application of Gypsum Board[1]

This standard is issued under the fixed designation C 514; the number immediately following the designation indicates the year of original adoption or, in the case of revision, the year of last revision. A number in parentheses indicates the year of last reapproval. A superscript epsilon (ε) indicates an editorial change since the last revision or reapproval.

This standard has been approved for use by agencies of the Department of Defense. Consult the DoD Index of Specifications and Standards for the specific year of issue which has been adopted by the Department of Defense.

1. Scope*

1.1 This specification covers requirements for steel wire nails suitable for use in the application of gypsum board.

NOTE 1—This specification does not necessarily cover nails for use where fire ratings are required. Consult manufacturers for independent data on assembly particulars, materials, and ratings.

1.2 The values stated in inch-pound units are to be regarded as the standard.

1.3 The following precautionary caveat pertains only to the test method portion, Sections 7 and 8, of this specification: *This standard does not purport to address all of the safety concerns, if any, associated with its use. It is the responsibility of the user of this standard to establish appropriate safety and health practices and determine the applicability of regulatory limitations prior to use.*

2. Referenced Document

2.1 *ASTM Standard:*

E 4 Practices for Force Verification of Testing Machines[2]

3. Materials

3.1 *Steel Wire*, used in the manufacture of nails, shall be of hard-drawn low or medium-low-carbon steel, entirely suitable for the purpose intended. Before fabrication it shall be sufficiently ductile to withstand cold-bending, without fracture, through 180° over a radius not greater than the diameter of the wire (see Section 8).

4. Physical Properties

4.1 *Ductility*—The nail shall be sufficiently ductile to withstand cold bending without fracture when tested according to Section 8.

4.2 *Withdrawal Resistance*—The average withdrawal resistance, both immediate and delayed, shall be at least equal to that provided by 12.5 gage (0.099 ± 0.003 in.) (2.515 ± 0.08 mm) bright, smooth shank, nails with a medium diamond point, when tested according to 8.2.

5. Dimensions and Permissible Variations

5.1 *Heads*—Shall be not less than 0.2375 in. (6.0 mm) and not more than 0.3875 in. (9.8 mm) in diameter, shall be not more than 1/64 in. (0.4 mm) thick at the peripheral edge, shall be uniformly tapered to a small fillet around the shank, shall be either flat or concave, and shall be free from protrusions and sharp, irregular edges.

5.2 *Shanks*—Diameter shall not vary more than ± 0.003 in. (0.08 mm) for shank diameters 0.076 in. (1.93 mm) or larger.

5.3 *Points*—The nails shall have medium to long diamond or needle points.

5.4 *Clearance*—Where a deformation process produces other than a smooth shank nail, a clearance area (round and smooth), measured from the top of the head to the deformed section, shall be provided equal to the nominal thickness of the gypsum board for which the nails are specified.

6. Workmanship, Finish, and Appearance

6.1 The nails shall be bright, or chemically treated, or coated with rust inhibiting material provided that such chemical treatment or rust inhibitor does not adversely affect the performance of the nail as specified in Section 4. The nails shall also be compatible with the joint compound and decoration. Nails shall be neatly formed and free from injurious defects or deformations.

7. Number of Tests

7.1 At least five nails from each lot of 100 individual containers shall be examined to determine conformance to the requirements of this specification.

8. Test Methods

8.1 *Bend Test:*

8.1.1 *Significance and Use*—This test method is performed to determine the ductility of a nail, and used to determine compliance with this specification.

8.1.2 *Procedure*—Hold the test nail in a vise and bend by means of a clamp or similar device attached to the free end of the nail through 90° over a radius not greater than the diameter of the nail.

8.1.3 *Precision and Bias*—Neither the precision nor the bias of the nail bend test have been determined.

8.2 *Withdrawal Resistance*

8.2.1 *Significance and Use*—This test method determines the resistance to withdrawal from a wood block of the test nail compared to a standard nail. It is used to determine

[1] This specification is under the jurisdiction of ASTM Committee C-11 on Gypsum and Related Building Materials and Systems and is the direct responsibility of Subcommittee C11.02 on Specifications and Test Methods for Accessories and Related Products.

Current edition approved Aug. 10, 1996. Published October 1996. Originally published as C 514 – 63 T. Last previous edition C 514 – 94.

[2] *Annual Book of ASTM Standards*, Vol 03.01.

* A Summary of Changes section appears at the end of this standard.

compliance with this specification.

8.2.2 *Apparatus:*

8.2.2.1 *Testing machine*—The testing machine shall be of any type that is of sufficient capacity and that is capable of applying a tensile load at the strain rate prescribed in 8.2.3.2. The load platform shall be of sufficient size to accept the block of wood described in 8.2.2.5. The testing machine shall be power operated and shall apply the load continuously and without shock.

8.2.2.2 The test apparatus shall be verified according to Practices E 4, Sections 16 through 18.

8.2.2.3 *Load Indicating Equipment*, capable of reading to 1 lbf (4.45 N) increments shall be provided for the test apparatus. A means shall be provided to indicate the maximum load attained during the test.

8.2.2.4 A means of securing the blocks of wood to the platform or lower cross head shall be provided so that the nails are aligned with the direction of pull. A device shall be attached to the upper cross head to grasp the nail head securely and hold it during the test for the purpose of pulling the nail from the block of wood.

8.2.2.5 *Wood Nailing Blocks*—Nominal 2 by 6 in. (38 by 140 mm) Douglas Fir, construction grade, containing at least 16 %, but not more than 19 %, free moisture as determined by a suitable moisture meter. Required wood blocks for a test series shall be cut from a single piece of lumber creating blocks free of knots.

8.2.3 *Procedure:*

8.2.3.1 Alternately use twenty of the standard 12.5 gage nails and twenty of the nails being tested and hammer drive to a depth of ⅞ in. (22.2 mm) into the longitudinal center line of one edge and one side face of the wood blocks. Space nails 3 in. (76 mm) apart. Determine the comparative performance as follows:

8.2.3.2 *Immediate*—Withdraw alternate samples of each type of nail from each face, ten of each nail, at the rate of 0.06 in. (1.5 mm)/min.

8.2.3.3 *Delayed*—Condition the lumber, into which the nails are driven, at 50 ± 2 % relative humidity and 70 ± 5°F (21.1 ± 3°C) to constant weight, after which withdraw the remaining nails.

8.2.4 *Precision and Bias*—Neither the precision nor the bias have been determined for the nail withdrawal test method.

9. Certification

9.1 When specified in the purchase order a producer's or supplier's certification shall be furnished to the purchaser that the material is in compliance with this specification.

10. Packaging and Package Marking

10.1 Unless otherwise specified, nails shall be packaged in substantial commercial containers of the type, size, and kind commonly used for the purpose, so constructed as to preserve the contents in good condition and to ensure acceptance and safe delivery by common or other carriers, at the lowest rate, to the point of delivery. In addition, the containers shall be so made that the contents can be partially removed without destroying the container's ability to serve as a receptacle for the remainder of the contents.

10.2 Unless otherwise specified, individual packages and shipping containers shall be marked with the type and length of nail and the name of the manufacturer or distributor. Individual packages shall also be marked with the name of the manufacturer or distributor and the net mass.

11. Keywords

11.1 gypsum board; nail; points; shanks; wire heads

SUMMARY OF CHANGES

Committee C-11 has identified the location of selected changes to this standard since the last issue (C 514-94) that may impact the use of this standard.

(*1*) Revised paragraph 5.1.

(*2*) Revised paragraph 5.2.

(*3*) Revised paragraph 5.4.

(*4*) Inserted a space between the words gypsum and board in the last sentence of paragraph 5.4.

(*5*) Deleted the phrase "may be" in 6.1.

(*6*) Deleted the word "approximately" in 8.2.3.1.

Designation: C 563 – 96

Standard Test Method for Optimum SO_3 in Hydraulic Cement Using 24-h Compressive Strength[1]

This standard is issued under the fixed designation C 563; the number immediately following the designation indicates the year of original adoption or, in the case of revision, the year of last revision. A number in parentheses indicates the year of last reapproval. A superscript epsilon (ε) indicates an editorial change since the last revision or reapproval.

1. Scope

1.1 This test method describes the determination of optimum SO_3 for maximum 24-h compressive strength by measuring the change in strength produced in hydraulic cement mortar as a result of substituting calcium sulfate for a portion of the cement. Usually, but not always, the SO_3 content that produces the highest 24-h strength at 23°C also produces approximately the lowest expansion in water and the lowest contraction in air at that temperature.

1.2 This test method refers to the sulfur trioxide (SO_3) content of the cement only. Slag cements and occasionally other hydraulic cements can contain sulfide or other forms of sulfur. The determination of SO_3 content by rapid methods may include these other forms, and may therefore produce a significant error. If a significant error occurs, analyze the cement for SO_3 content using the reference test method of Test Methods C 114 for sulfur trioxide.

1.3 Values stated as SI units are to be regarded as standard.

1.4 *This standard does not purport to address all of the safety concerns, if any, associated with its use. It is the responsibility of the user of this standard to establish appropriate safety and health practices and determine the applicability of regulatory limitations prior to use.*

2. Referenced Documents

2.1 *ASTM Standards:*

C 109/C 109M Test Method for Compressive Strength of Hydraulic Cement Mortars (Using 2-in. or 50-mm Cube Specimens)[2]

C 114 Test Methods for Chemical Analysis of Hydraulic Cement[2]

C 150 Specification for Portland Cement[2]

C 204 Test Method for Fineness of Hydraulic Cement by Air Permeability Apparatus[2]

C 305 Practice for Mechanical Mixing of Hydraulic Cement Pastes and Mortars of Plastic Consistency[2]

C 471 Test Methods for Chemical Analysis of Gypsum and Gypsum Products[2]

C 511 Specification for Moist Cabinets, Moist Rooms and Water Storage Tanks Used in the Testing of Hydraulic Cements and Concretes[2]

C 595 Specification for Blended Hydraulic Cements[2]

C 778 Specification for Standard Sand[2]

C 1157 Performance Specification for Blended Hydraulic Cement[2]

3. Significance and Use

3.1 The purpose of this test method is to estimate the SO_3 content for a hydraulic cement that gives maximum 24-h compressive strength.

3.2 The optimum SO_3 content for 24-h compressive strength is not necessarily the optimum SO_3 content for later ages or for volume change. Confirming tests can be made if the relationships between SO_3 and these properties are not known for the cement being tested.

3.3 This test method indicates optimum SO_3 content for cement in mortar made and cured at a standard temperature of 23 ± 1.7°C (73.4 ± 3°F). The optimum SO_3 increases with increasing temperature and may increase when water-reducing admixtures are used.

3.4 It should not be assumed that the optimum SO_3 estimated in this test method is the same SO_3 content for optimum performance of a concrete prepared from the cement.

3.5 The test method is applicable to cements specified in Specifications C 150, C 595, and C 1157.

4. Apparatus

4.1 Use the apparatus as specified in Test Method C 109.

5. Materials

5.1 *Calcium Sulfate*—Use calcium sulfate for addition to the cement that is either a high-grade natural gypsum having an SO_3 content of at least 46 %,[3] or the calcium sulfate from the source used for the intended plant production. Grind the calcium sulfate to 100 % passing the 75-μm (No. 200) sieve, and at least 800 m^2/kg Blaine fineness (Test Method C 204). If the SO_3 content of the calcium sulfate is unknown, analyze it in accordance with Test Methods C 471.

5.2 *Blended Standard Sand*—Use a blend of equal parts by mass of graded sand and 20–30 sand, both conforming to Specification C 778.

6. Test Specimens

6.1 Make six specimens from each batch of mortar. Make two rounds of three mixtures of mortar as described in 6.3. Make both rounds during the same day if possible. Make

[1] This test method is under the jurisdiction of ASTM Committee C-1 on Cement and is the direct responsibility of Subcommittee C01.28 on Sulfate Content.

Current edition approved Dec. 10, 1996. Published February 1997. Originally published as C 563 – 65. Last previous edition C 563 – 95.

[2] *Annual Book of ASTM Standards*, Vol 04.01.

[3] Terra Alba, available from the U.S. Gypsum Co., Southard, OK plant, meets these requirements for fineness and SO_3 content.

only complete rounds on any given day.

6.2 *Preparing Specimen Molds*—Prepare molds in accordance with the section on Preparation of Specimen Molds of Test Method C 109.

6.3 *Proportioning, Consistency, and Mixing of Mortar*—Proportion the dry material for the standard mortar as one part of cement to one part of blended sand by mass. Mix the following quantities of dry materials at one time for a 6-cube batch:

Mixture No. 1—940 g cement
470 g graded sand
470 g 20–30 sand

Mixture No. 2—930 g cement
10 g calcium sulfate
470 g graded sand
470 g 20–30 sand

Mixture No. 3—920 g cement
20 g calcium sulfate
470 g graded sand
470 g 20–30 sand

6.3.1 Use an amount of mixing water that is 30 % by mass of cement plus calcium sulfate for all types of hydraulic cement, except for Type III cement where the amount shall be 36 %.

6.3.2 Mix mechanically in accordance with the section on Procedure for Mixing Mortars of Practice C 305, except as follows:

6.3.2.1 Add the calcium sulfate to the water; then start the mixer and mix at slow speed (140 ± 5 rpm) for 15 s; then stop the mixer and add the cement to the water; then start the mixer and mix at slow speed (140 ± 5 rpm) for 30 s.

6.4 *Molding of Test Specimens*—Mold specimens in accordance with the Procedure Section of Test Method C 109.

6.5 *Storage of Test Specimens*—Immediately upon completion of molding, place the test specimens in the moist closet or moist room. Keep all test specimens, immediately after molding, in the molds on the base plates in the moist closet or moist room for 20 to 24 h with their upper surfaces exposed to the moist air but protected from dripping water. If the specimens are removed from the molds before 24 h, keep them on the shelves of the moist closet or moist room until they are 24-h old.

7. Temperature and Humidity

7.1 The temperature and humidity of the moist cabinet or moist room shall be as specified in Specification C 511.

7.2 The temperature and humidity of the mixing room shall be as specified in Test Method C 109.

8. Procedure

8.1 Test the specimens at the age of 24 ± ¼ h from the time the cement and water made contact during mixing, immediately after removal from the moist closet in accordance with Test Method C 109. If more than one specimen at a time is removed from the moist closet, keep these specimens covered with a damp cloth until time of testing. Note and observe precautions regarding testing and retesting given in the applicable sections of Test Method C 109.

9. Calculation

9.1 Calculate and record the strength of specimens as described in the section on Calculation in Test Method C 109.

9.1.1 Calculate the indicated optimum SO_3 content as follows:

$$G = [a/(a - b)]\ c + d + c/2$$

where:

G = optimum percent SO_3,
a = average strength both rounds of Mixture No. 2 minus Mixture No. 1,
b = average strength both rounds of Mixture No. 3 minus Mixture No. 2,
c = (percent SO_3 in calcium sulfate)/100, and
d = percent SO_3 in test cement.

10. Retest

10.1 Regard the results obtained as suspect and repeat the test under any of the following conditions:

10.1.1 If a and b both are positive and the ratio a/b is less than 2.00,

10.1.2 If a and b both are negative and the ratio a/b is greater than 0.500, or

10.1.3 If a is negative and b is positive.

11. Report

11.1 If repeat tests confirm that the ratio of a/b departs from the limiting values as stated above, report the SO_3 content of the cement as "above optimum" or "below optimum" as the data may indicate, but without stating the numerical value of optimum SO_3.

11.2 If the test data indicate that the cement is "below optimum" but do not permit a computation of the percentage of SO_3 required for optimum, an additional series of tests involving greater substitutions of gypsum for part of the cement should yield results that will permit an acceptable calculation of the percentage of SO_3 required for optimum. If the test data indicate that the cement is "above optimum" but do not permit a computation of the percentage of SO_3 required for optimum, in order to establish the required percentage of SO_3 for optimum, it is necessary to make tests of a new sample of similar cement but having an original SO_3 content lower than that of the original cement.

12. Precision and Bias

12.1 *Precision*—The coefficient of variation for this test method, using 6 cubes per batch and testing at 24-h age, is very similar to that for Test Method C 109 compression cubes tested at later ages. Refer to Test Method C 109 for suitable values.

12.2 *Bias*—Since there is no accepted reference material suitable for determining optimum SO_3, bias has not been determined.

13. Keywords

13.1 blended hydraulic cement; calcium sulfate; cement; compressive strength; gypsum; hydraulic cement; optimum sulfate content (of cement); portland cement; strength (of cement); sulfate content (of cement)

The American Society for Testing and Materials takes no position respecting the validity of any patent rights asserted in connection with any item mentioned in this standard. Users of this standard are expressly advised that determination of the validity of any such patent rights, and the risk of infringement of such rights, are entirely their own responsibility.

This standard is subject to revision at any time by the responsible technical committee and must be reviewed every five years and if not revised, either reapproved or withdrawn. Your comments are invited either for revision of this standard or for additional standards and should be addressed to ASTM Headquarters. Your comments will receive careful consideration at a meeting of the responsible technical committee, which you may attend. If you feel that your comments have not received a fair hearing you should make your views known to the ASTM Committee on Standards, 100 Barr Harbor Drive, West Conshohocken, PA 19428.

Designation: C 587 – 97

Standard Specification for Gypsum Veneer Plaster[1]

This standard is issued under the fixed designation C 587; the number immediately following the designation indicates the year of original adoption or, in the case of revision, the year of last revision. A number in parentheses indicates the year of last reapproval. A superscript epsilon (ε) indicates an editorial change since the last revision or reapproval.

This specification has been approved for use by agencies of the Department of Defense.

1. Scope

1.1 This specification covers calcined gypsum mixed at the mill with other ingredients to control working quality and setting time; specifically designed as a veneer plaster to be applied over gypsum base for veneer plasters, masonry or concrete surfaces to a maximum thickness of 1/4 in. (6.4 mm) providing a surface ready for decoration.

1.2 This specification covers test methods for determining the physical properties of gypsum veneer plasters and sets forth minimum requirements that must be met.

1.3 The values stated in inch-pound units are to be regarded as the standard. The SI (metric) values given in parentheses are approximate and are provided for information purposes only.

1.4 The text of this standard references notes and footnotes which provide explanatory material. These notes and footnotes shall not be considered as requirements of the standard.

1.5 The following safety hazards caveat applies only to the Test Methods portion, Section 5 of this specification: *This standard does not purport to address all of the safety concerns, if any, associated with its use. It is the responsibility of the user of this standard to establish appropriate safety and health practices and determine the applicability of regulatory limitations prior to use.*

2. Referenced Documents

2.1 *ASTM Standards:*

C 11 Terminology Relating to Gypsum and Related Building Materials and Systems[2]

C 473 Test Methods for Physical Testing of Gypsum Board Products and Gypsum Lath[2]

C 588 Specification for Gypsum Base for Veneer Plasters[2]

C 843 Specification for Application of Gypsum Veneer Plaster[2]

3. Terminology

3.1 *Definitions:*

3.1.1 Definitions of terms shall be in accordance with Terminology C 11 and Specification C 843.

3.2 *Definition of Term Specific to This Standard:*

3.2.1 *gypsum veneer plaster systems, n*—veneer plaster applied in accordance with Specification C 843 to gypsum base for veneer plasters.

4. Physical Properties

4.1 *Joint Strength*—Gypsum veneer plaster systems applied in accordance with Specification C 843 shall be tested in accordance with 5.7.4. The breaking load shall be not less than the parallel to surfacing loads specified in Specification C 588.

4.2 *Bond Strength*—The bond strength of the gypsum veneer plaster to the gypsum veneer base and, where applicable, between the base coat plaster and the finish coat shall not delaminate when tested in accordance with 5.7.1.

4.3 *Impact Strength*—Gypsum veneer plaster systems shall not crack or lose bond beyond the impact area when tested in accordance with 5.7.2.

4.4 *Flexure*—Gypsum veneer plaster systems shall exhibit crack resistance within the field of the gypsum base for veneer plasters such that radial or random cracking shall extend not more than 1/2 in. (13 mm) from the break line at failure when tested in accordance with 5.7.3. At rupture of the panel the veneer plaster shall be securely bonded to the face paper.

5. Test Methods

5.1 *Significance and Use*—These test methods provide procedures for evaluating the physical properties of gypsum veneer plaster. The degree of correlation between these test methods and service performance has not been determined.

5.2 *Materials*—Gypsum Base for Veneer Plaster, see Specification C 588.

5.3 *Sampling*—Sample not less than 1 % of the packages, but not less than 5 packages. Select packages to be sampled at random. Take samples both from the outer portion and the center of each package. Thoroughly mix the materials so obtained to provide a composite sample of not less than 15 lb (6.8 kg). Place this composite sample immediately in a clean, dry, airtight container for delivery to the laboratory.

5.4 *Test Unit Preparation*—From the sample, prepare panels of veneer plaster system and cut four 2 ft (600 mm) square, and three 12 by 16 in. (300 by 400 mm) specimens.

5.5 *Conditioning*—Condition specimens to constant weight at a temperature of 85 ± 15°F (29.5 ± 8.5°C) and relative humidity of 50 ± 2 % (see Note). Test specimens individually after constant weight is reached.

NOTE—Laboratory test results indicate that conditioning temperatures may range from 70 to 100°F (21 to 38°C) without significantly affecting the results. However, relative humidity is critical. Therefore, specimens shall be tested within 10 min upon removal from the conditioning chamber.

[1] This specification is under the jurisdiction of ASTM Committee C-11 on Gypsum and Related Building Materials and is the direct responsibility of Subcommittee C11.01 on Specifications and Test Methods for Gypsum Products.

Current edition approved Oct. 10, 1997. Published June 1998. Originally published as C 587 – 66. Last previous edition C 587 – 96.

[2] *Annual Book of ASTM Standards*, Vol 04.01.

5.6 *Number of Tests*—Run each test in triplicate.

5.7 *Test Procedure:*

5.7.1 *Bond Strength:*

5.7.1.1 Select three 2 ft (600 mm) square specimens. Raise the specimen and slam the back side (face not treated with veneer plaster) on a bare concrete floor repeatedly until the specimen is broken.

5.7.2 *Impact Strength*—Select one 2 ft (600 mm) square. Place the specimen horizontally, with the plaster surface face up, on a firm background. Drop a 1½ in. (38 mm) diameter polished steel ball that weighs 7.8 ± 0.1 oz (221 ± 3 g) three times onto different areas of the panel surface. Drop the ball from a height of 36 in. (915 mm).

5.7.3 *Flexure*—Select three 12 by 16 in. (300 by 400 mm) specimens. Subject each specimen to a continuous, midpoint, loading applied at a rate of 1 in./min ±5s (25 mm/min ±5s) downward to the back of the panel (plaster surface in tension) supported over a clear span of 14 in. (356 mm) until failure.

5.7.4 *Joint Strength:*

5.7.4.1 *Specimen preparation*—Cut two 8 by 12-in. (200 by 300 mm) pieces of gypsum base for veneer plaster for each specimen to be tested. One 12-in. edge of each piece shall be the paper-bound edge. Place the two pieces of gypsum base for veneer plaster face down with the paper-bound edges forming the joint. Apply a pregummed tape over the joint. (The tape will keep the joint from flexing during the specimen preparation.) Turn face up and apply joint tape and gypsum veneer plaster in accordance with Specification C 843.

5.7.4.2 *Procedure*—Condition the specimen and determine the flexural strength in accordance with Test Methods C 473. Apply the load parallel to and directly over the back of the joint.

5.8 *Precision and Bias*—Precision and bias have not been determined for these test methods.

6. Inspection

6.1 Inspection of the gypsum veneer plaster shall be agreed upon between the purchaser and the supplier as part of the purchase agreement.

7. Rejection and Rehearing

7.1 Rejection of gypsum veneer plaster that fails to conform to the requirements of this specification shall be reported to the producer or supplier promptly and in writing. The notice of rejection shall contain a statement documenting how the gypsum veneer plaster has failed to conform to the requirements of this specification.

8. Certification

8.1 When specified in the purchase agreement, a manufacturer's or supplier's report shall be furnished at the time of shipment certifying that the product is in compliance with this specification.

9. Packaging and Package Marking

9.1 Gypsum veneer plaster shall be dry and free from lumps, and shall be shipped in packages.

9.2 When shipped for resale, the following information shall be legibly marked on each package or on a tag of suitable size attached thereto:

9.2.1 Name of producer or supplier,

9.2.2 Brand,

9.2.3 Description, and

9.2.4 Net or gross weight, or both, of package.

10. Keywords

10.1 gypsum; gypsum veneer plaster; plaster; veneer plaster

Designation: C 588 – 95a

Standard Specification for Gypsum Base for Veneer Plasters[1]

This standard is issued under the fixed designation C 588; the number immediately following the designation indicates the year of original adoption or, in the case of revision, the year of last revision. A number in parentheses indicates the year of last reapproval. A superscript epsilon (ε) indicates an editorial change since the last revision or reapproval.

This specification has been approved for use by agencies of the Department of Defense. Consult the DoD Index of Specifications and Standards for the specific year of issue which has been adopted by the Department of Defense.

1. Scope*

1.1 This specification covers types of gypsum base that are designed to be used as a base for application of veneer plaster.

1.2 The values stated in inch-pound units are to be regarded as the standard.

2. Referenced Documents

2.1 *ASTM Standards:*

C 473 Test Methods for Physical Testing of Gypsum Board Products and Gypsum Lath[2]

C 645 Specification for Non-Load Bearing (Axial) Steel Studs, Runners (Tracks), and Rigid Furring Channels for Screw Application of Gypsum Board[2]

E 96 Test Methods for Water Vapor Transmission of Materials[3]

E 119 Test Methods for Fire Tests of Building Construction and Materials[4]

3. Materials and Manufacture

3.1 Gypsum base for veneer plaster shall consist of a noncombustible core, essentially gypsum, surfaced with paper bonded to the core.

3.2 Foil-backed gypsum base shall, in addition, have the back surface covered with aluminum foil.

3.3 Gypsum base for veneer plaster, Type X (special fire-resistant) designates gypsum base complying with this specification that provides not less than 1-h fire-resistance for boards ⅝ in. (15.9 mm) thick or ¾ h fire-resistance for boards ½ in. (12.7 mm) thick, applied parallel with and on each side of load bearing 2 by 4 wood studs spaced 16 in. (406 mm) on center with 6d coated nails, 1⅞ in. (48 mm) long, 0.0915 in. (2.3 mm) diameter shank, ¼ in. (6.4 mm) diameter heads, spaced 7 in. (178 mm) on center with gypsum base joints staggered 16 in. (406 mm) on each side of the partition and tested in accordance with Test Methods E 119.

NOTE 1—Consult manufacturers for independent test data on assembly details and fire resistance classifications for other types of construction. See official fire test reports for assembly particulars, materials, and classifications.

4. Physical Requirements

4.1 *Flexural Strength*—Specimens shall be tested in accordance with Test Methods C 473. When tested face up and when tested face down the specimens shall have an average breaking load of not less than the following:

	Method A		Method B	
Thickness, in. (mm)	Load, lbf (N) Bearing Edges Across Fiber of Surfacing	Load, lbf (N) Bearing Edges Parallel to Fiber of Surfacing	Load, lbf (N) Bearing Edges Across Fiber of Surfacing	Load, lbf (N) Bearing Edges Parallel to Fiber of Surfacing
⅜ (9.5)	80 (356)	30 (133)	77 (343)	26 (116)
½ (12.7)	110 (489)	40 (178)	107 (476)	36 (160)
⅝ (15.9)	150 (667)	50 (222)	147 (654)	46 (205)

4.2 *Humidified Deflection*—Specimens taken from the gypsum base for veneer plaster shall have an average deflection of no more than the following when tested in accordance with Test Methods C 473:

Thickness, in. (mm)	Humidified Deflection, Eighths of an inch (mm)
⅜ (9.5)	15 (48)
½ (12.7)	10 (32)
⅝ (15.9)	5 (16)

4.3 *Core, End, and Edge Hardness*—Specimens taken from the gypsum base for veneer plaster shall have an average hardness of not less than 15 lbf (67 N) when tested by Method A and 11 lbf (49 N) when tested by Method B for the core, ends, or edges when tested in accordance with Test Methods C 473.

4.4 *Nail-Pull Resistance*—Specimens taken from the gypsum base for veneer plaster shall have an average nail-pull resistance of not less than the following when tested in accordance with Test Methods C 473:

Thickness, in. (mm)	Method A Nail Pull Resistance, lbf (N)	Method B Nail Pull Resistance, lbf (N)
⅜ (9.5)	60 (267)	56 (249)
½ (12.7)	80 (356)	77 (343)
⅝ (15.9)	90 (400)	87 (387)

4.5 *Permeance*—Foil-backed gypsum base for veneer plaster shall have a permeance of not more than 0.30 perm for the condition of 50 % relative humidity on Side I (the

[1] This specification is under the jurisdiction of ASTM Committee C-11 on Gypsum and Related Building Materials and Systems, and is the direct responsibility of Subcommittee C11.01 on Specifications and Test Methods for Gypsum Products.

Current edition approved Dec. 10, 1995. Published February 1996. Originally published as C 588 – 66 T. Last previous edition C 588 – 95.

[2] *Annual Book of ASTM Standards*, Vol 04.01.

[3] *Annual Book of ASTM Standards*, Vol 04.06.

[4] *Annual Book of ASTM Standards*, Vol 04.07.

*** A Summary of Changes section appears at the end of this specification.**

face of the board) and 0 % relative humidity on Side II (the foil-covered back of the board) when tested in accordance with the Desiccant Procedure of Test Methods E 96.

5. Dimensions and Permissible Variations

5.1 *Thickness*—The nominal thickness of gypsum base for veneer plaster shall be 3/8, 1/2, or 5/8 in. (9.5, 12.7, or 15.9 mm), with permissible variations in the nominal thickness of ±1/64 in. (±0.4 mm), and with permissible local variations of ±1/32 in. (±0.8 mm) from the nominal thickness.

5.2 *Width*—The nominal width of gypsum base for veneer plaster shall be 48 in. (1.22 m) with a permissible variation of 3/32 in. (2.4 mm) under the specified width.

5.3 *Length*—The nominal length of gypsum base for veneer plaster shall be from 6 to 16 ft (1.8 to 4.9 m) with permissible variations of ±1/4 in. (±6.4 mm) from the specified length.

5.4 *Edges and Ends*—The edges and ends of gypsum base shall be straight and solid. Edges shall be square, rounded, or tapered. The average thickness of the tapered edge of gypsum base shall be at least 0.015 in. (0.38 mm) but not more than 0.075 in. (1.90 mm) less than the average thickness of the gypsum base as determined in accordance with Test Methods C 473. The corners shall be square with a permissible variation of ±1/8 in. (±3.2 mm) in the full width of the base.

6. Workmanship, Finish, and Appearance

6.1 The surface of gypsum base for veneer plaster shall be free of any imperfections or foreign substances that would render it unfit for the intended use.

7. Sampling

7.1 At least 0.25 % of the number of gypsum base for veneer plaster in a shipment, but not less than three boards, shall be so selected as to be representative of the shipment and shall constitute a sample for purpose of tests by the purchaser or user.

8. Inspection

8.1 Inspection of the gypsum base for veneer plaster shall be agreed upon between the purchaser and the supplier as part of the purchase agreement.

9. Rejection

9.1 Rejection of gypsum base for veneer plasters that fails to conform to the requirements of this specification shall be reported to the producer or supplier promptly and in writing. The notice of rejection shall contain a statement documenting how the gypsum base has failed to conform to the requirements of this specification.

10. Certification

10.1 When specified in the purchase agreement, a manufacturer's or supplier's report shall be furnished at the time of shipment certifying that the product is in compliance with this specification.

11. Packaging and Package Marking

11.1 Gypsum base for veneer plaster shall be shipped so as to be kept dry and free of moisture.

11.2 When shipped for resale, the name of the manufacturer or supplier and the brand shall be legibly marked on each board or package.

12. Keywords

12.1 gypsum base; gypsum board; veneer plaster

SUMMARY OF CHANGES

This section identifies the location of changes to this specification that have been incorporated since the last issue. Committee C-11 has highlighted those changes that affect the technical interpretation or use of this specification.

(1) Paragraph 9.1 was revised to change permissive language into mandatory language.
(2) A keywords section was added.

APPENDIX

(Nonmandatory Information)

This Appendix gives general information and also suggestions for inclusions to be made elsewhere by the specifier. They are not part of this specification.

X1. Alternate Definition for Type X

X1.1 Gypsum base for veneer plaster, type X (special fire- resistant) designates gypsum base for veneer plaster providing a greater fire resistance than regular gypsum base for veneer plaster of the same thickness. Type X (special fire-resistant) gypsum base for veneer plaster, when tested in accordance with Test Methods E 119, shall provide the following minimum fire resistance for the assemblies described.

X1.1.1 One hour for a 5/8 in. (15.9 mm) thickness applied to a partition in a single layer application on each side of 3 5/8 in. (92 mm) deep non-load bearing galvanized steel studs complying with Specification C 645 spaced 24 in. (610 mm) on center. The 5/8 in. (15.9 mm) thick gypsum base for veneer plaster 48 in. (1220 mm) wide shall be attached using 1 in. (25 mm) long drywall screws spaced 8 in. (203 mm) on center along the edges and ends, and 12 in. (305 mm) along intermediate studs. All joints shall be oriented parallel to and located over studs and staggered on opposite sides of the assembly. All joints shall be covered with joint reinforcing tape or mesh and covered with a coat of joint compound or veneer plaster. All screw heads shall be covered with joint compound or veneer plaster, and

X1.1.2 Two hours for a ½ in. (12.7 mm) thickness applied to a partition in a double layer application on each side of 2½ in. (64 mm) deep non-load bearing galvanized steel studs complying with Specification C 645 spaced 24 in. (610 mm) on center. The 48 in. (1220 mm) wide base layer shall be attached using 1 in. (25 mm) long drywall screws spaced 12 in. (305 mm) on center along board edges, ends and along intermediate studs. Joints shall be oriented parallel to and located over studs and staggered on opposite sides of the assembly. The 48 in. (1220 mm) wide face layer shall be attached using 1⅝ in. (41 mm) long drywall screws spaced 12 in. (305 mm) along board edges, ends and along intermediate studs. Joints shall be oriented parallel to and located over studs, offset 24 in. (610 mm) from the base layer joints, and staggered on opposite sides of the assembly. All joints in the face layer shall be covered with joint reinforcing tape or mesh and covered with a coat of joint compound or veneer plaster. All screw heads in the face layer shall be covered with joint compound or veneer plaster.

Standard Specification for Fly Ash and Other Pozzolans for Use With Lime[1]

This standard is issued under the fixed designation C 593; the number immediately following the designation indicates the year of original adoption or, in the case of revision, the year of last revision. A number in parentheses indicates the year of last reapproval. A superscript epsilon (ϵ) indicates an editorial change since the last revision or reapproval.

1. Scope

1.1 This specification covers fly ash and other pozzolans for use with lime in plastic mortars, nonplastic mixtures and other mixtures that affect lime pozzolanic reaction. Evaluation of pozzolans containing available lime, such as Class C fly ash, is given consideration. Pozzolans covered include artificial pozzolans such as fly ash, and natural pozzolans, such as diatomite and pumicite, in either raw or calcined state.

1.2 The following precautionary caveat pertains only to the test method portion, Sections 5 and 10 of this specification: *This standard does not purport to address all of the safety concerns, if any, associated with its use. It is the responsibility of the user of this standard to establish appropriate safety and health practices and determine the applicability of regulatory limitations prior to use.*

2. Referenced Documents

2.1 *ASTM Standards:*

C 25 Test Methods for Chemical Analysis of Limestone, Quicklime, and Hydrated Lime[2]

C 39 Test Method for Compressive Strength of Cylindrical Concrete Specimens[3]

C 50 Practice for Sampling, Inspection, Packing, and Marking of Lime and Limestone Products[2]

C 51 Terminology Relating to Lime and Limestone (As Used by the Industry)[2]

C 109/C 109M Test Method for Compressive Strength of Hydraulic Cement Mortars (Using 2-in. or 50-mm Cube Specimens)[2]

C 110 Test Methods for Physical Testing of Quicklime, Hydrated Lime, and Limestone[2]

C 207 Specification for Hydrated Lime for Masonry Purposes[2]

C 305 Practice for Mechanical Mixing of Hydraulic Cement Pastes and Mortars of Plastic Consistency[2]

C 311 Test Methods for Sampling and Testing Fly Ash or Natural Pozzolans for Use as a Mineral Admixture in Portland-Cement Concrete[3]

C 670 Practice for Preparing Precision and Bias Statements for Test Methods for Construction Materials[3]

C 821 Specification for Lime for Use With Pozzolans[2]

D 1557 Test Methods for Laboratory Compaction Characteristics of Soil Using Modified Effort (56,000 ft-lbf/ft^3 (2,700 kN-m/m^3))[4]

D 5239 Practice for Characterizing Fly Ash for Use in Soil Stabilization[5]

3. Terminology

3.1 *Descriptions of Terms:*

3.1.1 *calcined pozzolans*—materials that are produced by calcination of natural siliceous or alumino-siliceous earths, such calcination being for the purpose of activation of pozzolanic properties.

3.1.2 *fly ash*—the finely divided residue that results from the combustion of ground or powdered coal and is transported from the boiler by flue gases. Additional methods for characterization of fly ash can be found in Practice D 5239.

DISCUSSION—Section 12 provides guidance for determining the available lime index of fly ash, which may affect the desired proportions of fly ash with lime.

3.1.3 *lime*—all classes of quicklime and hydrated lime, both calcitic (high calcium) and dolomitic.

3.1.4 *natural pozzolans*—materials that, in the natural state, exhibit pozzolanic properties, such as some volcanic ash and lava deposits.

3.1.5 *pozzolan*—a siliceous or alumino-siliceous material that in itself possesses little or no cementitious value but that in finely divided form and in the presence of moisture will chemically react with alkali and alkaline earth hydroxides at ordinary temperatures to form or assist in forming compounds possessing cementitious properties.

4. Physical Properties

4.1 Pozzolans for use with lime in plastic mortars, when tested in accordance with the procedures of Sections 7 to 9, shall conform to the requirements prescribed in Table 1.

4.2 Pozzolans for use with lime in nonplastic mixtures shall conform to the requirements of Table 1, except the lime-pozzolan strength requirement, and in addition shall be tested in accordance with the procedures of Section 10 and shall conform to the following requirements:

Compressive strength, min, psi (MPa)	400 (2.8)
Vacuum saturation strength, min, psi (MPa)	400 (2.8)

NOTE 1—If the minimum value of the vacuum saturation strength specified in 4.2 of this specification is reduced, sufficient documentation shall be provided to the user to enable the determination of a satisfactory minimum residual strength for the given material in its intended use. Such documentation should include at least the following: (*1*) determination of a minimum residual strength requirement that will enable the material to perform its structural function in the pavement system; and

---

[1] This specification is under the jurisdiction of ASTM Committee C-7 on Lime and is the direct responsibility of Subcommittee C07.07 on Pozzolanic Materials.

Current edition approved Feb. 15, 1995. Published April 1995. Originally published as C 593 – 66 T. Last previous edition C 593 – 89 (1994)$^{\epsilon 1}$.

[2] *Annual Book of ASTM Standards,* Vol 04.01.

[3] *Annual Book of ASTM Standards,* Vol 04.02.

[4] *Annual Book of ASTM Standards,* Vol 04.08.

[5] *Annual Book of ASTM Standards,* Vol 04.09.

TABLE 1 Physical Requirements

Water-soluble fraction, max, %	10.0
Fineness, amount retained, when wet sieved:	
No. 30 (600-μm) sieve, max, %	2.0
No. 200 (75-μm) sieve, max, %	30.0
Lime-pozzolan strength, minimum compressive strength, psi (MPa):	
At 7 days, 130 ± 3°F (54 ± 2°C)	600 (4.1)
After additional 21 days, 73 ± 3°F (23 ± 2°C)	600 (4.1)

(*2*) a rational analysis, using actual climatic data, that will show the severity of exposure of the material to cyclic freeze-thaw action.

5. Sampling

5.1 Sample pozzolan in accordance with the applicable provisions of Test Methods C 311, except take one 10-lb (4.5-kg) sample from approximately each 400 tons (350 metric tons) of pozzolan.

5.2 The sampling procedures and techniques shall be consistent from original sample to project completion.

TEST METHODS

6. Significance and Use

6.1 This test method states various procedures that are recommended to quantify various aspects of the lime enhanced pozzolanic reaction. These tests are intended to qualify or quantify sources of fly ash and other pozzolans to meet specified job or project criteria.

7. Water-Soluble Fraction

7.1 *Procedure*—Place 10 g of a dried pozzolan sample (dried to constant weight in an oven at 221 to 230°F (105 to 110°C)) in a 200-mL Erlenmeyer flask and add 100 mL of distilled water at 73 ± 3°F (23 ± 2°C). Shake well by hand until no lumps can be observed; then with a mechanical shaker or stirring device, agitate at laboratory room temperature for a period of 1 h. Pour the material into a weighed Gooch or sintered-glass crucible, and wash all residue from the flask into the crucible with distilled water from a wash bottle. Wash the residue in the crucible free of adhering solution by repeated washings with distilled water. Dry the crucible to constant weight in an oven at 221°F (105°C).

7.2 *Calculation*—Calculate the percentage of water-soluble fraction by multiplying the loss in weight in grams by 10.

8. Fineness

8.1 Test in accordance with Test Methods C 110, except that the sample shall be 100 g of the dried pozzolan.

9. Lime-Pozzolan Strength Development

9.1 Test the pozzolan in accordance with the applicable portions of Test Method C 109 and Practice C 305, and in accordance with the following:

9.2 *Apparatus:*

9.2.1 *Oven,* closed, vapor-type.

9.3 *Materials:*

9.3.1 *Hydrated Lime*—Where possible, the lime shall be the same as that to be used on the job or shall meet the requirements of Specification C 207.

9.3.2 *Sand*—The sand shall be graded standard sand as required by Test Method C 109.

9.4 *Number of Test Specimens*—Three specimens shall be prepared for each age at which a strength test is desired.

9.5 *Proportioning, Consistency, and Mixing of Mortars*—Batches shall be of a size sufficient to make six specimens and shall consist of proportions of dry materials as follows:

Hydrated lime	180 g
Pozzolan (dry basis)	360 g
Graded standard sand	1480 g

9.5.1 The amount of mixing water, measured in millilitres, shall be such as to produce a flow of 65 to 75 as determined in accordance with 9.6, and shall be expressed as weight percent of the combined lime and pozzolan. The lime and pozzolan shall be blended together in a closed container. Mixing shall be done in accordance with the procedure described in Practice C 305, except that it shall be amended to read "Add the blended lime and pozzolan to the water and allow it to stand for 1 min. Then start the mixer and mix at slow speed (140 ± 5 rpm) for 30 s."

9.6 *Determination of Flow*—Determine the flow in accordance with Test Method C 109, except that the number of drops of the flow table shall be 10 drops in 6 s instead of 25 drops in 15 s. If the flow is less than the specified limit, the material used for the flow test may be returned to the mixing bowl and additional water added, the batch mixed for 1½ min, and a new flow taken. This operation may be repeated until a flow within the specified range is obtained. If the flow exceeds the range specified, discard the batch and give a new batch a new trial until a flow within range is obtained.

9.7 *Molding Test Specimens*—Immediately after the completion of the flow test, mold mortar specimens in accordance with Test Method C 109.

9.8 *Storage of Test Specimens*—When molding is completed, place the filled mold in the vapor immediately above water at 130 ± 3°F (54 ± 2°C) in a closed vapor oven with the top surface protected from the drip. Allow the specimens in the molds to remain in the vapor for a period of 7 days, after which remove them from the vapor and cool to 73 ± 3°F (23 ± 2°C) in air saturated sufficiently that no drying takes place during the cooling. When the specimens are cool, remove them from the molds, and store them at 73 ± 3°F (23 ± 2°C) at 95 to 100 % relative humidity until time of the compressive strength test.

10. Compressive Strength Development and Freeze-Thaw Resistance of Nonplastic Mixtures

10.1 *Materials:*

10.1.1 *Hydrated Lime*—Where possible, the lime shall be the same as that to be used on the job, or shall meet the applicable requirements of Specification C 207, Type N. Prior to usage, the lime shall be stored in a sealed container to prevent carbonation.

10.1.2 *Pozzolan*—The pozzolan used in this test shall be the same as intended for use on the job.

10.1.3 *Aggregate*—Where possible, the aggregate used in this test shall be the same as intended for use on the job. When using job aggregates, discard the material, if any, retained on the ¾-in. (19.0-mm) sieve. If the aggregate fraction between the ¾-in. and No. 4 (4.75-mm) sieve does not contain free surface moisture, then that fraction of the aggregate shall be soaked for 24 h and towel dried to obtain a saturated surface dry condition. If job aggregates are not available, graded standard sand as specified in Test Method C 109 shall be used.

10.2 *Number of Specimens*—Three specimens shall constitute one test for the compressive strength test with three additional specimens for the freeze-thaw test.

10.3 *Proportioning*—The proportion of dry materials by weight shall be, where possible, the same proportions as intended for use on the job. If graded standard sand is used as the aggregate, the proportions of dry materials by weight shall be as follows:

Hydrated lime	4 %
Pozzolan	24 %
Graded standard sand	72 %

The amount of mixing water shall be the optimum moisture content as determined by Method C of Test Methods D 1557, except that the 5-lift requirement is replaced with 3 lifts and Note 1 is not to be used. In determining the moisture-density relationship, mix dry materials in a Lancaster PC Mixer, or its equivalent, for 1 min, or until the mixture is uniform in color and texture, plus an additional 3 min after the water is added in order to obtain the first point on the moisture-density curve. The original sample may be reused for subsequent trials. The batch shall be mixed for an additional minute after the water has been added for each subsequent trial.

10.4 *Mixing and Molding Test Specimens*—After the optimum moisture content is obtained by the above procedure, a batch large enough to make three 4.0 by 4.6-in. (102 by 117-mm) cylinders (approximately 15 lb (7 kg)) shall be mixed in the following manner: Mix the dry materials in a Lancaster PC Mixer, or its equivalent, for 1 min or until the mixture is uniform in color and texture, followed by the addition of water that will give optimum moisture content and an additional 3 min of mixing. Mold the specimens immediately in accordance with Method C of Test Methods D 1557, except as previously noted. Each layer should be scarified to a depth of ¼ in. (6 mm) before the next layer is compacted in order to assure a good bond between the layers. Weigh a representative sample of the mixture, using a container with a tight lid to assure that no moisture is lost while determining the weight of the sample. Dry to constant weight and calculate the actual moisture content of the sample. After molding, weigh each sample in the mold to determine the uniformity of molded weights of the specimens and then carefully remove from the mold by the use of a sample extruded, such as a jack or lever frame.

10.5 *Curing of Test Specimens*—Immediately after the specimens are removed from the mold, reweigh the specimens and place in a sealed container (½-gal (2-L) or 1-gal (4-L) can with double friction lids) to prevent loss of moisture. Place the specimens in the sealed containers carefully in a room or cabinet with forced-air circulation maintained at 100 ± 3°F (38 ± 2°C) for a 7-day period. After this period, remove the specimens from the container, reweigh, and allow to cool to room temperature. Submerge the specimens for compressive strength testing in water for 4 h, remove, allow to drain on a nonabsorbent surface, and cap and test within 1 h of the time of removal from the water.

10.6 *Number of Test Specimens:*

10.6.1 Three specimens shall be tested in accordance with Test Method C 39; no l/d correction will be considered in the computation of the compressive strength.

10.6.2 Three specimens shall be tested for freeze-thaw resistance after being cured in accordance with 10.5 (without the 4-h soaking period) by means of the vacuum saturation strength testing procedure described in Section 11.

11. Vacuum Saturation Strength Testing Procedure

11.1 *Equipment:*

11.1.1 *Vacuum Saturation Chamber*—The vacuum saturation chamber is a 12-in. (305-mm) high by 12-in. inside diameter stainless steel cylindrical section welded to a ½-in. (12.7-mm) thick by 14-in. (356-mm) diameter stainless steel base plate. The wall thickness of the cylindrical section is ⅜ in. (9.5 mm). The lid of the vacuum saturation chamber is a poly(methylmethacrylate) (PMMA) plate 1 in. (13 mm) thick and 14 in. in diameter. Both the PMMA lid and top of the vacuum cylinder are grooved for a ¼-in. (6.4-mm) circular O-ring seal having an inside diameter of 12⅛ in. (308 mm). The lid is fastened to the chamber by six equally spaced threaded ¼-in. rods which pass along the outside wall of the cylindrical section and thread into the base plate.

11.1.1.1 A sketch of the vacuum saturation chamber described above is shown in Fig. 1. A vacuum saturation chamber of equivalent size and capability is permitted under this specification. Vacuum desiccators can also be used for this purpose.

11.1.1.2 A ¼-in. (6.4-mm) vacuum line connection is located 1 in. (13 mm) below the top of the vacuum chamber and a ⅜-in. (9.5-mm) water line connection with control valve is located at the base of the vacuum chamber. The vacuum line is connected to a commercial vacuum pump and the water line is connected to a reservoir of desired water. The vacuum is controlled by a pressure valve at the vacuum pump.

11.1.1.3 The specimen support plate inside of the chamber is constructed of ½-in. (12.7-mm) thick PMMA which is 11½ in. (292 mm) in diameter. The support plate sits on three 1½-in. (38.1-mm) long legs which elevate it off of the bottom of the chamber. The specimen support plate is perforated (approximately ten ⅛-in. (3.2-mm) diameter holes per square inch) so as to allow complete access of water to the specimens during saturation. For an equivalent size vacuum saturation chamber, a specimen support plate similar to that described above must be provided.

11.1.1.4 The vacuum saturation chamber must be of sufficient size to hold the same number of Proctor-sized specimens for vacuum saturation testing as the number of specimens tested for compressive strength.

11.1.2 *Vacuum System*—A system capable of maintaining a vacuum of 24 in. Hg (11.8 psi) for a minimum of 30 min is required.

11.2 *Procedure:*

11.2.1 At the end of the curing period, remove the specimens from the curing room and allow approximately 2 h to reach equilibrium with room temperature. The specimens should remain sealed in the containers during this 2-h equilibration period in order to prevent moisture loss.

11.2.2 Place the cured specimens in an upright position on the specimen support plate within the vacuum chamber. Place the lid on the vacuum chamber and evacuate the chamber to a pressure of 24 in. Hg gradually over a period of

FIG. 1 Pictorial View of Vacuum Saturation Equipment

not less than 45 s and hold for 30 min in order to remove air from the voids in the specimens. After the 30-min de-airing period, flood the vacuum chamber with water at room temperature to a depth sufficient to cover the specimens. Remove the vacuum and then soak the specimens for 1 h at atmospheric pressure.

11.2.3 At the end of the soaking period, remove the specimens from the water and allow to drain for approximately 2 min on a nonabsorptive surface. After the free surface water has drained, immediately test the specimens for unconfined compressive strength in accordance with Test Method C 39.

11.3 *Report*—Report of the compressive strength and vacuum saturation strength tests shall include the following:

11.3.1 Identification of each material used in the preparation of the specimens,

11.3.2 Percentage by dry weight of each of the constituents,

11.3.3 Actual percentage moisture content of mixture,

11.3.4 Actual dry unit weight of each specimen, nearest lb/ft^3 or g/cm^3,

11.3.5 Percentage of maximum dry unit weight of each specimen,

11.3.6 Cross-sectional area of each specimen, $in.^2$ or cm^2,

11.3.7 Maximum failure load of each specimen, lbf or N,

11.3.8 Compressive strength of each specimen, to nearest 5 psi or 50 kPa, and

11.3.9 Vacuum saturation strength of each specimen, to nearest 5 psi or 50 kPa.

11.3.10 The average compressive strength of the three specimens tested shall be designated as the test value for evaluation by this specification. The average vacuum saturation strength of the three specimens tested shall be designated as the test value for evaluation by this specification.

12. Available Lime Index of Fly Ash

12.1 The available lime index may be determined using the Available Lime Index Method of Test Methods C 25 (Hydrated Lime Procedure). See also Practice C 50, Terminology C 51, and Specification C 821.

12.2 The precision for pulverized coal fly ash, using the procedure of 12.1, is as follows:

12.2.1 The single operator standard deviation has been found to be 0.095 %.[6] Therefore, results of two properly conducted tests by the same operator on the same material should not differ by more than 0.268 %.[6]

12.2.2 The multilaboratory standard deviation has been found to be 0.264 %.[5] Therefore, results of two properly conducted tests from two different laboratories on identical samples should not differ by more than 0.75 %.[6]

13. Storage and Inspection

13.1 Pozzolans shall be stored in such a manner as to permit easy access for proper inspection and identification of each shipment. Reasonable facilities shall be provided the purchaser for careful sampling and inspection either at the source or at the site of the work, as may be specified by the purchaser.

14. Rejection

14.1 Pozzolan may be rejected if it fails to meet any of the requirements of this specification.

14.2 Packages varying more than 5 % from the stated weight may be rejected; and if the average weight of the packages in any shipment, as shown by weighing 50 packages taken at random, is less than that specified, the entire shipment may be rejected.

15. Packaging and Package Marking

15.1 When pozzolan is delivered in packages, the name and brand and the weight of material contained therein shall

---

[6] These numbers represent, respectively, the (1s) and (d2s) limits as described in Practice C 670.

be plainly marked on each package. Similar information shall be provided in shipping invoices accompanying the shipment of packaged or bulk pozzolans.

16. Keywords

16.1 fly ash; freeze-thaw resistance; lime; nonplastic mixture; pozzolan; plastic mortar; strength development

The American Society for Testing and Materials takes no position respecting the validity of any patent rights asserted in connection with any item mentioned in this standard. Users of this standard are expressly advised that determination of the validity of any such patent rights, and the risk of infringement of such rights, are entirely their own responsibility.

This standard is subject to revision at any time by the responsible technical committee and must be reviewed every five years and if not revised, either reapproved or withdrawn. Your comments are invited either for revision of this standard or for additional standards and should be addressed to ASTM Headquarters. Your comments will receive careful consideration at a meeting of the responsible technical committee, which you may attend. If you feel that your comments have not received a fair hearing you should make your views known to the ASTM Committee on Standards, 1916 Race St., Philadelphia, PA 19103.

Standard Specification for Blended Hydraulic Cements[1]

This standard is issued under the fixed designation C 595 ; the number immediately following the designation indicates the year of original adoption or, in the case of revision, the year of last revision. A number in parentheses indicates the year of last reapproval. A superscript epsilon (ε) indicates an editorial change since the last revision or reapproval.

This specification has been approved for use by agencies of the Department of Defense.

1. Scope

1.1 This specification pertains to five classes of blended hydraulic cements for both general and special applications, using slag or pozzolan, or both, with portland cement or portland cement clinker or slag with lime.

NOTE 1—This specification prescribes ingredients and proportions, with some performance requirements whereas Performance Specification C 1157 is a blended cement specification in which performance criteria alone govern the products and their acceptance.

1.2 For properties where values are given in both SI and non SI units, the values in SI units are to be regarded as the standard. Values in SI units shall be obtained by measurement in SI units or by appropriate conversion, using the Rules for Conversion and Rounding given in Standard IEEE/ASTM SI 10, of measurements made in other units.

1.3 The text of this standard refers to notes and footnotes which provide explanatory material. These notes and footnotes (excluding those in tables and figures) are not requirements of the standard.

2. Referenced Documents

2.1 *ASTM Standards:*

C 109/C 109M Test Method for Compressive Strength of Hydraulic Cement Mortars (Using 2-in. or 50-mm Cube Specimens)[2]

C 114 Test Methods for Chemical Analysis of Hydraulic Cement[2]

C 150 Specification for Portland Cement[2]

C 151 Test Method for Autoclave Expansion of Portland Cement[2]

C 157 Test Method for Length Change of Hardened Hydraulic-Cement Mortar and Concrete[3]

C 183 Practice for Sampling and the Amount of Testing of Hydraulic Cement[2]

C 185 Test Method for Air Content of Hydraulic Cement Mortar[2]

C 186 Test Method for Heat of Hydration of Hydraulic Cement[2]

C 187 Test Method for Normal Consistency of Hydraulic Cement[2]

C 188 Test Method for Density of Hydraulic Cement[2]

C 191 Test Method for Time of Setting of Hydraulic Cement by Vicat Needle[2]

C 204 Test Method for Fineness of Hydraulic Cement by Air Permeability Apparatus[2]

C 219 Terminology Relating to Hydraulic Cement[2]

C 226 Specification for Air-Entraining Additions for Use in the Manufacture of Air-Entraining Portland Cement[2]

C 227 Test Method for Potential Alkali Reactivity of Cement-Aggregate Combinations (Mortar-Bar Method)[3]

C 265 Test Method for Calcium Sulfate in Hydrated Portland Cement Mortar[2]

C 311 Test Methods for Sampling and Testing Fly Ash or Natural Pozzolans for Use as a Mineral Admixture in Portland-Cement Concrete[3]

C 430 Test Method for Fineness of Hydraulic Cement by the 45-μm (No. 325) Sieve[2]

C 465 Specification for Processing Additions for Use in the Manufacture of Hydraulic Cements[2]

C 563 Test Method for Optimum SO_3 in Portland Cement[2]

C 688 Specification for Functional Additions for Use in Hydraulic Cements[2]

C 821 Specification for Lime for Use with Pozzolans[2]

C 1012 Test Method for Length Change of Hydraulic-Cement Mortars Exposed to a Sulfate Solution[2]

C 1157M Performance Specification for Blended Hydraulic Cement[2]

IEEE/ASTM SI 10 Standard for Use of the International System of Units (SI): the Modern Metric System[4]

3. Terminology

3.1 *Definitions*—The terms used in this specification are defined in Terminology C 219.

4. Classification

4.1 This specification applies to the following types of blended cement that generally are intended for use as indicated.

4.1.1 Blended hydraulic cements for general concrete construction.

4.1.1.1 Type IS—Portland blast-furnace slag cement.

4.1.1.2 Type IP—Portland-pozzolan cement.

4.1.1.3 Type P—Portland-pozzolan cement for use when higher strengths at early ages are not required.

[1] This specification is under the jurisdiction of ASTM Committee C-1 on Cement and is the direct responsibility of Subcommittee C01.10 on Hydraulic Cement for General Concrete Construction.

Current edition approved Nov. 10, and Dec. 10, 1997, and Feb. 10, 1998. Published July 1998. Originally published as C 595 – 67 T. Last previous edition C 595M – 97.

[2] *Annual Book of ASTM Standards*, Vol 04.01.

[3] *Annual Book of ASTM Standards*, Vol 04.02.

[4] *Annual Book of ASTM Standards*, Vol 14.02.

4.1.1.4 Type I(PM)—Pozzolan—modified portland cement.

NOTE 2—Pozzolan-modified portland cement should not be used when special characteristics attributable to the larger quantities of pozzolan in portland-pozzolan cement are desired.

4.1.1.5 Type I(SM)—Slag-modified portland cement.

NOTE 3—Slag-modified portland cement should not be used when special characteristics attributable to the larger quantities of slag in portland blast-furnace slag cements are desired.

4.1.2 Type S—Slag cement for use in combination with portland cement in making concrete and in combination with hydrated lime in making masonry mortar.

4.2 *Special Properties:*

4.2.1 Air-entraining cement, when desired by the purchaser, shall be specified by adding the suffix (A) to any of the above types. The air-entraining option may be specified in combination with any of the other special properties.

4.2.2 Moderate sulfate resistance or moderate heat of hydration, or both, when desired by the purchaser, shall be specified by adding the suffix (MS) or (MH), respectively, to the type designation under 4.1.1.

5. Ordering Information

5.1 Orders for material under this specification shall include the following:

5.1.1 Specification number,

5.1.2 Type or types required,

5.1.3 Optional special properties required (see Note 2):

5.1.3.1 MS if moderate sulfate resistance is required;

5.1.3.2 MH if moderate heat of hydration is required;

5.1.3.3 LH if low heat of hydration is required, (Type P only);

5.1.3.4 A if air entraining is required;

5.1.3.5 Accelerating addition, if required;

5.1.3.6 Retarding addition, if required;

5.1.3.7 Water reducing addition, if required;

5.1.3.8 Water reducing and accelerating addition, if required; and

5.1.3.9 Water reducing and retarding addition, if required.

5.1.4 Certification, if desired (see Section 14).

NOTE 5—It is important to check for availability of various options. Some multiple options are mutually incompatible or unattainable.

6. Materials and Manufacture

6.1 *Portland Blast-Furnace Slag Cement*—The portland blast-furnace slag cement shall consist of an intimate and uniform blend (see Note 6) of portland cement and fine granulated blast-furnace slag produced either by intergrinding portland cement clinker and granulated blast-furnace slag, or by blending portland cement and finely ground granulated blast-furnace slag, or a combination of intergrinding and blending, in which the slag constituent is between 25 and 70 % of the mass of portland blast-furnace slag cement.

NOTE 6—The attainment of an intimate and uniform blend of two or more types of fine materials is difficult. Consequently, adequate equipment and controls must be provided by the manufacturer. The purchaser should assure himself of the adequacy of the blending operation.

6.2 *Air-Entraining Portland Blast-Furnace Slag Cement*—Air-entraining portland blast-furnace slag cement shall be portland blast-furnace slag cement to which sufficient air-entraining addition has been added so that the resulting product complies with the air content of mortar requirements.

6.3 *Slag-Modified Portland Cement*—Slag-modified portland cement shall be an intimate and uniform blend of portland cement and granulated blast-furnace slag produced either by intergrinding portland cement clinker and granulated blast-furnace slag (see Note 6), by blending portland cement and finely ground granulated blast-furnace slag, or a combination of intergrinding and blending in which the slag constituent is less than 25 % of the mass of the slag-modified portland cement.

6.4 *Air-Entraining Slag-Modified Portland Cement*—Air-entraining slag-modified portland cement shall be slag-modified portland cement to which sufficient air-entraining addition has been added so that the resulting product complies with the air content of mortar requirements.

6.5 *Blast-Furnace Slag*—Blast-Furnace slag shall be the nonmetallic product, consisting essentially of silicates and aluminosilicates of calcium and other bases, that is developed in a molten condition simultaneously with iron in a blast furnace.

6.6 *Granulated Blast-Furnace Slag*—Granulated blast-furnace slag shall be the glassy granular material formed when molten blast-furnace slag is rapidly chilled, as by immersion in water.

6.7 *Portland Cement*—See Terminology C 219. For purposes of this specification, portland cement meeting the requirements of Specification C 150 is suitable. Portland cement or other hydraulic materials, or both, containing high free lime may be used as long as the autoclave test limits for the blended cement are met.

6.8 *Portland Cement Clinker*—Portland cement clinker shall be partially fused clinker consisting primarily of hydraulic calcium silicates.

6.9 *Portland-Pozzolan Cement*—Portland-pozzolan cement shall be a hydraulic cement consisting of an intimate and uniform blend (see Note 6) of portland or portland blast-furnace slag cement and fine pozzolan produced either by intergrinding portland cement clinker and pozzolan, by blending portland cement or portland blast-furnace slag cement and finely divided pozzolan, or a combination of intergrinding and blending, in which the pozzolan constituent is between 15 and 40 mass % of the portland-pozzolan cement.

6.10 *Air-Entraining Portland-Pozzolan Cement*—Air-entraining portland-pozzolan cement shall be portland-pozzolan cement to which sufficient air-entraining addition has been added so that the resulting product complies with the air content of mortar requirements.

6.11 *Pozzolan-Modified Portland Cement*—Pozzolan-modified portland cement shall be an intimate and uniform blend (see Note 6) of portland cement or portland blast-furnace slag cement and fine pozzolan produced either by intergrinding portland cement clinker and pozzolan, by blending portland cement or portland blast-furnace slag cement and finely divided pozzolan, or a combination of intergrinding and blending, in which the pozzolan constituent is less than 15 mass % of the pozzolan-modified portland cement.

6.12 *Air-Entraining Pozzolan-Modified Portland Cement*—Air-entraining pozzolan-modified portland cement shall be pozzolan-modified portland cement to which sufficient air-entraining addition has been added so that the resulting product complies with the air content of mortar requirements.

6.13 *Pozzolan*—Pozzolan shall be a siliceous or siliceous and aluminous material, which in itself possesses little or no cementitious value but which will, in finely divided form and in the presence of moisture, chemically react with calcium hydroxide at ordinary temperatures to form compounds possessing cementitious properties.

6.14 *Slag Cement*—Slag cement shall be hydraulic cement consisting mostly of an intimate and uniform blend (see Note 6) of granulated blast-furnace slag and portland cement, or hydrated lime, or both, in which the slag constituent is at least 70 % of the mass of the slag cement.

6.15 *Air-Entraining Slag Cement*—Air-entraining slag cement shall be slag cement to which sufficient air-entraining addition has been added so that the resulting product complies with the air content of mortar requirements.

6.16 *Hydrated Lime*—Hydrated lime used as part of a blended cement shall meet the requirements of Specification C 821, except that when interground in the production process there shall be no minimum fineness requirement.

6.17 *Air-Entraining Addition*—When air-entraining cement is specified, an addition meeting the requirements of Specification C 226 shall be used.

6.18 When processing additions are used in the manufacture of cement, they shall have been shown to meet the requirements of Specification C 465 in the amounts used or greater, (see Section 14.2).

6.19 When functional additions (used at the sole option of the purchaser, and in amounts not to exceed 0.50 % by mass of the cement) are used they shall have been shown to meet the requirements of Specification C 688 when tested with the cement to be used, in the amount used or greater, (see Section 14.2 and Note 7).

NOTE 7—The 0.50 % by mass is an arbitrarily selected value.

6.20 *Other Additions*—The cement covered by this specification shall contain no additions except as provided for above except that water or calcium sulfate (see Terminology C 219), or both, may be added in amounts so that the limits shown in Table 1 for sulfate reported as SO_3 and loss on ignition are not exceeded.

7. Chemical Composition

7.1 Cement of the type specified shall conform to the applicable chemical requirements prescribed in Table 1.

7.2 If the purchaser has requested the manufacturer to state in writing the composition of the blended cement purchased, the composition of the cement furnished shall conform to that shown in the statement within the following tolerances (see Note 8).

	Tolerance, ± %
Silicon dioxide (SiO_2)	3
Aluminum oxide (Al_2O_3)	2
Calcium oxide (CaO)	3

NOTE 8—This means that if the manufacturer's statement of the composition says "SiO_2: 32 %," the cement when analyzed, shall be found to contain between 29 and 35 % SiO_2.

TABLE 1 Chemical Requirements

Cement Type	I(SM), I(SM)-A, IS, IS-A	S,SA	I(PM), I(PM)-A, P, PA, IP, IP-A
Magnesium oxide (MgO), max, %	. . .	. . .	6.0
Sulfur reported as sulfate (SO_3), max, %[B]	3.0	4.0	4.0
Sulfide sulfur (S), max, %	2.0	2.0	. . .
Insoluble residue, max, %	1.0	1.0	. . .
Loss on ignition, max, %	3.0	4.0	5.0
Water-soluble alkali, max, %	. . .	0.03[A]	. . .

[A] Applicable only when the cement is specified to be nonstaining to limestone. The amount and nature of the staining material in limestone vary with the stone. The alkali in any cement may, therefore, induce markedly different staining on different stone, even though the stone may have come apparently from the same source. The amount of alkali permitted by the specification should not cause stain unless stone high in staining material has been used, or unless insufficient means have been used to prevent infiltration of water into the masonry.

[B] When it has been demonstrated by Test Method C 563 that the optimum SO_3 exceeds a value 0.5 % less than the specification limit, an additional amount of SO_3 is permissible provided that, when the cement with the additional calcium sulfate is tested by Test Method C 265, the calcium sulfate in the hydrated mortar at 24 ± ¼ h, expressed as SO_3, does not exceed 0.50 g/L. When the manufacturer supplies cement under this provision, he will, upon request, supply supporting data to the purchaser.

8. Physical Properties

8.1 *Blended Cement*—Blended cement of the type specified shall conform to the applicable physical requirements prescribed in Table 2.

8.2 *Pozzolan or Slag*—Pozzolan or slag that is to be blended with cement shall be tested in the same state of subdivision as that in which it is to be blended. Pozzolan shall conform to the fineness requirement and the pozzolanic activity requirement of Table 3. Slag that is to be used for slag-modified portland cements shall conform to the slag activity requirement of Table 3. Such pozzolan or slag that is to be interground with portland cement clinker shall, before testing for conformance with requirements of Table 3, be ground in the laboratory to a fineness at which it is believed to be present in the finished cement. It is the manufacturer's responsibility to decide on the fineness at which the testing is to be carried out, and when requested to do so by a purchaser, to report the information upon which the decision was based.

8.3 Pozzolan for use in the manufacture of pozzolan-modified portland cement, Type I(PM) and I(PM)-A, shall meet the requirements of Table 3 when tested for mortar expansion of pozzolan as described in 10.1.13. If the alkali content of the clinker to be used for the production lots changes by more than 0.2 % total as equivalent Na_2O, calculated as $Na_2O + 0.658\ K_2O$, from that of the clinker with which the acceptance tests were carried out, the pozzolan shall be retested to show compliance with the requirements of Table 3.

9. Sampling

9.1 Sample the materials in accordance with the following methods:

9.1.1 *Sampling Blended Cements*—Practice C 183.

9.1.2 *Sampling Pozzolan*—Test Methods C 311. One 2 kg (4 lb) sample shall be taken from approximately each 360 Mg (400 tons) of pozzolan.

TABLE 2 Physical Requirements

Cement Type	I(SM), IS, I(PM), IP	I(SM)-A, IS-A, I(PM)-A, IP-A	IS(MS) IP(MS)	IS-A(MS) IP-A(MS)	S	SA	P	PA
Fineness	^{A}	^{A}	^{A}	^{A}	^{A}	^{A}	^{A}	^{A}
Autoclave expansion max, %	0.80	0.80	0.80	0.80	0.80	0.80	0.80	0.80
Autoclave contraction, max, %	0.20	0.20	0.20	0.20	0.20	0.20	0.20	0.20
Time of setting, Vicat test:^{C}								
Set, minutes, not less than	45	45	45	45	45	45	45	45
Set, h, not more than	7	7	7	7	7	7	7	7
Air content of mortar (Test Method C 185), volume %	12 max	19 ± 3	12 max	19 ± 3	12 max	19 ± 3	12 max	19 ± 3
Compressive strength, min, MPa (psi):								
3 days	13.0	10.0	11.0	9.0	...	...	...	...
	(1890)	(1450)	(1600)	(1310)	...	...	...	...
7 days	20.0	16.0	18.0	14.0	5.0	4.0	11.0	9.0
	(2900)	(2320)	(2610)	(2030)	(720)	(580)	(1600)	(1310)
28 days	25.0	20.0	25.0	20.0	11.0	9.0	21.0	18.0
	(3620)	(2900)	(3620)	(2900)	(1600)	(1310)	(3140)	(2610)
Heat of hydration:^{D}								
7 days, max, kJ/kg	290	290	290	290	...	...	250	250
(cal/g)	(70)	(70)	(70)	(70)	...	...	(60)	(60)
28 days, max, kJ/kg	330	330	330	330	...	...	290	290
(cal/g)	(80)	(80)	(80)	(80)	...	...	(70)	(70)
Water requirement, max weight % of cement	...	...	...	...	...	...	64	56
Drying shrinkage, max, %	...	...	...	...	...	...	0.15	0.15
Mortar expansion:^{E}								
At age of 14 days, max, %	0.020	0.020	0.020	0.020	0.020	0.020	0.020	0.020
At age of 8 weeks, max, %	0.060	0.060	0.060	0.060	0.060	0.060	0.060	0.060
Sulfate Resistance								
Expansion at 180 days, max, %	(0.10)^F	(0.10)^F	0.10	0.10			(0.10)^F	(0.10)^F

^{A} Both amount retained when wet sieved on 45-μm (No. 325) sieve and specific surface by air permeability apparatus, m^2/kg, shall be reported on all mill test reports requested under 14.4.

^{B} The specimens shall remain firm and hard and show no signs of distortion, cracking, checking, pitting, or disintegration when subjected to the autoclave expansion test.

^{C} Time of setting refers to initial setting time in Test Method C 191. The time of setting of cements containing a user-requested accelerating or retarding functional addition need not meet the limits of this table, but shall be stated by the manufacturer.

^{D} Applicable only when moderate (MH) or low (LH) heat of hydration is specified, in which case the strength requirements shall be 80 % of the values shown in the table.

^{E} The test for mortar expansion is an optional requirement to be applied only at the purchaser's request and should not be requested unless the cement will be used with alkali-reactive aggregate.

^{F} Optional sulfate resistance criterion that applies only if specifically invoked.

TABLE 3 Requirements for Pozzolan for Use in Blended Cements and for Slag for Use in Slag-Modified Portland Cements

Pozzolan	
Fineness:	
Amount retained when wet-sieved on 45-μm (No. 325) sieve, max, %	20.0
Alkali reactivity of pozzolan for use in Types I(PM) and I(PM)-A cements, six tests, mortar bar expansion at 91 days, max, %	0.05
Slag or pozzolan activity index:	
with portland cement, at 28 days, min, %	75

TABLE 4 Aggregate Grading Requirements for Mortar Expansion Test

Sieve Size		Weight %
Passing	Retained on	
4.75-mm (No. 4)	2.36-mm (No. 8)	10
2.36-mm (No. 8)	1.18-mm (No 16)	25
1.18-mm (No. 16)	600-μm (No. 30)	25
600-μm (No. 30)	300-μm (No. 50)	25
300-μm (No. 50)	150-μm (No. 100)	15

10. Test Methods

10.1 Determine the applicable properties enumerated in this specification in accordance with the following test methods:

10.1.1 *Chemical Analysis*—Test Methods C 114, with the special provisions noted therein applicable to blended cement analyses.

10.1.2 *Fineness by Sieving*—Test Method C 430.

10.1.3 *Fineness by Air-Permeability Apparatus*—Test Method C 204.

10.1.4 *Autoclave Expansion*—Test Method C 151, except that, in the case of slag cement, the test specimens shall remain in the moist cabinet for a period of 48 h before being measured for length, and the neat cement shall be mixed for not less than 3 min nor more than 3½ min.

10.1.5 *Time of Setting*—Test Method C 191.

10.1.6 *Air Content of Mortar*—Test Method C 185, using the actual specific gravity of the cement, if it differs from 3.15 by more than 0.05, in calculating the air content.

10.1.7 *Compressive Strength*—Test Method C 109/C 109M.

10.1.8 *Heat of Hydration*—Test Method C 186.

10.1.9 *Normal Consistency*—Test Method C 187, except that in the case of slag cement the paste shall be mixed for not less than 3 min nor more than 3½ min.

10.1.10 *Specific Gravity*—Test Method C 188.

10.1.11 *Water Requirement*—The mass of mixing water added to the six-cube batch in accordance with Test Method C 109, as a percentage of the total cementing ingredients.

10.1.12 *Mortar Expansion of Blended Cement*—Test Method C 227, using crushed Pyrex glass No. 7740[5] as aggregate and the grading prescribed in Table 4.

10.1.13 *Mortar Expansion of Pozzolan for Use in*

[5] Pyrex Glass No. 7740 is available as lump cullet from the Corning Glass Works, Corning, NY.

Pozzolan-Modified Portland Cement Types I(PM) and I(PM)-A—Using the pozzolan and the clinker or cement that are to be used together in the production of the blended cement, prepare pozzolan-modified portland cements containing 2.5, 5, 7.5, 10, 12.5, and 15 mass % of the pozzolan. These blends shall be tested in accordance with Test Method C 227 using a sand judged to be a nonreactive by the mortar bar test in Test Method C 227. The expansion of the mortar bars shall be measured at 91 days, and all the six blends shall meet the expansion requirement in Table 3.

10.1.14 *Drying Shrinkage*—Test Method C 157. Make three specimens using the proportion of dry materials of 1 part of cement to 2.75 parts of Test Method C 109 graded Ottawa sand. Use a curing period of 6 days and an air storage period of 28 days. Report the linear contraction during air storage based on an initial measurement after the 6-day water-curing period.

10.1.15 *Pozzolanic Activity Index with Portland Cement*—Test Methods C 311 – 85.[6]

10.1.16 *Slag Activity Test with Portland Cement*—Test Methods C 311, Section on Strength Activity with Portland Cement, except make three cube batches (half quantities) for the control mixture and test mixture and use for the test mixture 75 g of portland cement and an amount of slag determined by 175 g × density of slag ÷ density of cement. After removing the cubes from the molds at 20 to 24 h, store in airtight containers for 27 days at 38°C ± 1.7°C. Cool to 23°C ± 1.7°C before testing.

10.1.17 *Sulfate Resistance*—Test Method C 1012.

11. Testing Time Requirements

11.1 The following periods from time of sampling shall be allowed for the completion of testing:

3-day test	8 days
7-day test	12 days
14-day test	19 days
28-day test	33 days
8-week test	61 days

12. Inspection

12.1 Facilities shall be provided to the purchaser for careful inspection and sampling of the finished cement. Inspection and sampling of finished cement shall be at the mill or distribution site controlled by the manufacturer, or at any other location as agreed by the purchaser and seller.

12.2 The manufacturer shall provide suitable facilities to enable the inspector to check the relative masses of the constituents used, and the intergrinding or blending operation used to produce the cement. The plant facilities for intergrinding or blending and inspection shall be adequate to ensure compliance with the provisions of this specification.

13. Rejection

13.1 At the option of the purchaser, cement shall be rejected if it fails to meet any of the requirements of this specification applicable to the cement. Such rejection shall apply to an optional requirement only if that option has been invoked for the cement.

13.2 When the purchaser requires, cement in bulk storage for a period greater than six months shall be resampled and retested and, at the option of the purchaser, shall be rejected if it fails to meet any of the applicable requirements of this specification. Cement so rejected shall be the responsibility of the owner of record at the time of sampling for retest.

13.3 When the purchaser requires, packages more than 2 % below the mass marked thereon shall be rejected; or if the average mass of packages in any shipment, as shown by determining the mass of 50 packages taken at random, is less than that marked on the packages, the entire shipment, at the option of the purchaser, shall be rejected.

14. Certification

14.1 At the request of the purchaser, the manufacturer shall state in writing the source, amount, and composition of the essential constituents used in manufacture of the finished cement and the composition of the blended cement purchased.

14.2 At the request of the purchaser, the manufacturer shall state in writing the nature, amount, and identity of any processing, functional, or air-entraining addition used; and also, if requested, shall supply test data showing compliance of any such processing addition with the provisions of Specification C 465 and of any such functional addition with the provisions of Specification C 688, and of any such air-entraining addition with the provisions of Specification C 226.

14.3 In the case of portland-pozzolan cement or pozzolan-modified portland cement at the request of the purchaser, the manufacturer shall also state in writing that the amount of pozzolan in the finished cement will not vary more than ±5.0 mass % of the finished cement from lot to lot or within a lot.

14.4 Upon request of the purchaser in the contract or order, a manufacturer's certification shall be furnished indicating that the material was tested during production or transfer in accordance with this specification, that it complies with this specification, and a report of the test results shall be furnished at the time of shipment (to include both amount retained on the 45-μm (No. 325) sieve and specific surface by the air permeability method).

15. Packaging and Package Marking

15.1 When the cement is delivered in packages, the words, "portland blast-furnace slag cement," "portland-pozzolan cement," "pozzolan-modified portland cement," "slag-modified portland cement" or "slag cement," as appropriate; the type of cement, name and brand of the manufacturer, and the mass of the cement contained therein, shall be plainly marked on each package. When the cement contains a functional addition listed in 5.1.3.4 through 5.1.3.9, the type of functional addition shall be plainly marked on each package. Similar information shall be provided in the shipping documents accompanying the shipment of packaged or bulk cement. All packages shall be in good condition at the time of inspection.

16. Storage

16.1 The cement shall be stored in such a manner as to permit easy access for proper inspection and identification of

[6] The test, *Pozzolanic Activity Index with Portland Cement*, last appeared in Vol. 04.02 of the 1989 *Annual Book of ASTM Standards*.

each shipment, and in a suitable weathertight building that will protect the cement from dampness and minimize warehouse set.

17. Keywords

17.1 blended hydraulic cement; fly ash cement; hydraulic cement; portland blast-furnace slag cement; portland pozzolan cement; pozzolanic cement; slag cement; slag, granulated blast furnace

Designation: C 596 – 96

Standard Test Method for Drying Shrinkage of Mortar Containing Hydraulic Cement[1]

This standard is issued under the fixed designation C 596; the number immediately following the designation indicates the year of original adoption or, in the case of revision, the year of last revision. A number in parentheses indicates the year of last reapproval. A superscript epsilon (ε) indicates an editorial change since the last revision or reapproval.

1. Scope

1.1 This test method covers the determination of the change in length on drying of mortar bars containing hydraulic cement and graded standard sand.

1.2 The values stated in SI units are to be regarded as the standard. Values given in parentheses are for information only.

1.3 *This standard does not purport to address all of the safety concerns, if any, associated with its use. It is the responsibility of the user of this standard to establish appropriate safety and health practices and determine the applicability of regulatory limitations prior to use.*

2. Referenced Documents

2.1 *ASTM Standards:*

C 109/C 109M Test Method for Compressive Strength of Hydraulic Cement Mortars (Using 2-in. or 50-mm Cube Specimens)[3]

C 157 Test Method for Length Change of Hardened Hydraulic Cement Mortar and Concrete[2]

C 219 Terminology Relating to Hydraulic Cement[3]

C 305 Practice for Mechanical Mixing of Hydraulic Cement Pastes and Mortars of Plastic Consistency[3]

C 490 Practice for Use of Apparatus for the Determination of Length Change of Hardened Cement Paste, Mortar, and Concrete[3]

C 511 Specification for Moist Cabinets, Moist Rooms, and Water Storage Tanks Used in the Testing of Hydraulic Cements and Concretes[3]

C 778 Specification for Standard Sand[3]

C 1005 Specification for Weights and Weighing Devices for Use in the Physical Testing of Hydraulic Cements[3]

E 177 Practice for Use of the Terms Precision and Bias in ASTM Test Methods[4]

3. Terminology

3.1 The term "drying shrinkage" is defined as the decrease in length of the test specimen, where the decrease is caused by any factor other than externally applied forces under stated conditions of temperature, relative humidity and evaporation rate in the environment; the term includes the net effect of a variety of phenomena tending to bring about both increases and decreases in length during the period in which the test specimens under consideration are stored in the environment and in which a number of processes, including hydration of the cement, are taking place at a variety of rates.

3.2 Other terms used in this test method are defined in Terminology C 219.

4. Significance and Use

4.1 This test method establishes a selected set of conditions of temperature, relative humidity and rate of evaporation of the environment to which a mortar specimen of stated composition shall be subjected for a specified period of time during which its change in length is determined and designated "drying shrinkage."

4.2 The drying shrinkage of mortar as determined by this test method has a linear relation to the drying shrinkage of concrete made with the same cement and exposed to the same drying conditions.[5] Hence this test method may be used when it is desired to develop data on the effect of a hydraulic cement on the drying shrinkage of concrete made with that cement.

5. Apparatus

5.1 *Weighing Devices and Weights*—Weighing devices and weights used in determining the mass of materials for mortar mixtures shall conform to the requirements of Specification C 1005.

5.2 *Glass Graduates*—Glass graduates of suitable capacities shall conform to the requirements of Specification C 490.

5.3 *Molds*—Molds for test specimens shall provide for 25 by 25 by 285-mm or 1 by 1 by 11¼-in. prisms having an effective gage length of 250 mm or 10 in. respectively and shall conform to the requirements of Specification C 490.

5.4 *Trowel*—The trowel shall have a straight-edged steel blade 100 to 150 mm (4 to 6 in.) in length.

5.5 *Tamper*—The tamper shall conform to the requirements of Test Method C 157.

5.6 *Demolding Apparatus*—The demolding apparatus shall conform to the requirements of Test Method C 157.

5.7 *Length Comparator*—The length comparator shall

---

[1] This test method is under the jurisdiction of ASTM Committee C-1 on Cement and is the direct responsibility of Subcommittee C01.31 on Volume Change.

Current edition approved July 10, 1996. Published September 1996. Originally published as C 596 – 67 T. Last previous edition C 596 – 89.

[2] *Annual Book of ASTM Standards*, Vol 04.02.

[3] *Annual Book of ASTM Standards*, Vol 04.01.

[4] *Annual Book of ASTM Standards*, Vol 14.02.

[5] Data pertaining to the relation of the drying shrinkage of mortar test specimens and of the drying shrinkage of concrete test specimens as affected by the cement under specified laboratory test conditions, may be found in RR/120:C-1, available from ASTM Headquarters; in a report of the California Division of Highways. "Significance of the Test for Contraction of Mortar in Air with Respect to Performance of Cements in Concrete," Oct. 18, 1961; and in the paper by Mardulier, F. J., Schneider, A. M., and Stockett, A. L., "An Analysis of Drying Shrinkage. Data for Portland Cement Mortar and Concrete," *Journal of Materials*, JMLSA, Am. Soc. Testing Mats., Vol 2, No. 4, 1967, pp. 829–842. A relevant paper by H. T. Arni entitled "The Significance of the Correlation Coefficient for Analyzing Engineering Data" was published in *Materials Research and Standards*, MTRSA, Am. Soc. Testing Mats., Vol. 11, No. 5, 1971, pp. 16–19.

conform to the requirements of Specification C 490.

6. Temperature and Humidity

6.1 *Molding Room*—The temperature of the molding room, dry materials, and mixing water, and the relative humidity of the air in the molding room shall conform to the requirements of Specification C 490.

6.2 *Moist Storage Facility*—The temperature and humidity of the air in the moist storage facility shall conform to the requirements of Specification C 511.

6.3 *Drying Room and Controls*—The drying room and controls shall conform to the requirements of Test Method C 157.

7. Graded Standard Sand

7.1 The graded standard sand shall conform to Specification C 778.

8. Number of Test Specimens

8.1 Make at least four test specimens (Note 1).

NOTE 1—Although the number of test specimens may consist of four specimens from a single batch of mortar, it is preferable that twelve specimens be made with four specimens being made from each of three batches, with each batch being made on a different day.

9. Preparation of Molds

9.1 Prepare the specimen molds as required by Practice C 490.

10. Preparation of Test Specimens

10.1 *Mortar Proportions*—A batch of mortar shall consist of 750 g of cement, 1500 g of graded standard sand, and an amount of mixing water sufficient to produce a flow of 110 ± 5 %. The flow shall be determined in conformance with the procedure as described in Test Methods C 109/C 109M.

10.2 *Mixing Mortar*—Mix the mortar in a mechanical mixer as required by Practice C 305.

10.3 *Molding of Specimens*—Mold the specimens as required by Test Method C 157.

11. Curing, Storage, and Taking Comparator Readings of Test Specimens

11.1 Cure, store, and take comparator readings of the test specimens as required by Test Method C 157, except as follows:

11.1.1 Moist cure the specimens in the molds for 23½ ± ½ h. If the strength of the specimens is insufficient to allow proper removal from the mold at 24 h, moist cure in the mold for 47 ± ½ h.

11.1.2 Remove the specimens from the molds and cure in water for 48 h. If the specimens have been moist cured in the mold for 48 h, cure in water for 24 h.

12. Calculation

12.1 Calculate the length change of each specimen at each age of air drying by subtracting the initial comparator reading, taken after removal from water storage, from the comparator reading taken at each age of air drying and express as millionths and as the percent of the effective gage length. Do not prefix a shrinkage value with a minus sign.

12.2 Report the average change in unit length, expressed as millionths, and as a percent of the effective gage length of four specimens from the same batch of mortar as the drying shrinkage of the mortar. If any one test specimen is manifestly faulty, discard it (see 12.2.1). If more than one test specimen is discarded, do not report the results and repeat the test on a new batch of mortar. If more than one batch of mortar has been tested, report the average result of the individual batches. When the experimenter is clearly aware that a gross deviation from prescribed experimental procedure has taken place, discard the resultant observation, whether or not it agrees with the rest of the data and without recourse to statistical tests for outliers.

NOTE 2—An approximation of ultimate drying shrinkage of mortar may be determined by plotting the shrinkage values as a function of the reciprocal of the time. The time includes the moist curing period. An example of the method is given in Fig. 1 where a logarithmic scale is used for the drying shrinkage values and a reciprocal scale for the total time in weeks.

13. Precision and Bias

13.1 *Precision:*

13.1.1 The single-laboratory-operator multibatch-day precision has been found to be 0.0223 (1S%) and the single-operator multilaboratory-batch-day precision has been found to be 6.6 (R1S%) as defined in Practice E 177, therefore 95 % percent of the time two test results (each the average of four determinations of properly conducted tests) obtained by the same operator on different days should not differ by more than 70 millionths, and the test results (each the average of four determinations of properly conducted tests) obtained by two laboratories should not differ by more than 25 % of their mean.

13.1.2 *Repeatability*—Where a test result is the average of four specimens from a single batch, the difference between duplicate results by the same operator should not be considered suspect unless it exceeds 70 millionths. Where a test result is the average of twelve specimens, four specimens being tested from each of three batches, the difference between duplicate results by the same operator should not be considered suspect unless it exceeds 40 millionths.

13.1.3 *Reproducibility*—The average result of four or twelve specimens reported by one laboratory should not be considered suspect unless it differs from that of another laboratory by more than 25 % of their mean.

13.2 *Bias*—No justifiable statement can be made on the bias of this test method since no reference samples are available.

14. Keywords

14.1 drying shrinkage; hydraulic cement; method; mortar

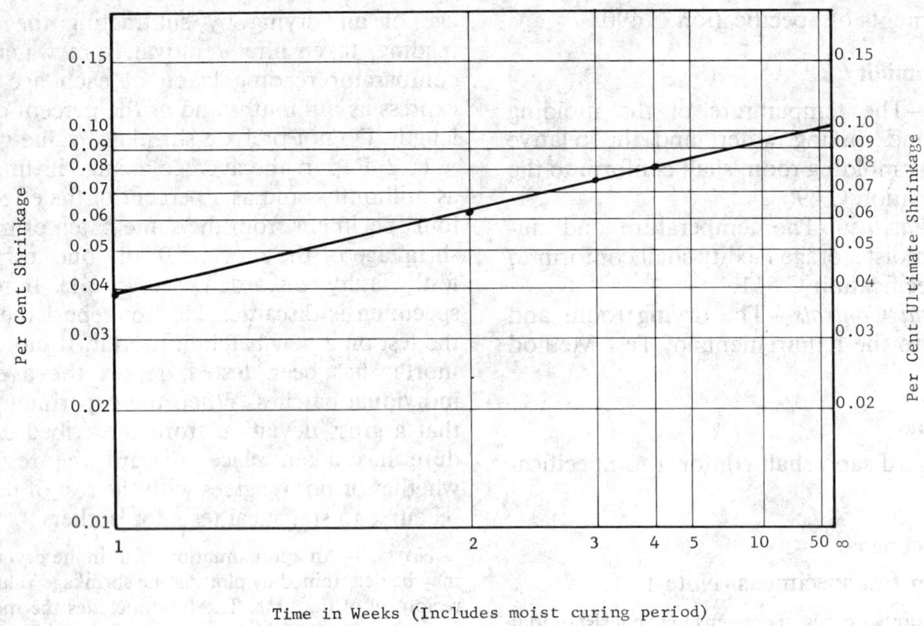

FIG. 1 Ultimate Drying Shrinkage of Mortar

Standard Specification for Agricultural Liming Materials[1]

This standard is issued under the fixed designation C 602; the number immediately following the designation indicates the year of original adoption or, in the case of revision, the year of last revision. A number in parentheses indicates the year of last reapproval. A superscript epsilon (ε) indicates an editorial change since the last revision or reapproval.

This standard has been approved for use by agencies of the Department of Defense. Consult the DoD Index of Specifications and Standards for the specific year of issue which has been adopted by the Department of Defense.

1. Scope

1.1 This specification covers agricultural liming materials, such as burnt lime (quicklime), hydrated lime, limestone, (calcitic and dolomitic), marl, shells, and byproducts including slag, and other materials.

1.2 The following precautionary caveat pertains only to the test method portion, Section 8: *This standard does not purport to address all of the safety concerns, if any, associated with its use. It is the responsibility of the user of this standard to establish appropriate safety and health practices and determine the applicability of regulatory limitations prior to use.*

2. Referenced Documents

2.1 *ASTM Standards:*

C 25 Test Methods for Chemical Analysis of Limestone, Quicklime, and Hydrated Lime[2]

C 125 Terminology Relating to Concrete and Concrete Aggregates[3]

D 3176 Practice for Ultimate Analysis of Coal and Coke[4]

E 11 Specification for Wire-Cloth Sieves for Testing Purposes[5]

3. Terminology

3.1 *Definitions:*

3.1.1 *agricultural liming material*—a product whose calcium and magnesium compounds are capable of neutralizing soil acidity.

3.1.2 *air-cooled blast-furnace slag and granulated blast-furnace slag*—air-cooled blast-furnace slag and granulated blast furnace slag as defined in Terminology C 125.

3.1.3 *calcium carbonate equivalent (C.C.E.)*—the acid-neutralizing capacity (of an agricultural liming material) of the material expressed as weight percent of calcium carbonate.

4. Chemical Classifications

4.1 Agricultural liming materials shall be classified in terms of calcium carbonate equivalent (C.C.E.), as shown in Table 1.

NOTE 1—Marl and some byproduct liming materials are used for neutralizing soil acidity, but due to their varying composition, their chemical limits are not included. In some economic circumstances limestone, slag, and shells of less than 80 % C.C.E. may be used.

5. Sieve Analysis Classifications for Agricultural Limestone

5.1 Agricultural limestone shall be classified according to the minimum percentages passing the No. 8 (2.36-mm) and No. 60 (250-μm) sieves conforming to Specification E 11, as shown in Table 2.

NOTE 2—These classifications apply where the agricultural limestone is obtained by the normal crushing procedure and the product contains the fines of fracture. In some economic circumstances, coarser products are used. The No. 60 (250-μm) sieve was selected because research has shown that this sieve gives a more accurate representation of the particle size distribution of most agricultural limestones presently produced than a finer or coarser sieve. The No. 8 (2.36-mm) sieve is used to control the upper limit on the amount of coarse limestone particles that may be in the product.

6. Sieve Analysis Classifications for Agricultural Slag

6.1 *Air-Cooled Blast-Furnace Slag*—Air-cooled blast-furnace slag shall be classified the same as agricultural limestone as shown in Section 5.

6.2 *Granulated Blast-Furnace Slag*—Granulated blast-furnace slag shall be classified in accordance with the minimum percentages passing the No. 8 (2.36-mm) and the No. 60 (250-μm) sieves.

TABLE 1 Agricultural Liming Materials

Material	Calcium Carbonate Equivalent (C.C.E.), percent
Burnt lime	not less than 140
Hydrated lime	not less than 110
Limestone	not less than 80
Slag	not less than 80
Shells	not less than 80

TABLE 2 Classification for Agricultural Limestone

Class Designation	Passing No. 8 (2.36-mm) Sieve, min, percent	Passing No. 60 (250-μm) Sieve, min, percent
S	100	100
T	99	75
O	95	55
N	90	40
E	80	25

[1] This specification is under the jurisdiction of ASTM Committee C-7 on Lime, and is the direct responsibility of Subcommittee C07.03 on Industrial Uses.

Current edition approved Nov. 10, 1995. Published February 1996. Originally published as C 602 – 67 T. Last previous edition C 602 – 95.

[2] *Annual Book of ASTM Standards*, Vol 04.01.

[3] *Annual Book of ASTM Standards*, Vol 04.02

[4] *Annual Book of ASTM Standards*, Vol 05.05.

[5] *Annual Book of ASTM Standards*, Vol 14.02.

7. Particle Size Requirements for Hydrated Lime and Burnt Lime

7.1 Hydrated lime and burnt lime for agricultural use shall be classified in accordance with the minimum percentages passing the No. 8 (2.36-mm) and No. 60 (250-μm) sieves, as follows:

Passing Sieve No.	Min. % Hydrated Lime	Min. % Burnt Lime
8 (2.36 mm)	100	95
60 (250 μm)	97	35

8. Sampling

8.1 Agricultural liming materials shall be sampled using the following equipment and procedures:

8.1.1 *Sampling Apparatus:*

8.1.1.1 *Sampling Tube, or Probe,* of thin brass or other noncorroding metal ¾ to 1 in. (19 to 25 mm) in diameter and 3 ft (915 mm) in length, fitted with a solid point and with an open side of which one edge is sharpened and flared so that upon turning, the liming material is forced into the tube.

8.1.1.2 *Soil Probe,* for sampling marl, 1 in. (25 mm) in diameter and 20 in. (508 mm) long with an open end and side.

8.1.1.3 *Pails,* galvanized iron, two, 2½-gal (0.5-L) capacity.

8.1.1.4 *Pointed Shovel.*

8.1.1.5 *Riffle with Pans,* as specified in Practice D 3176.

8.1.1.6 *Sample Containers,* airtight, vapor-proof, and moisture-resistant, such as polyethylene jars or metal cans, of 1-qt (946-cm^3) and 1-gal (3784-cm^3) capacities.

8.1.2 *Taking the Samples*—Samples shall be taken so as to represent an average of all parts of the lot, avoiding a disproportionate amount of surface material or of any modified or contaminated zone. Preferably, the sample shall be taken at the point of final weighing for sale.

8.1.2.1 *Burnt Lump Lime, in Bulk*—Collect a composite sample of not less than ten shovelsful per car, or shipment, taking each shovelful from different parts of the mass. Crush immediately to pass a 2-in. (51-mm) diameter circular opening, mix thoroughly and rapidly, quarter down to a 5-lb (2.3-kg) sample and place immediately in a labeled, dry sample container.

8.1.2.2 *Hydrated Lime and Ground Burnt Lime in Bags*—Select ten bags from different parts of each lot, or shipment, of 20 tons (18 metric tons), or less, and one additional bag for each additional 5 tons (4.5 metric tons). From each selected bag, withdraw a core from top to bottom with the sampling tube, place these subsamples in a pail and then mix them by passing through the riffle three times, recombining the splits between each pass. After the third pass, by means of the riffle, reduce the sample to 2 lb (0.92 kg) and place in a labeled, dry sample container. Mix and reduce the sample rapidly to prevent either loss or absorption of moisture.

8.1.2.3 *Ground Limestone, Slag, Shells, and Marl, in Bags*—Proceed as in 8.1.2.2.

8.1.2.4 *Ground Limestone, Ground Burnt Lime, Hydrated Lime, Slag, Shells, and Marl, in Bulk*—Take at least ten subsamples, representative of the lot, from the lot (car, truck, or stockpile), using the sampling tube. If possible, the tube should be inserted to its full length, given a one-half turn and then removed. If the material is too hard for use of the probe, dig holes distributed over the entire surface of the mass, to a depth of about 2 ft (610 mm) and from the sides of the hole take approximately 1-pt (473-cm^3) subsamples. Collect the subsamples in a pail and mix the sample by passing through the riffle three times, recombining the splits between passes. After the third pass, by means of the riffle, reduce the sample to 2 lb (0.92 kg) and place immediately in a labeled, dry sample container. Mix and reduce the sample rapidly to prevent loss or absorption of moisture. When the sample is too wet for adequate field mixing and reduction, the entire sample is dried in a laboratory dryer prior to mixing and reduction to size. Spread the sample in a dry, tared metal pan to a maximum depth of ½ in. (13 mm), weigh the pan and contents, dry, reweigh, and report the percentage lost. Mix the dry sample and reduce to size as specified above.

9. Chemical Methods

9.1 Reagent grade chemicals or equivalent and water purity shall be used as specified in Test Methods C 25.

9.2 The analytical sample for chemical methods relative to this standard shall be pulverized to pass a 60 mesh sieve.

9.3 The following chemical parameters are to be determined using the methods as specified in Test Methods C 25.

9.3.1 Calcium carbonate equivalent, and

9.3.2 *Total calcium and magnesium*—determination can be made either gravimetrically or by EDTA titration.

SIEVE ANALYSIS OF AGRICULTURAL LIMING MATERIALS

10. Preparation of Sample

10.1 Dry the sample to constant weight at 110 ± 5°C and store in an airtight container. Obtain the sample from a larger sample of the material to be tested by the use of a riffle or a sample splitter in accordance with Practice D 3176, or by the method of quartering. The sample for sieve analysis shall have a weight of 100 to 150 g.

10.2 Where limestone contains clay, some agglomeration of fine particles may occur. The agglomerates must be broken by rolling the dry sample with a hard rubber roller on a hard rubber mat or by some equally effective means that does not result in crushing the limestone.

11. Procedure

11.1 Weigh the dried sample to an accuracy of 0.1 g. Sieve the sample through an 8-in. (203-mm) diameter No. 8 (2.36-mm) sieve and an 8-in. diameter No. 60 (250-μm) sieve. The sieves shall conform to Specification E 11. Conduct the sieving operation by means of a lateral and vertical motion of this sieve, accompanied by a jarring action so as to keep the sample moving continuously over the surface of the sieve. Continue sieving until not more than 0.5 weight percent of the total sample passes either sieve during 1 min. The sieving operation may be performed either by hand or by a mechanical apparatus. Determine the weight of liming material passing each sieve on a scale or balance sensitive to at least 0.1 g. Calculate the amount of liming material passing the No. 8 (2.36-mm) and the No. 60 (250-μm) sieves to the nearest whole percentage of the total weight of the dry sample.

12. Report

12.1 Report the following results for agricultural liming materials:

12.1.1 *Percentage Calcium Carbonate Equivalent*—The percentage calcium carbonate equivalent (C.C.E.) shall be reported on the as-received basis. Calculate as follows:

$$\text{C.C.E. (as-received)} = [1-(\%\ \text{moisture} \div 100)] \times \text{C.C.E. (oven-dry)}$$

12.1.2 *Percentage Moisture*—Calculate as follows:

$$\text{Moisture, \%} = \frac{\text{Weight of sample (as received)} - \text{oven-dry weight} \times 100}{\text{Weight of sample (as-received)}}$$

12.1.3 *Percentage Calcium and Magnesium*—The percentages of elemental calcium and elemental magnesium on the basis of the as-received sample.

12.1.4 *Sieve Analysis*—The amount of dried sample passing the No. 8 (2.36-mm) and No. 60 (250-μm) sieves as a percentage of the total weight of the dried sample with the figures rounded to the nearest whole percentage point.

13. Keywords

13.1 agricultural liming materials; agricultural limestone; burnt lime; calcium carbonate equivalent; hydrated lime; lime byproducts; limestone; slag

Standard Specification for Water-Resistant Gypsum Backing Board[1]

This standard is issued under the fixed designation C 630/C 630M; the number immediately following the designation indicates the year of original adoption or, in the case of revision, the year of last revision. A number in parentheses indicates the year of last reapproval. A superscript epsilon (ε) indicates an editorial change since the last revision or reapproval.

1. Scope

1.1 This specification covers water-resistant gypsum backing board that is designed primarily to be used as a base for the application of ceramic or plastic tile on walls or ceilings. This product is also suitable for decoration.

NOTE 1—Specification C 840 contains application procedures for water-resistant gypsum backing board.

1.2 The values stated in either inch-pound units or SI (metric) units are to be regarded separately as standard. Within the text, the SI units are shown in brackets. The values stated in each system shall be used independent of the other. Values from the two systems shall not be combined.

1.3 The text of this standard references notes and footnotes which provide explanatory material. These notes and footnotes (excluding those in tables and figures) shall not be considered as requirements of the standard.

2. Referenced Documents

2.1 *ASTM Standards:*

C 11 Terminology Relating to Gypsum and Related Building Materials and Systems[2]

C 473 Test Methods for Physical Testing of Gypsum Board Products and Gypsum Lath[2]

C 645 Specification for Nonstructural Steel Framing Members[2]

C 840 Specification for Application and Finishing of Gypsum Board[2]

C 1264 Specification for Sampling, Inspection, Rejection, Certification, Packaging, Marking, Shipping, Handling, and Storage of Gypsum Board[2]

E 119 Test Methods for Fire Tests of Building Construction and Materials[3]

3. Terminology

3.1 Definitions used in this specification shall be in accordance with Terminology C 11.

4. Materials and Manufacture

4.1 Water-resistant gypsum backing board shall consist of a noncombustible water-resistant core, essentially gypsum, surfaced on both the face and back of the board with water-repellent paper bonded to the core.

4.2 Water-resistant gypsum backing board, type X (special fire resistant), designates water-resistant gypsum backing board complying with this specification that provides not less than 1-h fire-resistance for boards ⅝ in. [15.9 mm] thick or ¾-h fire-resistance for boards ½ in. [12.7 mm] thick, applied parallel with and on each side of load bearing 2 × 4 wood studs spaced 16 in. [406 mm] on center with 6d coated nails, 1⅞ in. [48 mm] long, 0.0915 in. [2.32 mm] diameter shank, ¼ in. [6.4 mm] diameter heads, spaced 7 in. [178 mm] on center with gypsum backing board joints staggered 16 in. [406 mm] on each side of the partition and tested in accordance with the requirements of Test Methods E 119.

NOTE 2—Consult producers for independent test data on assembly details and fire resistance classifications for other types of construction. See official fire test reports for assembly particulars, materials, and classifications.

5. Physical Properties

5.1 Specimens shall be taken from the samples obtained in accordance with Specification C 1264.

5.2 Specimens shall be tested in accordance with Test Methods C 473.

5.2.1 *Flexural Strength*—The specimens shall be tested face up and face down. The average breaking load shall be not less than the following:

	Method A		Method B	
Thickness, in. [mm]	Load, lbf [N] Bearing Edges Across Fiber of Surfacing	Load, lbf [N] Bearing Edges Parallel to Fiber of Surfacing	Load, lbf [N] Bearing Edges Across Fiber of Surfacing	Load, lbf [N] Bearing Edges Parallel to Fiber of Surfacing
½ [12.7]	110 [489]	40 [178]	107 [476]	36 [160]
⅝ [15.9]	150 [667]	50 [222]	147 [654]	46 [205]

5.2.2 *Humidified Deflection*—The specimens shall have an average deflection of not more than the following:

Thickness, in. [mm]	Humidified Deflection, Eighths of an in. [mm]
½ [12.7]	10 [32]
⅝ [15.9]	5 [16]

5.2.3 *Core, End, and Edge Hardness*—The specimens shall have an average hardness of not less than 15 lbf [67 N] when tested by Method A and 11 lbf [49 N] when tested by Method B.

5.2.4 *Nail Pull Resistance*—The specimen shall have an average nail-pull resistance of not less than the following:

Thickness, in. [mm]	Method A Nail Pull Resistance, lbf [N]	Method B Nail Pull Resistance, lbf [N]
½ [12.7]	80 [356]	77 [343]
⅝ [15.9]	90 [400]	87 [387]

5.2.5 *Water Resistance*—The three specimens selected

[1] This specification is under the jurisdiction of ASTM Committee C-11 on Gypsum and Related Building Materials and Systems and is the direct responsibility of Subcommittee C11.01 on Specifications and Test Methods for Gypsum Products.

Current edition approved Dec. 10, 1996. Published February 1997. Originally published as C 630 – 68T. Last previous edition C 630 – 96.

[2] *Annual Book of ASTM Standards*, Vol 04.01.

[3] *Annual Book of ASTM Standards*, Vol 04.07.

shall have an average water absorption of not more than 5 weight % after 2 h of immersion.

5.2.6 *Surface Water Absorption*—The specimens shall have an average surface water absorption of not more than 1.6 g after 2 h of elapsed time.

6. Dimensions and Permissible Variations

6.1 Specimens shall be taken from the samples obtained in accordance with Specification C 1264.

6.2 Thickness, width, length, tapered edge depth, and end squareness shall be determined in accordance with Test Methods C 473.

6.2.1 *Thickness*—The nominal thickness shall be ½ or ⅝ in. [12.7 or 15.9 mm] with permissible variations in the nominal thickness of 1⁄64 in. [0.4 mm] with permissible local variations of 1⁄32 in. [0.8 mm] from the nominal thickness.

6.2.2 *Width*—The nominal width shall be up to 48 in. [1220 mm], with widths up to 54 in. [1370 mm] permitted, with a permissible variation of 3⁄32 in. [3 mm] under the specified width.

6.2.3 *Length*—The nominal length and permissible variation shall be as follows:

Thickness, in. [mm]	Length, ft [mm]	Variation, in. [mm]
½ [12.7]	4 to 16 [1220 to 4880]	±¼ [6]
⅝ [15.9]	4 to 16 [1220 to 4880]	±¼ [6]

6.2.4 *Tapered Edge Depth*—The average thickness of the edge of recessed or tapered edge shall be not less than 0.015 in. [0.38 mm] but not more than 0.075 in. [1.90 mm] less than the average thickness of the water-resistant gypsum backing board.

6.2.5 *End Squareness*—Corners shall be square with a permissible variation of ⅛ in. [3 mm] in the full width of the board.

6.3 *Edges and Ends*—The edges and ends shall be straight and either square, recessed, beveled, featured, tapered, or featured and tapered.

7. Finish and Appearance

7.1 The surfaces of water-resistant gypsum backing board shall be true and free from imperfections that would render the board unfit for its designed use.

8. Sampling, Inspection, Rejection, Certification, Packaging, Marking, Shipping, Handling, and Storage

8.1 Sampling, Inspection, Rejection, Certification, Packaging, Marking, Shipping, Handling, and Storage of water-resistant gypsum backing board shall be in accordance with Specification C 1264.

9. Keywords

9.1 ceiling; gypsum; partitions; wall; water-resistant gypsum backing board; water-resistant gypsum backing board, type X

APPENDIX

(Nonmandatory Information)

This appendix gives general information and also suggestions for inclusions to be made elsewhere by the specifier. They are not part of this specification.

X1. ALTERNATE DEFINITION FOR TYPE X

X1.1 Water-resistant gypsum backing board, type X (special fire-resistant) designates water-resistant gypsum backing board providing a greater fire resistance than regular water-resistant gypsum backing board of the same thickness. Type X (special fire-resistant) water-resistant gypsum backing board, when tested in accordance with Test Methods E 119, shall provide the following minimum fire resistance for the assemblies described.

X1.1.1 One hour for a ⅝ in. [15.9 mm] thickness applied to a partition in a single layer application on each side of 3⅝ in. [92 mm] deep non-loadbearing galvanized steel studs complying with Specification C 645, spaced 24 in. [610 mm] on center. The ⅝ in. [15.9 mm] thick water-resistant gypsum backing board 48 in. [1220 mm] wide shall be attached using 1 in. [25 mm] long drywall screws spaced 8 in. [203 mm] on center along the edges and ends, and 12 in. [305 mm] along intermediate studs. All joints shall be oriented parallel to and located over studs and staggered on opposite sides of the assembly. All joints shall be filled with joint compound, covered with joint tape, and covered with another coat of joint compound. All screw heads shall be covered with joint compound, and

X1.1.2 Two hours for a ½ in. [12.7 mm] thickness applied to a partition in a double layer application on each side of 2½ in. [64 mm] deep non-loadbearing galvanized steel studs complying with Specification C 645, spaced 24 in. [610 mm] on center. The 48 in. [1220 mm] wide base layer shall be attached using 1 in. [25 mm] long drywall screws spaced 12 in. [305 mm] on center along board edges, ends, and along intermediate studs. Joints shall be oriented parallel to and located over studs and staggered on opposite sides of the assembly. The 48 in. [1220 mm] wide face layer shall be attached using 1⅝ in. [41 mm] long drywall screws spaced 12 in. [305 mm] along board edges, ends, and along intermediate studs. Joints shall be oriented parallel to and located over studs, offset 24 in. [610 mm] from the base layer joints, and staggered on opposite sides of the assembly. All joints in the face layer shall be filled with joint compound, covered with joint tape, and covered with another coat of joint compound. All screw heads in the face layer shall be covered with two coats of joint compound.

SUMMARY OF CHANGES

Committee C-11 has identified the location of selected changes to this standard since the last issue (C 630 – 96) that may impact the use of this standard.

(*1*) Revised the entire standard and placed into dual inch-pound and metric units.

(*2*) Added Terminology C 11 and Specification C 1264 to 2.1.

(*3*) Added Section 3.

(*4*) Reorganized and revised Sections 4, 5, 6, and 7.

(*5*) Removed previous sections 7, 8, 9, 10, and 11, as they are now covered by Specification C 1264.

Designation: C 631 – 95a

Standard Specification for Bonding Compounds for Interior Gypsum Plastering[1]

This standard is issued under the fixed designation C 631; the number immediately following the designation indicates the year of original adoption or, in the case of revision, the year of last revision. A number in parentheses indicates the year of last reapproval. A superscript epsilon (ϵ) indicates an editorial change since the last revision or reapproval.

1. Scope*

1.1 This specification covers minimum requirements for bonding compounds for adhering interior gypsum plaster to structurally sound surfaces.

1.2 This specification also covers test methods for determining performance requirements and physical properties.

1.3 The values stated in inch-pound units are to be regarded as the standard. The SI equivalents of inch-pound units are approximate.

1.4 The following safety hazards caveat pertains only to the test methods described in this specification: *This standard does not purport to address all of the safety concerns, if any, associated with its use. It is the responsibility of the user of this standard to establish appropriate safety and health practices and determine the applicability of regulatory limitations prior to use.*

2. Referenced Documents

2.1 *ASTM Standards:*

C 11 Terminology Relating to Gypsum and Related Building Materials and Systems[2]

C 28 Specification for Gypsum Plasters[2]

C 472 Test Methods for Physical Testing of Gypsum, Gypsum Plasters, and Gypsum Concrete[2]

C 511 Specification for Moist Cabinets, Moist Rooms and Water Storage Tanks Used in the Testing of Hydraulic Cements and Concretes[2]

3. Terminology

3.1 *Definitions*—For definitions relating to gypsum and related building materials and systems, see Terminology C 11.

4. Physical Properties

4.1 *General*—The bonding compound shall be a film forming, non-oxidizing, non-toxic, non-flammable composition suitable for application by brush, roller, and spray. The compound shall be tinted to show, by visual inspection, where it has been applied. The tinting material shall not show through the material being bonded. The surfaces to which bonding compounds are applied shall be relatively smooth, clean, and dry. Surface coatings shall be removed if their nature indicates an unsatisfactory condition. Surfaces shall be prepared in accordance with the manufacturer's directions.

4.2 *Consistency*—The bonding compound shall be of a homogeneous nature with no settling of the solids. The compound shall be free of foreign matter, and when applied in accordance with the manufacturer's directions, shall cover evenly.

4.3 *Film Characteristics*—A dry film of the bonding compound shall be uniform and smooth and shall remain flexible. The film shall not be noticeably affected by mild acid or alkali.

5. Performance Requirements

5.1 *Bonding Capability*—A dry film of bonding compound shall be capable of bonding gypsum plaster immediately after drying and 10 days after drying when tested as specified in Section 12.

5.2 *Degradation*—Bonding compound that separates shall be able to be re-mixed to a uniform consistency when tested as specified in Section 10.

5.3 *High Temperature Stability*—The bond strength shall be not less than 20 psi (138 kPa) when tested as specified in Section 9.

5.4 *Freeze-Thaw Stability*—The bond strength shall be not less than 20 psi (138 kPa) when tested as specified in Section 11.

5.5 *Bond Strength*—Bond strength of a fresh sample or a 30 day old sample shall be not less than 20 psi (138 kPa) when tested as specified in Section 12.

6. Sampling

6.1 Take a sample of not less than 1 kg from each shipment or consignment for analysis and tests. Except in special cases, take the sample from not less than three separate containers, chosen at random. In addition, take samples from containers that appear to be nonrepresentative, and test separately. Immediately place the samples in airtight glass containers and transport to the testing laboratory in these containers. Take precautions to reduce evaporation or drying to a minimum. Thoroughly mix the bonding compound in the container if there is a tendency for liquid phase separation.

[1] This specification is under the jurisdiction of ASTM Committee C-11 on Gypsum and Related Building Materials and Systems and is the direct responsibility of Subcommittee C11.02 on Specifications and Test Methods for Accessories and Related Products.

Current edition approved Oct. 10, 1995. Published December 1995. Originally published as C 631 – 88. Last previous edition C 631 – 95.

[2] *Annual Book of ASTM Standards*, Vol 04.01.

*** A Summary of Changes section appears at the end of this Specification.**

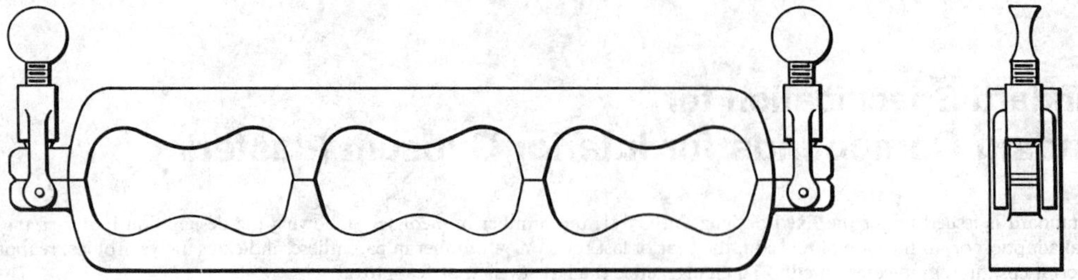

FIG. 1 Briquet Gang Mold

7. Apparatus

7.1 *Moist Cabinet*—Specification C 511.

7.2 *Oven*—A forced draft type oven, having a temperature controlled at 140 ± 5°F (60 ± 3°C) for high temperature tests and 100 ± 5°F (38 ± 3°C) for drying of the briquets.

7.3 *Freezer*—A freezer having a controlled temperature of −10 ± 2°F (−23 ± 1°C).

7.4 *Balance*—A balance capable of weighing at least 2500 g at a precision of 0.1 g.

7.5 *Timing Device*—A standard clock or watch with second hand.

7.6 *Tamper*—A tamper made up of a nonabsorptive, nonabrasive, non-brittle material and having a cross section of ½ in. by 1 in. (13 by 25 mm) and approximately 5 to 6 in. (130 to 150 mm) long. The tamping face shall be flat and at right angles to the length of the tamper.

7.7 *Straight Edge*—A steel straight edge not less than 4 in. (100 mm) long and not less than 1/16 in. (1.6 mm) nor more than ⅛ in. (3.2 mm) in thickness.

7.8 *Spatula*—A spatula with a metal blade 6 in. (150 mm) in length and ½ in. (13 mm) in width, with straight edges.

7.9 *Paint Brush*—A standard 1 in. (25 mm) wide brush, with natural or synthetic bristles.

7.10 *Testing Machine*—Any type that is of sufficient capacity and that is capable of applying the load continuously and without shock at the rate of 0.05 in. (1.27 mm)/min, with provision for adjustment of the rate of loading.

7.11 *Briquet Molds*—The molds for making test specimens shall be made of metal not attacked by the plaster mortar and shall have sufficient material in the sides to prevent spreading during molding. Gang molds, when used, shall be of the type shown in Fig. 1. The dimensions of the briquet molds shall conform to the following requirements: width of mold, between inside faces, at waist line of the briquet, 1 in. (25.4 mm) with permissible variations of ±0.01 in. (0.25 mm) for molds in use and ±0.005 in. (0.13 mm) for new molds; thickness of molds measured at the point of greatest thickness on either side of the mold at the waist line, 1 in. (25.4 mm) with permissible variations of +0.004 in. (0.10 mm) and −0.002 in. (0.05 mm) for new molds and −0.02 in. (0.5 mm) for molds in use. The inside faces at the waist line shall have a vertical slot 0.03 in. (0.75 mm) wide by 0.06 in. (1.5 mm) deep to accept the divider. The briquet specimens shall conform to the dimensional requirements shown in Fig. 2.

7.12 *Divider*—Sheet brass 0.02 in. (0.05 mm) thick, 1.1 in. (27.8 mm) wide by 1 in. (25.4 mm) high.

7.13 *Clips for Briquet Testing Machine*—The clips for holding the tension test specimen shall be in accordance with Fig. 3.

8. Conditioning

8.1 *Room Temperature and Humidity*—Maintain the air temperature in the vicinity of the mixing and testing area, 70 ± 5°F (21 ± 3°C). Maintain the relative humidity at 50 ± 2 %.

8.2 *Temperature of the Mixing Water*—72 ± 2°F (22 ± 1°C).

9. High Temperature Test

9.1 *Significance and Use*—This test method provides procedures for evaluating the bond strength of adhesives after being exposed to accelerated aging at high temperature.

9.2 *Specimen Preparation*—For each test specimen, transfer the bonding compound, taken from the sample obtained, into a standard 1 pt (400 mL) glass container.

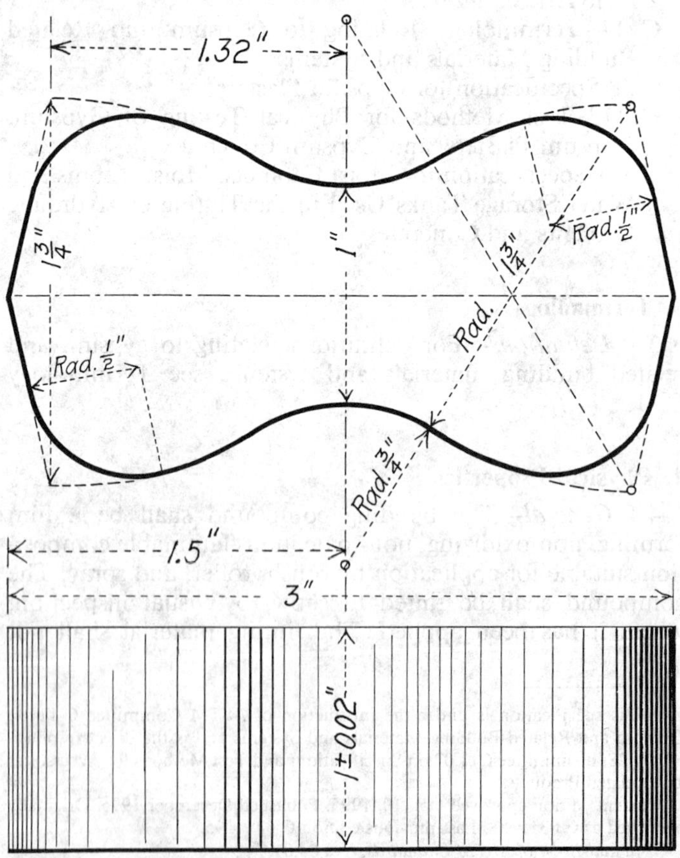

NOTE—1 in. = 25.4 mm

FIG. 2 Briquet Specimen for Tensile Strength Test

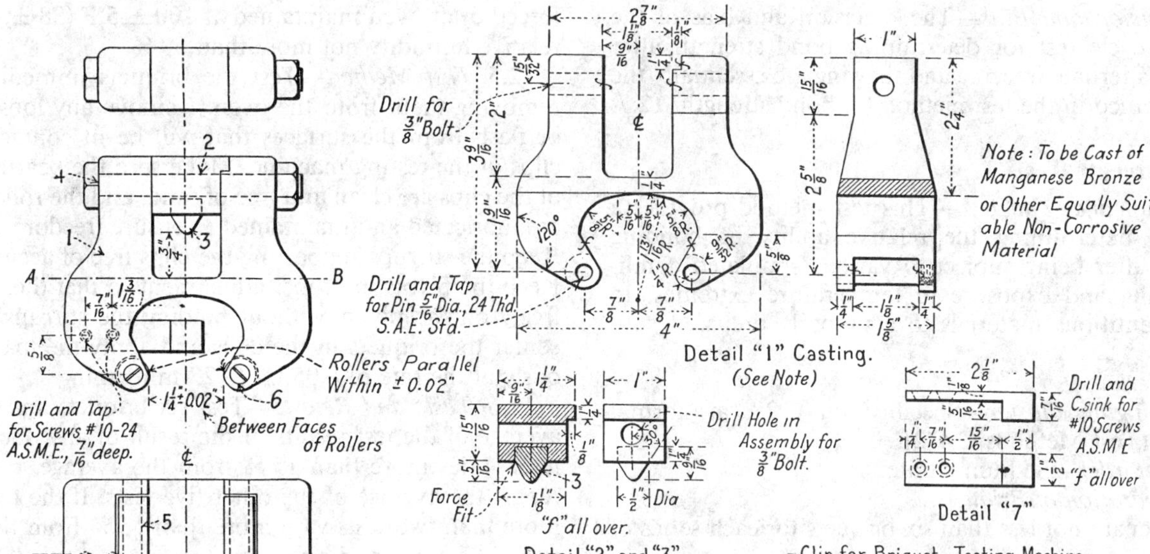

–Clip for Briquet-Testing Machine–

#	Name	Req'd.	Mat'l.	Remarks
1	Clip	2	*	* See Detail
2	Pivot Block	2	*	* Manganese Bronze
3	Pivot	2	Steel	Case-Hardened
4	Sq. H'd. Bolt, Hex. Nut	2	M.S.	$\frac{3}{8}$" Std.-$3\frac{1}{2}$" long
5	Roller	4	**	** Stainless Steel
6	Pin	4	**	Ditto
7	Guide	2	**	Ditto
8	Flat Head Mach. Scr.	4	Brss	#10-24 Th'd. A.S.M.E.

NOTE—1 in. = 25.4 mm

FIG. 3 Clips for Briquet Testing Machine

9.3 *Procedure*—Place the specimen in the oven for 30 days. Remove the container, allow the compound to cool to room temperature, mix the compound to a uniform consistency, and test as specified in Section 12.

9.4 *Test and Retests*—Test six briquets and report the average of the test results. If the result of any one of the six tests varied more than 15 % from the average, reject it and report the average of the other five tests. If the results from more than two tests vary more than 15 % from the average, reject the series and retest.

9.5 *Precision and Bias*—The precision and bias of the High Temperature Test for determining bond strength after exposure to high temperature is essentially the same as specified in the test method for bond strength, 12.7.

10. Degradation Test

10.1 *Significance and Use*—This test method simulates the effects of high temperature storage on the physical properties of bonding compound.

10.2 *Specimen Preparation*—For each test, transfer the bonding compound, from the sample obtained, into a standard 1 pt (400 mL) glass container.

10.3 *Procedure*—Place the specimen in the oven for 15 days. Remove the container from the oven and examine the contents for settling and separation. Place the container back in the oven for an additional 15 days. Remove the container and examine the contents again for settling and separation. Allow the compound to come to room temperature, and mix to a uniform consistency. Mix by hand, stirring with a wood, metal, or plastic paddle for not more than 5 min.

10.4 *Tests and Retests*—Bonding compound that fails to be remixed to a uniform consistency shall be considered to have failed the test.

10.5 *Report*—Determine the extent of settling and separation by measuring, and report as a percentage of the height of the specimen in the clear glass container, represented by the location of the liquid/solid interface.

10.6 *Precision and Bias*—No information is presented about either the precision or bias of the degradation test for measuring settling, separation, or the ability to be mixed to a uniform consistency since the test result is nonquantitative.

11. Freeze-Thaw Cycle Test

11.1 *Significance and Use*—This test method provides procedures for determining the bond strength of bonding compounds after being exposed to accelerated storage conditions of alternate freezing and thawing.

11.2 *Specimen Preparation*—For each test specimen, transfer the bonding compound, taken from the sample obtained, into a standard 1 pt (400 mL) glass container having a screw on or friction fit cover.

11.3 *Procedure*—Place a covered specimen in the freezer for 16 h. Remove the specimen and allow to thaw at room temperature for 8 h. Repeat this procedure for 5 cycles. After the fifth cycle, allow the compound to come to room temperature, and then test as specified in Section 12.

11.4 *Tests and Retests*—Test six briquets and report the average of the test results. If the result of any one of the six tests varies more than 15 % from the average, reject it and report the average of the other five tests. If the results from more than two tests vary more than 15 % from the average, reject the series and retest.

11.5 *Precision and Bias*—The precision and bias of the freeze-thaw cycle test for determining bond strength after exposure to alternate freezing and thawing are essentially the same as specified in the test method for bond strength, 12.7.

12. Bond Strength Test

12.1 *Significance and Use*—This test method provides a method for determining the relative ability of bonding compound, after being subject to various simulated conditions of aging and exposure to temperature extremes, to adhere cementitious materials to a properly prepared substrate.

12.2 *Materials:*

12.2.1 *Gauging Plaster*—Specification C 28, at normal consistency (Test Methods C 472).

12.2.2 *Retarder*—Sodium citrate.

12.3 *Preparation of Briquets:*

12.3.1 Prepare not less than six briquets for each series of tests.

12.3.2 Partition the briquet molds by placing the divider in the slots formed at the waist of the mold.

12.3.3 Add the plaster to the required amount of water and allow to soak for 30 s. Add approximately 0.1 g of retarder per 100 g of plaster to achieve a 2 to 3 h plaster set.

12.3.4 Prepare the molds by coating them with a thin film of mineral oil. Support the molds on a lightly oiled glass or metal plate. Protect the divider from contamination with oil (wrap in plastic food wrap) and carefully insert into the mold slots.

12.3.5 Cast the briquets immediately after completely mixing the mortar; half fill the molds and puddle the mortar uniformly with the tamper, taking care not to displace the divider. Slightly over fill the molds with additional mortar and repeat the puddling. After the mortar has set, cut off the excess to a plane surface flush with the top of the mold, using a broad knife or similar implement.

12.3.6 Immediately after molding, place the molds, on the base plate, into the moist cabinet for not less than 16 h.

12.3.7 Remove the briquets from the mold and store them at room temperature and humidity for not less than 48 h.

12.3.8 Examine the specimens, and discard any that have air holes or excessively rough surfaces.

12.4 *Procedure:*

12.4.1 Apply the bonding compound, in accordance with the producer's directions, to the flat waist surface of the briquet by means of a paint brush.

12.4.2 Remove the divider from the mold. Place the half briquet, coated with bonding compound, into the mold. Fill the remaining space in the mold with gypsum plaster, mixed and molded as specified in 12.2.

12.4.3 The time between coating the briquet with compound and casting the plaster against it shall be immediately after the compound has dried, or as recommended by the manufacturer, except when otherwise specified for a ten day waiting period.

12.4.4 Place the filled briquet mold in the moist cabinet for not less than 16 h. Remove the briquets from the molds as soon as they are thoroughly hardened, but retain them in the moist cabinet for not less than 16 h.

12.4.5 The briquets are then dried to constant weight in a forced draft oven maintained at 100 ± 5°F (38 ± 3°C) and a relative humidity not more than 50 %.

12.5 *Test Method*—Test the briquets immediately after removing them from the oven. Remove any loose particles or flash from the surfaces that will be in contact with the clips of the testing machine. Make sure the bearing surfaces of the clips are clean and free of sand, and the roller bearings are lubricated and maintained to ensure freedom of turning. Keep the stirrups supporting the clips free of accumulations. Keep the pivots in proper adjustment so that the clips swing freely on the pivots without binding the stirrups. Carefully center the briquets in the clips and apply the load continuously at the rate of 0.05 in. (1.27 mm)/min.

12.6 *Tests and Retests*—Test six briquets and report the average of the test results. If the result of any one of the six tests varies more than 15 % from the average, reject it and report the average of the other five tests. If the results from more than two tests vary more than 15 % from the average, reject the series and retest.

12.7 *Precision and Bias*—Precision and bias have not been determined for the test method specified.

13. Inspection

13.1 Inspection of the bonding compound shall be agreed upon between the producer or purchaser and the supplier as part of the purchase agreement.

14. Rejection

14.1 Rejection of bonding compound that fails to conform to the requirements of this specification shall be reported to the producer or supplier promptly and in writing. The notice of rejection shall contain a statement documenting how the product has failed to conform to the requirements of this specification.

15. Certification

15.1 When specified in the purchase agreement, the producer or supplier shall furnish a report certifying that, at the time of shipment, the bonding compound is in compliance with the requirements of this specification.

16. Packaging and Package Marking

16.1 *Packaging*—The bonding compound shall be packed in standard commercial containers. The containers shall be so constructed as to ensure acceptance by common or other carriers for safe transportation at the lowest rate to the point of delivery, unless otherwise specified in the purchase order.

16.2 *Marking*—Shipping containers shall be marked with the name of the bonding compound, the quantity contained therein, the name, brand, or trademark of the producer or supplier, the shelf life, the batch number, and the ASTM designation.

17. Keywords

17.1 bond; bonding compound; compound; plaster

SUMMARY OF CHANGES

This section identifies the location of changes to this specification that have been incorporated since the last issue. Committee C-11 has highlighted those changes that affect the technical interpretation of this specification.

(1) Mandatory language changes were made to Paragraphs 1.3, 5.2, and 10.4.

Designation: C 645 – 98

Standard Specification for Nonstructural Steel Framing Members[1]

This standard is issued under the fixed designation C 645; the number immediately following the designation indicates the year of original adoption or, in the case of revision, the year of last revision. A number in parentheses indicates the year of last reapproval. A superscript epsilon (ε) indicates an editorial change since the last revision or reapproval.

1. Scope*

1.1 This specification covers nonstructural steel framing members in interior construction assemblies.

1.2 The values stated in inch-pound units are to be regarded as the standard. The SI (metric) values given in parentheses are provided for information purposes only.

1.3 The following safety hazards caveat pertains only to the test methods portion, Sections 9 and 10, of this specification: *This standard does not purport to address all of the safety concerns, if any, associated with its use. It is the responsibility of the user of this standard to establish appropriate safety and health practices and determine the applicability of regulatory limitations prior to use.*

2. Referenced Documents

2.1 *ASTM Standards:*

A 653/A 653M Specification for Zinc-Coated (Galvanized) or Zinc-Iron Alloy-Coated (Galvanized) by the Hot-Dipped Process[3]

C 11 Terminology Relating to Gypsum and Related Building Materials and Systems[4]

C 36 Specification for Gypsum Wallboard[4]

C 1002 Specification for Steel Drill Screws for the Application of Gypsum Board or Metal Plaster Bases[4]

2.2 *AISI Standard:*

Specification for the Design of Cold-Formed Steel Structural Members[5]

3. Terminology

3.1 *Definitions*—Definitions shall be in accordance with Terminology C 11.

3.2 *Definitions of Terms Specific to This Standard:*

3.2.1 *members, n*—in screw application of gypsum board, studs, runners (track), rigid furring channels, main beams and cross furring members of grid suspension systems or other items manufactured in accordance with this specification.

3.2.2 *nonstructural wall stud, n*—a member in a steel framed wall system in which the loading does not exceed any of the following conditions: a transverse load of 20 lbf/ft (290 N/m) of member length; or an axial load, exclusive of sheathing, 200 lbf (890 N) per member.

4. Materials and Manufacture

4.1 Members shall be manufactured from steel that meets the mechanical requirements of the AISI Specification for the Design of Cold-Formed Steel Structural Members (1986 Edition and 1989 Addendum).

4.2 Members shall have a protective coating conforming to Specification A 653/A 653M – G 40 minimum or shall have a protective coating with an equivalent corrosion resistance.

4.3 Members shall be manufactured from steel having a minimum thickness, individual measurement of 0.0179 in. (0.455 mm) before application of protective coating.

5. Dimensions and Permissible Variations

5.1 Studs and rigid furring channels shall have a configuration and steel thickness such that the system in which they are used will carry the design transverse loads without exceeding either the allowable stress of the steel or the allowable design deflection. Main beams and cross furring of grid suspension systems shall be limited to a deflection of L/240. The manufacturer shall supply sufficient data for calculating design performance.

NOTE—Allowable deflection varies depending on the cladding used and architectural requirements. Detailed requirements shall be specified in application specifications.

5.1.2 Members, except main beams of grid suspension systems, shall be sufficiently rigid to permit penetration of the screw.

5.1.3 Minimum width of face to which gypsum board is screw-attached shall be not less than 1¼ in. (32 mm).

5.1.4 Minimum lip dimension shall be ⅛ in. (3.2 mm). See Appendix X1, Fig. X1.1.

5.2 Members shall be manufactured within the limits as shown in Table 1 and Fig. 1.

5.3 *Rigid Furring Channels*—Minimum depth shall be ¾ in. (19 mm). Minimum width of furring attachment flanges (see Fig. 3) shall be ½ in. (12.7 mm).

5.4 Grid suspension systems include main beams and cross

[1] This specification is under the jurisdiction of ASTM Committee C-11 on Gypsum and Related Building Materials and Systems and is the direct responsibility of Subcommittee C11.02 on Specifications and Test Methods for Accessories and Related Products.

Current edition approved May 10, 1998. Published December 1998. Originally published as C 645 – 70. Last previous edition C 645 – 97.

[2] *Annual Book of ASTM Standards*, Vol 01.03.

[3] *Annual Book of ASTM Standards*, Vol 01.06.

[4] *Annual Book of ASTM Standards*, Vol 04.01.

[5] Available from the American Iron and Steel Institute, 1000 16th St. N.W., Washington, DC 20036.

*** A Summary of Changes section appears at the end of this specification.**

TABLE 1 Manufacturing Tolerances

Dimension[A]	Item Checked	Drywall Studs, in. (mm)	Drywall Track, in. (mm)
A	length	+1/8 (3.18) −1/4 (6.35)	+1 (25.40) −1/4 (6.35)
B[B]	web width	+1/32 (0.79) −1/32 (0.79)	+1/8 (3.18) −0 (0)
C	flare overbend	+1/16 (1.59) −1/16 (1.59)	+0 (0) −3/16 (4.76)
D	hole center width	+1/8 (3.18) −1/8 (3.18)	NA
E	hole center length	+1/4 (6.35) −1/4 (6.35)	NA
F	crown	+1/8 (3.18) −1/8 (3.18)	+1/8 (3.18) −1/8 (3.18)
G	camber	1/32 per ft (0.79) 1/2 max (1.27)	1/32 per ft (0.79) 1/2 max (1.27)
H	bow	1/32 per ft (0.79) 1/2 max (1.27)	1/32 per ft (0.79) 1/2 max (1.27)
I	twist	1/32 per ft (0.79) 1/2 max (1.27)	1/32 per ft (0.79) 1/2 max (1.27)

[A] All measurements shall be taken not less than 1 ft (305 mm) from the end.
[B] Outside dimension for stud; inside for track.

furring members which mechanically interlock to form a modular supporting network. Length tolerance for grid suspension members shall be ±1/16 in. (1.59 mm).

5.5 Runners (track) shall be formed in a U-shaped configuration, having web depth compatible with those of the studs of the same nominal size. The runners (track) shall be designed such that when the studs are placed in both the top and bottom runners (track), they are held by friction. Minimum height of flanges shall be 1 in. (25 mm).

6. Edges

6.1 Members shall be manufactured in such a fashion as to minimize burrs and sharp edges.

7. Cutouts

7.1 Cutouts shall be permitted, but in no case shall they reduce the performance of the members in the gypsum board construction assembly below the specified performance requirements.

8. Sectional Properties

8.1 The sectional properties of members shall be computed in accordance with AISI Specifications for the Design of Cold-Formed Steel Structural Members (See Table 2 and Appendix X1).

9. Performance Requirements

9.1 *Penetration*—When tested in accordance with Section 10, members shall be capable of pulling the head of the screw below the surface of the gypsum board in less than 2 s without spin out.

10. Penetration Test

10.1 *Significance and Use*—This test method provides a

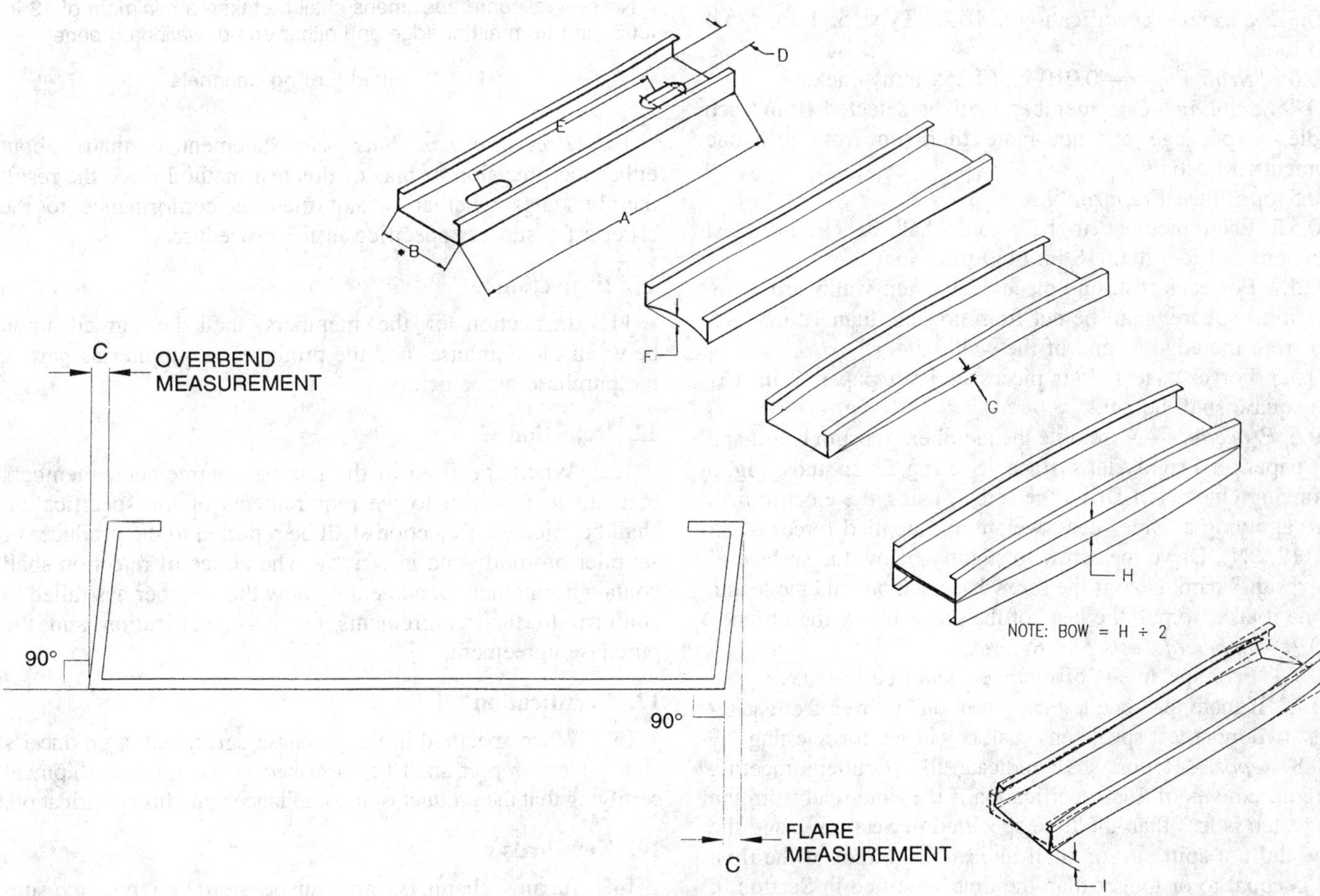

FIG. 1 Manufacturing Tolerances

TABLE 2 Minimum Section Properties for Various Drywall Studs[A,B,C]

Size, in. (mm)	Moment of Inertia,[D] I_x, in.^{4} (mm^4)	Section Modulus, S_x, in.^{3} (mm^3)
1⅝ (41.3)	0.028 (11500)	0.027 (483)
2½ (63.5)	0.075 (31100)	0.049 (800)
3½ (88.9)	0.168 (69900)	0.069 (1120)
3⅝ (92.1)	0.184 (76400)	0.071 (1170)
4 (101.6)	0.228 (98900)	0.078 (1290)

[A] Based on 0.0179 in. (0.46 mm) thickness.
[B] Properties of sections (moment of inertia and section modulus) are determined in accordance with 1986 AISI Specification for the Design of Cold-Formed Steel Structural Members, with 1989 Addendum.
[C] See Appendix for minimum drywall stud cross section.
[D] Based on Procedure 1 for deflection determination at the allowable moment in accordance with the 1986 AISI Specification For The Design of Cold-Formed Steel Structural Members, with 1989 Addendum.

procedure for evaluating the member's ability to pull the head of a screw below the surface of gypsum board. It shall be used to determine compliance with this specification. The degree of performance of this test method with service performance has not been determined.

10.2 *Apparatus shall satisfy the following:*

10.2.1 *Electric Drill,* capable of 2500 r/min (free spindle speed), supplied with a screw driving bit to fit the screw used in the test.

10.2.2 *Stop Watch,* capable of being read to the nearest 0.1 s.

10.3 *Materials shall satisfy the following:*

10.3.1 *Gypsum Wallboard*—Specification C 36, Type X, ⅝ in. (16 mm) thick.

10.3.2 *Screws*—Specification C 1002, Type S, 1 in. (25.4 mm) long.

10.3.3 *Kraft Paper*—0.010 in. (0.254 mm) thick.

10.4 *Sampling*—One member shall be selected from each bundle or package but not more than ten from any one shipment for testing.

10.5 *Specimen Preparation:*

10.5.1 Each member to be tested shall be cut into test specimens not less than 18 in. (460 mm) long.

10.5.2 For each test, one piece of gypsum wallboard, 6 in. (12.7 mm) square, shall be cut from not less than 12 in. (305 mm) from the edge or end of the wallboard.

10.5.3 For each test, four pieces of kraft paper, 2 in. (51 mm) square shall be cut.

10.6 *Procedure*—Assemble the member, gypsum board and kraft paper on a rigid, flat surface. (See Fig. 2 for studs, Fig. 3 for furring channels.) Drive the screw, using the electric drill while applying a force (dead weight and applied force) of 25 lbf (112.2 N). Drive the screw to slightly below the surface of the gypsum board. Note if the screw has spun out and the length of time it takes to pull the head of the screw below the surface.

10.7 *Number of Tests and Retests:*

10.7.1 Five specimens of members shall be tested.

10.7.2 If more than one test specimen fails to meet the requirements, two more test specimens shall be chosen for retesting.

10.8 *Report*—Report shall indicate all specimens meeting the requirements of this specification if the time read from the stop watch is less than the time specified in Section 9, and the screw did not spin out; or shall indicate all failing if the time read is equal to or longer than the time specified in Section 9, or the screw spun out.

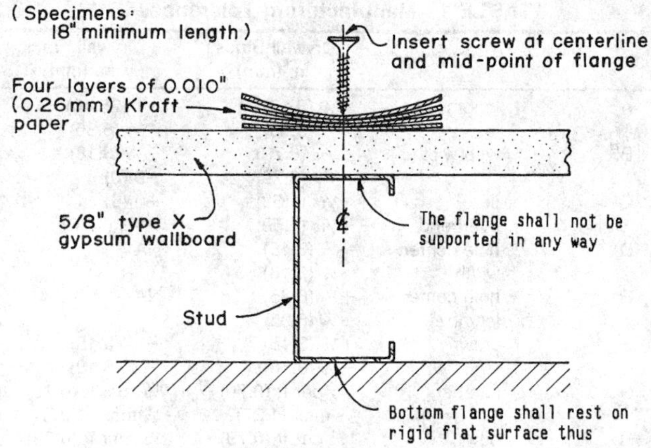

FIG. 2 Studs

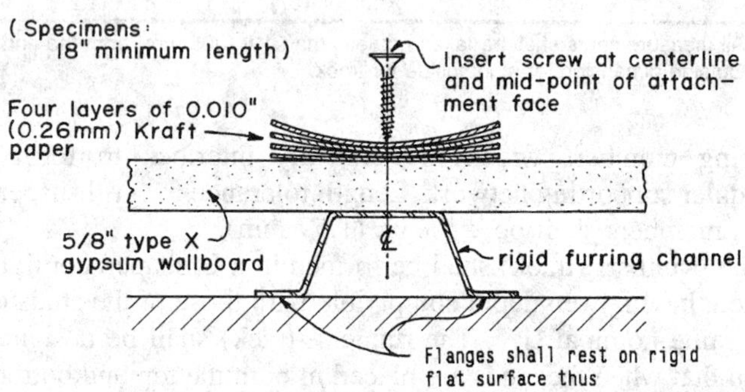

NOTE—Wallboard specimens shall be taken a minimum of 12 in. (305 mm) from either edge and either end of wallboard panel.

FIG. 3 Rigid Furring Channels

10.9 *Precision and Bias*—No statement is made about either the precision or bias of this test method since the result merely states whether or not there is conformance to the criteria for success specified in the procedure.

11. Inspection

11.1 Inspection of the members shall be agreed upon between the purchaser and the producer or supplier as part of the purchase agreement.

12. Rejection

12.1 When specified in the purchase agreement, members that fail to conform to the requirements of the specification shall be rejected. Rejection shall be reported to the producer or supplier promptly and in writing. The notice of rejection shall contain a statement documenting how the member has failed to conform to the requirements of this specification and the purchase agreement.

13. Certification

13.1 When specified in the purchase agreement, a producer's or supplier's report shall be furnished at the time of shipment certifying that the product is in compliance with this specification.

14. Keywords

14.1 furring channels; grid suspension systems; gypsum board; runners; screws; section properties; spinout; studs

APPENDIX

(Nonmandatory Information)

X1. MINIMUM DRYWALL STUD CROSS SECTION

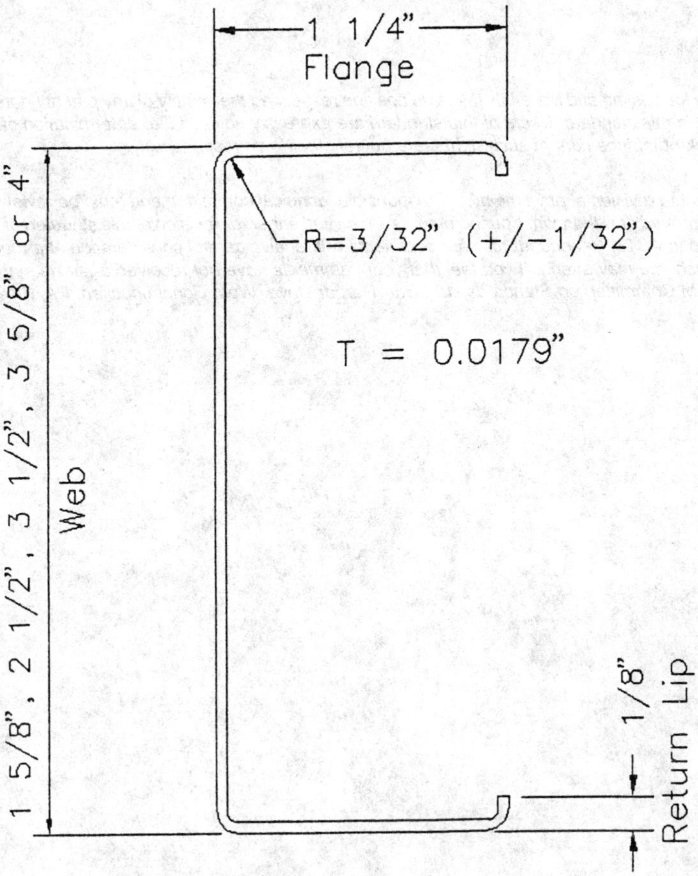

Fig. X1.1 Minimum Drywall Stud Cross Section

SUMMARY OF CHANGES

This section identifies the location of changes to this specification that have been incorporated since the last issue. Committee C-11 has highlighted those changes that affect the technical interpretation or use of this specification.

(*1*) Footnote A of Table 1 was revised.
(*2*) Revised paragraph 3.2.2.
(*3*) Added Specification C 36 to Section 2.
(*4*) Revised paragraphs 10.3.1 and 10.5.2.

Designation: C 688 – 96

Standard Specification for Functional Additions for Use in Hydraulic Cements[1]

This standard is issued under the fixed designation C 688; the number immediately following the designation indicates the year of original adoption or, in the case of revision, the year of last revision. A number in parentheses indicates the year of last reapproval. A superscript epsilon (ε) indicates an editorial change since the last revision or reapproval.

1. Scope

1.1 This specification covers methods to investigate the effectiveness of a material to beneficially change the properties of hydraulic cements when the material is interground with the clinker during manufacture of the cement.

1.2 The values stated in SI units are to be regarded as the standard.

1.3 The effect of additions in cement may markedly change properties other than those they are intended to modify. This specification is designed to test for such changes. Table 1 sets forth values for those properties of cement pastes and mortars that would permit a judgment of the changes effected by an addition. Likewise, Table 2 sets forth similar criteria for concrete. Certain additions may be found effective for more than one purpose as indicated in 3.1.4 and 3.1.5.

2. Referenced Documents

2.1 *ASTM Standards:*

C 39 Test Method for Compressive Strength of Cylindrical Concrete Specimens[2]

C 78 Test Method for Flexural Strength of Concrete (Using Simple Beam with Third-Point Loading)[2]

C 109/C 109M Test Method for Compressive Strength of Hydraulic Cement Mortars (Using 2-in. or 50-mm Cube Specimens)[3]

C 143 Test Method for Slump of Hydraulic Cement Concrete[2]

C 150 Specification for Portland Cement[3]

C 151 Test Method for Autoclave Expansion of Portland Cement[3]

C 157 Test Method for Length Change of Hardened Hydraulic Cement Mortar and Concrete[2]

C 187 Test Method for Normal Consistency of Hydraulic Cement[3]

C 219 Terminology Relating to Hydraulic Cement[3]

C 226 Specification for Air-Entraining Additions for Use in the Manufacture of Air-Entraining Portland Cement[3]

C 232 Test Method for Bleeding of Concrete[2]

C 234 Test Method for Comparing Concretes on the Basis of the Bond Developed with Reinforcing Steel[2]

C 266 Test Method for Time of Setting of Hydraulic-Cement Paste by Gillmore Needles[3]

C 403 Test Method for Time of Setting of Concrete Mixtures by Penetration Resistance[2]

C 451 Test Method for Early Stiffening of Portland Cement (Paste Method)[3]

C 465 Specification for Processing Additions for Use in the Manufacture of Hydraulic Cements[3]

C 595M Specification for Blended Hydraulic Cements[3]

C 596 Test Method for Drying Shrinkage of Mortar Containing Portland Cement[3]

C 666 Test Method for Resistance of Concrete to Rapid Freezing and Thawing[2]

C 845 Specification for Expansive Hydraulic Cement[3]

C 1157M Performance Specification for Blended Hydraulic Cement[3]

3. Terminology

3.1 *Definitions:*

3.1.1 *accelerating addition*—a functional addition that accelerates the setting or early strength, or both, of concrete and mortar.

3.1.2 *retarding addition*—a functional addition that retards the setting of concrete and mortar.

3.1.3 *set-control addition*—a functional addition composed essentially of calcium sulfate in any hydration state from $CaSO_4$ to $CaSO_4 \cdot 2H_2O$.

3.1.4 *water-reducing addition*—a functional addition used to reduce the quantity of mixing water required to produce concrete and mortar of a given consistency.

3.1.5 *water-reducing and accelerating addition*—a functional addition that reduces the quantity of mixing water required to produce concrete of a given consistency and that accelerates the setting or early strength development, or both, of concrete and mortar.

3.1.6 *water-reducing and retarding addition*—a functional addition that reduces the quantity of mixing water required to produce concrete and mortar of a given consistency and simultaneously retards the setting of concrete and mortar.

NOTE 1—This section is intended to provide a specification that may be applied to calcium sulfates as defined in Terminology C 219.

NOTE 2—It should be realized that some calcium sulfates, particularly some byproduct calcium sulfates, have produced cements with undesired set behavior after storage.

4. Ordering Information

4.1 The purchaser shall specify the type of functional addition desired.

5. Materials

5.1 *Cements*—The cements used in the evaluation of the addition shall be as described in Section 4.1 of Specification C 465 with the following exceptions:

---

[1] This specification is under the jurisdiction of ASTM Committee C-1 on Cement and is the direct responsibility of Subcommittee C01.20 on Additions.

Current edition approved May 10, 1996. Published July 1996. Originally published as C 688 – 71 T. Last previous edition C 688 – 95.

[2] *Annual Book of ASTM Standards*, Vol 04.02.

[3] *Annual Book of ASTM Standards*, Vol 04.01.

TABLE 1 Criteria for Evaluating Neat Cement and Mortar Containing Functional Cement Additions

Tests	Type of Addition					
	Water Reducing	Retarding	Accelerating	Water-Reducing and Retarding	Water-Reducing and Accelerating	Set-Control
Normal consistency, deviation from control, percentage points[A]	−1.0 min	+1.0 max	+1.0 max	−1.0 min	−1.0 mm	±1.0 max
Standard consistency (flow) deviation from control, percentage points[A]	−4.0 min	+2.0 max	+2.0 max	−4.0 min	−4.0 min	±2.0 max
Setting time, (Gillmore) deviation from control, h:min						
Initial						
At least	...	1:00 later[B]		1:00 later[B]		...
Not more than	1:00 earlier nor 1:30 later	3:30 later	1:30 earlier nor 1:30 later	3:30 later	1:30 earlier nor 1:30 later	1:00 earlier nor 1:30 later
Final						
Not more than	1:00 earlier nor 1:30 later	3:30 later	1:00 earlier nor 1:30 later	3:30 later	1:00 earlier nor 1:30 later	1:00 earlier nor 1:30 later
Compressive strength, min, percent of control:[C]						
1 day in moist air	110	90	125[D]	90	125	[E]
1 day in moist air, 2 days in water	110	90	125[D]	100	125	
1 day in moist air, 6 days in water	110	90	100	110	110	
1 day in moist air, 27 days in water	110	95	95	110	110	
Autoclave Expansion						
max increase in % change in length compared to control	0.10	0.10	0.10	0.10	0.10	0.10
Drying Shrinkage of Mortar						
max % change in length compared to control	0.020	0.020	0.030	0.020	0.020	0.020

[A] The minus sign indicates that the percentage of water required shall be less than that of the control cement by at least the indicated percentage points.
[B] Or 50 % later, whichever is the lesser.
[C] The compressive strength of mortar containing the test cement shall be not less than 95 % of that attained at any previous test age. The objective of this limit is to require that the strength of mortar containing the addition under test shall not decrease with age.
[D] The grand average of the 1, 3, 7, and 28-day strengths shall be not less than 95 % of the grand average for the corresponding control cement.
[E] In cases where the accelerated set time only is required, the strength can be reduced to 100 % of the control.

5.1.1 At least one of the Type I cements shall contain not less than 9 percent tricalcium aluminate (C_3A).

5.1.2 Disregard the last sentence of Section 4.1.5 of Specification C 465 and substitute the following: "Each control cement shall comply with all the requirements in the specification (C 150, C 595, C 845, and C 1157M) applicable to that type of cement. The method shall be adequate for the qualitative and quantitative determination of the addition in the finished cement, and shall be fully described in the report of the tests on the addition."

5.2 *Aggregates*—The aggregates used in the evaluation of the addition shall be in accordance with the Aggregates portion of the Materials section of Specification C 465, using proportions specified in Concrete Mixtures section of that specification.

6. General Requirements

6.1 The cement, mortar, and concrete in which each of the additions is used shall conform to the respective requirements prescribed in Tables 1 and 2, except that if the test cement fails to meet the requirements of Table 1, but possesses all the requirements listed in Table 2, then the requirements of Table 2 shall govern approval of the addition.

6.2 The trade name, source, and character of the material shall be specified and the means for the quantitative determination of the proposed addition in the finished cement, shall be furnished by the manufacturer or seller of the addition, and the results of such pertinent quantitative analysis shall form a part of the record of tests of the addition.

6.3 At the request of the purchaser of an addition for a specific functional purpose, the manufacturer shall state in writing the content of any substance which is known to or believed to impair other desirable properties of the cement or concrete.

6.4 The additions shall be evaluated by comparing cements containing the "test addition" to cements ground from similar clinkers from the same source. Those cements without test additions are referred to in this specification as "control cements."

6.4.1 *Special Provisions for Set-Control Additions Conforming to 3.1.3:*

6.4.1.1 The additions shall be evaluated by comparing cements containing the "test addition" to control cements ground from similar clinkers from the same source and under the same grinding conditions with the mill temperatures being within ±3°C (5°F) of the same value. The control cements shall be produced with calcium sulfate at any hydration state having a past record of satisfactory usage as an addition for portland cement.

6.4.1.2 Each test cement shall contain sufficient test addition $CaSO_4$ to produce a sulfur trioxide (SO_3) content that differs by no more than 0.24 percentage points from that of the control cement. The SO_3 shall be expressed as a percentage of the mass of the cement and reported to the nearest 0.01 %.

6.4.1.3 The test cements containing the additions shall conform to the requirements of 8.1 and Table 1. The cements need not be tested in concrete according to the requirements of 8.2 and Table 2.

6.5 The amount of addition to be interground with the cement for evaluation purposes shall be such as to produce the desired effects as listed in Table 1. When these tests show

TABLE 2 Criteria for Evaluating Concrete Containing Functional Cement Additions[A]

Tests	Type of Addition				
	Water Reducing	Retarding	Accelerating	Water-Reducing and Retarding	Water-Reducing and Accelerating
Water content, max, percent of control	95	. . .	. . .	95	95
Time of setting, (penetration) deviation from control, h:min:					
Initial					
At least	. . .	1:00 later[B]		1:00 later[B]	
Not more than	1:00 earlier nor 1:00 later	3:30 later	3:30 earlier	3:30 later	3:30 earlier
Final					
Not more than	1:00 earlier nor 1:30 later	3:30 later	. . .	3:30 later	. . .
Compressive strength, min, percent of control:[C]					
1 day[D]	110	85	125[E]	100	125
3 days	110	85	125[E]	100	125
7 days	110	90	100	110	110
28 days	110	90	95	110	110
3 months	100	90	95	100	100
1 year	100	90	95	100	100
Flexural strength, min, percent of control:[C]					
1 day[D]	100	85	110	90	110
3 days	100	85	110	100	110
7 days	100	90	100	100	100
28 days	100	90	90	100	100
3 months	100	90	90	100	100
1 year	100	90	90	100	100
Bond strength, percent of control:					
28 days	100	90	100	100	100
Volume change, expressed as change in length, max, increase over control, percentage points:					
28 days	0.010	0.010	0.030	0.010	0.010
3 months	0.010	0.010	0.030	0.010	0.010
1 year	0.010	0.010	0.030	0.010	0.010
Durability factor, min, percent of control[F]	80	80	80	80	80

[A] The values in the table are intended to allow for normal variation in test results. For example, the object of the 90 % compressive strength requirement for a retarding addition is to require a level of performance comparable to that of the reference concrete.

[B] Or 50 % later, whichever is the lesser.

[C] The compressive and flexural strength of the concrete containing the addition under test at any test age shall be not less than 90 % of that attained at any previous test age. The objective of this limit is to require that the compressive or flexural strength of the concrete containing the addition under test shall not decrease with age.

[D] One-day strengths applicable only to Type III cement. One-year strength tests shall be made; however, the addition may be approved after completion of the 3-month strength results, at the discretion of the purchaser.

[E] In cases where the accelerated set time only is required, the strength can be reduced to 100 % of the control.

[F] This requirement is applicable only when the addition is to be used in air-entrained concrete.

compliance with the requirements of this specification, the addition shall be used only in amounts up to the maximum amount tested in any one of the five cements used in the evaluation. For a cement requiring more than the established amount, a separate compliance test on this one cement can be made which will then establish the new maximum amount which can be used.

6.6 The cement produced with the functional additions shown in Section 3 shall comply with the specification for the respective cement and the ordinarily determined properties of cement paste, mortar, or concrete made with the cement containing the test addition shall not differ from these same properties of the cement paste, mortar, or concrete made with the control cement as indicated by the standard tests, except as provided in Tables 1 and 2.

6.7 Generally, the addition rates required for functional purposes are several times greater than those needed for processing purposes and the resultant increase in flowability, or lubricity, from these increased addition rates may cause sufficient reduction in mill retention time to affect significantly the particle size distribution of the treated cement. The test addition may, in instances where the full-scale tests have shown reduced mill retention time to be significant, be determined for acceptance purposes by making supplementary laboratory or pilot batch mill grinds.

6.8 Furthermore, the effect of the addition on the properties of the cement, mortar, or concrete shall be within the limits of Section 8 and Tables 1 and 2. In the event tests are conducted by making laboratory or pilot batch mill tests, this fact shall be entered in the report as specified in Section 9, and the specific tests shall be indicated.

6.9 After it has been documented once by the test results that a specific trade name and source of an addition have met the requirements of this specification, further tests shall be waived for subsequent shipments of the same product. Each shipment, however, shall be identified as to trade name, source, and date of shipment, and shall be essentially identical to nature and composition to that same trade name product which initially met the requirements of this specification.

7. Sampling

7.1 Sample the cements in accordance with the Sampling Cement section of Specification C 465. The quantity of sample shall be sufficient to make all the evaluation tests prescribed by this method. (A minimum of 300 kg (650 lb) per lot of cement is usually satisfactory.)

8. Test Methods

8.1 *Cements and Mortars*—Test the cements and mortars in accordance with the following ASTM test methods (see Table 1):

8.1.1 *Autoclave Expansion*—Test Method C 151.

8.1.2 *Drying Shrinkage of Mortar*—The length change of air-stored mortar bars made with the test cement shall not be more than 0.020 percentage points greater than that of similar mortar bars made with the control cement cured and tested according to Test Method C 596, except that for accelerators, an increased shrinkage of 0.03 percentage points will be allowed.

8.1.3 *Normal Consistency*—Test Method C 187.

8.1.4 *Standard Consistency (Flow)*—Test Method C 109.

8.1.5 *Setting Time*—Test Method C 266.

8.1.5.1 For set-control additions conforming to 3.1.3 repeat the setting time tests 3 months after grinding. Keep cement samples for these tests in sealed metal containers with minimum air contact.

8.1.6 *Compressive Strength*—For functional additions having no water-reducing effects, use Test Method C 109. For functional additions having some water-reducing effects, use Test Method C 109 except that mortars for both the test and the control specimens shall be made to a flow of 110 ± 5. In either case, there shall be six specimens for each condition of test.

8.1.6.1 It is required that cubes for companion cements be made and tested on the same days with storage of specimens side by side in the same section of the moist cabinet during the 24-h curing period. However, in water storage, store control and test specimens in separate tanks. If less than four specimens are left for each period, after discarding faulty cubes, make a retest.

NOTE 3—It is suggested that three batches of nine cubes each be made for each cement, then six cubes be picked at random for each age of test and the three excess cubes held to replace any faulty cubes found.

8.1.7 *False Set*—For set-control additions conforming to 3.6, perform false set tests initially and 3 months after grinding in accordance with Test Method C 451. The final penetration in percent shall be no less than 75 % as great as that with the control cement.

8.2 *Concretes*—Test the plastic and hardened concretes in accordance with the following ASTM methods (see Table 2):

8.2.1 *Consistency*—The consistency (slump) of the concrete produced with the test cement shall not differ from that of the control concrete by more than ±13 mm (±½ in.), when tested in accordance with Test Method C 143, and each shall have a slump of 64 ± 13 mm (2½ ± ½ in.).

8.2.2 *Water Content*—The maximum water content of the test concrete shall be as shown in Table 2 when compared with the control concrete of similar consistency tested as in 8.2.1.

8.2.3 *Time of Setting*—Test Method C 403.

8.2.4 *Compressive Strength*—Test Method C 39.

8.2.5 *Flexural Strength*—Test Method C 78.

8.2.6 *Bond to Steel*—Test Method C 234.

8.2.7 *Volume Change*—Test Method C 157.

8.2.8 *Resistance to Freezing and Thawing (Durability Factor)*—Test Method C 666. The air content of both the control and test concrete shall be 6 ± 0.5 % obtained by admixing appropriate quantities of any of the reference air entraining agents listed in the General Requirements of Specification C 226.

8.2.9 *Bleeding*—At the request of the purchaser, test the bleeding characteristics in accordance with Test Method C 232.

9. Report

9.1 The report shall conform to the Report section of Specification C 465, except that comparison of test results shall be made to the requirements prescribed in this specification instead of to the requirements of Specification C 465.

10. Keywords

10.1 additions; functional; hydraulic cement

Standard Specification for Limestone for Animal Feed Use[1]

This standard is issued under the fixed designation C 706; the number immediately following the designation indicates the year of original adoption or, in the case of revision, the year of last revision. A number in parentheses indicates the year of last reapproval. A superscript epsilon (ε) indicates an editorial change since the last revision or reapproval.

1. Scope

1.1 This specification covers limestone supplied for use as a mineral supplement in animal feeds.

NOTE—The calcium needed for animal nutrition is customarily supplied by limestone. Such limestone must be sufficiently fine to blend with mixed feeds and yet be free from excessive dusting.

1.2 *This standard does not purport to address all of the safety problems, if any, associated with its use. It is the responsibility of the user of this standard to establish appropriate safety and health practices and determine the applicability of regulatory limitations prior to use.*

2. Referenced Documents

2.1 *ASTM Standards:*

C 25 Test Methods for Chemical Analysis of Limestone, Quicklime, and Hydrated Lime[2]

C 50 Methods for Sampling, Inspection, Packing, and Marking of Lime and Limestone Products[2]

C 51 Terminology Relating to Lime and Limestone[2]

C 110 Test Methods for Physical Testing of Quicklime, Hydrated Lime, and Limestone[2]

E 11 Test Method for Wire-Cloth Sieves for Testing Purposes[3]

3. Terminology

3.1 Definitions of terms applying to this specification appear in Terminology C 51.

4. General Requirements

4.1 The product must be suitable for animal feeding.

5. Chemical Requirements

5.1 Limestone, as identified on feed labels, shall conform to the chemical requirements given in Table 1.

6. Physical Requirements

6.1 Screen analysis and other physical requirements shall be specified by the purchaser.

TABLE 1 Chemical Requirements

Limestone Classes (See C 51)	Calcium, min, percent (as $CaCO_3$)	Magnesium, max, percent (as $MgCO_3$)	Magnesium, min, percent (as $MgCO_3$)	Moisture, max, percent
High Calcium	95	5	. . .	0.5
Magnesium	60	35[A]	5	0.5
Dolomitic	49	46[A]	35	0.5
Ground Limestone	82	[B]	[B]	0.5

[A] The high magnesium content of these materials usually limits their use to special-purpose feeds.

[B] As specified by the purchaser.

7. Test Methods

7.1 *Chemical Analysis*—Unless otherwise agreed upon between the purchaser and the manufacturer, the chemical analysis of the limestone shall be used in accordance with Test Methods C 25.

7.2 *Fineness*—Unless otherwise agreed upon between the purchaser and the manufacturer, the sieve analysis of the material shall be determined in accordance with Test Methods C 110.

7.2.1 *Procedure*—If the entire sample is not to be dried, obtain lesser portions by riffling or quartering. Dry at between 115 to 120°C to a constant mass and cool to room temperature. Obtain a 90 to 120-g dry sample by riffling or quartering. If the material tends to cake, break up the agglomerates with a soft rubber pestle. Quantitatively transfer the weighed sample to an 8-in. diameter standard sieve or set of sieves (for example, Nos. 10, 20, 40, 60, 80, and 100 or other appropriate combination conforming to Test Method E 11). Conduct the sieving by lateral and vertical motion accompanied by a jarring action. Continue for a minimum of 5 min or until an additional 3 min of sieving time fails to change the results of any sieve fraction by 0.5 % of the total sample mass. Take care to avoid overloading any sieve when assaying closely sized materials. Determine the mass of each sieve fraction and report as the percentage of total sample mass.

8. Sampling, Inspection, etc.

8.1 The sampling, inspection, rejection, retesting, packing, and marking shall be conducted in accordance with Methods C 50.

9. Keywords

9.1 animal; feed; limestone; mineral; supplement

[1] This specification is under the jurisdiction of ASTM Committee C-7 on Lime and is the direct responsibility of Subcommittee C07.03 for Industrial Uses.

Current edition approved July 10, 1998. Published December 1998. Originally published as C 706 – 72. Last previous edition C 706 – 92.

[2] *Annual Book of ASTM Standards*, Vol 04.01.

[3] *Annual Book of ASTM Standards*, Vol 14.02.

Standard Specification for Limestone for Dusting of Coal Mines[1]

This standard is issued under the fixed designation C 737; the number immediately following the designation indicates the year of original adoption or, in the case of revision, the year of last revision. A number in parentheses indicates the year of last reapproval. A superscript epsilon (ε) indicates an editorial change since the last revision or reapproval.

1. Scope

1.1 This specification covers limestone suitable for use as dust in coal mines to prevent coal dust explosions.

NOTE—In coal mine operations, limestone is dusted onto coal exposures in sufficient amount so that not less than 65 % of all loose dust shall be limestone. With such a concentration of limestone, explosions cannot initiate or be propagated from nearby gas explosions. The limestone must be substantially dry in order to dust satisfactorily.

2. Referenced Documents

2.1 *ASTM Standards:*

C 25 Test Methods for Chemical Analysis of Limestone, Quicklime, and Hydrated Lime[2]

C 50 Methods for Sampling, Inspection, Packing, and Marking of Lime and Limestone Products[2]

C 110 Test Methods for Physical Testing of Quicklime, Hydrated Lime, and Limestone[2]

3. General Requirements

3.1 Either high calcium or dolomitic limestone may be furnished for this application.

4. Chemical Composition

4.1 Limestone shall conform to the following as to chemical composition:

Moisture (at point of manufacture), max, %	0.5
Silica, free and combined, max, %	4.0

5. Physical Properties

5.1 Limestone for this application shall have the following fineness:

Passing a No. 20 (850-μm) sieve, min, %	100
Passing a No. 200 (75-μm) sieve, min, %	70

6. Test Methods

6.1 *Chemical Analysis*—The chemical analysis of the limestone shall be made in accordance with Test Methods C 25.

6.2 *Fineness*—The sieve analysis of limestone shall be determined in accordance with Test Methods C 110.

7. Sampling, Inspection, etc.

7.1 The sampling, inspection, rejection, retesting, packing, and marking shall be conducted in accordance with Methods C 50.

8. Keywords

8.1 coal mines; dolomitic limestone; fineness; high calcium limestone; lime; limestone; limestone dust; mine explosions; moisture content; silica content

[1] This specification is under the jurisdiction of ASTM Committee C-7 on Lime and is the direct responsibility of Subcommittee C07.03 on Industrial Uses.

Current edition approved June 10, 1997. Published June 1998. Originally published as C 737–73. Last previous edition C 737–92.

[2] *Annual Book of ASTM Standards*, Vol 04.01.

Standard Specification for Installation of Steel Framing Members to Receive Screw-Attached Gypsum Panel Products[1]

This standard is issued under the fixed designation C 754; the number immediately following the designation indicates the year of original adoption or, in the case of revision, the year of last revision. A number in parentheses indicates the year of last reapproval. A superscript epsilon (ε) indicates an editorial change since the last revision or reapproval.

This standard has been approved for use by agencies of the Department of Defense.

1. Scope*

1.1 This specification covers the minimum requirements for the installation of interior nonstructural steel framing and furring members designed to receive screw-attached gypsum panel products. The steel framing and furring members covered in this specification are limited to those complying with Specification C 645.

1.2 Details of construction for a specific assembly to achieve the required fire resistance shall be obtained from reports of fire-resistance tests, engineering evaluations, or listings from recognized fire testing laboratories.

1.3 The values stated in inch-pound units are to be regarded as the standard. The SI (metric) values given in parentheses are approximate and are provided for information purposes only.

2. Referenced Documents

2.1 *ASTM Standards:*

A 366 Specification for Steel, Sheet, Carbon, Cold-Rolled, Commercial Quality[2]

A 641 Specification for Zinc-Coated (Galvanized) Carbon Steel Wire[3]

A 641M Specification for Zinc-Coated (Galvanized) Carbon Steel Wire (Metric)[3]

A 653/A 653M Specification for Steel Sheet, Zinc-Coated (Galvanized) or Zinc-Iron Alloy-Coated (Galvannealed) by the Hot-Dip Process[3]

C 11 Terminology Relating to Gypsum and Related Building Materials and Systems[4]

C 645 Specification for Non-Load (Axial) Bearing Steel Studs, Runners (Tracks), and Rigid Furring Channels for Screw Application of Gypsum Board[4]

E 119 Test Methods for Fire Tests of Building Construction and Materials[5]

3. Terminology

3.1 *Definitions*—Terms shall be as defined in Terminology C 11.

3.2 *Descriptions of Terms Specific to This Standard:*

3.2.1 *channel, n*—the material described in 4.3 to which furring members are attached.

3.2.2 *cross furring, n*—furring member attached perpendicular to main runners or framing members.

3.2.3 *cross furring member, n*—a member installed perpendicularly to the main beams designed to receive screw attached gypsum panel products.

3.2.4 *direct furring, n*—furring members attached directly to the structural members of the building.

3.2.5 *framing member, n*—metal studs, runners (track), and rigid furring channels designed to receive screw attached gypsum panel products.

3.2.6 *furred ceiling, n*—a ceiling in which the rigid furring channels and studs are attached directly to the structural members of the building.

3.2.7 *furring, v*—preparing a wall or ceiling with framing or furring members to provide a level surface or airspace.

3.2.8 *furring member, n*—metal studs, rigid furring channels, or channels used either as direct furring or as cross furring.

3.2.9 *grid suspension system, n*—a ceiling system composed of modular interlocking steel components designed to receive screw-attached gypsum panel products.

3.2.10 *main beam, n*—the main support member of a grid suspension system that receives cross furring members.

3.2.11 *main runner, n*—the channel or stud that is attached to or suspended from the structural members of the building.

3.2.12 *runner (track), n*—a member designed to receive the ends of metal studs, attached directly to the structural members of the building.

3.2.13 *suspended ceiling, n*—a ceiling in which the main runners and cross furring are suspended below the structural members of the building.

4. Materials and Manufacture

4.1 *Studs, Runners, Rigid Furring Channels, and Grid Suspension Systems*—see Specification C 645.

4.2 *Tie Wire and Hanger Wire*—Zinc-coated (galvanized) soft-annealed steel, or of a material and size having equivalent

[1] This specification is under the jurisdiction of ASTM Committee C-11 on Gypsum and Related Building Materials and Systems and is the direct responsibility of Subcommittee C11.03 on Specifications for Application of Gypsum and Other Products in Assemblies.

Current edition approved Oct. 10, 1998. Published March 1999. Originally published as C 754 – 74. Last previous edition C 754 – 98.

[2] *Annual Book of ASTM Standards*, Vol 01.03.

[3] *Annual Book of ASTM Standards*, Vol 01.06.

[4] *Annual Book of ASTM Standards*, Vol 04.01.

[5] *Annual Book of ASTM Standards*, Vol 04.07.

*** A Summary of Changes section appears at the end of this specification.**

corrosion resistance and strength. Wire diameters (uncoated) specified herein correspond with United States steel wire gage numbers as follows:

Diameter,[A] in.	mm	Wire Gage (U.S. Steel Wire Gage)
0.0348	0.88	No. 20
0.0410	1.04	No. 19
0.0475	1.21	No. 18
0.0540	1.37	No. 17
0.0625	1.59	No. 16
0.0800	2.03	No. 14
0.0915	2.32	No. 13
0.1055	2.68	No. 12
0.1205	3.06	No. 11
0.1350	3.43	No. 10
0.1483	3.77	No. 9
0.1620	4.12	No. 8

[A] Allowable variations in diameter shall be in accordance with tolerances as established in A 641 or A 641M.

4.3 *Channels*—Channels shall be cold-formed from steel with a minimum 33 000 psi (228 MPa) yield strength and 0.0538 in. (1.37 mm) minimum bare steel thickness. Channels shall have a protective coating conforming to Specification A 653–G 40 or shall have a protective coating with an equivalent corrosion resistance, and shall have the following minimum weights in lb per 1000 linear ft (kg/m):

Size, in.	(mm)	Weight lb/1000 ft	(kg/m)	Flange Width in.	(mm)
¾	(19)	277	(0.412)	½	(13)
1½	(38)	414	(0.616)	½	(13)
2	(51)	506	(0.753)	½	(13)
2½	(64)	597	(0.888)	½	(13)

4.4 *Grid Suspension System:*

4.4.1 *Main Beam*—Formed from cold-rolled steel "T" sections, indexed with slots to receive ends of cross furring members, and with stamped couplings at each end for the purpose of splicing.

4.4.2 *Cross Furring Members*—Formed from cold-rolled steel, designed to permit screw attachment of gypsum panel products, and formed with an end configuration that permits mechanical interlock with the indexed slots of the main beam.

4.5 *Rod and Flat Hangers*—Formed from steel conforming to Specification A 366. When specified, rod and flat hangers shall be protected with zinc coating or another equally rust-inhibiting coating.

5. Installation of Metal Framing

5.1 *Tolerances:*

5.1.1 Spacing of studs and furring members shall be not more than ±⅛ in. (3 mm) from the spacing shown in Tables 1 and 2. Any cumulative error shall be not more than ±⅛ in.

TABLE 1 Maximum Framing Spacing

NOTE—Where a conflict exists in spacing between base and face layers, the closer spacing shall govern.

Gypsum Panel Product Thickness: Base Layer, in. (mm)	Gypsum Panel Product Thickness: Face Layer, in. (mm)	Location	Application	Maximum Spacing, oc: One Layer Only, in. (mm)	Two Layers: Fasteners Only, in. (mm)	Two Layers: Adhesive Between Layers, in. (mm)
⅜ (9.5)	. . .	ceilings	perpendicular	16 (406)[A]	16 (406)[A]	16 (406)[A]
	⅜ (9.5)	ceilings	perpendicular	NA	16 (406)	16 (406)
	⅜ (9.5)	ceilings	parallel	NA	NR	16 (406)
½ (12.7)	. . .	ceilings	perpendicular	24 (610)[A]	24 (610)[A]	24 (610)[A]
	. . .	ceilings	parallel	16 (406)[A]	16 (406)[A]	16 (406)[A]
	⅜ (9.5)	ceilings	perpendicular	NA	16 (406)	24 (610)
	⅜ (9.5)	ceilings	parallel	NA	NR	24 (610)
	½ (12.7)	ceilings	perpendicular	NA	24 (610)	24 (610)
	½ (12.7)	ceilings	parallel	NA	16 (406)	24 (610)
⅝ (15.9)	. . .	ceilings	perpendicular	24 (610)[A]	24 (610)[A]	24 (610)[A]
	. . .	ceilings	parallel	16 (406)[A]	16 (406)[A]	16 (406)[A]
	⅜ (9.5)	ceilings	perpendicular	NA	16 (406)	24 (406)
	⅜ (9.5)	ceilings	parallel	NA	NR	24 (610)
	½ or ⅝ (12.7 or 15.9)	ceilings	perpendicular	NA	24 (610)	24 (610)
	½ or ⅝ (12.7 or 15.9)	ceilings	parallel	NA	16 (406)	24 (406)
¼ (6.4)	. . .	walls	parallel	NR	16 (406)[A]	16 (406)[A]
	⅜ (9.5)	walls	NR	NR	NR	NR
	½ or ⅝ (12.7 or 15.9)	walls	perpendicular or parallel	NA	16 (406)	16 (406)
⅜ (9.5)	. . .	walls	perpendicular or parallel	16 (406)[A]	16 (406)[A]	24 (610)[A]
	⅜ or ½ or ⅝ (9.5 or 12.7 or 15.9)	walls	perpendicular or parallel	NA	16 (406)	24 (610)
½ or ⅝ (12.7 or 15.9)	. . .	walls	perpendicular or parallel	24 (610)[A]	24 (610)[A]	24 (610)[A]
	⅜ or ½ or ⅝ (9.5 or 12.7 or 15.9)	walls	perpendicular or parallel	NA	24 (610)	24 (610)

Perpendicular—perpendicular to framing members
Parallel—parallel to framing members
NA—not applicable
NR—not recommended
oc—on center
[A] Denotes framing spacing for base layer in two-layer application.

TABLE 2 Spans and Spacings of Horizontal Furring Members

Type of Furring	Maximum[A] Spacing c to c, in. (mm)	Maximum Span, ft (mm)
Rigid Furring Channel	24 (610)	4 (1220)
1⅝ in. (41 mm) stud (erected with open side up and against support)	24 (610)	5 (1520)
2½ in. (64 mm) stud (erected with web vertical to support)[B]	24 (610)	6 (1830)
3⅝ in. (92 mm) stud (erected with web vertical to support)[B]	24 (610)	8 (2440)

c to c—center to center
[A] Consult Table 1 for maximum spacing as determined by gypsum panel product thickness.
[B] A 6 in. (150 mm) length of same size stud or track shall be nested to form a "box" at each saddle tie.

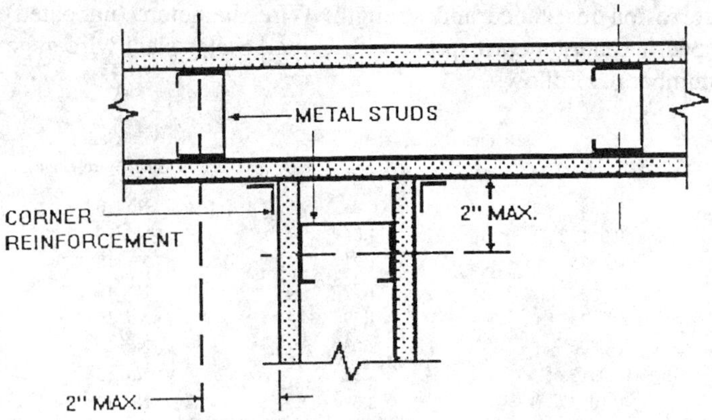

FIG. 1 Abutting Partition Detail

5.2 *Runner (Track) Installation:*

5.2.1 *General*—Runners shall be aligned accurately at the floor and ceiling and securely anchored approximately 2 in. (50 mm) from the runner ends, not more than 24 in. (610 mm) on center. Runners shall be secured with fasteners at partition corners. One runner shall extend to the end of the corner and the other runner shall butt to it and be gapped to allow clearance for the gypsum panel product thickness. Runners shall not be mitered.

5.2.2 *Runners to Concrete Slabs*—Shall be fastened with concrete stub nails, expansion anchors, shielded screws, or power-driven fasteners not exceeding 24 in. (610 mm) on center.

5.2.3 *Runners to Wood*—Shall be fastened with screws providing not less than 1 in. (25 mm) penetration or nails providing 1½ in. (38 mm) penetration into the wood.

5.2.4 *Runners to Suspended Ceilings*—Shall be fastened with "Molly"-type expandable fasteners, toggle bolts, clamps, or screws into channels, splines, "T" runners, or other members.

5.3 *Stud Installation:*

5.3.1 *Stud Height and Spacing Limitations:*

5.3.1.1 Maximum framing spacing determined by gypsum panel product thickness shall be in accordance with Table 1.

5.3.1.2 Stud heights shall be not greater than those shown in Tables 3 and 4.

5.3.1.3 Maximum stud heights for stud depths not shown in Tables 3 and 4 shall be based on design criteria of not less than 5 psf (240 Pa) load and deflection not greater than L/120.

5.3.2 *Location:*

5.3.2.1 Studs shall be positioned vertically, shall engage both the floor and ceiling runners, and shall be spaced not more than the maximum framing spacing allowed for the finish specified. Studs located adjacent to door and window frames, partition intersections, and corners shall be anchored to runner flanges by screws, or by crimping at each stud and runner flange.

5.3.2.2 At the junction of through and abutting partitions, a stud shall be located not more than 2 in. (50 mm) away from the intersection in the abutting partition from the through partition (see Fig. 1), and not more than 2 in. (50 mm) from partition corners and other construction. A stud shall be located adjacent to all door and borrowed light frames. Studs shall be securely anchored to the jamb anchor clips on each door frame or borrowed light frame by bolt or screw attachment. A header shall be formed over metal door and borrowed light frames with a cut-to-length section of runner placed horizontally with the flanges cut and web bent vertically at each end, and securely attached to the adjacent vertical studs. A cut-to-length stud shall be positioned at the location of vertical joints over the header extending to the ceiling runner. Additional cut to length studs required to comply with framing spacing in accordance with Table 1 shall also be added over the header, extending to the ceiling runner.

5.3.2.3 At partition corners, a stud shall be installed so that it forms the outside corner. Following application of a single layer of gypsum panel product to this stud, a second stud shall be installed in the abutting runner and the web shall be screw attached through the gypsum panel product into the flange of the first stud (see Fig. 2). A three-stud conventional corner shall be permitted (see Fig. 3).

5.4 *Chase Wall Partitions:*

5.4.1 A double row of runners and studs as specified in 5.2 and 5.3 shall be installed. Height shall be in accordance with 5.3.1.2.

5.4.2 Where a gypsum panel product is used as bracing between chase walls, a gap of not more than 20 in. (508 mm) between rows of studs shall be permitted.

5.4.3 Horizontal cross braces to opposite studs shall be installed not more than 4 ft (1220 mm) on center vertically. Horizontal cross braces shall be either of the following:

5.4.3.1 Gypsum panel product gussets 12 in. (305 mm) deep attached to the stud webs with three screws.

5.4.3.2 A stud or runner with the web screw attached to the wall stud web with not less than two screws.

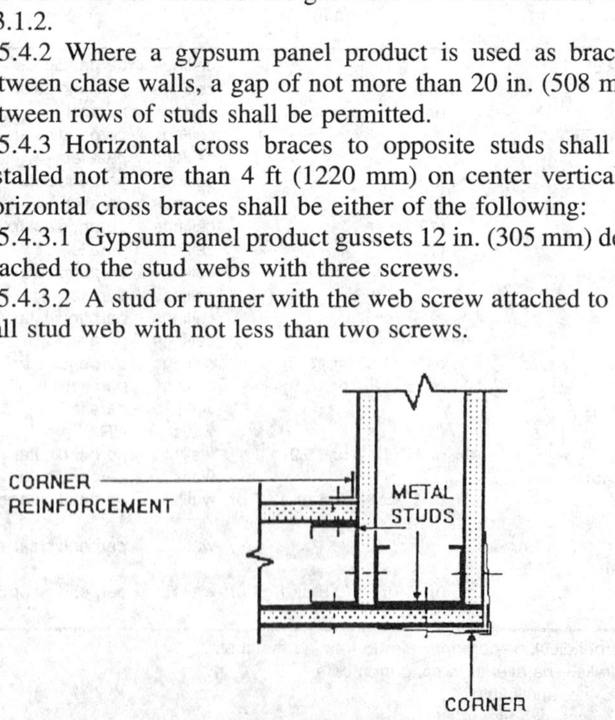

FIG. 2 Partition Corner Detail

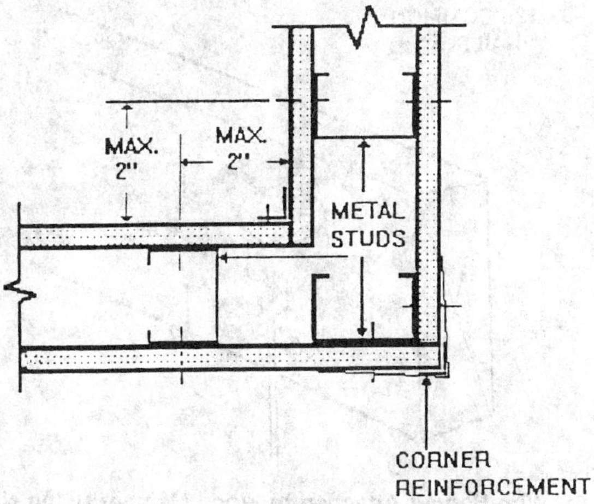

FIG. 3 Partition Corner Detail

5.5 *Rigid Furring Channel Installation, Direct Attachment to Masonry or Concrete*—The furring member shall be attached to masonry or concrete surfaces, either vertically or horizontally. Spacing shall be determined by gypsum panel product thickness in accordance with Table 1. For furring positioned horizontally, the center line of the furring members closest to the floor and ceiling shall be attached not more than 3 in. (76 mm) from the floor and ceiling lines. The furring member shall be secured with fasteners occurring on alternated flanges and spaced 24 in. (610 mm) on center.

5.6 *Wall Furring-Bracket System:*

5.6.1 Adjustable wall furring brackets with serrated edges facing upward shall be attached to masonry or concrete walls in the following spacing pattern: 48 in. (1220 mm) on center vertically, 6 in. (152 mm) maximum from floor and ceiling, 36 in. (910 mm) on center horizontally, 4 in. (100 mm) maximum from columns or other abutting construction, and as required above and below windows. Each bracket shall be fastened through the hole nearest to the serrated edges.

5.6.2 Channels ¾ in. (19 mm) shall be laid horizontally on the furring brackets so that the channel flanges engage the serrated edges of the bracket. Each channel shall be plumbed to align with ceiling and base channels. Channels shall be wire-tied to each bracket with a double strand of 16 gage or a triple strand of 18 gage tie wire. Each excess bracket length shall be bent down and inward toward the wall.

5.6.3 Rigid furring channels shall be positioned vertically with wing flanges against the channels with spacing determined by gypsum panel product thickness in accordance with Table 1. Each furring channel intersection shall be wire-tied with a double strand of 16-gage or a triple strand of 18-gage tie wire.

5.7 *Soffits*—Soffits shall be framed by attaching runners to ceilings and walls as specified in 5.2.1. Runners shall be used for backing of all outside corners. Hangers or spacers (cut-to-length pieces of stud), shall be provided from ceiling runner to outside corner and from outside corner to vertical surface. Where the hanger or spacer length is not more than the maximum framing spacing allowed in Table 1 for the gypsum panel product thickness specified, hangers or spacers shall be located not more than 4 ft (1220 mm) on centers. Where the hanger or spacer length is more than the maximum framing spacing allowed, they become the attachment means and shall be spaced in accordance with Table 1.

TABLE 3 Maximum Stud Height[A], ft-in. (mm), Single Layer ½ in. (12.7 mm) Thick Gypsum Board[B] on Each Side of Minimum 0.0179 in. (0.455 mm) Base Metal Thickness Steel Studs Spaced 24 in. (610 mm) o.c.[C, D]

Stud Depth, in. (mm) Industry Designator[E]	Deflection Limit	Lateral Pressure			
		5 psf (240 Pa)	7.5 psf (360 Pa)	10 psf (480 Pa)	15 psf (720 Pa)
1-⅝ (41.3) 162S125-18	L/120	9-9 (2970)	8-0 (2440)	[G]	[G]
	L/240	8-0 (2440)	[G]	[G]	[G]
	L/360	[G]	[G]	[G]	[G]
2-½ (63.5) 250S125-18	L/120	11-10 (3610)	9-8 (2950)	8-5 (2570)	[G]
	L/240	10-8 (3250)	9-3 (2820)	8-5 (2570)	[G]
	L/360	9-3 (2820)	8-1 (2460)	[G]	[G]
3-½ (88.9)[F] 350S125-18	L/120	13-9 (4190)	11-0 (3350)	9-5 (2870)	7-7 (2310)
	L/240	13-5 (4090)	11-0 (3350)	9-5 (2870)	7-7 (2310)
	L/360	11-8 (3560)	10-1 (3070)	9-1 (2770)	7-7 (2310)
4 (101.6) 400S125-18	L/120	15-1 (4600)	12-1 (3680)	10-5 (3180)	8-5 (2570)
	L/240	14-3 (4340)	12-1 (3680)	10-5 (3180)	8-5 (2570)
	L/360	12-4 (3760)	10-9 (3280)	9-9 (2970)	8-5 (2570)
6 (152.4) 600S125-18	L/120	16-9 (5110)	13-5 (4090)	11-5 (3480)	8-11 (2720)
	L/240	16-9 (5110)	13-5 (4090)	11-5 (3480)	8-11 (2720)
	L/360	16-9 (5110)	13-5 (4090)	11-5 (3480)	8-11 (2720)

[A] Based on tests conducted with gypsum board attached with screws spaced 12 in. (305 mm) o.c. to framing members.
[B] Maximum stud heights are also applicable to walls sheathed with gypsum board greater than ½ in. (12.7 mm) thick and multiple layers of gypsum board.
[C] Maximum stud heights are also applicable to walls with studs spaced less than 24 in. (610 mm) o.c.
[D] Runner flanges need not be fastened to studs except as required by paragraph 5.3.2.1.
[E] The *Industry Designator* defines the cold formed steel framing member.
Example: 350S125-18
350 designates the member web depth in 100ths of an inch, 350 = 3.50 in. (88.9 mm)
S designates the type of member, S = stud
125 designates the member flange width in 100ths of an inch, 125 = 1.25 in. (32 mm)
18 designates the member base metal thickness in mils, 18 = .0179 in. (0.455 mm)
[F] Also applicable to 3-⅝ in. (92.1 mm) stud depth, 362S125-18.
[G] Data not available.

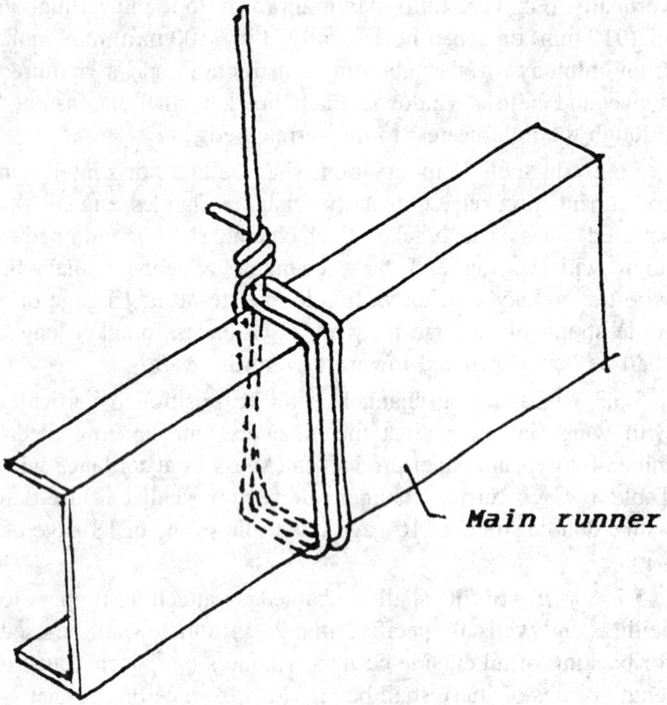

FIG. 4 Saddle Tie

6. Suspended and Furred Ceilings

6.1 *Hangers and Inserts for Suspended Ceilings:*

6.1.1 Sizes specified for wire, rods, and flats shall be in accordance with Table 5.

6.1.2 Hangers for supporting suspended ceilings shall be in accordance with Table 5.

6.1.3 Wire and rod hangers shall be saddle-tied around main runners and furring members so as to prevent turning or twisting of the member and to develop the full strength of the hangers (see Fig. 4).

6.1.4 Wire for attaching main runners and rigid furring channels directly to concrete beams and joists, to steel beams and joists, and to wood beams and joists shall be as specified in Table 3.

6.2 *Wood Construction*—Hangers for suspended ceilings under wood construction shall be as specified in Table 5, and shall be attached to supports by any of the following methods.

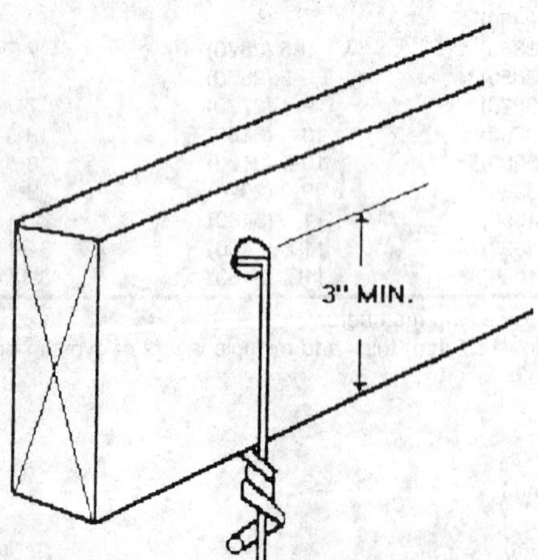

FIG. 5 Wire Hanger Attached to Wood Member Through a Drilled Hole

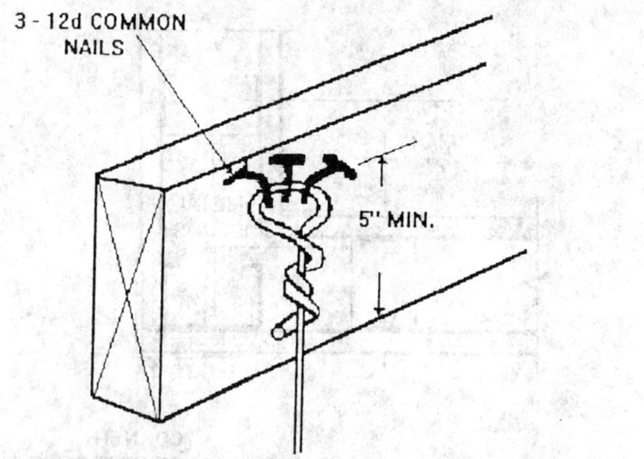

FIG. 6 Wire Hanger Attached to Wood Member Using Nails

6.2.1 A hole shall be drilled through the wood member not less than 3 in. (76 mm) above the bottom with the upper end of the wire hanger passed through the hole and twisted three times around itself (see Fig. 5).

6.2.2 Three 12-penny nails shall be driven, on a downward slant, into the sides of the wood member with not less than 1¼ in. (32 mm) penetration and not less than 5 in. (130 mm) from the bottom edges, and not more than 36 in. (910 mm) on center with the upper end of the wire hanger wrapped around the nails and twisted three times around itself (see Fig. 6).

6.2.3 A loop shall be formed on the upper end of the wire hanger and secured to the wood member by not less than four 1½ in. (38 mm), 9 gage diameter wire staples driven horizontally or on a downward slant into the sides of the wood members, three near the upper end of the loop, and the fourth to fasten the loose end (see Fig. 7).

6.2.4 Where supports for flooring are thicker than 1½ in. (38 mm) and are spaced more than 4 ft (1220 mm) on center, 1½ in. (38 mm) No. 1/0 (0.3065 in.) (7.78 mm) eye screws, or equivalent, spaced not greater than 3 ft (910 mm) on centers shall be screwed into the flooring with the upper end of the hanger inserted through the eye screws and twisted three times around itself.

6.2.5 *Flat Hangers*—Two holes shall be drilled in the upper end of the flat hangers and nailed to the sides of the wood

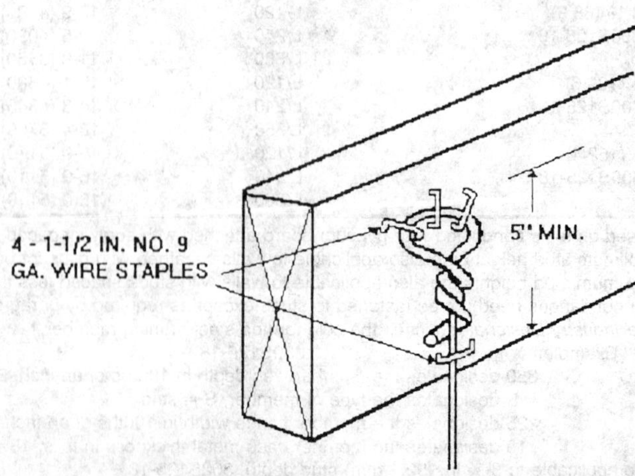

FIG. 7 Wire Hanger Attached to Wood Member Using Staples

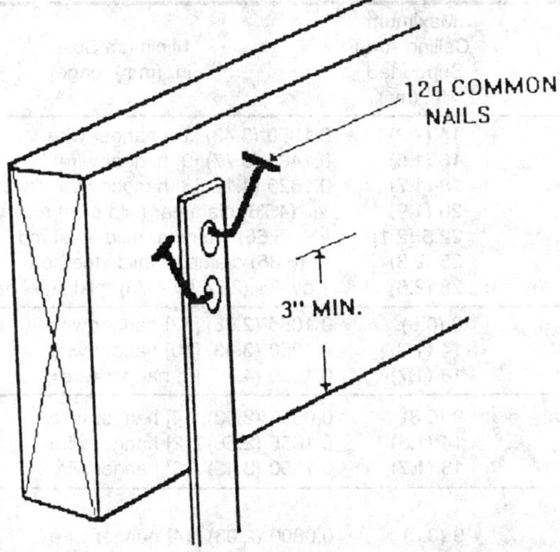

FIG. 8 Flat Hanger Attached to Wood Member Using Nails

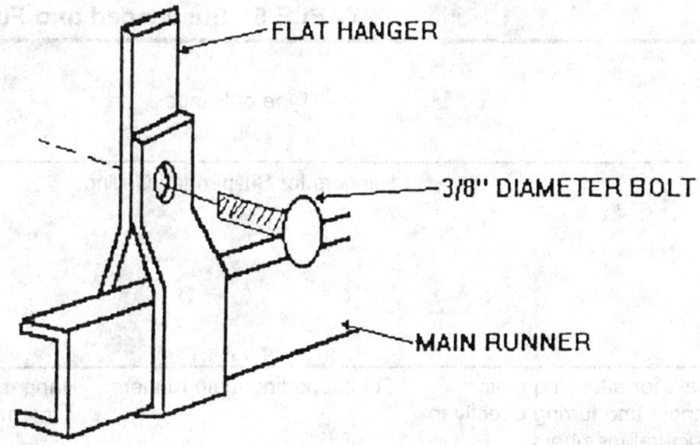

FIG. 9 Flat Hanger Attached to Main Runner Using Bolt

members with 12d penny nails driven through holes and clinched. Nails shall be not less than 3 in. (76 mm) above the bottom edge of the framing member (see Fig. 8).

6.3 *Rigid Furring Channels*—Rigid furring channels applied directly to wood framing shall be applied with screws providing 1-in. (25.4 mm) penetration or as otherwise specified by the manufacturer of the furring channels.

6.4 *Flat Hangers and Inserts:*

6.4.1 *Inserts*—Where 1 by 3⁄16 in. (25 by 4.7 mm) inserts are used, 7⁄16 in. (11 mm) diameter holes shall be punched on the center line at the lower ends of the insert to permit flat hangers to be bolted tightly to the inserts with 3⁄8 in. (10 mm) diameter bolts.

6.4.2 *Hangers*—Lower ends of the flat hangers shall be bolted to the main runners or shall be bent tightly around the members and bolted tightly to the main part of the hanger with 3⁄8 in. (9.5 mm) diameter bolts in a 7⁄16 in. (11 mm) diameter hole (see Fig. 9), or machine screws.

6.4.3 *General*—Holes in both inserts and hangers shall not be nearer to the ends than 3⁄8 in. (10 mm).

6.5 *Main Runners for Suspended Ceilings*—Channels for the various spacings of hangers shall be as specified in Table 6.

6.5.1 Ends and sides of main runners and cross furring shall not come in contact with abutting masonry or reinforced concrete walls or partitions. Clearance of not less than 1 in. (25 mm) from ends and 1⁄8 in. (3 mm) from sides shall be provided. A channel shall be located within 6 in. (150 mm) of the paralleling walls to support the ends of the cross furring. The ends at walls shall be supported by hangers located not more than 6 in. (150 mm) from such ends.

6.5.2 When main runners and cross furring are spliced, the

TABLE 4 Maximum Stud Height^A^, ft-in. (mm), Single Layer ½ in. (12.7 mm) Thick Gypsum Board^B^ on Each Side of Minimum 0.0329 in. (0.836 mm) Base Metal Thickness Steel Studs Spaced 24 in. (610 mm) o.c.^C,D^

Stud Depth, in. (mm) Industry Designator^E^	Deflection Limit	Lateral Pressure			
		5 psf (240 Pa)	7.5 psf (360 Pa)	10 psf (480 Pa)	15 psf (720 Pa)
1-⅝ (41.3) 162S125-33	L/120	11-2 (3400)	9-9 (2970)	8-11 (2720)	7-9 (2360)
	L/240	8-11 (2720)	7-9 (2360)	G	G
	L/360	7-9 (2360)	G	G	G
2-½ (63.5) 250S125-33	L/120	15-1 (4600)	13-2 (4010)	11-10 (3610)	10-2 (3100)
	L/240	11-10 (3610)	10-2 (3100)	9-2 (2790)	7-10 (2390)
	L/360	10-2 (3100)	8-9 (2670)	7-10 (2390)	G
3-½ (88.9)^F^ 350S125-33	L/120	19-1 (5820)	16-9 (5110)	15-2 (4620)	13-2 (4010)
	L/240	15-2 (4620)	13-2 (4010)	12-0 (3660)	10-5 (3180)
	L/360	13-2 (4010)	11-6 (3510)	10-5 (3180)	9-1 (2770)
4 (101.6) 400S125-33	L/120	20-11 (6380)	18-4 (5590)	16-8 (5080)	14-5 (4390)
	L/240	16-8 (5080)	14-5 (4390)	13-0 (3960)	11-3 (3430)
	L/360	14-5 (4390)	12-5 (3790)	11-3 (3430)	9-8 (2950)
6 (152.4) 600S125-33	L/120	27-6 (8380)	24-0 (7320)	19-1 (5820)	12-7 (3840)
	L/240	21-9 (6630)	19-0 (5790)	17-4 (5280)	12-7 (3840)
	L/360	19-0 (5790)	16-8 (5080)	15-1 (4600)	12-7 (3840)

^A^ Based on tests conducted with gypsum board attached with screws spaced 12 in. (305 mm) o.c. to framing members.
^B^ Maximum stud heights are also applicable to walls sheathed with gypsum board greater than ½ in. (12.7 mm) thick and multiple layers of gypsum board.
^C^ Maximum stud heights are also applicable to walls with studs spaced less than 24 in. (610 mm) o.c.
^D^ Runner flanges need not be fastened to studs except as required by paragraph 5.3.2.1.
^E^ The *Industry Designator* defines the cold formed steel framing member.
Example: 350S125-18
350 designates the member web depth in 100ths of an inch, 350 = 3.50 in. (88.9 mm)
S designates the type of member, S = stud
125 designates the member flange width in 100ths of an inch, 125 = 1.25 in. (32 mm)
18 designates the member base metal thickness in mils, 18 = .0179 in. (0.455 mm)
^F^ Also applicable to 3-⅝ in. (92.1 mm) stud depth, 362S125-33.
^G^ Data not available.

TABLE 5 Suspended and Furred Ceilings, Minimum Sizes for Hangers

Type of Hanger			Maximum Ceiling Area Supported, ft² (m²)	Minimum Size, in. (mm) [gage]
	Hangers for Suspended Ceilings		14 (1.3)	0.1350 (3.43) [10] hanger wire
			16 (1.5)	0.1483 (3.77) [9] hanger wire
			18 (1.7)	0.1620 (4.12) [8] hanger wire
			20 (1.9)	3⁄16 (4.76) diameter mild steel rod
			22.5 (2.1)	7⁄32 (5.56) diameter mild steel rod
			25 (2.3)	1⁄4 (6.35) diameter mild steel rod
			28 (2.6)	1 by 3⁄16 (25.4 by 4.76) mild steel flat
Hangers for attaching main runners and furring directly to structural members	For supporting main runners	Hangers between structural members[A]	9 (0.8)	0.1055 (2.68) [12] hanger wire
			14 (1.3)	0.1350 (3.43) [10] hanger wire
			18 (1.7)	0.1620 (4.12) [8] hanger wire
		Double-wire loops at structural member[A]	9 (0.8)	0.0800 (2.03) [14] hanger wire
			14 (1.3)	0.1055 (2.68) [12] hanger wire
			18 (1.7)	0.1350 (3.43) [10] hanger wire
	For supporting furring without a runner (wire loops at supports)	Types of support:		
		Concrete	9 (0.8)	0.0800 (2.03) [14] hanger wire
		Steel or Wood	9 (0.8)	0.0625 (1.59) [16] hanger wire (two loops)[B]

[A] Inserts, special clips or screws, or other devices of equal strength shall be permitted.

[B] Two loops of 0.0475 in. (1.21 mm), No. 18-gage galvanized wire shall be permitted to be substituted for each loop of 0.0625 in. (1.59 mm), No. 16-gage wire for attaching steel furring to steel or wood joists.

TABLE 6 Allowable Spans, ft-in. (mm) Cold-Rolled Channel Main Runners[A,B,C,D,E,F]

Member Size, In (mm)	Member Weight lb/1000 ft (kg/m)	Span Condition[G,H]	Total Uniform Load = 4 psf (0.19 kPa) Member Spacing, in. (mm) 24 (610)	36 (914)	48 (1220)	60 (1520)	72 (1830)
3⁄4	277	Simple	3–3 (990)	2–10 (860)	2–7 (790)	2–5 (740)	2–3 (690)
(19)	(0.412)	2 or More	4–1 (1240)	3–6 (1070)	3–3 (990)	3–0 (910)	2–10 (860)
1½	414	Simple	5–1 (1550)	4–5 (1350)	4–0 (1220)	3–9 (1140)	3–6 (1070)
(38)	(0.616)	2 or More	7–1 (2160)	6–2 (1880)	5–8 (173)	5–2 (1570)	4–10 (1470)
2	506	Simple	5–4 (1630)	4–8 (1420)	4–3 (1300)	4–0 (1220)	3–9 (1140)
(51)	(0.753)	2 or More	7–5 (2260)	6–6 (1980)	5–11 (1800)	5–6 (1680)	5–2 (1570)
2½	597	Simple	5–7 (1700)	4–11 (1500)	4–5 (1350)	4–2 (1270)	3–11 (1190)
(64)	(0.888)	2 or More	7–9 (2360)	6–9 (2060)	6–2 (1880)	5–9 (1750)	5–5 (1650)

Member Size, In (mm)	Member Weight lb/1000 ft (kg/m)	Span Condition[G,H]	Total Uniform Load = 4 psf (0.19 kPa) Member Spacing, in. (mm) 24 (610)	36 (914)	48 (1220)	60 (1520)	72 (1830)
3⁄4	277	Simple	2–10 (860)	2–6 (780)	2–3 (690)	2–1 (630)	2–0 (610)
(19)	(0.412)	2 or More	3–6 (1070)	3–1 (490)	2–10 (860)	2–7 (790)	2–5 (740)
1½	414	Simple	4–5 (1350)	3–11 (1190)	3–6 (1070)	3–3 (990)	3–1 (940)
(38)	(0.616)	2 or More	6–2 (1880)	5–5 (1650)	4–10 (1470)	4–6 (1370)	4–2 (1270)
2	506	Simple	4–8 (1420)	4–1 (1240)	3–9 (1140)	3–6 (1070)	3–3 (990)
(51)	(0.753)	2 or More	6–6 (1980)	5–8 (1730)	5–2 (1570)	4–10 (1470)	4–7 (1400)
2½	597	Simple	4–11 (1500)	4–3 (1300)	3–11 (1190)	3–8 (1120)	3–5 (1040)
(64)	(0.888)	2 or More	6–9 (2060)	5–11 (1800)	5–5 (1650)	5–0 (1520)	4–9 (1450)

[A] Bare metal thickness of cold-rolled main runners shall be not less than 0.0538 in. (1.367 mm).

[B] Inside corner radii shall not be greater than 1⁄8 in. (3.19 mm).

[C] Spans based on upper flange of main runners laterally unbraced.

[D] Maximum deflection limited to 1⁄360th of the span length.

[E] Steel yield stress, *Fy*, shall be not less than 33 000 psi (228 MPa).

[F] Tabulated spans apply only to main runners with webs oriented vertically.

[G] "2 or More" spans refers to two or more continuous, equal spans.

[H] For the "2 or More" span condition, listed spans represent the center-to-center distance between adjacent supports.

ends shall be overlapped not less than 12 in. (305 mm) with flanges interlocked and securely fastened near each end of the splice with double loops of No. 16 gage tie wire, minimum, or with screws.

6.6 *Furring Members for Suspended or Furred Ceilings:*

6.6.1 Furring members shall be spaced in accordance with Tables 2 and 6.

6.6.2 Furring members shall be securely fastened to main runners and structural supports with special clips, screws, or equivalent attachments. Where required, rigid furring channels shall be saddle-tied to main runners with No. 16 gage tie wire, a double strand of 18 gage tie wire, or with screws (see Fig. 10).

6.6.3 When furring members are spliced, the ends shall be overlapped not less than 8 in. (203 mm) and securely fastened near each end of the splice with double loops of No. 16 gage tie wire or with screws.

6.6.4 Flanges of rigid furring channels shall be interlocked.

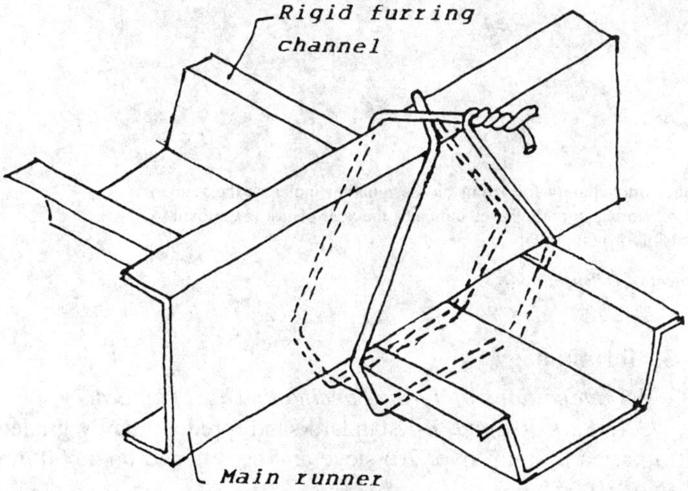

FIG. 10 Saddle Tie

Ends and sides of furring members shall not come in contact with abutting masonry or reinforced concrete walls or partitions.

6.7 *Grid Suspension System*—Main beams shall be suspended in parallel rows spliced together at their ends.

6.7.1 Hangers for supporting the main beams shall comply with the minimum size and ceiling area specified in Table 5.

6.7.2 Cross furring members of grid suspension systems shall interlock to the main beams in rows running perpendicular and spaced not to exceed maximums specified in Table 1. Cross furring members along the ceiling perimeter shall be supported by angle or channels attached to the wall.

7. Product Delivery, Identification, and Marking

7.1 All materials shall be delivered in original packages, containers, or bundles bearing the brand name and the name of the manufacturer or the supplier for whom the product is manufactured.

8. Product Storage

8.1 All materials shall be kept dry, preferably by being stored inside a building under a roof. Where necessary to store material outside, it shall be stacked off the ground, properly supported on a level platform, and fully protected from the weather.

9. Keywords

9.1 channel; framing; framing member; furring; grid suspension system; gypsum panel product; hangers; hanger wire; main runners; rigid furring channel; runner; stud; suspended ceiling; tie wire

SUMMARY OF CHANGES

This section identifies the location of changes to this specification that have been incorporated since the last issue. Committee C-11 has highlighted those changes that affect the technical interpretation or use of this specification.

(*1*) Change was made to the wording in 1.2.

Designation: C 778 – 99

Standard Specification for Standard Sand[1]

This standard is issued under the fixed designation C 778; the number immediately following the designation indicates the year of original adoption or, in the case of revision, the year of last revision. A number in parentheses indicates the year of last reapproval. A superscript epsilon (ε) indicates an editorial change since the last revision or reapproval.

This standard has been approved for use by agencies of the Department of Defense.

1. Scope

1.1 This specification covers standard sand for use in the testing of hydraulic cements.

1.2 The values stated in SI units are to be regarded as the standard.

1.3 Values in SI units shall be obtained by measurement in SI units or by appropriate conversion, using the Rules for Conversion and Rounding given in IEEE/ASTM SI 10, of measurements made in other units.

1.4 *This standard does not purport to address all of the safety concerns, if any, associated with its use. It is the responsibility of the user of this standard to establish appropriate safety and health practices and determine the applicability of regulatory limitations prior to use.*

2. Referenced Documents

2.1 *ASTM Standards:*

C 109/C 109M Test Method for Compressive Strength of Hydraulic Cement Mortars (Using 2-in. or [50-mm] Cube Specimens)[2]

C 127 Test Method for Specific Gravity and Absorption of Coarse Aggregate[3]

C 136 Test Method for Sieve Analysis of Fine and Coarse Aggregates[3]

C 150 Specification for Portland Cement[2]

C 185 Test Method for Air Content of Hydraulic Cement Mortar[2]

C 595 Specification for Blended Hydraulic Cements[2]

C 1005 Specification for Reference Masses and Devices for Determining Mass for Use in the Physical Testing of Hydraulic Cements[2]

E 11 Specification for Wire-Cloth Sieves for Testing Purposes[4]

IEEE/ASTM SI 10 Standard for Use of the International System of Units (SI): The Modern Metric System[4]

3. Terminology

3.1 *Definitions of Terms Specific to This Standard:*

3.1.1 *20–30 sand, n*—standard sand, predominantly graded to pass a 850-μm (No. 20) sieve and be retained on a 600-μm (No. 30) sieve.

3.1.2 *graded sand, n*—standard sand, predominantly graded between the 600-μm (No. 30) sieve and the 150-μm (No. 100) sieve.

3.1.3 *standard sand, n*—silica sand, composed almost entirely of naturally rounded grains of nearly pure quartz, used for preparing mortars in the testing of hydraulic cements.

4. Requirements

4.1 Sand shall meet the requirements of Table 1 with respect to grading, source of sand, and absence of undesirable air entraining characteristics.

METHODS OF SAMPLING AND TESTING STANDARD SANDS

5. Apparatus

5.1 *Sieves*—The sieves shall be standard 203-mm (8–in.) diameter, full-height, wire-cloth sieves, conforming to the requirements of Specification E 11, and of the following sizes:

1.18-mm (No. 16)	425-μm (No. 40)
850-μm (No. 20)	300-μm (No. 50)
600-μm (No. 30)	150-μm (No. 100)

5.2 *Sample Splitter*—The sample splitter shall be of the riffle type and shall have an even number of equal-width chutes that alternately discharge in opposite directions. The device shall have no fewer than eight chutes having a maximum opening no greater than 13 mm, and a minimum opening at least three times the diameter of the largest particle of sand in a sample to be split. It shall be equipped with a hopper or straightedged pan by which a sample may be fed to the chutes at a controlled rate, and two receptacles to hold the two halves of the sample following splitting. The length of the hopper or pan shall be approximately equal to the overall width of the assembly of chutes.

5.3 *Reference Masses and Devices for Determining Mass*—Reference masses and devices for determining mass shall conform to the requirements of Specification C 1005 as appropriate for the size of the sample the mass of which is to be determined.

---

[1] This specification is under the jurisdiction of ASTM Committee C-1 on Cement and is the direct responsibility of Subcommittee C01.95 on Coordination of Standards.

Current edition approved Jan. 10, 1999. Published May 1999. Originally published as C 778 – 74 T. Last previous edition C 778 – 97.

[2] *Annual Book of ASTM Standards*, Vol 04.01.

[3] *Annual Book of ASTM Standards*, Vol 04.02.

[4] *Annual Book of ASTM Standards*, Vol 14.02; and in the Related Materials Section of the *Annual Book of ASTM Standards*, Vol 04.01.

TABLE 1 Standard Sand Requirements

Characteristics	20-30 Sand	Graded Sand
Grading, percent passing sieve:		
1.18 mm (No. 16)	100	100
850 µm (No. 20)	85 to 100	
600 µm (No. 30)	0 to 5	96 to 100
425 µm (No. 40)		65 to 75
300 µm (No. 50)		20 to 30
150 µm (No. 100)		0 to 4
Difference in air content of mortars made with washed and unwashed sand, max, % air[A]	2.0	1.5[B]
Source of sand	Ottawa, IL or LeSuer, MN	Ottawa, IL

[A] This determination is needed when contamination of sand is suspected as discussed in 8.1.

[B] Compressive strength of Test Method C 109/C 109M mortar made with Specification C 150 or C 595 cement will be reduced approximately 4 % for each percent of air in the compacted cube. However, as many as three batches of washed and three of unwashed sand may be needed to reliably detect a difference in strength of 7 % between washed and unwashed sand mortars.

6. Sampling

6.1 *Procedure*—Pour an approximate quantity of sand from the bag or other container into the hopper or pan of the sample splitter, leveling the surface from end to end and from edge to edge so that approximately equal amounts will flow through each chute when the sand is introduced to the chutes. The rate at which the sand is introduced shall be such as to allow free flow through the chutes into the receptacles below. When all of the original material has been divided, the portion in one of the receptacles shall be reintroduced to the hopper. Repeat as many times as necessary to obtain a sample of the desired mass. As appropriate, the portion of the sample collected in the second receptacle each time may either be reserved for testing, for reduction in size for other tests, or discarded.

7. Sieve Analysis

7.1 *General*—Obtain approximately 700 g of sand from a full bag and reduce to approximately 100 g using the procedure described in the section on sampling.

7.2 *Sieving*—Sieve the sand by hand or mechanically as described in Test Method C 136.

7.3 *Reporting*—Report the results of the sieve analysis as total percentage passing each sieve.

8. Tests for Air-Entraining Potential of Sand

8.1 Both 20–30 and graded sands may contain small amounts of surfactants or other contaminants that will produce air voids in mortars made with the sand. When contamination is suspected, tests can be made using "washed" and "as received" sand to determine the effect on air content or strength (see Note 1).

NOTE 1—Often it may be desirable to wash these sands routinely to avoid potential problems. Additionally, the washing procedure outlined in 8.1.1 will tend to minimize segregation of sand that may occur when the sand is handled in a dry state.

8.1.1 Obtain a sample of at least 2800 g from a full bag of sand. Place the remainder of the sand from the bag in a concrete mixer of the drum type. Add sufficient water to cover the sand and mix for 2 min. Decant the water over a 75 µm sieve and drain for 2 min. Repeat the washing and decanting procedure four times. Tilt the mixer and drain the sand for 20 min or longer. Determine the mass of damp sand and place it in approximately batch-sized quantities in separate containers. Dry the sand to constant mass in an oven at 110°C. Cool and adjust the quantity of dry sand to the amount required for a test. Prepare two batches of each of "washed" and as "as received" sand for test following the procedures in Test Method C 185 for 20–30 sand or Test Method C 109/C 109M for graded sand.

8.1.2 Tests for air-entraining potential of a standard sand shall be made on a single day by a single operator.

8.2 *20–30 Sand*—Prepare two batches of mortar with washed sand and two with "as received" sand in accordance with Test Method C 185. Compare the average air content for washed and unwashed sand mortars for compliance with limit in Table 1.

8.3 *Graded Sand*—Prepare two batches of mortar with washed sand and two with "as received" sand in accordance with Test Method C 109/C 109M. When Test Method C 109/C 109M mortar cubes are removed from the molds, determine the density of a group of at least three cubes from each batch by weighing the cubes both in air and immersed in water. Follow procedures outlined in Test Method C 127 for bulk specific gravity (SSD). Weigh cubes to at least the nearest gram. Calculate the apparent air content as follows:

$$\text{apparent air content} = ((D_t - D_m)/D_t) \times 100 \quad (1)$$

where:

D_m = the measured density (bulk specific gravity (SSD)), and

D_t = the theoretical density of air-free mortar, g/cm^3.

Note that D_t is calculated as the total mass of the ingredients in a batch of mortar divided by the sum of the absolute volumes of the cement, sand, and water used in the batch. Average the apparent air content of the two batches made with washed sand and compare that average with the similar average for "as received" sand mortar to determine compliance with the limit in Table 1.

9. Rejection

9.1 *Standard Sand*—A bag of standard sand shall be rejected if the contents fail to meet one or more of the requirements of this specification. A shipment of standard sand shall be sampled by selecting every 15th bag by consecutive number beginning with the 5th bag of the shipment. Bags missing from the sequence shall be counted as if they were there. A shipment of standard sand may be accepted if the contents of the selected bags meet the requirements of this specification. If any of the selected bags fail to meet one or more of the requirements of this specification, the shipment shall be resampled by selecting every 15th bag by consecutive number beginning with the 10th bag of the shipment. A shipment of standard sand may be accepted if the contents of the selected bags on the resampling meet the requirements of this specification. If any of the selected bags in the resampling fail to meet one or more of the requirements of this specification, and the rejected bag is within a 15-bag sequence of a bag rejected in the first selection of bags, the entire sequence of bags between accepted tests shall be rejected. The entire shipment shall be rejected if half of the number of bags tested fail to meet the requirements of this specification, or if the shipment consists of five bags or less if one randomly selected bag fails the specification.

9.2 Table 2 shows for both sampling and resampling in Column 2 the number of bags to be tested based on the total number of consecutive bags in the shipment. Column 3 shows the identification of the bag to be selected for test based on the consecutive number of the bags.

TABLE 2 Sampling and Resampling of Standard Sand

Total No. of Consecutive Bags in Shipment	No. of Bags to Test	Test in Consecutive No., the Following Bags
Sampling:		
less than 5	1	random
5 to 19	1	5th
20 to 34	2	5th and 20th
35 to 49	3	5th, 20th, and 35th
50 to 64	4	5th, 20th, 35th, and 50th
Resampling:		
less than 5	none	reject shipment
5 to 19	1	10th
20 to 34	2	10th and 25th
35 to 49	3	10th, 25th, and 40th
50 to 64	4	10th, 25th, 40th, and 55th

NOTE 2—If, in a shipment of 30 bags, on first sampling, the 20th bag fails the specification, and on resampling, the 25th bag fails the specification, the entire shipment is rejected. If the shipment were 100 bags and on first sampling only the 20th bag fails, and on resampling only the 25th bag fails, only bags Nos. 11 to 34 would be rejected.

10. Packaging and Package Marking

10.1 *Packaging*—Standard sands shall be delivered in bags with impervious liners made of material that will not contaminate the sand.

10.2 *Marking*—Each bag shall be plainly marked as follows: ASTM C 778 20-30 Sand or ASTM C 778 Graded Sand as the case may be. The source of the sand and the weight of the sand contained therein shall be plainly indicated on each bag.

11. Keywords

11.1 20-30 sand; graded sand; hydraulic cement testing sand; standard sand

Designation: C 786 – 96

Standard Test Method for Fineness of Hydraulic Cement and Raw Materials by the 300-μm (No. 50), 150-μm (No. 100), and 75-μm (No. 200) Sieves by Wet Methods[1]

This standard is issued under the fixed designation C 786; the number immediately following the designation indicates the year of original adoption or, in the case of revision, the year of last revision. A number in parentheses indicates the year of last reapproval. A superscript epsilon (ε) indicates an editorial change since the last revision or reapproval.

1. Scope

1.1 This test method covers wet sieving techniques for determination of fineness of hydraulic cement and raw materials by means of the 300-μm (No. 50), the 150-μm (No. 100), and the 75-μm (No. 200) sieves.

1.2 The values stated in SI units are to be regarded as the standard.

1.3 *This standard does not purport to address all of the safety concerns, if any, associated with its use. It is the responsibility of the user of this standard to establish appropriate safety and health practices and determine the applicability of regulatory limitations prior to use.*

2. Referenced Documents

2.1 *ASTM Standards:*

C 114 Test Methods for Chemical Analysis of Hydraulic Cement[2]

C 184 Test Method for Fineness of Hydraulic Cement by the 150-μm (No. 100) and 75-μm (No. 200) Sieves[2]

C 430 Test Method for Fineness of Hydraulic Cement by the 45-μm (No. 325) Sieve[2]

E 11 Specification for Wire-Cloth Sieves for Testing Purposes[3]

3. Apparatus

3.1 *Wet Test Sieves*—Standard 300-μm (No. 50), 150-μm (No. 100), or 75-μm (No. 200) sieve cloth conforming to the requirements of Specification E 11, for standard sieves shall be woven from AISI Type 304 wire. The cloth shall be mounted in the frame without distortion, looseness, or wrinkling. Sieve frames are designated as 3 or 4-in. (76.2 or 101.6-mm) diameter type, as follows:

	Sieves	
	76 mm (3-in.) mm (in.)	102 mm (4-in.) mm (in.)
Diameter of frame	76 ± 6 (3.0 ± 0.25)	102 ± 6 (4.0 ± 0.25)
Depth of sieve from top of frame	83 ± 6 (3.25 ± 0.25)	108 ± 6 (4.25 ± 0.25)
Overall height	102 ± 6 (4.0 ± 0.25)	127 ± 6 (5.0 ± 0.25)

3.1.1 For a sieve fabricated by soldering the cloth to the frame, the joint shall be made smooth to prevent material from lodging in the joints between the sieve cloth and the frame. Two-piece sieves shall clamp tightly on the cloth to prevent particles from lodging in the joints between the sieve cloth and the frame, and shall have legs of sufficient length, 19-mm (0.75-in.) minimum, to allow air circulation beneath the sieve cloth.

3.2 *Spray Nozzle,* conforming to the requirements of Test Method C 430. Nozzles having an alternative design are acceptable if the sieve test results agree with those performed using a nozzle conforming to Test Method C 430.

3.3 *Pressure Gage,* conforming to the requirements of Test Method C 430.

3.4 *Balance,* analytical, accurate to within 0.005 g.

3.5 *Weights*—The weights used in fineness determinations shall conform to the requirements of Test Methods C 114.

3.6 *Brush*—A nylon or pure bristle brush will be required for use in cleaning the sieves. A 13-mm (0.5-in.) diameter round-style brush with a 229-mm (9-in.) handle is a convenient size.

NOTE 1: **Caution**—Do not use brass or steel-bristle brushes for cleaning sieves due to the possibility that the stiff bristle will part the wire weave, thereby altering the size of the openings and rendering the sieve useless. A 13-mm (½-in.) hog bristle stencil brush is also satisfactory for brushing sieves.

3.7 *Dry Test Sieves*—The standard samples for calibrating the wet test sieves must be standardized on 203-mm (8-in.) diameter sieves meeting the requirements of Test Method C 184. The 300-μm (No. 50) sieve shall meet the same requirements.

3.8 *NBS SRM No. 1004*—Glass Bead Standard.

4. Dry Sieve Standardization

4.1 *Correction Factors:*

4.1.1 Correction of the residues obtained on the 203-mm (8-in.) diameter 300-μm (No. 50) and 150-μm (No. 100) dry testing sieves is not required.

4.1.2 Where applicable, a correction factor for a 75-μm (No. 200) sieve shall be determined using the instructions given in Annex A1. A correction factor should be determined when accuracy is desired in order to compare results between laboratories.

4.2 *Standard Samples*—Each laboratory must prepare its own standard samples for wet-sieve tests for each of the sieve sizes used. Select standard samples at a level of fineness in the same range as that used in routine work. After the selected material is reserved, uniformly mix the gross sample by placing it on a sheet of rubber, oil cloth, or heavy

[1] This test method is under the jurisdiction of ASTM Committee C-1 on Cement and is the direct responsibility of Subcommittee C01.25 on Fineness.

Current edition approved May 10, 1996. Published July 1996. Originally published as C 786 – 74. Last previous edition C 786 – 94.

[2] *Annual Book of ASTM Standards,* Vol 04.01.

[3] *Annual Book of ASTM Standards,* Vol 14.02.

wrapping paper, depending on the sample size, and raising first one corner of the sheet and then the other so as to roll the sample over and over at least 100 times. Temporarily seal the prepared standard material in airtight containers during the standardization procedure prior to sealing small portions as standard samples in vials. Using the 203-mm (8-in.) diameter sieves from 3.7, perform the dry sieving tests, following the procedure of Test Method C 184. Repeat the test three times and use the average of the amounts passing, expressed as percent, as the standard value of the sample. Use this standard sample to calibrate the wet sieves. Place the entire sample in airtight vials as soon as possible to prevent changes due to humidity. Vials shall be prepared in denominations such as to contain approximately 50 g for standardizing the 300-µm (No. 50); 25 g for the 150-µm (No. 100); or 10 g for the 75-µm (No. 200) sieve.

5. Wet Sieve Calibration

5.1 Weigh the contents of the applicable size standard sample vial for the desired sieve determination on a balance of appropriate sensitivity to the nearest 0.01 g. Record the weight and transfer the sample quantitatively to a clean dry wet test 300-µm (No. 50), 150-µm (No. 100), or 75-µm (No. 200) sieve and proceed as directed in Section 6. The sieve correction factor is the difference between the test residue obtained and the residue value indicated by the standardization tests of Section 4, expressed as a percentage of the test residue. This factor is expressed as follows:

$$C = \frac{(Rs \times Wt/100) - Rt}{Rt} \times 100 \quad (1)$$

where:

C = sieve correction factor (which may be either plus or minus), %,
R_t = test residue from sample retained on sieve, g,
R_s = standard residue retained on sieve, %, and
W_t = weight of test sample, g.

5.1.1 Example of Determination of Wet Sieve Calibration:

% Residue on 150-µm (No. 100) sieve for standard sample, R_s	= 5.40%
Residue from test sample, R_t	= 1.25 g
Weight of sample, W_t	= 25.5 g

$$\text{Correction factor, } C, \% = \frac{(5.40 \times 25.5/100) - 1.25}{1.25} \times 100 = \pm 10.2$$

NOTE 2—The sieve correction is specified as a factor to be multiplied by the residue obtained, and therefore the amount to be added to or subtracted from the test residue in any given instance is proportional to the amount of the residue.

6. Procedure for Wet Sieving

6.1 Weigh the sample to the nearest 0.01 g using approximately 50 g for a 300-µm (No. 50), 25 g for a 150-µm (No. 100), or 10 g for a 75-µm (No. 200) determination. Record the weight and transfer the sample quantitatively to a clean dry sieve. Wet the sample thoroughly with a gentle stream of water. Remove the sieve from under the nozzle and adjust the pressure on the spray nozzle to 69 ± 4 kPa (10 ± 0.5 psi). Return the sieve to its position under the nozzle and wash for 1½ min, moving the sieve in the spray with a circular motion in a horizontal plane at the rate of one motion per second. Every portion of the screen should be sprayed during each circular motion of the sieve. Hold the sieve so that the bottom of the spray nozzle extends 13 mm (0.5 in.) below the top of the sieve frame. Immediately after removing the sieve from the spray, rinse once with about 50 cm^3 of distilled or deionized water using caution not to lose any of the residue. Gently blot the lower surface of the screen cloth with a damp, clean cloth. Dry the sieve and residue in an oven or over a hot plate (see Note 3), supporting the sieve in such a manner that air may pass freely beneath it. Cool the sieve; then brush the residue from the sieve, and weigh on a balance to the nearest 0.01 g (see Note 4).

NOTE 3—Care should be taken when heating the sieve, so that any solder that may have been used in assembling the sieve does not soften.

NOTE 4—Prior to each use, dip the sieve in dilute acetic acid (1+6) or dilute HCl (1+10) and immediately rinse it with distilled or deionized water to remove particles lodged in the meshes. Recalibrate the sieve after 25 determinations.

7. Calculation

7.1 Calculate the fineness of the material as follows:

$$R_c = (R_t/W) \times (100 + C) \quad (2)$$

$$F = 100 - R_c \quad (3)$$

where:

F = fineness expressed as the corrected percentage of sample passing the sieve,
R_c = corrected residue, %,
W = weight of sample used for test, g,
R_t = residue from the sample retained on the sieve, g, and
C = sieve correction factor (determined in accordance with Section 5, which may be either plus or minus), %.

7.1.1 Example for 150-µm (No. 100) Sieve Determination:

Sieve correction factor, C	= ±10.2 %
Exact weight of sample used for test, W	= 25.10 g
Residue from sample retained on sieve, R_t	= 1.42 g
Corrected residue,	
$R_c = (1.42/25.10) \times (100 + 10.2)$	= 6.2 %
$F = 100 - 6.2$	= 93.8 %

8. Precision and Bias

8.1 No precision data are available due to the limited use of this method. Therefore, users are advised to develop their own laboratory precision.

8.2 Since there is no accepted reference material suitable for determining any bias that may be associated with this test method, no statement on bias is being made.

9. Keywords

9.1 fineness; hydraulic cement and raw materials; sieve

ANNEX

(Mandatory Information)

A1. DRY SIEVE CORRECTION FACTORS

A1.1 300-µm (No. 50) and 150-µm (No. 100) 203-mm (8-in.) Diameter Sieves

A1.1.1 The particle size distribution curve of cement at the 300-µm (No. 50) and 150-µm (No. 100) is level in this region, and therefore it is felt that no sieve correction is necessary.

A1.2 75-µm (No. 200) 203-mm (8-in.) Diameter Sieve

A1.2.1 A 75-µm (No. 200) sieve may or may not need to be standardized, depending on the accuracy required. For internal laboratory use where changes in fineness are of more importance than the absolute value, it is not necessary to standardize the dry sieve. The sieve cloth, as a minimum, should conform to Specification E 11. A correction factor should be established where accuracy is desired in order to compare results between laboratories.

A1.2.2 To determine the percentage of test material passing through a nominal 75-µm (No. 200) sieve, two additional sieves are needed: a 90-µm (No. 170) and a 63-µm (No. 230) sieve. Using NBS SRM No 1004 (No. 140-No. 400 sieve), Glass Bead Standard, determine the effective openings of each of the three sieves.

NOTE A1—NBS SRM No. 1004 tends to blind the screen during use and the beads lodged in the meshes are difficult to remove. Tapping the frame of the sieve and gentle brushing of the cloth from the underside will aid in recovering all the beads and keep the bead loss down to approximately 50 mg.

A1.2.3 Select a test material. It is important that the sample for calibration be of the same typical fineness as the range in which the sieve is to be used. The use of the same test material selected for the 75-µm (No. 200) dry sieve calibration is encouraged (see 4.2). Determine the percent residue of the test material on each of the three sieves, following the procedure of Test Method C 184. Plot the average percent residues *versus* the effective opening. The percent residue at the nominal opening of the No. 200 sieve can be read off the plot and the difference between the actual obtained residue and the nominal residue at 75 µm may be used algebraically as a sieve correction factor for that sieve, only in the same general fineness area.

A1.3 Example—Determination of Sieve Correction factor for 75-µm (No. 200) Sieve

A1.3.1 Using the SRM 1004, the following was determined:

(*a*) Effective opening of nominal sieves being corrected:

75-µm (No. 200)	72 µm
90-µm (No. 170)	90 µm
63-µm (No. 230)	65 µm

(*b*) Test sample residue, % (to be plotted on graph paper):

at 90 µm	1.1 %
at 72 µm	2.3 %
at 65 µm	2.8 %

(*c*) From the above plot:

Residue at 75-µm = 2.1%

A1.3.2 Since the sieve is woven too tightly and retains more than it should if it were a 75-µm opening, the correction, C, to be applied to the residue in percent is −0.2%.

Designation: C 806 – 95

Standard Test Method for Restrained Expansion of Expansive Cement Mortar[1]

This standard is issued under the fixed designation C 806; the number immediately following the designation indicates the year of original adoption or, in the case of revision, the year of last revision. A number in parentheses indicates the year of last reapproval. A superscript epsilon (ε) indicates an editorial change since the last revision or reapproval.

1. Scope

1.1 This test method covers the determination of length changes of expansive cement mortar, while under restraint, due to the development of internal forces resulting from hydration of the cement.

1.2 The values stated in non-SI units are to be regarded as the standard.

1.3 *This standard does not purport to address all of the safety concerns, if any, associated with its use. It is the responsibility of the user of this standard to establish appropriate safety and health practices and determine the applicability of regulatory limitations prior to use.*

2. Referenced Documents

2.1 *ASTM Standards:*

A 307 Specification for Carbon Steel Bolts and Studs, 60 000 PSI Tensile Strength[2]

C 157 Test Method for Length Change of Hardened-Hydraulic-Cement Mortar and Concrete[3]

C 219 Terminology Relating to Hydraulic Cement[4]

C 305 Practice for Mechanical Mixing of Hydraulic Cement Pastes and Mortars of Plastic Consistency[4]

C 490 Practice for Use of Apparatus for the Determination of Length Change of Hardened Cement Paste, Mortar, and Concrete[4]

C 778 Specification for Standard Sand[4]

C 1005 Specification for Weights and Weighing Devices for Use in the Physical Testing of Hydraulic Cements[4]

E 11 Specification for Wire-Cloth Sieves for Testing Purposes[5]

F 606 Test Methods for Determining the Mechanical Properties of Externally and Internally Threaded Fasteners, Washers, and Rivets[2]

3. Terminology

3.1 Definitions of terms used in this test method are given in Terminology C 219.

4. Apparatus

4.1 *Molds*—The molds for casting test specimens, when used in conjunction with the restraining cage described in 4.2, shall provide for forming either 2 by 2 by 10-in. prisms having a 10-in. gage length, or 50 by 50 by 250-mm prisms having a 250-mm gage length. The molds shall otherwise conform to the requirements of Practice C 490, except that the gage stud holder, gage studs, and spacer screws described in that specification will not be used.

4.2 *Restraining Cage,* consisting of a threaded steel rod with steel end plates held in place by nuts as shown in Fig. 1. The rod shall be provided with cap nuts for prevention of corrosion. The rod shall conform to Specification A 307 Grade A except, it shall support 1180-lbf (5249 N) proof load when tested in accordance with Method 1 of Test Methods F 606.

4.3 *Length Comparator*—The length comparator for measuring length shall conform to and be used in accordance with the requirements of Practice C 490. If the terminals of the comparator are fitted with collars, they shall be such that the cap nuts on the ends of the threaded rod of the specimen do not rest on the collar during measuring.

4.4 *Tamper*—The tamper shall be made of nonabsorptive, nonabrasive material such as medium-hard rubber or seasoned oak wood, rendered nonabsorptive by immersion for 15 min in paraffin at approximately 392°F (200°C), and shall have a cross section of ½ by 1 in. (13 by 25 mm) and convenient length of about 6 in. (150 mm). The tamping face of the tamper shall be flat and at right angles to the length of the tamper.

4.5 *Apparatus for Demolding Specimens*—Apparatus similar to that described in the Apparatus section of Test Method C 157, may be found useful for demolding specimens.

4.6 *Weights and Weighing Devices*—Scales used shall comply with Specification C 1005.

NOTE 1—Generally defined, the sensibility reciprocal is the change in load required to change the position of rest of the indicating element or elements of a nonautomatic-indicating scale a definite amount at any load.

4.7 *Glass Graduates,* of suitable capacities (preferably large enough to measure the mixing water in a single operation) shall be made to deliver the indicated volume at 68°F (20°C). The permissible variation shall be ±2 mL. These graduates shall be subdivided to at least 5 mL, except that the graduation lines may be omitted for the lowest 10 mL for a 250-mL graduate and the lowest 25 mL for a 500-mL graduate. The main graduation lines shall be circles and shall be numbered. The least graduations shall extend at least one seventh of the way around, and intermediate graduations shall extend at least one fifth of the way around.

4.8 *Mixer, Bowl, and Paddle*—The mixer shall be an electrically driven mechanical mixer of the type equipped

---

[1] This test method is under the jurisdiction of ASTM Committee C-1 on Cement and is the direct responsibility of Subcommittee C01.13 on Special Cements.

Current edition approved Dec. 10, 1995. Published March 1996. Originally published as C 806 – 75. Last previous edition C 806 – 87 (1995)$^{\epsilon 1}$.

[2] *Annual Book of ASTM Standards,* Vol 15.08.

[3] *Annual Book of ASTM Standards,* Vol 04.02.

[4] *Annual Book of ASTM Standards,* Vol 04.01.

[5] *Annual Book of ASTM Standards,* Vol 14.02.

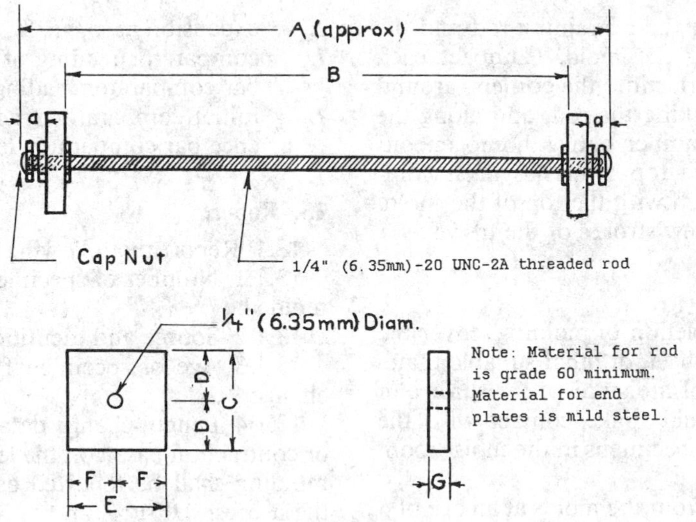

	Dimension, in. (mm)	Tolerance, in. (mm)
A	11¾ (300)	. . .
B	10 (250)	⅛ (3)
C	2 (50)	. . .
D	1 (25)	. . .
E	1 31/32 (49)	1/32 (1)
F	63/64 (24.5)	. . .
G	⅜ (10)	. . .

FIG. 1 Restraining Cage

with paddle and mixing bowl, as specified in the Apparatus section of Practice C 305.

4.9 *Trowel*—The trowel shall have a steel blade 4 to 6 in. (102 to 152 mm) in length, with straight edges.

5. Temperature and Humidity

5.1 The temperature of the air in the vicinity of the mixing operation, the dry materials, mixing water, molds, base plates, mixing bowl, moist closet or moist room, and storage water shall not vary from 73.4°F (23°C) by more than ±3°F (±1.7°C).

5.2 The relative humidity of the laboratory shall be not less than 50 %. The moist closet or moist room shall provide a relative humidity of not less that 95 %.

6. Graded Standard Sand

6.1 Sand used shall comply with the graded sand requirement of Specification C 778.

Passing Sieve	Percentage Retained
No. 100 (150-μm)	98 ± 2
No. 50 (300-μm)	72 ± 5
No. 30 (600-μm)	2 ± 2
No. 16 (1.18-mm)	none

NOTE 2—*Segregation of Graded Sand*—The graded standard sand should be handled in such a manner as to prevent segregation, since variations in the grading of the sand cause variations in the consistency of the mortar. In emptying sacks of sand into bins or in scooping sand out of bins or sacks, care should be exercised to prevent the formation of mounds of sand or craters in the sand, down the slopes of which the coarser particles will roll. Bins should be of sufficient size to permit these precautions. Devices for drawing the sand from bins by gravity should not be used.

7. Test Specimen

7.1 The test specimen shall be a 2 by 2 by 10-in. prism having a 10-in. gage length, or a 50 by 50 by 250-mm prism having a 250-mm gage length. At least two specimens shall be prepared for each test.

8. Preparation and Assembly of Specimen Molds and Restraining Cage

8.1 Install the cap nuts firmly in place on the ends of the restraining rods and measure the restraining rods for length while they are at a temperature of 73.4 ± 1°F (23.0 ± 0.5°C).

8.2 The edges of the end plates of the restraining cage that contact the sides and bottom of the mold shall be thinly covered with a heavy mineral oil or light cup grease such as petrolatum. After assembly of the molds, place a restraining cage in each compartment. Seal the contact lines between the outer sides and base plate of the molds with paraffin or micro-crystalline wax. Thinly cover the interior faces of the molds with mineral oil and remove any oil or grease from the restraining cage rod and end plates.

9. Proportioning and Mixing of Mortars

9.1 Proportion the standard mortar by using one part of cement to 2.75 parts of graded standard sand by mass. Mix 800 g of cement, 2200 g of graded standard sand, and 400 mL of water to produce a batch sufficient for two test specimens.

9.2 Mix in accordance with the procedure for mixing mortars given in Practice C 305. Upon completion of mixing, shake the mixing paddle into the mixing bowl to remove excess mortar.

10. Molding Specimens

10.1 Start molding specimens immediately after completion of mixing. Place the mortar in the mold in two layers with the first layer just covering the restraining rod and the second layer slightly overfilling the mold. Compact each layer by tamping. Work the mortar into the corners, around the restraining rod, without striking the rod, and along the surfaces of the mold with the tamper until a homogeneous specimen is obtained. After the top layer has been compacted, strike-off the mortar flush with the top of the mold, and smooth the surface with a few strokes of the trowel.

11. Curing of Specimens

11.1 Immediately after completion of molding, cover the specimens with a polyethylene sheet, or other suitable material for preventing loss or gain of moisture, at the surface of the specimens. The cover should make contact with the surface of the mortar. Place the specimens in the moist room or moist cabinet.

11.2 Remove the specimens from the molds at an age of 6 ± ¼ h after the addition of water to the cement during the mixing operation (Note 3). Particular care must be exercised in this operation in order not to damage the specimens because of their low strength at this age. It is preferable to use the demolding devices referred to in 4.5 for freeing specimens from the base and side plates.

NOTE 3—At this time comparator readings may be made. Although these readings may differ slightly from the initial reading made on the restraining rod prior to molding the specimen, it can serve as a check on the initial reading which should be used as the basis for calculating the length change.

11.3 Cure the specimens in lime-saturated water at 73.4 ± 3.0°F (23.0 ± 1.7°C) until they have reached an age of 7 days ± 1 h, including the period in the molds, except that the temperature of the curing water shall be 73.4 ± 1°F (23.0 ± 0.5°C) for at least the last 15 min. At the end of this period, remove the specimens from the water one at a time, wipe with a damp cloth, and immediately make comparator readings with the cap nuts in place.

11.4 Return the specimens to curing in lime-saturated water at 73.4 ± 3.0°F (23.0 ± 1.7°C) until they have reached an age of 28 days including the period in the molds. At the end of the curing period, make comparator readings immediately after the specimens have been subjected to about 15 min of storage in water at 73.4 ± 1.0°F (23.0 ±0.5°C) and in the same manner as described in 11.3.

12. Calculation

12.1 Calculate the expansion or shrinkage at any age as follows:

$$E_x = \frac{L_x - L_i}{10\ (250)} \times 100$$

where:

E_x = expansion at x age, %,

L_x = comparator reading of specimen at x age − reference bar comparator reading, in. (mm), and

L_i = initial comparator reading of restraining rod − reference bar comparator reading, in. (mm).

13. Report

13.1 Report the following information:

13.1.1 Number of specimens for each condition and date molded,

13.1.2 Source and identification of cement.

13.1.3 Age of specimen for each observation for length change,

13.1.4 Length-change data, expressed as linear expansion or contraction based on the length of the specimen at time of molding shall be reported as percent change dimension to the nearest 0.001 %, and

13.1.5 Any other pertinent information.

14. Precision and Bias[6]

14.1 *Precision:*

14.1.1 *Within Batch Precision*—The standard deviation of two bars made from the same batch and tested together was found to be 0.00276 %. Therefore two bars made from the same batch should not differ from each other by more than 0.008 % length change.[7]

14.1.2 *Single Operator Precision*—The single operator standard deviation has been found to be 0.0038 at seven and twenty eight days age, based on the average of two bars. Therefore the results of two properly conducted tests by the same operator on the same material should not differ by more than 0.011 % length change.[7]

14.1.3 *Multilaboratory Precision*—The multilaboratory standard deviation has been found to be 0.0093 at seven and twenty eight days age, based on the average of two bars. Therefore the results of two properly conducted tests on the same material in two different laboratories should not differ by more than 0.026 % length change.[7]

14.2 *Bias*—A bias statement cannot be prepared due to the fact that absolute standards are not available.

15. Keywords

15.1 cement mortar; expansive cement; restrained expansion

---

[6] Supporting data are available from ASTM Headquarters. Request RR: C01-1001.

[7] These numbers represent, respectively, the (1s) and (d2s) limits as described in Practice C 670.

Designation: C 807 – 99

Standard Test Method for Time of Setting of Hydraulic Cement Mortar by Modified Vicat Needle[1]

This standard is issued under the fixed designation C 807; the number immediately following the designation indicates the year of original adoption or, in the case of revision, the year of last revision. A number in parentheses indicates the year of last reapproval. A superscript epsilon (ϵ) indicates an editorial change since the last revision or reapproval.

1. Scope

1.1 This test method covers the determination of the time of setting of hydraulic cement mortar by means of the modified Vicat needle.

1.2 The values stated in SI units are to be regarded as standard. The values given in parentheses are for information only.

1.3 *This standard does not purport to address all of the safety concerns, if any, associated with its use. It is the responsibility of the user of this standard to establish appropriate safety and health practices and determine the applicability of regulatory limitations prior to use.* See Note 1 for a specific warning statement.

NOTE 1—**Warning:** Fresh hydraulic cementitious mixtures are caustic and may cause chemical burns to skin and tissue upon prolonged exposure. The use of gloves, protective clothing, and eye protection is recommended. Wash contact area with copious amounts of water after contact. Wash eyes for a minimum of 15 min. Avoid exposure of the body to clothing saturated with the liquid phase of the unhardened material. Remove contaminated clothing immediately after exposure.

2. Referenced Documents

2.1 *ASTM Standards:*

C 109/C 109M Test Method for Compressive Strength of Hydraulic Cement Mortars (Using 2-in. or 50-mm Cube Specimens)[2]

C 183 Practice for Sampling and the Amount of Testing of Hydraulic Cement[2]

C 187 Test Method for Normal Consistency of Hydraulic Cement[2]

C 219 Terminology Relating to Hydraulic Cement[2]

C 305 Practice for Mechanical Mixing of Hydraulic Cement Pastes and Mortars of Plastic Consistency[2]

C 490 Practice for Use of Apparatus for the Determination of Length Change of Hardened Cement Paste, Mortar, and Concrete[2]

C 511 Specification for Moist Cabinets, Moist Rooms, and Water Storage Tanks Used in the Testing of Hydraulic Cements and Concretes[2]

C 670 Practice for Preparing Precision and Bias Statements for Test Methods for Construction Materials[3]

C 778 Specification for Standard Sand[2]

C 845 Specification for Expansive Hydraulic Cement[2]

C 1005 Specification for Reference Masses and Devices for Determining Mass for Use in the Physical Testing of Hydraulic Cements[2]

D 1193 Specification for Reagent Water[4]

3. Terminology

3.1 Terms used in this standard are defined in Terminology C 219.

4. Summary of Test Method

4.1 A mortar is prepared with the cement to be tested, using stipulated quantities of cement and water, and sufficient standard sand to produce a required consistency as determined by a stipulated penetration using the plunger of the modified Vicat apparatus. Mortar of the proper consistency is then tested for time of setting, using the needle of the modified Vicat apparatus for the determination of a stipulated penetration. The time required to obtain the stipulated penetration of the modified Vicat needle is the time of setting.

5. Significance and Use

5.1 The purpose of this test method is to establish whether or not a cement complies with a specification limit on time of setting. It has been found to be particularly applicable for the determination of the setting time of expansive cements (see Specification C 845).

6. Apparatus

6.1 *Tamper and Trowel*, conforming to the requirements of Test Method C 109.

6.2 *Vicat Apparatus*, conforming to the requirements of Test Method C 187, with the following modifications:

6.2.1 For consistency determination, the plunger end of the movable rod shall be 17.5 ± 0.5 mm in diameter instead of 10 mm, and the total mass of the movable rod (including the 17.5-mm plunger) shall be 400 ± 0.5 g.

6.2.2 For time of setting determination, a 2 ± 0.05-mm

---

[1] This test method is under the jurisdiction of ASTM Committee C-1 on Cement and is the direct responsibility of Subcommittee C01.30 on Time of Set.

Current edition approved Jan. 10, 1999. Published May 1999. Originally published as C 807 – 75. Last previous edition C 807 – 89(1995)$^{\epsilon 1}$.

[2] *Annual Book of ASTM Standards*, Vol 04.01.

[3] *Annual Book of ASTM Standards*, Vol 04.02.

[4] *Annual Book of ASTM Standards*, Vol 11.01.

stainless steel needle shall be substituted for the 1-mm needle. The total mass of the movable rod with the 2-mm needle shall be 300 ± 0.5 g.

6.3 *Mold for Consistency and Setting Time*—A smooth cylindrical brass ring 76 ± 0.5 mm inside diameter (3-in. brass tubing) by 40 ± 1 mm deep, cemented to a plane, nonabsorptive plate, or smooth cylindrical brass cups having the above dimensions.

6.4 *Mixer, Bowl, Paddle, and Scraper*, conforming to the requirements of Practice C 305.

6.5 *Glass Graduates*, conforming to the requirements of Specification C 490.

6.6 *Masses and Mass Determining Devices*, conforming to the requirements of Specification C 1005.

7. Reagents and Materials

7.1 *Graded Standard Sand*, conforming to the requirements of Specification C 778.

7.2 *Mixing Water*—Potable water is satisfactory for routine tests. For all referee and cooperative tests, reagent water conforming to the requirements of Specification D 1193 for Type III or Type IV grade of reagent water shall be used.

8. Sampling

8.1 When the test is part of acceptance testing, sample the cement in accordance with Practice C 183.

9. Conditioning

9.1 The temperature of the room, dry materials, paddle, bowl, and molds shall be maintained between 20 and 27.5°C (68 and 81.5°F). The temperature of the mixing water, moist closet, or moist room shall not vary from 23°C (73.4°F) by more than ±1.7°C (3°F).

9.2 The relative humidity of the laboratory shall not be less than 50 %.

9.3 The moist closet or moist room shall conform to the requirements of Specification C 511.

10. Procedure

10.1 *Preparation of Mortar*:

10.1.1 Mix 750 g of cement, 375 mL of water, and the quantity of graded standard sand required to give the consistency specified below under" Test for Consistency" in accordance with the mortar mixing procedure described in Practice C 305.

10.1.2 Upon completion of mixing, place a layer of mortar about 20 mm in thickness in the mold and puddle with the tamper. Hold the tamper with the 25-mm side parallel to the radius of the mold for 14 strokes spaced equally around the outside of the mold. Apply an additional four strokes to the center of the specimen with the longitudinal axis of the tamper at the center of the mold and rotate the tamper 45° between each stroke. Then fill the mold to overflowing with mortar and puddle as specified for the first layer. Cut off the mortar to a plane surface flush with the top of the mold in two operations. First, with the trowel held at about a 20° angle, start at one side of the mold, 10 mm above the mold, and finish the stroke by cutting the mortar off the top of the mold at the opposite side. Second, starting from the opposite direction, cut the mortar flush with the top of the mold by drawing the straight edge of the trowel (held nearly perpendicular to the mold) with a sawing motion across the top of the mold.

10.2 *Test for Consistency*—Center the mortar confined in the mold under the movable rod of the modified Vicat apparatus described above. Bring the plunger in contact with the surface of the mortar and tighten the setscrew. Then set the movable indicator to the upper zero mark on the scale, or take an initial reading, and release the rod 35 to 50 s after starting to fill the mold. The mortar shall have the desired consistency when the rod settles to a point 20 ± 4 mm below the original surface 30 s after being released. Cover the bowl containing the remainder of the mortar with the lid while the consistency test is being made. Make trial mortars with varying percentages of standard sand until the desired consistency is obtained.

10.3 *Test for Time of Setting*—Mix the mortar remaining in the bowl at medium speed (285 ± 10 rpm) for 30 s after it has been found that the mortar has the proper consistency. Following the completion of the mixing, mold the test specimen as specified above. Immediately after molding, place the test specimen in the moist closet or moist room where it shall remain except when penetration tests are being made. Allow the time of setting specimen to remain in the moist closet or moist room for 30 min without being disturbed. Then, using the modified Vicat apparatus described above, bring the needle in contact with the surface of the mortar and tighten the setscrew. Set the movable indicator to the upper zero mark on the scale, or take an initial reading. Release the rod quickly by releasing the setscrew, and allow the needle to settle for 30 s. Determine the penetration of the needle at this time and every 30 min thereafter until the needle fails to penetrate to the bottom of the mold. Then, take penetration tests every 10 min until a penetration of 10 mm or less is obtained. No penetration test shall be made closer than 10 mm from any previous penetration test or from the inside edge of the mold. Record the results of all penetration tests, and, by interpolation, determine the time when a penetration of 10 mm is obtained. The difference, in minutes, between the time of contact of cement and water in the mortar mixing procedure and the time a penetration of 10 mm is obtained is the setting time.

10.4 *Precautions*—All of the apparatus shall be free of vibration during the penetration tests. Do not oil the plunger of the Vicat apparatus at any time, but keep the plunger and guide bearing clean at all times, and at the beginning of each day's test, rub a film of dry graphite on the plunger.

11. Report

11.1 The report shall indicate the time of setting in minutes.

12. Precision and Bias

12.1 The multilaboratory standard deviation has been found to be 15.2 min. Therefore, the results of two properly conducted tests from two different laboratories on samples of the same cement should not differ by more than 45 min.

13. Keywords

13.1 hydraulic-cement mortar; time of setting; Vicat needle

Designation: C 821 – 78 (Reapproved 1995)

Standard Specification for Lime for Use with Pozzolans[1]

This standard is issued under the fixed designation C 821; the number immediately following the designation indicates the year of original adoption or, in the case of revision, the year of last revision. A number in parentheses indicates the year of last reapproval. A superscript epsilon (ϵ) indicates an editorial change since the last revision or reapproval.

1. Scope

1.1 This specification covers all types of commercial hydrated lime such as: high-calcium, magnesium, or dolomitic-hydrated lime. By-product limes and slaked quick limes in dry, wet, or slurried form are also included.

1.2 *This standard does not purport to address all of the safety concerns, if any, associated with its use. It is the responsibility of the user of this standard to establish appropriate safety and health practices and determine the applicability of regulatory limitations prior to use.*

2. Referenced Documents

2.1 *ASTM Standards:*

C 25 Test Methods for Chemical Analysis of Limestone, Quicklime, and Hydrated Lime[2]

C 50 Practice for Sampling, Inspection, Packing, and Marking of Lime and Limestone Products[2]

C 204 Test Method for Fineness of Hydraulic Cement by Air Permeability Apparatus[2]

C 593 Specification for Fly Ash and Other Pozzolans for Use with Lime[2]

3. Terminology

3.1 *Definitions:*

3.1.1 *chemical factor*—a numerical value of the active constituents in the hydrated lime that react chemically with a pozzolan. The value is based on the combined calcium oxide equivalents of free calcium oxide, calcium hydroxide, and magnesium oxide. Calcium carbonate and magnesium hydroxide are excluded, since these compounds have been found to be nonreactive.

3.1.2 *pozzolanic receptivity index*—a performance factor incorporating both chemical quality and fineness, determined by Eq. 3.

4. Chemical and Physical Requirements

4.1 The lime shall conform to the requirements listed in Table 1, except as noted in Section 5.

5. Performance Requirements

5.1 If the chemical factor, Blaine fineness, or pozzolanic receptivity index is below the limits given in Table 1, the lime shall be required to meet the applicable performance requirements listed in Table 2. Where such lime complies with the limits listed in Table 2, it shall be considered acceptable for use with pozzolans. However, it may be necessary to increase the lime content in the lime-pozzolan mixtures in order to meet minimum design criteria.

6. Test Methods

6.1 *Chemical Factor:*

6.1.1 Determine calcium oxide (CaO), magnesium oxide (MgO), and carbon dioxide (CO_2) contents in accordance with Test Methods C 25. When the magnesium oxide content is greater than 5 %, determine the peak height of magnesium oxide and magnesium hydroxide using standard X-ray diffraction procedures and calculate the ratio (r) of magnesium oxide to hydroxide using Eq 1. When the magnesium oxide is less than 5 %, the value of r shall be 0.30.

$$r = \frac{\text{cps MgO}}{\text{cps MgO} + 2.0\,(\text{cps Mg(OH)}_2)} \tag{1}$$

where:

cps = counts per second of principal peaks corresponding to an interplanar spacing for MgO of 2.106 Å and for $Mg(OH)_2$ of 2.365 Å.

6.1.2 *Calculation of Chemical Factor*—Using the test results provided by 6.1.1 the chemical factor (C.F.) is determined by Eq 2:

$$\text{C.F.} = (\text{CaO}) - 1.27\,(\text{CO}_2) + 1.4\,r\,(\text{MgO}) \tag{2}$$

6.2 *Blaine Fineness*—Calculate in accordance with Method C 204, except use the specific gravity of the lime in calculating the trial weight of the sample, and the weight finally chosen must result in a firm bed as described in Test Method C 204. Use the specific gravity of the lime and finally chosen weight to determine the porosity of the bed in accordance with Eq 2 of Test Method C 204, and use this porosity (as well as specific gravity) in calculating the surface area.

6.3 *Pozzolanic Receptivity Index (PRI):*

$$\text{PRI} = \frac{\text{chemical factor} \times \text{Blaine fineness}}{10\ 000} \tag{3}$$

6.4 *Lime-Pozzolan Strength*—See Specification C 593, Section 7. The pozzolan used shall pass the requirements of Specification C 593 when tested with a lime having a pozzolanic receptivity index of 80 or higher.

6.5 *Lime-Pozzolan-Aggregate Strength, Freeze-Thaw Resistance*—See Specification C 593, Section 8. The pozzolan used shall pass the requirements of Specification C 593 when tested with a lime having a pozzolanic receptivity index of 80 or higher.

[1] This specification is under the jurisdiction of ASTM Committee C-7 on Lime, and is the direct responsibility of Subcommittee C07.07 on Pozzolanic Materials.

Current edition approved July 28, 1978. Published September 1978. Originally published as C 821 – 75 T. Last previous edition C 821 – 75 T.

[2] *Annual Book of ASTM Standards,* Vol 04.01.

TABLE 1 Chemical and Physical Requirements

Chemical factor, min	50
Blaine fineness, min, cm²/g	10 000
Pozzolanic receptivity index, min	100

TABLE 2 Performance Requirements

Plastic Compositions:		
Lime-pozzolan strength, min, psi (MPa)	600	(4.13)
Nonplastic Compositions:		
Lime-pozzolan-aggregate strength, min, psi (MPa)	400	(2.76)
Vacuum saturation strength, min, psi (MPa)	400	(2.8)

7. Sampling, Inspection, Packing, and Marking

7.1 The sampling, inspection, rejection, retesting, packing, and marking shall be conducted in accordance with Methods C 50.

8. Keywords

8.1 hydrated; lime; pozzolans

Designation: C 840 – 98

Standard Specification for Application and Finishing of Gypsum Board[1]

This standard is issued under the fixed designation C 840; the number immediately following the designation indicates the year of original adoption or, in the case of revision, the year of last revision. A number in parentheses indicates the year of last reapproval. A superscript epsilon (ε) indicates an editorial change since the last revision or reapproval.

This specification has been approved for use by agencies of the Department of Defense.

1. Scope

1.1 This specification covers the minimum requirements for the methods of application and finishing of gypsum board, including related items and accessories.

1.2 Where a fire resistance rating is required for a gypsum board assembly, details of construction shall be in accordance with reports of fire tests of assemblies that have met the requirements of the fire rating imposed.

1.2.1 Where this specification is more stringent (size or thickness of framing: size and spacing of fasteners) than the fire-rated construction, this specification shall govern.

1.3 Where sound control is required for a gypsum board assembly, details of construction shall be in accordance with reports of acoustical tests of assemblies that have met the required acoustical values.

1.4 Unheated spaces above gypsum board ceilings shall be properly ventilated (see Appendix X2).

1.5 The various application systems are located in the following sections:

I	Application of Single-Ply Gypsum Board to Wood Framing Members	8
II	Application of Two-Ply Gypsum Board to Wood Framing Members	9
III	Application of Gypsum Board by Adhesive Nail-On to Wood Framing Members	10
IV	Semi-Solid Gypsum Board Partitions	11
V	Solid Gypsum Board Partitions	12
VI	Application of Gypsum Board with Adhesives to Interior Masonry or Concrete Walls	13
VII	Application of Gypsum Board to Rigid Foam Insulation	14
VIII	Application of Gypsum Board to Steel Framing and Furring	15
IX	Arches and Bending Radii	16
X	Application of Gypsum Board to Receive Tile by Adhesive Application	17
XI	Exterior Application of Gypsum Wallboard and Exterior Gypsum Soffit Board	18
XII	Floating Interior Angles	19
XIII	Control (Expansion) Joints	20
XIV	Foil-Backed Gypsum Board	21

1.6 The values stated in inch-pound units are to be regarded as the standard. The SI (metric) values given in parentheses are approximate and are provided for information purposes only.

1.7 The text of this specification references notes and footnotes which provide explanatory material. These notes and footnotes (excluding those in tables and figures) shall not be considered as requirements of the specification.

1.8 The following precautionary caveat pertains only to Sections 6 through 23. *This standard does not purport to address all of the safety concerns, if any, associated with its use. It is the responsibility of the user of this standard to establish appropriate safety and health practices and determine the applicability of regulatory limitations prior to use.*

1.8.1 For specific precautionary statements, see 4.1.1, 4.1.2 and 22.5.

2. Referenced Documents

2.1 *ASTM Standards:*

C 11 Terminology Relating to Gypsum and Related Building Materials and Systems[2]

C 36 Specification for Gypsum Wallboard[2]

C 442 Specification for Gypsum Backing Board and Coreboard[2]

C 475 Specification for Joint Compound and Joint Tape for Finishing Gypsum Board[2]

C 514 Specification for Nails for the Application of Gypsum Board[2]

C 557 Specification for Adhesives for Fastening Gypsum Wallboard to Wood Framing[3]

C 630 Specification for Water-Resistant Gypsum Backing Board[2]

C 645 Specification for Non-Load (Axial) Bearing Steel Studs, Runners (Track), and Rigid Furring Channels for Screw Application of Gypsum Board[2]

C 754 Specification for Installation of Steel Framing Members to Receive Screw-Attached Gypsum Board[2]

C 931 Specification for Exterior Gypsum Soffit Board[2]

C 954 Specification for Steel Drill Screws for the Application of Gypsum Board or Metal Plaster Bases to Steel

[1] This specification is under the jurisdiction of ASTM Committee C-11 on Gypsum and Related Building Materials and Systems and is the direct responsibility of Subcommittee C11.03 on Specifications for Application of Gypsum and Other Products in Assemblies.

Current edition approved June 10, 1998. Published December 1998. Originally published as C 840 – 79. Last previous edition C 840 – 97.

[2] *Annual Book of ASTM Standards*, Vol 04.01.

[3] *Annual Book of ASTM Standards*, Vol 15.06.

Studs from 0.033 in. (0.84 mm) to 0.112 in. (2.84 mm) in Thickness[2]

C 1002 Specification for Steel Drill Screws for the Application of Gypsum Board or Metal Plaster Bases[2]

C 1007 Specification for the Installation of Load Bearing (Transverse and Axial) Steel Studs and Related Accessories[2]

C 1047 Specification for Accessories for Gypsum Wallboard and Gypsum Veneer Base[2]

2.2 *U.S. Department of Commerce Publication:*

PS20 American Softwood Lumber Standard[4]

2.3 *ANSI Standard:*

ANSI A 136.1 Organic Adhesives for Installation of Ceramic Tile, Type I and Type II[5]

3. Terminology

3.1 *Definitions*—Definitions shall be in accordance with Terminology C 11.

3.2 *Descriptions of Terms Specific to This Standard:*

3.2.1 *all-purpose compound, n*—a compound formulated and manufactured to serve as a taping or finishing compound, or both.

3.2.2 *decoration, n*—paint (including primers), texture, coatings, and coverings such as wallpaper and sheet plastic materials designed to conceal or protect the surface of the gypsum board (see Appendix X3).

3.2.3 *dry type, n*—a compound in powder form to be mixed with water before use.

3.2.4 *drying type, n*—a compound that hardens by the evaporation of the vehicle used to make the paste.

3.2.5 *face panel, n*—outside ply of multiple layer gypsum board assemblies.

3.2.6 *finished wallboard, n*—wallboard that has had the joints taped, has had the joints, fastener heads, and flanges of accessories concealed with joint compound, and has been sanded to prepare the surface to receive job applied decoration.

3.2.7 *finishing, n*—the preparation of gypsum board surfaces to receive the field application of decoration.

3.2.8 *finishing compound (sometimes called topping compound), n*—a compound used for successive coats over the taping or all-purpose compound that provides a smooth and level surface for the application of decoration.

3.2.9 *finishing of accessories, n*—the application of compound to flanges of accessories to create a monolithic surface.

3.2.10 *flange, n*—that part of an accessory extending over the face of the gypsum wallboard intended to become a part of the monolithic plane of the prepared surface.

3.2.11 *framing member, n*—that portion of the framing, furring, and blocking to which the gypsum board is attached.

3.2.12 *harden, v*—a condition reached when the compound has lost its plasticity to an arbitrary degree, measured in resistance to penetration or deformation.

3.2.13 *joint tape, n*—a strip of material designed to be embedded in compound to reinforce joints, cracks, or other small openings in or between gypsum boards.

3.2.14 *joint treatment, n*—application of joint tape and compound to the joint between gypsum boards.

3.2.15 *laminating compound, n*—a compound (abrasive) used to adhere gypsum board to gypsum board or other monolithic materials.

3.2.16 *parallel or vertical application, n*—gypsum board applied with the edges parallel to the framing member to which it is attached.

3.2.17 *perpendicular or horizontal application, n*—gypsum board application with the edges applied at right angles to the framing member to which it is attached.

3.2.18 *ready-mix type, n*—a factory-prepared compound ready to be used without the addition of water.

3.2.19 *required, adj*—pertaining to a mandatory obligation imposed by a force outside this standard, such as a building code, project specification, contract, or purchase order.

3.2.20 *setting type, n*—a compound that hardens by a chemical reaction and increases in straight through drying.

3.2.21 *skim coat, n*—a thin coat of skim coat compound applied over the finished wallboard surface.

3.2.22 *skim coat compound, n*—joint compound, or a compound specifically formulated and manufactured for use as a skim coat.

3.2.23 *specified, adj*—pertaining to a mandatory requirement of this standard or a referenced requirement (see 3.2.17).

3.2.24 *spotting fastener heads, n*—the application of compound to cover the fastener head to create a monolithic surface.

3.2.25 *taping compound (sometimes called embedding compound), n*—a compound used for embedding and bonding joint tape and for the first coat over the heads of fasteners and flanges of accessories.

3.2.26 *taping of joints, n*—the application of compound and joint reinforcing tape to the joints between adjoining gypsum boards.

3.2.27 *treated joint, n*—a joint between gypsum boards that has been covered with joint tape and joint compound as specified in 22.3.2.

DISCUSSION—Strip moldings or similar devices shall be permitted.

4. Environmental Conditions

4.1 *Application of Gypsum Board, Joint Treatment Materials, and Adhesives*—Room temperature shall be maintained at not less than 40°F (4°C) during application of gypsum board except when adhesive is used for the attachment of gypsum board. For the bonding of adhesive, joint treatment, texturing, and decoration, the room temperature shall be maintained at not less than 50°F (10°C) for 48 h prior to application and continuously thereafter until completely dry.

4.1.1 When a temporary heat source is used, the temperature shall not exceed 95°F (35°C) in any given room or area.

4.1.2 Adequate ventilation shall be maintained in the working area during installation and curing period.

4.2 Gypsum board shall be protected from direct exposure to rain, snow, sunlight, or other excessive weather conditions.

NOTE 1—Where manufacturers' recommendations differ from the above, follow their recommendations.

[4] Available from U.S. Government Printing Office, Washington, DC 20402.

[5] Available from American National Standards Institute, 11 West 42nd Street, 13th Floor, New York, NY 10036.

5. Materials and Manufacture

5.1 *Gypsum Boards*—A family of gypsum sheet products as defined in Terminology C 11.

5.1.1 *Type X (Special Fire-Resistant) Gypsum Wallboard, Gypsum Backing Board, Water-Resistant Gypsum Backing Board, or Exterior Gypsum Soffit Board*—Gypsum board that provides a greater degree of fire resistance than regular gypsum board as defined in Specifications C 36, C 442, C 630 and C 931.

5.1.2 *Foil-Backed Gypsum Wallboard or Gypsum Backing Board*—Regular or Type X gypsum board with foil laminated to the back surface. The foil is a vapor retarder.

5.1.3 *Predecorated Gypsum Board*—Gypsum board with a decorative wall covering or coating applied in-plant by the gypsum board manufacturer.

5.2 *Gypsum Wallboard*—See Specification C 36.

5.3 *Gypsum Backing Board and Coreboard*—See Specification C 442.

5.4 *Water-Resistant Gypsum Backing Board*—See Specification C 630.

5.5 *Exterior Gypsum Soffit Board*—See Specification C 931.

5.6 *Finishing Materials:*

5.6.1 *Compounds*—Taping compound, finishing compound, and all-purpose compound shall meet the requirements of Specification C 475.

5.6.2 Mix compounds in accordance with the manufacturers' directions.

5.6.3 *Joint Tape*—See Specification C 475.

5.7 *Fasteners:*

5.7.1 *Nails*—See Specification 514.

5.7.2 *Screws:*

5.7.2.1 See Specification C 1002 for screws for fastening gypsum board to wood members, steel members less than 0.03 in thickness, and to gypsum board.

5.7.2.2 See Specification C 754 and Specification C 954 for screws for fastening gypsum board to steel members from 0.033 to 0.112 in. in thickness.

5.7.3 *Staples*—No. 16 USS gage flattened galvanized wire staples with 7⁄16-in. (11.1-mm) wide crown outside measurement. Length of legs according to the following tables. Legs shall have divergent points.

NOTE 2—Use only for the base ply of two-ply gypsum board application.

Length of Legs,		Thickness of Gypsum Board,	
in.	mm	in.	mm
1	25.4	3⁄8	9.5
1 1⁄8	28.6	1⁄2	12.7
1 1⁄4	31.8	5⁄8	15.9

5.8 *Adhesives:*

5.8.1 *Fastening Gypsum Board to Wood Framing*—See Specification C 557.

5.8.2 *Fastening Gypsum Board to Steel Framing*—As specified by the manufacturer.

5.8.3 *Laminating Gypsum Board to Gypsum Board*—Laminating compounds, taping compound, or adhesive shall be as specified by the manufacturer.

NOTE 3—Adhesives for the installation of ceramic tile, see Appendix X4.

5.9 *Framing Members:*

5.9.1 Wood framing members shall conform to PS20, American Softwood Lumber Standards. The surface to which abutting edges or ends are attached shall be not less than 1 1⁄2 in. (38.1 mm) wide. For internal corners or angles, the bearing surface shall not be less than 3⁄4 in. (19.1 mm).

5.9.2 *Steel Studs, Furring Channels, and Runners,* Non-Load Bearing—See Specification C 645.

5.9.3 *Gypsum Studs*—Specification C 36 or C 442, not less than 1 in. (25.4 mm) thick by 6 in. (152 mm) wide. Studs shall be either solid or laminated.

5.10 *Accessories*—See Specification C 1047.

5.11 *Water*—Water shall be clean, fresh, and potable (suitable for domestic consumption).

5.12 *Face Panels*—Face panels shall be 1⁄2 in. (12.7 mm), 5⁄8 in. (15.9 mm), or multiple laminations of regular or Type X gypsum board.

5.13 *Core Board*—Core board shall be 3⁄4 in. (19.1 mm) or 1 in. (25.4 mm) either single thickness or multiple layers to the required thickness.

6. Substrate, Surface Preparation

6.1 The attachment surface of any framing member shall not vary more than 1⁄8 in. (3.2 mm) from the plane of the faces of adjacent framing members.

6.2 Wood framing shall be as straight and true as possible. Wood framing shall be securely attached following acceptable engineering practices and as required for the intended design.

NOTE 4—For installation of wood framing, see Appendix X5.

6.3 Metal framing members shall be of the proper size and design for their intended use and shall be installed in accordance with Specifications C 754 or C 1007 as required.

6.4 Masonry or concrete walls shall be dry, free of dust, oil, or form release agents, protrusions or voids, or foreign matter that would not permit proper bond for adhesively applied gypsum board.

6.5 All framing members and substrate shall be installed so that after the gypsum board has been applied, the finished surface will be in an even plane.

6.6 Gypsum board shall be kept free of any dirt, oil, or other foreign matter that cause a lack of bond. Foreign matter shall be removed.

6.7 All dents or gouges on the face of the gypsum board shall be brought up to a smooth level plane with the surface of the board.

6.8 Mechanical fasteners shall be set below the plane of the surface of the board.

6.9 All joints shall be true and even.

6.10 All gypsum board shall be tight against the framing member or substrate.

6.11 Deflection at design load of horizontal (ceiling) framing members supporting gypsum board shall be not more than L/240 of the span.

7. Application of Gypsum Board

7.1 *General:*

7.1.1 *Method of Cutting and Installation*—Gypsum board shall be cut by scoring and breaking or by sawing, working from the face side. When cutting by scoring, the face paper shall be cut with a sharp knife or other suitable tool. Gypsum board shall be broken by snapping the gypsum board in the reverse direction, or cut the back paper with a knife or suitable tool.

7.1.2 Cut edges and ends of the gypsum board shall be smoothed to obtain neat joints when installed. Holes for pipes, fixtures, or other small openings shall be scored on the back and the face in outline before removal or cut out with a saw or special tool designed for this purpose. Where gypsum board meets projecting surfaces, it shall be scribed and cut neatly.

7.1.3 When gypsum board is to be applied to both ceiling and walls, the gypsum board shall be applied first to the ceiling and then to the walls.

7.1.4 Where used at edges or ends, fasteners shall be spaced not more than 1 in. (25 mm) from edges and not less than ⅜ in. (9.5 mm) from edges and ends of gypsum board (except where floating angles are used). Perimeter fastening into the partition plate or sole at the top and bottom shall not be required except where the fire ratings, structural performance, or other special conditions require such fastening. While driving the fasteners, the gypsum board shall be held in firm contact with the underlying support. Application of fasteners shall proceed from the center or field of the gypsum board to the ends and edges.

7.1.5 Nails shall be driven with the heads slightly below the surface of the gypsum board, avoiding damage to the face and core of the board, such as breaking the paper or fracturing the core.

7.1.6 Screws shall be driven to provide screwhead penetration just below the gypsum board surface without breaking the surface paper of the gypsum board or stripping the framing member around the screw shank.

7.1.7 Staples shall be driven with the crown parallel to the framing members, in such a manner that the crown bears tightly against the gypsum board without cutting into the face paper.

7.1.7.1 Staple attachment shall be restricted to the base plies only of gypsum board in a multi-ply system.

7.1.8 The gypsum board shall be kept tight against the framing.

7.1.9 The external corners shall be protected with a metal corner bead or other suitable type of corner protection that shall be attached to supporting construction with fasteners or a crimping tool nominally 6 in. (152 mm) on centers (see Section 3 and Fig. 1).

7.2 Gypsum board shall not be used where it will be exposed to temperatures more than 125°F (52°C) for extended periods of time.

7.3 Gypsum board joints shall be located so that no joint will align with the edge of an opening unless control joints are to be installed at these locations.

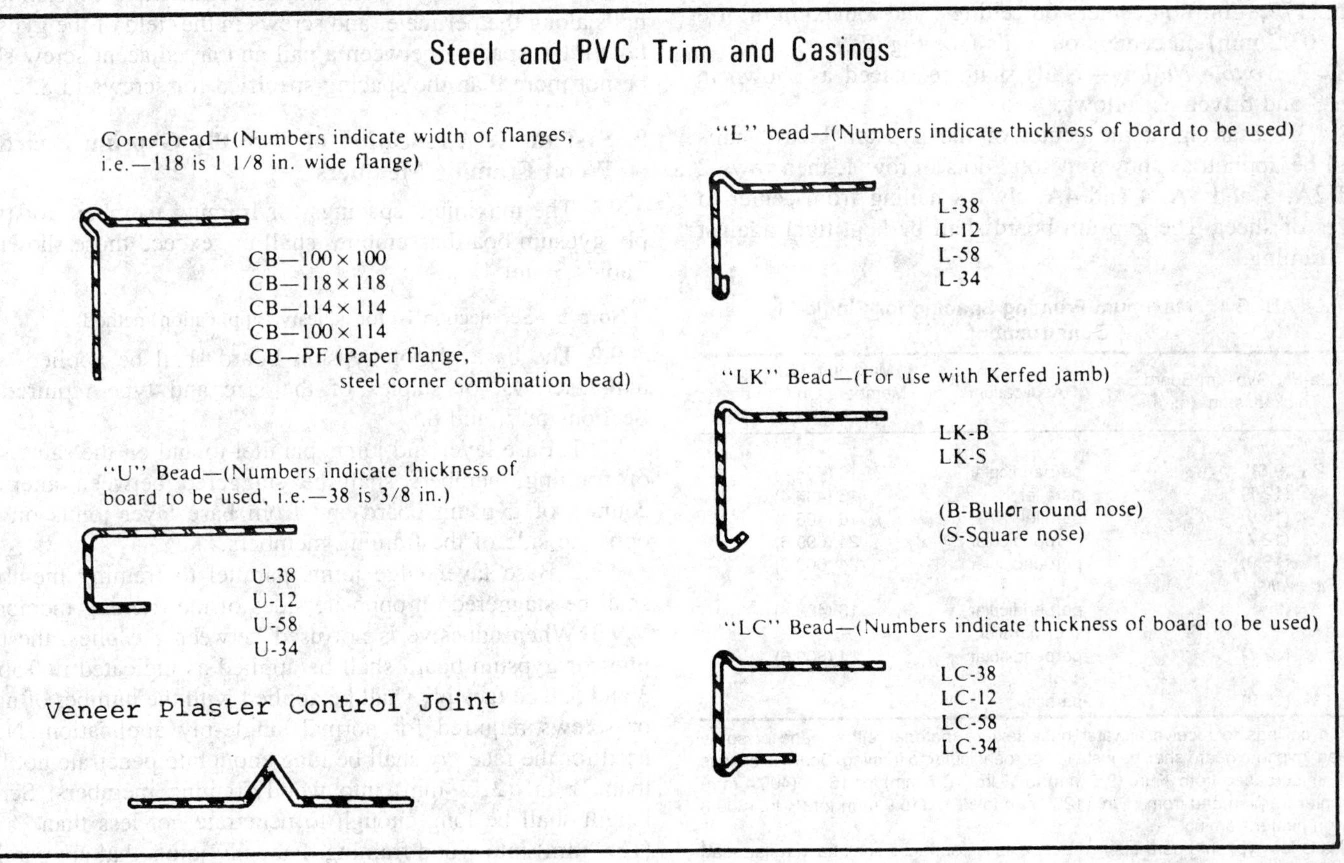

NOTE 1—All dimensions are in inch-pound units.

FIG. 1 Accessories

7.4 Joints shall be staggered, and joints on opposite sides of a partition shall not occur on the same stud.

7.5 Gypsum board used in building construction shall be not less than 8 in. (203 mm) from the finished grade in fully weather and water protected siding systems, and not less than 12 in. (305 mm) from the ground within properly drained and ventilated crawl spaces. Where ground moisture or humidity is extreme and/or continuous, the crawl space ground surface shall be covered with a vapor barrier.

8. System I: Application of Single-Ply Gypsum Board to Wood Framing Members

8.1 The maximum spacing for framing members for single-ply gypsum board assembly shall not exceed those shown in Table 1.

8.2 In single-ply installation, all ends and edges of gypsum board shall occur over framing members or other solid backing except where treated joints occur at right angles to framing or furring members.

8.3 End joints shall be staggered and joints on opposite sides of a partition shall be arranged to occur on alternate framing members.

8.4 Length of nails shall be as shown in Table 2.

8.4.1 Where a specific degree of fire resistance is required for gypsum board assemblies, nails of the same length, shank diameter, and head bearing area, as those described in the fire test report, shall be used.

8.4.2 *Single Nailing*—Nails shall be spaced a maximum of 7 in. (177.8 mm) on centers on ceilings, and a maximum of 8 in. (203.2 mm) on centers on walls (see Fig. 2).

8.4.3 *Double Nailing*—Nails shall be spaced as shown in Fig. 3 and driven as follows:

8.4.3.1 Starting at the center of the gypsum board, nails shall be applied as shown in solid dots in row 1, then rows 2 and 2A, 3 and 3A, 4 and 4A, always nailing from center to edges of sheet. The gypsum board shall be kept tight against the framing.

TABLE 1 Maximum Framing Spacing for Single-Ply Construction[A]

Single-Ply Gypsum Board Thickness, in. (mm)	Application[B]	Maximum Framing Members On Centers Spacing, in. (mm)
Ceilings:		
⅜ (9.5)[C]	perpendicular	16 (406.4)
½ (12.7)	parallel	16 (406.4)
⅝ (15.9)	parallel	16 (406.4)
½ (12.7)	perpendicular	24 (609.6)
⅝ (15.9)	perpendicular	24 (609.6)
Sidewalls:		
⅜ (9.5)	perpendicular or parallel	16 (406.4)
½ (12.7) or ⅝ (15.9)	perpendicular or parallel	24 (609.6)

[A] On ceilings to receive a water base texture material, either hand or spray applied, gypsum board shall be installed perpendicular to framing. Board thickness shall be increased from ⅜ in. (9.5 mm) to ½ in. (12.7 mm) for 16 in. (406.4 mm) on center framing and from ½ in. (12.7 mm) to ⅝ in. (15.9 mm) for 24 in. (609.6 mm) on center framing.

[B] Nails for gypsum board applied over existing surfaces shall have a flat head and diamond point, and shall penetrate not less than ⅞ in. (22.2 mm), nor more than 1 ¼ in. (31.8 mm) into the framing member.

[C] ⅜-in. (9.5-mm) single-ply gypsum board shall not be applied to ceilings where the gypsum board supports insulation.

TABLE 2 Nail Lengths for Single-Ply Gypsum Board Assembly

Gypsum Board Thickness, in. (mm)	Minimum Nail Length, in. (mm)
Application of gypsum board over existing surfaces:	Flat or concave head, diamond or needle point of such length as to provide not less than ⅞ in. (22.2 mm) nor more than 1¼ in. (31.8 mm) penetration into the following framing members:
⅜ (9.5)	1¼ (31.8)
½ (12.7)	1⅜ (34.9)
⅝ (15.9)	1½ (38.1)

8.4.3.2 Second nails, shown by circles, shall be applied in the same manner as the first nails, also starting at row 1.

8.4.3.3 As an alternative procedure, the second nail shall be applied immediately after all nails in each row are driven in accordance with 8.4.3.2.

8.4.3.4 Single nailing shall be used on the perimeter of the gypsum board, unless otherwise specified.

8.4.3.5 Nails shall be inspected for compliance with 7.1.5 after the second nails have been set.

8.5 *Spacing of Screws*—Screws shall be spaced not more than 12 in. (304.8 mm) on centers along framing members for ceilings and 16 in. (406.4 mm) on centers for walls where the framing members are 16 in. on centers. Screws shall be spaced not more than 12 in. on centers along framing members for ceilings and walls where the framing members are 24 in. (609.6 mm) on centers.

8.5.1 When using a combination of fasteners consisting of nails along the perimeter and screws in the field of the gypsum board, the spacing between a nail and an adjacent screw shall be not more than the spacing specified for screws in 8.5.

9. System II: Application of Two-Ply Gypsum Board to Wood Framing Members

9.1 The maximum spacings for framing members for two-ply gypsum board assemblies shall not exceed those shown in Tables 3 and 4.

NOTE 5—See Section 10 for adhesive application method.

9.2 The base ply of gypsum board shall be applied with nails, screws, or staples of the size and type required in Sections 5, 7, and 8.

9.2.1 Base layer end joints parallel to and on the same side of framing members shall be staggered between alternate courses of gypsum board and from base layer joints on the opposite side of the framing members.

9.2.2 Base layer edge joints parallel to framing members shall be staggered on opposite sides of the framing members.

9.3 When adhesive is not used between the plies, the two plies of gypsum board shall be applied as indicated in Tables 3 and 5. The face ply shall be applied with the number of nails or screws required for normal single-ply application. Nails used for the face ply shall be long enough to penetrate not less than ⅞ in. (22.2 mm) into wood framing members. Screw length shall be long enough to penetrate not less than ⅝ in. (15.9 mm) into wood framing. Face ply joints that are parallel to framing shall fall over framing members and be offset from the base ply joints when two plies of gypsum board are parallel.

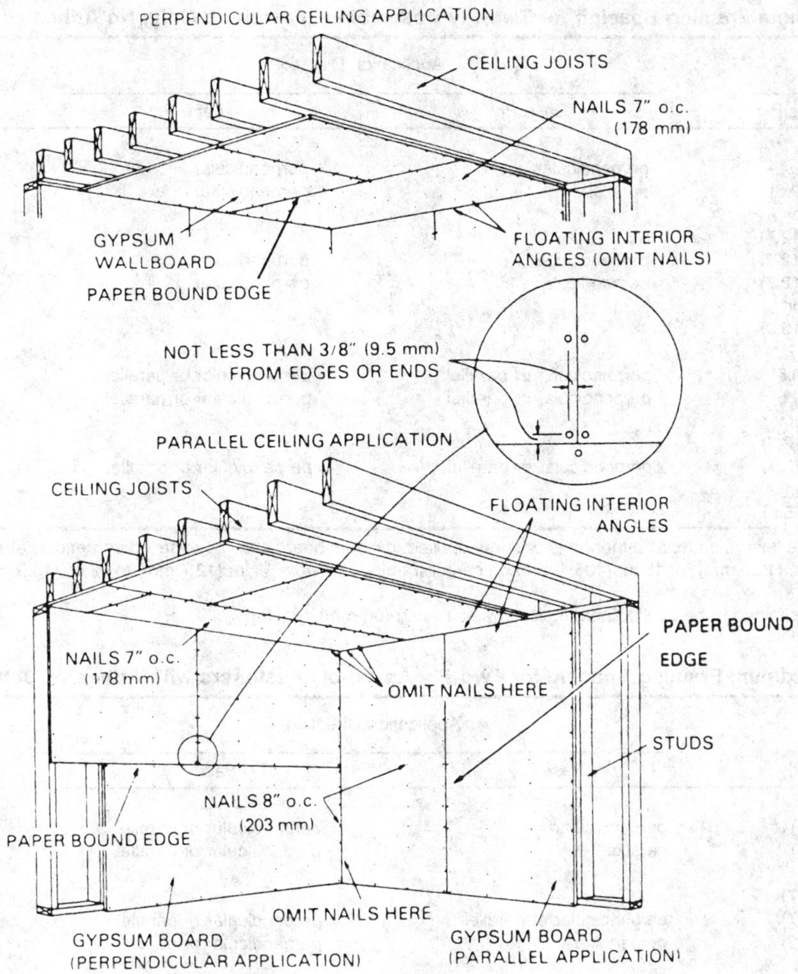

FIG. 2 Single Nailing

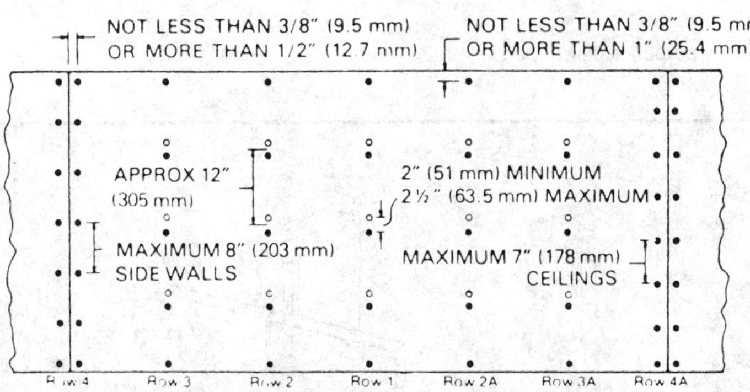

FIG. 3 Double Nailing

9.4 When an adhesive is used between the plies, the two plies shall be applied as indicated in Table 4. If the two plies are applied in parallel direction, the joints of the face ply shall be offset from the base ply.

9.4.1 Joints in the face ply need not occur over the framing members. The adhesive used between the two plies of gypsum board shall be uniformly applied over the back surface of the face ply of the gypsum board before it is erected or to the face surface of the base ply. The face ply of gypsum board shall be placed in position and fastened with nails or screws to hold gypsum board in place until the adhesive develops a bond.

9.4.2 Permanent fasteners shall be used around the perimeter 12 in. (304.8 mm) on centers and 16 in. (406.4 mm) on centers along framing members for the face ply of gypsum board applied on ceilings.

9.4.3 In place of nails and screws, the face ply of gypsum board applied on walls shall be held in position by shoring with props and headers, or other temporary support to ensure a pressure for bonding. Permanent fasteners shall be used on top and bottom of wall not more than 16 in. (406 mm) on centers. Nails or screws used to hold the gypsum board face ply shall be left in place and finished in the same manner as for single-ply gypsum board application (see Section 10).

9.5 Base layer joints and fasteners shall not be required to be taped or finished. Face layer joints and fasteners in multi-layer systems shall not be required to be finished unless required for appearance or decoration or as required for fire resistance.

10. System III: Application of Gypsum Board by Adhesive Nail-on to Wood Framing Members

10.1 Except as herein modified, application shall be in conformance with Section 8.

10.2 Surfaces of gypsum board and framing to receive the adhesive shall be free of dust, dirt, grease, or any other foreign matter that cause bond loss. Foreign matter shall be removed.

10.3 A bead of adhesive ⅜ in. (9.5 mm) in diameter shall be

TABLE 3 Maximum Framing Spacing for Two-Ply Assemblies, Fasteners Only, No Adhesive Between Plies[A]

Gypsum Board Thickness, in. (mm)		Application Direction		Maximum On Centers Spacing of Framing, in. (mm)
Base Ply	Face Ply	Base Ply	Face Ply	
Ceilings:				
⅜ (9.5)	⅜ (9.5)	perpendicular	perpendicular	16 (406.4)
½ (12.7)	⅜ (9.5) or ½ (12.7)	parallel	perpendicular	16 (406.4)
½ (12.7)	½ (12.7)	perpendicular	perpendicular	24 (609.6)
⅝ (15.9)[B]	½ (12.7) or ⅝ (15.9)	perpendicular	perpendicular	24 (609.6)
Sidewalls:				
⅜ (9.5)	⅜ (9.5)	perpendicular or parallel	perpendicular or parallel	16 (406.4)
½ (12.7)	⅜ (9.5) or ½ (12.7)	perpendicular or parallel	perpendicular or parallel	24 (609.6)
⅝ (15.9)	½ (12.7) or ⅝ (15.9)	perpendicular or parallel	perpendicular or parallel	24 (609.6)

[A] On ceilings to receive a water base texture material, either hand or spray applied, gypsum board shall be installed perpendicular to framing. Board thickness shall be increased from ⅜ in. (9.5 mm) to ½ in. (12.7 mm) for 16 in. (406.4 mm) on center framing and from ½ in. (12.7 mm) to ⅝ in. (15.9 mm) for 24 in. (609.6 mm) on center framing.

[B] ⅝-in. (15.9-mm) board shall be permitted to be applied perpendicularly at 16 in. (406 mm) spacing.

TABLE 4 Maximum Framing Spacing for Two-Ply Assembly Fasteners with Adhesive Between Plies[A]

Gypsum Board Thickness, in. (mm)		Application Direction		Maximum On Centers Spacing of Framing, in. (mm)
Base Ply	Face Ply	Base Ply	Face Ply	
Ceilings:				
⅜ (9.5)	⅜ (9.5)	perpendicular	perpendicular or parallel	16 (406.4)
½ (12.7)	⅜ (9.5) or ½ (12.7)	parallel	perpendicular or parallel	16 (406.4)
⅝ (15.9)	½ (12.7)	perpendicular or parallel	perpendicular or parallel	16 (406.4)
⅝ (15.9)	⅝ (15.9)	perpendicular	perpendicular or parallel	24 (609.6)
Sidewalls:				
⅜ (9.5)	⅜ (9.5)	perpendicular or parallel	perpendicular or parallel	24 (609.6)
½ (12.7)	⅜ (9.5) or ½ (12.7)	perpendicular or parallel	perpendicular or parallel	24 (609.6)
⅝ (15.9)	½ (12.7) or ⅝ (15.9)	perpendicular or parallel	perpendicular or parallel	24 (609.6)

[A] Adhesive between plies shall be dried or cured prior to joint treatment application.

TABLE 5 Fastener Spacing, Base-Ply of Two-Ply System, in. (mm) On Centers

	Nails	Screws	Staples
No adhesive between plies	24 (609.6)	24 (609.6)	16 (406.4)
Adhesive between plies	See 9.3	See 9.3	7 (177.8)

applied to the face of all wood framing members, except plates, that support the gypsum board. The adhesive shall be spread to an average width of ¾ in. (19 mm) and an average thickness of $\frac{1}{16}$ in. (1.6 mm). See Fig. 4 for application patterns.

10.3.1 Where a joining of two adjacent pieces of gypsum board occurs on a framing member, two parallel beads of adhesive shall be applied, one near each edge of the framing member.

10.4 Adhesive shall be applied to no greater area than can be covered with gypsum board within the "open time."

NOTE 6—"Open time" is the time period available for working with certain adhesives before they set in accordance with the adhesive manufacturer's specification.

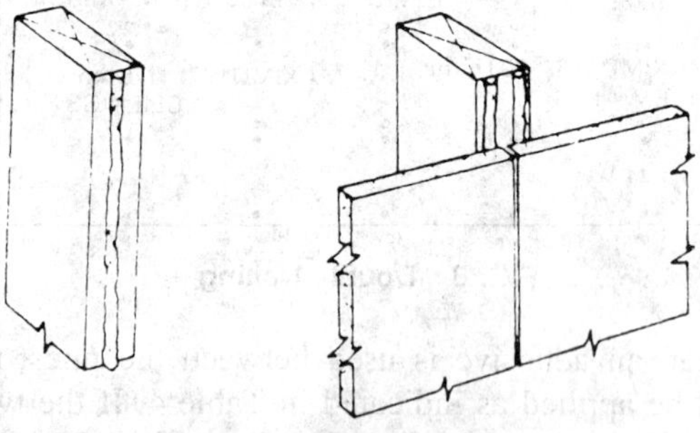

FIG. 4 Adhesive Application Patterns

10.5 *Fastener Spacing:*

10.5.1 If the properties of the adhesive ensure bridging between the gypsum board and the wood framing, nailing shall not be required in the field of the board for walls. In such cases, perimeter nailing, 16 in. (406.4 mm) on centers, shall be required.

10.5.2 When the properties of the adhesive are such that there is no positive bridging between the gypsum board and the wood framing, either temporary field nailing or temporary bracing shall be used to ensure contact between the gypsum board, the adhesive, and the wood framing, until the adhesive develops a bond.

10.5.3 Unless specified otherwise by the adhesive manufacturer, fastener spacing shall be in accordance with Table 6.

11. System IV: Semi-Solid Gypsum Board Partitions

11.1 *Installation:*

11.1.1 Runners or studs shall be installed where required to provide support at exterior walls, partition junctions, terminals, external corners, door frames and at other locations.

11.1.2 The gypsum board face panels and studs shall be positioned vertically.

11.1.3 The gypsum studs shall be laminated to face panels not more than 24 in. (609.6 mm) on centers and located at face panel vertical joints and at vertical centerline of panel.

11.1.4 Gypsum studs shall be laminated to face panels prior to erection or as erection of partition proceeds. A starter face panel shall be erected vertically at an intersecting wall. The starter panel shall be plumb and secured to the floor, the ceiling, and the vertical runners.

11.1.5 The next face panel adjacent to the starter panel shall be erected by butting its edge and end firmly to the starter panel and the ceiling. Erection of the face panels shall continue by laminating exposed faces of gypsum studs as work progresses.

11.1.6 Openings in partitions for doors and electrical outlets shall be carefully and accurately marked and cut.

11.1.7 Laminating compounds or adhesives shall be of a consistency and volume that will cover approximately three-fourths of the stud surface after lamination.

11.1.8 Type G screws shall be used to ensure a continuous bond between face panels and studs, and shall be spaced not more than 36 in. (914 mm) on centers.

11.1.9 Openings or changes in direction of partitions shall be reinforced with additional studs laminated in place at the following locations:

11.1.9.1 *External Corners*—Between the face panels in the corner opposite vertical runners.

11.1.9.2 *Abutting Walls*—Between the face panels of a partition to reinforce junction of an abutting wall.

11.1.9.3 *Door Openings*—Locate a vertical stud within 3 in. (76.2 mm) of door frame for reinforcement, and place a stud horizontally over the door header.

12. System V: Solid Gypsum Board Partitions

12.1 Non-load bearing solid partition consisting of gypsum coreboard faced on each side with gypsum board panels.

12.2 *Installation:*

12.2.1 Floor and ceiling runners shall be installed according to the layout and shall be secured not more than 24 in. (610 mm) on centers. Vertical runners shall be installed where specified in 11.1.1.

12.2.2 For partitions located parallel to and between ceiling members, steel or wood blocking not more than 24 in. (609.6 mm) on centers shall be provided to fasten ceiling runners prior to erection of ceiling.

12.2.3 Face panels shall be attached to runners at not more than 24 in. (609.6 mm) on centers.

12.2.4 Coreboard shall be installed prior to the installation of the face panel. Coreboard shall be attached not more than 24 in. (610 mm) on centers to the steel angles when steel angles are used as runners. When steel channels are used as runners to secure coreboard, attachment shall not be required.

NOTE 7—Combinations of wood and steel channels or angles will vary the installation procedure.

12.2.5 The adhesive shall be applied to the back surface of the face panels or the face of the coreboard as described for two-ply gypsum board construction (see Section 9).

12.2.6 The face panels shall be laminated to the core with sufficient pressure to ensure bonding. The joints of the face panels and core shall be staggered. To ensure the bond between face panels and core, fasteners shall be permitted.

13. System VI: Application of Gypsum Board with Adhesive to Interior Masonry or Concrete Walls

13.1 When applying gypsum board to monolithic concrete, brick, or concrete block, the adhesive shall be applied directly to the back of the gypsum board or on the wall in continuous beads not more than 12 in. (304.8 mm) on centers or daubs spaced not more than 12 in. (304.8 mm) on centers each way.

13.1.1 Beads shall be not less than ⅜ in. (9.6 mm) in diameter to provide a continuous bond between the gypsum board and the wall surface.

13.1.2 Daubs shall be 2 to 3 in. (50.8 to 76.2 mm) in diameter.

13.2 The gypsum board shall be positioned ⅛ in. (3.2 mm) from the floor and provide a tight fit at abutting edges or ends. The gypsum board shall not be slid on the adhesive. Mechanical fasteners, or temporary bracing, shall be used to support gypsum board until adhesive sets.

13.3 Delay the joint treatment until the gypsum board is firmly bonded.

13.4 Foil-backed gypsum board shall not be used for direct adhesive application.

13.5 Direct adhesive application shall be permitted only on interior masonry or concrete walls above grade, or the inside of exterior masonry cavity walls with 1-in. (25.4 mm) minimum width cavity between the inside and outside masonry for the full height of the above grade surface to receive gypsum board. Surfaces to which gypsum board is to be adhered shall be free of any foreign matter, projections, or depressions that will impair the bond.

TABLE 6 Fastener Spacing with Adhesive or Mastic Application and Supplemental Fastening

Framing Member Spacing, in. (mm) On Centers	Ceilings, in.		Partitions Load Bearing, in.		Partitions Nonload Bearing, in.	
	Nail	Screw	Nail	Screw	Nail	Screw
16 (406.4)	16	16	16	24	24	24
24 (609.6)	12	16	12	16	16	24

14. System VII: Application of Gypsum Board to Rigid Plastic Foam Insulation

14.1 *Application of Furring and Plastic Foam Insulation to Masonry and Concrete Walls:*

14.1.1 Rigid foam insulation shall be applied to the masonry or concrete in accordance with the foam manufacturer's specifications.

14.1.2 Furring strips or special metal furring members shall be attached by mechanical means to the masonry or concrete wall surface either before or after application of the insulation, depending on the system used. The furring members shall be installed in accordance with Table 1 and at gypsum board terminations above suspended ceilings, around doors, windows, or other openings, and for cabinet and fixture attachment.

14.2 The gypsum board shall be applied to furring as described in 8.4.1 or 8.5. The mechanical fasteners shall not penetrate completely to the masonry or concrete.

15. System VIII: Application of Gypsum Board to Steel Framing and Furring

15.1 Screw application shall be applied in accordance with 7.1.6.

15.1.1 Screws shall be of sufficient length so that the threaded portion shall penetrate not less than ⅜ in. (9 mm) into the framing members.

15.1.2 Where resilient furring channels are used, the screws used to attach gypsum board to the furring channels shall not contact the framing.

15.2 *Framing Spacing*—Maximum spacing of steel framing and furring for screw application shall be as specified in Table 1 for single-ply gypsum board and as specified in Tables 3 and 4 for two-ply gypsum board.

15.2.1 Installation of steel framing shall be in accordance with Specifications C 754 or C 1007.

15.3 *Screw Spacing:*

15.3.1 Screw spacing for single-ply gypsum board and face ply of two-ply gypsum board with no adhesive shall be in accordance with 8.5.

15.3.2 Screw spacing for parallel applied base-ply of two-ply gypsum board over steel framing with no adhesive between the plies shall be not more than 12 in. (304.8 mm) on centers along the edges of the gypsum board and 24 in. (609.6 mm) on centers into the stud or furring channel in the field of the gypsum board.

15.3.3 Screw spacing for a perpendicularly applied base-ply of two-ply gypsum board over steel framing with no adhesive between the plies shall be not less than one screw at each edge of the gypsum board at each framing member and one screw midway between the edges at each framing member.

15.3.4 Screw spacing for perpendicular or parallel applied base-ply of two-ply gypsum board over steel framing with adhesive between plies shall be not more than as specified for single-ply gypsum board in 8.5.

15.3.5 Screw spacing on ceilings for the face-ply of two-ply gypsum board over steel framing with adhesive between plies shall be the same as specified for the base-ply gypsum board in 15.3.2 and 15.3.3.

15.3.6 On wall surfaces with adhesive between the plies, the face-ply shall have only a sufficient number of screws to hold gypsum board in place.

16. System IX: Arches and Bending Radii

16.1 Where gypsum board is to be applied to the soffit of arches, it shall be carefully bent into place (see Table 7). If necessary, it first shall either be dampened or cut approximately 1 in. (25.4 mm) on centers on the back side. In the latter case, after the core has been broken at each cut, the gypsum board shall be applied to the curved framing member and fastened in place. At the arises of the arch (exterior or interior "corners" formed at the meeting of the adjoining angle surfaces), joint compound and joint tape or corner bead shall be applied. The tape or corner bead shall be snipped at intervals along one side to permit it to conform to the curved contour.

NOTE 8—To apply the board, place a stop at one end of the curve, then gently and gradually push on the other end of the board, forcing the center against the framing until the curve is complete.

NOTE 9—By thoroughly moistening the face and back paper and allowing the water to soak well into the core, the board will bend to still shorter radii. When the board thoroughly dries, it will regain its original hardness. Any subsequent joint treatment or decoration shall not be started until the gypsum board is thoroughly dry.

17. System X: Application of Gypsum Board to Receive Tile by Adhesive Application

17.1 Framing around tub enclosures and shower stalls shall allow sufficient room to permit the inside lip of the tub, prefabricated receptor, shower pan, or membrane to be applied as shown in Figs. 5, 6, and 7.

NOTE 10—This will necessitate furring out from the framing members the thickness of the gypsum board to be used (½ or ⅝ in. (12.7 or 15.9 mm)) less the thickness of the lip, on each wall abutting a tub receptor or subpan.

17.2 *Blocking or Backing:*

17.2.1 Where framing is spaced more than 16 in. (406 mm) o.c., blocking or backing shall be provided ¾–1¼ in. (20–30 mm) above the top of the tub or receptor and at gypsum board horizontal joints in the area to receive tile.

17.2.2 Interior angles shall be framed or blocked to provide solid backing for interior corners.

17.2.3 Blocking, headers, or supports shall be provided to support the tub and other plumbing fixtures and to receive soap dishes, grab bars, towel racks, and similar items.

17.3 *General:*

17.3.1 Water-resistant gypsum backing board shall be used as a base for adhesive application of ceramic or plastic wall tile in wet areas such as tub and shower enclosures. Either regular gypsum board or water-resistant gypsum backing board shall

TABLE 7 Bending Radii

Gypsum Board Thickness, in. (mm)	Bent Lengthwise, ft (m)	Bent Widthwise, ft (m)
½ (12.7)	10 (3.05)[A]	—
⅜ (9.5)	7 ½ (2.29)	25 (7.62)
¼ (6.4)	5 (1.52)	15 (4.57)

[A] Bending two ¼-in. (6.4-mm) pieces successively permits radii shown for ¼ in. (6.4 mm).

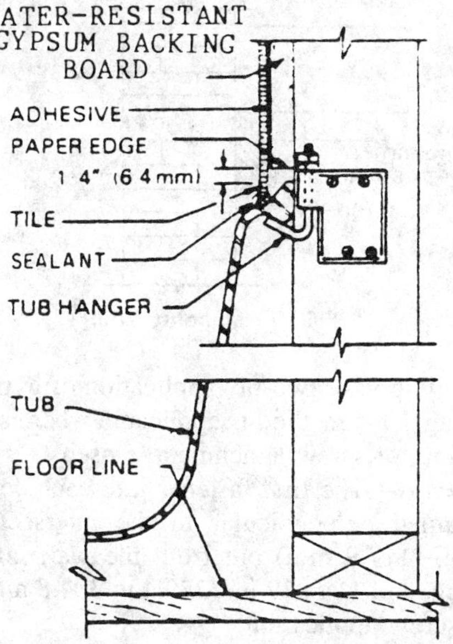

FIG. 5 Application of Gypsum Board at Bathtub Where WR Board Will Receive Ceramic Tile or Other Protective Covering

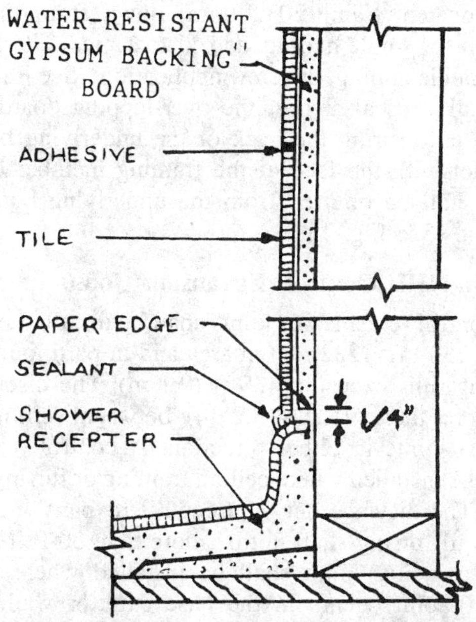

FIG. 6 Adhesive Application of Gypsum Board at Shower Where WR Board Will Receive Ceramic Tile or Other Protective Covering

be used as a base for tile in other areas. Water-resistant gypsum backing board and regular gypsum board shall not be used in extremely critical areas such as saunas, steam rooms, or gang shower rooms. Foil-backed gypsum board shall not be used as a base for tub or shower enclosures. Water-resistant gypsum board used as a base for tile or wall panels in tub and shower enclosures shall not be applied directly over a vapor retarder.

17.3.1.1 Gypsum board shall not be used where there will be direct exposure to water or continuous high humidity conditions such as are found in saunas, steam rooms, gang shower rooms, or indoor pools.

17.3.1.2 Foil-backed gypsum board shall not be used as a base for tub or shower enclosures.

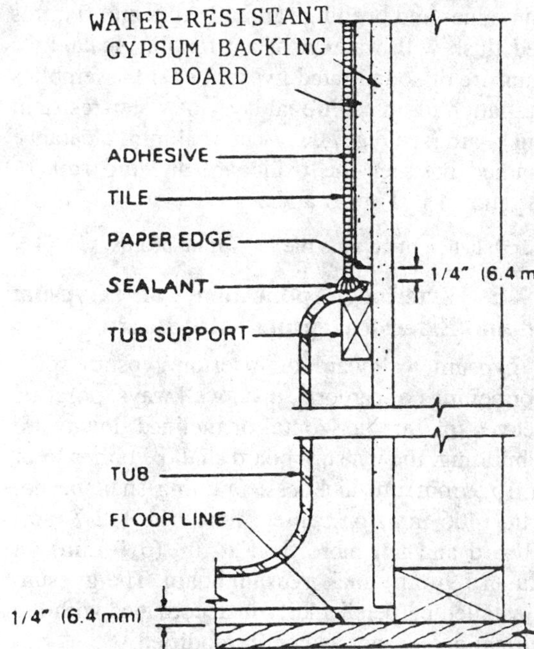

FIG. 7 Adhesive Application of Tile Over Gypsum Board

17.3.1.3 Water-resistant gypsum backing board used as a base for tile or wall panels in tub and shower enclosures shall not be applied directly over a vapor retarder.

17.3.1.4 Water-resistant gypsum backing board ½ in. (12.7 mm) thick shall be permitted to be used as a substrate for the application of ceramic or plastic tile on ceilings with framing spacing not more than 12 in. (305 mm) on centers.

17.3.1.5 Water-resistant gypsum backing board ⅝ in. (15.9 mm) thick shall be permitted to be used as a substrate for the application of ceramic or plastic tile on ceilings with framing spacing not more than 16 in. (406 mm) on centers.

NOTE 11—Asphalt-impregnated felt is not considered a vapor retarder.

17.3.2 All multiple plies of gypsum board in wet areas shall be attached with mechanical fasteners.

17.3.3 Waterproof receptors, pans, or sub-pans shall have an upstanding lip or flange that shall be a minimum of 1 in. (25.4 mm) higher than the water dam or threshold contained in the entry way to the shower.

17.4 *Installation:*

17.4.1 Water-resistant gypsum backing board shall be applied with the factory edge spaced not less than ¼ in. (6.4 mm) above the lip of the receptor, tub, or sub-pan. Shower pans, receptors, or tubs shall be installed prior to the erection of the water-resistant gypsum backing board.

17.4.2 Water-resistant gypsum backing board shall be attached by nails or screws spaced not more than 8 in. (203.2 mm) on centers. When ceramic tile more than ⅜ in. (9.5 mm) thick is to be applied, the nail or screw spacing shall be not more than 4 in. (101.6 mm) on centers.

17.4.3 Where treatment of water-resistant gypsum backing board joints under tile in wet areas (tub and shower enclosures) is required, the compound and application thereof, shall be as required by the joint compound manufacturer for this specific use, or shall be protected from penetration of moisture or water (see X4.2).

17.4.4 All cut edges and openings around pipes and fixtures shall be caulked flush with water-resistant flexible sealant.

17.4.5 Certain fire or sound-rated gypsum board assemblies require the installation of an additional layer of water-resistant gypsum backing board (see Fig. 7). A ¼ in. (6.4 mm) clearance shall be maintained between the wallboard and tub rim, or shower base, as shown in Figs. 6 and 7.

NOTE 12—For installation of ceramic tile to gypsum boards, see X4.3.

18. System XI: Exterior Application of Gypsum Wallboard and Exterior Gypsum Soffit Board

18.1 Where gypsum wallboard or exterior gypsum soffit board is used for ceilings of carports, open walkways, porches, and soffits of eaves that are horizontal or inclined downward away from the building, the gypsum board shall be either ½ or ⅝ in. (12.7 or 15.9 mm) in thickness. Framing shall be not more than 16 in. (406 mm) on centers for ½-in. (12.7-mm) thick gypsum board and not more than 24 in. (610 mm) on centers for ⅝-in. (15.9-mm) thick gypsum board. The gypsum board shall be installed perpendicularly in accordance with the foregoing specifications except as herein modified.

18.2 Suitable facia and molding shall be provided around the perimeter to protect the gypsum wallboard and exterior gypsum soffit board from direct exposure to water. Unless protected by metal or other water stops, the edges of the gypsum wallboard and exterior gypsum soffit board shall be placed not less than ¼ in. (6.4 mm) away from abutting vertical surfaces. (See Figs. 8, 9, and 10.) Joints and fastener heads shall be treated as specified in X3.6.

18.3 Exposed surface of gypsum board and exterior gypsum soffit board shall be treated as specified in X3.6.

18.4 Adequate ventilation shall be provided for the space immediately above installations (see Appendix X2).

19. System XII: Floating Interior Angles

19.1 Gypsum board shall be applied to ceiling first. (See Figs. 11, 12, and 13.)

19.1.1 To minimize the possibility of the fastener popping in areas adjacent to the wall and ceiling intersections, the floating angle method of application shall be permitted to be used for single-ply, or two-ply application of gypsum board to wood framing. This method is applicable where single nailing, double nailing, or screw attachment is used.

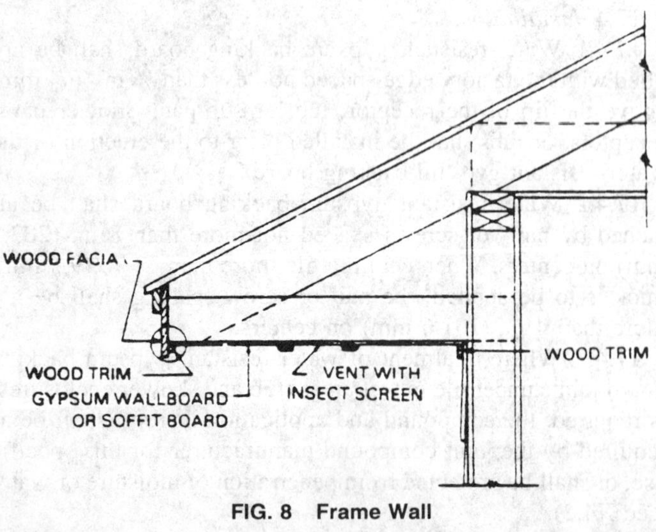

FIG. 8 Frame Wall

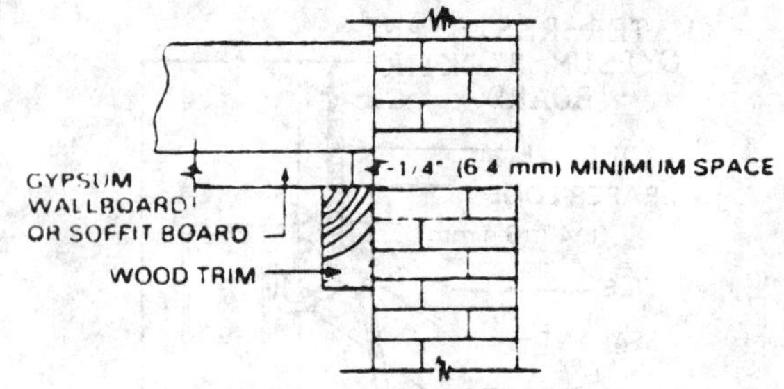

FIG. 9 Masonry Wall

19.2 *Ceilings*—The first fastener into each ceiling framing member framed perpendicular to the intersection shall be located 7 in. (117.8 mm) out from the wall intersection for single nailing, and 11 to 12 in. (279.4 to 304.8 mm) for double nailing, or screw application.

19.3 *Walls*—Gypsum board shall be applied on walls so as to provide support for the floated edges of the ceiling gypsum board. The top fastener into each vertical framing member shall be located 8 in. (203.2 mm) down from the ceiling intersection for single nailing, and 11 to 12 in. (279.4 to 304.8 mm) for double nailing, or screw application. (See Figs. 11 and 12.) At wall vertical angles, the over-lapping board shall be applied so as to bring the back of the underlying board into firm contact with the face of the framing member behind it. Fasteners shall be omitted from the underlying board at the intersection. (See Fig. 13.)

20. System XIII: Control (Expansion) Joints

20.1 Control (expansion) joints shall be installed in ceilings exceeding 2500 ft^2 (232 m^2) in area and in partition, wall and wall furring runs exceeding 30 ft (9.1 m). The distance shall not be more than 50 ft (15.2 m) between ceiling control (expansion) joints in either direction. A control (expansion) joint shall be installed where ceiling framing or furring changes direction. The distance between control (expansion) joints in walls or wall furring shall not be more than 30 ft. (9.1 m). A control (expansion) joint shall be installed where a control (expansion) joint occurs in the base exterior wall. Wall or partition height door frames shall be considered a control joint.

21. System XIV: Foil-Backed Gypsum Board

21.1 The application of foil-backed gypsum board shall conform to the specifications for the application of gypsum board. The reflective surface shall be placed against the face of the framing members. Foil-backed gypsum board shall not be used in the following areas:

21.1.1 As a backing material for tile in wet areas.

21.1.2 For the second ply on two-ply laminating systems.

21.1.3 For laminating directly to masonry or concrete.

21.1.4 In conjunction with electric heating cables.

22. Finishing of Gypsum Wallboard

22.1 *General:*

22.1.1 Compound for taping and finishing shall be either drying or setting types. Drying and setting type compounds

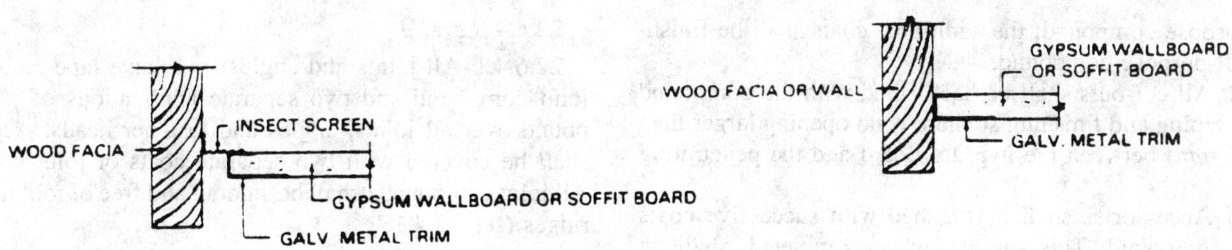

FIG. 10 Alternate Facia Details

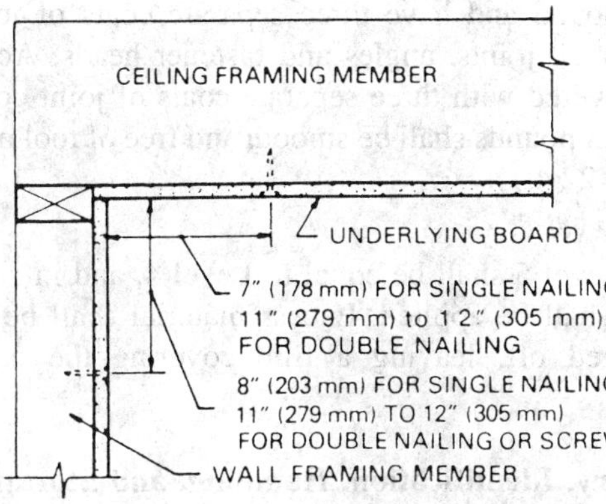

FIG. 11 Vertical Section, Ceiling Framing Member Perpendicular to Wall

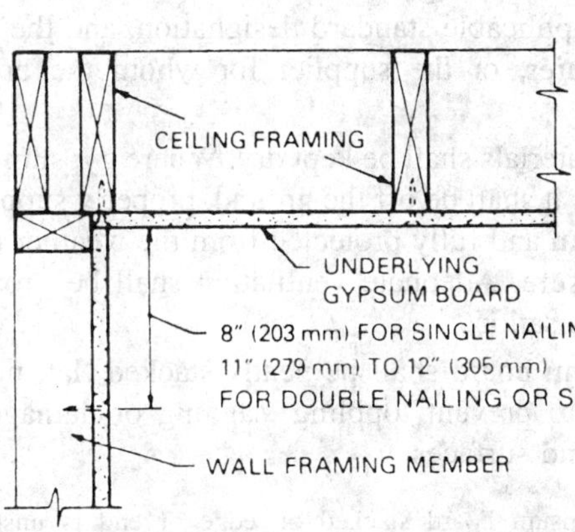

FIG. 12 Vertical Section, Ceiling Framing Parallel to Wall

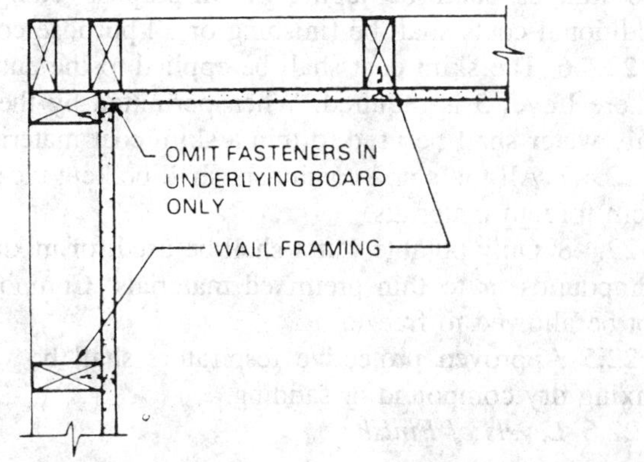

FIG. 13 Horizontal Section Through Interior Vertical Angle

shall not be mixed together unless specified otherwise by the joint compound manufacturer.

22.1.2 When applied, the compounds shall be of a chemical composition compatible with previous and successive coats.

22.1.3 No finishing operation shall be started until the interior temperature has been maintained at a minimum of 50°F (10°C) for a period of at least 48 h and thereafter until the compounds have completely dried.

22.1.4 When two-ply application is used with adhesives between plies, precautions shall be taken to ensure that the adhesive is thoroughly dried before any decorative finish is applied.

22.1.5 Adequate and continuous ventilation shall be provided to ensure proper drying and hardening of the compounds.

22.2 *Surface Preparation:*

22.2.1 Dirt, oil, and other materials that cause a lack of bond shall be removed from all surfaces to receive joint compound.

22.2.2 All dents, gouges, recesses, or other depressions shall be filled with compound to produce a monolithic surface.

22.3 *Taping:*

22.3.1 Joint tape and joint compounds shall be applied using tools designed for this work and the compounds being used. Such tools include hand-held broad knives, trowels or mechanical tools.

22.3.2 The tape shall be applied by applying joint compound to the joint (buttering), pressing in the tape and wiping off the excess compound, or by mechanical tools designed for this purpose. Sufficient compound shall be left under the tape to provide a bond of the tape to the gypsum board.

22.4 *Finishing:*

22.4.1 Finishing compound and all-purpose compound shall be applied with tools of sufficient width to extend a minimum of 3½ in. (89 mm) beyond both sides of the center of the joint tape. The compound shall be drawn down to a level plane. After the compound has dried, the surface shall be sanded (see 22.5) or wiped with a dampened sponge to eliminate high spots and excessive compound.

22.4.1.1 Non-setting type compounds shall be allowed to dry thoroughly between coats before sanding or the application of additional coats.

22.4.1.2 Additional coats of compounds shall be permitted to be applied over setting-type compound after it has set and before it dries completely.

22.4.2 Additional coats of finishing compound shall be applied with tools that will feather the material not less than 6 in. (152 mm) beyond both sides of the center of the joint tape. The compound shall be drawn down to a level plane, leaving a monolithic surface.

NOTE 13—A smooth surface is achieved by lightly sanding or wiping the joint compound with a dampened sponge. Do not raise the nap of the paper of the gypsum board (see 22.5). See Appendix X3.

22.4.3 Fastener heads shall be covered (spotted) with successive coats of joint compound. The first coat shall be taping or all-purpose compound; the additional coats shall be finishing or all-purpose compound.

22.4.4 All cut-outs shall be back filled with the compound used for taping and finishing so there is no opening larger than ¼ in. (6 mm) between the gypsum board and the penetrating element.

22.4.5 Accessories shall be finished with successive coats of joint compound. This surface, when completed, shall be flush with the surface of the gypsum board. The first coat over the flanges shall be taping or all-purpose compound; the additional coats shall be finishing or all-purpose compound.

22.4.6 The skim coat shall be applied to the entire surface where Level 5 is required. When permitted by the producer, only water shall be used to thin a skim coat material.

22.4.7 All tools and containers shall be kept clean and free from foreign materials.

22.4.8 Only potable water shall be used for mixing powder compounds or to thin premixed materials. Compounds shall not be allowed to freeze.

22.5 Approved protective respirators shall be worn when mixing dry compound or sanding.

22.6 *Levels of Finish:*

NOTE 14—The required level or extent of finishing of gypsum wallboard joints, fastener heads and overall surface can vary with the location in a structure and the intended type of decoration. This section describes various levels of finishing, that is, number of applications of joint compound, sanding or other finishing techniques; the recommendations of manufacturers may vary from what is specified herein and as such are not part of these specifications. The relationship of levels of finishing with location and intended decoration is described in Appendix X8.

22.6.1 *Level 0:*

22.6.1.1 No taping, finishing or corner beads are required.

22.6.2 *Level 1:*

22.6.2.1 All joints and angles shall have tape embedded in joint compound. Surface shall be free of excess joint compound. Tool marks and ridges shall be acceptable.

22.6.3 *Level 2:*

22.6.3.1 All joints and angles shall have tape embedded in joint compound and have one separate coat of joint compound over the tape and fastener heads. Accessories shall be covered by one coat of joint compound. Surface shall be free of excess joint compound. Tool marks and ridges are acceptable.

22.6.3.2 Joint compound applied over the body of the tape at the time of tape embedment shall be considered a separate coat of joint compound and shall satisfy the conditions of this level.

22.6.4 *Level 3:*

22.6.4.1 All joints and angles shall have tape embedded in joint compound and two separate applications of joint compound over all joints, angles and fastener heads. Accessories shall be covered with two separate coats of joint compound. All joint compounds shall be smooth and free of tool marks and ridges (see 22.4.1.1).

22.6.5 *Level 4:*

22.6.5.1 All joints and angles shall have tape embedded in joint compound and have three separate coats of joint compound over all joints, angles and fastener heads. Accessories shall be covered with three separate coats of joint compound. All joint compounds shall be smooth and free of tool marks and ridges (see 22.4.1).

22.6.6 *Level 5:*

22.6.1 Level 5 shall be equal to Level 4, and in addition, a skim coat shall be applied. Excess material shall be immediately sheared off, leaving a film covering the paper (see 22.4.1.1).

23. Delivery, Identification, Handling, and Storage

23.1 All materials shall be delivered in the original factory sealed, unopened packages, containers, or bundles bearing the brand name, applicable standard designation, and the name of the manufacturer, or the supplier for whom the product is manufactured.

23.2 All materials shall be kept dry. Where gypsum board is stored outside, it shall be off the ground, properly supported on a level platform and fully protected from the weather or direct sunlight exposure. Adequate ventilation shall be provided to prevent condensation.

23.3 Gypsum board shall be neatly stacked flat, not on its end or edge, to prevent toppling, sagging or damage to the ends, edges, and surfaces.

NOTE 15—Gypsum board stacked on edge or end is unstable and presents a serious hazard in the workplace should it accidentally topple.

23.4 Ready mixed and dry compounds which have been mixed with water shall be stored at temperatures above freezing or temperatures specified by the manufacturer.

24. Keywords

24.1 ceiling; gypsum; wallboard; walls

APPENDIXES

(Nonmandatory Information)

These Appendixes give general information and also suggestions for inclusions to be made elsewhere by the specifier. They are not a part of this specification.

X1. GENERAL INFORMATION

X1.1 Scaffolding shall be constructed and maintained in strict conformity with applicable laws and ordinances.

X1.2 This work shall be properly coordinated with the work of other trades.

X1.3 Gypsum board shall be protected from the elements of weather before, during, and after application.

X2. VENTILATION ABOVE GYPSUM BOARD CEILINGS

X2.1 Attics or similar unheated spaces above gypsum board ceilings shall be adequately ventilated by providing effective cross ventilation for all spaces between roof and top floor ceilings by screened louvers or other approved and acceptable means.

X2.2 The ratio of total net free ventilating area to ceiling area shall be not less than 1:150, except that the ratio may be 1:300 provided:

X2.2.1 A vapor barrier having a transmission rate not exceeding 1 perm is installed on the warm side of the ceiling, or

X2.2.2 At least 50 % of the required ventilating area is provided by ventilators located in the upper portion of the space to be ventilated (at least 3 ft (914.4 mm) above eave or cornice vents) with the balance of the required ventilation area provided by eave or cornice vents.

X2.3 Attic space that is accessible and suitable for future habitable rooms or walled-off storage space shall have at least 50 % of the required ventilating area located in the upper part of the ventilated space as near the high point of the roof as practicable and above the probable level of any future ceilings.

X3. JOB APPLIED DECORATION

X3.1 Prolonged exposure of gypsum board to sunlight may cause problems in decoration.

X3.2 With the joints and fastener head depressions treated as specified in Section 10, interior walls of gypsum board may be decorated in any of the popular variety of finishes, such as texture or stipple, flat paint or flat enamel paint, wall paper, or vinyl wall coverings.

X3.3 Because the porosity and texture of the gypsum board differs from that of the joint treatment, the surface shall be primed and sealed as may be required for the subsequent finish coats.

X3.4 In rooms where high humidity may be encountered, such as the kitchen, bath, or utility room, a flat or semigloss enamel finish is recommended.

X3.5 Care should be exercised in the selection of primer and sealer paints to make sure they will perform satisfactorily, and fulfill the following functions:

X3.5.1 Equalize variations of suction over the entire surface.

X3.5.2 Provide a bonding surface or "tooth" for the paint to be applied.

X3.5.3 Avoid nap raising.

X3.6 Before applying the sealer, remove all loose dirt and dust by brushing with a soft brush or by rubbing with a dry cloth. Be sure the joint treatment is thoroughly dry before any application of sealer or paint.

X3.7 In applying primers or sealers, apply sufficient quantity to assure that the surface is completely covered. Follow the manufacturer's printed directions and do not over thin. It is good practice to tint the sealer to approximately the shade of the finish coat. This will lead to better results in the finished job.

X3.8 In all cases where deep tones are to be used in the finish paint, best results will be achieved if the surface is first sealed. More than one coat of sealer may be necessary. Each coat must be thoroughly dry before applying another.

X3.9 Under normal atmospheric conditions, a waiting period of 12 to 18 h after application of primer-sealer should be observed before decoration is applied. In rainy, humid, and cold weather, a longer waiting period; sometimes as long as 36 to 48 h; may be necessary to make certain the sealer coat is absolutely dry.

X3.10 Exposed surfaces of gypsum board, as specified in Section 10 shall be painted with not less than two coats of exterior paint.

X3.11 Where semi-gloss or high-gloss paints are required or where severe lighting conditions occur, and prior to the application of a primer, a skim coat shall be applied to the entire surface of the wallboard, including the taped joints, covered fasteners, and accessories, to reduce the absorption and texture difference between the joint compound and the face of the wallboard.

X4. APPLICATION OF SURFACING MATERIALS AND CAULKING

X4.1 Caulk all cut edges and openings around pipes and fixtures flush with waterproof, nonhardening caulking compound or Type I adhesive complying with ANSI A136.1.

X4.2 Wall tile shall be used in combination with adhesive to protect the gypsum board and any water sensitive materials if present (such as joint compound) from penetration of moisture or water. Responsibility for performance of completed installation shall rest with the surface materials manufacturer or with the tile applicator.

X4.3 For application of ceramic, plastic, or metal wall tile, or plastic finished rigid wall panels, or other types of surfacing materials over gypsum board in wet or dry areas, the recommendations of the manufacturer of tile, wall panel, or other surfacing materials shall be followed.

X4.4 Apply the surfacing material down to the top surface or edge of the finished shower floor, return, or tub, or install so as to overlap the top lip of receptor, sub-pan, or tub. It shall completely cover the following areas:

X4.4.1 *Over Tubs without Showerheads*—Six in. (152.4 mm) above the rim of the tub.

X4.4.2 *Over Tubs with Showerheads*—A minimum of 5 ft (1.52 m) above the rim or 6 in. (152.4 mm) above the height of the showerhead, whichever is higher.

X4.4.3 *Shower Stalls*—A minimum of 6 ft (1.83 m) above the shower dam or 6 in. (152.4 mm) above the showerhead, whichever is higher.

X4.4.4 All gypsum board window sills and jambs in shower or tub enclosures shall be covered to a like height.

X4.4.5 The surfacing material shall be applied to the full specified height for a distance of at least 4 in. (101.6 mm) beyond the external face of the tub or receptor. Areas beyond an exterior corner are excluded.

X4.5 Where plastic finished rigid wall panels are used as a surfacing material, the following precautions shall be taken:

X4.5.1 The type and shape of moldings recommended by the manufacturer of the surfacing material shall be used. Recommended tub moldings shall be used at the base where the surfacing materials abut the tub, shower, floor, or curb. Such moldings shall be set in waterproof, nonhardening caulking compound.

X4.5.2 Joints shall be filled in such a manner as to leave no voids for water penetrations.

X4.5.3 A bead of adhesive shall be applied as a dam between the back surface of the finishing material and the tub or receptor to prevent any leakage of water at the joint.

X5. WOOD FRAMING REQUIREMENTS

X5.1 The following requirements should be included in the project specifications for framing and furring and are essential to provide a proper base to receive the gypsum board.

X5.2 All framing members to which gypsum board will be fastened shall be straight and true. Framing shall be in alignment and spaced not to exceed the maximum spacings shown in Table 1. Framing, bridging, and furring members shall be the proper grade for the intended use and members 2 by 4 in. nominal size or larger shall bear the grade mark of a recognized inspection agency. Framing, bridging, and furring shall be adequate to meet the design, or code loading, or both. Where there is an applicable local code, the framing shall be in accordance with the Manual for House Framing by the National Lumber Manufacturers. Supports shall be provided as necessary for the support of fixtures.

X5.3 When gypsum board is nailed to wood cross furring on ceilings, these furring members shall have a minimum cross section of 1½ by 1½ in. (38.1 by 38.1 mm) actual size and be spaced in accordance with 9.2 and 9.3. Where screw application is used, the furring member may be ¾ in. by 2½ in. (19.1 by 63.5 mm) actual size.

X5.4 Where wood furring is used over masonry or concrete, fasteners should be of a length that does not come into contact with the masonry surface.

X5.5 Insulating blankets or flanges of blankets shall not be applied over framing members that are to receive gypsum board.

X5.6 Foil-backed gypsum board may be used where a vapor retarder is required.

X6. PRECAUTIONS TO MINIMIZE POTENTIAL OF SAGGING

X6.1 Ensure framing spacing is adequate for thickness of board to be used. Ensure board is applied perpendicular to framing.

X6.2 Determine that excessive weight of insulation will not be added.

X6.3 Control the relative humidity within the structure by providing adequate ventilation before, during, and after board application. Watch for pouring of basement floors after board application.

X6.4 In cold weather, maintain inside temperature between

50°F (10°C) and 70°F (20°C). Where portable heaters are used, make sure to remove the extra humidity they produce.

X6.5 Ensure the gypsum board is thoroughly dry and at ambient temperature before application.

X6.6 Ensure joint treatment is thoroughly dry before applying any decoration.

X6.7 Ensure that primer and paint coats are dry before the application of successive coats.

X6.8 Where gypsum board supports ceiling insulation and is finished with a water-based texture material, either sprayed or hand applied, use only ⅝-in. (15.9-mm) gypsum board applied perpendicular to the framing members. Where severe adverse conditions are anticipated, reduce framing spacing to 16 in. (400 mm).

X7. ELECTRIC RADIANT HEATING SYSTEMS FOR GYPSUM BOARD CEILINGS

X7.1 The application of gypsum board under this specification shall be performed in accordance with Specification C 840, except as herein modified.

X7.2 Gypsum boards shall have no less thickness than that specified for perpendicular application for the ceiling framing spacing. Both layers shall be applied perpendicularly to framing members. Attachment shall be as required in Section 9.

X7.3 Electric heating cables shall be securely attached to gypsum board in accordance with the recommendations of the cable manufacturer and the National Electrical Code by other trades.

X7.4 Cables shall be parallel to and between framing members with at least 1¼-in. (31.8-mm) clearance from center of framing member on each side so that at least 2½-in. (63.5-mm) wide unobstructed strip is provided under each framing member.

X7.5 Cables shall cross framing members only at ends of ceiling 4 to 6 in. (101.6 to 152.4 mm) from the wall. There shall be at least a 4 in. (101.6 mm) space completely around the perimeter of each ceiling that is clear of cables. Cables shall be kept at least 8 in. (203.2 mm) clear of all openings.

X7.6 All inspections and testing of the heating system shall be completed before application of the noninsulating filler.

X7.7 Under no operating conditions shall the gypsum board core be exposed to a temperature exceeding 125°F (51.7°C) unless otherwise specifically recommended by the gypsum board manufacturer.

X7.8 Complete embedding of the heating cable is necessary. The area between the heating cables shall be filled with a nonshrinking, noninsulating filler that shall be leveled.

X7.9 If a face layer of gypsum board is to be applied then the nails or screws around the perimeter of the ceilings shall be spaced 8 to 10 in. (203.2 to 254.0 mm) away from the wall to prevent striking the heating elements where they cross the framing members.

X7.9.1 All joints and heads of fasteners shall be finished in accordance with the recommendations of the gypsum board manufacturer. Allow a minimum of one week with good drying conditions and two weeks in cold conditions before operating the heating system.

X7.10 If gypsum board is not used as a finish, a special spray texture for radiant heat ceilings may be applied.

X7.11 *Design Data:*

X7.11.1 It is the responsibility of the architect to prepare and correlate complete plans and specifications.

X7.11.2 It is the responsibility of the heating engineer to properly design the radiant heating system to work effectively with these materials.

X7.11.3 The data in Table X7.1 are recommended for use in calculating the heat transmission coefficient (U) for gypsum board constructions.

X7.12 *Decoration:*

X7.12.1 Decoration of the finished work, including sizing or sealing shall not proceed until the heating system has been test operated for a minimum of 24 h.

X7.12.2 The heating system should then be turned off and allowed to cool before decorating.

X7.12.3 Where texture finishes are used, such finishes must be allowed to thoroughly set or dry prior to the operation of the heating system.

TABLE X7.1 Thermal Conductivity Values

Thickness, in. (mm)	Conductance *C*		Resistance *R*	I/C
	Btu/h · ft² · °F	W/m²k	°F · h · ft²/Btu	k · m²/w
⅜ (9.5)	3.10	17.6	0.32	0.057
½ (12.7)	2.22	12.6	0.45	0.079
⅝ (15.9)	1.78	10.1	0.56	0.099
1 (25.4)	1.20	6.8	0.83	0.147

X8. LEVELS OF FINISH

NOTE X8.1 Joint compounds used to conceal the joints and fastener heads can vary in density and surface character from the adjacent wallboard surface. Terms such as "industry standard" or "workman-like finish" are often used but not specific and can lead to subjective interpretation.

The architect, general contractor and owners frequently anticipate or expect a higher level of finish than the sub-contractor can determine from the specifications that are furnished for the bid process.

Reasons for specifying levels are: (*1*) many building are designed with walls and partitions abutting window mullions, long hallways, or atriums with large surface areas flooded with artificial and natural lighting; (*2*) paints and painting methods have changed over the years with wider use of spray equipment and reduce solids in paint; and (*3*) differences in absorption and drying, shrinkage of the compound, tool marks and ridges in the compound, or scuffing or raising of the nap of the face paper may all, or in part, combine to emphasize visual differences.

It is recommended that the painting specification provide for proper treatment to ensure uniform absorption over the entire surface. Any special finish such as gloss, which demands a more stringent requirement than the standard surface provides, should be specifically pointed out in the design process.

X8.1 *Level 0:*

X8.1.1 As stated, there is no finish required. To be used in temporary construction or whenever the final finish level has not been determined.

X8.1.2 Also could be used where non-predecorated panels will be used in demountable type partitions which are to be painted as a final finish.

X8.2 *Level 1:*

X8.2.1 Frequently used in plenum areas above ceilings, in attics, in areas where the assembly would generally be concealed, or in building service corridors and other areas not normally open to public view. Some degree of sound and smoke control is provided; in some geographic locations referred to as fire-taping. Where a fire resistance rating is required for gypsum board assembly, details of construction shall be in accordance with reports of fire tests of assemblies that have met the requirements of the fire rating imposed.

X8.3 *Level 2:*

X8.3.1 May be used when Water Resistant Gypsum Backing Board, Specification C 630, is used as a substrate, or in garages, warehouse storage or other similar areas where surface appearance is not of primary concern.

X8.4 *Level 3:*

X8.4.1 Used in appearance areas that are to receive heavy texture (spray or hand applied) finishes before final painting. Should not be used under smooth painted surfaces.

X8.4.2 A high quality, high solids (undiluted) primer must be used to equalize absorption differences between the surface paper and the joint compound.

X8.5 *Level 4:*

X8.5.1 This method should be used where wallcoverings, flat paints, or light textures are specified.

X8.5.2 A high quality, high solids (undiluted) primer must be used to equalize the absorption of paper surface and the finished joints. Release agents for wall coverings are specifically formulated to minimize damage if coverings are subsequently removed.

X8.5.3 The weight texture and sheen level of the wallcovering selected should be taken into consideration when specifying wallcoverings over this level of drywall treatment. Joints and fasteners must be sufficiently concealed if the material is lightweight, contains limited patterns, has a glossy finish or any combination of these areas. Unbacked vinyls are not recommended.

X8.6 *Level 5:*

X8.6.1 The highest quality finish is the most effective method to minimize the possibility of highlighting or photographing of the treated joints and fasteners showing through the final decoration on all smooth-wall work.

X8.6.2 It is a requirement where glossy paints are specified or where severe lighting occurs.

NOTE X8.2—A high quality, high solids (undiluted) primer must be used prior to final decoration. A primer recommended by the paint or gypsum wallboard producer should be used where the specifications call for semi-gloss, gloss, or enamel paints.

Designation: C 841 – 98

Standard Specification for Installation of Interior Lathing and Furring[1]

This standard is issued under the fixed designation C 841; the number immediately following the designation indicates the year of original adoption or, in the case of revision, the year of last revision. A number in parentheses indicates the year of last reapproval. A superscript epsilon (ε) indicates an editorial change since the last revision or reapproval.

This standard has been approved for use by agencies of the Department of Defense.

1. Scope *

1.1 This specification covers the minimum requirements for, and the methods of application of, interior lathing and furring to receive gypsum plaster applied in accordance with Specification C 842. Other materials are not prohibited from being used provided that their physical characteristics and durability under conditions of usage are at least equal in performance to those described.

1.2 The values stated in inch-pound units are to be regarded as the standard. The metric values shown in parentheses are provided for information only.

1.3 Details of construction for a specific assembly to achieve the required fire resistance shall be obtained from reports of fire-resistance tests, engineering evaluations, or listings from recognized fire testing laboratories.

1.4 Where a specific degree of sound control is required for plastered assemblies and constructions, details of construction shall be obtained from reports of sound tests of assemblies that have met the requirements of Test Method E 492.

2. Referenced Documents

2.1 *ASTM Standards:*

A 641 Specification for Zinc-Coated (Galvanized) Carbon Steel Wire[2]

A 641M Specification for Zinc-Coated (Galvanized) Carbon Steel Wire [Metric][2]

C 11 Terminology Relating to Gypsum and Related Building Materials and Systems[3]

C 37 Specification for Gypsum Lath[3]

C 423 Test Method for Sound Absorption and Sound Absorption Coefficients by the Reverberation Room Method[4]

C 473 Test Methods for Physical Testing of Gypsum Board Products and Gypsum Lath[3]

C 514 Specification for Nails for the Application of Gypsum Board[3]

C 754 Specification for Installation of Steel Framing Members to Receive Screw-Attached Gypsum Panel Products[3]

C 842 Specification for Application of Interior Gypsum Plaster[3]

C 847 Specification for Metal Lath[3]

C 933 Specification for Welded Wire Lath[3]

C 954 Specification for Steel Drill Screws for the Application of Gypsum Board or Metal Plaster Bases to Steel Studs from 0.033 in. (0.84 mm) to 0.112 in. (2.84) in Thickness[3]

C 1002 Specification for Steel Drill Screws for the Application of Gypsum Board or Metal Plaster Bases[3]

C 1007 Specification for Installation of Load Bearing (transverse and Axial) Steel Studs and Related Accessories[3]

C 1032 Specification for Woven Wire Plaster Base[3]

D 1784 Specification for Rigid Poly(Vinyl Chloride) (PVC) Compounds and Chlorinated Poly(Vinyl Chloride) (CPVC) Compounds[5]

D 3678 Specification for Rigid Poly(Vinyl Chloride) (PVC) Interior-Profile Extrusions[6]

E 84 Test Method for Surface Burning Characteristics of Building Materials[7]

E 90 Test Method for Laboratory Measurement of Airborne Sound Transmission Loss of Building Partitions[4]

E 119 Test Methods for Fire Tests of Building Construction and Materials[7]

E 492 Method for Laboratory Measurement of Impact Sound Transmission Through Floor-Ceiling Assemblies Using the Tapping Machine[4]

2.2 *National Association of Architectural Metal Manufacturers Standard*

HMMA 840 Installation and Storage of Hollow Metal Doors and Frames[8]

3. Terminology

3.1 *Definitions*—For definitions relating to ceilings and walls see Terminology C 11.

3.2 *Definitions of Terms Specific to This Standard:*

[1] This specification is under the jurisdiction of ASTM Committee C-11 on Gypsum and Related Building Materials and Systems, and is the direct responsibility of Subcommittee C11.03 on Specifications for Application of Gypsum and Other Products in Assemblies.

Current edition approved Oct. 10, 1998. Published May 1999. Originally published as C 841 – 76. Last previous edition C 841 – 97.

[2] *Annual Book of ASTM Standards*, Vol 01.06.

[3] *Annual Book of ASTM Standards*, Vol 04.01.

[4] *Annual Book of ASTM Standards*, Vol 04.06.

[5] *Annual Book of ASTM Standards*, Vol 08.01.

[6] *Annual Book of ASTM Standards*, Vol 08.04.

[7] *Annual Book of ASTM Standards*, Vol 04.07.

[8] Available from NAAMM, 8 South Michigan Avenue, Suite 1000, Chicago, IL 60603.

***A Summary of Changes section appears at the end of this standard.**

3.2.1 *base coat, n*—the sum of the scratch and brown coats or the total coats in place prior to application of finish coats.

3.2.2 *contact ceiling, n*—a ceiling in which the lath is attached in direct contact with the construction above, without the use of main runners or cross furring.

3.2.3 *furred ceiling, n*—a ceiling in which the furring used for the support of the lath is attached directly to the structural members of the building.

3.2.4 *suspended ceiling, n*—a ceiling in which the main runners and cross furring are suspended below the structural members of the building.

3.2.5 *main runners, n*—the members that are attached to or suspended from the construction above for the support of cross furring.

3.2.6 *cross furring, n*—furring members that are attached at right angles to the underside of the main runners or construction above for support of the lath.

3.2.7 *furring , n*—spacer strips fastened to a wall, ceiling, or planar element that create an even surface for the application of metal plaster bases or gypsum lath.

3.2.8 *primary members, n*—the members (main runners or structural supports) to which the cross furring is attached.

3.2.9 *metal plaster bases, n*—expanded metal lath, sheet metal lath, welded or woven wire lath.

3.2.10 *face side (gypsum lath), n*—the side opposite the paper cover seam laps.

3.2.11 *surface transition, n*—the change in width of a surface, usually as a result of penetrations (doors, windows) or architectural demands.

4. Delivery of Materials

4.1 All materials shall be delivered in the original packages, containers, or bundles bearing the brand name and manufacturer's (or supplier's) identification.

5. Storage of Materials

5.1 All materials shall be kept dry, preferably by being stored inside. Where necessary to be stored outside, materials shall be stacked off the ground, supported on a level platform, and protected from the weather and surface contamination.

5.2 Materials shall be neatly stacked flat with care taken to avoid damage to edges, ends, or surfaces.

6. Materials

6.1 Wire Specification A 641 and A 641M, Class 1 coating (galvanized) soft temper.

6.1.1 *Tie Wire and Hanger Wire*—Zinc-coated (galvanized) soft-annealed steel, or a material having equivalent size, corrosion resistance and strength. Wire diameters (uncoated) specified herein correspond with United States steel wire gage numbers as follows:

Diameter, in.[A]	mm	Wire Gage (U.S. Steel Wire Gage)
0.0348	0.88	No. 20
0.0410	1.04	No. 19
0.0475	1.21	No. 18
0.0540	1.37	No. 17
0.0625	1.59	No. 16
0.0800	2.03	No. 14
0.0915	2.32	No. 13
0.1055	2.68	No. 12
0.1205	3.06	No. 11
0.1350	3.43	No. 10
0.1483	3.77	No. 9
0.1620	4.11	No. 8

[A]Allowable variations in diameter shall be in accordance with tolerances as established in A 641 or A 641M.

6.2 *Gypsum Lath*— Specification C 37.

6.3 *Expanded Metal Lath*— Specification C 847.

6.4 *Wire Lath*:

6.4.1 *Welded Wire Lath*, Specification C 933.

6.4.2 *Woven Wire Lath*, Specification C 1032.

6.5 *Fasteners*:

6.5.1 *Nails for Attaching Gypsum Lath*— Specification C 514. Nail size shall be in accordance with Table 1.

6.5.2 *Nails for Attaching Metal Plaster Base*—Galvanized roofing nails or common nails. length shall be as specified in 7.

6.5.3 *Screws*—Specifications C 954 or C 1002.

6.5.3.1 Screws for attaching gypsum lath shall be sized in accordance with Table 1.

6.5.3.2 Screws for attaching metal plaster base shall have a 7⁄16 in. (11.1 mm) diameter pan washer head and a 0.120 in. (3.0 mm) diameter shank. The length shall be sufficient enough to allow for the threaded portion of the screw to project not less than 3⁄8 in. (9.5 mm) through steel framing when the lath is installed.

6.5.4 *Staples*—Staples for attaching gypsum lath shall conform to the requirements of Table 1.

6.5.4.1 Staples for attaching metal plaster base shall be 0.062 in. (1.6 mm) diameter, round or flattened steel wire with 3⁄4 in. (19.0 mm) crown. Lengths shall be as specified for individual systems.

6.6 *Paper or Other Backing*—When used on metal plaster bases, over solid backing, backing shall be securely held in place by, or attached to the metal plaster base. Metal plaster base shall allow for 1⁄4 in. (6.4 mm) self furring. The backing shall permit embedment of plaster.

6.7 *Accessories*—Shall be steel, coated, with a protective rust inhibitive coating after fabrication, or fabricated from zinc-coated (galvanized) steel, aluminum, zinc alloy, or high impact plastic. Plastic accessories shall be nmanufactured in accordance with Specifcation D 1784 or Specification D 3678. Aluminum accessories shall be clear plastic coated. All accessories shall have perforated or expanded flanges or clips shaped so as to permit complete imbedment in the plaster, to provide means for secure alignment, and to secure attachment of the accessory to the underlying surface. Accessories shall be designed for the intended use and to receive or to permit application of the specified plaster thickness.

6.7.1 Thickness of base material for accessories shall be not less than as shown in the following:

Base Material (mm)	Thickness of Base Material, in.
Uncoated steel/zinc	0.014 (0.36)
PVC	0.030 (0.76)
Aluminum	0.050 (1.27)

6.7.2 *Control Joints*— Shall be formed by using a single

TABLE 1 Screws, Nails and Staples[A,B] for Attaching Gypsum Lath to Horizontal and Vertical Wood Supports

Maximum Spacing, in. (mm)					Minimum Gage Requirements, in. (mm)			
Width of Lath	Thickness of Lath	Distance Between Supports	Number of Attachments per Bearing	Approximate Spacing *c* to *c* of Attachments	Length of Leg	Depth of Support Penetration	Diameter of Flat Head or Blued Nails or Crown Width of Staples[A]	Gage of Shank of Nails or Staples[A]
16 (406)	⅜ (9.5)	16 (406)	4	5 (127)	1⅛ (28.6) 1 (25.4)[A]	¾(19) ⅝(15.9)[A]	19⁄64 (7.5) 7⁄16 (11)[A]	13 16[A]
24 (610)	⅜ (9.5)	16 (406)	6	4½ (114)	1⅛ (28.6) 1 (25.4)[A]	¾ (19) ⅝(15.9)[A]	19⁄64 (7.5) 7⁄16 (11)[A]	13 16[A]
16 (406)	½ (12.7)	24 (610)	4	5 (127)	1¼ (32) 1⅛ (28.6)[A]	¾(19) ⅝ (15.9)[A]	19⁄64 (7.5) 7⁄16(11)[A]	13 16[A]
24 (610)	½ (12.7)	24 (610)	6	4½ (114)	1¼ (32) 1⅛ (28.6)[A]	¾ (19) ⅝ (15.9)[A]	19⁄64 (7.5) 7⁄16(11)[A]	13 (330) 16[A]
48 (1219)	⅜ (9.5) ½ (12.7)	16 (406) 24 (610)	8	4½ (114)	1⅛ (28.6) 1 (25.4)[A] 1¼ (32) 1⅛ (28.6)[A]	¾ (19) ⅝ (15.9)[A] ¾ (19) ⅝(15.9)[A]		

[A] Galvanized staples.

[B] Screws for attaching gypsum lath to wood supports shall be Type W as described in Specification C 1002.

prefabricated member or fabricated by installing casing beads back to back with a flexible barrier membrane behind casing beads. The separation spacing shall be not less than ⅛ in. (3.2 mm) or as required by the anticipated thermal exposure range.

6.7.3 *Cornerite*—shall be not less than 1.75 lb/yd^2 (0.85 kg/m^2) expanded metal lath, galvanized, or given a protective coating of rust inhibitive paint after fabrication, or woven or welded fabric not less than 0.041 in. (1.04 mm) thick an weighing not less than 1.75 lb/yd^2 (0.85 kg/m^2), fabricated from galvanized wire. When shaped for angle reinforcing, cornerite shall have outstanding legs of not less than 2 in. (50.8 mm).

6.7.4 *Channels*—Shall be cold-rolled steel, free of rust with a coating of rust inhibitive paint or galvanized. Channels used in areas subject to corrosive action of salt air or other corrosive environments, shall be hot-dip galvanized. Weight of the channel shall be not less than as shown in the following:

Size		Weight	
in.	mm	lb/1000 lf	kg/m
¾	19	300	0.446
1 ½	38	475	0.707
2	50	590	0.818
2½	64	665	0.989

6.7.5 Rod and flat (strap) hangers shall be mild steel, zinc coated or protected with a rust-inhibiting paint.

6.7.6 *Clips*—Devices for attaching framing members to supports or to each other; for attaching lath to framing members; or for securing lath to lath, shall be formed of zinc-coated (galvanized) steel wire or sheet, and of sufficient size and strength to support the intended load

6.7.7 *Striplath*—Shall be not less than 1.75 lb/yd^2 (0.85 kg/m^2) expanded metal lath, or woven or welded fabric of not less than 0.041 in. (1.04 mm) thick weighing not less than 1.75 lb/yd^2 (0.85 kg/m^2) fabricated from galvanized steel.

7. Installation

7.1 *General Requirements for Application of Lath and Accessories*:

7.1.1 Lath shall be installed with the long dimension at right angles to supports, unless otherwise specified.

7.1.2 Corner beads shall be installed to protect all external corners and to establish grounds. Corner beads shall be attached to the substrate in such a manner as to ensure proper alignment during application of plaster.

7.1.3 Metal accessories shall be installed so that flanges and clips provided for their attachment are covered by the base coat plaster.

7.1.4 *Control Joints*—Shall be installed in ceilings greater than 2500 ft^2(232 m^2) in area and in partition, wall, and wall furring runs greater than 30 ft (9 m) in length. The distance between ceiling control joints shall not be more than 50 ft (15 m) in either direction. A control joint shall be installed where the ceiling framing or furring changes direction. The distance between control joints in walls or wall furring shall be not more than 30 ft (9 m). A control joint shall be installed where an expansion joint occurs in the base exterior wall or ceiling. A ceiling height door frame, installed in accordance with the spacing requirements set forth in this section, is not prohibited from being used as a control joint.

7.1.5 Where partition assemblies require the installation of control joints over the head of door frames to relieve stresses within the assemblies, such control joints shall be installed in the following manner:

7.1.5.1 Where control joints are located at the vertical center of the door opening two metal studs shall be installed not more than ½ in. (12.7 mm) apart over the center of the span at the head of the frame, with the lath installed so that a vertical joint occurs between the studs. A prefabricated control joint or casing beads shall be installed at the joint location.

7.1.5.2 Where control joints are located in line with the door jambs the procedure in 7.1.5.1 shall be followed with the exception that supporting members shall consist of two studs placed side-by-side.

7.1.6 Casing beads or other suitable accessories, shall be used to isolate non-load bearing partitions from load bearing members.

7.1.7 *Workmanship*—Metal construction furring and lathing shall be erected so that finished plaster surfaces will be true to line, level, plumb, square, curved, or as required to receive specified plaster thickness.

7.1.8 Except where frames are required to be fully grouted with plaster, lath shall be inserted into jambs as far as possible and notched to pass anchors.

7.2 *Application of Gypsum Lath to Supports*:

7.2.1 *General*—Gypsum lath shall be installed so that vertical joints occur not less than one full stud space from edges of openings in walls or partitions.

7.2.2 Gypsum lath shall be installed with the face side out and the edges and ends butted together. End joints shall fall on different supports in alternate courses, or the lath shall be installed so that end joints are continuous on one support. In the latter case, the continuous end joints shall be covered with 3-in. (76-mm) wide strips of metal lath or welded or woven wire fabric, and the long edge joints of lath in alternate courses shall be offset or staggered.

7.2.3 Cornerite shall be installed on gypsum lath at all internal angles, and lightly nailed or stapled to the gypsum lath and not to the framing members. Cornerite shall not be used in unrestrained construction, or where other provisions are made for the treatment of internal angles.

7.2.4 Where gypsum lath sheets intersect or where joint surfaces are to be plastered without lathing (such as masonry), a casing bead shall be installed at the intersection or joining.

7.2.5 Gypsum lath shall be reinforced at the corners of doors, windows, or other openings with diagonally installed strips of striplath or wire lath, not less than 6 in. (152 mm) wide by 12 in. (305 mm) long.

7.2.6 Metal lath stripping shall be used to cover chases and similar breaks in continuity or horizontal or vertical surfaces that are to receive plaster. Stripping shall be extended not less than 3 in. (76 mm) on all sides of the openings.

7.3 *Application of Gypsum Lath to Wood Supports*:

7.3.1 Except where required otherwise, for certain fire-resistant construction, gypsum lath shall be securely attached to wood supports using nails, screws, or staples conforming to Table 1.

7.3.2 Nails and screws shall be driven so that the face of the head is flush with the face of the gypsum lath, and not less than ⅜ in. (9.5 mm) from the edges of the lath. Staples shall be driven with the crown parallel to the nailing members. The crown shall bear lightly against, but not cut into, the face of the lath.

7.4 *Application of Gypsum Lath to Metal Supports*:

7.4.1 Gypsum lath shall be attached to horizontal or vertical metal supports with staples, screws, or nails, or a combination thereof.

7.4.2 Screws shall be driven so that the face of the head is flush with the face of the gypsum lath and not less than ⅜ in. (9.5 mm) from the edges of the lath.

7.5 *Studless Solid Partitions (Gypsum Lath and Plaster)*:

7.5.1 Gypsum lath shall be plain, ½ in. (12.7 mm) thick, 24 in. (610 mm) wide, and, when applied vertically, shall be not more than 12 ft (3.7 m) in length.

7.5.2 Wood floor runners shall be secured to the floor using anchors spaced not more than 24 in. (610 mm) on center. Anchors shall penetrate floor surface to a depth of not less than ⅝ in. (15.9 mm). The width of such runners shall correspond to the overall partition thickness and their upper surface shall be grooved parallel to the length of the runner in the center to a depth of not less than ½ in. (12.7 mm) and to a width to accommodate snugly the thickness of the lath.

7.5.3 Metal floor runnersshall be anchored to bases and clips at not more than 24 in. (610 mm) on center. Runners shall be designed to hold the lower edge of the lath securely in position.

7.5.4 Metal ceiling runners shall be aligned with floor runners to ensure plumb installation of the lath. Metal ceiling runners shall be firmly secured to the ceiling construction.

7.5.5 Gypsum lath shall be of such length as to allow not less than ¼ in. (6.4 mm) nor more than 1-in. (25.4-mm) top clearance in the ceiling runner. Erect gypsum lath vertically so as to engage the ends in or to the floor and ceiling runners.

7.5.6 the vertical edges of the lath shall be aligned with the vertical joints interlocked or mechanically fastened with type G screws spaced not more than 12 in. (305 mm) on centers.

7.6 *Application of Metal Plaster Bases to Framing Members*:

7.6.1 *General*—The spacing of supports for the type and weight of metal lath used shall be in accordance with the requirements of Table 2.

7.6.2 *Attachments for Metal Plaster Bases to Wood Framing Members*:

7.6.2.1 Diamond mesh expanded metal lath, flat rib expanded metal lath, and welded wire lath shall be attached to horizontal wood supports with 1½ in.(38 mm), 0.1205 in. (3.06 mm) shank, 7⁄16 in. (11 mm) diameter head, barbed, galvanized, or blued roofing nails driven flush with the lath. The metal lath shall be attached to vertical wood supports with 6d common nails, or 1-in. (25-mm) long roofing nails driven driven in accordance with Table 1 or 1-in. No. 14 gage wire staples driven flush with the lath.

7.6.2.2 ⅜-in. (9.5-mm) rib expanded metal lath shall be attached to horizontal and vertical wood supports with nails or staples not less than ⅜ in. (9.5 mm) longer than required (7.6.2.1) to provide not less than ⅜-in. (9.5-mm) longer than required (see 7.6.2.1) to provide penetration in accordance with Table 1.

7.6.2.3 When used on vertical wood supports, common nails shall be bent over to engage not less than three strands of lath. When used for the application of rib lath, common nails shall be bent over to engage not less than one rib of the lath.

7.6.2.4 Other methods of attachment shalll be used when it can be shown they provide a carrying strength not less than that described in 7.6.2.1-7.6.2.3 may be used.

7.6.3 *Attachments for Metal Plaster Bases to Metal Framing Members*:

TABLE 2 Types and Weights of Expanded Metal Lath, Sheet Lath, or Welded Wire Lath and Corresponding Maximum Permissible Spacing of Supports[A]

Type of Metal Plastering Base	Minimum Weight of Metal Base, lb/yd² (kg/m²)	Maximum Permissible Spacing of Supports Center for Center, in. (mm)				
		Walls (Partitions)			Ceilings	
		Wood Studs	Solid Partitions	Steel Studs, Wall Furring, etc.	Wood or Concrete	Metal
Expanded metal lath:						
Diamond mesh	2.5 (0.08)	16 (406)	16 (406)	16 (406)	12 (305)	12 (305)
	3.4 (0.12)	16 (406)	16 (406)	16 (406)	10 (254)	16 (406)
Flat rib	2.5 (0.08)	16 (406)	12 (305)	12 (305)	12 (305)	12 (305)
	2.75 (0.09)	16 (406)	16 (406)	16 (406)	16 (406)	16 (406)
	3.4 (0.12)	19 (483)	24 (610)	19 (483)	19 (483)	19 (483)
Flat rib (large opening)	1.8 (0.06)	24 (610)	24 (610)	24 (610)	16 (406)	16 (406)
⅜-in. (9.5-mm) rib						
	3.4 (0.12)	24 (610)	N/A	19 (482)	19 (482)	19 (482)
	4.0 (0.14)	24 (610)	N/A	24 (610)	24 (610)	24 (610)
Wire lath welded	1.4 (0.05)[B]	16 (406)	16 (406)	16 (406)	16 (406)	16 (406)
	1.95 (0.07)[C]	24 (610)	24 (610)	24 (610)	24 (610)	24 (610)

[A]Paper-backed lath shall be limited to lath having an absorbent and a perforated slotted paper separator only.
[B]Welded wire paper-backed lath, 16-gage, 0.0598 in., (1.52 mm).
[C]Welded wire, paper-backed lath, 16-gage face wire, 0.0598 in. (1.52 mm) face wire.

7.6.3.1 Metal plaster bases shall be attached to metal framing members with 0.048 in. (1.2 mm) diameter wire ties, clips, or modified truss (wafer-headed) screws installed so that the threaded portion of the screw will penetrate the metal supports not less than ⅜ in. (9.5 mm) and the head shall contact not less than three strands of lath.

7.6.3.2 Rib metal lath shall be attached to open web steel joists using single ties of galvanized, annealed steel wire, 0.048 in. (1.2 mm) in diameter, with the ends of each tie wire twisted together 1½ times.

7.6.3.3 Rib metal lath shall be installed to metal framing members with the point of the V rib in contact with the framing and attached to the framing by a screw driven through the rib so that the threaded portion of the screw penetrates the metal support not less than ⅜ in. (9.5 mm).

7.6.4 *Attachments for Metal Plaster Bases to Concrete Supports*—Rib metal lath shall be attached to concrete joists by loops of galvanized, annealed steel wire, not less than 0.076 in. (1.9 mm) in diameter, with the ends of each loop twisted together not less than 1 ½ times.

7.6.5 *Spacing of Attachments for Metal Plaster Bases*—Attachments for securing metal plaster bases to supports shall be spaced not more than 7 in. (178 mm) apart for diamond mesh and flat rib laths and at each rib for ⅜-in. (9.5-mm) rib lath.

7.6.6 *Lapping of Metal Plaster Bases*:

7.6.6.1 Side laps of metal plaster bases shall be secured to supports, and tied between supports at intervals not more than 9 in. (229 mm) on center.

7.6.6.2 Expanded metal lath shall be lapped ½ in. (12.7 mm) at edges, or the edge ribs shall be nested. Welded or woven wire lath shall be lapped one mesh at sides and ends. Expanded metal lath shall be lapped 1 in.(25 mm) at ends or ends shall be nested. When end laps occur between supports-,the ends of sheets of all metal plaster bases shall be tied together with 0.048 in. (1.2 mm), galvanized, annealed steel wire. Laps shall be located not more than 3 in. (76.2 mm) from the structural support. Lath shall be lapped not more than 3 in. (76.2 mm) in any direction.

7.6.6.3 Adjacent sheets of metal plaster base shall be installed so that metal laps onto metal and backing laps onto backing.

7.6.7 *Procedure for Application of Metal Plaster Bases to Ceiling Framing Members*:

7.6.7.1 Metal plaster bases shall be applied to ceilings first.

7.6.7.2 The ends of metal plaster bases shall be staggered.

7.6.7.3 Where furred or suspended ceilings greater than 50 ft (15 m) in either direction or 2500 ft² (232 m²) in area butt into or are penetrated by columns, walls, beams, or other elements, the following procedure shall be used: The sides and ends of the ceiling lath shall abut at the horizontal internal angles and terminate at a casing bead, control joint, or similar device designed to isolate the plaster membrane from the adjoining vertically oriented elements. Only casing beads shall be used at internal angles between ceilings, walls, or partitions. Lath shall not be continuous through control joints but shall be stopped and tied to the flanges of the control joint at each side.

7.6.8 Ends of sheets of diamond mesh metal lath and welded wire lath on partitions not abutting structural walls, columns, or floor-ceiling slabs shall be bent into or around vertical corners and continued on to not less than one support away from the corner. The ends of the sheets of lath shall be butted into corners when cornerite is applied over the abutting laths. Rib metal lath shall be butted into corners and cornerite shall be applied over the abutting laths.

7.6.9 The ends of metal plaster bases applied to load-bearing wall and partitions that abutt structural walls, columns, or floor-ceiling slabs shall be isolated from the adjacent surface at the interanl angle. Cornerite shall not be used at internal angles, only casing beads or control joint materials shall be used to isolate the lath and plaster from the abutting surface.

7.7 *Studless Solid Partitions (Metal Lath and Plaster)*:

7.7.1 Metal plaster bases used for studless solid partitions shall have the sides (long dimensions) of the sheet installed vertically. The plaster bases shall be wiretied to the ceiling runners using ties spaced not mroe than 8 in. (203 mm) on center and mechanically anchored to the floor runners or base.

7.7.1.1 At vertical internal and external corners, metal plaster bases shall be bent and returned 6 in. (152 mm) on abutting surfaces. Rib metal lath shall be lapped ½in. (12.7 mm) at sides, or the outside ribs shall be nested. Side laps shall be wire tied between supports at intervals not more than 9 in. (229 mm) on center.

7.8 *Suspended and Furred Ceilings (Metal Lath and Gypsum Lath Construction)*:

7.8.1 Hangers shall be of sufficient length to accommodate the ceiling construction.

7.8.1.1 Hangers (without inserts) shall be secured or otherwise embedded so as to develop full strength.

7.8.1.2 *Inserts*—Inserts for hangers shall be secured in concrete reinforced slabs so as to develop the full strength of the hangers that are attached to them.

7.8.2 *Hangers for Suspended Ceilings Under Wood Constructions*—Hangers shall conform with the requirements of Table 3 both as to size and maximum area to be supported, except as modified in this section. Hangers shall be attached to supports by one of the following methods:

7.8.2.1 Insert through holes drilled not less than 3 in. (76 mm) above the bottom of the joist with the upper end of the hanger twisted not less than three times around itself (see Fig. 1).

7.8.2.2 Twisted not less than three times around itself over not less than three 12d nails driven on a downward slant into the sides of joists not less than 5 in. (127 mm) from the bottom

TABLE 3 Suspended and Furred Ceilings, Minimum Sizes for Wire, Rod, and Rigid Hangers; Minimum Sizes and Maximum Spans and Spacings for Main Runners; and Minimum Sizes and Maximum Spans and Spacings for Cross Furring

NOTE 1— 1 in. = 25.4 mm; 1 ft² = 0.093 m²

HANGERS

	Maximum Ceiling Area Supported, ft²	Minimum Size of Hangers
Hangers for Suspended Ceilings	12.5	0.148 in. wire
	16	0.162 in. wire
	18	3/16-in. diameter, mild steel rod[A]
	20	7/32-in. diameter, mild steel rod[A]
	22.5	¼-in. diameter, mild steel rod[A]
	25.0	1 by 3/16-in. mild steel strap[B]
Attachments for Tying Runners and Furring Directly to Beams and Joists		
For Supporting Runners		
Single Hangers Between Beams[C]	8	0.106 in. wire
	12	0.135 wire
	16	0.162 wire
Double Wire Loops at Beams or Joists[C]	8	0.076 wire
	12	0.106 wire
	16	0.120 wire
For Supporting Furring Without Runners[C] (Wire Loops at Supports)		
Types of Support:		
Concrete	8	0.076 wire
Steel	8	0.062 (2 loops)[D]
Wood	8	0.062 (2 loops)[D]

SPANS AND SPACINGS OF MAIN RUNNERS[E,F]

Minimum Size and Type	Maximum Span Between Hangers or Supports, in.	Maximum Center to Center Spacing of Runners, in.
¾ in., 0.3 lb/ft, cold	24	36
1½ in. 0.475 lb/ft, cold-rolled channel	36	48
1½ in., 0.475 lb/ft, cold-rolled channel	42	42
1½ in., 0.475 lb/ft, cold-rolled channel	48	36
2 in., 0.59 lb/ft, cold-rolled channel	60	48
1½ by 1½by 3/16-in. angle	60	42

SPANS AND SPACINGS OF CROSS FURRING[E,F]

Minimum Size and Type	Maximum Span Between Runners or Supports, in.	Maximum Center to Center Spacing of Cross Furring Members, in.
¼-in. diameter pencil rods	24	12
3/8-in. diameter pencil rods	24	19
3/8-in. diameter pencil rods	30	12
¾ in., 0.3 lb/ft, cold	36	24
	42	19
	48	16

[A]All rod hangers shall be protected with a zinc cadmium coating.
[B]All flat hangers shall be protected with a zinc or cadmium coating or with a rust-inhibitive paint.
[C]Inserts, special clips, or other devices of equal strength are substitutes for those specified.
[D]Two loops of 0.048 in. wire are substitutes for each loop of 0.062 in. wire when used for attaching steel furring to steel or wood joists.
[E]Spans are based on webs of channels being erected and maintained in a vertical position.
[F] Other sections of cold-rolled members of equivalent beam strength are substitutes for those specified.

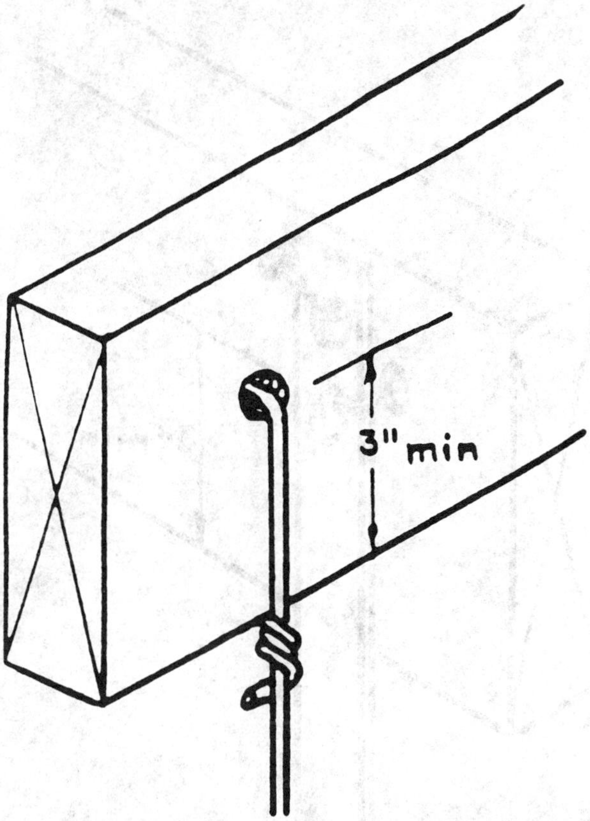

FIG. 1 Hanger Attached to Support Through a Drilled Hole

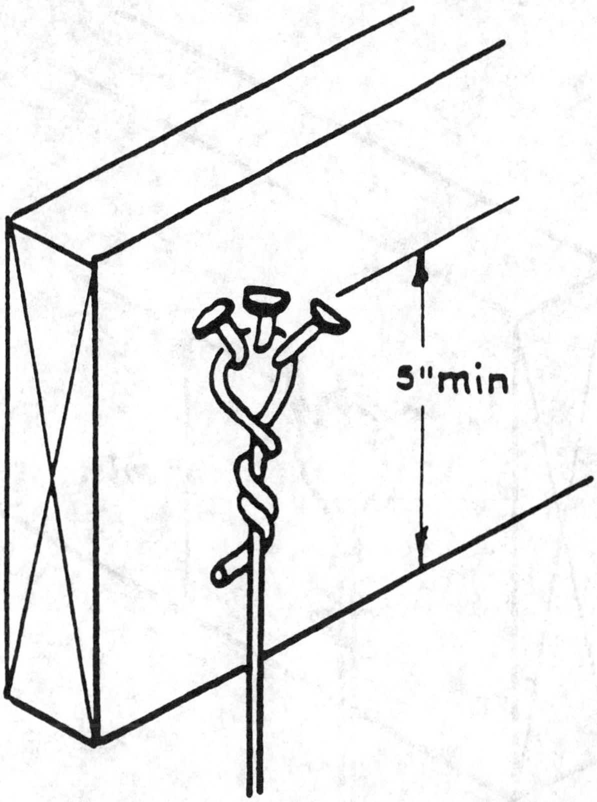

FIG. 2 Hanger Attached to Support Using Nails

edge of the joist and not more than 36 in. (914 mm) on center (see Fig. 2).

7.8.2.3 Formed into a loop to accommodate four 1 ½-in. (38.1-mm), 0.148-in. (3.8-mm) diameter staples driven horizontally or on a downward slant into sides of joists, placing three near the upper end of the loop and the fourth to fasten the loose end (see Fig. 3).

7.8.2.4 Where spacing of supports is more than 4 ft (1.2 m) on center, 1½-in. (38.1-mm) No. 0 screw eyes, or equivalent, spaced not more than 3 ft (0.9 m) on centers shall be screwed into supporting wood construction so that the supported area shall be not more than 9 ft^2(0.8 m^2). The upper end of the wire hanger shall be inserted through the screw eye and in each case twisted not less than three times around itself.

7.8.2.5 Flat hangers shall be secured with two 12d nails driven through holes drilled in the hanger and clinched not less than 3 in. (76 mm) above the bottom of the joists (see Fig. 4).

7.8.3 *Minimum Size for Hangers*—Thickness of wire hangers, diameter of rod hangers, and sizes of flat hangers shall be in accordance with Table 3.

7.8.4 Where 1 by 3⁄16-in. (25.4 by 4.8-mm) flat inserts and hangers are used, 7⁄16-in. (11-mm) diameter holes shall be punched on the center line at the lower end of inserts and upper end of hanger to permit the attachment of the hangers to the insert.

7.8.5 Holes in both inserts and hangers shall be not less than 3⁄8 in. (9.5 mm) from the ends.

7.8.6 Flat steel hangers shall be bolted to 1 by 3⁄16-in. (25.4 by 4.8-mm) inserts with 3⁄8-in. (9.5-mm) diameter round-head stove bolts.

7.8.7 The nuts of bolts shall be drawn up tight.

7.8.8 Wire or rod hangers shall be saddle-tied or attached to main runners so as to prevent turning or twisting of the runners and to develop the full strength of the hangers and the runners. Smooth or threaded rod hangers shall be attached to inserts and runners with special attachments designed for this application.

7.8.9 The lower ends of flat hangers shall be bolted to the main runners, or bent tightly around runners and carried up and above the runners and bolted to the main part of the hanger. Bolts shall be 3⁄8-in. (9.5-mm) diameter, stove bolts.

7.9 *Main Runners*:

7.9.1 Minimum sizes and maximum spans and spacings of main runners for the various spans between hangers or other supports shall be in accordance with Table 3.

7.9.2 A clearance of not less than 1 in. (25.4 mm) shall be provided between the ends of main runners and abutting masonry or reinforced concrete walls, partitions and columns. Where special conditions require main runners to penetrate the plane of abutting masonry or concrete construction, a clearance shall be provided within such constructions of not less than 1 in.(25 mm) from the ends and not less than ½ in. (12.7 mm) from the tops and sides of the runners.

7.9.3 A main runner shall be located not more than 6 in. (152 mm) from the paralleling wall to support the ends of the cross furring. The ends of main runners shall be supported by hangers located not more than 6 in. (152 mm) from the ends.

7.9.4 When main runners are spliced, the ends shall be overlapped not less than 12 in. (305 mm), with flanges of channels interlocked, and securely tied near each end of the splice with double loops of 0.062 in. (1.6 mm) galvanized wire

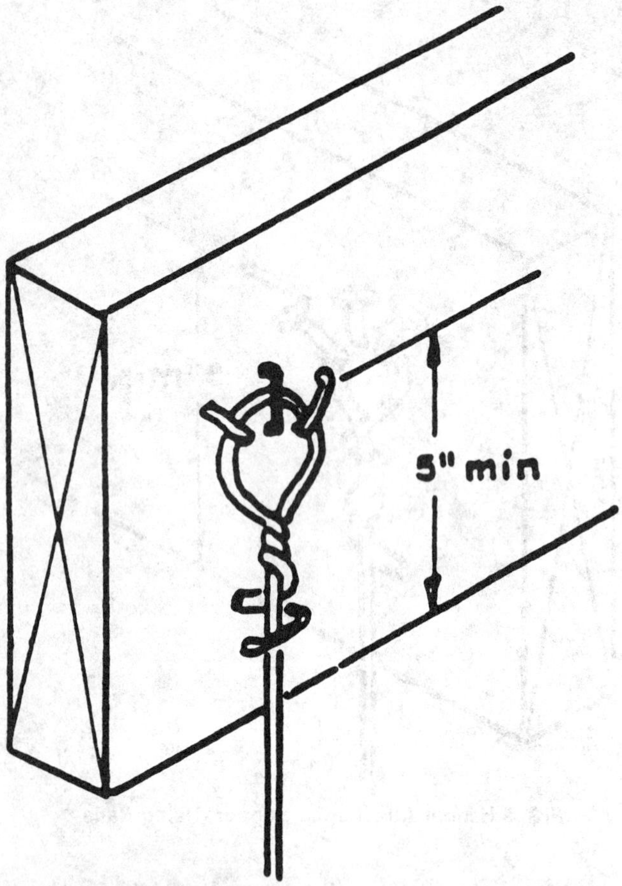

FIG. 3 Hanger Attached to Support Using Staples

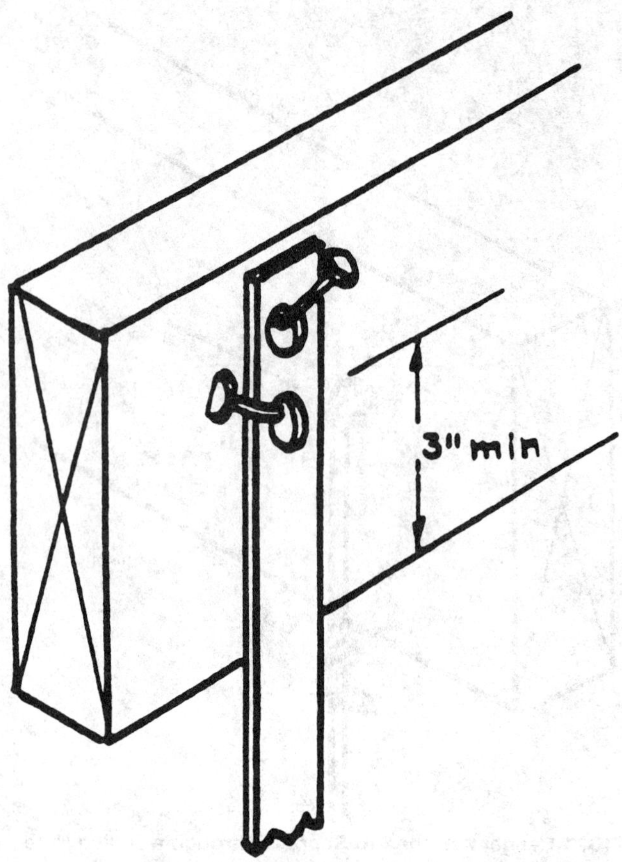

FIG. 4 Flat Hanger Attached to Support Using Nails

or double loops of twin strands of 0.048 in. (1.2 mm) galvanized wire.

7.10 *Cross Furring*:

7.10.1 Minimum size and maximum spans and spacings of various types of cross furring for various spans between main runners and supports shall be in accordance withTable 3.

7.10.2 Cross furring shall be securely saddle-tied to main runners with 0.062 in. (1.6 mm) galvanized wire, or a double strand of 0.048 in. (1.2 mm) galvanized wire or with special galvanized clips, or equivalent attachments.

7.10.3 When cross furring members are spliced, the ends shall be overlapped not less than 8 in. (203 mm), with flanges of channels interlocked, and securely tied near each end of the splice with double loops of either 0.062 in. (1.6 mm) galvanized wire or twin strands of 0.048 in. (1.2 mm) galvanized wire.

7.10.4 Cross furring shall not contact abutting masonry or reinforced concrete walls or partitions.

7.10.4.1 Where special conditions require cross furring to penetrate the place of the abutting masonry orconcrete construction, the applicable provisions of 7.9.2shall apply.

7.10.5 Main runners and cross furring shall be interrupted at control and expansion joints. The resulting intersections shall be constructed by splices, snugly tied, but loose enough to allow for lateral movement(telescoping) of the framing members and expansion and contraction of the plaster membrane.

7.11 *Metal Furring for Walls*:

7.11.1 Attachments for furring shall consist of nails driven securely into concrete or into masonry joints, short pieces of ¾ in. (19 mm) channels used as anchors driven into masonry joints, or other devices specifically designed as spacer elements. Furring shall be spaced horizontally not more than 2 ft (610 mm) on centers. Furring shall be spaced vertically in accordance with horizontal stiffener spacing so that it projects from the face of the wall enough to permit ties to be made.

7.11.2 Where dampproofing has been damaged in installation of attachments, the dampproofing shall be repaired with the same material before proceeding with the installation of the furring.

7.11.3 Horizontal stiffeners shall be not less than ¾-in. (19-mm) hot-rolled or cold-rolled channels, spaced not to exceed 60 in. (1524 mm) on center vertically, with the lower and upper channels not more than 6 in. (152 mm) from the floor and ceiling, and not less than ¼in. (6.4 mm) from the face of the wall. Stiffeners shall be securely tied to attachments with three loops of 0.048 (1.2 mm) wire.

7.11.4 Vertical members shall be not less than ¾-in. (19-mm) cold-rolled channels, in accordance with requirements in Table 2. Members ahll be saddle-tied to horizontal members with three loops of 0.048 in. (1.2 mm) galvanized soft-annealed wire, or equivalent devices, at each crossing, and securely anchored to the floor and ceiling constructions. If furring is not in contact with the wall, then channel braces shall be installed between horizontal stiffeners and the wall, spaced not more than 2 ft (600 mm) apart (horizontal spacing).

8. Keywords

8.1 furring; gypsum; lathing; plaster

ANNEXES

(Mandatory Information)

A1. GENERAL INFORMATION

The work shall include all labor, materials, services, equipment, and scaffolding required to complete the lathing and furring of the project in accordance with the drawings and specifications.

A1.1 *Floating Angle Construction*—Except where required otherwise for certain fire-resistive construction, nails or staples occurring less than 6 in. (152 mm) from internal angles shall not be prohibited from being omitted in the application of gypsum lath, provided each piece of lath is nailed to not less than two supports. Where lath abuts masonry, the angle shall not be "floated" and a plaster stop or casing bead shall be used.

A1.1.1 Gypsum lath, for floating angle construction, shall be applied to ceiling areas first, so that the ceiling lath will be supported in the angles by the upper edge of the top course of lath applied to walls or partitions.

A1.2 *Wood Furring*—For the attachment of wall finishes, wood furring shall be not less than ¾ in. (19 mm) by 1½in. (38.1 mm) (actual size) where applied to solid backing such as masonry. Wood furring attached to supports, spaced not more than 24 in. (610 mm) on centers, shall be not less than 1½by 1½ in.1

A2. TECHNICAL INFORMATION

A2.1 *Solid Plaster Partitions with Steel Studs:*

A2.1.1 Steel studs for solid plaster partitions shall be installed in accordance with Specifications C 754 and C 1007 and shall be of a size and number and located to provide backing at all corners. Steel studs shall be set to the required spacing, properly aligned, made plumb and true, securely anchored to the floor and ceiling construction, and temporarily braced, if necessary.

A2.1.2 Where studs do not fall at edges of openings, extra studs shall be added. Such studs shall be securely anchored to the door frames.

A2.1.3 Space above headers of openings extending to the ceiling shall be reinforced by installing studs 16 in. (406 mm) on center over the heads of the openings.

A2.2 *Hollow Metal Door Frames* for prefabricated steel stud, channel stud, lath and plaster hollow partitions.

A2.2.1 Door frames shall be installed in accordance with NAAMM Standard HMMA 840 .

A2.2.1.1 A vapor retarder having a transmission rate not exceeding 1 perm is installed on the warm side of the ceiling, or

A2.2.1.2 Not less than 50 % of the required ventilation area is provided by ventilators located in the upper portion of the space to be ventilated (not less than 3 ft (0.9 m) above eave or cornice vents) with the balance of the required ventilation area provided by eave or cornice vents.

A2.2.2 Attic space that is accessible and suitable for future habitable rooms or walled-off storage space shall have at least 50 % of the required ventilating area located in the upper part of the ventilated space as near the high point of the roof as practical and above the probable level of any future ceilings.

A3. DESIGN AND ERECTION OF HOLLOW DOOR METAL FRAMES FOR PREFABRICATED STEEL STUD AND CHANNEL STUD, LATH AND PLASTER HOLLOW PARTITIONS

A3.1 *Hollow Metal Frames* (for Prefabricated Steel Stud, Channel Stud, Lath and Plaster Hollow Partitions):

A3.1.1 Metal studs shall be installed at each jamb of the steel frame by wiretying, bolting, or welding them to jamb anchors in the frame. The edges of the studs shall be inserted into the notches or other devices provided in the jamb anchor insert clips.

A3.1.2 The first stud on either side of the frame shall be fastened to suitable anchorage at the top and bottom of the partition in accordance with the instructions of the stud manufacturer. A second metal stud shall be placed not more than 2 in. (50 mm) away from the first stud, using adequate means of attachments between the two; or, two metal studs shall be attached to jamb anchors and securely fastened back to back.

A3.1.3 Where metal door frames have an adjustable strut equal in stiffness to a stud, running from each jamb to the ceiling, one metal stud shall be installed at each jamb in the manner previously noted.

A3.1.4 At the head of the metal frame, a section of metal floor track shall be placed horizontally within the head section, and fastened to the jamb studs with wire ties, screws, or welds. Frames over 32 in. (813 mm) wide shall be provided with anchor insert clips spaced not over 24 in. (610 mm) on centers.

A3.1.5 Metal jack studs shall be installed over the head of the frame, spaced 16 in. (406 mm) on centers maximum, and shall be attached to the metal floor track. They shall be reinforced with a ¾-in. (19-mm) cold-rolled channel stiffener placed inside the partition and not more than 6 in. (152 mm) above the frame. Floor track is not required if one ¾-in. cold-rolled channel stiffener is placed horizontally on each interior face of the studs and not more than 6 in. above the frame. Channel stiffeners shall extend not less than two full stud spaces beyond the opening and shall be saddle-tied to each stud they cross.

APPENDIXES

X1. VENTILATION INFORMATION

X1.1 Ventilation above lath and plaster ceilings located under attics, roofs, or similar unheated spaces shall be adequately ventilated by providing effective cross ventilation for all spaces between roof and top floor ceilings by screened louvers or other approved or acceptable means.

X1.2 The ratio of total net free ventilating area to ceiling area shall not be less than 1:150 except that the ratio may be 1:300 provided that:

X1.2.1 A vapor retarder having a transmission rate not exceeding 1 perm is installed on the warm side of the ceiling, or

X1.2.2 Not less than 50 % of the required ventilation area is provided by ventilators located in the upper portion of the space to be ventilated (not less than 3 ft (0.9 m) above eave or cornice vents) with the balance of the required ventilation area provided by eave or cornice vents.

X1.3 Attic space that is accessible and suitable for future habitable rooms or walled-off storage space shall have not less than 50 % of the required ventilating area located in the upper part of the ventilated space as near the high point of the roof as practical and above the probable level of any future ceilings.

X2. GENERAL INFORMATION

X2.1 The work shall include all labor, materials, services, equipment, and scaffolding required to complete the lathing and furring of the project in accordance with the drawings and specifications.

X2.2 Scaffolding shall be constructed and maintained in strict conformity with applicable laws and ordinances.

X2.3 The work shall be coordinated with the work of other trades.

X2.4 Surfaces and openings shall be examined before furring or lathing is applied thereto, the proper authorities shall be notified, and unsatisfactory conditions shall be corrected prior to application of furring or lathing.

SUMMARY OF CHANGES

This section identifies the location of changes to this specification that have been incorporated since the last issue. Committee C-11 has highlighted those changes that affect the technical interpretation or use of this specification.

(1) This document has been editorially revised to be consistent with ASTM Form and Style.

(2) Figures 1 through 4 have been added to the document.

(3) References to ASTM Standards C 641M, C 754, C 847, C 933, C 1007, C 1032, D 1784, and D 3678, and NAAMA HMMA 840 have been added to Section 2 of the standard.

(4) Annex paragraphs A3.8.7, A3.8.7.1 and A3.8.7.2 have been moved to the body of the document as paragraph 7.1.5.

(5) Paragraph previously numbered 7.5.7 and 7.7.2 have been deleted.

(6) Annex paragraph A3.8.6 has been moved to the body of the document as paragraph 7.1.8.

(7) Paragraph previously numbered 7.12 has been moved to section 7.1.7.

(8) Tables 1, 2, and 3 have been revised.

(9) Annex sections A1.1, A1.4 through A1.6 and A2.1 through A2.1.2have been moved to the Appendix.

(10) Annex sections A2.2 and A3.1 through A3.6 have been eliminated.

(11) Annex sections A3.1 through A3.6 and A3,8 through A3.8.5 have been eliminated and replaced by reference.

(12) Annex sections A1.2 and A1.3 have been eliminated and replaced by sections 1.3 and 1.4.

Standard Specification for Application of Interior Gypsum Plaster[1]

This standard is issued under the fixed designation C 842; the number immediately following the designation indicates the year of original adoption or, in the case of revision, the year of last revision. A number in parentheses indicates the year of last reapproval. A superscript epsilon (ϵ) indicates an editorial change since the last revision or reapproval.

1. Scope *

1.1 This specification covers the minimum requirements for full-thickness (in accordance with Table 1) interior gypsum plastering on gypsum, metal, masonry, or monolithic concrete bases designed or prepared to receive gypsum plaster.

1.1.1 Gypsum plaster applied to the interior side of exterior walls of masonry or concrete shall be isolated from the exterior wall.

1.2 Gypsum plasters shall not be used in exterior locations or interior "wet" areas.

NOTE 1—To secure desirable results, this specification should be coordinated with the provisions of Specification C 841. General information concerning full thickness gypsum plastering is contained in Annex A2, Annex A3, and Annex A4.

1.3 Details of construction for a specific assembly to achieve the required fire resistance shall be obtained from reports of fire-resistance tests, engineering evaluations, or listings from recognized fire testing laboratories.

1.4 Details of construction to achieve required sound control shall be obtained from reports of tests conducted at recognized sound testing laboratories in accordance with the applicable sound tests of Test Methods E 90, C 423, or E 492.

1.5 The values stated in inch-pound units are to be regarded as the standard. The SI (metric) values given in parentheses are approximate and are provided for information purposes only.

1.6 The text of this standard references notes and footnotes which provide explanatory material. These notes and footnotes (excluding those in tables and figures) shall not be considered as requirements of the standard.

TABLE 1 Thickness of Plaster

Plaster Base	Thickness of Plaster Including Finish Coat, in. (mm)
Metal plaster base	⅝ (16) min
All other types of plaster base	½ (13) min
Unit masonry	⅝ (16) min
Monolithic concrete surfaces:[A]	
Vertical[B,C]	⅝ (16) min
Horizontal[C]	⅛ (3) to ⅜ (10)

[A] Base coat plastering of the same proportions as specified for unit masonry shall be permitted to be used over plain or reinforced monolithic concrete, provided the surface is first covered with a metal plaster base or first coated with a bonding compound.

[B] Finish coat plaster applied direct to a bonding compound over vertical monolithic concrete shall be not more than 3⁄16 in. (4.76 mm) in thickness. Where more than 3⁄16 in. of finish coat is required to bring such vertical surface to a true plane, a base coat of plaster shall first be applied to the bonding compound.

[C] Where horizontal or vertical monolithic concrete surfaces require more than ⅜ in. (9.52 mm) or ⅝ in. (15.87 mm) of plaster, respectively, to produce required lines or surfaces, metal plaster base shall be attached to the concrete before application of plaster. Where concrete surface requires the application of more than 1 in. (25.4 mm) of plaster to produce required lines or surfaces, lath shall be applied over furring secured to the concrete.

2. Referenced Documents

2.1 *ASTM Standards:*

C 5 Specification for Quicklime for Structural Purposes[2]

C 11 Terminology Relating to Gypsum and Related Building Materials and Systems[2]

C 28 Specification for Gypsum Plasters[2]

C 35 Specification for Inorganic Aggregates for Use in Gypsum Plaster[2]

C 59 Specification for Gypsum Casting and Molding Plaster[2]

C 61 Specification for Gypsum Keene's Cement[2]

C 206 Specification for Finishing Hydrated Lime[2]

C 423 Test Method for Sound Absorption and Sound Absorption Coefficients by the Reverberation Room Method[3]

C 631 Specification for Bonding Compounds for Interior Gypsum Plastering[2]

C 841 Specification for Installation of Interior Lathing and Furring[2]

E 90 Test Method for Laboratory Measurement of Airborne-Sound Transmission Loss of Building Partitions[3]

E 492 Test Method for Laboratory Measurement of Impact Sound Transmission Through Floor-Ceiling Assemblies Using the Tapping Machine[3]

3. Terminology

3.1 *Definitions:*

3.1.1 Definitions shall be in accordance with Terminology C 11.

3.2 *Definitions of Terms Specific to This Standard:*

3.2.1 *catfaces, n*—blemishes or rough depressions in the finish coat.

[1] This specification is under the jurisdiction of ASTM Committee C-11 on Gypsum and Related Building Materials and Systems and is the direct responsibility of Subcommittee C11.03 on Specifications for Application of Gypsum and Other Products in Assemblies.

Current edition approved Oct. 10, 1998. Published March 1999. Originally published as C 842 – 76. Last previous edition C 842 – 97a.

[2] *Annual Book of ASTM Standards*, Vol 04.01.

[3] *Annual Book of ASTM Standards*, Vol 04.06.

***A Summary of Changes section appears at the end of this standard.**

3.2.2 *coat, n*—a thickness or layer of plaster applied over a surface in a single application.

3.2.2.1 *base coat, n*—the sum of the scratch and brown coats or the total coats in place prior to application of finish coats.

3.2.2.2 *brown coat, n*—the second coat of plaster applied in three-coat work.

3.2.2.3 *finish coat, n*—the last coat of plaster applied in either two-coat or three-coat work.

3.2.2.4 *scratch coat, n*—the first coat of plaster applied over a lath or other substrate.

3.2.2.5 *three-coat work, n*—plaster applied in three successive coats, leaving time between coats for setting or drying, or both, of the plaster.

3.2.2.6 *two-coat work, n*—scratch and brown coats applied from the same mix with no time allowed for setting of the scratch coat before the brown coat is applied.

3.2.3 *gypsum ready mixed plaster, n*—calcined gypsum plaster, mixed at the mill with a mineral aggregate, designed to function as a base coat to receive various finish coats.

3.2.4 *metal bases, n*—expanded metal, welded or woven wire, or punched sheet metal plaster bases.

3.2.5 *smooth-trowel finish, n*—a finish resulting from steel troweling.

3.2.6 *texture finish, n*—a finish resulting from (*1*) trowel application followed by floating or texturing of the surface with any of a variety of tools using a minimum of water or (2) machine application which shall be permitted to be hand textured.

4. Delivery of Materials

4.1 All manufactured materials shall be delivered in the original packages, containers, or bundles bearing the brand name and manufacturer (or supplier) identification.

5. Protection of Materials

5.1 Plasters and other cementitious materials shall be kept dry until used; they shall be stored off the ground, under cover, and away from walls with condensation and other damp surfaces. Metal products shall be protected, while stored, against rusting.

6. Environmental Conditions

6.1 *Temperatures*—Where the ambient outdoor temperature at the building site is less than 55°F (13°C), a temperature of not less than 55°F (13°C) and not more than 80°F (27°C) shall be maintained continuously inside the building for a period of not less than one week prior to the application of plaster (Note 2), while the plastering is being applied, and for one week after the plaster has set, or until the plaster has dried. Heat shall be distributed evenly by using deflective or protective screens to prevent concentrated or uneven heat or cold on the plaster.

NOTE 2—The requirement should minimize the possibility of plaster cracking due to structural movements caused by thermal changes from outdoor temperature extremes during construction.

6.2 *Ventilation*—Sufficient ventilation shall be provided to remove excess water given off through the drying process. (See Appendix X2 for provisions for ventilating underside of roofs and glazed buildings with and without operable windows.)

7. Materials

7.1 *Gypsum Plasters*—The following plasters shall conform to Specification C 28:

7.1.1 Ready mixed.

7.1.2 Neat.

7.1.3 Wood-fibered.

7.1.4 Gauging for finish coat.

7.2 *Gypsum Casting and Molding Plasters*—Specification C 59.

7.3 *Gypsum Keene's Cement*—Specification C 61.

7.4 *Lime*:

7.4.1 *Finishing Hydrated Lime*—Specification C 206, Type S.

7.4.2 *Quicklime for Structural Purposes*—Specification C 5.

7.5 *Lime Putty*—Lime putty exceeding 8 weight % of unhydrated magnesium oxide shall not be used for finish coat plaster.

7.6 *Aggregates*:

7.6.1 *Aggregates for Base Coat Plaster*—Specification C 35.

7.6.2 *Aggregates for Finish Coat Plasters*—Specification C 35, except that gradation shall be within the limits specified in Table 2.

7.6.3 Sand for job mixed lime putty-gypsum gauged, sand float finish (see 9.6.6.1) shall be graded within the limits specified in Table 3.

7.7 *Water*—Water used in mixing and finishing plaster shall be clean, fresh, suitable for domestic water consumption, and free of such amounts of mineral or organic substances that affect the set, the plaster, or any metal in the system.

NOTE 3—Water containing salt or alum, or water in which tools have been washed, accelerates the "set" and may cause efflorescence. Water from stagnant pools and wells frequently contain organic or vegetable matter which may retard the "set," cause staining, or interfere with the bond.

7.8 *Bonding Compounds*—Specification C 631.

8. Surface Preparation

8.1 *Substrates*—Surfaces of substrates for the application of gypsum plaster shall be free of materials that will inhibit bond or adhesion, shall be straight, plumb, level, square, and true to required plan angles and curves. See A2.5.

8.1.1 All accessories shall be securely attached to the substrate and be installed to permit embedment of flanges.

8.1.2 *Gypsum and Metal Lath*—Shall have been installed in compliance with C841.

8.2 *Conditioning of Surfaces*—All depressions in masonry and concrete surfaces deeper than ⅛ in. (3.2 mm) shall be

TABLE 2 Aggregate for Finish Coat Plasters, Percentage Retained on Each Sieve, Cumulative

Sieve Size	Perlite, Natural and Manufactured Sand			
	Volume,%		Weight, %	
	max	min	max	min
No. 20 (850 μm)	0	...	0	...
No. 30 (600 μm)	10	...	0.5	...
No. 100 (150 μm)	100	40	100	40
No. 200 (75 μm)	100	70	100	70

TABLE 3 Sand for Job-Mixed Lime Putty-Gypsum Gaged Sand Float Finish, Percent Retained on Each Sieve by Weight, Cumulative

Sieve Size	max	min
No. 16 (1.18 mm)	0	0
No. 30 (600 μm)	50	20
No. 50 (300 μm)	70	50
No. 100 (150 μm)	100	80

brought flush to the surface with compatible materials prior to plaster application. Fins or protrusions extending more than 1⁄16 in. (1.6 mm) from the surface shall be removed. Protrusions less than 1⁄16 in. (1.6 mm) shall be feathered out with compatible materials prior to plaster application.

8.2.1 *Masonry*—All masonry surfaces shall be wetted immediately prior to the plaster application. No free water shall remain visible on the substrate surface.

8.2.1.1 Where the surface is too smooth to provide mechanical key, the surface shall be roughened or bonding compound shall be used. (See Annex A2.5 and A2.6.)

8.2.1.2 Where bond cannot be obtained over the entire surface by the methods specified in 8.2.1.1, self-furring metal lath shall be used in accordance with C841.

8.2.2 *Monolithic concrete*—bonding compounds shall be used prior to plastering all monolithic concrete surfaces. (See A2.6.)

8.2.3 Grounds, beads, and screeds shall be installed prior to the plastering as modified herein.

8.3 *Grouting*:

8.3.1 Metal bases for solid partitions shall be grouted with gypsum plaster leaving approximately 1⁄4 in. (6.4 mm) below the top edge of the base, and be formed with a center groove that will permit the lath to extend 3⁄4 in. (19 mm) below the top edge of the base, or plaster grout shall be placed after installation of lath.

8.3.2 *Metal Frames*—Hollow door and window frames shall be filled with gypsum plaster grout.

9. Mix Design

9.1 *Mixing, General*—All plaster mixing shall be done by mechanical means unless hand mixing is specified.

9.1.1 Retempered, partially set, frozen, caked or lumpy material shall not be used.

9.1.2 Each batch shall be mixed separately and mixers shall be free of all set and hardened materials prior to mixing each batch.

9.1.3 All tools shall be kept clean.

9.1.4 The setting time shall be permitted to be adjusted but in no case shall the setting time be more than 4 h.

9.2 *Hand Mixing*—When permitted, see Annex A3.1.

9.3 *Mechanical Mixing*—In accordance with manufacturer's printed directions. See Annex A3.2.

9.4 *Base Coat Proportions*:

9.4.1 *Gypsum Neat Plaster*—Proportions of sand, perlite, or vermiculite aggregate to 100 lb (45.4 kg) of gypsum neat plaster shall be not more than those specified in Table 4. (See

TABLE 4 Base Coat Proportions[A]

Plaster Base	Aggregates[B]		
	Sand		Perlite or Vermiculite[C]
	By Volume, ft³(m³), Damp and Loose	By Weight, lb (kg), Damp and Loose	By Volume, ft³(m³)
Over Gypsum Lath			
Two-coat work:			
Base coat	2½(0.071)	250 (113)	2 (0.06)
Three-coat work:			
Scratch coat	2 (0.056)	200 (91)	2 (0.06)
Brown coat	3 (0.085)	300 (136)	2 (0.06)
or			
Scratch and brown coats	2½(0.071)	250 (113)	...
Over Metal Lath			
Three-coat work:			
Scratch coat	2 (0.056)	200 (91)	2 (0.06)
Brown coat	3 (0.085)	300 (136)	2 (0.06)
or			
Scratch and brown coats	2½(0.071)	250 (113)	...
Over Unit Masonry (Note 6)			
Two-coat work:			
Base coat	3 (0.085)	300 (136)	3 (0.085)
Three-coat work:			
Scratch coat	3 (0.085)	300 (136)	3 (0.085)
Brown coat	3 (0.085)	300 (136)	3 (0.085)
Over Monolithic Concrete[D]			
For base coat proportions applicable to monolithic concrete, see 10.2.1.			

[A]The proportions in Table 3 are applicable for both hand and machine application of plaster. See plaster manufacturer's instructions for application of machine-applied plaster.

[B]Use of an accurate device to measure quantities, such as a measuring box or container of known capacity, is highly encouraged. Where such a device is not available, six No. 2, square-edge (not scoop) shovels, with a blade approximately 8½ in. (216 mm) wide and 11 in. (279 mm) long, with the maximum depth of sides not more than 1½in. (38 mm) higher than the face of the blade, and filled to an average depth of 4 in. (102 mm) of damp, loose sand, shall be considered as the approximate equivalent to 1 ft³ (0.028 m³).

[C]Where the plaster is 1 in. (25 mm) or more in total thickness, or where the finish coat is sand float, the proportions for the brown coat shall be permitted to be increased to 3 ft³ (0.085 m³).

[D]For use of bonding compounds for plastering on monolithic concrete, see 8.2.2, and Footnotes in Table 1 and A2.6.

A2.3 for equivalent measure for aggregates.)

9.4.2 *Gypsum Ready Mixed Plaster*—Shall be used with the addition of water only.

9.4.3 *Gypsum Wood-Fibered Plaster*—Proportion of damp, loose sand or perlite or vermiculite shall be not more than 1 ft^3 (0.028 m^3) to 100 lb (45.4 kg) of gypsum wood-fibered plaster.

9.5 *Preparation of Lime Putty*:

9.5.1 Lime putty shall be prepared from Type S hydrated lime or pulverized quicklime, in accordance with the manufacturer's printed directions.

9.5.2 *Hydrated Lime*—Type S lime shall only be used after soaking for the time period required in accordance with the manufacturer's printed directions.

9.6 *Finish Coat Proportions*:

9.6.1 Finish coats shall be either ready mixed or job mixed in accordance with Table 5 and Table 6.

9.6.1.1 Where thickness of the finish coat is more than ⅛ in. (3.2 mm), the proportion of the gypsum gauging plaster shall be increased to minimize shrinkage type cracks.

9.6.2 Troweled Finishes of lime putty gauged with gypsum gauging plaster or ready mixed gypsum plaster shall be proportioned in accordance with Table 5.

9.6.2.1 Where finish in 9.6.2 is to be applied over base coats containing perlite or vermiculite, the addition of not less than ½ ft^3(0.14 m^3) or not more than 1 ft^3 (0.028 m^3) of fine aggregate meeting the sieve analysis of Table 2 shall be added to the mix.

9.6.3 Troweled finishes of lime putty gauged with gypsum Keene's cement shall be specified as medium or hard and shall be proportioned in accordance with Table 5.

9.6.3.1 When finish in 9.6.3 is mechanically mixed, the water shall be placed in the mixer first, then the lime, the fine aggregate (if used), and finally the gypsum Keene's cement.

9.6.4 Troweled finishes of gypsum ready mixed plaster shall be mixed in accordance with Table 5.

TABLE 6 Trowel Finishes of Lime Putty Gaged With Gypsum Gaging Plaster Proportions

Proportions of Fine Aggregate[A,B,C]		
Not Less Than	Not More Than	Per
½ ft^3(0.014 m^3)	1 ft^3(0.028 m^3)	100 lb (45.4 kg) gypsum gauging
or ⅛ ft^3(0.0035 m^3)	¼ ft^3 (0.007 m^3)	50 lb (22.7 kg) dry hydrated lime
or 1 U.S. gal (3.78 L)	2 U.S. gal (7.57 L)	1 ft^3(0.028 m^3) lime putty
or 1 pt (0.473 L)	1 qt (0.946 L)	1 U.S. gal (3.78 L) lime putty

[A] Where such finish coats are to be applied over base coats containing perlite or vermiculite, the addition of not less than ½ ft^3 (0.014 m^3) of fine aggregate, meeting the sieve analysis of Table 1 shall be added.

[B] For any of the above proportions, up to ½ ft^3 (0.014 m^3) of fine aggregate in accordance with Table 1 shall be permitted to be added to the mix.

[C] The largest particle size of aggregate and its proportion ratio will determine the degree of coarseness of the sand finish. The specifier shall indicate the maximum size desired.

9.6.5 Float finishes of lime putty gauged with gypsum gauging plaster shall be proportioned in accordance with Table 5.

9.6.5.1 When finish in 9.6.6 is to be mixed with sand, the sand shall be graded within the limits shown for basecoats in Specification C 35 except that all of the sand shall pass a No. 8 (2.36 mm) sieve.

9.6.6 Float finishes of lime putty gauged with gypsum Keene's cement shall be proportioned in accordance with Table 5.

9.6.6.1 When the finish in 9.6.6 is to be mechanically mixed wet, follow procedures in 9.6.3.1. For mixing dry, add lime first, then gypsum Keene's cement, and sand. This dry mix shall be mixed to a uniform color and then add water to achieve the desired consistency.

9.6.7 Float finishes of job mixed gypsum and sand shall be proportioned in accordance with Table 5 and Table 6.

9.6.8 Float finishes of ready mixed plaster and sand shall be

TABLE 5 Proportion of Gypsum To Not More Than Lime/Aggregate, with Dry and Wet Equivalents

	Dry						Lime Putty Wet Equivalent		
	Weight, lb (kg)			Volume, ft^3(m^3)			ft^3(m^3)	U.S. gal (litres)	lb (kg)
	Gypsum	Lime	Aggregate	Gypsum	Lime	Aggregate			
Troweled Finishes:[A]									
Lime putty with:[B]									
Gypsum gauging	100 (45.4)	225 (102)	0[A]	1 (.028)	3 (.085)	0	6.75 (.191)	52.5 (199)	450 (204)
Gypsum Keene's cement:									
Medium	100 (45.4)	50 (22.7)	0[A]	1 (.028)	1 (.028)	0	1ln! (.032)	8¾ (33)	100 (45.4)
Hard	100 (45.4)	25 (11.3)	0[A]	1 (.028)	½ (.014)	0	ln%(.018)	4½ (17)	50 (22.7)
Ready-mixed gypsum plaster	100 (45.4)	0	0	1 (.028)	0	0	0	0	0
Gypsum vermiculite	100 (45.4)	0	7 to 15 (3.2 to 6.8)	1 (.028)	0	1 (.028)	0	0	0
Floated Finishes:									
Lime putty with:									
Gypsum gauging	100 (45.4)	225 (102)	200 (90.7)	1 (.028)	3 (.085)	2 (.057)	6.75 (.191)	52.5 (199)	450 (204)
Gypsum Keene's cement:									
Medium	150 (68.0)	100 (45.4)	450 (204)	1½(.042)	2 (.057)	4½(.127)	2¼ (.064)	17½ (66)	200 (90.7)
Ready-mixed gypsum plaster[C]	100 (45.4)	0	0	1 (.028)	0	0	0	0	0
Gypsum-vermiculite	100 (45.4)	0	7 to 15 (3.2 to 6.8)	1 (.028)	0	1 (.028)	0	0	0
Gypsum-sand (job-mixed)[D]	100 (45.4)	0	200 (90.7)	1 (.028)	0	2 (.057)	0	0	0

[A] See 9.6.2, 9.6.3, 9.6.4 and 9.6.5.

[B] If additional hardness of finish coat is desired, increased amounts of gypsum shall be used; however, hard finishes shall not be used over lightweight aggregate base coats.

[C] Mixed with water only, in accordance with manufacturers' printed directions.

[D] Gypsum shall be neat, unfibered plaster.

mixed with water only in accordance with the manufacturer's printed directions.

9.6.9 Float finishes of job mixed plaster shall be proportioned in accordance with Table 6.

9.6.10 Float or textured ready mixed colored plaster shall be mixed in accordance with the manufacturer's printed directions.

9.6.11 Float or textured job mixed color plaster shall be proportioned in accordance with Table 5 or Table 6.

9.6.12 Special finishes shall be proportioned in accordance with the applicable provision of Table 5 or Table 6.

10. Application

10.1 *General*—Plaster shall be applied by hand or machine as specified herein.

10.2 *Plaster Thickness*:

10.2.1 Gypsum plaster shall be applied to the thickness specified in Table 1. Plaster thickness shall be measured from the face plane of all plaster bases (Note 4).

NOTE 4—Installation of wood or metal grounds or plaster screeds will assist in achieving specified thicknesses.

10.3 *Application of Base Coats*:

10.3.1 *Two-Coat Work* (see 3.2.2.6 and Note 5)—The first coat shall be applied with sufficient material and applied with enough pressure to form a good bond to the substrate. The second coat shall be applied by doubling back before the first coat sets and with the plaster brought out to the grounds or screeds or specified thickness.

NOTE 5—Two coat work is not recommended over any type metal plaster base.

10.3.1.1 The plaster shall be straightened to a true plane without application of water leaving the surface porous and rough enough to provide a mechanical bond for the finish coat.

10.3.2 *Three-Coat Work* (see 3.2.2.5 and Appendix X3)—The first coat or scratch coat shall be applied with sufficient material and applied with enough pressure to form tight contact with and a good bond to solid plaster bases, or to form full keys through metal reinforcement. The first coat shall have sufficient depth of material leaving the surface sufficiently porous and rough (scratched or raked) to provide mechanical bond for the second or brown coat.

10.3.2.1 After the first or scratch coat has set, screeds shall be applied over the first coat prior to application of the second coat to ensure full plaster thickness. The second or brown coat shall be applied with sufficient material and pressure to ensure tight contact with the scratch coat and to bring the thickness of the second coat out to the grounds or specified thickness. The second coat shall straighten the surface to a true plane without application of water, and leave the surface sufficiently porous and rough enough to provide mechanical bond for the third or finish coat.

10.3.2.2 Metal base and metal frames for hollow partitions shall be plaster grouted prior to plastering, or shall be grouted solid between base or frame and plaster base at the time of plastering (see 8.3.1 and 8.3.2).

10.3.3 *Plastering on Monolithic Concrete*—All applicable requirements of 8.1, 8.1.1, 8.2, 8.2.1, 8.2.1.1, 8.2.2, A2.5, A2.6 and Table 1 footnotes shall be followed.

10.3.3.1 Walls and columns shall have a coat of dash-bond, bonding compound as provided in A2.5 and A2.6, or metal plaster base or shall have a two coat plaster application as provided in 10.3.1 with a proportion of gypsum neat plaster to aggregate of 100 lb (45.4 kg) of plaster to not more than 3 ft^3 (0.085 m^3) of aggregate.

10.3.3.2 Ceilings shall be prepared as in A2.5 and A2.6 and two coat work applied in accordance with 10.3.1.

10.3.4 *Solid Plaster Partitions with Steel Studs (Metal Lath and Plaster)*—Partitions shall be not less than 2 in. (51 mm) thick and shall have scratch, brown, and finish coats applied in accordance with 10.3.4.1-10.3.4.4. Where studs are temporarily braced, the bracing shall be maintained until the scratch coat on the lath side has set.

10.3.4.1 The scratch coat on the lath side shall be applied first, with sufficient material and pressure to form full keys and embed the lath and with sufficient depth of material to be scored to a rough surface immediately following its application.

10.3.4.2 The back-up coat on the channel side shall be applied in not less than two applications after the scratch coat on the lath side has set and partially dried. The first application shall completely cover the keys of the scratch coat. The second application shall bring the plaster out to the grounds in a true plane with the surface, left porous and sufficiently rough to provide mechanical bond for the finish coat. Extra water shall not be used in straightening the wall to a true plane.

10.3.4.3 The brown coat on the lath side shall be applied after the brown coat on the channel side has set, and left as specified for the finish coat.

10.3.4.4 Plaster shall extend to the floor, except that where plaster grouted combination metal bases and screeds are used, the plaster shall extend to the grout below the top of the base. Spaces between the grounds and all metal door frames shall be filled.

10.3.5 *Studless Solid Partitions (Metal Lath and Plaster)*—Partitions shall be not less than 2 in. (51 mm) thick, constructed using either rib lath or diamond mesh (flat expanded) lath, and with plaster applied in the same number of coats as for solid partitions with steel studs (see 10.3.4).

10.3.5.1 Where rib metal lath is used, the scratch coat shall be applied first to the flat side with temporary bracing on the rib side.

10.3.5.2 Where diamond mesh lath is used, the temporary bracing shall be attached to either side and the scratch coat applied on the opposite side.

10.3.5.3 Sequence application of brown coats in accordance with one of the following methods:

(*1*) Brown coat shall be applied over the face of the scratch coat followed by the brown coat on the opposite side of the partition, or

(*2*) Brown coat shall be applied on the rib side of the metal lath, or on the back side of the scratch coat followed by the brown coat applied over the face of the scratch coat.

10.3.6 *Studless Solid Partitions (Gypsum Lath and Plaster)*—Partitions shall be not less than 2 in. (51 mm) thick and shall be three-coat work on both sides.

10.3.6.1 Scratch coat shall be applied approximately ⅜ in.

(9.5 mm) thick with not less than 3⁄16 in. (4.8 mm) thickness at any point. The application shall be started at the side opposite the temporary bracing, then on the braced side without removing the bracing. The thickness shall be the same on both sides.

10.3.6.2 The brown coat shall be applied to the side opposite the bracing after the scratch coat has set and is partially dry. The bracing shall be removed when this brown coat has set and is partially dry. Both sides shall have been brought to 1⁄16 to 1⁄8 in. (1.6 to 3.2 mm) of the required thickness leaving a surface ready to receive a finish coat (see 10.3.4.2), which will bring the partition to an overall thickness of not less than 2 in. (51 mm).

10.3.7 *Studless Solid Partitions (Multiple-Thickness Gypsum Lath and Plaster)*—Partitions shall be not less than 2 in. (51 mm) thick with three coats on each side. Each side shall be not less than 1⁄2 in. (12.7 mm) thick.

10.3.8 *Gypsum Lath Ceilings Attached by Clips*—Plastered ceiling with clip attached gypsum lath shall be covered with three-coat work with the scratch coat not less than 3⁄16 to 1⁄4in. (4.8 to 6.4 mm) thick over the face of the gypsum lath. The total thickness of scratch and brown coats shall be not less than 7⁄16 in. (11 mm).

10.4 *Application of Finishes*:

10.4.1 Finish coats shall be applied to a partially dry base coat or to a thoroughly dry base coat that has been evenly wetted by brushing or spraying. No water shall be left standing on the surface prior to plaster application (Note 6).

NOTE 6—Precolored plaster is not recommended for smooth-trowel finishes. Where colored plaster is being applied, excessive water causes color variations. See A2.3 and A2.4.

10.4.2 Troweled finishes shall be applied by drawing the first application down tight to the entire base coat surface and doubling back to fill out to a true even surface of from 1⁄16 to 1⁄8 in. (1.6 to 3.2 mm) thick, free from catfaces and other blemishes or irregularities. Final troweling or texturing shall be done with a minimum of water after the finish coat has become firm.

10.4.2.1 Lime putty-gypsum gauged Keene's cement finish (see 9.6.3 and Note 7) shall be done in accordance with 10.4.2 except that the final troweling shall be continued until the finish has set.

NOTE 7—The thickness of this finish may be increased for decorative work as required.

10.4.2.2 Ready mixed gypsum troweled finishes (see 9.4.2) shall be applied in accordance with the printed directions of the manufacturer.

10.4.3 Textured/float finishes shall be applied in accordance with 10.4.2 except the float applied finish shall also meet the requirements of the specific texture required. The texture required shall be created by the type of float surface used (wood, carpet, cork, rubber, or other) and the length of time between trowel application and floating (take-up).

10.4.3.1 The mixes in 10.4.2.1 and 10.4.2.2 used for a float finish shall be applied in accordance with 10.4.3.

10.4.4 Texture or special finishes shall be applied in accordance with the printed directions of the manufacturer.

10.4.5 Job mixed finishes shall be applied in accordance with 10.4.2 or 10.4.3. The texture or special finish coat shall reduce the total thickness by not more than 1⁄16 in. (1.6 mm) while matching the approved sample as closely as possible.

10.4.5.1 Special finishes shall not reduce the combined thickness of the base coat and finish coat to less than that specified in Table 1. Inorganic coloring material added to the job mix shall be in accordance with A2.4.

10.4.5.2 Acoustical plaster and finish shall be applied in accordance with the printed directions of the manufacturer.

11. Keywords

11.1 aggregate; basecoat; ceiling; finish coat; gypsum; lath; plaster; texture

ANNEXES

(Mandatory Information)

A1. GENERAL INFORMATION

A1.1 Surfaces and openings shall be examined before plaster is applied thereto, the proper authorities shall be notified, and unsatisfactory conditions shall be corrected prior to application of plaster.

A2. TECHNICAL INFORMATION

A2.1 *Equivalent Measure for Aggregates*—Six No. 2 shovels of damp loose sand are approximately equivalent to 1 ft^3 (0.028 m^3) and weigh approximately 100 lb (45.4 kg). Perlite and vermiculite are normally packed in bags marked with the cubic foot (or cubic metre) content.

A2.2 *Job Addition of Fiber*—Job added fiber, where specified, shall be nonstaining natural or synthetic fiber, well-shredded and free from grease, oil, dirt, or other materials that adversely affect the strength, bond, or setting time of the plaster. Uniform quantities of the fiber, as specified, shall be added to the mixer, batch by batch, after mixing water, aggregate and plaster have been added. The type and amount of fiber to be used is dependent on job specifications or on that quantity of the selected fiber that provides the desired application characteristics. Where the type and quantity of fiber is not specified by contractual agreement, gradually increased or decreased quantities shall be used until the desired application characteristics are obtained, and that quantity used thereafter.

A2.3 *Ready Mixed Colored Finish Plasters*, shall be mixed and applied in strict accordance with the directions of the manufacturer. Where specified by contractual agreement, sample panels prepared in accordance with the job specifications shall be submitted to the architect or builder for approval of final color prior to the start of any job application.

A2.4 *Job Mixed Colored Finish Plasters*—Job mixed floated or textured colored finish plasters shall be prepared with inorganic or organic water-disperseable, nonbleeding, nonfading, lime-proof coloring materials, dry powdered, paste or liquid. Coloring materials shall be proportioned and thoroughly dispersed into the finish plaster in strict accordance with the directions of the manufacturer of the coloring material. In the absence of specific directions by the manufacturer of the coloring material, uniform quantities, mixer batch by mixer batch, shall be thoroughly dispersed into the mixing water. Uniform, identical proportioning by weight of coloring material, mixing water and dry finish plaster components, and uniform, identical mixing procedure, mixing time and application. Each unbroken area shall be completed in one continuous operation from angle to angle or natural interruption for best uniformity of color. Where specified by contractual agreement, sample panels prepared in accordance with the job specifications, including floating or texturing procedures, shall be submitted to the architect or builder for final color approval prior to the start of any job application. Floating or texturing procedure used for sample preparation shall be strictly followed for job application. The minimum quantity of coloring material necessary to obtain the specified or approved color shall be used, but in no case shall the proportion of coloring material exceed 8 weight % of the cementitious materials used.

A2.5 *Preparation:*

A2.5.1 All surfaces to be plastered shall be free from grease, oil, form release compounds, dirt, dust, paint, laitance, efflorescence or other foreign matter that would inhibit suction or mechanical keying, or cause staining.

A2.5.2 Where a cast-in-place or other monolithic wall is to be plastered and bond is dependent upon the ability of the surface to absorb water (suction) and surface roughness for a mechanical key, bond is obtained by roughening the surface to be plastered by sandblasting, wire brushing, acid wash, or chipping.

A2.5.3 Laitance or efflorescence on concrete and masonry shall be removed by washing first with a 10 % solution of commercial hydrochloric acid (muriatic acid) and water and then with clean water to remove all traces of acid. Grease, oil, and form release compounds shall be removed with cleaning agents compatible with surface and subsequent plastering.

A2.5.4 The use of dressed (smooth) lumber, metal, plastic, or plywood forms, in conjunction with the vibration of concrete, produce bonding surfaces so dense and smooth, that the mechanical key necessary for plaster to bond is absent. In addition, certain kinds of lumber, oiled forms, and exposed reinforcement metals cause staining of the plaster.

A2.6 Bonding compounds that are water re-emulsifiable shall not be used in areas of a building exposed to cyclic or continuous exposure to very humid or wet conditions, or where a dew point condition occurs on the plaster surface. (Dew point conditions frequently occur in commercial laundries, natatoriums, and other wet areas such as showers, hydrotherapy rooms, and where walls are to be washed with water or steam.)

A2.6.1 Applied bonding compound shall be protected from dirt and dust when it will not immediately be covered with plaster. Some bonding compounds require immediate application of plaster to be effective.

A3. MIXING

A3.1 *Hand Mixing*—Mixing boxes shall be watertight. Where mixing is done in a building, provide water-resistive protection under and around the mixing boxes and water containers.

A3.1.1 *Neat Plaster*

The aggregate shall be added to the plaster and mixed dry to a uniform color, pulled to one end of the box, hoed and chopped into water placed in the other end of the box until thoroughly mixed to the desired consistency.

A3.1.2 *Gypsum Ready Mixed Plaster*

Plaster shall be placed at one end of the box without adding aggregate, water shall be placed in the other end and plaster hoed and chopped into water until thoroughly mixed to the desired consistency.

A3.1.3 *Gypsum Wood-Fibered Plaster*

Where used without aggregate, the plaster shall be placed at one end of the box, hoed and chopped into water at the other end until thoroughly mixed to the desired consistency. Where aggregate is added, it shall be mixed in accordance with A3.1.1.

A3.2 *Mechanical Mixing:*

A3.2.1 *Gypsum Neat Plaster*

Aggregate shall be added at the job while the mixer is in continuous operation, as follows:

A3.2.1.1 Put in the approximate (less than total required) amount of water.

A3.2.1.2 If sand is used, add approximately half the sand. If vermiculite or perlite is used, add total amount required.

A3.2.1.3 Add all of the plaster.

A3.2.1.4 Add the remainder of the sand.

A3.2.1.5 Adding water as necessary, mix to the proper consistency. Do not overmix.

A3.2.1.6 Dump the entire batch and use. Clean the mixer for the next batch.

A3.2.2 *Gypsum Ready Mixed Plaster*

Aggregate shall be added at the producer's plant. While the mixer is in continuous operation, do as follows:

A3.2.2.1 Put in the approximate (slightly less than total required) amount of water.

A3.2.2.2 Add the ready mixed plaster.

A3.2.2.3 Adding water or plaster as necessary, mix to the desired consistency.

A3.2.2.4 Dump the entire batch and use. Clean the mixer for the next batch.

A3.2.3 *Gypsum Wood-Fibered Plaster*

Where used with aggregate added, the loading and mixing cycle shall be as prescribed in A3.2.1. Where used without aggregate, the loading and mixing cycle shall be as prescribed in A3.2.2.

A4. BASE COAT PROPORTIONS

A4.1 For job mixed plaster, proportions of sand, perlite, or vermiculite aggregate to 100 lb (45.4 kg) of gypsum neat plaster shall be not more than those specified in Tables 5 and 6.

A4.1.1 Gypsum ready mixed plaster shall be permitted to be used instead of job-prepared mixes of gypsum neat plaster and aggregate provided the proportion of aggregate to plaster is not more than that specified in Tables 5 and 6. Gypsum ready mixed plaster shall be used with the addition of water only, except as otherwise specified.

A4.1.2 *Gypsum Wood-Fibered Plaster*

The addition of extra aggregate to gypsum wood-fibered plaster is not recommended, but in no case shall the addition of sand, perlite, or vermiculite be more than 1 ft^3(0.028 m^3) to 100 lb (45.4 kg) of gypsum wood-fibered plaster.

APPENDIXES

(Nonmandatory Information)

X1. GENERAL INFORMATION

X1.1 The work includes all labor, materials, services, equipment, and scaffolding required to complete the plastering on the project in accordance with the drawings and specifications.

X1.2 Scaffolding shall be constructed and maintained in strict conformity with applicable laws and ordinances.

X1.3 The work shall be coordinated with the work of other trades.

X2. VENTILATION REQUIREMENTS

X2.1 Attics or similar unheated spaces above lath and plaster ceilings shall be adequately ventilated by providing effective cross ventilation for all spaces between roof and top floor ceilings by screened louvers or other approved or acceptable means.

X2.1.1 The ratio of total net free ventilating area to ceiling area shall be not less than 1:150, except that the ratio may be 1:300 provided:

X2.1.1.1 A vapor barrier having a transmission rate not exceeding 1 perm is installed on the warm side of the ceiling, or

X2.1.1.2 Not less than 50 % of the required ventilation area is provided by ventilators located in the upper portion of the space to be ventilated (not less than 3 ft (914.4 mm) above eave or cornice vents) with the balance of the required ventilation area provided by eave or cornice vents.

X2.1.2 Attic space that is accessible and suitable for future habitable rooms or walled-off storage space shall have not less than 50 % of the required ventilating area located in the upper part of the ventilated space as near the high point of the roof as practical and above the probable level of any future ceilings.

X2.2 In buildings with operable windows, ventilation shall be accomplished by keeping windows open approximately 2 in. (50 mm) top and bottom or side-pivoted windows approximately 4 in. (100 mm) to provide air circulation. In buildings without operable windows or having no windows, provisions shall be made to exhaust the moist air to the outside by mechanical means such as temporary circulators, exhaust fans, or the air conditioning system. A portion of the water used in mixing gypsum plaster is necessary for the chemical reaction that rehydrates and hardens the plaster. The excess amount required for mixing and application shall be exhausted from the plastered area after set has taken place. Free circulation of air is necessary to avoid "sweat-outs." Vent enclosed areas lacking natural ventilation by use of temporary mechanical means. If windows are not glazed and the building is subjected to hot dry winds, openings shall be screened with cheesecloth or similar materials. Free circulation of hot dry winds over the face of freshly applied plaster causes "dry-outs" by removing water from the plaster before the rehydration is completed.

X3. RECOMMENDED LOCATIONS OF THREE COAT WORK

X3.1 *Use Three-Coat Work on the Following Types of Construction:*

X3.1.1 Over gypsum lath attached to ceiling supports spaced more than 16 in. (406 mm) on center.

X3.1.2 Over gypsum lath attached to ceiling supports by clips.

X3.1.3 Over ⅜-in (9.5-mm) thick gypsum lath where a vapor retarder is used adjacent to the back of the lath.

X3.1.4 Over metal plaster bases.

SUMMARY OF CHANGES

This section identifies the location of changes to this specification that have been incorporated since the last issue.

(1) Paragraphs A1.2 and A1.3 were moved to 1.3 and 1.4. Changes were made to the wording of 1.3.

Standard Specification for Application of Gypsum Veneer Plaster[1]

This standard is issued under the fixed designation C 843; the number immediately following the designation indicates the year of original adoption or, in the case of revision, the year of last revision. A number in parentheses indicates the year of last reapproval. A superscript epsilon (ϵ) indicates an editorial change since the last revision or reapproval.

This specification has been approved for use by agencies of the Department of Defense.

1. Scope

1.1 This specification covers the minimum requirements for and methods of applying gypsum veneer plaster.

1.2 Where a specific degree of fire resistance is required for veneer plaster systems, applicable building code regulations shall be followed.

1.2.1 Details of construction for a specific assembly to achieve the required fire resistance shall be obtained from reports of fire-resistance tests, engineering evaluations, or listings from recognized fire testing laboratories.

1.3 Where a specific degree of sound control is required for veneer plaster assemblies and constructions, details of construction shall be in compliance with official reports of tests conducted in recognized sound testing laboratories in accordance with the applicable sound tests of Test Methods E 90, C 423, or E 492.

NOTE 1—To ensure desirable results, coordinate this specification with Specification C 844.

NOTE 2—General information regarding matters of a contractual nature concerning veneer plaster work is found in Appendix XI. Additional technical information related to veneer plastering is provided in Appendix X2.

1.4 The values stated in inch-pound units are to be regarded as the standard. The values given in parentheses are for information only.

1.5 The text of this specification references notes and footnotes which provide explanatory material. These notes and footnotes (excluding those in tables and figures) shall not be considered as requirements of the specification.

2. Referenced Documents

2.1 *ASTM Standards:*

C 11 Terminology Relating to Gypsum and Related Building Materials and Systems[2]

C 423 Test Method for Sound Absorption and Sound Absorption Coefficients by the Reverberation Room Method[3]

C 472 Test Methods for Physical Testing of Gypsum, Gypsum Plasters, and Gypsum Concrete[2]

C 587 Specification for Gypsum Veneer Plaster[2]

C 588 Specification for Gypsum Base for Veneer Plasters[2]

C 631 Specification for Bonding Compounds for Interior Gypsum Plastering[2]

C 844 Specification for Application of Gypsum Base to Receive Gypsum Veneer Plaster[2]

C 1047 Specification for Accessories for Gypsum Wallboard and Gypsum Veneer Base[2]

E 90 Test Method for Laboratory Measurement of Airborne Sound Transmission Loss of Building Partitions and Elements[3]

E 492 Test Method for Laboratory Measurement of Impact Sound Transmission Through Floor-Ceiling Assemblies Using the Tapping Machine[3]

3. Terminology

3.1 *Definitions:*

3.1.1 See Terminology C 11.

3.2 *Definitions of Terms Specific to This Standard:*

3.2.1 *base coat, n*—veneer plaster trowel or machine applied as the first coat of a two-component system.

3.2.2 *finish coat, n*—veneer finish plaster trowel or machine applied as the second coat over the base coat plaster in a two-component system.

3.2.2.1 *smooth-trowel finish, n*—a finish resulting from steel troweling with a minimum of water after the plaster has become firm. A smooth finish free of trowel marks, blemishes, or other imperfections.

3.2.2.2 *texture finish, n*—a finish resulting from (*1*) trowel application followed by floating or texturing of the surface with any of a variety of tools using a minimum of water or (*2*) machine application, left as applied or followed by hand texturing.

3.2.3 *joint-reinforcing embedment, n*—the cementitious material used to embed the strip material that is compatible with the veneer plaster used.

3.2.4 *one-component system, n*—a veneer plaster system designed for application directly over approved bases in a single plaster mix applied in a single coat or double-back operation with the same material.

3.2.5 *reinforced joint, n*—a joint between gypsum base that is reinforced with strip material embedded in a cementitious material.

3.2.6 *reinforcing, n*—joint reinforcing tape or mesh (strip material) that is applied over flat joints and interior angles.

3.2.7 *two-component system, n*—a veneer plaster system involving two separate materials mixed and applied separately

---

[1] This specification is under the jurisdiction of ASTM Committee C-11 on Gypsum and Related Building Materials and Systems and is the direct responsibility of Subcommittee C11.03 on Specifications for Application of Gypsum and Other Products in Assemblies.

Current edition approved Apr. 10, 1998. Published December 1998. Originally published as C 843 – 76. Last previous edition C 843 – 96.

[2] *Annual Book of ASTM Standards*, Vol 04.01.

[3] *Annual Book of ASTM Standards*, Vol 04.06.

for base coat and finish coat.

3.2.8 *veneer plaster system, n*—gypsum veneer plaster applied to (*1*) a gypsum base in accordance with 10.1 or (*2*) properly prepared masonry or monolithic concrete surfaces, neither side of which is exposed to moisture. The plaster shall be applied with one or more components not exceeding ¼ in. (6.4 mm) in total thickness.

4. Delivery of Materials

4.1 All manufactured materials shall be delivered in the original packages, containers, or cartons bearing the brand name and manufacturer identification.

5. Protection of Materials

5.1 Plasters and other cementitious materials shall be kept dry until ready to be used; they shall be kept off the ground, under cover, and away from damp walls and surfaces.

6. Environmental Conditions

6.1 *Temperature*—When the ambient outside temperature at the building site is less than 55°F (13°C), maintain a temperature in the building of not less than 55°F (13°C) or not more than 80°F (27°C) for not less than 24 h prior to the application of veneer plaster and for one week after the veneer plaster has set or until the veneer plaster has dried. Distribute heat evenly in all work areas, and use deflection or protective screens to prevent concentrated or uneven heat or cold on the veneer plaster.

NOTE 3—The requirements in 6.1 will minimize the cracking of veneer plaster due to structural movements caused by thermal changes from temperature extremes; and dry-outs due to hot spots caused by concentrated heat.

6.2 *Ventilation*—Ventilation shall be provided to remove excess water given off through the drying process.

7. Materials

7.1 *Gypsum Base for Veneer Plasters* (hereinafter referred to as "gypsum base") Specification C 588.

7.2 *Veneer Plasters*—Specification C 587.

7.3 *Reinforcing*—Noncorrosive strip providing the joint strength requirements of Specification C 587.

7.4 *Liquid Bonding Compounds*—Specification C 631 (see X2.3).

7.5 *Water*—Water shall be clean, fresh, and of potable water quality.

7.6 *Accessories*—Specification C 1047.

8. Surface Preparation

8.1 *Examination*—Carefully examine all surfaces, including but not limited to, unit masonry, monolithic concrete, gypsum bases of all types, and accessories to receive veneer plaster, before the veneer plaster is applied. Notify the proper authorities promptly of all unsatisfactory conditions. The examination shall determine whether the gypsum base has been exposed to excessive sunlight. Do not apply veneer plaster until after such unsatisfactory conditions are rectified to the satisfaction of the plastering contractor.

NOTE 4—The bond of alkaline veneer plaster to gypsum base may be impaired if the base is exposed to direct light or sunlight for extended periods.

8.1.1 Do not apply veneer plaster over any surfaces of unit masonry or concrete that have been coated with any bituminous compound or other waterproofing or dampproofing or form release agent.

8.2 *Conditioning of Surfaces:*

8.2.1 Immediately before the plaster is applied, wet down the masonry surfaces on which suction shall be reduced. Visible water shall not remain on the surface.

8.2.2 Carefully examine accessories, such as corner beads, control joints, casing beads, etc., to ensure that they are straight, curved, plumb, level, square, or true to the required angles and have been applied in a manner to ensure full coverage of flanges before the plaster is applied.

8.3 *Monolithic Concrete Surfaces*—Clean monolithic concrete surfaces of all dust, loose particles, and other foreign matter. Completely remove all grease, oil, noncompatible curing compounds, and form releasing agents. Remove all ridges and protrusions greater that ⅛ in. (3.2 mm) and fill all depressions greater than ¼ in. (6.4 mm) level with portland cement mortar and allow to set and dry. Any further preparation shall be as recommended by the veneer plaster manufacturer.

8.4 *Gypsum Base Joint Reinforcement*—Reinforce all interior angles and flat joints prior to application of the veneer plaster over the base.

8.4.1 *Interior Angles*—Position and secure reinforcement with staples (on 12-in. (305-mm) centers, one side only), veneer plaster, or other cementitious material compatible with the veneer plaster or by using self-adhering strip reinforcement. When stapling, staple along the ceiling edge only for wall-to-ceiling angles and along one edge for wall-to-wall angles. When securing reinforcement with veneer plaster, reinforcement shall be thoroughly embedded so that embedment material is both under and covering reinforcement.

8.4.2 *Flat Joints*—Center reinforcement over the joint line of the gypsum base and secure with staples or veneer plaster if not using self-adhering reinforcement. Keep reinforcement tight and flat against the gypsum base, and when stapling, position the staples no farther than 24 in. (610 mm) apart, staggered along each edge. Embed reinforcement and continue application of the veneer plaster over the field of the base.

NOTE 5: **Caution**—This method provides minimum reinforcement of the joints. To minimize incidence and severity of joint ridging and cracking, use one of the following methods (8.4.2.1, 8.4.2.2, or 8.4.2.3) in the absence of any specific method recommended by the veneer plaster manufacturer. Where paper tape and setting-type joint compounds are used, use 8.4.2.3.

8.4.2.1 *Reinforcement over Set Veneer Plaster*—Tightly trowel veneer plaster over the joint line leaving the plaster feathered out to a width of about 6 in. (152 mm) flush with the face of the gypsum base. Allow the plaster to set; then secure reinforcement over the joint line as described in 8.4.2.

8.4.2.2 *Reinforcement Secured and Embedded with Veneer Plaster*—Apply reinforcement over the joint line as described in 8.4.2. Tightly trowel the plaster over the reinforcement along the joint line to provide thorough embedment of the reinforcement. Allow the joint embedment to set before proceeding with general plastering.

8.4.2.3 *Reinforcement Embedded (no staples)*—Tightly trowel the embedment material to a depth of about 1⁄32 in. (0.8 mm), working the trowel in both directions along the joint line. Center the reinforcement over the joint line. Firmly and evenly press the reinforcement into the soft embedment material using a little soft material on the trowel to bury it completely, and leave the embedment material feathered out to a width of about 6 in. (152 mm) flush with the face of the gypsum base. Allow the embedment material to set before proceeding with general plastering.

9. Mixing

9.1 Do not use frozen, caked, or lumpy material. Do not retemper or use material that has partially set. Mix each batch separately. Clean mixers thoroughly after each batch so as not to accelerate the following batches. This can be done by spray hosing the mixer paddle and containers immediately after each batch. Water ratios and other techniques used for mixing shall conform to the manufacturer's recommendation for specific veneer plaster products. In the absence of explicit directions, follow the recommendations in X2.1.

9.2 *Setting Time*—Setting times are carefully controlled by the manufacturer. Do not make adjustments to the setting time on the job. If problems with setting develop under unusual job conditions, consult the manufacturer of the veneer plaster for recommendations.

9.2.1 Do not use gauging, molding, or casting plasters, lime, gypsum, Keene's cement, portland cement, etc. to adjust the veneer plaster setting time.

9.3 Occasionally, special textures are desired that require additional sand. Approval of the veneer plaster manufacturer shall be obtained prior to any such additions.

10. Application

10.1 *General*—Veneer plasters have widely differing working properties, physical characteristics, and limitations regarding their compatibility with one another and with various bases. Follow the particular methods, techniques, limitations, and procedures set forth by individual manufacturers for their product. In the absence of manufacturers' directions, follow the recommendations in X2.2.

10.2 *Plaster Thickness*—Measure the plaster thickness from the face of the base to which it is applied, exclusive of joint treatment. The minimum thickness shall be as recommended by the manufacturer of the plaster used but in no case less than shown as follows:

Base coats (trowel applied)	1⁄16 in. (1.6 mm) min
Base coats (spray applied)	1⁄16 in. (1.6 mm) min
Finishes (two-component systems)	1⁄32 in. (0.8 mm) to 1⁄16 in. (1.6 mm) min
One-component systems	1⁄16 in. (1.6 mm) min

10.3 Barriers shall be provided to prevent the free circulation of hot, dry winds over the face of freshly applied plaster.

11. Electric Radiant Heat Cable

11.1 Veneer plaster shall be applied to electric heat cable systems in accordance with the veneer plaster manufacturer's directions. In the absence of manufacturer's directions follow the recommendations in X2.5.

11.2 Veneer plaster shall be applied over heating systems in which the wire temperature will, at any time, be in excess of 125°F (52°C).

12. Keywords

12.1 electric heat cable; gypsum; plaster; veneer plaster

APPENDIXES

X1. GENERAL INFORMATION

X1.1 Construct and maintain scaffolding in strict conformity with applicable laws and ordinances so as not to interfere with or obstruct the work of others.

X1.2 Coordinate the work with the work of other trades.

X2. TECHNICAL INFORMATION

X2.1 *Mixing:*

X2.1.1 *Equipment*—A paddle-type agitator fitted to a ½-hp (373-W) heavy-duty, electric drill, rated from 900 to 1000 rpm (no load), and a clean drum of convenient size are recommended for rapid, efficient mixing of veneer plasters.

X2.1.2 *Procedure:*

X2.1.2.1 Put all but 1 or 2 qt (0.94 or 1.89 L) of the proper amount of water in the mixer for each bag of veneer plaster to be mixed.

X2.1.2.2 Add the veneer plaster to the water and immediately mix until uniformly wetted. Veneer plaster may be added while the agitator is turning.

X2.1.2.3 Continue mixing, adding water to obtain the desired lump-free mortar consistency.

X2.1.2.4 When mixing plaster for spray application, mix to a consistency so that 75 to 90 % of the batched mortar will pass a No. 8 (2.36-mm) sieve without shaking. Make sure the plaster is specifically designed for spray application to avoid quick set problems.

X2.1.2.5 Do not overmix. Three to four minutes is usually sufficient to attain fluidity.

X2.2 *Application:*

X2.2.1 *One-Component Plasters*—Apply the plaster with sufficient material and pressure to provide a good bond on the gypsum bases or other bases as approved by the veneer plaster manufacturer. Double back immediately with the same mixer batch of plaster to the desired thickness. Straighten to a true surface without application of water (water will cause blistering when applied at this time). Allow the material to "take-up" and texture with a float or sponge as desired. If a smooth finish is required, allow the plaster to become firm (so that water can be used without blistering) and trowel, using a minimum of water, to achieve a smooth finish, free of catfaces, trowel marks, blemishes, or other imperfections. Complete finishing before the veneer plaster sets.

X2.2.2 *Two-Component Plasters*—The finish component coat may be applied over set but still "green," partially dry, or dry base coat. Avoid application of a second component over a base coat that is damp on joints but dry in the field. It may result in photographing of the joints.

X2.2.3 *Base Coat-Hand Application*—Apply the base coat with sufficient material and pressure to provide a good bond on the gypsum bases or other bases as approved by the veneer plaster manufacturer and straighten to a true surface without application of water. Leave the surface sufficiently rough to provide a mechanical key for the finish coat.

X2.2.4 *Base Coat-Machine Application*—First spray flat joints, beads, and interior angles, and work the material with hand tools to cover the reinforcement, fill the beads, straighten the angles, and feather out the plaster over the joints. Proceed with full areas by spraying with a broad sweeping motion, holding the nozzle close enough to the surface to avoid excessive overspray and buildup in the angles. First apply a light coating while moving in one direction; then immediately double back over the same area, sweeping the nozzle in the opposite direction to obtain a level coat that requires no further working, and that is suitable to receive the finish component coat on setting.

X2.2.5 *Smooth-Trowel Finish*—Apply the plaster by troweling with firm pressure, then doubling back and filling out to a true, even surface. After the plaster has become firm, trowel it well with a minimum amount of water to a smooth finish, free of catfaces, blisters, trowel marks, blemishes, or other imperfections.

X2.2.6 *Texture Finishes*—Apply the plaster over the base coat by troweling thoroughly, building up to an even surface, and then floating with a variety of tools using a minimum of water. It may also be sprayed (depending on the type of texture desired) to a true, even surface free of slick spots or other blemishes.

X2.3 *Bonding Compounds*—Bonding compounds meeting Specification C 631 may be used to bond plaster to portland cement concrete and other sound surfaces. Such bonding agents may be used at the discretion of the specifier who may require performance records from the bonding agent manufacturer. Apply the bonding agent according to the bonding agent manufacturer's directions.

X2.4 *Heating and Ventilation*—A portion of the water used in mixing veneer plaster is necessary for the chemical reaction that rehydrates and hardens the plaster. The excess amount required for mixing and application must be removed from the plaster after set has taken place. Minimum circulation of air is desirable to prevent dry outs.

X2.4.1 Provide proper ventilation above veneer plaster ceilings by cross ventilating the space between the roof and top floor ceilings with screened louvers or other acceptable means. Provide a ratio of total net free ventilating area to the ceiling area of at least 1:150. The ratio may be as low as 1:300 if (*1*) foil-backed gypsum base or an approved vapor retarder is installed on the warm side of the ceiling above bases or (*2*) at least 50 % of the required ventilating area is provided by ventilators at least 3 ft (0.9 m) above the eave and cornice vents.

X2.4.2 For attic space accessible and suitable for future rooms or walled-off storage, locate at least 50 % of the required ventilating area as near the high point of the ventilated space as practicable and above the level of any future ceiling.

X2.4.3 Provide ventilation to remove excess moisture prior to, during, and after application of the gypsum base and plastering.

X2.4.4 Ventilate buildings with operable windows by keeping windows open approximately 2 in. (50 mm) top and bottom (or for side pivoted windows approximately 4 in. (100 mm)) to provide air circulation. Avoid rapid drying. During periods of low indoor humidity, minimum air circulation is desirable following plastering and until the plaster is dry.

X2.4.5 In buildings without operable windows or enclosed areas of buildings that lack natural ventilation, exhaust the moist air to the outside by mechanical means, such as temporary circulators, exhaust fans, or the air conditioning system.

X2.4.6 If glazed sashes are not in place and the building is subjected to hot, dry winds, screen the openings with cheesecloth or similar material. Hot, dry winds can cause "dry outs" by removing the water from the plaster before setting is complete.

X2.5 *Electric Radiant Heat Cable*—Apply veneer plaster over electric radiant heat cable according to this specification except as follows:

X2.5.1 Treat flat joints according to 8.4.2 before the electric heating cable is installed,

X2.5.2 Inspect and test the heating system before applying the veneer plaster,

X2.5.3 Apply veneer plaster in a tight scratch coat flush with the cable, working the trowel parallel to the heating cable. Immediately double back before the scratch has set and build up a thickness of veneer plaster over the cable to a minimum of 1⁄16 in. (1.6 mm) or a maximum of 1⁄8 in. (3.2 mm),

X2.5.4 Do not turn the heating system on until the veneer plaster has dried completely. Deactivate the heating system and provide supplemental heat as required while paint is being applied. Activating the electric heating cable while the veneer plaster is wet or while paint is being applied may produce photographing of the cables. If the room temperature is more than 15°F (8°C) below the desired temperature after decoration is complete, turn up the heat in 5°F (3°C) increments per 24 h period.

Standard Specification for Application of Gypsum Base to Receive Gypsum Veneer Plaster[1]

This standard is issued under the fixed designation C 844; the number immediately following the designation indicates the year of original adoption or, in the case of revision, the year of last revision. A number in parentheses indicates the year of last reapproval. A superscript epsilon (ε) indicates an editorial change since the last revision or reapproval.

This specification has been approved for use by agencies of the Department of Defense.

1. Scope

1.1 This specification covers the minimum requirements for, and methods of, application of gypsum veneer base for gypsum veneer plasters.

1.2 Gypsum veneer base shall not be used as a base for direct adhesive application of ceramic, metal, or plastic tile in areas exposed to abnormal moisture or humidity or temperature.

1.3 Details of construction for a specific assembly to achieve the required fire resistance shall be obtained from reports of fire-resistance tests, engineering evaluations, or listings from recognized fire testing laboratories.

1.4 Where a specific degree of sound control is required for veneer plaster assemblies and constructions, details of construction shall be in accordance with official reports of tests conducted in recognized sound testing laboratories in accordance with the applicable sound tests, Test Method C 423, Test Method E 90, or Test Method E 492.

1.5 Where this specification is more stringent (size or thickness of framing; spacing of fasteners) than the fire-rated construction, this specification shall govern; otherwise, the construction described in the fire test report shall govern.

1.6 Unheated spaces above gypsum veneer base ceilings shall be properly ventilated (see Appendix X3).

1.7 The values stated in inch-pound units are to be regarded as the standard. The SI (metric) values given in parentheses are approximate and are provided for information purposes only.

1.8 The text of this standard references notes and footnotes which provide explanatory material. These notes and footnotes (excluding those in tables and figures) shall not be considered as requirements of the standard.

2. Referenced Documents

2.1 *ASTM Standards:*

C 11 Terminology Relating to Gypsum and Related Building Materials and Systems[2]

C 423 Test Method for Sound Absorption and Sound Absorption Coefficients by the Reverberation Room Method[3]

C 514 Specification for Nails for the Application of Gypsum Board[2]

C 587 Specification for Gypsum Veneer Plaster[2]

C 588 Specification for Gypsum Base for Veneer Plasters[2]

C 754 Specification for Installation of Steel Framing Members to Receive Screw-Attached Gypsum Panel Products[2]

C 954 Specification for Steel Drill Screws for the Application of Gypsum Panel Products or Metal Plaster Bases to Steel Studs from 0.033 in. (0.84 mm) to 0.012 in. (2.84 mm) in Thickness[2]

C 955 Specification for Load-Bearing (Transverse and Axial) Steel Studs, Runners (Track), and Bracing or Bridging, for Screw Application of Gypsum Board and Metal Plaster Bases[2]

C 1002 Specification for Steel Drill Screws for the Application of Gypsum Board or Metal Plaster Bases[2]

C 1007 Specification for Installation of Load Bearing (Transverse and Axial) Steel Studs and Related Accessories[2]

C 1047 Specification for Accessories for Gypsum Wallboard and Gypsum Veneer Base[2]

E 90 Test Method for Laboratory Measurement of Airborne Sound Transmission Loss of Building Partitions and Elements[3]

E 492 Test Method for Laboratory Measurement of Impact Sound Transmission Through Floor-Ceiling Assemblies Using the Tapping Machine[3]

2.2 *American Lumber Softwood Standards:*

PS–20

3. Terminology

3.1 *Definitions*—Definitions shall be in accordance with Terminology C 11.

3.2 *Definitions of Terms Specific to this Standard:*

3.2.1 *base ply*—first ply of gypsum base in multiple ply application.

[1] This specification is under the jurisdiction of ASTM Committee C-11 on Gypsum and Related Building Material and Systems, and is the direct responsibility of Subcommittee C11.03 on Specifications for Application of Gypsum and Other Products in Assemblies.

Current edition approved Oct. 10, 1998. Published April 1999. Originally published as C 844 – 78. Last previous edition C 844 – 98.

[2] *Annual Book of ASTM Standards*, Vol 04.01.

[3] *Annual Book of ASTM Standards*, Vol 04.06.

3.2.2 *face ply*—outer or visible layer of gypsum base in multiple ply application.

3.2.3 *fastener, n*—nails, screws, or staples used for the application of the gypsum base or backing board.

3.2.4 *framing member, n*—that portion of the framing, furring, blocking, etc., to which the gypsum base is attached. Unless otherwise specified, the surface to which abutting edges or ends are attached shall be not less than 1ln$ in. (38 mm) wide for wood members, not less than 1¼ in. (32 mm) wide for steel members, and not less than 6 in. (152 mm) wide for gypsum studs. For internal corners or angles, the bearing surface shall be not less than ¾ in. (19 mm).

3.2.5 *parallel application, n*—gypsum base application where the gypsum base edge is applied parallel to the framing members.

3.2.6 *perpendicular application, n*—gypsum base application where the gypsum base edge is applied at right angles to the framing members.

4. Materials

4.1 Materials shall conform to the respective specifications and standards and to the requirements specified herein.

4.2 *Gypsum Base for Veneer Plasters* (hereinafter referred to as "gypsum base")—Specification C 588.

4.3 *Gypsum Backing Board or Coreboard*—Specification C 442.

4.4 *Special Fire-Retardant Gypsum Base*—Specification C 588, Type X.

4.5 *Foil-Backed Gypsum Base*—Specification C 588.

4.6 *Gypsum Veneer Plaster*—Specification C 587.

4.7 *Nails*—Specification C 514.

4.8 *Screws*:

4.8.1 *Non-load bearing steel*—Specification C 1002.

4.8.2 *Load bearing steel*—Specification C 954.

4.9 *Staples*—No. 16 USS gage 0.063 in. (1.6-mm) flattened galvanized wire with 7⁄16-in. (11.1-mm) wide-crown minimum outside measure and divergent points. Staples shall be used only for the base ply of two-ply application over wood framing.

4.10 *Steel Framing, Non-load bearing*—Specification C 645 for the specified design criteria.

4.11 *Accessories*—Specification C 1047.

4.12 *Adhesive*:

4.12.1 For application of gypsum base to steel framing, the adhesive shall be as recommended by the manufacturer of the gypsum base.

4.13 *Steel Framing Load Bearing*—Specification C 955.

4.14 *Wood Framing*—American Lumber Softwood Standard PS–20.

5. Delivery of Materials

5.1 All materials shall be delivered in the original packages, containers, or bundles bearing the brand name and manufacturer (or supplier) identification.

6. Shipping, Handling and Storage

6.1 Gypsum board shall be shipped so as to be kept dry.

6.2 Gypsum board shall be stored so as to be kept dry, preferably inside a building, under a roof. Gypsum board shall be neatly stacked flat with care taken to prevent sagging or damage to edges, ends, and surfaces.

6.3 Where necessary to store gypsum board outside, it shall be stacked flat, off the ground, supported on a level platform and fully protected from weather and direct sunlight exposure.

6.4 Prior to application, the building exterior shall be so enclosed to protect the gypsum base from exposure to weather and direct sunlight.

NOTE 1—The bond of alkaline veneer plaster to gypsum base will be impaired if the base is exposed to direct light or sunlight for extended periods.

7. Environmental Conditions

7.1 *Application of Gypsum Base and Adhesives*—A room temperature of not less than 40°F (4°C) shall be maintained during application of gypsum base except when adhesive is used for the attachment of gypsum base. For the bonding of adhesive, a room temperature of not less than 50°F (10°C) shall be maintained for 48 h prior to application and continuously thereafter until adhesive is completely dry. When a temporary heat source is used, the temperature shall not exceed 95°F (35°C) in any room or area. Adequate ventilation shall be maintained in the working area during the application and adhesive curing process.

8. Application of Gypsum Base

8.1 *General Requirements*:

8.1.1 Installation of steel framing shall be in accordance with applicable Specifications C 754 or C 1007 except where superceded by requirements of Table 1.

TABLE 1 Maximum Framing Spacing

Single-ply Base (Thickness), in. (mm)		Application to Framing	Maximum On Center Spacing of Framing, in. (mm)
Ceilings:			
	½ (12.7)	parallel	16 (406)
	½ (12.7)[A]	perpendicular	24 (610)
	⅝ (15.9)[B]	perpendicular	24 (610)
Sidewalls:			
	⅜ (9.5)[C]	perpendicular	16 (406)
	½ (12.7)	parallel	16 (406)
	½ (12.7)[A]	perpendicular	24 (610)
	⅝ (15.9)[B]	parallel or perpendicular	24 (610)

[A] ½-in. (12.7-mm) gypsum base applied perpendicular on 24-in. (610-mm) on centers framing shall be considered to be minimum construction.

[B] For ⅝-in. (15.9-mm) base, perpendicular on ceilings and either perpendicular or parallel on sidewalls at 24-in. (610-mm) on centers spacing, only two-component veneer plaster or special joint reinforcement, as recommended by the gypsum veneer plaster manufacturer, shall be used.

[C] ⅜-in. (9.5-mm) gypsum base shall be used over wood framing with two-component veneer plaster systems only.

NOTE 2—For general wood framing requirements, see Appendix X2.

8.1.2 *Method of Cutting*—The gypsum base shall be cut by scoring and breaking or by sawing, working from the face side. When cutting by scoring, the face shall be cut with a knife or other suitable tool and the gypsum base snapped back away from the cut face. The back shall be broken by snapping the gypsum base in the reverse direction, or by cutting with a knife or other suitable tool.

8.1.3 All cut edges and ends of the gypsum base shall be smoothed to obtain neat jointing. Cut-outs for pipes, fixtures, or other small openings on the face and back of the base shall be scribed in outline before removing or cut out with a saw or other suitable tool. The gypsum base shall be scribed where it meets projecting surfaces. Base shall be brought lightly into contact with adjacent panels but not forced into place. Abutting ends and edges shall fit neatly.

8.1.4 When gypsum base is to be applied to both ceiling and walls, the base shall be applied to the ceiling and then to the walls.

8.2 *Fastening, Mechanical*:

8.2.1 The gypsum base shall be held in firm contact with the underlying support while driving any fastener.

8.2.2 Fastener application shall proceed from the center of the field of the gypsum base to the ends and edges or shall begin along one edge and proceed toward the other edge.

8.2.3 Except where required for fire rating, structural performance, or other special considerations, fastening to top or bottom plates shall not be required.

8.2.4 Fasteners shall be spaced not less than ⅜ in. (9.5 mm) from the edges and ends of the gypsum base. Fasteners used at edges or at base ends in horizontal application shall not be more than 1 in. from edges or ends.

8.2.5 Nails and screws shall be driven to a point flush with the gypsum base surface without breaking the surface paper or damaging the surrounding gypsum core and without stripping the framing member around the screws.

8.2.6 Staples shall be driven so that both legs penetrate the support member and the crown bears tightly against the backing board, but does not cut into the face paper. The staple shall be driven with the crown parallel to framing members. Staples shall be used only for the first ply in two-ply gypsum base systems.

8.2.7 Penetration of mechanical fasteners into supporting framing members shall be not less than:

	Nails	Screws	Staples
Wood:	In& in. (22 mm)	In% in. (15 mm)	In% in. (15 mm), each leg
Steel:	...	In# in. (9 mm)	...

8.3 *Control Joints*

8.3.1 Control joints shall be installed in ceilings more than 2500 ft^2 (232 m^2) in area that are constructed with perimeter relief. The distance between control joints shall be not more than 50 ft (15 m) in either direction.

8.3.2 Ceilings of more than 900 ft^2 (85 m^2) in area constructed without perimeter relief shall have control joints installed with spacing not to exceed 30 ft (9 m) in either direction.

8.3.3 Control joints shall be installed where ceiling framing or furring changes direction or where construction joints occur in the base building construction. In partitions, walls, or wall furring, the distance between control joints shall be not more than 30 ft (9 m).

8.3.4 A ceiling height door frame, installed in accordance with the spacing requirements set forth in 8.3, shall be permitted to be used as a control joint.

9. Application of Single-Ply Gypsum Base

9.1 The maximum spacing of framing members for single-ply gypsum base construction shall be not more than those shown in Table 1.

9.2 In single-ply installation, all ends and edges of gypsum base shall occur over framing members or other solid backing except where treated joints occur at right angles to framing or furring members.

9.3 Joints on opposite sides of a partition shall be arranged to occur on different framing members.

9.4 *Fasteners (Single Ply Application)*:

9.4.1 *Nail Spacing*—Space nails a maximum of 7 in. (178 mm) on centers on the ceiling and a maximum of 8 in. (203 mm) on centers on the walls.

9.4.2 *Double Nailing*—Nails shall be spaced as shown in Fig. 1 and applied as follows:

9.4.2.1 Starting at the center of the base, nails shown as solid dots shall be applied by row starting with row 1, then rows 2 and 2A, 3 and 3A, 4 and 4A. Base shall be kept tight against the framing throughout the procedure.

9.4.2.2 Using the procedure in 9.4.2.1, second nails shown by circles shall be applied.

9.4.2.3 As an alternative procedure, second nails shall be

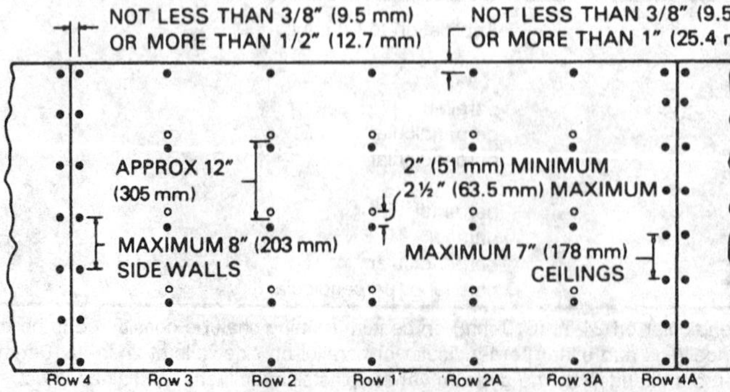

FIG. 1 Double Nailing

permitted to be applied immediately after all first nails in each individual row have been driven in accordance with 9.4.2.1.

9.4.2.4 Single nails shall be used on the perimeter of the base.

NOTE 3—It may be necessary to reset the first nails in each row after the second nails have been set.

9.4.3 Nails for single-ply application over wood members shall be as follows:

Gypsum Base Thickness, in. (mm)	Nails, in. (mm)
3⁄8 (9.5)	1¼ (32)
½ (12.7)	13⁄8 (35)
5⁄8 (15.9)	1½ (38)

9.4.4 Where a specific degree of fire resistance is required for gypsum board assemblies and constructions, nails of same or larger length, shank diameter, and head bearing area, as those described in the applicable fire test report shall be used.

9.4.5 *Screw Spacing*—Screws shall be spaced not more than 12 in. (305 mm) on center along framing members for wall or ceiling application.

9.5 *Floating Interior Angles*—To help minimize the possibility of cracking in areas adjacent to a wall and ceiling intersection, the floating angle method of application shall be permitted to be used. This method is applicable where single nailing, double nailing, or screw attachment is used. When using this method the base shall be applied to ceilings first.

9.5.1 *Ceilings*—Fasteners shall be spaced in accordance with 9.4 except that for single nailing the nail adjacent to the ceiling/wall intersection shall be located 7 in. (178 mm) away from the wall, or 11 to 12 in. (279 mm to 305 mm) away for double nailing or screw application.

9.5.2 *Walls*—The gypsum base shall be fastened to walls to provide a firm support for the floated edges of the ceiling gypsum base. The top attachment shall be located into each stud 8 in. (203 mm) down from the ceiling intersection for single nailing (Fig. 2), and 11 to 12 in. (279 to 305 mm) for double nailing (Fig. 1) or screw application (see Fig. 3 and Fig. 4). At sidewall vertical angles (Fig. 5), the overlapping base shall be applied so as to bring the back of the underlying board into firm contact with the face of the framing member behind it.

10. Adhesive-Mechanical Fastener Application

10.1 Except as modified herein, application shall be in accordance with Section 9.

10.2 Surfaces of gypsum base and framing to receive adhesive shall be free of dust, dirt, grease or any other foreign matter that could impair the bond.

10.3 A bead of adhesive 3⁄8 in. (9.5 mm) in diameter shall be applied to the face of all wood framing members, except plates, that support the gypsum base. As pressure is applied to the base the adhesive shall spread to an average width of ¾ in. (19 mm)

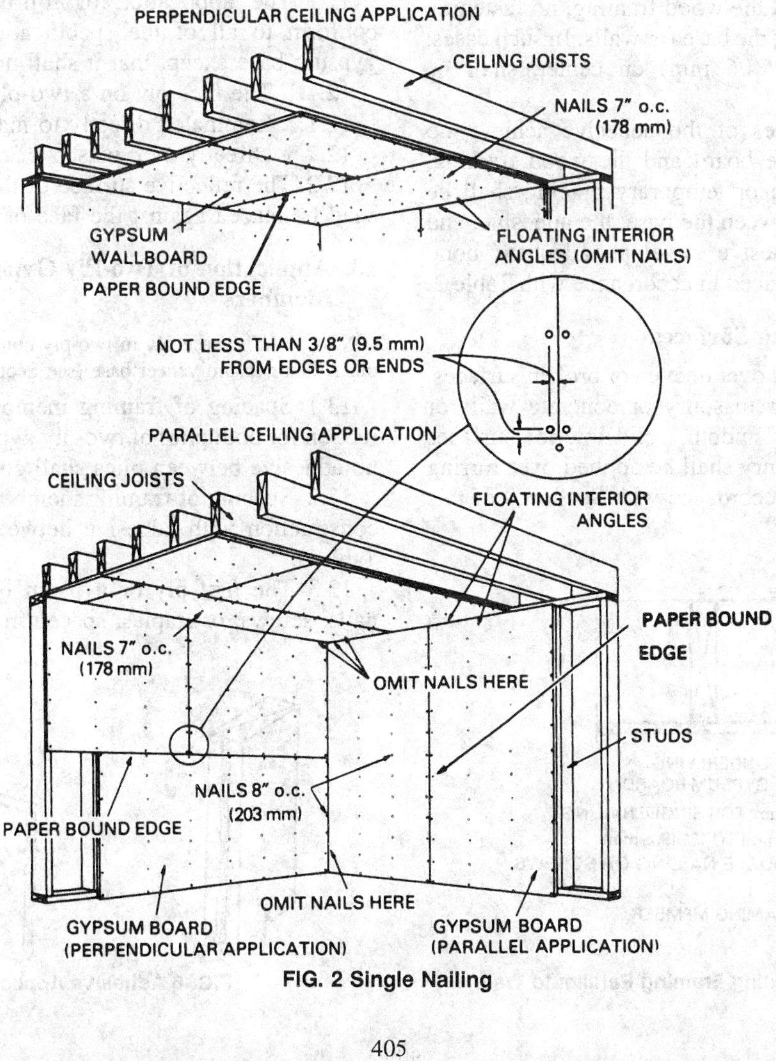

FIG. 2 Single Nailing

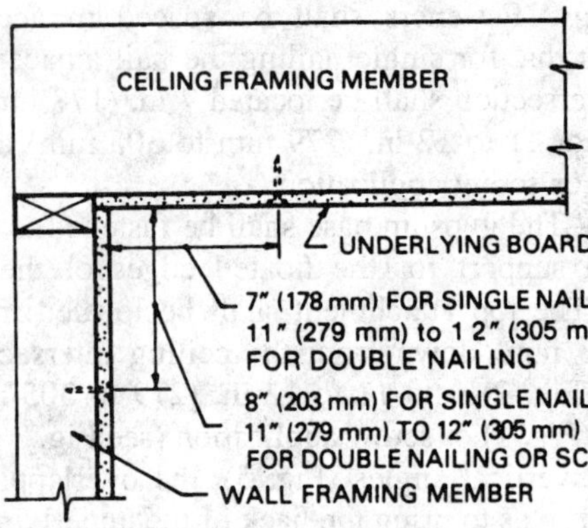

FIG. 3 Vertical Section, Ceiling Framing Perpendicular To Wall

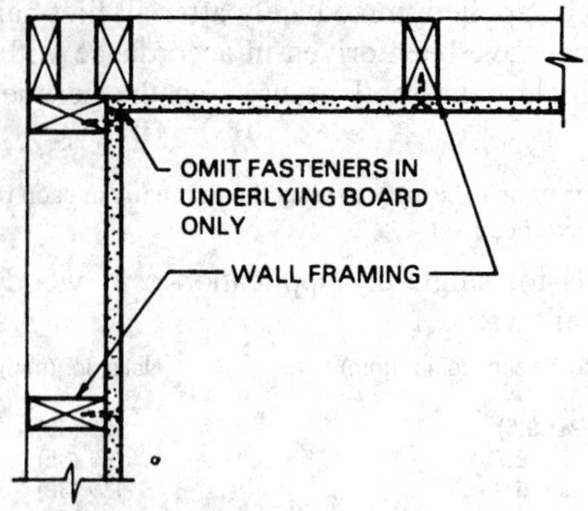

FIG. 5 Horizontal Section Through Interior Vertical Angle

approximately ⅛ in. (3.2 mm) thick. Application patterns shall be shown as in Fig. 6.

10.3.1 Where two adjacent pieces of gypsum base abut (ends or edges) on a framing member, two parallel beads of adhesive shall be applied to the framing member. Application patterns shall be as shown in Fig. 6.

10.4 Adhesive shall be applied to no greater area than can be covered with gypsum base within the "open time" of the adhesive.

10.5 *Fastener Spacing:*

10.5.1 Where the properties of the adhesive ensure bridging between the gypsum base and the wood framing, no fastening shall be required in the field of the base for walls. In such cases, perimeter fastening, 16 in. (406 mm) on center, shall be required.

10.5.2 Where the properties of the adhesive achieve no positive bridging between the board and the wood framing, either temporary field nailing or temporary bracing shall be required to ensure contact between the base, the adhesive, and the stud face until the adhesive develops adequate bond strength. Fasteners shall be spaced in accordance with Table 2.

11. Application over Existing Surfaces

11.1 Gypsum base installed over uneven or broken surfaces, on the interior of all exterior masonry or concrete walls or columns, or over reasonably smooth, solid interior surfaces such as concrete or unit masonry shall be applied over furring strips or furring channels in accordance with Section 9 of this specification.

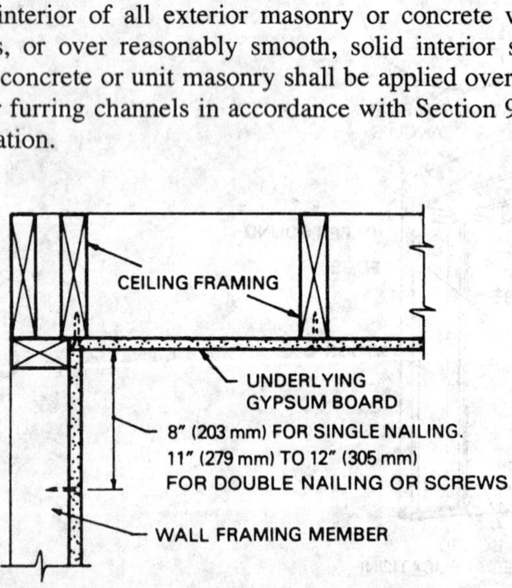

FIG. 4 Vertical Section, Ceiling Framing Parallel to Wall

11.2 Over reasonably smooth, sound, existing surfaces, such as gypsum wallboard or plaster in good condition, the gypsum base shall be applied by using fasteners of sufficient length to penetrate into the underlying framing members in accordance with 8.2.7. Fasteners shall be spaced in accordance with Sections 8 and 9 of this specification.

11.3 Fastener and adhesive attachment of gypsum base to existing surfaces shall be in accordance with Section 10 of this specification.

12. Foil-Backed Gypsum Base

12.1 The application of foil-backed gypsum base shall conform to all of the specifications for the application of gypsum base except that it shall not be used as follows:

12.1.1 The face ply on a two-ply laminating system.

12.1.2 Laminated directly to masonry.

12.1.3 Directly over plaster or other existing surfaces.

12.2 The reflective surface of the foil-backed gypsum base shall be placed against the face of the framing members.

13. Application of Two-Ply Gypsum Base to Framing Members

NOTE 4—The base ply in two-ply construction may be gypsum board other than gypsum veneer base (see Section 4).

13.1 Spacing of framing members and the application direction for both plies of two-ply gypsum base construction with no adhesive between plies shall comply with Table 1.

13.2 Spacing of framing members for two-ply gypsum base construction with adhesive between plies shall comply with Table 3.

13.3 The first ply of gypsum base shall be applied using nails, screws, or staples, spaced in accordance with Table 4.

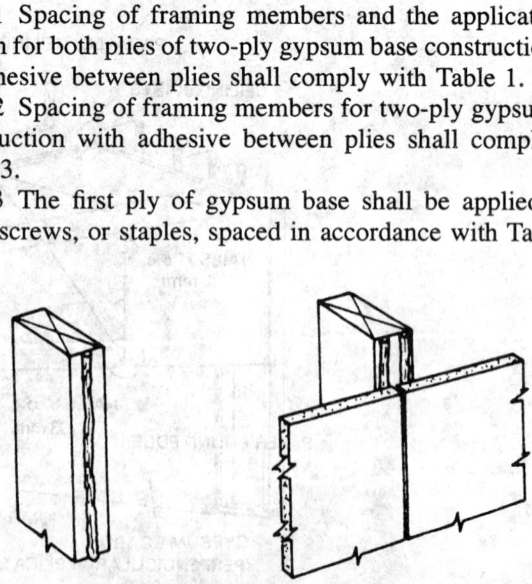

FIG. 6 Adhesive Application Patterns

TABLE 2 Fastener Spacing with Adhesive or Mastic Application and Supplemental Fastening, in. (mm) On Center

Framing Member Spacing	Ceilings		Partitions Load Bearing		Partitions Nonload-Bearing	
	Nail	Screw	Nail	Screw	Nail	Screw
16 (406)	16 (406)	16 (406)	16 (406)	24 (610)	24 (610)	24 (610)
24 (610)	12 (305)	16 (406)	12 (305)	16 (406)	16 (406)	24 (610)

13.4 Where no adhesive is used between plies, the two plies of gypsum base shall be applied either parallel or perpendicular to each other. The face ply shall be applied with the number of nails or screws required for normal single-ply application. Face ply joints that are parallel to framing shall fall over framing members and be offset from the base ply joints when two plies of gypsum base are installed parallel to each other.

13.5 Where a laminating adhesive is used between plies, joints in the face ply shall not be required to occur over framing members. Parallel joints of the two plies shall be staggered (offset) at least 8 in. (203 mm).

13.6 Where gypsum base is adhesively installed, the face ply shall be held firmly in place with temporary fasteners or shoring (supports, props, headers, etc.) in sufficient quantity to ensure pressure for bonding. When sufficient bond has developed, supports shall be removed and temporary fasteners removed or driven flush with the surface.

14. Solid Gypsum Base Partitions

14.1 *Face Ply*—shall be ½ in. (12.7 mm), ⅝ in. (15.9 mm), or multiple laminations of gypsum base of 4-ft (1200-mm) width.

14.2 *Core*—shall be 1-in. thick gypsum panel products, either single or multiple layer in 2-ft (610 mm) or 4-ft (1220 mm) width.

14.3 *Runners*—shall be metal not less than 0.0179 in. (0.046 mm) in thickness or construction grade wood.

14.4 Floor and ceiling runners shall be not less than 24 in. (610 mm) on center. Vertical runners shall be installed at required locations, such as exterior walls, exterior corners, and door frames. Partitions shall be secured at the floor and ceiling in accordance with the gypsum base manufacturer's details or as otherwise required.

14.5 Pressure shall be applied to gypsum base laminated to the backing board to ensure a secure, firm bond. Joints between the gypsum base and the backing board shall be staggered. Fasteners shall be used to ensure a continuous bond between the gypsum base and the backing board.

14.6 Veneer plaster shall not be applied until the adhesive is fully set and the gypsum base is firmly bonded.

15. Adhesive Application of Gypsum Base to Interior Masonry or Concrete Walls

15.1 Direct adhesive application shall be permitted only on interior masonry or concrete walls above grade, or the inside of exterior masonry cavity walls with a cavity not less than 1-in. (25.4-mm) in width between the inside and outside masonry for the full height of the above grade surface to receive gypsum base.

15.2 Foil-backed gypsum base shall not be used for direct adhesive application.

15.3 All surfaces to be adhered shall be free of any foreign matter that could impair the bond. Projections or elevations in the masonry shall be removed by chipping or wire brushing. Large hollows or depressions shall be face-plastered with portland cement and allowed to dry.

15.4 For application to monolithic concrete, porous brick, or concrete block, adhesive shall be applied to the back of the gypsum base or on the wall in continuous parallel beads spaced not more than 12 in. (305 mm) on center or daubs spaced not more than 12 in. on center each way.

15.4.1 Beads shall be not less than ⅜ in. (9.5 mm) in diameter to provide a continuous bond between the gypsum base and the wall surface.

15.4.2 Daubs shall be 2 to 3 in. (51 mm to 76 mm) in diameter.

NOTE 5—**Caution:** Be sure that the adhesive will bond to monolithic concrete. Form release compounds may cause bond failure.

15.5 *Gypsum Base Installation*—Gypsum base shall be positioned to provide a tight fit at abutting edges. The base shall not be slid into place. Moderate pressure shall be used to develop a full adhesive bond.

15.5.1 Veneer plaster shall not be applied until the adhesive is fully set and the gypsum base is firmly bonded.

15.6 *Gypsum Base Application over Rigid Plastic Foam*—

TABLE 3 Maximum Framing Spacing with Adhesive Between Plies

Gypsum Base Thickness, in. (mm)		Application Direction		Maximum On Center Spacing of Framing, in. (mm)
Ceilings				
Base	Face	Base	Face	
⅜ (9.5)	½ (12.7)	perpendicular	perpendicular or parallel	16 (406)
½ (12.7)	½ (12.7)	parallel	perpendicular or parallel	16 (406)
		perpendicular	perpendicular or parallel	24 (610)
⅝ (15.9)	½ (12.7)	perpendicular or parallel	perpendicular or parallel	24 (610)
⅝ (15.9)	⅝ (15.9)	perpendicular	perpendicular or parallel	24 (610)
Sidewalls				
⅜ (9.5)	½ (12.7)	perpendicular or parallel	perpendicular or parallel	16 (406)
½ (12.7)	½ (12.7)	perpendicular or parallel	perpendicular or parallel	24 (610)
⅝ (15.9)	½ or ⅝ (12.7 or 15.9)	perpendicular or parallel	perpendicular or parallel	24 (610)

TABLE 4 Attachment Spacing for Base Ply over Wood or Steel Framing

Fastener	Nails	Screws	Staples
Framing	wood	wood or metal	wood
No adhesive between plies	24 in. (610 mm) on centers	24 in. (610 mm) on centers	16 in. (406 mm) on centers
Adhesive between plies	Section 7.2.1	Section 7.2.2	7 in. (178 mm)

Gypsum base shall be permitted to be applied over rigid plastic foam insulation panels that have been applied to masonry or concrete walls.

15.6.1 Foam insulation shall be applied to the masonry or concrete in accordance with the foam manufacturer's specifications.

15.6.2 Furring strips or special metal furring members shall be attached using mechanical means to the masonry or concrete wall surface. This shall be done either before or after application of insulation, depending on the system used.

15.6.2.1 Furring members shall be installed in accordance with Table 1 of this specification and at the floor-wall angle, wall-ceiling angle (or at the termination of the gypsum base above suspended ceilings), around door, window, and other openings, and where required for cabinet or fixture attachment.

15.6.3 Gypsum base shall be applied to the furring as described in Section 9 of this specification. Mechanical fasteners shall be of such length that they shall not penetrate completely to the masonry or concrete.

16. Gypsum Base Application over Steel Framing and Furring

16.1 *Screw Application*—shall be as specified in Section 9 of this specification.

16.2 *Framing Spacing*—Spacing of steel framing and furring for screw application shall be in accordance with Table 1 for single-ply gypsum base and as specified in Table 3 for two-ply gypsum base.

16.3 *Screw Spacing*:

16.3.1 Screw spacing for single-ply gypsum base and face ply of two-ply gypsum base with no adhesive shall be as specified in 9.4.5 of this specification.

16.3.2 Screw spacing for parallel applied first-ply of two-ply gypsum base over steel framing on side walls with no adhesive between the plies shall be 12 in. (305 mm) on center along the edges of the gypsum base and at third points into each stud or furring channel in the field of the gypsum base.

16.3.3 Screw spacing for a perpendicular applied first-ply of two-ply gypsum base over steel framing on side walls with no adhesive between the plies shall be one screw at each edge at each stud or furring channel intersection and one screw midway between edges at each stud or furring channel.

16.3.4 Screw spacing for perpendicular or parallel applied first-ply of two-ply gypsum base over steel framing with adhesive between plies shall be as specified for single-ply gypsum base in 9.4.5 of this specification.

16.3.5 Screw spacing for the first ply of two-ply gypsum base over steel framing with adhesive between plies on ceilings shall be the same as specified for the face-ply of two-ply gypsum base in 13.4 of this specification. Only a sufficient number of screws to hold the gypsum base in place until the adhesive develops a bond shall be used.

17. Application to Arches

17.1 Gypsum base applied to the soffit of arches (see Table 5) shall be carefully bent into place. If necessary, the base shall first be either dampened or scored approximately 1 in. (25 mm) on center on the back side. In the latter case, after the core has been broken at each cut, the gypsum base shall be applied to the framing member and fastened in place. If the base is dampened, it shall be allowed to dry before plastering.

17.2 To apply the base, a stop shall be placed at one end of the curve and the base shall be gently and gradually pushed on the other end, forcing the center against the framing until the curve is complete.

NOTE 6—By moistening the face and back paper thoroughly and allowing the water to soak well into the core, the base may be bent to still shorter radii. When it dries thoroughly, the base will regain its original hardness.

18. Keywords

18.1 electric heat cable; gypsum; gypsum base; veneer plastic

TABLE 5 Bending Radii for Application to Arches

Gypsum Base Thickness, in. (mm)	Bent Lengthwise, ft (m)	Bent Widthwise, ft (m)
½ (12.7)	10 (3.05)[A]	–
⅜ (9.5)	7½ (2.29)	25 (7.62)

[A] Bending two ¼-in. (6.4-mm) pieces successively permits radii shown for ¼ in.

APPENDIXES

(Nonmandatory Information)

These Appendixes provide general information and also suggestions for inclusions to be made elsewhere by the specifier, and are not a part of this specification.

X1. GENERAL INFORMATION

X1.1 Scaffolding shall be constructed and maintained in strict conformity with applicable laws and ordinances and shall not interfere with or obstruct the work of others. The work shall be properly coordinated with the work of other trades.

X2. WOOD FRAMING REQUIREMENTS

X2.1 Requirements covering framing, furring, spacing, etc., are essential to provide a proper surface to receive the gypsum base. The following requirements should be included in the project specifications for framing and furring:

X2.1.1 All framing members to which gypsum base will be fastened shall be straight and true and spaced not more than the maximum spacings shown in Table 1. Wood framing, bridging, and furring members shall be the proper grade for the intended use and members 2 by 4 in. nominal size or larger shall bear the grade mark of a recognized inspection agency. Framing, bridging, and furring shall be adequate to carry the design or code loading, or both. In place of an applicable local code, the framing shall meet the minimum requirements of FHA for single or multifamily dwellings. The deflection of members supporting gypsum base shall be not more than L/360 of the span at full design load. Headers shall be provided as necessary for the support of fixtures.

X2.1.2 When the gypsum base is nailed to wood cross furring on ceilings, these members shall have a cross section of not less than 1½ by 1½ in. (38 by 38 mm) and shall be spaced in accordance with the requirements of Table 1. Where screw application is used, the furring member shall have a size of not less than ¾ by 2½ in. (19 by 64 mm) actual size. Lumber shall conform to American Softwood Lumber Standard PS-20.

X2.1.3 All vertical solid surfaces that are furred to receive gypsum base shall be furred with wood framing members not less than ¾ in. (19 mm) thick and 1½ in. (38 mm) wide. (Fastener penetration into the framing member shall be in accordance with 8.2.7 of this specification.)

X2.1.4 Insulating blankets or flanges of blankets shall not be applied over framing members that are to receive gypsum base. Foil-backed gypsum base may be used as a vapor retarder where required.

X3. VENTILATING ABOVE GYPSUM BASE CEILINGS

X3.1 Adequate ventilation of attics or similar unheated spaces above gypsum plaster systems is essential to the performance of these systems and shall be designed and provided by others per ASHRAE Fundamentals Handbook or applicable building code.

X4. CABLE ELECTRIC RADIANT HEATING SYSTEMS FOR CEILINGS

X4.1 The application of gypsum base under this specification shall be performed in accordance with the requirements of this specification except as follows:

X4.1.1 The gypsum base shall have a thickness of ½ in. (12.7 mm). The gypsum base shall be applied perpendicular to framing members by using attachment methods as required in Section 9 of this specification.

X4.1.2 Electric heating cables shall be securely attached to the gypsum base in accordance with the recommendations of the cable manufacturer or supplier.

X4.1.3 Cables shall run at right angles to the gypsum base paperbound edges. Cables shall not be positioned directly over joints that run parallel to the direction of cable run.

X4.1.4 Cables shall be positioned not less than 6 in. (152 mm) from all wall-ceiling intersections and openings.

X4.1.5 All inspections and testing of the heating system shall be completed before application of the veneer plaster.

X4.1.6 Under no conditions shall the heating cable have an operating temperature greater than 125°F (52°C).

Designation: C 845 – 96

Standard Specification for Expansive Hydraulic Cement[1]

This standard is issued under the fixed designation C 845; the number immediately following the designation indicates the year of original adoption or, in the case of revision, the year of last revision. A number in parentheses indicates the year of last reapproval. A superscript epsilon (ε) indicates an editorial change since the last revision or reapproval.

1. Scope

1.1 This specification covers hydraulic cements that expand during the early hardening period after setting.

2. Referenced Documents

2.1 *ASTM Standards:*

C 33 Specification for Concrete Aggregates[2]

C 109/C 109M Test Method for Compressive Strength of Hydraulic Cement Mortars (Using 2-in. or 50-mm Cube Specimens)[3]

C 114 Test Methods for Chemical Analysis of Hydraulic Cement[3]

C 183 Practice for Sampling and the Amount of Testing of Hydraulic Cement[3]

C 185 Test Method for Air Content of Hydraulic Cement Mortar[3]

C 188 Test Method for Density of Hydraulic Cement[3]

C 465 Specification for Processing Additions for Use in the Manufacture of Hydraulic Cements[3]

C 688 Specification for Functional Additions for Use in Hydraulic Cements[3]

C 806 Test Method for Restrained Expansion of Expansive Cement Mortar[3]

C 807 Test Method for Time of Setting of Hydraulic Cement Mortar by Modified Vicat Needle[3]

3. Terminology

3.1 *Definitions:*

3.1.1 *expansive cement (K)*—an expansive cement containing anhydrous calcium aluminosulfate ($4CaO \cdot 3Al_2O_3 \cdot SO_3$), calcium sulfate, and uncombined calcium oxide.

3.1.2 *expansive cement (M)*—an expansive cement containing calcium aluminate cement and calcium sulfate.

3.1.3 *expansive cement (S)*—an expansive cement containing tricalcium aluminate (C_3A) and calcium sulfate.

3.1.4 *expansive hydraulic cement*—a cement composed essentially of hydraulic calcium silicates, calcium aluminates and calcium sulfates, which, upon being mixed with water, forms a paste that increases significantly in volume during the early hardening period occurring after setting. This volume increase is due principally to the reaction of the calcium aluminates and calcium sulfates.

3.1.5 *shrinkage-compensating concrete*—a concrete made with an expansive cement in which the expansion, if restrained, induces compressive stresses that approximately offset tensile stresses induced by drying shrinkage.

4. Classification

4.1 The cement covered by this specification is suitable for use in shrinkage-compensating concrete and shall be designated as Type E-1.

NOTE 1—Three kinds of expansive cement are identified in Section 3 by the letters K, M, and S. The appropriate letter should be used as a suffix to the type designation when one of these kinds is desired. Expansive cement may also be prepared in other ways.

5. Ordering Information

5.1 When it is desired that an optional requirement apply or that the cement be of a particular kind, a specific request to this effect should be made by the purchaser.

6. Chemical Composition

6.1 Expansive cement shall conform to the requirements as to chemical composition prescribed in Table 1.

6.2 The purchaser has the option of specifying that the alkalies ($Na_2O + 0.658K_2O$) shall not exceed 0.60 % (See Table 2).

NOTE 2—This limit should be specified when the cement is to be used in concrete with aggregates that may be reactive. Reference should be made to Specification C 33 for suitable criteria of deleterious reactivity.

7. Physical Properties

7.1 Expansive cement shall conform to the physical requirements prescribed in Table 3.

8. Additions

8.1 The cement covered by this specification shall contain no additions except as follows:

8.1.1 Water, or a functional addition, or both, that has been shown to meet the requirements of Specification C 688, may be added in amounts such that the limits shown in Table 1 for loss on ignition shall not be exceeded.

8.1.2 At the option of the manufacturer, processing addi-

TABLE 1 Standard Chemical Requirements

	Composition, %
Magnesium oxide (MgO), max	6.0
Insoluble residue, max	1.0
Loss on ignition, max	4.0

TABLE 2 Optional Chemical Requirements

Alkalies ($Na_2O + 0.658 \times K_2O$) max, %	0.60

[1] This specification is under the jurisdiction of ASTM Committee C-1 on Cement and is the direct responsibility of Subcommittee C01.13 on Special Cements.

Current edition approved Jan. 10, 1996. Published March 1996. Originally published as C 845 – 76 T. Last previous edition C 845 – 90.

[2] *Annual Book of ASTM Standards*, Vol 04.02.

[3] *Annual Book of ASTM Standards*, Vol 04.01.

TABLE 3 Physical Requirements

Time of setting, min, minutes	75
Air content, max, vol %	12.0
Restrained expansion of mortar:	
7-day expansion:	
min, %	0.04
max, %	0.10
28-day, percentage of 7-day expansion, max	115
Compressive strength, min:	
7-day, psi(MPa)	2100(14.7)
28-day, psi(MPa)	3500(24.5)

tions may be used in the manufacture of the cement, provided such materials in the amounts used have been shown to be not harmful in accordance with Specification C 465.

9. Test Methods

9.1 The cement shall be sampled and the properties enumerated in this specification shall be determined in accordance with the following methods:

9.1.1 *Sampling*—Practice C 183.

9.1.2 *Chemical Analysis*—Test Methods C 114.

9.1.3 *Time of Setting*—Test Method C 807.

9.1.4 *Air Content*—Test Method C 185, using the actual specific gravity of the cement if it differs from 3.15 by more than 0.05 in calculating the air content.

9.1.5 *Restrained Expansion of Mortar*—Test Method C 806.

9.1.6 *Compressive Strength*—Test Method C 109/C 109M, except that a water-cement ratio of 0.500 shall be used, the specimens shall be covered with a polyethylene sheet or other suitable material for preventing loss or gain of moisture at the surface of the specimens during the moist storage period in the molds, and the specimens shall remain in the molds for 3 days.

9.1.7 *Density*—Test Method C 188.

10. Inspection

10.1 Every facility shall be provided the purchaser for careful inspection and sampling of the finished cement. The finished cement may be inspected and sampled either at the mill or at the site of the work, as may be specified by the purchaser. The following periods from time of sampling shall be allowed for completion of testing:

7-day test	12 days
28-day test	33 days

10.2 The manufacturer shall provide suitable facilities to enable the inspector to check the intergrinding or blending operation used to produce cement by either means. The plant facilities for intergrinding or blending and inspection shall be adequate to ensure compliance with the provisions of this specification.

11. Rejection

11.1 The cement may be rejected if it fails to meet any of the requirements of this specification.

11.2 Cement remaining in bulk storage at the mill prior to shipment for a period greater than 6 months or cement in bags in local storage in the hands of a vendor for more than 3 months, after completion of the tests, may be retested for expansion and may be rejected if it fails to conform to the requirements of this specification for expansion.

11.3 Packages more than 2 % below the net weight marked thereon, may be rejected. Also, if the average net weight of packages in any shipment, as shown by weighing 50 packages at random, is less than that marked on the packages, the entire shipment may be rejected.

12. Certification

12.1 At the request of the purchaser, the manufacturer shall state in writing the composition of a finished cement.

12.2 The composition of any cement shall conform to that shown in the statement of the manufacturer within the following tolerances:

	Tolerance, ± %
Silicon dioxide (SiO_2)	3
Aluminum oxide (Al_2O_3)	2
Calcium oxide (CaO)	3

12.3 At the request of the purchaser, the manufacturer shall state in writing the nature, amount, and identity of any processing or functional addition that may have been used. Also, if requested, the manufacturer shall supply test data showing compliance of any processing addition with the provisions of Specification C 465, and any functional addition with the provisions of Specification C 688.

12.4 Upon request of the purchaser in the contract or order, a manufacturer's certification that the material during production was tested in accordance with this specification together with a report of the test results shall be furnished at the time of shipment.

13. Packaging and Package Marking

13.1 When the cement is delivered in packages, the words "expansive cement," the type, the kind (if the cement is one of the three kinds defined in Section 3), the name and brand of the manufacturer, and the net weight shall be plainly marked on each package. Similar information shall be provided in the shipping documents accompanying the shipment of packaged or bulk cement. All packages shall be in good condition at the time of inspection.

14. Storage

14.1 The cement shall be stored in a manner such as to permit easy access for proper inspection and identification of each shipment and in a suitable weathertight building that will protect the cement from dampness and minimize warehouse set.

15. Keywords

15.1 calcium aluminate; calcium aluminosulfate; calcium oxide; calcium silicate; calcium sulfate; cement; expansive; hydraulic; shrinkage-compensated; tricalcium aluminate

Designation: C 847 – 95

Standard Specification for Metal Lath[1]

This standard is issued under the fixed designation C 847; the number immediately following the designation indicates the year of original adoption or, in the case of revision, the year of last revision. A number in parentheses indicates the year of last reapproval. A superscript epsilon (ε) indicates an editorial change since the last revision or reapproval.

This specification has been approved for use by agencies of the Department of Defense. Consult the DoD Index of Specifications and Standards for the specific year of issue which has been adopted by the Department of Defense.

1. Scope*

1.1 This specification covers sheet lath, expanded metal lath, diamond mesh, flat and self-furring, and rib metal lath, ⅛ and ⅜ in. (3.2 and 9.6 mm), all with or without backing and designed to be used as a base for gypsum or portland cement plaster.

1.2 The values stated in inch-pound units are to be regarded as the standard. The values in parentheses are for information only.

2. Referenced Documents

2.1 *ASTM Standards:*

A 366/A 366M Specification for Steel, Sheet, Carbon, Cold-Rolled, Commercial Quality[2]

A 653/A 653M Specification for Steel Sheet, Zinc-Coated (Galvanized) or Zinc-Iron Alloy-Coated (Galvannealed) by the Hot-Dip Process[3]

3. Material

3.1 Metal lath shall be fabricated from cold-rolled carbon steel sheet of commercial quality conforming to Specification A 366/A 366M. Galvanized metal lath shall have a G60 coating in accordance with Specification A 653/A 653M.

3.2 Backing shall be netting, film, paper, or felt attached to the lath sufficiently enough to prevent accidental removal during shipping, handling, or installation. Attachment of backing shall also allow lapping of metal to metal and backing to backing, 1 in. (25.4 mm) on the ends and ½ in. (12.7 mm) on the sides.

4. Dimensions, Mass, and Permissible Variations

4.1 *Thickness*—The nominal thickness of diamond mesh and flat rib metal lath shall be ⅛ in. (3.2 mm). The nominal thickness of other rib metal lath shall be as designated, ⅜ in. (9.6 mm). The nominal thickness of self-furring diamond mesh shall be 5⁄16 in. (7.9 mm).

4.2 *Width*—The nominal width of metal lath shall be 27 in. (686 mm).

4.3 *Length*—The nominal length of metal lath shall be 96 in. (2438 mm).

4.4 *Weight*—The nominal weight of metal lath shall be as follows:

4.4.1 *U.S. Nominal Weights:*

Type:	Weight, lb/yd^2 (kg/m^2)
Diamond mesh	2.5 (1.4); 3.4 (1.8)
Flat rib	1.8 (1.0); 2.75 (1.5); 3.4 (1.8)
⅜-in. rib	3.4 (1.8); 4.0 (2.1)
Sheet	4.5 (2.4)

4.4.2 *Canadian Nominal Weights:*

Type:	Weight, lb/yd^2 (kg/m^2)
Diamond mesh	2.5 (1.4); 3.0 (1.6); 3.4 (1.8)
Flat rib	1.8 (1.0); 2.5 (1.4); 3.0 (1.6)
⅜-in. rib	3.0 (1.6); 3.5 (1.9); 4.0 (2.1)

4.5 *Permissible Variations*—The permissible variations shall be as follows:

4.5.1 *Thickness,* ±1⁄64 in. (0.4 mm).

4.5.2 *Width,* ±3⁄16 in. (4.8 mm).

4.5.3 *Length,* −0 in., +2½ in. (12.7 mm).

4.5.4 *Weight,* ±10 %.

5. Finish

5.1 Metal lath shall be coated with a water barrier film such as asphalt or non-reemulsifiable water base paint unless fabricated from galvanized steel.

6. Inspection

6.1 Inspection of the material shall be agreed upon between the purchaser and supplier as part of the purchase contract.

7. Rejection and Rehearing

7.1 Any rejection shall be based upon the specific cause of failure to conform to the requirements of this specification, and shall be reported to the seller within 10 working days from the receipt of the shipment by the purchaser.

7.2 Claims for rehearing shall be valid only if made within 20 working days from receipt of notice of specific cause for rejection.

---

[1] This specification is under the jurisdiction of ASTM Committee C-11 on Gypsum and Related Building Materials and Systems and is the direct responsibility of Subcommittee C11.02 on Specifications and Test Methods for Accessories and Related Products.

Current edition approved Oct. 10, 1995. Published December 1995. Originally published as C 847 – 77. Last previous edition C 847 – 93.

[2] *Annual Book of ASTM Standards*, Vol 01.03.

[3] *Annual Book of ASTM Standards*, Vol 01.06.

*** A Summary of Changes section appears at the end of this specification.**

8. Certification

8.1 Where specified in the purchase order, a producer's or supplier's certification shall be furnished to the purchaser that the material was in accordance with this specification and has been found to meet the specified requirements.

9. Packaging and Package Marking

9.1 Metal lath shall be packaged ten sheets per bundle, and self-furring lath shall be five or ten sheets per bundle (option of manufacturer).

9.2 When shipped for resale, the name of the manufacturer or the supplier and the brand shall be legibly marked on each lath or package.

10. Keywords

10.1 gypsum cement plaster; lath; metal lath; portland cement plaster; rib lath; self-furring; sheet lath

SUMMARY OF CHANGES

This section identifies the location of changes to this specification that have been incorporated since the last issue. Committee C-11 has highlighted those changes that affect the technical interpretation or use of this specification.

(1) Paragraph 2.1 was revised.
(2) Paragraph 3.1 was revised.
(3) Keywords were added.

Designation: C 897 – 96

Standard Specification for Aggregate for Job-Mixed Portland Cement-Based Plasters[1]

This standard is issued under the fixed designation C 897; the number immediately following the designation indicates the year of original adoption or, in the case of revision, the year of last revision. A number in parentheses indicates the year of last reapproval. A superscript epsilon (ε) indicates an editorial change since the last revision or reapproval.

1. Scope*

1.1 This specification covers natural or manufactured aggregate for use in job-mixed base and finish-coat full thickness portland cement, portland cement-lime and modified portland cement plasters.

1.2 The values stated in inch-pound units are to be regarded as the standard. The SI (metric) values given in parentheses are approximate and are provided for information purposes only.

1.3 The text of this specification references notes and footnotes that provide explanatory material. These notes and footnotes (excluding those in tables and figures) shall not be considered as requirements of the specification.

2. Referenced Documents

2.1 *ASTM Standards:*

C 11 Terminology Relating to Gypsum and Related Building Materials and Systems[2]

C 29/C 29M Test Method for Unit Weight and Voids in Aggregate[3]

C 40 Test Method for Organic Impurities in Fine Aggregates for Concrete[3]

C 87 Test Method for Effect of Organic Impurities in Fine Aggregate on Strength of Mortar[3]

C 88 Test Method for Soundness of Aggregates by Use of Sodium Sulfate or Magnesium Sulfate[3]

C 123 Specification for Lightweight Pieces in Aggregate[3]

C 125 Terminology Relating to Concrete and Concrete Aggregates[3]

C 136 Test Method for Sieve Analysis of Fine and Coarse Aggregates[3]

C 142 Test Method for Clay Lumps and Friable Particles in Aggregates[3]

D 75 Practice for Sampling Aggregates[4]

3. Terminology

3.1 *Definitions*—Definitions used in this standard shall be in accordance with Terminologies C 11 and C 125.

3.2 *Definitions of Terms Specific to This Standard:*

3.2.1 *aggregate, n*—a granular material such as natural or manufactured sand used with a cementing medium to form plaster (stucco).

3.2.2 *manufactured sand*—the fine material resulting from the crushing and classification by screening, or otherwise, of rock, gravel, or blast furnace slag.

3.2.3 *natural sand*—the fine granular material resulting from the natural disintegration of rock.

4. Composition

4.1 *Deleterious Substances*—The amount of deleterious substances in aggregates, each determined on independent samples (see Test Method C 40) complying with the grading requirements of Section 6, shall not be more than the following:

Item	Maximum Permissible Weight, %
Friable particles	1.0
Light weight particles, floating on liquid having a specific gravity of 2.0	0.5

5. Physical Properties

5.1 *Soundness* (see 7.1.8):

5.1.1 Except as herein provided, aggregate subjected to five cycles of the soundness test shall show a loss, when weighed in accordance with the grading of a sampling complying with the limitations set forth in Section 6, not greater than 20 % when sodium sulfate is used, or 15 % when magnesium sulfate is used.

5.2.1 Aggregates failing to meet the requirements of 5.1.1 shall be accepted, provided there is evidence that plasters of comparable properties made from similar aggregates from the same source has been exposed to weathering, similar to that to be encountered, for a period of more than 5 years without appreciable disintegration.

6. Dimensions, Mass, and Permissible Variations

6.1 *Aggregate for Base Coat*—Aggregate for use in base coat portland cement-based plasters shall be graded as follows:

---

[1] This specification is under the jurisdiction of ASTM Committee C-11 on Gypsum and Related Building Materials and Systems, and is the direct responsibility of Subcommittee C 11.02 on Specifications and Test Methods for Accessories and Related Products.

Current edition approved Nov. 10, 1996. Published January 1997. Originally published as C 897 – 78. Last previous edition C 897 – 95a.

[2] *Annual Book of ASTM Standards*, Vol 04.01.

[3] *Annual Book of ASTM Standards*, Vol 04.02.

[4] *Annual Book of ASTM Standards*, Vol 04.03.

* A Summary of Changes section appears at the end of this specification.

U.S. Standard Sieve	Percent Retained (by Weight)			
	Natural Sand		Manufactured Sand	
	max	min	max	min
No. 4 (4.75 mm)	0	0	0	0
No. 8 (2.36 mm)	10	0	10	0
No. 16 (1.18 mm)	40	10	40	10
No. 30 (600 μm)	65	30	65	30
No. 50 (300 μm)	90	70	80	60
No. 100 (150 μm)	100	95	90	75
No. 200 (75 μm)	100	97	100	90

6.1.1 Not more than 50 % shall be retained between any two consecutive sieves shown in the above table, nor more than 25 % between No. 50 (300-μm) and No. 100 (150-μm) sieves. The amount of material finer than a No. 200 (75-μm) sieve shall not exceed 3 %.

6.1.2 The fineness modulus shall fall between 2.05 and 3.05. If the fineness modulus varies by more than 0.25 from the value assumed in selecting proportions for the plaster, the aggregate shall be rejected unless adjustments are made in proportions to compensate for change in grading.

NOTE 1—The gradation of aggregate used in portland cement-based plaster affects the workability, shrinkage characteristic porosity, and durability of the plaster.

6.2 *Aggregate for Finish Coat Plaster*—Aggregate for finish coat plaster shall all pass the No. 8 (2.36 mm) sieve and meet the remaining gradation requirements of 6.1, except where special texture finishes require gradation of aggregate outside of these limits.

6.2.1 When special texture finishes require gradation of the aggregate outside of the limits of 6.1, sample panels shall be required and evaluated in accordance with criteria defined in the purchase agreement.

NOTE 2—Special texture finishes may require modification of the plaster formulation and application to compensate for the effect of aggregate gradation.

7. Sampling and Testing

7.1 The aggregate shall be sampled and tested as specified in the following methods, except as otherwise specified in this specification:

7.1.1 *Sampling,* Practice D 75.

7.1.2 *Fineness Modulus,* Terminology C 125, is obtained by adding the total percentages shown by the sieve analysis to be retained on each of the sieves in 6.1 and dividing the sum by 100.

7.1.3 *Materials Finer than No. 200 (75-μm) Sieve,* Test Method C 87 by washing.

7.1.4 *Friable Particles,* Test Method C 142.

7.1.5 *Lightweight Constituents,* Specification C 123.

7.1.6 *Organic Impurities,* Test Method C 40.

7.1.7 *Effect of Organic Impurities on Strength,* Test Method C 87.

7.1.8 *Soundness,* Test Method C 88.

7.1.9 *Sieve Analysis,* Test Method C 136.

7.1.10 *Unit Weight,* Test Method C 29/C 29M.

8. Inspection

8.1 Inspection of the material shall be agreed upon between the purchaser and supplier as part of the purchase contract.

9. Rejection and Rehearing

9.1 Any rejection shall be based upon the specific cause of failure to conform to the requirements of this specification, and shall be reported to the seller within ten working days from the receipt of the shipment by the purchaser.

9.2 Claims for rehearing shall be valid only if made within 20 working days from receipt of notice of specific cause for rejection.

10. Certification

10.1 When specified in the purchase order a producer's or supplier's certification shall be furnished to the purchaser that the material was prepared or manufactured in accordance with this specification and has been found to meet the specified requirements.

11. Packaging and Package Marking

11.1 Where packaged, the materials shall be packaged in substantial commercial containers of the type, size, and kind commonly used for the purpose, so constructed as to preserve the contents in good condition, and to ensure acceptance and safe delivery by common or other carriers, at the lowest rate, to the point of delivery.

11.2 The containers shall be so made that the contents can be partially removed without destroying the container's ability to serve as a receptacle for the remainder of the contents.

11.3 The containers shall be labeled to show the content, weight, and the name of the manufacturer or supplier.

11.4 Where delivered in bulk quantities, the bill of lading shall show the quantity by weight, or cubic yards, and the name of the manufacturer or supplier.

11.5 Each package or container shall be marked with the name of manufacturer, description of contents, and amount.

12. Keywords

12.1 aggregate; fine aggregate; gradation; natural and manufactured sand; particle shape and size; plaster stucco; soundness

SUMMARY OF CHANGES

This section identifies the location of changes to this standard that have been incorporated since the last issue. Committee C-11 has highlighted those changes that affect the technical interpretation or use of this standard.

(1) Section 4 was deleted.

The American Society for Testing and Materials takes no position respecting the validity of any patent rights asserted in connection with any item mentioned in this standard. Users of this standard are expressly advised that determination of the validity of any such patent rights, and the risk of infringement of such rights, are entirely their own responsibility.

This standard is subject to revision at any time by the responsible technical committee and must be reviewed every five years and if not revised, either reapproved or withdrawn. Your comments are invited either for revision of this standard or for additional standards and should be addressed to ASTM Headquarters. Your comments will receive careful consideration at a meeting of the responsible technical committee, which you may attend. If you feel that your comments have not received a fair hearing you should make your views known to the ASTM Committee on Standards, 100 Barr Harbor Drive, West Conshohocken, PA 19428.

Designation: C 911 – 96

Standard Specification for Quicklime, Hydrated Lime, and Limestone for Chemical Uses[1]

This standard is issued under the fixed designation C 911; the number immediately following the designation indicates the year of original adoption or, in the case of revision, the year of last revision. A number in parentheses indicates the year of last reapproval. A superscript epsilon (ϵ) indicates an editorial change since the last revision or reapproval.

1. Scope

1.1 This specification covers lime and limestone products suitable for chemical uses as shown in Table 1. The type designations signify the following:

Limestone, high-calcium—CL
Limestone, dolomitic—DL
Limestone, magnesian—ML
Quicklime, high-calcium—CQ
Quicklime, dolomitic—DQ
Quicklime, magnesian—MQ
Hydrated lime, high-calcium—CH
Hydrated lime, dolomitic—DH
Hydrated lime, magnesian—MH
Hydrated lime, by-product—BH

NOTE 1—Lime is used in environmental systems when the control of factors influencing the environment such as gases, liquids and solids is required.

1.2 The buyer shall designate the use, as listed in Table 1, and may specify one or more of the type designations in 1.1.

2. Referenced Documents

2.1 *ASTM Standards:*

C 25 Test Methods for Chemical Analysis of Limestone, Quicklime, and Hydrated Lime[2]

C 50 Methods of Sampling, Inspection, Packing, and Marking of Lime and Limestone Products[2]

C 51 Terminology Relating to Lime and Limestone (as Used by the Industry)[2]

C 110 Test Methods for Physical Testing of Quicklime, Hydrated Lime, and Limestone[2]

C 400 Test Method for Quicklime and Hydrated Lime for Neutralization of Waste Acid[2]

3. General Requirements

3.1 Quicklime shall be reasonably free of unslakable residues and shall be capable of disintegrating in water to form a suspension of finely divided material. Rapidity of slaking is generally a desirable quality.

3.2 The amount of residue shall not exceed that stated in the physical requirements of Table 1 or as agreed upon between the manufacturer and the purchaser (the residue is the amount of material retained on the screen).

4. Chemical Composition and Physical Properties

4.1 The requirements for quicklime, hydrated lime, and limestone for the designated end uses are as shown in Table 1, and are on the basis of the weight of sample taken at the place of manufacture, except as noted in footnote *B* after the requirement. In this case, the requirement is on a moisture and carbon dioxide-free basis.

5. Sampling and Inspection

5.1 Conduct the sampling, inspection, rejection, retesting, packing, and marking in accordance with Methods C 50.

NOTE 2—Rags are cooked for the manufacture of paper in a digester under steam pressure with lime or with lime and soda ash. They are then washed to eliminate as much of the noncellulose material as possible. A standard composition without rejection limits is specified for the reason that lime of either higher or lower total oxides, available lime, calcium oxide, or calcium hydroxide than the standard, may safely be used under suitable conditions for the purpose herein specified, depending primarily upon economic considerations. In the present state of the art, it is believed that the more serviceable type of specification for the product herein specified is that which defines a reasonable standard rather than one that fixes actual rejection limits. It is generally recognized that, other things being equal, lime meeting this standard is preferable to lime that does not, and lime surpassing the standard should be considered of premium quality.

NOTE 3—Lime is used in the "milk of lime" or "tank" system of sulfite pulp manufacture for making the cooking liquor. The milk of lime is held in solution or suspension in a series of tanks equipped with suitable agitators. The sulfur dioxide (SO_2) is forced or drawn through these tanks successively. In some cases, the tanks are built on top of each other in the form of a tower. The contents of the first tank are drawn off when the liquor has reached a certain strength (3.5 to 6 % total SO_2) and the contents of the second and third tanks progress to the first and second tanks respectively. The third tank is again charged with fresh milk of lime. There are other systems of absorption that provide for continuous instead of intermittent operation. The function of the lime is to furnish the base for the formation of the bisulfites of calcium and magnesium.

NOTE 4—In the manufacture of silica brick, silica in the form of massive quartzite or quartz conglomerate is ground until the particles are less than ¼ in. (6 mm) in size. Lime in the form of either slaked or hydrated lime is then added in quantities varying from 1.5 to 3.0 % calcium oxide (CaO), with sufficient water to produce about 5 to 7 % moisture content, and the shapes are molded and dried. They are then burned in downdraft or tunnel kilns until most of the quartzite has been converted into tridymite or cristobalite.

NOTE 5—In the softening of water for municipal and industrial supplies, lime is used alone or with coagulant aids to produce a precipitate that assists in the clarification of water, removal of bacteria, and removal of hardness. Lime and soda ash may be used together for softening water. The lime serves as a chemical reagent in water softening. The only useful constituent of the lime is the calcium oxide content of the quicklime or hydrated lime capable of reacting with the other chemicals in the water or added to it. Inert material, besides reducing the value in proportion to its amount, also makes more sludge to be disposed of for a given amount of chemical action and thus reduces the capacity of the equipment in which it is used.

NOTE 6—In addition to water softening, lime is also used for silica removal from boiler feed-water, and for color removal and clarification of water for municipal and industrial supplies, for most such applications, high-calcium lime is preferred. However, for silica removal from boiler feedwater, dolomitic lime should be employed.

[1] This specification is under the jurisdiction of ASTM Committee C-7 on Lime and is the direct responsibility of Subcommittee C07.03 on Industrial Uses.

Current edition approved June 10, 1996. Published August 1996. Originally published as C 911 – 79. Last previous edition C 911 – 94.

[2] *Annual Book of ASTM Standards*, Vol 04.01.

TABLE 1 Limes and Limestones for Chemical Uses

Use	ASTM Specification[A]	Notes Referenced	Approved Types of Lime or Stone	Chemical Requirements, %: CaO min	CaO, +MgO min	SiO_2 +Fe_2O_3, +Al_2O_3, max	MgO, max	CO_2, max	SiO_2 + insoluble matter, max	Al_2O_3 + Fe_2O_3, max	Available CaO, min	Fe_2O_3, max	Chemical Other	Physical Requirements: Dry Sieve % −12 mesh min	Wet Sieve % − 30 mesh min	Wet Sieve % − 200 mesh min	Wet Sieve % − 325 mesh min
Cooking rags	C 45	2	CH	...	...	...	...	...	...	...	64.3[C]	...	...	...	...	...	...
Cooking rags	C 45	2	CQ	...	...	...	...	...	...	...	90.0[C]	...	...	...	...	...	...
Sulfite pulp	C 46	3	CQ, MQ	...	95.0[B]	3.0[B]	...	...	...	...	...	...	...	...	...	...	...
Sulfite pulp	C 46	3	CL, ML	...	95.0[B]	3.0[B]	...	...	...	...	...	...	...	...	...	...	...
Silica brick	C 49	4	CH	90[B]	...	...	2.5[B]	2.5	3.0[B]	1.5[B]	...	...	Free CaO ≤1.5	...	99	95	...
Silica brick	C 49	4	CQ	90[B]	...	...	2.5[B]	2.5	3.0[B]	1.5[B]	...	...		...	...	...	...
Water softening	C 53	5	CQ	...	...	...	...	...	...	...	90	...	...	...	...	...	...
Water softening	C 53	5	CH	...	...	...	...	...	...	...	68.1	...	...	...	...	...	...
Other water treatment uses	C 53	6	CQ, DQ, MQ	...	93.0	...	...	...	...	...	...	...	...	...	...	...	...
Other water treatment uses	C 53	6	CH, DH, MH	...	93.0[B]	...	...	...	...	...	...	...	...	...	...	...	...
Carbide	C 258		CQ	92[B]	...	...	1.75[B]	4.0	2.0[B]	1.0[B]	...	0.5	P ≤ 0.02[B] S 0.2[B]	...	...	...	...
Grease	C 259		CH	...	...	...	1.5	...	1.0	...	...	0.5	available Ca $(OH)_2$ ≥90	...	...	98	95
Calcium silicate product	C 415		CH	90[B]	...	...	1.3[B]	2.5	3.0[B]	1.5[B]	...	...	...	...	99	95	...
Calcium silicate product	C 415		CQ	90[B]	...	...	1.3[B]	2.5	3.0[B]	1.5[B]	...	...	...	99	...	...	...
Hypochlorite bleach	C 433	7	CH	...	...	...	...	...	...	...	68	0.3[B]	...	...	...	...	...
Hypochlorite bleach	C 433	7	CQ	...	...	...	...	...	...	...	90	...	...	...	...	...	...
Industrial waste treatment	C 826	8	CH DH MH	...	95.0[B]	...	...	5.0	...	...	...	...	BF > 0.72[D] BF > 0.81[D] BF > 0.74[D]	...	...	85	...
Industrial waste treatment	C 826	8, 9	BH	...	90.0[B]	...	...	...	...	...	...	...	...	...	...	70	
Industrial waste treatment	C 826	8	CQ DQ MQ	...	90.0	...	...	...	...	...	...	...	BF > 0.93[D] BF > 1.06[D] BF > 0.93[D]	...	...	85[E]	...
Industrial waste treatment	C 826	8, 10	CL DL ML	...	90[B]	...	...	...	...	...	...	...	BF > 0.45[D] BF > 0.56[D] BF > 0.48[D]	...	...	...	...

[A] These specifications have all been incorporated into the current issue of this specification and are now discontinued. The discontinued specifications are available through University Microfilms International, 300 North Zeebe Rd., Ann Arbor, MI 48106.

C 45 Specification for Quicklime and Hydrated Lime for Cooking of Rags in Paper Manufacture
C 46 Specification for Quicklime and Limestone for Sulfite Brick Manufacture
C 49 Specification for Quicklime and Hydrated Lime for Silica Brick Manufacture
C 53 Specification for Quicklime and Hydrated Lime for Water Treatment
C 258 Specification for Quicklime for Calcium Carbide Manufacture
C 259 Specification for Hydrated Lime for Grease Manufacture
C 415 Specification for Quicklime and Hydrated Lime for Calcium Silicate Products
C 433 Specification for Quicklime and Hydrated Lime for Hypochlorite Bleach Manufacture
C 826 Specification for Lime and Limestone Products for Industrial Waste Treatment

[B] On a nonvolatile basis.
[C] Standard composition.
[D] BF means Basicity Factor.
[E] Slaked basis.

NOTE 7—In manufacture of calcium hypochlorite bleach, lime hydrate in water suspension is reacted with chlorine. Lime hydrate suitable for this application should be rapidly reactive, low in sludge-forming impurities, and particularly low in iron oxide, that may catalyze bleach decomposition.

NOTE 8—When determining the quantity of liming material for the neutralization of liquid acidic wastes, use is made of an acid value–basicity factor relationship. Acid value and basicity factor test methods are in the Appendix. Calculate the acid value–basicity factor relationship as follows:

$$\text{Acid Value} \times \frac{56.08}{98.08} \times \frac{1}{\text{Basicity Factor}} = \text{grams of lime}$$

or limestone product per litre of acidic waste.

NOTE 9—When agreed upon between the manufacturer and the purchaser, the methods for neutralization coefficient as given in Test Method C 400.

NOTE 10—The chemical requirements for Type BH, shall be determined by agreement between the manufacturer and the purchaser.

NOTE 11—For limestone, sizing will affect the basicity factor results and size as shipped shall be determined by agreement between the manufacturer and the purchaser.

6. Test Methods

6.1 The chemical analyses shall be made in accordance with Test Methods C 25.

6.2 The physical tests shall be made in accordance with Test Methods C 110.

7. Keywords

7.1 calcium oxide; calcium silicate; dolomitic lime; hydrated lime; hypochlorite bleach; industrial waste; lime; limestone; quicklime; silica brick; sulfite pulp

APPENDIXES

(Nonmandatory Information)

X1. TEST METHOD FOR FREE CALCIUM OXIDE IN HIGH-CALCIUM HYDRATED LIME

X1.1 Scope

X1.1.1 This test method covers the determination of the amount of free calcium oxide (CaO) in high-calcium hydrated lime capable of being hydrated by steam at atmospheric pressure.

X1.2 Summary of Test Method

X1.2.1 This test method is based on the principle of gain in weight of CaO when it is hydrated to form calcium hydroxide ($Ca(OH)_2$).

X1.2.2 The sample is dried, subjected to steam at atmospheric pressure, again dried, and the gain in weight calculated to CaO.

X1.3 Apparatus

X1.3.1 *Drying Oven*, thermostatically controlled, CO_2-free atmosphere.

X1.3.2 *Analytical Balance.*

X1.3.3 *Steam Bath*, as shown in Fig. X1.1.

X1.4 Procedure

X1.4.1 Weigh a clean, dry, 10-mL Erlenmeyer flask on an analytical balance.

NOTE X1.1—Once started, complete the test without interruption.

X1.4.2 Add to the flask, by means of a widestem funnel, 3 to 5 g of the lime to be tested. Reweigh to get the exact sample weight.

NOTE X1.2—The flask should be kept stoppered at all times, except when weighing, drying, or steaming.

X1.4.3 Place the flask in the drying oven that has been previously heated to 120°C, and maintain at this temperature for 30 min. Remove, stopper, cool in a desiccator, and weigh.

NOTE X1.3—Oven atmosphere can be kept CO_2 free by placing therein a container of soda lime, quicklime, or other suitable CO_2-absorbing medium.

X1.4.4 Place the flask and the sample in a vigorously boiling steam bath and steam for 30 min. Protect the flask in the steam bath by a drip shield during the steaming period. Remove the flask to the oven, dry for 30 min at 120°C, stopper, cool in a desiccator, and weigh.

X1.4.5 Record all weighings to the fourth decimal place.

X1.5 Calculation

X1.5.1 Calculate the percentage of water and CaO as follows:

$$\text{Free water, \%} = [(W_1 - W_2)/S] \times 100$$

$$\text{Free CaO, \%} = \frac{(W_3 - W_2) \times 3.114}{S} \times 100$$

where:

W_1 = weight of sample and flask before drying,
W_2 = weight of sample and flask after drying,
W_3 = weight of sample and flask after steaming and drying, and
S = weight of sample.

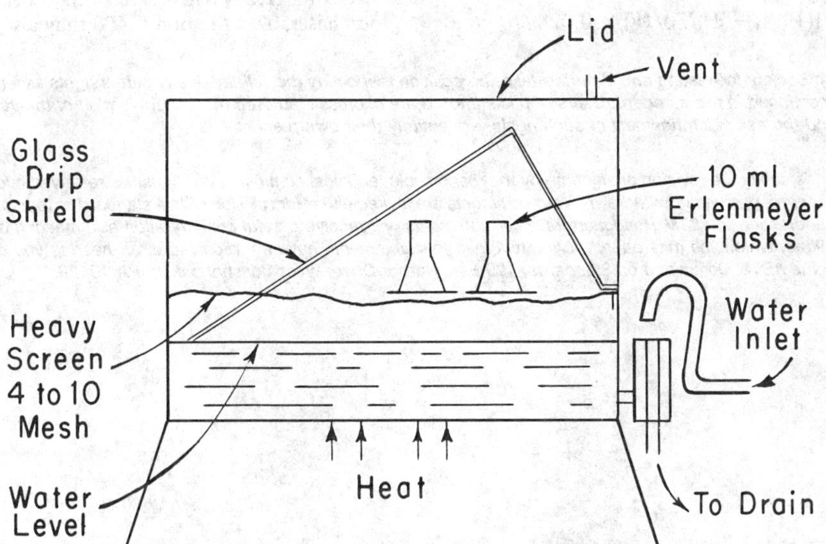

FIG. X1.1 Steam Chamber

X2. TEST METHOD FOR DETERMINING ACID VALUE–BASICITY FACTOR RELATIONSHIP

X2.1 Scope

X2.1.1 This test method covers determination of the quantity of liming material required for the neutralization of liquid acid wastes.

X2.2 Summary of Test Method

X2.2.1 The relationship is based on the basicity factor of the liming material and the acid value of the liquid acidic waste.

X2.2.2 The basicity factor of a liming material is a measure of the alkalinity that avoids dependence on chemical analysis. Basicity factor is grams of calcium oxide equivalent per gram of lime or limestone product and may be used for comparing the relative neutralizing values of the various liming materials for treating acid wastes.

X2.2.3 The acid value is an expression of grams of sulfuric acid equivalent per litre of acid waste.

X2.2.4 The determination is related on a stoichiometric basis to lime neutralization requirements of a liquid acid waste. Unless otherwise stated, neutralization is interpreted to occur at a pH of 8.40.

X2.3 Procedure

X2.3.1 *Basicity Factor*—Weigh 1 g ± 1 mg of the sample that has been prepared to pass a No. 100 (150-μm) sieve and transfer to a 500-mL Erlenmeyer flask containing approximately 20 mL of cold water. (For quicklime samples, heat to boiling to ensure complete slaking). Add from a buret 100 mL of 0.5000 *N* sulfuric acid and stopper with a two-hole rubber stopper. Place the flask on a hot plate and boil for 15 min. (Glass beads may be added to prevent bumping.) Remove the flask from the hot plate and cool in water. Add several drops of phenolphthalein indicator solution and titrate the excess acid with 0.5000 *N* potassium hydroxide solution. Calculate the basicity factor as follows:

$$\text{Basicity factor} = [(V_1N_1 - V_2N_2)/W] \times 0.028$$

where:

V_1 = H_2SO_4 (mL) required for titration of the sample,
N_1 = normality of H_2SO_4,
V_2 = KOH solution (mL) required for titration of the excess acid,
N_2 = normality of KOH solution, and
W = sample (g).

NOTE X2.1—A water-cooled condenser with a tapered, ground-glass joint and a magnetic stir plate may be used during the boiling step.

X2.3.2 *Acid Value*—Pipet 10 mL of the sample of acidic waste into a 250-mL Erlenmeyer flask. (This quantity may be increased or decreased depending on the strength of the acidic waste.) Add an excess of 25 to 30 mL of 0.5000 *N* potassium hydroxide solution and boil for 3 min. Filter through a small Büchner funnel with the aid of vacuum. Wash the residue on the funnel with water. Add several drops of phenolphthalein indicator solution to the filtrate and titrate the excess potassium hydroxide with 0.5000 *N* sulfuric acid solution. Calculate acid value as follows:

$$\text{Acid value} = [(V_1N_1 - V_2N_2)/V] \times 0.049$$

where:

V_1 = KOH solution (mL) required for titration of the sample,
N_1 = normality of KOH solution,
V_2 = H_2SO_4 (mL) required for titration of the excess KOH solution,
N_2 = normality of H_2SO_4, and
V = sample (mL).

X2.3.3 *Relationship of Acid Value to Basicity Factor*—Calculate the acid value–basicity factor relationship as follows:

$$\text{Acid value} \times \frac{56.08}{98.08} \times \frac{1}{\text{basicity factor}}$$

$$= \text{lime (g) or limestone product per litre of acidic waste.}$$

NOTE X2.2—When agreed upon between the manufacturer and the purchaser, Test Method C 400 may also be used.

Designation: C 917 – 98

Standard Test Method for Evaluation of Cement Strength Uniformity From a Single Source[1]

This standard is issued under the fixed designation C 917; the number immediately following the designation indicates the year of original adoption or, in the case of revision, the year of last revision. A number in parentheses indicates the year of last reapproval. A superscript epsilon (ε) indicates an editorial change since the last revision or reapproval.

1. Scope

1.1 This test method is intended for use in instances where the purchaser desires information on the strength uniformity of a hydraulic cement produced at a single source. It is intended that this method normally be used for the predominant cement manufactured at a cement plant. Guidelines for sampling, testing, presentation of results, and evaluation are given.

1.2 The values stated in inch-pound units are to be regarded as the standard.

1.3 *This standard does not purport to address all of the safety concerns, if any, associated with its use. It is the responsibility of the user of this standard to establish appropriate safety and health practices and determine the applicability of regulatory limitations prior to use.*

2. Referenced Documents

2.1 *ASTM Standards:*

C 109/C 109M Test Method for Compressive Strength of Hydraulic Cement Mortars (Using 2-in. or 50-mm Cube Specimens)[2]

C 150 Specification for Portland Cement[2]

C 219 Terminology Relating to Hydraulic Cement[2]

C 595M Specification for Blended Hydraulic Cements[2]

C 1157M Performance Specification for Blended Hydraulic Cement[2]

E 456 Terminology Relating to Quality and Statistics[3]

3. Terminology

3.1 *Definitions*—For definitions of terms relating to this test method refer to Terminology C 219 and Terminology E 456.

4. Significance and Use

4.1 This test method is designed to present in a standardized format information on the variability of strength of cement from a single source over a period of time. It can be applied to all hydraulic cements covered in Specifications C 150, C 595, and C 1157M.

NOTE 1—It should be recognized that concrete strength variability is influenced by other factors in addition to cement strength variability.

5. Sampling

5.1 All sampling shall be performed by quality control or testing personnel, or someone specifically trained for this purpose.

5.2 Take random grab samples from delivery units or during the loading or unloading process. Delivery units larger than 125 tons shall be sampled during loading or unloading. If samples are taken during loading or unloading, the two or more portions which are to be composited to make a sample shall be taken during the transfer to no more than 125 tons of cement. Identify samples by the date on which the cement they represent was shipped or received.

NOTE 2—Standard statistical procedures are recommended for ensuring that samples are selected by a random procedure. These procedures can be used to select the days within a month or within a week that samples will be taken. Then the delivery unit or the time of day can be chosen randomly.

5.3 If taken from a truck or rail car, take at least two separate 5-lb (approximately 2.3-kg) grab samples and thoroughly mix together to obtain a minimum 10-lb (4.5-kg) test sample. Sample only through hatches in the top of the unit. Remove approximately a 12-in. (300-mm) layer of cement. Make a hole before obtaining a sample to avoid collecting dust collector material that may be discharged into the delivery unit after the cement flow ceases.

5.4 If taken from another point in the loading or unloading process, the sample shall consist of a minimum of two separate 5-lb (approximately 2.3-kg) grab samples thoroughly mixed together or at least 10 lb (4.5 kg) as accumulated by a continuous sampler. Take care to avoid segregation and contamination of samples taken from screws, pneumatic systems, or air slides.

5.5 When samples are taken at the cement plant and shipments or rate of production of the cement exceeds 25 000 tons (23 000 Mg) per month, take samples at a rate of at least ten per month and at least two per week. When shipment or rate of production of the cement is less than 25 000 tons (23 000 Mg) per month, take samples at a rate of at least one per 2500 tons (2300 Mg). When samples are taken at the cement plant, in no instance shall samples be taken more frequently than one per 200 tons (180 Mg) of cement shipped

[1] This test method is under the jurisdiction of ASTM Committee C-1 on Cement and is the direct responsibility of Subcommittee C01.27 on Strength.

Current edition approved July 10, 1998. Published December 1998. Originally published as C 917 – 79. Last previous edition C 917 – 95a.

[2] *Annual Book of ASTM Standards,* Vol 04.01.

[3] *Annual Book of ASTM Standards,* Vol 14.02.

or received, except that sampling of consecutive shipments is permitted when they result from randomization.

6. Procedure

6.1 Test all samples for 7 and 28-day compressive strength in accordance with Test Method C 109 using three specimens for each test age. To be comparable, all tests used in a single evaluation must be made in a single laboratory, and preferably by the same laboratory operator.

6.1.1 When separate evaluations of a single source are made by two or more laboratories, additional tests of a standard cement or exchange of portions of the same sample of cement may be necessary to determine differences in testing that are likely to be obtained in the different laboratories. Five or more batches may be necessary to obtain a valid comparison between laboratories. Statistical techniques must be used to assess the validity of differences that might be obtained. Participation in the Cement Reference Sample Program of the CCRL by both laboratories will be helpful in resolving differences that are found.

6.1.2 When two laboratories exchange portions of the same sample and prepare single batches, results from the two laboratories shall not differ by more than 18.7 % of the average of the two laboratories (see Test Method C 109 multilaboratory d2s). If a larger number of samples are exchanged the difference in average strength shall not exceed $18.7/\sqrt{n}$ % of the overall average strength, where n is the number of samples exchanged and tested by each laboratory. A more precise calculation is outlined in Appendix X1.

6.2 Mix duplicate batches of mortar to determine the effect of testing variations on the uniformity of results made in a single laboratory. Make duplicate batches on a day different from the original batch of mortar.

6.2.1 When a uniformity testing program is started on shipments from a single source, make duplicate batches of mortar from every third cement sample. When duplicate tests have been made from a minimum of five cement samples, calculate the average range, $\bar{R}$, for the available duplicates, then calculate standard deviation and coefficient of variation for testing according to 7.1.3 and 7.1.4, respectively. Increase the number of duplicate batches used in the calculation until the results of ten cement samples are used in the calculation. After that time, use only the ten most recent results of duplicate testing in the calculation of the standard deviation and coefficient of variation for testing. See Table 1.

6.2.2 When at least ten sets of duplicate batches have been made and the coefficient of variation for testing is less than 4.0 %, the frequency of testing duplicate batches may be reduced to one out of ten consecutive cement samples. Resume testing one sample out of three if the coefficient of variation later exceeds 4.0 %. If the coefficient of variation for testing exceeds 5.5 %, the data are of questionable precision, and laboratory procedures and equipment should be thoroughly examined.

6.2.3 Use the results of duplicate tests indicating acceptable precision to estimate the single-laboratory testing variation for all other types of cement tested in that laboratory during the same period of time, provided that duplicate tests have been made on at least one sample per month.

7. Calculation

7.1 The calculations shall include the following:

7.1.1 *Average Strength:*

$$\bar{X} = \frac{X_1 + X_2 + \ldots + X_n}{n} \quad (1)$$

where:

$\bar{X}$ = average strength,

$X_1, X_2, \ldots, X_n$ = strength of individual tests, each of which is composed of the average of three cubes in accordance with Test Method C 109, and

n = number of individual samples.

7.1.2 *Total Standard Deviation:*

$$S_t = \sqrt{\frac{(X_1 - \bar{X})^2 + (X_2 - \bar{X})^2 + \ldots + (X_n - \bar{X})^2}{(n - 1)}} \quad (2)$$

where:

S_t = standard deviation, psi.

7.1.3 *Standard Deviation For Testing:*

$$S_e = 0.862\bar{R}$$

where:

S_e = standard deviation estimated from tests of duplicate batches mixed in a single laboratory,

R = range, the difference between the strengths of the duplicate batches from a single sample (all numbers are positive),

$\bar{R}$ = average of the individual ranges, R, for the preceding ten tests of duplicate batches. See 6.2.1 if fewer than ten ranges are available, and

0.862 = range coefficient for duplicate tests of the same sampleof cement.

7.1.4 *Coefficient of Variation for Testing:*

$$V_e = 100\, S_e / \bar{X}$$

where:

V_e = coefficients of variation estimated from tests of duplicate batches mixed in a single laboratory, and

$\bar{X}$ = average of the strengths of the duplicate batches from which $\bar{R}$ is calculated.

7.1.5 *Standard Deviation Corrected For Testing Variations:*

$$S_c = \sqrt{S_t^2 - S_e^2} \quad (3)$$

where:

S_c = net standard deviation of cement corrected for testing error,

S_t = total standard deviations for all tests included in the calculation, and

S_e = standard deviation of duplicate tests run on split sample to evaluate testing error.

The addition of the subscript 28 or 7 indicates the type of strength data used in the calculation.

NOTE 3—Values for averages and standard deviations can be calculated by other methods that are available in STP 15 D.[4] Electronic calculators are available for obtaining these statistics directly.

[4] *Manual on Presentation of Data and Control Chart Analysis, ASTM STP 15 D*, ASTM 1976.

TABLE 1 Calculation of Standard Deviation for Testing

Date	Sample Number	7-Day Data				Average^A		Range^B		$\bar{R}^C$		S_e^D		V_e^E	Note
		Test a		Test b											
		psi	(MPa)	psi	(MPa)	psi	(MPa)	psi	(MPa)	psi	(MPa)	psi	(MPa)		
01/06	3	4900	(33.7)	4960	(34.2)	4930	(34.0)	60	(0.41)	. . .	. . .	. . .	. . .	. . .	. . .
01/16	6	4580	(31.5)	4670	(32.2)	4625	(31.8)	90	(0.62)	. . .	. . .	. . .	. . .	. . .	. . .
01/30	9	4650	(32.0)	4850	(33.4)	4750	(32.7)	200	(1.37)	. . .	. . .	. . .	. . .	. . .	. . .
02/05	12	4400	(30.3)	4510	(31.1)	4455	(30.7)	110	(0.75)	. . .	. . .	. . .	. . .	. . .	. . .
02/13	15	4380	(30.2)	4300	(29.6)	4340	(29.9)	80	(0.55)	108	(0.74)	93	(0.64)	2.02 %	Av. 5
02/21	18	4700	(32.4)	4770	(32.8)	4735	(32.6)	70	(0.48)	102	(0.70)	88	(0.60)	1.89 %	Av. 6
03/04	21	4470	(30.8)	4610	(31.7)	4540	(31.3)	140	(0.96)	107	(0.73)	92	(0.64)	2.00 %	Av. 7
03/14	24	4030	(27.7)	3970	(27.3)	4000	(27.5)	60	(0.41)	101	(0.69)	87	(0.60)	1.92 %	Av. 8
03/19	27	4970	(34.2)	4820	(33.2)	4895	(33.7)	150	(1.03)	107	(0.73)	92	(0.63)	2.01 %	Av. 9
03/27	30	4550	(31.3)	4530	(31.2)	4540	(31.3)	20	(0.13)	96	(0.67)	84	(0.58)	1.84 %	Av. 10 (6)
04/30	40	4750	(32.7)	4920	(33.9)	4835	(33.3)	170	(1.17)	109	(0.75)	94	(0.65)	2.06 %	Av. last 10
05/31	50	5030	(34.6)	4820	(33.2)	4925	(33.9)	210	(1.44)	121	(0.83)	104	(0.72)	2.27 %	Av. last 10
06/29	60	4830	(33.3)	4720	(32.5)	4775	(32.9)	110	(0.75)	112	(0.77)	97	(0.67)	2.10 %	Av. last 10
07/28	70	4400	(30.3)	4460	(30.7)	4430	(30.5)	60	(0.41)	107	(0.73)	92	(0.64)	2.00 %	Av. last 10
08/30	80	4550	(31.3)	4460	(30.7)	4505	(31.0)	90	(0.62)	108	(0.74)	93	(0.64)	2.02 %	Av. last 10
09/25	90	4930	(34.0)	5000	(34.4)	4965	(34.2)	70	(0.48)	108	(0.74)	93	(0.64)	2.01 %	Av. last 10
10/26	100	4950	(34.1)	4820	(33.2)	4885	(33.6)	130	(0.89)	107	(0.73)	92	(0.64)	1.97 %	Av. last 10
11/25	110	4670	(32.2)	4720	(32.5)	4695	(32.3)	50	(0.34)	106	(0.73)	91	(0.63)	1.93 %	Av. last 10
12/21	120	4450	(30.6)	4520	(31.1)	4485	(30.9)	70	(0.48)	96	(0.67)	84	(0.58)	1.80 %	Av. last 10

Date	Sample Number	28-Day Data				Average^A		Range^B		$\bar{R}^C$		S_e^D		V_e^E	Note
		Test a		Test b											
		psi	(MPa)	psi	(MPa)	psi	(MPa)	psi	(MPa)	psi	(MPa)	psi	(MPa)		
01/06	3	6370	(43.9)	6620	(45.6)	6495	(44.7)	250	(1.72)	. . .	. . .	. . .	. . .	. . .	. . .
01/16	6	6250	(43.1)	6020	(41.5)	6135	(42.3)	230	(1.58)	. . .	. . .	. . .	. . .	. . .	. . .
01/30	9	6050	(41.7)	6120	(42.2)	6085	(41.9)	70	(0.48)	. . .	. . .	. . .	. . .	. . .	. . .
02/05	12	6020	(41.5)	6230	(42.9)	6125	(42.2)	210	(1.44)	. . .	. . .	. . .	. . .	. . .	. . .
02/13	15	5600	(38.6)	5420	(37.3)	5510	(38.0)	180	(1.24)	188	(1.29)	162	(1.12)	2.67 %	Av. 5
02/21	18	5500	(37.9)	5530	(38.1)	5515	(38.0)	30	(0.20)	162	(1.11)	139	(0.96)	2.33 %	Av. 6
03/04	21	6320	(43.5)	6280	(43.3)	6300	(43.4)	40	(0.27)	144	(0.99)	124	(0.86)	2.06 %	Av. 7
03/14	24	5920	(40.8)	6010	(41.4)	5965	(41.1)	90	(0.62)	138	(0.94)	119	(0.82)	1.97 %	Av. 8
03/19	27	6300	(43.4)	6050	(41.7)	6175	(42.5)	250	(1.72)	150	(1.03)	129	(0.89)	2.14 %	Av. 9
03/27	30	6350	(43.7)	6410	(44.2)	6380	(44.0)	60	(0.41)	141	(0.97)	122	(0.84)	2.00 %	Av. 10 (6)
04/30	40	6050	(41.7)	5940	(40.9)	5995	(41.3)	110	(0.75)	127	(0.87)	109	(0.75)	1.82 %	Av. last 10
05/31	50	6670	(46.0)	6530	(45.0)	6600	(45.5)	140	(0.96)	118	(0.81)	102	(0.70)	1.68 %	Av. last 10
06/29	60	6350	(43.7)	6190	(42.6)	6270	(43.2)	160	(1.10)	127	(0.87)	109	(0.75)	1.80 %	Av. last 10
07/28	70	6500	(44.8)	6300	(43.4)	6400	(44.1)	200	(1.37)	126	(0.86)	109	(0.75)	1.78 %	Av. last 10
08/30	80	6200	(42.7)	6150	(42.4)	6175	(42.5)	50	(0.34)	113	(0.77)	97	(0.67)	1.58 %	Av. last 10
09/25	90	6630	(45.7)	6540	(45.1)	6585	(45.4)	90	(0.62)	119	(0.82)	103	(0.71)	1.63 %	Av. last 10
10/26	100	6230	(42.9)	6010	(41.4)	6120	(42.2)	220	(1.51)	137	(0.94)	118	(0.81)	1.88 %	Av. last 10
11/25	110	5920	(40.8)	6020	(41.5)	5970	(41.1)	100	(0.68)	138	(0.95)	119	(0.82)	1.90 %	Av. last 10
12/21	120	. . .	. . .	. . .	. . .	. . .	. . .	. . .	. . .	. . .	. . .	. . .	. . .	. . .	. . .

Notes: Initially one out of three samples are tested in duplicate until at least ten duplicate test results are available.

^{A} Average of the test results a and b.

^{B} Absolute difference between tests a and b.

^{C} Average range is calculated for a minimum of 5 duplicate tests. Subsequently, ranges of the ten most recent duplicate tests are averaged.

^{D} Standard Deviation for testing is calculated as in 7.1.3: $S_e = 0.862 \times \bar{R}$.

^{E} Coefficient of Variation for testing is calculated as in 7.1.4: $V_e = 100\ S_e/\bar{X}$. Note that $\bar{X}$ is the average strength of the duplicate batches from which $\bar{R}$ is determined.

^{F} Frequency of duplicate tests can be reduced to one in ten samples when V_e is less than 4.0 % (see 6.2.2).

8. Report

8.1 Sufficient information shall be provided to identify the cement sampled including:

8.1.1 Name of manufacturer and location,

8.1.2 Type of cement or other identification,

8.1.3 Location of sampling,

8.1.4 Laboratory designation, and

8.1.5 Period of time represented by the report.

8.2 For ongoing programs the minimum period covered by the report shall include all strength tests made in the preceding three months, but in no instance less than that period of time necessary to include 28-day strength tests of 20 consecutive samples.

8.2.1 The report shall not cover a period of time greater than 12 months or tests of more than 120 samples.

8.3 The report of strength results shall be either in tabular form as shown in Table 2 or in graphical form as shown in Fig. 1, at the option of the reporting organization.

NOTE 4—For purposes of analyzing trends, the graphical presentation is to be preferred. Additionally, the average and standard deviation as calculated in Section 7 shall be shown.

8.4 Report the available 7 and 28-day compressive strength results on each sample including the date on which the sample was taken. Each value reported will be the average of tests of three cubes made from the same batch, except when one or more cubes are faulty. See Test Method C 109.

8.4.1 Report the results of tests of duplicate batches tested within the period covered by the report. When duplicate batches are made from a cement other than that being tested during the same period of time, by the same laboratory, these test results will not normally be reported on a regular basis, but results of such tests will be made available on request. However, report the standard deviation, S_e, and the coefficient of variation, V_e, of duplicate batches.

8.5 The report shall include the following values calculated from the reported data. Each cement sample shall be represented only by a single result at each age in these calculations. The second of a pair of duplicate batch test results shall not be included in overall calculations, but shall be used only to establish testing error.

8.5.1 Calculated from the reported 7-day strength data:

(1) $\bar{X}_7$, the average,

(2) S_{t7}, the total standard deviation,

(3) n_7, the number of samples tested,

(4) S_{c7}, the standard deviation corrected for testing, and

(5) $\bar{X}_{5-7}$, the moving averages of the five most recent 7-day results.

8.5.2 Calculated from the reported 28-day strength data:

(1) $\bar{X}_{28}$, the average,

(2) S_{t28}, the total standard deviation,

(3) n_{28}, the number of samples tested,

(4) S_{c28}, the standard deviation corrected for testing, and

(5) $\bar{X}_{5-28}$, the moving averages of the five most recent 28-day results.

8.5.3 The calculations in 8.5.1 and 8.5.2 shall not be made and reported until five results are available. The moving average of the five most recent results should be updated with each successive result by adding the new value in the calculation and deleting the oldest previous value (see Table 1).

8.5.4 Whenever the reporting agency concludes that a consistent change in strength producing properties has occurred, at its option, it may discontinue calculation until results

ABC Cement Inc., Qualitytown, NJ
Type I
Sampled on truck loading, Qualitytown
ABC Qualitytown Laboratory
Report date: January 5, 1992

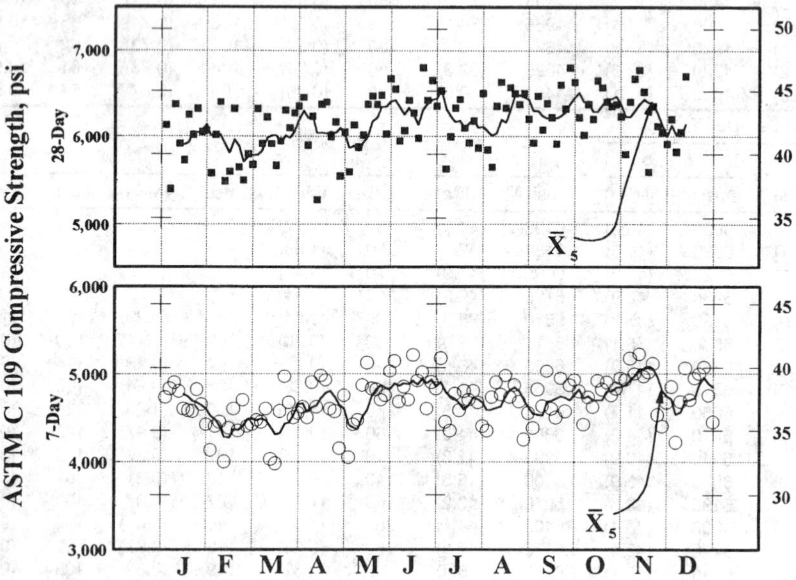

Test Data for cement produced from
January 2 to December 21, 1991

Average Strength, psi (MPa)

$\bar{X}_7$ = 4695 (32.3) n = 120

$\bar{X}_{28}$ = 6170 (42.5) n = 114

Average 5 most recent, psi (MPa)

$\bar{X}_{5-7}$ = 4846 (33.4)

$\bar{X}_{5-28}$ = 6140 (42.3)

Total Std. Dev., psi (MPa)

S_{t7} = 270 (1.85)

S_{t28} = 334 (2.30)

Testing Std. Dev., psi (MPa)

S_{e7} = 84 (0.58)

S_{e28} = 119 (0.82)

Corrected Std. Dev., psi (MPa)

S_{c7} = 256 (1.77)

S_{c28} = 312 (2.15)

FIG. 1 Uniformity Test Report

TABLE 2 Sample Uniformity Test Report

BC Cement Inc., Qualitytown, N.J.
Type I
Sampled on truck loading, Qualitytown
ABC Qualitytown Laboratory

Report Date: January 5, 1992

Item	7-Day	28-Day	7-Day	28-Day
Dates Represented:			(This space is provided for use if a break in the calculation is made. See 8.5.4 and 8.5.5.)	
From:	01/02/91	01/02/91		
To:	12/21/91	12/05/91		
Average Strength, psi (MPa), $\bar{X}$	4695 (32.3)	6170 (42.5)		
Total Standard Deviation, psi (MPa), S_t	270 (1.85)	334 (2.30)		
Number of Tests (n)	120	114		
Testing Standard Deviation, psi (MPa), S_e	84 (0.58)	119 (0.82)		
Number of Tests (n)	10	10		
Corrected Standard Deviation, psi (MPa), S_c	256 (1.77)	312 (2.15)		

Date Shipped	Sample No.[A]	7-Day Test a psi	7-Day Test a (MPa)	7-Day Average 5 psi	7-Day Average 5 (MPa)	28-Day Test a psi	28-Day Test a (MPa)	28-Day Average 5 psi	28-Day Average 5 (MPa)	Date Shipped	Sample No.	7-Day Test a psi	7-Day Test a (MPa)	7-Day Average 5 psi	7-Day Average 5 (MPa)	28-Day Test a psi	28-Day Test a (MPa)	28-Day Average 5 psi	28-Day Average 5 (MPa)
01/02	1	4730	(32.6)			6130	(42.3)			07/02	61	5180	(35.7)	4906	(33.8)	6530	(45.0)	6550	(45.2)
01/03	2	4830	(33.3)			5400	(37.2)			07/07	62	4450	(30.7)	4792	(33.0)	5630	(38.8)	6316	(43.6)
01/06	3a	4900	(33.8)			6370	(43.9)			07/09	63	4350	(30.0)	4742	(32.7)	6000	(41.4)	6232	(43.0)
01/08	4	4800	(33.1)			5920	(40.8)			07/11	64	4770	(32.9)	4716	(32.5)	6330	(43.7)	6168	(42.5)
01/14	5	4600	(31.7)	4772	(32.9)	5730	(39.5)	5910	(40.8)	07/12	65	4580	(31.6)	4666	(32.2)	6430	(44.3)	6184	(42.6)
01/16	6a	4580	(31.6)	4742	(32.7)	6250	(43.1)	5934	(40.9)	07/13	66	4800	(33.1)	4590	(31.7)	6100	(42.1)	6098	(42.1)
01/21	7	4570	(31.5)	4690	(32.3)	6020	(41.5)	6058	(41.8)	07/19	67	4700	(32.4)	4640	(32.0)	5930	(40.9)	6158	(42.5)
01/24	8	4820	(33.2)	4674	(32.2)	6320	(43.6)	6048	(41.7)	07/20	68	4800	(33.1)	4730	(32.6)	6180	(42.6)	6194	(42.7)
01/30	9a	4650	(32.1)	4644	(32.0)	6050	(41.7)	6074	(41.9)	07/24	69	4670	(32.2)	4710	(32.5)	5870	(40.5)	6102	(42.1)
01/31	10	4420	(30.5)	4608	(31.8)	6070	(41.9)	6142	(42.4)	07/28	70a	4400	(30.3)	4674	(32.2)	6500	(44.8)	6116	(42.2)
02/03	11	4130	(28.5)	4518	(31.2)	5580	(38.5)	6008	(41.4)	08/02	71	4350	(30.0)	4584	(31.6)	5850	(40.3)	6066	(41.8)
02/05	12a	4400	(30.3)	4484	(30.9)	6020	(41.5)	6008	(41.4)	08/05	72	4730	(32.6)	4590	(31.7)	5650	(39.0)	6010	(41.4)
02/06	13	4450	(30.7)	4410	(30.4)	6400	(44.1)	6024	(41.5)	08/07	73	4900	(33.8)	4610	(31.8)	6350	(43.8)	6044	(41.7)
02/07	14	4000	(27.6)	4280	(29.5)	5490	(37.9)	5912	(40.8)	08/09	74	4760	(32.8)	4628	(31.9)	6630	(45.7)	6196	(42.7)
02/13	15a	4380	(30.2)	4272	(29.5)	5600	(38.6)	5818	(40.1)	08/10	75	4980	(34.3)	4744	(32.7)	6330	(43.7)	6162	(42.5)
02/17	16	4600	(31.7)	4366	(30.1)	6320	(43.6)	5966	(41.1)	08/19	76	4730	(32.6)	4820	(33.2)	6570	(45.3)	6306	(43.5)
02/19	17	4350	(30.0)	4356	(30.0)	5650	(39.0)	5892	(40.6)	08/20	77	4870	(33.6)	4848	(33.4)	6500	(44.8)	6476	(44.7)
02/21	18a	4700	(32.4)	4406	(30.4)	5500	(37.9)	5712	(39.4)	08/24	78	4680	(32.3)	4804	(33.1)	6370	(43.9)	6480	(44.7)
02/25	19	4400	(30.3)	4486	(30.9)	5800	(40.0)	5774	(39.8)	08/25	79	4250	(29.3)	4702	(32.4)	6500	(44.8)	6454	(44.5)
02/27	20	4420	(30.5)	4494	(31.0)	5680	(39.2)	5790	(39.9)	08/30	80a	4550	(31.4)	4616	(31.8)	6200	(42.8)	6428	(44.3)
03/04	21a	4470	(30.8)	4468	(30.8)	6320	(43.6)	5790	(39.9)	09/01	81	4380	(30.2)	4546	(31.4)	5930	(40.9)	6300	(43.4)
03/07	22	4450	(30.7)	4488	(31.0)	5920	(40.8)	5844	(40.3)	09/05	82	4820	(33.2)	4536	(31.3)	6470	(44.6)	6294	(43.4)
03/12	23	4600	(31.7)	4468	(30.8)	6220	(42.9)	5988	(41.3)	09/08	83	4630	(31.9)	4526	(31.2)	6080	(41.9)	6236	(43.0)
03/14	24a	4030	(27.8)	4394	(30.3)	5920	(40.8)	6012	(41.5)	09/10	84	5030	(34.7)	4682	(32.3)	6380	(44.0)	6212	(42.8)
03/15	25	3980	(27.4)	4306	(29.7)	5670	(39.1)	6010	(41.4)	09/15	85	4600	(31.7)	4692	(32.4)	6220	(42.9)	6216	(42.9)
03/18	26	4570	(31.5)	4326	(29.8)	5950	(41.0)	5936	(40.9)	09/20	86	4500	(31.0)	4716	(32.5)	5920	(40.8)	6214	(42.9)
03/19	27a	4970	(34.3)	4430	(30.6)	6300	(43.4)	6012	(41.5)	09/21	87	4950	(34.1)	4742	(32.7)	6600	(45.5)	6240	(43.0)
03/22	28	4670	(32.2)	4444	(30.6)	6100	(42.1)	5988	(41.3)	09/23	88	4570	(31.5)	4730	(32.6)	6320	(43.6)	6288	(43.4)
03/25	29	4500	(31.0)	4538	(31.3)	6270	(43.2)	6058	(41.8)	09/24	89	4870	(33.6)	4698	(32.4)	6800	(46.9)	6372	(43.9)
03/27	30a	4550	(31.4)	4652	(32.1)	6350	(43.8)	6194	(42.7)	09/25	90a	4930	(34.0)	4764	(32.9)	6630	(45.7)	6454	(44.5)
04/02	31	4620	(31.9)	4662	(32.2)	6450	(44.5)	6294	(43.4)	10/01	91	4800	(33.1)	4824	(33.3)	6220	(42.9)	6514	(44.9)
04/05	32	4500	(31.0)	4568	(31.5)	5900	(40.7)	6214	(42.9)	10/02	92	4420	(30.5)	4718	(32.5)	6020	(41.5)	6398	(44.1)
04/10	33	4900	(33.8)	4614	(31.8)	6230	(43.0)	6240	(43.0)	10/05	93	4730	(32.6)	4750	(32.8)	6370	(43.9)	6408	(44.2)
04/12	34	4620	(31.9)	4638	(32.0)	5280	(36.4)	6042	(41.7)	10/07	94	4620	(31.9)	4700	(32.4)	6200	(42.8)	6288	(43.4)
04/14	35	4980	(34.3)	4724	(32.6)	6370	(43.9)	6046	(41.7)	10/08	95	4920	(33.9)	4698	(32.4)	6650	(45.9)	6292	(43.4)
04/16	36	4930	(34.0)	4786	(33.0)	6400	(44.1)	6036	(41.6)	10/13	96	4820	(33.2)	4702	(32.4)	6570	(45.3)	6362	(43.9)
04/18	37	4600	(31.7)	4806	(33.1)	6000	(41.4)	6056	(41.8)	10/15	97	4900	(33.8)	4798	(33.1)	6400	(44.1)	6438	(44.4)
04/23	38	4570	(31.5)	4740	(32.7)	6170	(42.6)	6044	(41.7)	10/19	98	4750	(32.8)	4802	(33.1)	6350	(43.8)	6434	(44.4)
04/25	39	4150	(28.6)	4646	(32.0)	5550	(38.3)	6098	(42.1)	10/21	99	4830	(33.3)	4844	(33.4)	6370	(43.9)	6468	(44.6)
04/30	40a	4750	(32.8)	4600	(31.7)	6050	(41.7)	6034	(41.6)	10/26	100a	4950	(34.1)	4850	(33.4)	6230	(43.0)	6384	(44.0)
05/05	41	4050	(27.9)	4424	(30.5)	5600	(38.6)	5874	(40.5)	10/29	101	4930	(34.0)	4872	(33.6)	5800	(40.0)	6230	(43.0)
05/07	42	4420	(30.5)	4388	(30.3)	6130	(42.3)	5900	(40.7)	11/02	102	5180	(35.7)	4928	(34.0)	6400	(44.1)	6230	(43.0)
05/09	43	4470	(30.8)	4368	(30.1)	5880	(40.6)	5842	(40.3)	11/04	103	5000	(34.5)	4978	(34.3)	6670	(46.0)	6294	(43.4)
05/15	44	4870	(33.6)	4512	(31.1)	6020	(41.5)	5936	(40.9)	11/06	104	5230	(36.1)	5058	(34.9)	6770	(46.7)	6374	(44.0)
05/17	45	5130	(35.4)	4588	(31.6)	6370	(43.9)	6000	(41.4)	11/09	105	4970	(34.3)	5062	(34.9)	6520	(45.0)	6432	(44.4)
05/19	46	4830	(33.3)	4744	(32.7)	6470	(44.6)	6174	(42.6)	11/12	106	5020	(34.6)	5080	(35.0)	5600	(38.6)	6392	(44.1)
05/21	47	4820	(33.2)	4824	(33.3)	6370	(43.9)	6222	(42.9)	11/15	107	5120	(35.3)	5068	(35.0)	6350	(43.8)	6382	(44.0)
05/27	48	4700	(32.4)	4870	(33.6)	6480	(44.7)	6342	(43.7)	11/17	108	4530	(31.2)	4974	(34.3)	6120	(42.2)	6272	(43.3)
05/29	49	4750	(32.8)	4846	(33.4)	6030	(41.6)	6344	(43.8)	11/19	109	4400	(30.3)	4808	(33.2)	6050	(41.7)	6128	(42.3)
05/31	50a	5030	(34.7)	4826	(33.3)	6670	(46.0)	6404	(44.2)	11/25	110a	4670	(32.2)	4748	(32.7)	5920	(40.8)	6008	(41.4)
06/08	51	5150	(35.5)	4890	(33.7)	6550	(45.2)	6420	(44.3)	11/26	111	4850	(33.4)	4714	(32.5)	6200	(42.8)	6128	(42.3)
06/10	52	4680	(32.3)	4862	(33.5)	5950	(41.0)	6336	(43.7)	11/30	112	4220	(29.1)	4534	(31.3)	5830	(40.2)	6024	(41.5)
06/12	53	4850	(33.4)	4892	(33.7)	6070	(41.9)	6254	(43.1)	12/04	113	4680	(32.3)	4564	(31.5)	6050	(41.7)	6010	(41.4)
06/16	54	4700	(32.4)	4882	(33.7)	6420	(44.3)	6332	(43.7)	12/05	114	5070	(35.0)	4698	(32.4)	6700	(46.2)	6140	(42.3)
06/18	55	5220	(36.0)	4920	(33.9)	6270	(43.2)	6252	(43.1)	12/10	115	4700	(32.4)	4704	(32.4)	...	...	...	...
06/21	56	4920	(33.9)	4874	(33.6)	5980	(41.2)	6138	(42.3)	12/12	116	4950	(34.1)	4724	(32.6)	...	...	...	...
06/23	57	5020	(34.6)	4942	(34.1)	6800	(46.9)	6308	(43.5)	12/14	117	5000	(34.5)	4880	(33.7)	...	...	...	...
06/25	58	4600	(31.7)	4892	(33.7)	6420	(44.3)	6378	(44.0)	12/16	118	5080	(35.0)	4960	(34.2)	...	...	...	...
06/27	59	4900	(33.8)	4932	(34.0)	6650	(45.9)	6424	(44.3)	12/19	119	4750	(32.8)	4896	(33.8)	...	...	...	...
06/29	60a	4830	(33.3)	4854	(33.5)	6350	(43.8)	6440	(44.4)	12/21	120a	4450	(30.7)	4846	(33.4)	...	...	...	...

[A] "a" is the first batch of two made from this sample. See Table 1.

from five additional samples of the cement have been obtained. In this instance, the values (8.5.1 and 8.5.2) calculated from the samples prior to the change shall also be reported (see 8.2.1).

8.5.5 When there is a break in the calculation, the sampling dates included in each set of calculated values (8.5.1 and 8.5.2) shall be clearly identified.

9. Keywords

9.1 cement; sampling; statistics; strength; uniformity

APPENDIXES

(Nonmandatory Information)

X1. COMPARISON OF RESULTS FROM TWO LABORATORIES USING TEST METHOD C 109

X1.1 Suppose n samples are split and tested by each Laboratory A and B with a single batch of mortar prepared on each sample in each laboratory.

X1.2 Compute the difference in results on each sample as $X_a - X_b$:

$$d_i = X_a - X_b \qquad \text{(X1.1)}$$

d_i values may be positive or negative.

X1.3 Compute the average of these differences:

$$\bar{D} = \frac{d_1 + d_2 + \ldots + d_n}{n} \qquad \text{(X1.2)}$$

where:

$\bar{D}$ = average difference in psi and d_i values may be positive or negative.

X1.4 Compute the standard deviation of these differences (7.1.2),

where:

S_d = standard deviation of the difference, psi.

X1.5 Compute the value $\frac{\bar{D}\sqrt{n}}{S_d}$ and compare to Student's t[5] for $n - 1$ at $\alpha = 0.05$, where t is from Student's t distribution.

X1.6 If $\left|\frac{\bar{D}\sqrt{n}}{S_d}\right| \geq t$, decide that averages in the two laboratories differ significantly; otherwise, decide that there is no reason to believe they differ.

[5] For additional information see Neville, A., and Kennedy, J., *Basic Statistical Methods for Engineers and Scientists,* (International Textbooks, Scranton, Pa. 1964), pp. 143–146.

$n - 1$	1	2	3	4	5	6	7	8	9
t[A]	12.7	4.30	3.18	2.78	2.57	2.45	2.36	2.31	2.26

[A] At $\alpha = 0.05$.

X2. WITHIN-PLANT VARIATIONS IN CEMENT STRENGTH

X2.1 With the assistance of the Cement and Concrete Reference Laboratory (CCRL), C 917 data for 1991 was collected from 87 plants in the United States and Canada.

X2.2 Information on standard deviations, coefficients of variation, and 7- and 28-day strengths is presented in Figs. X2.1 through X2.7.

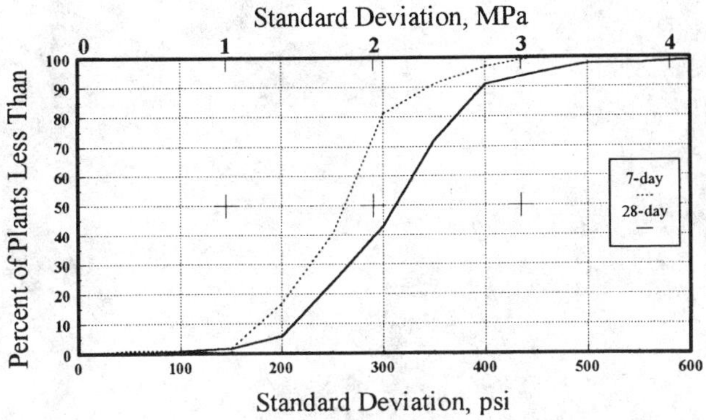

FIG. X2.1 Total Standard Deviation, St

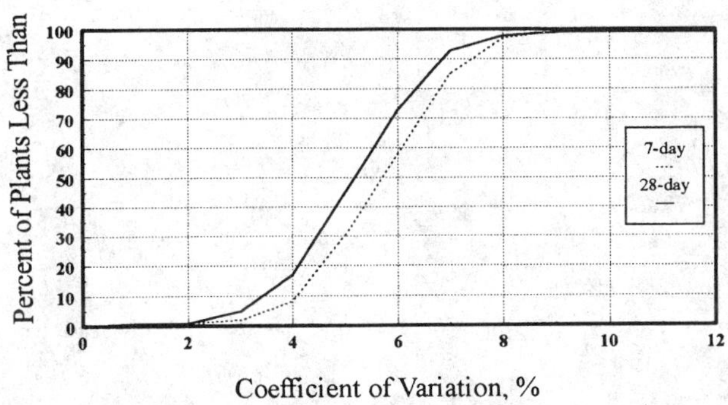

FIG. X2.2 Total Coefficient of Variation, Vt

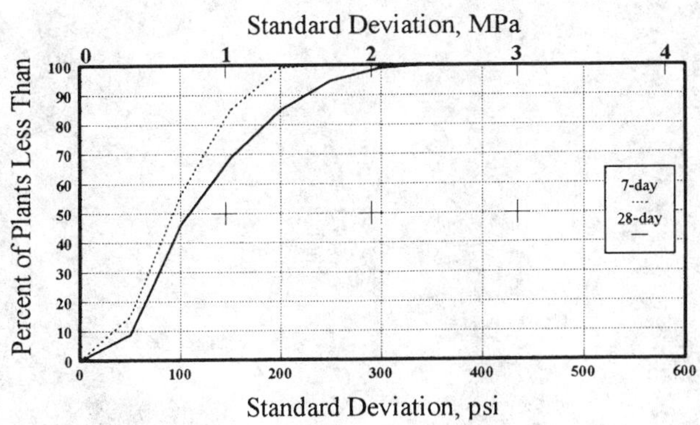

FIG. 1 Uniformity Test Report

FIG. X2.3 Standard Deviation from Duplicate Batches, Se

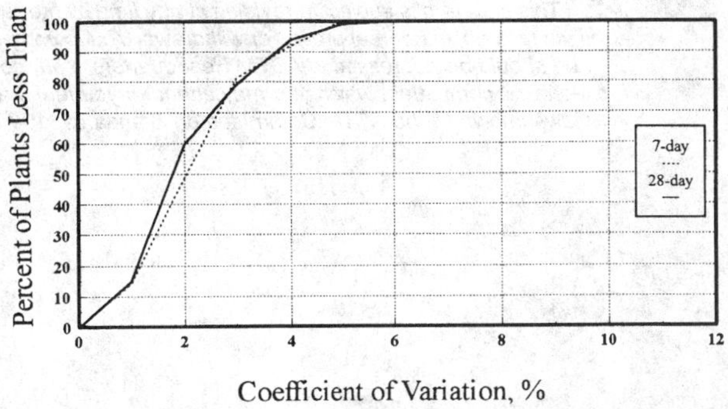

FIG. X2.4 Coefficient of Variation from Duplicate Batches, Ve

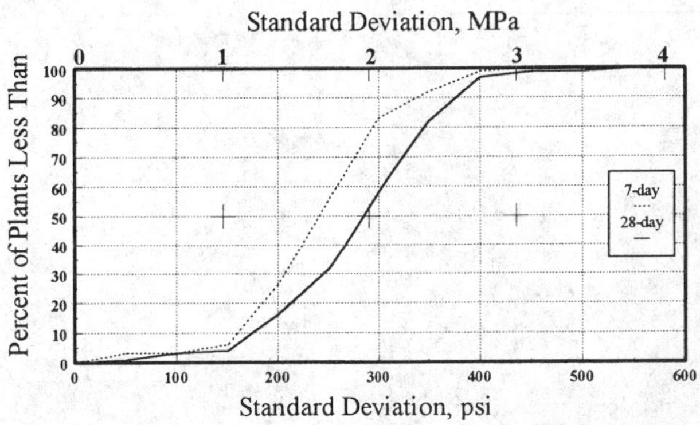

FIG. X2.5 Standard Deviation Corrected for Testing Error, Sc

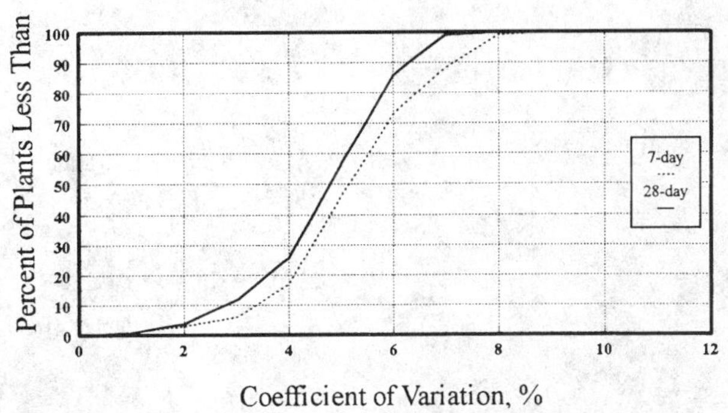

FIG. X2.6 Coefficient of Variation Corrected for Testing Error, Vc

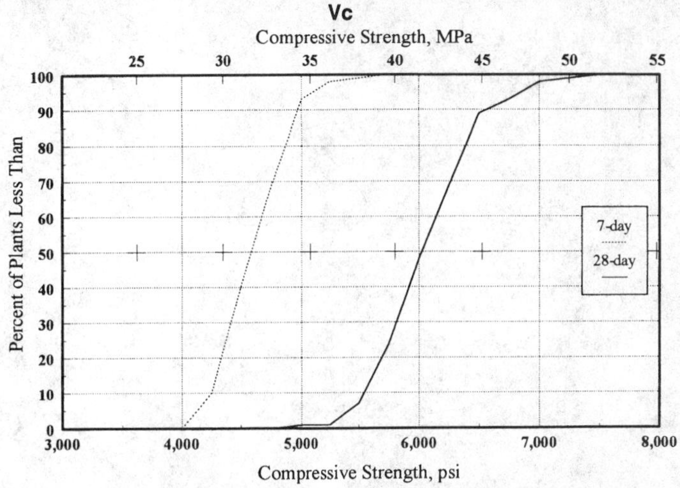

FIG. X2.7 Average Compressive Strength

Designation: C 926 – 98a

Standard Specification for Application of Portland Cement-Based Plaster[1]

This standard is issued under the fixed designation C 926; the number immediately following the designation indicates the year of original adoption or, in the case of revision, the year of last revision. A number in parentheses indicates the year of last reapproval. A superscript epsilon (ϵ) indicates an editorial change since the last revision or reapproval.

1. Scope

1.1 This specification covers the requirements for the application of full thickness portland cement-based plaster for exterior (stucco) and interior work.

1.2 This specification sets forth tables for proportioning of various plaster mixes and plaster thickness.

NOTE 1—General information will be found in Annex A1. Design considerations will be found in Annex A2.

1.3 The values stated in inch-pound units are to be regarded as the standard. The SI (metric) values given in parentheses are approximate and are provided for information purposes only.

1.4 The text of this specification references notes and footnotes that provide explanatory material. These notes and footnotes (excluding those in tables and figures) shall not be considered as requirements of the specification.

1.5 Details of construction for a specific assembly to achieve the required fire resistance shall be obtained from reports of fire-resistance tests, engineering evaluations, or listings from recognized fire testing laboratories.

2. Referenced Documents

2.1 *ASTM Standards:*

C 11 Terminology Relating to Gypsum and Related Building Materials and Systems[2]
C 25 Test Methods for Chemical Analysis of Limestone, Quicklime, and Hydrated Lime[2]
C 35 Specification for Inorganic Aggregates for Use in Gypsum Plaster[2]
C 91 Specification for Masonry Cement[2]
C 150 Specification for Portland Cement[2]
C 206 Specification for Finishing Hydrated Lime[2]
C 207 Specification for Hydrated Lime for Masonry Purposes[2]
C 219 Terminology Relating to Hydraulic Cement[2]
C 260 Specification for Air-Entraining Admixtures for Concrete[3]
C 595M Specification for Blended Hydraulic Cements[2,3]
C 631 Specification for Bonding Compounds for Interior Plastering[2]
C 897 Specification for Aggregate for Job-Mixed Portland Cement-Based Plasters[2]
C 932 Specification for Surface-Applied Bonding Agent for Exterior Plastering[2]
C 1063 Specification for Installation of Lathing and Furring to Receive Interior and Exterior Portland Cement-Based Plaster[2]
C 1116 Specification for Fiber-Reinforced Concrete and Shotcrete[3]
C 1328 Specification for Plastic (Stucco) Cement[2]
E 90 Test Method for Laboratory Measurement of Airborne-Sound Transmission Loss of Building Partitions[4]
E 119 Test Methods for Fire Tests of Building Construction and Materials[5]
E 492 Test Method for Laboratory Measurement of Impact Sound Transmission Through Floor-Ceiling Assemblies Using the Tapping Machine[4]

2.2 *ANSI Standard:*

A108.1 Specification for Installation of Ceramic Tile[6]

3. Terminology

3.1 Terms shall be defined as in Terminologies C 11 and C 219, except as modified herein.

3.2 *Descriptions of Terms Specific to This Standard:*

3.2.1 *accelerator*—an admixture that will shorten the setting time of plaster.

3.2.2 *admixture*—a material other than water, aggregate, or basic cementitious material added to the batch before or during job mixing.

3.2.3 *acid etching*—the cleansing and controlled erosion of a solid surface, using an acid wash.

3.2.4 *air entrainment*—the use of an air-entraining admixture or air-entraining cementitious material in a plaster mix to yield a controlled quantity of minute (typically between 10 and 1000 μm in diameter) disconnected air bubbles in the plaster (see *entrapped air*).

3.2.5 *backplaster*—plaster applied to the face of metal lath opposite a previously applied plaster.

[1] This specification is under the jurisdiction of ASTM Committee C-11 on Gypsum and Related Building Materials and Systems and is the direct responsibility of Subcommittee C11.03 on Specifications for Application of Gypsum and Other Products on Assemblies.

Current edition approved June 10, 1998. Published December 1998. Originally published as C 926 – 81. Last previous edition C 926 – 98a.

[2] *Annual Book of ASTM Standards,* Vol 04.01.

[3] *Annual Book of ASTM Standards,* Vol 04.02.

[4] *Annual Book of ASTM Standards,* Vol 04.06.

[5] *Annual Book of ASTM Standards,* Vol 04.07.

[6] Available from American National Standards Institute, 11 W. 42nd St., 13th Floor, New York, NY 10036.

3.2.6 *bond*—the state of adhesion between plaster coats or between plaster and plaster base.

3.2.7 *bonding compound or agent*—compounds surface applied or integrally mixed with plaster to improve the quality of bond between plaster and plaster base or between plaster coats.

3.2.8 *cementitious material*—a material that, when mixed with water and with or without aggregate, provides the plasticity and the cohesive and adhesive properties necessary for placement and the formation of a rigid mass.

3.2.9 *coat*—a thickness of plaster applied in a single operation.

3.2.9.1 *basecoat*—all plaster applied before the application of the finish coat.

3.2.9.2 *bedding coat*—a plaster coat that receives aggregate or other decorative material impinged into its surface before it sets.

3.2.9.3 *brown coat*—in three-coat work, the second coat, applied over the scratch coat. In two-coat work, brown coat refers to the double-up basecoat. In either use, the brown coat is the coat directly beneath the finish coat.

3.2.9.4 *dash-bond coat*—a thick wet mixture of portland cement and water, with or without aggregate, dashed onto the surface of a plaster base such as smooth monolithic concrete or concrete block surfaces to improve the mechanical key for subsequent plaster coats.

3.2.9.5 *double-up coat*—the brown-coat plaster applied to the scratch coat plaster before the scratch-coat plaster has set.

3.2.9.6 *finish coat*—the final layer of plaster applied over basecoat plaster.

3.2.9.7 *fog coat*—a light coat of cement and water, with or without aggregate or color pigment, applied by machine spray to improve color consistency.

3.2.9.8 *scratch coat*—the first coat of plaster applied to a plaster base.

3.2.9.9 *skim coat*—a thin finish coat applied to an existing plaster surface or other substrate to improve appearance.

3.2.9.10 *three-coat work*—application of plaster in three successive coats with time between coats for setting or drying, or both.

3.2.10 *cold joint ("joining" or "jointing")*—the juncture of fresh plaster application adjacent to set plaster, in the same plane.

3.2.11 *curing*—the act or processes of producing a moisture environment favorable to cement hydration, resulting in the setting or hardening of the plaster.

3.2.12 *entrapped air*—unintentional air voids in the plaster generally larger than 1 mm.

3.2.13 *factory prepared ("mill-mixed" or "ready mixed")*—pertaining to material combinations that have been formulated and dry-blended by the manufacturer, requiring only the addition of and mixing with water to produce plaster.

3.2.14 *fiber, natural or synthetic*—an elongated fiber or strand admixture added to plaster mix to improve cohesiveness or pumpability, or both.

3.2.15 *floating*—act of compacting and leveling brown-coat plaster to a reasonably true surface plane using a float tool or the act of bringing the aggregate to the surface of finish-coat plaster.

3.2.16 *key (also mechanical key)*—plaster that physically surrounds, penetrates, or deforms to lock onto the perforations or irregularities of the plaster base or previous coat of plaster.

3.2.17 *metal plaster base*—expanded metal lath, or welded or woven wire lath.

3.2.18 *plaster*—portland cement-based cementitious mixture (see *stucco*).

3.2.19 *required*—pertaining to a mandatory obligation imposed by a force outside of this specification, such as a building code, project specification, contract, or purchase order.

3.2.20 *rustication* (also "*break*")—an interruption or change in plane of a plastered surface.

3.2.21 *scoring* (also known as "*scratching*")—the grooving of the surface of an unset plaster coat to provide a key for a subsequent coat.

3.2.22 *set*—the chemical and physical change in plaster as it goes from a plastic, workable state to a rigid state.

3.2.23 *stucco*—portland cement-based plaster used on exterior locations.

3.2.24 *stucco finish*—a factory-prepared, dry blend of materials for finish coat applications.

3.2.25 *temper, v*—to mix or restore unset plaster with water to a workable consistency.

3.2.26 *texture*—any surface appearance as contrasted to a smooth surface.

4. Materials

4.1 Materials shall conform to the requirements of the referenced specifications and standards and to the requirements specified herein.

4.2 *Cement:*

4.2.1 *Portland Cement*—Specification C 150, Type I, II, and III, as specified. White where specified.

4.2.2 *Air-Entraining Portland Cement*—Specification C 150, type as specified. White where specified.

4.2.3 *Masonry Cement*—Specification C 91, Types N, S, and M. White where specified.

4.2.4 *Blended Hydraulic Cement*—Specification C 595, Type IP, I(PM), IS, and I(SM) as specified.

4.2.5 *Air-Entraining Blended Hydraulic Cement*—Specification C 595, Type IP-A, I(PM)-A, IS-A, and I(SM)-A as specified.

4.2.6 *Plastic Cement*—Plastic Cement shall meet the requirements of Specification C 1328, Standard Specification for Plastic (Stucco) Cement.

NOTE 2—Plastic cements are not available nationally.

4.3 *Type "S" Hydrated Lime*—A hydrated lime that contains not more than 8 % unhydrated oxides when tested in accordance with Test Methods C 25. See Specifications C 206 and C 207 for a complete description of a Type "S" hydrated lime.

4.4 *Aggregates:*

4.4.1 *Sand for Base Coats*—Specification C 897. Aggregate failing to meet gradation limits in Specification C 897 shall be permitted to be used, provided the plaster made with this sand has an acceptable demonstrated performance record in similar construction and climate conditions.

4.4.2 *Perlite*—Specification C 35.

4.4.3 *Sand for Job-Mixed Finish Coats*—Specification C 897.

4.5 *Water*—Water used in mixing, application, and finishing of plaster shall be clean, fresh, suitable for domestic water consumption, and free of such amounts of mineral or organic substances as would affect the set, the plaster, or any metal in the system.

4.6 *Admixtures*—See 3.2.2 and Annex A2.5.

4.7 *Fibers*—Specification C 1116 on alkali-resistant fibers, glass fibers, nylon, polypropylene or carbon fibers.

5. Requirements for Bases to Receive Portland Cement-Based Plaster

5.1 Metal bases and accessories used to receive plaster shall be installed in conformance with Specification C 1063, except as otherwise specified.

NOTE 3—All metal, or PVC, or CPVC plastic members should be free of deleterious amounts of rust, oil, or other foreign matter, which could cause bond failure or unsightly discoloration.

5.2 Surfaces of solid bases to receive plaster, such as masonry, stone, cast-in-place or precast concrete shall be straight and true within ¼ in. in 10 ft (2.1 mm/m) and shall be free of form oil or other elements, which would interfere with bonding. Form ties or other obstructions shall be removed or trimmed back even with the surface of the solid base.

5.2.1 Solid surfaces shall have the suction (ability to absorb water) or surface roughness, or both, to provide the bond required for the plaster.

5.2.2 Smooth or nonabsorbent solid surfaces, such as cast-in-place or precast concrete, shall be prepared to receive portland cement plaster by one of the following methods:

5.2.2.1 Sandblasting, wire brushing, acid etching, or chipping or a combination thereof,

5.2.2.2 Application of a dash-bond coat applied forcefully against the surface, left untroweled, undisturbed, and moist cured for at least 24 h, or

5.2.2.3 Application of a bonding compound suitable for exterior or interior exposure solid surfaces in accordance with the manufacturer's written directions.

5.2.3 Where bond cannot be obtained over the entire surface to receive plaster by one or more of the methods in 5.2.2, or where total plaster thickness will exceed the total thickness specified in Table 1 for types of solid bases, furred or self-furring metal plaster base shall be installed in accordance with Specification C 1063.

6. Plaster Proportions and Mixing

6.1 *Proportions:*

6.1.1 All portland cement plasters shall be mixed and proportioned in accordance with the following tables and accompanying requirements, using measuring devices of known volume with successive batches proportioned alike.

6.1.2 Plaster mix used shall be as designated and referenced to Table 2.

6.1.3 Base-coat proportions shall be as shown in Table 3 for the mix specified from Table 2.

6.1.3.1 *Measurement of Materials*—The method of measuring materials for the plaster shall be such that the specified proportions are controlled and accurately maintained. The weights per cubic foot of the materials are considered to be as follows:

Material	Weight, lb/ft^3 (kg/m^3)
Portland cement	94 (1505)
Blended cement	Weight printed on bag
Masonry or plastic cement	Weight printed on bag
Hydrated Lime	40 (640)
Lime Putty	80 (1280)
Sand, Damp and Loose (6.1.3.2)	80 (1280) of dry sand

6.1.3.2 For purposes of this specification, a weight of 80 lb (1280 kg) of oven-dried sand shall be used. This is, in most cases, equivalent to one cubic foot of loose, damp sand.

6.1.4 Finish-coat proportions for job-mixed finish coats shall be as specified in Table 4.

6.1.5 *Factory-Prepared Finish Coats*—see 3.2.14.

6.1.6 Dash-bond coat proportions shall be 1 volume part portland cement and not more than 2 volume parts of aggregate mixed to a consistency that will permit application as specified in 7.1.5.

6.1.7 Admixtures shall be proportioned, mixed, and applied in accordance with the printed directions of the manufacturer. (See A2.5.)

TABLE 1 Nominal Plaster Thickness[A] for Three- and Two-Coat Work, in. (mm)

BASE	Vertical				Horizontal			
	1st Coat	2nd Coat	3rd Coat[B]	Total	1st Coat	2nd Coat	3rd Coat[B]	Total
	Interior/Exterior							
Three-coat work:[C]								
Metal plaster base	⅜ (9.5)	⅜ (9.5)	⅛ (3)	⅞ (22)	¼ (6)	¼ (6)	⅛ (3)	⅝ (16)
Solid plaster base:								
Unit masonry	¼ (4)	¼ (6)	⅛ (3)	⅝ (16)	Use two-coat work			
Cast-in-place or precast concrete	¼ (6)	¼ (6)	⅛ (3)	⅝ (16)				⅜ (9.5), max
Metal plaster base over solid base	½ (12.5)	¼ (6)	⅛ (3)	⅞ (22)	½ (12.5)	¼ (6)	⅛ (3)	⅞ (22)
Two-coat work:								
Solid plaster base:								
Unit masonry	⅜ (9.5)	⅛ (3)		½ (12.5)				⅜ (9.5)
Cast-in-place or pre-cast concrete	¼ (6)	⅛ (3)		⅜ (9.5)				⅜ (9.5)

[A] Exclusive of texture.

[B] For solid plaster partitions, additional coats shall be applied to meet the finished thickness specified.

[C] For exposed aggregate finishes, the second (brown) coat shall become the "bedding" coat and shall be of sufficient thickness to receive and hold the aggregate.

TABLE 2 Plaster Bases—Permissible Mixes

NOTE—See Table 3 for plaster mix symbols

Property of Base	Mixes for Plaster Coats	
	First (Scratch)	Second (Brown)
Low absorption, such as dense, smooth clay tile, brick, or concrete	C	C, CL, M, or CM
	CM or MS	CM, MS, or M
	P	P
High Absorption, such as concrete masonry, absorptive brick, or tile	CL	CL
	M	M
	CM or MS	CM, MS, or M
	P	P
Metal plaster base	C	C, CL, M, CM, or MS
	CL	CL
	CM or MS	CM, MS, or M
	M	M
	CP	CP or P
	P	P

6.2 *Mixing:*

6.2.1 All plaster shall be prepared in a mechanical mixer, using sufficient water to produce a workable consistency and uniform color. (See X1.1.)

6.2.2 Base-coat plasters that have stiffened because of evaporation of water shall be permitted to be tempered one time only to restore the required consistency. Plaster not used within 1½ h from start of initial mixing shall be discarded.

NOTE 4—Severe hot, dry climate conditions accelerate the stiffening of plaster and require reduction of this limit. The use of cold waters will slow the stiffening process.

6.2.3 Finish-coat plaster shall not be tempered.

7. Application

7.1 *General:*

7.1.1 Portland cement plaster shall be applied by hand or machine to the nominal thickness specified in Table 1.

7.1.2 Plaster nominal thickness shall be measured from the back plane of the metal plaster base, exclusive of ribs or dimples, or from the face of the solid backing with or without metal plaster base, to the outer surface exclusive of texture variations.

7.1.3 Portland cement-based plaster shall be applied on furred metal plaster base when the surface of solid backing consists of gypsum board, gypsum plaster, wood, or rigid foam board-type products.

NOTE 5—On horizontal ceiling supports or roof soffits protected by a drip edge, gypsum board products shall be permitted to be used as backing for metal base to receive portland cement plaster.

7.1.4 Separation shall be provided where plaster abuts dissimilar construction materials or openings. (See A2.1.4.)

7.1.5 Each plaster coat shall be applied to an entire wall or ceiling panel without interruption to avoid cold joints and abrupt changes in the uniform appearance of succeeding coats. Wet plaster shall abut set plaster at naturally occurring interruptions in the plane of the plaster, such as corner angles, rustications, openings, and control joints where this is possible. Joinings, where necessary, shall be cut square and straight and not less than 6 in. (152 mm) away from a joining in the preceding coat.

7.1.6 Metal plaster base shall be covered with three-coat work with or without solid backing. The combined total nominal thickness shall be as shown in Table 1. A dash-bond coat shall not replace one of the specified number of coats.

7.1.7 Two-coat work shall be used only over solid bases meeting the requirements of 5.2. The combined total nominal thickness shall be as shown in Table 1. A dash-bond coat shall not replace one of the specified number of coats.

7.1.8 Backplaster where required, shall be applied only after the coat on the opposite side has set sufficiently to resist breaking or cracking the plaster keys.

7.1.9 Each coat shall be permitted to set before the next coat is applied. (See X1.4.2.)

7.1.10 Plaster coats that have become dry shall be evenly dampened with water prior to applying subsequent coats to obtain uniform suction. There shall be no visible water on the surface when plaster is applied.

7.2 *Plaster Application on Metal Plaster Bases:*

7.2.1 The first (scratch) coat shall be applied with sufficient material and pressure to form full keys through, and to embed the metal base, and with sufficient thickness of material over the metal to allow for scoring the surface.

7.2.1.1 As soon as the first (scratch) coat becomes firm, the entire surface shall be scored in one direction only. The vertical surfaces shall be scored horizontally.

7.2.1.2 The first (scratch) coat shall become sufficiently rigid to support the application of the second (brown) coat without damage to the monolithic continuity of the first (scratch) coat or its key.

7.2.2 The second (brown) coat shall be applied with sufficient material and pressure to ensure tight contact with the first (scratch) coat and to bring the combined thickness of the base coat to the nominal thickness shown in Table 4.

7.2.2.1 The surface of the second (brown) coat shall be brought to a true, even plane with a rod or straightedge, filling surface defects in plane with plaster. Dry rodding the surface of the brown coat shall be permitted.

7.2.2.2 The surface shall be floated uniformly to promote densification of the coat and to provide a surface receptive to bonding of the finish coat.

7.2.3 The third (finish) coat shall be applied with sufficient material and pressure to ensure tight contact with, and complete coverage of the base coat and to the nominal thickness shown in Table 1 and 7.3.1.1.

7.3 *Plaster Application on Solid Plaster Bases:*

7.3.1 High-suction bases shall be evenly dampened with clean water prior to the application of plaster. Do not dampen low-suction solid bases, such as dense concrete or smooth brick.

7.3.1.1 Where masonry or concrete surfaces vary in plane, plaster thickness required to produce level surfaces shall not be required to be uniform.

7.3.2 *Three-Coat Application on Solid Bases:*

7.3.2.1 The first (scratch) coat shall be applied with sufficient material and pressure to ensure tight contact and complete coverage of the solid base, to the nominal thickness shown in Table 1. As soon as the first (scratch) coat becomes firm, the entire surface shall be scored in one direction only. The vertical

TABLE 3 Base-Coat Proportions,[A] Parts by Volume[B]

Plaster Mix Symbols	Cementitious Materials					Volume of Aggregate per Sum of Separate Volumes of Cementitious Materials	
	Portland Cement or Blended Cement	Plastic Cement	Masonry Cement N	Masonry Cement M or S	Lime	1st Coat	2nd[C] Coat
C	1	...	...	...	0–¾	2½–4	3–5
CL	1	...	...	...	¾–1½	2½–4	3–5
M	...	...	1	...	...	2½–4	3–5
CM	1	...	1	...	...	2½–4	3–5
MS	...	...	...	1	...	2½–4	3–5
P	...	1	...	...	...	2½–4	3–5
CP	1	1	...	...	...	2½–4	3–5

[A] The mix proportions for plaster scratch and brown coats to receive ceramic tile shall be in accordance with the applicable requirements of ANSI A108.1 series applicable to specified method of setting time.

[B] Variations in lime, sand, and perlite contents are allowed due to variation in local sands and insulation and weight requirements. A higher lime content will generally support a higher aggregate content without loss of workability. The workability of the plaster mix will govern the amounts of lime, sand, or perlite.

[C] The same or greater sand proportion shall be used in the second coat than is used in the first coat.

TABLE 4 Job-Mixed Finish Coat Proportion Parts by Volume

Plaster Mix Symbols[A]	Cementitious Materials					Volume of Aggregate per Sum of Separate Volumes of Cementitious Materials[B]
	Portland Cement or Blended Cement	Plastic Cement	Masonry Cement[A] N	Masonry Cement[A] M or S	Lime	
F	1	...	...	...	¾–1½	1½–3
FL	1	...	...	...	1½–2	1½–3
FM	...	...	1	...	...	1½–3
FCM	1	...	1	...	...	1½–3
FMS	...	...	...	1	...	1½–3
FP	...	1	...	...	...	1½–3

[A] Additional portland cement is not required when Type S or M masonry cement is used.

[B] In areas not subject to impact, perlite aggregate shall be permitted to be used over base-coat plaster containing perlite aggregate.

surfaces shall be scored horizontally.

7.3.2.2 The second (brown) coat shall be applied using the same procedures specified in 7.2.2 and 7.2.2.1, bringing the surface to a true, even plane with a rod or straightedge, filling any defects in plane with plaster and darbying. The surface shall be floated uniformly to provide a surface receptive to the application of the third (finish) coat.

7.3.2.3 The third (finish) coat shall be applied as specified in 7.2.3.

7.3.3 *Two-Coat Application on Solid Plaster Bases:*

7.3.3.1 The first (scratch) coat shall be applied as specified in 7.3.2.1.

7.3.3.2 The second (finish) coat shall be applied as specified in 7.2.3.

7.4 *Finish-Coat Application:*

7.4.1 Job-mixed or factory-prepared finish coats shall be applied, by machine or by hand, as specified in 7.2.3.

7.4.2 The use of excessive water during the application and finishing of finish-coat plaster shall be avoided.

7.5 *Fog-Coat Application*—Job-mixed or factory-prepared fog coats shall be applied in accordance with the directions of the manufacturer.

8. Curing and Time Between Coats

8.1 Provide sufficient moisture in the plaster mix or by moist or fog curing to permit continuous hydration of the cementitious materials. The most effective procedure for curing and time between coats will depend on climatic and job conditions. (See X1.4.2.)

8.2 Sufficient time between coats shall be allowed to permit each coat to cure or develop enough rigidity to resist cracking or other physical damage when the next coat is applied. (See X1.4.2.)

9. Product Marking

9.1 Packaged materials shall be clearly marked or labeled to indicate product, brand name, the manufacturer, and the weight of the material contained therein. Similar information shall be provided in the shipping advices accompanying the shipment of bulk materials.

10. Delivery of Materials

10.1 Packaged materials shall be delivered in factory-sealed, unopened, and unbroken packages, containers, or bundles.

10.2 Bulk materials shall be delivered in clean transport vessels, free of contaminates.

11. Protection of Materials

11.1 Weather-sensitive materials shall be kept in a dry condition until ready for use. (See A2.4.)

11.2 Bulk materials shall be stored to prevent subsequent contamination and segregation.

12. Environmental Conditions

12.1 Portland cement-based plaster shall not be applied to frozen base or to a base containing frost. Plaster mixes shall not contain frozen ingredients. Plaster coats shall be protected from freezing for a period of not less than 24 h after set has occurred.

12.2 Portland cement plaster shall be protected from uneven and excessive evaporation during dry weather and from strong blasts of dry air.

12.3 *Plaster Application*—When artificial heat is required, heaters shall be located to prevent a concentration of heat on uncured plaster. Heaters shall be vented to the outside to prevent toxic fumes and other products of combustion from adhering to or penetrating plaster bases and plaster. Adequate ventilation shall be maintained in all areas, particularly in interior areas with little or no natural air movement.

12.3.1 Interior environment shall be maintained at a temperature above 40°F not less than 48 h prior to and during application of portland cement-based plaster. Interior temperature shall be maintained above 40°F until normal occupancy.

12.3.2 For exteriors, plaster shall be applied when the ambient temperature is higher than 40°F (4.4°C), unless the work area is enclosed and heat is provided as described in 12.3.

13. Keywords

13.1 brown coat; cementitious; exterior plaster; fog coat—bond; portland cement; scratch coat; stucco

ANNEXES

(Mandatory Information)

A1. GENERAL INFORMATION

A1.1 The work shall include all labor, materials, services, equipment, and scaffolding required to complete the plastering of the project in accordance with the drawings and specifications, except heat, electric power, and potable water.

A1.2 Where a specific degree of fire resistance is required for plastered assemblies and constructions, details of construction shall be in accordance with official reports of fire tests conducted by recognized testing laboratories, in accordance with Test Methods E 119.

A1.3 Where a specific degree of sound control is required for plastered assemblies and constructions, details of construction shall be in accordance with official reports of tests conducted by recognized testing laboratories, in accordance with applicable sound tests of Test Methods E 90 or E 492.

A1.4 Scaffolding shall be constructed and maintained in strict conformity with applicable laws and ordinances.

A1.5 Work schedules shall provide for completion of work affecting supports, framework or lath of a suspended ceiling (such as loading) before plastering work is accomplished.

A1.6 Surfaces and accessories to receive plaster shall be examined before plastering is applied thereto. The proper authorities shall be notified and unsatisfactory conditions shall be corrected prior to the application of plaster. Unsatisfactory conditions shall be corrected by the party responsible for such conditions.

A1.6.1 Metal plaster bases, backing, attachment, and accessories to receive plaster shall be examined to determine if the applicable requirements of Specification C 1063 have been met unless otherwise required by the contract specifications.

A1.6.2 The construction specifier shall describe, in the proper section of the contract specifications, the physical characteristics of solid surface bases to receive plaster. The plane tolerance shall be not more than ¼ in. in 10 ft (3.1 mm/m). The mortar joints shall be flush and not struck. Dissimilar materials such as ties, reinforcing steel, and so forth, shall be cut back ⅛ in. (3 mm) below the surface and treated with a corrosion-resistant coating. Masonry shall be solid at corners and where masonry changes thickness in a continuous construction. Form release compounds shall be compatible with plaster or be completely removed from surfaces to receive plaster. The plastering contractor shall use this portion of the construction specifications for acceptance or rejection of such surfaces.

A2. DESIGN CONSIDERATIONS

A2.1 Exterior plaster (stucco) is applied to outside surfaces of all types of structures to provide a durable, fire-resistant covering. Interior plaster is applied to inside surfaces that will be subjected to various exposures, such as abrasion, vibration, or to continuous or frequent moisture and wetting, or to freezing or thawing.

A2.1.1 Sufficient slope on faces of plastered surfaces shall be provided to prevent water, snow, or ice from accumulating or standing. Air-entrained portland cement plaster provides improved resistance to freeze/thaw deterioration. Resistance to rain penetration is improved where plaster has been adequately densified during application and properly cured. Plaster shall not, however, be considered to be "waterproof."

A2.1.2 The construction specifier shall describe, in the appropriate section of the contract specifications, the requirements for furnishing and application of flashing. Flashing shall be specified at openings, perimeters, and terminations to prevent water from getting behind plaster. Flashing shall be corrosion-resistant material. Aluminum flashing shall not be used. Flashing supplemented with sealant shall be permitted.

A2.1.3 Sealing or caulking of V-grooves, exposed ends, and edges of plaster panels exterior work to prevent entry of water shall be provided.

A2.1.4 To reduce spalling where interior plaster abuts openings, such as wood or metal door or window frames, or fascia boards, the edge of three-coat plaster shall be tooled through the second and finish coats to produce a continuous small V-joint of uniform depth and width. On two-coat work, the V-joint shall be tooled through the finish coat only.

A2.1.5 Provide in the appropriate project specification section that solid bases to receive plaster shall not be treated with bond breakers, parting compounds, form oil, or other material that will prevent or inhibit the bond of the plaster to the base.

A2.1.6 Maximum allowable deflection for vertical or horizontal framing for plaster, not including cladding, shall be L/360.

A2.2 *Provisions for Drainage Behind Exterior Plaster:*

A2.2.1 In multistory construction where lath and portland cement plaster exterior walls are continuous past a floor slab, tracks or plates and studs shall be offset to provide a space not less than ⅜ in. (9.5 mm) between the inner face of the exterior plaster and the edge of the floor slab.

A2.2.2 At the bottom of exterior walls where the wall is supported by a floor or foundation, a drip screed and through-wall flashing or weep holes or other effective means to drain away any water that may get behind the plaster shall be provided.

A2.2.3 Where vertical and horizontal exterior plaster surfaces meet, both surfaces shall be terminated with casing beads with the vertical surface extending at least ¼ in. (6 mm) below the intersecting horizontal plastered surface, thus providing a drip edge. The casing bead for the horizontal surface shall be terminated not less than ¼ in. from the back of the vertical surface to provide drainage.

A2.3 *Relief from Stresses:*

A2.3.1 For information on control joints and perimeter relief see the Installation Section of Specification C 1063.

A2.3.1.1 Control joints shall be cleaned and clear of plaster within the control area after plaster application and before final plaster set.

A2.3.1.2 Prefabricated control joint members shall be installed prior to the application of plaster; therefore, the decision to use them, the type selected, their location, and method of installation shall be determined and specified in project specification sections other than the section on plastering.

A2.3.1.3 The creation of a groove or cut in plaster only, shall be specified in the plastering section. Such groove shall not be considered a control or expansion joint.

A2.3.2 Where plaster and metal plaster base continues across the face of a concrete column, or other structural member, water-resistive building paper or felt shall be placed between the metal plaster base and the structural member (paper or plastic-backed metal plaster base shall be permitted). Where the width of the structural member exceeds the approved span capability of the metal plaster base, self-furring metal plaster base shall be used and sparingly scatter nailed to bring paper and metal base to general plane.

A2.3.3 Where dissimilar base materials abut and are to receive a continuous coat of plaster: (*1*) a two-piece expansion joint, casing beads back-to-back, or premanufactured control-expansion joint member shall be installed; or (*2*) the juncture shall be covered with a 6-in. (152-mm) wide strip of galvanized, self-furring metal plaster base extending 3 in. (76 mm) on either side of the juncture; or (*3*) where one of the bases is metal plaster base, self-furring metal plaster base shall be extended 4 in. (102 mm) onto the abutting base.

A2.4 *Weather-Sensitive Materials*—Water-sensitive materials shall be stored off the ground or floor and under cover, avoiding contact with damp floor or wall surfaces. Temperature-sensitive materials shall be protected from freezing. Bulk materials shall be stored in the area of intended use and caution shall be exercised to prevent contamination and segregation of bulk materials prior to use.

A2.5 *Admixtures*—Admixtures shall be proportioned and mixed in accordance with the published directions of the admixture manufacturer.

A2.5.1 The quantity of admixtures required to impart the desired performance is generally very small in relation to the quantities of the other mix ingredients. Batch-to-batch quantities shall be measured accurately.

A2.5.2 Air-entraining agents cause air to be incorporated in the plaster in the form of minute bubbles, usually to improve frost or freeze-thaw resistance, or workability of the plaster during application. Air-entraining agents for portland cement-based plaster shall meet the requirements of Specification C 260.

APPENDIX

(Nonmandatory Information)

X1. GENERAL INFORMATION

X1.1 *Additions*—Bonding compounds or agents may be pre-applied to a surface to receive plaster. In this usage it is not considered an admixture. Bonding compounds that are integrally mixed with plaster prior to its application are considered admixtures. Where exterior exposure and cyclic wetting are anticipated, the re-emulsification capability of the bonding material must be considered. Bonding agents are only as good as the material surface to which they are applied; therefore, form release materials must be removed from concrete or be compatible with the bonding material used. Bonding agents in plaster mixes may increase the cohesive properties of the plaster. Bonding agents should meet the requirements of Specifications C 631 or C 932.

X1.1.1 By the use of a suitable admixture or additive, it is possible to improve plaster's resistance to moisture movement. However, the use of the terms dampproofing or water proofing is misleading, and their use shall be discouraged.

X1.1.2 Natural or synthetic fibers, ½ to 2 in. (13 to 51 mm) in length and free of contaminates may be specified to improve resistance to cracking or to impart improved pumpability characteristics. The quantities per batch shall be in accordance with the published directions of the fiber manufacturer. No more than 2 lb (0.90 kg) of fiber should be used per cubic foot of cementitious material. Asbestos fibers should not be used. Alkaline-resistant glass fibers are recommended where glass fiber is used.

X1.1.3 Plasticizers containing hydrated lime putty, air-entraining agents, or approved fatteners to increase the workability of a portland cement plaster may be used. Plaster consistency and workability are affected by plasticizers that are beneficial in proper quantities from an economic standpoint, but in excess can be detrimental to the long-term performance of the plaster in place.

X1.1.4 Color material for integral mixing with plaster should not significantly alter the setting, strength development, or durability characteristics of the plaster. Natural or mineral pigments that are produced by physical processing of materials mined directly from the earth appear to offer the best long-term performance with respect to resistance to fading. Plaster color is determined by the natural color of the cementitious materials, aggregate, and any color pigment, and their proportions to each other. The use of white cement with the desired mineral oxide pigment color material may result in truer color.

X1.1.4.1 The uniformity of color cannot be guaranteed by the materials manufacturer of the component materials or by the applicating contractor. Color uniformity is affected by the uniformity of proportioning, thoroughness of mixing, cleanliness of equipment, application technique, and curing conditions and procedure, which are generally under the control of the applicator. Color uniformity is affected to an even greater degree by variations in thickness and differences in the suction of the base coat from one area or location to another, the type of finish selected, the migration of color pigments with moisture, and with job site climatic and environmental conditions. These factors are rarely under the control of the applicator.

X1.2 *Finish Coat Categories* (applicable to both natural and colored finishes):

X1.2.1 Texture, as a description of surface appearance, is identified generally with the method and tools used to achieve the finish. Texture can be varied by the size and shape of the aggregate used, the equipment or tools employed, the consistency of the finish coat mix, the condition of the base to which it is applied, and by subsequent decorative or protective treatment.

X1.2.2 There are many factors that affect the ultimate appearance of textured and integrally colored plaster. A suitably sized sample panel should be submitted for approval by the architect and the owner. Once approved, the sample should be maintained on site for reference and comparison.

X1.2.3 With the almost limitless variations possible for finish appearance or texture, the same term may not have the same meaning to the specifier, the contractor, and the actual applicator. The specifier is cautioned to use an approved range of sample panels. To provide some guidance, the following categories are generally understood and recognized to imply a particular method of application technique or resulting finished appearance:

X1.2.4 *Smooth Trowel*—Hand- or machine-applied plaster floated as smooth as possible and then steel-troweled. Steel troweling should be delayed as long as possible and used only to eliminate uneven points and to force aggregate particles into the plaster surface. Excessive troweling should be avoided.

X1.2.5 *Float*—A plaster devoid of coarse aggregate applied in a thin coat completely covering the base coat, followed by a second coat that is floated to a true plane surface yielding a relatively smooth to fine-textured finish, depending on the size of aggregate and technique used. It is also known as sand finish.

X1.2.6 *Trowel-Textured (such as Spanish Fan, Trowel Sweep, English Cottage)*—A freshly applied plaster coat is given various textures, designs, or stippled effects by hand troweling. The effects achieved may be individualized and may be difficult to duplicate by different applicators.

X1.2.7 *Rough-Textured (such as Rough Cast, Wet Dash, Scottish Harl)*—Coarse aggregate is mixed intimately with the plaster and is then propelled against the base coat by trowel or by hand tool. The aggregate is largely unexposed and deep textured.

X1.2.8 *Exposed Aggregate (also known at Marblecrete)*—Varying sizes of natural or manufactured stone, gravel, shell, or ceramic aggregates are embedded by hand or machine propulsion into a freshly applied finish "bedding" coat. The size of the aggregate determines the thickness of the "bedding" coat. It is generally thicker than a conventional finish coat.

X1.2.9 *Spray-Textured*—A machine-applied plaster coat directed over a previously applied thin smooth coat of the same mix. The texture achieved depends on the consistency of the sprayed mixture, moisture content of the base to which it is applied, the angle and distance of the nozzle to the surface, and the pressure of the machine.

X1.2.10 *Brush-Finish*—A method of surfacing or resurfacing new or existing plaster. The plaster is applied with a brush to a thickness of not less than 1⁄16 in. (1.6 mm). For an existing plaster surface the bond capability must be determined by test application or a bonding compound must be applied prior to the brush application.

X1.2.11 *Miscellaneous Types*—This finish coat category is somewhat similar to trowel-textured finishes, except that the freshly applied plaster is textured with a variety of instruments other than the trowel, such as swept with a broom or brush, corrugated by raking or combing, punched with pointed or blunt instrument, scored by aid of a straightedge into designs of simulated brick, block, stone, and so forth. A variation of texturing a finish coat involves waiting until it has partially set and then flattening by light troweling of the unevenly applied plaster or by simulating architectural terracotta.

X1.2.12 *Scraffitto*—A method of applying two or more successive coats of different colored plaster and then removing parts of the overlaid coats to reveal the underlaying coats, usually following a design or pattern. This is not generally considered a finish coat operation because of the number of thickness of coats.

X1.3 When specified as alternate for final coat, trowel- or plaster machine-applied textured acrylic finishes containing aggregate may be substituted for portland cement finish coats, provided brown coat is properly prepared and finish is applied according to the manufacturer's directions.

X1.3.1 *Staining of Plaster*—Staining and discoloration of plaster, caused by free water draining from one plane of plaster to another or from a dissimilar material onto a plaster surface, can be minimized by providing sufficient depth and angle for drip caps and the use of water-resistive surface coatings.

X1.3.2 Staining of plaster due to entrapment of moisture behind the plaster, can be avoided or minimized by providing an air space for ventilation between the back of the plaster and adjacent material. This type of staining may occur where insulation with or without vapor barrier, or other material containing asphaltic or coal tar derivatives, fireproofing salts, and so forth, can migrate with moisture movement to the finished plaster surface.

X1.3.3 Integrally colored plaster can be discolored or altered in shade if subjected to moisture, either from uncured base coats or external sources, such as rain, too soon after applications.

X1.4 *Installation Instructions:*

X1.4.1 Hand mixing should not be permitted, except as approved by the contract specifier.

X1.4.1.1 After all ingredients are in the mixer, mix the plaster for 3 to 5 min.

X1.4.1.2 The amount of water used in the plaster mix should be determined by the plasterer. Factors such as the suction of the base, or of the previous coat, water content of the aggregate, drying conditions, and finishing operations should be considered in determining water usage. Use of excessive water may result in dropouts, fall or slide off, excessive shrinkage, high porosity, and lower strength.

X1.4.2 *Time Between Coats and Curing for Portland Cement-Based Plaster:*

X1.4.2.1 The timing between coats will vary with climatic conditions and types of plaster base. Temperature and relative humidity extend or reduce the time between consecutive operations. Cold or wet weather lengthens and hot or dry weather shortens the time period. Moderate changes in temperature and relative humidity can be overcome by providing additional heating materials during cold weather and by reducing the absorption of the base by pre-wetting during hot or dry weather.

X1.4.2.2 In order to provide more intimate contact and bond between coats and to reduce rapid water loss, the second coat should be applied as soon as the first coat is sufficiently rigid to resist cracking, the pressures of the second coat application, and the leveling process.

X1.4.2.3 The amount of water and the timing for curing portland cement plaster will vary with the climatic conditions, the type of base, and use or nonuse of water-retentive admixtures.

X1.4.2.4 Some moisture must be retained in or added back to freshly applied portland cement-based plaster. If the relative humidity is relatively high (above 75 %), the frequency for rewetting a surface may be reduced. If it is hot, dry, and windy, the frequency of rewetting must be increased.

X1.4.2.5 Consider the physical characteristics of the structure as well as the previously mentioned conditions when selecting the method of curing. The method can be one or a combination of the following:

(*1*) Moist curing is accomplished by applying a fine fog spray of water as frequently as required, generally twice daily in the morning and evening. Care must be exercised to avoid erosion damage to portland cement-based plaster surfaces. Except for severe drying conditions, the wetting of finish coat should be avoided, that is, the base coat prior to application of the finish coat.

(*2*) Plastic film, when taped or weighted down around the perimeter of the plastered area, can provide a vapor barrier to retain the moisture between the membrane and plaster. Care must be exercised in placing the film: if too soon, the film may damage surface texture; if too late, the moisture may have already escaped.

(*3*) Canvas, cloth, or sheet material barriers can be erected to deflect sunlight and wind, both of which will reduce the rate of evaporation. If the humidity is very low, this option alone may not provide adequate protection.

Designation: C 931/C 931M – 98

Standard Specification for Exterior Gypsum Soffit Board[1]

This standard is issued under the fixed designation C931/931M; the number immediately following the designation indicates the year of original adoption or, in the case of revision, the year of last revision. A number in parentheses indicates the year of last reapproval. A superscript epsilon (ε) indicates an editorial change since the last revision or reapproval.

This specification has been approved for use by agencies of the Department of Defense.

1. Scope

1.1 This specification covers exterior gypsum soffit board, designed to be used for exterior soffits and carport ceilings that are completely protected from contact with liquid water.

1.2 The values stated in either inch-pound units or SI (metric) are to be regarded separately as the standard. Within the text, the SI units are shown in brackets. The values stated in each system shall be used independently of the other. Values from the two systems shall not be combined.

2. Referenced Documents

2.1 *ASTM Standards:*

C 473 Test Methods for Physical Testing of Gypsum Panel Products[2]

C 645 Specification for Nonstructural Steel Framing Members[2]

C 1264/C 1264M Specification for Sampling, Inspection, Rejection, Certification, Packaging, Marking, Shipping, Handling, and Storage[2]

E 84 Test Method for Surface Burning Characteristics of Building Materials[3]

E 119 Test Methods for Fire Tests of Building Construction and Materials[3]

3. Materials and Manufacture

3.1 Gypsum soffit board shall consist of an incombustible core, essentially gypsum, surfaced with paper bonded to the core.

3.2 Exterior gypsum soffit board, Type X (special fire-resistant) designates exterior gypsum soffit board complying with this specification that provides not less than 1-h fire-resistance for boards ⅝-in. (15.9-mm) thick or ¾-h fire-resistance for boards ½-in. (12.7-mm) thick, applied parallel with and on each side of load bearing 2 by 4 wood studs spaced 16 in. (406 mm) on center with 6d coated nails, 1⅞-in. (48-mm) long, 0.0915-in. (2.3-mm) diameter shank, ¼-in. (6.4-mm) diameter heads, spaced 7 in. (178 mm) on center with exterior gypsum soffit board joints staggered 16 in. (406 mm) on each side of the partition and tested in accordance with Test Methods E 119.

3.3 Gypsum soffit board shall have a maximum flame-spread classification of 25 when tested in accordance with the requirements of Test Method E 84.

4. Physical Properties

4.1 *Flexural Strength:*

4.1.1 Specimens shall be tested in accordance with Test Methods C 473. When tested face up and when tested face down the specimens shall have an average breaking load of not less than the following:

Thickness, in. (mm)	Method A Load, lbf (N) Bearing Edges Across Fiber of Surfacing	Method A Load, lbf (N) Bearing Edges Parallel to Fiber of Surfacing	Method B Load, lbf (N) Bearing Edges Across Fiber of Surfacing	Method B Load, lbf (N) Bearing Edges Parallel to Fiber of Surfacing
½ (12.7)	110 (489)	40 (178)	107 (476)	36 (160)
⅝ (15.9)	150 (668)	50 (222)	147 (654)	46 (205)

4.2 *Humidified Deflection:*

4.2.1 When tested in accordance with Test Methods C 473, specimens taken from the gypsum soffit board shall have an average deflection of no more than the following:

Thickness, in. (mm)	Humidified Deflection, Eighths of an in. (mm)
½ (12.7)	7 (22)
⅝ (15.9)	4 (13)

4.3 *Core, End, and Edge Hardness:*

4.3.1 When tested in accordance with Test Methods C 473, specimens taken from the gypsum soffit board shall have an average hardness of not less than 15 lbf (67 N) when tested by Method A and 11 lbf (49 N) when tested by Method B for the core, ends, or edges.

4.4 *Nail Pull Resistance:*

4.4.1 When tested in accordance with Test Methods C 473, specimens taken from the gypsum soffit board shall have an average nail pull resistance of not less than the following:

Thickness, in. (mm)	Method A Nail Pull Resistance, lbf (N)	Method B Nail Pull Resistance, lbf (N)
½ (12.7)	80 (356)	77 (343)
⅝ (15.9)	90 (400)	87 (387)

[1] This specification is under the jurisdiction of ASTM Committee C-11 on Gypsum and Related Building Materials and Systems and is the direct responsibility of Subcommittee C11.01 on Specifications and Test Methods for Gypsum Products.

Current edition approved April 10, 1998. Published December 1998. Originally published as C 930 – 80. Last previous edition C 930 – 97.

[2] *Annual Book of ASTM Standards,* Vol 04.01.

[3] *Annual Book of ASTM Standards,* Vol 04.07.

5. Dimensions and Permissible Variations

5.1 *Thickness, nominal*—½ or ⅝ in. (12.7 or 15.9 mm) with permissible variations in the nominal thickness of 1⁄64 in. (0.4 mm) and with permissible local variations of ±1⁄32 in. (0.8 mm) from the nominal thickness. Edges of ½ and ⅝ in. soffit board shall be either plain, recessed, beveled, or tapered to receive a joint reinforcing strip.

5.2 *Width, nominal*—up to 48 in. (1220 mm), with a permissible variation of 3⁄32 in. (3 mm) under the specified width.

5.3 *Length, nominal*—from 4 to 16 ft (1220 mm to 4880 mm), inclusive, with permissible variations of ±¼ in. (6 mm) from the specified length.

6. Workmanship, Finish, and Appearance

6.1 The surfaces of gypsum soffit board shall be true and free of imperfections that would render the board unfit for use with or without decoration. The edges shall be square with a permissible variation of ±⅛ in. (±3 mm) in the full width of the board.

7. Sampling, Inspection, Rejection, Certification, Packaging, Marking, Shipping, Handling, and Storage

7.1 Sampling, inspection, rejection, certification, packaging, marking, shipping, handling, and storage of exterior gypsum soffit board shall be in accordance with Specification C 1264/C 1264M.

APPENDIX

(Nonmandatory Information)

This Appendix gives general information and also suggestions for inclusions to be made elsewhere by the specifier. They are not part of this specification.

X1. Alternate Definition for Type X

X1.1 Exterior gypsum soffit board, Type X (special fire-resistant) designates exterior gypsum soffit board providing a greater fire resistance than regular exterior gypsum soffit board of the same thickness. Type X (special fire-resistant) exterior gypsum soffit board, when tested in accordance with Test Methods E 119, shall provide the following minimum fire resistance for the assemblies described.

X1.1.1 One hour for a ⅝-in. (15.9-mm) thickness applied to a partition in a single layer application on each side of 3⅝-in. (92-mm) deep non-loadbearing galvanized steel studs complying with Specification C 645, spaced 24 in. (610 mm) on center. The ⅝-in. (15.9-mm) thick exterior gypsum soffit board 48-in. (1220-mm) wide shall be attached using 1-in. (25-mm) long drywall screws spaced 8 in. (203 mm) on center along the edges and ends, and 12 in. (305 mm) along intermediate studs. All joints shall be oriented parallel to and located over studs and staggered on opposite sides of the assembly, and

X1.1.2 Two hours for a ½-in. (12.7-mm) thickness applied to a partition in a double layer application on each side of 2½-in. (64-mm) deep non-loadbearing galvanized steel studs complying with Specification C 645, spaced 24 in. (610 mm) on center. The 48-in. (1220-mm) wide base layer shall be attached using 1-in. (25-mm) long drywall screws spaced 12 in. (305 mm) on center along board edges, ends, and along intermediate studs. Joints shall be oriented parallel to and located over studs and staggered on opposite sides of the assembly. The 48-in. (1220-mm) wide face layer shall be attached using 1⅝-in. (41-mm) long drywall screws spaced 12 in. (305 mm) along board edges, ends and along intermediate studs. Joints shall be oriented parallel to and located over studs, offset 24 in. (610 mm) from the base layer joints, and staggered on opposite sides of the assembly.

Designation: C 932 – 98a

Standard Specification for Surface-Applied Bonding Compounds for Exterior Plastering[1]

This standard is issued under the fixed designation C 932; the number immediately following the designation indicates the year of original adoption or, in the case of revision, the year of last revision. A number in parentheses indicates the year of last reapproval. A superscript epsilon (ε) indicates an editorial change since the last revision or reapproval.

This standard has been approved for use by agencies of the Department of Defense.

1. Scope *

1.1 This specification covers minimum requirements for exterior surface-applied bonding compounds for improving the adhesion of cementitious material to concrete or other masonry surfaces or any structurally sound surfaces.

1.2 This specification also covers test methods for determining performance requirements and physical properties.

1.3 The values stated in inch-pound units are to be regarded as the standard. The SI metric values given in parentheses are approximate and are provided for information purposes only.

1.4 The following safety hazards caveat pertains only to the test methods described in Sections 9, 10, 11, and 12 in this specification: *This standard does not purport to address all of the safety concerns, if any, associated with its use. It is the responsibility of the user of this standard to establish appropriate safety and health practices and determine the applicability of regulatory limitations prior to use.*

2. Referenced Documents

2.1 *ASTM Standards:*

C 11 Terminology Relating to Gypsum and Related Building Materials and Systems[2]

C 109/C 109M Test Method for Compressive Strength of Hydraulic Cement Mortars (Using 2-in. or [50-mm] Cube Specimens[2]

C 150 Specification for Portland Cement[2]

C 305 Practice for Mechanical Mixing of Hydraulic Cement Pastes and Mortars of Plastic Consistency[2]

C 511 Specification for Moist Cabinets, Moist Rooms, and Water Storage Tanks Used in the Testing of Hydraulic Cements and Concretes[2]

C 778 Specification for Standard Sand[2]

3. Terminology

3.1 Definitions used in this specification shall be in accordance with Terminology C 11.

4. Physical Properties

4.1 *Surface Applied Bonding Compound*—A freeze-thaw stable composition, suitable for brush, roller, or spray application. It shall be tinted to show by visual inspection where it has been applied. The tint shall not bleed through the material being bonded.

4.2 *Consistency*—The bonding compound shall be free of foreign matter as determined by visual inspection and shall be of such uniform consistency that when applied in accordance with the producers directions by brush, roller, or spray to concrete, masonry, or other structurally sound surface, the bonding compound shall flow on evenly and dry uniformly.

4.3 *Film Characteristics*—The film-forming property shall be determined by visual inspection to determine the presence of a continuous film not broken by fisheyes, cracking, pull-back, or any other discontinuity in the film surface. It shall not be noticeably affected by alkaline surfaces or weak acids.

5. Performance Requirements

5.1 *Bonding Capability*—The bonding compound shall be capable of bonding cementitious materials when applied in accordance with the producer's directions and tested as specified in Section 12.

5.2 *Degradation*—Bonding compound which separates shall be able to be re-mixed to a uniform consistency when tested as specified in Section 10.

5.3 *High Temperature Test*—The bond strength shall be not less than 150 psi (1034 kPa) when tested as specified in Section 9.

5.4 *Freeze-Thaw Stability*—The bond strength shall be not less than 150 psi (1034 kPa) when tested as specified in Section 11.

5.5 *Tensile Bond Strength*—Bond strength of a fresh sample or a 6-month old sample of bonding compound shall have an average tensile strength of not less than 150 psi (1034 kPa) when tested as specified in Section 12.

6. Sampling

6.1 Take a sample of not less than 2.2 lb (1000 g) from each shipment or consignment for analysis and tests. Except in special cases, take the sample from not less than three separate containers, chosen at random. In addition, take samples from

[1] This specification is under the jurisdiction of ASTM Committee C-11 on Gypsum and Related Building Materials and Systems and is the direct responsibility of Subcommittee C11.02 on Specifications and Test Methods for Accessories and Related Products.

Current edition approved Oct. 10, 1998. Published March 1999. Originally published as C 932 – 80. Last previous edition C 932 – 98.

[2] *Annual Book of ASTM Standards*, Vol 04.01.

*A Summary of Changes section appears at the end of this standard.

containers that appear to be nonrepresentative, and test separately. Place the samples immediately in airtight glass containers and transport to the testing laboratory in these containers. Take precautions to reduce evaporation or drying to a minimum. Thoroughly mix the bonding compound in the container if there is a tendency for liquid phase separation.

7. Apparatus

7.1 *Moist Cabinet*—Specification C 511.

7.2 *Oven*—A forced draft type oven, having a temperature controlled at 140 ± 5°F (60 ± 3°C) for high temperature tests.

7.3 *Freezer*—A freezer having a controlled temperature of – 10 ± 2°F (–23 ± 1°C).

7.4 *Balance*—A balance capable of weighing not less than 5.5 lb (2500 g) at a precision of 0.0002 lb (0.1 g).

7.5 *Timing Device*—An instrument capable of reading to the nearest second.

7.6 *Tamper*—A tamper made up of a nonabsorptive, nonabrasive, non-brittle material and having a cross section of ½ in. by 1 in. (13 × 25 mm) and approximately 5 to 6 in. (130–150 mm) long. The tamping face shall be flat and at right angles to the length of the tamper.

7.7 *Straight Edge*—A steel straight edge not less than 4 in. (100 mm) long and not less than 1⁄16 in. (1.6 mm) nor more than ⅛ in. (3.2 mm) in thickness.

7.8 *Spatula*—A spatula with a metal blade 6 in. (150 mm) in length and ½in. (13 mm) in width, with straight edges.

7.9 *Paint Brush*—A standard 1 in. (25 mm) wide brush, with synthetic bristles.

7.10 *Testing Machine*—Any type that is of sufficient capacity and that is capable of applying the load continuously and without shock at the rate of 0.05 in. (1.27 mm) per minute, with provision for adjustment of the rate of loading.

7.11 *Briquet Molds*—The molds for making test specimens shall be made of metal not attacked by the mortar and shall have sufficient material in the sides to prevent spreading during molding. Gang molds, when used, shall be of the type shown in Fig. 1. The dimensions of the briquet molds shall conform to the following requirements: width of mold, between inside faces, at waist line of the briquet, 1 in. (25.4 mm) with permissible variations of ±0.01 in. (0.25 mm) for molds in use and ±0.005 in. (0.13 mm) for new molds; thickness of molds measured at the point of greatest thickness on either side of the mold at the waist line, 1 in. (25.4 mm) with permissible variations of + 0.004 in. (0.10 mm) and – 0.002 in. (0.05 mm) for new molds and – 0.02 in. (0.5 mm) for molds in use. The briquet specimens shall conform to the dimensional requirements shown in Fig. 2.

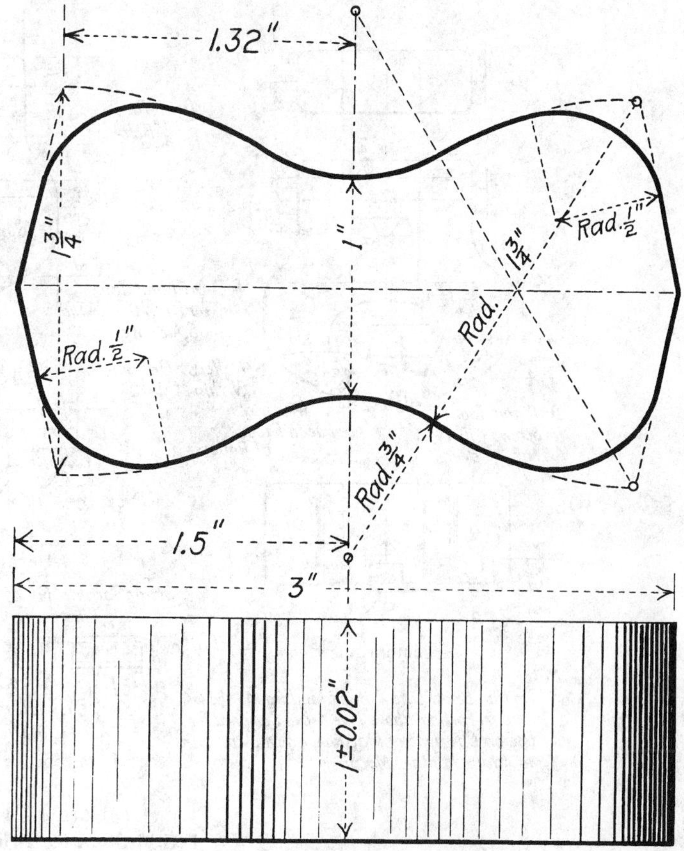

NOTE 1—1 in. = 25.4 mm.

FIG. 2 Briquet Specimen for Tensile Strength Test

7.12 *Saw*—A table saw equipped with an abrasive cutting blade or other blade suitable for cutting cementitious materials.

7.13 *Clips for Briquet Testing Machine*—The clips for holding the tension test specimen shall be in accordance with Fig. 3.

8. Conditioning

8.1 *Room Temperature and Humidity*—Maintain the air temperature in the vicinity of the mixing and testing area at 70 ± 5°F (21 ± 3°C). Maintain the relative humidity at 50 ± 2 %.

8.2 *Temperature of the Mixing Water*—72 ± 2°F (22 ± 1°C).

9. High Temperature Test

9.1 *Significance and Use*—This test method provides procedures for evaluating the adhesive strength of bonding compounds after being exposed to accelerated aging at high temperature.

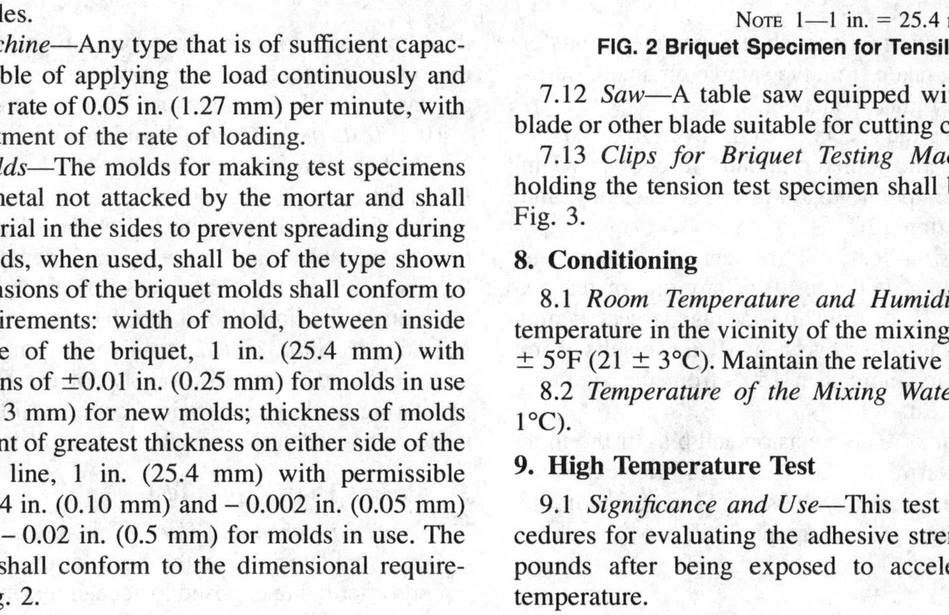

FIG. 1 Briquet Gang Mold

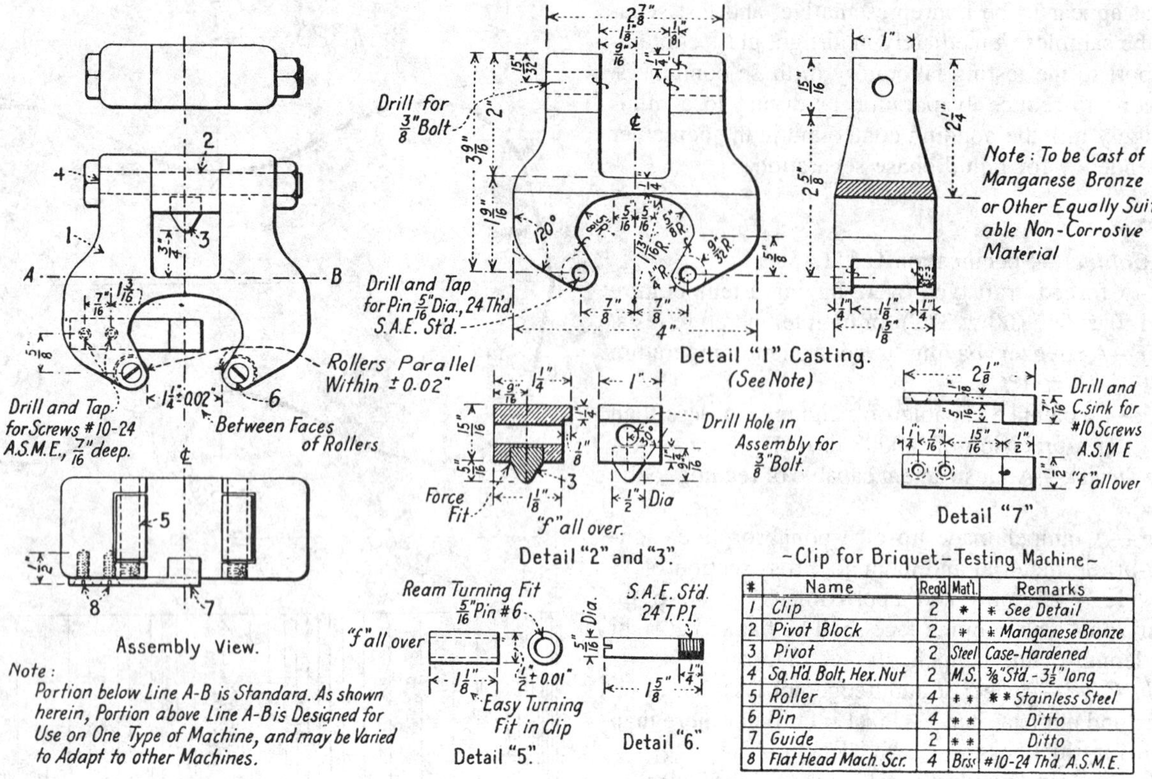

#	Name	Reqd	Mat'l.	Remarks
1	Clip	2	*	* See Detail
2	Pivot Block	2	*	* Manganese Bronze
3	Pivot	2	Steel	Case-Hardened
4	Sq. Hd. Bolt, Hex. Nut	2	M.S.	3/8" Std. - 3 1/2" long
5	Roller	4	**	** Stainless Steel
6	Pin	4	**	Ditto
7	Guide	2	**	Ditto
8	Flat Head Mach. Scr.	4	Brs.	#10-24 Th'd. A.S.M.E.

NOTE 1—1 in. = 25.4 mm.

FIG. 3 Clips for Briquet Testing Machine

9.2 *Specimen Preparation*—For each test specimen, transfer the bonding compound, taken from the sample obtained, into a standard 1 pt. (500 mL) glass container.

9.3 *Procedure*—Place the specimen in the oven for 30 days. Remove the container, allow the compound to cool to room temperature, mix the compound to a uniform consistency, and test as specified in Section 12.

9.4 *Tests and Retests*—Test six briquets and report the average of the test results. If the result of any one of the six tests varies more than 15 % from the average, reject it and report the average of the other five tests. If the results from more than two tests vary more than 15 % from the average, reject the series and retest.

9.5 *Precision and Bias*—The precision and bias of the high temperature test for determining bond strength after exposure to high temperature are essentially the same as specified in the test method for bond strength, paragraph 12.7.

10. Degradation Test

10.1 *Significance and Use*—This test method simulates the effects of high temperature storage on the physical properties of bonding compound.

10.2 *Specimen Preparation*—For each test, transfer the bonding compound, from the sample obtained, into a standard 1 pt. (500 mL) glass container.

10.3 *Procedure*—Place the specimen in the oven for 15 days. Remove the container from the oven and examine the contents for settling and separation. Place the container back in the oven for an additional 15 days. Remove the container and examine the contents again for settling and separation. Allow the compound to come to room temperature, and mix to a uniform consistency. Mix by hand stirring with a wood, metal, or plastic paddle for not more than five minutes.

10.4 *Tests and Retests*—Bonding compound that fails to be remixed to a uniform consistency shall be considered to have failed the test.

10.5 *Report*—Determine the extent of settling and separation by measuring and report as a percentage of the height of the specimen in the clear glass container, represented by the location of the liquid/solid interface.

10.6 *Precision and Bias*—No information is presented about either the precision or bias of the Degradation test for measuring settling, separation, or the ability to be mixed to a uniform consistency since the test result is non-quantitative.

11. Freeze-Thaw Cycle Test

11.1 *Significance and Use*—This test method provides procedures for determining the bond strength of bonding compounds after being exposed to accelerated storage conditions of alternate freezing and thawing.

11.2 *Specimen Preparation*—For each test specimen, transfer the bonding compound, taken from the sample obtained, into a standard 1 pt. (500 mL) glass container having a screw on or friction fit cover.

11.3 *Procedure*—Place a covered specimen in the freezer for 16 h. Remove the specimen and allow to thaw at room temperature for 8 h. Repeat this procedure for 5 cycles. After the fifth cycle, allow the compound to come to room temperature, then test as specified in Section 12.

11.4 *Tests and Retests*—Test six briquets and report the

average of the test results. If the result of any one of the six tests varies more than 15 % from the average, reject it and report the average of the other five tests. If the results from more than two tests vary more than 15 % from the average, reject the series and retest.

11.5 *Precision and Bias*—The precision and bias of the freeze-thaw cycle test for determining bond strength after exposure to alternate freezing and thawing are essentially the same as specified in the test method for bond strength, paragraph 12.7.

12. Bond Strength Test

12.1 *Significance and Use*—This test method provides a method for determining the ability of bonding compound, after being subject to various simulated conditions of aging and exposure to temperature extremes, to adhere cementitious materials to a properly prepared substrate.

12.2 *Materials*:

12.2.1 *Cement*—Specification C 150.

12.2.2 *Graded Standard Sand*—Specification C 778.

12.3 *Preparation of Briquets*:

12.3.1 Prepare not less than six briquets for each series of tests.

12.3.2 Prepare the proportions of the standard mortar in accordance with the procedure given in Test Method C 109/C 109M.

12.3.3 Mechanically mix the mortar in accordance with the procedure given in Practice C 305.

12.3.4 Prepare molds by coating the molds with a thin film of mineral oil. Support the molds on a lightly oiled glass or metal plate.

12.3.5 Cast the briquets immediately after completely mixing the mortar, half fill the molds and puddle the mortar uniformly with the tamper, taking care not to displace the divider. Slightly over fill the molds with additional mortar and repeat the puddling. After the mortar has set, cut off the excess to a plane surface flush with the top of the mold, using a broad knife or similar implement.

12.3.6 Immediately after molding, place the molds, on the base plate, into the moist cabinet for 48 h.

12.3.7 Remove the briquets from the mold and store them at room temperature and humidity for not less than 48 h.

12.3.8 Examine the specimens, discard any that have air holes or excessively rough surfaces.

12.3.9 Saw the specimens in half at the mid-point of the waist line, aligning the saw blade so that the entire saw kerf is taken out of one side of the briquet, and the remaining side maintains a 1 in. cross section. Discard the smaller portion of the briquet.

12.4 *Procedure*:

12.4.1 Thoroughly clean dust and loose particles from the flat waist surface of the specimen. Apply the bonding compound, in accordance with the producer's directions, to the flat waist surface of the briquet by means of a paint brush.

12.4.2 Place the half briquet, coated with bonding compound, into the mold. Fill the remaining space in the mold with mortar, mixed and molded as specified in 12.3.

12.4.3 The time between coating the briquet with compound and casting the mortar against it shall be immediately after the compound has dried, or as recommended by the manufacturer, except when otherwise specified for a ten day waiting period.

12.4.4 Place the filled briquet mold in the moist cabinet for 24 h, then remove the briquets from the molds.

12.5 *Test Method*:

12.5.1 Test the briquets at the following intervals:

12.5.1.1 Air cure three briquets for four days, soak in water for two days, air cure for 24 h, and test;

12.5.1.2 Air cure three briquets for seven days and test;

12.5.1.3 Air cure three briquets for 25 days, soak in water for two days, air cure 24 h, and test;

12.5.1.4 Air cure three briquets for 28 days and test.

12.5.2 Remove any loose particles or flash from the surfaces that will be in contact with the clips of the testing machine. Make sure the bearing surfaces of the clips are clean and free of sand, and the roller bearings are lubricated and maintained to ensure freedom of turning. Keep the stirrups supporting the clips free of accumulations. Keep the pivots in proper adjustment so that the clips swing freely on the pivots without binding the stirrups. Carefully center the briquets in the clips and apply the load continuously at the rate of 0.05 in. (1.27 mm)/min.

12.6 *Tests and Retests*—Test each group of three briquets for bond strength and report the average of the test results. If the result of any one of the six tests varies more than 15 % from the average, reject it and report the average of the other five tests. If the results from more than two tests vary more than 15 % from the average, reject the series and retest.

12.7 *Precision and Bias*—Precision and bias have not been determined for the test method specified.

13. Inspection

13.1 Inspection of the bonding compound shall be agreed upon between the producer or purchaser and the supplier as part of the purchase agreement.

14. Rejection

14.1 Rejection of bonding compound that fails to conform to the requirements of this specification shall be reported to the producer or supplier promptly and in writing. The notice of rejection shall contain a statement documenting how the product has failed to conform to the requirements of this specification.

15. Certification

15.1 When specified in the purchase agreement, the producer or supplier shall furnish a report certifying that, at the time of shipment, the bonding compound is in compliance with the requirements of this specification.

16. Package and Package Marking

16.1 *Packaging*—The bonding compound shall be packed in standard commercial containers. The containers shall be so constructed as to ensure acceptance by common or other carriers for safe transportation at the lowest rate to the point of delivery, unless otherwise specified in the purchase order.

16.2 *Marking*—Shipping containers shall be marked with the name of the bonding compound, the quantity contained therein, the name, brand, or trade mark of the producer or supplier, the shelf life, the batch number, and the ASTM designation.

17. Keywords

17.1 bond; bonding compound; mortar; surface-applied bonding compound

SUMMARY OF CHANGES

(1) Inch-pound units were substituted for SI units in 7.4.
(2) Test Method C 109/C 109M was substituted for specific mortar proportioning requirements in 12.3.2. Table 1 was deleted since it is no longer referenced in 12.3.2, and the information is covered in Test Method C 109/C 109M.
(3) Practice C 305, containing mechanical mixing instructions, was substituted for hand mixing requirements in 12.3.3.
(4) Test Method C 109/C 109M and Practice C 305 were placed in 2.1, and Test Method C 187 was deleted, as it is no longer referenced in 12.3.2.

Designation: C 933 – 96a

Standard Specification for Welded Wire Lath[1]

This standard is issued under the fixed designation C 933; the number immediately following the designation indicates the year of original adoption or, in the case of revision, the year of last revision. A number in parentheses indicates the year of last reapproval. A superscript epsilon (ϵ) indicates an editorial change since the last revision or reapproval.

1. Scope*

1.1 This specification covers welded wire lath, flat or self-furring, with or without backing, designed for use as a base to receive portland cement-based interior plaster and exterior stucco.

1.2 The values stated in SI units are to be regarded as the standard. The values given in parentheses are for information only.

2. Referenced Documents

2.1 *ASTM Standards:*

A 390 Specification for Zinc-Coated (Galvanized) Steel Poultry Netting (Hexagonal and Straight Line) and Woven Steel Poultry Fencing[2]

A 641 Specification for Zinc-Coated (Galvanized) Carbon Steel Wire[2]

C 11 Terminology Relating to Gypsum and Related Building Materials and Systems

2.2 *Federal Specifications:*[3]

UU-B-790 Building Paper, Vegetable Fiber: (Kraft, Waterproofed, Water Repellent and Fire Resistant)

3. Terminology

3.1 Definitions of terms used in this standard shall be in accordance with Terminology C 11.

4. Materials and Manufacture

4.1 Welded wire lath shall be fabricated from not less than 0.0625 in. (1.588 mm), cold-drawn, galvanized steel wire, conforming to Specification A 641.

4.1.1 The wire shall be zinc-coated (galvanized) in accordance with Specification A 390, Class 1.

4.1.2 The backing shall be paper or felt, conforming to Federal Specification UU-B-790. The backing shall be either absorptive or water resistant. The backing shall have a bursting strength of not less than that required to maintain integrity under normal hand- or machine-application pressures.

4.1.3 The backing shall be attached to the lath to prevent accidental removal during shipping, handling or installation. Attachment of the backing shall allow lapping of wire-to-wire and backing-to-backing of not less than one mesh at ends and edges and shall permit full embedment, in not less than 1/8 in. (3.2 mm) of plaster, of not less than one-half of the total length and width of the wire.

4.1.4 The thickness of the embedment of the lath and plaster shall be measured from the back plane of the back wire, exclusive of furring, to the backing or surface of the substrate.

5. Dimensions and Permissible Variations

5.1 *Openings and Stiffening:*

5.1.1 Lath shall be welded at all intersections of wire to form openings not more than 2 in. (51 mm) by 2 in. (51 mm). Lath shall be stiffened continuously and parallel to the long dimension of the lath at intervals of not more than 6 in. (152 mm) with wire not less than 0.0720 in. (1.829 mm) in diameter.

5.1.2 Self-furring crimps on self-furring lath shall project not less than 1/4 in. (6.4 mm) from the plane of the back of the lath.

5.2 *Thickness*—The nominal thickness shall be 1/8 in. (3.2 mm), exclusive of self-furring crimps, with a permissible variation of ±1/32 in. (0.8 mm).

5.3 *Width*—The nominal width shall be 28 in. (710 mm) or 38 in. (965 mm), with a permissible variation of ±3/4 in. (19 mm).

5.4 *Length*—The nominal length shall be 96 in. (2440 mm) or 104 in. (2640 mm) with a permissible variation of ±3/4 in. (19 mm).

5.5 *Weight*—The nominal weight, exclusive of backing, shall be from 1.14 to 1.83 lb/yd^2 (0.618 to 0.993 kg/m^2). Permissible variation in percentage over and under the nominal weights shall be in accordance with mill tolerances.

6. Sampling

6.1 When specified by the purchase agreement, samples of welded wire lath shall be taken at the place of manufacture or at the place of destination. If samples are taken at the destination, such samples shall be taken within 24 h of the receipt of the material.

6.2 At least 0.25 % of the number of sheets of welded wire lath in a shipment, but not less than three sheets, shall be so selected as to be representative of the shipment and shall constitute a sample for the purpose of tests by the purchaser or user.

[1] This specification is under the jurisdiction of ASTM Committee C-11 on Gypsum and Related Building Materials and Systems and is the direct responsibility of Subcommittee C11.02 on Specifications and Test Methods for Accessories and Related Products.

Current edition approved Nov. 10, 1996. Published January 1997. Originally published as C 933 – 80. Last previous edition C 933 – 96.

[2] *Annual Book of ASTM Standards*, Vol 01.06.

[3] Available from Standardization Documents Order Desk, Bldg. 4 Section D, 700 Robbins Ave., Philadelphia, PA 19111-5094, Attn: NPODS.

*** A Summary of Changes section appears at the end of this standard.**

7. Inspection

7.1 Inspection of the welded wire lath shall be agreed upon between the purchaser and the producer or supplier as part of the purchase agreement.

8. Rejection

8.1 Rejection of welded wire lath that fails to conform to the requirements of the specification shall be reported to the producer or supplier promptly and in writing. The notice of rejection shall contain a statement documenting how the welded wire lath has failed to conform to the requirements of the specification.

9. Packaging and Package Marking

9.1 Nominal 96 in. (2440 mm) lengths shall be packaged with 12 sheets [25 yd^2 (21 m^2)] per bundle; nominal 48 in. (1220 mm) lengths shall be packaged with 19 sheets [20 yd^2 (17 m^2)] per bundle.

9.2 Unless otherwise required by the purchase agreement, lath shall have legibly marked thereon the following: the product description; the name of the producer or seller; the brand name (if any); and the ASTM designation for the product.

9.2.1 Each bundle of lath with no backing shall be marked on metal or paper tags securely attached to the bundles.

9.2.2 Each sheet of lath with backing shall be marked on the backing material.

10. Keywords

10.1 lath; metal plaster base; plaster; self-furring lath; stucco; wire lath

SUMMARY OF CHANGES

Committee C-11 has identified the location of selected changes to this standard since the last issue C 933-96 that may impact the use of this standard.

(*1*) Paragraph 5.3 was revised.
(*2*) Paragraph 5.4 was revised.

Standard Specification for Steel Drill Screws for the Application of Gypsum Panel Products or Metal Plaster Bases to Steel Studs from 0.033 in. (0.84 mm) to 0.112 in. (2.84 mm) in Thickness[1]

This standard is issued under the fixed designation C 954; the number immediately following the designation indicates the year of original adoption or, in the case of revision, the year of last revision. A number in parentheses indicates the year of last reapproval. A superscript epsilon (ϵ) indicates an editorial change since the last revision or reapproval.

This specification has been approved for use by agencies of the Department of Defense.

1. Scope*

1.1 This specification covers minimum requirements for steel drill screws for use in fastening gypsum panel products or metal plaster bases to steel members from 0.033 in. (0.84 mm) to 0.112 in. (2.84 mm) in thickness.

1.2 This specification also covers physical properties and test methods for determining performance requirements.

1.3 The values stated in inch-pound units are to be regarded as the standard. The values given in parentheses are for information only.

1.4 The following safety hazards caveat pertains only to the test methods described in this specification. *This standard does not purport to address all of the safety concerns, if any, associated with its use. It is the responsibility of the user of this standard to establish appropriate safety and health practices and determine the applicability of regulatory limitations prior to use.*

2. Referenced Documents

2.1 *ASTM Standards:*

A 510 Specification for General Requirements for Wire Rods and Coarse Round Wire, Carbon Steel[2]

C 11 Terminology Relating to Gypsum and Related Building Materials and Systems[3]

C 36 Specification for Gypsum Wallboard[3]

3. Terminology

3.1 *Definitions:*

3.1.1 For definitions relating to gypsum and related building materials and systems, see Terminology C 11.

3.2 *Definition of Term Specific to This Standard:*

3.2.1 *steel drill screw*—a special self-tapping steel screw with a drilling tip configuration.

4. Materials

4.1 Screws shall be manufactured from Grade 1018 to 1022 steel wire manufactured in accordance with Specification A 510.

5. Physical Properties

5.1 *Metallurgical Requirements:*

5.1.1 The surface of the screw shall be carbonitrided to a depth of 0.004 to 0.008 in. (0.10 to 0.20 mm).

5.1.2 The core hardness shall be 32 to 40 HRB after being drawn at a minimum temperature of 625°F (330°C).

5.1.3 The screw shall have no band of free ferrite between the case and core nor shall the case contain appreciable amounts of retained austenite or other soft constituents.

5.1.4 Surface hardness shall be determined by a micro hardness instrument at "the root of the thread profile," as exposed by removal of enough material to form a flat surface along the length of the screw.

5.2 *Dimensions and Permissible Variations:*

5.2.1 *General:*

5.2.1.1 *Head Diameter*—The head of the screw shall not be out of round more than 0.021 in. (0.51 mm) and have the following shape and dimensions:

5.2.1.2 *Screw Diameter,* of the threaded portion shall be such that the screw will pass the performance tests in Section 6.

5.2.1.3 *Threads,* shall be of a single or double lead thread design.

5.2.1.4 *Points,* shall provide for self-drilling into steel studs from 0.333 in. (0.84 mm) to 0.112 in. (2.84 mm) in thickness and meet the performance tests in Section 6.

5.2.1.5 *Driving Recess,* shall be a No. 2 "Phillips" design with a minimum depth of 0.104 in. (2.64 mm) as determined with a "Phillips" penetration depth gage or a recess of equal performance.

5.2.1.6 *Length*—Nominal lengths shall be minimum length.

5.2.2 *Screws for Fastening Gypsum Board:*

5.2.2.1 Screw-head minimum diameter shall be 0.3145 in. (8.00 mm).

5.2.2.2 The top of the head shall be flat. The outer flange thickness shall be 0.025 ± 0.005 in. (0.64 ± 0.13 mm). The contour beneath the flange head shall be such that the screw

---

[1] This specification is under the jurisdiction of ASTM Committee C-11 on Gypsum and Related Building Materials and Systems and is the direct responsibility of Subcommittee C11.02 on Specifications and Test Methods for Accessories and Related Products.

Current edition approved May 10, 1998. Published December 1998. Originally published as C 954 – 81. Last previous edition C 954 – 96a.$^{\epsilon 1}$

[2] *Annual Book of ASTM Standards*, Vol 01.03.

[3] *Annual Book of ASTM Standards*, Vol 04.01.

* A Summary of Changes section appears at the end of this specification.

head shall be able to be driven so that the head of the screw rests immediately below the surface of the gypsum panel products.

5.2.2.3 The threads shall extend to the lower edge of the head contour.

5.2.3 *Screws for Fastening Metal Plaster Bases:*

5.2.3.1 Screw head minimum diameter shall be 0.437 in. (11.1 mm).

5.2.3.2 The top of the head shall be flat or contoured. The underside of the head shall be flat.

5.2.3.3 The threads shall extend to the underside of the head.

6. Performance Requirements

6.1 *Spin Out:*

6.1.1 Screws shall self-drill and drive into a stud 0.0598 in. (1.5 mm) thick with an approximate hardness of 65 HRB without spinout.

6.1.2 Screws shall self drill and drive into a stud 0.105 in. (2.7 mm) thick, with a hardness of 80 HRB without spinout.

6.2 *Performance:*

6.2.1 When tested in accordance with Section 10, screws shall meet the requirements as follows:

6.2.1.1 *Gypsum Panel Products*—Screw threads shall be adequate to pull the head of the screw below the surface of the gypsum panel through four layers of kraft paper placed over the gypsum panel products and a steel stud 0.033 in. (0.84 mm) thick when tested in accordance with 10.5.1.

6.2.1.2 *Metal Plaster Bases*—Screw threads shall be adequate to pull metal lath tight against the flange of a steel stud 0.033 in. (0.84 mm) thick so that when subjected to a steady pull, the lath will tear and not slip out from under the screw head when tested in accordance with 10.5.1.

7. Finish and Appearance

7.1 Screws shall be corrosion-resistant treated with a material which will not inhibit adhesion to joint compounds or plaster and will not bleed through decorative finishes.

7.2 Screws shall be straight, clean, smooth, neatly formed, and free of defects such as burrs and deformations.

8. Sampling

8.1 One screw from each of 5 containers in a lot of 16 000, but not less than 5 screws per 16 000 screws shall constitute a sample.

9. Number of Tests and Retests

9.1 Test the sample specified in 8.1 in accordance with Section 10. If any one screw of the sample fails to meet this specification, test a second sample using 25 screws. If two or more screws from the second sample fail, the represented lot of 16 000 screws is considered to have failed to meet this specification.

10. Test Method

10.1 *Summary of Test Method:*

10.1.1 This test method utilizes steel screws designed for use to hold gypsum panel products and metal plaster base materials applied with power-driven screw guns to steel studs.

10.2 *Significance and Use:*

10.2.1 This test method evaluates the ability of steel screws used to secure gypsum panel products and metal plaster bases to certain steel framing members.

10.2.2 The test shall be conducted in a laboratory or on-site.

10.3 *Apparatus:*

10.3.1 *Power-Driven Drill Screw Gun,* having the capability of turning 2500 r/min (free spindle speed).

10.3.2 A *Vice,* or similar device to support the stud during the test.

10.4 *Materials:*

10.4.1 *Kraft Paper,* 0.010 in. (0.25 mm) thick.

10.4.2 *Steel Stud*—One section of 0.033-in. thick stud (0.84-mm), 1 section of 0.0598-in. (1.5-mm) thick stud and with a hardness of 65 HRB, and 1 section of 0.105-in. (2.7-mm) thick stud and with a hardness of 80 HRB.

10.4.3 *Gypsum Wallboard*—Specification C 36, Type X, ⅝ in. (16 mm) thick.

10.4.4 *Screw Specimens,* to be tested.

10.4.5 *Diamond Mesh Metal Lath,* 2.5/lb/yd^2.

10.5 *Procedure for Performance Tests:*

10.5.1 Using the screw specimen, attach gypsum board or metal lath to center of a flange of a steel stud which has been securely and rigidly supported.

10.5.1.1 Grasp the metal lath with locking pliers and pull steadily in a plane parallel to the plane of the lath. Observe whether the lath tears before it slips out from under the screw head.

10.5.1.2 With the drill propelling the screw at 2500 r/min (free spindle speed), exert a total pressure of 30 lbf (135.5 N) (dead weight plus applied force).

10.6 *Precision and Bias*—No statement is made regarding the precision and bias of this test method, since the result of the test method is reported in nonnumerical terms.

11. Inspection

11.1 Inspection of steel drill screws shall be agreed upon by the purchaser and the producer or supplier as part of the purchase agreement.

12. Rejection and Rehearing

12.1 Rejection of steel drill screws that fail to conform to the requirements of this specification shall be reported to the producer or supplier promptly and in writing within 10 working days from receipt of shipment by the purchaser. Notice of rejection shall contain a specific statement as to the respects in which the screws have failed to conform to the requirements of this specification. In case of dissatisfaction with the results of the test, and at the request of the producer or supplier, such notice of rehearing shall be supported by results of a test conducted by a mutually agreeable independent laboratory.

13. Certification

13.1 When specified in the purchase agreement, the producer or supplier shall furnish a report certifying that, at time of shipment, the screws were in compliance with the requirements of this specification.

14. Packaging and Marking

14.1 Screws shall be packaged in substantial commercial containers constructed so as to preserve the contents in good condition and to ensure acceptance and safe delivery by common or other carriers, at the lowest rate, to the point of delivery.

14.2 The containers shall be so constructed that the contents can be partially removed without destroying the container's ability to serve as a receptacle for the remainder of the contents.

14.3 Individual packages and shipping containers shall be marked with the type and length of screw and the name of the manufacturer or distributor.

15. Keywords

15.1 steel drill screw

SUMMARY OF CHANGES

Committee C-11 has identified the location of selected changes to this standard since the last issue (C 954 – 96) that may impact the use of this standard.

(*1*) The term "gypsum board" was replaced with the term "gypsum panel products" in the title, scope, and paragraphs 5.2.2.2, 6.2.1.1, 10.1.1 and 10.2.1.

(*2*) Specification C 36 was added to Section 2.

(*3*) Paragraph 10.4.3 was revised.

Designation: C 955 – 98

Standard Specification for Load-Bearing (Transverse and Axial) Steel Studs, Runners (Tracks), and Bracing or Bridging for Screw Application of Gypsum Panel Products and Metal Plaster Bases[1]

This standard is issued under the fixed designation C 955; the number immediately following the designation indicates the year of original adoption or, in the case of revision, the year of last revision. A number in parentheses indicates the year of last reapproval. A superscript epsilon (ε) indicates an editorial change since the last revision or reapproval.

1. Scope*

1.1 This specification covers steel studs, runners (tracks), and bracing or bridging (with a base metal thickness range of not less than 0.0329 in. (0.836 mm)) for screw application of gypsum panel products and metal plaster bases in load-bearing (transverse and axial) construction assemblies. Steel of lesser thickness shall be permitted in additional engineered products.

1.2 The values stated in inch-pound units are to be regarded as the standard. The SI (metric) values given in parentheses are provided for information purposes only.

1.3 The following precautionary caveat pertains only to the test method portion, Section 7, of this specification: *This standard does not purport to address all of the safety concerns, if any, associated with its use. It is the responsibility of the user of this standard to establish appropriate safety and health practices and determine the applicability of regulatory limitations prior to use.*

2. Referenced Documents

2.1 *ASTM Standards:*

A 653/A 653M Specification for Zinc-Coated (Galvanized) or Zinc-Iron Alloy-Coated (Galvannealed) by the Hot-Dipped Process[2]

C 36 Specification for Gypsum Wallboard[3]

C 954 Specification for Steel Drill Screws for the Application of Gypsum Board or Metal Plaster Bases to Steel Studs from 0.33 in. (0.84 mm) to 0.112 in. (2.84 mm) in Thickness[3]

2.2 *AISI Standard:*

Specification for the Design of Cold-Formed Steel Structural Members[4]

3. Terminology

3.1 *Definitions*—Definitions shall be in accordance with Terminology C 11.

3.2 *Definitions of Terms Specific to This Standard:*

3.2.1 *members, n*—studs, runners, tracks, bracing, bridging, accessories or other items manufactured in accordance with this standard.

3.2.2 *structural member, n*—a member in a steel framed system in which the loading exceeds any of the following conditions: a transverse load of 20 lbf/ft (290 N/m) of member length, or an axial load, exclusive of sheathing, of 200 lbf (890 N) per member.

4. Materials and Manufacture

4.1 The mechanical properties of the steel shall meet the requirements of the AISI Specification for the Design of Cold-Formed Steel Structural Members.

4.2 The minimum steel thickness (base steel) shall be not less than 0.0329 in. (0.84 mm).

4.3 Individual measurements before the application of protective coating shall be not less than 95 % of the specified design thickness.

4.4 Members shall have a protective coating conforming to Specification A 653/A 653M, G60 minimum, or equivalent corrosion resistance, or shall have a rust-inhibitive coating providing equivalent corrosion resistance.

4.5 Edges of members shall be manufactured to minimize burrs and sharp edges.

4.6 Factory punch-outs in webs, when provided, shall not reduce the performance of the member in the assembly below the specified performance requirements.

4.7 The sectional properties of members shall be computed in accordance with the AISI Specification for the Design of Cold-Formed Steel Structural Members.

5. Performance Requirements

5.1 When tested in accordance with Section 7, members shall be capable of pulling the head of the screw below the surface of the gypsum panel product without spin out (see Fig. 1).

6. Dimensions and Permissible Variations

6.1 Data for calculating design performance shall be supplied by the manufacturer.

6.2 The width of the surface to which the sheathing board is attached shall be not less than 1¼ in. (32 mm).

6.3 Runners (track) shall be formed in a U-shaped configuration, having a depth compatible with that of the studs of the same nominal size.

[1] This specification is under the jurisdiction of ASTM Committee C-11 on Gypsum and Related Building Materials and Systems and is the direct responsibility of Subcommittee C11.02 on Specifications and Test Methods for Accessories and Related Products.

Current edition approved June 10, 1998. Published December 1998. Originally published as C 955 - 81. Last previous edition C 955 - 97a.

[2] *Annual Book of ASTM Standards,* Vol 01.06.

[3] *Annual Book of ASTM Standards,* Vol 04.01.

[4] Available from the American Iron and Steel Institute, 1000 16th St., N.W., Washington, DC 20036.

*** A Summary of Changes section appears at the end of this specification.**

TABLE 1 Manufacturing Tolerances

Dimension[A]	Item Checked	Structural Studs, in. (mm)	Structural Track, in. (mm)
A	length	+ 3/32 (2.38) − 3/32 (2.38)	+ 1/2 (1.27) − 1/4 (6.35)
B[B]	web width	+ 1/32 (0.79) − 1/32 (0.79)	+ 1/32 (0.79) + 1/8 (3.18)
C	flare overbend	+ 1/16 (1.59) − 1/16 (1.59)	+ 0 (0) − 3/32 (2.38)
D	hole center width	+ 1/16 (1.59) − 1/16 (1.59)	NA
E	hole center length	+ 1/4 (6.35) − 1/4 (6.35)	NA
F	crown	+ 1/16 (1.59) − 1/16 (1.59)	+ 1/16 (1.59) − 1/16 (1.59)
G	camber	1/32 per ft (0.79) 1/2 max (1.27)	1/32 per ft (0.79) 1/2 max (1.27)
H	bow	1/32 per ft (0.79) 1/2 max (1.27)	1/32 per ft (0.79) 1/2 max (1.27)
I	twist	1/32 per ft (0.79) 1/2 max (1.27)	1/32 per ft (0.79) 1/2 max (1.27)

[A] All measurements shall be taken not less than 1 ft (305 mm) from the end.
[B] Outside dimension for stud; inside for track.

6.4 Minimum height of runner (track) flanges shall be 3/4 in. (19 mm).

6.5 Members shall be manufactured within the limits as shown in Table 1 and Fig. 2.

6.6 Bracing and bridging shall have configuration and steel thickness to provide secondary support for the studs in accordance with the AISI Specification for the Design of Cold-Formed Steel Structural Members.

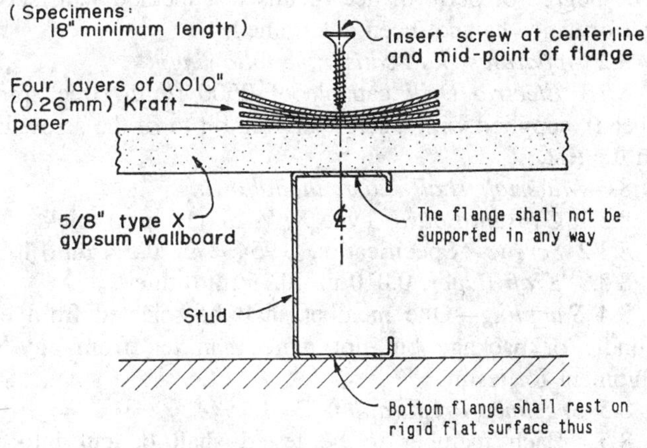

FIG. 1 Studs

7. Workmanship, Finish, and Appearance

7.1 The steel members shall be free of defects that interfere with the purpose for which they are intended.

8. Penetration Test

8.1 *Significance and Use*—This test method provides a procedure for evaluating the member's ability to pull the head of a screw below the surface of the gypsum panel product. It shall be used to determine compliance with this specification.

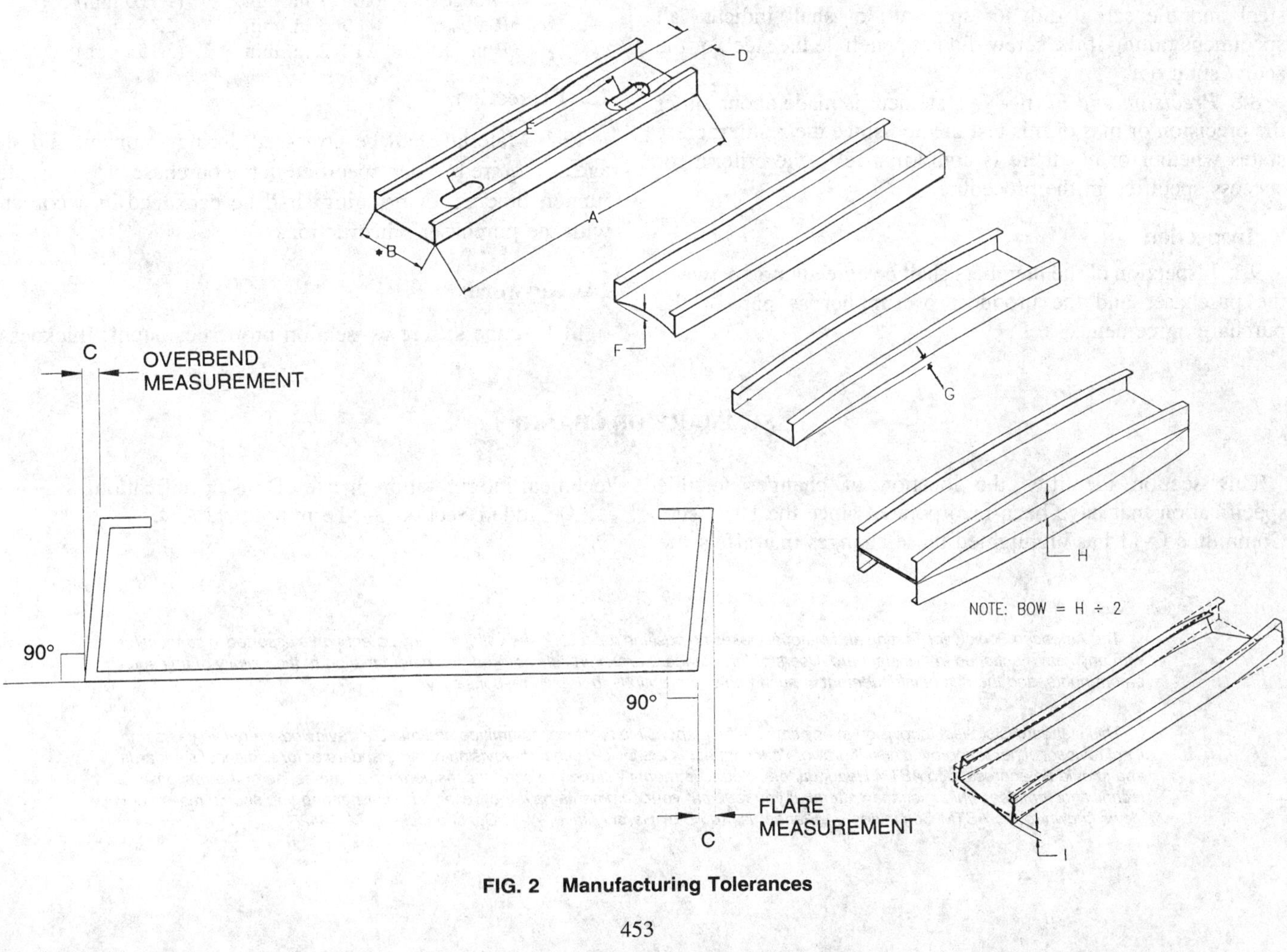

FIG. 2 Manufacturing Tolerances

The degree of performance of this test method with service performance has not been determined.

8.2 *Apparatus shall satisfy the following:*

8.2.1 *Electric Drill* capable of 2500 r/min (free spindle speed), supplied with a screw driving bit to fit the screw used in the test.

8.3 *Materials shall satisfy the following:*

8.3.1 *Gypsum Board,* Type X, ⅝ in. (16 mm) thick.

8.3.2 *Screws*—Specification C 954, 1 in. (25.4 mm) long.

8.3.3 *Kraft Paper,* 0.010 in. (0.25 mm) thick.

8.4 *Sampling*—One member shall be selected from each bundle or package but not more than ten from any one shipment for testing.

8.5 *Specimen Preparation:*

8.5.1 Each member to be tested shall be cut into test specimens not less than 18 in. (460 mm) long.

8.5.2 For each test, one piece of gypsum board, 6-in. (152-mm) square, shall be cut not less than 12 in. (305 mm) from the edge or end of the wallboard.

8.5.3 For each test, four pieces of kraft paper, 2-in. (51-mm) square shall be cut.

8.6 *Number of Tests and Retests:*

8.6.1 Five specimens of members shall be tested.

8.6.2 If more than one test specimen fails to meet the requirements, two more test specimens shall be chosen for retesting.

8.7 *Report*—Report shall indicate all specimens meeting the requirements of this specification if the screw penetrated the steel and the screw did not spin out; or shall indicate all specimens failing if the screw did not penetrate the steel, or the screw spun out.

8.8 *Precision and Bias*—No statement is made about either the precision or bias of this test method since the result merely states whether or not there is conformance to the criteria for success specified in the procedure.

9. Inspection

9.1 Inspection of the members shall be agreed upon between the purchaser and the producer or supplier as part of the purchase agreement.

10. Rejection

10.1 When specified in the purchase agreement, members that fail to conform to the requirements of this specification shall be rejected. Rejection shall be reported to the producer or supplier promptly and in writing. The notice of rejection shall contain a statement documenting how the member has failed to conform to the requirements of this specification and the purchase agreement.

11. Certification

11.1 When specified in the purchase agreement, a producer's or supplier's report shall be furnished at the time of shipment certifying that the product is in compliance with this specification.

12. Marking and Identification

12.1 Marking shall be on the material or a label, or a tag attached thereto. Marking shall include the order number, quantity, and description of size, configuration, metal thickness, and lengths.

12.2 Identification of product by base metal thickness as manufactured shall be by the following color code:

White	0.0329 in. min	(0.836 mm)
Yellow	0.0428 in. min	(1.087 mm)
Green	0.0538 in. min	(1.367 mm)
Orange	0.0677 in. min	(1.720 mm)
Red	0.0966 in. min	(2.454 mm)
Blue	0.1242 in. min	(3.155 mm)

13. Protection

13.1 Material shall be protected during shipment and storage as required. When specified in the purchase order, the stud, runner, bracing, or bridging shall be packaged in accordance with the purchaser's instructions.

14. Keywords

14.1 runners; screws; section properties; studs; thickness

SUMMARY OF CHANGES

This section identifies the location of changes to this specification that have been incorporated since the last issue. Committee C-11 has highlighted those changes that affect the technical interpretation or use of this specification.

(*1*) Added Section 3—Terminology.

Designation: C 956 – 97

Standard Specification for Installation of Cast-In-Place Reinforced Gypsum Concrete[1]

This standard is issued under the fixed designation C 956; the number immediately following the designation indicates the year of original adoption or, in the case of revision, the year of last revision. A number in parentheses indicates the year of last reapproval. A superscript epsilon (ε) indicates an editorial change since the last revision or reapproval.

1. Scope

1.1 This specification covers the minimum requirements for the installation of cast-in-place reinforced gypsum concrete over permanent formboard.

1.2 The values stated in inch-pound units are to be regarded as the standard. The values given in parentheses are for information only.

1.3 The text of this standard references notes which provide explanatory material. These notes shall not be considered requirements of the standard.

2. Referenced Documents

2.1 *ASTM Standards:*

A 82 Specification for Steel Wire, Plain, for Concrete Reinforcement[2]

A 185 Specification for Steel Welded Wire Fabric, Plain, for Concrete Reinforcement[2]

A 499 Specification for Steel Bars and Shapes, Carbon Rolled from "T" Rails[3]

A 525 Specification for General Requirements for Steel Sheet, Zinc-Coated (Galvanized) by the Hot-Dip Process[4]

A 568 Specification for Steel, Sheet, Carbon, and High-Strength, Low-Alloy, Hot-Rolled and Cold-Rolled, General Requirements for[5]

C 11 Terminology Relating to Gypsum and Related Building Materials and Systems[6]

C 317 Specification for Gypsum Concrete[6]

C 726 Specification for Mineral Fiber Roof Insulation Board[7]

E 72 Methods of Conducting Strength Tests of Panels for Building Construction[8]

2.2 *American Concrete Institute Standard:*

ACI 318 Building Code Requirements for Reinforced Concrete[9]

2.3 *American Welding Society Standard:*

D1.1 Structural Welding Code Steel[10]

3. Terminology

3.1 Definitions shall be in accordance with Terminology C 11.

3.2 *Definitions:*

3.2.1 *primary framing, n*—structural members provided to support the reinforced gypsum concrete roof deck assembly.

3.2.1.1 *purlin, n*—a secondary structural member that spans the primary framing members and provides support for subpurlins.

3.2.2 *subpurlin, n*—a steel member applied transversely to the primary framing and purlins to support the formboards and to transmit the dead and live loads from the gypsum concrete slab to the primary framing.

3.2.2.1 *bulb tee, n*—a subpurlin, hot-rolled formed steel, rail-shaped section.

3.2.2.2 *truss tee, n*—a subpurlin fabricated from steel wire and strip, tee-shaped section.

3.2.3 *cross tee, n*—a steel tee-shaped section used to support a formboard end at right angles to the subpurlins where a formboard end does not occur over the purlin or primary framing.

3.2.4 *formboard, n*—sheet material used as a permanent form to support the gypsum concrete.

3.2.5 *reinforcement, n*—steel wire mesh or fabric used within the gypsum concrete slab to provide longitudinal and transverse strength.

3.2.6 *double-pouring, n*—the application of gypsum concrete in more than one layer to complete the full slab thickness. See 9.7.4.3.

3.2.7 *ribbon-pouring or strip-pouring, n*—the application of narrow ribbons of gypsum concrete, about 10 to 12 in. (254 to 305 mm) wide by 1 to 1½ in. (25.4 to 38.1 mm) deep, across the width of the formboard at center of span between purlins.

4. Delivery of Materials

4.1 Materials shall be delivered in original packages, containers, or bundles bearing the brand name and name of producer or seller. Bulk materials shall be delivered with the

[1] This specification is under the jurisdiction of ASTM Committee C-11 on Gypsum and Related Building Materials and Systems and is the direct responsibility of Subcommittee C11.03 on Specifications for Application of Gypsum and Other Products in Assemblies.

Current edition approved April 10, 1997. Published June 1997. Originally published as C 956 – 81. Last previous edition C 956 – 96.

[2] *Annual Book of ASTM Standards*, Vol 01.04.

[3] *Annual Book of ASTM Standards*, Vol 01.05.

[4] *Annual Book of ASTM Standards*, Vol 01.06.

[5] *Annual Book of ASTM Standards*, Vol 01.03.

[6] *Annual Book of ASTM Standards*, Vol 04.01.

[7] *Annual Book of ASTM Standards*, Vol 04.06.

[8] *Annual Book of ASTM Standards*, Vol 04.07.

[9] Available from the American Concrete Institute, P.O. Box 19150, Detroit, MI 48219.

[10] Available from the American Welding Society, 550 N.W. La Jeune Rd., P.O. Box 351040, Miami, FL 33135.

brand name and name of the producer or seller shown on the accompanying bills of lading.

5. Storage of Materials

5.1 All materials shall be stored in a manner that prevents damage before use. When stored under tarpaulins, ventilation shall be provided to prevent moisture accumulation under the tarpaulin.

5.2 Formboard shall be stored flat and off the ground. Handling and stacking shall be done in such a manner to prevent damage to face, ends, and edges and keep dry until use.

5.3 When it is necessary to store gypsum concrete at the job site, the gypsum concrete shall be stored off the ground and kept dry until use.

6. Environmental Conditions

6.1 The minimum temperature at which gypsum concrete is mixed and placed is not specified. Gypsum concrete shall not be mixed with water containing ice crystals.

NOTE—An exothermic reaction during setting ensures complete hydration before freezing.

7. Materials

7.1 *Subpurlins:*

7.1.1 Subpurlins shall be designed to support live and dead loads of the roof deck.

7.1.2 Hot-rolled and cold-rolled steel shapes other than bulb tees or truss tees shall meet the requirements of this specification for subpurlins.

7.2 *Cross Tees,* shall be not less than 1¼ in. (31.75 mm) wide by ½ in. (12.7 mm) high, fabricated from not less than 26 gage zinc-coated steel conforming to Specifications A 525 or A 568.

7.3 *Formboard:*

7.3.1 *Mineral Fiber Formboard,* Specification C 726, not less than ¾ in. (19.0 mm) thick nor more than 3 in. (76 mm) thick. It shall sustain a uniform load of 20 lb/ft^2 (97 kg/m^2).

7.3.1.1 Mineral fiber formboard shall be tested in accordance with the transverse loading test of Methods E 72.

7.3.1.2 The test specimen shall be uniformly loaded and supported at both edges and ends. The test specimens shall be supported at the perimeter with a frame providing 1 in. (25.4 mm) bearing on ends and ½ in. (12.7 mm) bearing on edges and 7 in. (177.8 mm) nominal clearance from the bed of the apparatus.

7.3.1.3 For formboards ¾ in. (19.0 mm) thick, test specimens shall be 48 in. (1200 mm) long by 24 in. (600 mm) wide.

7.3.1.4 For formboards 1 in. (25.4 mm) thick or greater, test specimens shall be 48 in. (1200 mm) long by 32 in. (800 mm) wide.

7.4 *Reinforcement,* Specification A 82, shall be fabricated from zinc-coated (galvanized) welded or woven steel wire mesh or fabric having an effective cross-sectional area of not less than 0.026 $in.^2$ $(16.77\ mm)^2$ per foot of slab width. Reinforcement shall be free of rust, scale, or other materials that reduce bond to the gypsum concrete.

7.4.1 *Hexagonal Reinforcing Mesh,* shall be fabricated from 2 in. (50.8 mm) hexagonal mesh formed of not less than 0.0410 in. (1.04 mm) diameter wires with longitudinal wires not less than 0.0625 in. (1.59 mm) in diameter, spaced not more than 3 in. (76.2 mm) on centers. Steel wire for fabricating wire mesh and fabric—Specification A 82.

7.4.2 *Rectangular Reinforcing Fabric,* Specification A 185, shall be fabricated from longitudinal wires not less than 0.1055 in. (2.68 mm) in diameter spaced not more than 4 in. (102 mm) on centers and transverse wires not less than 0.0800 in. (2.03 mm) in diameter spaced not more than 8 in. (203 mm) on centers.

7.5 *Gypsum Concrete,* Specification C 317, Class A.

7.6 *Water,* shall be potable and free of substances that could adversely affect the gypsum concrete.

7.7 *Expansion Filler Strip,* shall be not less than ¾ in. (19 mm) thick and equal in height to the abutting gypsum concrete and of the type specified by the producer of the gypsum concrete.

8. Installation

8.1 *Installation of Subpurlins:*

8.1.1 *Installation, Normal:*

8.1.1.1 *Subpurlins*—The subpurlins shall be placed transversely to the purlins and primary framing and shall be spaced to suit the size of the formboards specified. Subpurlins shall be of sufficient length so that their joints occur over purlins or primary framing. Subpurlin joints shall be alternated so that the joints are staggered one purlin space.

8.1.1.2 Subpurlins shall be welded to purlins and steel primary framing with ⅛ in. (3.2 mm) fillet welds not less than ½ in. (12.7 mm) long on alternate sides of the subpurlin at each intersection with purlins. Except as otherwise specified, welding shall be in accordance with AWS D1.1. Where primary framing is of wood, nails not less than 16d shall be used on each side of the subpurlins at each intersection. The nails shall be bent over approximately ¾ in. (19 mm) to securely clinch the flange of the subpurlin.

8.1.2 *Installation Seismic:*

8.1.2.1 *Subpurlins*—The subpurlins shall be welded at each intersection with the purlins with ⅛-in. (3.2-mm) fillet welds not less than 1 in. (25.4 mm) long, on one side. Both sides of the subpurlin shall be welded with ⅛-in. fillet welds not less than 2 in. (50.8 mm) long where subpurlin joints occur over the purlin. Both sides of the subpurlin shall be welded with ⅛-in. fillet welds not less than 2 in. long at shear transfer points and at intersections with exterior frame.

8.1.3 *Maximum Spans*—Refer to X2.1.3.

8.2 *Cross Tees*—Cross tees shall be provided where formboard ends are supported by purlins or the primary framing.

8.3 *Formboards*—Formboards shall be located face down on the flanges of the supporting members and with the ends on top of the purlins.

8.3.1 Gypsum concrete shall not be poured over formboards which have been wetted by rain or snow.

8.3.2 The gypsum concrete shall be poured on the same day the formboards are installed.

8.4 *Reinforcement*—The mesh of wire fabric shall be located with the main longitudinal wires at right angles to the subpurlins and adjacent to the formboard at midspan between the purlins. Ends of the reinforcement shall be lapped not less than 4 in. (102 mm) or one mesh, whichever is greater.

8.4.1 Edges shall be butted, not lapped, unless required for fire rated or horizontal diaphragm construction. Reinforcement shall be cut to fit at walls, curbs, and openings, folded over on to itself, and extended into all areas when gypsum concrete is to be placed.

8.5 *Expansion Filler Strips*—Expansion filler strips shall be placed continuously against the walls, curbs, nailers, or other rigid surfaces that gypsum concrete will abut.

8.6 *Expansion Joints*—Expansion joints shall be located at expansion joints found in the main structure, at a maximum of 200 ft (61 m) on center in a direction parallel to the subpurlins on large buildings, at intersections of wings on L-, U-, T-, and H-shaped buildings and, wherever the roof framing changes direction. See Appendix X1.5.5.

8.7 *Gypsum Concrete Mixing and Placement:*

8.7.1 *General*—All equipment, including mixers, pumps, hoses, tools, and screeds shall be kept clean and free of set gypsum concrete throughout the placement operation.

8.7.2 *Proportioning*—The ratio of water to gypsum concrete shall be that specified by the producer of the gypsum concrete; additional water is prohibited. Volumetric or metering devices shall be used to ensure accurate measurement.

8.7.3 *Mixing*—Mixing shall be thorough but not overmixed. The slurry shall be discharged into placement equipment immediately after completion of mixing to avoid buildup of set material.

8.7.3.1 Calcium chloride or other admixtures shall not be added to the gypsum concrete.

8.7.4 *Placement:*

8.7.4.1 *General*—To avoid exposure to inclement weather and physical abuse, gypsum concrete shall be placed the same day as the installation of the formboard and reinforcement materials.

8.7.4.1.1 The deck shall be poured in sections of similar size.

8.7.4.2 *Screeds Grounds*—Screeds, such as metal bars or wood strips, shall be spaced approximately 10 ft (3050 mm) apart at right angles on top of the subpurlins at a height that will allow the gypsum concrete slurry to be leveled at the specified thickness.

8.7.4.2.1 Gypsum concrete cants, curbs, and drainage slopes shall be placed to the design thickness. Curbs around openings shall be permitted to be formed with the formboard.

8.7.4.3 The gypsum concrete shall be discharged into placement equipment immediately after completion of mixing.

8.7.4.3.1 The slurry shall be poured or discharged not more than 24 in. (600 mm) off the formboard to minimize impact and allowed to build to full thickness in one continuous operation. Double pouring shall not be permitted.

8.7.4.3.2 As soon as the slurry has stiffened enough to hold a full level, it shall be screeded to a true, even surface using a smooth, rigid straight-edge.

8.7.4.4 Place gypsum concrete cants, curbs, and drainage slopes to the design thickness. Curbs around openings shall be permitted to be formed with the formboard.

8.8 *Roof Overhangs*—Gypsum concrete shall not be placed on roof overhangs, eaves, or other similar locations, on formboards unless the underside is protected against direct wetting by a suitable facia drip cap overhang.

8.9 *Drying*—Construction moisture shall be removed from the building to permit drying of completed gypsum concrete slab, providing heat and ventilation where required. See Appendix X1.3.

8.10 *Protection of Completed Slab*—Gypsum concrete is not a finished traffic or weather protective surface. The top side of the completed gypsum concrete slab shall be protected with a permanent waterproof covering.

9. Keywords

9.1 calculation of diaphragm shear; gypsum concrete; gypsum roof deck

APPENDIXES

(Nonmandatory Information)

X1. GENERAL INFORMATION

X1.1 Limitations of Use

X1.1.1 *Roof Shapes and Designs*—Gypsum roof decks meeting this specification can be installed on flat, warped, sawtooth, curved, or pitched roofs.

NOTE X1.1—On flat roofs, a minimum pitch of 1/8 in./ft (10.4 mm/m) is recommeded.

X1.1.2 *Moisture Exposure*—Cast-in-place gypsum concrete is not recommended for general use in occupancies where it may be exposed to sustained relative humidity greater than 90 % or condensing moisture. The producer of the gypsum concrete should be consulted regarding limitations of use in high humidity occupancies.

X1.1.3 *Temperature Exposures*—Where intermittent or extended exposures to high temperatures are to be expected, such as in slabs directly exposed to radiant heat over furnace breachings, the gypsum concrete producer should be consulted for specific recommendations.

X1.1.4 *Acid Fumes*—If acid fumes are to be encountered, the producer of the gypsum concrete should be consulted for his specific recommendations.

X1.1.5 *Surface Protection*—Gypsum concrete decks shall not be used as a finished surface to bear traffic or other loads. Protective permanent waterproof coverings shall be installed as soon as possible after completion of the slab.

X1.1.6 *Welding Precautions*—Except as modified in the body of this specification, welding shall be in accordance with AWS D1.1. Welding will not be permitted when the temperature is lower than 0°F (−18°C), when steel surfaces are wet, or under other conditions not recommended by the American Welding Society.

X1.2 Support for Suspended Ceilings

X1.2.1 Suspended ceilings should be hung from the primary framing. Ceilings may be hung from the subpurlins. A design check shall be made in accordance with applicable building codes and regulations to verify that the structural adequacy of the subpurlins and the added weight of the ceiling will not cause the subpurlins to deflect more than that allowed by the roof deck or ceiling sign or that allowed by subpurlin design. Hangers should not be fastened into or through the gypsum concrete slab or the formboards. Similar precautions should be followed for the suspension of unit heaters, light troffers, and piping.

X1.3 Drying of Slabs

X1.3.1 Subsequent to the placing of gypsum concrete, ventilation accompanied by heat is recommended below the slabs to remove the excess construction moisture resulting from the slab pouring, placement of concrete floors, masonry plastering, and other similar construction work. Where natural ventilation is inadequate, mechanical ventilation is required to remove construction moisture from the building. When suspended ceilings are installed below the slabs, provisions should be made to permanently vent the enclosed space below the deck, particularly that space above a suspended ceiling, using sufficient gravity or mechanical ventilation to remove all construction moisture and excessive moisture vapor resulting from subsequent occupancy.

X1.4 Provisions for Expansion and Contraction

X1.4.1 Provision for relief of expansion and contraction should be considered for all cast-in-place roof decks. Wherever expansion joints have been provided in the main structure, they must also be provided in the roof deck and roof covering. The use of suitable expansion strips at the junction between roof slabs and parapets is recommended for further relief of expansion and contraction stresses in the slabs. Where subpurlins are used, edge expansion strips need be used only at walls where subpurlins abut walls at right angles.

X1.5 Roof Coverings

X1.5.1 Built-up or other suitable roof coverings should be specified over poured gypsum concrete roof slabs.

X1.5.2 The application of metal or other rigid types of roof covering over poured gypsum concrete decks is not generally recommended. If sheet metal roof coverings are required over portions of a poured gypsum concrete deck, their anchoring members must be attached either by the use of toggle bolts passing entirely through the slab and formboards or by suitable fastening to the primary members. Where shingle-type roof coverings are contemplated, the gypsum concrete producer's recommendations for attachment should be followed.

X1.5.3 Provisions for furnishing and installing roof sumps should be included in other sections of the specification.

X1.5.4 The application of roof covering should follow as promptly after the placing of gypsum concrete as possible. The top surface of the slab should be reasonably hard as indicated by the disappearance of visible moisture gloss from its surface.

X1.5.5 Expansion joints should be installed in the roofing wherever they are provided in the gypsum concrete and the main structure.

X1.6 Painting Undersides of Gypsum Concrete Roof Decks

X1.6.1 The underside of the gypsum deck is usually not further decorated. If painting is desired, it should be deferred until the formboard and the slab are thoroughly dry through their total thickness.

X1.7 Framing Requirements

X1.7.1 The primary framing to receive subpurlins and the primary framing spaced to directly receive the formboards for the gypsum concrete are not part of the gypsum concrete specifications. However, it is essential that the following provisions regarding such framing be incorporated in other sections.

X1.7.1.1 Spacing of primary structural supports to receive the subpurlins of the reinforced gypsum concrete assembly must be such that design stresses in the specified subpurlins are not exceeded.

X1.7.1.2 When subpurlins are not used, the following provisions should be made to anchor the slab to the primary framing members to resist uplift forces and movements due to temperature changes.

X1.7.1.2.1 Weld steel tee sections sized from 1 by 1 by $1/8$ in. (25.4 by 25.4 by 3.2 mm) to $1\frac{1}{2}$ by $1\frac{1}{2}$ by $3/16$ in. (38.1 by 38.1 by 4.7 mm) across the joists at approximately 32 in. (812.8 mm) to 48 in. (1219 mm) on center depending on type of formboard used. The reinforcing mat must then be wire tied to these tee sections.

X1.7.1.2.2 Wiring of formboard to the frame as a means of providing uplift resistance is not permissible.

X1.7.1.3 Heavy concentrated loads, such as water tanks, large fan bases, cooling towers, and flag poles, must not be imposed directly on the poured gypsum concrete roof deck. Details must provide for transmitting such loads directly to the walls, primary framing or structural supports.

X1.7.1.4 The architect or engineer should assign by specification the responsibility for furnishing and installing of roof nailers and for framing at openings and expansion joints, eaves and wall angles, and similar special items.

X1.7.1.5 It is important that the gypsum concrete roof deck be designed with sufficient slope to provide adequate drainage and eliminate the possibility of free-standing water.

X1.8 Evaluation and Inspection

X1.8.1 Metal framing members should be of the sizes specified and have a factory applied coating to prevent rust.

X1.8.1.1 Welds should meet the requirements of AWS D1.1 and be correctly located and of the size specified.

X1.8.2 Formboards should be of the type, size, and thickness specified and should be free of damage rendering them unfit for their intended use.

X1.8.3 Reinforcement should have main longitudinal strands laid at right angles to subpurlins and near formboard at midspan.

X1.8.3.1 Edges should be butted. Edges should not be lapped unless required for fire rated or horizontal diaphragm construction.

X1.8.3.2 Ends should be lapped a minimum of 4 in.

(101.6 mm). At slab edges, the fabric should be folded back to provide greater bond within the slab.

X1.8.4 Top surface of the gypsum concrete slab should be smooth and free of any protrusions.

X2. DESIGN

X2.1 General

X2.1.1 Reinforced gypsum concrete shall be designed to support the anticipated loads and to withstand the forces to which they may be subjected without exceeding the allowable stresses specified in X2.1.1.1. Except as otherwise specified, methods of design shall follow established principles of mechanics and principles of design for reinforced concrete in accordance with ACI 318.X2. The minimum ratio of area of reinforcement to area of portland cement concrete for shrinkage and temperature shall not apply to gypsum concrete.

X2.1.1.1 *Allowable Stresses*—not more than:

Compressive stress in bending	0.25 f_g (Note X2)
Axial compressive or bearing stress	0.20 f_g (Note X2)
Bond stress (plain reinforced bars)	0.02 f_g (Notes X2 and X3)
Bond stress (for deformed reinforced bars or electrically welded mesh)	0.03 f_g (Note X2)
Shear stress	0.02 f_g (Notes X2 and X3)

NOTE X2—f_g indicates the compressive strength of the gypsum concrete as determined in accordance with the requirements of Specification C 317, Class A.

NOTE X3—Gypsum concrete meets the bond and shear requirements of this section when reinforced with steel wire reinforcement conforming to the requirements of 7.4.

X2.1.2 *Minimum Thickness*—Not less than 2 in. (50.8 mm) measured at all points over the top surface of the formboard.

X2.1.3 *Maximum Spans*—Not more than 31 in. (787 mm) between supports when using subpurlins and formboards.

X2.1.4 *Maximum Spacing of Supports*—Not more than 36 in. (914 mm) for gypsum concrete slabs over formboards applied directly to primary framing members.

X2.2 *Seismic and Wind Conditions*—Where the gypsum concrete roof deck is used as a horizontal diaphragm to transmit seismic or wind loads, the subpurlins shall be installed in accordance with 9.1.2.

X2.2.1 *Diaphragm Shear*—Determine shear in poured gypsum concrete diaphragms as follows:

$$Q_e = [0.16\, f_g t C_1 + 1000\,(k_1 d_1 + k_2 d_2)] C_2 C_3$$

$$Q_m = [0.0000133\, f_g t C_1 + 0.175\,(k_1 d_1 + k_2 d_2)] C_2 C_3$$

where:

Q_e = allowable shear on diaphragm in lb per linear ft which includes a one-third increase for short-term loading,

Q_m = allowable shear on diaphragm in kN·m which includes a one-third increase for short-term loading,

f_g = oven-dry compressive strength of gypsum in lbf/in.^{2} (kPa) as determined by tests conforming to this specification,

C_1 = 1.0 for Class A gypsum concrete,

t = thickness of gypsum concrete between subpurlins in inches (mm). For the purpose of computing diaphragm shear values, t shall be not more than 4 in. (101.6 mm),

k_1 = number of mesh wires per ft (wires per m) passing over subpurlins,

d_1 = diameter of mesh wires passing over subpurlins in inches (mm) except hexagonal mesh,

k_2 = number of mesh wires per foot (wires per metre) parallel to subpurlins or 0.7 times the number of hexagonal wires,

d_2 = diameter in inches of mesh wires parallel to subpurlins or of hexagonal wires,

C_2 = 1.4 for Class A gypsum concrete using the truss tee and 1.0 for bulb tee, and

C_3 = 0.85 for perlited gypsum concrete, 1.0 for gypsum concrete made with wood chips or wood shavings.

The allowable shear values are shown in Tables X2.1 and X2.2.

NOTE X4—k_1 = X2.5 for 2 in. (50.8 mm) hexagonal mesh woven of No. 19 gage galvanized wire with additional longitudinal No. 16 gage galvanized wires spaced every 3 in. (75.8 mm) across the width of the mesh.

TABLE X2.1 Allowable Shear Values in Pounds per Foot Using Truss or Bulb Tee Subpurlins[A]

Concrete Thickness, in.	Truss Tee Subpurlins, Mesh Type: Galvanized 4 × 8 #12 × #14	Truss Tee Subpurlins, Mesh Type: Galvanized 6 × 6 #10 × #10	Truss Tee Subpurlins, Mesh Type: Hexagonal	Bulb Tee Subpurlins, Mesh Type: Galvanized 4 × 8 #12 × #14	Bulb Tee Subpurlins, Mesh Type: Galvanized 6 × 6 #10 × #10	Bulb Tee Subpurlins, Mesh Type: Hexagonal	Shear Values[B]
Perlited Gypsum Concrete—Class A (500 psi)							
2	718	831	878	513	593	627	
2.5	765	878	926	547	627	661	
3	813	926	974	581	661	695	
3.5	860	973	1021	615	695	729	
4	908	1021	1069	649	729	763	
Wood Chips or Wood Shavings Gypsum Concrete—Class A (500 psi)							
2	844	977	1033	603	698	738	840
2.5	900	1033	1089	643	738	778	
3	956	1089	1145	683	778	818	
3.5	1012	1145	1201	723	818	858	
4	1068	1201	1257	763	858	898	

[A] The tabulated shear values are for short-term loads due to wind or earthquake forces and are not permitted a one-third increase for duration of load.

[B] Recommended shear transfer provided such truss tees are embedded not less than ⅜ in. (9.5 mm) deep in gypsum concrete and are welded to the structural frame with not less than ⅛ in. (3.2 mm) fillet welds not less than 1½ in. (38.1 mm) long at 6 in. (152.4 mm) on centers both sides bottom flanges and the mesh is tied to top flange of such subpurlins.

TABLE X2.2 (Metric) Allowable Shear Values in kiloNewtons per Meter Using Truss or Bulb Tee Subpurlins (Metric)[A]

Concrete Thickness, mm	Truss Tee Subpurlins, Mesh Type: 4 × 8 #12 × #14	Truss Tee Subpurlins, Mesh Type: 6 × 6 #10 × #10	Truss Tee Subpurlins, Mesh Type: Hexagonal	Bulb Tee Subpurlins, Mesh Type: 4 × 8 #12 × #14	Bulb Tee Subpurlins, Mesh Type: 6 × 6 #10 × #10	Bulb Tee Subpurlins, Mesh Type: Hexagonal	Shear Values[B]
Perlited Gypsum Concrete—Class A (3450 kPa)							
50.8	10.5	12.1	12.8	7.47	8.65	9.15	
63.5	11.2	12.8	13.5	7.97	9.15	9.64	
76.2	11.8	13.5	14.2	8.46	9.64	10.1	
88.9	12.5	14.2	14.9	8.96	10.1	10.6	
101.6	13.2	14.9	15.6	9.45	10.6	11.1	
Wood Chips or Wood Shavings Gypsum Concrete—Class A (3450 kPa)							
50.8	12.3	14.2	15.1	8.79	10.2	10.8	12.3
63.5	13.1	15.1	15.9	9.37	10.8	11.3	
76.2	13.9	15.9	16.7	9.96	11.3	11.9	
88.9	14.8	16.7	17.5	10.5	11.9	12.5	
101.6	15.6	17.5	18.3	11.1	12.5	13.1	

[A] The tabulated shear values are for short-term loads due to wind or earthquake forces and are not permitted a one-third increase for duration of load.

[B] Recommended shear transfer provided such truss tees are embedded not less than 9.5 mm (⅜ in.) deep in gypsum concrete and are welded to the structural frame with not less than 3.2 mm (⅛ in.) fillet welds not less than 38.1 mm (1½ in.) long at 152.4 mm (6 in.) on centers both sides bottom flanges and the mesh is tied to top flange of such subpurlins.

Last ASTM Designation: C 959 – 91

Standard Specification for Radiant Heat Fill

This specification covers radiant heat fill used to cover electric heating cables applied over monolithic concrete or gypsum board conforming to Specification C 588 or C 36.

Formerly under the jurisdiction of ASTM Committee C-11 on Gypsum and Related Building Materials and Systems, this specification was withdrawn in May 1996.

Standard Specification for Predecorated Gypsum Board[1]

This standard is issued under the fixed designation C 960/C 960M; the number immediately following the designation indicates the year of original adoption or, in the case of revision, the year of last revision. A number in parentheses indicates the year of last reapproval. A superscript epsilon (ε) indicates an editorial change since the last revision or reapproval.

This specification has been approved for use by agencies of the Department of Defense.

1. Scope*

1.1 This specification covers predecorated gypsum board which is designed to be used as the finished surface for walls or ceilings.

1.2 The values stated in inch-pound units are to be regarded separately as the standard. Within the text, the SI units are shown in brackets. The values stated in each system shall be used independent of the other. Values from the two systems shall not be combined.

1.3 The following safety hazards caveat pertains only to the test method portion, Section 10 of this specification: *This standard does not purport to address all of the safety concerns, if any, associated with its use. It is the responsibility of the user of this standard to establish appropriate safety and health practices and determine the applicability of regulatory limitations prior to use.*

1.4 The text of this specification references notes and footnotes that provide explanatory material. These notes and footnotes (excluding those in tables and figures) shall not be considered as requirements of the standard.

2. Referenced Documents

2.1 *ASTM Standards:*

C 11 Terminology Relating to Gypsum and Related Building Materials and Systems[2]

C 36 Specification for Gypsum Wallboard[2]

C 645 Specification for Non-Load Bearing (Axial) Steel Studs, Runners (Tracks), and Rigid Furring Channels for Screw Application of Gypsum Board[2]

C 1264 Specification for Sampling, Inspection, Rejection, Certification, Packaging, Marking, Shipping, Handling and Storage of Gypsum Board[2]

E 84 Test Method for Surface Burning Characteristics of Building Materials[3]

E 119 Test Methods for Fire Tests of Building Construction and Materials[3]

3. Terminology

3.1 Definitions of terms shall be in accordance with Terminology C 11.

3.2 *Descriptions of Terms Specific to This Standard:*

3.2.1 *face, n*—the surface of the predecorated gypsum wallboard that is called the finish side.

4. Classification

4.1 Predecorated gypsum board is furnished in two classes as follows:

4.1.1 Class I predecorated gypsum board shall have a decorative sheet or film laminated to the paper surface.

4.1.2 Class II predecorated gypsum board shall have a decorative coating applied to the paper surface.

5. Materials and Manufacture

5.1 Predecorated gypsum board shall consist of a noncombustible core, essentially gypsum, surfaced with paper bonded to the core with the face covered with a decorative sheet, film, or coating.

5.2 Predecorated gypsum board, type X (special fire resistant) designates predecorated gypsum board complying with this specification that provides not less than 1 h fire-resistance for boards ⅝ in. [15.9 mm] thick or ¾ h fire-resistance for boards ½ in. [12.7 mm] thick, applied parallel with and on each side of load bearing 2 × 4 wood studs spaced 16 in. [406 mm] on center with 6d coated nails, 1⅞ in. [48 mm] long, 0.0915 in. [2.32 mm] diameter shank, ¼ in. [6.4 mm] diameter heads, spaced 7 in. [178 mm] on center with wallboard joints staggered 16 in. [406 mm] on each side of the partition and tested in accordance with the requirements of Test Methods E 119.

NOTE—Consult producers for independent test data on assembly details and fire resistance classifications for other types of construction. See official fire test reports for assembly particulars, materials, and classifications.

6. Physical Properties

6.1 Flexural strength, humidified deflection, core, end and edge hardness, and nail pull resistance shall be in accordance with Specification C 36.

6.2 No part of the test surface shall be disturbed where Class I predecorated gypsum board is tested by the pick (tape) test described in Section 10.

7. Dimensions and Permissible Variations

7.1 After decoration, thickness, width, length and end

---

[1] This specification is under the jurisdiction of ASTM Committee C-11 on Gypsum and Related Building Materials and Systems and is the direct responsibility of Subcommittee C11.01 on Specifications and Test Methods for Gypsum Products.

Current edition approved June 10, 1997. Published June 1998. Originally published as C 960 – 81. Last previous edition C 960 – 96.

[2] *Annual Book of ASTM Standards*, Vol 04.01.

[3] *Annual Book of ASTM Standards*, Vol 04.07.

*** A Summary of Changes section appears at the end of this specification.**

squareness shall be in accordance with Specification C 36.

7.1.1 Where decorative coating, sheet or film thicker than 0.020 in. [0.508 mm] is used, the dimension specification shall be waived when specified in the purchase order.

7.2 *Edges and Ends*—The edges and ends shall be straight. Edges shall be either square or featured.

8. Finish and Appearance

8.1 The surfaces of predecorated gypsum board shall be true and free from imperfections that would render it unfit for use.

9. Sampling, Inspection, Rejection, Certification, Packaging, Marking, Shipping, Handling and Storage

9.1 Sampling, inspection, rejection, certification, packaging, marking, shipping, handling, and storage shall be in accordance with Specification C 1264.

9.2 When required by the purchaser, the flame spread index, as tested in accordance with Test Method E 84, shall be legibly marked on each board, package, or pallet.

10. Test Method for Pick Resistance

10.1 *Significance and Use:*

10.1.1 This test method provides a procedure for measuring the adhesion and integrity of the decorative sheet or film laminated to the paper surface.

10.2 *Sampling and Test Specimens:*

10.2.1 Obtain samples in accordance with Section 9.

10.2.2 Prepare test specimens by cutting or marking full sheets of predecorated gypsum board into not less than 24 × 24 in. [600 × 600 mm] areas. Prepare not less than two specimens from each sheet of gypsum board in the sample.

10.3 *Procedure:*

10.3.1 Apply a ½ to 9⁄16 in. [12.7 to 14.3 mm] wide strip of transparent adhesive tape[4] approximately 3 in. [76 mm] in length to the test specimen face, leaving 1 in. [25 mm] free. Press the tape firmly to the test specimen face using thumb pressure for not less than 5 s. Grasp the free end of the tape and pull quickly at a 90° angle.

10.3.2 *Number of Tests*—Test not less than six specimens, three in the machine direction and three perpendicular to the machine direction.

10.4 *Precision and Bias*—Precision and bias of this test method have not been determined.

11. Keywords

11.1 ceiling; gypsum; partitions; predecorated gypsum board; predecorated gypsum board, type X; wall

[4] "Scotch" brand transparent tape No. 600, manufactured by 3M Co., St. Paul, MN 55101, or equivalent, has been found satisfactory for this purpose.

APPENDIX

(Nonmandatory Information)

This Appendix gives general information and also suggestions for inclusions to be made elsewhere by the specifier. They are not part of this specification.

X1. Alternate Definition for Type X

X1.1 Predecorated gypsum board, Type X (special fire-resistant) designates predecorated gypsum board providing a greater fire resistance than regular predecorated gypsum board of the same thickness. Type X (special fire-resistant) predecorated gypsum board, when tested in accordance with Test Methods E 119, shall provide the following minimum fire resistance for the assemblies described:

X1.1.1 One hour for a ⅝ in. [15.9 mm] thickness applied to a partition in a single layer application on each side of 3⅝ in. [92 mm] deep non-loadbearing galvanized steel studs complying with Specification C 645, spaced 24 in. [610 mm] on center. The ⅝ in. [15.9 mm] thick predecorated gypsum board 48 in. [1220 mm] wide shall be attached using 1 in. [25 mm] long drywall screws spaced 8 in. [203 mm] on center along the edges and ends, and 12 in. [305 mm] along intermediate studs. All joints shall be oriented parallel to and located over studs and staggered on opposite sides of the assembly, and

X1.1.2 Two hours for a ½ in. [12.7 mm] thickness applied to a partition in a double layer application on each side of 2½ in. [64 mm] deep non-loadbearing galvanized steel studs complying with Specification C 645, spaced 24 in. [610 mm] on center. The 48 in. [1220 mm] wide base layer shall be attached using 1 in. [25 mm] long drywall screws spaced 12 in. [305 mm] on center along board edges, ends, and along intermediate studs. Joints shall be oriented parallel to and located over studs and staggered on opposite sides of the assembly. The 48 in. [1220 mm] wide face layer shall be attached using 1⅝ in. [41 mm] long drywall screws spaced 12 in. [305 mm] along board edges, ends and along intermediate studs. Joints shall be oriented parallel to and located over studs, offset 24 in. [610 mm] from the base layer joints, and staggered on opposite sides of the assembly.

SUMMARY OF CHANGES

(1) Terminology for "decorative coating" and "decorative sheet or film" was deleted from Section 3.

(2) Paragraph 7.1.1 was revised to be more specific as to the thickness of decoration and refer it to the purchase order.

Standard Specification for Quicklime and Hydrated Lime for Soil Stabilization[1]

This standard is issued under the fixed designation C 977; the number immediately following the designation indicates the year of original adoption or, in the case of revision, the year of last revision. A number in parentheses indicates the year of last reapproval. A superscript epsilon (ε) indicates an editorial change since the last revision or reapproval.

This standard has been approved for use by agencies of the Department of Defense. Consult the DoD Index of Specifications and Standards for the specific year of issue which has been adopted by the Department of Defense.

1. Scope

1.1 This specification covers quicklime and hydrated lime, either high calcium, dolomitic, or magnesian lime, for use in stabilization of soils (See Note 2).

NOTE 1—Quicklime and hydrated lime act upon clay soils and may render such soils suitable for highway construction and for other load-bearing applications. In most cases, lime causes finely divided clay particles to agglomerate into coarser particles which improves load-bearing properties and subsequently the lime-treated soil hardens by chemical reaction.

NOTE 2—No attempt is made to present requirements for by-product lime, commercial lime slurry, etc. Specification requirements for these materials could be better determined on a local basis.

1.2 The following safety hazards caveat pertains only to the test method portion, Appendix X1, of this specification: *This standard does not purport to address all of the safety concerns, if any, associated with its use. It is the responsibility of the user of this standard to establish appropriate safety and health practices and determine the applicability of regulatory limitations prior to use.*

2. Referenced Documents

2.1 *ASTM Standards:*

C 25 Test Methods for Chemical Analysis of Limestone, Quicklime, and Hydrated Lime[2]

C 50 Practice for Sampling, Inspection, Packing, and Marking of Lime and Limestone Products[2]

C 51 Terminology Relating to Lime and Limestone (as Used by the Industry)[2]

C 110 Test Methods for Physical Testing of Quicklime, Hydrated Lime, and Limestone[2]

3. Chemical Composition

3.1 Unless otherwise specified, for definitions of terms used in this specification, refer to Terminology C 51.

3.2 Quicklime and hydrated lime for soil stabilization shall conform to the following chemical composition:

Calcium and Magnesium Oxides (on an LOI-free basis, minimum %)	90.0
Carbon Dioxide (taken at point of manufacture, maximum %)	5.0
Free Moisture (taken at point of manufacture, maximum %)	2.0

4. Physical Properties

4.1 *Hydrated Lime*, shall have not more than 3 % retained on a No. 30 (590-μm) sieve and not more than 25 % retained on a No. 200 (74-μm) sieve.

4.2 *Quicklime:*

4.2.1 *Fineness of Quicklime*—Quicklime shall all pass a 1.0-in. (24.4-mm) sieve.

4.2.2 Quicklime for soil stabilization shall have a temperature rise of a minimum of 30°C in 20 min as measured by Test Methods C 110.

4.2.3 *Residue of Quicklime*—Quicklime for soil stabilization shall have not more than 10 % residue as measured by Test Methods C 110.

5. Field Applications

5.1 When quicklimes are used, ensure that thorough mixing of the lime and soil is accomplished and all lime pebbles have been hydrated with additional water and distributed uniformly throughout the soil. There shall be no lime pebbles present before the compaction operation starts. Check by turning soil with a spade at representative intervals and inspect for visible lime pebbles. Care should be exercised on initial dry applications to minimize environmental dusting.

5.2 For hydrated limes, additional water shall be added to the lime-soil mixture to facilitate mixing and uniform distribution of the hydrated lime in the soil layer. There shall be no lime clumps present before the compaction operation starts. Check by turning soil with a spade at representative intervals and inspect for visible lime clumps. Care should be exercised on initial dry applications to minimize environmental dusting.[3]

6. Test Method

6.1 The chemical analysis of quicklime and hydrated lime shall be determined in accordance with Test Methods C 25 (See Appendix).

6.2 The fineness of hydrated lime shall be determined in accordance with the residue test for hydrated lime in Test Methods C 110.

7. Sampling, Inspection, Packing, and Marking

7.1 The sampling, inspection, rejection, retesting, packing, and marking shall be done in accordance with Practice C 50.

8. Keywords

8.1 highway construction; hydrated lime; lime treated soils; load-bearing; quicklime; soil stabilization

[1] This specification is under the jurisdiction of ASTM Committee C-7 on Lime and is the direct responsibility of Subcommittee C07.03 on Industrial Uses.

Current edition approved Aug. 15, 1995. Published October 1995. Originally published as C 977 – 83. Last previous edition C 977 – 92.

[2] *Annual Book of ASTM Standards*, Vol 04.01.

[3] Further information on soil stabilization construction technique is available from National Lime Association Bulletin No. 326, 3601 N. Fairfax Dr., Arlington, VA, 22201.

APPENDIX

(Nonmandatory Information)

X1. TEST METHOD FOR DETERMINING STABILIZATION ABILITY OF LIME

X1.1 This test method usually provides a lime-soil proportion for stabilization. It gives an indication whether the soil in question can be stabilized. For most stabilization work, the results of this test should be verified by performance tests in a soil laboratory.

X1.2 Air dry a sufficient quantity of the soil to be tested and screen through a No. 40 (425-μm) sieve. Store in a closed container to maintain uniform moisture. Weigh to the nearest 0.1 g a series of 20-g samples of soil and place in separate 150-mL plastic containers with water-tight lids.

X1.3 In the case of quicklime, rapidly crush the lime to pass a No. 6 (3.35-mm) sieve.

X1.4 Weigh, to the nearest 0.01 g, a series of quantities of lime equivalent to 2, 3, 4, 5, and 6 % of the soil sample.

X1.5 Add the lime quantity to the soil sample, mark the container with the appropriate percentage, and mix the dry contents thoroughly.

X1.6 Add the 100 mL of 70°F carbon dioxide-free distilled water or, if possible, 70°F actual water to be used on the job to each container of soil and lime. Seal with screw-cap lid and mix the three components by shaking the bottles. Shake each bottle for 30 s every ten min for 1 h. After 1 h, shake vigorously and transfer part of the slurry into a beaker. Measure the pH with a low-sodium error glass electrode (previously standardized to pH 12.45 with an agitated calcium hydroxide slurry). Record the pH reading for each mixture.

X1.7 If the pH readings are 12.40 or higher, the lowest percentage that gives a pH of 12.40 is the % required to stabilize the soil. If the pH readings do not go beyond pH of 12.30 and two percentages give this reading, the lowest % to give a pH of 12.30 is the % required to stabilize the soil. If the highest pH reading is a pH of 12.30 and only the highest percentage lime used gives a pH of 12.30, additional testing is required using higher percentages of lime.

Designation: C 1002 – 98

Standard Specification for Steel Drill Screws for the Application of Gypsum Panel Products or Metal Plaster Bases[1]

This standard is issued under the fixed designation C 1002; the number immediately following the designation indicates the year of original adoption or, in the case of revision, the year of last revision. A number in parentheses indicates the year of last reapproval. A superscript epsilon (ε) indicates an editorial change since the last revision or reapproval.

This standard has been approved for use by agencies of the Department of Defense.

1. Scope*

1.1 This specification covers steel drill screws for use in fastening gypsum panel products and metal plaster bases to light-gage steel and wood members and for fastening gypsum panel products to gypsum board.

1.2 This specification also covers test methods for determining performance requirements and physical properties.

1.3 The values stated in inch-pound units are to be regarded as the standard. The SI (metric) values given in parentheses are approximate and are for information only.

1.4 The following safety hazards caveat pertains only to the test methods described in this specification: *This standard does not purport to address all of the safety concerns, if any, associated with its use. It is the responsibility of the user of this standard to establish appropriate safety and health practices and determine the applicability of regulatory limitations prior to use.*

2. Referenced Documents

2.1 *ASTM Standards:*

A 548 Specification for Steel Wire, Carbon, Cold-Heading Quality for Tapping or Sheet Metal Screws[2]

C 11 Terminology Relating to Gypsum and Related Building Materials and Products[3]

C 36 Specification for Gypsum Wallboard[3]

C 645 Specification for Non-Load (Axial) Bearing Steel Studs, Runners (Track), and Rigid Furring Channels for Screw Application of Gypsum Board[3]

C 847 Specification for Metal Lath[3]

3. Terminology

3.1 *Definitions*—For definitions relating to gypsum and related building materials and systems, see Terminology C 11.

3.2 *Definitions of Terms Specific to This Standard:*

3.2.1 *light-gage steel, n*—steel less than 0.033 in. (0.84 mm) thick.

3.2.2 *self drill, v*—the ability to penetrate without benefit of a pre-drilled hole.

3.2.3 *spin-out, v—for a screw,* the continued rotation of a screw without further penetration into the substrate.

4. Classification

4.1 Steel drill screws covered by this specification are of three types:

4.1.1 *Type G,* for fastening gypsum board to gypsum board.

4.1.2 *Type S,* for fastening gypsum board or metal plaster bases to light-gage steel members.

4.1.3 *Type W,* for fastening gypsum board or metal plaster bases to wood members.

5. Materials

5.1 *Steel Wire,* for manufacturing screws, Specification A 548, Grades 1013 to 1022.

6. Physical Properties

6.1 *Hardness:*

6.1.1 The surface of Type S and Type W screws shall be case hardened to a depth of not less than 0.002 in. (0.05 mm) with case hardness not less than 45 HRC.

6.1.2 The surface of Type G screws is not required to be case hardened.

6.2 *Ductility*—The screws shall have sufficient ductility to be able to withstand a 15° bend without visible signs of fracture when tested as specified in 12.6.5.

7. Performance Requirements

7.1 *Type G Screws:*

7.1.1 Screws shall be able to self drill and drive into gypsum panel products.

7.1.2 Screw threads shall be adequate to pull the head of the screw below the surface of the gypsum panel product when tested as specified in 12.6.2.

7.2 *Type S Screws:*

7.2.1 *General*—Screws shall be able to self drill and drive into light-gage steel members in less than 2 s, when tested as specified in 12.6.3.1.

[1] This specification is under the jurisdiction of ASTM Committee C-11 on Gypsum and Related Building Materials and Systems and is the direct responsibility of Subcommittee C11.02 on Specifications and Test Methods for Accessories and Related Products.

Current edition approved May 10, 1998. Published December 1998. Originally published as C 1002 – 83. Last previous edition C 1002 – 96a.[ε1]

[2] *Annual Book of ASTM Standards,* Vol 01.03.

[3] *Annual Book of ASTM Standards,* Vol 04.01.

*** A Summary of Changes section appears at the end of this standard.**

7.2.2 *Screws for Fastening Gypsum Panel Products*—Screw threads shall be adequate to pull the head of the screw below the surface of the gypsum panel product, without spin-out, when tested as specified in 12.6.3.1.

7.2.3 *Screws for Fastening Metal Plaster Bases*—Screw threads shall be adequate to pull a metal plaster base tightly enough against the flange of a steel stud, without spin-out, so that when subjected to a steady pull the plaster base will tear before slipping out from under the screw head, when tested as specified in 12.6.3.2.

7.3 *Type W Screws:*

7.3.1 *General*—Screw threads shall be able to self drill and drive into wood members.

7.3.2 *Screws for Fastening Gypsum Panel Products*—Screw threads shall be adequate to pull the head of the screw below the surface of the gypsum panel product, without spin-out when tested as specified in 12.6.4.1.

7.3.3 *Screws for Fastening Metal Plaster Bases*—Screw threads shall be adequate to pull a metal plaster base tightly enough against the face of a wood stud, without spin-out, so that when subject to a steady pull, the plaster base will tear before slipping out from under the screw head, when tested as specified in 12.6.4.2.

8. Dimensions and Permissible Variations

8.1 *Head Diameter:*

8.1.1 *General*—Heads shall be permitted to be out of round not more than 0.02 in. (0.51 mm).

8.1.2 *Screws for Fastening Gypsum Board*—Not less than 0.315 in. (8 mm).

8.1.3 *Screws for Fastening Metal Plaster Bases*—Not less than 0.437 in. (11.1 mm).

8.2 *Head Contour:*

8.2.1 *Screws for Fastening Gypsum Panel Products*—The top of the head shall be flat. The outer flange thickness shall be 0.025 ± 0.005 in. (0.64 ± 0.13 mm). The contour beneath the flange head shall be such that the screw head shall be able to be driven so that the head of the screw rests immediately below the surface of the gypsum panel product.

8.2.2 *Screws for Fastening Metal Plaster Bases*—The top of the screw shall be either flat or contoured. The underside of the head shall be flat or near flat. The threads shall extend to the underside of the head.

8.3 *Driving Recess:*

8.3.1 *Screws for Fastening Gypsum Board*—No. 2 "Phillips" design, depth of recess, not less than 0.104 in. (2.64 mm).

8.3.2 *Screws for Fastening Metal Plaster Bases*—No. 2 "Phillips" design, depth of recess, not less than 0.077 in. (1.95 mm).

8.3.3 Determine depth of recess with a Phillips penetration depth gage.

8.3.4 Other types of screw-driving recesses having performance values equal to those of the recesses specified are acceptable.

8.4 *Shank Diameter*—Not less than No. 6, with major diameter not less than 0.136 in. (3.45 mm).

8.5 *Length:*

8.5.1 *Nominal Length*—The nominal length shall be the minimum length.

8.5.2 Type W screws shall be long enough to penetrate wood members not less than ⅝ in. (15.9 mm).

8.6 *Threads*—Either single or multiple thread design.

8.7 *Points*—Self-drilling design.

9. Finish and Appearance

9.1 *Form*—The form shall be straight and neatly formed.

9.2 *Threads*—The threads shall be clean and smooth.

9.3 *Finish*—Screws shall have a corrosion-resistant treatment. The treatment shall not inhibit adhesion to finishing compounds nor bleed through field-applied decoration.

10. Sampling

10.1 Obtain not less than one specimen of each type of screw to be tested from each of five containers, with not less than five screws per 16 000 screws, to constitute a lot for testing.

11. Number of Tests and Retests

11.1 If any of the first lot fails the test, test 25 more screws.

11.2 If two or more screws of the second lot fail the second test, then the represented lot fails to meet the specified requirements.

12. Test Methods

12.1 The following test methods set forth procedures used to determine the ability of the screws to self drill through gypsum panel and metal plaster base and into the backup material without spin-out.

12.1.1 The test methods can also be used to determine whether or not alternate screw driving recesses will have performance values equal to those of the "Phillips" recess specified.

12.2 *Significance and Use*—The test methods provide procedures for evaluating the physical properties and performance requirements of steel drill screws. The degree of correlation between these tests and service performance has not been determined.

12.3 *Apparatus:*

12.3.1 *Power Drill*—Standard manufacture electric screw gun, turning at 2500 rpm having a bit suitable for the driving recess of the screw being tested.

12.3.2 *Timing Device*—Standard stopwatch, calibrated in ⅒ s.

12.3.3 *Vice*—Standard machinists vice, not less than 3 in. (76 mm).

12.3.4 *Clamp*—Locking pliers ("Vice-Grip"), 6 in. (152 mm).

12.4 *Materials:*

12.4.1 *Gypsum Wallboard*—Specification C 36, Type X, ⅝ in. (16 mm) thick.

12.4.2 *Metal Lath*—As in Specification C 847, diamond mesh, weight 2.5 lb/yd^2 (1.4 kg/m^2).

12.4.3 *Steel Stud*—As in Specification C 645, 0.0179 in. (0.455 mm) thick, hardness not less than 52 HRB, size 3⅝ in. (92.1 mm).

12.4.4 *Wood Stud*—Douglas fir, construction grade, 2 by 4 or 2 by 6, nominal, containing not less than 16 % nor more than 19 % free moisture as determined by a suitable moisture meter.

12.4.5 *Kraft Paper*—0.01 in. (0.25 mm) thick, 2 in. (51 mm) square.

12.5 *Specimen Preparation:*

12.5.1 Cut 6-in. (152-mm) specimens from a sheet of gypsum wallboard not less than 12 in. (305 mm) away from the edge or end.

12.5.1.1 Weigh the specimens to within 1 g, and then condition at a temperature of 85 ± 15°F (29.5 ± 5.5°C) in an atmosphere having a relative humidity of 50 ± 2 %. Individually test the specimens after constant weight is reached.

12.5.2 Cut 6 in. (152 mm) square specimens from a sheet of metal lath not less than 4 in. (102 mm) away from the edge or end.

12.5.3 Cut 12 in. (305 mm) long specimens from a length of a steel stud.

12.5.4 Cut 12 in. (305 mm) long specimens from a length of wood stud.

12.6 *Procedure:*

12.6.1 *General:*

12.6.1.1 When driving Type G screw specimens, the total pressure shall be not more than 20 lbf (89.0 N).

12.6.1.2 When driving Types S and W screw specimens, the total pressure shall be 25 lbf (111 N).

12.6.1.3 Drive the screws not closer than ⅜ in. (9.5 mm) from the edges of the gypsum wallboard and not closer than 2 in. (50 mm) from the edges of metal lath.

12.6.2 *Type G Screw Test*—Rigidly support the two specimens of gypsum wallboard sandwiched together. With a power drill, drive the screw specimen into the gypsum boards, so that the head of the screw is below the surface of the wallboard.

12.6.3 *Type S Screw Test:*

12.6.3.1 *Screws for Fastening Gypsum Board*—Rigidly support a specimen of gypsum wallboard over a steel stud specimen. With a power drill, drive the screw specimen through four layers of Kraft paper, through the wallboard and into the steel stud. Observe the length of time it takes to drive the screw to the depth required to meet the performance requirements specified in 7.2.2 and observe if there was spin-out.

12.6.3.2 *Screws for Fastening Metal Lath*—Rigidly support a specimen of metal lath over a steel stud specimen. With a power drill, drive the screw specimen through the metal lath, and into the steel stud. Observe the length of time it takes to drive the screw and observe if there was spin-out. Grasp the metal lath with the locking pliers and pull steadily in a plane parallel to the plane of the lath. Observe whether the lath tears before it slips out from under the screw head.

12.6.4 *Type W Screw Test:*

12.6.4.1 *Screws for Fastening Gypsum Board*—Rigidly support a specimen of gypsum wallboard over a wood stud specimen. With a power drill, drive the screw specimen through four layers of Kraft paper, through the wallboard and into the wood stud.

12.6.4.2 *Screws for Fastening Metal Lath*—Rigidly support a specimen of metal lath over a wood stud specimen. With a power drill, drive the screw specimen through the metal lath, and into the wood stud. Grasp the metal lath with the locking pliers and pull steadily in a direction parallel to the plane of the lath. Observe whether the lath tears before it slips out from under the screw head.

12.6.5 *Ductility Test*—Hold one end of a screw specimen in a vice. Hold the other end with the locking pliers and bend the screw to an angle of 15°. Observe for signs of fracture in the screw shank.

12.7 *Precision and Bias*—No estimate of the precision or bias of these methods is available at this time.

13. Inspection

13.1 Inspection of the screws shall be agreed upon between the purchaser and the producer or the supplier as part of the purchase agreement.

14. Rejection

14.1 Rejection of screws that fail to conform to the requirements of this specification shall be reported to the producer or supplier promptly and in writing. The notice of rejection shall contain a statement documenting how the screws have failed to conform to the specification requirements.

15. Certification

15.1 When specified in the purchase agreement, a producer's or supplier's report shall be furnished at the time of shipment, certifying that the screws are in compliance with this specification.

16. Packaging and Marking

16.1 *Packaging:*

16.1.1 Screws shall be packaged in substantial commercial shipping containers, constructed so as to preserve the contents in good condition and to ensure acceptance and safe delivery by common or other carriers, at the lowest rate, to the point of delivery.

16.1.2 Individual packages shall be so constructed that the contents shall be able to be partially removed without destroying the container's ability to serve as a receptacle for the remainder of the contents.

16.2 *Marking*—Individual packages and shipping containers shall be marked with the type, size, use, and quantity of screws contained therein, the name, brand, or trademark of the producer or supplier, and the ASTM designation.

17. Keywords

17.1 drill screw; gypsum board; lath; metal plaster base; self drill

SUMMARY OF CHANGES

Committee C-11 has identified the location of selected changes to this standard since the last issue (C 1002 – 93) that may impact the use of this standard.

(*1*) The term "gypsum board" was replaced with the term "gypsum panel products" in the title, scope, and paragraphs 7.1.1, 7.1.2, 7.2.2, 7.3.2, 8.2.1, and 12.1.

(*2*) The term "gypsum wallboard" was added to paragraphs 12.4.1, 12.5.1, 12.6.1.3, 12.6.2, 12.6.3.1, and 12.6.4.1.

Designation: C 1005 – 97

Standard Specification for Reference Masses and Devices for Determining Mass for Use in the Physical Testing of Hydraulic Cements[1]

This standard is issued under the fixed designation C 1005; the number immediately following the designation indicates the year of original adoption or, in the case of revision, the year of last revision. A number in parentheses indicates the year of last reapproval. A superscript epsilon (ε) indicates an editorial change since the last revision or reapproval.

1. Scope

1.1 This specification covers the minimum requirements for scales, balances, and reference masses used in the physical testing of hydraulic cements.

1.2 Requirements for analytical reference masses and balances are not included in this specification, but are to be found in Test Methods C 114. The use of restrictive terminology, classes, ranges, etc. has been intentionally avoided to allow the use of this specification by other standards-writing bodies with similar requirements for reference masses and devices for determining mass, if desired.

1.3 These requirements are not sufficiently descriptive to be used as the sole specifications for the purchase of reference masses or devices for determining mass without amplification.

1.4 Values in SI units shall be obtained by measurement in SI units or by appropriate conversion, using the Rules for Conversion and Rounding given in Standard IEEE/ASTM SI 10, of measurement made in other units.

2. Referenced Documents

2.1 *ASTM Standards:*

C 114 Test Methods for Chemical Analysis of Hydraulic Cement[2]

C 185 Test Method for Air Content of Hydraulic Cement Mortar[2]

E 617 Specification for Laboratory Weights and Precision Mass Standards[3]

IEEE/ASTM SI 10 Standard for Use of the International System of Units (SI): The Modern Metric System[3]

2.2 *NIST Standards*:[4]

NIST Circular 547 Precision Laboratory Standards of Mass and Laboratory Weights

NIST Handbook 44 Specification Tolerances and Other Technical Requirements for Weighing and Measuring Devices

2.3 *International Organization for Legal Metrology (OIML)*:[5]

International Recommendation No. 1 Cylindrical Weights from One Gram to Ten Kilograms, Medium Accuracy Class

International Recommendation No. 20 Weights of Accuracy Classes

3. Terminology

3.1 *Definitions:*

3.1.1 *accuracy*—degree of conformity of the indication of a device for determining mass to the true value of an applied mass.

3.1.2 *balance*—a device for determining mass that compares a mass with standard masses.

3.1.3 *precision*—reproducibility of the readings of a device for determining mass when a given test mass is applied.

3.1.4 *range of mass determination*—the range of indications of a device from the minimum test load plus any aids or containers on the load-receiving elements to the maximum test load with the same aids or containers, as specified by the particular method of test being used.

3.1.5 *readability*—the smallest fraction of a division to which the index scale of a device for determining mass can be read with ease either by estimation or by use of a vernier.

3.1.6 *reference mass*—a piece of material of known specified mass for use in comparing or measuring the mass of other masses. (Note 1.)

3.1.6.1 *Discussion*—The definition of "reference mass" was formerly incorrectly designated "weight"; weight is a force (see Standard IEEE/ASTM SI 10).

3.1.7 *scale*—a device for determining mass having a load-receiving element and an index scale (possibly in combination with the use of internal masses or proportional masses), almost always calibrated to indicate mass; generally of lesser accuracy than a balance.

3.1.8 *sensitivity*—the minimum change of applied mass required to perceptibly move the indicating element of a device for determining mass.

3.1.9 *sensitivity requirement*—the minimum change in position of rest of the indicating element of a device for determining mass in response to a specified change of mass load on the load-receiving element.

3.1.10 *test load*—the mass to be determined in a single determination, exclusive of aids or containers, as specified in the test method being followed.

3.1.11 *tolerance*—precision and accuracy criteria for reference masses or devices for determining mass.

acceptance tolerance—the maximum permissible deviation from correct indication for new or newly reconditioned or adjusted reference masses or devices for determining mass; equal to ½ the maintenance tolerance.

---

[1] This specification is under the jurisdiction of ASTM Committee C-1 on Cement and is the direct responsibility of Subcommittee C01.95 on Coordination of Standards.

Current edition approved May 10, 1997. Published June 1998. Originally published as C 1005 – 83. Last previous edition C 1005 – 96.

[2] *Annual Book of ASTM Standards*, Vol 04.01.

[3] *Annual Book of ASTM Standards*, Vol 14.02.

[4] Available from National Institute of Standards and Technology (NIST), Gaithersburg, MD 20899.

[5] Available from Office of International Standards, Room A413 Admin., National Institute of Standards and Technology, Washington, DC 20234.

TABLE 1 Tolerances for Reference Masses[A,B]

Mass, g	Acceptance Tolerance, ±, mg	Maintenance Tolerance, ±, mg
10 000	1 000	2 000
5 000	500	1 000
3 000	300	600
2 000	200	400
1 000	100	200
500	50	100
300	30	60
200	20	40
100	10	20
50	7	14
30	5	10
20	3	6
10	2	4
5	2	4
3	2	4
2	2	4
1	2	4

[A] Values from Table X7.1 of Specification E 617 for Class 6 metric reference masses.

[B] For reference masses with unlisted nominal values, the tolerance shall be that for the next lower denomination listed.

adjustment tolerance—acceptance tolerance.

maintenance tolerance—maximum permissible deviation from correct indication for reference masses and devices for determining mass in service.

3.1.12 *total load*—the sum of the masses of materials and aids or containers applied to the load-receiving element at any one time.

4. Reference Masses

4.1 Reference masses for use in the methods of physical testing of hydraulic cement and related and similar materials shall conform at least to the requirements of Class 6 reference masses in Specification E 617, except that the maintenance tolerances given in Specification E 617 for Class 6 reference masses are to be considered acceptance tolerances for purposes of this specification and the maintenance tolerances twice these values. See Table 1. (Note 2.)

NOTE 2—Former NBS Class P and F reference masses meet this specification as do OIML Class M1 reference masses. Former NBS Class T reference masses of 100 mg and less, Class C reference masses of 1 kg and greater, and OIML Class M2 reference masses of 200 g and greater meet the specification.

4.2 A mass used strictly as a mass, whose actual value need not be known, need not meet the requirements of 4.1. (Note 3.)

NOTE 3—A tare mass for a container used in mass determination is in this category. The actual value of its mass need not be known, but it must be equal in mass to that of the container.

5. Requirements for Scales and Balances

5.1 *Capacity*—The capacity of a device for mass determination shall be at least equal to the maximum total load to be applied to the load-receiving element at any one time. (Note 4.)

NOTE 4—In general, the capacity of a device for mass determination should not greatly exceed twice the maximum total load required for a given test because of the reduced accuracy, readability, and sensitivity with higher capacity weighing devices. There are exceptions to this generality as, for example, the high quality substitution balances.

5.2 *Precision and Accuracy*—The maintenance tolerance for a device for mass determination shall be no greater than 0.05 % of the test load throughout the range. The device for mass determination shall be capable of reproducing readings with the same test mass at least to within the acceptance tolerance. (Note 5.)

NOTE 5—Devices in service prior to January 1983 that conform to the requirements of the section on Scales of Test Method C 185 are considered to meet these requirements. See Note 4.

5.3 *Readability*—The index scale of a device for mass determination shall be easily readable to one-half the acceptance tolerance within its range; that is, readable to at least 0.0125 % of the test load.

5.4 *Sensitivity Requirement*—The indicating element of the device for determination of mass shall show a change of at least one of the smallest divisions of the index scale for a change of 0.1 % of the test load within the range. The device shall show perceptible movement of the indicating element (sensitivity) with a change in test load of no more than ¼ of this value.

6. Rejection

6.1 New or newly reconditioned or adjusted reference masses and devices for determination of mass shall not be placed in service if they do not meet the acceptance tolerances specified in this standard.

6.2 Reference masses and devices for determination of mass in service shall be checked annually and shall be removed from service if they exceed the maintenance tolerances specified in this standard.

7. Keywords

7.1 devices for determining mass; reference masses

Designation: C 1007 – 98$^{\epsilon 1}$

Standard Specification for Installation of Load Bearing (Transverse and Axial) Steel Studs and Related Accessories[1]

This standard is issued under the fixed designation C 1007; the number immediately following the designation indicates the year of original adoption or, in the case of revision, the year of last revision. A number in parentheses indicates the year of last reapproval. A superscript epsilon (ε) indicates an editorial change since the last revision or reapproval.

ε[1] NOTE—Mandatory language was updated in Section A2 of the Annex.

1. Scope *

1.1 This specification covers the installation and erection requirements for load bearing (transverse and axial) steel studs and related accessories 0.0329 in. (0.836 mm) to 0.1120 in. (2.845 mm) thick.

1.2 Values stated in inch-pound units are to be regarded as the standard. The SI (metric) values given in parentheses are approximate and are provided for information purposes only.

1.3 The text of this standard references notes and footnotes which provide explanatory material. These notes and footnotes (excluding those in tables and figures) shall not be considered as requirements of the standard.

2. Referenced Documents

2.1 *ASTM Standards:*

C 11 Terminology Relating to Gypsum and Related Building Materials and Systems[2]

C 754 Specification for Installation of Steel Framing Members to Receive Screw-Attached Gypsum Board[2]

C 840 Specification for Application and Finishing of Gypsum Board[2]

C 841 Specification for Installation of Interior Lathing and Furring[2]

C 844 Specification for Application of Gypsum Base to Receive Gypsum Veneer Plaster[2]

C 847 Specification for Metal Lath[2]

C 954 Specification for Steel Drill Screws for the Application of Gypsum Board or Metal Plaster Bases to Steel Studs from 0.033 in. (0.84 mm) to 0.112 in. (2.84 mm) in Thickness[2]

C 955 Specification for Load-Bearing (Transverse and Axial) Steel Studs, Runners (Track), and Bracing or Bridging for Screw Application of Gypsum Board and Metal Plaster Bases[2]

C 1063 Specification for Installation of Lathing and Furring to Receive Interior and Exterior Portland Cement-Based Plaster

2.2 *American Iron and Steel Institute Publication (AISI):*[3]

Cold-Formed Steel Design Manual: Specifications for the Design of Cold-Formed Steel Structural Members, 1986 Revision and 1989 addendum

2.3 *American Welding Society Documents (AWS):*[4]

D1.3 Specification for Welding Sheet Steel in Structures

2.4 *Military Specification:*[5]

MIL-P-21035 Paint, High Zinc Dust Content, Galvanizing Repair

2.5 *Federal Specification:*[5]

FF-P-395 Pin, Drive, Guided and Pin Drive, Power Actuated Fasteners for Power Actuated and Hand Actuated Fastening Tools

3. Terminology

3.1 Definitions shall be in accordance with Terminology C 11.

3.2 *Definitions of Terms Specific to This Standard:*

3.2.1 *axial loads, n*—those loads applied to a member in the same plane as its major centroidal axis.

3.2.2 *cripple studs, n*—less than full height studs above a header or below a sill.

3.2.3 *framing members, n*—studs joist, runners (tracks), bridging and bracing and related accessories.

3.2.4 *header, n*—the stud assembly or the track assembly, or both, placed at a right angle to supporting studs that creates the top of a framed opening.

3.2.5 *jack stud, n*—a framing member which provides support at the end of a header.

3.2.6 *panelized construction, n*—fabrication of framing members into an assembly prior to erection.

3.2.7 *sill, n*—the stud assembly or the track assembly, or both, placed at a right angle to supporting studs that creates the bottom of a framed opening.

[1] This specification is under the jurisdiction of ASTM Committee C-11 on Gypsum and Related Building Materials and Systems and is the direct responsibility of Subcommittee C11.03 on Specifications for Application of Gypsum and Other Products in Assemblies.

Current edition approved June 10, 1998. Published December 1998. Originally published as C 1007 – 83. Last previous edition C 1007 – 97.

[2] *Annual Book of ASTM Standards*, Vol 04.01.

[3] Available from the American Iron and Steel Institute, 1000 16th St., N.W., Washington, DC 20036.

[4] Available from American Welding Society, 550 N.W. La Jeune Rd., Miami, FL 33135.

[5] Available from Standardization Documents Order Desk, Bldg. 4 Section D, 700 Robbins Ave., Philadelphia, PA 19111-5094, Attn: NPODS.

***A Summary of Changes section appears at the end of this standard.**

3.2.8 *transverse loads, n*—those loads applied to a member in a plane perpendicular to its major centroidal axis.

4. Storage of Materials

4.1 Products shall be protected from adverse weather and job site conditions that will cause any physical damage.

4.2 Materials shall be stored on a flat plane.

4.3 Any damaged materials shall be removed from the job site immediately.

5. Materials

5.1 *Framing members*—Specification C 955.

5.2 *Zinc-rich paint*—MIL-P-21035.

5.3 *Steel drill screws*—Specification C 954. Screws shall have rust inhibitive coating suitable for the intended installation.

5.4 *Power actuated drive pins*—FF-P-395.

6. Fastenings and Attachments

6.1 Anchorage of the tracks to the structure shall be with methods designed for the specific application of sheet steel to that surface. Size, penetration, type, and spacing shall be determined by design.

6.2 Welds shall conform to the requirements of AWS D1.1, and AISI Manual, section E 2. Welds shall be butt, fillet, spot, or groove type, the appropriateness of which shall be determined by, and within, the design calculations. All welds shall be touched-up using zinc rich paint for galvanized members, and paint similar to that used by the manufacturer for painted members.

6.3 Steel drill screws shall be of the minimum diameter indicated by the design of that particular attachment detail. Penetration through joined materials shall not be less than three exposed threads.

6.4 Wire tying in structural applications shall not be permitted.

7. Tolerances

7.1 Vertical alignment (plumbness) of studs shall be within 1/960 (ln! in. in 10 ft 0 in.) of the span.

7.2 Horizontal alignment (levelness) of walls shall be within 1/960 (ln! in. in 10 ft 0 in.) of their respective lengths.

7.3 Spacing of the framing members shall not be more than ±1⁄8 in. (3 mm) from the designed spacing, provided that the cumulative error does not exceed the requirements of the finishing materials.

7.4 Squareness of the prefabricated panels shall be not more than 1⁄8 in. (3 mm) out of square within the length of that panel.

INSTALLATION

8. General Installation

8.1 Methods of construction shall be either stickbuilt or panelized.

8.2 Stud to track connections shall be designed to meet or exceed the design loads required.

8.3 Allowance for vertical deflection of the structure shall be as required by the designer.

8.4 Transversely loaded studs shall not be required to sit squarely in tracks, but shall be attached to the tracks.

8.5 Axially loaded studs shall be fabricated with the studs seated squarely, within 1⁄16in. (2 mm) of the web portion of the top and bottom tracks and shall be attached to the tracks.

8.6 *Cutting and Splicing of Members*:

8.6.1 Cutting of steel framing members shall be accomplished with a saw or shear.

8.6.2 Torch cutting of framing members shall not be permitted.

8.6.3 Splicing of axially loaded framing members shall not be permitted.

8.6.4 Cutting of flanges in stud and joist framing members shall not be permitted.

8.6.5 Cutting of additional holes other than those provided by the manufacturer in framing members shall not be permitted.

8.7 Temporary bracing shall be provided and left in place until the work is permanently stabilized.

8.8 Bridging shall be of the size and type shown on the shop drawings and as called for in the design calculations. For further description of the shop drawings see Annex A3.

8.9 *Header and Jack Stud Installation*:

8.9.1 Headers shall be installed in all openings in axially loaded walls that are larger than the stud spacing in that wall.

8.9.1.1 Headers shall be installed so that they lie wholly within the width of the stud wall.

8.9.1.2 Insulation (equal to the job requirements) shall be placed in all jamb and header type conditions that will be in accessible after their installation into the wall.

8.9.2 Jack studs shall be securely connected to the header and must seat squarely in the lower track of the wall in accordance with 8.4.

8.9.3 If the header is designed to be installed immediatley above the wall opening the cripple studs that occur over the header shall be designed to carry all superimposed loads.

8.10 Wall track (runners) shall not be used to support any load unless specifically designed for that purpose.

8.11 All axially loaded members shall be aligned vertically to allow for full transfer of the loads down to the foundation. Vertical alignment shall be maintained at floor/wall intersections.

8.12 All steel bared by fabrication and erection procedures shall be touched up using zinc-rich paint on galvanized steel framing members and paint equal to that used by the manufacturer on painted steel framing members.

8.13 Bearing shall be provided under tracks to provide for load transfer in axially loaded assembles.

9. Panelized Construction

9.1 Panels shall be designed to resist all construction and handling loads as well as live loads.

9.2 Handling and lifting of prefabricated panels shall not cause permanent distortion in any member or collateral material.

9.3 All stud-to-track connections shall be installed prior to hoisting the panel.

9.4 Where splicing of the track is necessary between stud spacings, a piece of stud not less than 12 in. (305 mm) long shall be placed in the track fastened with not less than two screws per flange to each piece of track or welds shall be provided as required.

NOTE 1—Claddings may be permitted to be attached prior to panel installation.

9.5 Attachment of the panel to the structure shall be as shown on the design drawings.

9.6 All panels shall be aligned to provide continuity of any wall/floor surface.

10. Stick Built (Non-Panelized) Members

10.1 Tracks shall be aligned accurately at supporting structure and shall be fastened to the structure.

10.2 Track intersections shall butt evenly.

10.3 Doors, windows, and other items installed in the wall shall be securely anchored to the wall by means of clips, angles, screws, bolts, etc.

11. Keywords

11.1 axial; loadbearing; runners; steel; studs; tracks; transverse

ANNEXES

(Mandatory Information)

A1. DESIGN REQUIREMENTS

A1.1 Physical properties and allowable load capabilities of materials shall be developed in accordance with AISI Design Manual, as provided by Specification C 955.

A1.2 Studs shall be spaced to suit the design requirements and limitations of collateral materials.

A1.3 All applicable loads, strains, deflections, shear and web crippling, and bending of all sections shall be examined.

A1.4 All connections (member-to-member and structure-to-structure) shall be carefully examined taking into account all relevant physical strengths and properties.

A1.5 Selected walls shall be designed to provide frame stability and lateral load resistance. Diagonal steel strapping or other engineered methods shall be used to provide frame stability and lateral load resistance. Diagonal steel strapping shall be used to transfer lateral loads to the structure and foundation. Additional studs shall be provided to resist the vertical component of the loads from the diagonal bracing.

A1.6 Wall bridging shall be designed to provide resistance to minor axis bending and rotation of wall studs.

A2. INSTALLATION REQUIREMENTS

A2.1 Gypsum board shall be attached to the steel studs in accordance with the appropriate specification.

A2.2 Metal plaster bases shall be attached in accordance with Specification C 841 or Specification C 1063, except screw heads shall be of a size and type suitable for positive (no movement) attachment.

A2.3 When diaphragm rated components are substituted for bridging, they shall be installed prior to loading of the wall. If components are installed on one side of the wall only, then the other stud flanges shall be bridged with suitable bridging. Bridging is not required to be removed when diaphragm rated components are installed.

A2.4 Care shall be taken to allow for additional studs at panel intersections, corners, doors, window, control joints, etc.

A2.5 Provision for structure movement shall be provided as indicated and necessary by design or code requirements.

A2.6 Properly designed splices, cutting of flanges, and holes other than those provided by the manufacturer shall be approved in writing by the engineer of record.

A2.7 Care shall be taken to properly distribute construction materials so that they do not exceed the project design loads.

A3. SUBMITTALS

A3.1 The following items shall be considered for approval prior to delivery of materials to the site:

A3.1.1 *Shop Drawings*—Drawings shall illustrate materials; shop coatings; steel thickness; details of fabrication; details of attachment to adjoining work; size, location, spacing of fasteners for attaching framing to itself; details of attachment to the structure; accessories and their installation; and critical installation procedures. Drawings shall include plans, elevations, sections and details.

A3.1.2 *Samples*—Samples shall be representative pieces of all framing component parts and accessories.

A3.1.3 *Certification*—Certification shall be a statement from the manufacturer certifying that the materials conform to the appropriate requirements as outlined in the contract documents.

A3.1.4 *Calculations*—Engineering calculations shall be prepared verifying the assembly's ability to meet or exceed design requirements as required by local codes and authorities.

A3.1.5 *Descriptive Literature*—Manufacturer's literature shall contain product and installation specifications and details.

SUMMARY OF CHANGES

This section identifies the location of changes to this specification that have been incorporated since the last issue. Committee C-11 has highlighted those changes that affect the technical interpretation of use of these test methods.

(*1*) Revised paragraph A2.1.

Designation: C 1012 – 95a

Standard Test Method for Length Change of Hydraulic-Cement Mortars Exposed to a Sulfate Solution[1]

This standard is issued under the fixed designation C 1012; the number immediately following the designation indicates the year of original adoption or, in the case of revision, the year of last revision. A number in parentheses indicates the year of last reapproval. A superscript epsilon (ε) indicates an editorial change since the last revision or reapproval.

1. Scope

1.1 This test method covers the determination of length change of mortar bars immersed in a sulfate solution. Mortar bars made using mortar described in Test Method C 109/C 109M are cured until they attain a compressive strength of 20.0 ± 1.0 MPa (3000 ± 150 psi), as measured using cubes made of the same mortar, before the bars are immersed.

1.2 The values stated in SI units are to be regarded as the standard. The values shown in parentheses are for information purposes only.

1.3 *This standard does not purport to address all of the safety concerns, if any, associated with its use. It is the responsibility of the user of this standard to establish appropriate safety and health practices and determine the applicability of regulatory limitations prior to use.*

2. Referenced Documents

2.1 *ASTM Standards:*

C 109/C 109M Test Method for Compressive Strength of Hydraulic Cement Mortars (Using 50-mm (2-in.) Cube Specimens)[2]

C 150 Specification for Portland Cement[3]

C 157 Test Method for Length Change of Hardened Hydraulic-Cement Mortar and Concrete[3]

C 215 Test Method for Fundamental Transverse, Longitudinal, and Torsional Frequencies of Concrete Specimens[3]

C 305 Practice for Mechanical Mixing of Hydraulic Cement Pastes and Mortars of Plastic Consistency[2]

C 348 Test Method for Flexural Strength of Hydraulic Cement Mortars[2]

C 349 Test Method for Compressive Strength of Hydraulic Cement Mortars (Using Portions of Prisms Broken in Flexure)[2]

C 452 Test Method for Potential Expansion of Portland Cement Mortars Exposed to Sulfate[2]

C 490 Practice for Use of Apparatus for the Determination of Length Change of Hardened Cement Paste, Mortar, and Concrete[2]

C 595 Specification for Blended Hydraulic Cements[2]

C 597 Test Method for Pulse Velocity Through Concrete[3]

C 618 Specification for Fly Ash and Raw or Calcined Natural Pozzolan for Use as a Mineral Admixture in Portland Cement Concrete[3]

C 684 Test Method for Making, Accelerated Curing, and Testing of Concrete Compression Test Specimens[3]

C 778 Specification for Standard Sand[2]

C 917 Test Method for Evaluation of Cement Strength Uniformity from a Single Source[2]

C 989 Specification for Ground Iron Blast-Furnace Slag for Use in Concrete and Mortars[3]

D 1193 Specification for Reagent Water[4]

E 18 Test Methods for Rockwell Hardness and Rockwell Superficial Hardness of Metallic Materials[5]

3. Significance and Use

3.1 This test method provides a means of assessing the sulfate resistance of concretes and mortars made using portland cement, blends of portland cement with pozzolans or slags, and blended hydraulic cements. Method C 452 is suitable for evaluating portland cements but not blended cements or blends of portland cement with pozzolans or slags.

3.2 The standard exposure solution used in this test method, unless otherwise directed, contains 352 moles of Na_2SO_4 per m^3 (50 g/L). Other sulfate concentrations or other sulfates such as $MgSO_4$ may be used to simulate the environmental exposure of interest. Further discussion of these and other technical issues is given in the Appendix.

4. Apparatus

4.1 *Mixer*, conforming to the requirements of Practice C 305.

4.2 *Cube Molds*, conforming to the requirements of Test Method C 109.

4.3 *Bar Molds*, conforming to the requirements of Specification C 490.

4.4 *Comparator*, conforming to the requirements of Specification C 490.

4.5 *Containers*—The containers in which the bars are immersed may be plastic, glass, ceramic, or metal. A means for supporting the bars shall be included so that no end or side of a bar rests against the container. A lid shall be included that can be sealed on so that the sulfate solution cannot evaporate.

4.6 *Curing Tank*, conforming to the requirements of Test Method C 684.

---

[1] This test method is under the jurisdiction of ASTM Committee C-1 on Cement and is the direct responsibility of Subcommittee C01.29 on Sulfate Resistance.

Current edition approved Oct. 10, 1995. Published December 1995. Originally published as C 1012 – 1984. Last previous edition C 1012 – 95.

[2] *Annual Book of ASTM Standards*, Vol 04.01.

[3] *Annual Book of ASTM Standards*, Vol 04.02.

[4] *Annual Book of ASTM Standards*, Vol 11.01.

[5] *Annual Book of ASTM Standards*, Vol 03.01.

5. Reagents and Materials

5.1 *Purity of Reagents*—USP or technical grade chemicals may be used, provided it is established that any reagent used is of sufficiently high purity to permit its use without lessening the accuracy of the determination. When tests are made that are expected to produce results that are close to an acceptance-rejection value, it is recommended that reagent grade chemicals be used. Such chemicals shall conform to the specifications of the Committee on Analytical Reagents for the American Chemical Society where such specifications are available.[6]

5.2 *Purity of Water*—Unless otherwise indicated, references to water shall be understood to mean reagent water conforming to Type IV of Specification D 1193.

5.3 *Sodium Sulfate* (Na_2SO_4)—The water content shall be checked by loss on ignition each time the solution is prepared. Any anhydrous or hydrated sodium sulfate may be used if the water content of the salt is checked by loss on ignition and proper corrections made to account for the specified sulfate concentration.

5.4 *Sulfate Solution*—Each litre of solution shall contain 50.0 g of Na_2SO_4 dissolved in 900 mL of water, and shall be diluted with additional distilled or deionized water to obtain 1.0 L of solution. Mix the solution on the day before use, cover, and store at 23 ± 1.7°C (73.4 ± 3°F). Determine the pH of the solution before use; reject the solution if the pH range is outside 6.0 to 8.0. The volume proportion of sulfate solution to mortar bars in a storage container shall be 4 ± 0.5 volumes of solution to 1 volume of mortar bars.

NOTE 1—The volume of a mortar bar may be taken as 184 mL (11.25 in.^{3}).

5.5 *Materials:*

5.5.1 *Graded Standard Sand*, as specified in Specification C 778.

5.5.2 *Stainless Steel Gage Studs*, as specified in Specification C 490.

6. Preparing Mortars

6.1 Make mortars as described in Test Method C 109, that is, 1 part cement to 2.75 parts of sand by mass. The water-cement ratio by mass shall be 0.485 for all non-air-entraining portland cements and 0.460 for all air-entraining portland cements. The water-cement ratio for non-air-entraining portland-pozzolan (IP) and portland-blast furnace slag (IS) cements shall be 0.485. For blends of portland cement with a pozzolan or slag, the water-cement ratio shall develop a flow within ±5 of that of the portland-cement mortar at a water-cement ratio of 0.485.

7. Specimen Molds

7.1 Prepare the specimen molds in accordance with the requirements of Specification C 490 except, the interior surfaces of the mold shall be covered with a release agent. A release agent will be acceptable if it serves as a parting agent without affecting the setting of the cement and without leaving any residue that will inhibit the penetration of water into the specimen.

NOTE 2—TFE-fluorocarbon tape complies with the requirements for a mold release agent.

8. Procedure

8.1 *Molding and Initial Curing of Specimens*—Mold the test bars in accordance with Test Method C 157. Mold the cubes in accordance with Test Method C 109. A set of specimens to test one cement will consist of 6 bars and up to 21 cubes (Note 3). Immediately after molding, cover the molds with a rigid steel, glass, or plastic plate, seal the plate to the mold so as to be watertight, and place the mold in the curing tank in water at 35 ± 3°C (95 ± 5°F) for 23½ h ± 30 min, as in Test Method C 684, Procedure A (Warm-Water Method). Place the sealed molds in the curing tank with the bottom of the bars as cast down, that is, in the same relative position in which the bars were cast. At 23½ h ± 30 min, remove molds from tank and demold the specimens.

NOTE 3—The set of cubes will consist of 21 cubes to be tested as described herein when significant information on the strength development rate is not available. When information is available (as for example, from the use of the procedures of Test Method C 917) that would justify making fewer cubes, only those needed to confirm the time the mortar achieves 20.0 ± 1.0 MPa (3000 ± 150 psi) will be needed.

8.2 *Subsequent Curing and Preparation for Test*—After demolding, store all bars and cubes, except the two to be broken, in a curing tank of saturated limewater at 23 ± 1.7°C (73.4 ± 3°F). Break two cubes in compression in accordance with Test Method C 109 after demolding when the specimens have cooled to ambient temperature under moist cloths. If the mean strength of the two cubes is 20 MPa (2850 psi) or more, observe and record comparator readings in accordance with Specification C 490 and as prescribed in the section on Measurements of Length Change and place all the bars in the sulfate solution. If 20 MPa (2850 psi) is not achieved, store the demolded cubes and mortar bars in the curing tank and test additional cubes. Predict from the first two cubes when a compressive strength of at least 20 MPa (2850 psi) will be reached. Verify the prediction, and at that time observe and record comparator readings and place all the bars in the sulfate solution (Note 4). This measurement is designated as the initial length. The storage temperature and test temperature shall be 23 ± 1.7°C (73.4 ± 3°F).

NOTE 4—If the value for strength at 24 h is less than 20.0 MPa (2850 psi) and additional testing on the same day is not possible, or, is unlikely to yield a value over 20.0 MPa (2850 psi) and the strength is over 21 MPa (3150 psi) when tested early the next day, it is not necessary to remake the batch.

8.3 *Measurements of Length Change*—At 1, 2, 3, 4, 8, 13, and 15 weeks after the bars are placed in the sulfate solution, test them for length change using the length comparator in accordance with Specification C 490. At 15 weeks the data should be reviewed. If slight, gradual, and uniform length change is taking place, the next measurements should be made at 4, 6, 9, and 12 months. When the expansion is

[6] *Reagent Chemicals, American Chemical Society Specifications*, American Chemical Society, Washington, DC. For suggestions on the testing of reagents not listed by the American Chemical Society, see *Analar Standards for Laboratory Chemicals*, BDH Ltd., Poole, Dorset, U.K., and the *United States Pharmacopeia and National Formulary*, U.S. Pharmaceutical Convention, Inc. (USPC), Rockville, MD.

changing rapidly at any period in the test, the interval between readings should be adjusted so that it is short enough to permit observing and reporting the behavior of the bars.

8.3.1 *Details of Measurement of Bars for Length Change:*

8.3.1.1 Clean the hole in the base of the comparator into which the gage stud on the lower end of the bar fits (this hole tends to collect water and sand and should be cleaned after every reading). Read and record the comparator indication of the length of the reference bar. Take one bar out of immersion, blot the pins, put the bar in the comparator, read, and record the indication. Return the bar to immersion and clean the hole in the base of the comparator. Take out the second bar and treat it in a like manner. Return the second bar to immersion, record the reading, and clean the hole in the base of the comparator. Continue the procedure until all bars have been read, returned to immersion, and the readings recorded, cleaning the hole in the bottom of the comparator each time. After reading the last bar, clean the hole in the comparator base and read and record the reference-bar indication.

8.3.1.2 When, based on cube strength of mortar, the bars are initially ready to be stored in sulfate solution, or after they have been removed from sulfate solution storage for length change test, they shall be placed in storage in fresh sulfate solution of a known pH of 6-8. At subsequent readings for length change, proceed as described above; cleaning the socket in the base of the comparator before reading the reference bar initially and after reading each mortar bar. Record reference bar and mortar bar readings. Also, read reference bar and record reading after the last bar. Blot only around the pins (Note 5). After reading each bar return it to used solution. Prior to replacing the solution, rinse the container once with water, pouring out water and debris. Replace the frame holding the bars in the container, fill the container with new solution of known pH of 6-8 to immerse bars, and secure on the container lid.

NOTE 5—The purpose of the minimal blotting of the pins and no blotting of the bars is to avoid drying and shrinkage of the bars. It has been observed that if the pins are blotted, and the bar placed in the comparator and the dial read, and the bar is then wiped gently with a dry cloth, the bar will shrink measurably. Therefore, drying should be minimized.

8.3.2 *Examination of Specimens After Measuring Length Change*—When the bars seem to have behaved in an unusual way or when the test is part of a research study, the specimens shall be tested for warping by being placed on a plane surface so that the ends are curved down and the maximum bowing measured. Cracking (presence, location, type); surface deposits, mottling, exudations (nature, thickness, type) shall be noted.

8.4 *Tolerance on Time*—All references to elapsed time in 8.3 are intended to have a tolerance of ±2 %.

9. Calculation

9.1 Calculate the length change at any age as follows:

$$\Delta L = \frac{L_x - L_i}{Lg} \times 100$$

where:

ΔL = change in length at x age, %,

L_x = comparator reading of specimen at x age—reference bar comparator reading at x age, and

L_i = initial comparator reading of specimen-reference bar comparator reading, at the same time

Lg = nominal gage length, or 250 mm (10 in.) as applicable. (See C 490).

9.2 Calculate length change values for each bar to the nearest 0.001 % and report averages to the nearest 0.01 %.

TABLE 1 Maximum Permissible Range of Values

Remaining No. of Specimens	Blended Cements	Portland Cement
3	0.034	0.010
4	0.037	0.011
5	0.039	0.012
6	0.041	0.012

10. Report

10.1 Report type of cement, and, if blending material is used, its identification and amount and whether cement and blending material, if used, meet the applicable specifications. Report the initial comparator reading of each bar at the time of immersion in sulfate solution; this is the base value for calculation of length changes. Report the subsequent length changes in percent of base value to the nearest 0.001 % for individual bars and the nearest 0.01 % for averages. Comment on the visual condition of bars at end of the test. With each report of mean length change of bars at a particular age, report number of bars (n), standard deviation of length change, and coefficient of variation of length change. Data from at least three bars must be available at any age to constitute a valid test at that age. Depending upon how many remaining specimens there are, the maximum permissible range of the values must not exceed the values of length change in percent in Table 1.

10.2 Report any variation from the procedure outlined in the text as it relates to solution composition, concentration, or temperature, mortar proportions, age, or maturity, and the like.

11. Precision and Bias

11.1 The precision of this test method has been evaluated by cooperative testing and found to vary with the type of cement studied, hence separate precision statements are given.[7]

11.1.1 *Blended Cements*—The single-operator standard deviation has been found to be 0.010 % for expansions between 0.04 % and 0.07 %. Therefore, results of two properly conducted tests by the same operator on the same blended cement Type IP or IS, should not differ from each other by more than 0.028 %. The multi-laboratory standard deviation has been found to be 0.020 % for expansions between 0.04 %, and 0.07 %. Therefore, results of two properly conducted tests on the same blended cement Type IP or IS in two different laboratories, should not differ from each other by more than 0.056 %.

11.1.2 *Type II Cement*—The single-operator standard deviation has been found to be 0.005 %, for expansions

[7] A research report is available from ASTM Headquarters. Request RR: C01 - 1003.

between 0.04 % and 0.07 %. Therefore, results of two properly conducted tests by the same operator on the same cement, should not differ from each other by more than 0.014 %. The multi-laboratory standard deviation has been found to be 0.020 % for expansions between 0.04 % and 0.07 %. Therefore, results of two properly conducted tests on the same cement in two different laboratories, should not differ from each other by more than 0.056 %.

11.1.3 *Type V Cement*—The single-operator standard deviation has been found to be 0.003 % for expansions between 0.04 % and 0.07 %. Therefore, results of two properly conducted tests by the same operator on the same cement, should not differ from each other by more than 0.009 %. The multilaboratory standard deviation has been found to be 0.010 % for expansions between 0.04 % and 0.07 %. Therefore, results of two properly conducted tests on the same cement in two different laboratories should not differ from each other by more than 0.028 %.

11.2 *Bias*—Since there is no accepted reference material suitable for determining the bias for the procedure in this test method, no statement on bias is being made.

12. Keywords

12.1 acceptability; expansion; fly ash; mortars; pozzolans; precision; search report; slag; sulfate attack; sulfate-resisting cements; tests

APPENDIX

(Nonmandatory Information)

X1. Technical Background

X1.1 Test Method C 1012 was developed to explore the suitability of a test method for sulfate resistance based on length-change measurements of molded prisms immersed in sulfate solution. This approach was taken after it was established that Test Method C 452, in which the sulfate is added to the mortar as it is mixed, and the prisms stored in fresh water, was not able properly to evaluate sulfate resistance of blended cements and blends of portland cement with pozzolan or slag.

X1.2 This test method has been subjected to two rounds of cooperative testing. The first program involved five cements (Types I, II, and V of Specification C 150, and Types IS and IP of Specification C 595). The test solution contained both sodium sulfate and magnesium sulfate, each in the amount of 0.176 mol/L. A report of this work is available. A paper giving results using this test method has been published **(1)**.[8] The second program involved eight blends using a Type I and a Type II cement with 50 and 70 % slag, 25 % Class F, and 35 % Class C fly ash (ASTM C 618). Two test solutions were used: one contained 0.303 mol/L sodium sulfate and 0.049 % magnesium sulfate; the other contained 0.352 mol/L sodium sulfate only.

X1.3 This test method involves length change of hydraulic-cement mortar bars immersed in a sulfate solution. Mortar bars are placed in the sulfate solution after companion mortar cubes have reached a compressive strength of 20.0 ± 1.0 MPa (3000 ± 150 psi). Attempts to use Test Method C 452 for blended cements and blends of portland cement and slag (Specification C 989) or pozzolan (Specification C 618) with portland cement failed to yield results that correlate with field and laboratory experience where the mortar or concrete developed some maturity prior to being exposed to external sulfate solution **(2)**. Since some blended cements and some blends of slag and pozzolan with portland cement have been successfully used where sulfate resistance was needed, they are tested under circumstances which permit comparisons to be based on results obtained when all the sulfate exposures are begun at equivalent strength since, in practice, it is likely that the concrete will be at about the same strength when sulfate attack begins regardless of the type of cementitious medium employed. This test method is also applicable for use in evaluating portland cements.

X1.4 The solution of 0.176 mol/L of each of the two sulfates (magnesium and sodium) studied in the first cooperative program, is believed likely to adversely affect blends or blended cements containing slag, due to a magnesium ion **(3, 4)** attack, this attack being unrelated to sulfate attack. Therefore, this solution should not be used to evaluate such systems unless they are to be exposed in service to attack in which the magnesium ion (Mg^{++}) is present in amounts approaching that of the test solution (4200 ppm).

X1.5 This test method does not simulate the mechanism of sulfate attack by solutions of sulfate compositions other than that used. If evaluation of behavior due to exposure to a given sulfate solution is desired, that solution should be used.

X1.6 The user may therefore modify this test method to use any relevant compositions or concentrations of aggressive solutions, other ages or degrees of maturity as the basis for beginning the exposure, mortars of different proportions, or different or additional means of assessing the influence of the aggressive solution on the specimens. Much work was done using mortar bars of lower strength (higher porosity) in previous cooperative tests sponsored by Committee C-1 **(5–7)**. A procedure in which sodium sulfate is kept at constant alkalinity/acidity has been proposed **(8)**. Other methods of assessing the influence of the aggressive solutions that have been used or proposed include (*a*) change in pulse velocity (Test Method C 597), (*b*) change in resonant frequency (Test Method C 215), (*c*) change in compressive strength (Test Methods C 109 and C 349), (*d*) change in flexural strength (Test Method C 348), (*e*) change in mass, and (*f*) change in hardness (Test Methods E 18).

X1.7 Work reported by Polivka and Brown **(9)** in 1958

[8] The boldface numbers in parentheses refer to the list of references at the end of this test method.

and by Mehta and Polivka in 1975 **(10)** included tests of concrete exposed to a mixed sulfate solution containing 5 % each of sodium and magnesium sulfate. Other relevant work is cited in the references **(1 to 19)**.

X1.8 The second cooperative study used both a mixed sodium/magnesium sulfate solution and a sodium sulfate solution. Precision of the results and the ranking of the cements did not differ between the solutions. Hence, the simpler test, using sodium sulfate, was selected as the standard.

REFERENCES

(1) Rosnerl, J. C., Chehovits, J. G., and Wharburton, R. G., "Sulfate Resistance of Mortars Using Fly Ash as a Partial Replacement for Portland Cement," *Proceedings*, Sixth International Conference on Utilization of Fly Ash, Reno, March 1982.

(2) Mather, Bryant, "Laboratory Tests of Portland Blast-Furnace Slag Cements," *Journal of the American Concrete Institute*, Proceedings, Vol 54, 1957, pp 205–232.

(3) Biczok, I. *Concrete Corrosion, Concrete Protection,* Chemical Publishing Company, New York, 1967, p 178.

(4) Miller, D. G., and Snyder, C. G., "Report on Comparative Short-Time Tests for Sulfate Resistance of 121 Commercial Cements," Report of Committee C-1 on Cement, Appendix III, *Proceedings*, ASTM, Vol 45, 1945, pp 165–194.

(5) Wolochow, D., "Determination of the Sulfate Resistance of Portland Cement," Report of Committee C-1 on Cement, Appendix, *Proceedings*, ASTM, Vol 52, 1952, pp. 250–363.

(6) Wolochow, D., "A Lean Mortar Bar Expansion Test for Sulfate Resistance of Portland Cements," Appendix A, *Proceedings*, ASTM, Vol 52, 1952, pp 264–265.

(7) Mather, Katharine, "Tests and Evaluation of Portland and Blended Cements for Resistance to Sulfate Attack," *ASTM STP 663*, 1978, pp 74–86.

(8) Mehta, P. K., "Evaluation of Sulfate-Resisting Cements by a New Test Method," *Proceedings of the American Concrete Institute*, Vol 72, Oct. 1975, pp 573–575.

(9) Polivka, M., and Brown, E. H., "Influence of Various Factors on Sulfate Resistance of Concretes Containing Pozzolan," *Proceedings*, ASTM, Vol 58, 1958, pp 1077–1100.

(10) Mehta, P. K., and Polivka, M., "Sulfate Resistance of Expansive Cement Concretes," American Concrete Inst SP-47, 1975, pp 367–379.

(11) Regourd, M., "The Action of Sea Water on Cements," *Annales de L'Institut Technique du Batiment et des Travaux Publics*, Vol 329, 1975, pp 86–102.

(12) Mehta, P. K., and Haynes, H. H., "Durability of Concrete in Sea Water," *Proceedings of the American Society for Civil Engineers*, Vol 101, No. ST 8, 1975, pp 1679–1686.

(13) Campus, F., "Essais de resistance des Mortiers et Betons à la Mer (1934–1964)," Silicates Industriel, Vol 28, 1963, pp 79–88.

(14) Bakker, R., "On the Cause of Increased Resistance of Concrete Made From Blast Furnace Cement to the Alkali Silica Reaction and to Sulfate Corrosion," (English Translation of doctoral thesis at RWTH 1980), Maastricht, 1981, 144 pp.

(15) Miller, D. G., and Manson, P. W., "Tests of 106 Commercial Cements for Sulfate Resistance," *Proceedings*, ASTM, Vol 40, 1940, pp 988–1001.

(16) Lea, F. M., *The Chemistry of Cement and Concrete*, Third Edition, Chemical Publishing Co., Inc., New York, NY, 1970, pp 727.

(17) Brown, P. W., "An Evaluation of the Sulfate Resistance of Cements in a Controlled Environment," Cement and Concrete Research. Vol 11, 1981, pp 719–727.

(18) Patzias, Terry, "Evaluation of Sulfate Resistance of Hydraulic-Cement Mortars by the ASTM C 1012 Test Method. *Concrete Durability*, Katharine and Bryant Mather International Conference, American Concrete Institute SP-100, Vol 2, 1987, pp 2103–2120.

(19) Patzias, T., "The Development of ASTM C 1012 with Recommended Acceptance Limits for Sulfate Resistance of Hydraulic Cement," Cement, Concrete, and Aggregates, CCA GDP, Vol 13, No. 1, Summer 1991, pp. 50–57.

Designation: C 1032 – 96

Standard Specification for Woven Wire Plaster Base[1]

This standard is issued under the fixed designation C 1032; the number immediately following the designation indicates the year of original adoption or, in the case of revision, the year of last revision. A number in parentheses indicates the year of last reapproval. A superscript epsilon (ε) indicates an editorial change since the last revision or reapproval.

1. Scope*

1.1 This specification covers woven wire plaster base, flat or self-furring, with or without stiffener wires, and with or without backing, designed for use as a base to receive gypsum or portland cement based plaster.

1.2 The values stated in inch-pound units are to be regarded as the standard. The SI equivalents of inch-pound units are approximate.

1.3 The text of this standard references notes and footnotes which provide explanatory material. These notes and footnotes (excluding those in tables and figures) shall not be considered as requirements of the standard.

2. Referenced Documents

2.1 *ASTM Standards:*

A 641 Specification for Zinc-Coated (Galvanized) Carbon Steel Wire[2]

C 11 Terminology Relating to Gypsum and Related Building Materials and Systems[3]

2.2 *Federal Specifications:*

UU-B-790a Building Paper, Vegetable Fiber: Kraft, Waterproofed, Water Repellent and Fire Resistant[4]

UU-P-31 B/Gen. Paper; General Specifications and Methods of Testing[4]

UU-P-31 B/171 Tensile Breaking Strength (Dry)[4]

3. Terminology

3.1 *Definitions*—Definitions shall be in accordance with Terminology C 11.

4. Material

4.1 In accordance with Specification A 641 and Table 1.

4.2 *Kraft Building Paper*, Federal Specification UU-B-790a.

5. Physical Properties

5.1 *Backing:*

5.1.1 Factory attached backing shall provide a one mesh wire to wire lap joint at one end and one edge of each flat sheet or roll. Attachment of the backing shall allow lapping of wire-to-wire and backing-to-backing of one mesh at ends and edges and shall permit full embedment in not less than ¼ in. (6.4 mm) of plaster of not less than one half the total length and width of the wire.

5.1.2 Backing shall have a minimum tensile breaking strength (dry) of 20 lbf/in. (3.5 kN/m) width, as tested in accordance with Federal Specification UU-P-31 B, Method 171.

6. Dimensions, Weight, and Permissible Variations

6.1 *Openings:*

6.1.1 Woven wire plaster base shall be fabricated to provide the size of openings in accordance with Table 1, except that any single opening shall be not more than 3.0 in.^{2} (1935.5 mm^2).

6.1.2 Wire diameter for opening size shall be in accordance with Table 1.

6.2 Diameter tolerance for galvanized wire shall be in accordance with Table 3 of Specification A 641.

6.3 Weight for woven wire plaster base shall be in accordance with Table 2.

6.4 When included as part of the woven wire plaster base, factory fabricated stiffener wires shall have a diameter not less than that specified in Table 1 and shall be continuous, parallel to the long dimension of the plaster base, and shall occur at not more than 6-in. (154.4-mm) intervals.

6.5 Factory fabricated crimps for self-furring woven wire plaster base shall provide a nominal ¼-in. (6.4-mm) separation between the back plane of the plaster base and the surface of solid backing to permit embedment in both directions.

6.6 Width of woven wire plaster base shall be not less than 36 in. (914 mm).

6.7 Length of woven wire plaster base shall be in accordance with Table 3.

7. Inspection

7.1 Inspection of the material shall be agreed upon between the purchaser and the producer or supplier as part of the purchase agreement.

8. Rejection and Rehearing

8.1 Material that fails to conform to the requirements of this specification shall be reported to the producer or supplier promptly and in writing. The notice of rejection shall contain a statement documenting how the product has failed to conform to the requirements of the specification.

---

[1] This specification is under the jurisdiction of ASTM Committee C-11 on Gypsum and Related Building Materials and Systems and is the direct responsibility of Subcommittee C11.02 on Specifications and Test Methods for Accessories and Related Products.

Current edition approved May 10, 1996. Published July 1996. Originally published as C 1032 – 85. Last previous edition C 1032 – 95$^{\epsilon 1}$.

[2] *Annual Book of ASTM Standards*, Vol 01.06.

[3] *Annual Book of ASTM Standards*, Vol 04.01.

[4] Available from U.S. Government Printing Office, Washington, DC 20402.

*** A Summary of Changes section appears at the end of this specification.**

TABLE 1 Minimum Wire Diameter for Size of Opening

Opening Size, in. (mm)	Wire Diameter, in. (mm)
1 (25.4)	0.035 (0.89)
1½ (38.0)	0.0510 (1.295)

TABLE 2 Minimum Weight for Size of Opening[A]

Opening Size, in. (mm)	Weight per Square Yard, lb/yd^2 (kg/m^2)
1 (25.4)	0.86 (0.466)
1½ (38.0)	1.40 (0.760)

[A] The weight of attached backing is not included.

TABLE 3 Minimum Length of Flat Sheets and Rolls[A]

Flat Sheets	100 linear in. (2.540 m)
Rolls	150 linear ft. (45.72 m)
Rolls, with backing	100 linear ft. (30.48 m)

[A] Allowable variations shall be not more than −¾ in. (−19.0 mm) in width or length.

9. Certification

9.1 When specified in the purchase order or contract, the producer or supplier shall furnish a report certifying that at the time of shipment, the materials were in compliance with the requirements of this specification.

10. Packaging and Package Marking

10.1 *Packaging*—Woven wire plaster base shall be packaged in bundles, rolls or pallets.

10.2 *Marking*—When shipped for sale or resale, the name of the producer or the supplier and the brand and description of the contents shall be legibly marked on each pallet, bundle or roll. The quantity for bundles shall be square yards and number of sheet per bundle. The quantity for pallets, bundles or rolls shall indicate the mesh-opening size and wire thickness.

11. Keywords

11.1 backing; base; cement plaster; flat; gypsum plaster; plaster; self-furning; woven wire

SUMMARY OF CHANGES

This section identifies the location of changes to this specification that have been incorporated since the last issue. Committee C-11 has highlighted those changes that affect the technical interpretation or use of this specification.

(1) Added paragraph 1.3.

Designation: C 1038 – 95

Standard Test Method for Expansion of Portland Cement Mortar Bars Stored in Water[1]

This standard is issued under the fixed designation C 1038; the number immediately following the designation indicates the year of original adoption or, in the case of revision, the year of last revision. A number in parentheses indicates the year of last reapproval. A superscript epsilon (ε) indicates an editorial change since the last revision or reapproval.

1. Scope

1.1 This test method covers the determination of the expansion of mortar bars made from portland cement, of which sulfate is an integral part. This test method applies only to portland cements.

1.2 The values stated in SI units are to be regarded as the standard. Values in parentheses are for information only.

1.3 *This standard does not purport to address all of the safety concerns, if any, associated with its use. It is the responsibility of the user of this standard to establish appropriate safety and health practices and determine the applicability of regulatory limitations prior to use.*

2. Referenced Documents

2.1 *ASTM Standards:*

C 109/C 109M Test Method for Compressive Strength of Hydraulic Cement Mortars (Using 2-in. or 50-mm Cube Specimens)[2]

C 305 Practice for Mechanical Mixing of Hydraulic Cement Pastes and Mortars of Plastic Consistency[2]

C 490 Practice for Use of Apparatus for the Determination of Length Change of Hardened Cement Paste, Mortar, and Concrete[2]

C 511 Specification for Moist Cabinets, Moist Rooms, and Water Storage Tanks Used in the Testing of Hydraulic Cements and Concretes[2]

C 778 Specification for Standard Sand[2]

C 1005 Specification for Weights and Weighing Devices for Use in the Physical Testing of Hydraulic Cements[2]

D 1193 Specification for Reagent Water[3]

2.2 *National Standard of Canada:*

CAN 3-A5-M83 Portland Cements[4]

3. Significance and Use

3.1 This test method, when applied to portland cement containing calcium sulfate, determines the amount of expansion of a mortar bar when it is stored in water. The amount of mortar bar expansion is related to the amount of calcium sulfate in the cement; expansion becomes excessive when the cement contains too much calcium sulfate.

3.2 When desired, cement specifications may limit the amount of calcium sulfate contained in a portland cement by requiring that the amount of expansion in water not exceed a specified value.

NOTE 1—An expansion limit of 0.020 % in fourteen days of water immersion is in use in Canadian Standards document CAN 3-A5-M83.

4. Apparatus

4.1 *Weights and Weighing Devices,* conforming to the requirements of Specification C 1005. The weighing device shall be evaluated for precision and accuracy at a total load of 2 kg.

4.2 *Glass Graduates, Molds, and Length Comparator,* conforming to the requirements of Specification C 490.

4.3 *Moist Cabinet or Room,* conforming to the requirements of Specification C 511.

4.4 *Mixer, Bowl, and Paddle,* conforming to the requirements of Practice C 305.

4.5 *Trowel and Tamper,* conforming to the requirements of Test Method C 109.

5. Temperature and Humidity

5.1 *Molding Room, Dry Materials, and Mixing Water*—The temperature of the molding room, dry materials, and mixing water shall be maintained between 20 and 28°C (68 and 82.4°F) and the relative humidity of the molding room shall not be less than 50 %.

6. Reagents and Materials

6.1 *Mixing Water*—Potable water is satisfactory for routine tests. For all cooperative tests and in case of dispute, reagent water conforming to Type III B of Specification D 1193 shall be used.

6.2 *Graded Sand*—The graded sand for making the test specimens shall conform to the requirements for graded standard sand in Specification C 778.

7. Procedure

7.1 *Number and Dimensions of Test Specimens*—Make four 25 by 25 by 285-mm (1 by 1 by 11¼-in.) test specimens for each cement.

NOTE 2—In routine tests, 25 by 25 by 160-mm (1 by 1 by 6¼-in.) specimens may be used; however, in case of dispute, results obtained with 25 by 25 by 285-mm (1 by 1 by 11¼-in.) specimens shall govern.

7.2 *Specimen Molds*—Prepare the specimen molds in accordance with the requirements of Specification C 490, except the interior surfaces of the mold shall be covered with a release agent. A release agent will be acceptable if it serves as a parting agent without affecting the setting of the cement and without leaving any residue which will inhibit the penetration of water into the specimen.

---

[1] This test method is under the jurisdiction of ASTM Committee C-1 on Cement and is the direct responsibility of Subcommittee C01.28 on Sulfate Content.

Current edition approved Dec. 10, 1995. Published March 1996. Originally published as C 1038 – 85. Last previous edition C 1038 – 89.

[2] *Annual Book of ASTM Standards,* Vol 04.01.

[3] *Annual Book of ASTM Standards,* Vol 11.01.

[4] Available from Canada Standards Association, 178 Rexdale Blvd., Rexdale, Ontario Canada M9W 143.

NOTE 3—TFE-fluorocarbon tape complies with the requirements for a mold release agent.

7.3 *Proportioning, Consistency, and Mixing of Mortar:*

7.3.1 The quantity of dry materials required to prepare four 25 by 25 by 285-mm (1 by 1 by 11¼-in.) specimens is 500 g of cement and 1375 g of graded sand. The amount of mixing water is 243 mL for all non-air-entraining portland cements and 230 mL for all air-entraining portland cements.

7.3.2 Mix the mortar in accordance with the procedure described in Practice C 305.

7.4 *Molding Test Specimens*—Immediately upon completion of mixing, remove the paddle and the bowl from the mixer and shake the excess mortar from the paddle into the bowl. Fill the mold in two layers, compacting each layer with the tamper. Work the mortar into the corners, around the gage studs, and along the surfaces of the mold with the tamper until a homogeneous specimen is obtained. After the top layer has been compacted, cut the mortar off flush with the top of the mold and smooth the surface with a few strokes of the trowel.

8. Storage of Test Specimens

8.1 *Initial Storage*—Immediately upon completion of molding, place the test specimens in the moist cabinet or moist room. Keep all test specimens in the molds in the moist cabinet or moist room for 22 to 23 h with upper surfaces exposed to the moist air but protected from dripping water. After this time, remove the specimens from the molds, properly identify, and place in water maintained at 23 ± 2°C (73.4 ± 3.6°F) for at least 30 min prior to making the initial measurement.

8.2 *Subsequent Storage*—After the initial measurement, store the test specimens in saturated lime water in a tank in the moist cabinet or moist room. The specimens shall be covered with at least 5-mm (¼-in.) of water during storage. Keep the storage water clean by changing as necessary.

9. Length Measurement

9.1 Remove the specimens from water storage, one at a time, and wipe with a damp cloth before measuring. Measure the specimens for length by means of the length comparator.

9.2 Make the first reading at the age of 24 h ± 15 min from the time the cement and water are mixed together. Measure the specimens again at the age of 14 days.

NOTE 4—Additional information of value may be obtained by making measurements of the specimens at ages earlier and later than 14 days.

10. Calculation and Report

10.1 Calculate the difference in length of the specimen at 24 h and at 14 days to the nearest 0.001 % of the effective gage length and report as the expansion of the specimen at that period. Report the average of the four specimens.

11. Precision and Bias

11.1 *Precision*—The following precision statements are applicable when replica samples are made using the same cement.

11.1.1 The single-operator standard deviation has been found to be 0.00165 % for expansions between 0.0075 and 0.011 %. Therefore, results of two properly conducted tests by the same operator on the same material should not differ from each other by more than 0.005 %.

11.1.2 The multi-laboratory standard deviation has been found to be 0.00287 % for expansions between 0.0075 and 0.0115 %. Therefore, results of two properly conducted tests on the same material in two different laboratories should not differ from each by more than 0.008 %.

11.2 *Bias*—Since there is no accepted reference material suitable for determining the bias for the procedure in Test Method C 1038 for measuring the expansion of the bar specimens, no statement on bias is being made.

12. Keywords

12.1 calcium sulfate; hydraulic cement; sulfate content

Designation: C 1047 – 98

Standard Specification for Accessories for Gypsum Wallboard and Gypsum Veneer Base[1]

This standard is issued under the fixed designation C 1047; the number immediately following the designation indicates the year of original adoption or, in the case of revision, the year of last revision. A number in parentheses indicates the year of last reapproval. A superscript epsilon (ε) indicates an editorial change since the last revision or reapproval.

1. Scope*

1.1 This specification covers accessories used in conjunction with assemblies of gypsum wallboard and gypsum veneer plaster to protect edges and corners and to provide architectural features (see Fig. 1).

1.2 The values stated in inch-pound units are to be regarded as the standard. The SI units in parentheses are provided for information purposes only. The SI equivalents of inch-pound units may be approximate.

2. Referenced Documents

2.1 *ASTM Standards:*

A 463 Specification for Steel Sheet, Aluminum-Coated, by Hot-Dip Process
A 591/A 591M Specification for Steel Sheet, Electrolytic Zinc-Coated, for Light Coating Mass Applications[2]
A 653/A 653M Specification for Steel Sheet, Zinc-Coated (Galvanized) or Zinc-Iron Alloy-Coated (Galvannealed) by the Hot-Dip Process[2]
B 69 Specification for Rolled Zinc[3]
B 117 Practice of Salt Spray (Fog) Apparatus[4]
C 475 Specification for Joint Compound and Joint Tape for Finishing Gypsum Board[5]
C 587 Specification for Gypsum Veneer Plaster[5]
D 1788 Specification for Rigid Acrylonitrile-Butadiene-Styrene (ABS) Plastics[6]
D 2092 Guide for Preparation of Zinc-Coated (Galvanized) Steel Surfaces for Painting[7]
D 3678 Specification for Rigid Poly(Vinyl Chloride) (PVC) Interior Profile Extrusions[8]

3. Terminology

3.1 *Definition:*

3.1.1 *flange*—that part of the accessory extending out on the face of the gypsum board.

3.2 *Definitions of Terms Specific to This Standard:*

3.2.1 *accessories*—cornerbeads, edge trims, and control joints, such as casing beads, bull noses, and stops.

3.2.2 *control joint*—a formed product used for designed or required separations between adjacent surfaces of gypsum boards or gypsum veneer base.

3.2.3 *cornerbead*—a formed metal, plastic, or metal and paper angle for outside corners of gypsum boards or gypsum veneer base.

3.2.4 *edge trim*—typically "J"- or "L"-shaped strip, as shown in Fig. 1, formed of metal, plastic, or metal and paper to cover exposed ends or edges of gypsum board or gypsum veneer base.

4. Materials and Manufacture

4.1 Steel accessories and steel components of accessories manufactured from steel and paper in combination shall be manufactured from zinc-coated cold-roll (coil or cut length) sheet steel not less than 0.012-in. (0.30-mm) thick before application of coating.

4.1.1 Sheet steel, zinc-coated by the hot-dip process, shall be in accordance with Specification A 653/A 653M, minimized spangle, minimum G-30 coating.

4.1.2 Sheet steel, zinc-coated by the electrolytic process shall be in accordance with Specification A 591/A 591M, minimum coating on surface side to be 0.000140 in. (0.00356 mm), minimum coating on reverse side to be 0.00010 in. (0.00254 mm).

4.1.3 Sheet steel, aluminum-coated, shall be in accordance with Specification A 463 minimum T1-40 coating.

4.1.4 Phosphatizing (as specified in Method A of Guide D 2092) or other surface treatments may be used to ensure compatibility and bond as specified in Section 5.

4.2 Zinc accessories shall be manufactured from rolled zinc in accordance with Specification B 69, Type I, not less than 0.012-in. (0.305-mm) thick.

4.3 Plastic for accessories shall be manufactured from rigid PVC or ABS plastic not less than 0.012 in. (0.305 mm).

4.3.1 PVC, Specification D 3678, Class II or III.

4.3.2 ABS, Specification D 1788 - 81, all limits 5-3-3-1-1.

[1] This specification is under the jurisdiction of ASTM Committee C-11 on Gypsum and Related Building Materials and Systems and is the direct responsibility of Subcommittee C11.02 on Specifications and Test Methods for Accessories and Related Products.

Current edition approved April 10, 1998. Published December 1998. Originally published as C 1047 – 92. Last previous edition C 1047 – 95.

[2] *Annual Book of ASTM Standards,* Vol 01.06.
[3] *Annual Book of ASTM Standards,* Vol 02.04.
[4] *Annual Book of ASTM Standards,* Vol 03.02.
[5] *Annual Book of ASTM Standards,* Vol 04.01.
[6] *Discontinued;* see 1989 *Annual Book of ASTM Standards,* Vol 08.02.
[7] *Annual Book of ASTM Standards,* Vol 06.02.
[8] *Annual Book of ASTM Standards,* Vol 08.04.

*** A Summary of Changes section appears at the end of this specification.**

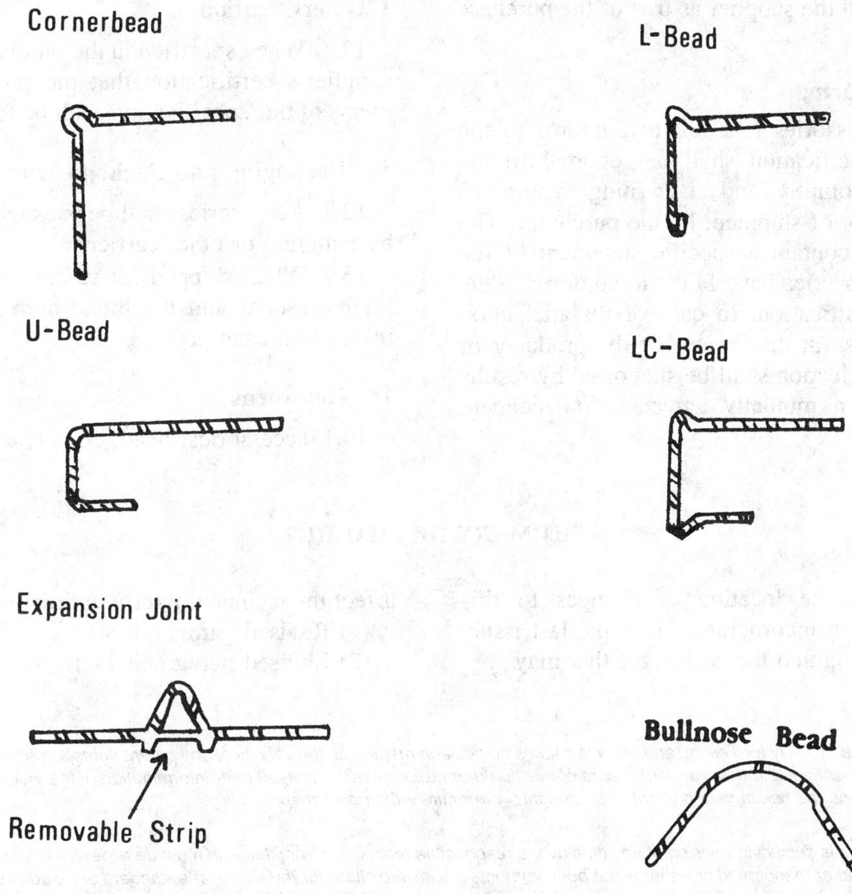

NOTE—Flanges may have punchouts, may be knurled, mesh, deformed, expanded or otherwise shaped for intended use.

FIG. 1 Accessories for Gypsum Wallboard and Gypsum Veneer Base

4.4 Paper components of accessories manufactured from steel and paper in combination shall comply with requirements for thickness, tensile strength, dimensional stability, and bond of joint tape to joint compound as specified in Specification C 475.

5. Physical Properties

5.1 *Compatibility and Bond*—Accessories shall be compatible with and provide a surface bond to the materials specified in Specifications C 475 and C 587.

5.2 *Test Performance*—Steel accessories and steel components of accessories shall not show any red oxidation when tested as specified in Practice B 117 for 192 h.

6. Dimensions and Permissible Variations

6.1 Cornerbeads shall have an interior angle between the flanges no greater than 89°.

6.2 Accessories shall be free of twist or camber of a degree that will prevent their use in the assembly for which they are intended.

6.3 Length tolerances shall be ±3⁄16 in. (±4.8 mm).

6.4 The size of the edge trim shall suit the thickness of the gypsum board used.

6.5 The minimum width of the flange shall be 7⁄8 ± 1⁄32 in. (22 ± 0.8 mm).

7. Appearance

7.1 Edges of accessories shall be free of burrs and sharp edges.

8. Configuration

8.1 Flanges of accessories, when required to be attached to framing members, shall have punch outs to accommodate fastening to the framing members. When required to meet performance requirements, the flanges shall be either knurled, made of mesh, deformed, expanded, or otherwise shaped to meet the performance requirements and intended use. Figure 1 depicts the most commonly used shapes.

NOTE—Other types of accessories designed for special finish application (bull-nose, cove, and so forth) shall be manufactured in accordance with the general requirements of this specification.

9. Sampling

9.1 At least 0.25 % of the number of each type of accessory in a shipment, but not less than three pieces, shall be selected to represent the shipment and shall constitute a sample for the purpose of tests by the purchaser or user.

10. Inspection

10.1 Inspection of the accessories shall be agreed upon

between the purchaser and the supplier as part of the purchase order.

11. Rejection and Rehearing

11.1 Rejection of accessories that fail to conform to the requirements of this specification shall be reported to the producer or supplier promptly and in writing within 10 working days from receipt of shipment by the purchaser. The notice of rejection shall contain a specific statement of the respect in which the accessories have failed to conform to the requirements of this specification. In case of dissatisfaction with the results of the test, at the request of the producer or supplier, such notice of rejection shall be supported by results of a test conducted by a mutually agreeable independent laboratory.

12. Certification

12.1 When specified in the purchase order, a producer's or supplier's certification that the product meets the requirements of this specification shall be furnished to the purchaser.

13. Packaging and Package Marking

13.1 Accessories shall be packaged to ensure safe delivery by common or other carriers.

13.2 When shipped for resale, the name of the manufacturer or seller and the brand and contents shall be legibly marked on each package.

14. Keywords

14.1 accessories; bead; corner; cornerbead; edge

SUMMARY OF CHANGES

This section identifies the location of changes to this specification that have been incorporated since the last issue. Committee C-11 has highlighted those changes that may affect the technical interpretation or use of this specification.

(*1*) Revised paragraph 8.1.

(*2*) Revised paragraph 11.1.

Designation: C 1063 – 99

Standard Specification for Installation of Lathing and Furring to Receive Interior and Exterior Portland Cement-Based Plaster[1]

This standard is issued under the fixed designation C 1063; the number immediately following the designation indicates the year of original adoption or, in the case of revision, the year of last revision. A number in parentheses indicates the year of last reapproval. A superscript epsilon (ε) indicates an editorial change since the last revision or reapproval.

1. Scope

1.1 This specification covers the minimum requirements for lathing and furring for the application of exterior and interior portland cement-based plaster as in Specification C 926.

1.2 Where a fire resistance rating is required for plastered assemblies and constructions, details of construction shall be in accordance with reports of fire tests of assemblies that have met the requirements of the fire rating imposed.

1.3 Where a specific degree of sound control is required for plastered assemblies and constructions, details of construction shall be in accordance with official reports of tests conducted in recognized testing laboratories in accordance with the applicable requirements of Test Method E 90.

1.4 The values stated in inch-pound units are to be regarded as the standard. The SI (metric) values given in parentheses are approximate and are provided for information purposes only.

2. Referenced Documents

2.1 *ASTM Standards:*

A 526/A 526M Specification for Steel Sheet, Zinc-Coated (Galvanized) by the Hot-Dip Process, Commercial Quality[2]

A 641/A 641M Specification for Zinc-Coated (Galvanized) Carbon Steel Wire[3]

B 69 Specification for Rolled Zinc[4]

B 221 Specification for Aluminum-Alloy Extruded Bars, Rods, Wire, Profiles, and Tubes[5]

C 11 Terminology Relating to Gypsum and Related Building Materials and Systems[6]

C 841 Specification for Installation of Interior Lathing and Furring[6]

C 847 Specification for Metal Lath[6]

C 926 Specification for Application of Portland Cement-Based Plaster[6]

C 933 Specification for Welded Wire Lath[6]

C 954 Specification for Steel Drill Screws for the Application of Gypsum Panel Products or Metal Plastic Bases to Steel Studs from 0.033 in. (0.84 mm) to 0.112 in. (2.84 mm) in Thickness[6]

C 1002 Specification for Steel Drill Screws for the Application of Gypsum Panel Products or Metal Plaster Bases[5]

C 1032 Specification for Woven Wire Plaster Base[6]

D 1784 Specification for Rigid Poly(Vinyl Chloride) (PVC) Compounds and Chlorinated Poly(Vinyl Chloride) (CPVC) Compounds[7]

D 4216 Specification for Rigid Poly(Vinyl Chloride) (PVC) and Related Plastic Building Products Compounds[8]

E 90 Test Method for Laboratory Measurement of Airborne-Sound Transmission Loss of Building Partitions[9]

3. Terminology

3.1 *Definitions*—For definitions relating to Ceilings and Walls, see Terminology C 11.

3.2 *Definitions of Terms Specific to This Standard:*

3.2.1 *hangers*—wires or steel rods or straps used to support main runners for suspended ceilings beneath floor or roof constructions.

3.2.2 *inserts*—devices embedded in concrete structural members to provide a loop or opening for attachment of hangers.

3.2.3 *saddle tie*—see Figs. 1 and 2.

3.2.4 *self-furring*—a metal plaster base manufactured with evenly-spaced indentations that hold the body of the lath approximately ¼ in. (6.4 mm) away from solid surfaces to which it is applied.

3.2.5 *framing member, n*—studs, joist, runners (track), bridging, bracing, and related accessories manufactured or supplied in wood or light gage steel.

4. Delivery of Materials

4.1 All materials shall be delivered in the original packages, containers, or bundles bearing the brand-name and manufacturer's (or supplier's) identification.

5. Storage of Materials

5.1 All materials shall be kept dry. Materials shall be stacked off the ground, supported on a level platform, and protected from the weather and surface contamination.

5.2 Materials shall be neatly stacked with care taken to avoid damage to edges, ends, or surfaces.

[1] This specification is under the jurisdiction of ASTM Committee C-11 on Gypsum and Related Building Materials and Systems and is the direct responsibility of Subcommittee C11.03 on Specifications for Application of Gypsum and Other Products in Assemblies.

Current edition approved June 10, 1999. Published August 1999. Originally published as C 1063 – 86. Last previous edition C 1063 – 98.

[2] Discontinued; see 1994 *Annual Book of ASTM Standards*, Vol 01.06.

[3] *Annual Book of ASTM Standards*, Vol 01.06.

[4] *Annual Book of ASTM Standards*, Vol 02.04.

[5] *Annual Book of ASTM Standards*, Vol 02.02.

[6] *Annual Book of ASTM Standards*, Vol 04.01.

[7] *Annual Book of ASTM Standards*, Vol 08.01.

[8] *Annual Book of ASTM Standards*, Vol 08.02.

[9] *Annual Book of ASTM Standards*, Vol 04.06.

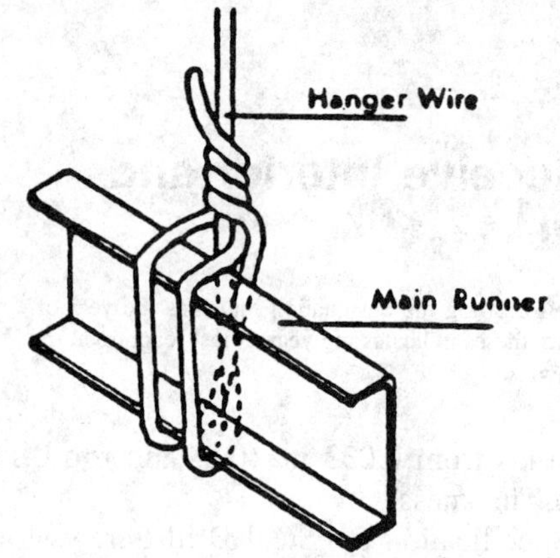

FIG. 1 Saddle Tie

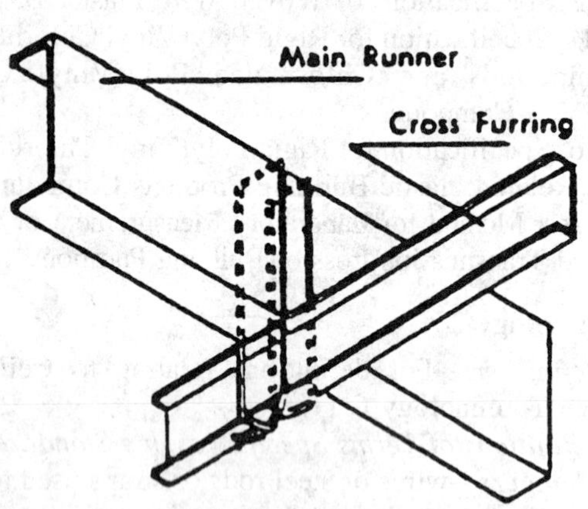

FIG. 2 Saddle Tie

5.3 Paper backed metal plaster bases shall be handled carefully in delivery, storage, and erection to prevent puncturing or removal of paper.

6. Materials

6.1 *Metal Plaster Bases:*

6.1.1 *Expanded Metal Lath*—Specification C 847, galvanized.

6.1.2 *Wire Laths:*

6.1.2.1 *Welded Wire Lath*—Specification C 933.

6.1.2.2 *Woven Wire Lath*—Specification C 1032.

6.1.2.3 *Paper Backed Plaster Bases*—Specification C 847.

6.2 *Accessories:*

6.2.1 *General*—All accessories shall have perforated or expanded flanges or clips shaped to permit complete embedment in the plaster, to provide means for accurate alignment, and to secure attachment of the accessory to the underlying surface. Accessories shall be designed to receive application of the specified plaster thickness.

6.2.2 Accessories shall be fabricated from Zinc Alloy (99 % pure zinc), galvanized (zinc coated) steel, rigid PVC or CPVC plastic, or anodized aluminum alloy. See Specification B 221. (See Table 1 for minimum allowable thicknesses).

NOTE 1—The selection of an appropriate type of material for accessories shall be determined by applicable surrounding climatic and environmental conditions specific to the project location, such as salt air, industrial pollution, high moisture, or humidity.

6.2.3 *Steel*—Specification A 526 and shall have a G60 coating.

TABLE 1 Minimum Thickness of Accessories

Accessory	Base Material, in. (mm)		
	Steel	Zinc Alloy	P.V.C.
Corner Beads	0.0172 (0.44)	0.0207 (0.53)	0.035 (0.89)
Casing Beads	0.0172 (0.44)	0.0207 (0.53)	0.035 (0.89)
Weep Screeds	0.0172 (0.44)	0.0207 (0.53)	0.050 (1.27)
Control Joints	0.0172 (0.44)	0.018 (0.46)	0.050 (1.27)

6.2.4 *PVC Plastic*—Specification D 1784 or D 4216.

6.2.5 *Zinc Alloy*—Specification B 69, 99 % pure zinc.

6.2.6 Thickness of base material shall be as shown in Table 1.

6.2.7 *Cornerite*—1.75 lb/yd^2 (0.059 kg/m^2), galvanized expanded metal lath, galvanized, 1.7 lb/yd^2 (0.057 kg/m^2) galvanized woven or welded wire fabric of 0.0410 in. (1.04 mm) wire. When shaped for angle reinforcing, it shall have outstanding flanges (legs) of not less than 2 in. (51 mm).

6.3 *Channels*—Shall be cold-formed from steel with minimum 33 000 psi (228 MPa) yield strength and 0.0538 in. (1.37 mm) minimum bare steel thickness. Channel shall have a protective coating conforming to Specification A 653–G60, or have a protective coating with an equivalent corrosion resistance for exterior applications, or shall be coated with a rust inhibitive paint, for interior applications, and shall have the following minimum weights in pounds per 1000 linear feet (kg/m).

Sizes, in. (mm)	Weight, lb/1000 ft (kg/m)	Flange Width, in. (mm)
¾ (19)	277 (0.412)	½ (13)
1½ (38)	414 (0.616)	½ (13)
2 (51)	506 (0.753)	½ (13)
2½ (64)	597 (0.888)	½ (13)

NOTE 2—Channels used in areas subject to corrosive action of salt air shall be hot-dipped galvanized, G60 coating.

6.3.1 *External Corner Reinforcement*—Expanded lath, welded wire, or woven wire mesh bent to approximately 90° used to reinforce portland cement stucco at external corners. This accessory shall be fully embedded in the stucco.

6.3.2 *Foundation Weep Screed*—Accessory used to terminate portland cement based stucco at the bottom of exterior walls. This accessory shall have a sloped, solid, or perforated, ground or screed flange to facilitate the removal of moisture from the wall cavity and a vertical attachment flange not less than 3½ in. (89 mm) long.

6.4 *Wire*—As specified in Specification A 641 with a Class I zinc coated (galvanized), soft temper steel. Wire diameters (uncoated) specified herein correspond with United States Steel Wire Gage numbers as follows:

Wire Gage (US Steel Wire Gage)	Diameter (in.)	mm
No. 20	0.0348	.88
No. 19	0.0410	1.04
No. 18	0.0475	1.21
No. 17	0.0540	1.37
No. 16	0.0625	1.59
No. 14	0.0800	2.03
No. 13	0.0915	2.32
No. 12	0.1055	2.68
No. 11	0.1205	3.06
No. 10	0.1350	3.43
No. 9	0.1483	3.77
No. 8	0.1620	4.12

6.5 *Rod and Strap Hangers*—Mild steel, zinc or cadmium plated, or protected with a rust-inhibiting paint.

6.6 *Clips*—Form from steel wire, Specification A 641 zinc-coated (galvanized), Specification A 641 or steel sheet, Specification A 526, depending on use and manufacturer's requirements.

6.7 *Fasteners:*

6.7.1 *Nails*—For attaching metal plaster bases to wood supports, 0.1205-in. 11 gage (3.06-mm) diameter, 7⁄16-in. (11.1-mm) head, barbed, galvanized roofing nails or galvanized common nails.

6.7.1.1 Nails for attaching metal plaster bases to solid substrates shall be not less than ¾ in. (19 mm) long.

6.7.2 Screws for attaching metal plaster base shall be fabricated in accordance with either Specification C 954 or C 1002 and shall have a 7⁄16 in. (11.1 mm) diameter pan wafer head and a 0.120 in. (3.0 mm) diameter shank. Screws used for attachment to metal framing members shall be self-drilling and self-tapping. Screws used for attachment to wood framing members shall be sharp-point.

7. Installation

7.1 *Workmanship*—Metal furring and lathing shall be erected so that the finished plaster surfaces are true to line (allowable tolerance of ¼ in. [6.4 mm] in 10 ft [3.05 m]), level, plumb, square, or curved as required to receive the specified plaster thickness.

7.2 *Hangers and Inserts:*

7.2.1 Hangers shall be of ample length and shall conform to the requirements of Table 2 both as to size and maximum area to be supported, except as modified in this section.

7.2.2 When 1 by 3⁄16-in. (25 by 4.8-mm) flat inserts and hangers are used, 7⁄16-in. (11.1-mm) diameter holes shall be provided on the center line at the lower end of the insert and upper end of the hanger to permit the attachment of the hanger to the insert. The edge of the holes in both the inserts and the hangers shall be not less than ⅜ in. (9.5 mm) from the ends.

7.2.3 In concrete, hangers shall be attached to inserts embedded in the concrete or to other attachment devices designed for this purpose and able to develop full strength of the hanger.

7.2.4 Flat, steel hangers shall be bolted to 1 by 3⁄16-in. (25 by 4.8-mm) inserts with ⅜-in. (9.5-mm) diameter round-head stove bolts. (See Fig. 3.)

7.2.5 The nuts of the bolts shall be drawn up tight.

NOTE 3—Hangers required to withstand upward wind pressures shall be of a type to resist compression. Struts of formed channels shall be permitted.

7.3 *Installation of Hangers for Suspended Ceilings Under Wood Constructions*—Hangers shall be attached to supports by any of the following methods:

7.3.1 A hole shall be drilled through the wood member not less than 3 in. (76 mm) above the bottom with the upper end of the wire hanger passed through the hole and twisted three times around itself. See Fig. 4.

7.3.2 Three 12d nails shall be driven, on a downward slant, into the sides of the wood member with not less than 1¼ in. (31.8 mm) penetration and not less than 5 in. (127 mm) from the bottom edges, and not more than 36 in. (914 mm) on the center with the upper end of the wire hanger wrapped around the nails and twisted three times around itself. See Fig. 5.

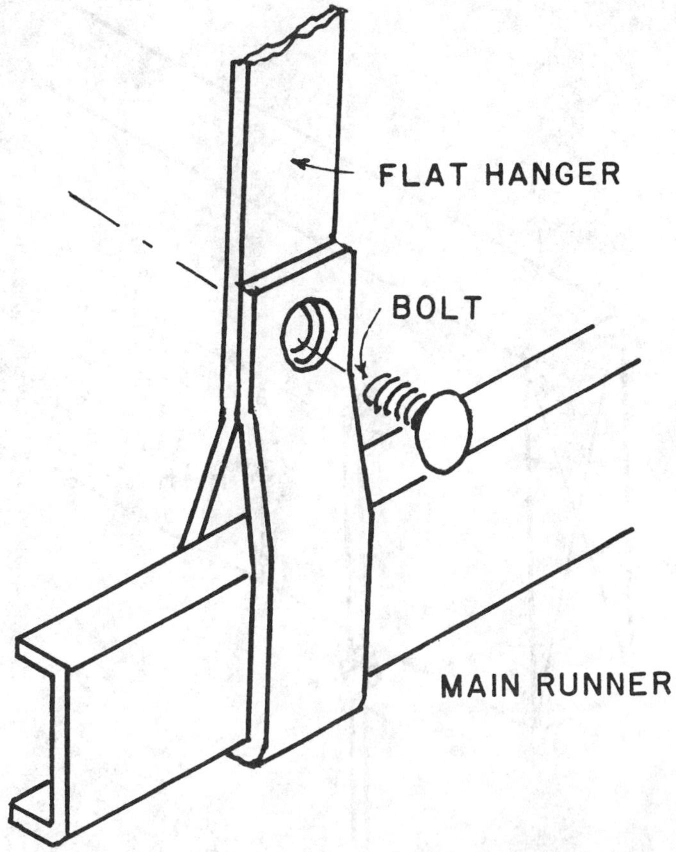

FIG. 3 Flat Hanger Attached to Main Runner Using Round-Head Stove Bolt

7.3.3 A loop shall be formed in the upper end of the wire hanger and secured to the wood member by four 1½-in. (38.1-mm), not less than 9 gage, 0.1483-in. (3.77-mm) diameter wire staples driven horizontally or on a downward slant into the sides of the wood members, three near the upper end of the loop and the fourth to fasten the loose end. See Fig. 6.

7.3.4 Where supports for flooring are thicker than 1½ in. (38.1 mm) and are spaced more than 4 ft (1.2 m) on center, 1½ in. (38.1-mm) No. 1/0 (0.3065 in.) (7.78 mm) eye screws, or equivalent, spaced not more than 3 ft (0.9 m) on centers shall be screwed into the flooring supports with the upper end of the wire hanger inserted through the eye screws and twisted three times around itself.

7.3.5 Two holes shall be drilled in the upper end of the flat hangers and nailed to the sides of the wood members with 12d nails driven through the holes and clinched. Nails shall be not less than 3 in. (76 mm) above the bottom edge of the framing member. See Fig. 7.

7.4 *Attachment of Hangers to Main Runners:*

7.4.1 Wire hangers shall be saddle-tied to the runners. See Fig. 1.

7.4.2 Smooth or threaded rod hangers shall be fastened to the runners with special attachments appropriate to the design.

7.4.3 The lower ends of flat hangers shall be bolted to the main runners, or bent tightly around the runners and carried up

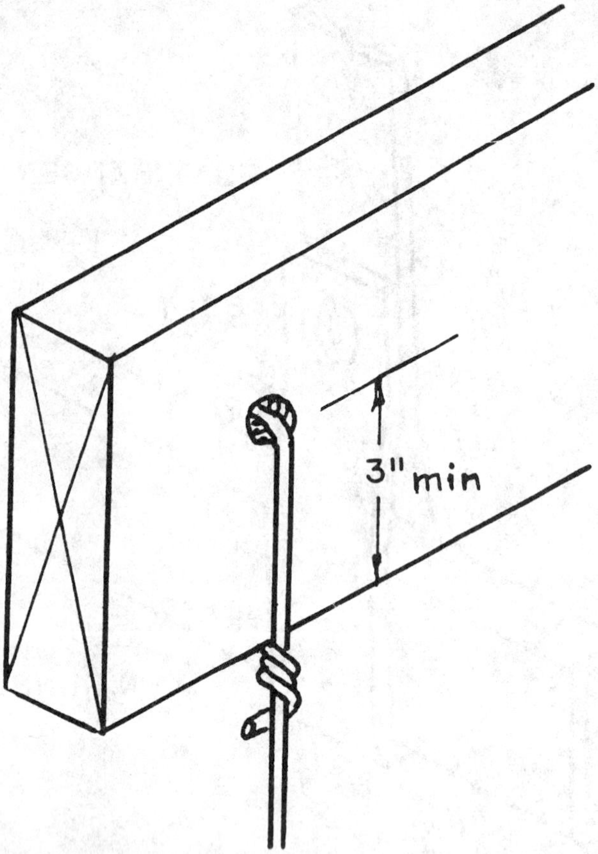

FIG. 4 Hanger Attached to Support Through a Drilled Hole

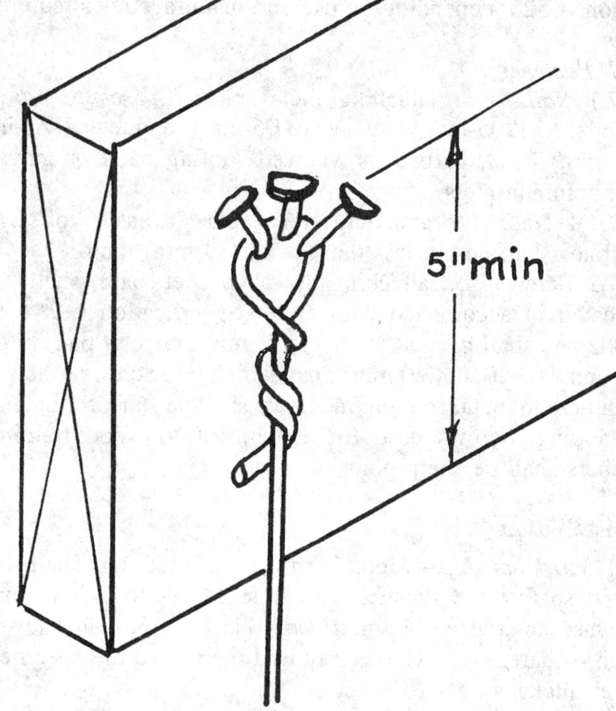

FIG. 5 Hanger Attached to Support Using Nails

and above the runners and bolt to the main part of the hanger. Bolts shall be ⅜-in. (9.5-mm) diameter, round-head stove bolts. See Fig. 3.

7.5 *Installation of Main Runners:*

7.5.1 Minimum sizes and maximum spans and spacings of main runners for the various spans between hangers or other supports shall be in accordance with the requirements of Table 2.

7.5.2 A clearance of not less than 1 in. (25 mm) shall be maintained between the ends of the main runners and the abutting masonry or the concrete walls, partitions, and columns. Where special conditions require that main runners let into abutting masonry or concrete construction, within such constructions maintain a clearance of not less than 1 in. (25 mm) from the ends and not less than ¼ in. (6.4 mm) from the tops and sides of the runners.

7.5.3 A main runner shall be located within 6 in. (152 mm) of the paralleling walls to support the ends of the cross furring. The ends of main runners shall be supported by hangers located not more than 6 in. (152 mm) from the ends.

7.5.4 Where main runners are spliced, the ends shall be overlapped not less than 12 in. (305 mm) with flanges of channels interlocked and securely tied near each end of the splice with double loops of 0.0625 in. (1.59 mm) or double loops of twin strands of 0.0475-in. (1.21-mm) galvanized wire. However, when the splice occurs at an expansion/control joint, the channel shall be nested and loosely tied to hold together but still allow movement.

7.5.5 Hanger wires shall hang straight down. If an obstacle prevents this, a trapeze type device shall be used to allow hanger wires to hang straight.

7.6 *Installation of Cross Furring:*

7.6.1 Minimum size and maximum spans and spacings of various types of cross furring for various spans between main runners and supports shall conform to the requirements of Table 2.

7.6.2 Cross furring shall be saddle-tied to main runners with 0.0625-in. 16 gage (1.59-mm) galvanized wire, or a double strand of 0.0475-in. 18 gage (1.21-mm) galvanized wire or with special galvanized clips, or equivalent attachments. (See Fig. 2.)

7.6.3 Where cross furring members are spliced, the ends shall be overlapped not less than 8 in. (203 mm), with flanges of channels interlocked, and securely tied near each end of the splice with double loops of 0.0625-in. 16 gage (1.59-mm) galvanized wire or twin strands of 0.0475-in. 18 gage (1.21-mm) galvanized wire.

7.6.4 Cross furring shall not come into contact with abutting masonry or reinforced concrete walls or partitions except, where special conditions require that cross furring be let into abutting masonry or concrete construction, the applicable provisions of 7.5.2 shall apply.

7.6.5 Main runners and cross running shall be interrupted at control (expansion) joints. However, when the splice occurs at an expansion/control joint, the channel shall be nested and loosely tied to hold together but still allow movement.

7.7 *Metal Furring for Walls:*

7.7.1 Attachments for furring shall be concrete nails driven

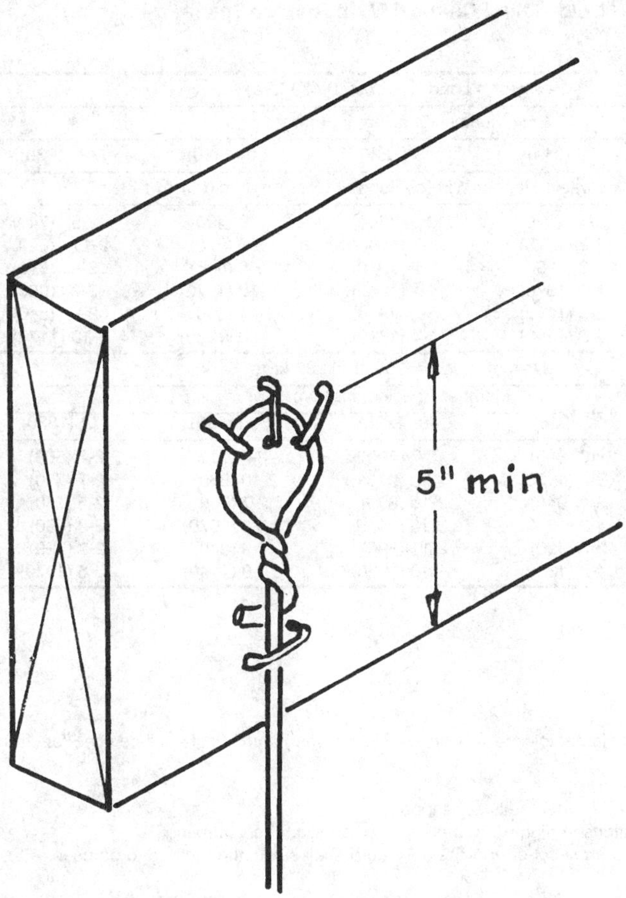

FIG. 6 Hanger Attached to Support Using Staples

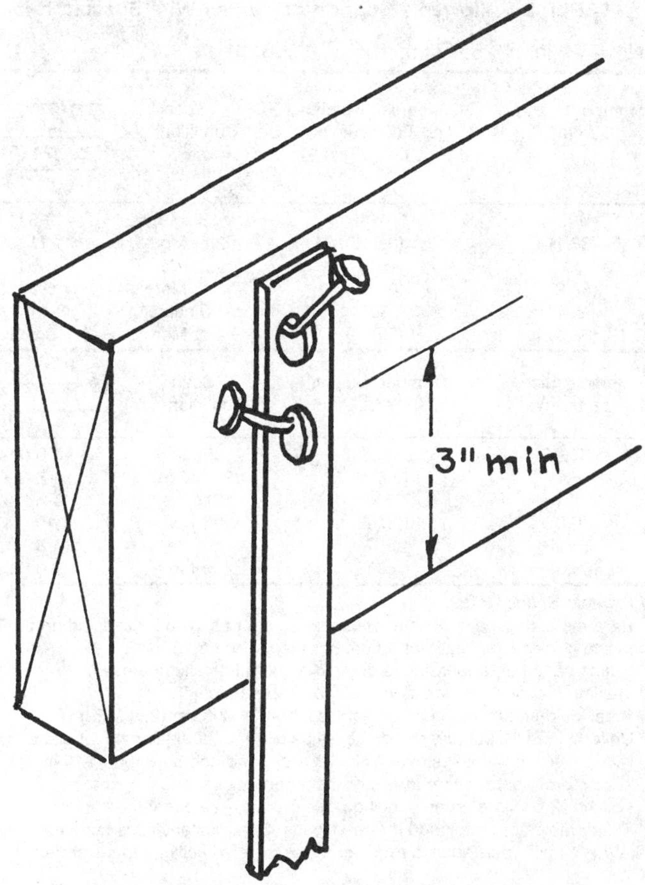

FIG. 7 Flat Hanger Attached to Support Using Nails

securely into concrete or into masonry joints, short pieces of ¾-in. (19.1-mm) channels used as anchors driven into masonry joints, or other devices specifically designed as spacer elements, spaced horizontally not more than 2 ft (0.6 m) on centers. They shall be spaced vertically in accordance with horizontal stiffener spacing so that they project from the face of the wall in order for ties to be made.

7.7.2 Horizontal stiffeners shall be not less than ¾ in. (19.5-mm) cold-rolled channels, spaced not to exceed 54 in. (1372 mm) on centers vertically, with the lower and upper channels not more than 6 in. (152 mm) from the ends of vertical members and not less than ¼ in. (6.4 mm) clear from the wall face, securely tied to attachments with three loops of galvanized, soft-annealed wire, or equivalent devices. Approved furring is not prohibited from use in this application.

7.7.3 Vertical members shall be not less than ¾ in. (19.5-mm) cold-rolled channels in accordance with the requirements of Table 3. Vertical members shall be saddle-tied to horizontal stiffeners with three loops of 0.0475-in. (1.21-mm) galvanized soft-annealed wire, or equivalent devices, at each crossing, and securely anchored to the floor and ceiling constructions. Where furring is not in contact with the wall, channel braces shall be installed between horizontal stiffeners and the wall, spaced horizontally not more than 2 ft (600 mm) on centers.

7.7.4 Where dampproofing has been damaged during installation of attachments, the dampproofing shall be repaired with the same material before proceeding with the installation of the furring.

7.8 *Lapping of Metal Plaster Bases:*

7.8.1 Side laps of metal plaster bases shall be secured to framing members. They shall be tied between supports with 0.0475-in. (1.21-mm) wire at intervals not more than 9 in. (229 mm).

7.8.2 Metal lath shall be lapped ½ in. (12.7 mm) at the sides, or nest the edge ribs. Wire lath shall be lapped one mesh at the sides and the ends. Lap metal lath 1 in. (25 mm) at ends. Where end laps occur between the framing members, the ends of the sheets of all metal plaster bases shall be laced or wire tied with 0.0475-in. (1.21-mm) galvanized, annealed steel wire.

7.8.3 Where metal plaster base with backing is used, the vertical and horizontal lap joints shall be backing on backing and metal on metal.

7.8.3.1 Backing shall be lapped not less than 2 in. (50 mm). On walls, the backing shall be lapped so water will flow to the exterior. Backing shall not be placed between plaster base (lath) and flanges of accessories. Metal lath to metal flange contact shall be required to ensure that flanges are mechanically locked together.

7.9 *Spacing of Attachments for Metal Plaster Bases*—Attachments for securing metal plaster bases to framing members shall be spaced not more than 7 in. (178 mm) apart for diamond mesh and flat rib laths and at each rib for ⅜ in. (9.5-mm) rib lath.

7.10 *Application of Metal Plaster Bases:*

7.10.1 *General:*

TABLE 2 Allowable Support or Hanger Wire Spacing ft-in. (mm) and Cold-Rolled Channel Main Runner Spans, ft-in. (mm)[1–10]

NOTE—1 in. = 25.4 mm; 1 ft^2 = 0.093 m^2

Member Size, in. (mm)	Member Weight, lb/1000 ft (kg/m)	Span Condition[9,10]	Uniform Load = 12 psf (0.479 kPa)				
			Member Spacing, in. (mm)				
			24 (610)	36 (914)	48 (1220)	60 (1520)	72 (1830)
			Allowable Hanger Wire or Support Spacing, ft-in. (mm)				
1½	414	Single	3-6 (1070)	3-1 (940)	2-9 (840)	2-9 (790)	2-5 (740)
(38.1)	(0.615)	2 or More	4-11 (1500)	4-2 (1270)	3-7 (1090)	3-2 (970)	2-11 (890)
2	506	Single	3-9 (1140)	3-3 (990)	3-0 (910)	2-9 (840)	2-8 (810)
(50.8)	(0.753)	2 or More	5-2 (1570)	4-6 (1370)	4-1 (1240)	3-10 (1170)	3-7 (1090)
2½	597	Single	3-11 (1190)	3-5 (1040)	3-2 (970)	2-11 (890)	2-9 (840)
(63.5)	(0.888)	2 or More	5-5 (1650)	4-9 (1450)	4-4 (1320)	4.0 (1220)	3-10 (1170)
Member Size, in. (mm)	Member Weight, lb/1000 ft (kg/m)	Span Condition[9,10]	Uniform Load = 15 psf (0.287 kPa)				
			Member Spacing, in. (mm)				
			24 (610)	36 (914)	48 (1220)	60 (1520)	72 (1830)
1½	414	Single	3-3 (990)	2-10 (860)	2-7 (790)	2-4 (710)	2-2 (660)
(38.1)	(0.616)	2 or More	4-6 (1370)	3-8 (1120)	3-2 (970)	2-10 (860)	2-7 (790)
2	506	Single	3-6 (1070)	3-1 (940)	2-10 (880)	2-7 (790)	2-5 (740)
(50.8)	(0.753)	2 or More	4-10 (1470)	4-3 (1300)	3-10 (1170)	3-6 (1070)	3-3 (990)
2½	597	Single	3-8 (1120)	3-3 (990)	2-11 (890)	2-9 (840)	2-7 (790)
(63.5)	(0.888)	2 or More	5-0 (1520)	4-5 (1350)	4-0 (1220)	3.9 (1140)	3-6 (1070)

Allowable Spans Notes:

[1] Bare metal thickness of cold-rolled main runners shall be not less than 0.0538 in. (1.367 mm).

[2] Inside corner radii shall not be greater than ⅛ in. (3.19 mm).

[3] Spans based on upper flange of main runners laterally unbraced.

[4] Maximum deflection limited to 1/360 of the span length.

[5] Steel yield stress, Fy, shall be not less than 33 000 psi (228 MPa).

[6] Uniform load 12 psf (dry density) shall be used for portland cement plaster ceilings with plaster thicknesses up to ⅞ in. (22 mm) and 15 psf shall be used for ceilings with plaster thicknesses over ⅞ in. (22 mm) and not more than 1-¼ in. (32 mm).

[7] "2 or More" spans refers to two or more continuous, equal spans.

[8] For the "2 or More" span condition, listed spans represent the center-to-center distance between adjacent supports.

[9] These tables are designed for dead loads. Specific conditions such as exterior installations in high wind areas require additional engineering.

[10] Where uplift resistance is required for suspended ceilings to resist negative forces, the architect or engineer of record shall select the method to be used.

7.10.1.1 The spacing of framing members for the type and weight of metal plaster base used shall conform to the requirements of Table 3. Metal plaster bases shall be attached to framing members at not more than 7 in. (178 mm) along framing members except for ⅜-in. (9.5-mm) rib metal lath that shall be attached at each rib.

7.10.1.2 Lath shall be applied with the long dimension at right angles to the supports, unless otherwise specified.

7.10.1.3 Ends of adjoining plaster bases shall be staggered.

7.10.1.4 Lath shall not be continuous through control joints but shall be stopped and tied at each side.

7.10.1.5 Where furred or suspended ceilings butt into or are penetrated by columns, walls, beams, or other elements, the edges and ends of the ceiling lath shall be terminated at the horizontal internal angles with a casing bead, control joint, or similar device designed to keep the edges and ends of the ceiling lath and plaster free of the adjoining vertically oriented, or penetrating elements. Cornerite shall not be used at these locations. A clearance of not less than ⅜ in. (9.5 mm) shall be maintained between the bead and all such elements.

7.10.1.6 Where load bearing walls or partitions butt into structural walls, columns, or floor or roof slabs, the sides or ends of the wall or partition lath shall be terminated at the internal angles with a casing bead, expansion/control joint, or similar device designed to keep the sides and ends of the wall or partition lath free of the adjoining elements. Cornerite shall not be used at these internal angles. A clearance of not less than ⅜ in. (9.5 mm) shall be maintained from all abutting walls, columns, or other vertical elements.

7.10.2 *Attachments for Metal Plaster Bases to Wood Framing Members:*

7.10.2.1 Lath shall be attached to framing members with attachments spaced not more than 7 in. (178 mm) on centers along supports.

7.10.2.2 Diamond-mesh expanded metal lath, flat-rib expanded metal lath, and wire lath shall be attached to horizontal wood framing members with 1½-in. (38.1-mm) roofing nails driven flush with the plaster base and attached to vertical wood framing members with 6d common nails, or 1-in. (25-mm) roofing nails driven to a penetration of not less than ¾ in. (19.1 mm), or 1-in. (25-mm) wire staples driven flush with the plaster base. Staples shall have crowns not less than ¾ in. (19.05 mm) and shall engage not less than three strands of lath and penetrate the wood framing members not less than ¾ in. (19.05 mm). When metal lath is applied over sheathing, use fasteners that will penetrate the structural members not less than ¾ in. (19 mm).

7.10.2.3 Expanded ⅜ in. (9.5 mm) rib lath shall be attached to horizontal and vertical wood framing members with nails or staples to provide not less than 1¾-in. (44.5-mm) penetration into horizontal wood framing members, and ¾-in. (19.1-mm) penetration into vertical wood framing members.

7.10.2.4 Common nails shall be bent over to engage not less than three strands of lath, or be bent over a rib when rib lath is installed.

7.10.2.5 Screws used to attach metal plaster base to horizontal and vertical wood framing members shall penetrate not less than ⅝ in. (15.9 mm) into the member when the lath is

TABLE 3 Types and Weights of Metal Plaster Bases and Corresponding Maximum Permissible Spacing of Supports

Type of Metal Plaster Base	Minimum Weight of Metal Plaster Base, lb/yd^2 (kg/m^2)	Maximum Permissible Spacing of Supports Center to Center, in. (mm)				
		Walls (Partitions)			Ceilings	
		Wood Studs or Furring	Solid Partitions^{A}	Steel Studs or Furring	Wood or Concrete	Metal
U.S. Nominal Weights:						
Diamond Mesh^{D}	2.5 (1.4)	16 (406)^{C}	16 (406)	16 (406)^{C}	12 (305)	12 (305)
	3.4 (108)	16 (406)^{C}	16 (406)	16 (406)^{C}	16 (406)	16 (406)
Flat Rib	2.75 (1.5)	16 (406)	16 (406)	16 (406)	16 (406)	16 (406)
	3.4 (1.8)	19 (482)	24 (610)	19 (482)	19 (482)	19 (482)
Flat Rib (large opening)	1.8 (0.95)	24 (610)	24 (610)	24 (610)	16 (406)	16 (406)
⅜ in. Rib	3.4 (1.8)	24 (610)	N/A^{B}	24 (610)	24 (610)	24 (610)
	4.0 (2.1)	24 (610)	N/A	24 (610)	24 (610)	24 (610)
¾ in. Rib	5.4 (2.9)	24 (610)	N/A	24 (610)	36 (914)	36 (914)
Welded Wire^{D}	1.4 (0.8)	16 (406)	16 (406)	16 (406)	16 (406)	16 (406)
	1.95 (1.1)	24 (610)	24 (610)	24 (610)	24 (610)	24 (610)
Woven Wire^{D}	1.1 (0.6)	24 (610)	16 (406)	16 (406)	16 (406)	24 (610)
	1.4 (0.6)	24 (610)	16 (406)	16 (406)	24 (610)	16 (406)
Canadian Nominal Weights:						
Diamond Mesh^{D}	2.5 (1.4)	16 (406)	12 (305)	12 (305)	12 (305)	12 (305)
	3.0 (1.6)	16 (406)	12 (305)	12 (305)	12 (305)	12 (305)
	3.4 (1.8)	16 (406)	16 (406)	16 (406)	16 (406)	16 (406)
Flat Rib	2.5 (1.4)	16 (406)	12 (305)	12 (305)	12 (305)	12 (305)
	3.0 (1.6)	16 (406)	16 (406)	16 (406)	16 (406)	13½ (343)
⅜ in. Rib	3.0 (1.6)	19 (482)	N/A	16 (406)	16 (406)	16 (406)
	3.5 (1.9)	24 (610)	N/A	19 (482)	19 (482)	19 (482)
	4.0 (2.1)	24 (610)	N/A	24 (610)	24 (610)	24 (610)

^{A} Where plywood is used for sheathing, a minimum of ⅛ in. (3.2 mm) separation shall be provided between adjoining sheets to allow for expansion.

^{B} Not applicable.

^{C} These spacings are based on unsheathed walls. Where self-furring lath is placed over sheathing or a solid surface, the permissible spacing of supports shall be no more than 24 in. (610 mm).

^{D} Metal plaster bases shall be furred away from vertical supports or solid surfaces at least ¼ in. Self-furring lath meets furring requirements; except, furring of expanded metal lath is not required on supports having a bearing surface of 1⅝ in. or less.

installed and shall engage not less than three strands of lath. When installing rib lath, the screw shall pass through, but not deform, the rib.

7.10.3 *Attachments for Metal Plaster Bases to Metal Framing Members:*

7.10.3.1 Except as described in 7.10.3.2, all metal plaster bases shall be securely attached to metal framing members with 0.0475-in. 18 gage (1.21-mm) wire ties, clips or by other means of attachment which afford carrying strength and resistance to corrosion equal to or superior to that of the wire.

7.10.3.2 Rib metal lath shall be attached to open-web steel joists by single ties of galvanized, annealed steel wire, not less than 0.0475 in. (1.21 mm), with the ends of each tie twisted together 1½ times.

7.10.3.3 Screws used to attach metal plaster base to metal framing members shall project not less than ⅜ in. (9.5 mm) through the metal framing member when the lath is installed and shall engage not less than three strands of lath. When installing rib lath, the screw shall pass through, but not deform, the rib.

7.10.4 *Attachments for Metal Plaster Bases to Concrete Joists*—Rib metal lath shall be attached to concrete joists by loops of 0.0800-in. (2.03-mm) galvanized, annealed steel wire, with the ends of each loop twisted together.

7.10.5 Metal plaster bases shall be attached to masonry or concrete with power or powder actuated fasteners or a combination of power or powder actuated fasteners and hardened concrete stub nails. One power or powder actuated fastener shall be located at each corner and one at the mid point of the long dimension adjacent to the edge of the metal plaster base sheet. The balance of the sheet shall be fastened with power or powder actuated fasteners or hardened concrete stub nails. The fasteners shall be installed in rows not more than 16 in. (406 mm) on center and spaced vertically along each row not more 7 in. (178 mm) on center. All fasteners shall be corrosion resistant and shall be not less than ¾ in. (19 mm) long, with heads not less than ⅜ in. (9.5 mm) wide.

7.11 *Application of Accessories:*

7.11.1 *General*—All metal accessories shall be installed in such a manner than flanges and clips provided for their attachment are completely embedded in the plaster.

7.11.1.1 Accessories shall be attached to substrate in such a manner as to ensure proper alignment during application of plaster. Flanges of accessories shall be secured to the face of the plaster base at not more than 7 in. (178 mm) intervals along supports.

7.11.2 *Corner Beads*—Corner beads shall be installed to protect all external corners and to establish grounds.

7.11.2.1 *External Corner Reinforcement*—External corner reinforcement shall be installed to reinforce all external corners where corner bead is not used. Where no external corner reinforcement or corner bead is used, lath shall be furred out and carried around corners not less than one support on frame construction.

7.11.3 *Casing Beads*—Nonload-bearing members shall be isolated from load-bearing members, and all penetrating elements, with casing beads or other suitable means, to avoid transfer of structural loads, and to separate from dissimilar materials.

7.11.4 *Control Joints-General*—Control joints shall be formed by using a single prefabricated member or fabricated by installing casing beads back to back with a flexible barrier membrane behind the casing beads. The separation spacing shall be not less than ⅛ in. (3.2 mm) or as required by the anticipated thermal exposure range and shall be in conformance with 7.10.1.4.

7.11.4.1 *Control Joints*—Control (expansion and contraction) joints shall be installed in walls to delineate areas not more than 144 ft^2 (13.4 m^2) and to delineate areas not more than 100 ft^2 (9.30 m^2) for all horizontal applications, that is, ceilings, curves, or angle type structures.

7.11.4.2 The distance between control joints shall not exceed 18 ft (5.5 m) in either direction or a length-to-width ratio of 2½ to 1. A control joint shall be installed where the ceiling framing or furring changes direction.

7.11.4.3 A control joint shall be installed where an expansion joint occurs in the base exterior wall.

7.11.4.4 Wall or partition height door frames shall be considered as control joints.

7.11.5 *Foundation Weep Screed*—Foundation weep screed shall be installed at the bottom of all steel or wood framed exterior walls to receive lath and plaster. Place the bottom edge of the foundation weep screed not less than 1 in. (25 mm) below the joint formed by the foundation and framing. The nose of the screed shall be placed not less than 4 in. (102 mm) above raw earth or 2 in. (51 mm) above paved surfaces. The weather resistive barrier and lath shall entirely cover the vertical attachment flange and terminate at the top edge of the nose or ground flange.

8. Keywords

8.1 ceiling; expansion control joints; lath; plaster; screed; suspended ceiling; walls

Designation: C 1097 – 95

Standard Specification for Hydrated Lime for Use in Asphaltic-Concrete Mixtures[1]

This standard is issued under the fixed designation C 1097; the number immediately following the designation indicates the year of original adoption or, in the case of revision, the year of last revision. A number in parentheses indicates the year of last reapproval. A superscript epsilon (ϵ) indicates an editorial change since the last revision or reapproval.

1. Scope

1.1 This specification covers high calcium, dolomitic and magnesian-hydrated lime for use in asphaltic/concrete mixtures.

NOTE 1—Hydrated lime, either calcitic, dolomitic, or magnesian, improves bonding of bitumen and aggregates which reduces susceptibility to moisture damage, reduces age hardening by chemically stabilizing polar compounds found in asphalts, and increases initial stiffness of asphalt mixtures.

NOTE 2—No attempt is made to present requirements for any by-product lime.

1.2 *This standard does not purport to address the safety concerns associated with its use. It is the responsibility of the user of this standard to establish appropriate safety and health practices and determine the applicability of regulatory limitations prior to use.*

2. Referenced Documents

2.1 *ASTM Standards:*

C 25 Test Methods of Chemical Analysis of Limestone, Quicklime, and Hydrated Lime[2]

C 50 Practice for Sampling, Inspection, Packing, and Marking Lime and Limestone Products[2]

C 51 Terminology Relating to Lime and Limestone (As Used by the Industry)[2]

C 110 Test Methods for Physical Testing of Quicklime, Hydrated Lime, and Limestone[2]

3. Chemical Requirements

3.1 Unless otherwise specified, for definitions of terms used in this specification, refer to Terminology C 51.

3.2 Hydrated lime for use in bituminous paving mixtures shall conform to the following chemical composition:

Calcium and Magnesium Oxides (on an LOI-free basis, minimum %)	90.0
Carbon Dioxide (taken at point of manufacture, maximum %)	5.0
Unhydrated Calcium and Magnesium Oxides (maximum %)	5.0
Free Moisture of Dry Hydrates (taken at point of manufacture, maximum %)	2.0

4. Physical Requirements

4.1 Hydrated lime, either dry or slurry form, shall not have more than 3.0 % retained on a No. 30 (590 microns) sieve and not more than 30 % retained on a No. 200 (74 microns) sieve.

5. Test Method

5.1 The chemical analysis of the hydrated lime shall be determined in accordance with Test Methods C 25.

5.2 The fineness of hydrated lime shall be determined in accordance with the residue test for hydrated lime and Test Methods C 110.

NOTE 3—Some hydrated limes may require a full 30-min wash time.

6. Sampling, Inspection, Packing and Marking

6.1 The sampling, inspection, rejection, retesting, packing, and marking shall be done in accordance with Practice C 50.

7. Keywords

7.1 Asphaltic; concrete; hydrated; lime; mixtures

[1] This specification is under the jurisdiction of ASTM Committee C-7 on Lime and is the direct responsibility of Subcommittee C07.03 on Industrial Uses.

Current edition approved Aug. 15, 1995. Published October 1995. Originally published as C 1097 – 90. Last previous edition C 1097 – 90.

[2] *Annual Book of ASTM Standards*, Vol 04.01.

Standard Performance Specification for Hydraulic Cement[1]

This standard is issued under the fixed designation C 1157; the number immediately following the designation indicates the year of original adoption or, in the case of revision, the year of last revision. A number in parentheses indicates the year of last reapproval. A superscript epsilon (ε) indicates an editorial change since the last revision or reapproval.

1. Scope

1.1 This specification covers hydraulic cements for both general and special applications. This is a specification giving performance requirements. There are no restrictions on the composition of the cement or its constituents.

1.2 The specification classifies cements by type based on specific requirements for general use, high early strength, resistance to attack by sulfates, and heat of hydration. Optional requirements are provided for the property of low reactivity with alkali-reactive aggregates.

1.3 For properties where values are given in both acceptable metric and inch-pound units, the values in SI units are to be regarded as the standard. Values in SI units shall be obtained by measurement in SI units or by appropriate conversion, using the Rules for Conversion and Rounding given in Standard IEEE/ASTM SI 10, of measurements made in other units.

1.4 The text of this standard refers to notes and footnotes which provide explanatory material. These notes and footnotes (excluding those in tables and figures) are not requirements of the standard.

1.5 *This standard does not purport to address all of the safety concerns, if any, associated with its use. It is the responsibility of the user of this standard to establish appropriate safety and health practices and determine the applicability of regulatory limitations prior to use.*

NOTE 1—Specification C 150 is a prescription specification for portland cement and specification C 595 is a specification for blended hydraulic cement where the products that may be furnished are of restricted nature and proportions and where the products must meet prescriptive requirements as to chemical and physical properties.

2. Referenced Documents

2.1 *ASTM Standards:*

C 109 Test Method for Compressive Strength of Hydraulic Cement Mortars (Using 2-in. or 50-mm Cube Specimens)[2]

C 114 Test Methods for Chemical Analysis of Hydraulic Cement[2]

C 150 Specification for Portland Cement[2]

C 151 Test Method for Autoclave Expansion of Portland Cement[2]

C 183 Practice for Sampling and the Amount of Testing of Hydraulic Cement[2]

C 185 Test Method for Air Content of Hydraulic Cement Mortar[2]

C 186 Test Method for Heat of Hydration of Hydraulic Cement[2]

C 188 Test Method for Density of Hydraulic Cement[2]

C 191 Test Method for Time of Setting of Hydraulic Cement by Vicat Needle[2]

C 204 Test Method for Fineness of Hydraulic Cement by Air Permeability Apparatus[2]

C 219 Terminology Relating to Hydraulic Cement[2]

C 226 Specification for Air-Entraining Additions for Use in the Manufacture of Air-Entraining Hydraulic Cement[2]

C 227 Test Method for Potential Alkali Reactivity of Cement-Aggregate Combinations (Mortar-Bar Method)[3]

C 311 Test Methods for Sampling and Testing Fly Ash or Natural Pozzolans for Use as a Mineral Admixture in Portland-Cement Concrete[3]

C 430 Test Method for Fineness of Hydraulic Cement by the 45-μm (No. 325) Sieve[2]

C 441 Test Method for Effectiveness of Mineral Admixtures or Ground Blast-Furnace Slag in Preventing Excessive Expansion of Concrete Due to the Alkali-Silica Reaction[3]

C 451 Test Method for Early Stiffening of Hydraulic Cement (Paste Method)[2]

C 465 Specification for Processing Additions for Use in the Manufacture of Hydraulic Cements[2]

C 595 Specification for Blended Hydraulic Cements[2]

C 688 Specification for Functional Additions for Use in Hydraulic Cements[2]

C 1012 Test Method for Length Change of Hydraulic-Cement Mortars Exposed to a Sulfate Solution[2]

C 1038 Test Method for Expansion of Portland Cement Mortar Bars Stored in Water[2]

IEEE/ASTM SI 10 Standard for Use of the International System of Units (SI): the Modern Metric System[4]

3. Terminology

3.1 *Definition:*

---

[1] This specification is under the jurisdiction of ASTM Committee C-1 on Cement and is the direct responsibility of Subcommittee C01.10 on Hydraulic Cements for General Concrete Construction.

Current edition approved Nov. 10, 1998. Published March 1999. Originally published as C 1157 – 92. Last previous edition C 1157 – 98.

[2] *Annual Book of ASTM Standards*, Vol 04.01.

[3] *Annual Book of ASTM Standards*, Vol 04.02.

[4] *Annual Book of ASTM Standards*, Vol 14.02.

3.1.1 Terms used in this specification are defined in Terminology C 219.

3.1.2 *blended hydraulic cement*—a hydraulic cement consisting of two or more inorganic constituents which contribute to the strength-gaining properties of the cement, with or without other constituents, processing additions, and functional additions.

3.1.2.1 Blended hydraulic cements are made by intergrinding or other blending processes.

4. Nomenclature

4.1 As selected by the producer, designate cements under this specification in accordance with the following names. The nomenclature does not establish or require any limits on the composition or properties of a cement. Its only purpose is to provide a uniform means for designating the cement after its ingredients and proportions have been selected by the manufacturer, thereby providing general information about the cement when the manufacturer so chooses.

4.2 *Hydraulic Cement*—any cement meeting the requirements of this specification is a hydraulic cement. At the option of the producer, use this designation in addition to, or in lieu of, any other proper designation.

4.3 Portland cement consists of portland cement clinker to which, at the option of the manufacturer, one or more of the following additions, and no others, shall have been made during manufacture.

4.3.1 Calcium sulfate set control addition meeting the requirements of Specification C 688.

4.3.2 Water, including that present in clinker and other additions, in amounts such that loss-on-ignition does not exceed 3.0 % by mass of the cement.

4.3.3 Processing additions up to the maximum amount demonstrated to meet the requirements of Specification C 465 for portland cements.

4.3.4 *Air Entraining Portland Cement* shall contain an interground air entraining addition conforming to Specification C 226.

4.4 Modified portland cements contain up to 15 % of a mineral additive, and, as selected by the producer, are designated "[additive]-modified portland cement" in accordance with the identity of the mineral addition used. [Additive]-modified cements are blended cements.

4.5 Blended hydraulic cements contain more than 15 % of a mineral additive and, as selected by the producer, are named in accordance with the two or, if more appropriate, three constituents present in greatest amount, listed in decreasing order of abundance by mass.

5. Classification and Use

5.1 The types of hydraulic cement covered by this specification are given in 5.2.1-5.2.6 and are classified in accordance with specific properties.

5.2 Cements conforming to this specification shall be designated in accordance with the nomenclature with special characteristics indicated by type in accordance with the types in 5.2.1-5.2.6. When the type is not specified, the requirement of type GU shall apply.

5.2.1 *Type GU*— Hydraulic cement for general construction. Use when one or more of the special types are not required.

5.2.2 *Type HE*—High Early Strength.

5.2.3 *Type MS*—Moderate Sulfate Resistance.

5.2.4 *Type HS*—High Sulfate Resistance.

5.2.5 *Type MH*—Moderate Heat of Hydration.

5.2.6 *Type LH*—Low Heat of Hydration.

5.3 *Additional Option*—The following additional purchase option applies for any of the principal types listed. When this option is invoked, its letter designation and title shall follow immediately after the respective letter designation and title of the principal type.

5.3.1 *Option R*—Low Reactivity with Alkali-Reactive Aggregates. When tested for potential activity with alkali-reactive aggregates, the cement shall meet the requirement (Table 1) for low reactivity with alkali-reactive aggregates.

6. Ordering Information

6.1 Orders for cement meeting the requirements of this specification shall include:

6.1.1 The specification designation ASTM C 1157,

6.1.2 The specification date, if other than the latest published version,

6.1.3 The quantity of cement desired,

6.1.4 The nomenclature (see Section 4) and the type of cement desired (if no type is specified, the cement shipped shall meet the requirements for Type GU).

6.1.5 *Strength Range at Desired Age*—When cement is required, that is, within a particular range of strength, specify cement as being within one Strength Range from Table 2 at one age selected from among 3, 7, and 28 days, except that Type HE may be specified within one Strength Range selected from among 1, 3, 7, or 28 days. If no Strength Range is specified, only the minimum strengths of all the ranges in Table 1, as established in Table 2, shall apply.

6.1.6 A statement that the low reactivity with alkali-reactive aggregates option is invoked, when such is desired, and

6.1.7 A request for the manufacturer's certification, if desired.

7. Chemical Composition

7.1 The chemical composition for the cement is not specified. However, the cement and the individual constituents interground or blended to produce it shall be analyzed.

8. Physical Properties

8.1 Cement of the type specified shall conform to all of the applicable physical requirements of Table 1.

8.2 When optional requirements for early stiffening or 28-day strength are specified, the cement shall conform to the applicable optional limits of Table 1.

8.3 Pozzolan to be used as an ingredient of a cement shall not cause unacceptable expansion due to an alkali-aggregate reaction. To determine that a pozzolan meets this requirement, it shall be tested for alkali-reactivity potential. If the mortar expansion exceeds 0.05 % at the 91-day age for any of the three pozzolan-containing compositions, the pozzolan shall be considered potentially reactive and unacceptable under this specification. If the average total alkali content of any lot of the

TABLE 1 Standard Physical Requirements

Cement Type	GU	HE	MS	HS	MH	LH
Fineness	A	A	A	A	A	A
Autoclave Length Change, max,%	0.80	0.80	0.80	0.80	0.80	0.80
Time of Setting, Vicat Test[B]						
Initial, not less than, min	45	45	45	45	45	45
Initial, not more than, min	420	420	420	420	420	420
Air Content of Mortar Volume, %	C	C	C	C	C	C
Strength Range[D]						
1 day	...	10	...	...	...	...
3 days	10	17	10	5	5	...
7 days	17	...	17	10	10	5
28 days	...	...	...	17	...	17
Heat of Hydration						
7 days, max, kJ/kg (kcal/kg)	...	...	...	...	290 (70)	250 (60)
28 days, max, kJ/kg (kcal/kg)	...	...	...	...	...	290 (70)
Mortar Bar Expansion						
14 days, % max	0.020	0.020	0.020	0.020	0.020	0.020
Sulfate Expansion (Sulfate Resistance)[E]						
6 months, max, %	...	...	0.10	0.05	...	...
1 year, max, %	...	...	...	0.10	...	...
Option R—Low Reactivity with Alkali-Reactive Aggregates[F]						
Expansion						
14 days, max, %	0.020	0.020	0.020	0.020	0.020	0.020
56 days, max, %	0.060	0.060	0.060	0.060	0.060	0.060
Optional Physical Requirements						
Early Stiffening, Final Penetration, min,%	50	50	50	50	50	50
Compressive Strength,[G] *min, MPa (psi),* 28 days	28.0 (4060)	...	28.0 (4060)	...	22.0 (3190)	...

[A]Both amount retained when wet sieved on the 45-μm (No. 325) sieve and specific surface area by air permeability apparatus in m^2/kg shall be reported on all certificates of test results requested from the manufacturer.

[B]Time of setting refers to initial setting time in Test Method C 191.

[C]Air content shall be reported on all certificates of test results requested from the manufacturer. A given value in mortar does not necessarily assure that the desired air content will be obtained in concrete.

[D]Lowest Strength Range whose minimum shall apply at the specified age unless a higher Strength Range is specified by the purchaser. See Table 2 for the applicable strength limits.

[E]In the testing of HS cement, testing at one year shall not be required when the cement meets the 6 month limit. An HS cement failing the 6 month limit shall not be rejected unless it also fails the one year limit.

[F]Compliance with this requirement should not be requested unless the cement will be used with alkali-reactive aggregate.

[G]When 28-day strengths are specified, sufficient time must be allowed for completion of the test. When required on a certificate of test results, special arrangements for storage of the cement pending completion of the test may be required.

TABLE 2 Strength Range Limits

Strength Range	5	10	17	25	35	45
Compressive Strength, min, Mpa	5	10	17	25	35	45
(psi)	(725)	(1450)	(2465)	(3625)	(5075)	(6525)
Compressive Strength, max, Mpa	15	20	30	40	60	...
(psi)	(2175)	(2900)	(4350)	(5800)	(8700)	

cement, expressed as Na_2O equivalent (% $Na_2O + 0.658 \times \%K_2O$), changes by more than 0.10 % by mass from that with which the alkali-reactivity potential tests were carried out, the test for alkali-reactivity of the pozzolan shall be rerun.

8.4 If processing additions are used in the manufacture of a cement, they shall conform to the requirements of Specification C 465.

8.5 If functional additions of the type defined in Specifications C 226 or C 688 are used in a cement, they shall conform to the requirements of the applicable specification.

9. Sampling

9.1 When the purchaser requires that the cement be sampled and tested to verify compliance with this specification, sampling shall be in accordance with Practice C 183 (Note 2). Adequate facilities shall be provided for sampling the finished cement. The finished cement shall be sampled at the mill or at the job site as may be specified by the purchaser.

NOTE 2—Practice C 183 is not designed for manufacturing quality control and is not required for manufacturer's certification.

10. Test Methods

10.1 When testing a cement for compliance with this specification, use the following methods, with modifications or exceptions as indicated.

10.2 *Chemical Analysis*—The cement shall be subject to chemical analysis using Test Methods C 114 for major and minor oxides present in greatest quantity that together, including loss-on-ignition, constitute at least 98 % of the total mass of the cement and for insoluble residue. Each constituent included in a cement in amounts greater than 1 % by mass shall be subjected to chemical analysis using the appropriate methods, such as Test Methods C 311 for pozzolan or other methods consistent with Test Methods C 114.

10.2.1 *Sulfate*—Any appropriate method for determining SO_3 shall be used. If the reference method of Test Methods C 114 is used, the sample need not be completely decomposed as long as excess acid is used.

10.2.2 *Loss-on-Ignition*—For pozzolan, use the method for portland cement in Test Methods C 114, except ignite the sample at 700 to 800°C using an uncovered porcelain crucible. For cements containing sulfides or other constituents which gain weight on heating, use the slag cement procedure in Test Methods C 114.

10.2.3 *Sodium and Potassium Oxides*—For pozzolans or pozzolan-containing cements, the procedure for solution of cement in the procedure for total alkali given in Test Methods C 114 often will not dissolve all of the alkalies. Complete dissolution of the alkalies is essential for accurate analysis. For cements, igniting the sample prior to dissolving will often make the alkalies soluble in acid. Other approaches are igniting with lime, hydrated lime, or calcium carbonate or fusion with lithium borate prior to dissolving, especially for pozzolans. A blank must be run.

10.3 *Fineness*:

10.3.1 Fineness shall be determined in accordance with Test Method C 204.

10.3.2 Amount retained on the 45-μm (No. 325) sieve is to be determined using Test Method C 430.

10.4 Autoclave expansion shall be tested in accordance with Test Method C 151, except that for cements with 1-day compressive strengths below 3.4 MPa (500 psi), the neat cement paste shall be mixed for not less than 3, nor more than 3½ min, and the test specimen shall remain in the moist cabinet for a period of 48 h before being demolded and measured for length.

10.5 Time of setting shall be determined using Test Method C 191. Only time of initial setting is required.

10.6 Air content of mortar shall be tested in accordance with Test Method C 185 using the actual density of the cement, as determined by Test Method C 188, if it differs from 3.15 Mg/m^3 by more than 0.05 Mg/m^3.

10.7 Compressive strength shall be determined in accordance with Test Method C 109. Tests shall be run at each age specified in Table 1 and, if optional requirements are selected, at the age specified in the optional requirements of Table 1. The mixing water requirement in Test Method C 109 shall be adjusted to produce a flow of 110 ± 5 for blended hydraulic cements.

10.8 Heat of hydration shall be tested using Test Method C 186.

10.9 Sulfate resistance shall be determined using Test Method C 1012.

10.10 Reactivity of cement with alkali-reactive aggregate (Option R) should be tested using Test Method C 227 using crushed borosilicate glass, as described in Test Method C 441, as aggregate.

10.11 Alkali-reactivity potential for pozzolan shall be determined using Test Method C 227. Use all constituents, except the pozzolan, at the same relative proportions as they are used in the cement. Prepare blends containing the pozzolan at 0.0 %, the percentage intended for use in the cement, the intended level +5 %, and the intended level −5 %. Test these blends in accordance with Test Method C 227 using sand judged to be non-reactive by Test Method C 227.

10.12 Early stiffening shall be determined by Test Method C 451.

10.13 Mortar bar expansion shall be determined by Test Method C 1038 using a 14-day immersion period.

11. Testing Time Requirements

11.1 The testing time requirements of Practice C 183 shall apply.

11.2 The purchaser shall make necessary arrangements for shipment of samples to the testing laboratory. The time required for transport to the laboratory shall be added to the minimum time intervals allowed from receipt of the samples by the testing laboratory.

11.3 The minimum time allowed from receipt of samples by the testing laboratory to the report of test results shall be determined by the required age of specimen at the time of testing plus seven days.

11.4 A written report of results of applicable tests shall be available within not more than three days of the interval indicated in 11.3.

11.5 Retesting or additional testing shall be without regard to these time limits. Such testing shall not provide the basis for initial acceptance or rejection of the cement.

NOTE 3—Aging effects on small samples of cement stored for long periods of time can produce test results that are not representative of the fresh cement nor of cement stored in large quantity for equal periods of time.

12. Testing by the Manufacturer

12.1 Representative samples of cement shall be taken during production or transfer and tested for compliance with the specification. Such testing shall include a chemical analysis and determination of fineness. Location and frequency of sampling are at the discretion of the manufacturer. Sampling and testing shall be either part of, or in addition to, the manufacturer's normal quality control.

12.2 *Special Testing*:

12.2.1 The following requirements for testing apply only to tests for sulfate resistance and low reactivity with alkali-reactive aggregates.

12.2.1.1 For sulfate resistance of Types MS and HS and for Option R of any cement, tests for compliance with the applicable requirements shall be made at least once every 12 months. Retesting is required if the analysis of any oxide in any of the constituents, or the cement as a whole, departs by more than the values given below from values previously shown to satisfy the applicable requirements.

	Variation from composition for which compliance has been shown	
	Cement or any constituent making up 10 % or more of the cement	Constituents making up less than 10 % of the cement
Change in constituent composition, any oxide, Mass %	±3	±5

Retesting is also required if the amount of a constituent making up 10 % or more of the cement is changed by 5 % or more by mass of the cement, or if a constituent making up less than 10 % of the cement or an addition is changed by 50 % or more of the amount previously present.

12.2.1.2 At the request of the purchaser, the manufacturer's data on compliance with sulfate resistance for Types MS or HS or Option R for any cement shall be made available when such requirements are called for in the cement specification. At the option of the purchaser, the manufacturer's data on compliance may be used in lieu of additional sampling and testing for acceptance or rejection of cement.

13. Inspection

13.1 Inspection of the material shall be as agreed upon between the purchaser and the seller as part of the purchase contract.

13.2 All packages shall be in good condition at the time of inspection.

14. Rejection

14.1 At the option of the purchaser, material that fails to conform to the applicable requirements of this specification for the type specified shall be rejected. Rejection shall be reported to the manufacturer or supplier promptly and in writing, stating the specific reasons for rejection.

14.2 At the option of the purchaser, packages more than 2 % below the mass marked thereon shall be rejected, and if the average mass of packages in any shipment, as shown by determining the mass of 50 packages taken at random, is less than that marked on the packages, the entire shipment shall be rejected.

15. Certification

15.1 When specified in the purchase order or contract, the manufacturer's certification shall be furnished at the time of shipment stating the results of tests, including chemical analysis, made on cement sampled during production or transfer and certifying that the applicable requirements of this specification have been met.

15.2 At the option of the purchaser, bulk cement remaining in storage for more than six months after testing or packaged cement remaining in the possession of a vendor for more than three months shall be sampled and retested and rejected if it fails to meet any of the applicable requirements of this specification.

15.3 *Sulfate Resistance (Types MS and HS) and Low Reactivity with Aggregates (Option R)*—When specified, results from the tests demonstrating compliance with this specification shall be available for inspection and the manufacturer's report shall state that the applicable requirements of this specification for those properties have been met.

16. Package Marking

16.1 When the cement is delivered in packages, words stating an appropriate cement nomenclature selected by the producer, the type of cement and, if specified, compliance with Option R, the name and brand of manufacturer, the mass of cement contained therein, and a list of constituents, using generic names in decreasing order of abundance, shall be plainly marked on each package. Similar information shall be provided with the manufacturer's certification accompanying the shipment of packaged or bulk cement.

NOTE 4—The following list contains suggested generic names for some possible constituents of blended cements. The list is representative and is not inclusive.

Class of Constituent	Examples of Generic Terms
Cement	Portland cement, portland cement clinker.
Calcium Component	Calcium carbonate, limestone, lime, hydrated lime.
Pozzolan	Class F fly ash, Class C fly ash, uncalcined natural pozzolan, calcined natural pozzolan, silica fume.
Slag	Granulated iron blast-furnace slag.
Class of Constituent	Examples of Generic Terms
Additions	Calcium sulfate (if added separately, in amounts greater than normal in portland cement, or with a component other than portland cement), water reducer, accelerator, retarder, water-reducing retarder, air-entraining addition, processing addition.

17. Storage

17.1 The cement shall be stored in such a manner as to permit reasonable access for proper inspection and identification of each shipment and in a suitable weather-tight building, container, or package that will protect the cement from dampness and minimize warehouse set.

18. Keywords

18.1 blended hydraulic cement; fly ash; hydraulic cement; pozzolan; silica fume; slag; granulated blast furnace

Standard Practice for Evaluation of Limestone or Lime Uniformity From a Single Source[1]

This standard is issued under the fixed designation C 1164; the number immediately following the designation indicates the year of original adoption or, in the case of revision, the year of last revision. A number in parentheses indicates the year of last reapproval. A superscript epsilon (ε) indicates an editorial change since the last revision or reapproval.

ε1 Note—Section 10, Keywords were added editorially in November 1997.

1. Scope

1.1 This practice is intended for use in instances where the purchaser desires information on the uniformity of limestone or lime produced at a single source. It is intended that this test method normally be used for the predominant material manufactured at a plant. Guidelines for sampling, testing and presentation of results (Table 1) are given.

1.2 *This standard does not purport to address all of the safety concerns, if any, associated with its use. It is the responsibility of the user of this standard to establish appropriate safety and health practices and determine the applicability of regulatory limitations prior to use.*

2. Referenced Documents

2.1 *ASTM Standards:*

C 25 Test Methods for Chemical Analysis of Limestone, Quicklime, and Hydrated Lime[2]

C 50 Methods of Sampling, Inspection, Packing, and Marking of Lime and Limestone Products[2]

C 51 Terminology Relating to Lime and Limestone (As Used by the Industry)[2]

C 110 Test Methods for Physical Testing of Quicklime, Hydrated Lime, and Limestone[2]

C 141 Specification for Hydraulic Hydrated Lime for Structural Purposes[2]

3. Terminology

3.1 *Definitions:*

3.1.1 For definitions of terms relating to this practice refer to Terminology C 51.

4. Significance and Use

4.1 This practice is designed to present in a standardized format information on the variability of limestone or lime from a single source over a period of time. It can be applied to all materials covered in Test Methods C 25 and C 110, and Specification C 141.

TABLE 1 High Calcium Quicklime January 1989

Note 1—This data is presented only as a sample of the correct method for presentation. Other tests can be substituted and others omitted.

Day	Test			
	Acid Insoluble	% CaO	% MgO	Available CaO
1	0.66	95.85	2.27	94.90
2	0.84	95.03	2.87	94.08
3	1.24	95.85	1.57	94.90
4	1.22	96.67	1.49	95.71
5	1.22	96.26	1.32	95.31
8	0.68	95.85	2.06	94.69
9	0.96	95.54	2.27	94.89
10	1.26	96.15	1.74	95.20
11	1.04	96.04	2.00	95.11
13	1.26	96.26	1.51	95.31
15	1.34	95.23	2.09	94.28
16	1.28	95.44	2.34	94.69
19	0.68	96.15	1.46	95.20
20	1.42	95.85	1.27	94.90
22	1.06	95.85	1.86	94.90
23	0.58	96.15	2.09	95.31
24	0.84	95.03	2.61	94.08
26	1.28	95.64	1.39	94.69
27	1.48	95.54	1.58	94.39
29	0.98	94.82	2.71	93.67
30	1.24	94.41	3.42	93.67
Avg	1.07	95.70	2.00	94.76
Range	0.90	2.26	2.15	2.04
Std Dev	0.26	0.54	0.56	0.54

5. Sampling

5.1 The sampling shall be done in accordance with Methods C 50.

5.2 All sampling shall be performed by personnel qualified by specific training for this purpose.

5.3 Data points shall be an average of three or more separate determinations relative to a specified unit of time. For example: (*1*) *n* determinations for daily average, (*2*) daily determinations for weekly average.

5.4 The samples shall be identified by the date or dates that the represented material was obtained.

6. Procedure

6.1 The minimum amount of testing (see 5.3) and the type of test(s) to be performed should be agreed upon by the purchaser and producer.

Note 1—A single daily analysis does not necessarily characterize a

[1] This practice is under the jurisdiction of ASTM Committee C-7 on Lime and is the direct responsibility of Subcommittee C07.11 on Statistics.

Current edition approved Feb. 15, 1992. Published April 1992. Originally published as C 1164 – 90. Last previous edition C 1164 – 91.

[2] *Annual Book of ASTM Standards*, Vol 04.01.

product. A minimum of three analyses is more reasonable.

6.2 It is desirable that all tests used in a single evaluation be performed in the same laboratory, preferably by the same person, in order for the results to be comparable.

NOTE 2—In order to obtain comparable results when exchanging samples between laboratories, methods used in preparation, handling and storage of those samples must be uniform.

6.2.1 When separate evaluations of a single source are made by two or more laboratories, additional tests of a standard material or exchanged portions of the same sample may be necessary to determine differences in testing that are likely to be obtained in the different laboratories. Three or more batches may be necessary to obtain comparison between laboratories.

6.2.2 Where a within-laboratory testing standard deviation history has not been established, duplicate tests made from a single sample are required to determine the effect of testing variation on the uniformity of results made in a single laboratory.

7. Test Method

7.1 The chemical analysis of quicklime, hydrated lime, and limestone shall be determined in accordance with Test Methods C 25.

7.2 The physical analysis of quicklime, hydrated lime, and limestone shall be determined in accordance with Test Methods C 110.

7.3 The analysis of hydraulic hydrated lime shall be determined in accordance with Specification C 141.

8. Calculation

8.1 Average $\bar{X}$:

$$\bar{X} = \frac{(X_1 + X_2 + \ldots + X_n)}{n} \quad (1)$$

where:

$X_1, X_2, \ldots, X_n$ = the individual test results in accordance with Test Methods C 25 or C 110, or Specification C 141, and

n = the number of individual samples.

8.2 Total Standard Deviation:

$$s = \sqrt{\frac{(X_1 - \bar{X})^2 + (X_2 - \bar{X})^2 + \ldots + (X_n - \bar{X})^2}{(n - 1)}} \quad (2)$$

9. Report

9.1 Report the following information. Sufficient information shall be provided to identify the sample including:

9.1.1 Name of manufacturer and location,

9.1.2 Type of material or other identification,

9.1.3 All sampling and analytical practices used are to be noted (see Note 3), and

9.1.4 Period of time represented by the report.

9.2 The report shall not cover a period of time greater than one year.

NOTE 3—Laboratory designations are lost or are meaningless when averaging a number of determinations on individual samples. The purchaser needs to know the sampling and analytical practices used in generating a report to properly evaluate the information they receive concerning a specific product.

10. Keywords

10.1 hydrated lime; limestone; product uniformity; quicklime; sampling

Standard Specification for Glass Mat Gypsum Substrate for Use as Sheathing[1]

This standard is issued under the fixed designation C 1177/C 1177M; the number immediately following the designation indicates the year of original adoption or, in the case of revision, the year of last revision. A number in parentheses indicates the year of last reapproval. A superscript epsilon (ε) indicates an editorial change since the last revision or reapproval.

1. Scope*

1.1 This specification covers glass mat gypsum substrate, which is designed to be used as an exterior substrate for a weather barrier.

1.2 The values stated in either inch-pound units or SI (metric) are to be regarded separately as the standard. Within the text, the SI units are shown in brackets. The values stated in each system shall be used independently of the other. Values from the two systems shall not be combined.

2. Referenced Documents

2.1 *ASTM Standards:*

C 11 Terminology Relating to Gypsum and Related Building Materials and Systems[2]

C 473 Test Methods for Physical Testing of Gypsum Board Products and Gypsum Lath[2]

C 645 Specification for Nonstructural Steel Framing Members[2]

C 1264 Specification for Sampling, Inspection, Rejection, Certification, Packaging, Marking, Shipping, Handling, and Storage of Gypsum Board.[2]

E 119 Test Methods for Fire Tests of Building Construction and Materials[3]

3. Terminology

3.1 Definitions used in this standard shall be in accordance with Terminology C 11.

3.2 *Definitions of Terms Specific to This Standard:*

3.2.1 *glass mat, n*—a mat of glass fibers with or without a binder.

3.2.2 *edge, n*—the bound edge as manufactured.

3.2.3 *weather barrier, n*—a material that prevents the passage of rain, sleet and snow.

4. Materials and Manufacture

4.1 *Glass Mat Gypsum Substrate* shall consist of a noncombustible water-resistant core, essentially gypsum, surfaced with glass mat partially, or completely embedded in the core.

4.2 *Glass Mat Gypsum Substrate, type X (special fire-resistant)* designates glass mat gypsum substrate complying with this specification that provide not less than 1 h fire resistance for substrate ⅝ in. [15.9 mm] thick or ¾ h fire resistance for substrate ½ in. [12.7 mm] thick, applied parallel with and on each side of load bearing 2 by 4 wood studs spaced 16 in. [406 mm] on center with 6d coated nails, 1⅞ in. [48 mm] long, 0.0915 in. [2.3 mm] diameter shank, ¼ in. [6.4 mm] diameter heads, spaced 7 in. [178 mm] on center with glass mat gypsum substrate joints staggered 16 in. [406 mm] on each side of the partition and tested in accordance with Test Methods E 119.

NOTE—Consult manufacturers for independent test data on assembly details and fire resistance classifications for other types of construction. See official fire test reports for assembly particulars, materials, and classifications.

5. Physical Properties

5.1 Specimens shall be taken from the samples obtained in accordance with 8.1.

5.2 Specimens shall be tested in accordance with Methods C 473.

5.2.1 *Flexural Strength*—The specimens shall be tested face up and face down. The average breaking load shall not be less than the following:

Thickness, in. [mm]	Method B Load, lbf [N] Bearing edges perpendicular to the substrate edge	Load, lbf [N] Bearing edges parallel to the substrate edge
½ [12.7]	100 [445]	80 [356]
⅝ [15.9]	140 [623]	100 [445]

5.2.2 *Humidified Deflection*—Specimens shall have an average deflection of not more than the following:

Thickness, in. [mm]	Humidified deflection, eighths of an in. [mm]
½ [12.7]	2 [6]
⅝ [15.9]	1 [3]

5.2.3 *Core, End and Edge Hardness*—Specimens shall have an average hardness of not less than 15 lbf [67 N] when tested in accordance with Method B:

5.2.4 *Nail-Pull Resistance*—Specimens shall have an average nail-pull resistance of not less than the following when tested in accordance with Method B:

Thickness, in. [mm]	Nail pull resistance, lbf [N]
½ [12.7]	80 [356]
⅝ [15.9]	90 [400]

5.2.5 *Water Resistance*—The three specimens selected shall have an average water absorption of not more than 10 weight % after a 2-h immersion.

[1] This specification is under the jurisdiction of ASTM Committee C-11 on Gypsum and Related Building Materials and Systems and is the direct responsibility of Subcommittee C11.01 on Specifications and Test Methods for Gypsum Products.

Current edition approved Dec. 10, 1996. Published February 1997. Originally published as C 1177 – 91. Last previous edition C 1177 – 95.

[2] *Annual Book of ASTM Standards*, Vol 04.01.

[3] *Annual Book of ASTM Standards*, Vol 04.07.

*** A Summary of Changes section appears at the end of this standard.**

6. Dimensions and Permissible Variations

6.1 Specimens shall be taken from the samples obtained in accordance with 8.1.

6.2 *Thickness, width, length and end squareness* shall be determined in accordance with Test Methods C 473.

6.2.1 *Thickness*—The nominal thickness shall be ½ or ⅝ in. [12.7 or 15.9 mm], with permissible variations in the nominal thickness of ±1⁄32 in. [±0.8 mm], and with permissible local variation of ±1⁄16 in. [±1.6 mm] from the nominal thickness.

6.2.2 *Width*—The nominal width shall be up to 48 in. [1220 mm], with permissible variations of ⅛ in. [3 mm] under the specified width.

6.2.3 The nominal length permissible variation shall be as follows:

Thickness, in. [mm]	Length ft [mm]	Variation in. [mm]
½ [12.7]	6 to 12 [1840 to 3660]	±¼ [6]
⅝ [15.9]	6 to 12 [1840 to 3660]	±¼ [6]

6.2.4 *End Squareness*—Corners shall be square with a permissible variation of ±1.8 in. [±3 mm] in the full width of the substrate.

6.2.5 *Edges and Ends*—The edges and ends shall be straight and either square, recessed, beveled, featured, tapered, or featured and tapered.

7. Finish, and Appearance

7.1 Glass mat gypsum substrate shall be free of any cracks and imperfections that would render it unfit for its designed use.

8. Sampling, Inspection, Rejection, Certification, Packaging, Marking, Shipping, Handling, and Storage

8.1 Sampling, inspection, rejection, certification, packaging, marking, shipping, handling, and storage shall be in accordance with C 1264.

9. Keywords

9.1 core; exterior substrate; fire resistant classification; flexural strength; glass mat; glass mat gypsum substrate; gypsum; hardness; humidified deflection; nail pull resistance; noncombustible; sheathing; substrate; type X; water resistance; weather barrier

APPENDIX

(Nonmandatory Information)

This Appendix gives general information and also suggestions for inclusions to be made elsewhere by the specifier. They are not part of this specification.

X1. ALTERNATE DEFINITION FOR TYPE X

X1.1 Glass mat gypsum substrate, type X (special fire-resistant) designates glass mat gypsum substrate providing a greater fire resistance than regular glass mat gypsum substrate of the same thickness. Type X (special fire-resistant) glass mat gypsum substrate, when tested in accordance with Test Methods E 119, shall provide the following minimum fire resistance for the assemblies described:

X1.1.1 One hour for a ⅝ in. [15.9 mm] thickness applied to a partition in a single layer application on each side of 3⅝ in. [92 mm] deep non-loadbearing galvanized steel studs complying with Specification C 645 spaced 24 in. [610 mm] on center. The ⅝ in. [15.9 mm] thick glass mat gypsum substrate 48 in. [1220 mm] wide shall be attached using 1 in. [25 mm] long drywall screws spaced 8 in. [203 mm] on center along the edges and ends, and 12 in. [305 mm] along intermediate studs. All joints shall be oriented parallel to and located over studs and staggered on opposite sides of the assembly; and

X1.1.2 Two hours for a ½ in. [12.7 mm] thickness applied to a partition in a double layer application on each side of 2½ in. [64 mm] deep non-loadbearing galvanized steel studs complying with Specification C 645 spaced 24 in. [610 mm] on center. The base layer 48 in. [1220 mm] wide shall be attached using 1 in. [25 mm] long drywall screws spaced 12 in. [305 mm] on center along substrate edges, ends and along intermediate studs. Joints shall be oriented parallel to and located over studs and staggered on opposite sides of the assembly. The face layer 48 in. [1220 mm] wide shall be attached using 1⅝ in. [41 mm] long drywall screws spaced 12 in. [305 mm] along substrate edges, ends and along intermediate studs. Joints shall be oriented parallel to and located over studs, offset 24 in. [610 mm] from the base layer joints, and staggered on opposite sides of the assembly.

SUMMARY OF CHANGES

Committee C-11 has identified the location of selected changes to this standard since the last issue (C 1177-95) that may impact the use of this standard.

(*1*) The term "board" was changed to "substrate" in Sections 1, 4, 7, 8, and 9.

(*2*) The term "panel" was changed to "substrate" in Section 4 and the Appendix.

(*3*) Section 1, 3, 5, and 6 were revised to comply with standard format.

(*4*) Specification C 1264 was added to Section 2.

(*5*) Paragraph 5.2.3 and 5.2.4 were revised to designate Method B.

(*6*) Reference to Specification C 1264 was added to Section 8.

(*7*) Section 9, 10, 11, and 12 were deleted.

Standard Specification for Glass Mat Water-Resistant Gypsum Backing Panel[1]

This standard is issued under the fixed designation C 1178/C 1178M; the number immediately following the designation indicates the year of original adoption or, in the case of revision, the year of last revision. A number in parentheses indicates the year of last reapproval. A superscript epsilon (ϵ) indicates an editorial change since the last revision or reapproval.

1. Scope*

1.1 This specification covers glass mat water-resistant gypsum backing panel designed for use on ceilings and walls in bath and shower areas as a base for the application of ceramic or plastic tile.

1.2 The values stated in either inch-pound units or SI (metric) are to be regarded separately as the standard. Within the text, the SI units are shown in brackets. The values stated in each system shall be used independently of the other. Values from the two systems shall not be combined.

2. Referenced Documents

2.1 *ASTM Standards:*

C 11 Terminology Relating to Gypsum and Related Building Materials and Systems[2]

C 473 Test Methods for Physical Testing of Gypsum Board Products and Gypsum Lath[2]

C 645 Specification for Nonstructural Steel Framing Members[2]

C 1264 Specification for Sampling, Inspection, Rejection, Certification, Packaging, Marking, Shipping, Handling, and Storage of Gypsum Board[2]

E 119 Test Methods for Fire Tests of Building Construction and Materials[3]

3. Terminology

3.1 Definitions used in this specification shall be in accordance with Terminology C 11.

3.2 *Definitions of Terms Specific to this Standard:*

3.2.1 *glass mat, n*—a mat of glass fibers with or without a binder.

3.2.2 *edge, n*—the bound edge as manufactured.

3.2.3 *face, n*—the coated surface.

4. Materials and Manufacture

4.1 *Glass mat water-resistant gypsum backing panel* shall consist of a noncombustible water-resistant core, essentially gypsum, surfaced with glass mat, partially or completely embedded in the core, and with a water resistant coating on one surface.

4.2 Glass mat water-resistant gypsum panels, type X (special fire-resistant) designates glass mat gypsum panels complying with this specification that provide not less than 1-h fire-resistance for boards ⅝ in. [15.9 mm] thick or ¾-h fire-resistance for panels ½ in. [12.7 mm] thick, applied parallel with and on each side of load bearing 2 by 4 wood studs spaced 16 in. [406 mm] on center with 6d coated nails, 1⅞ in. [48-mm] long, 0.0915 in. [2.3 mm] diameter shank, ¼ in. [6.4 mm] diameter heads, spaced 7 in. [178 mm] on center with glass mat gypsum panel joints staggered 16 in. [406 mm] on each side of the partition and tested in accordance with Test Methods E 119.

NOTE 1—Consult manufacturers for independent test data on assembly details and fire resistance classifications for other types of construction. See official fire test reports for assembly particulars, materials, and classifications.

5. Physical Properties

5.1 Specimens shall be taken from the samples obtained in accordance with 8.1.

5.2 Specimens shall be tested in accordance with Test Methods C 473.

5.2.1 *Flexural Strength*—The specimens shall be tested face up and face down. The average breaking load shall not be less than the following:

Thickness, in. [mm]	Method B Load, lbf [N]: Bearing edges perpendicular to the board edge	Bearing edges parallel to the board edge
½ [12.7]	100 [445]	80 [356]
⅝ [15.9]	140 [623]	100 [445]

5.2.2 *Humidified Deflection*—Specimens shall have an average deflection of not more than the following:

Thickness, in. [mm]	Humidified deflection, eighths of an in. [mm]
½ [12.7]	2 [6]
⅝ [15.9]	1 [3]

5.2.3 *Core, End and Edge Hardness*—Specimens shall have an average hardness of not less than 15 lbf [67 N] when tested in accordance with Method B.

5.2.4 *Nail Pull Resistance*—Specimens shall have an average nail pull resistance of not less than the following when tested in accordance with Method B:

Thickness, in. [mm]	Nail pull resistance, lbf [N]
½ [12.7]	70 [311]
⅝ [15.9]	90 [400]

5.2.5 *Water Resistance*—The three specimens selected shall have an average water absorption of not more than 5 weight % after 2-h immersion.

[1] This specification is under the jurisdiction of ASTM Committee C-11 on Gypsum and Related Building Materials and Systems and is the direct responsibility of Subcommittee C11.01 on Specifications and Test Methods for Gypsum Products.

Current edition approved Dec. 10, 1996. Published February 1997. Originally published as C 1178 – 91. Last previous edition C 1178 – 95.

[2] *Annual Book of ASTM Standards*, Vol 04.01.

[3] *Annual Book of ASTM Standards*, Vol 04.07.

*** A Summary of Changes section appears at the end of this standard.**

5.2.6 *Surface Water Absorption*—Specimens shall have an average surface water absorption of the face side of the panel not more than 0.5 g after 2 h of elapsed time.

6. Dimensions and Permissible Variations

6.1 Specimens shall be taken from the samples obtained in accordance with 8.1.

6.2 Thickness, width, and length shall be determined in accordance with Test Methods C 473.

6.2.1 *Thickness*—The nominal thickness shall be ½ in. or ⅝ in. [12.7 or 15.9 mm], with permissible variations in the nominal thickness of ±1⁄64 in. [±0.4 mm] and with permissible local variations of ±1⁄32 in. [±0.8 mm].

6.2.2 *Width*—The nominal width shall be up to 48 in. [1220 mm], with a permissible variation of 3⁄32 in. [3 mm] under the specified width.

6.2.3 *Length*—The nominal length and permissible variations shall be as follows:

Thickness in. [mm]	Length ft [mm]	Variation in. [mm]
½ [12.7]	5 to 12 [1520 to 3660]	+¼ in. [6 mm]
⅝ [15.9]	5 to 12 [1520 to 3660]	+¼ in. [6 mm]

6.2.4 *End Squareness*—Corners shall be square with a permissible variation of ±¼ in. [±6 mm] in the full width of the panel.

6.2.5 *Edges and Ends*—The edges and ends shall be straight and solid.

7. Finish and Appearance

7.1 Glass mat water-resistant gypsum backing panel shall have surfaces true and free of imperfections that would render the panel unfit for its designed use.

8. Sampling, Inspection, Rejection, Certification, Packaging, Marking, Shipping, Handling, and Storage

8.1 Sampling, Inspection, Rejection, Certification, Packaging, Marking, Shipping, Handling, and Storage shall be in accordance with Specification C 1264.

9. Keywords

9.1 bath; ceilings; ceramic or plastic tile; construction; core; end and edge hardness; fire resistant; flexural strength; glass mat; gypsum; humidified deflection; immersion; nail pull resistance; shower; surface water absorption; type X; walls; water absorption; water resistance; water resistant

APPENDIX

(Nonmandatory Information)

This Appendix gives general information and also suggestions for inclusions to be made elsewhere by the specifier. They are not part of this specification.

X1. ALTERNATE DEFINITION FOR TYPE X

X1.1 Glass mat water-resistant gypsum panels, type X (special fire-resistant) designates glass mat gypsum panels providing a greater fire resistance than regular glass mat water-resistant gypsum panels of the same thickness. Type X (special fire-resistant) glass mat water-resistant gypsum panels, when tested in accordance with Test Methods E 119, shall provide the following minimum fire resistance for the assemblies described:

X1.1.1 One hour for a ⅝ in. [15.9 mm] thickness applied to a partition in a single layer application on each side of 3⅝ in. [92 mm] deep non-load bearing galvanized steel studs complying with Specification C 645 spaced 24 in. [610 mm] on center. Thc ⅝ in. [15.9 mm] thick glass mat water-resistant gypsum panels 48 in. [1220 mm] wide shall be attached using 1 in. [25 mm] long drywall screws spaced 8 in. [203 mm] on center along the edges and ends, and 12 in. [305 mm] along intermediate studs. All joints shall be oriented parallel to and located over studs and staggered on opposite sides of the assembly; and

X1.1.2 Two hours for a ½ in. [12.7 mm] thickness applied to a partition in a double layer application on each side of 2½ in. [64 mm] deep non-load bearing galvanized steel studs complying with Specification C 645 spaced 24 in. [610 mm] on center. The base layer 48 in. [1220 mm] wide shall be attached using 1 in. [25 mm] long drywall screws spaced 12 in. [305 mm] on center along panel edges, ends and along intermediate studs. Joints shall be oriented parallel to and located over studs and staggered on opposite sides of the assembly. The face layer 48 in. [1220 mm] wide shall be attached using 1⅝ in. [41 mm] long drywall screws spaced 12 in. [305 mm] along panel edges, ends and along intermediate studs. Joints shall be oriented parallel to and located over studs, offset 24 in. [610 mm] from the base layer joints, and staggered on opposite sides of the assembly.

SUMMARY OF CHANGES

Committee C-11 has identified the location of selected changes to this standard since the last issue (C 1178 – 95) that may impact the use of this standard.

(1) The term "board" was changed to "panel" in the title and sections 1, 4, and 7.

(2) Sections 1, 3, 5, and 6 were revised.

(3) Specification C 1264 was added to 2.1.

(4) Paragraphs 5.2.3 and 5.2.4 were revised to designate Method B.

(5) Paragraph 6.2.4 was added.

(6) Section 8 was revised.

(7) Sections 9, 10, 11, and 12 were deleted.

Standard Practice for Evaluation of Laboratories Testing Hydraulic Cement[1]

This standard is issued under the fixed designation C 1222; the number immediately following the designation indicates the year of original adoption or, in the case of revision, the year of last revision. A number in parentheses indicates the year of last reapproval. A superscript epsilon (ε) indicates an editorial change since the last revision or reapproval.

1. Scope

1.1 This practice covers the technical training and experience of laboratory testing personnel and identifies the minimum technical requirements for laboratory equipment used in testing of hydraulic cement as prescribed by ASTM.

1.2 This practice provides minimum criteria for evaluating the capability of a laboratory to perform chemical or physical tests listed in the various specifications on hydraulic cement (Note 1).

NOTE 1—Relevant hydraulic cement specifications are C 91, C 150, C 595, C 845, and C 1157.

1.3 Values in SI units shall be obtained by measurement in SI units or by appropriate conversion, using the Rules for Conversion and Rounding given in Standard IEEE/ASTM SI 10, of measurement made in other units.

1.4 *This standard does not purport to address all of the safety concerns, if any, associated with its use. It is the responsibility of the user of this standard to establish appropriate safety and health practices and determine the applicability of regulatory limitations prior to use.*

2. Referenced Documents

2.1 *ASTM Standards:*

C 91 Specification for Masonry Cement[2]
C 109/C 109M Test Method for Compressive Strength of Hydraulic Cement Mortars (Using 2-in. or 50-mm Cube Specimens)[2]
C 114 Test Methods for Chemical Analysis of Hydraulic Cement[2]
C 115 Test Method for Fineness of Portland Cement by the Turbidimeter[2]
C 125 Terminology Relating to Concrete and Concrete Aggregates[3]
C 150 Specification for Portland Cement[2]
C 151 Test Method for Autoclave Expansion of Portland Cement[2]
C 185 Test Method for Air Content of Hydraulic Cement Mortar[2]
C 187 Test Method for Normal Consistency of Hydraulic Cement[2]
C 191 Test Method for Time of Setting of Hydraulic Cement by Vicat Needle[2]
C 204 Test Method for Fineness of Hydraulic Cement by Air Permeability Apparatus[2]
C 219 Terminology Relating to Hydraulic Cement[2]
C 230 Specification for Flow Table for Use in Tests of Hydraulic Cement[2]
C 266 Test Method for Time of Setting of Hydraulic-Cement Paste by Gillmore Needles[2]
C 305 Practice for Mechanical Mixing of Hydraulic Cement Pastes and Mortars of Plastic Consistency[2]
C 451 Test Method for Early Stiffening of Portland Cement (Paste Method)[2]
C 595 Specification for Blended Hydraulic Cements[2]
C 778 Specification for Standard Sand[2]
C 845 Specification for Expansive Hydraulic Cement[2]
C 1005 Specification for Weights and Weighing Devices for Use in the Physical Testing of Hydraulic Cements[2]
C 1157 Performance Specification for Blended Hydraulic Cement[2]
E 4 Practices for Force Verification of Testing Machines[4]
IEEE/ASTM SI 10 Standard for Use of the International System of Units (SI): The Modern Metric System[5]

2.2 *Other Documents:*

ACI 116R Cement and Concrete Terminology[6]
ASTM Manual of Cement Testing[7]

3. Terminology

3.1 *Definitions:*

3.1.1 *inspection, n*—a process of measuring, examining, testing, gaging, or using other procedures to ascertain the quality or state, detect errors or defects, or otherwise appraise materials, products, services, systems, or environments when compared to preestablished criteria.

3.1.2 Additional definitions may be found in Terminology C 125 and C 219, Practices E 4, and ACI 116R.

3.2 *Descriptions of Terms Specific to This Standard:*

3.2.1 *evaluation authority, n*—an independent entity, apart from the organization being evaluated, that can provide an unbiased evaluation of the organization and shall have the capability to assess the technical activities of testing laboratories (see Discussion).

3.2.1.1 *Discussion*—One such evaluation authority is the Cement and Concrete Reference Laboratory (CCRL).[8] Laboratory inspection is broadened into accreditation programs by such independent authorities as the National Voluntary

[1] This practice is under the jurisdiction of ASTM Committee C-1 on Cement and is the direct responsibility of Subcommittee C01.95 on Coordination of Standards.

Current edition approved May 10, 1997. Published June 1998. Originally published as P 227 in 1991. Last previous edition C 1222 – 96.

[2] *Annual Book of ASTM Standards*, Vol 04.01.

[3] *Annual Book of ASTM Standards*, Vol 04.02.

[4] *Annual Book of ASTM Standards*, Vol 03.01.

[5] *Annual Book of ASTM Standards*, Vol 14.02.

[6] Available from American Concrete Institute, P.O. Box 19150, Detroit, MI 48219.

[7] *Annual Book of ASTM Standards*, Vol 04.01.

[8] CCRL, National Institute of Standards and Technology, Bldg. 226, Rm. A 365, Gaithersburg, MD 20899.

Laboratory Accreditation Program (NVLAP),[9] American Association for Laboratory Accreditation (A2LA)[10] AASHTO Accreditation Program (AAP),[11] and others established.

3.2.2 *laboratory technician, n*—an employee of the laboratory who is assigned to perform the actual testing operations primarily conducted in the laboratory.

3.2.3 *quality systems, n*—those internal procedures and practices that a laboratory utilizes to ensure continued compliance with applicable testing standards.

3.2.4 *subcontracting, n*—employing another organization to provide testing services that the laboratory contracted to provide.

3.2.5 *testing laboratory, n*—an organization that measures, examines, performs tests, or otherwise determines the characteristics or performance of materials or products. This may include organizations that offer commercial testing services, an in-house quality control function, an academic institution, or any other organization providing the specified testing services.

4. Significance and Use

4.1 The testing of hydraulic cement is an important element in obtaining quality construction. A testing laboratory must be selected with care.

4.2 A testing laboratory shall be deemed qualified to perform and report the results of its tests if the laboratory meets the requirements of this practice.

4.3 This practice provides guidance for evaluating the organization, personnel, facilities, and quality systems of the laboratory. This practice may be supplemented by criteria and requirements for particular projects.

5. Documentation of Organization

5.1 The following information shall be readily available for review:

5.1.1 A description of the organization, including the complete legal name and address of the main office and each laboratory location, names and positions of the principal officers and the individual in charge of the laboratory,

5.1.2 A description of the organization management structure, and

5.1.3 A listing of the range of services offered.

6. Human Resources

6.1 The manager of the laboratory shall be a chemist, materials analyst, or an engineer and a full-time employee of the organization having at least three years supervisory experience in the testing of hydraulic cement; however, a person with equivalent science-oriented education or experience having satisfactorily directed testing of hydraulic cement is acceptable.

7. Testing and Additional Requirements

7.1 *Testing Requirements:*

7.1.1 The organization shall have the capability of performing tests associated with its range of services (Note 2). The laboratory shall have the facilities and equipment required for preparing, storing, conditioning, and testing specimens.

NOTE 2—The range of services of a testing laboratory may involve either chemical testing or physical testing, or both. The laboratory need not perform all tests listed in the applicable ASTM specification, but rather have the required equipment and demonstrate the ability to perform the procedures within its reported range of services.

7.1.2 The laboratory shall use the latest version of each referenced ASTM standard within one year of its publication in the Annual Book of ASTM Standards except where an earlier version is specifically required.

7.1.3 Laboratory personnel shall have convenient access to applicable standards.

7.2 *Additional Requirements*—The following requirements apply to only those tests performed by the laboratory. They are intended to supplement certain provisions of the standard test methods:

7.2.1 Water storage recording thermometers shall be checked for accuracy at least every six months by comparing their output with that of a mercury in glass thermometer, graduated in 1°F (0.5°C) divisions, placed in the water adjacent to the probe of the recording thermometer. Adjustments shall be made if differences in the observed readings exceed 2°F (1°C).

7.2.2 General-purpose weighing devices and weights shall be checked at least annually and shall conform to the requirements of Specification C 1005.

7.2.3 Analytical balances and reference masses shall be checked at least annually and shall conform to the requirements of Test Methods C 114.

7.2.4 The 45-μm (No. 325) sieve fineness nozzles shall be checked at least every six months using the flow rate test procedure described in the ASTM Manual of Cement Testing.

7.2.5 Flow tables shall be checked at least every 2½ years using the calibration material described in Specification C 230.

7.2.6 Compression machines shall be verified, in accordance with Practices E 4 at least annually to determine if indicated loads, with and without the maximum load indicator (when so equipped), are accurate to ±1.0 %.

7.2.7 Compression machine bearing blocks shall be checked for planeness to 0.001 in. at least annually using a straightedge and 0.001-in. feeler stock and shall be refinished if found to be out of tolerance.

7.2.8 Air content measures (400 mL) shall be calibrated at least every 2½ years following the procedures described in Test Method C 185.

7.2.9 Wagner turbidimeter apparatus shall be calibrated at least every six months using the procedures described in Test Method C 115.

7.2.10 Each lot of standard sand shall be checked upon receipt to determine if it conforms to the requirements of Specification C 778.

7.2.11 Autoclave apparatus shall be maintained and tested in accordance with Test Method C 151.

---

[9] NVLAP, National Institute of Standards and Technology, Building 411, Room A 124, Gaithersburg, MD 20899.

[10] American Association for Laboratory Accreditation, Quince Orchard, Gaithersburg, MD 20878.

[11] American Association of State Highway and Transportation Officials (AASHTO), 444 N. Capital St., NW, Suite 225, Washington, DC 20001.

7.2.12 Cube molds and tampers shall be checked for conformance to the design and dimensional requirements of Test Method C 109/C 109M at least every 2½ years.

7.2.13 Vicat apparatus and vicat ring shall be inspected and checked for conformance to Test Methods C 187, C 191, and C 451 at least every 2½ years.

7.2.14 Gillmore test apparatus shall be inspected and checked for conformance to the requirements of Test Method C 266 at least every 2½ years.

7.2.15 Mechanical mixing apparatus shall be inspected and checked for conformance to the requirements of Practice C 305 at least every 2½ years.

7.2.16 Air-permeability apparatus and related accessory equipment shall be calibrated and checked for conformance to Test Method C 204 at least every 2½ years.

7.2.16.1 Check determinations shall be made against standard or secondary standard reference samples every six months. If the results exceed the precision of the test method, the apparatus shall be recalibrated.

7.2.17 Water-retention apparatus and related equipment shall be checked for conformance to Specification C 91 at least every 2½ years.

8. Subcontracting

8.1 Subcontracting for the performance of part or all of a test method is permitted. When subcontracting is used the quality of the work provided must be ensured. The requirements for evaluation differ depending on the permanence of the subcontracting.

8.2 *Permanent Subcontracting*—When all or a portion of a test method is subcontracted on a permanent basis then the subcontracted organization must be subjected to separate evaluation according to this practice for the methods subcontracted. The laboratory shall retain a copy of this evaluation for review by the evaluation authority.

8.3 *Temporary Subcontracting*—When all or a portion of a test method is subcontracted on a temporary basis then the laboratory shall ensure the quality of the subcontracted work. The level and type of assurance will depend on the extent of the subcontrating (Note 3). The laboratory shall retain records of this assurance for review by the evaluation authority.

NOTE 3—Some examples of assurance for different levels of temporary subcontracting are: (*1*) a review of calibrations records when the use of a piece of equipment is subcontracted; and (*2*) a satisfactory rating on a proficiency sample when a method is subcontracted. Other means of assurance are test results from a standard reference material or, in the case of chemical testing records of qualification testing.

9. Quality System

9.1 *Quality System Requirements*—The laboratory shall establish a quality system for ensuring the quality of services offered.

9.1.1 The laboratory shall maintain written documentation for the following:

9.1.1.1 Procedures for sampling, handling, and testing the hydraulic cement,

9.1.1.2 Procedures for training and evaluating laboratory personnel,

9.1.1.3 Procedures for calibrating and maintaining test equipment, and

9.1.1.4 Procedures for handling technical complaints from clients about results, procedures, or methods used.

9.1.2 The laboratory shall maintain current standard test methods and other pertinent reference material.

9.1.3 The laboratory shall participate in the appropriate CCRL Cement Proficiency Sample Program (Note 4).

NOTE 4—The CCRL offers proficiency sample programs for physical and chemical testing of portland and blended cements and physical testing of masonry cement. The laboratory should participate in the program compatible with its scope.

10. Laboratory Records and Reports

10.1 *Quality System Records*—The laboratory shall maintain records indicating compliance with the procedures described in their quality system documentation. Records shall be maintained for at least five years.The records shall include at least the following:

10.1.1 Results of all calibrations and verifications performed in the laboratory,

10.1.2 Traceability or accuracy of any in-house calibration or verification equipment used,

10.1.3 An inventory and description of all required test equipment,

10.1.4 Records on laboratory personnel that document work experience, education, on-the-job training, and action taken to ensure continued competence in performing the required test methods,

10.1.5 Records of audits and inspections by outside agencies and all reports or certifications, with applicable dates, of any evaluations or accreditations issued by an evaluating authority, and

10.1.6 Records of participation in the appropriate CCRL Cement Proficiency Sample Program.

10.2 *Test Report Records*—The laboratory shall maintain a system of records that permits verification of any test report issued. Each test report and related records shall be retained for at least three years. Corrections or additions to reports shall clearly reference the report being amended. Include in the test report or the records the following information:

10.2.1 Name and address of the laboratory,

10.2.2 Identification of the report and the date issued,

10.2.3 Name of the client, if applicable,

10.2.4 Project identification, if applicable,

10.2.5 Sample identification,

10.2.6 Date of receipt of test samples,

10.2.7 Date(s) of test performance,

10.2.8 Identification of the standard test method used, and a notation of all known deviations from the test method observed during the testing of the sample being reported (Note 5).

NOTE 5—Deviation from standard test methods may adversely affect results.

10.2.9 Identification of any standard test methods subcontracted and name of the organization that provided such services,

10.2.10 Test results and other pertinent data required by this practice,

10.2.11 Identification of the person performing the tests, and

10.2.12 Name of the person in charge of the laboratory or their designee.

11. Laboratory Evaluation

11.1 *Laboratory Evaluation*—The laboratory shall have its facilities, personnel, equipment, and procedures inspected at intervals of approximately two years by an evaluation authority (see Discussion 3.2.1.1) to confirm its ability to perform the required tests.

11.2 The evaluation authority, who performs the inspection of the laboratory, shall report to the manager of the laboratory on all applicable requirements of this practice including deficiencies that existed during the inspection. The laboratory manager is responsible for the correction of any deficiencies.

11.3 The manager of the laboratory shall report corrective actions taken on all inspection deficiencies when required by the evaluating authority for accreditation.

11.4 The evaluation authority providing the accreditation shall provide to the laboratory a statement of conformance when the laboratory evaluation report, including satisfactory correction of the deficiencies, has been accepted. The statement of conformance shall be evidence of the ability of the laboratory to conduct tests on cements in accordance with relevant standards. The report and statement of conformance shall be in such form that they may be used for public disclosure at the discretion of the laboratory.

12. Keywords

12.1 hydraulic cement; laboratory evaluation

Designation: C 1264/C 1264M – 95

Standard Specification for Sampling, Inspection, Rejection, Certification, Packaging, Marking, Shipping, Handling, and Storage of Gypsum Board[1]

This standard is issued under the fixed designation C 1264/C 1264M; the number immediately following the designation indicates the year of original adoption or, in the case of revision, the year of last revision. A number in parentheses indicates the year of last reapproval. A superscript epsilon (ε) indicates an editorial change since the last revision or reapproval.

1. Scope

1.1 This specification covers sampling; inspection; rejection; certification; packaging and marking; and shipping, handling, and storage of gypsum board.

1.2 The values stated in either inch-pound units or SI units are to be regarded separately as standard. Within the text, the SI units are shown in brackets. The values stated in each system are not exact equivalents; therefore, each system must be used independently of the other. Combining values from the two systems may result in nonconformance with the specification.

2. Sampling

2.1 When specified by the purchase agreement, samples of gypsum board shall be taken at the place of manufacture or at the destination. If the samples are taken at the destination, such samples shall be taken within 24 h of the receipt of the material.

2.2 At least 0.25 % of the number of gypsum boards in a shipment, but not less than three boards, shall be so selected as to be representative of the shipment and shall constitute a sample for the purpose of tests by the purchaser or user.

3. Inspection

3.1 Inspection of the gypsum board shall be agreed upon between the purchaser and the producer or supplier as part of the purchase agreement.

4. Rejection

4.1 Rejection of gypsum board that fails to conform to the requirements of the specification shall be reported to the producer or supplier promptly and in writing. The notice of rejection shall contain a statement documenting how the gypsum board has failed to conform to the requirements of the specification.

5. Certification

5.1 When specified in the purchase agreement, a producer's or supplier's report shall be furnished at the time of shipment certifying that the product is in compliance with the ASTM specification referencing this specification.

6. Packaging and Marking

6.1 Unless otherwise specified in the purchase agreement, each board or package shall have legibly marked thereon the following: the thickness; the name of the producer or supplier; the brand name, if any; and the ASTM specification for the product.

7. Shipping, Handling, and Storage

7.1 Gypsum board shall be shipped so as to be kept dry.

7.2 Gypsum board shall be stored so as to be kept dry, preferably inside a building, under a roof. Gypsum board shall be neatly stacked flat with care taken to prevent sagging or damage to edges, ends, and surfaces.

7.3 Where necessary to store gypsum board outside, it shall be stacked flat, off the ground, supported on a level platform, and fully protected from weather and direct sunlight exposure.

8. Keywords

8.1 certification; gypsum; gypsum board; handling; inspection; marking; packaging; rejection; sampling; shipping; storage

[1] This specification is under the jurisdiction of ASTM Committee C-11 on Gypsum and Related Building Materials and Systems and is the direct responsibility of Subcommittee C11.01 on Specifications and Test Methods for Gypsum Products.

Current edition approved July 15, 1995. Published October 1995. Originally published as C 1264 – 94. Last previous edition C 1264 – 94.

Designation: C 1271 – 94a

Standard Test Method for X-ray Spectrometric Analysis of Lime and Limestone[1]

This standard is issued under the fixed designation C 1271; the number immediately following the designation indicates the year of original adoption or, in the case of revision, the year of last revision. A number in parentheses indicates the year of last reapproval. A superscript epsilon (ϵ) indicates an editorial change since the last revision or reapproval.

1. Scope

1.1 This test method covers the X-ray emission spectrometric analysis of limestone, quicklime, hydrated lime, and hydraulic lime using wavelength dispersive instruments.

1.2 The values stated in SI units are to be regarded as the standard.

1.3 *This standard does not purport to address all of the safety concerns, if any, associated with its use. It is the responsibility of the user of this standard to establish appropriate safety and health practices and determine the applicability of regulatory limitations prior to use.* Specific precautionary statements are given in Section 10.

2. Referenced Documents

2.1 *ASTM Standards:*

C 25 Test Methods for Chemical Analysis of Limestone, Quicklime, and Hydrated Lime[2]

C 50 Methods for Sampling, Inspection, Packing, and Marking of Lime and Limestone Products[2]

C 51 Terminology Relating to Lime and Limestone (As Used by the Industry)[2]

E 50 Practices for Apparatus, Reagents, and Safety Precautions for Chemical Analysis of Metals[3]

E 135 Terminology Relating to Analytical Chemistry for Metals, Ores, and Related Materials[3]

E 305 Practice for Establishing and Controlling Spectrochemical Analytical Curves[3]

E 456 Terminology Relating to Quality and Statistics[4]

E 691 Practice for Conducting an Interlaboratory Study to Determine the Precision of a Test Method[4]

E 1060 Practice for Interlaboratory Testing of Spectrochemical Methods of Analysis[5]

E 1361 Guide for Correction of Interelement Effects in X-Ray Spectrometric Analysis[5]

2.2 *NIST Documents:*[6]

ANSI-NIST Handbook 114

NIST Standards Catalog

3. Terminology

3.1 *Definitions:*

3.1.1 *emission spectroscopy*—unless otherwise specified, for definitions of terms used in this test method pertaining to emission spectroscopy, refer to Terminology E 135.

3.1.2 *lime*—unless otherwise specified, for definitions of terms used in this test method pertaining to lime, refer to Terminology C 51.

3.1.3 *statistical*—unless otherwise specified, for definitions of terms used in this test method pertaining to statistics, refer to Terminology E 456.

4. Summary of Test Method

4.1 A briquetted powder specimen or a fused-glass disk specimen is irradiated by a high-energy X-ray beam. The secondary X rays produced are dispersed by means of crystals, and the intensities are measured by suitable detectors at selected wavelengths. Data are collected based on the time required to reach a fixed number of counts, total count for a fixed time, or integration of voltage for a fixed time. Concentrations of the elements are determined by relating the measured radiation of unknown specimens to analytical curves prepared from reference materials of known composition.

5. Significance and Use

5.1 This procedure is suitable for manufacturing control and verifying that the product meets specifications. It provides rapid, multi-element determinations with sufficient accuracy to ensure product quality and minimize production delays. The analytical performance data included may be used as a benchmark to determine whether similar X-ray spectrometers provide equivalent precision and accuracy, or whether the performance of a particular X-ray spectrometer has changed.

6. Interferences

6.1 Interelement effects or matrix effects may exist for some of the elements listed. One way to compensate for these effects is to prepare a series of calibration curves to cover the designated concentration ranges. The composition of the specimen being analyzed must match closely the composition of the reference materials used to prepare the calibration curve. Mathematical corrections, derived from empirical relationships or fundamental parameter calculations, may be used alternatively. Any of these are acceptable, providing that the analytical accuracy required by this test method is achieved.

6.2 Calcium is the primary component of the matrix analyzed by this test method. The analyst must be aware of all possible interferences and matrix effects of this element. Orders of all wavelengths may cause interference, for example, the effect of CaKβ1 (2nd order) on PKα1.

6.3 Contamination from the grinding apparatus is an

---

[1] This test method is under the jurisdiction of ASTM Committee C-7 on Lime and is the direct responsibility of Subcommittee C07.05 on Chemical Tests.

Current edition approved Sept. 15, 1994. Published November 1994. Originally published as C 1271 – 94. Last previous edition C 1271 – 94.

[2] *Annual Book of ASTM Standards*, Vol 04.01.

[3] *Annual Book of ASTM Standards*, Vol 03.05.

[4] *Annual Book of ASTM Standards*, Vol 14.02.

[5] *Annual Book of ASTM Standards*, Vol 03.06.

[6] Available from National Institute for Standards and Technology (NIST), Gaithersburg, MD 20899.

ever-present source of interference of which the analyst must always be cognizant. For example, stainless steel will impart iron, chrome, and nickel to the prepared sample.

6.4 Spectral interferences from the X-ray tube may also occur, for example, line overlap from CrKβ1 on MnKα1 caused by a chrome target tube.

7. Apparatus

7.1 *Specimen Preparation Equipment:*

7.1.1 *Jaw Crusher*, for initial crushing of lumps.

7.1.2 *Plate Grinder or Pulverizer*, with one static and one rotating disk for further grinding.

7.1.3 *Rotary Disk Mill or Shatter Box*, with hardened grinding containers and timer control for final grinding.

7.2 *Briquetting Equipment:*

7.2.1 *Briquetting Press*, capable of providing pressures up to 550 MPa (80 000 psi). The press shall be equipped with a mold assembly that provides a briquet compatible with the X-ray specimen holder.

7.3 *Fusion Equipment:*

7.3.1 *Furnace or Gas Burners*, with a timer, capable of heating the sample and flux to at least 1000°C and homogenizing the melt.

7.3.2 *Fusion Crucibles:*

7.3.2.1 *Vitreous Carbon or Graphite*, 20 to 30-mL capacity, with a flat bottom 30 to 35 mm in diameter.

7.3.2.2 *Platinum/Gold*, 95 % platinum/5 % gold alloy, 30 to 35-mL capacity.

7.3.2.3 *Platinum/Gold Casting Dish*, 95 % platinum/5 % gold alloy, 30 to 35-mL capacity, with a flat bottom 30 to 35 mm in diameter.

7.3.3 *Polishing Wheel*, suitable for polishing the fusion disk to obtain a uniform surface for irradiation. The analyst should be aware at all times of possible contamination from the polishing media used for surfacing the disk (6.3).

7.4 *Excitation Source:*

7.4.1 *X-Ray Tube Power Supply*, providing constant potential or rectified power of sufficient energy to produce secondary radiation of the specimen for the elements specified. The generator may be equipped with a line voltage regulator and current stabilizer.

7.4.2 *X-Ray Tubes*, with targets of various high-purity elements, capable of continuous operation at required potentials and currents, and that will excite the elements to be determined.

7.5 *Spectrometer*, designed for X-ray emission analysis and equipped with specimen holders and a specimen chamber. The chamber may contain a specimen spinner, and it must be equipped for vacuum operation.

7.5.1 *Analyzing Crystals*—Flat or curved crystals with optimized capability for diffraction of the wavelengths of interest.

7.5.2 *Collimator*, for limiting the characteristic X rays to a parallel bundle when flat crystals are used in the instrument. A collimator is not necessary for curved crystal optics.

7.5.3 *Detectors*, sealed or gas flow, proportional type, Geiger counters, scintillation counters or equivalent.

7.5.4 *Vacuum System*, providing for the determination of elements whose radiation is absorbed by air (atomic number below 20). The system shall consist of at least one vacuum pump, gage, and electrical controls to provide automatic pumpdown of the optical path and maintain a controlled pressure, usually 13 Pa or less.

7.6 *Measuring System*, consisting of electronic circuits capable of counting or integrating pulses received from the detectors. The counts, count rate, or integrated voltages may be displayed on meters, recorders, digital counters, or voltmeters. The counts, count rates, or integrated voltages can also be presented to a computer or programmable calculator for conversion to percent concentration. Pulse height analyzers may be required to provide more accurate measurements for some measurements.

8. Reagents and Materials

8.1 *Purity and Concentration of Reagents*—The purity and concentration of chemical reagents shall conform to the requirements prescribed in Practices E 50.

8.2 *Binders*—Various binders have been used successfully to prepare briquettes suitable for presentation to the instrument. As a general rule, the binder should not contain an element that will be determined. In addition, the sample to binder ratio must be present in the analytical sample as in the reference materials that were used to establish the calibration.

8.3 *Fluxes*—Various fluxes have been used successfully to prepare fusion disks. The flux must be capable of dissolving or dispersing the analyzed elements in an homogeneous and reproducible fashion in the melt. The prepared disks must then be suitable for presentation to the instrument.

8.4 *Detector Gas (P-10)*, consisting of a mixture of 90 % argon and 10 % methane for use with gas flow proportional counters only.

9. Reference Materials

9.1 Certified reference materials are available from the National Institute for Standards and Technology (NIST) and other international certification agencies. Refer to Test Methods C 25 for a current listing.

9.2 Reference materials with matrixes similar to that of the test specimen and containing varying amounts of the elements to be determined may be used provided that they have been analyzed in accordance with ASTM standard test methods.

9.3 Standard additions method or spiked samples may also be used to create standards for which appropriate reference materials are not available for an element desired to be analyzed. The matrix material must match that of the test specimen.

10. Safety Precautions

10.1 Occupational health and safety standards for ionizing radiation shall be observed at all X-ray emission spectrometer installations. It is also recommended that operating and maintenance personnel follow the guidelines of safe operating procedures given in current handbooks and publications from the original equipment manufacturer, NIST, U.S. Government Printing Office, or similar handbooks on radiation safety. See NIST Standards Catalog and ANSI-NIST Handbook 114.

10.2 *Personnel*—X-ray equipment shall be used only under the guidance and supervision of a responsible, qualified person.

10.3 *Monitoring Devices*—Either film badges or dosimeters shall be worn by all operating and maintenance personnel. Periodic radiation surveys of the equipment for leaks and excessive scattered radiation shall be made by a qualified person using an ionization chamber detector to meet local, state, and federal radiation standards. The personal film badge survey record, the radiation survey record, and an equipment maintenance record shall be available upon request.

10.4 Special precautions for operators and maintenance personnel shall be posted at the equipment site.

10.5 Radiation caution signs shall be posted near the X-ray equipment and at all entrances to the radiation area.

10.6 Fail-Safe "X-ray on" warning lights shall be used on the equipment.

11. Sampling

11.1 Gross sample is to be taken in accordance with Methods C 50.

11.2 Sample preparation is to be performed in accordance with Methods C 50 and Test Methods C 25.

11.3 Special preparation may be required to grind the analytical sample finer than is required in 11.2.

12. Preparation of Reference Materials and Test Specimens

12.1 Treat reference materials and specimens exactly the same way throughout the procedure. Either briquetted powder or fused disk specimens may be used.

13. Preparation of Apparatus

13.1 Prepare and operate the spectrometer in accordance with the manufacturer's instructions, using the parameters given in Table 1. Once established, control all instrument parameters closely to ensure repeatable analyses.

14. Calibration and Standardization

14.1 *Calibration (Preparation of Analytical Curves)*—Select a sufficient number of reference materials to cover the concentration ranges of the elements sought. Average duplicate intensity measurements of each reference material, and establish a calibration curve for each element in accordance with Practice E 305. Establish the curve by a least squares or multiple regression fit of the X-ray intensity measurements versus the corresponding weight percent concentrations of the element in the reference materials.

TABLE 1 Analytical Lines and Crystals

Element	Wavelength, Å	Crystals
Sodium	11.91	TLAP, multilayer
Magnesium	9.889	ADP, TLAP, RAP, multilayer
Aluminum	8.338	PET, EDdT, RAP
Silicon	7.126	PET, EDdT, RAP, InSb
Phosphorus	6.155	Ge, PET
Sulfur	5.373	Ge, PET, NaCl
Chlorine	4.729	PET, Ge
Potassium	3.742	LiF 200, PET
Calcium	3.359	LiF 200, PET
Titanium	2.750	LiF 200, LiF 220
Manganese	2.103	LiF 200, LiF 220
Iron	1.937	LiF 200, LiF 220
Strontium	0.8766	LiF 200, LiF 220
Barium	2.775	LiF 200, LiF 220
Lead	1.175	LiF 200, LiF 220

14.2 *Standardization (Analytical Curve Adjustment)*—Verify that the calibration curve for each element has not drifted using a control reference material. Drift is indicated by a change in counts that exceeds the normal variation of counts for that element. A change of 1 % or greater usually signifies that the calibration curve for that element has drifted and should thus be adjusted. Re-measure the reference materials to verify that acceptable analyses are obtained after the adjustments are made.

15. Procedure

15.1 *Excitation*—Introduce the specimen into the specimen chamber, using care not to contaminate the surface to be excited. Produce the secondary fluorescence using the equipment used in 7.4.

15.2 *Radiation Measurements*—Obtain and record the counting rate measurement for each element. Either fixed count or fixed time modes may be used. A predetermined minimum count rate is obtainable only after unacceptable long counting times for lime materials, with the exception of the major oxides (Si, Fe, Al, Ca, and Mg). A minimum counting time of 60 s is acceptable.

15.3 *Spectral Interferences*—Some X-ray spectrometers will not resolve radiation completely from several line overlaps. Care must therefore be exercised in these cases. Alternate wavelengths without interference may be used. Otherwise, mathematical calculations must be used to correct for the interferences.

NOTE 1—If a goniometer is used, the setting for the 2 θ peak must be determined experimentally within each laboratory. Periodic checks to verify the setting are advisable.

15.4 *Replicate Measurements*—Make a minimum of two measurements on each test specimen. The performance of an X-ray spectrometer is not improved significantly by making multiple measurements on the same surface of the same specimen; however, for those elements not having sufficient counts in the predetermined time, the statistics near the detection limit will improve with multiple exposures. Confidence in the accuracy of analysis will also improve by analyzing more than one specimen.

16. Calculation of Results

16.1 Determine the concentration of the various elements using the radiation measurements for the test specimen and the appropriate analytical curves.

16.2 If mathematical corrections must be made to correct the concentrations for interelement effects, any one of a number of correction procedures may be used. Refer to the equipment manufacturer's manual for the instrument being used. Computers are normally interfaced to the X-ray spectrometers in laboratories using such correction systems. Refer to Guide E 1361 for a guide to interelement corrections.

NOTE 2—With computer-controlled X-ray spectrometers that convert output signals automatically to percent concentrations and make interelement corrections, calculated results are usually printed without operator involvement.

17. Precision and Bias

17.1 *Precision*—An acceptable number of laboratories tested this test method in accordance with Practice E 1060.

Precision data are provided in the annex.

17.1.1 *Repeatability (R1)*—Calculations for within-laboratory precision are listed in the annex.

17.1.2 *Reproducibility (R2)*—Calculations for between-laboratory precision are listed in the annex.

17.2 *Bias (Accuracy)*—Bias data calculated in accordance with Practice E 1060 are included in the annex.

18. Keywords

18.1 chemical analysis; hydrated lime; hydraulic lime; limestone; quicklime; wavelength dispersive; X-ray spectrometric

ANNEX

(Mandatory Information)

A1. Annex of Test Methods

A1.1 *Intent*—It is the intent of ASTM Task Group C07.05.03 to produce test methods for insertion into the annex of this test method.

A1.2 *Requirements:*

A1.2.1 *Method*—Universal application to current models of X-ray spectrometers is to be strived for.

A1.2.2 *Provisions*—Only deviations, special provisions, and interferences not referred to in the test method must be written into the specific test method.

A1.2.3 *Precision and Bias*—Test methods to be inserted in the annex will not be incorporated until a round robin study has been made generating acceptable precision and bias statements. Practice E 691 will be followed for the round robin guidelines.

Designation: C 1278/C 1278M – 96

Standard Specification for Fiber-Reinforced Gypsum Panels[1]

This standard is issued under the fixed designation C 1278/C 1278M; the number immediately following the designation indicates the year of original adoption or, in the case of revision, the year of last revision. A number in parentheses indicates the year of last reapproval. A superscript epsilon (ϵ) indicates an editorial change since the last revision or reapproval.

1. Scope

1.1 This specification covers fiber-reinforced gypsum panels designed to be used for walls, ceilings, or partitions and affords a suitable surface to receive decoration.

1.2 The values stated in either inch-pound units or SI (metric) are to be regarded separately as the standard. Within the text, the SI units are shown in brackets. The values stated in each system shall be used independently of the other. Values from the two systems shall not be combined.

1.3 The text of this specification references notes and footnotes that provide explanatory material. These notes and footnotes shall not be considered as requirements of the standard.

2. Referenced Documents

2.1 *ASTM Standards:*

C 11 Terminology Relating to Gypsum and Related Building Materials and Systems[2]

C 22 Specification for Gypsum[2]

C 473 Test Methods for Physical Testing of Gypsum Board Products and Gypsum Lath[2]

C 645 Specification for Nonstructural Steel Framing Members[2]

C 1264 Specification for Sampling, Inspection, Rejection, Certification, Packaging, Marking, Shipping, Handling, and Storage of Gypsum Board[2]

E 84 Test Method for Surface Burning Characteristics of Building Materials[3]

E 96 Test Methods for Water Vapor Transmission of Materials[4]

E 119 Test Methods for Fire Tests of Building Construction and Materials[3]

3. Terminology

3.1 *Definitions*—Definitions of terms shall be in accordance with Terminology C 11.

4. Materials and Manufacture

4.1 Fiber-reinforced gypsum panels shall consist essentially of gypsum with fibers dispersed throughout.

4.1.1 *Gypsum*—Specification C 22.

4.1.2 *Fibers*—Cellulose, derived from paper, paperboard stock, wood, or other organic material.

4.2 The back surface of foil-backed fiber-reinforced gypsum panels shall in addition be covered with aluminum foil.

4.3 Fiber-reinforced gypsum panels, type X (special fire-resistant) designates fiber-reinforced gypsum panels complying with this specification that provide not less than 1-h fire resistance for panels ⅝-in. [15.9-mm] thick or ¾-h fire-resistance for panels ½-in. [12.7-mm] thick, applied parallel with and on each side of load bearing 2 by 4 wood studs spaced 16 in. [406 mm] on center with 6d coated nails, 1⅞-in. [48-mm] long, 0.0915-in. [2.3-mm] diameter shank, ¼-in. [6.4-mm] diameter heads, spaced 7 in. [178 mm] on center with fiber-reinforced gypsum panel joints staggered 16 in. [406 mm] on each side of the partition and tested in accordance with Test Methods E 119.

NOTE 1—Consult producers for independent test data on assembly details and fire resistance classifications for other types of construction. See official fire test reports for assembly particulars, materials, and classification.

4.4 Fiber-reinforced gypsum panels shall have a flame spread index of not more than 25 when tested in accordance with Test Method E 84.

5. Mechanical Properties

5.1 Specimens shall be taken from the samples obtained in accordance with Specification C 1264.

5.2 Specimens shall be tested in accordance with Test Methods C 473.

5.2.1 *Flexural Strength*—The specimens shall have an average breaking load of not less than the following when tested face up or face down:

Thickness, in. [mm]	Breaking Load, lbf [N], Bearing Edges Parallel or Perpendicular to Panel Length, Method B
¼ [6.4]	50 [222]
5/16 [7.9]	65 [289]
⅜ [9.5]	75 [334]
½ [12.7]	110 [490]
⅝ [15.9]	155 [689]

5.2.2 *Humidified Deflection*—The specimens shall have an average deflection of not more than the following:

Thickness, in. [mm]	Humidified Deflection, Eighths of an in. [mm]
¼ [6.4]	not applicable
5/16 [7.9]	not applicable
⅜ [9.5]	not applicable
½ [12.7]	3 [10]
⅝ [15.9]	2 [6]

5.2.3 *Core, End and Edge Hardness*—The specimens shall have an average hardness of not less than 40 lbf [178 N] when tested in accordance with Method B.

5.2.4 *Nail Pull Resistance*—The specimens shall have an average nail pull resistance of not less than the following:

---

[1] This specification is under the jurisdiction of ASTM Committee C-11 on Gypsum and Related Building Materials and is the direct responsibility of Subcommittee C11.01 on Specifications and Test Methods for Gypsum Products.

Current edition approved Nov. 10, 1996. Published January 1997. Originally published as C 1278 – 94. Last previous edition C 1278 – 95.

[2] *Annual Book of ASTM Standards*, Vol 04.01.

[3] *Annual Book of ASTM Standards*, Vol 04.07.

[4] *Annual Book of ASTM Standards*, Vol 04.06.

Thickness, in. [mm]	Nail Pull Resistance, lbf [N], Method B
1/4 [6.4]	40 [178]
5/16 [7.9]	50 [222]
3/8 [9.5]	75 [334]
1/2 [12.7]	120 [534]
5/8 [15.9]	145 [645]

5.3 *Foil-Backed Fiber-Reinforced Gypsum Panels:*

5.3.1 Foil-backed fiber-reinforced gypsum panels shall meet all of the requirements for fiber-reinforced gypsum panels. In addition, aluminum foil shall be bonded to the back surface.

5.3.2 When tested in accordance with Test Methods E 96, the permeance of foil-backed fiber-reinforced gypsum panels shall be not more than 0.30 perm [17 ng/Pa·s·m^2] (Desiccant Method) for the condition of 50 % relative humidity on the face of the panel and 0 % relative humidity on the foil-covered back side of the panel.

6. Dimensions and Permissible Variations

6.1 Specimens shall be taken from the samples obtained in accordance with Specification C 1264.

6.2 Thickness, width, length, and end squareness shall be determined in accordance with Test Methods C 473.

6.2.1 *Thickness*—The nominal thickness shall be 1/4, 5/16, 3/8, 1/2, or 5/8 in. [6.4, 7.9, 9.5, 12.7, or 15.9 mm], with permissible variations in the nominal thickness of ±1/64 in. [0.4 mm] and permissible local variations of ±1/32 in. [0.8 mm] from the nominal thickness.

6.2.2 *Width*—The nominal width shall be 24 to 96 in. [610 to 2440 mm], with permissible variation of 3/32 in. [3 mm] under the specified width.

6.2.3 *Length*—The nominal length shall be from 6 to 24 ft [1830 to 7315 mm], with permissible variations of ±1/4 in. [6 mm] from the specified length.

6.2.4 The average thickness of the edge of recessed or tapered-edge shall be not less than 0.015 in. [0.38 mm] but not more than 0.075 in. [1.90 mm] less than the average thickness of the fiber-reinforced gypsum panel.

6.4 *Squareness:*

6.4.1 Corners of panels not more than 54 in. [1370 mm] in width shall be square with a permissible variation of ±1/8 in. [3 mm] in the full width of the panel.

6.4.2 Corners of panels more than 54 in. [1370 mm] in width shall be square with a permissible variation of ±5/16 in. [8 mm] in the full width of the panel.

7. Workmanship, Finish, and Appearance

7.1 The surfaces of the fiber-reinforced gypsum panels shall be true and free from imperfections that would render the panels unfit for use with or without decoration.

8. Sampling, Inspection, Rejection, Certification, Packaging, Marking, Shipping, Handling, and Storage

8.1 Shall be in accordance with Specification C 1264.

9. Keywords

9.1 ceiling; fiber-reinforced gypsum panels; foil-backed; FRGP; gypsum; partition; type X; wall

APPENDIX

(Nonmandatory Information)

This Appendix gives general information and also suggestions for inclusions to be made elsewhere by the specifier. They are not part of this specification.

X1. ALTERNATE DEFINITION FOR TYPE X

X1.1 Fiber-reinforced gypsum panels, type X (special fire-resistant) designates fiber-reinforced gypsum panels providing a greater fire resistance than regular fiber-reinforced gypsum panels of the same thickness. Type X (special fire-resistant) fiber-reinforced gypsum panels, when tested in accordance with Test Methods E 119, shall provide the following minimum fire resistance for the assemblies described:

X1.1.1 One hour for a 5/8-in. [15.9-mm] thickness applied to a partition in a single layer application on each side of 3 5/8-in. [92-mm] deep non-loadbearing galvanized steel studs complying with Specification C 645 spaced 24 in. [610 mm] on center. The 5/8-in. [15.9-mm] thick fiber-reinforced gypsum panels 48-in. [1220-mm] wide shall be attached using 1-in. [25-mm] long drywall screws spaced 8 in. [203 mm] on center along the edges and ends, and 12 in. [305 mm] along intermediate studs. All joints shall be oriented parallel to and located over studs and staggered on opposite sides of the assembly, and

X1.1.2 Two hours for a 1/2-in. [12.7-mm] thickness applied to a partition in a double layer application on each side of 2 1/2-in. [64-mm] deep non-loadbearing galvanized steel studs complying with Specification C 645 spaced 24 in. [610 mm] on center. The base layer 48-in. [1220-mm] wide shall be attached using 1-in. [25-mm] long drywall screws spaced 12 in. [305 mm] on center along panel edges, ends and along intermediate studs. Joints shall be oriented parallel to and located over studs and staggered on opposite sides of the assembly. The face layer 48-in. [1220-mm] wide shall be attached using 1 5/8-in. [41-mm] long drywall screws spaced 12 in. [305 mm] along panel edges, ends and along intermediate studs. Joints shall be oriented parallel to and located over studs, offset 24 in. [610 mm] from the base layer joints, and staggered on opposite sides of the assembly.

Designation: C 1280 – 98

Standard Specification for Application of Gypsum Sheathing[1]

This standard is issued under the fixed designation C 1280; the number immediately following the designation indicates the year of original adoption or, in the case of revision, the year of last revision. A number in parentheses indicates the year of last reapproval. A superscript epsilon (ϵ) indicates an editorial change since the last revision or reapproval.

1. Scope

1.1 This specification covers the minimum requirements for and methods of application of gypsum sheathing for use as a substrate for exterior wall cladding.

1.2 Details of construction for a specific assembly to achieve the required fire resistance shall be obtained from reports of fire-resistance tests, engineering evaluations, or listings from recognized fire testing laboratories.

1.2.1 This specification shall govern where it is more stringent (size or thickness of framing and size and spacing of fasteners) than the fire-rated construction.

1.3 Where sound control is required for a gypsum sheathing assembly, the details of construction shall be in accordance with the acoustical test report of an assembly that has met the required acoustical value(s).

1.3.1 This specification shall govern where it is more stringent (size or thickness of framing and size and spacing of fasteners) than the sound-rated construction.

1.4 Where resistance to racking loads or shear is required for a gypsum sheathing assembly, the details of construction shall be in accordance with the racking or shear test report of an assembly that has met the required racking or shear value(s).

1.4.1 This specification shall govern where it is more stringent (size or thickness of framing and size and spacing of fasteners) than the racking or shear-tested construction.

1.5 The values stated in inch-pound units are to be regarded as the standard. The SI (metric) values given in parentheses are approximate and are provided for information purposes only.

1.6 The text of this standard references footnotes which provide explanatory material. These footnotes shall not be considered as requirements of the standard.

1.7 *This standard does not purport to address all of the safety concerns, if any, associated with its use. It is the responsibility of the user of this standard to establish appropriate safety and health practices and determine the applicability of regulatory limitations prior to use.*

2. Referenced Documents

2.1 *ASTM Standards:*

C 11 Terminology Relating to Gypsum and Related Building Materials and Systems[2]

C 79 Specification for Treated Core and Nontreated Core Gypsum Sheathing Board[2]

C 954 Specification for Steel Drill Screws for the Application of Gypsum Panel Products or Metal Plaster Bases to Steel Studs from 0.033 in. (0.84 mm) to 0.112 in. (2.84 mm) in Thickness[2] C 955 Specification for Load-Bearing (Transverse and Axial) Steel Studs, Runners (Tracks), and Bracing or Bridging for Screw Application of Gypsum Panel Products and Metal Plaster Bases[2]

C 1002 Specification for Steel Drill Screws for the Application of Gypsum Panel Products or Metal Plaster Bases[2]

C 1007 Specification for the Installation of Load Bearing (Transverse and Axial) Steel Studs and Related Accessories[2]

C 1063 Specification for Installation of Lathing and Furring to Receive Interior and Exterior Portland Cement-Based Plaster[2]

2.2 *U.S. Department of Commerce Publication:*

PS20 American Softwood Lumber Standard[3]

3. Terminology

3.1 *Definitions*—Definitions shall be in accordance with Terminology C 11.

3.2 *Definitions of Terms Specific to This Standard:*

3.2.1 *exterior cladding, n*—a permanent material or system that impedes the transmission of environmental elements to the sheathing.

3.2.2 *fastener, n*—nails, staples, or screws used for application of the sheathing.

3.2.3 *framing member, n*—studs, headers, bracing, and blocking that serve to receive the sheathing.

3.2.4 *horizontal application, n*—a synonym for *perpendicular application.*

3.2.5 *parallel application, n*—gypsum sheathing applied with the edges parallel to the framing members; a synonym for *vertical application.*

---

[1] This specification is under the jurisdiction of ASTM Committee C-11 on Gypsum and Related Building Materials and Systems and is the direct responsibility of Subcommittee C11.03 on Specifications for Application of Gypsum and Other Products in Assemblies.

Current edition approved Apr. 10, 1998. Published September 1998. Originally published as C 1280 – 94. Last previous edition C 1280 – 97.

[2] *Annual Book of ASTM Standards,* Vol 04.01.

[3] Available from Standardization Documents Order Desk, Bldg. 4 Section D, 700 Robbins Ave., Philadelphia, PA 19111-5094, Attn: NPODS.

3.2.6 *perpendicular application, n*—gypsum sheathing applied with the edges at right angles to the framing members; a synonym for *horizontal application.*

3.2.7 *require, v*—to mandate by a force outside this specification, such as a building code, project specification, contract, or purchase order.

3.2.8 *specified, adj*—pertaining to a mandatory requirement of this specification or a referenced requirement.

3.2.9 *specify, v*—to mandate by an obligation of this standard or a referenced document.

3.2.10 *vertical application, n*—synonym for *parallel application.*

3.2.11 *weather resistive barrier, n*—a temporarily exposed protective membrane that is intended to impede the penetration of environmental elements until the installation of a permanent exterior cladding.

4. Exposure After Installation

4.1 *Core-Treated Gypsum Sheathing Board*—Core-treated gypsum sheathing board is a substrate that shall be covered by an exterior cladding or other weather-resistive barrier and is not intended for long-term exposure. It shall not be exposed to the elements for more than 30 days after it has been installed. The core-treated gypsum sheathing shall be covered with a weather-resistive barrier within 30 days if the exposure time will be more than 30 days.

NOTE—Some building codes require an additional weather-resistive barrier. The exterior face paper of the sheathing shall be dry prior to application of the additional weather-resistive barrier.

4.2 *Untreated-Core Gypsum Sheathing Board*—Untreated-core gypsum sheathing board should be covered immediately with a weather-resistive barrier.

5. Materials and Manufacture

5.1 *Gypsum Sheathing*—Specification C 79.

5.1.1 *Core-Treated Gypsum Sheathing*—Gypsum sheathing with a water-resistive material incorporated in the core as defined in Specification C 79.

5.1.2 *Type X (Special Fire-Resistant) Gypsum Sheathing*—Gypsum sheathing that provides a greater degree of fire resistance than regular gypsum sheathing as defined in Specification C 79.

5.2 *Fasteners*—Fasteners shall be as described in 5.2.1 through 5.2.3. The fastener length shall be not less than that specified in Table 1.

5.2.1 *Nails*—Nails shall be not less than 12-gage galvanized, 7⁄16-in. (11.1-mm) diameter head.

TABLE 1 Minimum Fastener Lengths

Framing Type	Fastener Type	Sheathing Thickness, in. (mm)	Minimum Fastener Length, in. (mm)
Wood	nails	½ (12.7)	1½ (38)
		⅝ (15.9)	1¾ (45)
	screws	½ (12.7)	1¼ (32)
		⅝ (15.9)	1¼ (32)
	staples	½ (12.7)	1½ (38)
		⅝ (15.9)	1¾ (45)
Steel	screws	½ (12.7)	1 (25)
		⅝ (15.9)	1¼ (32)

5.2.2 *Screws:*

5.2.2.1 Screws for fastening gypsum sheathing to wood framing members, and to steel framing members less than 0.033 in. (0.84 mm) in thickness, shall meet the requirements of Specification C 1002.

5.2.2.2 Screws for fastening gypsum sheathing to steel framing members from 0.033 to 0.112 in. (0.84 to 2.84 mm) in thickness shall meet the requirements of Specification C 954.

5.2.3 *Staples*—Staples shall be of galvanized steel, not less than 16-gage, 7⁄16-in. (11.1-mm) wide crown outside measurement. Legs shall have divergent points.

5.3 *Framing Members:*

5.3.1 *Wood Framing*—Wood framing members shall conform to PS20.

5.3.2 *Steel Framing*—Load-bearing steel framing shall meet the requirements of Specification C 955.

6. Substrate, Surface Preparation

6.1 The maximum spacing for framing members for gypsum sheathing shall be not more than 24 in. (610 mm) on center on walls.

6.1.1 Wood framing shall be straight and true, attached securely following accepted engineering practices and as required for the intended design. The surfaces to which abutting edges or ends of gypsum sheathing are to be attached shall be not less than 1½-in. (38-mm) wide. The bearing surface shall be not less than ¾ in. (19 mm) for internal corners or angles.

6.1.2 Load-bearing metal framing members shall be of the proper size and design for their intended use and shall be installed in accordance with Specification C 1007.

7. Framing Alignment

7.1 Framing members shall be installed so that the surface will be in an even plane, unless otherwise specified, after the gypsum sheathing has been applied.

8. Cutting and Application of Gypsum Sheathing

8.1 *Cutting Gypsum Sheathing:*

8.1.1 The gypsum sheathing shall be cut by scoring and breaking or by sawing, working from the face side. The face paper shall be cut with a sharp knife or other suitable tool when cutting by scoring. The gypsum sheathing shall be broken by snapping the gypsum sheathing in the reverse direction or by cutting the back paper with a knife or suitable tool.

8.1.2 The cut edges and ends of gypsum sheathing shall be trimmed to obtain neat fitting joints when installed.

8.1.3 Holes for pipes, fixtures, or other small openings shall be scored on the face paper before removing the cutout with a saw or special tool designed for this purpose.

8.1.4 Where gypsum sheathing meets projecting surfaces, the gypsum sheathing shall be scribed and cut neatly.

8.2 *Application of Gypsum Sheathing:*

8.2.1 Gypsum sheathing shall be fitted snugly around all window and door openings.

8.2.1.1 Sheathing joints shall be offset a minimum of 4 in. (100 mm) from the edge of any opening.

8.2.1.2 Gypsum sheathing shall be flashed at openings so that water intrusion will not contact the sheathing.

8.2.2 All vertical end and edge joints shall abut over the centers of framing members and shall be offset a minimum of one framing space between adjacent rows of gypsum sheathing.

8.2.3 *Control Joints:*

8.2.3.1 Control joints shall be provided in wall runs exceeding 30 ft (9100 mm) in length. The distance between control joints shall be not more than 30 ft (9100 mm).

8.2.3.2 Control joints shall be installed in walls wherever there is a building construction joint or where required.

8.2.3.3 The location of the control joints shall be as required by either the building design or the exterior cladding specified.

8.2.3.4 Control joints, other accessories, and metal plaster base shall be fastened through gypsum sheathing to framing members.

8.2.4 Gypsum sheathing, 4-ft (1220-mm) wide, shall be permitted to be applied either parallel or perpendicular to framing members.

8.2.4.1 Square edge gypsum sheathing applied perpendicular to framing members shall be covered with a weather-resistive barrier, or horizontal joints shall be sealed.

8.2.5 Tongue and groove gypsum sheathing shall be applied with the long dimension perpendicular to framing members, with the grooved edges down, interlocking the tongue of the panel below without forcing.

8.2.6 Fasteners shall be spaced not more than 8 in. (203 mm) on center along vertical ends or edges and intermediate supports. The length of fasteners shall be as indicated in Table 1.

8.2.6.1 Fasteners shall be located not less than ⅜ in. (9.5 mm) from the ends and edges of the gypsum sheathing.

8.2.6.2 Gypsum sheathing shall be held tight against the framing when driving fasteners.

8.2.6.3 Fasteners shall be driven so that the heads are at or slightly below the surface of the gypsum sheathing without breaking the face paper or fracturing the core.

8.2.6.4 Screws shall be driven without stripping the framing member around the screw shank.

8.2.6.5 Staples shall be driven with the crown parallel to the framing member.

8.2.7 *Application of Gypsum Sheathing Under Self-Furred Metal Lath*—Where fire resistance or shear resistance is not required, and when metal lath and portland cement plaster are to be applied as the exterior cladding over gypsum sheathing installed over framing members spaced not more than 16 in. (406 mm) on center, and the metal lath shall be installed in accordance with Specification C 1063 within 24 h after the gypsum sheathing, the gypsum sheathing shall be permitted to be applied as specified in 8.2.7.1 and 8.2.7.2, and fastening of the gypsum sheathing shall be completed with the attachment of the self-furred metal lath.

8.2.7.1 For 2-ft (610-mm) wide sheathing installed in accordance with 8.2.5, vertical ends of gypsum sheathing shall be fastened in accordance with 8.2.6. Gypsum sheathing shall be permitted to be fastened in place at each intermediate stud with a minimum of two fasteners per board spaced not more than 2 in. (50 mm) from opposite edges.

8.2.7.2 For 4-ft (1220-mm) wide sheathing installed in accordance with 8.2.4, vertical ends or edges of gypsum sheathing shall be fastened in accordance with 8.2.6. Gypsum sheathing shall be permitted to be fastened in place at each intermediate stud with fasteners spaced not more than 24 in. (610 mm) on center.

9. Keywords

9.1 exterior walls; gypsum; sheathing

Standard Test Method for Major and Trace Elements in Limestone and Lime by Inductively Coupled Plasma-Atomic Emission Spectroscopy (ICP) and Atomic Absorption (AA)[1]

This standard is issued under the fixed designation C 1301; the number immediately following the designation indicates the year of original adoption or, in the case of revision, the year of last revision. A number in parentheses indicates the year of last reapproval. A superscript epsilon (ε) indicates an editorial change since the last revision or reapproval.

1. Scope

1.1 The following test method covers the use of inductively coupled plasma-atomic emission spectroscopy (ICP) and atomic absorption spectroscopy (AA) in the analysis of major and trace elements in limestone and lime (calcined limestone).

1.2 Table 1 lists some of the elements that can be analyzed by this test method and the preferred wavelengths. Also see U.S. EPA Method 200.7 and 200.9.

1.3 *This standard does not purport to address all of the safety concerns, if any, associated with its use. It is the responsibility of the user of this standard to establish appropriate safety and health practices and determine the applicability of regulatory limitations prior to use.*

2. Referenced Documents

2.1 *ASTM Standards:*

C 51 Terminology Relating to Lime and Limestone (As Used by the Industry)[2]

D 1193 Specification for Reagent Water[3]

E 135 Terminology Relating to Analytical Chemistry for Metals, Ores, and Related Materials[4]

E 863 Practice for Describing Flame Atomic Absorption Spectroscopy Equipment[5]

E 1479 Practice for Describing and Specifying Inductively-Coupled Plasma Optical Emission Spectrometers[5]

2.2 *U.S. EPA Standards:*

Methods for the Determination of Metals in Environmental Samples; U.S. EPA Methods 200.2, 200.7 and 200.9; Smoley, C. K., 1992[6]

Method 6010, Inductively Coupled Plasma Method, SW-846, Test Methods for Evaluating Solid Waste[7]

3. Terminology

3.1 *Definitions*—Definitions for terms used in this test method can be found in Terminologies C 51 and E 135.

3.2 *Additional Definitions:*

3.2.1 *total recoverable, n*—trace element concentration in an unfiltered sample after heating in acid.

3.2.2 *total digestion, n*—complete digestion of a sample, including silica and silicate minerals, using the fusion-flux method.

TABLE 1 Elements and Some Suggested Wavelengths[A]

Major Elements	ICP Wavelength, nm	AA Wavelength, nm
Calcium	317.933 (315.887)[B]	422.7
Magnesium	279.079 (285.213)	285.2
Silicon	251.611 (288.160)	251.6
Aluminum	308.215 (309.271)	309.3
Iron	259.940	248.3
Manganese	257.610	279.5
Sodium	588.995 (589.59)	589.0
Potassium	766.491	766.5
Phosphorus	214.914 (213.618)	. . .[C]
Strontium	421.552	460.7
Trace Elements	**ICP Wavelength, nm**	**AA Wavelength, nm**
Antimony	206.833	217.6
Arsenic	193.696	193.7
Barium	455.403 (493.409)	553.6
Beryllium	313.042	234.9
Boron	249.773	249.8
Cadmium	226.502 (228.80)	228.8
Chromium	267.716 (205.552)	357.9
Cobalt	228.616	240.7 (242.5)
Copper	324.754	324.8
Lead	220.353	217.0 (283.3)
Molybdenum	202.030 (203.844)	313.3
Nickel	231.604 (221.647)	232.0
Selenium	196.090	196.0
Silver	328.068	328.1
Sulfur	180.731 (180.669)	. . .[C]
Thallium	190.864	276.8
Tin	189.989	235.5 (286.3)
Vanadium	292.402	318.4
Zinc	213.856 (202.551)	213.9

[A] The suggested wavelengths may vary for your particular instrument.
[B] Numbers in parentheses are alternate wavelengths.
[C] Not recommended or not used.

4. Summary of Test Method

4.1 A sample, digested by either fusion or acid, is atomized and passed into an excitation medium (a plasma in the case of ICP; a flame in the case of AA). The resulting ions are analyzed by atomic spectroscopy. Elemental concentrations are determined by graphically relating the emission/absorption at specific wavelengths for an unknown sample to analytical curves made from reference standards of known composition.

5. Significance and Use

5.1 The presence and concentration of elements in lime and limestone is important in determining product quality

[1] This test method is under the jurisdiction of ASTM Committee C-7 on Lime and is the direct responsibility of Subcommittee C07.05 on Chemical Tests.

Current edition approved Aug. 15, 1995. Published October 1995.

[2] *Annual Book of ASTM Standards*, Vol 04.01.

[3] *Annual Book of ASTM Standards*, Vol 11.01.

[4] *Annual Book of ASTM Standards*, Vol 03.05.

[5] *Annual Book of ASTM Standards*, Vol 03.06.

[6] Available from CRC Press, 2000 Corporate Blvd., N. W., Boca Raton, FL 33431.

[7] Available from U.S. Government Printing Office, Washington, DC 20402.

and its suitability for various uses. This test method provides a means of measuring the major and trace element concentration in lime and limestone.

6. Interferences

6.1 *Chemical*—Chemical interferences, most common in AA, arise from the formation of molecular compounds that cause absorbances at the wavelength of interest. This molecular band spectral overlap can be minimized by buffering the sample with matrix modifiers (a Lanthanum additive, for example), using standard additions techniques, matrix matching or by careful selection of operating conditions (for example, using a hotter nitrous oxide/acetylene flame, selecting an alternate wavelength).

6.2 *Physical*—Physical interferences are the result of the inconsistancies in the introduction of the sample into the instrument, namely the transport and atomization/nebulization of the sample. These inconsistancies are a function of changing viscosity and surface tension, and are found primarily in samples of high-dissolved solids or high-acid concentrations. Physical interferences can be reduced by diluting the sample and by the use of a peristaltic pump.

6.3 *Spectral*—Spectral interference, most common in ICP, consists of overlapping and unresolved peaks. Computer software, along with the analysis of the suspected interfering element, can compensate for this effect. Using an alternate wavelength is also a solution. Another spectral interference is caused by background, both stray light and continuous spectrum (continuous argon spectrum, for example). Background correction adjacent to the analyte line will correct background spectral interference.

7. Apparatus

7.1 *Spectrometer.*

7.1.1 *Inductively Coupled Plasma Emission Spectrometer (ICP)*—Either a scanning sequential or multielement simultaneous type ICP, with resolution appropriate for the elements to be analyzed. The optical path may be in air, vacuum or an inert gas. A detailed description of an ICP is given in Practice E 1479.

7.1.2 *Atomic Absorption Spectrometer (AA)*—An atomic absorption spectrometer consisting of single or double beam optics, a monochromator, photomultiplier detector, adjustable slits, a wavelength range from 190 to 800 nm, and provisions for interfacing with either a strip chart recorder or a computer. A simultaneous background correction system is also recommended. A detailed description of an AA is given in Practice E 863.

7.1.2.1 *Hollow Cathode Lamps*—Single hollow cathode lamps, one for each element. Multielement hollow cathode lamps can be used but spectral interferences are possible.

8. Reagents

8.1 *Purity of Reagents*—Reagents should conform to the specifications of the Committee on Analytical Reagents of the American Chemical Society as a minimum when such specifications are available.[8] The high sensitivity of both the ICP and AA may require reagents of high purity. It is recommended that the reagents be of sufficiently high purity so as not to lessen the accuracy of the determination.

8.2 *Purity of Water*—At minimum, water should conform to Type II of Specification D 1193.

8.3 *Stock Solutions*—Standard stock solutions may be purchased or prepared from high purity metals or metal salts (Method 6010, SW-846; EPA Method 200.7 and 200.9). Salts should be dried at 105°C for 1 h, unless otherwise specified.

8.4 *Multielement Calibration Standards*—ICP calibration is most often performed using multielement calibration standards prepared from single element stock solutions. Prior to preparing the mixed standards, each stock solution should be analyzed separately to determine possible spectral interference or the presence of impurities. Standards are combined in such a way that they are chemically compatible (no precipitation occurs) and do not cause spectral interferences. An example of multielement combinations is given in EPA Method 200.7.

8.5 *Interference Check Sample*—Interference check samples are made from single element stock solutions at a concentration level equal to that of the samples to be analyzed.

8.6 *Calibration Blank*—A calibration blank is prepared at the same acid strength as that of the samples to be analyzed; usually 5 or 10 %. To prepare a 10 % nitric acid calibration blank, add one volume of nitric acid to nine volumes of water. This same blank can be used as the rinse solution for flushing the system between standards and samples.

8.7 *Reagent Blank*—The reagent blank contains all the reagents in the same concentrations (including nitric acid) as the samples to be analyzed. The reagent blank is carried through the same processes as a sample for analysis.

8.8 *Nitric Acid*—High purity nitric acid is recommended.

8.9 *Lithium Tetraborate (Li2B4O7) Powder or Lithium Metaborate (LiBO2)) Powder.*

8.10 *Non-Wetting Agent*—Saturated solution of Hydrogen Bromide (HBr), Potassium Bromide (KBr) or Potassium Iodide (KI) as a non-wetting agent to prevent the flux from sticking to the crucible.

8.11 *Lanthanum Chloride (LaCl3) Powder*—Lanthanum is added to samples for AA analysis as a releasing agent (for Calcium) and ionization suppressant (for Aluminum). When added to the sample solution, Lanthanum will preferentially react with potential interferents and "release" the analyte. In addition, the Lanthanum will preferentially ionize relative to Aluminum, aiding in the number of ground state Aluminum atoms. A typical Lanthanum additive is prepared by dissolving 175 g LaCl3 in 1 L of water (equivalent to 100 g/L Lanthanum).

[8] *Reagent Chemicals, American Chemical Society Specifications,* American Chemical Society, Washington, DC. For suggestions on the testing of reagents not listed by the American Chemical Society, see *Analar Standards for Laboratory Chemicals,* BDH Ltd., Poole, Dorset, U.K., and the *United States Pharmacopeia and National Formulary,* U.S. Pharmaceutical Convention, Inc. (USPC), Rockville, MD.

9. Preparation of Apparatus

9.1 Prepare and operate the spectrometer in accordance with the manufacturer's instructions. The present method assumes that good operating procedures are followed. Design differences between spectrometers make it impractical to specify the required steps in detail here.

10. Calibration and Standardization

10.1 Allow a warm-up time of at least 30 min. Operate the spectrometer according to the operation manual for the instrument.

10.2 Calibrate the instrument by aspirating the blank and standards. A 10 % by volume HNO3 rinse solution is aspirated for a minimum of 60 s between each standard. Most new systems are controlled by computer. The computer will establish the slope, intercept and correlation coefficients for each element. Some suggested wavelengths are given in Table 1 and EPA Methods 200.2, 200.7, and 200.9.

10.3 A peristaltic pump is recommended for aspirating standards and samples. The peristaltic pump will reduce physical interferences caused by changes in specimen viscosity and concentration (transport processes).

11. Sample Preparation

11.1 *Major Elements*—Samples for major element analysis are prepared for total digestion using lithium tetraborate or lithium metaborate as a flux. Major elements include Calcium, Magnesium, Silicon, Aluminum, Iron, Manganese, Sodium and Potassium. Trace elements such as Lead, Arsenic, Selenium and Antimony will partially volatilize using this fusion method and it is therefore not recommended for trace element analysis.

11.1.1 Take a representative minus 100 mesh sample split and dry at 105°C for 2 h.

11.1.2 Weigh 0.25 g of dried sample in a graphite or platinum crucible. Then weigh 1.00 g of lithium metaborate in the crucible. Add a few drops of non-wetting agent if needed. Mix the sample and lithium metaborate (the flux) well. Cover the mixed sample-lithium metaborate with an additional 0.50 g of lithium metaborate. This will give a total sample-flux ratio of 1:6.

11.1.3 Place a lid (optional) on the crucible prepared in 11.1.2 and place in a muffle furnace at 1000°C for 30 min. Gently agitate the molten contents of the crucible at least once during the 30° min heating.

11.1.4 Add 12.5 mL of concentrated nitric acid and 40 mL of water to a clean 250 mL wide-mouth plastic bottle.

11.1.5 When the 30 min heating in 11.1.3 is complete, quickly pour molten contents of the crucible into the plastic bottle described in 11.1.4. The water will bubble and sizzle. Quickly put the lid on the plastic bottle and shake. To aid in digestion place the bottle in a warm ultrasonic bath.

11.1.6 The contents of the plastic bottle can either be quantitatively transferred to a 250 mL volumetric flask and diluted to volume or diluted to volume by weight (that is, 1 mL = 1 g) in the same 250 mL plastic bottle it was digested in. Keep in mind, however, that the standards need to be made in the same manner as the samples. Add 10 mL Lanthanum additive to samples for AA analysis (10 mL addition is part of the dilution to volume). Filtering is not necessary.

11.2 *Trace Elements*—Samples for trace element analysis are prepared using hydrochloric acid (1:5 or 1+4) and nitric acid (1:2 or 1+1). The trace elements concentrations determined by this method are termed "total recoverable" (that is, components not digested in hot acid are not recovered). Trace elements include, but are not limited to, Antimony, Arsenic, Barium, Beryllium, Boron, Cadmium, Chromium, Cobalt, Copper, Lead, Molybdenum, Nickel, Selenium, Silver, Strontium, Thallium, Tin, Vanadium, and Zinc.

11.2.1 Take a representative minus 100 mesh sample split and dry at 105°C for 2 h.

11.2.2 Weigh 1.00 g of dried sample and transfer to a clean beaker.

11.2.3 Add 10 mL of dilute hydrochloric acid (see 11.2) and 4 mL of dilute nitric acid (see 11.2). In the case of limestone, add the acids slowly so that the powdered sample will not splatter on the sides of the beaker during it's effervescent reaction with the acid. Cover the sample (a ribbed watch glass is best) and place on a hot plate. Heat at approximately 85°C for 30 min. Boiling should be kept to a minimum. After 30 min allow the sample to cool. Then quantitatively transfer the liquid and any undissolved residue to a 100 mL volumetric flask. Dilute to volume.

12. Procedure

12.1 Aspirate the specimens prepared in Section 11 into the AA or ICP using the same conditions used during calibration.

12.2 Analyze the instrument check standards, blanks and internal control samples at a 10 % or better frequency. The results on the instrument check standards are to be within 10 %, and the internal control specimens should fall within established limits of deviation. If the results exceed these limits, investigate the cause and take corrective action.

12.3 It is recommended that 10 % of the samples analyzed be duplicates. Process duplicates through the same dissolution and analysis procedures as any other sample. The results should fall within established limits of deviation. If the results exceed these limits, investigate the cause and take corrective action. The applicability of duplicate analysis data as a measure of analytical and sampling reproducibility assumes specimen homogeneity.

12.4 Analyze an interference check specimen at the beginning and the end of each specimen run or a minimum of twice per 8 h work shift, whichever is more frequent. This check specimen should contain, in relatively high concentration, those elements which are expected to be present at significant levels in the samples and which are known interfering species. All other elements should be present at relatively low levels in order to assess the quality of interference corrections.

12.5 Use the background and interference corrected data to calculate the concentration of each element. This calculation, including the dilution factor, is performed by the computer.

13. Precision and Bias

13.1 Participation in a round robin is anticipated.

14. Keywords

14.1 atomic absorption; atomic emission spectroscopy; digestion; dissolution; inductively coupled plasma; lime; limestone; spectrometer; trace elements

Designation: C 1318 – 95

Standard Test Method for Determination of Total Neutralizing Capability and Dissolved Calcium and Magnesium Oxide in Lime for Flue Gas Desulfurization (FGD)[1]

This standard is issued under the fixed designation C 1318; the number immediately following the designation indicates the year of original adoption or, in the case of revision, the year of last revision. A number in parentheses indicates the year of last reapproval. A superscript epsilon (ε) indicates an editorial change since the last revision or reapproval.

1. Scope

1.1 This test method covers analysis of magnesian, dolomitic and high-calcium limes for total neutralizing capability and dissolved major oxides. Dissolved calcium and magnesium are the major species that neutralize acid under the conditions of the test.

1.2 The test conditions are chosen to measure the acid-neutralizing capacity of both calcium hydroxide and magnesium hydroxide contained in slaked lime. By controlling the neutralization pH at 6, magnesium hydroxide and magnesium oxide are titrated in addition to calcium hydroxide fraction.

1.3 This test method also determines the fraction of Mg ions present in the lime that will dissolve under lime flue gas desulfurization (FGD) conditions. Because the Mg^{++} ion alters FGD performance, it is important to know its concentration.

1.4 *This standard does not purport to address all of the safety concerns, if any, associated with its use. It is the responsibility of the user of this standard to establish appropriate safety and health practices and determine the applicability of regulatory limitations prior to use.*

2. Referenced Documents

2.1 *ASTM Standards:*

C 25 Test Methods for Chemical Analysis of Limestone, Quicklime, and Hydrated Lime[2]

C 50 Practice for Sampling, Inspection, Packing and Marking of Lime and Limestone Products[2]

C 51 Terminology Relating to Lime and Limestone (as Used by the Industry)[2]

3. Terminology

3.1 *Definitions*—Unless otherwise specified, for definitions of terms used in these test methods refer to Terminology C 51.

4. Summary of Test Method

4.1 Lime is slaked by boiling and is reacted with acid at a pH and residence time similar to those found in full-scale FGD reaction tanks.

4.2 A sample of lime is titrated with 1N hydrochloric acid, maintaining a pH of 6 for 30 min. After 30 min, the acid consumption is recorded. The total neutralizing capacity is calculated from the acid consumption and reported as CaO. Dissolved magnesium is determined by atomic absorption spectrometry or by EDTA titration and reported as percent dissolved Magnesium Oxide (as MgO).

5. Significance and Use

5.1 There are existing lime-based flue gas desulfurization units in operation that require a method to measure the oxides available for sulfur dioxide absorption. Dissolved magnesium oxide varies among limes depending on the limestone sources and calcination conditions.

6. Interferences

6.1 Any substance reacting with acid under the conditions of the test will contribute to the total oxide and dissolved oxide values.

6.2 Magnesium in forms other than MgO, which dissolve under test conditions, may affect the dissolved MgO and total oxide value.

7. Apparatus

7.1 *Digital Readout pH Meter*, with combination electrode readable to 0.01 pH units, or an autotitrator with an automatic temperature compensator capable of titrating to a preset endpoint (Note 1).

NOTE 1—Use of an automatic titrator is recommended.

7.2 *Stirrers and Stir Bars.*

7.3 *Atomic Absorption Spectrometer*, if not using EDTA titration.

7.4 *Filtering Funnel*, Type AE glass fiber paper, volumetric flasks (size as needed), 500 mL volumetric flasks, and various Class A pipettes.

8. Reagents

8.1 *1.0 Normal Hydrochloric Acid*, standardize using procedures in Test Methods C 25, Section 28.

8.2 *Distilled Water*, CO_2 free.

8.3 *Calcium and Magnesium Standard Solutions* (commercially available or use methods in Test Methods C 25, Section 31).

8.4 *EDTA Standard Solution* (0.4 %), standardize according to Test Methods C 25, Section 31.5, Standardization of EDTA.

9. Sample Preparation

9.1 The sampling and grinding of any lime sample shall be carried out as rapidly as possible, so that the absorption of

[1] This test method is under the jurisdiction of ASTM Committee C-7 on Lime and is the direct responsibility of Subcommittee C07.05 on Chemical Tests.

Current edition approved Nov. 10, 1995. Published February 1996.

[2] *Annual Book of ASTM Standards*, Vol 04.01.

moisture and carbon dioxide is held to a minimum.

9.2 The sample as received at the laboratory shall be thoroughly mixed and a representative sample with minimum weight of 100 g shall be taken and pulverized to pass a No. 50 mesh sieve for analysis.

9.3 The prepared sample shall be stored in a tightly capped container.

10. Procedure

10.1 *Slaking of Lime Sample:*

10.1.1 Weigh rapidly 1.402 g of the finely pulverized sample and brush carefully into a 250 mL Erlenmeyer flask containing about 50 mL of CO_2 free water and immediately stopper the flask.

10.1.2 Remove the stopper. Swirl the flask, place on a hot plate and boil actively 1 min for complete slaking. Remove from the hot plate, stopper the flask loosely and place in a cold-water bath to cool to room temperature.

10.2 *Titration of Sample:*

10.2.1 Set the pH electrode, burette tip, and temperature sensing device in place, while maintaining agitation.

10.2.2 Begin the titration by adding standard 1.0 normal hydrochloric acid solution. Titrate to and maintain a pH of 6 within 0.4 pH units for 30 min (Note 2). Time begins from initial addition of acid.

NOTE 2—When doing a manual titration, the increment of acid addition may require the use of partial drops (suspend a small amount of titrant on the buret tip and wash into the titration flask with CO_2 free distilled water) to avoid exceeding the 0.4 unit limit of pH 6.

10.2.3 Record the amount of acid consumed after 30 min.

10.2.4 Filter the solution and rinse the flask thoroughly with CO_2 free distilled water.

10.2.5 Carefully transfer the solution to an appropriately sized volumetric flask, add approximately 10 mL of 1:1 hydrochloric acid, and dilute to volume.

10.3 *Determination of Dissolved Oxides:*

10.3.1 *Atomic Absorption:*

10.3.1.1 From the diluted solution make the required dilutions and determine the dissolved magnesium by standard atomic absorption spectroscopy techniques.

10.3.1.2 Calculate the dissolved magnesium as MgO according to 11.2.

10.3.2 *EDTA Titration:*

10.3.2.1 From the diluted solution pipette 20 mL of sample into a 500 mL Erlenmeyer flask and dilute with CO_2 free distilled water to an approximate volume of 100 mL.

10.3.2.2 Determine the total dissolved calcium according to Test Methods C 25, Paragraph 31.6.4. Record mL of EDTA required for titration.

10.3.2.3 Repeat 10.3.2 and determine the total dissolved oxides according to Test Methods C 25, Paragraph 31.6.5. Record mL of EDTA required for this titration.

11. Calculation of Results

11.1 *Percent Total Neutralizing Capability (TNC):*

$$\%\ \text{TNC (as CaO)} = \frac{A \times B \times C_1}{W} \tag{1}$$

where:
A = mL HCl used in 10.2.3,
B = normality HCl,
C_1 = chemical factor = 2.804 (mequi of CaO/10), and
W = sample weight, g.

11.2 *Atomic Absorption:*

11.2.1 *Dilution Factor:*

$$\text{Dilution factor} = \text{Initial Volume (see 10.2.5)/Pipette volume for final dilution} \tag{2}$$

11.2.2 *Percent Dissolved Magnesium:*

$$\%\ \text{Dissolved Mg}^{++}\ \text{(as MgO)} = \frac{A \times B \times C_2 \times D}{W \times 1000} \tag{3}$$

where:
A = concentration of magnesium from AA, ppm Mg^{++},
B = dilution factor,
C_2 = chemical factor = 1.658 (MgO/Mg^{++}),
D = final dilution volume, mL, and
W = sample weight, g.

11.3 *EDTA Titration:*

11.3.1 *Dilution Factor:*

$$\text{Dilution factor} = \text{Initial Volume (see 10.2.5)/titration volume (20 mL)} \tag{4}$$

11.3.2 *Percent Dissolved Calcium:*

$$\%\ \text{Dissolved Ca}^{++}\ \text{(as CaO)} = \frac{A \times B \times C_3 \times D}{W \times 10} \tag{5}$$

where:
A = mL of EDTA standard solution used,
B = dilution factor,
C_3 = chemical factor = 1.399 (CaO/Ca^{++}),
D = EDTA titer, mg Ca^{++}/mL EDTA, and
W = sample weight, g.

11.3.3 *Percent Total Dissolved Oxides (TDO):*

$$\%\ \text{TDO (as CaO)} = \frac{A \times B \times C_3 \times D}{W \times 10} \tag{6}$$

where:
A = mL of EDTA standard solution used,
B = dilution factor,
C_3 = chemical factor = 1.399 (CaO/Ca^{++}),
D = EDTA titer, mg Ca^{++}/mL EDTA, and
W = sample weight, g.

11.3.4 *Percent Dissolved Magnesium:*

$$\%\ \text{Dissolved Mg}^{++}\ \text{(as MgO)} = (A - B) \times C_4 \tag{7}$$

where:
A = % total dissolved oxides,
B = % dissolved calcium, and
C_4 = chemical factor = 0.7188 (MgO/CaO).

12. Precision and Bias

12.1 The precision and bias of this test method have not been determined.

12.2 When sufficient data has been obtained and analyzed, a statement of precision will be provided.

12.3 The user is cautioned to verify by the use of reference materials, if available, that the bias of the test method is adequate for the contemplated use.

13. Keywords

13.1 dissolved calcium; dissolved magnesium; flue gas desulfurization (FGD); lime; pH determination; total neutralizing capability

Standard Specification for Plastic (Stucco) Cement[1]

This standard is issued under the fixed designation C 1328; the number immediately following the designation indicates the year of original adoption or, in the case of revision, the year of last revision. A number in parentheses indicates the year of last reapproval. A superscript epsilon (ε) indicates an editorial change since the last revision or reapproval.

1. Scope

1.1 This specification covers two types of plastic cement for use in portland cement-based plasters for exterior (stucco) and interior application.

1.2 The values stated in SI units are to be regarded as the standard. Values in SI units shall be obtained by measurement in SI units or by appropriate conversion, using the Rules for Conversion and Rounding given in IEEE/ASTM SI 10, of measurements made in other units.

1.3 The text of this standard refers to notes and footnotes that provide explanatory material. These notes and footnotes (excluding those in tables and figures) shall not be considered as requirements of the standard.

1.4 The following safety hazards caveat pertains only to Sections 15 and 16. *This standard does not purport to address all of the safety concerns, if any, associated with its use. It is the responsibility of the user of this standard to establish appropriate safety and health practices and determine the applicability of regulatory limitations prior to use.*

2. Referenced Documents

2.1 *ASTM Standards:*

C 91 Specification for Masonry Cement[2]

C 109/C 109M Test Method for Compressive Strength of Hydraulic Cement Mortars (Using 2-in. or 50-mm Cube Specimens)[2]

C 151 Test Method for Autoclave Expansion of Portland Cement[2]

C 183 Practice for Sampling and the Amount of Testing of Hydraulic Cement[2]

C 185 Test Method for Air Content of Hydraulic Cement Mortar[2]

C 187 Test Method for Normal Consistency of Hydraulic Cement[2]

C 188 Test Method for Density of Hydraulic Cement[2]

C 219 Terminology Relating to Hydraulic Cement[2]

C 230 Specification for Flow Table for Use in Tests of Hydraulic Cement[2]

C 266 Test Method for Time of Setting of Hydraulic-Cement Paste by Gillmore Needles[2]

C 305 Practice for Mechanical Mixing of Hydraulic Cement Pastes and Mortars of Plastic Consistency[2]

C 430 Test Method for Fineness of Hydraulic Cement by the 45-μm (No. 325) Sieve[2]

C 511 Specification for Moist Cabinets, Moist Rooms, and Water Storage Tanks Used in the Testing of Hydraulic Cements and Concretes[2]

C 778 Specification for Standard Sand[2]

C 926 Specification for Application of Portland Cement-Based Plaster[2]

IEEE/ASTM SI 10 Standard for Use of the International System of Units (SI) (the Modernized Metric System)[3]

3. Terminology

3.1 *Definitions*—Terms used in this specification are defined in Terminology C 219 and Specification C 926.

3.2 *Definition of Term Specific to This Standard:*

3.2.1 *plastic cement*—a hydraulic cement, primarily used in portland cement-based plastering construction, consisting of a mixture of portland or blended hydraulic cement and plasticizing materials (such as limestone or hydrated or hydraulic lime), together with other materials introduced to enhance one or more properties such as setting time, workability, water retention, and durability.

3.2.1.1 *Discussion*—The term "plastic" does not refer to the inclusion of one or more organic components in the cement. The cement is predominantly inorganic in chemical composition. The term "plastic" refers to the ability of the cement to impart to the plaster a high degree of workability, and for the plaster to remain workable or plastic for a period of time so that, after initial application and floating on the wall, it can be reworked to obtain both densification and desired texture.

4. Physical Properties

4.1 Plastic cement shall conform to the applicable requirements prescribed in Table 1.

5. Sampling

5.1 When the purchaser desires that the cement be sampled and tested to verify compliance with this specification, sampling and testing should be performed in accordance with Practice C 183.

5.2 Practice C 183 is not designed for manufacturing quality control and is not required for manufacturer's certification.

6. Temperature and Humidity

6.1 The temperature and relative humidity of the air in the vicinity of the mixing slab and dry materials, molds, base plates, and mixing bowl shall conform to the requirements of Test Method C 109/C 109M.

6.2 The moist cabinet or moist room shall conform to the

---

[1] This specification is under the jurisdiction of ASTM Committee C-1 on Cement and is the direct responsibility of Subcommittee C01.11 on Masonry Cement.

Current edition approved Feb. 10, 1998. Published June 1998. Originally published as C 1328 – 96. Last previous edition C 1328 – 97a.

[2] *Annual Book of ASTM Standards*, Vol 04.01.

[3] *Annual Book of ASTM Standards*, Vol 14.02.

TABLE 1 Physical Requirements

Plastic Cement Type	S	M
Fineness, residue on a 45-μm (No. 325) sieve, max, %	24	24
Autoclave expansion, max, %	1.0	1.0
Time of setting, Gillmore method:		
Initial set, min, not less than	90	90
Final set, min, not more than	1440	1440
Compressive strength (average of three cubes):		
The compressive strength of mortar cubes, composed of 1 part cement and 3 parts blended sand (half graded standard sand and half standard 20–30 sand) by volume, prepared and tested in accordance with this specification, shall be equal to or higher than the values specified for the ages indicated below:		
7 days, MPa (psi)	9.0 (1300)	12.4 (1800)
28 days, MPa (psi)	14.5 (2100)	20.0 (2900)
Air content of mortar:		
Min, volume %	8	8
Max, volume %	20	20
Water retention value, min, %, of original flow	70	70

requirements of Specification C 511.

7. Fineness

7.1 Determine the residue on the 45-μm (No. 325) sieve in accordance with Test Method C 430.

8. Normal Consistency

8.1 Determine normal consistency by the Vicat apparatus in accordance with Test Method C 187.

9. Autoclave Expansion

9.1 Determine the autoclave expansion in accordance with Test Method C 151. After molding, store the bars in the moist cabinet or room for 48 h ± 30 min before removal from the molds for measurement and testing in the autoclave. Calculate the difference in length of the test specimen before and after autoclaving to the nearest 0.01 % of the effective gage length, and report as the autoclave expansion of the plastic cement.

10. Time of Setting

10.1 Determine the time of setting by the Gillmore needle method in accordance with Test Method C 266.

11. Density

11.1 Determine the density of the plastic cement in accordance with Test Method C 188, using kerosine as the liquid. Use the density so determined in the calculation of the air content of the plaster.

12. Apparatus for Plaster Tests

12.1 *Weights and Weighing Devices, Sieves, Glass Graduates, Specimen (Cube) Molds, Tamper, Trowel, and Testing Machine*, as described in Test Method C 109/C 109M.

12.2 *Flow Table*, conforming to the requirements prescribed in Specification C 230.

12.3 *Mixing Apparatus*, conforming to the requirements prescribed in Practice C 305.

12.4 *Measure, Straightedge, Tapping Stick, and Spoon*, conforming to the requirements given in Test Method C 185.

12.5 *Specimen (Cube) Molds*—Molds shall be prepared in accordance with Test Method C 109/C 109M.

13. Blended Sand

13.1 The sand shall be a blend of equal parts by weight of graded standard sand and standard 20–30 sand conforming to Specification C 778. The fineness of each sand may be checked by using either of the procedures described in Specification C 778.

14. Preparation of Plaster

14.1 *Proportions for Plaster*—Plaster for air entrainment, compressive strength, and water retention tests shall be proportioned to contain 1620 g of sand and a mass of cement, in grams, equal to 15 times the net mass in kilograms printed on the bag. The sand shall consist of 810 g of graded standard sand and 810 g of 20–30 standard sand (Note 1). The quantity of water, measured in millilitres shall be such as to produce a flow of 110 ± 5, as determined by the flow table.

NOTE 1—If the net mass printed on the bag is 35 kg, the mortar shall contain 525 g of cement. The specified mortar proportions are approximately the 1:3 nominal proportions by volume, commonly specified for construction, as indicated in the following calculation:

If a bag of mortar cement contained 28 L of cement, and 28 L of loose damp sand contained 36 kg of dry sand,

$$A = 1620 \times (C/B) \quad (1)$$

where:

A = number of grams of cement to be used in the mortar with 1620 g of sand,

B = 3 × 36 = 108 kg, the mass of dry sand in 84 (or 3 × 28) L of loose damp sand, and

C = mass per bag, kg.

14.2 *Mixing of Plasters*—Mix the plaster in accordance with Practice C 305.

14.3 *Determination of Flow*—Determine the flow in accordance with Test Method C 109/C 109M.

15. Air Entrainment

15.1 *Procedure*—If the plaster has the correct flow, use a separate portion of the plaster for the determination of entrained air. Determine the mass of 400 mL of plaster in accordance with Test Method C 185.

15.2 *Calculation*—Calculate the air content of the plaster, and report it to the nearest 1 % as follows:

$$D = (W_1 + W_2 + V_w)/[(W_1/S_1) + (W_2/S_2) + V_w]$$
$$A = 100 - (W_m/4D) \quad (2)$$

where:

D = density of air-free plaster, g/cm³,

W_1 = mass of cement, g,

W_2 = mass of sand, g,
V_w = mL-g of water used,
S_1 = density of cement, g/cm^3,
S_2 = density of standard sand, 2.65 g/cm^3,
A = volume percent of entrained air, and
W_m = mass of 400 mL of plaster, g.

16. Compressive Strength

16.1 *Test Specimens:*

16.1.1 *Molding*—Immediately after determining the flow and mass of 400 mL of plaster, return all of the plaster to the mixing bowl and remix for 15 s at the medium speed. Then mold the test specimens in accordance with Test Method C 109/C 109M, except that the elapsed time for mixing plaster, determining flow, determining air entrainment, and starting the molding of cubes shall be within 8 min.

16.1.2 *Storage*—Immediately after molding, store all test specimens in the molds on plane plates in a moist cabinet or moist room for 48 to 52 h in such a manner that the upper surfaces shall be exposed to the moist air. Then remove the cubes from the molds, and place them in the moist cabinet or moist room for 5 days in such a manner as to allow free circulation of air around at least five faces of the specimens. At the age of 7 days, immerse the cubes for the 28-day tests in saturated lime water in storage tanks of noncorrodible materials.

16.2 *Procedure:*

16.2.1 Test the cube specimens immediately after their removal from the moist cabinet or moist room for 7-day specimens, and immediately after their removal from storage water for all other specimens. If more than one specimen at a time is removed from the moist cabinet or moist room for 7-day tests, cover these cubes with a damp cloth until the time of testing.

16.2.2 The remainder of the testing procedure shall conform to Test Method C 109/C 109M.

17. Water Retention

17.1 To determine water retention, use the test method and calculation described in the section on water retention in Specification C 91, except that wherever the term "mortar" is used, read "plaster."

18. Storage

18.1 The cement shall be stored in such a manner as to permit easy access for the proper inspection and identification of each shipment, and in a suitable weathertight building that will protect the cement from dampness and minimize warehouse set.

19. Inspection

19.1 Adequate facilities shall be provided to the purchaser for the necessary inspection and sampling.

19.2 All packages shall be in good condition at the time of inspection.

20. Rejection

20.1 At the option of the purchaser, the cement shall be rejected if it fails to meet any of the requirements of this specification.

20.2 At the option of the purchaser, packages more than 2 % below the mass marked thereon shall be rejected. At the option of the purchaser, the entire shipment represented shall be rejected if the average mass of packages in any shipment as shown by weighing 50 packages taken at random is less than that marked on the packages.

20.3 At the option of the purchaser, cement remaining in storage prior to shipment for a period greater than six months after testing shall be retested and, at the option of the purchaser, shall be rejected if it fails to meet any of the requirements of this specification.

21. Manufacturer's Certification

21.1 Upon request of the purchaser in the contract or order, a manufacturer's report shall be furnished at the time of shipment stating the results of the tests made on samples of the material taken during production or transfer and certifying that the applicable requirements of this specification have been met.

22. Packaging and Package Marking

22.1 When plastic cement is delivered in packages, the brand, name of the manufacturer, net mass of the package in kg (see Note 2) shall be indicated plainly thereon. Similar information shall be provided in the shipping documents accompanying the shipment of plastic cement in bulk.

NOTE 2—To facilitate the change to SI units, a standard SI package size of 35 Kg for Type S and 42 Kg for Type M will provide convenient mass increments reasonably similar to the traditional 77– and 94–lb packages.

23. Keywords

23.1 plaster; plastic cement; portland cement-based plaster; stucco

Designation: C 1329 – 98

Standard Specification for Mortar Cement[1]

This standard is issued under the fixed designation C 1329; the number immediately following the designation indicates the year of original adoption or, in the case of revision, the year of last revision. A number in parentheses indicates the year of last reapproval. A superscript epsilon (ϵ) indicates an editorial change since the last revision or reapproval.

1. Scope

1.1 This specification covers three types of mortar cement for use where mortar for masonry is required.

1.2 The values stated in SI units are to be regarded as the standard. Values in SI units shall be obtained by measurement in SI units or by appropriate conversion, using the Rules for Conversion and Rounding given in IEEE/ASTM SI 10, of measurements made in other units.

1.3 The text of this standard refers to notes and footnotes which provide explanatory material. These notes and footnotes (excluding those in tables and figures) shall not be considered as requirements of the standard.

1.4 The following safety hazards caveat pertains only to Annex A1. *This standard does not purport to address all of the safety concerns, if any, associated with its use. It is the responsibility of the user of this standard to establish appropriate safety and health practices and determine the applicability of regulatory limitations prior to use.*

2. Referenced Documents

2.1 *ASTM Standards:*

C 91 Specification for Masonry Cement[2]

C 109/C 109M Test Method for Compressive Strength of Hydraulic Cement Mortars (Using 2-in. or 50-mm Cube Specimens)[2]

C 128 Test Method for Specific Gravity and Absorption of Fine Aggregate[3]

C 140 Test Methods of Sampling and Testing Concrete Masonry Units[4]

C 151 Test Method for Autoclave Expansion of Portland Cement[2]

C 183 Practice for Sampling and the Amount of Testing of Hydraulic Cement[2]

C 185 Test Method for Air Content of Hydraulic Cement Mortar[2]

C 187 Test Method for Normal Consistency of Hydraulic Cement[2]

C 188 Test Method for Density of Hydraulic Cement[2]

C 219 Terminology Relating to Hydraulic Cement[2]

C 230 Specification for Flow Table for Use in Tests of Hydraulic Cement[2]

C 266 Test Method for Time of Setting of Hydraulic-Cement Paste by Gillmore Needles[2]

C 305 Practice for Mechanical Mixing of Hydraulic Cement Pastes and Mortars of Plastic Consistency[2]

C 430 Test Method for Fineness of Hydraulic Cement by the 45-μm (No. 325) Sieve[2]

C 511 Specification for Moist Cabinets, Moist Rooms, and Water Storage Tanks Used in the Testing of Hydraulic Cements and Concretes[2]

C 778 Specification for Standard Sand[2]

C 780 Test Method for Preconstruction and Construction Evaluation of Mortars for Plain and Reinforced Unit Masonry[4]

C 1072 Test Method for Measurement of Masonry Flexural Bond Strength[4]

IEEE/ASTM SI 10 Standard for Use of the International System of Units (SI): The Modern Metric System[5]

3. Terminology

3.1 *Definitions*—Terms used in this specification are defined in Terminology C 219.

3.2 *Definition of Term Specific to This Standard:*

3.2.1 *mortar cement*—a hydraulic cement, primarily used in masonry construction, consisting of a mixture of portland blended hydraulic cement and plasticizing materials (such as limestone or hydrated or hydraulic lime), together with other materials introduced to enhance one or more properties such as setting time, workability, water retention, and durability.

3.2.1.1 *Discussion*—Mortar cement is similar to masonry cement in use and function. However, this specification requires lower air contents and includes a flexural bond strength requirement.

4. Classification

4.1 Mortar cements are classified as Type N, S, or M, according to the physical requirements prescribed in Table 1.

5. Physical Properties

5.1 Mortar cement shall conform to the applicable requirements prescribed in Table 1 for its classification.

6. Sampling

6.1 When the purchaser desires that the cement be sampled and tested to verify compliance with this specification, sampling and testing shall be performed in accordance with Practice C 183.

6.2 Practice C 183 is not designed for manufacturing quality control and is not required for manufacturer's certification.

---

[1] This specification is under the jurisdiction of ASTM Committee C-1 on Cement and is the direct responsibility of Subcommittee C01.11 on Masonry Cement.

Current edition approved March 10, 1998. Published June 1998. Originally published as C 1329 – 96. Last previous edition C 1329 – 97.

[2] *Annual Book of ASTM Standards*, Vol 04.01.

[3] *Annual Book of ASTM Standards*, Vol 04.02.

[4] *Annual Book of ASTM Standards*, Vol 04.05.

[5] *Annual Book of ASTM Standards*, Vol 14.02.

TABLE 1 Physical Requirements

Mortar Cement Type	N	S	M
Fineness, residue on a 45-µm (No. 325) sieve, max, %	24	24	24
Autoclave expansion, max, %	1.0	1.0	1.0
Time of setting, Gillmore method:			
Initial set, min, not less than	120	90	90
Final set, min, not more than	1440	1440	1440
Compressive strength (average of three cubes):			
The compressive strength of mortar cubes, composed of 1 part cement and 3 parts blended sand (half graded standard sand and half standard 20–30 sand) by volume, prepared and tested in accordance with this specification, shall be equal to or higher than the values specified for the ages indicated below:			
7 days, MPa (psi)	3.5 (500)	9.0 (1300)	12.4 (1800)
28 days, MPa (psi)	6.2 (900)	14.5 (2100)	20.0 (2900)
Flexural bond strength			
28 days, min, MPa (psi)	0.5 (70)	0.7 (100)	0.8 (115)
Air content of mortar:			
Min, volume %	8	8	8
Max, volume %	16	14	14
Water retention value, min, %, of original flow	70	70	70

7. Temperature and Humidity

7.1 The temperature and relative humidity of the air in the vicinity of the mixing slab and dry materials, molds, base plates, and mixing bowl shall conform to the requirements of Test Method C 109/C 109M.

7.2 The moist cabinet or moist room shall conform to the requirements of Specification C 511.

8. Fineness

8.1 Determine the residue on the 45-µm (No. 325) sieve in accordance with Test Method C 430.

9. Normal Consistency

9.1 Determine the normal consistency by the Vicat apparatus in accordance with Test Method C 187.

10. Autoclave Expansion

10.1 Determine the autoclave expansion in accordance with Test Method C 151. After molding, store the bars in the moist cabinet or room for 48 h ± 30 min before removal from the molds for measurement and testing in the autoclave. Calculate the difference in length of the test specimen before and after autoclaving to the nearest 0.01 % of the effective gage length, and report as the autoclave expansion of the mortar cement.

11. Time of Setting

11.1 Determine the time of setting by the Gillmore needle method in accordance with Test Method C 266.

12. Density

12.1 Determine the density of the mortar cement in accordance with Test Method C 188, using kerosine as the liquid. Use the density so determined in the calculation of the air content of the mortars.

13. Apparatus for Mortar Tests

13.1 *Weights and Weighing Devices, Sieves, Glass Graduates, Specimen (Cube) Molds, Tamper, Trowel, and Testing Machine*, as described in Test Method C 109/C 109M.

13.2 *Flow Table*, conforming to the requirements prescribed in Specification C 230.

13.3 *Mixing Apparatus*, conforming to the requirements prescribed in Practice C 305.

13.4 *Measure, Straightedge, Tapping Stick, and Spoon*, conforming to the requirements given in Test Method C 185.

14. Preparation of Mortar

14.1 The sand shall be a blend of equal parts by weight of graded standard sand and 20–30 standard sand conforming to Specification C 778. The fineness of each sand may be checked by using either of the procedures described in Specification C 778.

14.2 *Proportions for Mortar*—Mortar for air entrainment, compressive strength, and water retention tests shall be proportioned to contain 1620 g of sand and a mass of cement, in grams, equal to fifteen times the net mass in kilograms printed on the bag. The sand shall consist of 810 g of graded standard sand and 810 g of 20–30 standard sand (Note 1). The quantity of water, measured in millilitres, shall be such as to produce a flow of 110 ± 5 as determined by the flow table.

NOTE 1—If the net mass printed on the bag is 32 kg, the mortar shall contain 480 g of cement. The specified mortar proportions are approximately the 1:3 nominal proportions by volume, commonly specified for construction, as indicated in the following calculation:

If a bag of mortar cement contained 28 L of cement, and 28 L of loose damp sand contained 36 kg of dry sand,

$$A = 1620 \times (C/B) \tag{1}$$

where:

A = number of grams of cement to be used in the mortar with 1620 g of sand,

B = 3 × 36 = 108 kg, the mass of dry sand in 84 (or 3 × 28) L of loose damp sand, and

C = mass per bag, kg.

14.3 *Mixing of Mortars*—Mix the mortar in accordance with Practice C 305.

14.4 *Determination of Flow*—Determine the flow in accordance with Test Method C 109/C 109M.

15. Air Entrainment

15.1 *Procedure*—If the mortar has the correct flow, use a separate portion of the mortar for the determination of entrained air. Determine the mass of 400 mL of mortar in accordance with Test Method C 185.

15.2 *Calculation*—Calculate the air content of the mortar, and report it to the nearest 1 % as follows:

$$D = (W_1 + W_2 + V_w)/[(W_1/S_1) + (W_2/S_2) + V_w] \quad A = 100 - (W_m/4D) \qquad (2)$$

where:

D = density of air-free mortar, g/cm^3,
W_1 = mass of cement, g,
W_2 = mass of sand, g,
V_w = mL-g of water used,
S_1 = density of cement, g/cm^3,
S_2 = density of standard sand, 2.65 g/cm^3,
A = volume percent of entrained air, and
W_m = mass of 400 mL of mortar, g.

16. Compressive Strength

16.1 *Test Specimens:*

16.1.1 *Molding*—Immediately after determining the flow and mass of 400 mL of mortar, return all of the mortar to the mixing bowl and remix for 15 s at the medium speed. Then mold the test specimens in accordance with Test Method C 109/C 109M, except that the elapsed time for mixing mortar, determining flow, determining air entrainment, and starting the molding of cubes shall be within 8 min.

16.1.2 *Storage*—Immediately after molding, store all test specimens in the molds on plane plates in a moist cabinet or moist room for 48 to 52 h in such a manner that the upper surfaces shall be exposed to the moist air. Then remove the cubes from the molds, and place them in the moist cabinet or moist room for 5 days in such a manner as to allow free circulation of air around at least five faces of the specimens. At the age of 7 days, immerse the cubes for the 28-day tests in saturated lime water in storage tanks of noncorrodible materials.

16.2 *Procedure:*

16.2.1 Test the cube specimens immediately after their removal from the moist cabinet or moist room for 7-day specimens, and immediately after their removal from storage water for all other specimens. If more than one specimen at a time is removed from the moist cabinet or moist room for 7-day tests, cover these cubes with a damp cloth until the time of testing.

16.2.2 The remainder of the testing procedure shall conform to Test Method C 109/C 109M.

17. Water Retention

17.1 Water retention shall be determined in accordance with the procedures described in Specification C 91.

18. Flexural Bond Strength

18.1 Flexural bond strength shall be determined in accordance with the procedure set forth in Annex A1 of this specification.

19. Storage

19.1 The cement shall be stored in such a manner as to permit easy access for proper inspection and identification of each shipment, and in a suitable weathertight building that will protect the cement from dampness and minimize warehouse set.

20. Inspection

20.1 Adequate facilities shall be provided to the purchaser for the necessary inspection and sampling.

20.2 All packages shall be in good condition at the time of inspection.

21. Rejection

21.1 At the option of the purchaser, the cement shall be rejected if it fails to meet any of the requirements of this specification.

21.2 At the option of the purchaser, packages more than 2 % below the mass marked thereon shall be rejected. At the option of the purchaser, the entire shipment shall be rejected if the average mass of packages in any shipment, as shown by weighing 50 packages taken at random, is less than that marked on the packages.

21.3 At the option of the purchaser, cement remaining in storage prior to shipment for a period greater than six months after testing shall be retested and, at the option of the purchaser, shall be rejected if it fails to meet any of the requirements of this specification.

22. Manufacturer's Certification

22.1 Upon request of the purchaser in the contract or order, a manufacturer's report shall be furnished at the time of shipment stating the results of the tests made on samples of the material taken during production or transfer and certifying that the applicable requirements of this specification have been met.

23. Packaging and Package Marking

23.1 When mortar cement is delivered in packages, the brand, name of the manufacturer, type of mortar cement, and net mass of the package in kilograms (Note 2) shall be indicated plainly thereon. Similar information shall be provided in the shipping documents accompanying the shipment of mortar cement in bulk.

NOTE 2—To facilitate the change to SI units, a standard metric package size of 32 kg for Type N, 34 kg for Type S, and 36 kg for Type M will provide convenient mass increments reasonably similar to the traditional 70–, 75–, and 80–lb packages.

24. Keywords

24.1 flexural bond strength; masonry; mortar; mortar cement

ANNEX

(Mandatory Information)

A1. TEST METHOD FOR DETERMINATION OF FLEXURAL BOND STRENGTH

A1.1 Scope

A1.1.1 This test method covers a laboratory procedure for determination of the flexural bond strength of masonry assemblies made using mortar-cement mortars and standard masonry units.

A1.2 Summary of Test Method

A1.2.1 This test method includes the following: requirements for standard concrete masonry units used in this test procedure; procedures for proportioning and mixing test mortars used in the fabrication of test prisms; procedures for fabricating, curing, and testing prism specimens; and reporting requirements.

A1.3 Significance and Use

A1.3.1 This test method provides a means for the determination of flexural bond strength values to be used to determine compliance with the bond-strength requirements of this specification.

A1.3.2 The flexural bond strength determined by this test method should not be interpreted as the flexural bond strength of a masonry wall constructed using the same mortar cement. Nor should it be interpreted as an indication of the extent of bond for purposes of water permeance evaluation.

A1.4 Apparatus

A1.4.1 *Prism Fabrication Jig*, as shown in Fig. A1.1.

A1.4.2 *Mortar Joint Template*, as shown in Fig. A1.2.

A1.4.3 *Drop Hammer*, as shown in Fig. A1.3.

A1.4.4 *Mechanical Drum Mortar Mixer.*

A1.4.5 *Flow Table, Flow Mold, and Caliper*, conforming

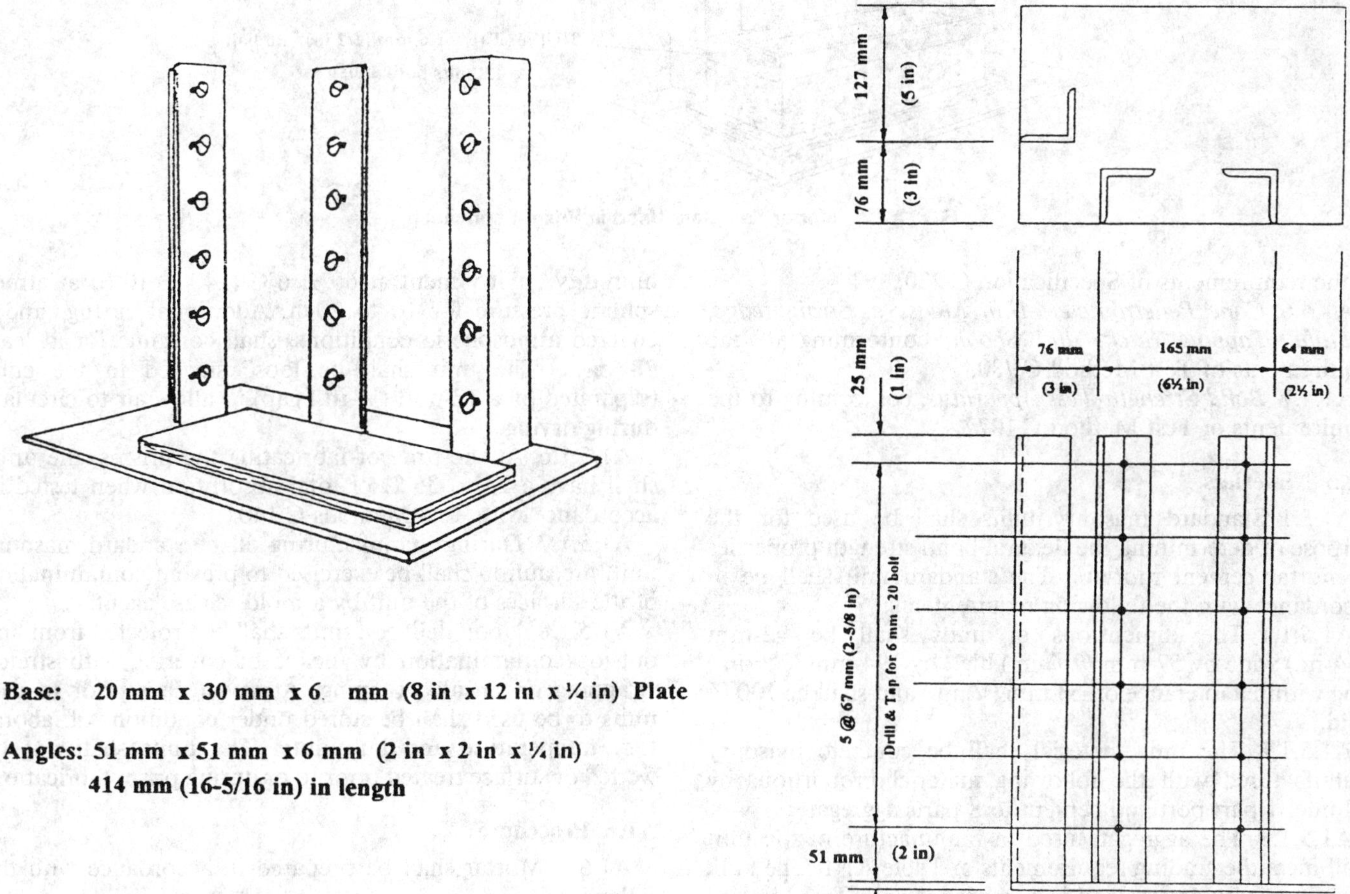

NOTE—Base: 20 by 30 by 6.0-mm (8 by 12 by ¼-in.) plate. Angles: 51 by 51 by 6 mm (2 by 2 by ¼ in.), 414 mm (16⁵⁄₁₆ in.) in length.

FIG. A1.1 Jig Used in Prism Fabrication

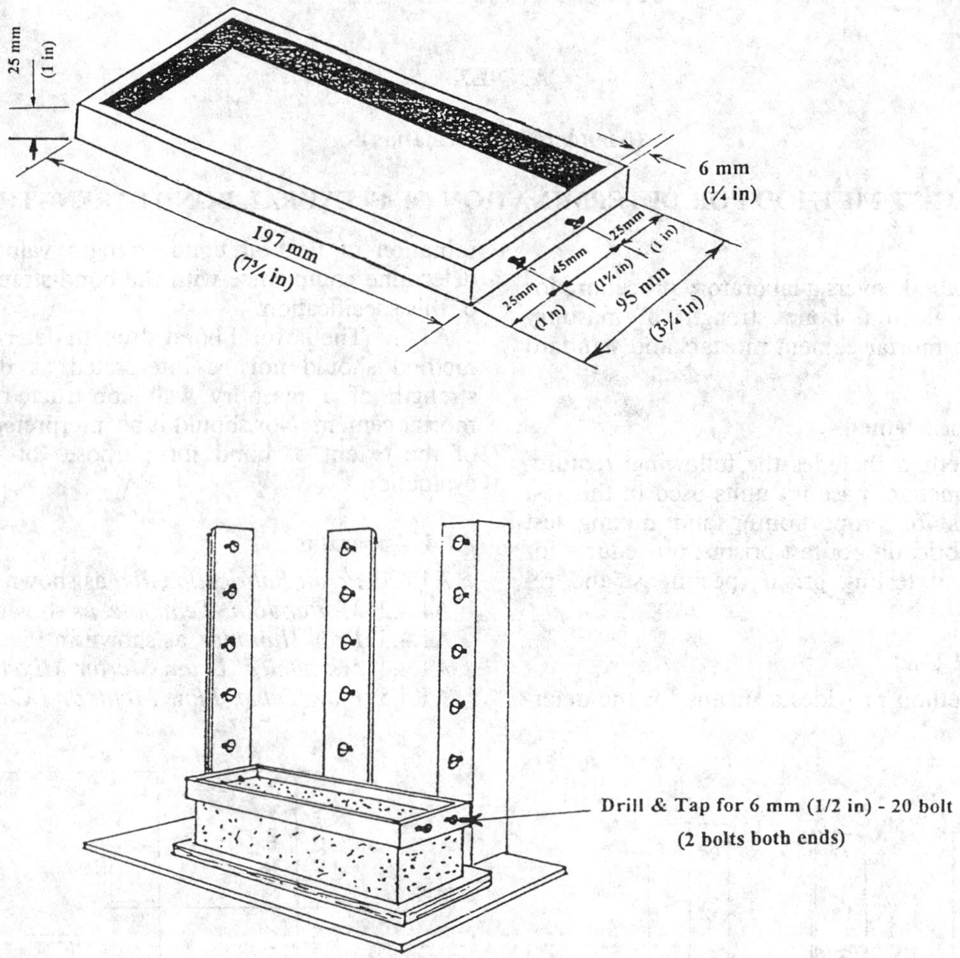

FIG. A1.2 Mortar Template Used in Prism Fabrication

to the requirements of Specification C 230.

A1.4.6 *Cone Penetrometer, Unit Measure, Straightedge, Spatula, Tapping Stick, and Spoon*, conforming to the requirements of Test Method C 780.

A1.4.7 *Bond Strength Test Apparatus*, conforming to the requirements of Test Method C 1072.

A1.5 Materials

A1.5.1 Standard masonry units shall be used for the purpose of determining the flexural bond strength properties of mortar cement mortars. The standard unit shall be in accordance with the following requirements:

A1.5.1.1 The dimensions of units shall be 92-mm (3⅝-in.) wide by 57-mm (2¼-in.) high by 194-mm (7⅝-in.) long within a tolerance of ±3 mm (⅛ in.) and shall be 100 % solid.

A1.5.1.2 The unit material shall be concrete masonry manufactured with the following material proportions by volume: 1 part portland cement to 8 parts aggregate.

A1.5.1.3 The aggregate used in manufacture of the unit shall meet the grading requirements of Table A1.1. The bulk specific gravity determined in accordance with Test Method C 128 shall be not less than 2.6 and not more than 2.7.

A1.5.1.4 The density of the unit shall be 2000 to 2160 kg/m^3 (125 to 135 lb/ft^3).

A1.5.1.5 The unit shall be cured in a 100 % relative humidity environment at 60 ± 6°C (140 ± 10°F) at atmospheric pressure for 10 to 20 h. Additional curing, under covered atmospheric conditions, shall continue for at least 28 days. The unit shall be loose stacked in the cube (separated by a 12-mm (½-in.) gap) to allow air to circulate during drying.

A1.5.1.6 At the time of fabricating the prisms, the units shall have a 25 to 35 % of total absorption when tested in accordance with Test Methods C 140.

A1.5.1.7 During manufacturing of the standard masonry unit, precaution shall be exercised to prevent contamination of the surfaces of the unit by a mold release agent.

A1.5.1.8 Upon delivery, units shall be protected from any outside contamination by means of covering with stretch wrapping or shrink wrapping. At seven days prior to use, units to be used shall be stored under conditions of laboratory temperature and humiduty. The units shall not be wetted or surface treated prior to or during prism fabrication.

A1.6 Procedure

A1.6.1 Mortar shall be prepared in accordance with the following:

A1.6.1.1 Sand shall be a blend of an equal mass of graded standard sand and standard 20–30 µm sand conforming to Specification C 778. Mortar shall be proportioned to contain a mass of cement in kg equal to 0.20 times the net mass in kg

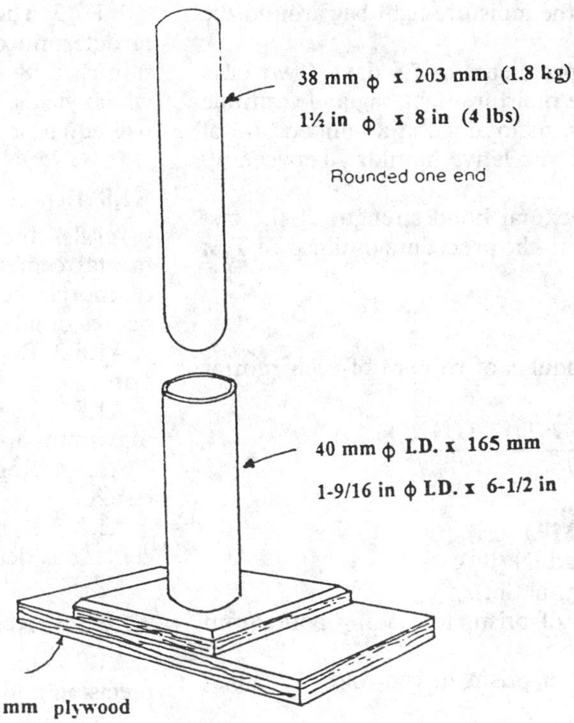

FIG. A1.3 Drop Hammer Used in Prism Fabrication

TABLE A1.1 Grading of Aggregate Used in the Manufacture of Standard Concrete Masonry Units

Grading	Percent Retained by Mass
9.53 mm (3/8 in.) sieve	0
4.75 mm (No. 4) sieve	0 to 5
2.36 mm (No. 8) sieve	20 to 30
1.18 mm (No. 16) sieve	20 to 30
600 μm (No. 30) sieve	15 to 25
300 μm (No. 50) sieve	5 to 15
150 μm (No. 100) sieve	5 to 10
Pan	5 to 10

printed on the bag and 22.0 kg of sand.

A1.6.1.2 Mortar materials shall be mixed in a paddle-type mortar mixer of no less than 18 L (0.6 ft^3) capacity for 5 min.

A1.6.1.3 Mortar flow shall be determined in accordance with Test Method C 109/C 109M. Water shall be adjusted until flow of 125 ± 5 is achieved. The flow shall be determined on mortar mixed 3 min after all of the ingredients are added to the mixer. If the flow meets the required consistency, the batch (including material used to determine flow) shall be mixed for the remaining 2 mins. If the flow is less than 120, additional water may be added to the batch (including material used to determine flow) and mixed for 1 min. If the flow then meets the required consistency, the batch (including material used to determine flow) shall be mixed for the remaining minute. If after the one-time addition of water the flow does not meet the required consistency, the batch shall be discarded. Whenever the measured flow exceeds 130 the batch shall be discarded.

A1.6.1.4 The initial cone penetration shall be determined in accordance with Test Method C 780, immediately after mixing the mortar. Mortar shall not be used when the cone penetration is less than 80 % of the initial cone penetration value.

A1.6.2 Test specimens shall consist of a set of prisms such that 30 mortar joints are fabricated for testing. Prism specimens shall be not more than six units in height.

A1.6.3 Fabricate the test specimens as follows:

A1.6.3.1 Each prism shall be built in an opened moisture-tight bag that is large enough to enclose and seal the completed prism. Set the first unit of each prism on a 12-mm (½-in.) plywood pallet in an alignment jig, as shown in Fig. A1.1.

A1.6.3.2 Place the mortar template shown in Fig. A1.2 on the unit so that the mortar bed depth prior to compaction is 12 mm (½ in.). Place the mortar in the template, and strike off excess mortar with a straight edge.

A1.6.3.3 Remove the template, and immediately place the next unit on the mortar bed in contact with the three alignment bolts for that course using a bulls-eye level to ensure uniform initial contact of the unit surface and bed mortar. Carefully position the drop hammer apparatus shown in Fig. A1.3 on top of the unit, and drop its 1.8 kg (4 lb), round end down, once from a height of 38 mm (1.5 in.).

A1.6.3.4 Repeat the steps given in A1.6.3.2 and A1.6.3.3 until the prisms are complete.

A1.6.3.5 Joints shall be cut flush after the prism is completely built. The joints shall not be tooled.

A1.6.3.6 Place two masonry units of the type used to construct the prism upon the top course 1 h ± 15 min after the completion of construction.

A1.6.3.7 Identify all test specimens using a water-resistant marker.

A1.6.3.8 Draw and seal the moisture-tight bag around the prism.

A1.6.3.9 Cure all test specimens for 28 days. Two days prior to testing, remove the moisture-tight bag and continue curing in the laboratory air, maintained at a temperature of 24 ± 9°C (75 ± 16°F), with a relative humidity between 30 and 70 %.

A1.6.4 Determine the flexural bond strength of the test specimens in accordance with the procedure outlined in Test Method C 1072.

A1.7 Calculation

A1.7.1 Calculate the modulus of rupture of each mortar joint as follows:

$$f_r = \frac{6(PL + P_1L_1)}{bd^2} - \frac{(P + P_1)}{bd}$$

where:

f_r = modulus of rupture, MPa (psi),
P = maximum applied load, N (lbf),
P_1 = weight of loading arm, N (lbf),
L = distance from center of prism to loading point, mm (in.),
L_1 = distance from center of prism to centroid of loading arm, mm (in.),
b = average width of cross section of failure surface, mm (in.), and
d = average thickness of cross section of failure surface, mm (in.).

A1.7.2 The flexural bond strength of mortar cement shall be determined as the average modulus of rupture of 30 joints minus 1.28 times the standard deviation of the sample, which yields a value that a mortar joint's modulus of rupture will equal or exceed 9 out of 10 times.

A1.8 Report

A1.8.1 Include the name of the manufacturer of the mortar cement being evaluated, place of manufacture, type of mortar cement, date of testing, laboratory name, and names of laboratory staff in the report.

A1.8.2 Report the mortar flow and initial cone penetration.

A1.8.3 Identify the prism and joint tested. Report the maximum applied test load and the modulus of rupture.

A1.8.4 Report the standard deviation for all 30 joints tested.

A1.8.5 Report the flexural bond strength of the mortar cement as determined in accordance with A1.7.2.

A1.9 Precision and Bias

A1.9.1 Insufficient data are currently available for a precision and bias statement.

A1.10 Keywords

A1.10.1 flexural bond strength; mortar cement; standard concrete masonry unit

Designation: C 1355/C 1355M – 96

Standard Specification for Glass Fiber Reinforced Gypsum Composites[1]

This standard is issued under the fixed designation C 1355/C 1355M; the number immediately following the designation indicates the year of original adoption or, in the case of revision, the year of last revision. A number in parentheses indicates the year of last reapproval. A superscript epsilon (ε) indicates an editorial change since the last revision or reapproval.

1. Scope

1.1 This specification covers glass fiber reinforced gypsum (GRG) composites having minimum properties and quality suitable to allow the production of GRG parts for non-loading bearing, thin section, ornamental shapes for architectural embellishment of interior building construction.

1.2 The values stated in either inch-pound units or SI (metric) are to be regarded separately as standard. Within the text, the SI units are shown in brackets. The values stated in each system shall be independent of the other. Values from the two systems shall not be combined.

1.3 The text of this standard references footnotes which provide explanatory material. These footnotes shall not be considered as requirements of the standard.

2. Referenced Documents

2.1 *ASTM Standards:*

C 11 Terminology Relating to Gypsum and Related Building Materials and Systems[2]

C 472 Test Methods for Physical Testing of Gypsum, Gypsum Plasters, and Gypsum Concrete[2]

C 473 Test Methods for Physical Testing of Gypsum Board Products and Gypsum Lath[2]

C 947 Test Method for Flexural Properties of Thin-Section Glass Fiber Reinforced Concrete (Using Simple Beam with Third-Point Loading)[3]

D 256 Test Methods for Determining the Pendulum Impact Resistance of Specimens of Plastics[4]

D 578 Specification for Glass Fiber Strands[5]

D 696 Test Method Coefficient of Linear Thermal Expansion of Plastics Between −30°C and 30°C[4]

D 2583 Test Method for Indentation Hardness of Rigid Plastics by Means of a Barcol Impresser[6]

E 84 Test Method for the Surface Burning Characteristics of Building Materials[7]

E 136 Test Method for the Behavior of Materials in a Vertical Tube Furnace at 750°C[7]

3. Terminology

3.1 *Definitions*—Definitions of terms shall be in accordance with Terminology C 11.

3.2 *Definitions of Terms Specific to This Standard:*

3.2.1 *GRG composite, n*—a thin section laminate made from the combination of alpha gypsum cement, glass fiber, additives and water.

3.2.2 *GRG part, n*—an individual molded component used as architectural embellishment.

4. Materials and Manufacture

4.1 *Materials:*

4.1.1 *Alpha Gypsum Cement*—Alpha-calcium sulfate hemihydrate-plaster which is noncombustible, has a low water demand, neutral (pH) or low alkalinity, and a purity of not less than 90 % by weight of $CaSO_4 \cdot 2H_2O$.

4.1.2 *"E" Glass Fiber*—Chopped glass fiber strands or continuous strand mats of calcia-alumina silicate glasses conforming to Specification D 578.

4.1.3 *Water*—Potable water.

4.1.4 *Additives*—In accordance with the alpha gypsum cement manufacturer's specifications.

4.2 *Manufacturer's Certification of Raw Materials:*

4.2.1 *Alpha Gypsum Cement*—Each lot of alpha gypsum cement shall be certified to be in compliance with 4.1 and 5.1.

4.2.2 *Glass Fiber Reinforcement*—Each lot of glass fiber reinforcement shall be certified to be in compliance with Specification D 578.

4.3 *Composite Preparation:*

4.3.1 Prepare a flat, rectangular composite of sufficient size to obtain specimens required in 5.2. The composite prepared shall be the same formula as the intended GRG part.

5. Mechanical Properties

5.1 *Neat Alpha Gypsum Cement:*

5.1.1 Normal Consistency shall be not more than 30 when tested in accordance with Test Methods C 472.

5.1.2 Compressive Strength shall be not less than 6000 psi (41.4 MPa) when tested in accordance with Test Methods C 472.

5.1.3 Hardness shall be not less than 33 Barcol when tested in accordance with Test Method D 2583.

5.2 *Composite:*

5.2.1 *Specimen Preparation*

5.2.1.1 *Flexural Strength*—Average Flexural Yield (FY) shall be not less than 1200 psi (8.3 MPa) and an average Flexural Ultimate (FU) shall be not less than 2500 psi (17.2 MPa) when tested in accordance with Test Method C 947, using a crosshead speed of 0.08 in./min (2 mm/min). Specimens shall be conditioned to a constant weight in

---

[1] This specification is under the jurisdiction of ASTM Committee C-11 on Gypsum and Related Building Materials and Systems and is the direct responsibility of Subcommittee C11.01 on Specifications and Test Methods for Gypsum Products.

Current edition approved Nov. 10, 1996. Published January 1997.

[2] *Annual Book of ASTM Standards*, Vol 04.01.

[3] *Annual Book of ASTM Standards*, Vol 04.05.

[4] *Annual Book of ASTM Standards*, Vol 08.01.

[5] *Annual Book of ASTM Standards*, Vol 07.01.

[6] *Annual Book of ASTM Standards*, Vol 08.02.

[7] *Annual Book of ASTM Standards*, Vol 04.07.

accordance with Test Methods C 473.

5.2.1.2 *Impact Resistance*—Average impact resistance shall be not less than 3 ft-lb (4 j) when tested in accordance with Test Methods D 256, Test Method A (Izod-unnotched method).

5.2.1.3 *Hardness*—Average hardness shall be not less than 50 Barcol when tested in accordance with Test Methods D 2583. Specimens shall be conditioned according to Test Methods C 473.

5.2.1.4 *Coefficient of Linear Thermal Expansion*—Coefficient of linear thermal expansion shall be not more than 11.1 $\times 10^{-6}$ in./in.°F (20.0 $\times 10^{-6}$ mm/mm°C) when tested in accordance with Test Method D 696.

5.2.1.5 *Humidified Deflection*—Average humidified deflection shall be not more than 1/8 in. (3 mm) when tested on a 3/16 in. specimen (5 mm) in accordance with Test Methods C 473.

5.2.1.6 *Surface Burning Characteristics*—Flame spread index shall be 0 and smoke developed index shall be less than 10 when tested in accordance with Test Method E 84.

5.2.1.7 *Behavior at 750°C*—Shall pass all criteria when tested in accordance with Test Method E 136.

5.2.1.8 *Nail Pull Resistance*—Shall be not less than 110 lbf (489 N) when tested on a 3/16 in. specimen (5 mm) in accordance with Test Methods C 473 (Method B).

6. Samples Taken at the Point of Manufacture

6.1 When specified by the purchase agreement, samples of the GRG composite shall be taken at the place of manufacture and tested for compliance with 5.2.

7. Number of Tests and Retests

7.1 The number of tests and type of tests to be performed shall be part of the purchase agreement. Data from prior tests showing compliance with Section 5 are permitted to be supplied by the producer showing compliance with the specified requirements.

8. Inspection

8.1 Inspection of the GRG composite shall be agreed upon between the purchaser and the producer or the supplier as part of the purchase agreement.

9. Rejection and Rehearing

9.1 GRG composite that fails to conform to the requirements of this specification shall be reported to the producer or the supplier promptly and in writing. In case of dissatisfaction with the results of the test, the producer or supplier, shall be permitted to make a claim for a hearing to retest.

10. Certification

10.1 When specified in the purchase agreement, a producer's or supplier's report certifying that the GRG composite is in compliance with this specification, shall be furnished at the time of shipment.

11. Quality Assurance

11.1 Testing for compliance with this specification is required whenever glass fiber reinforced gypsum formulae are initially established.

11.2 Additional testing for compliance with this specification is required whenever constituent materials are added or deleted.

11.3 Each producer's test specimens shall be representative of actual production procedures and materials.

12. Keywords

12.1 alpha gypsum cement; glass fiber reinforced gypsum; glass fiber reinforced gypsum composite; GRG; GRG composite

Designation: C 1356M – 96
METRIC

Standard Test Method for Quantitative Determination of Phases in Portland Cement Clinker by Microscopical Point-Count Procedure [Metric][1]

This standard is issued under the fixed designation C 1356M; the number immediately following the designation indicates the year of original adoption or, in the case of revision, the year of last revision. A number in parentheses indicates the year of last reapproval. A superscript epsilon (ε) indicates an editorial change since the last revision or reapproval.

1. Scope

1.1 This test method covers a systematic procedure for measuring the percentage volume of the phases in portland cement clinker by microscopy.

1.2 *This standard does not purport to address all of the safety concerns, if any, associated with its use. It is the responsibility of the user of this standard to establish appropriate safety and health practices and determine the applicability of regulatory limitations prior to use.*

1.3 The values stated in SI units are to be regarded as the standard.

2. Referenced Documents

2.1 *ASTM Standards:*

C 150 Specification for Portland Cement[2]

C 670 Practice for Preparing Precision and Bias Statements for Test Methods for Construction Materials[3]

D 75 Practice for Sampling Aggregates[4]

D 3665 Practice for Random Sampling of Construction Materials[4]

3. Terminology

3.1 *Definitions:*

3.1.1 *clinker phase, n*—a physically and chemically distinct optically identifiable portion of the clinker sample, including both principal phases (alite, belite, aluminate, and ferrite), minor phases (for example, free lime, periclase, and alkali sulfates), and voids.

3.1.1.1 *Discussion*—Voids, though not a phase in the sense of being a crystalline compound, are a distinct, identifiable portion of a clinker microstructure.

3.1.2 *voids, n*—isolated or interconnected open areas in the clinker, also called pores.

3.2 *Principal Clinker Phases:*[5]

3.2.1 *alite, n*—crystalline tricalcium silicate (C_3S), modified in composition and crystal structure by incorporation of foreign ions; the crystals are pseudo-hexagonal with well-defined faces, though less regular shapes commonly occur.

3.2.2 *aluminate, n*—tricalcium aluminate (C_3A) modified in composition and crystal structure by incorporation of a substantial proportion of foreign ions; aluminate forms cubic crystals when relatively pure, and forms identifiable elongated crystals commonly called "alkali aluminate" when in solid solution with significant amounts of potassium or sodium, or both.

3.2.3 *belite, n*—crystalline dicalcium silicate (C_2S), modified in composition and crystal structure by incorporation of foreign ions; belite usually occurs as rounded crystals marked by striations formed by cross sections of lamellae, and may occur as single crystals or in clusters.

3.2.4 *ferrite, n*—a solid solution of approximate composition tetracalcium aluminoferrite (C_4AF) modified in composition by variation in the Al/Fe ratio and by substantial incorporation of foreign ions; ferrite is characterized by high reflectivity in polished sections and is normally the only strongly colored compound among the principal clinker phases.

3.2.4.1 *Discussion*—Aluminate and ferrite form most of the interstitial material between the silicate crystals and, under certain conditions of cooling, may not be easily identifiable or resolved by ordinary light microscopy.

3.3 *Minor Clinker Phases:*

3.3.1 *alkali sulfates, n*—sodium sulfate, potassium sulfate, and double sulfates such as calcium langbeinite (K_2SO_4–$2CaSO_4$).

3.3.2 *free lime, n*—calcium oxide (C) found mostly as round crystals.

3.3.3 *periclase, n*—crystalline form of free magnesium oxide (M), that has not been taken up in solid solution with other phases.

4. Summary of Test Method

4.1 The test method consists of the preparation and microscopical examination of a specimen produced by encapsulating clinker in a mounting medium and sectioning the specimen so as to expose the interior of particles for visual examination. Polishing the section surface and treating it with etchants to highlight specific phases complete the preparation. During microscopical examination phases are identified and their proportions determined by a point-count procedure. In this procedure, the specimen is moved in uniform increments on a microscope stage, and phases falling under the cross hairs of the eyepiece are identified and counted **(1–5)**.

5. Significance and Use

5.1 This test method provides a relatively simple and reliable microscopical means of measuring the phase abundance of portland cement clinker (Note 1). Microscopical point counting provides a direct measure of the clinker phase

---

[1] This test method is under the jurisdiction of ASTM Committee C-1 on Cement and is the direct responsibility of Subcommittee C01.23 on Compositional Analysis.

Current edition approved Dec. 10, 1996. Published March 1997.

[2] *Annual Book of ASTM Standards*, Vol 04.01.

[3] *Annual Book of ASTM Standards*, Vol 04.02.

[4] *Annual Book of ASTM Standards*, Vol 04.03.

[5] C = CaO, S = SiO_2, A = Al_2O_3, F = Fe_2O_3, $\overline{S}$ = SO_3, M = MgO, N = Na_2O, and K = K_2O in cement chemistry notation.

composition in contrast to the calculated Bogue phase composition (Note 2).

NOTE 1—This test method utilizes a reflected light microscope. Related methods such as transmitted light microscopy, scanning electron microscopy, and automated imaging techniques may also be used for clinker analysis but are not presently included in this test method.

NOTE 2—This test method allows direct determination of the proportion of each individual phase in portland-cement clinker. This test method is intended to provide an alternative to the indirect estimation of phase proportion using the equations in Specification C 150 (footnote C in Table 1 and footnote B in Table 2).

5.2 This test method assumes the operator is qualified to operate a reflected light microscope and the required accessories, is able to correctly prepare polished sections and use necessary etchants, and is able to correctly identify the constituent phases.

5.3 This test method may be used as part of a quality control program in cement manufacturing as well as a troubleshooting tool. Microscopic characterization of clinker phases may also aid in correlating cement properties and cement performance in concrete, to the extent that properties and performance are a function of phase composition.

6. Apparatus

6.1 Reflected light microscope.

6.2 Mechanical stage with stepping increments ranging from 0.05 to 2.0 mm (to enable analysis of clinkers of different average crystal sizes) and vernier scales graduated in both X and Y directions.

6.3 Microscope objectives of magnification 5×, 10×, 20×, and 40× or other magnifications suitable for the task.

NOTE 3—The use of reflected light with oil immersion is optional. It is highly recommended for study of finely crystalline aluminate and ferrite which typically form the ground mass in which the silicates occur. Reflected light objective lenses with magnification up to 100× designed for use in oil-immersion are required.

6.4 Assorted eyepieces (5×, 10×, 20×) which when combined with the objectives described in 6.3 will provide magnifications up to 800×.

6.5 Eyepiece reticles (graticulae) with a linear grid pattern containing 9, 16, or 25 intersections.

6.6 Eyepiece micrometer for measuring dimensions of the object under investigation and calibrated for each magnification.

6.7 Stage micrometer for the calibration of the eyepiece micrometer.

6.8 Light source that provides uniform and consistent illumination of the field and light of constant intensity.

6.9 Counting (tallying) device capable of recording up to ten categories of data.

6.10 Crushing device capable of reducing sample particle size to between 1 and 4 mm.

6.11 Riffle sample splitter to reduce sample from initial volume to approximately 100 g.

6.12 Wire cloth sieves with openings suitable for sieving the entire clinker sample to broadly define the model size class, and sieves with 1-mm and 4-mm square openings to concentrate particles of recommended size for specimen preparation.

6.13 Vacuum impregnation device to force epoxy into clinker voids. (Vacuum bell jar or desiccator connected to a vacuum pump.)

6.14 Curing oven, hot plate, slide warmer, or ultraviolet light may be used to accelerate the epoxy hardening.

6.15 Thin, diamond-rimmed wafering saw for sectioning the encapsulated clinker.

6.16 Glass grinding (lapping) plates (300 mm × 300 mm × 5 mm) required only if the mechanical system is not equipped to handle the final grinding with alumina powder.

6.17 Ultrasonic cleaning device (optional) to clean the sample prior to, between, and after polishing steps.

7. Reagents and Materials

7.1 Consumable grinding (lapping) and polishing supplies. After the encapsulated specimen has been cut with the saw, all or most of the following grinding and polishing steps are required: 120-, 320-, and 600-grit silicon carbide grinding papers or equivalent and 5 µm, 0.3 µm, and 0.05 µm alumina polishing powders or their equivalent. Diamond grinding discs, silicon carbide paper, or polishing cloths and alumina polishing powder may be used. Various types of polishing cloths may be used to produce a nearly flat clinker surface or a relief surface to aid in identification of periclase **(1)**.[6]

7.2 Sample cups (with volumes ranging from 10 to 20 mL) to contain epoxy-clinker mix during hardening.

7.3 Epoxy resin and hardener for encapsulation of the clinker. Low viscosity resin will facilitate penetration into clinker voids. When hardened it should have an abrasive resistance close to that of the clinker to minimize relief during polishing. It should be resistant to substances used for washing and etching.

7.4 Isopropyl alcohol (2-propanol) for washing the specimen and for use in the ultrasonic cleaner. Propylene glycol is suitable as a lubricant for the saw blade.

7.5 Immersion oil with an index of refraction of 1.51 if reflected light immersion-oil technique is used.

7.6 Etching material to highlight different phases for counting. (See Appendix X3.)

8. Sampling

8.1 Take samples of portland-cement clinker in accordance with the applicable provisions of Practices D 75 and D 3665 so as to be representative of the quantity of material with which testing is concerned (see Appendix).

8.2 Sieve the initial sample to obtain clinker particles representing approximately 70 % of the clinker particle size distribution, centered about the mode. This particle size interval represents a size range of approximately two standard deviations, one on each side of the mode, and is herein defined as the "bulk mode" (M_b), the combined material between the extremes of the particle size distribution. State the sieve sizes used. If either extreme of the particle size distribution is to be studied, the selected portion shall be identified as non-modal, and the percentages retained or passing standard sieves shall be stated. The recommended size fraction for microscopical analysis is 2 to 4 mm.

[6] The boldface numbers in parentheses refer to the list of references at the end of this standard.

Therefore, that portion of the initial clinker sample representing the bulk mode shall be crushed, sieved, and riffled to provide approximately 100 g. Whole clinkers may be encapsulated to study the phase distribution within clinker nodules.

9. Preparation of Sample Specimen

9.1 Polished sections shall be polished to a fineness such that grinding pits and scratches have been eliminated (see Appendix).

9.2 Etching of the clinker surface may be used to facilitate identification of clinker phases **(1)**; additional information may be found in the Appendix.

10. Counting Procedure

10.1 Choose the microscope magnification such that adjacent reticle grid points do not fall on the same crystal, except for a few unusually large crystals. Magnification from 200× to 500× will accommodate most clinkers. Reticles with multiple grid points are recommended. Single crosshair reticles are not suitable.

10.2 Choose the stepping interval such that an entirely different field of view is observed after each step.

10.3 Attach a mechanical stage to the microscope. The mechanical stage may be an electrically driven specimen carriage connected to an automatic electronic counter or a simple hand-operated carriage and counter. Place a small amount of soft modeling clay on a standard, petrographic glass slide (27 mm × 46 mm) and level the sample thereon. Either use a commercial leveling device or use a small spirit level to adjust the mount while pressing it firmly into the clay. A tissue paper between the polished section and the leveling device or spirit level prevents surface scratches. Place the glass slide with attached clay and leveled mount on the mechanical stage. Check the accuracy of leveling by observing the focus at several points on the polished surface. Move to a starting point (the initial field of view from which data will be taken) at the edge of the mount and record the position in the X-Y coordinate system, using the graduated scales on the mechanical stage.

10.4 Use a tally sheet or a counting device to record the phases as required. The observer should not keep a mental tally of any data because of possible bias. Identify and record each phase under the grid intersections. In some cases, the aluminate and ferrite quantities may be combined and labeled "matrix". Move the mechanical stage a distance of one stepping interval in the chosen X or Y direction to bring another field into view. The phases under the grid points are identified, counted, and the mechanical stage advanced one stepping interval to an adjacent field of view. This procedure is continued until a range of 3000-4000 points are recorded.

At the clinker periphery, some of the reticle points may fall on the encapsulating epoxy that surrounds the clinker particle. Count only points within the clinker, and disregard the reticle points over epoxy at the clinker periphery. Thus, the clinker void space (porosity), if determined in the point count, does not include cavities on the surface of the clinker particle. As one steps over irrelevant areas (such as epoxy exterior to the clinker or severally damaged portions of the polished surface that obscure the phase identification) the count is temporarily suspended until a suitable clinker surface again falls under the reticle grid points. Artifacts (for example, blot marks, residual liquids) on the section surface are not to be counted. If the identity of the phase is obscured by the area formed by the grid intersection, one should consistently use a specified corner of the intersection where the phase can be clearly observed.

When the edge of the mount is reached at the end of a line of traverse, the mount is translated one stepping interval perpendicular to that line and counting continues in the opposite direction, or the mount is repositioned to a point at the original starting boundary to keep the same direction of travel. As one progresses along the lines of traverse the data are accumulated until 3000–4000 points (N) are recorded.

11. Calculation of Results

11.1 Calculate the volume fraction of each phase in the sample by dividing the number of points counted for that phase by the total number of points counted. This number multiplied by 100 yields the phase content in volume per cent.

11.2 The number of points to be counted is to be between 3000 and 4000 **(2)**. The absolute error at the 96 % confidence interval has been shown to be:

$$\delta = 2.0235 \sqrt{\frac{P(100 - P)}{N}} \tag{1}$$

where:

δ = absolute measuring error in percentage for a given constituent,

P = phase percentage, and

N = total number of points counted.

Therefore for any error percentage deemed acceptable and expected value of P, the required number of points, N, can be calculated:

$$N = \left(\frac{2.0235}{\delta}\right)^2 \times (100 - P)P \tag{2}$$

If an estimate of P is not available, the above equation takes on a maximum when $P = 50$. Using this value will provide a conservative estimate for the number of points required.

11.3 To express phase abundance in terms of mass fraction, multiply phase volume fractions by their respective mass density (see Table 1) and normalize the totals to 100 %.

12. Reporting Results

12.1 The report of the point count results to microscopically determine the phase percentages in portland cement clinker should include the following information:

TABLE 1 Mass Densities in g/cm³ for Clinker Phases (7)

Phase	Density
alite	3.18
belite	3.31
aluminate	3.03
ferrite	3.73
free lime	3.35
periclase	3.58
alkali sulfate	2.66

12.1.1 Identify the source of the samples, including sampling location and type of sample.

12.1.2 State how the sample was treated. If the sample was sieved, give the sieve designation size for the final sample. Indicate the type of each etchant used and etch time.

12.1.3 State the number of data points and percentages of each determined phase.

13. Precision and Bias

13.1 *Within-Laboratory Precision*—Single operator values of the one-sigma limit in percent (1S %), defined in accordance with Practice C 670 determined for volume fraction determination by point count analysis (see Note 4) have been found to be 0.71 % (1S) **(6)**. Therefore, results of two properly conducted tests by the same operator on the same material should not differ by more than 2.58 % (D2S) from the average of the two results.[7]

NOTE 4—These levels of precision are based on data from the NIST RM Clinker certificate data. A bias statement is not available as these data are used to establish the phase abundance compositions of the RM clinkers.

13.2 A multi-laboratory round robin is planned to generate data for use in a precision and bias statement in accordance with Practice C 670.

14. Keywords

14.1 alite; aluminate; belite; ferrite; periclase; portland-cement clinker; reflected light microscopy

[7] Data pertaining to this precision statement are on file at ASTM Headquarters as report RR:C01-1005.

APPENDIX

(Nonmandatory Information)

X1. SAMPLING

X1.1 The task of obtaining a representative clinker sample for microscopical examination from the stream of material leaving the kiln presents special problems, and the solution depends greatly on local conditions. When sampling from a conveyor belt, an entire cross section of the material should be taken. When the sample is collected from material in free fall, for example at the end of the conveyor belt or at the discharge of a bucket elevator, all the particles falling during the sampling operation shall be included, and sampling devices and containers shall be of sufficient volume to contain the resulting amount. One litre is recommended for the initial clinker sample.

X1.2 If continuous sampling of a clinker flow is impractical, samples representing more extended periods of operation shall be composites of two of more grab samples obtained according to X1.1. The period of time covered and the frequency of sampling will depend on local conditions and the purpose of the test, but the procedure for obtaining each individual grab sample shall be identical.

X1.3 Material judged atypical for the period, the process, or the equipment under study (kiln dust, kiln coating, etc.) may be removed prior to microscopical examination, but the removal of such material should be noted.

X2. SAMPLE IMPREGNATION AND POLISHING

X2.1 Label a sample cup with the sample identification (a tag may also be embedded in the epoxy) and coat the interior with a mold release agent. Place 10 to 20 g of the clinker sample in the cup and add low-viscosity epoxy resin previously mixed with the hardener. Place the sample cup under the vacuum jar and operate the vacuum pump to obtain adequate penetration of the resin. Break the vacuum several times to permit atmospheric pressure to force the epoxy into the material.

X2.2 Cure the epoxy as recommended by the manufacturer in an oven at a temperature not higher than 70°C. Some epoxies will harden rapidly at room temperature or under ultraviolet light.

X2.3 Remove the specimen from the cup and, if necessary, label with a diamond engraver or indelible pen.

X2.4 Cut the encapsulation with the diamond-rimmed wafering saw to expose the clinker particle cross sections. Preparation liquids retained in clinker voids may interfere with etching and staining; therefore, a second application of vacuum impregnated epoxy on the polished surface may be necessary. A cyanoacrylate resin, smeared over the polished surface, may be used. The encapsulation is then reground and polished.

X2.5 Grind the exposed surface by lapping with each of the selected grinding media (see 7.1), beginning with the coarsest, for 2 to 4 min each. Use propylene glycol, isopropanol, or other liquids as lubricants during grinding. After each step, wash the surface with a forceful spray of isopropyl alcohol. Do not use water in cutting, grinding, polishing, or cleaning the encapsulated clinkers.

X2.6 Polish the surface with each of the chosen polishing media, beginning with the coarsest, for 2 to 4 min. Wash the surface after each polishing step with a forceful isopropyl alcohol spray. Clean the specimen ultrasonically after the last wash, using isopropyl alcohol as the cleaning fluid. Rinse again with the alcohol spray. Dry the surface with a forced warm air current. Blotting is not recommended because of the production of blot marks.

X3. ETCHES AND STAINS

X3.1 Certain liquids or vapors applied to the polished surface will produce a preferential phase coloration. One versatile liquid is nital (1 mL nitric acid in 99 mL isopropyl alcohol). Other commonly used stains and etchants are hydrofluoric acid vapor, potassium hydroxide, distilled water, aqueous ammonium chloride, dimethyl ammonium citrate, and others. Details of various etches and stains, immersion times, and techniques are given in Ref **(1)**. The choice of etchant, etch time, and temperature of etchant can affect results. Therefore, these choices should be consistent for all specimens prepared for point counting.

NOTE X1.1: **Precaution**—Appropriate safety precautions must be observed when preparing and using stains and etchants. Review all Material Safety Data Sheets and always use suitable personal protective equipment including eye protection, apron, and gloves.

X3.2 For aluminate and free-lime coloration, the application of distilled water at 40°C for 5 s with the specimen at room temperature, or 0.1 molar KOH for 20 to 30 s is recommended. The polished surface can also be placed on a water-saturated, clean, low-nap polishing cloth for 2 to 3 s and then quickly washed with a forceful isopropyl alcohol spray. This technique produces a structure etch on the silicates. The polishing cloth may be attached to a rotating wheel or may be stationary on a glass plate.

X3.3 For a clear differentiation of the silicates, alite and belite, a 6- to 8-s immersion in nital is recommended, quickly followed by an alcohol spray wash. One may choose to carry out a high magnification point count (550× or more) on the water-stained matrix, followed by a relatively low-magnification point count (approximately 300×) of the silicates on a nital-etched surface.

REFERENCES

(1) Campbell, D. H., *Microscopical Examination and Interpretation of Portland Cement and Clinker*, Portland Cement Association, Skokie, Illinois, 1986, 128 pp.

(2) Hofmanner, F., *Microstructure of Portland Cement Clinker*, Holderbank Management & Consulting Ltd., Switzerland, 1973, 48 pp.

(3) Fundal, E., *Microscopy of Cement Raw Mix and Clinker*, F. L. Smidth, Review 25, 1980, 15 pp.

(4) Chayes, F., *Petrographic Modal Analysis, An Elementary Statistical Appraisal*, John Wiley and Sons, Inc., New York, 1956, 113 pp.

(5) Brown, L. S., *Microscopical Study of Clinkers*, in Long-Time Study of Cement Performance in Concrete, Portland Cement Association R&D Bulletin 26, 1948, pp. 877–923.

(6) Kanare, H. M., *Production of Portland Cement Clinker Phase Abundance Standard Reference Materials*, Construction Technology Laboratories Final Report, Project No. CRA012-840, 1987, 32 pp.

(7) ICDD/JCPDS Powder Diffraction File, International Centre for Diffraction Data, Newton Square, PA.

Standard Test Method for Determination of the Proportion of Phases in Portland Cement and Portland-Cement Clinker Using X-Ray Powder Diffraction Analysis[1]

This standard is issued under the fixed designation C 1365; the number immediately following the designation indicates the year of original adoption or, in the case of revision, the year of last revision. A number in parentheses indicates the year of last reapproval. A superscript epsilon (ϵ) indicates an editorial change since the last revision or reapproval.

1. Scope

1.1 This test method covers direct determination of the proportion by mass of individual phases in portland cement or portland-cement clinker using quantitative X-ray (QXRD) analysis. The following phases are covered by this standard: aluminate (tricalcium aluminate,[2] C_3A), ferrite (tetracalcium aluminoferrite, C_4AF), and periclase (magnesium oxide, M) (see Note 1).

NOTE 1—In the future, Subcommittee C1.23 plans to add additional phases: alite (tricalcium silicate, C_3S), belite (dicalcium silicate, C_2S), gypsum ($C\bar{S}H_2$), hemihydrate ($C\bar{S}H½$), and anhydrite ($C\bar{S}$).

1.2 This test method specifies certain general aspects of the analytical procedure, but does not specify detailed aspects. A recommended procedure is described, but not specified. Regardless of the procedure selected, the user must demonstrate by analysis of reference materials (RM's) that the particular analytical procedure selected for this purpose qualifies (that is, provides acceptable precision and bias) (see Note 2). The recommended procedure is the one used in the round-robin analysis to determine the precision and bias levels of this test method.

NOTE 2—A similar approach was used in the performance requirements for alternative methods for chemical analysis in Test Methods C 114.

1.3 The values stated in SI units shall be regarded as the standard.

1.4 *This standard does not purport to address all of the safety concerns, if any, associated with its use. It is the responsibility of the user of this standard to establish appropriate safety and health practices and determine the applicability of regulatory limitations prior to use.* For specific hazards, see Section 9.

2. Referenced Documents

2.1 *ASTM Standards:*

C 114 Test Methods for Chemical Analysis of Hydraulic Cement[3]

C 150 Specification for Portland Cement[3]

C 183 Practice for Sampling and the Amount of Testing of Hydraulic Cement[3]

C 219 Terminology Relating to Hydraulic Cement[3]

C 670 Practice for Preparing Precision and Bias Statements for Test Methods for Construction Materials[4]

E 29 Practice for Using Significant Digits in Test Data to Determine Conformance with Specifications[5]

E 691 Practice for Conducting an Interlaboratory Study to Determine the Precision of a Test Method[5]

3. Terminology

3.1 *Definitions are in accordance with Terminology C 219.*

3.2 *Phases* (***1***):[6]

3.2.1 *alite, n*—tricalcium silicate (C_3S) modified in composition and crystal structure by incorporation of foreign ions; occurs typically between 30 to 70 % (by mass) of the Portland-cement clinker; and is normally either the M_1 or M_3 crystal polymorph, each of which is monoclinic.

3.2.2 *aluminate, n*—tricalcium aluminate (C_3A) modified in composition and sometimes in crystal structure by incorporation of a substantial proportion of foreign ions; occurs as 2–15 % (by mass) of the Portland-cement clinker; is normally cubic when relatively pure and orthorhombic when in solid solution with significant amounts of sodium, though tetragonal aluminate containing a substantial amount of potassium has been reported (**2**).

3.2.3 *belite, n*—dicalcium silicate (C_2S) modified in composition and crystal structure by incorporation of foreign ions; occurs typically as 15 to 45 % (by mass) of the Portland-cement clinker as normally the β polymorph, which is monoclinic. In lesser amounts, other polymorphs can be present.

3.2.4 *ferrite, n*—tetracalcium aluminoferrite solid solution of approximate composition $C_2(A,F)$ modified in composition by variation in the Al/Fe ratio and by substantial incorporation of foreign ions as $C_4A_XF_{2-X}$ where $O < x < 1.4$; constituting 5 to 15 % (by mass) of a portland-cement clinker; and is orthorhombic.

3.2.5 *periclase, n*—free magnesium oxide (M); cubic.

3.3 *Definitions of Terms Specific to This Standard:*

3.3.1 *standardization, n*—process of determining the relationship between XRD intensity and phase proportion for one or more phases.

3.3.1.1 *Discussion*—In the literature of X-ray powder diffraction analysis, this process has been commonly referred to

---

[1] This test method is under the jurisdiction of ASTM Committee C-1 on Cement and is the direct responsibility of Subcommittee C01.23 on Compositional Analysis.

Current edition approved July 10, 1998. Published December 1998.

[2] When expressing chemical formulae, C = CaO, S-SiO_2, A = Al_2O_3, F = Fe_2O_3, M = MgO, $\bar{S}$ = SO_3, and H = H_2O.

[3] *Annual Book of ASTM Standards,* Vol 04.01.

[4] *Annual Book of ASTM Standards,* Vol 04.02.

[5] *Annual Book of ASTM Standards,* Vol 14.02.

[6] The boldface numbers in parentheses refer to the list of references at the end of this test method.

as calibration; however, we have determined that standardization is a more accurate term.

3.3.2 *diffractometer, n*—the instrument, an X-ray powder diffractometer, for determining the X-ray diffraction pattern of a crystalline powder.

3.3.3 *phase, n*—a homogeneous, physically distinct, and mechanically separable portion of a material, identifiable by its chemical composition and crystal structure.

3.3.3.1 Phases in portland-cement clinker that are included in this test method are two major phases (aluminate and ferrite) and one minor phase (periclase).

3.3.4 *qualification, n*—process by which a QXRD procedure is shown to be valid.

3.3.5 *Reference Material (RM) Clinkers, n*—three cement clinkers with known proportions of each major phase, available as RM's, which shall be used in demonstrating qualification of a particular procedure for determination of the phases in portland cement or portland-cement clinker.[7]

3.3.6 *Standard Reference Material (SRM), n*—a material whose properties (in this case XRD peak position or intensity, or both) are known and certified.[8]

3.3.7 *X-ray diffraction (XRD), n*—the process by which X-rays are coherently scattered by electrons in a crystalline material.

4. Background

4.1 This test method assumes general knowledge concerning the composition of portland cement and portland-cement clinker. Necessary background information may be obtained from a number of references **(1,3)**.

4.2 This test method also assumes general expertise in XRD and QXRD analysis. Important background information may be obtained from a number of references **(4,5,6,7)**.

4.3 QXRD analysis is difficult and time-consuming and QXRD of portland cement and portland-cement clinker is especially difficult due to the complex and variable nature of the individual clinker phases.

5. Summary

5.1 A QXRD test procedure includes some or all of the following:

5.1.1 Specimen preparation,

5.1.2 Standardization,

5.1.3 Use of an internal or external standard (to correct for various effects on intensity besides phase proportion),

5.1.4 Analysis of the sample (in which the intensity of selected XRD peaks is measured), and

5.1.5 Calculation of the proportion of each phase.

5.2 This test method does not specify details of the procedure. The user must demonstrate by analysis of reference materials that the particular analytical procedure selected for this purpose provides acceptable levels of precision and bias. A recommended procedure (the procedure used to determine the acceptable levels of precision and bias) is given in the Appendix.

6. Significance and Use

6.1 This test method covers direct determination of the proportion of some individual phases in portland cement or portland-cement clinker. Thus it provides an alternative to the indirect estimation of phase proportion using the equations in Specification C 150 (see Note C of Table 1 and Note B of Table 2 of C 150).

6.2 This test method assumes that the operator is qualified to operate an X-ray diffractometer and to interpret X-ray diffraction spectra.

6.3 This test method may be used as part of a quality control program in cement manufacturing.

6.4 This test method may be used in predicting properties and performance of hydrated cement and concrete, to the extent that properties and performance are a function of phase composition.

6.5 QXRD provides a bulk analysis (that is, the weighted average composition of several grams of material). Results may not agree precisely with results of microscopical methods.

7. Apparatus

7.1 *X-Ray Diffractometer*—The X-ray diffractometer allows measurement of the X-ray diffraction pattern from which the crystalline phases within the sample may be qualitatively identified and the proportion of each phase may be quantitatively determined. X-ray diffractometers are manufactured commercially and a number of instruments are available. The suitability of the diffractometer for this test method shall be established using the qualification procedure outlined in this test method.

8. Materials

8.1 *Standardization Phases*—The use of standardization phases is recommended. These phases must usually be synthesized **(8)**.

8.2 *RM Clinkers*—The use of RM clinkers is required to qualify the QXRD procedure.

8.3 *Internal Standard*—The use of an internal standard is recommended. Suitable materials include chemical reagents (see 8.4) or SRM's (see the Appendix).

8.4 *Reagent Chemicals*—Reagent grade chemicals, if used either as an internal standard or during chemical extraction of the calcium silicate phases, shall meet the specifications of the Committee on Analytical Reagents of the American Chemical Society where such specifications are available.[9] Other grades may be used, provided it is first ascertained that the chemical is sufficiently pure to permit its use without lessening the accuracy of the determination.

9. Hazards

9.1 The need for careful and safe operation of an X-ray

[7] In this test method, RM clinkers refer specifically to RM 8486, RM 8487, and RM 8488. These are available from the Standard Reference Material Program, National Institute of Standards and Technology.

[8] SRM's are available from the Standard Reference Material Program, National Institute of Standards and Technology.

[9] *Reagent Chemicals, American Chemical Society Specifications*, American Chemical Society, Washington, DC. For suggestions on the testing of reagents not listed by the American Chemical Society, see *Analar Standards for Laboratory Chemicals*, BDH Ltd., Poole, Dorset, U.K., and the *United States Pharmacopeia and National Formulary*, U.S. Pharmacopeial Convention, Inc. (USPC), Rockville, MD.

diffractometer cannot be overemphasized. X-rays are particularly hazardous. An X-ray diffractometer must be operated safely to avoid serious injury or death. Furthermore, the X-rays are generated by high voltages, perhaps as high as 55 kV peak, requiring care to avoid serious electric shock. Klug and Alexander (**5**, pp. 58 – 60) state, "*The responsibility for safe operation rests directly on the individual operator*" (italics are theirs).

10. Sampling and Sample Preparation

10.1 Take samples of portland cement in accordance with the applicable provisions of Practice C 183. Take samples of portland-cement clinker so as to be representative of the material being tested.

10.2 Prepare samples as required for the specific analytical procedure (see Appendix).

11. Qualification and Assessment

11.1 *Qualification of Test Procedure:*

11.1.1 When analytical data obtained in accordance with this test method are required, any QXRD test procedure that meets the requirements described in this section may be used.

11.1.2 Prior to use for analysis of portland cement or portland-cement clinker, each QXRD test procedure must be qualified for such analysis. The laboratory shall maintain records that include a description of the QXRD procedure and the qualification data (or, if applicable, requalification data). These records shall be made available to the purchaser if requested in the contract or order.

11.1.3 If more than one X-ray diffractometer is used in a specific laboratory for the same analysis, even if the instruments are substantially identical, each shall be qualified separately.

11.1.4 If more than one procedure is used to mount specimens for QXRD, the use of each procedure shall constitute a separate test procedure and each procedure shall be qualified separately.

11.1.5 Qualification shall consist of analyzing the three RM clinkers (see Note 3) for the proportions of C_3A (cubic and orthorhombic), C_4AF, and M using the desired QXRD procedure (see Note 4).

NOTE 3—Prior to qualification, it may be convenient to carry out a preliminary assessment in which one or more mixtures of synthetic phases is analyzed. Such a preliminary assessment should produce no more than the permissible variation described in 11.2.

NOTE 4—It is recommended that at least two replicate analyses be carried out.

11.2 *Permissible Variation:*

11.2.1 The values of permissible variation were computed from the within-laboratory standard deviation values obtained in the cooperative standardization and round robin analysis of mixtures of synthetic phases (see 14.2.1).

11.2.2 If replicate analyses are carried out, results should differ from each other by no more than the value shown in Table 1 for the particular number of replicates.

11.2.3 The mean result shall differ from the known value by no more than the value shown in Table 2 for the particular number of replicates.

11.2.4 *Known Values*—The known values of each phase in

TABLE 1 Permissible Maximum Difference Between Replicate Values (percent of clinker)

Phase	Number of Replicates	Permissible Difference[A]
Orthorhombic C_3A	2	1.06
	3	1.25
	4	1.37
	5	1.48
Cubic C_3A	2	1.04
	3	1.22
	4	1.33
	5	1.44
C_4AF	2	1.68
	3	1.98
	4	2.16
	5	2.34
M	2	0.50
	3	0.59
	4	0.65
	5	0.70

[A] As described in Practice C 670.

the RM clinkers provided by NIST were determined using quantitative optical microscopy and differ somewhat from those obtained using QXRD. For the purpose of this test method, the known values of the RM clinkers shall be the values in the QXRD round robin analysis (see 14.2.2). These values are listed in Table 3.

11.3 *Partial Results:*

11.3.1 QXRD procedures that provide acceptable results for some phases but not for others shall be used only for those phases for which acceptable results are obtained. However, it is not expected that a QXRD procedure would provide acceptable results for some phases and not for others, and such a result may indicate that the procedure is not, in fact, valid.

11.4 *Assessing the Diffractometer:*

11.4.1 The procedures described in the Annex shall be used to assess the diffractometer. Note that assessment is different from qualification or requalification.

11.4.2 The diffractometer shall be assessed each day that this test method is used.

11.4.3 The diffractometer shall be assessed after any substantial modification in the instrument (see Note 5).

NOTE 5—Substantial modification of the diffractometer includes changing the X-ray tube, changing slits, adding or removing a monochromator, and realigning the diffractometer.

11.4.4 QXRD procedure shall be assessed upon receipt of evidence that the test procedure is not providing data in accordance with the permissible variation.

11.5 *Requalification of QXRD Procedure:*

11.5.1 If assessment shows that the X-ray diffractometer is not properly aligned (as discussed in the Annex), it shall be realigned following the manufacturer's instructions. When subsequent assessment shows that the X-ray diffractometer is properly aligned (or was not properly aligned when the QXRD procedure was previously qualified), qualification of the QXRD procedure shall be repeated.

12. Recommended Procedure

12.1 For required analytical data see Section 11 and a recommended QXRD procedure described in Appendix X1.

13. Report

13.1 *Report the Following Information:*

TABLE 2 Permissible Maximum Difference Between Mean Value and Known Value (percent of clinker)

Phase	Number of Replicates	Permissible Difference[A]
Orthorhombic C_3A	1	0.74
	2	0.53
	3	0.43
	4	0.37
	5	0.33
Cubic C_3A	1	0.73
	2	0.51
	3	0.42
	4	0.36
	5	0.32
C_4AF	1	1.18
	2	0.83
	3	0.68
	4	0.59
	5	0.53
M	1	0.35
	2	0.25
	3	0.20
	4	0.18
	5	0.16

[A] Computed from within-laboratory standard deviation using 95 % confidence interval and 30 df.

TABLE 3 Phase Proportions in RM Clinkers (determined by QXRD) (percent of clinker)

Clinker	Phase	Proportion
RM 8486	orthorhombic C_3A	2.63
	cubic C_3A	1.07
	C_4AF	10.61
	M	3.40
RM 8487	orthorhombic C_3A	1.18
	cubic C_3A	11.02
	C_4AF	3.11
	M	0.48
RM 8488	orthorhombic C_3A	3.21
	cubic C_3A	1.95
	C_4AF	11.46
	M	0.18

13.1.1 The phase and its proportion, and

13.2 Round figures to the number of significant places required in the report only after calculations are completed, in order to keep the final results substantially free of calculation errors. Follow the rounding procedure outlined in Practice E 29 or Test Methods C 114.

13.3 Note in the report that this test method was used.

14. Precision and Bias

14.1 *Analysis*—A cooperative standardization of mixtures of synthetic phases and a round-robin analysis[9] of the RM clinkers have been carried out **(8,9,10)** following the experimental procedures described in the Appendix.

14.1.1 A cooperative standardization and round robin analysis was undertaken to determine precision and bias levels for analysis of mixtures of synthetic phases. Results were analyzed statistically according to Practice E 691 to determine precision and bias levels.

14.1.2 A round robin was undertaken to determine precision and bias levels for analysis of portland cement or portland-cement clinker (see Note 6) using the RM clinkers. Results were analyzed statistically according to Practice E 691 to determine precision and bias levels.

TABLE 4 Statistical Results for Mixtures of Synthetic Standardization Phases (percent of clinker)

Phase	Repeatability Standard Deviation	Reproducibility Standard Deviation
orthorhombic C_3A	0.38	0.83
cubic C_3A	0.37	0.78
C_4AF	0.60	0.64
M	0.18	0.19

TABLE 5 Statistical Results for RM Clinkers (percent of clinker)

Phase	Repeatability Standard Deviation	Reproducibility Standard Deviation
orthorhombic C_3A	0.30	1.05
cubic C_3A	0.37	0.84
C_4AF	0.59	1.61
M	0.07	0.56

NOTE 6—Analysis of clinker is likely to include variance in addition to that found in analysis of mixtures of synthetic phases.

14.1.3 The precision and bias values are all expressed as percentage points by mass relative to the total clinker.

14.2 *Precision:*

14.2.1 *Mixtures of Synthetic Phases*—Ten laboratories participated in the cooperative standardization and round-robin analysis of mixtures of synthetic phases. The within-laboratory standard deviation for all phases averaged 0.39 percentage point, and the between-laboratory standard deviation for all phases averaged 0.61 percentage point. Values for each phase are given in Table 4.

14.2.2 *Clinkers*—Five laboratories participated in the round-robin analysis of the RM clinkers. The within-laboratory standard deviation for all phases averaged 0.33 percentage point and the between-laboratory standard deviation for all phases averaged 1.01 percentage point. Values for each phase are given in Table 5.

14.3 *Bias*—In the cooperative standardization and round-robin analysis of mixtures of synthetic phases the measured values differ from the known values by less than the reproducibility standard deviation, so there appears to be no bias in these results.

14.4 *Discussion*—These precision and bias levels appear reasonable. Taylor **(1)** concluded that the four major phases in portland-cement clinker may be determined using QXRD with an absolute accuracy of 2 to 5 percentage points (by mass) for alite and belite and 1 to 2 percentage points (by mass) for aluminate and ferrite. A bias may occur if the chemical composition or crystal structure of a phase being analyzed is different from that of the phase used for standardization. For example, a greater bias in ferrite determination may occur if the A/F ratio of that phase in the unknown is quite different from that of the standardization mixtures.

15. Keywords

15.1 aluminate; clinker; diffractometer; ferrite; periclase; phase analysis; portland cement; quantitative X-ray diffraction analysis; X-ray diffraction

ANNEX

(Mandatory Information)

A1. ASSESSING THE X-RAY DIFFRACTOMETER

A1.1 *Introduction*—A QXRD analysis of portland cement and portland-cement clinker is made particularly difficult by the fact that individual clinker phases used for standardization are not stable over long periods of time because they hydrate easily, and are not easily synthesized. Thus it is difficult to assess standardizations directly by reanalysis of one or more standardization specimen. In addition, it is not desirable to repeat the standardization unless absolutely necessary. A more reasonable strategy is to use an external standard to assess the diffractometer and to decide when it is necessary to requalify a particular procedure. This Annex provides a procedure for assessing the diffractometer to assure the validity of the QXRD procedure over a long period of time (several years or longer).

A1.2 *Overview:*

A1.2.1 As long as certain aspects of the procedure are not changed, the relationship between peak intensity (relative to the intensity of an internal standard peak) and proportion of the phase (relative to the proportion of the internal standard) is assumed to be universal (that is, valid over an indefinite period of time, even after changes in the diffractometer such as realignment and replacement of the X-ray tube, and transferable from one diffractometer to another).

A1.2.2 The requirements for the QXRD standardization to be universal are: (*1*) specimens are free from preferred orientation, primary extinction, and microabsorption, (*2*) the irradiated volume of the specimen is constant and independent of scattering angle, (*3*) monochromator polarization effects are corrected, (*4*) integrated peak intensity is used, (*5*) when using an internal standard, standardization and analyses are carried out with an internal standard from the same lot because differences in the particle size distribution between lots of the same material can cause significant difference in peak intensity, and (*6*) standardization and analyses are carried out with the diffractometer in proper geometric alignment.

A1.2.3 If analyses are carried out using only the instrument on which the standardization was carried out, then it is necessary only that preferred orientation, extinction, microabsorption, irradiated volume, and integrated peak intensity are reproducible. In that case, the standardization is valid (though not universal, in that it cannot be transferred from one diffractometer to another) as long as methods of specimen preparation, specimen mounting, and data collection are suitable and are not changed. For example, thus the use of a variable divergence slit is acceptable, because it provides reproducible irradiated volume.

A1.3 Terminology

A1.3.1 *extinction*—a decrease in intensity during diffraction due to interference by successive crystal planes.

A1.3.1.1 *Discussion*—Extinction is affected by the crystallite size and is negligible for specimens ground to a particle diameter of 5 or 10 μm.

A1.3.2 *irradiated volume*—the volume of specimen that produces XRD signal.

A1.3.2.1 *Discussion*—Irradiated volume is constant from specimen to specimen as long as the proper geometric alignment is maintained and the specimen is sufficiently thick.

A1.3.3 *microabsorption*—an increase or decrease in intensity produced by a combination of phases that differ in absorption coefficient.

A1.3.3.1 *Discussion*—Microabsorption is affected by the extent to which the absorption coefficients differ and by the crystallite size. For phases whose mass absorption coefficients differ by less than 100, microabsorption is not significant for specimens ground to a particle diameter of <10 μm.

A1.3.4 *peak intensity measurements*—the integrated intensity of the particular diffraction peak.

A1.3.5 *preferred orientation*—the nonrandom orientation of grains relative to the specimen surface.

A1.3.5.1 *Discussion*—Preferred orientation causes major changes in intensity of certain XRD lines, and therefore may be a source of error in QXRD analysis. Preferred orientation is not thought to be a major problem with portland-cement clinker phases because they do not typically cleave along crystallographic directions. Preferred orientation is reduced (but not prevented) by prolonged grinding. The mounting procedure must be one that reduces preferred orientation, such as the procedure as described by Klug and Alexander (**5**, pp. 372–374).

A1.4 Alignment

A1.4.1 Loss of proper alignment causes systematic variations in peak intensity with 2θ angle, thus rendering the QXRD procedure invalid.

A1.4.2 In order to assess alignment, an external standard shall be analyzed each day that this test method is used. Measurements shall include peak position, intensity, and resolution (that is, peak width or the ratio of the peak to valley intensity of partially overlapping peaks) of two or more peaks at widely separated 2θ angles. Suitable external standards include polished specimens of novaculite quartz or silicon, or SRM 1976.[8]

A1.4.3 Proper alignment is indicated by all of the following: (*1*) correct peak position, (*2*) suitable peak intensity, (*3*) suitable ratio of peak intensity of one or more peaks, and (*4*) suitable peak resolution. These must all be determined for suitably intense peaks. The correct peak position is within $0.01°2\theta$ (Cu Kα) of its nominal value; for the (101) line of novaculite quartz, this value is $26.64°2\theta$ (Cu Kα). Suitable peak intensity depends on many aspects besides alignment and therefore must be determined for a particular diffractometer based on experience; 1000 counts per second per mA is a reasonable expected value for the (101) line of novaculite quartz. Suitable peak intensity ratio is within 5 % of the nominal value. Suitable peak resolution must likewise be determined for a particular diffractometer based on experience. A reasonable indication is provided by clear separation of the five quartz peaks [(122) α_1, (122) α_2, (203) α_1, (203) α_2 plus (301) α_1, and (301) α_2] that appear at about 68°2θ (Cu Kα) (**5**, p. 392–394). Another indication is provided by resolution of the (110) Kα1–2 doublet of tungsten that appears at about 40.4°2θ (Cu Kα); the valley between these peaks must be no greater than 0.5 times the intensity of the α_2 peak (**5**, p. 282–285).

A1.4.4 NIST SRM 1976 may be used for instrument sensitivity assessment (**10**). Certified relative intensities of diffraction peaks, by both peak height or peak area, may be used to assess and correct for instrument bias. Plotting the ratios of the observed to certificate relative intensities will allow assessment of instrument performance relative to a diffractometer deemed to be "in control." If the plot of intensity ratios is patternless and falls within the control limits the diffractometer may be considered "in control."

A1.4.5 When a diffractometer is found to not be properly aligned, then it must be realigned according to the manufacturer's instructions.

APPENDIX

(Nonmandatory Information)

X1. QXRD PROCEDURES USED TO DETERMINE ACCEPTABLE LEVELS OF PRECISION AND BIAS

X1.1 *Introduction*—During development of this test method, a round-robin analysis of the RM clinkers was carried out to determine acceptable levels of precision and bias for the test method (**8,9,10**). Details of the QXRD procedure were specified during this analysis. Although this test method does not include detailed QXRD procedures, the procedures used in the cooperative standardization and round robin are recommended.

X1.2 *Overview*—This QXRD procedure has a number of critical features: standardization using mixtures of synthetic phases, use of an internal standard, intensity measurement of a specified peak for each phase, measurement of peak area to determine intensity, grinding samples until at least 90 % (by mass) is smaller than 10 μm in diameter, and the use of chemical extractions in the analytical scheme (see Note X1.1).

NOTE X1.1—Alternative QXRD procedures may provide satisfactory precision and bias and thus be acceptable under this test method. For example, acceptable procedures need not include a standardization, but may instead involve calculating the relationships between intensity and proportion. An external standard may be used rather than an internal standard. Intensity of each phase may be measured using many peaks or even the entire pattern. Chemical extractions need not be used in the analytical scheme. The particular procedure described in this appendix simply constitutes a recommendation.

X1.3 *Standardization Mixtures:*

X1.3.1 Three standardization mixtures were prepared using cubic C_3A, orthorhombic C_3A, C_4AF, and M. These individual phases were obtained from the Construction Technology Laboratories, and procedures used to prepare these individual phases and standardization mixtures were described by Struble and Kanare (**8**).

TABLE X1.1 Proportion of Phases in Standardization Mixtures

Phase	Percent of Mixture			Percent of Clinker		
	A	B	C	A	B	C
Orthorhombic C_3A	0	51	7	0.00	8.67	1.19
Cubic C_3A	65	11	29	11.05	1.87	4.93
C_4AF	35	11	60	5.95	1.87	10.20
M	0	27	4	0.00	4.59	0.68

X1.3.2 The proportions in each mixture are listed in Table X1.1. The phases in this standardization are the principal phases in a clinker that has been chemically extracted using maleic acid or salicylic acid (see X1.3.3).

X1.3.3 Standardization phases were ground such that at least 95 % (by mass) was finer than 10 μm and not more than 5 % were finer than 1 μm in diameter. This particle size was selected to minimize microabsorption preferred orientation and to avoid loss of X-ray intensity due to grinding to particle sizes <1 μm.

X1.4 *Internal Standard:*

X1.4.1 Participants were required to use an internal standard—a pure, stable material, 95 % of which is finer than 10 μm. One recommended material was SRM 640a (silicon).[8] However, it should be noted that silicon intensity may be affected by particle size distribution. Also recommended were any one of the materials in SRM 674.[8] Another suitable material is SRM 676.[8]

X1.4.2 Standardization mixtures were mixed with the internal standard material in recommended proportions of 0.1200 g internal standard per 1.0000 g standardization or unknown mixture, corresponding to 0.0200 g internal standard per 1.0000 g clinker. These proportions assume that the standardization mixture and unknown mixtures represent 17 % (by mass) of a clinker.

X1.5 *Blending*—Components must be blended to provide homogeneous specimens. The recommended procedure was to blend each standardization mixture for approximately 10 minutes using a vibratory-type mill with approximately 5 mL of a nonaqueous solvent (see Note 14) for each gram of powder and with appropriate grinding media.

NOTE 14—Suitable solvents include cyclohexane or alcohol (ethanol, methanol, or 2-propanol). Alcohol often contains sufficient water to cause hydration, so this water must be removed (**9**).

X1.6 *Diffraction Procedures*—Participants were instructed to use their normal procedures for preparing specimens and for collecting diffraction patterns. Specimen preparation procedures should minimize preferred orientation. Procedures to collect diffraction patterns should provide reproducible peak intensity measurements (for example, slow scanning speeds and moderately large receiving-slit widths).

X1.7 *Peak Selection*—Participants were required to measure intensity of the peaks specified in Table X1.2 plus one or more peaks for the internal standard, though they were encouraged to measure additional peaks, as many as possible for each phase. Analysis using a manual diffractometer is probably limited to the few peaks in Table X1.2, whereas analysis using a computer-controlled diffractometer can utilize many peaks for each phase.

X1.8 *C_3A Correction*—The intensity of the (113) peak of orthorhombic C3A must be corrected for any contribution from the (023) peak of cubic C3A. This (023) intensity contribution is calculated by multiplying the (213) peak by a correction factor, which is the intensity ratio of the (023) peak to the (213) peak for cubic C3A using Sample *A* (which contains no orthorhombic C_3A).

X1.9 *Intensity Measurement*—Integrated intensity above the background must be measured for each peak. There are a number of procedures for measuring integrated intensity, and no specific procedure is required for this standardization.

X1.10 *Standardization Curve*—Once the standardization mixtures have been analyzed, standardization curves relating the measured intensity of each peak, relative to the intensity of the internal standard peak, to the known proportion of the phase must be developed. These curves may be prepared graphically or using accepted statistical procedures to determine the best-fitting curve. When an internal standard is used, it is assumed that the peak intensity is linearly related to the phase proportion. However under certain circumstances it may be found that the relationship is not linear, but is better described by some other mathematical function.

X1.11 *Analysis of Unknown Mixtures*—Two additional mixtures of the standardization phases were analyzed using the same QXRD procedures to provide a preliminary assessment of the standardization.

X1.12 *Analysis of Cement or Clinker*—The following procedures were followed during analysis of the RM clinkers and are recommended for analysis of any portland cement or portland-cement clinker.

TABLE X1.2 Peaks Recommended for QXRD Analysis

Phase	hkl	*d*-spacing (nm)	Position (°2θ) using Cu Kα
Orthorhombic C_3A[A]	(113)	0.4210	21.1
Cubic C_3A	(023)	0.4236	21.0
	(213)	0.4083	21.8
C_4AF	(130)	0.3654	24.3
M	(200)	0.2106	42.9

[A] Note that the intensity of the (113) peak of orthorhombic C_3A must be corrected for any contribution from the (023) peak of cubic C_3A. The intensity of the (023) peak is calculated by multiplying the (213) peak by a correction factor. The correction factor is determined by measuring the intensity ration of the (023) peak to the (213) peak in a sample containing cubic C_3A and no orthorhombic C_3A.

X1.12.1 *Grinding*—The clinker samples were ground such that at least 95 % was finer than 10 μm and not more than 5 % was finer than 1 μm. The following procedure has been found to be suitable: grind using a tungsten carbide ring and puck mill, with approximately 5 drops ethylene glycol as a grinding aid, for 4 to 6 min. It is necessary to start with enough sample to provide sufficient powder for the specimen mount, considering that extraction of the calcium silicates will leave ca. 15 % to 20 % (by mass) for XRD analysis.

X1.12.2 *Ignition*—In the case of portland cement, gypsum ($C\bar{S}H_2$) and hemihydrate ($C\bar{S}H_{1/2}$) must be converted to anhydrite ($C\bar{S}$) before QXRD analysis to eliminate interfering XRD peaks. This conversion can be carried out by igniting the cement for 30 min at 500°C.

X1.12.3 *Blending with Internal Standard*—The ground sample was blended with the internal standard using the same procedure as used during the standardization. The recommended proportion was 0.0200 g internal standard per gram of clinker to approximate the 12 % (by mass) recommended for the standardization mixtures.

X1.12.4 *Extraction*—The calcium silicates were extracted chemically using salicylic acid or maleic acid and methanol using procedures described by Struble **(11)**. This extraction was not quantitative; because the internal standard was added relative to unextracted clinker, the phase contents were determined relative to the unextracted clinker.

X1.12.5 *XRD Analysis*—The clinkers were analyzed using the same procedures (for example, collection of the diffraction pattern, measurement of peak intensities, and calculation of phase proportions) that were used in the standardization.

REFERENCES

(1) Taylor, H. F. W., *Cement Chemistry,* Academic Press, San Diego, CA, 1990.

(2) Gebhardt, R. F., "KC_8A_3: X-Ray Diffraction Characterization of the Compound and Comparison with NC_8A_3 and C_3A," *Cement, Concrete and Aggregates,* Vol 9, No. 2, 1987, pp. 117–128.

(3) Lea, F. M., *The Chemistry of Cement and Concrete,* 3rd ed., Arnold, London, 1970, 727 pp.

(4) Cullity, B. D., *Elements of X-Ray Diffraction,* Addison-Wesley Publishing Company, Inc., Reading, MA, 1978.

(5) Klug, H. P., and Alexander, L. E., *X-Ray Diffraction Procedures for Polycrystalline and Amorphous Materials,* 2nd ed, John Wiley & Sons, New York, NY, 1974.

(6) Bish, D. L., and Post, J. E., eds, *Modern Powder Diffraction, Reviews in Mineralogy,* Vol 20, Mineralogical Society of America, Washington, DC, 1989.

(7) *Methods and Practices in X-Ray Powder Diffraction,* 3rd ed, International Centre for Diffraction Data, Swarthmore, PA, 1989.

(8) Struble, L., and Kanare, H., "Cooperative Calibration and Phases, Report 2," *Structural Research Series No. 556,* University of Illinois at Urbana-Champaign, 1990.

(9) Struble, L., "Quantitative Phase Analysis of Clinker Using X-Ray Diffraction," *Cement, Concrete and Aggregates,* Vol 13, No. 2, 1991, pp. 97–102.

(10) Certificate of Analysis, "Standard Reference Material 1976, Instrument Sensitivity Standard for X-ray Powder Diffraction," *Standard Materials Reference Program,* National Institute of Standards and Technology, Gaithersburg, MD.

(11) Struble, L., "The Effect Of Water On Maleic Acid And Salicylic Acid Extractions," *Cement and Concrete Research,* Vol 15, 1985, pp. 631–636.

Standard Specification for Molded Glass Fiber Reinforced Gypsum Parts[1]

This standard is issued under the fixed designation C 1381; the number immediately following the designation indicates the year of original adoption or, in the case of revision, the year of last revision. A number in parentheses indicates the year of last reapproval. A superscript epsilon (ε) indicates an editorial change since the last revision or reapproval.

1. Scope

1.1 This specification covers molded glass fiber reinforced gypsum (GRG) parts, which are nonload-bearing, thin-shell, ornamental shapes for architectural embellishment of interior building construction.

1.2 The values stated in either inch-pound or SI units are to be regarded separately as standard. Within the text, the SI units are shown in parentheses. The values stated in each system shall be used independent of the other. Values from the two systems shall not be combined.

1.3 The information presented refers to the molded surface of the part.

1.4 The text of this standard references footnotes, which provide explanatory material. These footnotes shall not be considered as requirements of the standard.

2. Referenced Documents

2.1 *ASTM Standards:*

C 11 Terminology Relating to Gypsum and Related Building Materials and Systems[2]

C 1355/C 1355M Specification for Glass Fiber Reinforced Gypsum (GRG) Composites[2]

3. Terminology

3.1 *Definitions*—Definitions of terms shall be in accordance with Terminology C 11.

3.2 *Definitions of Terms Specific to This Standard:*

3.2.1 *control thickness, n*—the thickness of the GRG part where abutting other GRG parts, other products or at fastening points.

3.2.2 *embedments, n*—materials encapsulated into the part for the purpose of suspension, attachment, and stiffening.

3.2.3 *finish surface, n*—a surface against a mold.

3.2.4 *GRG part, n*—an individual molded component used as architectural embellishment.

3.2.5 *GRG product, n*—combination of GRG parts used as architectural embellishment.

3.2.6 *mock-up, n*—a presentation assembly of GRG parts defined by the purchase agreement.

3.2.7 *part thickness, n*—thickness of a part at any measurement point.

4. Materials and Manufacture

4.1 *Materials*—GRG parts shall be produced using the same materials used to make the composites in conformance with Specification C 1355/C 1355M.

4.2 *Manufacture*:

4.2.1 *GRG* parts shall be manufactured from the mixed slurry of alpha gypsum cement, potable water and additives, when used, and not less than 5 % by weight glass fiber reinforcement using either the hand lay-up or the spray-up process in a mold.

4.2.2 *GRG* parts shall be produced according to appıoved shop drawings.

4.2.3 Shop drawings shall specify part dimensions, quantities, recommended attachment points and methods, reinforcements, embedments, and tolerance.

5. Mechanical Properties

5.1 *GRG* parts shall be made from materials to make composites having properties in accordance with Specification C 1355/C 1355M.

6. Dimensions and Permissible Variations of GRG Parts

6.1 *Part Thickness*—The thickness of the GRG part shall be not less than 3⁄16 in. (5 mm).

6.2 *Control Thickness*, shall be not less than 5⁄16in. (8 mm).

6.3 *Straightness*—At any point on a line generally along a plane, an edge, or a surface, the molded side of the part shall not vary from straight ± 1⁄8 in. in 8 linear ft (1 mm in 750 mm).

6.4 *Dimensions*, as specified in approved shop drawings, pertaining to the individual part.

6.4.1 *Overall Width*, ± 1⁄8 in. (3 mm).

6.4.2 *Dimensions Within the Overall Width*, ± 1⁄16 in. (2 mm).

6.4.3 *Overall Length*, ± 1⁄8 in. (3 mm).

6.4.4 *Dimensions Within the Overall Length*, ± 1⁄16 in. (2 mm).

7. Workmanship, Finish and Appearance

7.1 *GRG* parts shall be produced to quality, tolerances, and surface standards set within this specification.

7.2 The standard finished surface shall be suitable to receive flat paint.

7.3 Finishes with sheen, direct light or special lighting conditions require additional coats of primer, sanding and finishing by the installer or painter.

---

[1] This specification is under the jurisdiction of ASTM Committee C-11 on Gypsum and Related Building Materials and Systems and is the direct responsibility of Subcommittee C11.01 on Specifications and Test Methods for Gypsum Products.

Current edition approved Dec. 10, 1997. Published June 1998.

[2] *Annual Book of ASTM Standards*, Vol 04.01.

8. Sampling

8.1 When specified by the purchase agreement, samples of the GRG part shall be taken at the place of manufacture for measuring in accordance with Section 6.

8.2 *Mock-Up*—When specified by the purchase agreement, a mock-up shall be supplied.

9. Inspection

9.1 Inspection of the GRG parts shall be agreed upon between the purchaser and the supplier as part of the purchase agreement.

10. Rejection and Rehearing

10.1 Rejection of GRG parts that fail to conform to the requirements of this specification shall be reported in writing to the producer or supplier within 14 calendar days from receipt of parts and prior to installation.

11. Certification

11.1 When specified in the purchase agreement, the purchaser shall be furnished certification that samples representing each lot have been either tested or inspected as directed in this specification and the requirements have been met. When specified in the purchase agreement, a report of test results shall be furnished.

12. Quality Assurance

12.1 Each manufacturer's measuring of samples shall be representative of actual production procedures and materials. The GRG manufacturer shall certify that the materials are the same as those used to produce the composite according to C 1355/C 1355M.

13. Marking

13.1 Each GRG part shall be marked clearly according to the project shop drawings.

14. Packaging

14.1 Packaging shall ensure that the GRG parts arrive at the jobsite in good condition when handled and shipped according to industry operating standards.

15. Keywords

15.1 glass fiber reinforced gypsum; glass fiber reinforced gypsum composite; glass fiber reinforced gypsum parts; GRG; GRG composites; GRG parts

Specification for Gypsum Ceiling Board[1]

This standard is issued under the fixed designation C 1395 / C 1395M; the number immediately following the designation indicates the year of original adoption or, in the case of revision, the year of last revision. A number in parentheses indicates the year of last reapproval. A superscript epsilon (ε) indicates an editorial change since the last revision or reapproval.

1. Scope

1.1 This specification covers ½ in. [12.7 mm] thick gypsum ceiling board designed for use on interior ceilings with framing spaced not more than 24 in. [610 mm] on center and that affords a surface suitable to receive water–based texture and other decoration. This product is also suitable for use on interior walls.

1.2 The values stated in either inch–pound units or SI (metric) are to be regarded separately as standard. Within the text, the SI units are shown in brackets. The values stated in each system shall be used independent of the other. Values from the two systems shall not be combined.

2. Referenced Documents

2.1 *ASTM Standards:*

C 11 Terminology Relating to Gypsum and Related Building Materials and Systems[2]

C 473 Test Methods for Physical Testing of Gypsum Panel Products[2]

C 645 Specification for Nonstructural Steel Framing Members[2]

C 1264 Specification for Sampling, Inspection, Rejection, Certification, Packaging, Marking, Shipping, Handling, and Storage of Gypsum Board[2]

E 84 Test Methods for Surface Burning Characteristics of Building Materials[3]

E 119 Test Methods for Fire Tests of Building Construction and Materials[3]

3. Terminology

3.1 Definitions used in this specification shall be in accordance with Terminology C 11.

4. Materials and Manufacture

4.1 Gypsum ceiling board shall consist of a noncombustible core, essentially gypsum, surfaced with paper bonded to the core.

4.2 *Gypsum ceiling board, type X (special fire-resistant),* designates gypsum ceiling board complying with this specification that provides not less than ¾ h fire–resistance for boards applied parallel with and on each side of load bearing 2 × 4 wood studs spaced 16 in. [406 mm] on centers with 6d coated nails, 1⅞ in. [48 mm] long, 0.0915 in. [2.3 mm] diameter shank, ¼ in. [6.4 mm] diameter heads, spaced 7 in. [178 mm] on centers with gypsum board joints staggered 16 in. [406 mm] on each side of the partition and tested in accordance with Test Methods E 119.

NOTE 1—Consult producers for independent test data on assembly details and fire resistance classifications for other types of construction. See official fire test reports for assembly particulars, materials, and classifications.

4.3 Gypsum ceiling board shall have a flame spread index of not more than 25 when tested in accordance with Test Method E 84.

5. Physical Properties

5.1 Specimens shall be taken from the samples obtained in accordance with Specification C 1264.

5.1.1 Specimens shall be tested in accordance with Test Methods C 473.

5.1.2 *Flexural Strength*—The specimens shall be tested face up and face down. The average breaking load shall be not less than that given in Table 1.

5.1.3 *Humidified Deflection*—The specimens shall have an average deflection of not more than 5⁄16 in. [8 mm].

5.1.4 *Core, End, and Edge Hardness*—The specimens shall have an average hardness of not less than 15 lbf [67 N] when tested by Method A and 11 lbf [49 N] when tested by Method B.

5.1.5 *Nail Pull Resistance*—The specimens shall have an average nail–pull resistance of not less 80 lbf [356 N] when tested by Method A and 77 lbf [343 N] when tested by Method B.

6. Dimensions and Permissible Variations

6.1 Specimens shall be taken from the samples obtained in accordance with Specification C 1264.

6.2 Thickness, width, length, and end squareness shall be determined in accordance with Test Methods C 473.

6.2.1 *Thickness*—The nominal thickness shall be ½ in. [12.7 mm] with permissible variations in the nominal thickness of ±1⁄64 in. [0.4 mm] with permissible local variations of ±1⁄32 in. [0.8 mm] from the nominal thickness.

6.2.2 *Width*—The width shall be up to 48 in. [1220 mm]

---

[1] This test method is under the jurisdiction of ASTM C-11 on Gypsum and Related Building Materials and Systems and is the direct responsibility of Subcommittee C11.01 on Specifications and Test Methods for Gypsum Products.

Current edition approved July 10, 1998. Published December 1998.

[2] *Annual Book of ASTM Standards*, Vol 04.01.

[3] *Annual Book of ASTM Standards*, Vol 04.07.

TABLE 1 Flexural Strength

METHOD A		METHOD B	
Load, lbf [N] Bearing Edges Across Fiber of Surfacing	Load, lbf [N] Bearing Edges Parallel to Fiber of Surfacing	Load, lbf [N] Bearing Edges Across Fiber of Surfacing	Load, lbf [N] Bearing Edges Parallel to Fiber of Surfacing
110 [489]	40 [178]	107 [476]	36 [160]

with a permissible variation of 3⁄32 in. [3 mm] under the specified width.

6.2.3 *Length*—The permissible variation in length shall be ± 1⁄4in. [6 mm].

6.2.4 *Tapered Edge Depth*—The average thickness of the edge of recessed or tapered edge shall be not less than 0.020 in. [0.51 mm] but not more than 0.090 in. [2.29 mm] less than the average thickness of the gypsum ceiling board.

6.2.5 *End Squareness*—Corners shall be square with a permissible variation of ± 1⁄8 in. [±3 mm] in the full width of the board.

6.3 *Edges and Ends*—The edges and ends shall be straight and either square, recessed, beveled, featured, tapered, or featured and tapered.

7. Finish and Appearance

7.1 The surfaces of gypsum ceiling board shall be true and free from imperfections that would render it unfit for use with or without decoration.

8. Sampling, Inspection, Rejection, Certification, Packaging, Marking, Shipping, Handling, and Storage

8.1 Sampling, Inspection, Rejection, Certification, Packaging, Marking, Shipping, Handling, and Storage of gypsum ceiling board shall be in accordance with Specification C 1264.

9. Keywords

9.1 ceiling; drywall; gypsum; gypsum board; gypsum ceiling board; gypsum wallboard; type X

APPENDIX

(Nonmandatory Information)

This Appendix gives general information and also suggests for inclusions to be made elsewhere by the specifier. It is not part of this specification.

X1. Alternate Definition for Type X

X1.1 Gypsum ceiling board, type X (special fire-resistant) designates gypsum ceiling board providing a greater fire resistance than regular gypsum ceiling board of the same thickness. Type X (special fire–resistant) gypsum ceiling board, when tested in accordance with Test Methods E 119, shall provide the following minimum fire resistance ratings for the assembly described.

X1.1.1 2-h for a 1⁄2 in. [12.7 mm] thickness applied to a partition in a double layer application an each side of 2-1⁄2 in. [64 mm] deep non–loadbearing galvanized steel studs complying with Specification C 645 spaced 24 in. [610 mm] on center. The base layer 48 in. [1220 mm] wide shall be attached using 1 in. [25 mm] long drywall screws spaced 12 in. [305 mm] on center along board edges, ends, and along intermediate studs. Joints shall be oriented parallel to and located over studs and staggered on opposite sides of the assembly. The face layer 48 in. [1220 mm] wide shall be attached using 1-5⁄8 in. [41 mm] long drywall screws spaced 12 in. [305 mm] along board edges, ends, and along intermediate studs. Joints shall be oriented parallel to and located over studs, offset 24 in. [610 mm] from the base layer joints, and staggered on opposite sides of the assembly. All joints in the face layer shall be filled with joint compound, covered with joint tape and covered with an additional coat of joint compound. All screw heads in the face layer shall be covered with two coats of joint compound.

Standard Specification for Gypsum Board[1]

This standard is issued under the fixed designation C 1396 / C 1396M; the number immediately following the designation indicates the year of original adoption or, in the case of revision, the year of last revision. A number in parentheses indicates the year of last reapproval. A superscript epsilon (ϵ) indicates an editorial change since the last revision or reapproval.

1. Scope

1.1 This specification covers the gypsum boards described in 1.1.1-1.1.8.

1.1.1 *Gypsum wallboard*, designed for use on walls, ceilings, or partitions and that affords a surface suitable to receive decoration.

1.1.2 *Predecorated gypsum board*, designed for use as the finished surfacing for walls, ceilings, or partitions.

1.1.3 *Gypsum backing board* and *coreboard*, designed for use as a base in multilayer systems or as a gypsum stud or core in semisolid or solid gypsum board partitions.

NOTE 1—The terms backing board and coreboard refer to different end uses. The term gypsum backing board, as used in this specification shall include gypsum coreboard unless otherwise stated.

1.1.4 *Water-resistant gypsum backing board*, designed primarily to be used as a base for the application of ceramic or plastic tile on walls or ceilings. This product is also suitable for decoration.

1.1.5 *Exterior gypsum soffit board*, designed for use on exterior soffits and carport ceilings that are completely protected from contact with liquid water.

1.1.6 *Gypsum sheathing board*, designed for use as sheathing on buildings.

1.1.7 *Gypsum base for veneer plaster*, designed for use as a base for the application of gypsum veneer plaster.

1.1.8 *Gypsum lath*, designed for use as a base for the application of gypsum plaster.

1.2 Specifications applicable to all gypsum boards are located in Section 1-4 and 12-14. Specifications applicable to specific gypsum boards are located in the following sections:

	Section
Gypsum wallboard, predecorated gypsum board	5
Gypsum backing board, gypsum coreboard	6
Water-resistant gypsum backing board	7
Exterior gypsum soffit board	8
Gypsum sheathing board	9
Gypsum base for veneer plaster	10
Gypsum lath	11

1.3 The values stated in either inch-pound units or SI (metric) are to be regarded separately as standard. Within the text, the SI units are shown in brackets. The values stated in each system shall be used independent of the other. Values from the two systems shall not be combined.

2. Referenced Documents

2.1 *ASTM Standards:*

C 11 Terminology Relating to Gypsum and Related Building Materials and Systems[2]

C 473 Test Methods for Physical Testing of Gypsum Panel Products [2]

C 645 Specification for Nonstructural Steel Framing Members[2]

C 1264/C 1264M Specification for Sampling, Inspection, Rejection, Certification, Packaging, Marking, Shipping, Handling, and Storage of Gypsum Board[2]

E 84 Test Method for Surface Burning Characteristics of Building Materials[3]

E 96 Test Methods for Water Vapor Transmission of Materials[4]

E 119 Test Methods for Fire Tests of Building Construction and Materials[3]

3. Terminology

3.1 Definitions used in this specification shall be in accordance with Terminology C 11.

3.2 *Definitions of Terms Specific to This Standard:*

3.2.1 *decorative coating, n*—a paint or other liquid material with or without aggregate.

3.2.2 *decorative sheet or film, n*—a plastic film, either backed or unbacked; plastic sheet; or combination of these.

3.2.3 *edge, n*—the bound edge as manufactured.

3.2.4 *gypsum backing board, n*—a ¼ to ⅝-in. [6.4 to 15.9-mm] gypsum board used as a backing for gypsum wallboard, acoustical tile, or other dry cladding.

3.2.5 *gypsum coreboard, n*—a ¾ to 1-in. [19.0 to 25.4-mm] gypsum board used as a gypsum stud or core in semisolid or solid gypsum board partitions.

4. Materials and Manufacture

4.1 Gypsum board shall consist of a noncombustible core, essentially gypsum, surfaced with paper bonded to the core.

4.1.1 Gypsum sheathing board shall consist of a noncombustible core, essentially gypsum, surfaced on both the face

[1] This specification is under the jurisdiction of ASTM Committee C-11 on Gypsum and Related Building Materials and Systems and is the direct responsibility of Subcommittee C11.01 on Specifications and Test Methods for Gypsum Products.

Current edition approved July 10, 1998. Published December 1998.

[2] *Annual Book of ASTM Standards*, Vol 04.01.

[3] *Annual Book of ASTM Standards*, Vol 04.07.

[4] *Annual Book of ASTM Standards*, Vol 04.06.

and back with water-repellant paper bonded to the core.

4.1.1.1 Water-resistant core (core treated) gypsum sheathing board shall have a water-resistant material incorporated in the core.

4.1.2 Gypsum coreboard shall be either a single board or composed of two factory-laminated gypsum boards to provide up to 1 in. [25.4 mm] total nominal thickness.

4.1.3 Water-resistant gypsum backing board shall consist of a noncombustible water-resistant core, essentially gypsum, surfaced on both the back and face with water-repellant paper bonded to the core.

4.1.4 Predecorated gypsum board shall consist of a noncombustible core, essentially gypsum, surfaced with paper bonded to the core with the face covered with a decorative sheet, film, or coating.

4.1.4.1 Class I predecorated gypsum board shall have a decorative sheet or film laminated to the face.

4.1.4.2 Class II predecorated gypsum board shall have a decorative coating applied to the face.

4.2 The back surface of foil-backed gypsum board shall in addition be covered with aluminum foil.

4.3 *Gypsum Board, Type X* (Special Fire-Resistant):

4.3.1 Gypsum Board, Type X, designated gypsum board, except gypsum lath, complying with this specification that provides not less than 1 h fire-resistance for boards ⅝ in. [15.9 mm] thick or ¾ h fire-resistance for boards ½ in. [12.7 mm] thick, applied parallel with and on each side of load bearing 2 by 4 wood studs spaced 16 in. [406 mm] on centers with 6d coated nails, 1⅞ in. [48 mm] long, 0.0915-in. [2.3-mm] diameter shank, ¼-in. [6.4-mm] diameter heads, spaced 7 in. [178 mm] on centers with gypsum board joints staggered 16 in. [406 mm] on each side of the partition and tested in accordance with Test Methods E 119.

4.3.2 Gypsum Lath, type X designates gypsum lath complying with this specification that provides not less than a 1-h fire-resistance classification for gypsum lath ⅜in. [9.5 mm] thick, when applied at right angles to and on each side of loadbearing 2 × 4 wood studs spaced 16 in. [406 mm] on centers with blue lath nails spaced 5 in. [127 mm] on centers over which is applied ½ in. [12.7 mm] 1:2 gypsum sand plaster and tested in accordance with the requirements of Test Method E 119.

NOTE 2—The definitions for type X given in 4.3.1 and 4.3.2 are intended only as criteria to establish a definition for qualifying gypsum board as meeting the requirements for type X. These defining tests are neither intended to indicate a preferred application nor to limit the use of the product in other fire-resistance rated assemblies.

NOTE 3—Consult producers for independent test data on assembly details and fire-resistance classifications for other types of construction. See official fire test reports for assembly particulars, materials, and classifications.

4.4 Gypsum wallboard, gypsum backing board, and exterior gypsum soffit board shall have a flame spread index of not more than 25 when tested in accordance with Test Method E 84.

4.5 *Physical Properties, Dimensions, and Permissible Variations of Gypsum Board*:

4.5.1 Specimens shall be taken from the samples obtained in accordance with Specification C 1264/C 1264M.

4.5.1.1 Specimens shall be tested in accordance with Test Methods C 473.

4.5.2 *Core, End, and Edge Hardness*—The specimens shall have an average hardness of not less than 15 lbf [67 N] when tested by Method A and 11 lbf [49 N] when tested by Method B.

4.5.3 *Edges and Ends*—The edges and ends of gypsum board shall be straight and solid.

4.5.4 *Length*—The permissible variation in length shall be ±¼ in. [±6 mm].

4.5.5 *Tapered Edge Depth*—The average thickness of the edge of recessed or tapered edge gypsum board shall be not less than 0.020 in. [0.38 mm] but not more than 0.090 in. [1.90 mm] less than the average thickness of the gypsum board.

5. Gypsum Wallboard and Predecorated Gypsum Board

5.1 *Physical Properties of Gypsum Wallboard and Predecorated Gypsum Board*:

5.1.1 *Flexural Strength*—The specimens shall be tested face up and face down. The average breaking load shall be not less than the following:

Thickness, in. [mm]	Method A Bearing Edges Perpendicular to Panel Length, lbf [N]	Method A Bearing Edges Parallel to Panel Length, lbf [N]	Method B Bearing Edges Perpendicular to Panel Length, lbf [N]	Method B Bearing Edges Parallel to Panel Length, lbf [N]
¼ [6.4]	50 [222]	20 [89]	46 [205]	16 [71]
5⁄16 [7.9]	65 [289]	25 [111]	62 [276]	21 [93]
⅜ [9.5]	80 [356]	30 [133]	77 [343]	26 [116]
½ [12.7]	110 [489]	40 [178]	107 [476]	36 [160]
⅝ [15.9]	150 [667]	50 [222]	147 [654]	46 [205]
¾ [19.0]	170 [756]	60 [267]	167 [743]	56 [249]

5.1.2 *Humidified Deflection*—The specimens shall have an average deflection of not more than the following:

Thickness, in. [mm]	Deflection, eighths of an inch [mm]
¼ [6.4]	not applicable
5⁄16 [7.9]	not applicable
⅜ [9.5]	15 [48]
½ [12.7]	10 [32]
⅝ [15.9]	5 [16]
¾ [19.0]	5 [16]

5.1.3 *Nail Pull Resistance*—The specimens shall have an average nail-pull resistance of not less than the following:

Thickness, in. [mm]	Method A, lbf [N]	Method B, lbf [N]
¼ [6.4]	40 [178]	36 [160]
5⁄16 [7.9]	50 [222]	46 [205]
⅜ [9.5]	60 [267]	56 [249]
½ [12.7]	80 [356]	77 [343]
⅝ [15.9]	90 [400]	87 [387]
¾ [19.0]	100 [445]	97 [432]

5.1.4 *Foil-Backed Gypsum Wallboard and Foil-Backed Predecorated Gypsum Board*:

5.1.4.1 Foil-backed gypsum wallboard and foil-backed predecorated gypsum board shall meet all of the requirements for gypsum wallboard and predecorated gypsum board.

5.1.4.2 When tested in accordance with Test Methods E 96, the permeance of foil-backed gypsum wallboard and foil-backed predecorated gypsum board shall be not more than 0.30 perm [17 ng/Pa·s·m^2] (Desiccant Method) for the condition of

50 % relative humidity on the face of the board, and 0 % relative humidity on the back (foil-covered) side of the board.

5.2 *Dimensions and Permissible Variations of Gypsum Wallboard and Predecorated Gypsum Board*:

5.2.1 *Thickness*—The nominal thickness shall be from ¼to ¾ in. [6.4 to 19.0 mm] with permissible variations in the thickness of ±1⁄64 in. [±0.4 mm], and with permissible local variations of ±1⁄32 in. [±0.8 mm].

5.2.2 *Width*—The width shall be up to 48 in. [1220 mm], with widths up to 54 in. [1370 mm] permitted, with a permissible variation of 3⁄32in. [3 mm] under the specified width.

5.2.3 *End Squareness*—Corners shall be square with a permissible variation of ±⅛ in. [±3 mm] in the full width of the board.

5.3 *Edges*:

5.3.1 Edges of gypsum wallboard shall be either square, recessed, beveled, featured, tapered, or featured and tapered.

5.3.2 Edges of predecorated gypsum board shall be either square or featured.

6. Gypsum Backing Board and Gypsum Coreboard

6.1 *Physical Properties of Gypsum Backing Board and Gypsum Coreboard*:

6.1.1 *Flexural Strength*—The specimens shall be tested face up and face down. The average breaking load shall be not less than the following:

Thickness, in. [mm]	Method A Bearing Edges Perpendicular to Panel Length, lbf [N]	Method A Bearing Edges Parallel to Panel Length, lbf [N]	Method B Bearing Edges Perpendicular to Panel Length, lbf [N]	Method B Bearing Edges Parallel to Panel Length, lbf [N]
¼ [6.4]	50 [222]	20 [89]	46 [205]	16 [71]
⅜ [9.5]	80 [356]	30 [133]	77 [343]	26 [116]
½ [12.7]	110 [489]	40 [178]	107 [476]	36 [160]
⅝ [15.9]	140 [622]	50 [222]	137 [609]	46 [205]
¾ [19.0]	170 [756]	60 [267]	167 [743]	56 [249]
1 [25.4]	230 [1023]	80 [356]	228 [1014]	77 [343]

6.1.2 *Humidified Deflection*—The specimens shall have an average deflection of not more than the following:

Thickness, in. [mm]	Deflection, eighths of an inch [mm]
¼ [6.4]	not applicable
⅜ [9.5]	15 [48]
½ [12.7]	10 [32]
⅝ [15.9]	5 [16]
¾ [19.0]	not applicable
1 [25.4]	not applicable

6.1.3 *Nail Pull Resistance*—The specimens shall have an average nail-pull resistance of not less than the following:

Thickness, in. [mm]	Method A, lbf [N]	Method B, lbf [N]
¼ [6.4]	40 [178]	36 [160]
⅜ [9.5]	60 [267]	56 [249]
½ [12.7]	80 [356]	77 [343]
⅝ [15.9]	90 [400]	87 [387]
¾ [19.0]	not required	not required
1 [25.4]	not required	not required

6.1.4 *Foil-Backed Gypsum Backing Board and Foil-Backed Gypsum Coreboard*:

6.1.4.1 Foil-backed gypsum backing board and foil-backed gypsum coreboard shall meet all of the requirements for gypsum backing board and gypsum coreboard.

6.1.4.2 When tested in accordance with Test Methods E 96, the permeance of foil-backed gypsum backing board and foil-backed gypsum coreboard shall be not more than 0.30 perm [17 ng/Pa·s·m^2] (Desiccant Method) for the condition of 50 % relative humidity on the face of the board, and 0 % relative humidity on the back (foil-covered) side of the board.

6.2 *Dimensions and Permissible Variations of Gypsum Backing Board and Gypsum Coreboard*:

6.2.1 *Thickness*—The nominal thickness shall be from ¼ to 1 in. [6.4 to 25.4 mm] with permissible variations in the thickness of ±1⁄32 in. [±0.8 mm], and with permissible local variations of ±1⁄16 in. [±1.6 mm].

6.2.2 *Width*—The width shall be from 16 to 48 in. [406 to 1220 mm] with a permissible variation of ⅛ in. [3.2 mm] under the specified width.

6.2.3 *End Squareness*—Corners shall be square with a permissible variation of ±⅛ in. [±3 mm] in the full width of the board.

6.3 *Edges*—Edges shall be either square, round, V-tongue and groove, or featured.

7. Water-Resistant Gypsum Backing Board

7.1 *Physical Properties of Water-Resistant Gypsum Backing Board*:

7.1.1 *Flexural Strength*—The specimens shall be tested face up and face down. The average breaking load shall be not less than the following:

Thickness, in. [mm]	Method A Bearing Edges Perpendicular to Panel Length, lbf [N]	Method A Bearing Edges Parallel to Panel Length, lbf [N]	Method B Bearing Edges Perpendicular to Panel Length, lbf [N]	Method B Bearing Edges Parallel to Panel Length, lbf [N]
½ [12.7]	110 [489]	40 [178]	107 [476]	36 [160]
⅝ [15.9]	150 [667]	50 [222]	147 [654]	46 [205]

7.1.2 *Humidified Deflection*—The specimens shall have an average deflection of not more than the following:

Thickness, in. [mm]	Deflection, eighths of an inch [mm]
½ [12.7]	10 [32]
⅝ [15.9]	5 [16]

7.1.3 *Nail Pull Resistance*—The specimens shall have an average nail-pull resistance of not less than the following:

Thickness, in. [mm]	Method A, lbf [N]	Method B, lbf [N]
½ [12.7]	80 [356]	77 [343]
⅝ [15.9]	90 [400]	87 [387]

7.1.4 *Water Resistance*—The specimens shall have an average water absorption of not more than 5 weight % after 2-h immersion.

7.1.5 *Surface Water Absorption*—The specimens shall have an average surface water absorption of not more than 1.6 g after 2 h of elapsed time.

7.2 *Dimensions and Permissible Variations of Water-Resistant Gypsum Backing Board*:

7.2.1 *Thickness*—The nominal thickness shall be ½ or ⅝ in. [12.7 or 15.9 mm] with permissible variations in the thickness of ±1⁄64 in. [±0.4 mm], and with permissible local variations of ± 1⁄32 in. [±0.8 mm].

7.2.2 *Width*—The width shall be up to 48 in. [1220 mm], with widths up to 54 in. [1370 mm] permitted, with a permissible variation of 3⁄32 in. [3 mm] under the specified width.

7.2.3 *End Squareness*—Corners shall be square with a permissible variation of ± 1⁄8 in. [±3 mm] in the full width of the board.

7.3 *Edges*—Edges shall be either square, recessed, beveled, featured, tapered, or featured and tapered.

8. Exterior Gypsum Soffit Board

8.1 *Physical Properties of Exterior Gypsum Soffit Board*:

8.1.1 *Flexural Strength*—The specimens shall be tested face up and face down. The average breaking load shall be not less than the following:

Thickness, in. [mm]	Method A Bearing Edges Perpendicular to Panel Length, lbf [N]	Method A Bearing Edges Parallel to Panel Length, lbf [N]	Method B Bearing Edges Perpendicular to Panel Length, lbf [N]	Method B Bearing Edges Parallel to Panel Length, lbf [N]
½ [12.7]	110 [489]	40 [178]	107 [476]	36 [160]
⅝ [15.9]	150 [667]	50 [222]	147 [654]	46 [205]

8.1.2 *Humidified Deflection*—The specimens shall have an average deflection of not more than the following:

Thickness, in. [mm]	Deflection, eighths of an inch [mm]
½ [12.7]	7 [22]
⅝ [15.9]	4 [13]

8.1.3 *Nail Pull Resistance*—The specimens shall have an average nail-pull resistance of not less than the following:

Thickness, in. [mm]	Method A, lbf [N]	Method B, lbf [N]
½ [12.7]	80 [356]	77 [343]
⅝ [15.9]	90 [400]	87 [387]

8.2 *Dimensions and Permissible Variations of Exterior Gypsum Soffit Board*:

8.2.1 *Thickness*—The nominal thickness shall be ½ or ⅝ in. [12.7 or 15.9 mm] with permissible variations in the thickness of ± 1⁄64 in. [±0.4 mm], and with permissible local variations of ± 1⁄32 in. [±0.8 mm].

8.2.2 *Width*—The width shall be up to 48 in. [1220 mm] with a permissible variation of 3⁄32 in. [3 mm] under the specified width.

8.2.3 *End Squareness*—Corners shall be square with a permissible variation of ± 1⁄8 in. [±3 mm] in the full width of the board.

9. Treated Core and Nontreated Core Gypsum Sheathing Board

9.1 *Physical Properties of Treated Core and Nontreated Core Gypsum Sheathing Board*

9.1.1 *Flexural Strength*—The specimens shall be tested face up and face down. The average breaking load shall be not less than the following:

Thickness, in. [mm]	Method A Bearing Edges Perpendicular to Panel Length, lbf [N]	Method A Bearing Edges Parallel to Panel Length, lbf [N]	Method B Bearing Edges Perpendicular to Panel Length, lbf [N]	Method B Bearing Edges Parallel to Panel Length, lbf [N]
⅜ [9.5]	80 [356]	30 [133]	77 [343]	26 [116]
4⁄10 [10.2]	88 [391]	32 [142]	85 [378]	28 [125]
½ [12.7]	110 [489]	40 [178]	107 [476]	36 [160]
⅝ [15.9]	150 [667]	50 [222]	147 [654]	46 [205]

9.1.2 *Humidified Deflection*—The specimens shall have an average deflection of not more than the following:

Thickness, in. [mm]	Deflection, eighths of an inch [m]
⅜ [9.5]	15 [48]
4⁄10 [10.2]	12 [38]
½ [12.7]	10 [32]
⅝ [15.9]	5 [16]

9.1.3 *Nail-Pull Resistance*—The specimens shall have an average nail-pull resistance of not less than the following:

Thickness, in. [mm]	Method A, lbf [N]	Method B, lbf [N]
⅜ [9.5]	60 [267]	56 [249]
4⁄10 [10.2]	70 [312]	67 [298]
½ [12.7]	80 [356]	77 [343]
⅝ [15.9]	90 [400]	87 [387]

9.1.4 *Surface Water Absorption*—The specimens shall have an average surface water absorption of not more than 1.6 g after 2 h of elapsed time.

9.2 *Water Resistance of Treated Core Gypsum Sheathing Board*—The specimens shall have an average water absorption of not more than 10 weight % after 2-h immersion.

9.3 *Dimensions and Permissible Variations of Treated Core and Nontreated Core Gypsum Sheathing Board*

9.3.1 *Thickness*—The nominal thickness shall be from ⅜to ⅝ in. [9.5 to 15.9 mm] with permissible variations in the thickness of ± 1⁄32 in. [±0.8 mm], and with permissible local variations of ± 1⁄16 in. [±1.6 mm].

9.3.2 *Width*—The nominal width shall be 24 or 48 in. [610 or 1220 mm] with a permissible variation of ⅛ in. [3.2 mm] under the specified width.

9.3.3 *End Squareness*—Corners shall be square with a permissible variation of ± ⅛ in. [±3 mm] in the full width of the board.

9.4 *Edges*—Edges shall be either square or V-tongue and groove.

10. Gypsum Base for Veneer Plaster

10.1 *Physical Properties of Gypsum Base for Veneer Plaster*:

10.1.1 *Flexural Strength*—The specimens shall be tested face up and face down. The average breaking load shall be not less than the following:

Thickness, in. [mm]	Method A Bearing Edges Perpendicular to Panel Length, lbf [N]	Method A Bearing Edges Parallel to Panel Length, lbf [N]	Method B Bearing Edges Perpendicular to Panel Length, lbf [N]	Method B Bearing Edges Parallel to Panel Length, lbf [N]
⅜ [9.5]	80 [356]	30 [133]	77 [343]	26 [116]
½ [12.7]	110 [489]	40 [178]	107 [476]	36 [160]
⅝ [15.9]	150 [667]	50 [222]	147 [654]	46 [205]

10.1.2 *Humidified Deflection*—The specimens shall have an average deflection of not more than the following:

Thickness, in. [mm]	Deflection, eighths of an inch [m]
3⁄8 [9.5]	15 [48]
1⁄2 [12.7]	10 [32]
5⁄8 [15.9]	5 [16]

10.1.3 *Nail Pull Resistance*—The specimens shall have an average nail-pull resistance of not less than the following:

Thickness, in. [mm]	Method A, lbf [N]	Method B, lbf [N]
3⁄8 [9.5]	60 [267]	56 [249]
1⁄2 [12.7]	80 [356]	77 [343]
5⁄8 [15.9]	90 [400]	87 [387]

10.1.4 *Foil-Backed Gypsum Base for Veneer Plaster*:

10.1.4.1 Foil-backed gypsum base for veneer plaster shall meet all of the requirements for gypsum base for veneer plaster.

10.1.4.2 When tested in accordance with Test Methods E 96, the permeance of foil-backed gypsum base for veneer plaster shall be not more than 0.30 perm [17 ng/Pa·s·m^2] (Desiccant Method) for the condition of 50 % relative humidity on the face of the board, and 0 % relative humidity on the back (foil-covered) side of the board.

10.2 *Dimensions and Permissible Variations of Gypsum Base for Veneer Plaster*:

10.2.1 *Thickness*—The nominal thickness shall be from 3⁄8to 5⁄8 in. [9.5 to 15.9 mm] with permissible variations in the thickness of ±1⁄64 in. [±0.4 mm], and with permissible local variations of ±1⁄32 in. [±0.8 mm].

10.2.2 *Width*—The width shall be 48 in. [1220 mm] with a permissible variation of 3⁄32 in. [3 mm] under the specified width.

10.2.3 *End Squareness*—Corners shall be square with a permissible variation of ±1⁄8 in. [±3 mm] in the full width of the board.

10.3 *Edges*—Edges shall be either square, rounded, or tapered.

11. Gypsum Lath

11.1 *Physical Properties of Gypsum Lath*:

11.1.1 *Flexural Strength*—The specimens shall be tested face up and face down. The average breaking load shall be not less than the following:

Thickness, in. [mm]	Method A Bearing Edges Perpendicular to Panel Length, lbf [N]	Method A Bearing Edges Parallel to Panel Length, lbf [N]	Method B Bearing Edges Perpendicular to Panel Length, lbf [N]	Method B Bearing Edges Parallel to Panel Length, lbf [N]
3⁄8 [9.5]	60 [267]	25 [111]	56 [249]	21 [93]
1⁄2 [12.7]	100 [445]	35 [156]	97 [431]	31 [138]

11.1.2 *Humidified Deflection*—The specimens shall have an average deflection of not more than the following:

Thickness, in. [mm]	Deflection, eighths of an inch [m]
3⁄8 [9.5]	15 [48]
1⁄2 [12.7]	10 [32]

11.1.3 *Nail Pull Resistance*—The specimens shall have an average nail-pull resistance of not less than the following:

Thickness, in. [mm]	Method A, lbf [N]	Method B, lbf [N]
3⁄8 [9.5]	50 [222]	46 [205]
1⁄2 [12.7]	70 [312]	66 [294]

11.1.4 *Foil-Backed Gypsum Lath*:

11.1.4.1 Foil-backed gypsum lath shall meet all of the requirements for gypsum lath.

11.1.4.2 When tested in accordance with Test Methods E 96, the permeance of foil-backed gypsum lath shall be not more than 0.30 perm [17 ng/Pa·s·m^2] (Desiccant Method) for the condition of 50 % relative humidity on the face of the board, and 0 % relative humidity on the back (foil-covered) side of the board.

11.2 *Dimensions and Permissible Variations of Gypsum Lath*:

11.2.1 *Thickness*—The nominal thickness shall be 3⁄8 or 1⁄2 in. [9.5 or 12.7 mm] with permissible variations in the thickness of ±1⁄32 in. [±0.8 mm], and with permissible local variations of ±1⁄16 in. [±1.6 mm].

11.2.2 *Width*—The nominal width shall be up to 48 in. [1220 mm] with a permissible variation of 3⁄16 in. [4.8 mm] under and 1⁄8 in. [3.2 mm] over the specified width.

11.2.3 *End Squareness*—Corners shall be square with a permissible variation of ±1⁄4 in. [±6.4 mm] in the full width of the board.

11.3 *Edges*—Edges shall be either round or square.

12. Finish and Appearance

12.1 The surfaces of gypsum board shall be true and free from imperfections that would render it unfit for use with or without decoration.

13. Sampling, Inspection, Rejection, Certification, Packaging, Marking, Shipping, Handling, and Storage

13.1 Sampling, inspection, rejection, certification, packaging, marking, shipping, handling, and storage of gypsum board shall be in accordance with Specification C 1264/C 1264M.

13.2 Treated core gypsum sheathing shall be marked legibly on the face or the back as treated core.

14. Keywords

14.1 ceiling; coreboard; drywall; exterior gypsum soffit board; foil-backed; gypsum; gypsum backing board; gypsum base; gypsum board; gypsum coreboard; gypsum lath; gypsum sheathing; gypsum sheathing board; gypsum wallboard; partitions; plaster; predecorated gypsum board; sheathing; Type X; veneer base; veneer plaster; wall; wallboard; water-resistant gypsum board

APPENDIX

(Nonmandatory Information)

This appendix gives general information and also suggestions for inclusions to be made elsewhere by the specifier. They are not part of this specification.

X1. Alternate Definition for Type X Gypsum Board, Except Gypsum Lath

X1.1 Gypsum board, Type X (special fire-resistant) designates gypsum board, except gypsum lath, providing a greater fire-resistance than regular gypsum board of the same thickness. Type X (special fire-resistant) gypsum board, when tested in accordance with Test Methods E 119, shall provide the following minimum fire resistance for the assemblies described:

X1.1.1 One hour for a ⅝-in. [15.9-mm] thickness applied to a partition in a single-layer application on each side of 3⅝-in. [92-mm] deep non-loadbearing galvanized steel studs complying with Specification C 645, spaced 24 in. [610 mm] on center. The ⅝-in. [15.9-mm] thick gypsum board 48 in. [1220 mm] wide shall be attached using 1-in. [25-mm] long drywall screws spaced 8 in. [203 mm] on center along the edges and ends, and 12 in. [305 mm] along intermediate studs. All joints shall be oriented parallel to and located over studs and staggered on opposite sides of the assembly, and

X1.1.2 Two hours for a ½-in. [12.7-mm] thickness applied to a partition in a double-layer application on each side of 2½-in. [64-mm] deep non-loadbearing galvanized steel studs complying with Specification C 645, spaced 24 in. [610 mm] on center. The 48-in. [1220-mm] wide base layer shall be attached using 1-in. [25-mm] long drywall screws spaced 12 in. [305 mm] on center along board edges, ends, and along intermediate studs. Joints shall be oriented parallel to and located over studs and staggered on opposite sides of the assembly. The 48-in. [1220-m] wide face layer shall be attached using 1⅝-in. [41-mm] long drywall screws spaced 12 in. [305 mm] along board edges, ends, and along intermediate studs. Joints shall be oriented parallel to and located over studs, offset 24 in. [610 mm] from the base layer joints, and staggered on opposite sides of the assembly.

NOTE X1.1—The definitions for type X given in X1.1.1 and X1.1.2 are intended only as criteria to establish a definition for qualifying gypsum board as meeting the requirements for type X. These defining tests are neither intended to indicate a preferred application nor to limit the use of the product in other fire resistance rated assemblies.

Designation: C 1397 – 98

Standard Practice for Application of Class PB Exterior Insulation and Finish Systems[1]

This standard is issued under the fixed designation C 1397; the number immediately following the designation indicates the year of original adoption or, in the case of revision, the year of last revision. A number in parentheses indicates the year of last reapproval. A superscript epsilon (ε) indicates an editorial change since the last revision or reapproval.

1. Scope

1.1 This practice covers the minimum requirements and procedures for field or prefabricated application of Class PB Exterior Insulation and Finish Systems (EIFS). Class PB EIFS are defined as systems applied over various types of insulation board, in which the base coat ranges from not less than 1⁄16 in. (1.6 mm) to 1⁄4 in. (6.4 mm) in dry thickness, depending upon the number of nonmetallic reinforcing mesh layers encapsulated in the base coat. The base coat is then covered with a finish coat of various thickness in a variety of textures and colors.

1.2 The values stated in inch-pound units are to be regarded as the standard. The metric values given in parentheses are approximate and are provided for information purposes only.

1.3 *This standard may involve hazardous materials, operations, and equipment. This standard does not purport to address all of the safety concerns, if any, associated with its use. It is the responsibility of the user of this standard to establish appropriate safety and health practices and determine the applicability of regulatory limitations prior to use.*

2. Referenced Documents

2.1 *ASTM Standards:*

C 11 Standard Terminology Relating to Gypsum and Related Building Materials and Systems[2]

C 79/C 79M Standard Specification for Treated Core and Nontreated Core Gypsum Sheathing Board[2]

C 1063 Standard Specification for Installation of Lathing and Furring to Receive Interior and Exterior Portland Cement-Based Plaster[2]

C 1177 Standard Specification for Glass Mat Gypsum Substrate for Use as Sheathing[2]

C 1186 Standard Specification for Flat Non-Asbestos Fiber-Cement Sheets[3]

3. Terminology

3.1 Definitions used in this standard shall be in accordance with Terminology C 11.

3.2 *Definitions of Terms Specific to This Standard:*

3.2.1 *accessories, n*—preformed metal, fiberglass or plastic members for use to form corners, edges, control joints, or decorative effects.

3.2.2 *aesthetic joint, n*—a deprecated term. See Aesthetic Reveal.

3.2.3 *aesthetic reveal, n*—a groove cut into the insulation board which serves the function of decoration and/or to provide a starting or stopping point for finish coat application.

3.2.4 *back wrapping, n*—a deprecated term. See Wrap.

3.2.5 *base coat, n*—the initial wet state material, either factory or field-mixed, used to encapsulate the nonmetallic reinforcing mesh or fasten the insulation board to the substrate.

3.2.6 *cold joint, n*—the visible junction in a finish coat.

3.2.7 *cure, v*—to develop the ultimate properties of a wet state material by a chemical process.

3.2.8 *dry, v*—to develop the ultimate properties of a wet state material solely by evaporation of volatile ingredients.

3.2.9 *durability, n*—the capability of a building, assembly, component, product, or construction to maintain serviceability over not less than a specified time.

3.2.10 *embed, v*—to encapsulate the nonmetallic reinforcing mesh in the base coat.

3.2.11 *expansion joint, n*—a structural separation between building elements that allows independent movement without damage to the assembly.

3.2.12 *exterior insulation and finish system, n*—(EIFS)—nonload bearing, exterior wall cladding system that consists of an insulation board attached either adhesively, mechanically, or both to the substrate; an integrally reinforced base coat; and a textured protective finish coat.

3.2.13 *factory mix, v*—a material that is prepared at the point of manufacture and is ready to use without the addition of other materials, except possibly water to adjust consistency.

3.2.14 *flash set (quick set), n*—the early hardening or stiffness in the working characteristics of a Portland-cement paste, mortar, or concrete, usually with the evolution of considerable heat. Stiffness cannot be dispelled nor the plasticity regained by further mixing without addition of water; also known as "quick set."

3.2.15 *field mix, n*—a material that is mixed in the field with other components and/or water.

3.2.16 *finish coat, n*—the final wet state material, which

---

[1] This practice is under the jurisdiction of ASTM Committee C-11 on Gypsum and Related Building Materials and Systems and is the direct responsibility of Subcommittee C11.05 on Application of Exterior Insulating and Finish Systems and Related Products.

Current edition approved Aug. 10, 1998. Published October 1998.

[2] *Annual Book of ASTM Standards*, Vol 04.01.

[3] *Annual Book of ASTM Standards*, Vol 04.05.

provides color and texture, applied over the reinforced base coat.

3.2.17 *framing member*, *n*—studs, joist, runners (tracks), bridging, bracing, and related accessories manufactured or supplied in wood or hot or cold formed steel.

3.2.18 *initial grab*, *n*—the ability of a wet state material to remain in place initially after it has been applied.

3.2.19 *initial set*, *n*—a time related set caused by the hydration process.

3.2.20 *lamina*, *n*—composite of base coat, reinforcement, and finish coat.

3.2.21 *mechanical fastener*, *n*—corrosion resistant component used to attach the insulation board to the substrate or framing member.

3.2.22 *nonmetallic reinforcing mesh*, *n*—a fiberglass component of the EIFS encapsulated in the base coat to strengthen the system.

NOTE 1—Nonmetallic reinforcing mesh is available in various weights to achieve different levels of impact resistance.

3.2.23 *pot life*, *n*—the duration of time that the wet state material remains workable after it has been mixed.

3.2.24 *primers*, *n*—liquid coatings applied to improve the adhesion of the EIFS to the substrate or of the finish to the base coat.

NOTE 2—Primers are sometimes applied to improve the water resistance of cementitious base coats.

3.2.25 *reinforced base coat*, *n*—base coat that has been reinforced with a nonmetallic reinforcing mesh.

3.2.26 *running bond*, *n*—pattern used when installing the thermal insulation board, to offset the vertical insulation board joints from joints in previous rows of insulation board.

3.2.27 *substrate*, *n*—surface to which the EIFS is applied.

3.2.28 *surface sealer*, *n*—material used to enhance weather resistance.

3.2.29 *temper*, *v*—to bring to a workable state by adding water.

3.2.30 *texture*, *n*—any surface appearance as contrasted to a smooth surface.

3.2.31 *thermal insulation board*, *n*—a system component of a specific type and density that functions to reduce heat flow through the wall and serves as the surface to receive the base coat.

3.2.32 *wet edge*, *n*—the leading edge of a continuously applied wet state material.

3.2.33 *wet state materials*, *n*—the adhesive, base coat and finish coat components applied in liquid or semi-liquid state.

3.2.34 *wrap*, *v*—to protect the exposed edges of thermal insulation board.

4. Significance and Use

4.1 This practice provides minimum requirements for the application of Class PB EIFS. The requirements for materials, mixtures, and details shall be contained in the project plans and specifications.

5. Delivery of Materials

5.1 All materials shall be delivered in packages, containers, or bundles with the identification and markings intact.

6. Inspection

6.1 Inspection of the materials shall be agreed upon between the purchaser and the supplier as part of the purchase agreement.

7. Rejection

7.1 Materials that are damaged, frozen or in any way defective shall not be used. Rejection of materials shall be promptly reported verbally to the producer and immediately reported in writing. The notice of rejection shall contain a statement documenting the basis for material rejection.

8. Certification

8.1 When specified in the contract documents, the producer shall furnish a report certifying that the materials are in conformance with product and material standards and contract documents.

8.2 The insulation boards shall have been subjected to a third party quality control inspection and shall be marked as approved for use within the EIFS and in accordance with applicable building code.

8.3 Prior to installation, the insulation board shall be inspected for conformance with contract documents. Nonconforming insulation board shall not be used.

8.4 The insulation board producer shall furnish, for each shipment, a written certificate of conformance with the EIFS producer's specifications.

9. Storage of Materials

9.1 All materials shall be kept dry by storage under cover and protected from the weather. When outside storage is required, materials shall be stacked off the ground, supported on a level platform and protected from the weather, surface contamination or physical damage in accordance with the EIFS producer's written instructions.

9.2 Materials shall be protected from exposure to direct sunlight and temperatures less than 40°F (4°C).

9.3 Portland Cement shall be kept dry until ready for use. It shall be kept off the ground, under cover and away from damp walls and surfaces.

9.4 Insulation board shall be stacked flat with care taken to avoid damage to edges, ends, or surfaces; or exposure to direct sunlight (ultraviolet radiation).

10. Environmental Conditions

10.1 *Cold Weather Conditions*:

10.1.1 Wet materials shall not be applied when the temperature is less than 40°F (4°C) unless temporary heat and enclosures are provided to maintain minimum 40°F (4°C) for a minimum period of not less than 24 h before, during, and after application (see Appendix X1.9.3).

10.1.2 Materials shall not be applied to a base containing frost. Substrate surface temperature shall be not less than 40°F (4°C). Mixtures for application shall not contain any frozen ingredients.

11. Assessment of Condition of Substrates to Receive Exterior Insulation and Finish Systems

11.1 The substrate shall be as required by the project plans and specifications and as recommended by the EIFS producer for the particular system specified.

11.2 The substrate shall be inspected by the applicator and general contractor to ensure that it meets the project plans and specifications and the requirements of 11.3-11.8.

11.3 Sheathing materials shall be inspected to ensure that:

11.3.1 The specified sheathing thickness has been installed for the stud spacing used.

11.3.2 Fastener type and fastener spacing are as specified.

11.3.3 All Specification C 79/C 79M gypsum sheathing boards shall have a water-resistant core.

11.3.4 Tongue and groove water-resistant core gypsum sheathing boards conforming to Specification C 79/C 79M shall be installed with the tongue edge oriented to the top.

11.3.5 Gypsum sheathing is protected from moisture at all penetrations and terminations.

11.3.6 Glass mat gypsum sheathing complying with Specification C 1177 shall be installed in accordance with the sheet producer's written installation instructions, which include details of framing type and spacing, fastener type and spacing, and sheet orientation and spacing.

11.3.7 Plywood shall be not less than ½ in. (12.7 mm) nominal in thickness, exterior or exposure 1 wood-based panel sheathing. The plywood shall be installed with a ⅛ in. (3.2 mm) space between sheets.

11.3.8 Fiber-cement sheets complying with Specification C 1186, Type A, shall be installed in accordance with the sheet producer's written installation instructions, which include details of stud type and spacing, fastener type and spacing, and sheet spacing recommendations.

11.4 *Alignment*—All substrate surfaces shall be straight and true within ¼ in. in 10 ft (2 mm/m). More stringent requirements by the EIFS producer shall supersede the above stated requirements.

11.5 *Suitability for Use*—The substrate surface shall be firm, sound, and undamaged in order to receive the EIFS.

11.5.1 Loose, spalling or crumbling concrete or masonry shall be removed by sandblasting or other appropriate means. If the surface condition of the substrate is at all in doubt, an alternative attachment surface or method shall be provided.

11.5.2 Broken, cracked or delaminated substrate boards shall be replaced or restored to a condition equal to adjacent undamaged boards.

11.5.3 All rotted, decayed, or delaminated wood and plywood sheathing shall be replaced.

11.5.4 Paper-faced gypsum sheathing that has been installed and exposed to the elements for more than 30 days shall be checked at not less than two locations, or every 5000 ft^2 (465 m^2), whichever is greater, for integrity of the surface. The procedures for evaluating the gypsum sheathing are found in Appendix X2.

NOTE 3—Gypsum sheathing conforming to Specification C 79 is sometimes not recommended for adhesive attachment by the gypsum sheathing producer. Consult with the gypsum board producer before use.

11.6 *Cleaning*—The surface of all substrates shall be clean and free from any foreign materials such as form release agents, curing compounds, dust, dirt, frost, oil or grease, efflorescence and laitance.

11.6.1 All sheathing board substrates shall have all loose dirt and dust removed by cleaning methods appropriate for the job and job conditions.

11.6.2 Efflorescence and laitance on concrete, masonry, stucco or clay tile substrates shall be removed prior to EIFS application. Concrete masonry shall be cleaned by light sandblasting, pressure washing or brushing. Heavy deposits shall be removed through use of hand or power impact tools followed by washing with an appropriate cleaner. Light deposits shall be removed by washing with an appropriate cleaner. All loose particles and cleaner residue shall be removed by washing with clean, potable water. The surface shall be allowed to dry.

11.6.3 Existing paint on concrete or masonry surfaces shall be removed by sandblasting or grit blasting. If removal of existing paint is not practical, an alternate attachment surface or method shall be provided.

11.7 *Dryness*—No visible free water.

11.7.1 Newly constructed concrete or masonry surfaces shall be allowed to cure not less than 28 days prior to application of the EIFS. Repaired areas on existing (aged) walls shall meet the same 28-day curing time.

11.7.2 Sheathing must be dry prior to EIFS application.

11.8 *Metal Lath or Furring and Accessories*—If metal lath or furring and accessories are used, they shall be installed in conformance with Specification C 1063, except butt-lapped, or as otherwise noted. Metal lath shall be 3.4 lb/yd^2 (1.3 kg/m^2), self furring, galvanized.

11.8.1 These members shall be properly attached, straight, and true unless required by the system design.

11.8.2 All metal members shall be free of rust, oil, or other foreign matter or contaminants, which cause bond failure or unsightly discoloration.

12. Insulation Board Installation

12.1 Method of attachment shall be approved by the system producer and be in accordance with applicable building codes.

12.2 *Adhesive Attachment Method*—The adhesive shall be applied to the back of the insulation board by one to the following methods:

12.2.1 *Notched Trowel*:

12.2.1.1 When applied, and before the insulation board is placed, the height of the adhesive measured from the surface of the insulation board shall not be less than ¼in. (6.4 mm) for factory mixed adhesive and ⅜ in. (9.5 mm) for field mixed adhesives.

12.2.2 *Ribbon and Dab*:

12.2.2.1 The adhesive shall be applied to the entire perimeter of the insulation board in a ribbon fashion that is not less than 2 in. (51 mm) wide by not less than ⅜ in. (9.5 mm) thick. Dabs not less than 4 in. (102 mm) in diameter by not less than ⅜ in. (9.5 mm) thick shall be applied not more than 8 in. (204 mm) on center over the remainder of the board.

13. Adhesive and Mechanical Attachment Method

13.1 Adhesive shall be applied using the ribbon and dab method, the notched trowel method, or in accordance with system producer's current published instructions.

13.2 Mechanical fasteners shall be installed into the framing or nailable substrate.

13.3 Fasteners shall penetrate not less than 5⁄16 in. (8 mm) into steel framing members, 1 in. (25 mm) into wood framing members, and 1 in. (25 mm) into masonry substrates.

13.4 Mechanical fasteners shall be corrosion resistant. Fastener type and spacing shall be in accordance with system producer's current published instructions.

13.5 Selection and frequency of fasteners and fastener type will vary depending on substrate type, insulation board thickness and design wind load and whether used in combination with an adhesive attachment. Therefore fastener type, patterns, and use shall be in accordance with the EIFS producer's written instructions and shall be included in the project plans and specifications.

14. Mechanical Attachment Method

14.1 Mechanical fasteners shall be installed into the framing or nailable substrate.

14.2 Mechanical fasteners shall be corrosion resistant. Fastener type and spacing shall be in accordance with system producer's current published instructions.

NOTE 4—Specific consideration of building code wind-load requirements shall be given for mechanically fastened systems.

15. General Requirements

15.1 Insulation boards shall be placed from a level base line with vertical joints staggered in a running bond pattern and butted tightly.

15.2 Insulation board joints at all inside and outside corners shall be staggered and interlocked.

15.3 Insulation boards must abut without gaps, eliminating space for adhesive or base coat intrusion between the board edges and ends.

15.4 Window corners and similar reentrant corners shall be cut out of a single insulation board. Board joints shall not be aligned at corners with head, sill or jambs of windows, doors, similar openings, and aesthetic reveals.

15.5 *Adhesively Attached*:

15.5.1 The insulation board thickness shall be not less than ¾ in. (19.1 mm) and not more than allowed by building codes.

15.5.2 Insulation board joints shall be offset not less than 6 in. (150 mm) from horizontal sheathing board joints.

15.6 *Mechanically Fastened*:

15.6.1 The insulation board thickness shall be not less than 1 in. (25 mm) or not more than allowed by building codes.

15.6.2 Insulation board joints shall be offset from horizontal sheathing board joints. All vertical insulation board joints shall fall upon a framing member.

15.7 When attachment is by adhesive or adhesive and mechanical fasteners, each board shall be installed by sliding it into place until it abuts the adjoining board tightly. Pressure shall be applied over the entire surface of the board to achieve uniform contact and an overall level surface. The insulation board shall be occasionally checked for proper contact with the substrate by removing a piece of insulation board. Proper contact has been achieved when approximately equal portions of the adhesive remain on both the substrate and the insulation board when the board is removed from the wall. The insulation board shall remain undisturbed for no less than 24 h prior to proceeding with the installation of mechanical fasteners or the base coat. Cool, damp weather conditions extend minimum curing or drying time.

15.8 When the boards are attached only with mechanical fasteners, they shall be affixed to a nailable substrate or structural members with the proper fasteners, ensuring that all boards abut tightly. Fastening patterns shall be in accordance with the EIFS producer's recommendation. When installed, the portion of the fastener left flush with the face of the insulation board shall not fracture the insulation board.

16. Aesthetic Reveals

16.1 Aesthetic reveals shall be cut into the insulation board prior to applying the reinforcing mesh.

16.2 The insulation board thickness in the bottom of a reveal shall be not less than ¾ in. (19.1 mm).

16.3 Aesthetic reveals shall be cut true and straight.

16.4 Horizontal aesthetic reveal configuration shall provide for outward positive drainage.

16.5 Aesthetic reveals shall not occur at the abutment of two pieces of insulation board.

17. Reinforced Base Coat Application

17.1 Inspect the insulation board surface to ensure that it is clean, dry, free of all foreign materials, and damage of any type. Insulation board planar irregularities of more than 1⁄16 in. (1.6 mm) shall be corrected. All board joints shall be tightly abutted or shall be filled with an insulating material. The entire wall area shall be sanded or rasped to minimize any irregularities.

17.2 *Base Coat Preparation*—All materials requiring field preparation shall be mixed in accordance with the EIFS producer's specifications.

17.3 *Base Coat Application*—The prepared base coat shall be uniformly spread over the entire surface of the insulation board.

17.4 *Nonmetallic Reinforcing Mesh*—The single layer reinforcing mesh shall be fully encapsulated in the field of the wall, at corners, edges, and joints. Trowel from the center to the edge of the reinforcing mesh to avoid wrinkles. The single layer reinforcing mesh shall be continuous at all corners.

17.4.1 The surface shall be free of voids, projections, trowel marks and other surface irregularities. The base coat thickness shall be not less than 1⁄16 in. (1.6 mm) dry as measured from the surface of the insulation board.

NOTE 5—The recommended method is to apply the base coat in two applications.

17.5 *Reinforcing Mesh Overlap*—All reinforcing mesh edges shall be overlapped not less than 2½ in. (64 mm).

17.6 *Impact Layers*—When required, higher impact performance shall be achieved with multiple layers of reinforcing mesh or by incorporating heavier weight reinforcing mesh. All areas requiring higher impact performance shall be detailed on the project plans and specifications. When overlapping of high impact mesh is not required, a complete second layer of reinforcing mesh shall be applied over the layer of high impact mesh.

17.7 *Corners*—Reinforcing mesh shall not be lapped within 8 in. (204 mm) of any corner.

17.8 *Wall Penetrations*—All edges of the insulation board at penetrations of the EIFS, such as at windows, doors, HVAC, pipes, ducts, electrical boxes, and at the base of the wall shall be wrapped with either the base coat and reinforcing mesh or trim, or as specified by the project plans and specifications. Wall openings such as doors, windows, and HVAC shall be diagonally reinforced at corners with mesh not less than 9 in. (230 mm) by 12 in. (305 mm).

17.8.1 The corners of returns to windows, doors, and similar wall openings shall have reinforcing mesh the full width of the return and extending not less than 2½in. (64 mm) on both sides of the corner.

17.9 *Expansion Joints*—All edges of the insulation board at expansion joints shall be wrapped with either the base coat and reinforcing mesh or trim as specified by the EIFS producer for the particular EIFS. The reinforcing mesh shall be fully encapsulated.

17.10 *Aesthetic Reveal*—The reinforcing mesh shall be continuous and care shall be taken to ensure that the reinforcing mesh is fully encapsulated into the reveal and that the reinforcing mesh shall not be cut during application of the base coat.

17.11 *Reinforced Base Coat*—The reinforcing mesh shall be fully encapsulated in the base coat throughout the field of the wall, at corners, edges, and joints. The surface shall be free of voids, projections, trowel marks and other surface irregularities.

17.12 *Special Insulation Shapes*—When special insulation shapes are installed over EIFS, fiberglass reinforcing mesh encapsulated in the base coat shall be used.

18. Finish Coat Application

18.1 The reinforced base coat shall be allowed to dry and harden for not less than 24 h prior to finish coat application.

18.2 *Material Preparation*—The finish coat shall be mixed in accordance with the EIFS producer's specifications. Only clean potable water shall be added to adjust workability. Always add the same amount of water to each container within a given lot of finish.

18.3 *Execution*—Finish shall be applied with a stainless steel trowel or appropriate equipment as specified by the EIFS producer. Tools and equipment shall be kept clean at all times. Finish shall be applied continuously to a natural break such as corners, joints or tape line. Apply finish to the base coat maintaining a wet edge. Sufficient manpower and scaffolding shall be provided to continuously finish a distinct wall area. Scaffolding shall be spaced a distance from the wall, consistent with safety standards, that will allow uniform texturing of the finish without staging marks. On hot windy days, the wall shall be permitted to be fogged with clean, potable water to cool the wall. Work shall precede the sun whenever possible. Water shall not be added to the finish once it is applied to the wall. Use the same tool and hand motion and match the texture of the surrounding area. The finish coat thickness shall be as specified by the EIFS producer. Protect all finish work from damage until fully dried.

18.4 *Joints*—The application of finish into the joints shall be as required by the project plans and specifications.

19. Curing Time Between Application Steps

19.1 Manufacturer's recommendations for climatic conditions, mixing, and application practices shall be followed to allow respective coats sufficient time to cure or dry before any subsequent coats are applied. Each application step shall be permitted to cure or dry to prevent degradation to the system integrity with subsequent steps.

19.2 Auxiliary protection (tenting, supplemental heat, shading) from adverse environmental conditions shall be utilized to minimize deviations from standard, expected cure or dry times. (See Appendix X1.9.3.)

19.3 All substrates shall be sound and cured or dry before continuing on to the next application step.

20. Cleanup

20.1 Adjacent materials, surfaces, and the work area shall be cleaned of foreign materials resulting from the work.

20.2 All excess EIFS materials shall be removed from the job site.

21. Keywords

21.1 class PB; EIFS; exterior insulation and finish system

APPENDIXES

(Nonmandatory Information)

These Appendixes give general information and also suggestions for inclusions to be made elsewhere by the specifier. They are not part of this practice.

X1. GENERAL IFORMATION

X1.1 *General:*

X1.1.1 The work includes all labor, materials, services, equipment and scaffolding required to complete the EIFS installation on the project in accordance with the project drawings and specifications, except electrical power and potable water.

X1.1.2 Where a specific fire resistance rating is required for the EIFS assemblies and construction, details of construction shall be in accordance with official reports of tests conducted by recognized testing laboratories and a part of approved plans and specifications.

X1.1.3 The installation of the EIFS is to be coordinated by

the owner's authorized agent with work of other trades.

X1.1.4 Surfaces and accessories to receive EIFS components shall be examined and accepted as provided in 11.2 before the EIFS components are installed thereto. The proper authorities shall be notified about unacceptable conditions. The unsatisfactory conditions shall be corrected prior to the application of the EIFS components. Unsatisfactory conditions shall be corrected by the party or parties responsible for such conditions.

X1.1.5 The construction specifier shall describe, in the proper section of the project specifications, the physical characteristics of the surfaces to receive the EIFS.

X1.2 *Design Considerations:*

X1.2.1 EIFS are attached to outside surfaces of all types of structures to provide weather resistant, energy efficient, and durable claddings. The claddings will be subjected to various exposures such as abrasion, vibration, moisture, freezing, thawing, negative and positive wind loading, ultraviolet light, acid rain, thermally induced expansion and contraction, and impact.

X1.2.2 Provide sufficient slope on faces of EIFS surfaces to prevent water, snow or ice from accumulating or standing. A minimum 6 in. (152 mm) in 12 in. (305 mm) (1:2) pitch is recommended. Length of sloped surfaces shall not be more than 12 in. (305 mm).

X1.2.3 Provide parapet cap made only from corrosion resistant materials. The parapet cap shall extend not less than 2½ in. (64 mm) over the faces of the EIFS, not including wood blocking and shall include a drip edge detail. Detailed requirements for furnishing and installing the parapet cap shall be in the appropriate section of the project specifications. Parapet caps shall be clearly shown in large scale details. Parapet caps shall be installed as soon as practical after the installation of the EIFS.

X1.2.4 Provide for sealing of expansion joints, panel joints, system terminations and all interfaces of the EIFS with other building components.

X1.2.5 At soffits, window and door heads provide for a drip edge.

X1.2.6 Windows installed in EIFS walls shall have metal sill flashing.

X1.2.7 The specifier shall evaluate the potential of the EIFS surface being subjected to abnormal stresses or impacts.

X1.2.8 Wall assemblies have properties that in certain applications make a dew point analysis critical to ensure successful performance of the wall system. The specifier shall evaluate these conditions.

X1.2.9 Sheathing must be continuous over floor slabs, intersecting walls, columns, beams and similar constructions. The EIFS shall not be installed over areas without sheathing.

X1.2.10 Where the sheathing is interrupted or out of plane, corrections must be made prior to EIFS application. If the sheathing cannot be continuous, the installation of expansion joints shall be considered.

X1.3 *Relief from Stress:*

X1.3.1 Where the EIFS and dissimilar types of substrates, materials and building components interface, provide for appropriate size and location of expansion joints.

X1.3.2 Where roof elevations change, provide for appropriate size and location of expansion joints.

X1.4 *Bonding Agents:*

X1.4.1 Bonding compounds or agents are sometimes applied to a substrate surface to receive base coat adhesive. Follow the EIFS producer's instructions and specifications for their use.

X1.5 *Adhesives:*

X1.5.1 The types of adhesives and the methods of installation used will vary depending on the substrate to which the insulation boards will be attached. Therefore, the adhesive type and application method used shall be in accordance with the EIFS producer's written instructions and applicable building code requirements.

X1.6 *Mechanical Fasteners:*

X1.6.1 The types of mechanical fasteners used and the methods of installation will vary depending on the substrate to which the insulation boards are to be fastened.

X1.6.2 Fastener washers shall be installed flush with the insulation board surface.

X1.7 *Reinforced Base Coats:*

X1.7.1 Base coat thickness is permitted to exceed 1⁄16 in. (1.6 mm) dry depending on the type of base coat used and the impact resistance specified.

X1.7.2 A small amount of base coat shrinkage is normal. Due to color variations, it shall not be unusual to see a very slight telegraphing of the reinforcing mesh pattern after the base coat cures. Bare mesh shall not be visible.

X1.7.3 The producers instructions shall be followed to ensure that the base coat cement is properly proportioned. The proper portioning of cement based material is important. Excess cement may cause the base coat to become hard and stiff as the cement hydrates over time.

X1.7.4 The minimum amount of mix water to form a workable mix shall be used. Excess mix water will cause the base coat to be porous and less weather resistant. Never add more water than recommended by the producer. Measure water in containers of known volume.

X1.7.5 Mix materials only at low speed with paddles designed to reduce air entrapment.

X1.7.6 Generally the second base coat layer is applied as soon as the first layer has dried firmly enough. Depending upon material composition and weather conditions, this sometimes occurs within the first 4 to 8 h after the completion of the application of the initial base coat, or, in some cases, it will require waiting until the following day.

X1.7.7 Trowel aesthetic reveals and corners with a trowel shaped to fit the profiles.

X1.8 *Finish Textures:*

X1.8.1 Textures, as a description of surface appearance, generally are identified with the method and tools used to achieve the finish appearance. Texture is sometimes varied by the size and shape of the aggregate contained in the finish, equipment or tools employed, the consistency of the finish mix, the condition of the base to which it is applied and by decorative and protective treatments.

X1.8.2 With almost limitless variations possible for finish appearance or texture, the same term sometimes does not have the same meaning to the specifier, contractor or applicator. Sample panels of sufficient size including components and features specified shall be constructed using the methods, tools and techniques to be used in the actual construction. It is recommended that in the case of prefabrication, two panels be constructed. One panel remains at the jobsite. One panel remains at the fabrication site. The samples shall be approved by the architect and owner in writing.

X1.8.3 To provide some guidance, the following categories are generally understood and recognized to imply a particular method of application or finished appearance.

X1.8.3.1 *Sand Finish*—A finish material devoid of coarse aggregate applied in a thin coat completely covering the base coat and floated to a true plane, yielding a relatively fine-textured finish.

X1.8.3.2 *Trowel Texture*—A freshly applied finish is given various textures, designs or stippled effects by hand motion and trowel used. The effects achieved are individualized and it is sometimes difficult for different applicators to duplicate.

X1.8.3.3 *Standard Finish*—A finish with various size aggregates, applied in a thin coat completely covering the base coat and floated in a random pattern. The applicators are cautioned to use the same type texturing tools to help ensure a consistent texture.

X1.8.3.4 *Exposed Aggregate*—Varying sizes of natural or manufactured stone, gravel, shell or ceramic aggregates are embedded by hand or by machine propulsion into a freshly applied" bed coat". The size of the aggregate shall not be more than ½ in. (12.7 mm) in diameter. The size of the aggregate determines the thickness of the bedding coat. The bedding coat shall be no less than one-half the thickness of the aggregate to be embedded.

X1.8.3.5 *Spray Texture*—A machine applied finish material. The texture achieved depends on many factors some of which are the consistency of the sprayed mixture, the angle and distance of the spray nozzle to the surface, the pressure of the machine, orifice size used, operator expertise, climatic conditions (wind specifically), and base coat conditions.

X1.8.3.6 *Miscellaneous Types*—Similar to trowel textures except that the freshly applied finish is textured with instruments other than a trowel, such as swept with a broom or brush, or corrugated by raking or combing. A variation of texturing a finish involves waiting until the applied finish has partially dried and then lightly troweling the unevenly applied finish. The effects achieved are sometimes difficult for different applicators to duplicate.

X1.9 *Conditions for EIFS Application:*

X1.9.1 The amount of water used to mix materials is to be within the allowable range published by the EIFS producer. Factors such as the drying conditions and finishing operations must be considered in determining water usage. The applicator is cautioned that the use of excessive water results in fall off or slide off, excessive shrinkage, higher porosity, lower strength, and inconsistent color.

X1.9.2 Timing between coats will vary with climatic conditions and types of materials used. Temperature and relative humidity extend or reduce the curing and drying time. Cold or wet weather lengthens, and hot or dry weather shortens the drying period. Moderate changes in temperature and relative humidity are overcome by providing additional heating during cold weather and by pre-wetting the base during hot, dry weather.

X1.9.3 Heated enclosures require adequate air circulation to prevent hot or cold spots that allow too rapid drying or freezing. When using heated enclosures it is important to control the relative humidity as well as the temperature. Maintain relative humidity at not more than 90 %.

X1.9.4 Staining and discoloration of finishes or other building components that are caused by free water draining from one plane to another, dissimilar materials, or other building components, are minimized by providing sufficient depth and angle for drip caps and drip edges.

X2. EVALUATION OF GYPSUM SHEATHING FOR SURFACE ACCEPTABILITY

X2.1 The following procedure is valid only if, at the time of the test, the gypsum sheathing has not been wetted or exposed to high humidity in the previous 24 h.

X2.2 Using a sharp knife or razor blade, cut a 3-in. (76-mm) long diagonal in the middle of the gypsum sheathing. The cut must be sufficient to cut through the paper face and just into the underlying gypsum. Make a 3-in. (76-mm) cut at a right angle to the first, forming an "X".

X2.3 Using the sharp edge of the cutting tool, peel the paper face (in one quadrant) back approximately ¼ in. (6 mm) from the intersection of the two lines.

X2.4 Firmly holding this peeled edge between the thumb and forefinger, lift the paper face and continue removing the facing material.

X2.5 If the paper splits cohesively so that part is still adhering to the underlying gypsum, the gypsum sheathing has not been overly exposed to the weather or moisture. If the paper lifts and directly exposes the gypsum, the gypsum sheathing has been exposed to the weather beyond the required limit and shall be replaced.

X3. SEALANT SYSTEMS

X3.1 The size and location of joints and selection of the sealant to be used is the responsibility of the design professional and shall be consistent with project conditions and guidelines of the EIFS producer.

X3.2 The sealant system must be tested for compatibility with the specified EIFS by an independent laboratory and found to be compatible with the EIFS.

X3.3 The sealant system is recommended by the EIFS and sealant producer. Non-recommended sealant systems must be jointly approved by the architect and the sealant producer.

X3.4 Base the joint design and surface preparation on the sealant producer's recommendations and project conditions.

X3.5 Allow the EIFS material to cure prior to sealant system installation.

RELATED DOCUMENTS

The following are ASTM Standards

Specification C 150 for Portland Cement, Vol 04.01

Specification C 514 for Nails for the Application of Gypsum Wall Board, Vol 04.01

Specification C 522 for Cellular Glass Thermal Insulation, Vol 04.06

Specification C 578 for Preformed, Cellular Polystyrene Thermal Insulation, Vol 04.06

Specification C 612 for Mineral Fiber Block and Board Thermal Insulation, Vol 04.06

Specification C 630 for Water-Resistant Gypsum Backing Board, Vol 04.01

Specification C 645 for Nonstructural Steel Framing Members, Vol 04.01

Specification C 754 for Installation of Steel Framing Members to Receive Screw-Attached Gypsum, Vol 04.01

Specification C 847 for Metal Lath, Vol 04.01

Specification C 920 for Elastomeric Joint Sealants, Vol 04.07

Specification C 932 for Surface-Applied Bonding Agents for Exterior Plastering, Vol 04.01

Specification C 954 for Steel Drill Screws for the Application of Gypsum Board or Metal Plaster Bases to Steel Studs from 0.033 in. (0.84 mm) to 0.112 in. (2.84 mm) in Thickness, Vol 04.01

Specification C 955 for Load-Bearing (Transverse and Axial) Steel Studs, Runners (Track), and Bracing or Bridging, for Screw Application of Gypsum Board and Metal Plaster Bases, Vol 04.01

Specification C 1002 for Steel Drill Screws for the Application of Gypsum Board or Metal Plaster Bases, Vol 04.01

Specification C 1007 for Installation of Load-Bearing (Transverse and Axial) Steel Studs and Related Accessories, Vol 04.01

Provisional Standard Test Method for Length Change of Mortar Bars Caused by Reaction Between Hydraulic Cement and Alkali-Reactive Aggregate[1]

This standard is issued under the fixed designation PS 81; the number immediately following the designation indicates the year of original adoption.

ϵ^1 NOTE—Footnotes Six and Seven were added editorially in May 1999.

1. Scope

1.1 This provisional test method covers the potential of hydraulic cement to cause excessive expansion through reaction between alkalies (sodium and potassium) in the cement and alkali-silica reactive aggregate (see Note 1). The determination is based on the change in length of mortar bars containing the cement in combination with a standard alkali-reactive aggregate during storage under prescribed conditions of test.

NOTE 1—.Although the reaction is called "alkali-silica reaction", it in fact is a sequence of processes initiated by reaction between thermodynamically metastable silica in the aggregate and hydroxide ion (OH^-) in the pore fluid, the concentration of which is controlled largely by the concentration of alkali ions (Na^+ and K^+), usually provided by the cement.

1.2 Two types of alkali reactivity of aggregates are recognized: alkali-silica reaction, that involves certain siliceous rocks, minerals, and natural or artificial glasses, and alkali-carbonate reaction, that involves dolomite in certain calcitic dolomites and dolomitic limestones. This provisional test method addresses only alkali-silica reaction.

1.3 Provisional standards[2] achieve limited concensus through approval of the sponsoring subcommittee.

1.4 This provisional test method was approved as a provisional standard because it is critical in the development of a performance specification for Portland Cement, a high-priority effort in C-1. This provisional test method has only achieved sub conssessus and will be published for two years only.

1.5 The values stated in SI units are to be regarded as the standard.

1.6 *This standard does not purport to address all of the safety concerns, if any, associated with its use. It is the responsibility of the user of this standard to establish appropriate safety and health practices and determine the applicability of regulatory limitations prior to use.*

2. Referenced Documents

2.1 *ASTM Standards:*

C 109/C 109M Test Method for Compressive Strength of Hydraulic Cement Mortars (Using 2-in. or [50-mm] Cube Specimens)[3]

C 114 Test Methods for Chemical Analysis of Hydraulic Cement[3]

C 150 Specification for Portland Cement[3]

C 227 Test Method for Potential Alkali Reactivity of Cement-Aggregate Combinations (Mortar-Bar Method)[4]

C 305 Practice for Mechanical Mixing of Hydraulic Cement Pastes and Mortars of Plastic Consistency[3]

C 490 Practice for Use of Apparatus for the Determination of Length Change of Hardened Cement Paste, Mortar, and Concrete[3,4]

C 511 Specification for Moist Cabinets, Moist Rooms, and Water Storage Tanks Used in the Testing of Hydraulic Cements and Concretes[4]

C 670 Practice for Preparing Precision Statements for Test Methods of Construction Materials[3,4]

C 778 Specification for Standard Sand[3]

C 856 Practice for Petrographic Examination of Hardened Concrete[4]

C 1005 Specification for Reference Masses and Devices for Determining Mass for Use in the Physical Testing of Hydraulic Cements[3]

E 11 Specification for Wire Cloth and Sieves for Testing Purposes[4]

3. Significance and Use

3.1 This test method may be used to evaluate the potential of a hydraulic cement to cause excessive expansion when combined with alkali-silica reactive aggregate.

3.2 When increases in length are produced in excess of 0.25 at 14 d or 0.30 at 28 d, it is recommended that supplementary information be developed to confirm that expansion is actually due to alkali-silica reaction. Sources of such supplementary information include: determination of alkali content of the cement (see Test Methods C 114), and examination of the specimens according to 9.3 and 9.4 after the final length change measurement to establish that alkali-silica reaction has taken place.

---

[1] This provisional test method is under the jurisdiction of ASTM Committee C-1 on Cement and is the direct responsibility of Subcommittee C01.31 on Volume Change.

Current edition approved May 22, 1998. Published May 1999. Originally published as PS 81–98. Last previous edition PS 81–98.

[2] Provisional standards exist for two years subsequent to the approval date.

[3] *Annual Book of ASTM Standards*, Vol 04.01

[4] *Annual Book of ASTM Standards*, Vol 04.02.

3.3 Excessive expansion may occur rarely in the test for reasons other than alkali-silica expansion. When it is suspected that this may occur or has occurred, parallel tests are suggested using this procedure with a relatively non-reactive aggregate.

4. Apparatus

4.1 *Devices for Determining Mass*—See Specification C 1005.

4.2 *Sieves*—See Specification E 11.

4.3 *Mixer, Paddle, and Mixing Bowl*—See Test Method C 305.

4.4 *Tamper and Trowel*—See Test Method C 109.

4.5 *Molds*—See Practice C 490.

4.6 *Storage Container*—Covered containers for storing the test specimens shall be constructed of material that is resistant to corrosion under the test conditions. Containers shall be at least 150 mm inside diameter and 250 mm inside height. The containers shall provide continuous air flow, as discussed in 6.3.4. A satisfactory container is shown in Fig. 1. The containers shall provide a seal between the cover and the container wall (See Note 2).

NOTE 2—If the cover is grooved to fit the container wall, as shown in Fig. 1, the seal may be achieved using a coating of silicone sealant in the groove of the bottom cover and silicone grease in the groove of the top cover. Alternatively if the cover is designed to fit over the outside of the top portion of the container wall, then the seal may be achieved by a double wrap of vinyl tape overlaying the cover and the container wall along its full circumference.

4.7 *Mortar-Bar Support Rack*, for storing the bars vertically. The rack shall fit into the storage container and allow for air flow from the bottom to the top of the storage container. The rack shall be constructed of a material that is resistant to corrosion under the test conditions. A satisfactory rack is shown in Fig. 2. The rack described in Test Method C 227 is satisfactory.

4.8 *Length Comparator*, conforming to Practice C 490.

5. Materials

5.1 *Reactive Aggregate*, glass tubing, Vycor brand, No. 7713[5], crushed and graded according to Table 1 (See Note 3). Separate the crushed glass into sieve fractions and wash each fraction to remove adhering dust and fine particles (See Note 4). To wash, immerse the fraction in water, agitate, then retrieve the aggregate using the retaining sieve. Dry each fraction and either use immediately or store in a clean container provided with a tight-fitting cover.

NOTE 3—If it is desired to test a relatively non-reactive aggregate, material of the specified grading can be produced by using equal amounts of graded sand and 20–30 sand (See Specification C 778).

NOTE 4—Aggregate often is contaminated with metallic iron during the crushing operation. If such contamination is observed, remove this using a magnet.

5.2 *Cement*—Pass the cement through an 850-µm sieve in order to mix the sample, break up lumps, and remove foreign material. Discard the foreign materials and hardened lumps that do not break up on sieving or brushing. Store the cement in airtight moisture-proof containers to prevent aeration or absorption of moisture.

6. Temperature and Humidity

6.1 *Preparation of Test Specimens*—The temperature of the molding room and dry materials shall be maintained between 20 and 27.5°C. The relative humidity shall be not less than 50 %. The temperature of the mixing water shall be 23.0 ± 2.0°C.

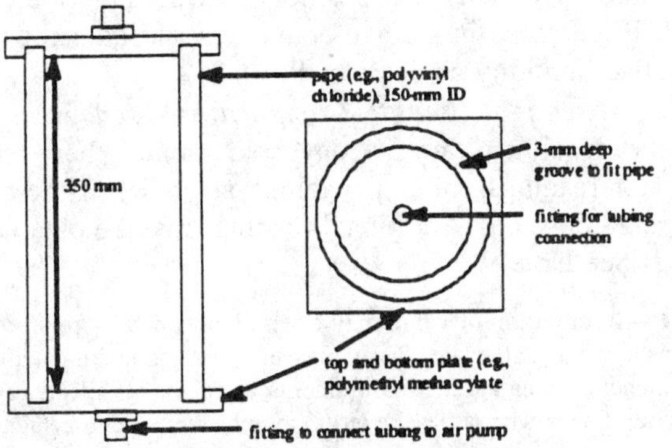

FIG. 1 Satisfactory Storage Container

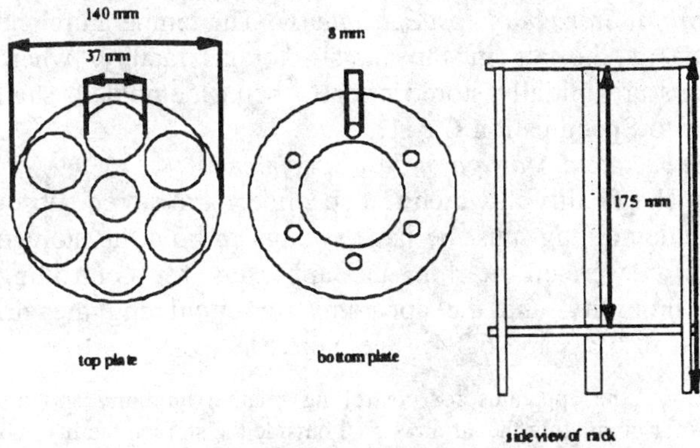

FIG. 2 Satisfactory Storage Rack

[5] The sole source of supply of material suitable for this provisional standard known to the committee at this time is the Cement and Concrete Reference Laboratory, National Institute of Standards and Technology, Gaithersburg, MD 20899. If you are aware of alternative suppliers, please provide this information to ASTM Headquarters. Your comments will receive careful consideration at a meeting of the responsible technical committee,[1] which you may attend.

TABLE 1 Grading

Sieve Size		% by Mass
Passing	Retained	
. . .	1.18 mm	0.0
1.18 mm	850 mm	7.5
850 mm	600 mm	42.5
600 mm	425 mm	17.5
425 mm	300 mm	17.5
300 mm	150 mm	15.0
150 mm	. . .	0.0

6.2 *Initial Storage of Test Specimens*—The temperature and humidity of the air in the moist storage facility, where specimens are initially stored prior to being demolded, shall conform to Specification C 511.

6.3 *Subsequent Storage of Test Specimens*:

6.3.1 The facility in which test specimens are stored subsequent to demolding must be large enough to hold the storage container, described in 4.6., the apparatus for controlling relative humidity, and the apparatus for circulating the air, described in 6.3.3 and 6.3.4, (See Note 5).

NOTE 5—If the apparatus for controlling relative humidity and the apparatus for circulating the air are stored outside the storage facility and air is allowed to cool during its circulation, then water will condense in the air lines, causing air flow to be blocked.

6.3.2 Maintain the temperature of the cabinet in which specimens are stored after being demolded at 37.5 ± 0.5°C.

6.3.3 Maintain the relative humidity inside the container in which specimens are stored subsequent to demolding at 96 to 99 % (See Note 6).

NOTE 6—If the relative humidity is reduced, measured expansion values are reduced. If the relative humidity reaches 100 %, then water condenses on the bars, causing alkalies to leach from the bars and possibly reducing expansion.

6.3.4 Circulate air through the storage container throughout the storage period. The flow rate shall be approximately 1 mL/s (See Note 7).

NOTE 7—The air flow reduces the tendency of water to condense on the bars, known to be a problem at this high relative humidity. While the precise flow rate is not important, it is important that the flow rate be similar through each container so relative humidity does not vary from container to container. It is useful to monitor flow rate to assure that similar rates are achieved.

6.3.5 Satisfactory procedures for storage of test specimens are illustrated in the appendix, and additional details are available elsewhere.[6]

6.4 *Measurement of Test Specimens*:

6.4.1 Temperature of the facility in which length-comparator readings are made shall conform to one of the following: a molding room as described in 6.1.1 or the storage facility as described above. It has been found that length-comparator measurements are not significantly affected by measurement temperature as long as bars are at the same temperature as the facility in which they are measured.[7]

6.4.2 If the measurement temperature is higher than the initial storage temperature, bars shall be stored at the measurement temperature for 2 h prior to the initial length-comparator measurement. The relative humidity during this equilibration period shall conform to that of the storage facility as described above, but no air flow is required.

6.4.3 If the measurement temperature is lower than the storage temperature, bars shall be stored at the measurement temperature for 2 h prior to each subsequent length-comparator measurement. The relative humidity during this equilibration period shall conform to that of the storage facility as described above, but no air flow is required.

7. Procedure

7.1 *Preparation of Test Specimens*:

7.1.1 *Number of Specimens*—Make at least three specimens using each test cement.

7.1.2 *Proportioning of Mortar*—Use 500 g of cement, 250 g of water, and 650 g of reactive aggregate (See Note 8).

NOTE 8—If it is desired to test a relatively nonreactive aggregate using standard sand as described in Note 3, a satisfactory mortar can be made using 400 g of cement, 200 g of water, and 828 g of fine aggregate. The aggregate/cement ratio by volume is greater for the unreactive mortar than for the reactive mortar to provide similar workabilities. Otherwise workability would be different due to the differences in density (2.20 Mg/m^3 for the reactive aggregate and 2.65 Mg/m^3 for the unreactive aggregate) and shape and texture (highly angular and elongated for the reactive aggregate and smooth and spherical for the unreactive aggregate). The proportions provide an aggregate/cement ratio by volume of 1.86 for the reactive mortar and 2.67 for the unreactive mortar.

7.1.3 *Mixing of Mortar*— Mix the mortar in accordance with Test Method C 305.

7.1.4 *Molding Test Specimens*—Immediately after mixing, start molding the specimens. Fill the molds with two approximately equal layers, each layer being compacted with the tamper. Work the mortar into the corners, around the gage studs, and along the surfaces of the mold with the tamper until a homogeneous specimen is obtained. After the top layer has been compacted, cut off the mortar flush with the top of the mold and smooth the surface with a few strokes of the trowel.

7.2 *Storage and Measurement of Test Specimens*:

7.2.1 *Initial Storage*— Immediately after each mold has been filled, place the mold in the moist room. The specimens shall remain in the molds, stored in the moist cabinet, for 24 ± 2 h. At the end of this period, remove the bars from the molds and, while protecting them from loss of moisture, properly identify bars and apply a mark so that each end of the bar can be distinguished from the other.

7.2.2 *Initial Length-Comparator Measurement*—Obtain the initial length-comparator reading in accordance with Practice C 490. The temperature of the facility in which this reading is obtained is described in 6.4.

7.2.3 *Subsequent Storage*—Stand the specimens in storage rack so the specimens are positioned with their long axes vertical. Place racks in storage containers and adjust air flow and relative humidity as described in 6.3

7.2.4 *Subsequent Length-Comparator Measurement*—When the specimens are 14 days old, obtain their length-comparator readings. Obtain readings again when the specimens are 28 days old. Additional readings may be obtained as required (See Note 9).

NOTE 9—Information of value may often be obtained by returning the specimens to storage after the 28-day measurement and making additional measurements at later ages. Measurements at ages of 2,3,4,6,9, and 12 months and, if necessary, at least every 6 months thereafter are suggested.

7.2.5 Make all length-comparator measurements according

[6] Pade, C., and Struble, L., "Mortar-Bar Expansion Due to Alkali-Silica Reaction," Institute for Standards Research, 1996. A copy of this ISR report is available from ASTM Headquarters as ISR: # C01.001.

[7] Struble, L. J., "Measuring Potential Expansion Due to Alkali-Silica Reaction in Concrete, Phase II — Follow-Up Studies," Institute for Standards Research, 1995. A copy of this ISR report is available from ASTM Headquarters as ISR: # C 01.001.

to Practice C 490 following the procedure for bars stored moist.

7.2.6 After the final length measurement, examine specimens in the manner described in Section 9.

7.2.7 Make all specimens placed in a given container on the same day and place them in the container at the same time, so that all specimens in a given container are due for measurement at the same time.

7.3 Always place the specimens in the comparator with the same end uppermost. After measurement, replace the specimens in the container in an inverted position as compared with the previous period.

8. Calculation

8.1 Calculate length-change values according to Practice C 490.

9. Examination at End of Test

9.1 After the final length change measurements have been made, test each specimen for warping and examine each specimen for visual evidence of reaction.

9.2 *Warping*—Determine the warping, if any, by placing the specimen on a plane surface and measuring the maximum separation between the specimen and the surface. Place the specimen so that its ends curve down to the reference surface and measure separation to the nearest 0.5 mm.

9.3 *Visual Examination*—Items to be noted in the visual examination include:

9.3.1 Presence, location and type of pattern of any cracking,

9.3.2 Surface mottling and surficial deposits or exudations (their nature, thickness, and continuity), and

9.4 When it is believed that additional important features are likely to be revealed thereby, make a further examination of the interior of the specimen to establish whether alkali-silica reaction has taken place, according to the relevant sections of Practice C 856. Note such items as the presence of gel-filled pores and fissures, evidence of reaction with aggregate particles, and properties of the reaction products.

10. Report

10.1 Report the following information:

10.1.1 The type of cement tested and its total alkali content, expressed as eq. Na_2O, determined according to the relevant portion of Test Methods C 114,

10.1.2 The average increase in length of the mortar bars at 14 and 28 days, in percent,

10.1.3 Any deviation from the test procedure, including the aggregate grading when it differs from that given in Table 1, and

10.1.4 Any significant features revealed by examination of specimens after the test as described in Section 9.

11. Precision and Bias

11.1 Interlaboratory test data[7] were developed in accordance with Practice C 670. Seven sets of data were analyzed and the average length change was 0.523 %.

11.2 *Precision*:

11.2.1 The single-operator standard deviation was found to be 0.08. Therefore, results of two properly conducted tests by the same operator on the same material should not differ by more than 0.24. These numbers represent, respectively, the (1s) and (d2s) limits as described in Practice C 670.

11.2.2 The multi-laboratory standard deviation was found to be 0.08. Therefore, results of two properly conducted tests from two different laboratories on the same material should not differ by more than 0.24. These numbers represent respectively the (1s) and (d2s) limits as described in Practice C 670.

11.3 *Bias*—This procedure has no bias because the values measured are defined in terms of the test method.

12. Keywords

12.1 alkalies; alkali-silica reaction; cement; length change

APPENDIXES

(Nonmandatory Information)

X1. SATISFACTORY PROCEDURES FOR STORAGE OF TEST SPECIMENS

X1.1 This appendix illustrates satisfactory experimental procedures for storage of test specimens. The procedures provide circulation through the storage container of air at a relative humidity of 96 to 99 %.

X1.2 *Differential Temperature Method*—Saturate the air with water at a temperature 0.5°C lower than the temperature used for storage of test (Fig. X1.1).

X1.2.1 This provisional test method requires a water bath or some other means to control temperature that cools as well as heats, and maintains temperature within ± 0.05°C.

X1.2.2 When temperature is increased, relative humidity decreases. If air is saturated with water at a temperature of 37.0°C and then warmed to 37.5°C, its relative humidity

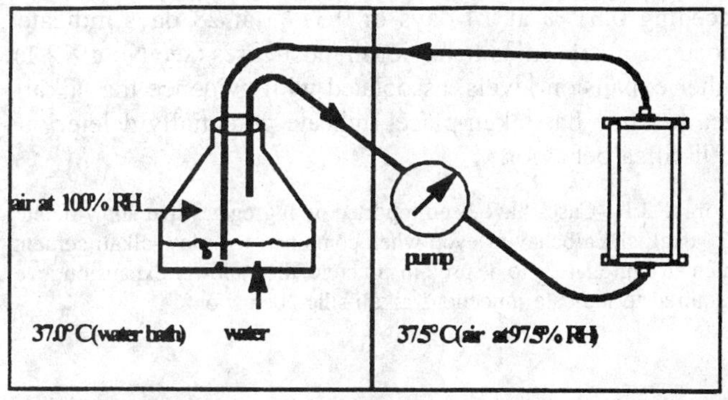

FIG. X1.1 Experimental Procedure for Differential Temperature Method

decreases from 100 % to 97.5 %.

X1.2.3 The procedure allows visual observation that the flow rate is within the required range by counting bubbles in the water reservoir.

X1.3 *Saturated Salt Solution Method*—Equilibrate the air with a saturated aqueous solution of reagent-grade K_2SO_4 at the storage temperature (See Note X1.1 and Note X1.2) (See Fig. X1.2).

NOTE X1.1—When using the experimental procedure in Fig. X1.2, it is important to periodically monitor airflow. When airflow stops, K_2SO_4 may precipitate in the air line, blocking subsequent flow of air.

NOTE X1.2—At 37.5°C, this method provides a relative humidity of 96.4 % (estimated from published data).[8]

X1.4 *Air Flow*—A satisfactory air flow may be achieved using a speed adjustable peristaltic pump equipped with a pump head for each container. An example is the Masterflex 7568-00 pump drive and Masterflex 7024-20 pump head.[9] Using this pump, a setting of 6 on the knob scale adjustable between 0 and 10 was found to be adequate for obtaining a constant humidity.

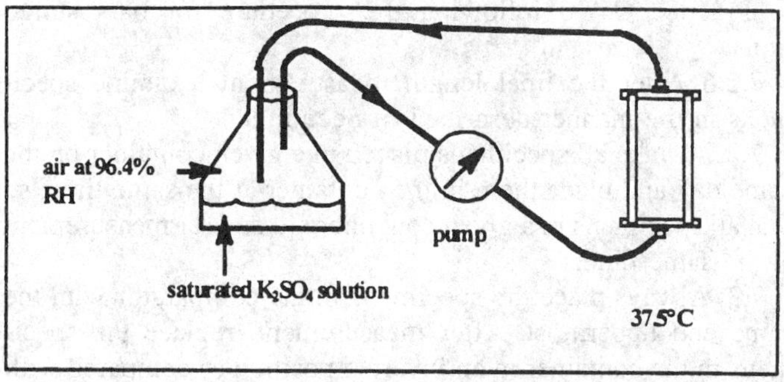

FIG. X1.2 Experimental Procedure for Saturated Salt Solution Method

[8] Young, J.F., "Humidity Control in the Laboratory Using Salt Solutions– A Review," *J. Appl. Chem.*, Vol 17, 1967, pp. 241-245.

[9] The sole source of supply of the apparatus known to the committee at this time is Cole-Parmer Instrument Company, Chicago, IL. If you are aware of alternative suppliers, please provide this information to ASTM Headquarters. Your comments will receive careful consideration at a meeting of the responsible technical committee,[1] which you may attend.

X2. INTERPRETATION OF TEST RESULTS

X2.1 The following expansion limits are based on research results and on Specification C 150.

X2.2 The optional requirement in Specification C 150 for low-alkali cement is that the total alkali content (expressed as equivalent sodium oxide, calculated as % Na_2O + 0.658 % K_2O) not exceed 0.60 %. This optional chemical requirement should be invoked when the cement is to be used with aggregate that is potentially reactive, and no other provisions have been made to protect the concrete from deleterious expansion.

X2.3 Tests using portland cements having various alkali contents combined with Vycor glass have shown that the expansion level associated with an alkali content of 0.60 % is 0.31 % at 14 days and 0.37 % at 28 days (See Fig. X2.1).

X2.4 Based on these results, an expansion level not exceeding 0.31 % at 14 days or 0.37 % at 28 days indicates innocuous alkali-silica behavior in most cases (see Note X2.1). Higher expansion levels, associated with evidence that alkali-silica reaction has taken place, indicate potentially deleterious alkali-silica behavior.

NOTE X2.1—Cases have been reported of aggregates that show deleterious alkali-silica behavior even when combined with low-alkali cement. If such an aggregate is to be used in concrete, then a lower expansion level is required to indicate innocuous alkali-silica behavior.

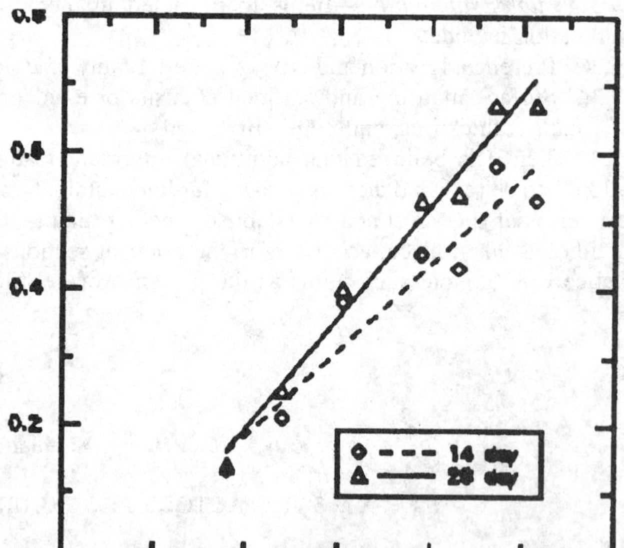

NOTE—The lines were determined using linear regression analysis; the equation for the 14-day expansion is E=0.63 A-0.07 and R^2 is 0.91; the equation for the 28-day expansion is E=0.81 A-0.13 and R^2 is 0.98; E is % expansion and A is % alkali.

FIG. X2.1 Expansion Versus Cement Alkali Content

RELATED MATERIAL

Any documents included in this section that are marked "proposed" are published for information only. They have received the approval of the sponsoring technical committee for publication as "proposed" but have not been officially accepted by the Society. Comments are solicited and should be addressed to ASTM, 100 Barr Harbor Drive, West Conshohocken, PA 19428.

MANUAL OF CEMENT TESTING[1]

INTRODUCTION

This Manual of Cement Testing is intended to supplement–not in any way to supersede–the various ASTM standard methods of sampling and physical testing of hydraulic cement. This manual was prepared by Committee C-1 on Cement and has been accepted by the Society for publication as information only. The manual is not a part of the ASTM methods.

The ASTM standard methods of testing cement are essential adjuncts to the specifications, because faulty test methods may lead to erroneous conclusions as to the quality of the cement.

Precautionary statements are found in many of the test methods. Before using a test method for cement testing, the precautionary statement should be noted and clearly understood.

The approved test methods are frequently revised, sometimes as often as twice a year. The cement tester must be sure the latest revision of the test method is used. A test method or other ASTM standard is revised and promulgated in the following manner. After a document, based on technical merit, has passed the ASTM Society letter ballot, which is published in the monthly issue of the ASTM *Standardization News*, it goes to the Committee on Standards. This standing committee of the ASTM Board of Directors determines if the procedural requirements of the Society have been met. If favorable action is taken, the document is approved as a standard. This is the date when the new standard is officially in effect. An announcement in a later issue of the ASTM *Standardization News* notifies individuals of its approval and the availability of the standard in published form. The revised test method should then be used when testing is specified to be performed by the latest test method.

Existing methods of testing cement are, at best, crude tools, and it must not be expected that exact agreement will be secured in tests by different operators, or by the same operator at different times, even when the tests are made on the same sample and under the same conditions. Many steps are involved in making cement tests, and the results are sometimes greatly influenced by seemingly minor variations in procedure.

It is the purpose of this manual to emphasize those factors which may affect results of tests, and to call attention to less apparent influences which are important but which are sometimes overlooked.

The ASTM specifications for cement are intended to set forth minimum requirements that must be met. Cement testing laboratories should have on hand copies of current ASTM specifications for cements, as well as the standard methods of sampling and testing cement. The cement testers should be thoroughly familiar with these specifications and methods, and should use every care to follow the standards strictly and maintain their testing equipment in good condition. If careless methods and faulty equipment are used, and short-cuts made, the results may lead to the rejection of a good product, or, on the other hand, to the acceptance of a product that may not conform to the specification requirements.

It is suggested that the different cement testers in a given city or district arrange to carry out tests on the same cement and exchange results. To be of the greatest value, such tests should be made not only by each operator in his laboratory but also by all operators in a single laboratory where the equipment and test conditions are identical. Each operator should repeat his tests on several different days to see how closely he can check his own results. This procedure will do more than anything else to create an interest in standard methods and will be found effective in eliminating erroneous practices and in calling attention to defective apparatus. A suggested course of training for cement testers is given in Section 55.

The subjects covered in this manual appear in the following order:

	Sections
Sampling	1 to 2
Fineness by the 150-µm (No. 100) and 75-µm (No. 200) Sieves	3 to 5
Fineness by the Turbidimeter	6 and 7
Fineness by the 45-µm (No. 325) Sieve	8
Fineness by the Air Permeability Test	9
Temperature and Relative Humidity, Storage Equipment	10 to 13
Mixing Cement Pastes and Mortars	14 to 24
Normal Consistency	25 to 27
Autoclave Expansion	28 to 32
Time of Setting	33 to 36
Strength Tests	37 and 38
Tensile Strength	39 to 42
Compressive Strength	43 to 45
Flow Table	46
Air Content of Mortar	47 to 51
Standard Sand	52
Calibration and Care of Apparatus	53
Cement and Concrete Reference Laboratory	54
Training of Cement Testers	55
Cement Tests for Research Purposes	56
Safety in the Cement Testing Laboratory	57
Chemical Analysis of Hydraulic Cement	58
Compositional Analysis of Hydraulic Cement	59

SAMPLING

(See Practice C 183)[2]

1. Treatment of Samples

1.1 Test samples must be forwarded to the laboratory promptly in airtight, moistureproof containers. Ordinary paper or cloth sacks must not be used. Poly(vinyl chloride)

[1] This manual is under the jurisdiction of ASTM Committee C-1 on Cement and is the direct responsibility of Subcommittee C01.97 on Manual for Cement Testing.

Published as information, June, 1932; revised, 1944, 1946, 1950, 1952, 1954, 1955, 1956, 1958, 1960, 1968, 1986, 1989, 1990 and 1992; editorially corrected 1970, 1971, 1972, 1975, 1977, 1978, 1981, and 1986.

[2] *Annual Book of ASTM Standards*, Vol 04.01.

sample containers, upon occasion, have been found to affect the air entraining potential of a cement sample. The same problem might be experienced with containers made from other plastics. These containers should be used with caution. Each sample should be marked plainly and identified completely on the outside of the body of the container as to source and date of sampling. Do not mark the lids of containers, since the lids may be readily interchanged and lead to confusion. Also, a complete identification should be enclosed in the container. As it is important to keep variable influences at a minimum, the sample should be kept in airtight, moisture-proof containers from the time it is taken until it is tested or made into specimens. Reserve portions of samples should be likewise protected as long as they are kept.

1.2 Table 1 of Practice C 183 prescribes the number of tests per lot for most properties. After a quality history, as described in 9.5.1 of the practice, has been established, testing may be carried out at the reduced rate as long as no pairs of results produce an average in the critical range. The critical range is so defined that when the average of a pair of tests is outside the range, there is a 95% probability that if a large number of samples were tested, fewer than 5% would fail the specification limit. It is based on the average range of results of pairs of samples accumulated during the development of the quality history. Where the specification limit is a minimum value, the critical limit is placed above the specification limit by an amount equal to the average range multiplied by 2.49. The critical range includes all possible test results with values below the critical limit. Where the specification limit is a maximum value, the critical limit is below the specification limit, and the critical range includes all results above the critical limit. The critical limit is intended to ensure that the mean of the population of test results is removed from the specification limit by at least 1.645 standard deviations, that being the value corresponding to a failure rate of 5 %. Since the standard deviation of a sampling distribution of averages of pairs is equal to the population standard deviation divided by the square root of 2, the critical limit, which is the 95% confidence limit, must be removed from the desired position mean by a distance equal to $1.645/\sqrt{2}$ or 1.163 standard deviation. Thus, the distance between the critical limit and specification limit must be the sum of 1.645 and 1.163, or 2.808 standard deviations. Since, for a sample size of 2 the standard deviation is exactly equal to 0.886 range, the distance between the critical limit and the specification may be expressed as 2.49 average ranges.

1.3 Chapter 8 of The American Concrete Institute document, Guide to Selection and Use of Hydraulic Cements, ACI 225.1 R,[3] is a review of sampling and testing procedures of hydraulic cement that are commonly found in the industry. The various aspects of sampling, testing, and reporting of data are discussed and may provide useful information on the several quality control and acceptance procedures that are commonly used.[4]

[3] American Concrete Institute, P.O. Box 19150, Detroit, MI 48219.

[4] For additional information see the paper by Dywer, J. R., and Young, R. N., "Inspection of Portland Cement," *Concrete*, Vol 21, August and September 1992, pp. 56 and 95. This paper has been filed at ASTM Headquarters and may be obtained by requesting RR:C1-1002.

2. Preparation of Samples for Test

2.1 Prior to testing, a sample should be thoroughly mixed and passed through the 850-μm (No. 20) sieve or any other sieve having approximately 20 openings per inch, and brought to room temperature. All foreign matter and lumps that do not pulverize easily in the fingers must be discarded.

FINENESS BY THE 150-μm (No. 100) AND THE 75-μm (No. 200) SIEVES

(See Test Method C 184)[2]

3. Accuracy of Sieve Tests

3.1 The requirement for 150 μm (No. 100) and 75 μm (No. 200) sieves has been deleted from the ASTM specification for portland cement. These fineness tests may be useful for research or special purposes.

3.2 In order to make fineness determinations satisfactorily, the operator must possess considerable skill, which can be acquired only by practice. With proper care, an experienced operator should check himself within 1 %, and generally within 0.5 %, on portions of the same thoroughly mixed sample on the same sieve in duplicate tests. In general, it may be stated that if an accuracy of 1 % or better is required, it is necessary that standard calibrated sieves be used.

3.3 It is first necessary to remove the flour from the sample before beginning the sieve test itself. This is most easily done by suspending the sieve between the fingers of each hand and shaking it loosely back and forth, alternately striking the palms of the hands. As the sifting progresses it will be noted that the residue on the screen assumes a darker color because the lighter colored flour has been removed. This procedure also prevents blinding the screen with flour so that a cleaner and more decisive end point can be reached promptly. Five to ten min of sifting will clean the sample very adequately, at which time it will be noted that the material passing through the sieve also becomes darker. Before beginning the 1-min shaking tests, it is important to see that no flour has blinded any of the screen openings. When performing the 1-min shaking test, each operator should experiment to determine the angle from the horizontal at which the sieve is to be held. When the proper angle has been reached, the residue distributes itself across the screen during each stroke so that most effective screening action is obtained. It is also important that the sieve is moved against the hand held in the upright position. Some operators take liberties with the test and strike the sieve with the palm of the open hand. Screening is not complete when this is done. The weight of residue will tend to be high. The rate at which the sieve strikes the palm is important. Take a full 10 s for each 25 strokes in each of the six positions. Do not try to do the 125 strokes and the five rotations in 1 min. When this is done the stroke rate is speeded up and there is a tendency to strike the palm less hard, which again tends toward high residues.

4. Design of Sieves

4.1 It is essential that all sieves used in the tests of fine materials be well designed and constructed and that they be provided with closely fitting covers and pans in order to avoid the loss of small quantities of material during the sieving operation.

5. Care of Sieves

5.1 Sieves should be examined frequently for defects in the cloth. Holes or breaks are sometimes indicated by very noticeable irregularities in the end-point test weighings. Holes are sometimes found in the soldered joint between the sieve and frame. Soldering must be done with great care so that the wire cloth is not injured by excessive heat or by harmful fluxes. Sieves should be kept thoroughly clean and dry. The battering of edges of sieve frames and pans should be avoided, for this will interfere with the proper fitting together of pan, sieve, and cover. The brush used for cleaning the sieve should be stiff enough to clean cloth, but not so heavy as to injure it. Be careful that the brush is not so held that the metal binding will drag across the sieve cloth and press objectionable ridges in it.

5.2 New sieves should be cleaned with a suitable solvent to remove any grease before making tests. Alcohol is not recommended because it may attack the lacquer on the sieve frames.

FINENESS BY THE TURBIDIMETER

(See Test Method C 115)[2]

6. Difficulties

6.1 Some operators have found it difficult to oscillate the tank the full 180° about a horizontal axis through the center of the tank. Complete rotation is necessary for proper agitation. A design of a rotatable holder is available on request from the Cement and Concrete Reference Laboratory at the National Institute of Standards and Technology, Gaithersburg, MD 20899.

6.2 Occasionally, a standard fineness sample may be unsuitable for calibration purposes. Whenever an unexpected difficulty is encountered, a new sample should be opened. When opening a sample, contamination with glass chips should be avoided. One satisfactory method of opening a vial consists of inverting the vial and pressing the sides in the empty section against a thin curved high-resistance wire mounted across the terminals of a small transformer.

6.3 The derivation of the Wagner equation is based on a light intensity (I_o) of 100 microamperes (μA). For some unexplained reasons, the light intensities arrived at by calibrating with NIST SRM 114 have shown a great increase with some I_o's being 150 or more. In addition to the calibration being time consuming and laborious, test results are subject to increasingly larger errors as the I_o deviates further from 100 μA.

6.4 The development work in the Fineness Subcommittee of ASTM Committee C-1 showed that the revised method showed a large reduction in the scatter of results of participating laboratories in the CCRL Proficiency Sample Program. The work also indicated that Wagner fineness results are independent of sample size when the I_o is set at 100 μA.

7. Care of Apparatus

7.1 Great care must be taken to keep the various parts of the apparatus clean, and to maintain good electrical contacts. This last precaution especially relates to the electric bulb, rheostat, and the wire leads to the shelf. It is particularly important to keep the water cell clean. At times the cell glasses will become dirty. A wavy film of growth will sometimes develop in the cell, causing decided fluctuations in the intensity of the transmitted light when the water is disturbed by the movement of the shelf. Keep the cell completely filled with distilled water. A drop or two of potassium dichromate solution added to the water in the cell will serve to inhibit possible algae growth but will not be enough to discolor the water.

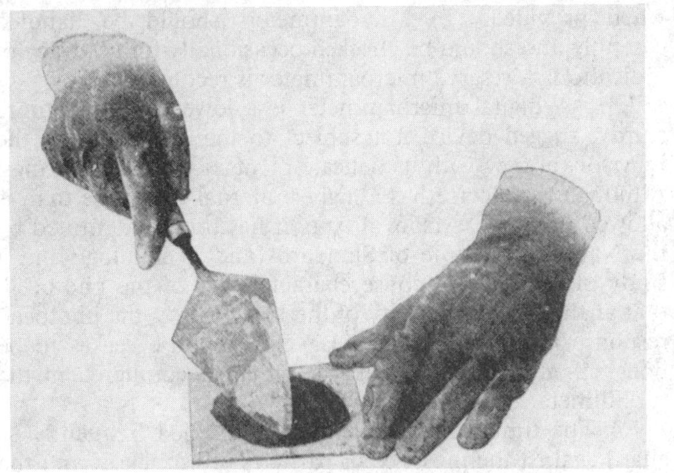

FIG. 1 Correct Method of Forming Cement Pat

7.2 The tank support must be so adjusted that, when using the specified volume of suspension the top of the suspension will be at the level of the zero mark on the graduated scale of the turbidimeter case, and the tank will be in a perpendicular position. If a new tank is put into use, the adjustment of the tank support should be checked for elevation.

7.3 D'Arsonval microammeters are very sensitive instruments, and are necessarily made up of very delicate parts. (*1*) They should be used with great care since pivots, jewels, and springs are subject to damage due to rough handling. (*2*) Microammeters should not be placed where they will be exposed to vibration. One should not hammer on a bench or table on which a microammeter has been placed, since one severe shock may permanently damage pivots and jewels. (*3*) Care should be taken that static charges do not exist on the face or case of the instrument while in use. Such a static charge may result from rubbing the glass face with a cleaning cloth, and may be eliminated by breathing on the affected part. Errors may result if the instrument is read while placed on a table top made of magnetic material or near some strong external magnetic field. This has been done inadvertently when a magnifier was used, and the microammeter set on the metal plate which was the magnifier stand. (*4*) Slight friction in the moving system may be eliminated by lightly tapping the case before taking a reading. Hard tapping defeats its purpose and may damage pivots and jewels. (*5*) When not in use, instruments should be stored in cases free from dust, acid fumes, excessive heat, strong magnetic fields, moisture and vibration. Binding posts should be tightened—this prevents their loss and helps to keep the contact surfaces clean. (*6*) Terminals of a microammeter should be shorted while the instrument is in transit, and locking devices for moving systems should be used for their assigned purposes when provided. (*7*) Microammeters should be handled carefully and should be checked occasionally for accuracy of indication. A reserve microammeter is recommended.

7.3.1 A digital microammeter is a lower cost and more sturdy, rugged device not subject to the difficulties of the D'Arsonval type with its delicate pivots, jewels, and springs. Although this device has a higher internal resistance than is allowed for the D'Arsonval type, it has been determined by the National Institute of Standards and Technology and a study of the light response characteristics of the photocell, that at the lumen intensity of the light source, the photocell response is linear in that area. The response ceases to be linear at much higher light levels than encountered in the turbidimeter.

7.4 The timing buret should be examined frequently to guard against the presence of particles of lint, glass, and the like in the capillary. A filter made of 45-μm (No. 325) wire cloth is recommended for the top of the buret, and a cover should be placed over the top of the buret when it is not in use. The dimensional requirements for the buret are shown in Fig. 1 of Test Method C 115, for Fineness of Portland Cement by the Turbidimeter.[2]

7.4.1 If a marked buret is not available an unmarked one may be used by placing a tape lengthwise along the barrel of the buret where the markings usually appear. Drainage times may be calculated if the viscosity of the kerosine is known, or the drainage times may be measured with a stop watch by timing the flow from a buret known to be correct. Lines are then marked on the tape corresponding to the measured drainage times.

7.5 The stirring brush should be examined for dimensions and condition. Although there is no specified distance between the brush and the inside of the test tube, the distance is critical for stirring action. Restandardization is necessary when a new brush is installed or a different test tube is used for agitating the sample. The contour of the bottom of the stirring brush should match the contour of the test tube. Many stirring brushes when received do not meet this requirement and it may be necessary to trim the bristles in order to achieve this condition.

7.6 Many of the difficulties experienced with the turbidimeter stem from electrical troubles. (*1*) Corrosion can occur in electrical contacts and wires of rheostats. Some turbidimeters ground the light to the movable shelf through the reflector housing and the wire from the power source is terminated with a spade connector fastened to the bottom of the movable shelf. This can be a hidden source of corrosion and looseness. (*2*) Connecting wires develop hidden breaks under constant flexing from raising and lowering the shelf. (*3*) Looseness in connecting wires, the light bulb and its socket, and the reflector in its holder can cause flickering or a reduction in light intensity. (*4*) Storage batteries should be kept well charged. (*5*) A light bulb, after prolonged usage, can become darkened and emit light of a different color or intensity which changes the instrument characteristics. (*6*) Voltage regulators, which some laboratories use to replace storage batteries, are subject to malfunctioning and failures.

7.7 Kerosine should not be used more than once. There are some brands of specially processed kerosines on the market. These are usually water white and have an odor differing markedly from the characteristic odor of customary market grade kerosine. **Caution**—Caution should be used in substituting specialties for the market grade of kerosine since differences in viscosity and density will influence results at other than the calibration value. If difficulty is encountered in calibration, a new supply of kerosine has been known to solve the problem.

7.8 It should be remembered that the basic premise of the Wagner turbidimeter is that the standard light intensity is adjusted to a value of 100 μA. Current standardizing procedures "force" the I_o regardless of condition of the photocell or light bulb or a suspending liquid of incorrect density or viscosity. By setting the light intensity near 50 μA with the filter out of position and then measuring the resulting intensity through the filter, the ratio can be used to calculate the theoretical I_o through the filter. If the theoretical I_o differs too greatly from the I_o determined in the standardization procedure, there is the possibility that considerable errors could creep into the calculation of fineness for high-fineness materials. The digital microammeter with its range of 199.1 μA allows the theoretical I_o to be measured directly without using the above ratio method.

7.9 Linearity of response of the photocell is also basic to proper operation. As photocells age their response to higher intensity of light may decrease. When a new cell is installed, set the intensity of light with the filter removed at a value near 50. Also measure the intensity with the filter in place and record both values. After a cell is in use, to see if the response has changed, set the intensity through the filter at the observed value. After the filter is removed, a substantial decrease in value from the original reading indicates that the cell should be replaced. This drop in response, although not influencing results at or near the calibration value, will show a trend toward calculated high values for high fineness products.

7.10 The parabolic reflector should remain clean, bright, and highly polished. Cleaning may be accomplished by using lens tissue or a very soft, lint-free cloth. Avoid touching the reflector surfaces with the fingers. The lens over the reflector assembly should be clean and free from fingerprints. A lamp, after being placed in the socket, should be wiped clean.

7.11 When the lamp filament is at the focus of the mirror, the rays of light will be approximately parallel and the response of the photocell should be greatest. Turning the cell in the hood could vary the response. Rotate to obtain maximum response. Rotating the bulb will also aid in "peaking-in" before locking the bulb in position. When this has been accomplished, a piece of paper placed in front of the hood of the photocell will show an image of the slot in the shield. The edges of the spot will be reasonably straight and the image will have a reasonably uniform distribution of light.

7.12 The turbidimeter should not be located near a source of variable heat such as a radiator, hot plate, or autoclave. Good laboratory planning should include the turbidimeter location in the constant temperature room.

7.13 No light-colored wiring or shiny pieces of metal should be used in the interior of the turbidimeter cabinet. Black gaskets should be used to seal the ends of the water cell.

7.14 Both linseed oil and oleic acid change with age or by exposure to heat and light through oxidation. If difficulties arise in calibration, a fresh supply of dispersing agent should be obtained. Dropping bottles commonly used to dispense the agent should be kept free from the thickened jelly-like oxidation products. Many laboratories use a small 1 or 2-oz dropping bottle and keep a primary supply in a container which is not opened frequently.

7.15 Scratches on the face of the tank may affect transmission of light. The tank should be kept clean. An oily film accumulates on the interior tank faces. Removal is facilitated with a lint-free wiper, or a synthetic rubber "squeegee." One product used successfully is a piece of lintless foam on a stick.[5]

FINENESS BY THE 45-μm (No. 325) SIEVE

(See Test Method C 430)[2]

8. Spray Nozzle

8.1 Care of the spray nozzle is important. In areas where considerable lime is present in the water, the holes of the spray nozzle have a tendency to become plugged with lime which will alter the flow rate. When cleaning a nozzle, wires should not be used due to the possibility of enlarging the holes. Dilute hydrochloric or acetic acid should be used to avoid chemical attack on the brass from which the nozzles are usually constructed.

8.2 A spray nozzle is deemed acceptable if the flow rate is between 1500 and 3000 g/min at 10 psi (69 kPa). No single hole in the perforated plate should be large enough to accept a No. 74 drill (0.0225 in. or 0.57 mm). **Caution**—Caution should be used when checking hole diameters with a twist drill because all holes in the plate, except the center one, are drilled on an angle of either 5 or 10°, and if the drill is not inserted at the proper angle there is a possibility of enlarging the hole with the drill which is made from a much harder material.

8.3 *Care of Apparatus*—Ultrasonic means for cleaning sieves is practical, convenient, and thorough. Small ultrasonic cleaners having a capacity of about 1 pt (500 cm^3) and rated approximately 40 W, similar to those used for cleaning watch parts or drafting pens, are priced reasonably. Trial and error methods must be used to determine the proper cleaning time and the best cleaning solution. A typical cleaning cycle for a sieve used on testing cement is 15 s using a dilute HCl (1 + 10) solution. Other detergents may be used. Remove the screen and rinse immediately with distilled or deionized water. More cleaning time may be necessary but this is a matter of individual requirements. Two-piece sieves appear to clean better if the sieve is disassembled and the screen cleaned separately. This is not necessary each time, however. Some laboratories have large ultrasonic cleaners for cleaning cube molds, etc. To use this apparatus it is suggested that the sieve be immersed in the cleaning solution in a large beaker which in turn is placed in the large ultrasonic tank filled with the usual cleaner.

8.4 *45-μm (No. 325) Sieve Cloth*–After considerable research into the reasons for high correction factors for the 45-μm sieves available, it was determined by Subcommittee C01.25 on Fineness that the limits in Specification E 11, for Wire-Cloth Sieves for Testing Purposes,[2] are not tight enough to ensure a low correction even if Specification E 11 is met. Accordingly, if an operator has trouble in obtaining a reasonable sieve correction of 10 % or less, the reliability of the sieve should be checked using the "picket fence" approach discussed in Appendix X1.4 of Specification E 11. In order to check a sieve, a 350 line to the inch interference grating such as furnished by W. S. Tyler can be used. A piece of cloth with 45-μm openings and average wire diameter will have 339 openings. If the number of openings is within ± 10 of the nominal 339, a sieve correction of 10 % or less should be obtained.

FINENESS BY THE AIR PERMEABILITY TEST

(See Test Method C 204)[2]

9. Air Permeability Fineness Test

9.1 Details of this procedure are well covered in ASTM Test Method C 204, for Fineness of Portland Cement by Air Permeability Apparatus.[2] The following remarks supplement those instructions:

9.1.1 The U-tube manometer should be filled to the lowest mark with the manometer fluid. The liquid should be at this midpoint after any liquid on the side of the manometer has had sufficient time to drain. Immediately after a test, the liquid will be slightly below the mark, due to the fact that a small portion remains on the glass. This will not affect the results of a subsequent test. Slight variations in the height of this liquid may be noted at different temperatures. Do not attempt to adjust the quantity of liquid for different temperatures. Silicone stopcock lubricants should not be used with this apparatus. Each new lot of filter paper disks should be carefully checked for conformance with specification requirements.

9.1.2 When preparing the bed of cement in the cell the side of the cell should be tapped sufficiently to level the bed. This leveling should be confirmed by visual inspection, prior to placing a filter paper disc on top of the cement.

9.2 The latest revision of Test Method C 204 substitutes the quantity $(b - \epsilon)$ for the quantity $(1 - \epsilon)$ in the denominator of equations 5, 6, 7 and 8 of Section 6 and the value of $(b - \epsilon_s)$ or $(b_s - \epsilon_s)$ for $(1 - \epsilon_s)$ in the numerator of the same equations. The purpose of these changes is to eliminate inadequacies in the former porosity function. Data indicate that differences in the fineness test results with changes in porosity are considerable when using the former function. The new function nearly eliminates the differences caused by changes in porosity.

9.2.1 The value of the constant, b, for a material other than portland cement can be determined from air permeability tests at different porosities (4 or 5 should be ample) on that material. Plot the value of $\sqrt{\epsilon^3 T}$ on the horizontal axis against ϵ on the vertical axis. The porosity value, b, is the value of ϵ where a straight line drawn through the points intersects the zero $\sqrt{\epsilon^3 T}$ value.

9.2.2 Although the 0.9 value has been determined for portland cement and new values determined by experiment for other materials, work in ASTM Subcommittee C01.25 on Fineness demonstrated that for blended cements, granulated blast-furnace slags, silica flour, and fly ashes, that the constant of 0.9 in the porosity function in all cases gave

---

[5] This item is marketed by Texwipe, 51 Prospect Place, Hillsdale, NJ 07642.

better correlation than the use of the unity constant.

TEMPERATURE AND RELATIVE HUMIDITY, STORAGE EQUIPMENT

10. Laboratory

10.1 The ASTM standard test methods require that the temperature of the cement testing laboratory and the dry materials shall be not less than 20°C (68°F) nor more than 27.5°C (81.5°F). The temperature of the mixing table and apparatus should be within these limits. The temperature of the mixing water must be maintained at 23 ± 1.7°C (73.4 ± 3°F). The relative humidity of the laboratory shall be not less than 50 %. In general, increase in temperature is accompanied by decrease in setting time and usually by accelerated increase in strength. Drafts of air and low relative humidity will cause rapid evaporation of water from the mixes. Numerous laboratories have installed automatically controlled equipment for close control of room conditions. Because the temperature and relative humidity of the laboratory have a great effect upon the behavior of the mixes, it is essential that they be controlled within close limits.

10.2 The laboratory temperature and humidity requirements have been developed for technical reasons and are an integral part of the testing system, and are not for the comfort of the cement tester.

11. Moist Closet and Moist Room

11.1 A well-made moist closet or moist room is essential in portland cement testing in order to maintain both the required temperature of 23 ± 1.7°C (73.4 ± 3°F) and a relative humidity of at least 95 %. Recirculating water sprayed through ducts, or water curtains flowing down inner walls, as well as fog sprays, have been used for maintaining the required humidity in cabinets and moist rooms. When fog sprays are used, precautions should be taken to prevent water falling upon the surfaces of the freshly molded specimens. A recording thermometer (see Section 13), preferably of the rapid-air wet and dry-bulb type, is recommended as a means of indicating the performance of the cabinet or moist room. Pans of water in the top of the closet, and pieces of wet wicking on inner surfaces of walls, may prove helpful in maintaining the necessary humidity by bringing large areas of wet surfaces into contact with the air. Numerous cabinets are available completely equipped with automatically controlled conditioning units.

11.2 It is important to remember that the moist cabinets and moist rooms should be well insulated if they are exposed to surrounding air of substantially different temperature. Plenty of water does not necessarily mean high relative humidity in a cabinet. If the necessary cooling of a poorly insulated cabinet is done by cold water, it may mean that the differential between the air and water temperatures will be such that the relative humidity will be less than the minimum relative humidity allowed. Refrigerator cooling coils are best immersed in the conditioning water system, because when they are exposed they tend to remove moisture from the air. Recirculation of water, when employed, must be sufficiently rapid to ensure that there is no great temperature differential between the water and the cabinet air.

11.3 The doors of the moist cabinet should be well-fitting and should be kept closed except during the actual operation of introducing or withdrawing the test specimens. It is desirable to have separate compartments for the time-of-set test specimens and the specimens for other tests, because of the necessity of frequently opening and closing the doors to make observations on the time-of-set test specimens. The shelves on which freshly molded specimens are placed should be level.

11.4 Specimens in storage should both look moist and feel moist.

12. Storage Tanks

12.1 Storage tanks for test specimens are required to be of noncorroding material, and may or may not contain racks for holding the test specimens. Test specimens should be completely immersed during the water storage period. In many instances, and particularly during the summer, the temperature of the tap water exceeds the allowable maximum. The use of some form of temperature control device to keep the storage water temperature within the specified range (23 ± 1.7°C) (73.4 ± 3°F) at all times is recommended, and is necessary in most locations. Placing the storage tanks in temperature-controlled moist cabinets or moist rooms sometimes solves the problem, provided the feed water to the tanks is maintained within required temperature limits. If the tanks are located in a conditioned laboratory room, tank covers must be provided in order to lessen the humidifying effect which the large area of water surface would have on the room air and also to avoid the lowering of the tank water temperature below the room temperature because of evaporation.

12.2 Storage water shall be saturated lime water. A small cloth bag of high-calcium hydrated lime placed in each tank should keep the lime water in a saturated condition. This water should not be used as mix water in any test procedure. Storage water and tanks should be clean and free of any substance that might possibly affect the test specimens. The water should be changed as required. Continuous running water, or demineralized water, has been known to lower strength results due to excessive leaching and should not be used in storage tanks. Test specimens containing materials possessing accelerating or retarding effects should not be placed in the same storage tank with other test specimens. When the storage tanks are independent of moist closets or moist rooms, it is recommended that the tanks be equipped with a recording thermometer with its bulb in the storage water.

13. Recording Thermometers and Hygrometers

13.1 Recording thermometers are recommended for the laboratory air, moist cabinet, and moist room, and also for the storage tanks if the latter are separate from the moist storage equipment. Recording hygrometers are advisable for the laboratory. Recording wet- and dry-bulb thermometers are well suited for the moist cabinet and moist room, but they should be of the "ventilated" type, that is, provided with rapidly circulating air over the bulbs. Recording instruments for moist cabinets and moist rooms should be of the distant bulb type, so that the recording mechanism and chart are not in the moist air.

MIXING CEMENT PASTES AND MORTARS

(See Test Method C 1005[2] and Specification E 617[6])

14. Weights and Weighing Devices

14.1 The balance and the weights used in weighing the materials for the mixes should conform to the requirements of the respective standards, and should be kept clean and in good order at all times. Occasional check tests should be made, including the checking of any graduated beam on the scale. Appropriate accuracy requirements for graduated beams or dials on scales can be found in Test Method C 1005. Special single weights of sizes suitable to certain tests, such as 250, 300, 500, 750, and 900 g, may be used, both for convenience and to minimize the probability of errors in weighing the batches. Appropriate tolerances for weights between 1 and 1000 g, but not listed in the test methods, may be interpolated from the tables given in the methods. Weights should be checked occasionally and should be kept in blocks or other suitable containers to minimize wear. Pockets of weight containers should be kept free of sand and dirt. Balances should not be loaded beyond their capacity, which should be marked on the balance. Tops of balance tables should be plane to guard against balances easily sliding out of position.

15. Glass Graduates

15.1 The mixing water should be measured in a glass graduate of the capacity and type specified in the applicable test method. If not definitely specified, the capacity should not be more than twice the volume of water required in the mix. A 200-mL cylindrical graduate, reading in one direction only, is particularly recommended as suitable for use in the tests for normal consistency, tensile strength, time of setting, and autoclave expansion. Graduates should be checked to see that they conform to the requirements for accuracy of delivery. In some cases a very large error has been found—as much as several millilitres. Graduates that have been tested and found satisfactory should be marked appropriately. Rubber guards around the tops of the graduates will greatly aid in reducing breakage.

16. Glass Burets

16.1 Some laboratories have found burets of the dispensing type, which discharge rapidly, very satisfactory for measuring the mixing water. However, such burets should be so mounted as to discharge directly into the mix, and not into an intermediate container which involves possible loss of water from the mix. The use of rubber delivery tubes with pinchcocks, which can readily shift from one position to another, may introduce appreciable error.

17. Testing Graduates and Burets

17.1 The National Institute of Standards and Technology *Circular C434,* "Testing of Glass Volumetric Apparatus," gives information on specifications, special requirements, tolerances, and other matters relating to the testing of glass graduates and burets. The following information will be helpful to operators in calibrating water-measuring apparatus:

17.2 A temperature of 20°C (68°F) is regarded as standard for testing glass volumetric apparatus. The relation of height to diameter should be such that the graduation marks are not less than 1 mm apart and also that the graduated height is at least five times the inside diameter. Graduations should read in one direction only and should be of uniform width, continuous, finely but distinctly etched or engraved, and perpendicular to the axis of the cylinder. The main graduation lines of cylinders should extend entirely around the circumference and should be numbered. The least graduations should extend at least one seventh of the way around, and the intermediate graduations should extend at least one fifth of the way around. The numbers indicating the capacity of the graduate should be placed immediately above the marks to which they refer.

17.3 Graduates should be calibrated with water to deliver the indicated volume at 20°C (68°F). In all apparatus where the volume is limited by a meniscus, the reading is made on the lowest point of the meniscus. Apparatus must be sufficiently clean to permit uniform wetting of the surface, and the entire interior of the vessel should be wetted in filling. Apparatus must be allowed to drain a sufficient time before readings are taken.

17.4 In testing graduates and burets, the capacity of any interval may be determined by weighing the water delivered by the interval in question. The water is delivered into a weighing flask of convenient size provided with a stopper. The temperature of the water is observed either immediately before or immediately after delivery from the apparatus.

17.5 In making the weighings, it is both convenient and accurate to use the method of substitution. By this method a constant tare is kept on one pan of the balance, while on the other pan is placed the object to be weighed and with it sufficient weights to secure equilibrium.

17.6 Methods based on the transfer of water from a standard container, such as a buret, into the instrument to be tested can be utilized for items such as cylinders.

18. Gloves

18.1 Rubber gloves should always be used when handling cement pastes and mortars. Gloves should be well-fitting and comfortable, and flexible enough to permit easy movement of the fingers.

19. Mixing Water

19.1 In general, water fit for drinking is satisfactory for cement testing. In case of doubt, distilled water may be used for comparison or reference. The amount of water present in a mix is of great importance, particularly with regard to the setting time and strength. Water much in excess of the required quantity will prolong the setting time and lower the strength. Less than the proper amount of water may shorten the setting time and result in strength variations from the normal, the latter sometimes being due to the difference in molding properties of the mortar. The amount of water used is expressed in millilitres, is usually measured by volume, and must be carefully measured.

[6] *Annual Book of ASTM Standards,* Vol 14.02.

20. Time of Mixing

20.1 A timer, preferably one with a sweep-second hand, should be so located that the operator can conveniently observe the time.

21. Size of Batch

21.1 The size of the batch is closely related to the possible thoroughness and uniformity of mixing, other conditions being the same, and is therefore not without its effects on the results of the tests. Small batches may lose relatively large amounts of water by evaporation. On the other hand, with the same amount of time and mixing, much larger batches may not be so thoroughly or so uniformly mixed. Moreover, the longer time required to mold the larger batches into specimens may result in considerable evaporation of water, stiffening of the cement through setting action, and an accompanying decrease in the ease of molding and in the uniformity of the results.

22. Mechanical Mixing (See Practice C 305)[2]

22.1 *Mixer*—The mechanical mixer should be inspected frequently, and kept clean and oiled. Care should be taken to prevent leakage of oil from the gear housing falling into the bowl during mixing operations. This will change the consistency and greatly affect the tests. Before the bowl is raised to the mixing position and the mixing operation started, a check should be made to ensure that both the paddle and bowl are firmly seated. Care should be taken in removing the paddle from the mixer, as the bowl can be easily dented if the paddle is dropped. If dents appear in the bottom of the bowl, they should be removed. Filling the underside of the raised portion of the bowl bottom with solder reduces denting, thus increasing the useful life of the bowl.

22.2 *Bowl Clearances*—Proper paddle to bowl clearances are necessary to make the tests accurate and reproducible. When the clearance is too great, nonuniform mixing results. When the clearance is too tight, grinding action reduces the size of the standard sand, which changes consistency and results. The tight clearance also causes excessive wear of the paddle and the bowl, necessitating frequent replacement. Small adjustments can be made by use of the adjustment bracket specified in Practice C 305. Major adjustments can be made by raising or lowering the motor housing.

22.2.1 There are several important factors in monitoring the specified clearances between the paddle and the mixing bowl. The assembly, shaft and paddle should be visually inspected for wear each time clearances are checked or adjusted.

22.2.1.1 The bushing, which holds the paddle shaft in place in the paddle gear assembly, must be properly snug during operation. By moving the paddle-shaft assembly by hand, the condition of the bushing can be determined. This hand movement of the paddle-shaft assembly should only be checked with the power cord to the mixer disconnected. Remove the bowl from the mixer and place the paddle on the shaft. Try to move the paddle by hand with a push-pull motion toward the center of the assembly. Movement or eccentric "play" in the assembly above the shaft indicates a worn bushing.

22.2.1.2 If the shaft remains still, while the paddle visibly moves during the hand movement, then three possibilities should be considered: the shaft has become worn and unable to properly support the paddle during mixing; the pin holding the paddle in position on the shaft has become worn and is no longer able to retain the paddle in the correct position to achieve the specified clearance; or the bore hole in the paddle has become enlarged and the paddle will no longer fit snugly on the shaft.

22.2.1.3 Another potential problem is the spring on the paddle shaft which holds the paddle in place during mixing. The spring may be damaged over a period of normal use. This seldom occurs but should not be overlooked. A damaged spring will be especially noticeable during the mixing procedure.

22.2.2 It is recommended that a set of tolerance gages be available for checking the critical bowl to paddle clearance. These gages can be made from the appropriate size drill rods. Long-shanked Allen wrenches of the proper size can also be used. Three sizes are required as described in the Apparatus section of Practice C 305.

22.3 *Markings of Bowls and Paddles*—Since most bowls and paddles are not quite identical, once a set has been adjusted for clearance, it should be marked for identification.

22.4 *Procedure*—The following are some of the conditions that can cause error:

22.4.1 Incomplete drying of the bowl and paddle between tests,

22.4.2 Excessive heating or cooling of the bowl and paddle during washing, which will affect consistency,

22.4.3 Failure to follow correct sequence of mixing operation, and

22.4.4 Loss of material during addition. (The introduction of sand into the mixer bowl can be facilitated by the use of a funnel.)

22.5 In any case requiring a remixing interval, any mortar adhering to the side of the bowl shall be quickly scraped down into the batch with the scraper prior to remixing.

23. Hand Mixing

23.1 The requirements for hand mixing have been deleted from the ASTM specifications for portland cement. The procedures may be useful for research and special purposes.

23.2 *Mixing Slab*—The height of the mixing table slab above the floor may have a great influence on the efficiency of hand-mixing and molding operations. A height of about 36 in. is correct for the average operator. For unusually tall or unusually short operators, the height of the table top should be adjusted accordingly. The slab should be plane and made of nonabsorbent material. Steel plate at least ¼ in. in thickness, or plate glass about ¾ to 1 in. in thickness, is recommended for the tops of mixing tables. The mixing table should not be subjected to undue exposure to sun and drafts, or be placed too near radiators, hot plates, steam apparatus, etc. The slab must be kept clean and free of cement and oil. Glass plates are frequently chipped and steel plates nicked by the practice of tapping the trowel corners edgewise and sharply downward on the slab to remove mortar from the trowel. Placing unusually hot or cold objects on glass mixing slabs may crack the glass.

24. Mixing of Cement, Sand, and Water

24.1 In some cases the specifications require that the

batch must be thoroughly mixed dry, on the slab, prior to the introduction of the water. The materials should not be scattered widely, but should be neatly and thoroughly mixed by turning with the trowel and by rubbing them lightly on the mixing table with the hand until the mass shows by its uniform appearance that the ingredients are well mixed. This method will cause less segregation than rolling the dry materials on a paper. A crater which will hold the entire desired quantity of mixing water without overflow or loss should be then formed in the dry materials.

24.2 The mixing water must be added exactly as required by the test method. When mixed by hand on the slab, the whole of the mixing water must be poured at one time into the crater in the dry materials. The dry materials should then be turned, as quickly as possible, from the outside into the crater by means of a trowel, covering all the water in not more than 30 s. It should be noted that this 30-s interval for covering is the maximum time allowed for this operation. A skilled operator will usually finish covering in 15 or 20 s, and then the timing of the 30 s for absorption should be started immediately. The absorption time is always 30 s, regardless of the time required to complete the covering operation. For example, if the covering is completed in 15 s, the total elapsed time for the two operations, covering and absorption, should be 15 + 30, or 45 s.

24.3 The amount of water in a mix may be unintentionally increased by using wet gloves, or by pouring the dry materials onto a wet slab. Loss of water may be somewhat avoided by mixing the materials and the water so as to spread as little as possible of the mix over the surfaces of the gloves or slab.

NORMAL CONSISTENCY

(See Test Method C 187)[2]

25. Vicat Apparatus

25.1 The Vicat apparatus should be inspected frequently and kept clean, free of rust, and oiled. The plunger and needle ends should be free of oil or grease during a test. The weight and diameter of the plunger assembly should be checked to determine whether they conform to the requirements of ASTM Test Method C 187, for Normal Consistency of Hydraulic Cement.[2] The plunger should be removed from the frame and weighed with the proper assembly of parts. Some Vicat apparatus are equipped with separate weights for use with needle or plunger. The plunger and needle diameters should be measured with a micrometer at the end and at several points over their penetration portions. The plunger and needle should be longer than the height of the mold, and must be straight with end surfaces flat and perpendicular to the axis. The edges should be true and sharp and the surfaces free of rust. The plunger should be clean, lightly oiled at contact with the frame, and should slide freely through the guides. If plunger and needle assembly require considerable adjustment, weights should be rechecked after such adjustments have been made.

25.2 The top and bottom inside diameters of molds, as well as height of mold, should be measured at several points.

26. Test Method

26.1 For details of mixing in the determination of normal consistency, see Test Method C 187. If a trial paste is too dry, it should not be remixed with more water but should be discarded, and a new sample should be taken in order to avoid error resulting from evaporation and other causes. The tossing referred to in Section 6.2 of Test Method C 187 is comparable to tossing a ball from one hand to another with the free travel of the ball being approximately 6 in. (150 mm). The tossing of the paste from hand to hand is considered to be one of the most effective methods to form the paste into a spherical shape, enabling easy insertion into the Vicat ring, with a minimum amount of additional sample manipulation.

26.2 After the correct Normal Consistency has been determined, the test specimen may be used for the Vicat Time of Set test, the remaining paste may be used for the Autoclave test and the Gillmore Time of Set test.

27. Care in Determination

27.1 The accuracy of the normal consistency determination depends upon the proper mixing and placing of the cement paste in the mold, removal of the surplus with as little disturbance of the mass as possible, freedom of the apparatus from vibration, and careful manipulation throughout the test. Care should be taken to center the test specimen under the plunger. Take initial reading and release plunger immediately. The time for operations performed between the end of mixing and release of the plunger shall not exceed 30 s. Care must also be taken to avoid shaking the Vicat apparatus when releasing the plunger and to avoid leaning upon or shaking the table during the test. The operator must possess considerable skill and dexterity in order to make this test satisfactorily. It is a very important determination, however, because upon it depends the percentage of water that is to be used for certain other determinations.

AUTOCLAVE EXPANSION

(See Test Method C 151 and Specification C 490)[2]

28. Autoclave

28.1 If the autoclave does not meet the specified heating rate, one of the following may be the cause:

28.1.1 Low input voltage, or

28.1.2 Need of a new heating element.

28.2 If the autoclave does not meet the specified cooling rate, the cooling can be increased by providing more ventilation for the autoclave and by allowing a fan to blow on the autoclave after the heat is shut off. An accumulation of dirt in the space between the pressure chamber and the outer shell may retard cooling.

29. Use of Thermometers

29.1 The section on Safety Precautions (6.2) of Test Method C 151 for Autoclave Expansion of Portland Cement, requires that a thermometer always be used, together with the pressure gage, so as to provide a means of detecting any failure of the pressure gage to act properly and to indicate any unusual conditions. The thermometer is to be placed in the thermometer well which is specified in Test Method C 151. Neither mercury nor asbestos shall be used in the well due to health hazards. Oil is not recommended for use in the well. If a laboratory considers that conditions require the use of some liquid in the well, use might be made of one of the liquid organic phosphates of the type used as plasticizers in the plastic industry. Dow Plasticizer No. 5 has been found satisfactory for temperatures above that prescribed for the autoclave test in Test Method C 151.

30. Reference Bar

30.1 The steel reference bar mentioned in Specification C 490 should be used at least at the beginning and end of the readings made within a half day. Such use of the reference bar may seem like an unnecessary refinement for routine tests, particularly in those cases where test specimens are measured in temperature-controlled rooms. Nevertheless the use of the reference bar will guard against errors caused by unsuspected changes in position of the dial, and will permit measurements to be made even when the comparator may have undergone adjustment or replacement during the autoclave test. After the bar is put into use, the ends should not be rubbed with abrasives, but should merely be wiped with a cloth. The ends should be protected against corrosion or injury, and the length of the bar should be carefully determined occasionally to detect any appreciable change. The bar should be so marked that the same end can always be kept uppermost when the bar is used. The central 4 in. (100 mm) should be covered with a rubber insulating grip.

31. Rupture Disk and Vent Valve

31.1 In Test Method C 151 a rupture disk is the pressure relief safety device specified for use with the autoclave used in this method. State and local regulations should be checked regarding the use of rupture disks for this purpose. Section 4.5.1 of C 151 describes the autoclave rupture disk specified in 4.5.

31.1.1 Care should be taken in selecting the disk material. Of the materials available for making disks, Inconel has the flattest temperature/pressure curve retaining approximately 93 % at 420°F (216°C) of the rated burst pressure at 68°F (20°C). The holder material for the disk may be either carbon steel or stainless steel. Disks may have a relief area of 0.5 to 1 in. A 1-in. relief is considered more than adequate. The manufacturer should be consulted about the particular application.

31.1.2 Rupture disks cannot be tested without destructive bursting. A periodic inspection/replacement is recommended at least yearly as a method of preventive maintenance. If there is any reason to suspect corrosion or the autoclave is used heavily, a more cautious approach should be taken by replacing the rupture disk every six months until a history is established. Then, if there are no problems, an annual replacement may be used.

31.1.3 There are applications in industry where a rupture disk is placed in line ahead of the relief valve as protection against corrosion. Although this arrangement does not appear in Test Method C 151, some manufacturers recommend the practice to ensure an operable valve at all times while reducing maintenance.

31.1.4 The American Institute of Chemical Engineers in their publication 54-B, "Selection of Rupture Discs for Emergency Relief of Chemical Process Equipment," describes the sizing, selection, and installation of rupture disks. This publication is recommended for those designing or operating a high-pressure steam vessel used in cement testing.

31.1.5 Rupture disks are actually differential pressure devices. If a disk is rated to burst at 100 psig, it will burst at 100 psig process pressure, only if the downstream pressure is 0 psig. If the disk is back loaded to 50 psig, it will not rupture until the process reaches 150 psig. It is important, therefore, to ensure that no pressure pockets can build in the piping between double-disk assemblies or disk-relief valve assemblies. This is generally accomplished by drilling the hold-down flange and installing an excess flow valve, pressure gage, or combination of the two. There is an ASME code requirement covering this matter.

31.2 During the early portion of the heating period, the vent valve is left open to permit air to escape from the chamber. A rubber tube, one end of which is attached to the vent and the other end of which is submerged in a beaker of water, will serve to indicate when the escape of the air is completed. A few drops of kerosine placed in the vent valve about once a week will aid in keeping the needle clean and in good working condition.

31.3 **Safety Precuations**—Experience has prompted Committee C-1 to express the desire that publicity be given to the safety precautions that should be observed when operating cement testing autoclaves (see Section 6 of Test Method C 151). Safety Precautions in Test Method C 151 will alert the operator to potential safety hazards.

32. Cooling of Test Specimens After Autoclaving

32.1 Test Method C 151 specifies that upon removal from the autoclave at the end of the test period, the test specimens should be placed in water the temperature of which exceeds 194°F (90°C), after which the water in the bath is to be cooled to the measurement temperature at a uniform rate by the addition of cold water. A cylindrical container approximately 8 in. (200 mm) in diameter and 16 in. (400 mm) in height, equipped with a bail, makes a very convenient chamber for the cooling operation. In use, the container is filled with water and placed on a hot plate at the end of the 3-h autoclaving period. Later, when the autoclave is opened, the rack and bars as a single unit may be lifted from the autoclave by means of a suitable metal hook and immersed in the near-boiling water in the container. The container may then be transferred to a sink, and cold water introduced through a rubber tube connected to a cold water spigot. The rate of cooling may be controlled by regulation of the flow of cold water through the tube.

32.2 It is to be noted that the temperature of heated tap water is seldom greater than 194°F (90°C).

TIME OF SETTING

(See Test Methods C 191 and C 266)[2]

33. Vicat and Gillmore Needles

33.1 The cement specifications permit the use of either the Vicat apparatus or the Gillmore needle, as specified. Weights of needle assemblies, and the straightness and diameters of needles should be checked frequently to determine whether they conform to the respective requirements of the applicable ASTM test methods; see also Section 25 of this manual. It is sometimes found that even new needles are not standard either as to weight or diameter.

34. Method of Making Gillmore Specimens

34.1 The method of making Gillmore test specimens is described in Test Method C 266.[2] The shape and size of the pats should be as specified in Test Method C 266, and the pats should be formed as shown in Fig. 1 of this manual. Clean, dry, unoiled, plane glass plates should be used. There should be just enough troweling to give the proper shape and a smooth surface. The specified thickness should be observed closely. An overtroweled surface or a rough surface may materially affect the apparent time of set. The top of the pat should be brought to a smooth horizontal surface by a single stroke of the trowel. The flat surface of the pat should be parallel to the surface of the glass.

35. Storage of Specimens

35.1 Specimens should be stored in the moist closet or moist room during the determination of time of setting. It is usually necessary to take the specimens from the moist storage when a determination is being made. However, they should not be left in the air of the laboratory but returned to the moist closet as soon as possible after the trial test has been made.

36. Test Method

36.1 The surface of the test specimen must be well lighted throughout the test. The needle must be held vertical and applied perpendicular to the surface without shock. This is especially difficult in the case of the Gillmore needle, and great care is necessary. The use of a frame, as illustrated in Fig. 1(*b*) of Test Method C 266, will assist in this regard. In determining the end point, the needle, if not applied perpendicular to the surface, will give a crescent-shaped mark some time after the setting time should have been recorded.

STRENGTH TESTS (GENERAL)

37. Storage of Specimens in Moist Closet or Moist Room

37.1 It is intended that the molds containing test specimens should be stored so that the upper surfaces of the specimens are freely and uniformly exposed to the atmosphere of the moist closet or room. Molds should not be piled one upon the other in such a way that the plates of the upper molds exclude air from the specimens beneath. The specimens should be kept in moist storage until they have attained the age of 24 h, when they are to be placed in the storage tanks. In no case should they be removed trom the molds until they are at least 20-h old. If removed from the molds before they are 24-h old they should be replaced on the shelves of the closet until the expiration of the 24-h period. The removal from the molds should be done with care and with as little disturbance as possible, as specimens at the end of 24 h are not very hard and rough treatment may produce injuries which, although not apparent, will nevertheless affect the strengths.

37.2 When the freshly molded specimens are placed in the moist storage, each mold should be accompanied by an identification tag. On no account should identification depend on the position of a mold on the shelves or its proximity to some identified mold.

37.3 When the specimens are removed from the moist storage, it will be found very desirable to mark the reference numbers of the bottom faces of the specimens, after the plates have been removed but while the specimens are still in the molds. A graphite crayon, when used on these oily bottom surfaces which have been in contact with the mold plates, will form very distinct and durable markings. Colored crayons are not recommended because their marks may quickly fade under the circumstances. The practice of using a sharp tool to scratch identification marks on the surfaces of mortar specimens is objectionable.

38. Precautions to be Observed in Testing Specimens

38.1 Specimens, except those to be tested at 24 h, must be tested as soon as removed from the storage water. Specimens to be tested at 24 h should be taken directly from moist storage and tested at once without previous immersion in water. When several 24-h specimens are to be tested at one time, it is recommended that they be removed from moist storage, placed in a pan, and kept covered with a moist cloth until ready for test. Indications of injuries, defects, or unusual appearances should be noted. When two or more specimens, other than 24-h specimens, are to be tested at one time, they should be removed from the tanks, placed immediately in a pan of water, and kept there until actually tested. As it is important to prevent changes in temperature, it is recommended that specimens be covered with water from the storage tank.

TENSILE STRENGTH

(See Test Method C 190)[7]

39. Briquet Molds

39.1 The requirements for tensile strength have been deleted from the ASTM specifications for portland cement. This procedure[7] may be useful for research and special purposes.

39.2 Briquet molds must be of the form specified in Figs. 1 and 3 of ASTM Test Method C 190, for Tensile Strength of Hydraulic Cement Mortars,[8] and should meet the requirements for all dimensions given in Section 4.4 of Test Method C 190. Molds should be frequently checked to determine whether they conform to the tolerances specified. A three-gang mold is the most convenient to handle. Clamps should be of a type which will close the joints in the molds completely so as to ensure briquets of specified width (see Fig. 1 of Test Method C 190). There should be enough metal in the sides of the molds to prevent distortion during use. The molds, well cleaned, should be examined to see whether the halves are properly mated and the clips are of such type and condition as will hold the halves tightly together. The widths at waist line should be measured, preferably and most readily, with "go and no-go" gages.

39.3 Compare the molds with the standard outline of

---

[7] Those interested in this procedure should refer to Test Method C 190, 1990 *Annual Book of ASTM Standards*, Vol 04.01.

[8] Discontinued; see *1990 Annual Book of ASTM Standards*, Vol 04.01.

briquet. Considerable difference in outline near ends may be encountered, which is important in case weights of briquets are being compared. The mold outlines should conform closely with the standard at the points of contact with the rollers of the testing machine grips. As these contact points will be about ½ in. (13 mm) above and below the center line of briquets, the briquet mold outline should compare closely with the standard in these regions.

39.4 The greatest thickness of each half of new molds should be measured at points between inner and outer edges at the waistline of the briquet, opposite each briquet space.

39.5 The molds should be cleaned soon after the hardened mortar test specimens have been removed; otherwise, the cleaning will be more difficult. Brushes having bristles made of brass, fiber, or other material softer than the molds are recommended for use in cleaning briquet molds.

40. Testing Machines

40.1 The type of testing machine should be such as to conform strictly with the requirements of Test Method C 190 as regards rate and uniformity of application of the load and other details. The clips should be provided with rollers ½ in. (13 mm) in diameter, with full bearing of 1 in. (25 mm) on the briquet, and mounted so as to permit easy turning during the test. (See Fig. 2 of Test Method C 190.) Many laboratories do not observe the latter detail sufficiently, with the result that eccentric strains may be produced in the briquet and erratic results may be secured. Moreover, the distance between the rollers in the clips should not vary appreciably, and the clips should be so designed as to keep the rollers in parallel positions during the test. The bearings of the rollers should be well lubricated. Petrolatum or cup grease will be found quite effective. Grit in the bearings or lack of lubrication often results in a jerky or nonuniform operation of a machine, and may produce undue stresses in a specimen. Long use will sometimes so roughen or deform the rollers that they must be replaced. Daily balancing of the beam is a good habit to acquire. The testing machine should be thoroughly cleaned before examination. Knife-edges and bearings should be inspected. Stirrups and knife-edge bearings should be free of nicks, dents, or worn spots. Knife-edges should be straight, sharp, and tightly held in beams. The various parts of the machine should be adjusted so that the beam balances and swings freely at zero load. The method of balancing the machine will vary according to the type and model. Individual parts of the machine should be calibrated as follows:

40.2 Check grips with respect to condition, lubrication, dimensions, and spacing of rollers. (See Fig. 2 of Test Method C 190 for details of important dimensions.) The rollers should be 0.50 ± 0.01 in. (12.7 ± 0.3 mm) in diameter and should be approximately 1⅛ in. (28.6 mm) in length. The distance between the roller faces should be 1.25 ± 0.02 in. (31.8 ± 0.5 mm), and the rollers should be parallel within 0.02 in. and should be in the same horizontal plane. The bore of the rollers should be straight and the rollers should bear on the pins throughout their length.

40.3 Check pivots with respect to center of briquets when in grips, particularly with respect to the front and back of specimen.

40.4 Make at least three determinations of the rate of shot flow, using not less than 30-s intervals and timing by the second hand of a watch. The result should conform to the tolerances of the specifications. Inability to secure sufficient flow rate may be due to dirty or badly worn shot or to stoppage in outlets. A small quantity of flake graphite will assist in preserving the shot and facilitating flow.

40.5 Load tests may be made by means of a lever, or by dead weights placed upon a support suspended from the upper grip of the machine either by a saddle straddling the base of the machine or by a rod. In the latter method a metal briquet is placed in the upper grip and a rod, passing through the base of the machine, is screwed into this briquet. A support sufficiently large to hold test weights is attached to the lower end of the rods. It is necessary to remove the lower grip and "take-up" gearing to accommodate the rod.

40.6 A small, good-grade spirit level, firmly secured to the top of a test lever near the fulcrum before balancing the beam, may prove preferable to the separate reference pointer. The weight of the level must be included as part of the weight of the lever assembly.

40.7 Loads should be applied in increments of not more than 100 lbf (445 N) and the weight of shot required to counterbalance each load carefully determined. Make several determinations of the amount of shot required to balance the machine at each loading. After removing the loading lever or other device, ascertain what load indications are obtained with the amount of shot required for each of the loadings used. If the machine has both a spring scale and graduated beam device, record the readings by both devices. If the indicated loads are not within the tolerance, examine the various weights of the machine for discrepancies. If the error is constant for the various loadings, look for alterations in the machine, defective counterweights, and see whether proper balance was obtained at zero load. If the error varies, examine bearings, rider weights, and spring scale weighing device. The latter may be tested for irregularity in operation or graduation of dial by applying small increments of load and determining whether the indicated dial readings are proportional to the load applied. The weight of the test lever, suspension devices, etc., must be taken into account when calculating the test load. The dead weights should be accurately known and not assumed to be exactly as indicated.

40.8 If the testing machine cannot be corrected within the specified tolerances by minor adjustments, it is recommended that it be returned to the manufacturer for proper adjustments.

41. Molding of Briquets

41.1 The mold plates should be of sufficient thickness to prevent bending during molding and handling of the briquets.

41.2 It is recommended that the thumb pressures be applied in groups of two per briquet on each of six trips along the mold. Care should be taken that the thumb pressures are applied entirely to the surface of the mortar, and not to the surface of the mold. The specified thumb pressure of between 15 and 20 lbf (67 and 89 N) may be measured by molding the briquets on a platform scale or on other suitable apparatus that will indicate the pressure exerted. In the smoothing operation, the mortar should be finished flush with the top surface of the mold, using the flat bottom surface of the trowel to smooth the mortar, and exerting not more than 4 lbf (18 N) pressure. The trowel should be held so that the long axis of the blade is at a right angle to the length of the mold. Sand grains should not be allowed to remain on the face of the mold to which the plate is to be applied, as this may result in thick specimens and also may cause a loss of water between the mold and plate. Care should be taken to see that the briquets when furnished are exactly flush with the surface of the molds. The joints of the molds should be tightly closed in order to ensure specimens of correct width. Many operators find that a slightly curved trowel is of great asssistance in making briquets because of the ease with which it can be used to clean off the mold faces and smooth the mortar within the molds. Only mineral oil should be used in oiling molds and plates.

42. Precautions to be Observed in Testing Briquets

42.1 Briquets must be carefully centered in the clips; otherwise large variations in the stresses and consequently in the results are probable. Briquets should be inserted in clips so that the smooth surfaces that have been in contact with the mold plate are turned toward the back of the clips and placed in contact with the guide bar at the back. Bearing surfaces of the briquets and of the rollers should be clean, smooth, and free of sand or grit, and the rollers should be well lubricated.

42.2 After the briquet is properly centered in the clips, just enough load should be applied by hand, before starting the mechanically applied load, to take up the play of the machine and prevent jerky application of the load. Very weak briquets may not permit taking up the play of the machine, and must be handled carefully. Adjustments of the load control mechanism are often needed and at least daily trials should be made before tests are commenced, to determine whether the load is being applied at the specified rate of 600 ± 25 lbf (2670 ± 111 N) per min. A quantity of flake graphite added to the shot hopper will aid the smooth and uniform flow of the shot.

COMPRESSIVE STRENGTH

(See Test Method C 109)[2]

43. Cube Molds

43.1 Particularly important is the requirement for planeness of mold faces. When sealing the molds to the plates, care should be taken to seal the joints and openings at the ends of the molds. The mold plates should be of sufficient thickness to prevent bending during molding and subsequent handling of the filled molds. When cleaning cube molds, care should be taken to preserve the planeness of the interior faces. It is recommended that steel brushes not be used on any softer metal molds for which brushes with brass or fiber bristles should be used. Molds should be cleaned promptly after cubes are removed. There should be no excess of grease or oil on the interior faces or in the bottoms of the molds after they have been sealed to the plates.

44. Compression Testing Machine

44.1 The machine should be periodically tested to determine whether the error in load indication, over the working range involved in testing the 2-in. (50-mm) cubes, is within the 1 % allowed by Test Method C 109. Compression testing machines should be frequently inspected, and the following are among the items that should be checked (not all of these items are applicable to all machines):

44.1.1 *All Machines*—smooth application of load, surface of lower platen protected by use of lower bearing block, power switch within easy reach of machine operator, availability of manufacturer's literature and drawings relative to the machine, information as to time and results of the last verification of the machine, and proper fuses on power line.

44.1.2 *Screw Loaded, Beam and Poise Machines*—weighing levers in proper positions and free-moving parts of machine, proper adjustment of counterweights on weighing beam, loading screws well lubricated and straight, ability to balance machine at zero load, and clutches and gear shifts in good working order.

44.1.3 *Hydraulic Machines*—freedom of movement of gage hands, proper "gag" in machine fitted with hydraulic capsule weighing system (gap should be measured at least once each month), ample oil in hydraulic loading system, hydraulic systems free of air, and proper operation of automatic cutoffs.

44.2 Compression machines should be regularly lubricated. Lack of proper lubrication of loading screws may result in "jumpy" application of load. A mixture of grease and flake graphite has been used successfully for this purpose. This mixture should be well brushed into the threads, the head of the machine being moved up and down during the lubricating process.

45. Bearing Blocks

45.1 Particular attention should be given to the planeness of faces of the bearing blocks, their diameter, and the proper centering and freedom of movement of the upper, spherically seated bearing block.

FLOW TABLE

(See Specification C 230)[2]

46. Flow Table

46.1 The flow table top should be maintained in good condition. Avoid defacing the top with metal tampers or other hand implements. The top must be clean, dry, and free of grease or oil when a flow test is made. In addition to maintaining the top in good condition, the table top must be level. This task can be accomplished by placing 4-in. square metal shims of various thickness between the cork pads and the floor. The leveling should be checked at several points of reference on the table top. Due to possible compression of the cork pads, leveling should be checked periodically and adjusted, if necessary. The table should be operated a dozen or more times just prior to use if it has not been operated for some hours. The vertical shaft should be cleaned frequently and reoiled with a light oil. Oil or dirt should not be present between the contact faces of the table top and frame. Oil on the cam face will lessen wear and promote smoothness of operation. It is important to keep reducing gears of the motor drive mechanism packed with suitable grease. The flow mold should rest evenly on the table top. A kitchen spoon or similar tool is useful in transferring mortar from the mixing bowl to the flow mold.

46.2 Specification C 230 states that the performance of a flow table is to be judged by the results obtained from flow tests made using a suitable calibration material. If the results of these tests are considered to be unsatisfactory, a thorough examination of the apparatus should be made. Some important points other than those mentioned above to be considered during this examination are as follows:

46.2.1 Flow tables that do not conform to the general design requirements of C 230 are unreliable.

46.2.2 A pedestal must contain dense concrete in good condition. Pedestals, when new, should be moist-cured 28 days before use.

46.2.3 Flow tables located on flexible floors, such as old wooden floors, frequently fail to give satisfactory service.

46.2.4 The use of shims between table base and plate, or plate and pedestal, is undesirable.

46.2.5 Excessive rotation of the table top is undesirable. This problem may be caused by a convexity on the end of the vertical shaft.

46.2.6 The cam should meet the shaft smoothly, and the shaft should drop without hitting the cam.

46.2.7 The accuracy of the flow caliper, the weight of the table top, the height of drop, the rate of drop, and the planeness of the contact surfaces have an influence on test results.

46.2.8 Clearances between shaft and bore must be within specified limits both at the top and bottom.

46.2.9 Full contact between table base and metal plate is necessary. Check for the following:

46.2.9.1 Oil, rust, dirt, paint, etc., in the joint between the base and the plate,

46.2.9.2 Planeness of bottom surface of table base and top surface of plate, and

46.2.9.3 Nonplaneness in vicinity of tapped holes in the metal plate.

46.2.10 Full contact between the metal plate and the pedestal is necessary. The plate should not overhang the pedestal and there should be no visible spaces between them.

AIR CONTENT OF MORTAR

(See Test Method C 185)[2]

47. Test Procedure

47.1 The procedure for determining the air content of hydraulic cement mortar, Test Method C 185, was changed in 1988. The new procedure requires tamping each layer with a tamper rather than spading with a spatula. The new test method will produce air content values about 1 % lower than the old method.

48. Equipment

48.1 The measure should be kept scrupulously clean since accumulations of hardened cement or mortar on the inside of the measure will change the calibrated volume and tare weight leading to erroneous results. Accurate calibration is essential since any error will be reflected directly in the determinations of air content. Care should be used in handling the measure to avoid dents or other damage that will change the volume. Some measures are two piece, having a bottom that is press fit or soldered into the measure. This bottom should be checked periodically for a secure fit since any movement will change the volume of the measure.

48.2 The straightedge, tamper, spoon, and tapping stick shall conform to the requirements in the Test Method.

49. Preparation and Testing of the Mortar

49.1 Because of the nature of this test, the accurate batching of all materials and the cleanliness and grading uniformity of the standard sand are of great importance in producing accurate results. This is also true with respect to the mixing procedure, flow determination, and the procedure for filling and weighing the measure. Close adherence to the timing and temperature requirements of the method are of extreme importance for the same reasons. It is because of the importance of each of these operations that the procedure for each is described in more explicit detail in the test method than is normal for some other less critical test methods. Departures from the stated procedures will give inaccurate and noncomparable results.

50. Causes of Variations Attributable to Test Procedure

50.1 *Use of an Incorrect Amount of Cement, Sand, or Water*—If water requirements for all cements seem to be consistently in error, the flow table and mechanical mixer should be thoroughly checked for conformance with the requirements of applicable specifications.

50.2 *Failure to Observe Specified Time Limits on Test Operations*—The desired accuracy cannot be obtained if the time limits on various test operations are disregarded.

50.3 *Failure to Use Specified Equipment in Performance of Test*—Particular attention is invited to the advantages of using the specified spoon, tamper, tapping stick, and straightedge in the filling of the 400-mL measure.

50.4 *Failure to Place Mortar Gently in the Measure*—The technique of throwing mortar into the measure does not conform with standard practice.

50.5 *Improper Handling of the 400-mL Measure*—Shaking or jolting of the measure during the filling operation must be avoided.

50.6 *Variations in Strike-Off Procedure*—The use of too few sawing strokes with the straightedge results in a tearing of the top surface of the mortar in the measure and the use of too many results in the formation of a ridged surface. A total of 10 to 12 strokes in each direction is recommended.

50.7 *Variations in Weighings*—In many instances, bench weights will give more uniform results than the beam and poise with which many balances are now equipped.

51. Air Entraining Potential of Testing Sands

51.1 Difficulties have been experienced with Test Method C 185 in some laboratories due to the variable air entraining potential of the testing sands. These difficulties have usually taken the form of values for air content that are abnormally high and may be greater than the specified maxima in specifications for hydraulic cements. The air entraining potential of the testing sands may be reduced by washing by decantation with potable water until a clear supernatant liquid is obtained and then drying the sand before using in the test procedure.

STANDARD SAND

(See Test Method C 778)[2]

52. Gradation

52.1 *20–30 Sand*—A natural silica sand from Ottawa, IL or Lesueur, MN, used primarily in testing for the percentage of entrained air in hydraulic cements.

52.2 *Graded Sand*—A natural silica sand from Ottawa, IL used primarily in testing hydraulic cements for strengths.

52.3 A number of other types of sand are prepared by the processing plants. Therefore, it is advisable for every laboratory to check each new shipment to ascertain that sand conforming to the requirements of the cement testing sand for 20–30 sands or graded sands has been received.

CALIBRATION AND CARE OF APPARATUS

53. Calibration and Care of Apparatus

53.1 It is important that the apparatus used in cement testing be kept in good condition at all times. Apparatus should be inspected at frequent intervals to see that it is functioning properly and that it conforms to the requirements of the standard method involved. In such inspections, the laboratory should refer to the related paragraphs in this manual, and also to the latest applicable ASTM standard test methods. The good appearance of laboratory equipment is conducive to good work and therefore should not be neglected. Periodic inspection of equipment is recommended.

CEMENT AND CONCRETE REFERENCE LABORATORY

54. Cement and Concrete Reference Laboratory

54.1 Since 1929, there has been maintained at the National Institute of Standards and Technology a Research Associate Program currently known as the Cement and Concrete Reference Laboratory. The CCRL is a cooperative project of the government and the American Society for Testing and Materials, sponsored by ASTM Committees C-1 on Cement and C-9 on Concrete and Concrete Aggregates. Its primary activities are inspection of laboratories and distribution of comparative test samples. At the request of the supervisor of a cement testing laboratory, the CCRL will schedule an on-site inspection of the procedures and equipment used by the laboratory in performing the standard physical tests for hydraulic cement, or register the laboratory for participation in a cement testing proficiency sample program covering the physical or chemical tests, or both for portland, masonry, and blended hydraulic cements. Charges for these services are made in accordance with fee schedules established by the sponsors. Inquiries should be addressed to the Cement and Concrete Reference Laboratory, National Institute of Standards and Technology, Building 226, Room A365, Gaithersburg, MD 20899.

54.2 In addition to this service, it is suggested that each laboratory check its own apparatus and methods at regular intervals. Correspondence relative to standard test methods and apparatus will be welcomed by the Cement and Concrete Reference Laboratory.

TRAINING OF CEMENT TESTERS

55. Outline of Instruction

55.1 Considerable experience is necessary before an operator is able to secure consistent results in repeated tests on the same cement sample. Therefore, before assuming the responsibility of making tests which involve the acceptance of cement, the operator should be thoroughly trained in all details of the standard tests. The following outline of instruction is suggested:

55.1.1 Study thoroughly the various ASTM cement specifications and test methods, and also this manual. Training video tapes on various cement tests are available from the Portland Cement Association, Skokie, Illinois. Before using any training aide it should be thoroughly checked for agreement with current ASTM specifications and test methods.

55.1.2 Make normal consistency tests on several samples from the same lot of cement (at least 20 samples) until in successive tests the water requirements agree within 0.5 %. This will give practice in mixing and in preparing the sample for the determination of time of setting by the Vicat method. The normal consistency test is very important, for the quantity of mixing water used in some of the other tests depends on this determination.

55.1.3 Practice making pats until test specimens of the specified dimensions can be secured with a minimum of effort. Not less than 20 to 30 trials should be made. The top surface of Gillmore pats should be horizontal.

55.1.4 Make five test specimens each for time-of-setting tests by the Vicat and the Gillmore methods and determine the setting time by each method. Between intervals consumed in making observations of time of setting, the operator may continue his practice on other tests.

55.1.5 Make Blaine air-permeability fineness tests on several samples of each of a number of cements of different finenesses until the results of duplicate determinations on any given cement, made on separate beds, agree well within 2 % of each other.

55.1.6 Make Wagner turbidimeter fineness tests on several samples of each of a number of cements of different finenesses. Reference to some of the preceding comparative work in his own laboratory should readily give an operator a good idea of the proficiency which he has attained.

55.1.7 Make compressive strength tests for four ages (1, 3, 7, and 28 days) on each of at least ten samples from the same lot of cement.

55.1.8 Autoclave tests should be made on five different days of at least three samples from the same lot of cement.

55.1.9 Make several air content determinations on a sample of non-air-entrained and on a sample of air-entrained cement, varying the water-to-cement ratio to produce flows within the range from 80 to 95 %.

55.1.10 Having attained a reasonable proficiency in a given test, the new operator should make parallel determinations with a skilled operator on five different days on five samples of cement. The samples should be from different lots of cement, preferably including different brands and types. Comparison between the skilled operator's results and his own will furnish the new operator with a good measure of his ability to make the various tests. Attention is again called to the advantages to be gained by regular, frequent interchange of samples for test by different laboratories.

CEMENT TESTS FOR RESEARCH PURPOSES

56. Research Tests

56.1 In research and particularly important tests, reliance should not be placed on single tests for fineness, normal consistency, time of setting, soundness of pats, and autoclave expansion. The results should be based on at least two, and preferably three or more, tests.

56.2 When making mortar tests of cement for research, the specified methods of mixing, molding, storing, etc., should be carefully followed. In order to obtain desirably representative results in research or unusually important work, it is not sufficient to test only three strength specimens of a kind made from the same batch. At least five test specimens should be made for each mix, consistency, age, curing conditions, etc., on different days. For example, if it is desired to test specimens at ages of 1, 3, and 7 days, 3 months, and 1 year, five specimens may be taken from a batch, one for test at each of the above ages. Additional batches should be mixed on the different days so that the results of tests of the independent specimens may be averaged for the final result for each age. By following this procedure, accidental variations in the tests are distributed so that no one set of specimens is unduly affected.

56.3 Experience and tests have shown the tensile strength of briquets, Test Method C 190, is not a satisfactory indicator of the concrete-making properties of cements. Test Method C 190 is used primarily by those interested in research on methods of determining tensile strength of hydraulic cements. For research in the area of strength and strength potential, it is suggested that compressive strength tests be made on both concrete and standard mortar. For the mortar tests, Test Method C 109 should be followed. (See Section 39 on Tensile Strength in this manual.)

SAFETY

57. Safety in the Cement Testing Laboratory

57.1 Safety must be one of the first considerations when designing a laboratory, training technicians, and in the daily operation of the laboratory. A modern laboratory often utilizes intricate combinations of energy, materials, and apparatus. Hazardous chemicals, radioactive materials, X rays, high-voltage equipment, and compressed gases have been introduced into the cement laboratory in recent years. Reliable equipment and prime operating techniques are imperative for the safe and efficient performance of laboratory duties.

57.2 Existing testing facilities should be inspected to be certain that proper work areas and environment conditions are adequate for existing equipment. New apparatus should be installed in the laboratory only after necessary modifications have been made to the laboratory to accommodate the demands of the new equipment. A systematic maintenance and safety inspection system should be developed and utilized in each testing facility. The maintenance and operating instructions provided by the equipment manufacturer and current operating procedures should be readily available for each item of equipment. A record of repairs, adjustments, and service should be kept for each piece of apparatus.

57.3 Good safety practices must be insisted upon in every laboratory. Safety must be included in training programs for new personnel. Some of the ASTM test methods include safety requirements as part of the test method. These instructions and requirements should be rigidly followed.

57.4 In all cases the proper OSHA, MSHA, state, and local safety regulations must be followed. These are often minimum requirements in some areas, and should not prevent the cement testing laboratory from using initiative and resourcefulness in developing safer working environments.

57.5 The laboratory is often looked upon as a resource center for providing technical information on the handling, storage, and use of certain chemicals and materials that require special attention. Resource information on hazardous materials likely to be found in the facility must be readily available. A list of agencies and individuals who could provide specific and precise information on hazardous materials and conditions must be on hand for emergencies. Most laboratories and testing facilities are members of societies and trade associations such as the Portland Cement Association for cement producer member companies who have safety departments and established safety programs. The cement testing laboratory should avail themselves of these resource centers and facilities.

57.6 A bibliography of books and publications that have been found to be useful in the cement testing laboratory is provided at the end of this manual as a guide for selecting reference materials on laboratory safety.

57.7 The standards described in this manual may involve the use of hazardous materials, operations, and equipment. It is the responsibility of whoever uses these standards to establish appropriate safety practices and to determine the applicability of regulatory limitations prior to use.

57.8 **WARNING**—Chemical abuse of substances is a factor in the health, safety, and welfare of a worker in a cement testing laboratory. There are several chemicals and materials common to the laboratory and routinely used that can alter human behavior in an undesirable manner when misused intentionally or accidentally consumed. These substances must be identified so that a system for regulation and control can be provided. Access to certain substances must be restricted. A periodic inventory is required to monitor the accidental loss or intentional misuse of the materials. A substance abuse program shall be included in the health and safety program of the laboratory.

CHEMICAL ANALYSIS OF HYDRAULIC CEMENT

(See Test Methods C 114)[2]

58. Chemical Analysis

58.1 Method of Analysis:

58.1.1 The scope of Test Methods C 114 says, in essence, that any method may be used for analysis of hydraulic cement provided that it is demonstrated that the methods, as used by the individual analysis, achieve required levels of precision and bias. The required way to accomplish that demonstration is spelled out in Section 3 of the standard. It requires testing of the methods used with the current series of

Certified National Institute of Standards and Technology (NIST) Standard Reference Materials (SRM) Cements.

58.2 Requalification of a method (3.3.6 of Test Methods C 114) is required at least once every two years (3.3.6.1) or when there is evidence that results of the required precision and bias are not being obtained (3.3.6.2 and 3.3.6.3).

58.3 Material in Section 4 of Test Methods C 114, General, applies to all methods. It provides general information on precautions, interferences and limitations, apparatus and materials, reagents, and sample preparation. General procedures such as calculation and rounding of figures, and the recommended order of reporting analysis are found in this Section.

58.4 Material in Section 5 and succeeding sections of Test Methods C 114 are, as noted in the Scope, not required for analysis of cements. They are in the books solely to serve as a guide for those who desire to use them. They are the classical wet chemical methods for analysis of cement proven over many years to be capable of providing precise and accurate results when used with proper care. They require no expensive instruments and can be run successfully by almost any normally equipped chemical laboratory. Practice and experience with these methods is very important since many of the procedures are technique dependent.

COMPOSITIONAL ANALYSIS OF HYDRAULIC CEMENT

(See Specification C 150)[2]

59. Compositional Analysis

59.1 Possible confusion exists between Specification C 150 and Test Methods C 114 in regard to calculation of the Bogue potential compounds required in Specification C 150 for determination of conformance to specifications.

59.2 Specification C 150 requires, at the present time, use of the traditional value $R_2O_3 - Fe_2O_3$ to be called "Al_2O_3" or "alumina" for purposes of calculating the potential compounds. Specification C 150 also states that Test Methods C 114 must be used. Since Test Methods C 114 allows any method of demonstrated precision and bias to be used, any method may be used which legitimately obtains the value $R_2O_3 - Fe_2O_3$ for the purpose of calculating the potential compounds. This means that when traditional wet chemistry is used, the value $R_2O_3 - Fe_2O_3$ may be used directly as obtained. If this is the case, and that result is reported as Al_2O_3 on the report of analysis, it must be noted that that value is "$R_2O_3 - Fe_2O_3$".

59.3 When methods which determine true Al_2O_3 are used, values for minor oxides such as TiO_2 and P_2O_5 must be added to true Al_2O_3 in order to obtain the "Al_2O_3" ($R_2O_3 - Fe_2O_3$) desired for calculation of potential compounds. While the values for TiO_2 and P_2O_5 (and in some cases, ZnO) may be determined on the individual sample, this will usually not be necessary. With rare exception, cements from a given source will have nearly constant values for the minor oxides traditionally included in "Al_2O_3". Long and short-term differences normally are on the order of a few hundredths of a percent by weight of the total sample. These differences are insignificant when compared with the precision and bias of the Al_2O_3 determination which is on the order of 0.2 %.

59.4 When historical data is available as to the levels of the minor oxides in cements from a given source, those values may be used to correct true Al_2O_3 to "compound Al_2O_3" except when there is good reason not to do so.

BIBLIOGRAPHY

Kayser, D. S., "Selection of Rupture Discs for Emergency Relief of Chemical Process Equipment," Paper 54b, 64th Annual Meeting, San Francisco, Calif., Nov. 28, 1971, American Institute of Chemical Engineers, New York, NY

"Safety Precautions for Oxygen, Nitrogen, Argon, Helium, Carbon Dioxide, Hydrogen, Acetylene, Ethylene Oxide, and Stimulant Mixtures," F34996, Linde Division, Union Carbide, Buffalo, NY

Radiation Safety for X-Ray Diffraction and Fluorescence Analysis Equipment, NBS Handbook III, National Bureau of Standards

Allison, W. W., *Compressed Gases,* Recommended Safety Guide, SC-M-68-378, Sandia Office of Industrial Cooperation, Albuquerque, N. Mex.

Matheson Gas Data Book, Matheson Gas Products, East Rutherford, NJ

The Safe Handling of Compressed Gases, Compressed Gas Assn., New York, NY

Hazardous Chemicals and Reactions, National Fire Protection Assn., Boston, MA

Sax, N. I., *Dangerous Properties of Industrial Materials,* 3rd edition, Reinhold Publishing Co., New York, NY, 1978.

Steere, N. V., *Handbook of Laboratory Safety,* The Chemical Rubber Co., Cleveland, Ohio.

Safety in a Chemistry Laboratory, SC-R-66-959, Sandia Office of Industrial Cooperation, Albuquerque, N. Mex.

Guide for Safety in the Chemical Laboratory, 2nd Edition, Safety and Fire Protection Committee of the Manufacturing Chemists Assn., D. Van Nostrand Co., Princeton, NJ, 1972.

SI QUICK REFERENCE GUIDE:
International System of Units (SI)
*The Modernized Metric System**

UNITS

The International System of Units (SI) is based on seven fundamental (base) units:

Base Units

Quantity	*Name*	*Symbol*
length	metre	m
mass	kilogram	kg
time	second	s
electric current	ampere	A
thermodynamic temperature	kelvin	K
amount of substance	mole	mol
luminous intensity	candela	cd

and a number of derived units which are combinations of base units and which may have special names and symbols:

Examples of Derived Units

Quantity	*Expression*	*Name*	*Symbol*
acceleration			
angular	rad/s^2		
linear	m/s^2		
angle			
plane	dimensionless	radian	rad
solid	dimensionless	steradian	sr
area	m^2		
Celsius temperature	K	degree Celsius	°C
density			
heat flux	W/m^2		
mass	kg/m^3		
current	A/m^2		
energy, enthalpy			
work, heat	N·m	joule	J
specific	J/kg		
entropy			
heat capacity	J/K		
specific	J/(kg·K)		
flow, mass	kg/s		
flow, volume	m^3/s		
force	$kg \cdot m/s^2$	newton	N
frequency			
periodic	1/s	hertz	Hz
rotating	rev/s		
inductance	Wb/A	henry	H
magnetic flux	V·s	weber	Wb
mass flow	kg/s		
moment of a force	N·m		
potential,electric	W/A	volt	V
power, radiant flux	J/s	watt	W
pressure, stress	N/m^2	pascal	Pa
resistance, electric	V/A	ohm	Ω
thermal conductivity	W/(m·K)		
velocity			
angular	rad/s		
linear	m/s		
viscosity			
dynamic (absolute)(μ)	Pa·s		
kinematic (ν)	m^2/s		
volume	m^3		
volume, specific	m^3/kg		

*For complete information see *IEEE/ASTM SI-10.*

SYMBOLS

Symbol	Name	Quantity	Formula
A	ampere	electric current	base unit
Bq	becquerel	activity (of a radio nuclide)	1/s
C	coulomb	electric charge	A·s
°C	degree Celsius	temperature interval	°C = K
cd	candela	luminous intensity	base unit
F	farad	electric capacitance	C/V
Gy	gray	absorbed dose	J/kg
g	gram	mass	kg/1000
H	henry	inductance	Wb/A
Hz	hertz	frequency	1/s
ha	hectare*	area	10 000 m^2
J	joule	energy, work, heat	N·m
K	kelvin	temperature	base unit
kg	kilogram	mass	base unit
L	litre	volume	m^3/1000
lm	lumen	luminous flux	cd·sr
lx	lux	illuminance	lm/m^2
m	metre	length	base unit
mol	mole	amount of substance	base unit
N	newton	force	kg·m/s^2
Ω	ohm	electric resistance	V/A
Pa	pascal	pressure, stress	N/m^2
rad	radian	plane angle	m/m (dimensionless)
S	siemens	electric conductance	A/V
Sv	sievert	dose equivalent	J/kg
s	second	time	base unit
sr	steradian	solid angle	m^2/m^2 (dimensionless)
T	tesla	magnetic flux density	Wb/m^2
t	tonne, metric ton	mass	1000 kg; Mg
V	volt	electric potential	W/A
W	watt	power, radiant flux	J/s
Wb	weber	magnetic flux	V·s

*allowed with SI

Use of Symbols

The correct use of symbols is important because an incorrect symbol may change the meaning of a quantity. Some SI symbols are listed in the Symbol table.

SI has no abbreviations—only symbols. Therefore, no periods follow a symbol except at the end of a sentence.

Examples: A, *not* amp; s, *not* sec; SI, *not* S.I.

Symbols appear in lower case unless the unit name has been taken from a proper name. In this case the first letter of the symbol is capitalized.

Examples: m, metre; Pa, pascal; W, watt

Exception: L, litre

Symbols and prefixes are printed in upright (roman) type regardless of the type style in surrounding text.

Example: *. . . a distance of* 73 km *between . . .*

Unit symbols are the same whether singular or plural.

Examples: 1 mm, 100 mm; 1 kg, 65 kg

Leave a space between the value and the symbol.

Examples: 115 W, *not* 115W; 0.75 L, *not* 0.75L
88 °C, not 88°C or 88° C

Exception: No space is left between the numerical value and symbol for degree of plane angle.

Examples: 73°, not 73 °

Note: Symbol for coulomb is C; for degree Celsius it is °C

Do not mix symbols and names in the same expression.

Examples: radians per second or rad/s
not radians/second; not radians/s
m/s or metres per second,
not metres/second; *not* metres/s
J/kg or joules per kilogram,
not joules/kilogram; *not* joules/kg

Symbol for product—use the raised dot (·)

Examples: N·m; mPa·s; W/(m^2·K)

Symbol for quotient—use one of the following forms:

Examples: m/s or $\frac{m}{s}$ or use negative exponent

Note: Use only one solidus (/) per expression and parentheses to avoid any ambiguity.

Do not use modifying terms such as electrical, alternating current, etc.

Examples: kPa (gage); MW (e); V (ac)

When a prefix ends with a vowel and the unit name begins with a vowel, retain and pronounce both vowels.

Example: kiloampere
Exceptions: hectare; kilohm; megohm

When compound units are formed by multiplication, leave a space between units that are multiplied.

Examples: newton metre, *not* newton-metre; volt ampere, *not* volt-ampere

Use the modifier squared or cubed after the unit name.

Example: metre per second squared
Exception: For area or volume the modifier may be placed before the units.
Example: square millimetre; cubic metre

When compound units are formed by division, use the word *per*, not a solidus (/).

Examples: metre per second, *not* metre / second; watt per square metre, *not* watt / square meter

Do not use modifying terms such as electrical, alternating current, etc. after the symbol.

Examples: kPa (gage); MW (e); V (ac)

SELECTED CONVERSION FACTORS

CAUTION: These conversion values are rounded to three or four significant figures, which is sufficiently accurate for most applications. When making conversions, remember that a converted value is no more precise than the original value. Round off the final value to the same number of significant figures as those in the original value. See ANSI SI 10 for additional conversions with more significant figures.

Multiply	*By*	*To Obtain*
acre	0.4047	ha
atmosphere, standard	*101.325	kPa
bar	*100	kPa
barrel (42 US gal, petroleum)	159	L
Btu, (International Table)	1.055	kJ
Btu / lb·°F (specific heat, c^P)	4.184	kJ/(kg·K)
bushel	0.03524	m³
calorie, kilogram (kilocalorie)	4.187	kJ
candle, candlepower	*1.0	cd
centipoise, dynamic vicosity, μ	*1.00	mPa·s
centistokes, kinematic viscosity, ν	*1.00	mm²/s
ft	*0.3048	m
ft	*304.8	mm
ft / min, fpm	*0.00508	m/s
ft / s, fps	*0.3048	m/s
ft of water	2.99	kPa
ft²	0.09290	m²
ft²/s, kinematic viscosity, ν	92 900	mm²/s
ft³	28.32	L
ft³	0.02832	m³
ft³/h, cfh	7.866	mL/s
ft³/min, cfm	0.4719	L/s
ft³/s, cfs	28.32	L/s
footcandle	10.76	lx
ft·lb_f (torque or moment)	1.36	N·m
ft·lb_f (work)	1.36	J
ft·lb_f / lb (specific energy)	2.99	J/kg
ft·lb_f / min (power)	0.0226	W
gallon, US (*231 in³)	3.785	L
gph	1.05	mL/s
gpm	0.0631	L/s
gpm / ft²	0.6791	L/(s·m²)
gr / gal	17.1	g/m³
horsepower (550 ft·lb_f / s)	0.746	kW
inch	*25.4	mm
in of mercury (60°F)	3.377	kPa
in of water (60°F)	248.8	Pa
in·lb_f (torque or moment)	113	mN·m
in²	645	mm²
in³ (volume)	16.4	mL
in³ (section modulus)	16 400	mm³
in⁴ (section moment)	416 200	mm⁴
km / h	0.278	m/s
kWh	*3.60	MJ
kip / in² (ksi)	6.895	MPa
litre	*0.001	m³
micron (μm) of mercury (60°F)	133	mPa
mil (0.001 in.)	*25.4	μm
mile	1.61	km
mile, nautical	1.85	km
mph	1.61	km/h
mph	0.447	m/s
millibar	*0.100	kPa
mm of mercury (60°F)	0.133	kPa
mm of water (60°F)	9.80	Pa
ounce (mass, avoirdupois)	28.35	g
ounce (force of thrust)	0.278	N
ounce (liquid, US)	29.6	mL
ounce (avoirdupois) per gallon	7.49	kg/m³
pint (liquid, US)	473	mL
pound		
lb_m (mass)	0.4536	kg
lb_m (mass)	453.6	g
lb_f (force or thrust)	4.45	N
lb_m / ft (uniform load)	1.49	kg/m
lb_m / (ft·h) (dynamic viscosity, μ)	0.413	mPa·s
lb_m / (ft·s) (dynamic viscosity, μ)	1490	mPa·s
lb_f·s / ft² (dynamic viscosity, μ)	47 880	mPa·s
lb_m / min	0.00756	kg/s
lb_m / h	0.126	g/s
lb_f / ft²	47.9	Pa
lb_m / ft²	4.88	kg/m²
lb_m / ft³ (density, ρ)	16.0	kg/m³
lb_m / gallon	120	kg/m³
ppm (by mass)	*1.00	mg/kg
psi	6.895	kPa
quad (10¹⁵ Btu)	1.06	EJ
quart (liquid, US)	0.946	L
rpm	0.105	rad/s
tablespoon (approx.)	15	mL
teaspoon (approx.)	5	mL
therm (100,000 Btu)	105.5	MJ
ton, short (2000 lb)	0.907	Mg; t (tonne)
yd	*0.9144	m
yd²	0.836	m²
yd³	0.7646	m³

*Conversion factor is exact.

Note: In this list the kelvin (K) expresses temperature intervals. The degree Celsius symbol (°C) may be used for this purpose as well.

PREFIXES

Most prefixes indicate orders of magnitude in steps of 1000 and provide a convenient way to express large and small numbers and to eliminate nonsignificant digits and leading zeros in decimal fractions.

Examples: 64 000 watts is the same as 64 kilowatts*
0.057 metre is the same as 57 millimetres
16 000 metres is the same as 16 kilometres*
*except for intended accuracy

Prefix	*Symbol*	*Represents*
yotta	Y	10^{24}
zetta	Z	10^{21}
exa	E	10^{18}
peta	P	10^{15}
tera	T	10^{12}
giga	G	10^{9}
mega	M	10^{6}
kilo	k	10^{3}
hecto	h*	10^{2}
deka	da*	10^{1}
deci	d*	10^{-1}
centi	c*	10^{-2}
milli	m	10^{-3}
micro	µ	10^{-6}
nano	n	10^{-9}
pico	p	10^{-12}
femto	f	10^{-15}
atto	a	10^{-18}
zepto	z	10^{-21}
yocto	y	10^{-24}

To realize the full benefit of the prefixes when expressing a quantity by numerical value, choose a prefix so that the number lies between 0.1 and 1000. For simplicity, give preference to prefixes representing 1000 raised to an integral power (i.e., µm, mm, km).

**Exceptions*:

In expressing area and volume, the prefixes hecto, deka, deci, and centi may by required; for example, cubic decimetre (L), square hectometre (hectare), cubic centimetre.

Tables of values of the same quantity.

Comparison of values.

For certain quantities in particular applications. For example, the millimetre is used for linear dimensions in architectural and engineering drawings even when the values lie far outside the range of 0.1 mm to 1000 mm; the centimetre is usually used for anatomical measurements and clothing sizes.

Compound units. A compound unit is a derived unit expressed with two or more units. The prefix is attached to a unit in the numerator.

Examples: V/m *not* mV/mm
MJ/kg *not* kJ/g

Compound prefixes formed by a combination of two or more prefixes are not used. Use only one prefix.

Examples: 2 nm *not* 2 mµm
6 m^3 *not* 6 kL
6 MPa *not* 6 kkPa

Exponential Powers. An exponent attached to a symbol containing a prefix indicates that the multiple (of the unit with its prefix) is raised to the power of 10 expressed by the exponent.

Examples: 1 $mm^3 = (10^{-3}\ m)^3 = 10^{-9}\ m^3$
1 $ns^{-1} = (10^{-9}\ s)^{-1} = 10^{9}\ s^{-1}$
1 $mm^2/s = (10^{-3}\ m)^2/s = 10^{-6}\ m^2/s$

NUMBERS

International practice separates the digits of large numbers into groups of three, counting from the decimal to the left and to the right, and inserts a space to separate the groups. In numbers of four digits, the space is not necessary except for uniformity in tables.

Examples: 6.358 568; 85 365; 51 845 953; 88 000; 0.246 113 562; 7 258

Small Numbers. When writing numbers less than one, always put a zero before the decimal marker.

Example: 0.046

Decimal Marker. The recommended decimal marker is a dot on the line (period). (In some countries, a comma is used as the decimal marker.)

Because **billion** means a million million in most countries but a thousand million in the United States, avoid using billion in technical writing.

DO'S AND DON'TS

The units in the international system of units are called SI units—*not* Metric Units and *not* SI Metric Units.

Non-SI units in the US are called Inch-Pound units (I-P units)—*not* conventional units, *not* U.S. customary units, *not* English units, and *not* Imperial units.)

Treat all spelled out names as nouns. Therefore, do not capitalize the first letter of a unit except at the beginning of a sentence or in capitalized material such as a title.

Examples: watt; pascal; ampere; volt; newton; kelvin
Exception: Always capitalize the first letter of Celsius.

Do not begin a sentence with a unit symbol—either rearrange the words or write the unit name in full.

Use plurals for spelled out words when required by the rules of grammar.

Examples: metre—metres; henry—henries; kilogram—kilograms; kelvin—kelvins
Irregular: hertz—hertz; lux—lux; siemens—siemens

Do not put a space or hyphen between the prefix and unit name.

Examples: kilometre *not* kilo metre or kilo-metre; milliwatt *not* milli watt or milli-watt

METRIC PRACTICE GUIDE FOR CEMENT

Inch-Pound Units	Factor	SI Units
inch	× 25.40	= millimetre (mm)
inch^2	× 6.452	= centimetre^2 (cm^2)
foot^3	× 0.02832	= metre^3 (m^3)
pound per foot^3	× 16.02	= kilogram per cubic metre (kg/m^3)
pound per yard^3	× 0.5933	= kilogram per cubic metre (kg/m^3)
quart (U.S. fluid)	× 0.9464	= litre decimetre^3 (dm^3)
pound	× 0.4536	= kilogram (kg)
ton	× 0.9072	= megagram (Mg)
psi	× 6.895	= kilopascal (kPa)
psi	× 0.006895	= megapascal (MPa)
grain	× 0.06480	= gram (g)
horsepower	× 745.7	= watt (W)
centimetre^2 per gram (cm^2/g)	× 0.1000	= metre^2 per kilogram (m^2/kg)
calorie per gram	× 4.184	= kilojoule per kilogram (kJ/kg)
ounce (U.S. fluid)	× 29.57	= cubic centimetre (cm^3) (millilitre)
pound-force (lbf avoirdupois)	× 4.448	= newton (N)
pound-force (lbf avoirdupois)	× 0.004448	= kilonewton (kN)
temperature, degree Fahrenheit (°F)	(°F − 32)/1.8	= temperature, degree Celsius (°C)
temperature interval (°F)	× 5/9	= temperature interval (°C)

NOTE—When converting from alternative designation of sieves in inch-pound units, the proper quantity in SI units is that given in Table 1 of Specification E 11, Wire-Cloth Sieves for Testing Purposes, and not the quantity calculated from the alternative designation. (For example, the "1-in." sieve is the 25.0-mm sieve, not the 25.4-mm sieve).

ANNUAL BOOK of ASTM STANDARDS

Index

Volume 04.01

Cement; Lime; Gypsum

This index covers the standards and related material appearing in this volume(s) or section only. The boldface references represent the ASTM designations. For multiple volume or section indexes, the volume in which the standard appears is included in parentheses. A triangle (Δ) preceding the ASTM designation denotes that Adjunct Material for the standard is available separately; the Adjunct No. is given in the standard. A Combined Index, covering the standards appearing in all volumes of the *1999 Annual Book of ASTM Standards*, is issued as Volume 00.01.

Alphabetization in the index is letter-for-letter, with no consideration to punctuation or word division. Initial prepositions of (indented) subentries are ignored for alphabetization.

In the preparation of indexes, every attempt has been made to index standards on three levels: (1) by main subject, using general and specific search terms; (2) by tests or other significant sections of ASTM standards; and (3) by cross-references to locate main subject entry terms. *(See also* references are abbreviated as *Sa* and appear under main entry terms.) The following examples illustrate ASTM's method of indexing.

INDEX TERMS FOR SPECIFICATIONS

Steel pipe—carbon steel
Seamless and Welded Steel Pipe for Low-Temperature Service, Specification for, **A 333/333M (01.01)**

Steel tube—ferritic stainless steel
Seamless and Welded Ferritic/Austenitic Stainless Steel Tubing for General Service, Specification for, **A 789/789M (01.01)**

INDEX TERMS FOR TESTS

Gold—electrodeposited coating
Installing Corrugated Aluminum Structural Plate Pipe for Culverts and Sewers, Practice for, **B 789 (02.02)**

Effective elastic parameter (E_{eff})
Determining the Effective Elastic Parameter for X-Ray Diffraction Measurements of Residual Stress, Test Method for, **E 1426 (03.01)**

CROSS-REFERENCES

Water—drinking
See **Drinking water**

PDB (pressure design basis)
See **Pressure design basis (PDB)**
Sa **Pressure testing**

A

Acceptance criteria/testing
Sampling and the Amount of Testing of Hydraulic Cement, Practice for, **C 183**
Accessories
Accessories for Gypsum Wallboard and Gypsum Veneer Base, Specification for, **C 1047**
Accountability
See **Quality control**
Acidity, alkalinity, pH
Physical Testing of Quicklime, Hydrated Lime, and Limestone, Test Methods for, **C 110**
Acid pickling residue
See **Residue**
Active humidification systems
See **Humidity**
Additions to cement
Air-Entraining Additions for Use in the Manufacture of Air-Entraining Hydraulic Cement, Specification for, **C 226**
Functional Additions for Use in Hydraulic Cements, Specification for, **C 688**
Processing Additions for Use in the Manufacture of Hydraulic Cements, Specification for, **C 465**
Admixtures—concrete
See **Additions to cement**
Admixtures—soil
Sa **Soil stabilization**
Quicklime and Hydrated Lime for Soil Stabilization, Specification for, **C 977**
Aggregate
Gypsum Concrete, Specification for, **C 317**
Inorganic Aggregates for Use in Gypsum Plaster, Specification for, **C 35**
Aggregate—natural mineral aggregate
See **Mineral aggregate**
Agricultural liming materials
Sa **Lime and limestone products/testing**
Agricultural Liming Materials, Specification for, **C 602**
Limestone for Animal Feed Use, Specification for, **C 706**
Agricultural residue
See **Residue**
Air-entraining portland cement
Air-Entraining Additions for Use in the Manufacture of Air-Entraining Hydraulic Cement, Specification for, **C 226**
Portland Cement, Specification for, **C 150**
Air entrainment
Air Content of Hydraulic Cement Mortar, Test Method for, **C 185**
Air-Entraining Additions for Use in the Manufacture of Air-Entraining Hydraulic Cement, Specification for, **C 226**
Masonry Cement, Specification for, **C 91**
Physical Testing of Quicklime, Hydrated Lime, and Limestone, Test Methods for, **C 110**
Air permeability method
Fineness of Hydraulic Cement by Air Permeability Apparatus, Test Method for, **C 204**
Alcohol wash method
Physical Testing of Gypsum, Gypsum Plasters and Gypsum Concrete, Test Methods for, **C 472**
Alite
Quantitative Determination of Phases in Portland Cement Clinker by Microscopical Point Count Procedure, Test Method for, **C 1356**
Alkalies content—cement
Chemical Analysis of Hydraulic Cement, Test Methods for, **C 114**
Masonry Cement, Specification for, **C 91**
Alkalinity
See **Acidity, alkalinity, pH**
Alkali reactive aggregate
Length Change of Mortar Bars Caused by Reaction Between Hydraulic Cement and Alkali-Reactive Aggregate, Provisional Test Method for, **PS 81**
Alkali reactivity
Length Change of Mortar Bars Caused by Reaction Between Hydraulic Cement and Alkali-Reactive Aggregate, Provisional Test Method for, **PS 81**
All-day testing
See **Twenty-four hour testing**
Alpha gypsum cement
Glass Fiber Reinforced Gypsum Composite, Specification for, **C 1355/C1355M**
Alumina
See **Aluminum oxide (Al_2O_3) content**
Aluminate
Determination of the Proportion of Phases in Portland Cement and Portland Cement Clinker Using X-Ray Powder Diffraction Analysis, Test Method for, **C 1365**
Quantitative Determination of Phases in Portland Cement Clinker by Microscopical Point Count Procedure, Test Method for, **C 1356**
Aluminum oxide (Al_2O_3) content
Chemical Analysis of Gypsum and Gypsum Products [Metric], Test Methods for, **C 471M**
Chemical Analysis of Hydraulic Cement, Test Methods for, **C 114**
Chemical Analysis of Limestone, Quicklime, and Hydrated Lime, Test Methods for, **C 25**
Ammonium acetate method
Chemical Analysis of Gypsum and Gypsum Products [Metric], Test Methods for, **C 471M**
Ammonium hydroxide content
Chemical Analysis of Hydraulic Cement, Test Methods for, **C 114**
Anhydrous gypsum
Gypsum Keene's Cement, Specification for, **C 61**
Animal feeds
Limestone for Animal Feed Use, Specification for, **C 706**
Appearance of materials
Sa **Optical materials/properties/tests**
Bleeding of Cement Pastes and Mortars, Test Method for, **C 243**
Artificial pozzolans
See **Pozzolans**
Aseptic packaging
See **Packaging materials**
Asphaltic concrete mixtures
Hydrated Lime for Use in Asphaltic-Concrete Mixtures, Specification for, **C 1097**
ASTM Committee C-1 on Cement
Manual of Cement Testing, (Related Material)
Atomic absorption
Major and Trace Elements in Limestone and Lime by Inductively Coupled Plasma-Atomic Emission Spectroscopy (ICP) and Atomic Absoprtion (AA), Test Method for, **C 1301**
Atomic absorption spectrophotometry/spectroscopy
See **Spectrophotometry—atomic absorption**
Atomic emission spectrometry
Major and Trace Elements in Limestone and Lime by Inductively Coupled Plasma-Atomic Emission Spectroscopy (ICP) and Atomic Absoprtion (AA), Test Method for, **C 1301**
Autoclave expansion
Sa **Expansion—cement/concrete**
Autoclave Expansion of Portland Cement, Test Method for, **C 151**
Physical Testing of Quicklime, Hydrated Lime, and Limestone, Test Methods for, **C 110**

Available lime index
Chemical Analysis of Limestone, Quicklime, and Hydrated Lime, Test Methods for, **C 25**
Fly Ash and Other Pozzolans for Use with Lime, Specification for, **C 593**

B

Balances
Reference Masses and Devices for Determining Mass for Use in the Physical Testing of Hydraulic Cements, Specification for, **C 1005**
Ball-pan hardness
See **Hardness tests**
Band-pass spectrophotometers
See **Spectrophotometry**
Basicity
See **Acidity, alkalinity, pH**
Bathrooms
Glass Mat Water-Resistant Gypsum Backing Board, Specification for, **C 1178/C1178M**
Bed moisture
See **Moisture analysis**
Belite
Quantitative Determination of Phases in Portland Cement Clinker by Microscopical Point Count Procedure, Test Method for, **C 1356**
Bending moment
See **Stiffness**
Bituminous coal
See **Coal products**
Blaine fineness test
Fineness of Hydraulic Cement by Air Permeability Apparatus, Test Method for, **C 204**
Bleeding
Bleeding of Cement Pastes and Mortars, Test Method for, **C 243**
Blended hydraulic cement
Blended Hydraulic Cement, Performance Specification for, **C 1157**
Blended hydraulic cements
Blended Hydraulic Cements, Specifications for, **C 595**
Heat of Hydration of Hydraulic Cement, Test Method for, **C 186**
Bonding
Surface-Applied Bonding Compounds for Exterior Plastering, Specification for, **C 932**
Bonding compounds
Bonding Compounds for Interior Gypsum Plastering, Specification for, **C 631**
Bond strength
Joint Treatment Materials for Gypsum Board Construction, Test Methods for, **C 474**
Brightness
Physical Testing of Quicklime, Hydrated Lime, and Limestone, Test Methods for, **C 110**
Brinell hardness
See **Hardness tests**
Brittleness
See **Stiffness**
Building constructions/materials—partitions
See **Partitions**
Burnt lime
Sa **Lime and limestone products/testing**
Agricultural Liming Materials, Specification for, **C 602**
Lime for Use with Pozzolans, Specification for, **C 821**

C

Cabinets
Moist Cabinets, Moist Rooms and Water Storage Tanks Used in the Testing of Hydraulic Cements and Concretes, Specification for, **C 511**
CA-DIR spectrophotometers
See **Spectrophotometry**
Calcined gypsum
Sa **Gypsum products**
Gypsum Concrete, Specification for, **C 317**
Calcium
Determination of Total Neutralizing Capability and Dissolved Calcium and Magnesium Oxide in Lime for Flue Gas Desulfurization, Test Method for, **C 1318**
Calcium carbonate ($CaCO_3$) equivalent
Agricultural Liming Materials, Specification for, **C 602**
Calcium content
Agricultural Liming Materials, Specification for, **C 602**
Calcium oxide (CaO) content
Chemical Analysis of Gypsum and Gypsum Products [Metric], Test Methods for, **C 471M**
Chemical Analysis of Hydraulic Cement, Test Methods for, **C 114**
Chemical Analysis of Limestone, Quicklime, and Hydrated Lime, Test Methods for, **C 25**
Calcium sulfate ($CaSO_4$)
Calcium Sulfate in Hydrated Portland Cement Mortar, Test Method for, **C 265**
Chemical Analysis of Gypsum and Gypsum Products [Metric], Test Methods for, **C 471M**
Canadian cube de-molding procedure
Compressive Strength of Hydraulic Cement Mortars (Using 2-in. or 50-mm Cube Specimens), Test Method for, **C 109/C109M**
Carbon dioxide content
Chemical Analysis of Gypsum and Gypsum Products [Metric], Test Methods for, **C 471M**
Chemical Analysis of Limestone, Quicklime, and Hydrated Lime, Test Methods for, **C 25**
Carbon residue (petroleum products)
See **Residue**
Cargo residue/waste
See **Residue**
Cast-in-place gypsum concrete
See **Gypsum products**
Ceiling board
Gypsum Ceiling Board, Specification for, **C 1395**
Ceilings
Fiber-Reinforced Gypsum Panels, Specification for, **C 1278/C1278M**
Glass Mat Water-Resistant Gypsum Backing Board, Specification for, **C 1178/C1178M**
Predecorated Gypsum Board, Specification for, **C 960/C960M**
Cement
Compressive Strength of Hydraulic-Cement Mortars (Using Portions of Prisms Broken in Flexure), Test Method for, **C 349**
Evaluation of Laboratories Testing Hydraulic Cement, Practice for, **C 1222**
Flow Table for Use in Tests of Hydraulic Cement, Specification for, **C 230**
Hydraulic Cement, Terminology Relating to, **C 219**
Manual of Cement Testing, (Related Material)
Metric Practice Guide for Cement, (Related Material)
Plastic (Stucco) Cement, Specification for, **C 1328**
Cement—additions
Air-Entraining Additions for Use in the Manufacture of Air-Entraining Hydraulic Cement, Specification for, **C 226**

Cement—additions
Functional Additions for Use in Hydraulic Cements, Specification for, **C 688**
Processing Additions for Use in the Manufacture of Hydraulic Cements, Specification for, **C 465**
Cement—hydraulic
See **Hydraulic cement concrete**
Cement—masonry
Masonry Cement, Specification for, **C 91**
Cement—plasters
Installation of Lathing and Furring to Receive Interior and Exterior Portland Cement-Based Plaster, Specification for, **C 1063**
Surface-Applied Bonding Compounds for Exterior Plastering, Specification for, **C 932**
Cement—portland (air-entraining)
Air-Entraining Additions for Use in the Manufacture of Air-Entraining Hydraulic Cement, Specification for, **C 226**
Portland Cement, Specification for, **C 150**
Cement mortars
Air Content of Hydraulic Cement Mortar, Test Method for, **C 185**
Bleeding of Cement Pastes and Mortars, Test Method for, **C 243**
Calcium Sulfate in Hydrated Portland Cement Mortar, Test Method for, **C 265**
Compressive Strength of Hydraulic Cement Mortars (Using 2-in. or 50-mm Cube Specimens), Test Method for, **C 109/C109M**
Compressive Strength of Hydraulic-Cement Mortars (Using Portions of Prisms Broken in Flexure), Test Method for, **C 349**
Drying Shrinkage of Mortar Containing Portland Cement, Test Method for, **C 596**
Early Stiffening of Portland Cement (Mortar Method), Test Method for, **C 359**
Expansion of Portland Cement Mortar Bars Stored in Water, Test Method for, **C 1038**
Flexural Strength of Hydraulic Cement Mortars, Test Method for, **C 348**
Length Change of Hydraulic-Cement Mortars Exposed to a Sulfate Solution, Test Method for, **C 1012**
Mechanical Mixing of Hydraulic Cement Pastes and Mortars of Plastic Consistency, Practice for, **C 305**
Restrained Expansion of Expansive Cement Mortar, Test Method for, **C 806**
Time of Setting of Hydraulic Cement Mortar by Modified Vicat Needle, Test Method for, **C 807**
Use of Apparatus for the Determination of Length Change of Hardened Cement Paste, Mortar, and Concrete, Practice for, **C 490**
Cement pastes
Bleeding of Cement Pastes and Mortars, Test Method for, **C 243**
Mechanical Mixing of Hydraulic Cement Pastes and Mortars of Plastic Consistency, Practice for, **C 305**
Use of Apparatus for the Determination of Length Change of Hardened Cement Paste, Mortar, and Concrete, Practice for, **C 490**
Cement testing
Manual of Cement Testing, (Related Material)
Certification
Sampling, Inspection, Rejection, Certification, Packaging, Marking, Shipping, Handling and Storage of Gypsum Board, Specification for, **C 1264/C1264M**
Change in length
Autoclave Expansion of Portland Cement, Test Method for, **C 151**
Drying Shrinkage of Mortar Containing Portland Cement, Test Method for, **C 596**
Expansion of Portland Cement Mortar Bars Stored in Water, Test Method for, **C 1038**
Length Change of Hydraulic-Cement Mortars Exposed to a Sulfate Solution, Test Method for, **C 1012**
Physical Testing of Quicklime, Hydrated Lime, and Limestone, Test Methods for, **C 110**
Restrained Expansion of Expansive Cement Mortar, Test Method for, **C 806**
Use of Apparatus for the Determination of Length Change of Hardened Cement Paste, Mortar, and Concrete, Practice for, **C 490**
Chemical analysis—cement/concrete/aggregate
Chemical Analysis of Hydraulic Cement, Test Methods for, **C 114**
X-Ray Spectrometric Analysis of Lime and Limestone, Test Method for, **C 1271**
Chemical composition
See **Composition analysis**
Chloride content
Chemical Analysis of Gypsum and Gypsum Products [Metric], Test Methods for, **C 471M**
Chloroform-soluble organic substances content
Chemical Analysis of Hydraulic Cement, Test Methods for, **C 114**
Class PB exterior insulation/finish system
Application of Class PB Exterior Insulation and Finish Systems, Practice for, **C 1397**
Clinker
Determination of the Proportion of Phases in Portland Cement and Portland Cement Clinker Using X-Ray Powder Diffraction Analysis, Test Method for, **C 1365**
Coal mine dusting
Limestone for Dusting of Coal Mines, Specification for, **C 737**
Coal products
Limestone for Dusting of Coal Mines, Specification for, **C 737**
Commercial hydrated lime
Lime for Use with Pozzolans, Specification for, **C 821**
Composites
Glass Fiber Reinforced Gypsum Composite, Specification for, **C 1355/C1355M**
Composition analysis
Evaluation of Limestone or Lime Uniformity from a Single Source, Practice for, **C 1164**
Compression testing—cement/concrete
Compressive Strength of Hydraulic Cement Mortars (Using 2-in. or 50-mm Cube Specimens), Test Method for, **C 109/C109M**
Compressive Strength of Hydraulic-Cement Mortars (Using Portions of Prisms Broken in Flexure), Test Method for, **C 349**
Hydraulic Hydrated Lime for Structural Purposes, Specification for, **C 141**
Masonry Cement, Specification for, **C 91**
Optimum SO_3 in Hydraulic Portland Cement Using 24-h Compressive Strength, Test Method for, **C 563**
Physical Testing of Gypsum, Gypsum Plasters and Gypsum Concrete, Test Methods for, **C 472**
Computer-assisted spectrophotometry
See **Spectrophotometry**
Concrete
Hydrated Lime for Use in Asphaltic-Concrete Mixtures, Specification for, **C 1097**
Use of Apparatus for the Determination of Length Change of Hardened Cement Paste, Mortar, and Concrete, Practice for, **C 490**
Concrete—bonding
Surface-Applied Bonding Compounds for Exterior Plastering, Specification for, **C 932**
Concrete—compressive strength
See **Compression testing—cement/concrete**
Concrete—freshly-mixed
Sa **Hydraulic cement concrete**
Moist Cabinets, Moist Rooms and Water Storage Tanks Used in

the Testing of Hydraulic Cements and Concretes, Specification for, **C 511**

Concrete—gypsum
Sa **Gypsum products**
Gypsum Concrete, Specification for, **C 317**
Installation of Cast-in-Place Reinforced Gypsum Concrete, Specification for, **C 956**
Physical Testing of Gypsum, Gypsum Plasters and Gypsum Concrete, Test Methods for, **C 472**

Concrete—hardened
Use of Apparatus for the Determination of Length Change of Hardened Cement Paste, Mortar, and Concrete, Practice for, **C 490**

Concrete—test specimens
Moist Cabinets, Moist Rooms and Water Storage Tanks Used in the Testing of Hydraulic Cements and Concretes, Specification for, **C 511**

Condensation
See **Humidity**

Consistency
Joint Treatment Materials for Gypsum Board Construction, Test Methods for, **C 474**
Normal Consistency of Hydraulic Cement, Test Method for, **C 187**
Physical Testing of Gypsum, Gypsum Plasters and Gypsum Concrete, Test Methods for, **C 472**
Physical Testing of Quicklime, Hydrated Lime, and Limestone, Test Methods for, **C 110**

Control joint
Accessories for Gypsum Wallboard and Gypsum Veneer Base, Specification for, **C 1047**

Controlled-potential coulometry
See **Coulometry**

Conversion units/factors
SI Quick Reference Guide: International System of Units (SI)*(The Modern Metric System)*, **IEEE/ASTM SI-10** (Related Material)

Coreboard
Sa **Gypsum products—board/wallboard**
Gypsum Board, Specification for, **C 1396**

Core hardness
Sa **Gypsum products—board/wallboard**
Physical Testing of Gypsum Panel Products, Test Methods for, **C 473**

Corner bead
Accessories for Gypsum Wallboard and Gypsum Veneer Base, Specification for, **C 1047**

Coulometry
Chemical Analysis of Gypsum and Gypsum Products [Metric], Test Methods for, **C 471M**

Countersunk head cap screws
See **Steel bolting materials—screws**

Cracking
Joint Treatment Materials for Gypsum Board Construction, Test Methods for, **C 474**

Crushed bituminous coal
See **Coal products**

Crushed lime
Sa **Quicklime**
Quicklime for Structural Purposes, Specification for, **C 5**

D

Data analysis—coal/coke
See **Coal products**

Deflection—building constructions/materials
Physical Testing of Gypsum Panel Products, Test Methods for, **C 473**

Degree of set
See **Setting time**

Density
Density of Hydraulic Cement, Test Method for, **C 188**

Depth (recessed/tapered-edge)
Physical Testing of Gypsum Panel Products, Test Methods for, **C 473**

Derived units
SI Quick Reference Guide: International System of Units (SI)*(The Modern Metric System)*, **IEEE/ASTM SI-10** (Related Material)

Desulfurization systems
Determination of Total Neutralizing Capability and Dissolved Calcium and Magnesium Oxide in Lime for Flue Gas Desulfurization, Test Method for, **C 1318**

Diffraction intensity (of X-rays)
See **X-ray diffraction**

Diffractometer
Determination of the Proportion of Phases in Portland Cement and Portland Cement Clinker Using X-Ray Powder Diffraction Analysis, Test Method for, **C 1365**

Digestion
Major and Trace Elements in Limestone and Lime by Inductively Coupled Plasma-Atomic Emission Spectroscopy (ICP) and Atomic Absoprtion (AA), Test Method for, **C 1301**

Dimensional stability
Joint Treatment Materials for Gypsum Board Construction, Test Methods for, **C 474**

Direct aspiration spectrophotometric procedure
See **Spectrophotometry**

Direct photometric method
See **Photometry**
See **Spectrophotometry**

Dissolution
Major and Trace Elements in Limestone and Lime by Inductively Coupled Plasma-Atomic Emission Spectroscopy (ICP) and Atomic Absoprtion (AA), Test Method for, **C 1301**

Dissolved calcium and magnesium
Determination of Total Neutralizing Capability and Dissolved Calcium and Magnesium Oxide in Lime for Flue Gas Desulfurization, Test Method for, **C 1318**

Dolomitic limestone
Sa **Lime and limestone products/testing—hydrated**
Determination of Total Neutralizing Capability and Dissolved Calcium and Magnesium Oxide in Lime for Flue Gas Desulfurization, Test Method for, **C 1318**
Lime for Use with Pozzolans, Specification for, **C 821**

Drill screws
Sa **Screws**
Steel Drill Screws for the Application of Gypsum Panel Products or Metal Plaster Bases, Specification for, **C 1002**
Steel Drill Screws for the Application of Gypsum Panel Products or Metal Plaster Bases to Steel Studs from 0.033 in. (0.84 mm) to 0.112 in. (2.84 mm) in Thickness, Specification for, **C 954**

Drill screws (steel)
See **Steel bolting materials—screws**

Dry brightness
Physical Testing of Quicklime, Hydrated Lime, and Limestone, Test Methods for, **C 110**

Dry film thickness
See **Thickness**

Drying shrinkage
Drying Shrinkage of Mortar Containing Portland Cement, Test Method for, **C 596**

Drywall
Sa **Gypsum products—board/wallboard**
Gypsum Board, Specification for, **C 1396**
Gypsum Ceiling Board, Specification for, **C 1395**

Dusting
Limestone for Dusting of Coal Mines, Specification for, **C 737**

E

Early stiffening
Early Stiffening of Hydraulic Cement (Paste Method), Test Method for, **C 451**
Early Stiffening of Portland Cement (Mortar Method), Test Method for, **C 359**
Edge hardness
Physical Testing of Gypsum Panel Products, Test Methods for, **C 473**
Edge trim
Accessories for Gypsum Wallboard and Gypsum Veneer Base, Specification for, **C 1047**
Emission spectrometry
See **Spectrometry**
End hardness
Physical Testing of Gypsum Panel Products, Test Methods for, **C 473**
Equilibrium moisture
See **Moisture analysis**
Evaporation residue
See **Residue**
Expansion—cement/concrete
Autoclave Expansion of Portland Cement, Test Method for, **C 151**
Expansion of Portland Cement Mortar Bars Stored in Water, Test Method for, **C 1038**
Physical Testing of Quicklime, Hydrated Lime, and Limestone, Test Methods for, **C 110**
Potential Expansion of Portland-Cement Mortars Exposed to Sulfate, Test Method for, **C 452**
Restrained Expansion of Expansive Cement Mortar, Test Method for, **C 806**
Expansion joints
Accessories for Gypsum Wallboard and Gypsum Veneer Base, Specification for, **C 1047**
Expansive cement
Expansive Hydraulic Cement, Specification for, **C 845**
Exterior gypsum soffit board
Gypsum Board, Specification for, **C 1396**
Exterior insulation/finish system (EIFS)
Application of Class PB Exterior Insulation and Finish Systems, Practice for, **C 1397**
Exterior lathing/furring
See **Gypsum products—lath**
Exterior walls
Application of Gypsum Sheathing Board, Specification for, **C 1280**

F

Fasteners (metal)
See **Nails**
Ferrite
Determination of the Proportion of Phases in Portland Cement and Portland Cement Clinker Using X-Ray Powder Diffraction Analysis, Test Method for, **C 1365**
Quantitative Determination of Phases in Portland Cement Clinker by Microscopical Point Count Procedure, Test Method for, **C 1356**
Ferrous iron content
Chemical Analysis of Limestone, Quicklime, and Hydrated Lime, Test Methods for, **C 25**
Ferrous oxide content
See **Iron oxide content**
FGD systems
See **Flue gas desulfurization (FGD)**
Fine-coal fractions
See **Coal products**
Fineness
Physical Testing of Gypsum, Gypsum Plasters and Gypsum Concrete, Test Methods for, **C 472**
Fineness—hydraulic cement (blended/portland)
Fineness of Hydraulic Cement by Air Permeability Apparatus, Test Method for, **C 204**
Fineness of Hydraulic Cement by the 150-µm (No. 100) and 75-µm (No. 200) Sieves, Test Method for, **C 184**
Fineness of Hydraulic Cement by the 300-µm (No. 50), 150-µm (No. 100), and 75-µm (No. 200) Sieves by Wet Methods, Test Method for, **C 786**
Fineness of Hydraulic Cement by the 45-µm (No. 325) Sieve, Test Method for, **C 430**
Fineness of Portland Cement by the Turbidimeter, Test Method for, **C 115**
Hydraulic Hydrated Lime for Structural Purposes, Specification for, **C 141**
Fixed wavelength photometric detectors
See **Photometry**
Flexural strength
Flexural Strength of Hydraulic Cement Mortars, Test Method for, **C 348**
Physical Testing of Gypsum Panel Products, Test Methods for, **C 473**
Floor/roof assemblies (of gypsum board)
See **Gypsum products**
Flow and flow rate—cement/concrete
Flow Table for Use in Tests of Hydraulic Cement, Specification for, **C 230**
Flue gas desulfurization (FGD)
Determination of Total Neutralizing Capability and Dissolved Calcium and Magnesium Oxide in Lime for Flue Gas Desulfurization, Test Method for, **C 1318**
Fly ash and other pozzolans
See **Pozzolans**
Foil-backed gypsum board
Gypsum Board, Specification for, **C 1396**
Formboard
Sa **Gypsum products—board/wallboard**
Gypsum Formboard, Specification for, **C 318/C318M**
Free acid content
See **Acidity, alkalinity, pH**
Free silica content
See **Silica content**
Free water content
Chemical Analysis of Gypsum and Gypsum Products [Metric], Test Methods for, **C 471M**
Physical Testing of Gypsum, Gypsum Plasters and Gypsum Concrete, Test Methods for, **C 472**
Freshly-mixed concrete
Moist Cabinets, Moist Rooms and Water Storage Tanks Used in the Testing of Hydraulic Cements and Concretes, Specification for, **C 511**
FRGP (fiber-reinforced gypsum panels)
Fiber-Reinforced Gypsum Panels, Specification for, **C 1278/C1278M**
Furnace spectrophotometric procedure
See **Spectrophotometry**
Furring channels
Non-Load Bearing (Axial) Steel Studs, Runners (Tracks), and Rigid Furring Channels for Screw Application of Gypsum Board, Specification for, **C 645**

G

Gas composition
See **Composition analysis**
Gillmore needles
Time of Setting of Hydraulic-Cement Paste by Gillmore Needles, Test Method for, **C 266**
Glass-fiber-reinforced gypsum materials
Glass Fiber Reinforced Gypsum Composite, Specification for, **C 1355/C1355M**
Molded Glass Fiber Reinforced Gypsum Parts, Specification for, **C 1381**
Glass mat gypsum board
Sa **Gypsum products**
Glass Mat Gypsum Substrate for Use as Sheathing, Specification for, **C 1177/C1177M**
Glass Mat Water-Resistant Gypsum Backing Board, Specification for, **C 1178/C1178M**
Granular blast furnace
Blended Hydraulic Cement, Performance Specification for, **C 1157**
Granular lime
Sa **Quicklime**
Quicklime for Structural Purposes, Specification for, **C 5**
GRG composite
Glass Fiber Reinforced Gypsum Composite, Specification for, **C 1355/C1355M**
Grid suspension systems
Non-Load Bearing (Axial) Steel Studs, Runners (Tracks), and Rigid Furring Channels for Screw Application of Gypsum Board, Specification for, **C 645**
Ground lime
See **Quicklime**
Gypsum board panels
See **Gypsum products—board/wallboard**
Gypsum concrete
Gypsum Concrete, Specification for, **C 317**
Installation of Cast-in-Place Reinforced Gypsum Concrete, Specification for, **C 956**
Physical Testing of Gypsum, Gypsum Plasters and Gypsum Concrete, Test Methods for, **C 472**
Gypsum Keene's cement
Gypsum Keene's Cement, Specification for, **C 61**
Gypsum plaster
Application of Gypsum Base to Receive Gypsum Veneer Plaster, Specification for, **C 844**
Application of Gypsum Veneer Plaster, Specification for, **C 843**
Application of Interior Gypsum Plaster, Specification for, **C 842**
Application of Portland Cement-Based Plaster, Specification for, **C 926**
Bonding Compounds for Interior Gypsum Plastering, Specification for, **C 631**
Chemical Analysis of Gypsum and Gypsum Products [Metric], Test Methods for, **C 471M**
Gypsum Base for Veneer Plasters, Specification for, **C 588**
Gypsum Casting and Molding Plaster, Specification for, **C 59**
Gypsum Keene's Cement, Specification for, **C 61**
Gypsum Plasters, Specification for, **C 28**
Gypsum Veneer Plaster, Specification for, **C 587**
Inorganic Aggregates for Use in Gypsum Plaster, Specification for, **C 35**
Installation of Lathing and Furring to Receive Interior and Exterior Portland Cement-Based Plaster, Specification for, **C 1063**
Metal Lath, Specification for, **C 847**
Physical Testing of Gypsum, Gypsum Plasters and Gypsum Concrete, Test Methods for, **C 472**
Surface-Applied Bonding Compounds for Exterior Plastering, Specification for, **C 932**
Woven Wire Plaster Base, Specification for, **C 1032**
Gypsum products
Chemical Analysis of Gypsum and Gypsum Products [Metric], Test Methods for, **C 471M**
Fiber-Reinforced Gypsum Panels, Specification for, **C 1278/C1278M**
Gypsum, Specification for, **C 22/C22M**
Gypsum and Related Building Materials and Systems, Terminology Relating to, **C 11**
Molded Glass Fiber Reinforced Gypsum Parts, Specification for, **C 1381**
Gypsum products—board/wallboard
Sa **Installation—gypsum products**
Accessories for Gypsum Wallboard and Gypsum Veneer Base, Specification for, **C 1047**
Application and Finishing of Gypsum Board, Specification for, **C 840**
Application of Gypsum Sheathing Board, Specification for, **C 1280**
Exterior Gypsum Soffit Board, Specification for, **C 931/C931M**
Glass Mat Gypsum Substrate for Use as Sheathing, Specification for, **C 1177/C1177M**
Glass Mat Water-Resistant Gypsum Backing Board, Specification for, **C 1178/C1178M**
Gypsum Backing Board and Coreboard, Specification for, **C 442**
Gypsum Base for Veneer Plasters, Specification for, **C 588**
Gypsum Board, Specification for, **C 1396**
Gypsum Ceiling Board, Specification for, **C 1395**
Gypsum Formboard, Specification for, **C 318/C318M**
Gypsum Wallboard, Specification for, **C 36**
Installation of Load Bearing (Transverse and Axial) Steel Studs and Related Accessories, Specification for, **C 1007**
Installation of Steel Framing Members to Receive Screw-Attached Gypsum Panel Products, Specification for, **C 754**
Joint Compound and Joint Tape for Finishing Gypsum Board, Specification for, **C 475**
Joint Treatment Materials for Gypsum Board Construction, Test Methods for, **C 474**
Load-Bearing (Transverse and Axial) Steel Studs, Runners (Tracks), and Bracing or Bridging for Screw Application of Gypsum Board and Metal Plaster Bases, Specification for, **C 955**
Nails for the Application of Gypsum Board, Specification for, **C 514**
Non-Load Bearing (Axial) Steel Studs, Runners (Tracks), and Rigid Furring Channels for Screw Application of Gypsum Board, Specification for, **C 645**
Physical Testing of Gypsum Panel Products, Test Methods for, **C 473**
Predecorated Gypsum Board, Specification for, **C 960/C960M**
Sampling, Inspection, Rejection, Certification, Packaging, Marking, Shipping, Handling and Storage of Gypsum Board, Specification for, **C 1264/C1264M**
Steel Drill Screws for the Application of Gypsum Panel Products or Metal Plaster Bases, Specification for, **C 1002**
Steel Drill Screws for the Application of Gypsum Panel Products or Metal Plaster Bases to Steel Studs from 0.033 in. (0.84 mm) to 0.112 in. (2.84 mm) in Thickness, Specification for, **C 954**
Treated Core and Nontreated Core Gypsum Sheathing Board, Specification for, **C 79/C79M**
Water-Resistant Gypsum Backing Board, Specification for, **C 630/C630M**
Gypsum products—lath
Gypsum Board, Specification for, **C 1396**
Gypsum Lath, Specification for, **C 37**
Installation of Interior Lathing and Furring, Specification for, **C 841**
Installation of Lathing and Furring to Receive Interior and Exte-

Gypsum products—lath
rior Portland Cement-Based Plaster, Specification for, **C 1063**
Metal Lath, Specification for, **C 847**
Physical Testing of Gypsum Panel Products, Test Methods for, **C 473**
Welded Wire Lath, Specification for, **C 933**

H

Handling material
Sampling, Inspection, Rejection, Certification, Packaging, Marking, Shipping, Handling and Storage of Gypsum Board, Specification for, **C 1264/C1264M**
Hardness tests
Physical Testing of Gypsum Panel Products, Test Methods for, **C 473**
Head cap screws
See **Screws**
Heat of hydration
Heat of Hydration of Hydraulic Cement, Test Method for, **C 186**
Hexagon socket screw
See **Steel bolting materials—screws**
High-calcium limes
Sa **Lime and limestone products/testing—hydrated**
Determination of Total Neutralizing Capability and Dissolved Calcium and Magnesium Oxide in Lime for Flue Gas Desulfurization, Test Method for, **C 1318**
High-carbonate coals
See **Coal products**
High-humidity exposure tests
See **Humidity**
Humidity
Physical Testing of Gypsum Panel Products, Test Methods for, **C 473**
Hydrated lime
Sa **Lime and limestone products/testing**
Agricultural Liming Materials, Specification for, **C 602**
Chemical Analysis of Limestone, Quicklime, and Hydrated Lime, Test Methods for, **C 25**
Finishing Hydrated Lime, Specification for, **C 206**
Hydrated Lime for Masonry Purposes, Specification for, **C 207**
Hydrated Lime for Use in Asphaltic-Concrete Mixtures, Specification for, **C 1097**
Hydraulic Hydrated Lime for Structural Purposes, Specification for, **C 141**
Lime for Use with Pozzolans, Specification for, **C 821**
Physical Testing of Quicklime, Hydrated Lime, and Limestone, Test Methods for, **C 110**
Quicklime and Hydrated Lime for Neutralization of Waste Acid, Test Methods for, **C 400**
Quicklime and Hydrated Lime for Soil Stabilization, Specification for, **C 977**
Quicklime, Hydrated Lime, and Limestone for Chemical Uses, Specification for, **C 911**
X-Ray Spectrometric Analysis of Lime and Limestone, Test Method for, **C 1271**
Hydraulic cement concrete
Air Content of Hydraulic Cement Mortar, Test Method for, **C 185**
Blended Hydraulic Cement, Performance Specification for, **C 1157**
Blended Hydraulic Cements, Specifications for, **C 595**
Calcium Sulfate in Hydrated Portland Cement Mortar, Test Method for, **C 265**
Chemical Analysis of Hydraulic Cement, Test Methods for, **C 114**
Compressive Strength of Hydraulic Cement Mortars (Using 2-in. or 50-mm Cube Specimens), Test Method for, **C 109/C109M**
Compressive Strength of Hydraulic-Cement Mortars (Using Portions of Prisms Broken in Flexure), Test Method for, **C 349**
Density of Hydraulic Cement, Test Method for, **C 188**
Evaluation of Cement Strength Uniformity From a Single Source, Test Method for, **C 917**
Evaluation of Laboratories Testing Hydraulic Cement, Practice for, **C 1222**
Expansive Hydraulic Cement, Specification for, **C 845**
Fineness of Hydraulic Cement by Air Permeability Apparatus, Test Method for, **C 204**
Fineness of Hydraulic Cement by the 150-μm (No. 100) and 75-μm (No. 200) Sieves, Test Method for, **C 184**
Fineness of Hydraulic Cement by the 300-μm (No. 50), 150-μm (No. 100), and 75-μm (No. 200) Sieves by Wet Methods, Test Method for, **C 786**
Fineness of Hydraulic Cement by the 45-μm (No. 325) Sieve, Test Method for, **C 430**
Flexural Strength of Hydraulic Cement Mortars, Test Method for, **C 348**
Flow Table for Use in Tests of Hydraulic Cement, Specification for, **C 230**
Functional Additions for Use in Hydraulic Cements, Specification for, **C 688**
Heat of Hydration of Hydraulic Cement, Test Method for, **C 186**
Hydraulic Cement, Terminology Relating to, **C 219**
Length Change of Hydraulic-Cement Mortars Exposed to a Sulfate Solution, Test Method for, **C 1012**
Length Change of Mortar Bars Caused by Reaction Between Hydraulic Cement and Alkali-Reactive Aggregate, Provisional Test Method for, **PS 81**
Moist Cabinets, Moist Rooms and Water Storage Tanks Used in the Testing of Hydraulic Cements and Concretes, Specification for, **C 511**
Normal Consistency of Hydraulic Cement, Test Method for, **C 187**
Optimum SO_3 in Hydraulic Portland Cement Using 24-h Compressive Strength, Test Method for, **C 563**
Portland Cement, Specification for, **C 150**
Processing Additions for Use in the Manufacture of Hydraulic Cements, Specification for, **C 465**
Reference Masses and Devices for Determining Mass for Use in the Physical Testing of Hydraulic Cements, Specification for, **C 1005**
Restrained Expansion of Expansive Cement Mortar, Test Method for, **C 806**
Sampling and the Amount of Testing of Hydraulic Cement, Practice for, **C 183**
Time of Setting of Hydraulic Cement by Vicat Needle, Test Method for, **C 191**
Time of Setting of Hydraulic-Cement Paste by Gillmore Needles, Test Method for, **C 266**
Hydraulic cement mortar
Compressive Strength of Hydraulic-Cement Mortars (Using Portions of Prisms Broken in Flexure), Test Method for, **C 349**
Flexural Strength of Hydraulic Cement Mortars, Test Method for, **C 348**
Time of Setting of Hydraulic Cement Mortar by Modified Vicat Needle, Test Method for, **C 807**
Hydraulic-cement strength
Compressive Strength of Hydraulic-Cement Mortars (Using Portions of Prisms Broken in Flexure), Test Method for, **C 349**
Flexural Strength of Hydraulic Cement Mortars, Test Method for, **C 348**
Hydraulic cement testing sand
Standard Sand, Specification for, **C 778**
Hydraulic lime
Sa **Lime and limestone products/testing**
X-Ray Spectrometric Analysis of Lime and Limestone, Test Method for, **C 1271**
Hydrogen ion concentration
See **Acidity, alkalinity, pH**

Hydroxyl (hydroxide ion) content
See **Acidity, alkalinity, pH**

I

ICP-AES (inductively-coupled plasma atomic emission spectrometry)
Major and Trace Elements in Limestone and Lime by Inductively Coupled Plasma-Atomic Emission Spectroscopy (ICP) and Atomic Absoprtion (AA), Test Method for, **C 1301**
Ignition loss
Chemical Analysis of Hydraulic Cement, Test Methods for, **C 114**
Chemical Analysis of Limestone, Quicklime, and Hydrated Lime, Test Methods for, **C 25**
Inductively-coupled plasma (ICP)
Major and Trace Elements in Limestone and Lime by Inductively Coupled Plasma-Atomic Emission Spectroscopy (ICP) and Atomic Absoprtion (AA), Test Method for, **C 1301**
Industrial limes
See **Lime and limestone products/testing**
Industrial materials/applications—water treatment
Quicklime, Hydrated Lime, and Limestone for Chemical Uses, Specification for, **C 911**
Inorganic aggregate
Inorganic Aggregates for Use in Gypsum Plaster, Specification for, **C 35**
Insoluble matter content
Chemical Analysis of Hydraulic Cement, Test Methods for, **C 114**
Chemical Analysis of Limestone, Quicklime, and Hydrated Lime, Test Methods for, **C 25**
Insoluble residue content
Chemical Analysis of Gypsum and Gypsum Products [Metric], Test Methods for, **C 471M**
Chemical Analysis of Hydraulic Cement, Test Methods for, **C 114**
Inspection
Sampling, Inspection, Packing, and Marking of Lime and Limestone Products, Practice for, **C 50**
Sampling, Inspection, Rejection, Certification, Packaging, Marking, Shipping, Handling and Storage of Gypsum Board, Specification for, **C 1264/C1264M**
Installation—gypsum products
Application and Finishing of Gypsum Board, Specification for, **C 840**
Application of Class PB Exterior Insulation and Finish Systems, Practice for, **C 1397**
Application of Gypsum Base to Receive Gypsum Veneer Plaster, Specification for, **C 844**
Application of Gypsum Sheathing Board, Specification for, **C 1280**
Application of Gypsum Veneer Plaster, Specification for, **C 843**
Application of Interior Gypsum Plaster, Specification for, **C 842**
Application of Portland Cement-Based Plaster, Specification for, **C 926**
Bonding Compounds for Interior Gypsum Plastering, Specification for, **C 631**
Gypsum Backing Board and Coreboard, Specification for, **C 442**
Installation of Cast-in-Place Reinforced Gypsum Concrete, Specification for, **C 956**
Installation of Interior Lathing and Furring, Specification for, **C 841**
Installation of Lathing and Furring to Receive Interior and Exterior Portland Cement-Based Plaster, Specification for, **C 1063**
Installation of Load Bearing (Transverse and Axial) Steel Studs and Related Accessories, Specification for, **C 1007**
Installation of Steel Framing Members to Receive Screw-Attached Gypsum Panel Products, Specification for, **C 754**
Load-Bearing (Transverse and Axial) Steel Studs, Runners (Tracks), and Bracing or Bridging for Screw Application of Gypsum Board and Metal Plaster Bases, Specification for, **C 955**
Metal Lath, Specification for, **C 847**
Nails for the Application of Gypsum Board, Specification for, **C 514**
Non-Load Bearing (Axial) Steel Studs, Runners (Tracks), and Rigid Furring Channels for Screw Application of Gypsum Board, Specification for, **C 645**
Steel Drill Screws for the Application of Gypsum Panel Products or Metal Plaster Bases, Specification for, **C 1002**
Steel Drill Screws for the Application of Gypsum Panel Products or Metal Plaster Bases to Steel Studs from 0.033 in. (0.84 mm) to 0.112 in. (2.84 mm) in Thickness, Specification for, **C 954**
Insulation
Application of Class PB Exterior Insulation and Finish Systems, Practice for, **C 1397**
Interior lathing and furring
See **Gypsum products—lath**
Interlaboratory testing—construction materials
Evaluation of Laboratories Testing Hydraulic Cement, Practice for, **C 1222**
Iron content
Chemical Analysis of Gypsum and Gypsum Products [Metric], Test Methods for, **C 471M**
Chemical Analysis of Limestone, Quicklime, and Hydrated Lime, Test Methods for, **C 25**
Iron oxide content
Chemical Analysis of Hydraulic Cement, Test Methods for, **C 114**
Chemical Analysis of Limestone, Quicklime, and Hydrated Lime, Test Methods for, **C 25**

J

Job-mixed portland cement plaster
Aggregate for Job-Mixed Portland Cement-Based Plasters, Specification for, **C 897**
Joint compound
Joint Compound and Joint Tape for Finishing Gypsum Board, Specification for, **C 475**
Joint treatment materials
Joint Compound and Joint Tape for Finishing Gypsum Board, Specification for, **C 475**
Joint Treatment Materials for Gypsum Board Construction, Test Methods for, **C 474**

K

Keene's cement
Sa **Gypsum products—board/wallboard**
Gypsum Keene's Cement, Specification for, **C 61**

L

Laboratory
Evaluation of Laboratories Testing Hydraulic Cement, Practice for, **C 1222**
Laboratory apparatus—scales
Reference Masses and Devices for Determining Mass for Use in the Physical Testing of Hydraulic Cements, Specification for, **C 1005**
Lath
See **Gypsum products—lath**
Length
Physical Testing of Gypsum Panel Products, Test Methods for, **C 473**

Length change
Sa **Change in length**
Length Change of Mortar Bars Caused by Reaction Between Hydraulic Cement and Alkali-Reactive Aggregate, Provisional Test Method for, **PS 81**
Lignite
See **Coal products**
Lime and limestone products/testing
Agricultural Liming Materials, Specification for, **C 602**
Determination of Total Neutralizing Capability and Dissolved Calcium and Magnesium Oxide in Lime for Flue Gas Desulfurization, Test Method for, **C 1318**
Fly Ash and Other Pozzolans for Use with Lime, Specification for, **C 593**
Major and Trace Elements in Limestone and Lime by Inductively Coupled Plasma-Atomic Emission Spectroscopy (ICP) and Atomic Absoprtion (AA), Test Method for, **C 1301**
Sampling, Inspection, Packing, and Marking of Lime and Limestone Products, Practice for, **C 50**
Lime and limestone products/testing—hydrated
Agricultural Liming Materials, Specification for, **C 602**
Chemical Analysis of Limestone, Quicklime, and Hydrated Lime, Test Methods for, **C 25**
Finishing Hydrated Lime, Specification for, **C 206**
Hydrated Lime for Masonry Purposes, Specification for, **C 207**
Hydrated Lime for Use in Asphaltic-Concrete Mixtures, Specification for, **C 1097**
Hydraulic Hydrated Lime for Structural Purposes, Specification for, **C 141**
Lime for Use with Pozzolans, Specification for, **C 821**
Physical Testing of Quicklime, Hydrated Lime, and Limestone, Test Methods for, **C 110**
Quicklime and Hydrated Lime for Neutralization of Waste Acid, Test Methods for, **C 400**
Quicklime and Hydrated Lime for Soil Stabilization, Specification for, **C 977**
Quicklime, Hydrated Lime, and Limestone for Chemical Uses, Specification for, **C 911**
X-Ray Spectrometric Analysis of Lime and Limestone, Test Method for, **C 1271**
Lime and limestone products/testing—limestone
Chemical Analysis of Limestone, Quicklime, and Hydrated Lime, Test Methods for, **C 25**
Evaluation of Limestone or Lime Uniformity from a Single Source, Practice for, **C 1164**
Lime and Limestone (as used by the industry), Terminology Relating to, **C 51**
Limestone for Animal Feed Use, Specification for, **C 706**
Limestone for Dusting of Coal Mines, Specification for, **C 737**
Physical Testing of Quicklime, Hydrated Lime, and Limestone, Test Methods for, **C 110**
Quicklime, Hydrated Lime, and Limestone for Chemical Uses, Specification for, **C 911**
X-Ray Spectrometric Analysis of Lime and Limestone, Test Method for, **C 1271**
Lime and limestone products/testing—quicklime
See **Quicklime**
Lime byproducts
Agricultural Liming Materials, Specification for, **C 602**
Lime putty
Physical Testing of Quicklime, Hydrated Lime, and Limestone, Test Methods for, **C 110**
Liquid air-entraining admixtures
See **Additions to cement**
Liquid density
See **Density**
Livestock food products
Limestone for Animal Feed Use, Specification for, **C 706**
L-/LC-LK-bead
Accessories for Gypsum Wallboard and Gypsum Veneer Base, Specification for, **C 1047**
Load-bearing steel studs
Installation of Load Bearing (Transverse and Axial) Steel Studs and Related Accessories, Specification for, **C 1007**
Long-span slabs
See **Gypsum products**
Loss on ignition
See **Ignition loss**
Lump lime
Sa **Quicklime**
Quicklime for Structural Purposes, Specification for, **C 5**

M

Magnesian lime (ML)
Sa **Lime and limestone products/testing**
Determination of Total Neutralizing Capability and Dissolved Calcium and Magnesium Oxide in Lime for Flue Gas Desulfurization, Test Method for, **C 1318**
Quicklime, Hydrated Lime, and Limestone for Chemical Uses, Specification for, **C 911**
Magnesium
Determination of Total Neutralizing Capability and Dissolved Calcium and Magnesium Oxide in Lime for Flue Gas Desulfurization, Test Method for, **C 1318**
Magnesium content
Agricultural Liming Materials, Specification for, **C 602**
Magnesium oxide (MgO) content
Chemical Analysis of Gypsum and Gypsum Products [Metric], Test Methods for, **C 471M**
Chemical Analysis of Hydraulic Cement, Test Methods for, **C 114**
Chemical Analysis of Limestone, Quicklime, and Hydrated Lime, Test Methods for, **C 25**
Manganese content
Chemical Analysis of Limestone, Quicklime, and Hydrated Lime, Test Methods for, **C 25**
Manganic oxide content
Chemical Analysis of Hydraulic Cement, Test Methods for, **C 114**
Manuals
Manual of Cement Testing, (Related Material)
Marking
Sampling, Inspection, Packing, and Marking of Lime and Limestone Products, Practice for, **C 50**
Sampling, Inspection, Rejection, Certification, Packaging, Marking, Shipping, Handling and Storage of Gypsum Board, Specification for, **C 1264/C1264M**
Masonry mortar
Masonry Cement, Specification for, **C 91**
Masonry mortar and grout
Hydrated Lime for Masonry Purposes, Specification for, **C 207**
Masonry Cement, Specification for, **C 91**
Material handling
See **Handling material**
Mechanical mixing
Mechanical Mixing of Hydraulic Cement Pastes and Mortars of Plastic Consistency, Practice for, **C 305**
Mercaptan sulfur content
See **Sulfur content**
Metal lath
Sa **Gypsum products**
Installation of Interior Lathing and Furring, Specification for, **C 841**
Installation of Lathing and Furring to Receive Interior and Exterior Portland Cement-Based Plaster, Specification for, **C 1063**

Metal Lath, Specification for, **C 847**
Welded Wire Lath, Specification for, **C 933**
Metric practice—SI (International System of Units)
Sa **SI (International System of Units)**
Metric Practice Guide for Cement, (Related Material)
SI Quick Reference Guide: International System of Units (SI)*(The Modern Metric System)*, **IEEE/ASTM SI-10** (Related Material)
Microphotometers
See **Photometry**
See **Spectrophotometry**
Mineral aggregate
Sa **Aggregate**
Aggregate for Job-Mixed Portland Cement-Based Plasters, Specification for, **C 897**
Mixing
Mechanical Mixing of Hydraulic Cement Pastes and Mortars of Plastic Consistency, Practice for, **C 305**
Modified Vicat needle
Time of Setting of Hydraulic Cement Mortar by Modified Vicat Needle, Test Method for, **C 807**
Moist cabinets and rooms
Moist Cabinets, Moist Rooms and Water Storage Tanks Used in the Testing of Hydraulic Cements and Concretes, Specification for, **C 511**
Moisture analysis
Chemical Analysis of Limestone, Quicklime, and Hydrated Lime, Test Methods for, **C 25**
Moisture loss
See **Water retention**
Mortar—cement mortar
Air Content of Hydraulic Cement Mortar, Test Method for, **C 185**
Bleeding of Cement Pastes and Mortars, Test Method for, **C 243**
Compressive Strength of Hydraulic Cement Mortars (Using 2-in. or 50-mm Cube Specimens), Test Method for, **C 109/C109M**
Compressive Strength of Hydraulic-Cement Mortars (Using Portions of Prisms Broken in Flexure), Test Method for, **C 349**
Drying Shrinkage of Mortar Containing Portland Cement, Test Method for, **C 596**
Flexural Strength of Hydraulic Cement Mortars, Test Method for, **C 348**
Length Change of Hydraulic-Cement Mortars Exposed to a Sulfate Solution, Test Method for, **C 1012**
Mechanical Mixing of Hydraulic Cement Pastes and Mortars of Plastic Consistency, Practice for, **C 305**
Mortar Cement, Specification for, **C 1329**
Restrained Expansion of Expansive Cement Mortar, Test Method for, **C 806**
Time of Setting of Hydraulic Cement Mortar by Modified Vicat Needle, Test Method for, **C 807**
Use of Apparatus for the Determination of Length Change of Hardened Cement Paste, Mortar, and Concrete, Practice for, **C 490**
Mortar—cement mortar (portland)
Early Stiffening of Hydraulic Cement (Paste Method), Test Method for, **C 451**
Early Stiffening of Portland Cement (Mortar Method), Test Method for, **C 359**
Potential Expansion of Portland-Cement Mortars Exposed to Sulfate, Test Method for, **C 452**
Mortar bars
Expansion of Portland Cement Mortar Bars Stored in Water, Test Method for, **C 1038**
Length Change of Mortar Bars Caused by Reaction Between Hydraulic Cement and Alkali-Reactive Aggregate, Provisional Test Method for, **PS 81**
Mortar cement
Mortar Cement, Specification for, **C 1329**
Mortar method
Early Stiffening of Portland Cement (Mortar Method), Test Method for, **C 359**
Mortar strength
Compressive Strength of Hydraulic-Cement Mortars (Using Portions of Prisms Broken in Flexure), Test Method for, **C 349**
Flexural Strength of Hydraulic Cement Mortars, Test Method for, **C 348**

N

Nail pull resistance
Nails for the Application of Gypsum Board, Specification for, **C 514**
Physical Testing of Gypsum Panel Products, Test Methods for, **C 473**
Nails
Nails for the Application of Gypsum Board, Specification for, **C 514**
Physical Testing of Gypsum Panel Products, Test Methods for, **C 473**
Narrow-band-pass instrumentation
See **Spectrophotometry**
Natural cement
See **Cement**
See **Lime and limestone products/testing**
Natural mineral aggregate
See **Mineral aggregate**
Natural moisture method
See **Moisture analysis**
Natural pozzolans
See **Pozzolans**
Neutralization
Quicklime and Hydrated Lime for Neutralization of Waste Acid, Test Methods for, **C 400**
Neutralization numbers
See **Acidity, alkalinity, pH**
Neutralizing capability
Determination of Total Neutralizing Capability and Dissolved Calcium and Magnesium Oxide in Lime for Flue Gas Desulfurization, Test Method for, **C 1318**
Node-to-node bond strength
See **Bond strength**
Nominal thickness
Physical Testing of Gypsum Panel Products, Test Methods for, **C 473**
Nondispersive spectrometry
See **Spectrometry**
Nonload-bearing steel studs
Non-Load Bearing (Axial) Steel Studs, Runners (Tracks), and Rigid Furring Channels for Screw Application of Gypsum Board, Specification for, **C 645**
Nontreated core gypsum sheathing board
Treated Core and Nontreated Core Gypsum Sheathing Board, Specification for, **C 79/C79M**
Nonvolatile acidity
See **Acidity, alkalinity, pH**
Normal finishing hydrated lime
See **Lime and limestone products/testing—hydrated**

O

Object-color specimens
See **Spectrophotometry**
Operable partition
See **Partitions**

Optical materials/properties/tests
Physical Testing of Quicklime, Hydrated Lime, and Limestone, Test Methods for, **C 110**
Optimum moisture content
See **Moisture analysis**
Optional methods
Chemical Analysis of Hydraulic Cement, Test Methods for, **C 114**
Organic matter content
Chemical Analysis of Hydraulic Cement, Test Methods for, **C 114**
Oxide analysis
Determination of Total Neutralizing Capability and Dissolved Calcium and Magnesium Oxide in Lime for Flue Gas Desulfurization, Test Method for, **C 1318**

P

Packaging materials
Sampling, Inspection, Packing, and Marking of Lime and Limestone Products, Practice for, **C 50**
Sampling, Inspection, Rejection, Certification, Packaging, Marking, Shipping, Handling and Storage of Gypsum Board, Specification for, **C 1264/C1264M**
Partitions
Fiber-Reinforced Gypsum Panels, Specification for, **C 1278/C1278M**
Installation of Load Bearing (Transverse and Axial) Steel Studs and Related Accessories, Specification for, **C 1007**
Predecorated Gypsum Board, Specification for, **C 960/C960M**
Paste method
Early Stiffening of Hydraulic Cement (Paste Method), Test Method for, **C 451**
Pebble lime
Sa **Quicklime**
Quicklime for Structural Purposes, Specification for, **C 5**
Performance—spectrochemical equipment
See **Spectrometry**
See **Spectrophotometry**
Periclase
Determination of the Proportion of Phases in Portland Cement and Portland Cement Clinker Using X-Ray Powder Diffraction Analysis, Test Method for, **C 1365**
Quantitative Determination of Phases in Portland Cement Clinker by Microscopical Point Count Procedure, Test Method for, **C 1356**
Perlite
Inorganic Aggregates for Use in Gypsum Plaster, Specification for, **C 35**
pH
Sa **Acidity, alkalinity, pH**
Determination of Total Neutralizing Capability and Dissolved Calcium and Magnesium Oxide in Lime for Flue Gas Desulfurization, Test Method for, **C 1318**
Phosphorus content
Chemical Analysis of Limestone, Quicklime, and Hydrated Lime, Test Methods for, **C 25**
Phosphorus pentoxide content
Chemical Analysis of Hydraulic Cement, Test Methods for, **C 114**
Photometry
Sa **Spectrophotometry**
Chemical Analysis of Gypsum and Gypsum Products [Metric], Test Methods for, **C 471M**
Plaster
Application of Gypsum Base to Receive Gypsum Veneer Plaster, Specification for, **C 844**
Application of Gypsum Veneer Plaster, Specification for, **C 843**
Application of Interior Gypsum Plaster, Specification for, **C 842**
Application of Portland Cement-Based Plaster, Specification for, **C 926**
Bonding Compounds for Interior Gypsum Plastering, Specification for, **C 631**
Chemical Analysis of Gypsum and Gypsum Products [Metric], Test Methods for, **C 471M**
Gypsum Base for Veneer Plasters, Specification for, **C 588**
Gypsum Casting and Molding Plaster, Specification for, **C 59**
Gypsum Keene's Cement, Specification for, **C 61**
Gypsum Plasters, Specification for, **C 28**
Gypsum Veneer Plaster, Specification for, **C 587**
Inorganic Aggregates for Use in Gypsum Plaster, Specification for, **C 35**
Installation of Lathing and Furring to Receive Interior and Exterior Portland Cement-Based Plaster, Specification for, **C 1063**
Metal Lath, Specification for, **C 847**
Physical Testing of Gypsum, Gypsum Plasters and Gypsum Concrete, Test Methods for, **C 472**
Plastic (Stucco) Cement, Specification for, **C 1328**
Surface-Applied Bonding Compounds for Exterior Plastering, Specification for, **C 932**
Woven Wire Plaster Base, Specification for, **C 1032**
Plastic cement
Plastic (Stucco) Cement, Specification for, **C 1328**
Plasticity index
Physical Testing of Quicklime, Hydrated Lime, and Limestone, Test Methods for, **C 110**
Points
Nails for the Application of Gypsum Board, Specification for, **C 514**
Popping and pitting
Physical Testing of Quicklime, Hydrated Lime, and Limestone, Test Methods for, **C 110**
Portland cement
Aggregate for Job-Mixed Portland Cement-Based Plasters, Specification for, **C 897**
Application of Portland Cement-Based Plaster, Specification for, **C 926**
Autoclave Expansion of Portland Cement, Test Method for, **C 151**
Chemical Analysis of Hydraulic Cement, Test Methods for, **C 114**
Compressive Strength of Hydraulic Cement Mortars (Using 2-in. or 50-mm Cube Specimens), Test Method for, **C 109/C109M**
Density of Hydraulic Cement, Test Method for, **C 188**
Determination of the Proportion of Phases in Portland Cement and Portland Cement Clinker Using X-Ray Powder Diffraction Analysis, Test Method for, **C 1365**
Drying Shrinkage of Mortar Containing Portland Cement, Test Method for, **C 596**
Early Stiffening of Hydraulic Cement (Paste Method), Test Method for, **C 451**
Early Stiffening of Portland Cement (Mortar Method), Test Method for, **C 359**
Evaluation of Cement Strength Uniformity From a Single Source, Test Method for, **C 917**
Fineness of Hydraulic Cement by Air Permeability Apparatus, Test Method for, **C 204**
Fineness of Hydraulic Cement by the 150-μm (No. 100) and 75-μm (No. 200) Sieves, Test Method for, **C 184**
Fineness of Hydraulic Cement by the 300-μm (No. 50), 150-μm (No. 100), and 75-μm (No. 200) Sieves by Wet Methods, Test Method for, **C 786**
Fineness of Hydraulic Cement by the 45-μm (No. 325) Sieve, Test Method for, **C 430**
Fineness of Portland Cement by the Turbidimeter, Test Method for, **C 115**
Heat of Hydration of Hydraulic Cement, Test Method for, **C 186**

Hydraulic Cement, Terminology Relating to, **C 219**
Installation of Lathing and Furring to Receive Interior and Exterior Portland Cement-Based Plaster, Specification for, **C 1063**
Normal Consistency of Hydraulic Cement, Test Method for, **C 187**
Optimum SO_3 in Hydraulic Portland Cement Using 24-h Compressive Strength, Test Method for, **C 563**
Portland Cement, Specification for, **C 150**
Potential Expansion of Portland-Cement Mortars Exposed to Sulfate, Test Method for, **C 452**
Quantitative Determination of Phases in Portland Cement Clinker by Microscopical Point Count Procedure, Test Method for, **C 1356**

Portland cement-based plaster
Plastic (Stucco) Cement, Specification for, **C 1328**

Portland-cement clinker
Determination of the Proportion of Phases in Portland Cement and Portland Cement Clinker Using X-Ray Powder Diffraction Analysis, Test Method for, **C 1365**
Quantitative Determination of Phases in Portland Cement Clinker by Microscopical Point Count Procedure, Test Method for, **C 1356**

Portland cement concrete (PCC)
See **Concrete**

Portland-pozzolan cement
See **Hydraulic cement concrete**

Potassium oxide (K_2O) content
Chemical Analysis of Hydraulic Cement, Test Methods for, **C 114**

Potential alkali reactivity
See **Alkali reactivity**

Potential volume change
See **Volume change**

Pozzolans
Sa **Concrete**
Sa **Lime and limestone products/testing**
Blended Hydraulic Cement, Performance Specification for, **C 1157**
Fly Ash and Other Pozzolans for Use with Lime, Specification for, **C 593**
Lime for Use with Pozzolans, Specification for, **C 821**

Precast gypsum products
See **Gypsum products**

Predecorated gypsum board
Sa **Gypsum products—board/wallboard**
Gypsum Board, Specification for, **C 1396**
Predecorated Gypsum Board, Specification for, **C 960/C960M**

Prefixes
SI Quick Reference Guide: International System of Units (SI)*(The Modern Metric System)*, **IEEE/ASTM SI-10** (Related Material)

Primary beam filter
See **Spectrometry**

Processing additions (for cement)
Sa **Additions to cement**
Processing Additions for Use in the Manufacture of Hydraulic Cements, Specification for, **C 465**

Pulverized coal
See **Coal products**

Pulverized lime
Sa **Lime and limestone products/testing**
Quicklime for Structural Purposes, Specification for, **C 5**

Putrefaction
Joint Treatment Materials for Gypsum Board Construction, Test Methods for, **C 474**

Q

Quality control
Sampling and the Amount of Testing of Hydraulic Cement, Practice for, **C 183**

Quantitative analysis/measurement
Determination of the Proportion of Phases in Portland Cement and Portland Cement Clinker Using X-Ray Powder Diffraction Analysis, Test Method for, **C 1365**
Quantitative Determination of Phases in Portland Cement Clinker by Microscopical Point Count Procedure, Test Method for, **C 1356**

Quicklime
Sa **Lime and limestone products/testing**
Agricultural Liming Materials, Specification for, **C 602**
Chemical Analysis of Limestone, Quicklime, and Hydrated Lime, Test Methods for, **C 25**
Physical Testing of Quicklime, Hydrated Lime, and Limestone, Test Methods for, **C 110**
Quicklime and Hydrated Lime for Neutralization of Waste Acid, Test Methods for, **C 400**
Quicklime and Hydrated Lime for Soil Stabilization, Specification for, **C 977**
Quicklime for Structural Purposes, Specification for, **C 5**
Quicklime, Hydrated Lime, and Limestone for Chemical Uses, Specification for, **C 911**
X-Ray Spectrometric Analysis of Lime and Limestone, Test Method for, **C 1271**

QXRD (quantitative x-ray diffraction) analysis
Determination of the Proportion of Phases in Portland Cement and Portland Cement Clinker Using X-Ray Powder Diffraction Analysis, Test Method for, **C 1365**

R

Rapid methods
Chemical Analysis of Hydraulic Cement, Test Methods for, **C 114**

Raw coal
See **Coal products**

Ready-mixed concrete
See **Concrete**

Real density (RD)
See **Density**

Recessed depth (of gypsum boards)
Sa **Gypsum products**
Physical Testing of Gypsum Panel Products, Test Methods for, **C 473**

Referee methods
Chemical Analysis of Hydraulic Cement, Test Methods for, **C 114**

Reference masses
Reference Masses and Devices for Determining Mass for Use in the Physical Testing of Hydraulic Cements, Specification for, **C 1005**

Reinforced class PB exterior insulation
See **Class PB exterior insulation/finish system**

Rejection
Sampling, Inspection, Rejection, Certification, Packaging, Marking, Shipping, Handling and Storage of Gypsum Board, Specification for, **C 1264/C1264M**

Relative density
See **Density**

Relative humidity (RH)
See **Humidity**

Residual moisture
See **Moisture analysis**

Residue
Chemical Analysis of Gypsum and Gypsum Products [Metric], Test Methods for, **C 471M**
Chemical Analysis of Hydraulic Cement, Test Methods for, **C 114**
Physical Testing of Quicklime, Hydrated Lime, and Limestone, Test Methods for, **C 110**
Resistance
See **Water resistance**
Restrained expansion
Restrained Expansion of Expansive Cement Mortar, Test Method for, **C 806**
Rigid furring channels (for gypsum products)
Sa **Gypsum products**
Non-Load Bearing (Axial) Steel Studs, Runners (Tracks), and Rigid Furring Channels for Screw Application of Gypsum Board, Specification for, **C 645**
Rockwell hardness
See **Hardness tests**
Roll stability
See **Stability**
Roof decks
Gypsum Formboard, Specification for, **C 318/C318M**
Roofing and waterproofing materials—mineral aggregate
See **Mineral aggregate**
Runners
Installation of Load Bearing (Transverse and Axial) Steel Studs and Related Accessories, Specification for, **C 1007**
Runners (track)
Load-Bearing (Transverse and Axial) Steel Studs, Runners (Tracks), and Bracing or Bridging for Screw Application of Gypsum Board and Metal Plaster Bases, Specification for, **C 955**
Non-Load Bearing (Axial) Steel Studs, Runners (Tracks), and Rigid Furring Channels for Screw Application of Gypsum Board, Specification for, **C 645**

S

Sampling—cement/cement products
Agricultural Liming Materials, Specification for, **C 602**
Sampling and the Amount of Testing of Hydraulic Cement, Practice for, **C 183**
Sampling, Inspection, Packing, and Marking of Lime and Limestone Products, Practice for, **C 50**
Standard Sand, Specification for, **C 778**
Sand
Chemical Analysis of Gypsum and Gypsum Products [Metric], Test Methods for, **C 471M**
Inorganic Aggregates for Use in Gypsum Plaster, Specification for, **C 35**
Standard Sand, Specification for, **C 778**
Sand-lime brick
See **Gypsum products**
See **Lime and limestone products/testing**
Scales
Reference Masses and Devices for Determining Mass for Use in the Physical Testing of Hydraulic Cements, Specification for, **C 1005**
Screw applications (for gypsum boards)
See **Gypsum products**
See **Installation—gypsum products**
Screws
Steel Drill Screws for the Application of Gypsum Panel Products or Metal Plaster Bases, Specification for, **C 1002**
Steel Drill Screws for the Application of Gypsum Panel Products or Metal Plaster Bases to Steel Studs from 0.033 in. (0.84 mm) to 0.112 in. (2.84 mm) in Thickness, Specification for, **C 954**
Screws—gypsum board applications
Installation of Steel Framing Members to Receive Screw-Attached Gypsum Panel Products, Specification for, **C 754**
Load-Bearing (Transverse and Axial) Steel Studs, Runners (Tracks), and Bracing or Bridging for Screw Application of Gypsum Board and Metal Plaster Bases, Specification for, **C 955**
Non-Load Bearing (Axial) Steel Studs, Runners (Tracks), and Rigid Furring Channels for Screw Application of Gypsum Board, Specification for, **C 645**
Steel Drill Screws for the Application of Gypsum Panel Products or Metal Plaster Bases, Specification for, **C 1002**
Steel Drill Screws for the Application of Gypsum Panel Products or Metal Plaster Bases to Steel Studs from 0.033 in. (0.84 mm) to 0.112 in. (2.84 mm) in Thickness, Specification for, **C 954**
Section properties
Non-Load Bearing (Axial) Steel Studs, Runners (Tracks), and Rigid Furring Channels for Screw Application of Gypsum Board, Specification for, **C 645**
Non-Load Bearing (Axial) Steel Studs, Runners (Tracks), and Rigid Furring Channels for Screw Application of Gypsum Board, Specification for, **C 645**
Sediment content
See **Residue**
Selected conversion factors
SI Quick Reference Guide: International System of Units (SI)*(The Modern Metric System)*, **IEEE/ASTM SI-10** (Related Material)
Sequential spectrometer
See **Spectrometry**
Setting time
Physical Testing of Gypsum, Gypsum Plasters and Gypsum Concrete, Test Methods for, **C 472**
Time of Setting of Hydraulic Cement Mortar by Modified Vicat Needle, Test Method for, **C 807**
Time of Setting of Hydraulic Cement by Vicat Needle, Test Method for, **C 191**
Time of Setting of Hydraulic-Cement Paste by Gillmore Needles, Test Method for, **C 266**
Settlement
Physical Testing of Quicklime, Hydrated Lime, and Limestone, Test Methods for, **C 110**
Sewerage treatment limes
Sa **Lime and limestone products/testing**
Quicklime, Hydrated Lime, and Limestone for Chemical Uses, Specification for, **C 911**
Shanks
Nails for the Application of Gypsum Board, Specification for, **C 514**
Sheaths/sheathing
Application of Gypsum Sheathing Board, Specification for, **C 1280**
Glass Mat Gypsum Substrate for Use as Sheathing, Specification for, **C 1177/C1177M**
Gypsum Board, Specification for, **C 1396**
Shipping
Sampling, Inspection, Rejection, Certification, Packaging, Marking, Shipping, Handling and Storage of Gypsum Board, Specification for, **C 1264/C1264M**
Shotcrete
See **Concrete**
Shrinkage
Joint Treatment Materials for Gypsum Board Construction, Test Methods for, **C 474**
Shrinkage—drying
Drying Shrinkage of Mortar Containing Portland Cement, Test Method for, **C 596**

Shrinkage-compensating concrete
Expansive Hydraulic Cement, Specification for, **C 845**
Sieve analysis—building materials
Agricultural Liming Materials, Specification for, **C 602**
Fineness of Hydraulic Cement by the 150-μm (No. 100) and 75-μm (No. 200) Sieves, Test Method for, **C 184**
Fineness of Hydraulic Cement by the 300-μm (No. 50), 150-μm (No. 100), and 75-μm (No. 200) Sieves by Wet Methods, Test Method for, **C 786**
Fineness of Hydraulic Cement by the 45-μm (No. 325) Sieve, Test Method for, **C 430**
Physical Testing of Quicklime, Hydrated Lime, and Limestone, Test Methods for, **C 110**
SI (International System of Units)
Metric Practice Guide for Cement, (Related Material)
SI Quick Reference Guide: International System of Units (SI)*(The Modern Metric System),* **IEEE/ASTM SI-10** (Related Material)
Silica content
Chemical Analysis of Limestone, Quicklime, and Hydrated Lime, Test Methods for, **C 25**
Silica fume
Blended Hydraulic Cement, Performance Specification for, **C 1157**
Silicon dioxide (SiO_2) content
Chemical Analysis of Gypsum and Gypsum Products [Metric], Test Methods for, **C 471M**
Chemical Analysis of Hydraulic Cement, Test Methods for, **C 114**
Chemical Analysis of Limestone, Quicklime, and Hydrated Lime, Test Methods for, **C 25**
Simultaneous spectrometer
See **Spectrometry**
Slag
Agricultural Liming Materials, Specification for, **C 602**
Blended Hydraulic Cement, Performance Specification for, **C 1157**
Slag—cement
See **Hydraulic cement concrete**
Slaking rate
Physical Testing of Quicklime, Hydrated Lime, and Limestone, Test Methods for, **C 110**
Slimes
See **Residue**
Slit width (of spectrophotometers)
See **Spectrophotometry**
Socket set screws (SSS)
See **Screws**
Sodium chloride content
Chemical Analysis of Gypsum and Gypsum Products [Metric], Test Methods for, **C 471M**
Sodium content
Chemical Analysis of Gypsum and Gypsum Products [Metric], Test Methods for, **C 471M**
Sodium oxide (Na_2O) content
Chemical Analysis of Hydraulic Cement, Test Methods for, **C 114**
Soffit board
Sa **Gypsum products**
Exterior Gypsum Soffit Board, Specification for, **C 931/C931M**
Soil
Quicklime and Hydrated Lime for Soil Stabilization, Specification for, **C 977**
Soil stabilization
Quicklime and Hydrated Lime for Soil Stabilization, Specification for, **C 977**
Soluble chloride
See **Chloride content**
Specimen handling
See **Handling material**
Spectrochemical analysis
See **Spectrometry**
See **Spectrophotometry**
See **X-ray emission spectrometry**
Spectrometry
Major and Trace Elements in Limestone and Lime by Inductively Coupled Plasma-Atomic Emission Spectroscopy (ICP) and Atomic Absoprtion (AA), Test Method for, **C 1301**
Spectrophotometry
X-Ray Spectrometric Analysis of Lime and Limestone, Test Method for, **C 1271**
Spectrophotometry—atomic absorption
Chemical Analysis of Gypsum and Gypsum Products [Metric], Test Methods for, **C 471M**
Spectrophotometry—flame photometry
Chemical Analysis of Gypsum and Gypsum Products [Metric], Test Methods for, **C 471M**
Chemical Analysis of Hydraulic Cement, Test Methods for, **C 114**
Spinout
Non-Load Bearing (Axial) Steel Studs, Runners (Tracks), and Rigid Furring Channels for Screw Application of Gypsum Board, Specification for, **C 645**
SRE (stray radiant energy)
See **Spectrophotometry**
Stability
Quicklime and Hydrated Lime for Soil Stabilization, Specification for, **C 977**
Stability (dimensional)
See **Dimensional stability**
Standard sand
Sa **Sand**
Standard Sand, Specification for, **C 778**
Steel
Accessories for Gypsum Wallboard and Gypsum Veneer Base, Specification for, **C 1047**
Steel bolting materials
See **Nails**
Steel bolting materials—screws
Steel Drill Screws for the Application of Gypsum Panel Products or Metal Plaster Bases, Specification for, **C 1002**
Steel Drill Screws for the Application of Gypsum Panel Products or Metal Plaster Bases to Steel Studs from 0.033 in. (0.84 mm) to 0.112 in. (2.84 mm) in Thickness, Specification for, **C 954**
Steel bolting materials—studs
Installation of Load Bearing (Transverse and Axial) Steel Studs and Related Accessories, Specification for, **C 1007**
Installation of Steel Framing Members to Receive Screw-Attached Gypsum Panel Products, Specification for, **C 754**
Load-Bearing (Transverse and Axial) Steel Studs, Runners (Tracks), and Bracing or Bridging for Screw Application of Gypsum Board and Metal Plaster Bases, Specification for, **C 955**
Non-Load Bearing (Axial) Steel Studs, Runners (Tracks), and Rigid Furring Channels for Screw Application of Gypsum Board, Specification for, **C 645**
Steel screws
See **Steel bolting materials—screws**
Stiffness
Early Stiffening of Hydraulic Cement (Paste Method), Test Method for, **C 451**
Early Stiffening of Portland Cement (Mortar Method), Test Method for, **C 359**
Storage
Sampling, Inspection, Rejection, Certification, Packaging, Marking, Shipping, Handling and Storage of Gypsum Board,

Storage
Specification for, **C 1264/C1264M**
Storage tanks
Moist Cabinets, Moist Rooms and Water Storage Tanks Used in the Testing of Hydraulic Cements and Concretes, Specification for, **C 511**
Strength
Sa **Bond strength**
Compressive Strength of Hydraulic Cement Mortars (Using 2-in. or 50-mm Cube Specimens), Test Method for, **C 109/C109M**
Evaluation of Cement Strength Uniformity From a Single Source, Test Method for, **C 917**
Strength (of cement)
Compressive Strength of Hydraulic-Cement Mortars (Using Portions of Prisms Broken in Flexure), Test Method for, **C 349**
Flexural Strength of Hydraulic Cement Mortars, Test Method for, **C 348**
Strontium oxide content
Chemical Analysis of Limestone, Quicklime, and Hydrated Lime, Test Methods for, **C 25**
Structural building materials
Hydraulic Hydrated Lime for Structural Purposes, Specification for, **C 141**
Quicklime for Structural Purposes, Specification for, **C 5**
Structural lime
Sa **Lime and limestone products/testing**
Hydraulic Hydrated Lime for Structural Purposes, Specification for, **C 141**
Quicklime for Structural Purposes, Specification for, **C 5**
Stucco
Plastic (Stucco) Cement, Specification for, **C 1328**
Stucco work
Application of Portland Cement-Based Plaster, Specification for, **C 926**
Studs—gypsum board applications
Sa **Gypsum products**
Installation of Load Bearing (Transverse and Axial) Steel Studs and Related Accessories, Specification for, **C 1007**
Installation of Steel Framing Members to Receive Screw-Attached Gypsum Panel Products, Specification for, **C 754**
Load-Bearing (Transverse and Axial) Steel Studs, Runners (Tracks), and Bracing or Bridging for Screw Application of Gypsum Board and Metal Plaster Bases, Specification for, **C 955**
Non-Load Bearing (Axial) Steel Studs, Runners (Tracks), and Rigid Furring Channels for Screw Application of Gypsum Board, Specification for, **C 645**
Sub-bituminous coal
See **Coal products**
Substrates
Glass Mat Gypsum Substrate for Use as Sheathing, Specification for, **C 1177/C1177M**
Sulfate resistance
Length Change of Hydraulic-Cement Mortars Exposed to a Sulfate Solution, Test Method for, **C 1012**
Potential Expansion of Portland-Cement Mortars Exposed to Sulfate, Test Method for, **C 452**
Sulfide content
Chemical Analysis of Hydraulic Cement, Test Methods for, **C 114**
Sulfur content
Chemical Analysis of Limestone, Quicklime, and Hydrated Lime, Test Methods for, **C 25**
Sulfur trioxide (SO_3) content
Calcium Sulfate in Hydrated Portland Cement Mortar, Test Method for, **C 265**
Chemical Analysis of Gypsum and Gypsum Products [Metric], Test Methods for, **C 471M**
Chemical Analysis of Hydraulic Cement, Test Methods for, **C 114**
Chemical Analysis of Limestone, Quicklime, and Hydrated Lime, Test Methods for, **C 25**
Optimum SO_3 in Hydraulic Portland Cement Using 24-h Compressive Strength, Test Method for, **C 563**
Surface-applied bonding compound
Surface-Applied Bonding Compounds for Exterior Plastering, Specification for, **C 932**
Surface hardening/hardness
See **Hardness tests**
Surface water resistance
See **Water resistance**

T

Tallow content
Chemical Analysis of Hydraulic Cement, Test Methods for, **C 114**
Tapered-edge depth (of gypsum boards)
Sa **Gypsum products**
Physical Testing of Gypsum Panel Products, Test Methods for, **C 473**
Tension (tensile) properties/tests—building materials/applications
Joint Treatment Materials for Gypsum Board Construction, Test Methods for, **C 474**
Terminology—cement/concrete
Gypsum and Related Building Materials and Systems, Terminology Relating to, **C 11**
Hydraulic Cement, Terminology Relating to, **C 219**
Lime and Limestone (as used by the industry), Terminology Relating to, **C 51**
Terra alba
See **Calcium sulfate ($CaSO_4$)**
Testing and inspection agencies—cement/concrete
Evaluation of Laboratories Testing Hydraulic Cement, Practice for, **C 1222**
Moist Cabinets, Moist Rooms and Water Storage Tanks Used in the Testing of Hydraulic Cements and Concretes, Specification for, **C 511**
Testing methods
Sampling and the Amount of Testing of Hydraulic Cement, Practice for, **C 183**
Thickness
Non-Load Bearing (Axial) Steel Studs, Runners (Tracks), and Rigid Furring Channels for Screw Application of Gypsum Board, Specification for, **C 645**
Tile—gypsum
See **Gypsum products**
Time of setting
See **Setting time**
Time tests
See **Twenty-four hour testing**
Titanium dioxide content
Chemical Analysis of Hydraulic Cement, Test Methods for, **C 114**
Titratable acidity
See **Acidity, alkalinity, pH**
Topping compound
Joint Treatment Materials for Gypsum Board Construction, Test Methods for, **C 474**
Total acid/base number
See **Acidity, alkalinity, pH**
Total moisture content
See **Moisture analysis**
Total neutralizing capability
Determination of Total Neutralizing Capability and Dissolved Calcium and Magnesium Oxide in Lime for Flue Gas Desulfurization, Test Method for, **C 1318**

Trace elements
Major and Trace Elements in Limestone and Lime by Inductively Coupled Plasma-Atomic Emission Spectroscopy (ICP) and Atomic Absoprtion (AA), Test Method for, **C 1301**
Tracks (runners)
Sa **Gypsum products**
Load-Bearing (Transverse and Axial) Steel Studs, Runners (Tracks), and Bracing or Bridging for Screw Application of Gypsum Board and Metal Plaster Bases, Specification for, **C 955**
Non-Load Bearing (Axial) Steel Studs, Runners (Tracks), and Rigid Furring Channels for Screw Application of Gypsum Board, Specification for, **C 645**
Transverse load-bearing steel studs
Installation of Load Bearing (Transverse and Axial) Steel Studs and Related Accessories, Specification for, **C 1007**
Treated core gypsum sheathing board
Treated Core and Nontreated Core Gypsum Sheathing Board, Specification for, **C 79/C79M**
Turbidimeter method
Fineness of Portland Cement by the Turbidimeter, Test Method for, **C 115**
Twenty-four hour testing
Optimum SO_3 in Hydraulic Portland Cement Using 24-h Compressive Strength, Test Method for, **C 563**

U

U-bend specimens
Accessories for Gypsum Wallboard and Gypsum Veneer Base, Specification for, **C 1047**
Unhydrated oxides
Chemical Analysis of Limestone, Quicklime, and Hydrated Lime, Test Methods for, **C 25**
Uniformity
Evaluation of Cement Strength Uniformity From a Single Source, Test Method for, **C 917**
Evaluation of Limestone or Lime Uniformity from a Single Source, Practice for, **C 1164**
Unsulfonated residue
See **Residue**

V

Veneer plaster
Sa **Plaster**
Gypsum Base for Veneer Plasters, Specification for, **C 588**
Vermiculite
Inorganic Aggregates for Use in Gypsum Plaster, Specification for, **C 35**
Vicat needle
Time of Setting of Hydraulic Cement Mortar by Modified Vicat Needle, Test Method for, **C 807**
Vicat needles
Time of Setting of Hydraulic Cement by Vicat Needle, Test Method for, **C 191**
Vinsol resin content
Chemical Analysis of Hydraulic Cement, Test Methods for, **C 114**
Volume change
Autoclave Expansion of Portland Cement, Test Method for, **C 151**
Expansion of Portland Cement Mortar Bars Stored in Water, Test Method for, **C 1038**
Length Change of Hydraulic-Cement Mortars Exposed to a Sulfate Solution, Test Method for, **C 1012**
Restrained Expansion of Expansive Cement Mortar, Test Method for, **C 806**
Use of Apparatus for the Determination of Length Change of Hardened Cement Paste, Mortar, and Concrete, Practice for, **C 490**
Volume shrinkage
Drying Shrinkage of Mortar Containing Portland Cement, Test Method for, **C 596**

W

Wagner turbidimeter
Fineness of Portland Cement by the Turbidimeter, Test Method for, **C 115**
Wallboard
Sa **Gypsum products—board/wallboard**
Gypsum Board, Specification for, **C 1396**
Gypsum Ceiling Board, Specification for, **C 1395**
Wall panels
Glass Mat Water-Resistant Gypsum Backing Board, Specification for, **C 1178/C1178M**
Waste acid solutions
Quicklime and Hydrated Lime for Neutralization of Waste Acid, Test Methods for, **C 400**
Water absorption
See **Moisture analysis**
Water content
See **Moisture analysis**
Water content—gypsum products (free/combined)
Chemical Analysis of Gypsum and Gypsum Products [Metric], Test Methods for, **C 471M**
Physical Testing of Gypsum, Gypsum Plasters and Gypsum Concrete, Test Methods for, **C 472**
Water resistance
Physical Testing of Gypsum Panel Products, Test Methods for, **C 473**
Water-resistant core gypsum sheathing board
Treated Core and Nontreated Core Gypsum Sheathing Board, Specification for, **C 79/C79M**
Water-resistant gypsum board
Gypsum Board, Specification for, **C 1396**
Water retention
Masonry Cement, Specification for, **C 91**
Physical Testing of Quicklime, Hydrated Lime, and Limestone, Test Methods for, **C 110**
Water-soluble acidity or alkalinity
See **Acidity, alkalinity, pH**
Water-soluble-matter content
Chemical Analysis of Hydraulic Cement, Test Methods for, **C 114**
Wavelength dispersive spectrophotometry
X-Ray Spectrometric Analysis of Lime and Limestone, Test Method for, **C 1271**
Webster hardness
See **Hardness tests**
Weight
Physical Testing of Gypsum Panel Products, Test Methods for, **C 473**
Reference Masses and Devices for Determining Mass for Use in the Physical Testing of Hydraulic Cements, Specification for, **C 1005**
Welding/welds—wire
Sa **Gypsum products—lath**
Welded Wire Lath, Specification for, **C 933**
Width
Physical Testing of Gypsum Panel Products, Test Methods for, **C 473**
Wire heads
Nails for the Application of Gypsum Board, Specification for, **C 514**
Wire lath
See **Gypsum products—lath**

Withdrawal resistance
Nails for the Application of Gypsum Board, Specification for, **C 514**
Wood-fiber gypsum plaster
Chemical Analysis of Gypsum and Gypsum Products [Metric], Test Methods for, **C 471M**
Wood screws
See **Steel bolting materials—screws**

X

X-ray diffraction
Determination of the Proportion of Phases in Portland Cement and Portland Cement Clinker Using X-Ray Powder Diffraction Analysis, Test Method for, **C 1365**
X-ray emission spectrometry
X-Ray Spectrometric Analysis of Lime and Limestone, Test Method for, **C 1271**

100 Barr Harbor Drive
West Conshohocken, PA 19428-2959
Phone: (610) 832-9693
FAX: (610) 832-9667

European Office
27-29 Knowl Piece, Wilbury Way
Hitchin, Herts SG4 OSX, England
Phone: 1462-437933
FAX: 1462-433678

1999 INDIVIDUAL MEMBERSHIP

SEE FEE AND BENEFITS ON REVERSE SIDE

VISIT OUR WEB SITE AT
http://www.astm.org

APPLICATION IS MADE FOR MEMBERSHIP IN ASTM:

INDIVIDUAL MEMBERSHIP AN INDIVIDUAL MEMBER SHALL PAY AN ANNUAL FEE AND MAY PARTICIPATE ON ASTM TECHNICAL COMMITTEES REPRESENTING HIS OR HER EMPLOYER UNLESS UNASSOCIATED WITH ANY ORGANIZATION.

PLEASE PRINT ALL INFORMATION CLEARLY. DO NOT EXCEED THE CAPACITY OF EACH LINE.

LAST NAME ____ FIRST NAME ____ INITIAL ____

AFFILIATION ____

FACILITY ____

STREET ____ ☐ CHECK HERE IF HOME ADDRESS

P.O. BOX ____

CITY ____ STATE ____ ZIP ____ — ____

COUNTRY ____

PHONE ( ) — ____ EXTENSION ____

FAX ( ) — ____

E-MAIL ____

JOB TITLE ____

NOTE: IF YOUR AFFILIATION IS A SUBSIDIARY, PLEASE IDENTIFY THE PARENT ORGANIZATION.

COMMITTEE APPLICATION REQUEST

THIS APPLICATION IS FOR SOCIETY MEMBERSHIP ONLY. A TECHNICAL COMMITTEE REQUIRES A SEPARATE APPLICATION. PLEASE SEND APPLICATIONS FOR THE FOLLOWING COMMITTEE(S):

☐ ENCLOSED IS PAYMENT.
ALL CHECKS MADE PAYABLE TO ASTM
IN U.S. FUNDS ON U.S. BANKS.

☐ EFT: FIRST UNION NATIONAL BANK
123 S. BROAD STREET, PHILADELPHIA, PA 19109
ACCOUNT 2030000863723
ACH/EFT 031000503 WIRES 031201467

OPTIONAL METHODS OF PAYMENT

AMOUNT $____

☐ AMERICAN EXPRESS ☐ MASTER CARD ☐ VISA

ACCOUNT NO. (ALL DIGITS) ____ EXPIRATION DATE ____ SIGNATURE ____

MAIL TO: ASTM ATTN: MEMBER SERVICES

FEE AND BENEFITS FOR 1999

Annual Administrative Fee	\$65.00*†
Participation on ASTM Technical Committees	Yes
Reduced Symposium Registration Fees	Yes
Complete Set of Book of ASTM Standards (Vols. 1-72)	\$5,200.00 Less 10% (Prepaid-North America)† \$5,720.00 Less 10% (Prepaid-All Other Countries)
Complete Sections at Member Price	No Limit
Individual Book of Standards at Member Price	No Limit
Other Publications at Member Price	No Limit
Annual Subscription to **ASTM STANDARDIZATION NEWS**	1 Free
Advance Information about Standards & Activities	Yes

Membership year is January 1 to December 31. Fees are payable in advance and are not prorated. Membership will become effective upon payment of fees.

* Administrative fee payments made to ASTM are not tax deductible as charitable contributions for federal income tax purposes.

† Canadian Members – Please add 7% GST – Registration Number R129162244.

NOTE: Membership benefits are subject to change.

Printed in the USA.